AF323154

MATERIALS AND PROCESS AFFORDABILITY
KEYS TO THE FUTURE

Dedicated to the Memory of
Dr. Charles L. Hamermesh
SAMPE Technical Director
1925-1998

Dr. Hamermesh was very active in directing SAMPE's future growth as a premier technical organization, giving guidance in the planning of informative technical programs for SAMPE conferences. He was Editor of the SAMPE Journal. He also promoted active co-location activities with other technical societies. He was SAMPE's Technical Director since 1990.

Dr. Hamermesh had recently been selected as a SAMPE Fellow for his outstanding contribution to SAMPE and the materials and processing industry over the last 39 years.

He was with North American/Rockwell from 1959-1985 and was also a Consultant to the industry since retiring from Rockwell in 1985.

"You will be missed, Charlie, not only by your immediate family, but also by your SAMPE family and your numerous colleagues and friends within the industry."

SOCIETY FOR THE ADVANCEMENT OF MATERIAL AND PROCESS ENGINEERING

43rd International SAMPE Symposium and Exhibition

VOLUME 43
Book 1 of 2 Books

MATERIALS AND PROCESS AFFORDABILITY KEYS TO THE FUTURE

Edited by
Dr. Howard S. Kliger
Benjamin M. Rasmussen
Dr. Louis A. Pilato
Tia Benson Tolle

Anaheim Convention Center
Anaheim, California

May 31-June 4, 1998

INTERNATIONAL OFFICERS

President	Dr. A. Brent Strong
Executive Vice President	John Willis
Senior Vice President	Allan B. Goldberg
Vice President	Allen P. Penton
Vice President-Europe	Jean-Claude Mezzadri
Vice President-Asia	Nobuo Ohashi
Secretary	Dr. Scott W. Beckwith
Treasurer	Clark W. Johnson
Immediate Past President	Ronald B. Keyson

OFFICERS OF SPONSORING CHAPTERS
(New Jersey Chapter)

Chairman	Gail D. DiSalvo
First Vice Chairman	Ash Chopra
Second Vice Chairman	David F. Vincenti
Secretary	Tom B. Borne
Treasurer	John F. Osterndorf
Director	Dr. Louis A. Pilato
Director	Dr. Howard S. Kliger

(Midwest Chapter)

Chairman	Tia Benson Tolle
First Vice Chairman	Dr. Allan S. Crasto
Second Vice Chairman	Dr. Donald A. Klosterman
Secretary	Dr. David P. Anderson
Treasurer	Dr. Anthony E. Saliba
Director	Jim T. Johnson
Director	Robert J. Gran

(Foothill Chapter)

Chairman	Susan Ruth
Vice Chairman	Winston C. Mih
Secretary	John D. Harper, Jr.
Treasurer	Steven M. Breitbart
Director	Rudolph B. Baggett
Director	Gerald L. Horstman

43RD INTERNATIONAL SAMPE SYMPOSIUM COMMITTEE

Howard S. Kliger
General Co-Chairman
HSK Associates

Benjamin M. Rasmussen
General Co-Chairman
BMR Associates

Tia Benson Tolle
Technical Program Co-Chairman
Air Force Research Laboratory

Louis A. Pilato
Technical Program Co-Chairman
Bakelite AG

John Russell
Finance Co-Chairman
Air Force Research Laboratory

John Osterndorf
Finance Co-Chairman
U.S. Army Picatinny Arsenal

43RD INTERNATIONAL SAMPE SYMPOSIUM COMMITTEE

Ray Wegman
Arrangements Co-Chairman
Adhesion Associates

Allan Crasto
Arrangements Co-Chairman
University of Dayton Research Institute

Karla Strong
Arrangements Co-Chairman
University of Dayton Research Institute

Bob Gran
Publicity Co-Chairman
Universal Technology Corporation

Gail DiSalvo
Publicity Co-Chairman
GSD Consultings

Kathy Froelich
Publicity Co-Chairman
Huls America

43RD INTERNATIONAL SAMPE SYMPOSIUM COMMITTEE

Dick Walther
Audio Visual Co-Chairman
Walther Consulting

Dave Anderson
Audio Visual Co-Chairman
University of Dayton Research Institute

Susan Ruth
Volunteers Chairman
The Aerospace Corporation

Charles Browning
Advisor
Air Force Research Laboratory

George Schmitt
Advisor
Air Force Research Laboratory

Charles Hurley
Advisor
University of Dayton Research Institute (retired)

PREFACE

The New Jersey, Midwest and Foothill Chapters of SAMPE are honored to host the 43rd International SAMPE Symposium and Exhibition.

The theme of this Symposium is "Materials and Process Affordability--Keys to the Future". The technology developments in materials and processes have provided significant advances worldwide in improving performance and affordability of products for aircraft, aerospace, automotive, industrial, infrastructure, sporting goods, and other application areas.

This symposium was designed to foster the goals of SAMPE through communication and educational technology transfer. It was focused on the needs of industry to expand its horizons and to provide a platform for non-traditional groups to meet and interact with our members. In addition to the traditional program, SAMPE partnered with the Composites Fabricators Association, the Interagency Working Group on Fire and Materials, the National Hydrogen Association, the U.S. Air Force Office of Scientific Research, and the U.S. Air Force Manufacturing Problem Prevention Program Team to provide expanded forums for interaction.

The technical program presented a diversity of affordable technologies and applications present in today's materials and processes arena. It was structured to present an integrated program of technical sessions, plenaries, panels, tutorials, and workshops. We were particularly pleased with the two plenary offerings, which are not part of this publication. One addressed the current worldwide shortage of carbon fiber and the other was a first public presentation of the material and process applications for the F-22 Air Superiority Fighter.

Special thanks for the development of the program are extended to the authors, session chairmen, speakers, and panelists. The help of the committee chairmen, committee workers, and the SAMPE International Business Office has been invaluable.

Howard S. Kliger
General Co-Chairman

Benjamin M. Rasmussen
General Co-Chairman

Tia Benson Tolle
Program Co-Chairman

Louis A. Pilato
Program Co-Chairman

CONTENTS

Processing
*Chairmen: Brian Rice, University of Dayton Research Institute
Dr. John Russell, Air Force Research Laboratory*

Sporting Goods

Chairmen: David Reppetto, Wilson Sporting Goods
Frederick F. Saremi, Newport Adhesives and Composites

PAPERS, AUTHORS, SESSIONS AND SESSION CHAIRMEN

LOW TEMPERATURE CURING MATERIALS
THE NEXT GENERATION

K. Jackson
The Advanced Composites Group, Inc.
Tulsa, OK

ABSTRACT

Low Temperature curing preimpregnated composite materials were developed in the early 1980's and were used for a number of applications where accuracy and low cost were considered important. In the early 1990's the same materials were used for the first time for aerospace structures , offering the same promise of lower overall cost through processing flexibility . In the intervening years the mechanical performance and handlability of these types of materials have been improved in research and development programs dedicated to this goal without sacrificing the processing flexibility that made them so attractive at their inception.
The latest generation of low temperature structural materials show a 17% improvement in RT/dry compression strength compared with first generation LTM 45-EL. The same system shows an improvement in hot/wet performance and void contents < 1% for oven/vacuum bag cured unidirectional tape and woven laminates.
An improved understanding of the physical mechanisms that prevail during the oven/vacuum bag cure cycle have lead to lower void level laminates.

KEYWORDS; Composites, Prepregs , Low temperature cure.

1. INTRODUCTION

Almost since the beginning of modern day composites in the mid 1960's with the advent of carbon fiber, polymer based composite materials have been available in a preimpregnated (prepreg.) format. In this form the partly reacted (B-staged) resin is impregnated into the desired reinforcement , protected by paper or plastic sheeting and stored ready for use in roll form. Prepregs have proved to be a convenient building block from which composite structures can be readily laid up and cured to produce a desired structure. Balanced against the structural performance of the resulting structure is the cost. The cost is divided between the cost of the material (the prepreg.) , any non-recurring cost such as tooling and the labor necessary to build the part. The well documented superior specific mechanical performance of advanced composites has established their desirability in a wide range of applications and it has only really been their relatively high cost which has prevented their use in preference to 'conventional engineering' materials in all areas where they are technically suited. This has led understandably to investigations aimed at cost reduction without performance reduction. Some parties have moved away from prepreg. composite materials and expended their energy on techniques such as resin film infusion (RFI) , resin transfer molding (RTM) , filament winding etc. All have their cost merits which are typically related to size and geometry. Conventional epoxy prepregs. particularly for aerospace use fall into two broad processing bands ; those curing at 120° C and those curing at 175° C. Systems with these cure characteristics once established were replicated and data on the performance of the resulting cured materials and structures was determined and a number of the systems have since become 'qualified'. Working with a known quantity is always preferable to delving into the unknown especially when there are constraints to minimise cost and risk in a program. Low temperature curing prepreg. materials in the mid 1980's were a new concept that offered the potential for overall cost reduction for advanced composite parts but with very little supportive mechanical data to characterize their performance as compared to what were then well established high temperature cured composite materials.

1.1 Cost Effective Composite Structures. Low temperature curing prepreg. materials are no less expensive than conventional 175°C curing epoxy prepregs. They both consist of a base epoxy resin , a blend of curing agents and maybe some additional agents to affect flow or toughen the cured composite. All prepreg composite parts need to be molded against a tool surface. If the prepreg is required to be cured at 175°C then the mold tool material must maintain its integrity at that temperature. Typically materials operating at these higher temperatures are more expensive. If the number of parts to taken from an given tool is large then the amortized cost of the tooling is low. For new designs and prototypes the number of parts produced are frequently small in number and the proportional cost of the tooling is high. A study of costs for aerospace composite parts showed that tooling can account for 20% of the program cost for a production run of 250 parts. For a prototype program looking at only two or three parts tooling can account for 70% of the total program cost. If the tooling cost in prototype and development programs can be reduced it can make a significant impact on the economics of modern day composite part development. Adopting low temperature curing prepreg materials in the place of conventional higher temperature curing materials does allow the use of lower cost tooling materials and hence reduces the overall cost of the program.

1.2 Mechanical Performance . If lower temperature curing prepreg. materials returned the same mechanical performance as the established conventional systems and at lower overall cost they would have a larger proportion of the current market for such products. Toughness when expressed as a retention of compression strength after low energy impact is a key design parameter for aerospace primary structure. First generation low temperature curing epoxy matrices do not return compression after impact (CAI) figures commonplace in toughened 175°C cure systems. Conversely , the hot/wet compression performance of early low temperature cure materials were found to exceed those of established conventional 175°C systems. Programs have looked to improve the toughness of these systems without compromising any of their better qualities.

1.3 Logistics and Handling . Systems that will cure between room temperature and 80°C are generally regarded as low temperature curing systems. A material that will cure at room temperature will naturally have a limited out life – usable life outside the freezer. For the most reactive systems the room temperature out life can be as short as two days. For an industry used to six months out life at room temperature this has been considered as a potential problem when selecting these types of materials particularly for large or complex structures. Experience has shown that the limited out times rarely cause a problem and frequently improve materials handling practices across the board with greater attention paid to the age and state of all uncured prepreg materials. To help with the much larger structures a range of low temperature curing materials have been developed which require slightly higher initial cure temperatures but with correspondingly longer out time at room temperature.

1.4 Milestones in the Development of LTM Aerospace Structural Materials.
In 1975 the original low temperature curing system LTM 10 was developed for use in motor racing structures. Developments of the same system are used to this day for track side repairs and for higher service temperature component parts. During the 1980's the LTM tooling system was widely established in Europe and North America. In parallel toughened low temperature curing prepreg materials were developed and used to manufacture marine and communications structures. These toughened systems were limited to maximum dry glass transistion temperatures of around 125°C , to low for many aerospace applications. In 1991 McDonnell Douglas , via the Low Cost Composite Processing program (LCCP) evaluated LTM 10 and LTM 22 for prototype structure manufacture. The LTM 22 system is a first generation toughened low temperature curing epoxy. In 1992 McDonnell Douglas manufactured an Astroquartz III/ LTM 10 radome which was subsequently fitted to the leading edge of a F15EF and test flown. At this time the decision was made to improve on the temperature capability of the LTM 20 series without compromising on their toughness – a marriage between the LTM 20 and LTM 10 series of materials. The LTM 45 prepreg resin system was the result of this work a system with improved toughness when compared to LTM 10 but with a comparable upper service temperature. Between 1993 and 1995 Lockheed Martin Skunk Works used LTM45 for the DarkStar Tier III UAV fuselage. Using LTM45 allowed them to use low cost tooling materials to manufacture the initial prototype fuselage. Made in two halves similar in shape to a clam shell the fuselage was a sandwich construction with woven carbon skins either side of a Nomex honeycomb core. The DarkStar was a joint program with the major design and build packages divided between Lockheed Martin and Boeing. Lockheed Martin made the fuselage and made good use of low temperature curing component materials. Boeing were responsible for the wings which they designed and built using a conventional aerospace prepreg curing at 175°C. Significantly the wing skins were molded on LTM12 carbon tools which reduce the cost of conventional composite tooling by requiring only two processing steps between master model and part. Both McDonnell

Douglas and Lockheed Martin had determined independently that the LTM 40 and LTM 10 series materials were displaying good hot/wet property retention. In 1995 Aurora Flight Sciences and NASA selected LTM 25 for the high altitude experimental UAV Theseus. In the same year the Israeli UAV Hermes 450 selected LTM26-EL (an extended out life version of LTM 26) for the manufacture of the fuselage. In 1996 McDonnell Douglas and NASA selected LTM 10 as the skinning material for the X-36 highly agile scaled demonstrator.

The potential use of low temperature curing structural materials has not been lost on the Space industry. In 1993 a low temperature curing cyanate ester prepreg material was developed and given the designation LTM 110. Satellite structures have a high demands on stability and cyanate ester based materials are widely used because of their reported low microcracking and low moisture uptake related distortion. The additional feature of low temperature cure effectively reduces the internal strains present from the initial cure . In 1996 NASA approved LTM 110 for use aboard the Space Shuttle. Many of the top floor mounted structures on the GPS III satellite designed and built by Boeing Space Systems used glass and astroquartz LTM 110 materials in their construction. In 1996 Boeing qualified the use of LTM45-EL for use on the Delta Launcher. In 1997 Orbital Sciences and NASA selected LTM 45-EL for the X-34 Reusable Launch Vehicle.

In all of the above listed applications the potential for program cost savings was balanced against any reductions in mechanical performance compared with conventional 175°C cured prepregs. The most widely used system at this time is LTM45-EL which exhibits the characteristics listed in table 1.

Table 1

Out Life at 21° C (70°F)	3-4 days	5-6 days
Minimum Initial Cure Temperature	50° C	50° C
Overnight Initial Cure Temperature	60° C	60° C
Tg After 140°F (60°C) Cure	75° C	75° C
Tg After 350°F (177°C) Postcure	180° C	180° C
Tg After 390°F (199°C) Postcure	>200°C	> 200° C
Oven/Vacuum Bag Moldability	Excellent	Excellent
Autoclave/Press Moldability	Excellent	Excellent
0°F (-18°C) Storage Life	>6 Months	>6 Months
Honeycomb Bondability	Adhesive Required	Adhesive Required
Foam Core Bondability	Yes	Yes
Solvent Content	Nil	Nil

With LTM45-EL as a baseline improvements in the following specific areas were targeted ;

1. Improvements in the oven/vacuum bag laminate quality for non-optimized layups. Earlier work had noted significant variations in the void content of laminates with the same resin and cure processing but different fiber architecture.

2. Improvements in the mechanical performance particularly hot/wet compression.

3. Improvements in toughness related properties e.g. compression after impact.

1.5 Out of Autoclave Processing. LTM45-EL is a resin system capable of being processed in the autoclave or under oven/ vacuum bag curing only. Autoclave processing at low temeperature but high pressure still allows for the use of lower cost tooling materials – the most signficant area for cost reduction in a typical low volume composite structure program. However there remain many instances when it is advantageous to consider out of autoclave processing – for very large structures, for prototype structures and for material characterization programs. These structures demand high quality , low void level laminates like the ones readily achievable when autoclave processing. Greater care and a fuller understanding of the constituent materials is required when only oven/vacuum bag curing to achieve the very optimum results.

2.0 EXPERIMENTAL

2.1 The Significance of Rheology. The greatest potential use for low temperature curing materials is for structures manufactured out of the autoclave. In looking to improve on the mechanical performance of the LTM45-EL system it was important that this is not done at the expense of the handling and processing characteristics where they are key to manufacturing cost savings. Early work on the first generation of low temperature molding materials had indicated that significant variations were possible in the void contents of laminates with different lay-ups despite identical cure processing. Void contents were high enough with some reinforcement configurations to be in the regime where a clear correlation was possible between void content and compression strength. *Typical compression failure in a laminate will start from localized fiber buckling at points within the laminate lacking lateral support – such as the volume local to a void(s).* A key aspect of this development work was to formulate a resin and impregnation process that would work with a wide range of fabrics and cross-plied tape layup configurations. The resulting system was designated LTM45-1. One of the key physical differences between this formulation and LTM45-EL is the room temperature viscosity the comparative values for which are shown in table 2. The viscosity of LTM45-1 is several orders of magnitude higher than LTM45-EL at room temperature while at a typical processing temperature of 60°C the viscosity's are of a similar order. Table 2 also includes comparative void content results for comparative LTM45-EL and LTM45-1 panels manufactured under oven and vacuum bag only processing conditions.

Table 2

Resin Type	Isothermal Viscosity (Poise)		Dynamic Viscosity (Poise)		Voidage % v/v, 0°/90° Unidirectional laminate, oven/vac bag cured	0° Compression Strength Mpa (Ksi)	Comments
	21 C	60 C	21 C	60 C	60 C		
LTM45EL	332K	265	332K	250	1% – 3%	1241 (180)	Baseline system
LTM45-1	2.02M	444	2.02M	310	0 – 0.29%	1476 (214)	Blending of reactive toughener – 4 HR @ 80 C. Outlife > 6 days at 21°C

The panels were laid up from cross plied tape ,a configuration which is most difficult to manufacture void free in LTM45-EL without the assistance of augmented pressure. The LTM45-1 panels returned consistent mean void level values of less than 0.5% while the LTM45-EL panels had mean void levels ranging up to 3%. The lower void content LTM45-1 panels returned a mean compression strength some 17% higher than that achieved with the baseline LTM45-EL panels.

LTM 45-1 is handled and processed like any other LTM resin system requiring an initial low temperature cure between 20°C and 80°C followed after demolding by a freestanding postcure. The postcure is required to complete the reaction and develop the glass transistion temperature Tg. As with all LTM resin systems the Tg is intended to step ahead of the postcure temperature.
The initial cure used for the manufacture of the test panels consisted of a 0.5°C /minute ramp from room temperature to 60°C followed by a dwell for 16 hours. The minimum vacuum permitted on cure was 25" Hg. The panels were postcured freestanding in an oven ramping from room temperature to 177° C at 20°C /hour. The oven temperature was held at 177°C for 2 hours before cooling.

2.2 Flow Mechanism during Cure. Modeling the behavior of any resin under any given set of curing conditions can be quite complex. In general for most composite structures the aim is to produce a cured continuum of resin with an even distribution of a known amount of reinforcing fiber. There is a window of time from the start of the cure up to the gel point (when the resin starts to resist applied shearing forces) when entrapped air not previously removed by debulking can be removed from the curing laminate. Under the augmented pressure in an autoclave much of the air can be driven into solution, however this is not the case when curing with an oven and vacuum bag. There needs to be a physical escape path and sufficient time before gelation with the limited force available to permit the removal of entrapped air. In LTM 45-1 the higher initial viscosity in the early part of the cure keeps open a path along the fiber tows and between the prepreg plies which allows the air to be evacuated. In systems with lower initial viscosities these pathways are closed off early in the cure cycle and any remaining air must travel through the resin as bubble in solution. If the resin system drops dramatically in viscosity and remains in that state for a sometime before gelation it is viable method for air removal but one which is more prone to being effected by the format of the reinforcement. Certain fabrics and stacking sequences return better results than others and this was clearly seen in the experimental work on LTM45-EL where cross-plied tape and some satin weave reinforcements returned significantly higher void contents (up to 3%) compared to twill weaves and laminates containing both tape and fabric reinforcement (0.5%). When developing a cure cycle for a composite component it is quite common to incorporate an intermediate dwell to improve on the surface finish of the part. This is effectively giving the laminate enough time to flow and remove any entrapped air which if successful we can see as bubbles in the cured resin flash. This mechanism is affected by the size and geometry of the part and the format of the reinforcement. The principal mechanism at work with LTM45-1 and rheologically similar systems for air removal will be less affected by size and shape and offer greater potential for the consistent high quality larger out of autoclave structures expected in the future.

2.3 The Effect of Out Time on the Quality of Out of Autoclave Cured Laminates. From
the foregoing it will be evident that the change in the rheology of the resin with time
during cure is crucial to achieving low voidage in the final laminate. Low temperature
curing resin systems are inherently highly reactive this being the characteristic which is
key to saving cost in low volume and prototype applications. A corollary to high reactivity
is a relatively short out life at ambient temperatures. LTM 45-EL has a work life at 21°C of
5 to 6 days. During this time the material remains pliable and can be cut, flexed and fitted
to the desired shape of the part under construction. Beyond its work life the material
becomes 'boardy' and difficult if not impossible to laminate with. In this state if subject to
an elevated temperature and consolidated with a minimum of a vacuum bag it may be
possible for the resin to flow and cure. If the material is left at room temperature (21°C)
for an extended period of time no flow or acceptable cure will be possible even with the
application of heat and pressure. These three states in-life and usable, out of life and
curable if already laid-up and unusable are common to all resins and prepregs with the
distinction for low temperature curing systems that the time-scales between states are much
reduced. Experiments were conducted to establish if the quality of the laminates in terms
of void content were significantly effected by using in-life 'fresh' material (0 days at RT)
and in life 'aged' material (4 days at RT). A series of large panels were made and for a
typical cure cycle (ramp to 60°C and hold) a variation in the mean void content between
'fresh' and 'aged' material was evident.

For cross plied tape laminates the fresh material returned mean void levels of 0.54% while
the 4 day aged material dropped to a mean void level of only 0.07%. The older material
will have a higher initial resin viscosity and this may be explained allow air removal under
the applied vacuum for a longer period of the cure compared with the fresher material.
Despite the difference both mean void levels are respectable and in a practical laminate it
would be unlikely that any difference in the shear and compression strengths would be
detectable.

Experiments conducted on 370 gsm 6k 5 harness satin carbon fabric reinforced mimicked
the results found with the unidirectional tape cross plied laminates **except** the variation in
void level between 'fresh' and 'aged' material was much greater. The fresh material had a
mean void content of 2.73% after oven/ vacuum bag cure at 60°C while the four day aged
material had a mean void content of only 0.38%.
 This particular fabric style was used in early years when developing the LTM45-EL resin
system. At that time it proved difficult to consistently achieve low void contents with this
fabric style while similar areal weight twill weaves did not prove to be a problem. The
results of the experiment described above present us with a possible explanation. In
conclusion the 5HS fabric architecture is such that to achieve void levels well below 1%
under oven /vacuum bag only processing when curing at 60°C a higher initial resin
viscosity is required. The wider more general point is that the result of this experiment
emphasizes that greater knowledge of both resin and reinforcement is required when
considering oven/ vacuum bag processing to produce high quality laminates. It is worth
noting that all of the systems , LTM 45, LTM45-EL and LTM45-1 when subject to an
autoclave cure at 60°C returned void free laminates.

2.4 Variations in theLevel of Impregnation. Clearly from the observations above
variations in the resin rheology with time have a dramatic effect on the quality of the
resulting laminate. We can conclude that each reinforcement/lay up may need a slightly
different dynamic rheology profile or a different vacuum bag cure cycle to achieve
repeatable optimum results. Tailoring each resin system rheology to each reinforcement
and/or developing an optimum vacuum only cure cycle would be a costly exercise. One

practical solution to the problem is to consider partial impregnation on reinforcements which trials have shown require a longer time at higher viscosity to completely evacuate any entrapped air. Trials were conducted on the 5 harness satin weave which when fully impregnated with 'fresh' material produced a consistently high void level. The prepreg used for these trials were coated with a resin film on one side only. The other side being free of resin presents a multitude of pathways for the air to escape by evacuation when the vacuum is applied. Under the applied vacuum the resin completes the impregnation of the reinforcement and in the early stages of the cure any trapped air is pulled from the laminate along the dry fiber tows ahead of the impregnating resin. Results with the 5HS with 'fresh' material which had previously achieved void levels of nearly 3% were reduced on average to one tenth this figure – 0.33%. This result was encouraging as it demonstrated that optimum vacuum bag processing could be a achieved with two different reinforcements by changing the manufacturing parameters of the prepreg. rather than the resin formulation.

The trial was completed by molding a series of panels of 'aged' partially impregnated material. The result was a sharp rise in the mean void level to around 1% possibly indicating that the optimum processing window had been exceeded in this case by the resin not achieving the minimum viscosity required to flow adequately to wet out the fibers.

2.5 Hot/Wet Mechanical Performance. LTM materials typically exhibit good strength retention when tested hot after wet conditioning. LTM45-1 was no different in this respect showing a marginal increase in interlaminar shear strength when compared to LTM45-EL and tested at both room temperature and 120°C. Oven / vacuum bag cured LTM45-1 has very similar performance to autoclave cured 3501-6 with slightly better hot/wet strength retention.

2.6 Toughness . An attempt was made to improve on the toughness of LTM45-1 without detracting from its high service temperature performance. Toughness was evaluated by determining the compression after impact performance using the SACMA method. The compression strength was determined after impacting the laminates with an energy of 6.6 Joules/mm. All of the unidirectional tape laminates were reinforced with HTA carbon fiber and again autoclave cured 3501-6 was used as a comparative control. CAI results for LTM45-1 and 3501-6 were of the same order of magnitude ranging between 140 to 150 MPa. Improvements in toughness remains a key area for future development.

4. CONCLUSIONS

4.1 To manufacture the highest quality out of autoclave laminates the prepreg material specification must address format as well as the constituents of the prepreg material.

4.2 Void contents <<1% are achievable for oven/vacuum bag cured unidirectional tape or woven laminates.

4.3 LTM45-1 has improved on the RT/dry compression strength of LTM45-EL by 17%.

4.4 LTM45-1 shows improvements in hot/wet mechanical performance

4.5 LTM45-1 mechanical properties and toughness are comparable with or better than those of first generation 175 °C curing structural systems.

Cycom® X5215 -- An Epoxy Prepreg that Cures Void Free out of Autoclave at Low Temperature

Guo Feng Xu, Linas Repecka, Jack Boyd
CytecFiberite Engineered Materials, Inc.
1440 N. Kraemer Blvd, Anaheim, CA 92806
(714)666-4370

ABSTRACT

It is difficult to achieve void free composite parts in a low temperature (65 °C) and vacuum bag only cure process. The mechanical properties of non-autoclaved, low temperature cured composites are usually inferior to high pressure, 177 °C, autoclave cured composites. This paper presents a newly developed material, CYCOM® X5215, which cures void free at low temperature using vacuum bag only processing. A typical cure cycle for this prepreg is a cure at 65 °C (150 °F) under vacuum bag only pressure for 14 hours followed by a freestanding postcure at 177 °C (350 °F) for 2 hours. Both cured UD tape and woven laminates show very low void content (<1%). This composite also possesses mechanical properties equivalent to epoxy/carbon fiber composite materials cured at 177 °C in an autoclave. Details such as void performance, mechanical properties, bagging procedure, handling and out life are discussed.

KEY WORDS: Vacuum Bag, Low Temperature, Composites

1.0   INTRODUCTION

Composite laminates are widely used in the aerospace industry. The advantages of these materials are their high strength and low weight. However, the manufacturing cost of composite parts is high. Standard manufacturing processes require a high pressure, high temperature autoclave, and high temperature tooling that are expensive. This is because composites cured under these conditions are normally void free. The high pressure minimizes the amount and size of any trapped air, resulting in low void parts. Further, high temperature and high pressure processing allows use of the widest variety of epoxy resins.

It is often desired to manufacture a part without using an autoclave. For example, the size of the part may be too big for an autoclave, or in low rate manufacture, low cost tools made of wood or low Tg polymers would be desirable.

When these tools are used, composite parts can only be cured at low temperature and low pressure. For these reasons, a vacuum bag only low temperature curable prepreg would fill an important need in the industry.

Cycom®X5215 was developed to meet this need. It is curable at low temperature and under vacuum bag pressure. A typical cure cycle is 14 hours at 65 °C in a vacuum bag followed by a freestanding postcure at 177 °C for 2 hours. Also it has the mechanical performance of a first generation 177 °C cure epoxy, such as Rigidite®5239-1 or 5208. Cycom®X5215 is a no bleed system and is available as unitape or woven carbon fiber prepregs. Both woven and UD tape prepregs cure void free (<1%) and the out life is a minimum of 10 days. This prepreg can be used for affordable tooling, rapid prototyping and low rate manufacture.

2.0   EXPERIMENTAL

2.1   Materials
Cycom®X5215 is an epoxy prepreg. G30-500 carbon fiber from TOHO Japan was used. Unidirectional tape prepreg has areal fiber weight of 145 g/m^2 and resin content of 36% by weight. Woven material is G30-500-5HS-6K with areal fiber weight 373 g/m^2 and resin content of 40%.

2.2   Laminates Manufacture Procedure

Bagging Procedure
The material is cured using a standard vacuum bag procedure (see Figure 1).

Cure Cycle
All the laminates are cured using an initial cure on the tool followed by a free standing postcure. These cure conditions are described as follows:

<u>Initial Cure</u>
Under vacuum (absolute pressure ≤12 mmHg), increase the temperature at 2 °C/min to 65 °C. Cure under vacuum at 65 °C for 14 hours. Decrease temperature at 3 °C/min to room temperature.

<u>Freestanding Postcure</u>
Place the sample in the oven, increase the temperature at 2 °C/min to 177 °C. Cure at 177 °C for 2 hours. Decrease temperature at 3 °C/min to room temperature.

All cures were carried out in an ordinary air circulating oven.

2.3   Testing

Void Performance Evaluation
The void content of the cured laminates was examined using C-scan. A lead calibration standard was used.

Void performance was also examined using photomicrographs. The samples were from the center area of the cured panels. This is because the center area is furthest from the edge breather, so if air gets trapped in the panel, the voids will be greatest in the center. The panel cross sections were polished and examined with a microscope.

Tg Measurement
Glass transition temperature (Tg) was measured using DuPont 983 Dynamic Mechanical Analyzer. The Tg data reported here is based on the tangent intercept from the storage modulus curve.

Compression Strength After Impact (CAI) Test
A Dynatup Model 8200 instrumented impact machine was equipped with a 1.58 cm diameter hemispherical tip impactor weighing 5.155 Kg which was set at a height to achieve a target impact energy level of 6.675 KJ/m (1,500 in-lb/in) of thickness. The test coupon was prepared in accordance with SACMA SRM2R-94 and placed on an impact support base which is made of steel. After the impact, the residual compression strength was measured.

Short Beam Shear Test
A three point loading test fixture with 0.633 cm diameter loading nose and 0.316 cm diameter support noses was used. The test was carried out using ASTM D2344 method. Coupon size is 2.53 cm x 0.633 cm (1"x0.25") with span of 4:1 panel thickness. For UD tape prepreg, panel was laid up using $[0]_{16}$ configuration. For cloth prepreg, panel was laid up using $[0]_8$ configuration.

0°-Compression Strength Test
0°-Compression was carried out using ASTM D695. For both UD tape and woven prepregs, panels are laid up using $[0]_8$ configuration. Metalbond™ 1515 adhesive was used to bond the tabs to the laminates.

Resin Viscosity Measurement
Resin viscosity was measured using a rheometrics RDA instrument.

3.0   RESULTS AND DISCUSSION

Cycom®X5215 was developed to cure at 65 °C/14 hours, and then have enough green strength to free stand postcure at 177 °C/2 hours. The cure is done under a vacuum bag and in an air circulating oven. This allows the use of low cost polyurethane or wood tools.

3.1   Tg and Degree of Cure
Table 1 shows the Tg and degree of cure of Cycom®X5215 after initial cure and postcure cycle. As can be seen, after the initial cure, the material has 61% cure and Tg of 63 °C. This level of cure offers sufficient mechanical strength for a freestanding postcure. After the postcure, the laminates did not show any distortion, and the Tg reached 191 °C.

Table 1. Tg and Degree of Cure of Cycom®X5215.

Property	Neat Resin
Time to double viscosity at 24 °C	>10 days
Degree of cure after 14 hours at 65 °C	61%
Degree of cure after postcure	90%
Tg after cure at 65 °C for 14 hours	63 °C
Tg after postcure at 177 °C for 2 hours (dry)	191 °C
Tg after postcure at 177 °C for 2 hours (wet)*	163 °C
* Wet denotes the coupon exposed to boiling water for 48 hours.	

3.2. Void Performance

In vacuum bag pressure, low temperature cured composite laminates, high void content is always a problem. In the case of no bleed cure, the low void part manufacture becomes even more difficult. These voids are caused by trapped air in both intralaminar and interlaminar areas. Using a standard prepreg, the void content level is above 3% or even higher in a vacuum bag only process. Figure 2 shows the results when a standard prepreg was cured in a vacuum bag only process. The void content is higher than 5%.

Cycom®X5215 cured under the same conditions has a very low void content (<1%). The void content was evaluated on 61 cm x 61 cm size panels. For the unidirectional tape panels, the lay-up configuration was $[0, 90]_{10S}$. This lay-up configuration is considered as the least nested of any laminate construction, and therefore presents the greatest challenge to make a void free part.

C-scan examination showed both UD tape and cloth panels were void free. The low void content was further shown by the cross-section micrograph. Since the middle area of the panel is normally the worst, the cross section area for this study was from the middle portion of the panel. Figure 3a shows the micrograph of the cross section of cloth panel. It can be seen that the panel is void free. Figure 3b is the cross section micrograph of unidirectional tape panel. It too is void free.

3.3 Out-life

Low temperature cure prepregs normally do not have long out-life. The prepreg can lose tack and the resin viscosity can increase at room temperature. High resin viscosity can prevent the prepreg from consolidating, resulting in voids.

Cycom®X5215 resin stability at room temperature was tested. After the resin was staged at room temperature for various lengths of time, the viscosity was measured. Figure 4 shows the resin viscosity at 65 °C, where part consolidation and initial cure takes place, does increase with time at room temperature.

However, the viscosity has less than doubled in 11 days.

The out-life cannot be assessed by viscosity and tack alone. Panels must be built and the void performance determined.

The prepreg out-life was assessed by stagging both UD tape and woven prepreg at room temperature (below 25°C) for 10 days. The tack was virtually unchanged. Panels were then laid-up and cured using the staged prepreg and the standard cure cycle. The panels were 61 cm x 61 cm with a configuration of $[0,90]_{10S}$ for tape, and $[0, 45, 0, 45, \underline{0}]_S$ configuration for fabric. The C-scan and photomicrograph tests showed the panels to be void free.

3.4 Mechanical properties

The mechanical properties of Cycom®X5215 are equal to first generation 177 °C cure epoxy/carbon fiber prepregs. The tensile, compression, SBS and Tg all fall within a range typical for a 177 °C system.

The CAI value of Cycom®X5215 was tested side by side with Cycom®5239-1 which is a typical 177 °C cure first generaton autoclave cure epoxy/carbon fiber prepreg system of CytecFiberite. Under the same testing conditions, both prepregs show the equivalent CAI (Figure 5), both on UD tape and woven.

4.0 SUMMARY

CytecFiberite has developed Cycom®X5215 which is designed for 65 °C cure on low temperature tools using a vacuum bag only. The mechanical properties after a 177 °C post cure are equivalent to first generation, 177 °C cure epoxies. For a low temperature cure material, the prepreg has an outstanding out life of 10 days. The parts manufactured under vacuum bag only, no bleed conditions are virtually void free. This prepreg offers the designer the capability of rapidly and inexpensively developing prototype parts that have all the mechanical properties of a standard epoxy composite.

BIOGRAPHY

Guo Feng Xu is currently a chemist at CytecFiberite. He received his Ph.D in Organic/Polymer Chemistry from Texas A&M University in College Station, TX in 1994. He also served as a Post Doctoral Fellow at Southern Methodist University in Dallas, TX before he joined CytecFiberite.

Linas Repecka is a Lead Engineer for manufacturing improvement and development for CytecFiberite. He has a B.S. in Chemical Engineering from the University of California at Berkeley and has over 15 years of experience in improving and developing new materials and processes for the manufacture and application of advanced composites and structural adhesives. He is a long time active member of SAMPE and has served in several positions including Chairman of the Orange County Chapter.

Jack Boyd is currently a Senior Chemist at CytecFiberite. Previously he worked as an R&D Chemist with Occidental Research Corporation. He received his Ph.D in Organic Chemistry from the University of Utah in 1977 and then served as a Post Doctoral Fellow at the University of California, Los Angeles.

Test	*Initial Requirement	UD Tape	*Initial Requirement	Cloth
Tg (DMA), °C (°F)				
(Dry)	177 (350)	192 (377)	177 (350)	192 (377)
[a](Wet)	121 (250)	163 (325)	121 (250)	166 (330)
Tensile Strength GPa (Ksi), RTD	1.379 (200.0)	2.048 (297.0)	0.517 (75.0)	0.655 (95.0)
[b]SBS , MPa (Ksi)				
RTD	103 (15.0)	119 (17.3)	68.9 (10.0)	72.4 (10.5)
121 °C (Dry)	62 (9.0)	74.5 (10.8)	55 (8.0)	54 (7.9)
[c]121 °C (wet)	52 (7.5)	55 (8.0)	38 (5.5)	42 (6.1)
[d]0° Compression, GPa (Ksi)				
RTD	1.413 (205.0)	1.631 (236.6)	0.6895 (100.0)	0.674 (97.8)
121 °C (Dry)	1.000 (145.0)	1.532 (222.2)	0.586 (85.0)	0.606 (87.9)
[e]CAI, MPa, (Ksi) 6.675 KJ/m (1,500 in-lb/in) impact energy		126 (18.4)		161 (23.3)

*This initial requirement is generated based on the typical 177 °C (350 °F) cure epoxy composites.

[a]Wet denotes 48 hours exposure in boiling water.

[b]Lay-up configuration of UD tape is $[0]_{16}$. Lay-up configuration of cloth prepreg is $[0]_8$.

[c]Wet denotes 24 hours exposure in boiling water.

[d]Lay-up configuration of UD tape is $[0]_8$. Lay-up configuration of woven prepreg is $[0]_8$.

[e]Lay-up configuration of UD tape is $[45, 0, -45, 90]_{4S}$. Lay-up configuration of woven prepreg is $[45, 0]_{3S}$.

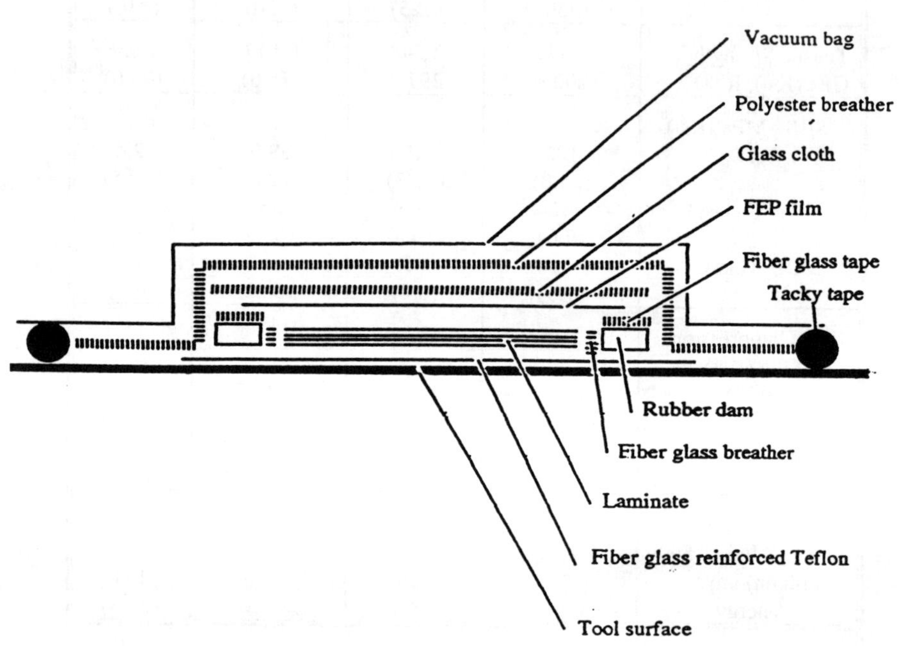

Bagging Scheme

Figure 1. Bagging procedure for Cycom®X5215.

Figure 2. Cross-section micrograph of a standard low temperature cure epoxy UD tape laminate.

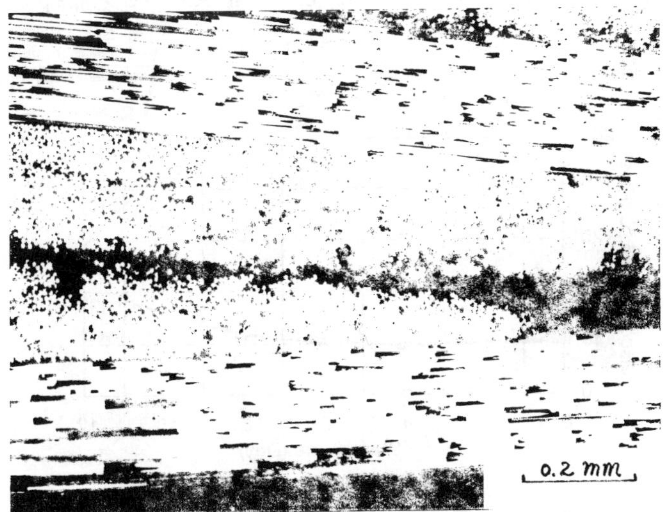

Figure 3a. Cross-section micrograph of Cycom®X5215 woven prepreg laminate.

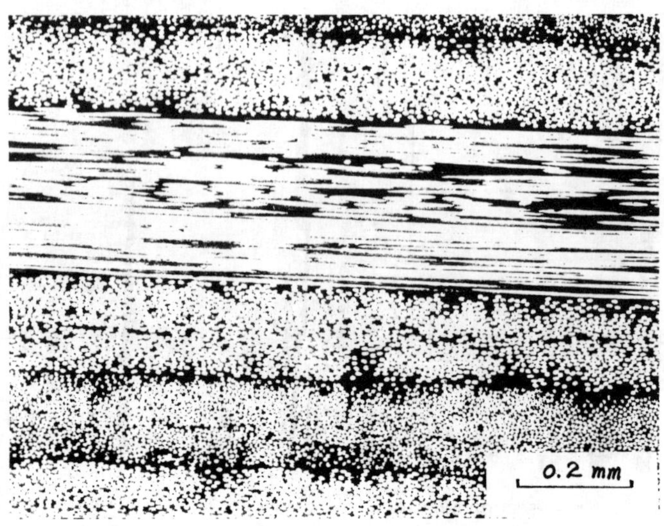

Figure 3b. Cross-section micrograph of Cycom®X5215 UD tape prepreg laminate.

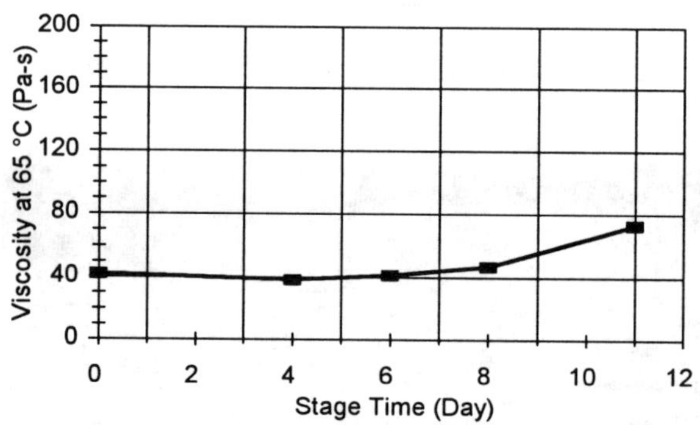

Figure 4. Viscosity of Cycom®X5215 resin at 65 °C changes with the stage at room temperature.

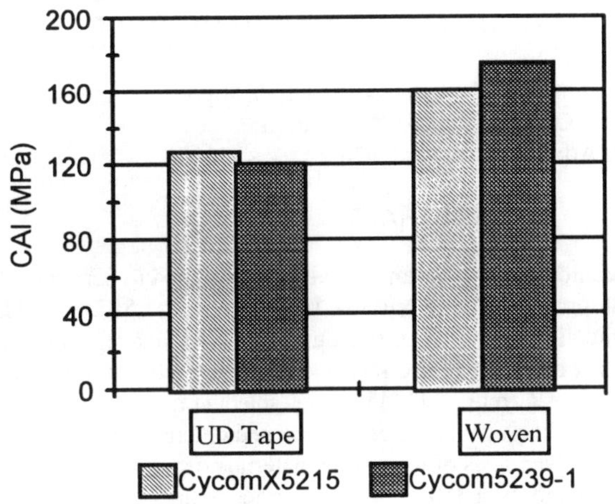

Figure 5. CAI of Cycom®X5215 equivalent to first generation epoxy.

MATERIALS CHARACTERIZATION AND JOINT TESTING ON THE X-34 REUSABLE LAUNCH VEHICLE

Dr. Thomas L. Dragone & Patrick A. Hipp
Orbital Sciences Corporation
21700 Atlantic Boulevard, Dulles, VA 20166

ABSTRACT

The X-34 Reusable Launch Vehicle is an experimental vehicle designed to demonstrate technologies vital to the development of future reusable launch vehicles. Stringent demands are placed on this all-composite airframe for cost, schedule and performance. Balancing these demands has resulted in the use of the LTM45EL resin system, a low-cost toughened epoxy produced by Advanced Composites Group. LTM45EL was selected from a range of composite prepreg materials including cyanate esters, toughened epoxies and non-toughened epoxies because it provided the advantages of lower raw material cost and better processing flexibility. In addition, it also maintains a 350°F *(176°C)* service temperature (necessary for a Mach 8 reentry) and sufficient structural properties for use in this application. The material has been fully characterized at room and elevated temperatures for strength and stiffness in tension, compression, shear and flexure, as well as for bearing and sandwich properties. Full qualification test results will be presented as well as characterization of the strength of bolted and bonded joint assemblies. The X-34 design utilizes the material in both multi-part bolted assemblies as well as single part bonded assemblies to provide an optimal combination of low-cost initial fabrication with long-term low-cost operability.

KEY WORDS: Composite Structures, Aerospace / Aircraft, Mechanical Properties

1. BACKGROUND

The X-34 Reusable Launch Vehicle (RLV) Technology Demonstration program is a joint industry/government project to develop, test, and operate a small, fully-reusable vehicle with the objective of demonstrating technologies and operating concepts applicable to future RLV systems. Orbital Sciences Corporation is developing the X-34 (shown in Figure 1) as an air-launched, liquid-fueled vehicle that draws heavily on Orbital's Pegasus and Taurus heritage and incorporates many RLV technologies including an all-composite primary airframe structure, composite propellant tanks, SIRCA leading edge tiles, and autonomous flight control with safe abort capabilities. The X-34 is carried uprange by Orbital's Pegasus Carrier Aircraft, can fly trajectories that reach Mach 8 and 250,000 feet. lands horizontally on a conventional runway, and is quickly readied for subsequent flights using aircraft-style ground operations. The first flight structure is being fabricated and assembled at Orbital's Dulles Integration and Test Facility. System testing will begin later this year with captive carry flight tests commencing on the L1011 Carrier Aircraft in mid 1998 and powered flight of the X-34 scheduled for early 1999.

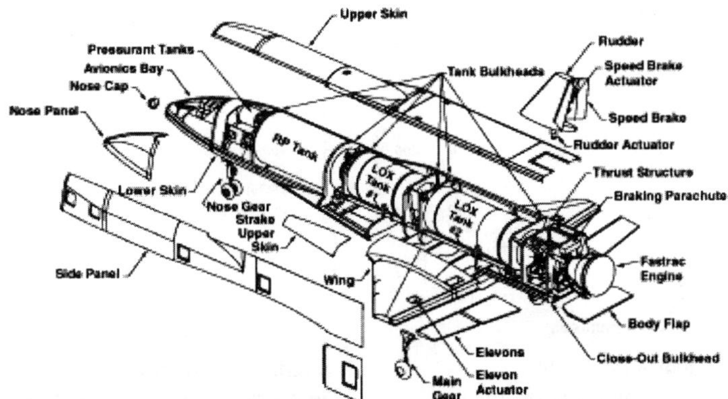

Figure 1 - X-34 expanded view indicating primary vehicle components.

2. MATERIALS REQUIREMENTS

Critical requirements for the selection of an airframe material for the X-34 RLV included:
- Use of graphite /epoxy composite material,
- 350°F service temperature capability for limited duration,
- Limited multiple use (100 mission design life),
- Processing flexibility,
- Low cost.

The use of a graphite/epoxy composite material in the primary airframe (as opposed to aluminum or titanium) was strongly encouraged by NASA as it sees these materials as necessary to achieve the vehicle mass fraction needed for future full-scale RLVs. Nearly 95% of the X-34 airframe is designed using graphite/epoxy. Since the X-34 is an experimental vehicle, there is no specific requirement for using a particular resin or fiber system that has an extensive flight heritage.

The thermal environment for the X-34 vehicle structure is somewhat unique in that the ceramic thermal blankets and tiles that cover the vehicle create a significant lag between the time when the peak heat flux is applied to the vehicle and the time when the structure reaches its maximum temperature of 350°F. Figure 2 indicates that the peak structural loads on the vehicle occur shortly after drop from the L1011 Carrier Aircraft, when the vehicle is cold-soaked below freezing as a result of the extended captive carry flight to the launch point. Maximum heating and secondary loads occur during entry when the structure has warmed to nearly room temperature again. This heat soaks into the structure and causes the peak structure temperature to occur several minutes later during terminal maneuvers, when the loading on the structure is much lower. Therefore, the structure material must be capable of withstanding temperatures of 350°F, but full retention of properties at temperature is not required. The limited design life of the X-34 vehicle (100 missions) means that fatigue properties are not a driving factor in material selection.

The 350°F temperature capability appears to be a local optimum in terms of overall vehicle design. In general, as the structure temperature capability is increased, the thickness and weight of the ceramic blankets and tiles that insulate the structure is correspondingly decreased. However, because of the heat soakback phenomenon described above, there is no simple linear correlation between the structure temperature capability and the thermal

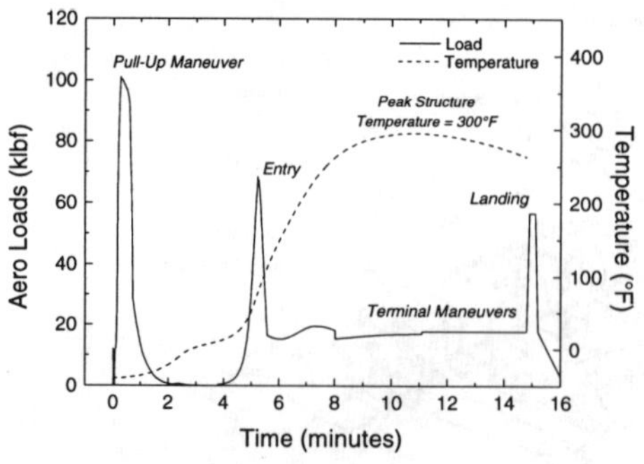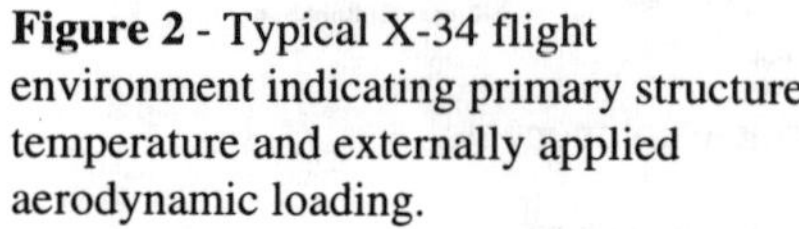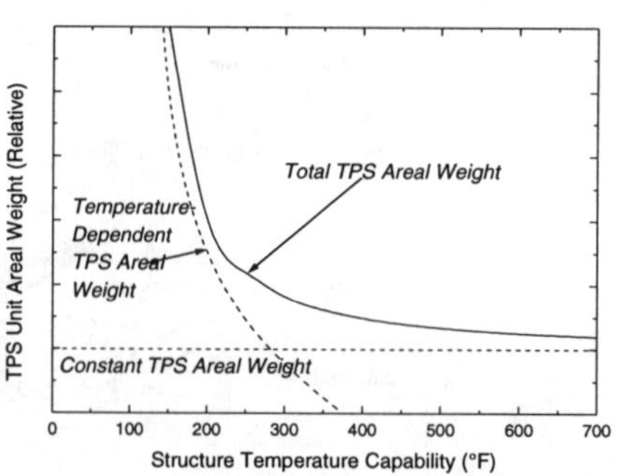

Figure 2 - Typical X-34 flight environment indicating primary structure temperature and externally applied aerodynamic loading.

Figure 3 - Effect of increasing structure temperature capability on total TPS weight indicating that 350°F is optimal.

protection system (TPS) weight. Rather, for the X-34 flight environment, there is a significant "knee" in the curve defining that relationship at about 350°F as shown schematically in Figure 3. Above this value, structure temperature capability does not decrease TPS weight because of certain constant "overhead" weights that cannot be reduced (i.e. stitching, protective coatings, and adhesive). Therefore, higher performance materials, such as bismaleimides or polyimides, are not appropriate for this vehicle.

The processing flexibility requirement flows from the rapid integration schedule requirements of the program. In order to meet this schedule, parallel fabrication of the vehicle has been coordinated between three separate composite structure vendors, for the fuselage, the wing and the control surfaces. The fabricators have different facilities and experience with composite manufacturing. Therefore, the material chosen must be capable of producing adequate mechanical properties by vacuum bag / oven curing as well as autoclave curing.

The program goal of low-cost access to space was one of the most important design drivers. In order to achieve the desired structure cost, all the design and manufacturing parameters needed to be considered, including raw material cost, tooling, processing, and assembly. Low material cost is defined as delivered composite part cost, including both the raw material cost as well as the associated costs of the tooling, processing and assembly. Since the X-34 is not an orbital payload delivery system, it does not have the extreme mass requirements of many other spacecraft; therefore high price, high performance materials are not required.

3. MATERIAL SELECTION

3.1 Resin Screening   An extensive screening evaluation was conducted early in the program to evaluate various matrix resin candidates against the material and process requirements for the X-34 vehicle. Candidate materials included cyanate esters (Fiberite 954-2A, YLA RS12, and Bryte BTCy-1), 350°F curing toughened epoxies (Fiberite 977 and 970) and low temperature curing toughened epoxies (Advanced Composite Group LTM45EL). Screening tests included uniaxial tension and compression strength and stiffness as well as shear strength and stiffness. These tests were conducted at room temperature and 350°F for the cyanate esters and the LTM45EL. Data was only available at room temperature and 250°F *(121°C)* for the toughened epoxies.

Some of the screening test results are shown in Figure 4. The material samples used in the screening tests had slightly different fiber properties, due to the availability of prepreg samples at the time the tests were conducted. For comparison purposes, the strength values have been normalized to the same nominal intermediate modulus M35J fiber properties: 50 Msi *(344 GPa)* fiber stiffness and 683ksi *(4710 MPa)* fiber strength. For a given material type (cyanate ester, for example) there were only minor differences between material suppliers in comparison to the differences between material types. Therefore, only representative properties of one material within a given material type are shown.

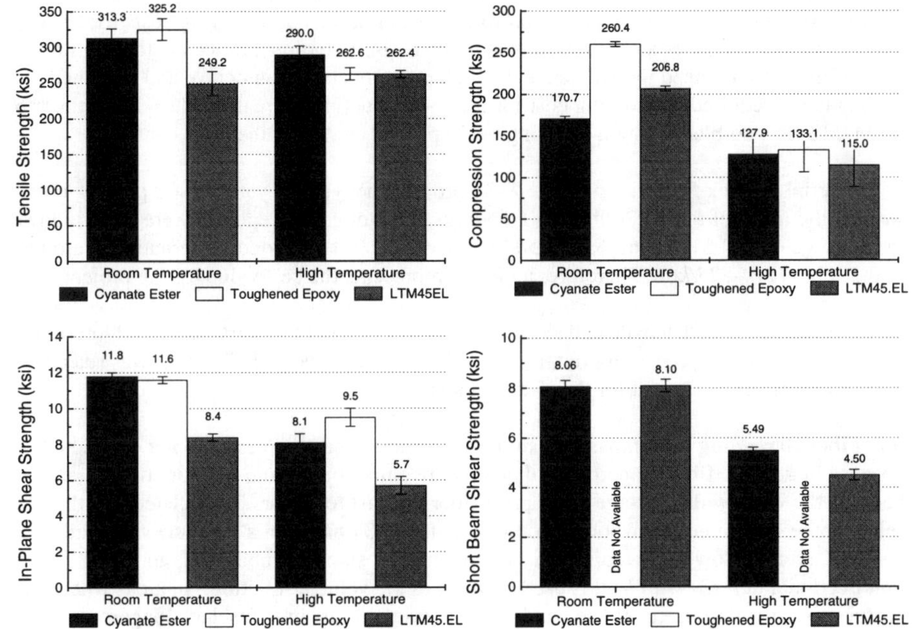

Figure 4 - Screening test results comparing cyanate esters to toughened epoxies to LTM45.

All materials showed relatively good retention of fiber stiffness with less than 3% difference in tensile or compressive moduli between material types. Therefore, stiffness was not a discriminator in material selection.

At room temperature, both the cyanate ester and the toughened epoxy composites showed substantially higher (as much as 30%) tensile and in-plane shear strength than the low-cost LTM45EL. On the other hand, LTM45EL had much better compressive strength properties than the cyanate esters tested, and this proved to be much more important in the selection process since a greater portion of the X-34 vehicle is sized by compressive strength (upper wing spar caps and upper fuselage longerons, for example). Room temperature interlaminar strength (as indicated by the short beam shear strength results) were comparable between the cyanate esters and the LTM45EL.

As expected, the cyanate esters showed the greatest retention of strength at high temperature. The toughened epoxies also had good retention of properties to 250°F, but fell off significantly above that mark. Therefore only the 250°F results are shown. LTM45EL showed acceptable retention of properties at 350°F, losing only 45% in compression, 30% in shear, and actually increasing in tensile strength by 5%. This has been attributed to additional cross-linking in the matrix at the test condition that leads to higher matrix properties and better fiber-matrix bonding.

While LTM45EL did not meet the mechanical performance of some of the other prepreg materials screened, it did offer significant processing advantages. Unlike most 350°F curing resin systems that require an autoclave to achieve the desired quality and performance, the viscosity profile and reaction rates for the LTM45EL system do not require a high pressure curing process or a tight temperature control. In general, the non-recurring costs for prototype and low volume vehicles are closely linked to the required tooling . Since the tooling for LTM45EL does not need to experience the high pressures of the autoclave and hardened tooling is not required, the tooling cost for the X-34 structure is greatly reduced. Furthermore, LTM45EL is a no-bleed system allowing for improved control of the resin content.

For material prepreg forms used on the X-34 project, the typical cyanate ester prepreg costs were in the range of $90-$120/lbm *($41-$54/kg)*. The toughened epoxies were slightly lower in the range of $70-$100/lbm *($32-$45/kg)*. LTM45EL is the low-cost alternative at around $30-50/lbm *($14-$23/kg)*. Therefore, it clearly represents the best value for the project.

Based on its acceptable tensile and shear properties, comparable compressive and interlaminar properties, extensive processing flexibility, and exceptional value, LTM45EL was selected as the composite resin material for the X-34 vehicle.

3.2 Fiber Screening   Comparative studies of various fiber sources were not conducted because in general, fiber properties and quality are comparable from various sources, and because the worldwide shortage of graphite fiber the past few years has dictated that the fiber selection be determined by whoever could meet the X-34 program's schedule and availability needs. Given the low cost requirements on the project, standard modulus graphite unidirectional tape (34-700WD) is used on most of the vehicle structure. In areas where stiffness is critical to meet vehicle bending stiffness requirements (fuselage longerons and wing spar caps) intermediate modulus graphite unidirectional tape (T800WD) is used. For areas that are bolted or highly curved, standard modulus 2 x 2 twill fabric (CF302) is used.

4. MATERIALS USAGE

The low temperature initial cure and oven processing capability of the X-34 composite material has broadened the design options and has fostered more creative engineering solutions. Composite parts for the X-34 structure are provided by three separate composite fabricators: Vermont Composites (fuselage), Aurora Flight Sciences (wing) and R-Cubed Composites (rudder and elevons). The processing methods used to fabricate the X-34 structure was highly dependent on the structural design, the composite fabricator's experience and the available facilities. As a result , both oven and autoclave curing were extensively employed. Furthermore, the processing flexibility of LTM45EL allows for the largest structures to be cured in a single step in an oven instead of requiring a costly autoclave or performance-degrading splice joints. As a by-product of the low temperature initial cure of this material, the resin must be very reactive, and therefore has a limited 5-day out-life. This limitation has resulted in the requirement for very careful planning for the larger parts of the X-34 vehicle (the wing skins in particular) to accomplish all the required layup operations before the resin had reacted sufficiently to lose its tack and preclude good consolidation.

In all cases, part quality was dependent on the debulking process that were necessitated by the resin formulation. During hand layup of the larger structures, the entrapment of air between successive layers became a critical issue. Several methods of removing the air pocket were attempted with limited degrees of success. The only method of eliminating all of the trapped air was through debulking at regular intervals. If too many layers were applied without a debulk it became difficult to eliminate all of the pockets on the large parts.

In general, the manufacturer's recommended cure profile was used during part fabrication [1]. A slow temperature heat-up ramp and a long hold are used with constant vacuum for the oven process as shown in Figure 5. Similarly, autoclave cures used the cure profile shown in Figure 6. Both the autoclave and oven processes provided high quality, thin laminates. The only thick (0.60" or 1.5cm) oven-cured details were the wing spar caps. These were mostly comprised of unidirectional tape that consolidated well under vacuum pressure. Although the initial cure for the composite material can be achieved at a low temperature, an elevated cure/postcure must be utilized to raise the resin's glass transition temperature and obtain the higher operating temperatures. The post-cure can follow the initial cure immediately (as shown in Figures 5 & 6) or may be done at a later time in a free-standing condition since the part is fully gelled after the initial cure.

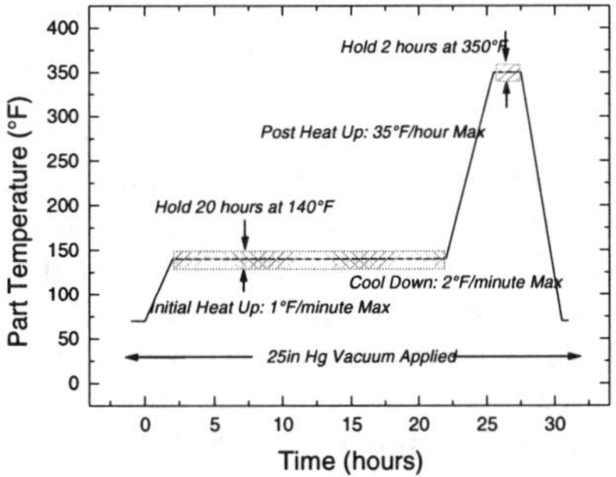

Figure 5 - Typical vacuum bag / oven cure cycle and post-cure heat treatment.

Figure 6 - Typical autoclave cure cycle and post-cure heat treatment..

The LTM45EL resin system provides the flexibility for part fabrication to be implemented using a variety of tooling techniques. For instance, the tooling for the wing skins was fabricated using graphite/epoxy prepreg material. Since the wing structure is of considerable size, (28' x 14' or 10m x 5m) and includes complex contours, matching the part and tool coefficients of thermal expansion was necessary. The composite face sheet of the tooling provided a sufficiently smooth outer mold line and provided adequate durability for the limited number of part being produced. The wing skin layup process is shown in Figures 7 and 8. The X-34 fuselage employed a combination of wood and composite tooling. On the other hand, the control surfaces were fabricated using aluminum and steel tools. In order to simplify the tools and lower tooling costs, hybrid tools were utilized. Split lines in the tools were masked using fairing compounds. The initial cure tooling for the wing structures and the control surfaces were also used for post curing. The mahogany wood tools used to fabricate the fuselage details did not have high temperature capability and as a result, a free-standing post-cure was employed. The hybrid tool approach is possible because of the flexible processing of LTM45EL. As a result, the cure profiles were tailored according to the type of tooling and the cure methods used for the individual structural details.

Figure 7 - X-34 wing lower skin lamination in progress.

Figure 8 - Lower wing skin core inspection prior to laminating the inner skin laminate.

As a further cost reduction method, extensive co-curing and co-bonding has been implemented through out the X-34 structure. A room temperature curing adhesive was selected which has an operating temperature to 350°F. The utilization of these joining techniques has been well documented [2] in reducing tooling cost and providing a more structural and higher quality joint. The wing and control surfaces predominantly use bonded joints. By using the parts themselves as the tool surface , in contrast to secondary bonding which requires a near perfect fit, co-curing/bonding provides a controlled bond line without prestressing the structure. As a result, secondary bonding is only used during the final close out of the wing and when planar surfaces are joined. However, extensive verification of the mating surfaces during the assembly sequences was employed to assure that a more controlled bondline was achieved during the secondary bonding. Even though the co-cured/bonded joint design has been identified as being more efficient, assembly and access requirements have forced the fuselage structure to have a large percentage of mechanical joints. Wherever possible, the mating surfaces were tooled to aid in dimensional control. Furthermore, the assembly plan incorporated the use of liquid shims for the majority of the bolted joints.

In order to reduce the number of structural parts in the vehicle, and consequently the number of joints, large sandwich structure skins were utilized. This design philosophy was also adopted to minimize the tooling cost and design time for a prototype vehicle as noted by Vosteen [3]. In general, standard sandwich construction techniques and guidelines were followed. Since core migration was a concern, the honeycomb panels containing transition ramps were fabricated with a ramp angle of less than twenty degrees. Similar fabrication techniques were documented by Hart-Smith [4]. For many of the vehicles internal structures, flat, honeycomb sandwich billets were fabricated and routed to shape. The exposed core of these panels were potted with epoxy filler material, as were many of the hardpoint regions used for attachments. Only in the most highly loaded areas were specially designed reinforcements incorporated.

The X-34 program has truly utilized the full flexibility of the LTM45EL resin system as indicated in Table 1. Various combinations of tooling approaches, cure cycles, post cure supports, and sandwich construction have been utilized by the three X-34 structure fabricators. When all these component parts have been delivered, Orbital will have a substantial database of experience, performance and cost from which to make conclusions about the true cost savings of using LTM45EL.

Table 1 - Composite Materials Usage on the X-34 RLV by Major Component

	FUSELAGE	WING	RUDDER*
Fabricator	Vermont Composites	Aurora Flight Sciences	R-Cubed Composites
Composite Parts	~100	~20	~10
Construction	Bonded / Bolted	Bonded	Bonded
Process	Autoclave	Oven	Autoclave
Honeycomb Cure	3-Step	1-Step	3-Step
Post Cure	Separate	Combined	Separate
	Free-Standing	On-Tool	On-Tool
Tooling	Steel / Wood	Composite	Aluminum

Typical of Elevons and Body Flap

5. MATERIALS CHARACTERIZATION

5.1 Laminate Qualification Testing   Materials qualification testing was conducted to evaluate LTM45EL's properties for use in design of the vehicle structure. Standard test methods were used to characterize tension (ASTM D3039), compression (ASTM D3410), in-plane shear (SACMA 8-77), short beam shear (ASTM D2344) and flexural properties (ASTM D790). The results are presented in Tables 2, 3, and 4 for the three material forms are used in the vehicle: standard modulus (SM) unidirectional tape (34-700WD), intermediate modulus (IM) unidirectional tape (t800HB), and SM fabric (CF302).

The unidirectional tape prepreg has 59% fiber volume fraction and a 145 gm/m^2 areal weight while the fabric has 50% fiber volume fraction and 199 gm/m^2 areal weight. All specimens were cured at 180°F fir 6 hours followed by a post-cure at 350°F for 2 hours following the recommended practices for this material as provided by Advanced Composites Group (ACG) [1]. Testing was conducted at room temperature in the as fabricated condition and at 350°F in the as-fabricated condition. No special specimen pre-conditioning was performed. The 350°F dry condition was determined to be appropriate for the X-34 vehicle since flight tests will be conducted at White Sands Missile Range and the vehicle will be housed in an environmentally controlled hangar. For comparative purposes, short beam shear testing was conducted at 350°F in the fully saturated condition (160°F *(71°C)* 85% RH to full saturation) to estimate the materials further degradation due to moisture ingression. Transverse and flexure properties were only measured at room temperature. Ten samples were tested for each condition noted in the table. Mean values and standard deviations are listed along with B-Basis design values for the strengths [5]. For design purposes, average moduli are used.

Table 2 - Qualification Test Results for Standard Modulus Tape LTM45EL Laminates

| | | | Standard Modulus Unidirectional Tape | | | | | |
| | | | RT | | | 350°F | | |
Direction	Property	Units	Mean		CV	Mean		CV
0° Tension	Strength	ksi *(MPa)*	192.9	*(1330)*	8.36%	230.0	*(1586)*	6.67%
	Modulus	Msi *(GPa)*	17.3	*(119)*		25.8	*(178)*	
0° Compression	Strength	ksi *(MPa)*	170.1	*(1173)*	5.88%	70.4	*(485)*	8.06%
	Modulus	Msi *(GPa)*	16.6	*(114)*		24.8	*(171)*	
90° Tension	Strength	ksi *(MPa)*	4.1	*(28.1)*	12.01%			
	Modulus	Msi *(GPa)*	1.2	*(8.07)*				
90° Compression	Strength	ksi *(MPa)*	13.7	*(94.5)*	16.99%			
	Modulus	Msi *(GPa)*	1.3	*(8.83)*				
±45° Shear	Strength	ksi *(MPa)*	10.8	*(74.5)*	3.80%	10.92	*(75.3)*	2.11%
	Modulus	Msi *(GPa)*	0.6	*(4.00)*		0.57	*(3.93)*	
Short Beam Shear	Dry	ksi *(MPa)*	11.4	*(78.9)*	4.28%	6.32	*(43.6)*	5.06%
	Wet	ksi *(MPa)*				5.8	*(40.4)*	7.71%
Flexure	Strength	ksi *(MPa)*	190.7	*(1315)*	3.73%			
	Modulus	Msi *(GPa)*	16.6	*(114)*				

Table 3 - Qualification Test Results for Intermediate Modulus Tape LTM45EL Laminates

Direction	Property	Units	RT Mean		RT CV	350°F Mean		350°F CV
			Intermediate Modulus Unidirectional Tape					
0° Tension	Strength	ksi *(MPa)*	300.0	*(2068)*	7.65%	295.8	*(2039)*	5.63%
	Modulus	Msi *(GPa)*	21.9	*(151)*		29.7	*(205)*	
0° Compression	Strength	ksi *(MPa)*	144.1	*(994)*	9.53%	72.4	*(499)*	7.41%
	Modulus	Msi *(GPa)*	19.6	*(135)*		20.8	*(143)*	
90° Tension	Strength	ksi *(MPa)*	3.9	*(26.7)*	9.30%			
	Modulus	Msi *(GPa)*	1.1	*(7.52)*				
90° Compression	Strength	ksi *(MPa)*	16.8	*(115.8)*	14.89%			
	Modulus	Msi *(GPa)*	1.9	*(12.8)*				
±45° Shear	Strength	ksi *(MPa)*	7.97	*(55.0)*	2.46%	7.58	*(52.3)*	4.35%
	Modulus	Msi *(GPa)*	0.45	*(3.10)*		0.32	*(2.22)*	
Short Beam Shear	Dry	ksi *(MPa)*						
	Wet	ksi *(MPa)*						
Flexure	Strength	ksi *(MPa)*	175.1	*(1207)*	4.15%			
	Modulus	Msi *(GPa)*	19.4	*(134)*				

Table 4 - Qualification Test Results for Standard Modulus Fabric LTM45EL Laminates

Direction	Property	Units	RT Mean		RT CV	350°F Mean		350°F CV
			Standard Modulus Bidirectional Fabric					
0° Tension	Strength	ksi *(MPa)*	60.8	*(419)*	4.66%	101.5	*(700)*	3.94%
	Modulus	Msi *(GPa)*	7.3	*(50.0)*		14.8	*(102)*	
0° Compression	Strength	ksi *(MPa)*	81.6	*(562)*	10.42%	40.2	*(277)*	9.11%
	Modulus	Msi *(GPa)*	8.2	*(57.0)*		8.2	*(57.0)*	
90° Tension	Strength	ksi *(MPa)*						
	Modulus	Msi *(GPa)*						
90° Compression	Strength	ksi *(MPa)*						
	Modulus	Msi *(GPa)*						
±45° Shear	Strength	ksi *(MPa)*	12.0	*(82.5)*	11.11%	8.14	*(56.1)*	10.20%
	Modulus	Msi *(GPa)*	0.56	*(3.86)*		0.41	*(2.80)*	
Short Beam Shear	Dry	ksi *(MPa)*	6.98	*(48.1)*	5.30%	4.38	*(30.2)*	4.79%
	Wet	ksi *(MPa)*				2.63	*(18.1)*	11.03%
Flexure	Strength	ksi *(MPa)*	78.3	*(540)*	3.46%			
	Modulus	Msi *(GPa)*	16.6	*(114)*				

5.2 Discussion and Comparison with Other Data   In general, the results were as expected based on the screening tests described above and on the available product literature [1]. A few anomalous tests were noted. The high temperature longitudinal tension strengths and moduli measured in the test appear to be higher than would seem reasonable. It is conceivable that exposure of this resin to high temperature results in further cross-linking of the molecular chains and therefore higher resin strengths and stiffnesses. Such an effect has also been noted in this material in the data from the Low Cost Composite Processing (LCCP) Program [6]. However, this effect is only expected to be on the order of a few percent, not the 20% noted in the test. Since all 10 samples in all three material types exhibited this characteristic, some systematic error must have induced this effect, and the high temperature tension data should be considered suspect.

The room temperature fabric tension strengths were somewhat lower than expected based on the previous screening tests and data from the LCCP Program [6]. Initial qualification tests done on a Toray fiber woven into a 2x2 twill pattern indicated a mean warp tensile strength of over 83 ksi *(572 MPa).* As the prepreg was about to be produced for the X-34 , it was determined that the fiber supplier could not provide sufficient quantities of that fiber to meet the needs of the program. Therefore, the fabric was changed to a Soficar fiber with the same nominal tensile strength (530 ksi or 3650 MPa). All other factors remained the same including weave style, tow size, fiber sizing, and resin content. However, the new fabric was tested to have only a 61 ksi *(420 MPa)* tensile strength. Further discussions with the fiber manufacturer indicated that in processing, this fiber is twisted and then untwisted before weaving. It is assumed that the twisting process induces surface flaws that act as initiation sites for fiber cracks that reduce the tensile strength of the laminate. Since no other fiber could be procured in time, the vehicle was designed around this low value for fabric tensile strength.

The fabric tension values are also anomalous. The high temperature results indicate that strength increases over 60% seemingly beyond the realm of plausibility. Comparative data from the LCCP program indicate a 10% decrease in properties with temperature. Apparently, the test results at room temperature are suspect and nee to be adjusted.

Tension and compression properties of the IM tape material compare favorably with the data published for IM fibers in LTM25, a closely related resin formulation, also supplied by ACG [7]. Strength values from the X-34 data are about 10% higher in tension (attributable to a slightly higher fiber strength) but about 15% lower in compression, perhaps attributable to autoclave curing of the NASA specimens compared with oven curing of the X-34 material.

5.3 Sandwich Testing   Since much of the X-34 vehicle structure is fabricated from honeycomb sandwich panels, flatwise tensile tests were conducted to measure the adhesive bond between the 5052 Aluminum honeycomb core (1/8" cell size, 0.0007" cell wall) and the laminate skins. ACG's LTA45 film adhesive (with 120 style fiberglass scrim cloth) was used for the bonding since it is compatible with the laminate resin system. The results of these tests are summarized in Table 5. Both co-bonding of pre-cured skins and co-curing of skins and core are used in the vehicle with no statistical difference in flatwise strength. Considering the data as a single set, the average flatwise strength measured was 545psi *(3.75 MPa)*.

Table 5 - Flatwise Tensile Test Results for LTM45EL Sandwich Panels

		Mean Strength		
Process	**N**	*psi*	*(MPa)*	**CV**
Co-Bond	14	516.6	3.56	8.92%
Co-Cure	16	569.4	3.93	14.17%
All Samples	30	544.8	3.75	13.03%

6. JOINT TESTING

6.1 Bolted Joints   While much of the X-34 vehicle structure is bonded in order to reduce the overall complexity of the vehicle and the overall part count, accessibility requirements, particularly in the fuselage, dictate that some of the structural panels be bolted together over each other and into the fuselage bulkheads. Prudent design practice [8] indicates that bolted joints in composites should be fabricated with a substantial amount of fabric in a near quasi-isotropic laminate. However, since the fuselage design relies on its skin panels for vehicle pitch bending stiffness, these panels are strongly reinforced with high modulus unidirectional tape. Therefore, bearing tests were conducted to determine the effects of laminate sequence and material type. Results were also obtained for tension versus compression bearing, 1/4" versus 3/8" pin diameter and autoclave versus oven cured material.

Pin bearing tests were conducted in accordance with ASTM 953. The baseline condition was a 1/4" thick quasi-isotropic fabric laminate ($[\pm45\ 0/90]_{NS}$ with 25%/50%/25% 0°/45°/90° plies) that was vacuum bag/oven-cured and tested in tension with a 1/4" diameter pin. Edge distance was greater then 4 hole diameters in all cases. The effect of layup was investigated by varying the proportions of 0°, 45°, and 90° plies within the laminate. Ten specimens were tested for the baseline condition as well as the compression condition and the 3/8" pin bearing condition to determine B-Basis allowables. Three specimens were tested for all other conditions to observe general trends. Ultimate strengths were determined from the maximum load achieved prior to failure:

$$\sigma_{Brg,Ult} = P_{max} / (d\,t)$$

where Pmax is the maximum load, d is the pin diameter, and t is the specimen thickness.

Bearing results showing the effects of geometry and material type are shown in Figures 9 and 10. Compression strengths were nearly 50% higher than tension strengths because of the added support and constraint of the material away from the edge of the specimen. 3/8" ultimate bearing strength was 12% lower than the 1/4" bearing strength; however, the 3/8" pin has 50% more bearing area so can support 17% higher ultimate load. Autoclave-cured laminates were marginally stronger (3%) than vacuum bag/oven-cured laminates, but this difference is within the level of experimental error and is not considered significant. Laminates fabricated exclusively with unidirectional tape had ultimate bearing strengths that were 28% lower than those fabricated with fabric. Ply sequence had only a minor effect on the ultimate strength. The weakest laminate (37.5/25/37.5) was only 10% weaker than the quasi-isotropic laminate. In general, strength increases with the %45° plies up to about 50%.

A single bearing test series on a specimen that represents an actual laminate sequence from the fuselage design (combining SM tape, IM tape and SM fabric) resulted in an ultimate bearing strength of 83.5 ksi *(576 MPa)*, indicating that the addition of higher strength, higher modulus T800HB fibers increases the bearing strength and offsets the weakening effect of the tape over the fabric. Further investigation is needed to characterize the bearing strength of these hybrid laminates combining different fiber types.

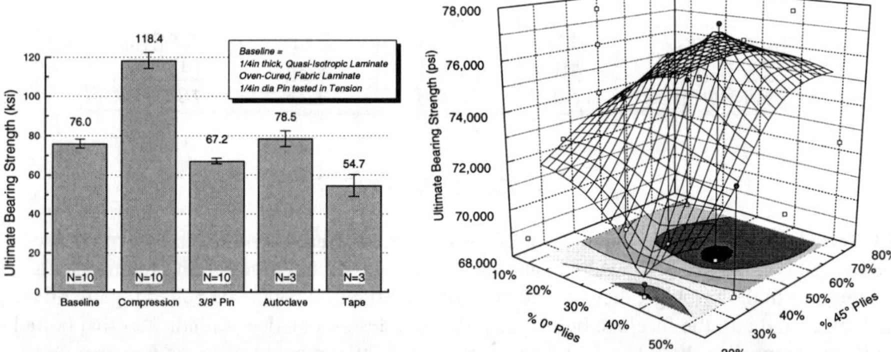

Figure 9 - Geometry effects on the bearing strength of LTM45 laminates.	**Figure 10** - Material effects on the bearing strength of LTM45 laminates.

6.2 Bonded Joints   The accessibility requirements on the X-34 wing and control surfaces (elevons, tail and body flap) are significantly less than those on the fuselage, so nearly all of the internal joints on these structures are bonded in order to reduce vehicle complexity, improve reliability, and reduce overall part count. The typical bonded joint, shown in Figure 11, consists of a sandwich panel web attached to a laminated section of the skin. Bonded interfaces exist between the U-Clip (closing out the web) and the skin laminate, and between the L-Clips and the skin laminate. For the control surfaces, these structural components are pre-cured and all interfaces are bonded with Hysol EA9394 paste adhesive. For the wing, these same structural components are co-cured using LS416 laminating resin (supplied by ACG, also compatible with the LTM45 prepreg resin) with an EA9394 paste adhesive bond between the U-clip and the skin.

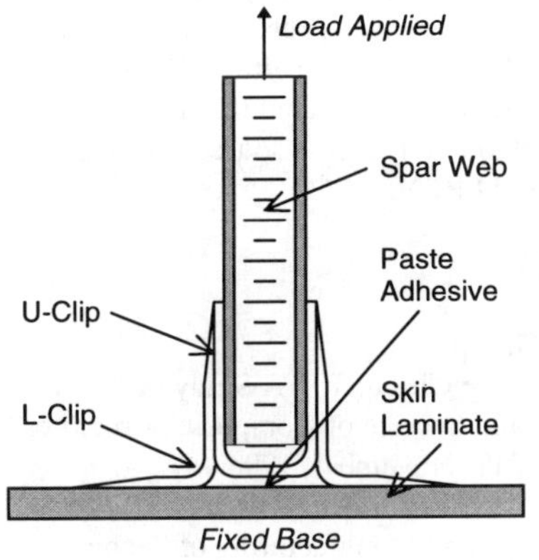

Figure 11 - Typical bonded joint geometry showing U-Clips and L-Clips.

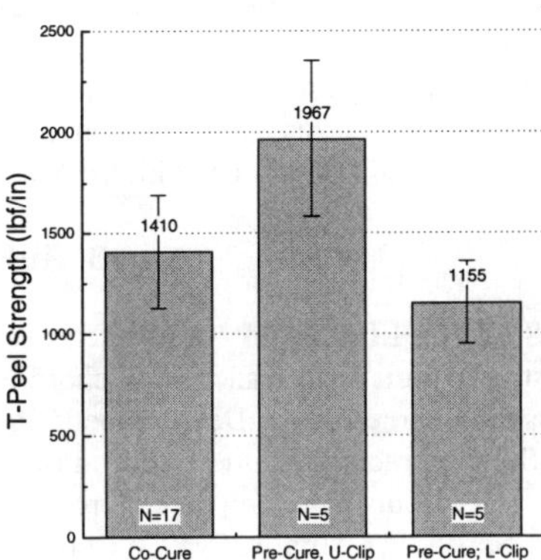

Figure 12 - Effects of bond process on the ultimate strength of bonded joints.

To assess the structural integrity of this type of bonded joint, T-peel tests were conducted on 1" deep specimens that have the cross section indicated in Figure 11. The joint specimen was typical of the joint in the regions of the structure that were most highly loaded, following the design verification approach advocated by Whitehead and Deo [9]. The skin laminate was clamped to the test rig just outboard of the L-clips, and load was applied to the web until failure occurred. Several different design features were varied in the test matrix including the effectiveness of the U-clips in the pre-cured configuration and the relative thicknesses of the U-Clips, L-Clips and web skin laminates in the co-cured configuration. Three to six specimens of each configuration were tested to observe overall trends in the strength characteristics.

Results of the bonded joint testing are shown in Figure 12. For the co-cured geometry, failure occurred as a result of an interface crack that initiated under the web either at the bondline or a single ply into the skin laminate. No statistically significant differences between the various specimen configurations could be detected, indicating that the interface failure mode was not affected by changes in the U-clip and L-Clip geometry. For the pre-cured geometry, the use of a U-Clip increases the T-peel strength by 70% over the L-clip only geometry. Apparently, the U-Clip is effective in spreading the load from the web skins more uniformly into the skin laminate than the single L-Clip on each side. The pre-cured U-Clip configuration also exhibited failure by interface cracks that started at the bondline under the web or in the skin laminate, but at 39.5% higher load. It is possible that the greater strength of the pre-cured configuration is directly due to more uniform consolidation and properties of the clip details resulting in more uniform load distribution at the interface. It is also possible that the greater strength could be partly attributed to higher skin laminate interlaminar strength since the skin laminates were autoclave cured while the co-cured specimens were vacuum bag/oven-cured.

7. CONCLUSIONS

LTM45EL is an affordable and flexible prepreg material that is uniquely suited to application in the primary airframe of the X-34 Reusable Launch Vehicle. Its low temperature initial cure results in simple, low-cost tooling ideally suited for this limited production experimental vehicle. Its high glass transition temperature and good retention of fiber properties also validate its use on this high temperature, high performance vehicle. Problems in the testing precluded the possibility of completely characterizing the material properties. Ongoing testing is addressing the gaps in the data. However, sufficient data was collected for laminate properties and joint properties such that, combined with other available sources of data, the material can be confidently used in the design of the X-34 RLV.

8. REFERENCES

1. PDS LTM45EL/5.24.95 - Advanced Composites Group.
2. Vosteen, Louis F. & Hadcock, Richard N., "Composite Chronicles: A Study of the Lessons Learned in the Development, Production, and Service of Composite Structures"; NASA Contractor Report 4620 Contract NAS1-19317, November, 1994.
3. Vosteen , Louis F. "Composite Aircraft Structures", ACEE Project Office.
4. Hart-Smith, L.J. "Innovative Concepts for the Design and Manufacturing of Secondary Composite Aircraft Structures" 5th Australian Aeronautical Conference, Melbourne, Australia, September 13-15, 1993.
5. Silverman, Edward M. ed. "Composite Spacecraft Structures Design Guide", NASA Contractor Report CR 4708 under Contract NAS1-19319, March 1996.
6. Materials Directorate, Wright Laboratory, LCCP 3rd Semiannual Technical Report, April 1994.
7. Cruz, Juan R., Shah, C.H., & Postyn, A.S., "Properties of Two Carbon Composite Materials Using LTM25 Epoxy Resin." NASA Technical Memorandum 110286, November, 1996.
8. Niu, Michael C.Y., Composite Airframe Structures. Conmilit Press Ltd. Hong Kong, 1992.
9. Whitehead, R.S. and Deo, R.B, "A Building Block Approach to Design Verification of Primary Composite Structures"; AIAA 83-0947

ACKNOWLEDGMENTS

The authors would like to thank the X-34 Program Office at the NASA Marshal Space Flight Center for funding this work. Thanks also to Mr. Steve Larby and Mr. Kevin Jackson of the Advanced Composites Group for their many helpful discussions.

PUSHING THE LIMITS OF VARTM

Thomas D. Juska
Newport News Shipbuilding, Newport News, VA 23607
H. Benson Dexter
NASA Langley Research Center, Hampton, VA 23681
William H. Seemann, III
Seemann Composites, Inc., Gulfport, MS 39505

ABSTRACT

Vacuum Assisted Resin Transfer Molding (VARTM) has found widest application in the marine industry. The process is usually used with conventional marine construction materials, specifically, polyester or vinyl ester resin, glass fabric, and PVC or balsa core. The primary reason for the increasing acceptance of VARTM is that it minimizes styrene emissions. An additional advantage is that VARTM can attain fiber contents only slightly lower than those of prepreg fabrics. Since typical vinyl esters have elastic moduli (3 GPa) and failure strains (5%) equivalent to those of many epoxy resins, the properties of room temperature cured VARTM laminates are, except for fracture toughness, equivalent to those of glass fabric/epoxy prepreg cured at 120 °C. Since VARTM has proven to be a low cost method capable of fabricating large structures, investigations have been made into increasing the application of this process. Herein the results of three such efforts are reported: increasing the fracture toughness of glass fabric laminates, increasing the hot/wet strength of carbon laminates, and the development of process modifications which allow the fabrication of large structures in one stage. The fracture toughness of room temperature cured glass fabric reinforced VARTM laminates was shown by the Naval Surface Warfare Center, Carderock Division (NSWC-CD) to be significantly improved with a toughened vinyl ester formulated by Seemann Composites, Inc. Measured values of Mode I fracture toughness during crack propagation were 1.9-2.6 kJ/m^2. An investigation into maximizing the hot/wet compression strength of carbon laminates was recently initiated at NASA/Langley for possible applications of VARTM in aerospace structures. Improvements are continually being made in the VARTM process; in particular, the Seemann Composites Resin Infusion Molding Process (SCRIMP) was used to fabricate the sandwich hull of an 27m motor yacht in one shot.

KEY WORDS: Composites, Vacuum Infiltration, Mechanical Properties

1. INTRODUCTION

Large composite structures are fairly common in the marine industry, such as private motor yachts. The vast majority of these structures are fabricated by hand lay-up. One advantage with hand lay-up is that the laminate thickness is produced in multiple stages and builds gradually, so there are essentially no limits to the size and thickness of structures which can be made with this process.

Fabricating a large structure with VARTM, however, introduces several challenges. Large quantities of catalyzed resin must be handled, and the gel time of the resin must be extended to allow time for infusion. If the thickness is built up in more than one stage, the surface of the first stage must be adequately prepared, which generally requires sanding or sandblasting. (Surface preparation is not necessary between the multiple stages of a structure made by hand lay-up due to the primary bonding which occurs to air-inhibited surfaces.)

There are various VARTM processes in use today. Most are proprietary, and the methods used are not discussed in even general terms. Evaluations of VARTM at the Naval Surface Warfare Center, Carderock Division (NSWC-CD) and NASA Langley Research Center have predominantly been limited to SCRIMP since it is patented and publicly described in detail.

SCRIMP has found widespread application. The process was invented by a boat builder, and the majority of structures fabricated with SCRIMP are boats. It has also had a significant impact on military marine structures. NSWC-CD evaluated a variety of prototype structures made with SCRIMP, including a deckhouse module with a stiffened-skin construction, and a 13.4m balsa-cored Advanced Materiel Transporter (1). However, it has also found application in civil engineering projects such as bridge deck prototypes, and in the transportation industry for rail cars (2).

In conventional SCRIMP, the distribution apparatus is positioned over one or both sides of the part, and the resin is drawn under vacuum into the distribution media, through which it infuses into the fabric. The structure is usually infused with a single batch of catalyzed resin. For thick parts, use of retarders, or other compositional changes which effect a long working life resin, are sufficient to complete the infusion. Similarly, large flat panels are readily fabricated by suitable arrangement of the distribution media, and if necessary, resin working life.

Fabrication of a large, thick part becomes challenging, however, if the resin must move a significant vertical distance. In any vacuum-assisted infusion, the flow rate decreases as the resin advances through the dry fabric, since the same pressure drop must move an increasing volume of resin. Application of Darcy's Law to the flow of resin in one direction through dry fabric predicts, in Poiseuille's Equation, that the time required for infusion is proportional to the square of the flow length (3). The rate of infusion is further reduced up vertical surfaces since gravity works against the pressure drop. (In theory, resin should be introduced at the top of the part, but in practice, for complex shaped parts with variable thickness, infusion from the top down is difficult to control.)

To meet the challenge of fabricating large, thick, complex-shaped parts with VARTM, Seemann Composites developed a modification of the SCRIMP process. Instead of infusing with a long working life resin, in which curing reactions do not commence until all of the fabric is impregnated, they developed a method using fast-curing resins. With this method, the resin cures during the infusion. As the resin cures progressively, additional resin is introduced into the part, which keeps the flow length short and fairly constant. Seemann's "progressive gelation" process requires well-characterized resin formulations, which is made difficult from mass effects on cure rate and temperature effects on viscosity. And it should be realized that placement of the distribution media cannot be done by "trial and error" when making a large, thick part given the obvious costs involved. But with experience the variables can be controlled. North End Marine used SCRIMP progressive gelation to fabricate the sandwich hull of an 27m motor yacht, which had 2.54 cm thick outer and 1.27 cm thick inner skins, in one shot, and required 22 drums of resin.

2.0 HIGH FRACTURE TOUGHNESS GLASS FABRIC LAMINATES

2.1 Marine Applications Conventional marine laminates are made by hand lay-up with orthophthalic polyester resin and woven glass fabric. Orthophthalic polyester, also called general purpose polyester, dominated the resin market for commercial boats until about a decade ago, but now there are numerous options. The major problem encountered with ortho resins is unpredictable blistering, which possibly happens because they are made in a single cook process and the resins contain various impurities (4). Polyesters based on dicyclopentadiene (DCPD) are in common usage in the Glass Reinforced Plastics (GRP) industry, usually blended with other polyesters or vinyl esters. DCPD polyesters are less expensive than ortho resins, cure rapidly and completely (are not air inhibited) and have the best cosmetics because they shrink less. The major concern with DCPD polyesters is secondary bond strength (5). As a result of the "air-drying" nature of DCPDs, the secondary bonding window is narrow. Subsequent laminations must not be delayed more than a few hours. Isophthalic polyesters, which use isophthalic acid in place of phthalic anhydride as the saturated acid, are superior resins. They cost about 20% more than orthos based on current pricing, but have improved corrosion and blister resistance and higher heat distortion temperatures. Isophthalic polyesters are available with a fairly wide range of mechanical properties as a result of the variations possible in oligomer structure. Vinyl esters are a type of polyester, with which they have much in common: both are unsaturated oligomers dissolved in styrene and are cured with similar catalyst systems. There are, however, major differences in the oligomer chemistry which sets vinyl esters apart from polyesters.

Vinyl esters have supplanted polyesters in many applications, particularly those requiring corrosion resistance. Vinyl esters generally have higher failure strains than polyesters, which results in superior laminate mechanical properties, impact damage resistance, and fatigue life. The advantage they have over epoxies is that they cure at room temperature but still have heat distortion temperatures which approach 90 °C without a post-cure (6). There are six types of vinyl ester, described briefly in Table 1.

TABLE 1. TYPES OF VINYL ESTER RESIN

Vinyl Ester	Description
General purpose	Methacrylated bisphenol A epoxy
Fire retardant	Contains brominated components
Novolac	High service temperature
Rubber Modified	High tensile elongation
Toughened w/o rubber	High tensile elongation
Low volatile content	Reduced emissions

Resin selection in boatbuilding is rarely driven by laminate mechanical properties. However, for Naval applications, the various resin, fabric, and resin content options were evaluated to maximize mechanical properties (7). It was found that increasing resin failure strain increases laminate tensile, compressive, and flexural strength, but above about 4% strain, further increases in resin elongation do not improve mechanical properties. The exception to this rule is fracture toughness.

2.2 Fracture Toughness Testing The most commonly used procedure for calculation of Mode I fracture toughness is the ASTM D 5528 method. Typically, specimens are fabricated with a release film inserted at the mid-plane to facilitate measurement of the strain energy required to initiate and propagate a crack under Mode I loading. A modification of the ASTM D 5528 methodology was used in this study. The ASTM procedure allows calculation of the initiation value of fracture toughness, but for this study, average propagation values were also computed. Propagation values were determined by measuring the strain energy release rate after the crack had propagated 1.25, 2.5, and 3.8 cm from the tip of the insert, and reporting the average of the three values. Propagation values were measured because the resistance to delamination growth, which is important for large structures susceptible to impact, is for some glass fabric laminates substantially higher than the resistance to the onset of a delamination. Fracture toughness increases with crack length (due to fiber bridging) until the length of the bridged zone attains a steady state value, which occurs after about 1.25 cm of crack growth.

A second change to the ASTM procedure was that the beam theory expression for strain energy release rate was not corrected for rotation of the beam at the delamination front, as required in ASTM D 5528. Most fiberglass laminates delaminate by stick-slip, so it is usually not possible to perform the compliance corrections. Mode I fracture toughness is reported as: $G_{Ic} = 3P\delta/2ba$, where P is the load, δ is the crack opening displacement, b is the specimen width, and a is the crack length.

2.3 Fracture Toughness Results Mode I fracture toughness measurements were made on several types of marine construction laminates fabricated by VARTM, the values of which are given in Table 2. The reported toughness values are typical for the given class of resin.

TABLE 2. FRACTURE TOUGHNESS OF MARINE LAMINATES

RESIN	MODE I FRACTURE TOUGHNESS (KJ/M^2)	
	INITIATION	PROPAGATION
orthophthalic polyester	0.18	
isophthalic polyester	0.26	
modified isophthalic polyester	0.44-0.53	
DCPD polyester	0.18	
general purpose vinyl ester	0.53	
untoughened epoxy RTM resin	0.53	
fire retardant vinyl ester	0.53	1.23
novolac vinyl ester	0.26	0.79
low VOC vinyl ester	0.35	1.05
toughened vinyl ester*	1.05	1.93

* not commercially available

Orthophthalic and DCPD polyesters have relatively low resistance to delamination. Isophthalic polyesters have somewhat higher values of fracture toughness, and modified isophthalic resins can be as tough as vinyl esters. (Best results were achieved by replacing the propylene glycol by diethylene glycol (8)). Vinyl esters and RTM epoxies have about the same resistance to delamination. Several vinyl esters were tested, and all had a fracture toughness of about 0.53 kJ/m^2. Only one low VOC vinyl ester was evaluated, but it is likely that reducing the styrene content will in general have an adverse effect on delamination resistance. Similarly, only one novolac was tested, but these high crosslink density, high service temperature materials are expected to be lower in toughness than vinyl esters derived from bisphenol A epoxy.

2.4 High Fracture Toughness Vinyl Esters   Specific attempts to formulate a high fracture toughness vinyl ester have probably not been made by vendors of these resins because there is not a substantial market for such a material. In Table 1, it is noted that two types of toughened vinyl esters are available, rubber-modified resins and grades toughened without rubber. However, these materials were formulated for the corrosion protection industry for their superior resistance to microcracking due to thermal cycling or for enhanced adhesion to steel. Laminates made with one of the rubber-modified resins, Dow Derakane 8084, was tested for Mode I fracture toughness and found to have about the same value as general purpose vinyl ester. The rubber particles in Derakane 8084 are only about 0.05 μm in diameter (9), whereas the optimum size for toughening is generally believed to be 1-5 μm (10).

Two vinyl esters were evaluated which were formulated to have high fracture toughness. One was developed by Seemann Composites, Inc. (11), and the other by BFGoodrich Aerospace (12). These two materials were used to generate the last row of data in Table 2. Interestingly, woven roving reinforced laminates of both resins had initiation values of 1.05 kJ/m^2 and propagation values of 1.93 kJ/m^2. Neat resin castings of the Seemann Composites material was observed to stress-whiten extensively upon deformation, which indicates that the resin is capable of substantial plastic deformation.

VARTM is compatible with virtually any fabric, but it is relatively restrictive with respect to the resin characteristics. This is particularly true when fabrication temperature is limited to ambient. The high fracture toughness achieved by room temperature curing VARTM laminates with the Seemann Composites and BFGoodrich vinyl esters is a significant advancement in low cost composites technology. Prior to the development of these two resins, this level of toughness could only be attained with elevated temperature curing epoxy prepreg laminates.

3. HIGH STRENGTH CARBON LAMINATES

3.1 Aerospace Applications   The NASA Langley Research Center is involved in technology to support the development of an all-composite wing for commercial transports. The material concept being developed differs markedly from that used in aircraft to date. For example, the tail section of the Boeing 777 is fabricated in an autoclave with a thermoplastic toughened 177°C curing epoxy prepreg carbon tape. The technology currently being considered for the wing uses Kevlar-stitched, multi-axial warp knit carbon preforms. Stiffeners, intercostal clips, and spar caps are stitched to the skin and the structure is cured in an autoclave with the resin film infusion (RFI) process.

In Stitched/RFI, the resin is degassed and cast into tiles. These tiles are placed under the preform, and the assembly is vacuum bagged. Machined aluminum blocks are included in the lay-up, under the bag, to keep the stiffeners vertical and parallel and to achieve the required dimensional tolerances. The resin currently used is Hexcel's 3501-6.

In a recently initiated effort, NASA Langley is evaluating lower cost alternatives to Stitched/RFI. The research was motivated in part by the cost of the materials and tooling involved in RFI, but also by the fact that the autoclave limits the size of the structure which can be fabricated. The initial evaluation of lower cost alternatives is focused on conventional VARTM, in particular the SCRIMP process. SCRIMP is currently used predominantly for fabrication of commercial and military marine structures, which are composed of glass fabrics and room temperature curing resins, so most of the knowledge base was developed to support this technology. There have been few applications to date of VARTM in the aerospace industry, and those to our knowledge have been limited to secondary structures made with the SCRIMP process.

3.2 Summary of the Challenge   The study summarized herein was an initial attempt at maximizing the performance of VARTM laminates and structures. The goal of this effort was to evaluate the feasibility of fabricating aerospace structures with VARTM whose mechanical performance and quality are comparable to that achieved by autoclave-consolidated stitched/RFI structures. This would require comparable fiber and void content, base laminate strength and stiffness, and environmental durability. There are additional quality requirements on stiffened panels, such as tolerances on parallelism and verticality of the stiffeners. The focus of this initial study was low cost tooling concepts and fabricability. However, a significant effort was made to use high temperature curing resins since these materials will probably be necessary to meet the performance requirements.

3.3 Resin Requirements for High Performance VARTM   Attempts were made to formulate resins which would result in laminates whose properties are comparable to those of the incumbent Stitched/RFI composites. The properties of interest are compression strength, both room temperature (RT) dry and 82°C wet, open hole compression strength, both RT dry and hot/wet, and compression-after-impact strength. The difficulty with achieving this goal is that high performance resins, such as 3501-6, are not compatible with VARTM.

It is important to be specific about the characteristics of a resin which make it "high performance". It is fairly well established that composite compression strength is directly proportional to resin Young's modulus (13), all other factors being equal. Therefore, the initial attempts at maximizing the properties of VARTM laminates were focused on attaining as high a resin modulus as possible. The laminates must also retain properties at elevated temperatures (82 °C) in the moisture saturated condition. But since resin components which result in high modulus also provide high values of glass transition temperature (Tg), no specific attempts were made to formulate for a high Tg. And although increasing resin failure strain generally increases the fracture toughness of a laminate, as with the thermoplastic toughened Toray 3900-2 prepreg material used in the Boeing 777 tail section, no attempts were made to formulate for high resin failure strain since through-the-thickness stitching in the fabric is highly effective for delamination resistance (14).

For this first attempt, resins were formulated which could infuse at 150 °F and gel at less than 120 °C, which had the following additional characteristics:

1. Low viscosity (< 200 cps at 66 °C)
2. Long working life (> 1 hour at 66 °C)
3. High resin modulus (> 4.2 GPa)
4. High wet Tg (> 177 °C)

3.4 Resin Formulations   Two classes of resin were evaluated: anhydride-cured epoxy and amine-cured epoxy. The challenge is to produce a 4.2 GPa modulus resin which cures at low temperatures. There are two ways to build modulus in polymers, one being to increase crosslink density and the other to increase the fraction of aromatic carbon. Usually both are present, but relatively speaking, anhydride-cured epoxies owe their high modulus to crosslink density, and high performance amine-cured epoxies, like 3501-6, to aromatic carbon. For the purposes of this effort, anhydrides might be considered to have more promise because the chemical reactions which result in crosslinks with an aliphatic anhydride can occur at lower temperatures than those with aromatic amine.

3.4.1 Epoxy   The epoxy selected for this initial study was Ciba Geigy MY 722. This is a low viscosity version of MY 720, the predominant epoxy used in 3501-6 and other high performance formulations. MY 722 is tetrafunctional. Tetrafunctional epoxies results in higher crosslink density than difunctional epoxies, and as a result, produce higher values of Tg and modulus than the difunctional resins which compose conventional RTM and filament winding formulations. There are other tetrafunctional epoxies which could have been selected, such as Shell Chemical's EPON HPT 1077 (15), but the limited scope of this initial effort did not allow a comparison of the various tetrafunctional epoxy options.

3.4.2 Anhydride Curing Agents   Epoxy laminates in common usage are typically cured with amines or anhydrides. Anhydride curing agents are widely used in filament winding operations, but are rarely used in prepregs because they are difficult to stage, i.e., achieve a controlled increase in degree of conversion. For this reason, anhydrides are not as well known in the advanced composites community as amines. There are, however, several advantages to anhydrides. The main advantage is that anhydrides will crosslink with the pendant hydroxyls which form during the initial epoxy:anhydride reaction. Reaction with hydroxyls is beneficial for two reasons: 1) It results in a higher crosslink density compared to amine-cured epoxies, and 2) Reaction with hydroxyls reduces the amount of water the resin will absorb. The pendant hydroxyls in amine-cured resins cause large water absorption, as high as 5-6%, and reductions in Tg as high as 40 °C (16). It is expected that water conditioned anhydride-cured epoxy laminates will retain a higher fraction of their RTD strength than amine cured resins.

There are two potential problems with anhydride curing agents. One is that the resulting failure strain of the neat resin is relatively low. However, given the resistance to delamination afforded by the stitching in the fabric, high resin elongation may not be required. Also, they are water sensitive *prior* to cure.

From the above considerations, an anhydride cured epoxy, LaRC-RTMx was formulated as below:

LaRC-RTMx

Component	Relative Amount
MY 722	75
EPON 828	25
HY 906	125
AC-8	4

MY 722 and HY 906 are available from Ciba Geigy distributors. HY 906 is nadic methyl anhydride, and is available from a wide variety of sources. Any difunctional epoxy is suitable to replace EPON 828, such as Shell Chemical EPON 826, Dow DER 332 or Ciba Geigy Araldite GY 6010, but if Shell DPL 862 (or other Bisphenol F epoxies) is used the reactivity of the resulting resin may be reduced somewhat over LaRC-RTMx. Anhydrides and Chemicals, Inc., AC-8 was selected as an accelerator. AC-8 is a liquid zinc salt, which reportedly converts epoxy more gradually and completely than imidazoles or tertiary amines (17).

This formulation uses a high viscosity epoxy mixture, 13,250 cps at room temperature, and a low viscosity anhydride. It is stable at room temperature: after 27 hours, the viscosity only increased from 1375 cps to 1975 cps, and after one week at room temperature the viscosity was about 20,000 cps.

The viscosity was measured as a function of time at various temperatures. At 66 °C, the viscosity was constant at 75 cps for 45 minutes, and was only 90 cps after 1 hour 30 minutes. At 80°C the viscosity doubled after 10 minutes, and gelled in about 20 minutes. At 107 °C, the viscosity doubled in 1 minute and reached the gel point in about 8 minutes. At 120 °C the resin gels in about 3 minutes. These data are shown in Figure 1.

Figure 1. LaRC- RTMx Viscosity Vs Time

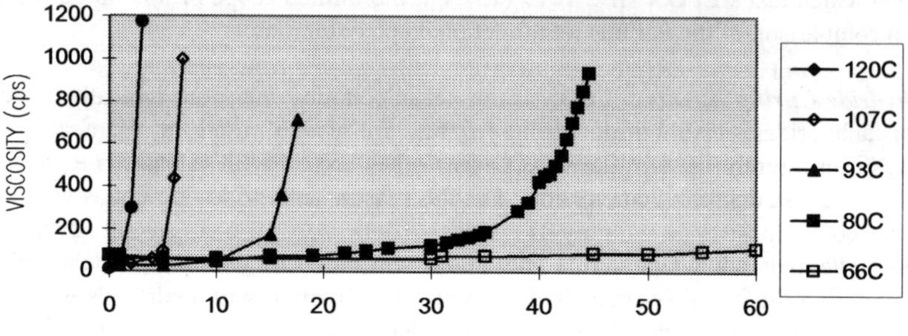

It is clear from Figure 1 why epoxy:anhydride prepregs are rare; it is difficult to B-stage the resins since they do not increase molecular weight gradually, like epoxy:amines.

3.4.3 Amines   Since 3501-6 is an amine-cured epoxy, the approach was to formulate a close match and still be compatible with VARTM. The curing agent in 3501-6 is a tetrafunctional aromatic amine, diaminodiphenyl sulfone (DDS). Since DDS is a solid at room temperature, and making adducts is somewhat involved, a liquid tetrafunctional aromatic amine, diethyltoluenediamine (DETDA) was selected.   The material actually used was Shell Chemical's Curing Agent W. (Curing Agent W has the same hydrogen equivalent weight as DETDA, and product literature (18) describes it as an aromatic amine.) There are other tetrafunctional aromatic amines available, such as Ciba's XU HY 350, which are not DETDA. As with the epoxy resins evaluated, the limited scope of this initial effort did not allow a comparison of the various tetrafunctional aromatic amine options.

MY 722:W was used in the first attempt at formulating an RTM version of 3501-6. Unfortunately it has a very high cure temperature, and did not advance at all after several hours at 120 °C. A 75:25 mixture of epoxies, MY 722:EPON 828 was evaluated, but the EPON 828 did not lower the reaction temperature (with curing agent W) significantly. A catalyst was then added to the formulation, which did increase the reactivity at lower temperatures, but not to the extent that the resin would gel in a reasonable time at 93 °C. The formulation eventually developed for evaluation was LaRC-RTMd, which used a curing agent composed of a 50:50 mixture of aromatic and cycloaliphatic amines, given below. (Amicure 101 is about a 50:50 mixture of DETDA:PACM (19), and is available from Air Products and Chemicals, Inc.) LaRC-RTMd has the cure kinetics shown in Figure 2.

LaRC-RTMd

Component	Relative Amount
MY 722	75
EPON 828	25
Amicure 101	36

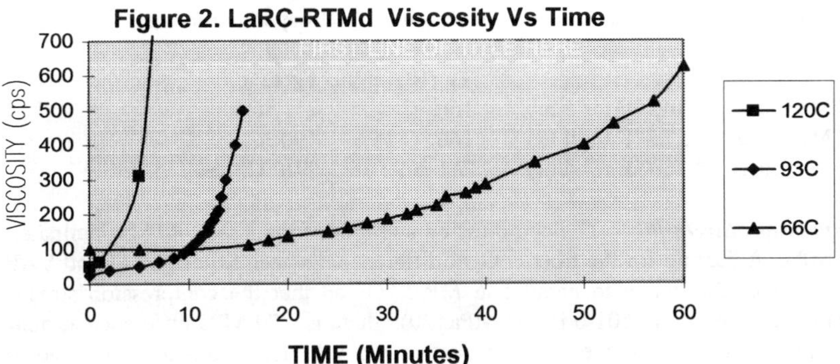

3.5 Fabrication   Both RTMd and RTMx were infused at 66 °C, into 5 stack 12"x 12" preforms, which were vacuum bagged inside a walk-in oven. After bagging, the resin components were put inside the oven with the preform and the temperature was raised to 66 °C. For RTMx, all the components were mixed in a single pot except for the AC-8 accelerator. For RTMd, the epoxy mixture and the Amicure 101 were placed in separate pots. (It is not necessary to mix the epoxy components, which are viscous at room temperature. After 1 hour at 66 °C the viscosity of the epoxy mixture decreases to about 100 cps, at which time they are readily mixed.) After the resins reach 66 °C the respective formulations are completed in the oven, which are then used to impregnate the preform with the SCRIMP process.

The cure cycles were developed from the data in Figures 1 and 2. After infusion at 66 °C, the RTMd panel was raised to 93 °C and held for 1 hour, then to 135 °C for 1 hour, followed by a slow cool. RTMx was raised to 120 °C and held for 2 hours, followed by 2 hours at 150 °C.

3.5.1 Fabrication Quality   Two quality problems were encountered, both of which were due to the absence of autoclave pressure: voids, and low fiber content compared to Stitched/RFI laminates. The voids are due to volatilization of resin components. There are ways to reduce, and possibly eliminate void formation, such as advancing the resin at 66 °C and/or decreasing the pressure drop after infusion. Advancing the resin at the infusion temperature may also increase fiber content by eliminating thermal expansion of liquid resin after impregnation.

The fiber volume fraction of the stitched carbon laminates, as evidenced by the stack thickness, is unacceptably low. VARTM results in a stack thickness of 0.152 cm, whereas autoclave consolidations produces a 0.140 mil thickness. There is at least one possible method to increase fiber content with VARTM, which is to stitch the preform to as close to net thickness as possible. One of the fabrication advantages VARTM resins have over prepreg resins is their low viscosity, which allows them to wet-out tightly stitched preforms. The tight stitching/low viscosity approach can further be used to advantage by decreasing the vacuum (after infusion) to lower levels than is normally done, to prevent resin volatilization.

41

3.6 Laminate Properties   At the time of this writing, minimal testing has been conducted on the two RTM neat resins and laminates.   Initial compression strength measurements were made, and the dry values of resin Tg were measured.

3.6.1 Glass Transition Temperature (Tg)   The dry values of resin Tg were measured with DMA, with the data given below. A broad transition of RTMx is evident, which is typical of anhydride cured epoxy. The Tg has not been measured after the 177 °C postcure. The upper use temperature of both resins are probably acceptable for aerospace structures.

	Tg (E')	Tg(E")
RTMx (cured @ 300 °F for 5 hr)	166 °C	211 °C
RTMd (cured @ 275 °F for 4 hr)	166 °C	190 °C
RTMd (cured @ 350 °F for 4 hr)	211 °C	240 °C

3.6.2 Compression Strength   The compression strength of the LaRC-RTMx laminates was about 517 MPa. Adjusting for the fiber content difference between Stitched/RFI and VARTM laminates increases this value to about 586 MPa. Given that the compression strength of Stitched/RFI laminates with 3501-6 is 655 MPa (20), there is a 70 MPa difference attributable to lower resin modulus and higher void content of the VARTM laminates compared to the autoclave cured laminates.

3.7 Three Stringer Panel Fabrication   As mentioned, the focus of this initial evaluation of VARTM for aerospace structures was a fabricability study. Four identical three-stringer stitched carbon preforms were prepared at NASA/Langley, three of which were delivered to Seemann Composites. The fourth duplicate preform was infused and autoclave cured with the stitched/RFI process at NASA. From this cured stiffened panel, Seemann Composites prepared a reusable silicone rubber vacuum bag with resin distribution media molded into the bag.

VARTM has an advantage over Stitched/RFI in that the resin can be infused across the top of the part. To date, one of the three available preforms was infused in this manner with the SCRIMP process under the silicone bag. A method was devised to maintain verticality and parallelism of the stiffeners. Investigations of the quality of the first three stringer panel made by VARTM are underway at the time of this writing. The investigation will include dimensions, ultrasonic inspection, and void content measurements.

4.0 CONCLUDING REMARKS

4.1 Marine Applications   Vacuum Assisted Resin Transfer Molding (VARTM) has found widest application in the marine industry, where the primary reason for the increasing acceptance is that it minimizes styrene emissions. Advances in vinyl ester resin chemistry have resulted in room temperature cured VARTM laminates with high fracture toughness, equivalent to those of glass fabric/epoxy prepreg cured at 120 °C.   Improvements are continually being made in the VARTM process; in particular, the Seemann Composites Resin Infusion Molding Process (SCRIMP) was used to fabricate the sandwich hull of a 27m motor yacht in one shot.

4.2 Aerospace Applications Mechanical properties of the VARTM laminates developed in this screening effort were lower than those of Stitched/RFI laminates. There are three apparent reasons for the property differences: 1) lower fiber contents, 2) higher void contents, and 3) lower resin modulus. As mentioned, all three material variables can be improved upon, and the maximum attainable properties of VARTM laminates remains to be determined .

4.2.1 Fiber Content The fiber volume fraction of the stitched carbon laminates made by VARTM, as evidenced by the stack thickness, was unacceptably high. Investigations will be made into increasing fiber content by stitching the preform to as close to the net thickness as possible.

4.2.2 Void Content Void content can be decreased most effectively by developing resins which gel at low temperatures to prevent resin volatilization, without sacrificing resin modulus or maximum Tg.

4.2.3 Resin Properties Epoxy:anhydride formulations appear to have promise for low cost, high performance aerospace RTM structures. The material evaluated in this study can be modified to increase compression strength. Specifically, the difunctional components can be removed, and part of the difunctional anhydride can be replaced with higher performance tetrafunctional anhydride. These changes, however, may increase the cure temperature, and therefore exacerbate void formation. The compression strength of the epoxy:amine formulation was significantly less than expected. The 25% difunctional epoxy and 50% cycloaliphatic amine used evidently resulted in substantial property reduction over purely tetrafunctional aromatic components. However, a tetrafunctional aromatic formulation could not be cured without an excessive void content, indicating that sophisticated catalyst systems will be required to cure these resins at sufficiently low temperatures to prevent resin volatilization.

5. REFERENCES

1. NSWC-CD, Bethesda, Md., Low Cost, High Quality Composite Ship Structures Technology Demonstrated, Research Release, May 1993.
2. Reinforced Plastics, November 1995, p.48
3. K. Watahiki, M. Ichimura, H. Goto, and Y. Abe, 49th Annual Conference, Composites Institute, The Society of Plastics Industry, Inc., Feb 1994.
4. J.S. Ghotra and G. Pritchard, Developments in Reinforced Plastics-3, Elsevier, London, p.63.
5. B. Pfund, Professional Boatbuilder, October/November, (1992).
6. T. Juska and J.S. Mayes, CARDIVNSWC-SSM-64-94/18, March (1995).
7. T. Juska, J.S. Mayes, and W.H. Seemann, CARDIVNSWC SSM-64-93/04, August (1993).
8. B. Bogner, Amoco Chemical Co., Technical Service Report 92-144(R), July (1992).
9. BFGoodrich Data Sheet AB-259, February (1992).
10. W.D. Bascom and D.L. Hunston, Treatise on Adhesion and Adhesives, Marcel Dekker, vol. 6, 1989, p. 137.
11. Seemann Composites, Inc., Gulfport, MS.
12. BFGoodrich, Engineered Polymer Products, Jacksonville, FL.
13. R.J. Palmer, NASA Contractor Report 165677, March (1981).
14. R. Palmer, Fiber-Tex 1992, NASA Conference Publication 3211, p.316.
15. Shell Chemical Co., Technical Bulletin SC:1357-93.
16. T. Juska, D. Loup, and S. Mayes, NSWCCD-TR-65-96/23, September 1996.
17. P. Rhodes, Anhydrides and Chemicals, personal communication.
18. Shell Chemical Co., Technical Bulletin SC:1183-90.
19. Air Products and Chemicals, personal communication.
20. S. Hinrichs, R. Palmer, A. Ghumman, J. Deaton, K. Furrow, and L.C. Dickinson, NASA Conference Publication 3294, Volume I, Part 2, May 1995, p. 712.

43rd International SAMPE Symposium
May 31-June 4, 1998

A HYBRID METHODOLOGY FOR QUANTIFYING THE COST AND PRODUCTIVITY OF MANUFACTURING SCENARIOS FOR THE AUTOMATED FIBER PLACEMENT OF AIRCRAFT SKINS

Scott Jones, Riccardo Raciti, and Kevin Stolfo
University of Delaware, Department of Business and Accounting

Karl V. Steiner
University of Delaware, Center for Composite Materials

Mark Lamontia
Cytec Fiberite, Inc., Newark, DE

ABSTRACT

As new composites manufacturing processes become technically feasible, one of the first questions to be answered prior to investing into process development and deployment is the cost competitiveness with other existing manufacturing methodologies. While there are considerable analytical test methods available for assuring the mechanical performance of composite components, only a few cost analysis tools for composites manufacturing are available. In particular, there exists a need predict and analyze the cost-benefit for new and developing processes, since process economics is perhaps the biggest obstacle to the practical application of many composite materials. The estimation of the cost of composite structures is complicated by the interrelation of process parameters within each step, and the general lack of historical protocol since such information would not yet be present for a developing process. This paper describes selected results from the first part of a multi-year research effort aimed at establishing a robust cost and productivity model applicable to economics analysis of automated fiber placement of thermoset and thermoplastic composites for aerospace structures, including alternative curing procedures. The objective of this effort is not only to predict specific costs for a composite structure, but to be able to estimate the partial derivative of cost with respect to changes in process features. This objective helps determine the potential payoff of alternate R&D investment scenarios.

The concept of activity-based cost modeling is presented and scenarios are detailed that outline the influence of selected cost factors on the productivity and component cost for the automated fiber placement of a "model" aircraft skin.

1. SIMULATION OF PROCESS ECONOMICS

Modeling process economics for a developing process can be challenging since many of the scenarios to be considered are not yet reduced to practice or even perhaps to hardware, and there is often no exact historical analogy from which to draw the specific parameters of a manufacturing process that would be needed to calculate cost. Furthermore, even if such cost data scenarios existed for some for the process steps, for most composite components there are now several alternative manufacturing processes and consequently the modeling procedure must be capable of rapidly considering alternative scenarios. Consequently, at best it may be difficult to focus process development because a considerable amount of uncertainty exists. At worst, R&D resources are not directed at those process and equipment features that have potential for high payoff. A good example of this situation and the motivation for this work involves the choices an organization has to make when developing a number of new techniques to fabricate cost-effective composite aircraft skins.

The methodology outlined in this paper involves the use of a commercially available simulation system known as Simprocess® and incorporates activity-based cost estimation techniques for existing and developing processes such as, in this case, the automated fiber placement process. Through interviews, experiments, and research of archival sources, several manufacturing scenarios to simulate the fiber placement of generic geometries such as wing skins, aircraft fuselages, or rocket motor cases are collected and integrated into the cost model. These scenarios include information about related process parameters such as fiber placement head or gantry cost and depreciation, required support services such as tool preparation, autoclave cycles, quality control, and the manpower necessary to operate the equipment for each scenario. The work presented in this paper draws heavily on prior studies conducted at the University of Delaware [1-3]. While some of the earlier cost modeling work [1,2] was focused on a different manufacturing process, resin transfer molding, the focus of the more recent work [3] has been on the fiber placement process. The cost model is perturbed to enable the evaluation of alternative scenarios such as variations in tape width, placement velocities, and number of personnel available for the production.

The general role of a cost-benefit simulation is to identify and assist in the appropriate allocation of resources to increase the production rate for given resources. The primary goal of the simulation is to calculate the hourly cost and total time needed for both personnel and equipment resources to increase system throughput for the resources available. To produce a large composites structure, such as an aircraft wing skin, a certain investment in scale is required. The organization which can achieve the greatest productivity for this investment will achieve the lowest cost and, for any given price, the greatest profit.   Therefore, contemporary economic modeling, especially that involving theoretical processes, needs to rely on process simulation. The idea of focusing on throughput has been popularized by Goldratt [4]. The methodology used in this research adopts the notion that the focus should be on throughput, measuring cost with an activity-based model. The goal of activity-based costing is to model the causal relationships among process activities and the commensurate consumption of resources. The result is a robust model that is useful not only for estimating production cost, but in achieving optimal resource utilization. The question of optimal resource specification is important since the actual production scale required to produce the particular component may not be well understood.

2. ACTIVITY-BASED COST MODELS

There are many examples in the literature of cost research models that use Monte Carlo simulation as the basis for estimating costs [e.g., 5,6]. In all cases, their advantage is that the stochastic nature of manufacturing processes can be considered in developing cost projections. Simulations are also permit sensitivity analysis on various process parameters. In effect, the user of the simulation can obtain the cost response to a range of input parameter values.

In this work, Simprocess® is used to conduct Monte Carlo simulations of the specific production scenarios. It is a hierarchical process simulation tool that can account for random variations, waiting time, and bottlenecks in a specified production setting. The simulation attaches activity-based costs to the production entities as they pass from the initial start to the finished product. The modeling research is thus free from the burden of developing algorithms and accumulation of costs, and can focus on the following elements:

- Identify activity sequences (materials requisition, tool preparation, fiber placement, consolidation cycle in autoclave, part removal, post inspection);
- Develop activity execution times and related functions (time needed for placement based on placement velocity, tape width, and part complexity);
- Identify resources consumed at each activity (cost of tape to be placed, personnel needs at each location, power, Nitrogen for the autoclave, and other support costs);
- Develop or estimate resource costs (development or acquisition cost for placement head, tools, etc. and estimated useable life cycle to determine depreciation);
- Establish experiment specifications (select generic geometries of typical aircraft component that lend themselves to be produced with the fiber placement process).

Extant studies tend to concentrate on comparisons between processes for a given part, including production volume and automation. Volume is found to be the major determinant of cost-effectiveness[7,9]. Research on thermoplastic composite systems indicates that geometric constraints and the processing required to achieve these shapes play an important part in assessing the cost[8]. Research also indicates that larger integrated parts tend to be cost-effective[10]. Existing cost models use volume-based costing. Volume-based costing applies overhead based on some measure of volume (e.g., units produced, labor hours, or labor dollars). Many prior studies note the inequities of assigning overhead based on a volume basis [7,10]. High overhead rates are a signal that volume-based methods are distorting product costs.

Activity-based costing is an alternative method and is better suited to calculating composite costs. A large part of a product's cost is created by activities that are unrelated to the production volume. The activities themselves drive cost every time they are performed, regardless of production volume. A purchase of one unit of a material, for example, is a very similar effort to a purchase of one hundred units. The cost of the purchase is therefore related to the complexity of the purchase, not the volume purchased. Activity-based costing takes this into account, and assigns resource costs to the activities that cause the resources to be consumed. Activities are then associated with products, and resource costs can thus be accurately linked to products in proportion to the activities generated in the manufacturing process. Figure 1 provides a schematic comparison of the volume-based and activity-based costing methods.

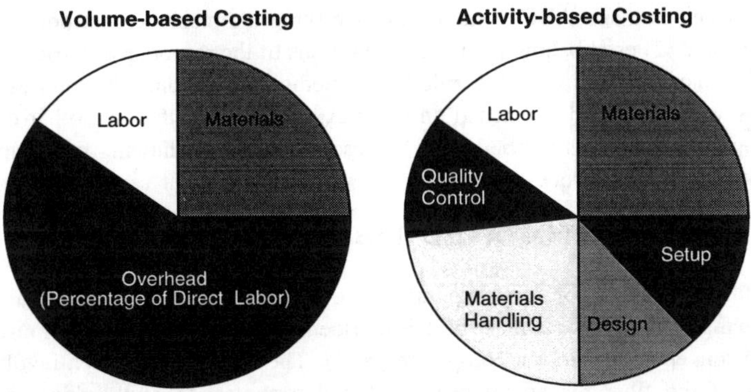

Fig. 1: Compared to volume-based costing, activity-based costing removes the direct link between labor cost and overhead leading to a more detailed breakdown of the cost-driving forces in a manufacturing process.

Though Figure 1 depicts volume- and activity-based pie charts of the same size, in practice there is often no resemblance between the total cost and relative proportions. In fact, many costs excluded from a volume-based cost allocation will actually be included in an activity-based computation. The activity-based costing computation is made according to the following procedure:

- Specify the major activities performed in a process;
- Identify the resources and cost of those resources which are consumed by the activities;
- Attribute activity consumption to individual products proportional to the products' demands for these activities;
- Spread attributed costs among all units of the products that are produced.

Activity-based costing (ABC) was popularized by Johnson and Kaplan [11] and takes two major factors into account, which are not always considered in traditional systems: volume diversity and complexity. A more complex product will consume more resources, which will be taken into account by the fact that the number of activities increases and therefore so do the costs that are added to this product. A major requirement to use an ABC system most effectively is the quite time consuming step to initially record all activities and properly observe them. Accordingly, the development of model parameters and cost distributions can be challenging and costly. Activity-based costing computes the cost of a product based on the resources consumed during the production cycle [12]. ABC is typically developed for strategic purposes - for example, to help companies understand how they can become competitive with a certain process or product. Overall, ABC displays a more realistic picture than volume-based costing of the factors that comprise the costs of a process. It places emphasis on processes and activities and encourages rethinking of what causes costs to be incurred. ABC enables cost estimation in the design stage of a product since it provides a description and understanding of the manufacturing process design itself [2].

Perhaps the most difficult aspect of cost analysis in the design of products made from advanced composites is that relevant costs have to be modeled in the absence of historical cost information and with an incomplete specification of the process. This poses at least two challenges in developing and ABC system: (1) specifying the resource cost pools associated with activities, and (2) establishing parametric functions of the response of these cost pool to changes in the quantity of activities. A rule-based method for dividing the total process into specific steps was developed and based on prior experience with this and other composites manufacturing systems. This method shall be described in the following paragraph for the automated fiber placement process.

3. AUTOMATED FIBER PLACEMENT

Automated fiber placement of very large structure has become the de-facto manufacturing process for wing skins – for example the tail of Boeing's new 777 commercial airplane was fiber placed using epoxy thermoset composite parts. The authors have been involved in the development of the fiber placement process for thermoplastics and lightly cross-linked thermoplastic tapes [13, 14]. This process is contemplated for fabrication of wing skins for a future high-speed civil transport. In that case, a heated deposition head partially or fully consolidates a skin laminate, and the autoclave is used to complete laminate consolidation or secondary bonding of stiffeners. The material may be fully dry thermoplastic and use an in-situ placement process to consolidate the part, or it may be a solvated thermoplastic with tack and drapability, where conventional "epoxy" tape placement heads and machines are used. Both processes are followed by an autoclave cycle. The question at hand is: which method yields the lowest cost?

In the fiber placement process, various bandwidths can be used along with different placement velocities. Clearly, while heads placing at higher velocities increase throughput, but at substantial increased development costs. Is the extra development cost worth the time and cost? Specifically, bandwidth can vary from one to three inches based on the number of tapes placed simultaneously and on the complexity of the head, since each tape requires independent cut and refeed mechanisms and tape tensioning to allow for close control of the tape when placing over or into highly contoured structures. Thus, the ability to place wider tapes may be desirable for rapid placement of material, but there are physical limits due to the need for high agility and conforming consolidation shoes and a cost limitation due to the dramatically rising increase in head cost to achieve these objectives.

Similarly, one might desire increased placement speed leading to increased materials throughput, but again one will face a cost penalty in increased placement head cost for better heater systems and a more robust design to assure high and consistent quality of the placed tape along the wing skin.

4. COST MODEL DEVELOPMENT FOR THE PROCESS

An activity-based cost model of the automated thermoplastic fiber placement process was developed to accurately model the placement itself and to combine it with the needed support processes. This ABC model was then used to evaluate the influence of numerous cost drivers of the process. Figure 2 provides an overview of the ATP cost model components in the simulation. Each of the boxes in the simulation represents individual process steps and includes the resources such as labor, equipment, supplies, maintenance, etc., required for each subprocess.

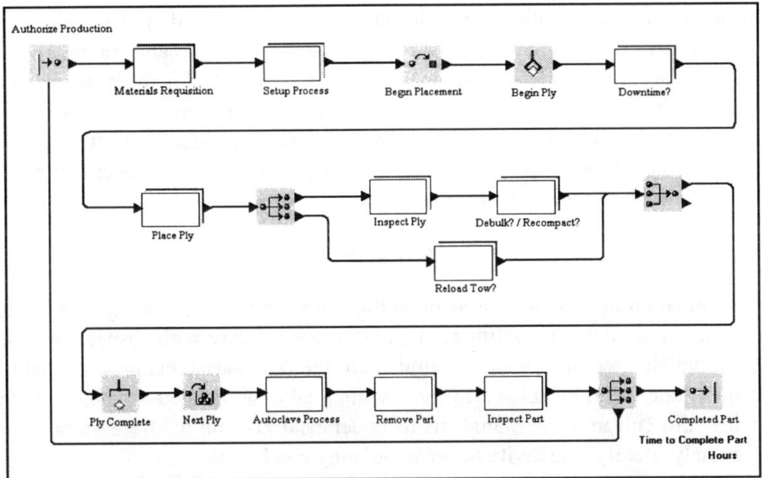

Fig. 2: Structure of the cost model for the Automated Fiber Placement process. The boxes represent individual process steps and include the required resources.

The hierarchical nature of the simulation tool allows for additional detail of the manufacturing process to be added as modeling efforts evolve towards more complex analyses. As an example, the actual fiber placement process is captured in a more detailed subprocess as shown in Figure 3.

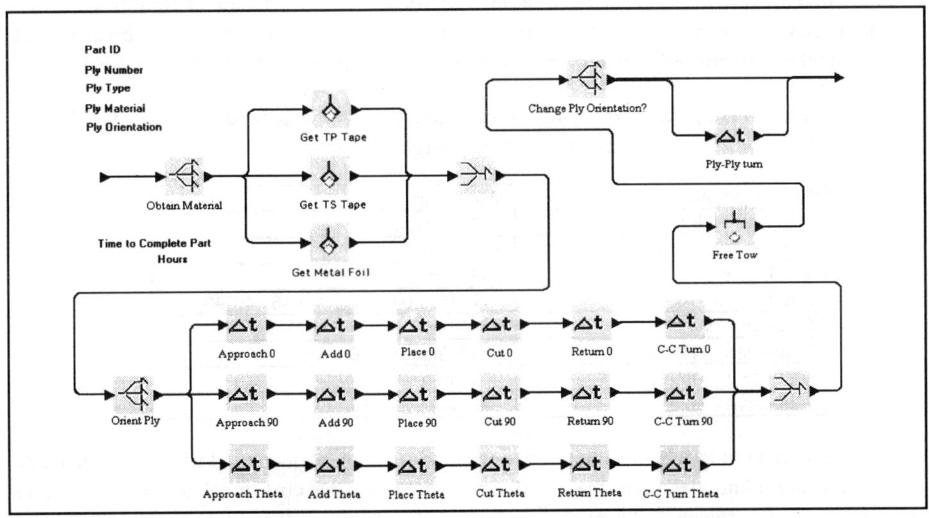

Fig. 3: Subprocess for placing plies during Automated Fiber Placement. The subprocess simulates the placement of each ply including cut and restart steps to estimate the total time required for part production.

Based on an externally generated placement pattern featuring such details as number of plies, fiber orientation of each ply, and material selection for each ply, the process simulator will follow the placement pattern of an actual part. The simulation also considers such process-specific features such as available tape length on the spool, and planned downtime for maintenance or service or unexpected downtime due to equipment malfunction. The simplified fiber placement model is designed to act as a substitute for more detailed types of models available in CAD based tools. With the simulator, the model is able to consider various cases with details such as alternative ply stacking sequences, the presence of stiffeners or inserts, and multiple defect rates. Elements of the fiber placement model have been validated through experiments on an actual fiber placement head [3].

5. CASE STUDY

The role of simulation in cost analysis shall be illustrated with the following case study. The intention of the simulation is to optimize the placement velocity with respect to cost of the final part. In the absence of a dynamic simulation, the cost variations in relationship to the fiber placement rate could be calculated by assuming all other related variables as constant. However, it is difficult in a static analysis to understand how all of these variables interact and to adequately specify sensitivity to differentiating conditions.

Table 1 lists the parameters of a generic part to be manufactured with thermoplastic fiber placement. The selected dimensions of 15 ft by 60 ft are considered large for a composite part to be manufactured and a special gantry system would have to be installed to successfully produce such a part. To simplify the simulation in this case study, no stiffeners are attached to the skin. A set of production scenarios is conducted in which the fiber placement rate is incremented by 30 inches per minute starting from a lower limit of 120 in/min to up to 480 in/min. In order to accomplish the increase in placement velocity, it is assumed that there will be linear increase in equipment cost by a factor of four from a 120 in/min placement speed to a 480 in/min speed. This cost includes the research and development efforts related to achieving the necessary upgrades on the placement head.

Table 1: Placement parameters selected for a case study to evaluate cost drivers for fiber placement of a very large wing skin.

Part Geometry	Wing Skin
Materials Selection	IM7 / Polyimide
Dimensions	15 ft x 60 ft
Thickness	24 layers
Layup Sequence	$[45,-45,0,0,0,90,0,0,0,45,-45,90]_s$
Placement Width	3 inches
Placement Velocity	120 in/min -> 480 in/min
Velocity Increments	30 in/min
Autoclave Cycle	19 hours

The question at hand is whether the added productivity due to increased placement velocity can justify the additional investment. Figure 4 presents the cost breakdown per part in relationship to the selected placement velocities. As expected, the cost per part decreases with increasing placement velocity and the relative contribution of the cost components remains relatively constant. The equipment component in Figure 4 includes the placement gantry, fiber placement head, autoclave, and the quality control facilities. The fiber placement, consolidation, and quality control components include primarily the personnel needed to operate the facilities. The facility component includes the rent or depreciation for a manufacturing site. This item is assumed to remain constant and increased production will therefore lead to a reduction in this cost component per part. As can be expected, the relative cost related to the fiber placement process which is comprised of the labor associated with the placement step will decrease with increased productivity in this step.

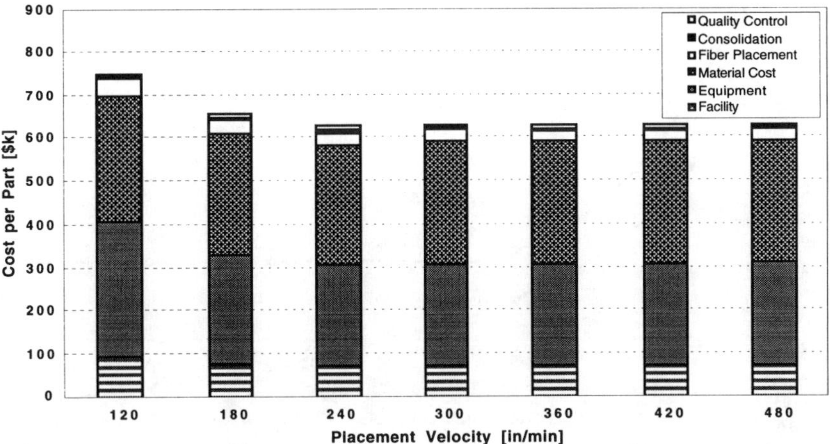

Fig. 4: A comparison of production cost components for 15 by 60 ft, 24 ply fiber-placed aircraft skin does not a continuous reduction in cost per part at placement velocities above 240 in/min.

As shown in Figure 4, increases in placement velocity beyond about 240 in/min do not result in a decreased cost per unit. One could therefore conclude that there is no economic justification to efforts to increase velocity beyond 240 in/min. What is hidden in this analysis is the effect of a constraining resource. As noted above, the simulation procedure is able to track resource utilization and hence identify potential production bottlenecks. A closer evaluation of the case study shows that the most constraining resource to improving productivity beyond velocities of 240 in/min is that of available support technicians.

In addition to the two operators required to run the fiber placement machine and several additional operators for the autoclave and the NDE facility, the production scenario includes one support technician per shift to transport materials to the work cells, assist in the setup of tools, etc., and handle the movement of components between the individual work cells.

To evaluate the influence of this support technician on the entire process, the analysis was perturbed and run after adding a second technician at a cost of $60 per hour. The effect on production time per part and the related productivity is shown in Figure 5. As expected, the addition of a second technician will reduce the production time. The reduction in completion time for each part is between 5 and 9%. This holds true for velocity up to 240 in/min. In order to achieve similar productivity improvements one would have to increase the placement velocity by 20 to 25% while holding the number of technicians constant. However, the introduction of a second technician permits productivity gains to be realized well beyond 240 in/min.

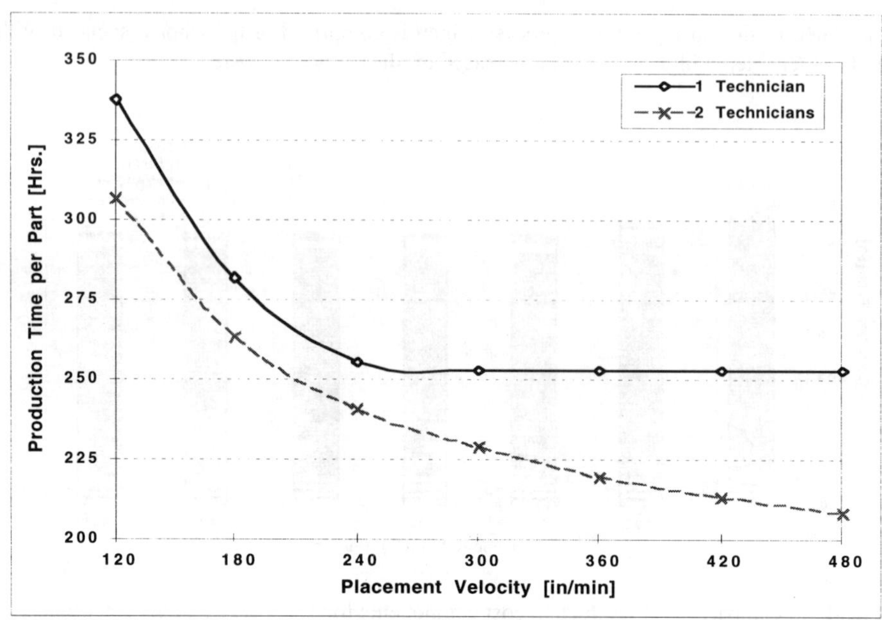

Fig. 5: The addition of a second technician to assist in the delivery of materials and movement of components between work cells will lead to increased productivity above placement speeds of 240 in/min.

What may not be quite as obvious as the reduction in production time with the addition of an second technician is the influence this added resource has on the variability and predictability of the process cost. Figure 6 provides a chart of the standard deviation of production time obtained from runs for each production scenario where between 25 and 42 parts are made within a selected production scenario. The number of produced parts at a given production scenario is linked to the productivity for certain scenarios. While there is a considerable increase in variability for a single technician, this effect is drastically reduced when adding the second resource. This can be attributed to eliminating bottlenecks in the process that may occur at random, such as unexpected downtime on the fiber placement workcell due to interruption in material supply or finished part transportation. With a single technician available, these events may greatly influence the smoothness of the operation and create a backup in the entire production. The addition of the second technician shows a more uniform process which helps reduce costs through reduction in waiting time, better scheduling, and quality.

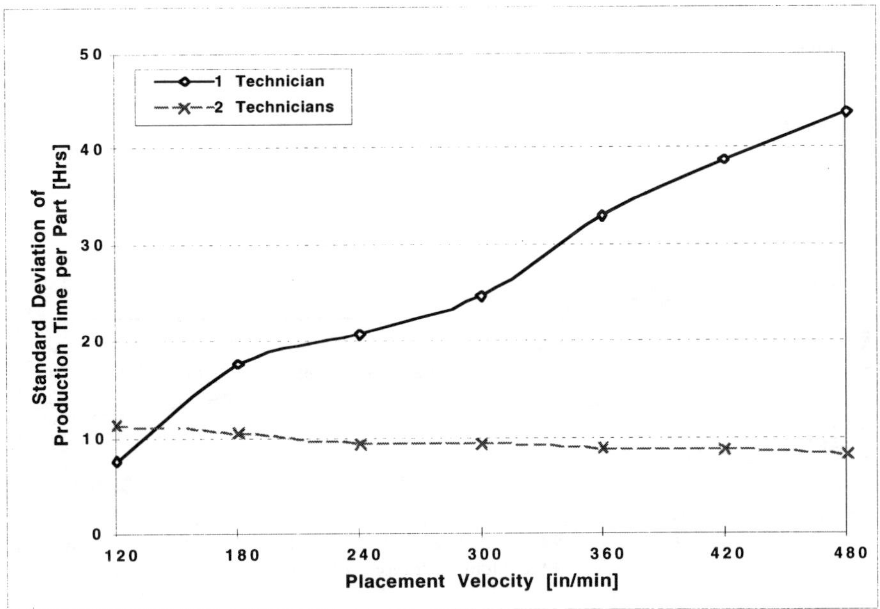

Fig. 6: A comparison of the standard deviations in productivity related to the number of technicians available as simulated during twelve different runs. The graph shows how an additional technician makes productivity more predictable due to a more even work load.

Figure 7 provides a chart of the total cost per part for increasing placement velocities for one and two technicians. Placement velocities at this rate seem at this time outside the actual performance envelope of commercially available tape placement machines but should be considered in this case study. As the placement velocity increases from 120 in/min, the cost per part for two technicians decreases rapidly for both one- and two-technician scenarios. At placement velocity of 240 in/min, there appears to be a clear cost advantage to having two technicians on the shop floor, as the two total cost curves begin to diverge. There is a minimum cost for the one technician at a placement rate of about 240 in/min, beyond which productivity gain ceases and the cost plateaus. A similar limit can be anticipated for the two-technicians scenario at a theoretical placement velocity above 480 in/min.

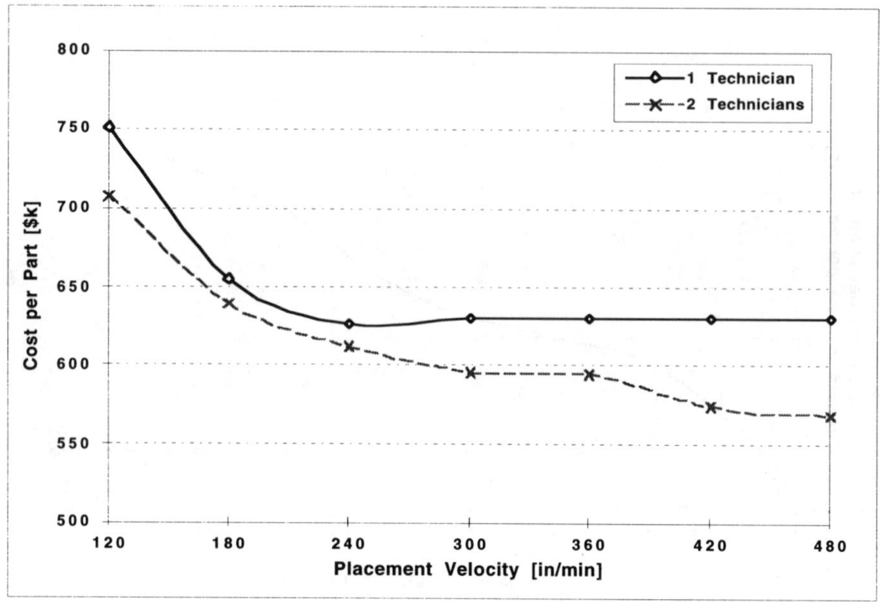

Fig. 7: A comparison of the production cost per part in relationship to the number of technicians on the process highlights crossovers and minimum values for productivity.

It should be noted that the costs per part shown are not intended to be taken literally but rather to be seen as generic values to determine specific correlations between different resources in the manufacturing cycle. Although the ongoing work has been successful in providing a solid foundation for the cost analysis of the fiber placement process, more detailed evaluations and additional information about the cost drivers of the fiber placement process and its secondary supporting processes are required to closer correlate the simulation with actual manufacturing costs.

6. CONCLUSIONS

An activity-based cost model of the automated fiber placement process has been developed and initial experiments have been conducted to correlate to the ongoing development with actual production data. Some of the cost drivers in the fiber placement production process were located, and process-specific features such as planned downtime for maintenance or service or unexpected downtime due to equipment malfunction have been integrated in the process simulator. Additional research will be required continue this evaluation in more detail. To this end, other fiber placement options with and without autoclave or post consolidation cycles are being introduced.

It was found that productivity of the process cannot be determined by looking at placement velocity in isolation. Investment in research to develop greater placement velocity heads will not result in the expected productivity increases unless other constraining resources are also targeted. In this case, the effect of relaxing the technician constraint not only results in a decrease in cycle time, but also leads to a considerable reduction of variability in the production.

This result shows the value of the activity-based costing approach to determining the optimum use of resources such as investment or personnel to arrive at the optimum cost structure for a new and developing manufacturing process.

7. REFERENCES

[1] Wright, D. L., "Activity-Based Analysis of the Unit-Level Complexity of Resign Transfer Molded Composites," CCM Report 94-25, 1994.

[2] Eggert, A., "Activity-Based Cost Analysis of Manufacturing Processes of the Composite Armored Vehicle," CCM Report 96-15, 1996.

[3] Pachler, T., "Activity-Based Cost Estimation for the Automated Fiber Placement Process," CCM Report 98-02, 1998.

[4] Goldratt, E., *The Goal*, 1992.

[5] Edwards, W.C. and J. F. Wong, "A Computer Model to Estimate Capital and Operating Costs," Cost Engineering, Vol. 29, No 10, 1987.

[6] Pope, R.D. and R.E. Just, "Empirical Estimation of ex ante Cost Functions," Journal of Econometrics 72, 1996.

[7] Krolewski, S., "Economic Comparison of Advanced Composite Technologies," 34th International SAMPE Symposium, 1989.

[8] Foley, M.F. and E. Bernardon, "Thermoplastic Composite Manufacturing Cost Analysis for the Design of Cost Effective Automated Systems," SAMPE Journal, Vol. 26, No. 4, 1990.

[9] Foley, M.F., "Techno-Economic Automated Composite Manufacturing Techniques," SAMPE Quarterly, January 1991.

[10] Gutowski, T., R. Henderson, and C. Shipp, "Manufacturing Costs for Advanced Composites Aerospace Parts," SAMPE Journal, Vol. 27, No. 3, 1991.

[11] Johnson, H.T. and R.S. Kaplan, "Relevance Lost: The Rise and Fall of Management Accounting," Harvard Business School Press, Boston, MA, 1987.

[12] Thompson, B.W., "Expanding the Benefits of Activity-Based Costing," APICS, Vol. 6, No. 9, 1996.

[13] "Rapid Placement Technology for Polymer Matrix Composites," RapTech-PMC, Final Report to Advanced Research Projects Agency, MDA 972-92-J-1026, 1994.

[14] "Rapid Placement Technology for Affordable Composites Manufacturing," RapTech-ACM, Final Report to Army Research Office, DAAH04-94-G-0285, 1998.

Affordable Composite Structure
for Next Generation Fighters[*]

Larry Bersuch
Ross Benson
Steve Owens
Lockheed Martin Tactical Aircraft Systems
Fort Worth, Texas

ABSTRACT

Over the past twenty years, composite materials have been increasingly added to the mixture of structural materials that comprise modern combat aircraft. Composite materials have literally bought their way onto the airframe in order to meet stringent weight and performance requirements. However, the "fly-away" part costs have remained high relative to comparable metallic structure. With the enormous emphasis on reducing weapons systems acquisition cost, the buzzword throughout government and aerospace industry is "affordability." Novel applications of composite materials are being pursued as a means to dramatically reduce airframe manufacturing and assembly costs. Carbon composite textile preforms are being incorporated in airframe structures to manage out-of-plane loads. These innovative design concepts are true technology "enablers" necessary to meet affordability goals. Preforms are being incorporated with net shape two-dimensional (2-D) composite material forms using manufacturing processes such as resin transfer molding (RTM), electron beam cure, diaphragm forming, fiber placement, and cocuring/cobonding. This results in the elimination of machined metal load fittings, fasteners, and substructure members saving weight and cost. Three-dimensional (3-D) woven preforms, when cocured into primary wing and fuselage laminate structure, offers improved damage tolerance and ballistic survivability attributes. To achieve these benefits 3-D textiles and structural joints must be characterized through the development of design and manufacturing processes. In addition, Z-direction reinforcement can be achieved through in-process fiber insertion with processes such as ultrasonic Z-fiber insertion, stitching, and short fiber additions to adhesives.

Lockheed Martin has been actively developing 2-D/3-D composite structures under various internal and contracted development efforts, such as the Robust Composite Sandwich Structures (ROCSS), Low Cost Composite Substructure (LCCS), and the Composite Affordability Initiative (CAI). The purpose is to develop 2-D/3-D composite structures through a building block development program for application on the Joint Strike Fighter (JSF) and future post JSF air vehicles. This paper is directed at design for manufacturing of 2-D/3-D composite structures to best exploit the future benefits they exhibit.

KEYWORDS: Composite materials, 3-D preforms, Z-fiber reinforcement, graphite/epoxy, failure criteria, design concepts, and low cost processing of 3-D composite materials.

INTRODUCTION

The application of 3-D woven textiles and Z-fiber reinforcement materials in structural concepts enables effective management of out-of-plane loads. Three-dimensional woven textile composites when integrated with 2-D laminates offer future composite airframes both breakthrough performance improvement and cost reduction. Composite usage in state-of-the-art fighters has typically been limited to under thirty percent by vehicle weight. The primary barriers to increased composite utilization have been high fabrication and assembly cost, structural integrity concerns, lack of high fidelity design tools, poor ballistic survivability, and poor interlaminar mechanical properties. Application of 3-D woven textile composites with other interlaminar reinforcement techniques, such as Z-fiber reinforcement, offer airframe structural design

---

concepts that can eliminate these barriers. This paper examines innovative design concepts and issues related to cost effective manufacture of composite structures.

DISCUSSION

Innovative structural concepts based on complex, yet low cost, woven fiber textile reinforcements are currently under development. Figure 1 shows a collage of these technologies. Intersecting structure with continuous fiber reinforcement through-the-intersection and woven 3-D fiber preforms for highly loaded composite wing carry-through bulkheads are being investigated. The development of Z-direction or through-the-thickness (Z-fiber) reinforcement to greatly increase the interlaminar strength and toughness of laminated composite structures is enabling the local tailoring of out-of-plane strength. Emphasis is being placed on coupling complex textile based fiber architecture with vacuum assist RTM, resin film infusion, and electron beam curing materials. Electron beam curing technologies are being pursued to enable the use of extremely low cost tooling materials. This enables designers to consider innovative structural arrangements that were previously prohibitive in cost because of the complex tooling needed for autoclave cure. These different manufacturing and design technologies are discussed in the following sections.

3-D Woven Composite Technologies

Carbon fiber composite textiles have found significant airframe structural applications. Inclusion of 3-D woven and braided textile "preforms" at joints in 2-D composite laminates dramatically improves out-of-plane load capability. Attributes of the textile manufacturing process are summarized below. Woven preforms can be produced by a number of techniques, including the following:

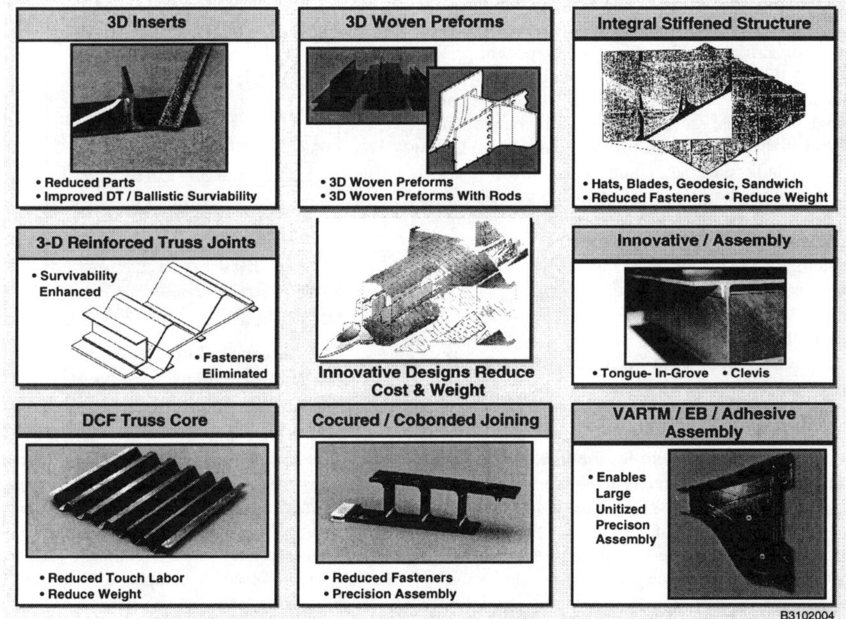

Figure 1 3-D Woven Preforms, Z-Spiking, and Low Energy Processing Will Enhance Joint Strength and Reduce Costs

- **Orthogonal pattern** – Z-fibers are pulled through the warp (0 degree direction) and fill (90 degree direction) fiber, intersecting the layers at a 90 degree angle.

- **Through-the-thickness angular interlock** – The Z-direction fibers are woven through the layers of the fabric, intersecting the layers at an angle. These Z-fibers may be the same type and filament count as the warp and fill fibers. In some cases a finer fiber is preferred for through-thickness reinforcement because it allows tight packing of the warp and fill fibers.

- **Layer-to-layer interlock** – Warp fibers are woven into adjacent layers within the weave at designated intervals. This process improves interlaminar shear strength but does not provide true through-thickness reinforcement. The adjacent interlocked fibers provides through thickness strength improvement.

To make 3-D woven preforms more affordable will require standardization of shapes (T, cruciform, special shapes) in stepped sizes for various loads. These could be standardized in a manner similar to aluminum extrusions. To reduce storage costs, preforms can be impregnated with resin prior to the cocure process. Reducing the cost of 3-D preform weaving will require automation of the more complex shaped preforms. Additional technology development is needed in the textile manufacturing arena to add bias (+/- 45 degree) plies.

Z-Fiber Reinforcement. The "Z-fiber reinforcement" process, as patented and marketed by Aztex, Inc., utilizes a novel ultrasonic insertion technique to drive carbon and titanium pins into preimpregnated or staged composite details. The Z-fiber reinforcement process is illustrated in Figure 2. A foam block containing Z-fibers is placed on the component prior to curing at the specific locations where enhanced damage tolerance and/or through-thickness reinforcement is desired. The most cost effective Z-fibers are small, 0.010-inch diameter carbon fiber pultruded rods. During cure or debulk, the foam compacts and the Z-fibers are forced through the prepreg composite materials. Thus, this process is commonly referred to as "Z-spiking." A GLCC program is developing the manufacturing technologies for factory implementation of the Z-fiber process.

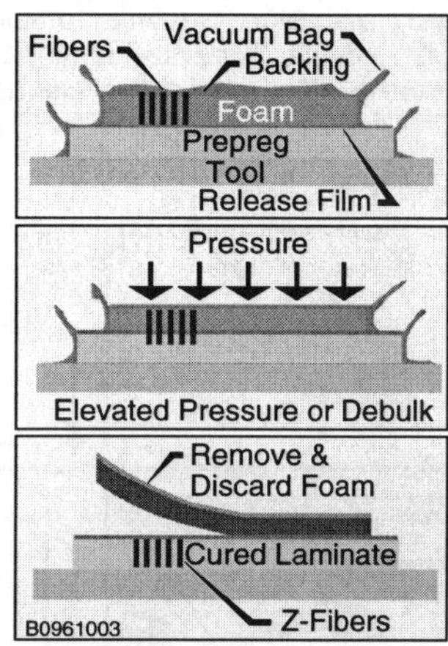

Figure 2 The "Z-Fiber Reinforcement" Concept

The Z-fiber reinforcement process can be used in damage-prone areas such as blade- and hat-stiffened laminates, in sandwich joints where 3-D woven preforms and 2-D materials intersect, and as reinforcements to prevent delamination growth. Z-fiber reinforcement provides damage tolerance and survivability advantages in composite structures. However, this must be balanced with local decreases in in-plane properties compared to 2-D components with no through-thickness reinforcement. Knockdowns in tensile and compressive strengths have been determined to range from zero up to 40% in local Z-fiber reinforced areas. The knockdowns are dependent on the areal density of the Z-fiber reinforcement and the percentage of in-plane fibers. Z-fiber reinforcement can provide a 30-to-50 fold improvement in interlaminar fracture toughness. This translates into a significant improvement in damage tolerance and ballistic survivability when applied in cocured composite structure. In addition, the Z-fibers can eliminate the need for radius blocks and fasteners at stiffener ends. These improvements can prevent catastrophic failure from propagation of a manufacturing flaw or in-service damage event.

3-D Composite Application for Bulkheads and Frames

Composite utilization in the heaviest parts of the airframe such as highly loaded fuselage bulkheads and moderately loaded frames has been virtually non-existent because of the poor interlaminar properties of conventional laminated or two-dimensional reinforced composites. Because of these low interlaminar properties, composites have not been used in structures subjected to high out-of-plane or complex loads without requiring costly integration of metallic fittings and fasteners. This also often results in an increase in the amount of composite material required to accommodate the low bearing strength of typical polymer matrix composite structures.

The development of composite structures that can react out-of-plane loads will greatly reduce the reliance on bolted joints and will offer improved structural efficiency. One of the primary keys to lowering the cost of composite structures is to minimize touch labor, part count, and fastener count through unitized structural assemblies. Figure 3 shows a moderately loaded composite bulkhead that primarily reacts engine hammershock and fuselage bending loads. The composite bulkhead meets the structural load requirements of the corresponding F-22 aluminum part with a 10% weight savings. In this bulkhead, the 3-D preforms were staged and cocured with conventional laminated composite details during final cure. The staged details are net shape prior to the final cocure assembly. The 3-D textile preform was manufactured by Techniweave for this part. The preform was incorporated to react large punch loads introduced by the opening of the attached weapons bay door actuator. The insert data show that the 3-D inserts carry approximately twice the load of the baseline uni-tape insert. The cruciform data demonstrate that the bonded joint with embedded cruciform carries three times the load of the baseline bonded joint.

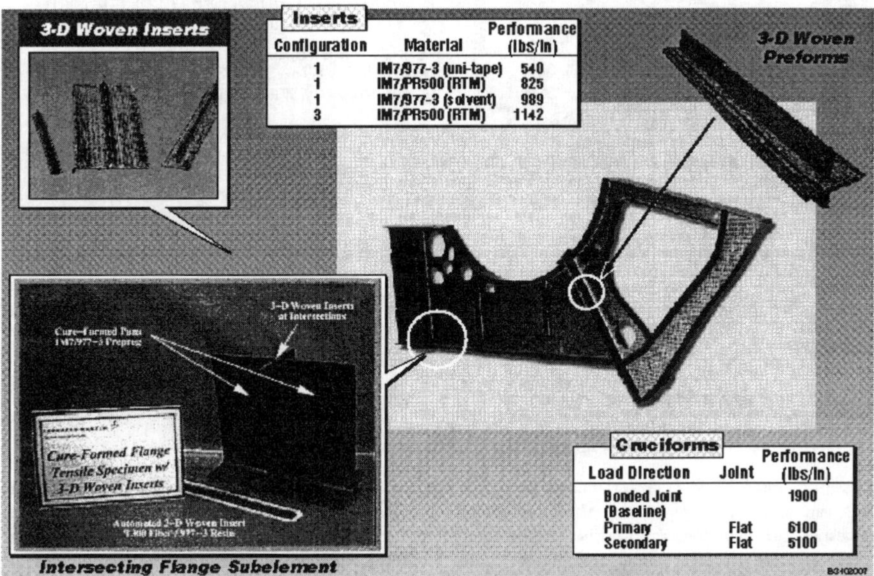

Inserts		
Configuration	Material	Performance (lbs/ln)
1	IM7/977-3 (uni-tape)	540
1	IM7/PR500 (RTM)	825
1	IM7/977-3 (solvent)	989
3	IM7/PR500 (RTM)	1142

Cruciforms		
Load Direction	Joint	Performance (lbs/ln)
Bonded Joint (Baseline)		1900
Primary	Flat	6100
Secondary	Flat	5100

Figure 3 Generic F-22 Laminated Composite Bulkhead with Embedded 3-D Preforms

Current composite structure designs require labor intensive fabrication and assembly which account for up to 60% of the total cost. The use of integral stiffened skin structure and application of 3-D composites at out-of-plane intersecting structures can allow unitization, i.e., reduced fastener and part count. In addition, the long term support costs will be reduced since the majority of repair actions in fielded fighter air vehicles are conducted on mechanically fastened joints which will be reduced through the use of unitized structure. Support costs can be reduced by deletion of fastener inventories, reducing fuel seal problems, improved fatigue life, and reduced weight.

3-D Composite Application for Sandwich Structure

Sandwich structure is the most weight efficient design for a large percentage of an aircraft structure because of buckling stability requirements. Unfortunately, use of sandwich structure is limited on state-of-the-art aircraft due to the high manufacturing and supportability costs associated with metallic honeycomb structure on earlier aircraft. Figure 4 shows the B-58, which used a high percentage of aluminum sandwich construction. The illustration shows a typical wing joint in which the sandwich was full depth through the joint with a metallic insert (internal slug) for joining the skin to the spars. The spars were spaced fifteen inches apart to reduce the parts count. The mission requirements of the B-58 dictated low structural weight fraction (0.24) and good thermal and sonic fatigue properties; this drove the design to sandwich construction.

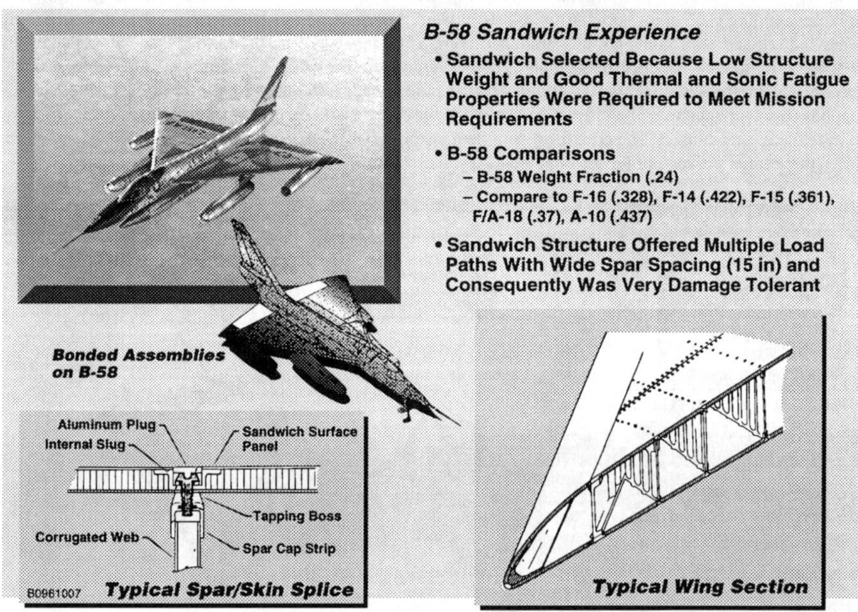

Figure 4 B-58 The Bonded Bomber

These structures, however, experienced high maintenance costs due to moisture retention, severe corrosion, and damage tolerance problems. The emergence of composite materials, novel core concepts, tougher adhesives, textile preforms and Z-fiber reinforcement have now created a new opportunity to take advantage of the superior structural efficiency of sandwich and monolithic integral stiffened construction without the penalty of maintainability problems. These emerging technologies apply non-corrosive honeycombs, utilize foam filling to reduce moisture retention and improve damage tolerance, and employ design concepts that eliminate leak paths.

The use of noncorrosive materials is one step to applying honeycomb core sandwich structure for eliminating supportability problems. For strength to weight reasons the preferable material is titanium.

Another benefit of titanium is it does not allow the migration of moisture or fuel from damaged skin areas because each cell is sealed with metal walls.

Using innovative bonding approaches with 3-D woven preforms, sandwich structure can be attached to other structure or other sandwich structure with simplified assembly techniques. Using several different shaped 3-D preforms, a variety of joints can be built for assembling sandwich structure. Also, several bonding techniques (cocuring, cobonding, and secondary bonding) and Z-direction reinforcements can be employed to enhance properties and/or simplify assembly methods when used with preforms. As a result, sandwich structure is now an option in areas not traditionally considered. The ROCSS first generation 3-D woven preforms have been further developed to enhance the joint efficiency use for applications in monolithic (non-sandwich) and sandwich joints.

Structural concepts applying 3-D woven materials for stiffening laminate structure and for usage at intersecting substructure joints are under development. Using "π" or "T" shaped 3-D woven preforms as a portion of a hat or blade, durable high-pull-off-load stiffeners can be bonded to panels. Depending on tooling and fabrication desires, these stiffeners can be cocured, cobonded, secondarily bonded, or some combination of the three bonding methods.

As manufacturing techniques improve and new material forms become available, such as short discontinuous fiber composite materials, stiffening concepts like waffle grid patterns have become viable. Like honeycomb, the thin-wall waffle grid (TWWG) uses large percentage bond areas with significant lightweight shear load paths. The primary advantage of TWWG over core is that it can offer more depth than sandwich for improved efficiency without sacrificing fuel volume. Also, TWWG can be secondarily bonded with some adhesives without autoclaving.

Current high-payoff applications feature the use of integral stiffened or sandwich skins in highly loaded fuselage and wing structure. The unitized center fuselage shown in Figure 5 can make extensive use of advanced composite integral stiffened structure and 3-D composite preforms and Z-spiking in the joint

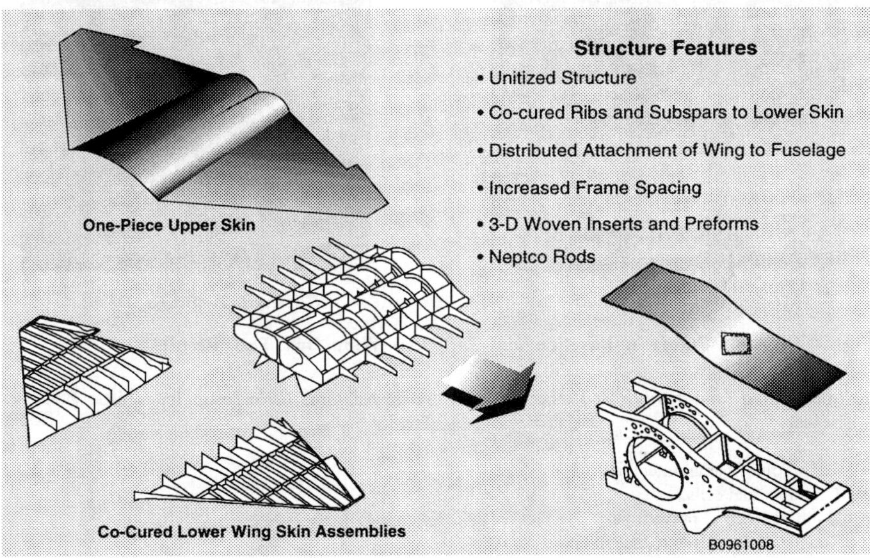

Figure 5 Unitized Center Fuselage Application of Advanced Sandwich Structure for a 21st Century Fighter

intersections. Extensive integration of the wing and fuselage will significantly reduce weight, fabrication costs, and maintenance costs. Parts count is reduced by cocuring (or cobonding) the lower wing skin with ribs and subspars, and building the upper skin in one piece, tip-to-tip. Longeron, spar, and bulkhead caps are embedded in the skins to provide unbroken primary load paths. Reducing parts count will lead to a more cost effective airframe and the unitized, efficient structural approach will reduce airframe weight fraction.

Ballistic Survivable Composite Structure

Testing has been conducted at Wright Laboratory to determine the response of cocured wingbox composite designs with Z-spiked joints to high explosive incendiary ballistic threats. This testing evaluated the merits of a cocured lower skin to I-section spars concept with Z-spiking in the joint interface against a conventional composite skin with aluminum substructure design. The ballistic tests were conducted under fully simulated combat conditions which included airflow, ballistic impact, structural load, hydraulic ram and blast from a high explosive incendiary projectile.

The composite cocured design allowed the joint to rotate as the explosive pressures impinged on the skin/spar substructure and placed the entire fuel cell in tension. The cocured design enabled the joint to act as a fully integrated assembly. This was unlike the skin of the mechanically fastened baseline design that rotated independently from the substructure. By designing the cocured wingbox with laminate plies blended from the skin into the substructure the Z-spiked spars were able to rotate and the skin did not fail at the fastened joints as in the baseline tests. The skin was loaded as a tension membrane that initiated a spanwise crack across the impacted base of the test section. Figure 6 shows the results of the ballistic test for the baseline fastened wingbox and for the cocured Z-spiked wingbox.

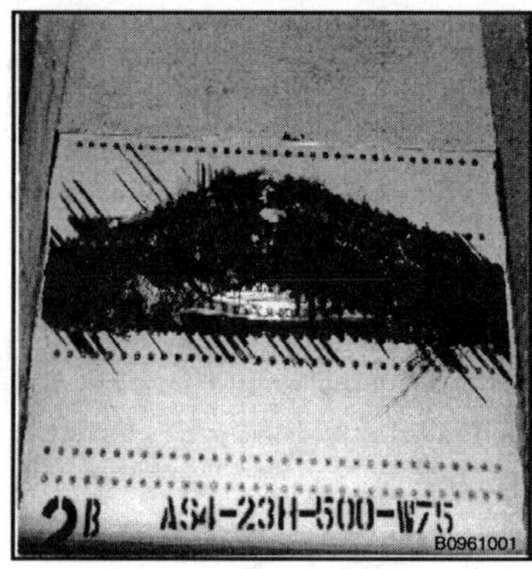
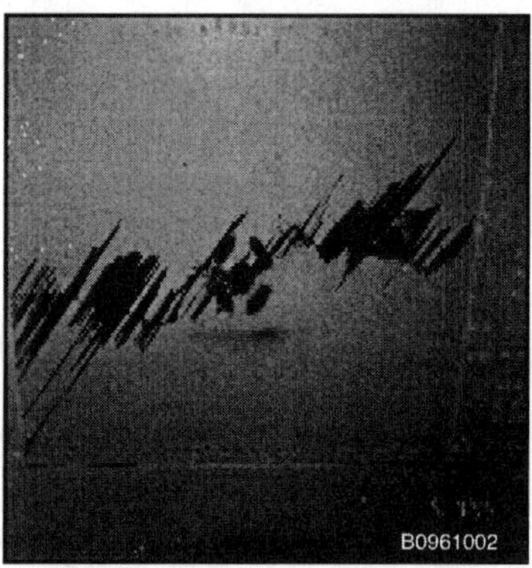

<table>
<tr><td align="center">Mechanically Fastened</td><td align="center">Cocured Z-spiked Design</td></tr>
</table>

Figure 6 Baseline and Z-Spiked Wing Box Test Results

Lessons learned from this box and other ballistic tests reveal some basic design-for-ballistic guidelines. These include:

- Tailor joint fiber architecture.
- Decouple fuel cells.
- Z-pin or stitch to prevent peel.
- Toughen materials to reduce damage.
- Hybridization – buffer strips, coweaving, tough resins.
- Use buffer strips.

62

- Establish zones of controlled failure through stiffness tailoring. Want energy to escape through skins and load to redistribute around joint.
- Provide weak point in the skin at desired failure locations.
- Use soft skins with hard buildups at spars – wide spar caps are good – soften skins to carry aerodynamic loads and torsion only.
- Make joints elastic – avoid stiff joints like sinewave spars.

Current evaluations are assessing these approaches.

ANALYSIS AND CERTIFICATION

The certification issues and design methodologies for 3-D composites center around the measurable quality of the as-manufactured structure, and the resulting allowables. The quality of a part is typically measured by ultrasonic and x-ray non-destructive inspection (NDI) techniques. The NDI of 3-D composite joints will be less critical because of the increased strength and inherent damage tolerance. The ability to resolve flaws in complex 3-D joints (such as those using inserts, preforms, Z-fibers, and cocured assemblies) will to a large extent drive the manner in which allowables will be calculated for a particular joint technology. Analysis tool and methodology development, physical tests, and design and certification advisors will all reflect the flaw resolution limitations imposed by quality inspection capabilities. Though advanced inspection techniques such laser ultrasonics can be readily applied to inspect 3-D composites with a high degree of accuracy, structural load, stress, and strain allowables development needs to reflect manufacturing variability so that flaws produced by typical manufacturing capability are accounted for in the allowables determination. Allowables development will also be based on the manufacturing process (e.g., RTM, resin film infusion, and electron beam) and the knowledge gained from testing 3-D joints manufactured with these technologies.

The many advantages of 3-D composites have already been discussed above. However, there is much work to be done to provide a production structural certification capability to allow the structural designer to take advantage of all these promises.

The certification of structures containing 3-D preforms will center around the development of design specific mechanical properties, allowables, and analysis methods development. The calculation of woven properties, particularly at intersections, has proven to be very difficult. To develop a-priori certification techniques the properties and allowables for the woven or braided inserts alone (not in a joint) must be developed, along with basic joint allowables. This is particularly necessary when calculating margins-of-safety for the insert, and when specifying material response properties for detailed finite element analyses. As previously noted, mechanical allowables will be processing-, structure-, and materials-dependent. Because of the lack of maturity in complex 3-D composite material response prediction, a comprehensive testing and analysis correlation will be needed in the near-term.

The methodology and criteria must be updated to account for the effects of manufacturing defects and delaminations in Z-fiber reinforced structures if the advantages of the delamination arrestment capabilities are to be realized. Static test programs to date have shown that significant gains in delamination arrestment are possible. However, a comprehensive test program to document the behavior of post-delaminated 3-D composite reinforced joints and components must include not only static strength but service life testing for multiple lifetimes under realistic loading spectrums. Once the joint stiffness behavior of post-delaminated 3-D composite reinforcement structure has been completely characterized, the possibility of changing aircraft certification criteria to allow flight of locally disbonded structures will become a possibility. This will make possible the potential weight and safety benefits promised by the early structural tests.

Joints containing cocured or cobonded 3-D composites preforms can exhibit a failure mode characterized by initial cracking followed by significant additional load carrying capability prior to rupture, as shown in Figure 7 by the graphite T-Blade test data. The initial failure in this specimen was a delamination in the wrap plies at 480 lb/in which did not grow. The ultimate failure which occurred at 911 lb/in was a delamination failure of the skin laminate at the 3-D preform interface. From the testing performed thus far,

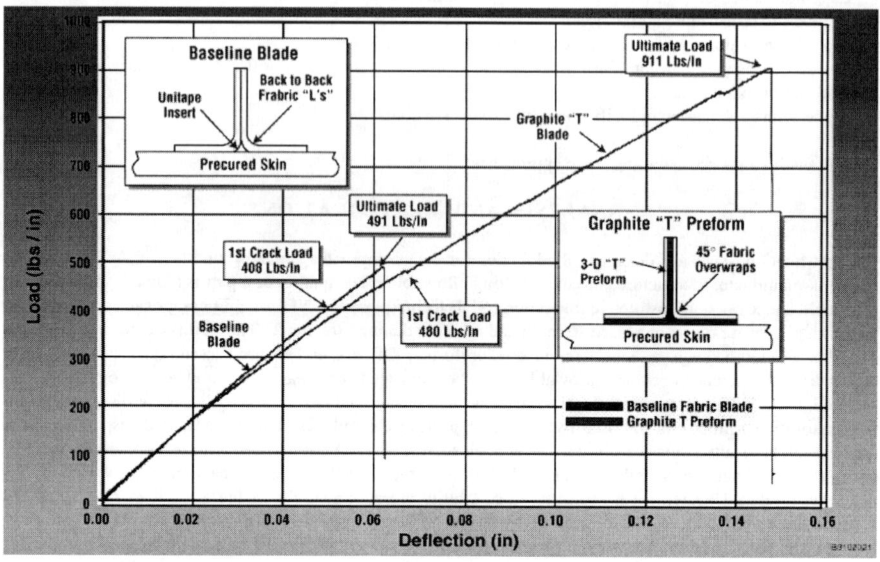

Figure 7 Blade-Stiffened Panel Test Results With Embedded 3-D "T" Preform Versus Baseline

there appears to be little propensity for the crack (usually a localized delamination in the attached skin, noodle of the 3-D preform, or overwrap plies), to propagate under cyclic loading at or below design limit load (DLL). Failure criteria for joints constructed of 3-D preforms and/or Z-fiber reinforcement could be sized using criteria shown in Figure 8.

The justification for the above failure criteria hinges upon the development of new 3-D composite joints (with 3-D preforms, and/or Z-direction reinforcement) which exhibit progressive failure under cyclic load as opposed to the often classical explosive, catastrophic failure characteristic of 2-D composite joints as exhibited by the baseline blade shown in Figure 7. The initial failure occurred in the unitape radius insert at 408 lb/in followed shortly by catastrophic ultimate failure at 491 lb/in.

There are many gaps in the capabilities of 3-D composite joint analysis tools and methodologies. As previously mentioned, tools to reliably calculate material properties will provide the cornerstone that will enable detail design and structural certification. These toolsets are being co-developed by the airframe industry under the Composites Affordability Initiative (CAI) program. Building upon this methodology will be the application of finite element analysis (FEA) techniques for certification. Both 2-D (plane stress/strain and shell elements) and 3-D (solid elements) FEA analysis techniques must be developed and verified by test comparisons.

Testing will perform a crucial role in the development of 3-D structural certification tools and methodology. Each of the aspects of 3-D materials and joints must be characterized both within a structure and individually. Examples already discussed include woven inserts and Z-fiber reinforcement. One of the

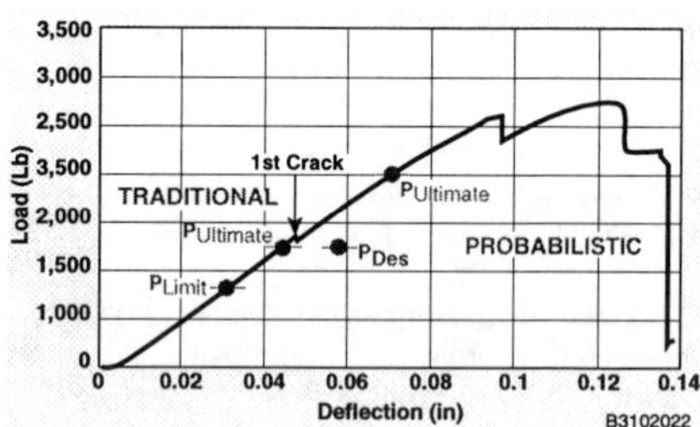

Figure 8 New Failure Criteria Will Exploit Damage Tolerance Characteristics

most difficult aspects of testing is the accurate determination of out-of- plane properties for 3-D laminates. This is only made more complicated by the extremely complex geometries of fiber placement in the woven and braided intersections. A comprehensive program to test these properties combined with the development of mathematical material response property calculation tools will provide tools capable of accurate predictions of structural strength and margins-of-safety. Once these tools have been validated, future mechanical tests can be minimized.

It is clear from the above discussions that the design and certification of 3-D woven insert joints will require the application of a large body of knowledge gained from test and analysis. Durability and damage tolerance approaches will have to be a mix of experience-based heuristics and analyses until much more sophisticated mathematical models are available. The complex body of knowledge for 3-D composite joint structural certification, design technology, and methodology is very suitable to automation in either a procedural or rule based advisor system. The advisor will enable information and skill sharing in the integrated design-analysis environment in today's product development approach by minimizing the chance that one of the many aspects of 3-D composite joint design and certification are not considered.

SUMMARY AND CONCLUSIONS

The application of 3-D preforms and Z-fiber reinforcement in continuous fiber reinforced composite materials offer future aircraft significant improvements in aero-performance, maintainability, and survivability over current in-service composite structures. Incorporation of 3-D textile preforms in 2-D laminates makes it feasible to use polymeric composite materials for structural applications that were previously not possible from a cost and performance perspective. To obtain these benefits will require continued maturation of the processing and manufacturing technologies required to (1) fabricate net shape preforms and (2) cure these preforms into precision affordable assemblies. Composite certification and design methodology development is also underway, however it will require continued attention to be brought to full maturity.

Z-fiber reinforcement offers improvements in out-of-plane joint strength, damage tolerance, and ballistic survivability compared to current cocure structures. The Z-fiber manufacturing process is readily adaptable to current cocure tooling approaches. The cost of in-process Z-fiber insertion will be offset by elimination of fasteners currently employed for joint strength or to meet damage tolerance or ballistic survivability requirements.

Emerging 3-D composite and innovative structural joint concepts will offer future generation aircraft both weight reduction and cost reduction through unitization and through significant improvements in load carrying capability and survivability.

REFERENCES

1. Burgess, K., Paradis, S., "Application of 3-D Woven Preforms to Laminated Composites," Paper No. 95-3890, Proceedings of the 1st AIAA Aircraft Engineering, Technology, and Operations Congress, September 1995.

2. Fisher, K., "3D Reinforcements: on the Verge?," High Performance Composites Magazine, March/April 1996.

3. Gause, L. W., Alper, J. M., "Braided to Net Section Graphite/Epoxy Composite Shapes," Journal of Composites Technology & Research, 1988.

4. Baron, W., "Survivability of Multispar Cocured Composite Wing Structure," 37th Structural Dynamics and Materials Conference, April 1995.

5. Bersuch, L., Hunten, K., Baron, W., Tuss, J., "3-D Composites in Primary Aircraft Structure Joints," Paper No. AGARD-CP-590, 83rd AGARD Conference, 1996.

43rd International SAMPE Symposium
May 31-June 4, 1998

METHODS TO STREAMLINE RESEARCH IMPLEMENTATION IN PRODUCTION OPERATIONS

Tom DeMint and Richard Struve
Boeing Commercial Aircraft Group
Seattle, WA 98124

ABSTRACT

As companies begin to enter the aerospace market, we see established companies beginning to focus on their core abilities. This can result in scrutinizing of manufacturing research budgets, increasing the demand for applied research. The ability to phase new manufacturing processes quickly and effectively into production thus becomes an important competitive advantage. The process of applying manufacturing research and development in the aerospace industry, though often tedious and costly, can be streamlined using available tools in the industry. Front end tools for characterizing technical and economic feasibility can be used to launch a promising concept. Further down the road, statistical methodologies can be applied to harden a process for a factory environment. These tools are demonstrated in a case study for a new commercial aircraft interior manufacturing process. The application of these methods will demonstrate how they can be used to minimize the development time for a robust production process.

KEY WORDS: Process Capability, Aircraft Interiors, Thermoplastic Sandwich Panels

1. INTRODUCTION

To remain competitive, manufacturing organizations are continuously looking for ways to reduce the cost of building products and reduce the time required to build the product and introduce new products to the market. If the current manufacturing methods are not suited to this strategy, then a completely new way of doing business may be needed. Even the most promising "breakthrough" initiatives can run amuck if the approach is not well planned and executed. Involving all the affected organizations and establishing a sound technical approach can lead to a successful implementation of research initiatives.

The authors were, among other team members, involved in developing and implementing a new manufacturing process for building aircraft interiors. In doing so, a methodology was used for bringing a robust manufacturing process from concept through implementation in a logical fashion that allows progress to be quantified. The methodology proceeds in four stages: Concept Demonstration, Feasibility Study, Implementation, and Production. This methodology is described in the context of the following case study.

1.1 Concept Origin: A Research Initiative to Reduce Overall Cost   Boeing resources were teamed up to reduce the cost of airplane interiors. Initially, Payloads Engineering, Materiel, Boeing Materials Technology, and Operations Technology teamed up to develop new technologies for airplane interiors. Each organization assigned one person to the team. The objective for this work was to find technologies that could reduce the overall cost of interiors; the design, fabrication, and assembly costs all-inclusive. Suppliers to Boeing were invited to participate. Several new materials, new design approaches, and new fabrication processes were proposed.

1.2 Our Direction, Strategically, Tactically   The research approach was to find technologies that contributed advantages on multiple fronts; to reduce design and planning time; to reduce fabrication time; to reduce assembly time; to reduce overall flow time in the interiors fabrication factory. We focused on technologies that either eliminated or combined current production processes. Also, we looked to reduce the total part count within assemblies.

1.3 How TSP was Found   Several suppliers proposed Thermoplastic Sandwich Panels (TSP) as a means to achieve the objective. Thermoplastic sandwich panels are defined as panels having lightweight cores with fiber-reinforced thermoplastic facesheets (skins) on both sides of the core. Such panels were supposed to offer advantages in design, fabrication, and assembly (1).

1.4 Why TSP Appeared to Fit Our Interests   Thermoplastic sandwich panels (TSP) appeared, on a speculative level, to have cost advantages over incumbent designs and incumbent production processes. Incumbent designs employ thermoset sandwich panels. The thermoset raw materials require refrigeration, while thermoplastic raw material does not. Some incumbent designs employ panel bending to achieve formed shapes, but such bent panels are laboriously fit and faired due to inaccuracy in the incumbent panel bending process. TSP can also be bent or folded to make formed shapes (1). Bent TSP panels produced by suppliers did not appear to require any fit/fair rework.

Some incumbent designs join multiple flat panels together into assemblies. Reducing the total number of parts within an assembly generally leads to overall cost reduction (2). Despite the potential for reducing part count, conventional bend forming is not used within incumbent designs to reduce the number of parts in an assembly; this is because of the inherent problems of fit/fair and inaccuracy related to the conventional bend forming process. Bend-formed TSP panels, however, could replace multiple part assemblies if the TSP bend formed panels met the overall design requirements. On the basis of these speculations, TSP bend formed parts, and the TSP bend forming process were chosen for research. The primary questions to be answered by this research were, 1) Are bend formed TSP panels capable of meeting the design requirements? and 2) Are assemblies made with bend formed TSP panels less expensive?

2. Concept Demonstration: Determining Requirements and Evaluating TSP vs. Requirements

The steps taken to evaluate the feasibility of TSP started with determining the typical design requirements for aircraft interior panels. The performance of TSP panels was then evaluated and compared to each typical requirement. Concurrently, an economic comparison was made between an incumbent design and the same assembly made with TSP.

A variety of opinions existed about requirements for aircraft interiors. The research team initially set its own requirements and targets for TSP. Due to the various technical and business disciplines represented by team members, a wide range of requirements was established, including design performance requirements and targets, fabrication requirements, and cost targets. A requirement was set if some characteristic of the panel or assembly had a hard, unalterable, limit. For example, self extinguishing, fireworthy, aircraft interiors material is mandated by a hard requirement, the Federal Aviation Regulation FAR23. A target value was set if a characteristic of the panel was desirable, but the exact performance limit, a limit beyond which the panel or assembly would certainly fail in-use or in-factory, was unknown. In the absence of a known performance limit, the performance value of the incumbent design was chosen as the target for the TSP design. It was important to distinguish between requirement characteristics and target characteristics. By doing so, characteristics of incumbent designs which may be excessive could be still used to represent target values. If TSP failed to meet such a target, then the TSP would require further analysis to ensure its fitness for use. It was thought this approach would avoid disqualifying TSP if it turned out some characteristics of TSP are satisfactory for its fitness for use, but potentially lower than comparable characteristics of the incumbent design. The exact numerical values for many characteristics of Boeing incumbent designs are considered proprietary, and are not presented here. Table 1 indicates the properties and targets that were initially compared to measurements taken for TSP.

2.1 Panel Bending Process Development   Corner strength is the maximum load a bend-formed panel can withstand. The inferior value of TSP corner strength indicated in Table 1 was cause for investigating the bending process. TSP bending is generally accomplished by the steps illustrated in Figure 1. A proprietary method was developed that increased corner strength from 2.4 kg/cm. to 6.1 kg/cm. The corner strength measurements for this research were carried out using the test rig in Figure 2.

The panel bending process was developed on a small scale prototype. The machine was entirely hand operated. It required several people at a time to coordinate movements of the heater, the panel, and the folding blade. The folding blade is the component that performs the actual folding operation, applying a load to the folded panel. This load was manually exerted on the folding blade by a lever attached to the end of the blade. This arrangement lead to torsional deflection in the blade. Torsional deflection of the folding blade limited the capacity of this machine. Even though panels as wide as 43 cm could be bent (folded) on this machine, its capacity was limited to 15 cm wide panels. Wider than this, and the resulting panels were found to have measurable variation in fold angle due to the twisting deflection of the folding blade.

2.2 Initial Scouting Experiments and Economic Studies   Scouting experiments were run to evaluate the performance of TSP against requirements and targets. In general, only five samples were run for each requirement or target. Also, the scouting experiments include only one of the many possible configurations of facesheet type and core type. This singular panel configuration was 2 ply 7781 fiberglass/thermoplastic for each facesheet, and 1.36 kg Nomex®/phenolic honeycomb core. Even though aircraft interiors production currently involves many different configurations, the scouting experiments focused on this single construction. Table 1 includes results for a number of Manufacturing Requirements; for example, Compatibility with mechanical trimming process. To improve our confidence in the results, we worked with production personnel and production equipment/processes to fabricate the samples for evaluating manufacturing performance of TSP. The TSP panel bending process itself was the only exception to this initial involvement of production personnel.

Table 1. Results of Scouting Experiments.

Property	Test Method	Units of Measure	Requirement (R) or Target (T)	Result for TSP: Superior (+) Acceptable (~) Unacceptable (!)
Sandwich tensile strength	MIL-STD-401B	psi	T	+
Sandwich compressive Strength	MIL-STD-401B	psi	T	~
Sandwich shear strength	MIL-STD-401B	psi	T	~
Sandwich flexural strength	BMS 8-226G	psi	T	+
Facesheet peel strength	BMS 8-226G	lbf/3inch width	T	+
Flammability	BSS 7230	in sec	R	+
Smoke generation (Optical smoke density)	BSS 7238	Ds at 4minutes	R	~
Toxic gas generation	BSS 7239	ppm	R	+
Heat release, flaming mode	BSS 7322	$kW\text{-}min/m^2$ kW/m^2	R	+
Flame spread index	BSS 7304	-	R	+
Burnthrough resistance	BSS 7306	-	R	~
Panel weight	MIL-STD-401B	lbm/ft^3	T	~
Decorative Laminate peel strength	ASTM D 903	lb.	T	~
Fastener insert pullout strength	BMS 5-28	lb.	T	~
Folded Panel corner strength	WR-920337 WR-9203706	lb.	T	!
Decorative edge trim adhesive bonding	WR-9300060	lb.	T	~
Panel folding process dimensional accuracy	Trials by research team	angle in degrees	T	~
Compatibility with mechanical trimming and drilling process	Trials by production personnel	trim and drill edge quality	R	~
Economics		direct labor raw material flow time part count	T	+

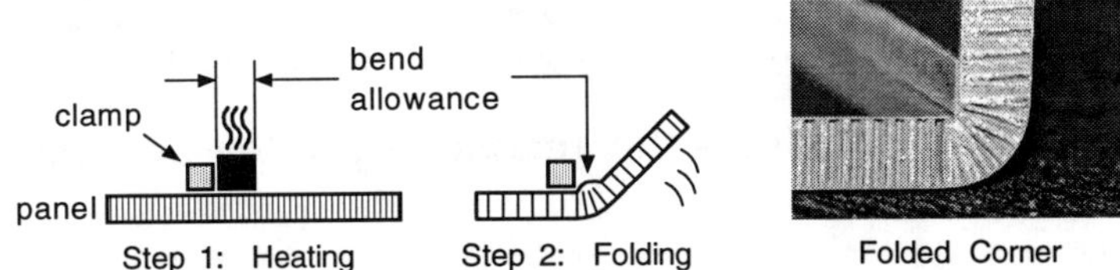

Figure 1. Basic steps of TSP thermofolding (panel bending) process.

2.3 Initial Economic Study Based on Only One Part Type   The initial economic study focused on a single candidate part. The part is a floor-mounted stowage box from the 747-400. The part uses similar processes found in the fabrication and assembly of closets and other stowage bins. Because of this similarity, we reasoned that the results of this study could be used to identify other application opportunities. The box has three major sub-assemblies, the front panel, the back panel and the lid assembly (Figure 3). The incumbent design called for two 90° bends to fabricate the u-shaped front panel. The initial economic study looked only at raw material costs, direct labor, flow time and part count reduction. The economic study did not include other part types for which TSP could be applied. We decided that if TSP could pass our economic target for this single candidate, then we could recommend that the scope of the study be increased.

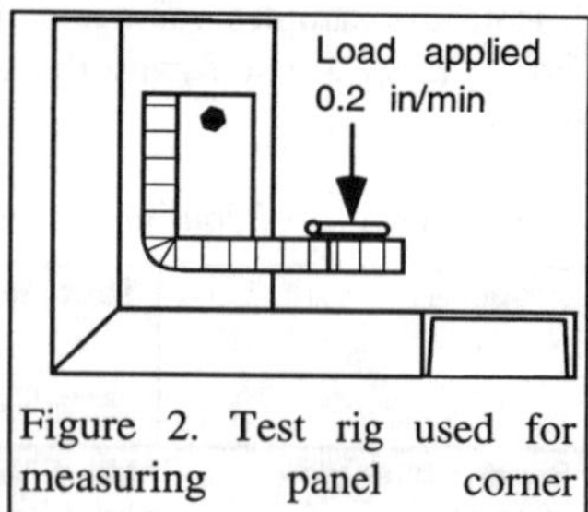

Figure 2. Test rig used for measuring panel corner strength.

2.4 Initial Go Decision and New Feasibility Study   After the corner strength was improved, the targets and the requirements shown in Table 1 were either met, or surpassed. The research up to this point indicated that TSP would work, and it would be beneficial, but we could only be confident about this for a single part type, and a single panel configuration. Up to this point, the investment of resources in studying TSP was kept to a minimum by limiting the scope to one part type, one panel type, and minimal sample numbers. To explore the feasibility of TSP in applications across many different part types, more research was required. Based on these initial results, the research team determined that enough was known about TSP to warrant a deeper and broader study on the feasibility of TSP.

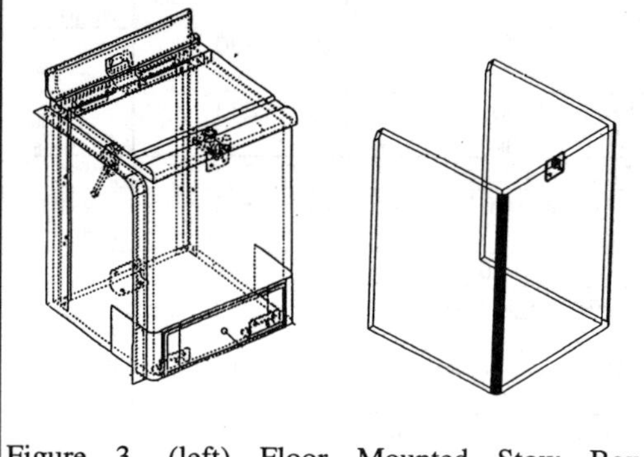

Figure 3. (left) Floor Mounted Stow Box Assembly, (right) front panel only.

The initial results, along with recommendations to start a new feasibility study, were presented throughout the Interiors Responsibility Center (Boeing's business unit for design, fabrication and assembly of aircraft interiors). These presentations were made to non-management planners, NC programmers, and design, process and tooling engineers. This proved useful in gathering ideas for suggested TSP applications across many different part types. Finally, the initial results were presented to Interiors Responsibility Center management. A new feasibility study was authorized to study TSP, to determine the entire scope of its applications, to estimate its impact if applied to such a wide extent, and to involve a wider set of skills on the study team so that a more thorough set of targets and requirements could be evaluated.

3. Feasibility Study: Validate the Technology's Key Components at Full Scale

One of the most significant aspects of the new feasibility study was its objective to evaluate TSP at full scale for a wide range of applications. As shown in Figure 4, the major areas of effort for the feasibility study proceeded concurrently with the idea of using the experience gained in fabricating full-scale parts to estimate the eventual impact of TSP on the business and people at the Interiors Responsibility Center.

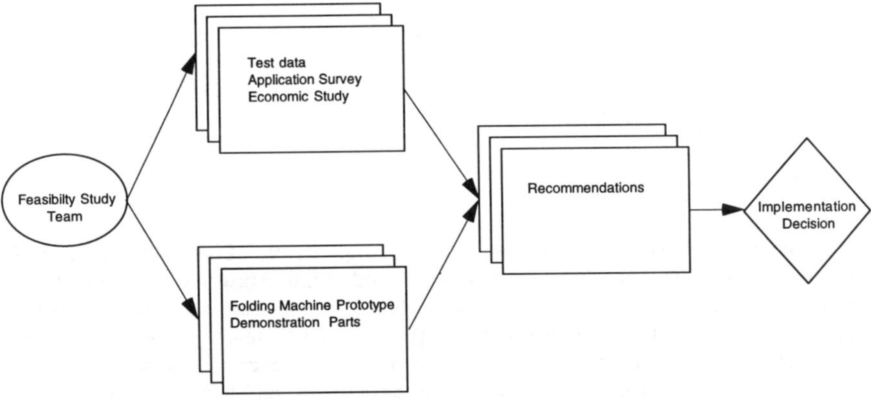

Figure 4. Activities in the Feasibility Study.

3.1 Mechanized Folding Machine Prototype Parts chosen for feasibility study are shown Figure 5 through Figure 8. The fold line length (panel width) for some of these parts greatly exceeded the capacity of the manually operated panel bending machine. The force exerted by the machine and the machine stiffness needed for TSP panel bending demanded that we develop a mechanized folding machine. Such a machine was developed, giving a capacity to fold panels up to 305 cm wide. The scope of the feasibility study included evaluating the accuracy and repeatability of folds produced with the folding machine. Also, the folding machine, though regarded as a prototype, was used as a reference for estimating the capital cost of a production TSP folding machine.

Figure 5. Overhead stowbin back shell, 777.

Figure 6. Emergency equipment rack, 737.

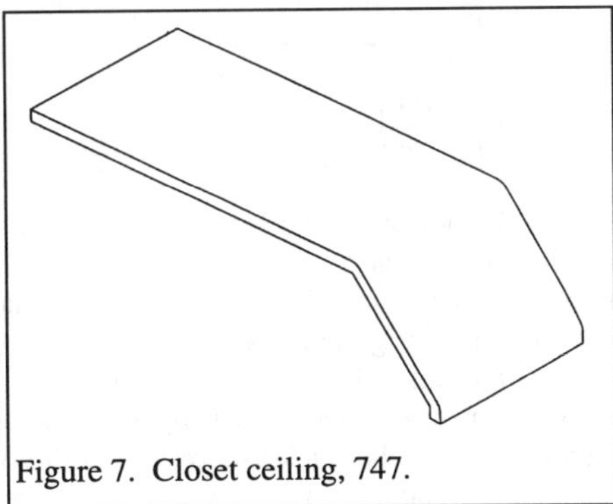

Figure 7. Closet ceiling, 747.

71

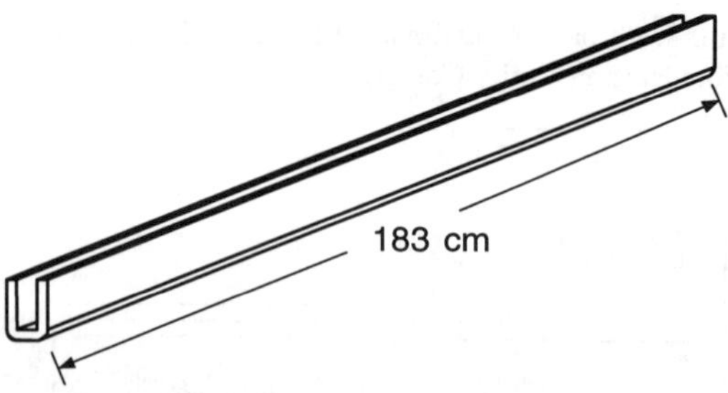

Figure 8. Air conditioning plenum, 767.

3.2 Involvement of Design Engineering, Tool Engineering, Planning, Industrial Engineering, Process Engineering, Production, Production Materiel, and Facilities Engineering   The feasibility study formally included people from all areas of design, fabrication, and assembly. This approach was thought to enable the team to look at the issues surrounding TSP from the widest possible perspective; to uncover all the technical and economic questions that needed to be answered before we could confidently state if TSP was feasible for aircraft interiors production. All team members were asked to estimate how TSP would impact their activities on a production basis. For example, would their effort designing, fabricating, assembling or supporting production be affected by using TSP? How would each person have to react to implement TSP? Would new equipment be required, new documentation, new safety measures, etc.? If some impacts were uncertain in the opinion of a team member, the team took on the responsibility of developing enough new information about TSP so these impacts could be evaluated.

All the initial study methods were repeated, this time on the new part types and panel types selected for the feasibility study (see Table 2). Also, as the team had hoped, new issues were uncovered due to the broader cross section of people on the team. The following are all examples of new issues that emerged because of the new people who came onto the team: panel flatness and thickness variation, raw material handling and storage methods, compatibility of repair and folding processes, and structural adhesive bonding. As in the initial study, the number of samples was typically limited to five replicates.

Table 2. Types of parts and panel configurations in the feasibility study.

Part reference	Facesheets	Core
Figure 3	2 ply, 7781 glass	1.27 cm x 1.36 kg
Figure 5	2 ply, 3K-70-PW graphite	0.95 cm x 1.36 kg
Figure 6	2 ply, 7781 glass	0.95 cm x 1.36 kg
Figure 7	2 ply, 7781 glass	2.54 cm x 1.36 kg
Figure 8	1 ply, 3K-70-PW graphite	0.64 cm x 1.36 kg

3.3 Application Base Determined   One of the biggest challenges to estimating the impact of TSP was estimating the total number of TSP applications. TSP would not be implemented overnight, so how could we estimate which future designs would someday call for TSP? We reasoned that design engineers for today's parts were best able to judge the requirements for tomorrow's parts. For this reason, the responsibility was largely on the design engineering organization to estimate which parts would someday be TSP. It was therefore considered appropriate for the leader of the feasibility team to be a design engineer. A survey was conducted among design engineers and planners in all the functional groups: stowbins, closets, partitions, stowage, and ceilings. Each person was asked to look at their area's incumbent designs, and determine which could be good TSP designs. Results of this survey were compiled into a list of potential applications for TSP. This list was used as the application base for financial evaluations. By multiplying the current production rate of incumbent designs by the quantity of those designs per plane, we came up with a yearly estimate of maximum TSP application.

3.4 Financial Benefit Estimated for Phased Implementation on New Airplane Designs
We reasoned that future production rates would be approximately equal to current production rates. We also reasoned that, as new airplane models are produced, TSP would be implemented in new designs that replace candidates on the earlier and similar airplane models. This mode of implementation avoids redesign costs that arise from substituting TSP into old designs. As new airplane models are designed, TSP could be introduced without cost of design change. By estimating the phased introduction schedule of new airplane models, we developed a phased rate of annual TSP application.

The results of this application estimate were used to evaluate the required quantity and the required capacity of TSP folding machines for future production. We found that one machine would be sufficient for up to five years, and then a second unit would cover all remaining applications out to ten years. All of the implementation costs (non-recurring) were estimated, as were the recurring benefits of TSP. The return on investment was estimated at 33%.

3.5 Implementation Decision   The feasibility study results were presented to Interior Responsibility Center management along with the recommendation to implement TSP on new airplane designs. Also, presented were the teams findings on financial and technical risk. Schedule and production rate uncertainty was discussed as a risk; potentially a downside risk if TSP implementation proceeded slower than anticipated. The uncertain ability of TSP to enable new designs that were, until now, impractical with incumbent material/production systems was also discussed; potentially an upside risk because such new designs were not accounted for in the application base. The recommendation to implement TSP in aircraft interiors production was accepted. The feasibility study had determined, evaluated, and recommended what activities were required to implement TSP. It evaluated those activities to estimate the cost of implementing TSP, and to layout a template for the eventual implementation team to follow. Following this recommendation, Interiors Responsibility Center Management authorized a team to go and carry out these implementation activities. Up to this point, the Interiors Responsibility Center management was our internal customer. From this point on they will be referred to as the internal customer.

After the internal customer has bought into the concept and agreed to scale it for production, the task of preparing the process for production began. The following steps describe the method for bringing a process from feasibility to robustness.

4. Implementation

4.1 Define the "Product" A flexible process such as thermofolding can produce products of many shapes, as shown in Figure 5 through Figure 8. For initial capability studies of a new process being developed for production, we must choose one item to study which economically represents all or most of the marketable features of the product, while representing the natural variation of those features as well. We therefore chose the item illustrated below for the following reasons.

1. It *economically* represents all of the marketable features of the product (defined below as "key characteristics").
2. Its simple shape allows easy measurement of the key characteristics, also described below.

3. It spans two of the five folding machine zones (see Figure 22), which should yield an adequate sample of the natural variation of the key characteristics across zones.

4.2 Define "Key Characteristics" of the Product (For what features of the product is the paying customer willing to pay?)   The important features of the product must be defined. These are the "key characteristics" of the product, the features important to the customer buying the product, features for which the customer is willing to pay and, when improved, increase customer satisfaction, thus adding value. Table 3 lists the key characteristics for folded panels, and they are described in Figure 9.

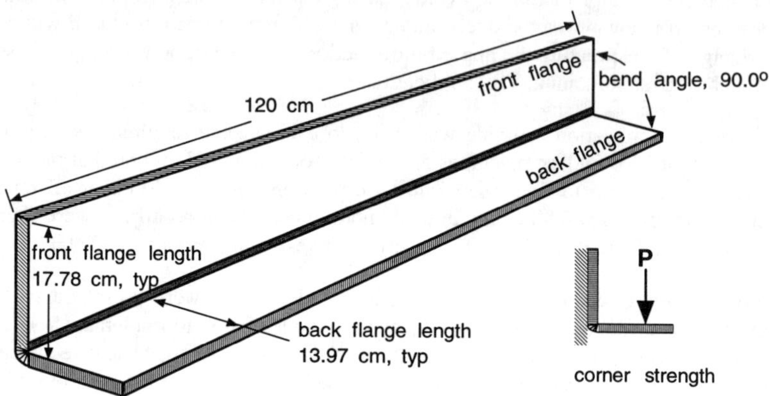

Figure 9. Product and corresponding key characteristics

Each of these characteristics are imparted by the folding process and affect the fit and function of the parts during assembly and service. Other characteristics of the panel are also key, such as thickness, but these are imparted at other production stages besides folding.

Table 3. Key Characteristics of Folded Panels
1. Back Flange Length
2. Front Flange Length
3. Bend Angle
4. Corner Strength

4.3 Define Key Characteristic Measurement Methods (How do we quantify these?) Once the key characteristics are identified, a method to quantify them is required. The measurement should, ideally, quantify the level to which the product feature meets the intended function. It should meet the following criteria for characterizing key characteristics.

- It is a relatively simple test, economical and quickly run;
- It tests elements which contribute to the product's performance;
- Is sensitive to the differences within the characteristic themselves which affect performance;
- Provides relatively consistent results (does not introduce too much noise into the measurements).

The first three items we wish to measure from Table 3 are relatively straightforward. They are the geometric features of the product, which must be within certain limits for the product to fit and function properly, and represent the dimensional accuracy of the process. A caliper can be used to measure the front and back flange length. Bend Angle, defined as the included angle between the front and back flange, can be measured using a digital protractor. The test shown in Figure 2 was retained for measuring corner strength (3).

4.4 Define Desired Ranges for the Key Characteristics   Again, for the geometric properties above, the desired range of linear dimensions is already defined as drawing tolerances, usually $\pm.08$ cm. This may not be the true dimensional range required for the part to function, but it is generally accepted and a good starting point. Angular tolerance is usually defined on drawings a bit more liberally as $\pm3°$; however, this may conflict with the previous dimensional requirement, depending on the flange length. (A 30 cm flange end can range .5 cm over 6°.) An angular tolerance of $\pm.5°$ is less incongruous with the linear tolerance.

Regarding the corner strength measurement, no desired range has yet been defined from a functional standpoint, i.e., the function of the product has not yet been defined in terms of this test. Therefore the desired range of this property was defined in terms of the baseline hand-folding process (4). L-shaped coupons were fabricated using 1.3 cm thick core with 2-ply fiberglass face sheets and tested using the incumbent process to get an initial target for the thermofolding process. This target was initially measured at 2.26 kg/cm.

4.5 Sampling Process Capability Now that the measurement tools are in place and the tests are considered reliable, it is time to get a "feel" for the distribution of the data and sample the process capability.   Fifty one panels, representing a range of panel constructions (2 core thicknesses, 2 fiber types, and 2 skin thicknesses) were folded, and data for each of the key characteristics listed above was collected.   The distribution of the Back Flange length is shown here.

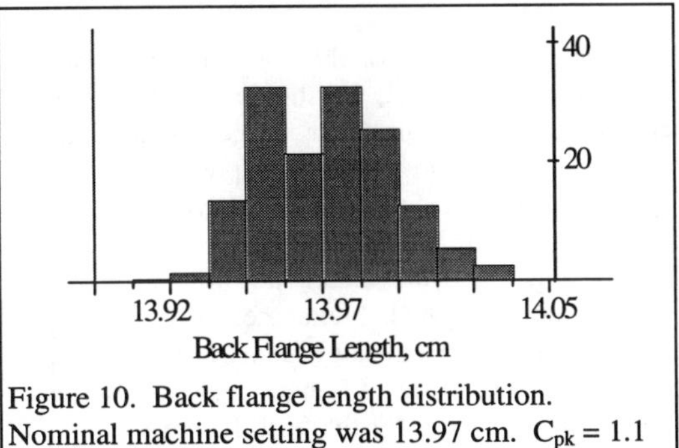

Figure 10. Back flange length distribution. Nominal machine setting was 13.97 cm. $C_{pk} = 1.1$

The nominal machine setting for the back flange was 13.97 cm; however, the clamp tooling built in an offset of .12 cm (see Figure 11). Since we can account for this offset in the folding process by adjusting the machine setting, we can remove its effect from the distribution and examine the distribution around the nominal setting, as shown above. The vertical lines represent the process specification limits ($\pm.076$ cm from nominal), and the vertical axis represents the number of observations.

Since we now know the observed back flange mean and the spec limits, we can now calculate the process capability index, or C_{pk} as defined here.

$$C_{pk} = \min\left(\frac{\hat{\mu} - LSL}{3\hat{\sigma}} , \frac{USL - \hat{\mu}}{3\hat{\sigma}} \right)$$

where,
$\hat{\mu}$ = observed mean of the distribution

$\hat{\sigma}$ = observed standard deviation of the distribution
LSL = lower spec limit
USL = upper spec limit .

The process capability index not only gives a ratio of the desired process spread to the actual process spread, but also gives an indication of how close the process mean is to the desired target.

Figure 11 illustrates the dimensional changes that occur in the panel during the folding process. One can see the that front flange length is affected by the panel thickness, i.e. given the same initial panel length and back flange length, a thinner panel will have a shorter "bend allowance" and therefore a longer front flange than a thicker panel. These are geometric properties and can be expresses as follows.

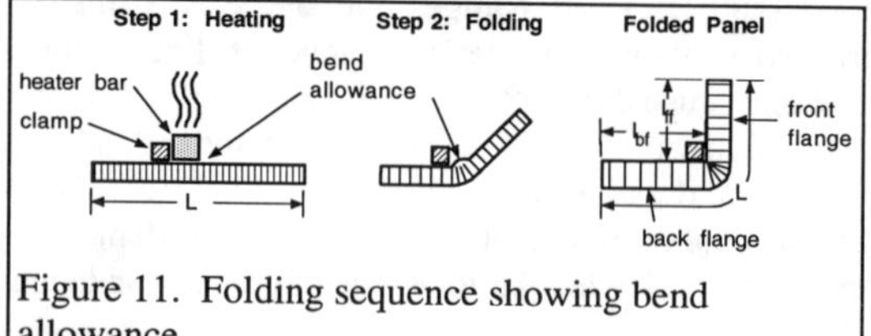

Figure 11. Folding sequence showing bend allowance

$$l_i = l_{bf} + l_{ff} + l_{ba}$$

where,

l_i = initial length of the panel,

l_{bf} = length of the back flange,

l_{ff} = length of the front flange,

l_{ba} = length of the bend allowance.

Since we must design for this by using the bend allowance to lay out the flat pattern of the panel prior to folding it, we should remove this effect from the data to focus on the process capability regarding the front flange distribution. We can calculate the effect of the panel thickness on the front flange and predict the bend allowance for the two different thicknesses we are studying. Subtracting the difference in the bend allowance between the two panel thicknesses from the data for the .64cm-thick panels, we can plot the front flange length distribution as we did for the back flange, and plot the capability, as follows.

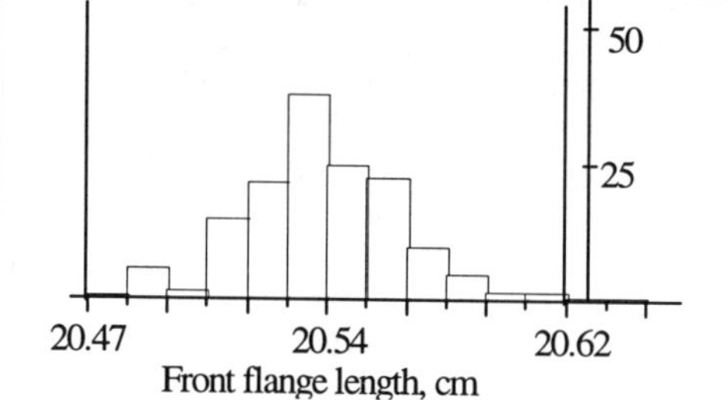

Figure 12. Front flange length distribution after accounting for the bend allowance. C_{pk} = .9

$$l_{BAthick} - l_{BAthin} = (l_i - l_{bf} - l_{ff})_{thick}$$
$$- (l_i - l_{bf} - l_{ff})_{thin}$$
$$= l_{ff\,thin} - l_{ff\,thick}$$

Therefore, $l_{ff} = \begin{cases} l_{ff}, \text{ for } 2.54cm \text{ panels} \\ l_{ff} - (\hat{\mu}_{ff\,thin\,panels} - \hat{\mu}_{ff\,thick\,panels}), \text{ for } .64cm \text{ panels} \end{cases}$

The bend angle target of 90° is the same for all constructions, so it can be plotted in the same way as the front flange distribution, is shown here. As we can see, it is not as well-behaved as the flange length distributions. It is not normally distributed, and its initial C_{pk} value is less than desirable. The spec limits are centered around the mean of the distribution (90.2°), and some of the data spills out beyond the

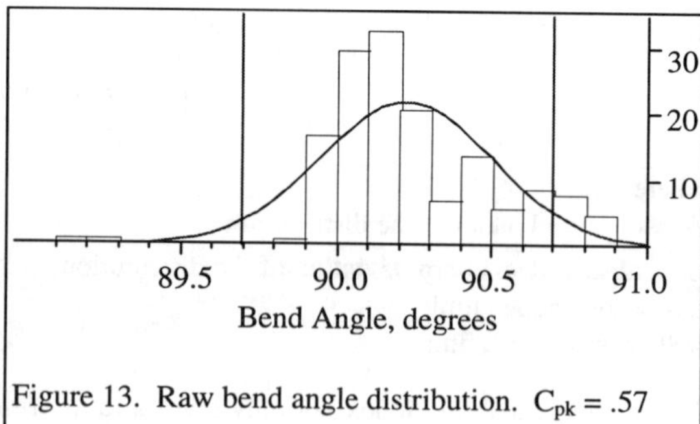

Figure 13. Raw bend angle distribution. C_{pk} = .57

desired limits. The distribution suggests an effect of one of the panel construction attributes. This can be investigated by applying statistical modeling methods such as ANOVA (5), which includes the effects of the panel construction, such as core thickness (.64 cm and 2.54 cm), and the fiber used in the panel skin (fiberglass and graphite). In doing so we find that the

76

model, does, in fact, account for a significant, .5° effect of the core thickness on the bend angle, as shown here.

Since we can account for this in production by adjusting the machines folding angle set point according to the panel thickness, we can filter this effect from the bend angle data. When the data is adjusted for this effect and replotted, the distribution appears less skewed, as shown here.

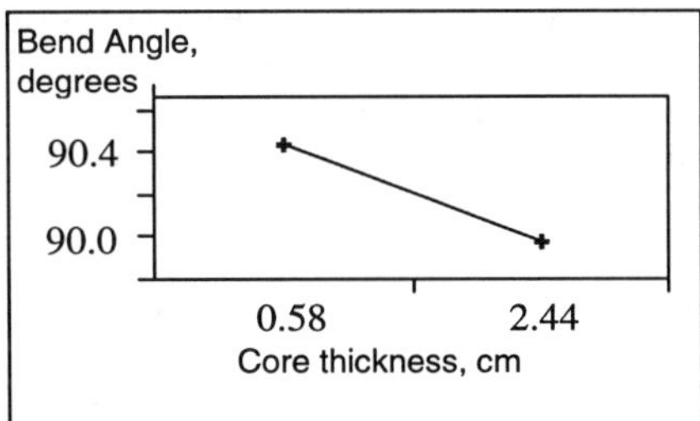

Figure 14. Effect of panel thickness on bend angle

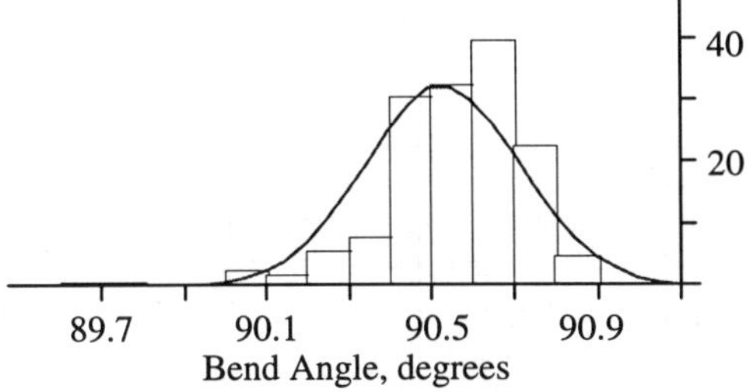

Figure 15. Bend angle distribution after accounting for the panel thickness effect. C_{pk}=.83

The raw corner strength distribution is shown here .The desired, baseline corner strength was previously introduced for L-shaped coupons built with .5"-thick core and 2-ply fiberglass skins. Since the data above is for .25"- and 1"-thick panels only, we do not have any data to directly compare to the baseline process. Therefore another experiment was conducted using .5"- and 1"-thick panels and a variety of constructions. The corner strength distribution for this new set of data is shown below.

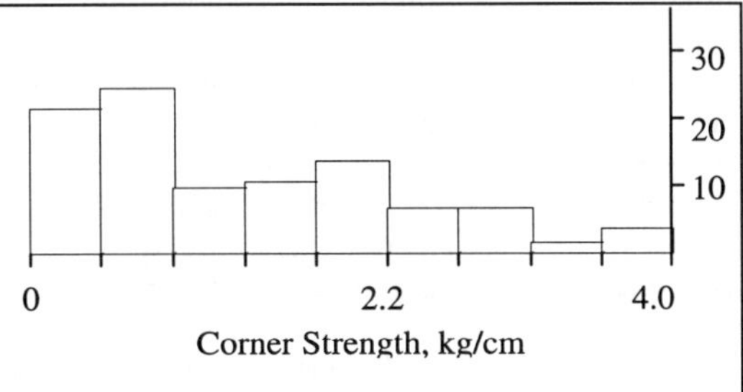

Figure 16. Raw corner strength distribution for .25"- and 1"-thick panels

The shaded portions represent the corner strength that corresponds to the 1.27cm-thick, 2-ply fiberglass construction that we wish to compare to the baseline. As one can see the values are somewhat below the baseline corner strength target of 2.2 kg/cm. Increasing the thermofolded corner strength can be the topic of further investigation; however, we are still interested in sampling the process capability. Even though we are somewhat below the corner strength target, we can sample the "potential" process capability, C_p, defined as follows.

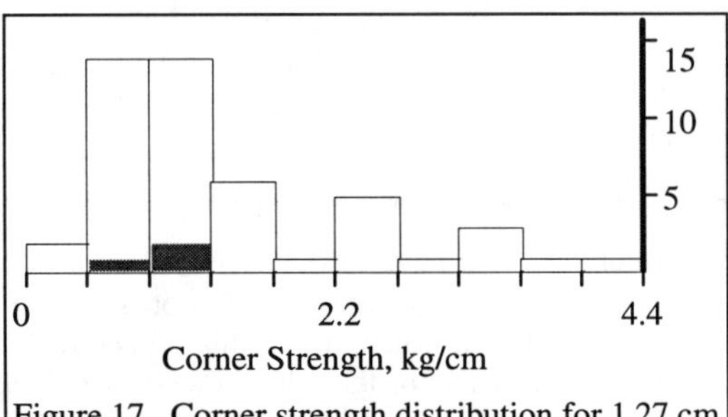

Figure 17. Corner strength distribution for 1.27 cm and 2.54 cm thick panels

From the corner strength distributions above we can see that it is the key characteristic which is most affected by the panel construction, and is therefore the most widely distributed. As we did for the front flange and bend angle, we can develop an ANOVA model to account for the effects of the panel

$$C_p = \frac{desired\ process\ spread}{actual\ process\ spread}$$

$$= \frac{Upper\ Spec\ Limit - Lower\ Spec\ Limit}{6\hat{\sigma}}$$

construction on the corner strength. After doing so , we can then predict the corner strength for each construction type and examine the distribution of the difference between the expected and actual, observed data, or the *residuals.*

residual = observed value - predicted value

This is shown in Figure 18 for the new set of data from Figure 17.

4.6 Benchmark or Explore?   Table 4 shows product rejection rate for various C_p values. A rejection rate of 3.6% seems reasonable for a new process, so we choose a lower limit of .7 for the process capability. We can now rank the key characteristics according to relative robustness, that is, according to the ability of the process to control them within the desired ranges. We must now decide if these estimated folding process capabilities are acceptable. From Table 5 we can see that the corner strength is the only key characteristics which does not meet this criteria. Until now we have been merely characterizing the process as it emerged from its initial development. We therefore wish to tune the folding process to improve the corner strength.

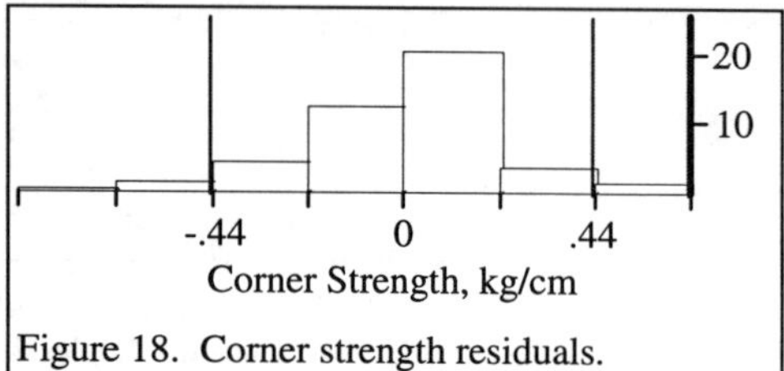

Figure 18. Corner strength residuals. Estimated process capability C_p = .6

Table 4. C_p values and corresponding rejection rates (6)

C_p	Percentage rejected
.3	36.8%
.4	23.0%
.5	13.4%
.6	7.2%
.7	3.6%
.8	1.6%
.9	.7%
1.0	.3%

Table 5. Folding Key Characteristics Ranked by Estimated Process Capability

Key Characteristic	Standard Deviation	Tolerance	Process Capability, C_{pk}
Back Flange, l_{bf}, cm	.023	$\pm.076$	1.1
Front Flange, l_{bf}, cm	.028	$\pm.076$	.9
Bend Angle, α, degrees	.2	$\pm.5°$	.83
Corner Strength, kg/cm	.24	$\pm.44$	.6

4.7 Explore   We have now discovered that, of the key characteristics, only the corner strength does not achieve the desired process capability. By exploring some potential key factors of the process which affect this property, we may find a way to improve it without adversely affecting any of the other key characteristics. Since the thermofolding process is, essentially, a reinforced thermoplastic welding process, we suspect that the amount of heat applied to the fold during the folding process can affect the corner strength. The factors which affect the heat flow during the process are the heater bar temperature and soak time (time of contact between the heater bar and the panel). The effect of these factors on the corner strength, however, has not been quantified. After performing some preliminary tests to explore the effects of varying these factors on the corner strength, we do, in fact, find an alternate folding cycle which appears to improve the corner strength. We can execute a designed experiment and use ANOVA to verify the effect of the new cycle on the corner strength. In doing so we find an improvement in the corner strength of approximately 40% over the original folding cycle.

Shown here in the shaded portion is the distribution of the corner strength data from the previous folding cycle (Figure 17), and the distribution of the corner strength data of the new cycle is superimposed. We can apply the ANOVA model that we used on the previous corner strength data to model the effects from the new cycle, plot the residuals and sample the capability, as shown here.

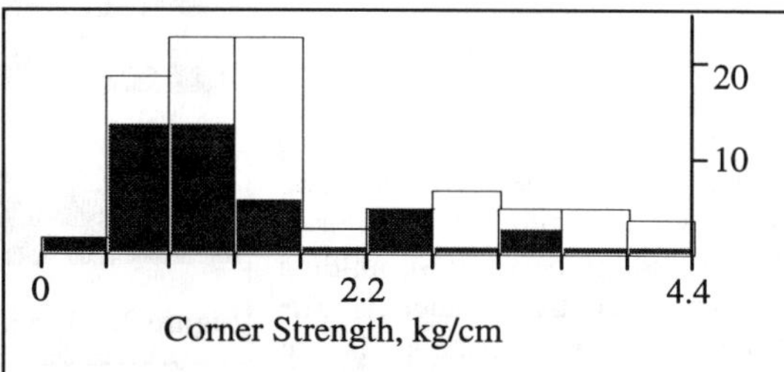

Figure 19. Corner strength distributions for the original (shaded) and improved thermofolding cycle

From the residual distribution we can see that the C_p has not increased, i.e., the variation in the corner strength data for which the ANOVA model does not account (noise), has not changed between cycles. Thus the new folding cycle only improves the process mean for the corner strength but does not reduce the variation. However, this data samples variation across

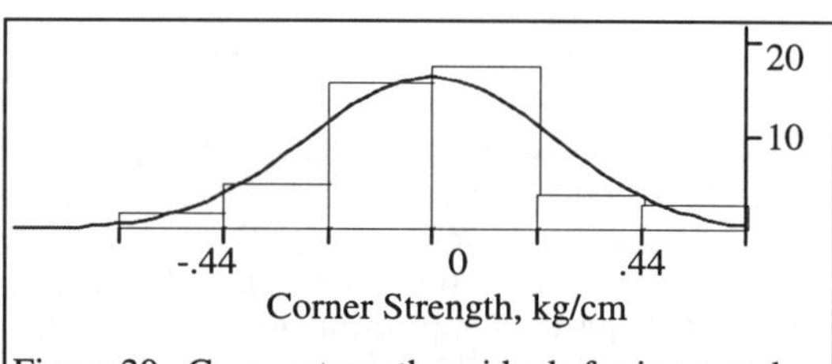

Figure 20. Corner strength residuals for improved folding cycle. $C_p = .57$

various constructions (graphite and glass skin materials using 1- and 2-ply face sheets). Therefore it may be conservative compared to sampling process capability for only one construction. Sampling the capability for one construction should represent a production scenario more closely than sampling several constructions simultaneously. We could choose one construction which has displayed the most variation and sample representative process data. Another approach would be to choose the construction which is most likely to be applied in production and sample data from that population. For this process, the variation of the corner strength is roughly the same for the different constructions, so we will sample data from the construction which is most likely to be fabricated in production, the 1.27cm-thick, 2-ply fiberglass panels. The distribution is shown here.

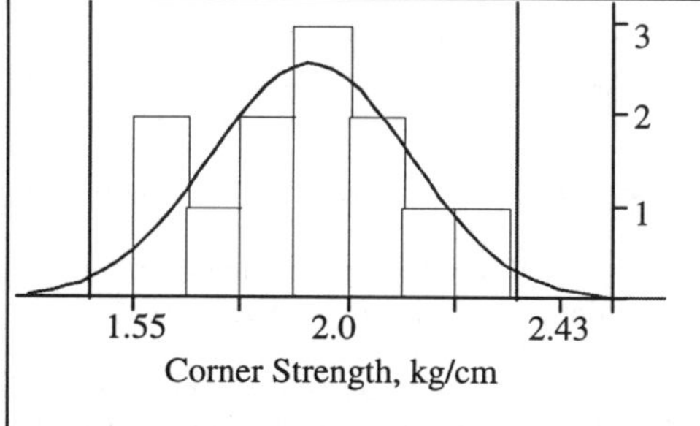

Figure 21. Corner strength distribution for process control panels. $\hat{\sigma} = .2$ kg/cm, $C_p = .7$

This sample of 12 data points for the 2-ply, 1.27cm-thick construction suggests that the new cycle may meet the minimum process capability criteria of .7, and for now we will accept this level of capability and proceed with the implementation.

4.8   Implementation   Wrap-up
The final implementation step is to show that panels produced on the production equipment will meet the desired process capability. This involves producing a quantity of products in a production scenario which sample a representative amount of process variation, and measuring the process capability. The prototype machine was installed on the shop floor, shown in Figure 22. As this is written we are in the stage of building the "process control" panels. The purpose of these panels is to

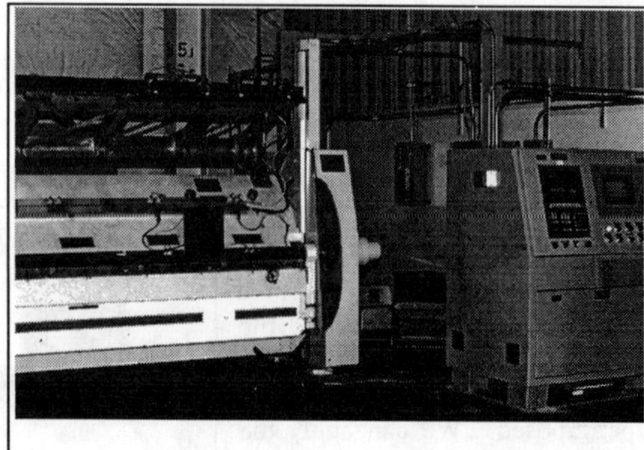

Figure 22. Production thermofolding equipment

continuously sample the process capability, as described in this section, throughout production. 12 "process control" panels of each viable construction will be fabricated and folded, and the capability of all key characteristics listed in Table 3 will be calculated. This will complete the final phase of Implementation.

5. Conclusion

We have proposed and demonstrated a viable approach for introducing a new technology into the aircraft interior production line. The concept was demonstrated, and this justified launching a full scale feasibility study. Figure 4 summarizes this feasibility study, which used strategies to establish the full scale technical and economic feasibility of the concept. The feasibility study justified implementation. Implementation used statistical tools to benchmark the process capability. This drove out the product features which did not meet the initial process capability criteria. More advanced statistics were used to further tune the process to become more robust. Final production readiness will be determined by meeting statistical criteria which show that the process has stabilized and is capable of economically producing production quantities.

6. References

1. "Origami-Technology; Creative Manufacturing of Advanced Composite Parts," W. Van Druemel, published brochure of Ten Cate Advanced Composites BV, the Netherlands, 1991.

2. *Product Design for Assembly,* G. Boothroyd, 1989.

3. "Forming of Thermoplastic Sandwich Panels," Boeing Process Specification BAC 5344, Boeing Materials Technology, August 1997

4. "Process for Potting Inserts and Edges into Sandwich Panels," Boeing Process Specification BAC 5538, Boeing Materials Technology, 10 February, 1997

5. *Statistics for Experimenters,* George Box, William Hunter, J. Stuart Hunter, John Wiley &Sons, New York, 1978

6. "Advanced Statistical Process Control," Boeing Publication No. ACC 7015, Organization & Management Development Training, 16 July, 1990

Composite Properties of Polyimide Resins Made From "Salt-Like" Solution Precursors[1]

Roberto J. Cano, Erik S. Weiser, and Terry L. St. Clair
NASA Langley Research Center
Hampton, VA 23681-0001

Yoshiaki Echigo and Hisayasu Kaneshiro
Unitika Ltd.
Kyoto, Japan 611

ABSTRACT

Recent work in high temperature materials at NASA Langley Research Center (LaRC™) have led to the development of new polyimide resin systems with very attractive properties. The majority of the work done with these resin systems has concentrated on determining engineering mechanical properties of composites prepared from a poly(amide acid) precursor.

Three NASA Langley-developed polyimide matrix resins, LaRC™-IA, LaRC™-IAX, and LaRC™-8515, were produced via a "salt-like" process developed by Unitika Ltd. The "salt-like" solutions (sixty-five percent solids in NMP) were prepregged onto Hexcel IM7 carbon fiber using the NASA LaRC™ Multipurpose Tape Machine. Process parameters were determined and composite panels fabricated. Mechanical properties are presented for these three intermediate modulus carbon fiber/polyimide matrix composites and compared to existing data on the same polyimide resin systems and IM7 carbon fiber manufactured via poly(amide acid) solutions (thirty-five percent solids in NMP). This work studies the effects of varying the synthetic route on the processing and mechanical properties of polyimide composites.

KEY WORDS: Composites, Polyimides, Prepreg, and Processing

---

1. INTRODUCTION

Polyimides are attractive for aerospace applications because of their excellent thermooxidative stability and high mechanical properties. Controlled molecular weight, low melt viscosity polyimides are easily processed via thermoplastic forming techniques which makes them attractive matrix materials for advanced composite applications. Polyimides provide an excellent combination of high glass transition temperatures, excellent solvent resistance and high mechanical properties.[1-5]

NASA Langley Research Center has developed several polyimides that can be melt processed into various useful forms as coatings, adhesives, composite matrix resins and films.[1-10] These polyimides are prepared from various aromatic diamines and dianhydrides in several different solvents. The use of phthalic anhydride as an endcapping agent is to control the molecular weight of the polymer and, in turn, to make it easier to process in molten form.

Current technology for making prepreg and composites from polyimides utilizes solutions of the poly(amide) acids which are processed into prepreg with various reinforcing fibers.[11-12] In general, poly(amide acid) solutions are prepared at solid contents of 25 to 35% by weight with resulting Brookfield viscosities at ~20°C of 15,000 to 35,000 centipoise (cp). Therefore, processing these types of solutions requires overcoming significant problems such as solvent management and good fiber wet out from the high viscosity solutions. The resultant prepregs require residual solvent contents of 20 to 25 % by weight (of which ~2-3% is water from the imidization reaction) for adequate tack and drape. This residual solvent must then be removed during the composite cure cycle. Again, this poses a significant solvent removal and recovery problem, especially in large parts. Typically, solvent removal is accomplished by holding the material at an intermediate drying temperature, above the boiling temperature of the solvent, for extended periods of time (i.e. one to two hours) under full vacuum, 30" Hg. The material is then ramped to its final cure temperature at high pressure (~1.4 MPa) to ensure complete fiber wet-out and full consolidation.

In the present work, preimpregnated tape was fabricated from "salt-like" solutions of high temperature polyimides prepared by Unitika Ltd., Japan. These low viscosity (5,000 to 9,700 cp at 20°C), high solids content (65% by weight) solutions were solution coated onto reinforcing fiber to produce high quality prepreg with excellent tack and drape at 12-15% residual solvent (of which ~4-6% is water from the imidization reaction). Composites from this prepreg were of high quality, required significantly lower intermediate drying temperature utilizing only partial vacuum and a lower final pressure.

Mechanical properties are presented for the three intermediate modulus carbon fiber/polyimide matrix composites (IM7/ LaRC™-IA, IM7/ LaRC™-IAX, and IM7/ LaRC™-8515) and compared to existing data on the same polyimide resin systems with IM7 carbon fiber manufactured via poly(amide acid) solutions.

2. MATERIALS[2]

IM7 12K, unsized carbon fiber from Hexcel, Salt Lake City, UT was used exclusively in this work. Solutions of LaRC™-IA, -IAX, and -8515 were made by Unitika Ltd., Kyoto, Japan and prepared as described in Section 3.1. Prepreg was made at NASA LaRC™ on the Multipurpose Prepregger as described in Section 3.2. Composite panels were manufactured at NASA LaRC™ as described in Section 3.3. Test coupons were machined and tested as described in Section 3.4.

3. EXPERIMENTAL

3.1 "Salt-like" Solutions For example, LaRC™-IA polyimide "salt-like" solutions were formed from the reaction of oxydiphthalic anhydride (ODPA) dissolved in a mixture of NMP and MeOH at room temperature. This solution is treated at 60°C for 3 hours in order to convert the ODPA into ODP-dimethyl ester. The resulting solution, phthalic acid (PA) and 3,4'-oxydianiline is added to the ODP-dimethyl ester and stirred for 2 hours to yield a homogeneous polyimide "salt-like" resin. The chemical structure of the three polyimides utilized in this work are shown in Figure 1 while the properties of the three "salt-like" solutions are shown in Table 1.

3.2 Prepreg Fabrication Polyimide "salt-like" solutions were placed in the resin dip tank of the LaRC Multipurpose Prepreg Machine (Figures 2 and 3). The LaRC™-IA, -IAX and -8515 solutions had solids contents of 65% resin by weight in NMP with Brookfield viscosities ranging from 4955 to 9660 cp (Table 1). The hotplates were set at 149°C while the oven was set at 138°C. Nip stations 2, 3, and 4 utilized only contact pressure and were set at 93°C. The metering bar gap was set at ~0.4064 mm and the line speed was set between 0.55 and 0.76 m/min. The comb was adjusted to attain a fiber aerial weight of 145 g/m². Seventy ends of 12k unsized Hexcel IM7 carbon fiber were utilized. Prepreg tapes with FAW's of 148 to 150 g/m², ~35 wt. % dry resin content, and a wet volatiles content of 13 to 15 wt.% were produced. The 20.6 centimeter wide prepreg tape made from each material was of high quality with excellent wet out, tack and drape. A total of 107 linear meters of each material was produced. The properties of the each prepreg system are presented in Table 2.

3.3 Composite Panel Fabrication LaRC™-8515, -IA, and -IAX "salt-like" prepregs were processed into several different composite laminates by the process illustrated in Figure 4. As shown in Figure 5, one ply of Kapton™ film and one ply of 0.0635 mm Teflon™ bleeder /

---

[2] The use of trademarks or names of manufacturers in this report is for accurate reporting and does not constitute an official endorsement, either expressed or implied, of such products or manufacturers by NASA.

breather cloth was placed on either side of several plies of unidirectional "salt-like" prepreg, which was placed in a closed mold and processed in a hydraulic vacuum press. The prepreg was heated to 121°C under isobaric conditions and held for one hour. During the one hour hold, the residual solvent within the material was removed. After 30 minutes at 121°C, 508 mm of Hg of vacuum was applied. Once the 121°C hold was completed, the material was heated to 225°C and full vacuum was applied. During the one and half hour hold at 225°C, the material becomes fully imidized. After the hold at 225°C, the polyimide was heated to the final cure temperature of 380°C, 0.6895 MPa was applied and the material was held at this temperature and pressure for 30 minutes. During this hold, the polyimides become amorphous and reach minimum viscosity which allows for proper consolidation. Once the hold has been completed, the material was cooled to ambient temperature under full vacuum and 0.6895 MPa pressure. At ambient temperature, the mold was removed from the hydraulic vacuum press and the laminate was released from the mold. Upon visual and ultrasonic examination, laminates were determined to be of excellent quality.

3.4 Composite Panel Mechanical Testing Specimens were machined for short beam shear (SBS) testing which is a flexural test used to qualitatively assess the interlaminar strength of a composite. This test is designed to force an interlaminar failure along the centerline of the composite beam. The SBS test utilizes a three point bending fixture and the span is determined from the distance between the two base points in the fixture. The test specimen of approximate length and width of 1.9 cm and 0.64 cm, respectively, were tested in accordance with ASTM D2344 at both room temperature (RT) and 177°C. The load is applied at a rate of 0.13 cm/min.

The transverse flexural test, ASTM D790, is very similar to the SBS test, except in this case a tension failure is forced to occur along one free face while the other is put in compression. The same fixture can be used, but the specimens differ in size in order to affect this tension failure. These test specimens were 7.0 cm long and 1.27 cm wide, while test temperatures were RT and 177°C. In this test there is an initial load of 0.34 MPa, and the load is again applied at the rate of 0.13 cm/min. until the initial failure occurs.

4. RESULTS AND DISCUSSION

4.1 Unidirectional Prepreg Tape Unidirectional prepreg was readily made with the "salt-like" solutions of LaRC™-IA, -IAX, and -8515. The low viscosity of these solutions resulted in excellent wet out of the fiber bundles and prepreg tape with excellent tack and drape with significantly less residual solvent than prepreg from poly(amide acid) solutions of polyimides (13-15 wt.% versus 20-25 wt.%).

4.2 Composite Panels Unidirectional panels fabricated according to the procedure describe earlier were determined to be of very good quality by both visual and ultrasonic inspection. Panels fabricated with poly(amide) acid solutions of the same resin systems required more pressure (1.4 MPa) to adequately consolidate the laminates. Therefore, "salt-like" solutions allow for the fabrication of quality panels at lower pressures than the same resins fabricated via poly(amide) acid solutions. The need for less residual solvent in the prepreg translate into less solvent removal during cure.

4.3 Mechanical Properties   Short beam shear and 0° and 90° flexural strength and modulus were determined for each "salt-like" solution composite. Mechanical properties for composites fabricated from "salt-like" solutions are presented in Figures 6-9 and Tables 3-5. For comparison, data obtained from composites fabricated from poly(amide) acid solutions [13-15] are also presented in these figures. As shown in Figure 6, IM7/LaRC™-8515, IM7/LaRC™-IA, and IM7/LaRC™-IAX composites fabricated via the "salt-like" solution demonstrated superior or comparable 0° flexural properties at both RT and 177°C. Similar results were obtained for 0° flexural modulus properties as shown in Figure 7.   IM7/ LaRC™-8515 composites also demonstrated improved or comparable 90° flexural strengths as shown in Figure 8.   However, IM7/LaRC™-IA and IM7/LaRC™-IAX composites did not.   SBS strengths were not as good as those determined for the poly(amide acid) solutions of all three materials at elevated temperature. Only IM7/ LaRC™-IA composites from "salt-like" solutions demonstrated improved SBS properties at RT.   IM7/ LaRC™-8515 and IM7/ LaRC™-IAX composites resulted in lower RT SBS strengths compared to their equivalent poly(amide acid) composites.

In general, material properties are considered good if the 0° flexure moduli are between 110-131 GPa and RT short beam shear strengths are above 97 MPa [16]. Although the mechanical data for the "salt-like" solution materials were not all equivalent to the poly(amide acid) materials, determined test values are considered acceptable by the above criteria.   The three "salt-like" solutions resulted in composites with 0° flexural moduli above 131 GPa at both RT and 177°C and RT SBS values above 103 MPa.

The poor SBS properties of the "salt-like" IM7/LaRC™-8515 composites may have been due to insufficient molecular weight build-up. The 8515 "salt-like" solution composites had a cured T_g of around 235°C as measured by DSC. This is significantly lower than the amide acid solution composite T_g of ~250-260°C. The lower T_g would indicate a lower molecular weight which would translate into poor SBS properties. These properties should improve with an increased dwell time at the final cure temperature which would allow for further molecular weight build-up to occur. Although, no apparent difference in T_g was observed between the "salt-like" and amide acid solution composites of IA and IAX (T_g ~235°C), a lack of molecular weight build-up may have resulted in the difference observed in the data. The lower 90° flexure, SBS, and elevated temperature 0° flexural of the IA and IAX "salt-like" solution composites may be improved by a longer dwell at 380°C allowing for more molecular weight growth.

5. CONCLUSIONS

"Salt-like" solutions of the NASA Langley-developed polyimide matrix resins, LaRC™-IA, LaRC™-IAX, and LaRC™-8515 were produced by Unitika, Ltd and solution prepregged onto Hexcel IM7 carbon fiber by NASA LaRC™. Using these unidirectional prepregs which contained ~12-15 percent N-methylpyrrolidinone solvent, processing parameters were developed to fabricate fully consolidated high quality flat panels. RT 0° flexural composite mechanical data for each "salt-like" polyimide prepreg was superior or comparable to the same material synthesized via the poly(amide) acid route. SBS and 90° flexural properties, however, were not equivalent. Overall, processing of polyimide composites from a "salt-like" precursor was readily achievable at lower pressures and with less residual solvent and resulted in composites of good quality with good mechanical properties.

6. REFERENCES

1. P. M. Hergenrother, R. G. Bryant, B. J. Jensen, J. G. Smith, Jr. and S. P. Wilkinson, Soc.. Adv. Matl. Proc. Eng. Series, 39(1), 961 (1994).

2. R. G. Bryant, B. J. Jensen, J. G. Smith, Jr. and P. M. Hergenrother, Ibid.,(Closed Papers), 39, 273(1994).

3. C. W. Paul R. A. Schultz and S. P. Fenelli, U. S. Pat 5,138,028 (1992) [to National Starch and Chemical Co.].

4. C. W. Paul, R. A. Schultz and S. P. Fenelli in Advances in Polyimide Science and Technology, C. Feger, M. M. Khojasteh and M. S. Htoo, ed. Technomic Pub. Co., N.Y., 1993, p 220.

5. R. G. Bryant, B. J. Jensen and P. M. Hergenrother, Polym. Prepr., 34(1), 566 (1993).

6. B. J. Jensen, R. G. Bryant and S. P. Wilkinson, ibid., 35(1), 539 (1994).

7. S. J. Havens, R. G. Bryant, B. J. Jensen, and P. M. Hergenrother, ibid., 35(1), 553 (1994).

8. J. G. Smith, Jr. and P. M. Hergenrother, ibid., 35(1), 353 (1994).

9. J. G. Smith, Jr. and P. M. Hergenrother, Polymer, 35(22), 4857 (1994).

10. P. M. Hergenrother, R. G. Bryant, B. J. Jensen, and S. J. Havens, J. of Polymer Science: Polymer Chemistry Edition, 32 , 3061 (1994).

11. R. J. Cano, N. J. Johnston, and J. M. Marchello., Solution Prepreg Quality Control, Soc. Adv. Matl. Proc. Eng. Series, 40, 583, (1995).

12. R. J. Cano, M. Rommel, J. A. Hinkley, E. D. Estes, Fiber Study Involving a Polyimide Matrix, Soc.. Adv. Matl. Proc. Eng. Series, 41, 1047, (1996).

13. B. J. Jensen, T. H. Hou, and S. P. Wilkinson, Adhesive and Composite Properties of LaRC™-8515 Polyimide, High Perfromance Polymers, 7, 11-21, (1995)

14. T. H. Hou, N. J. Johnston, and T. L. St. Clair, IM7/LaRC™-IA Polyimide Composites, High Perfromance Polymers, 7, 105-124, (1995)

15. T. H. Hou, N. J. Johnston, E. S. Weiser, and J. M. Marchello, Processing and Properties of IM7 Composites Made From LARC™-IAX Polyimide Powders, <u>Soc. Adv. Matl. Proc. Eng. Series,</u>27, 135, (1995).

16. Leif A. Carlsson and R. B. Pipes, ed., <u>Composite Materials Series</u>, <u>Thermoplastic Composite Materials</u>, Vol. 7, Elsevier Science Publishing Company Inc., New York, 1991, pp. 47.

Table 1. "Salt-like" Solution Properties.

Resin System	Solids in NMP, wt. percent	Solution Brookfield viscosity, cp @ °C
LaRC™ IA	65	6780 @ 23
LaRC™ IAX	65	4955 @ 22
LaRC™ 8515	65	9660 @ 18

Table 2. Unidirectional IM7/ "Salt-like" Solution Prepreg Characteristics.

Resin	FAW, g/m2	Solvent Content , wt. %, wet	Solids Content, wt. %, dry
LaRC™ IA	150	13.2	34.0
LaRC™ IAX	150	13.8	37.5
LaRC™ 8515	148	15.0	35.7

Table 3. Mechanical Tests and Properties for LaRC™ 8515 / IM-7 "Salt-Like" Composites.

Mechanical Test	Test Temp., °C	No. spec. at each Temp.	Failure Load (kg)	Failure Stress (MPa)	Modulus (GPa)
SBS	RT, 177	10 10	352.7 ± 12.2 135.6 ± 9.9	117.8 ± 3.8 45.4 ± 3.0	N/A
0° Flexural	RT, 177	6 9	79.9 ± 4.4 57.4 ± 2.5	1725.1 ± 84.1 1299.0 ± 101.4	146.9 ± 3.4 137.2 ± 9.7
90° Flexural	RT, 177	5 5	26.4 ± 1.9 14.2 ± 3.3	117.9 ± 6.1 58.6 ± 13.1	4.62 ± 0.28 4.34 ± 0.28

Table 4. Mechanical Tests and Properties for LaRC™ IAX / IM-7 "Salt-Like" Composites.

Mechanic al Test	Test Temp., °C	No. spec. at each Temp.	Failure Load (kg)	Failure Stress (MPa)	Modulus (GPa)
SBS	RT, 177	10 12	313.9 ± 14.1 118.8 ± 3.6	103.3 ± 4.6 39.1 ± 1.2	N/A
0° Flexural	RT, 177	5 4	67.3 ± 2.9 45.8 ± 1.5	1525.1 ± 131.0 984.6 ± 91.7	145.5 ± 12.4 135.1 ± 9.7
90° Flexural	RT, 177	6 6	31.71 ± 2.54 16.83 ± 1.32	121.34 ± 11.72 66.88 ± 4.83	4.34 ± 0.28 3.03 ± 0.28

Table 5. Mechanical Tests and Properties for LaRC™ IA / IM-7 "Salt-Like" Composites.

Mechanic al Test	Test Temp., °C	No. spec. at each Temp.	Failure Load (kg)	Failure Stress (MPa)	Modulus (GPa)
SBS	RT, 177	10 12	380.2 ± 16.3 126.1 ± 3.2	125.1 ± 5.4 41.4 ± 1.0	N/A
0° Flexural	RT, 177	5 5	73.4 ± 3.4 47.1 ± 3.5	1589.9 ± 84.8 1065.9 ± 79.3	137.2 ± 9.7 131.0 ± 6.9
90° Flexural	RT, 177	5 5	35.74 ± 7.03 19.01 ± 1.59	133.1 ±17.2 73.1 ± 5.5	4.27 ± 0.34 3.44 ± 0.29

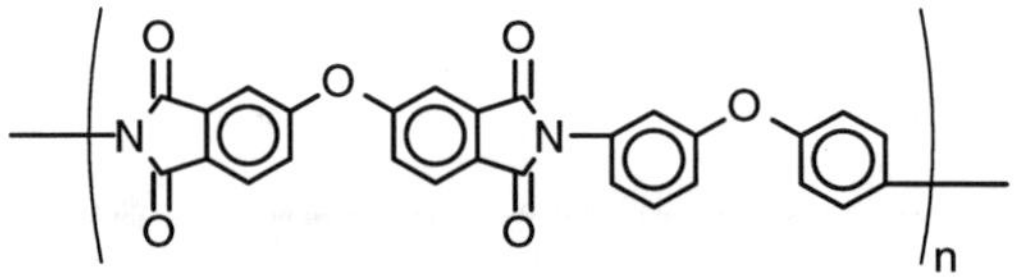

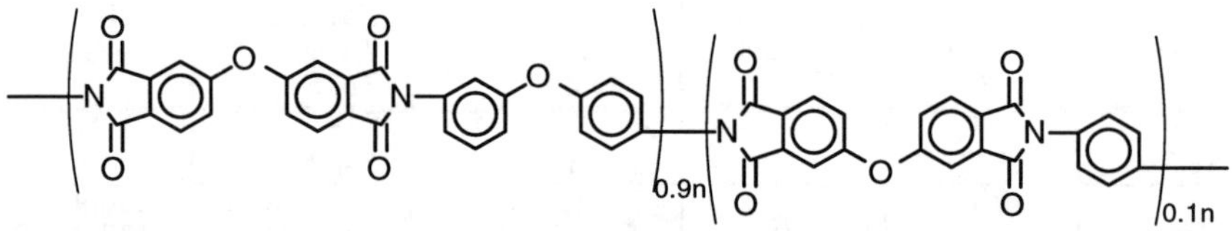

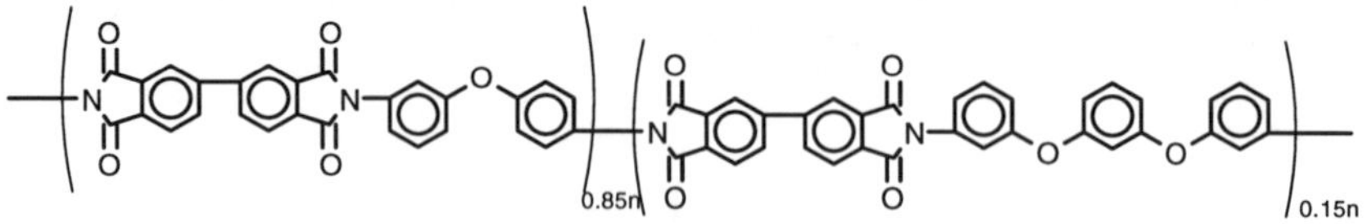

Figure 1. Molecular structure of LaRC™-IA, LaRC™-IAX, and LaRC™-8515.

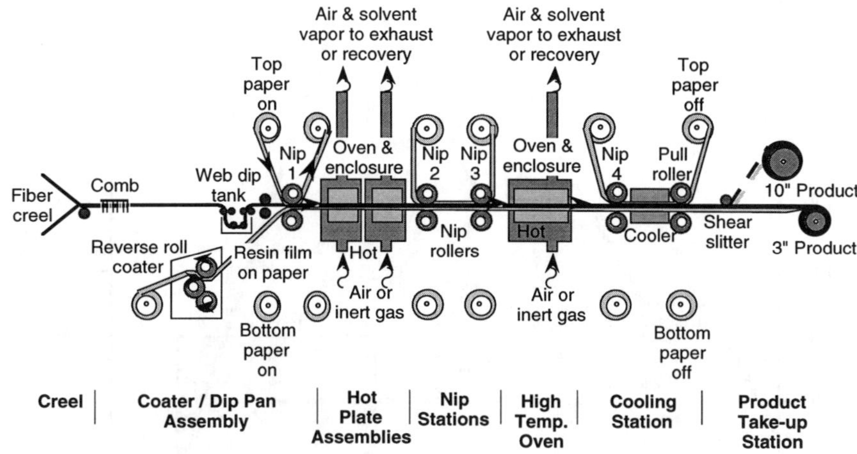

Figure 2. Schematic Diagram of the Tape Machine Modular Components.

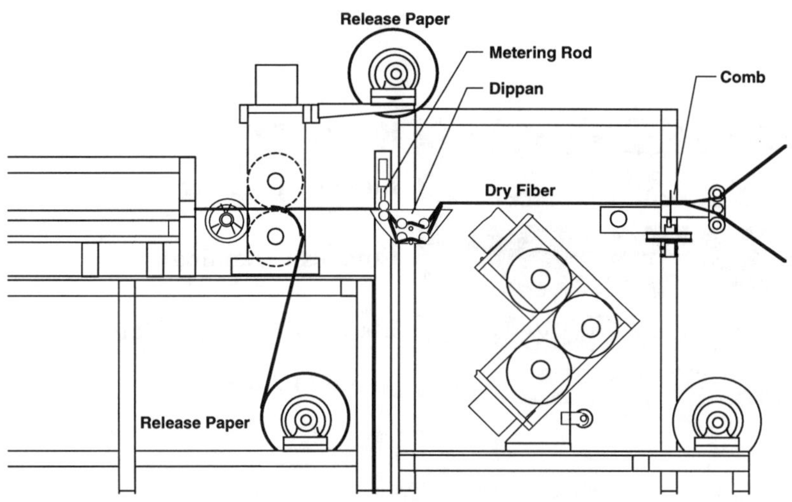

Figure 3. Solution Prepregging Using the Dip Tank Method.

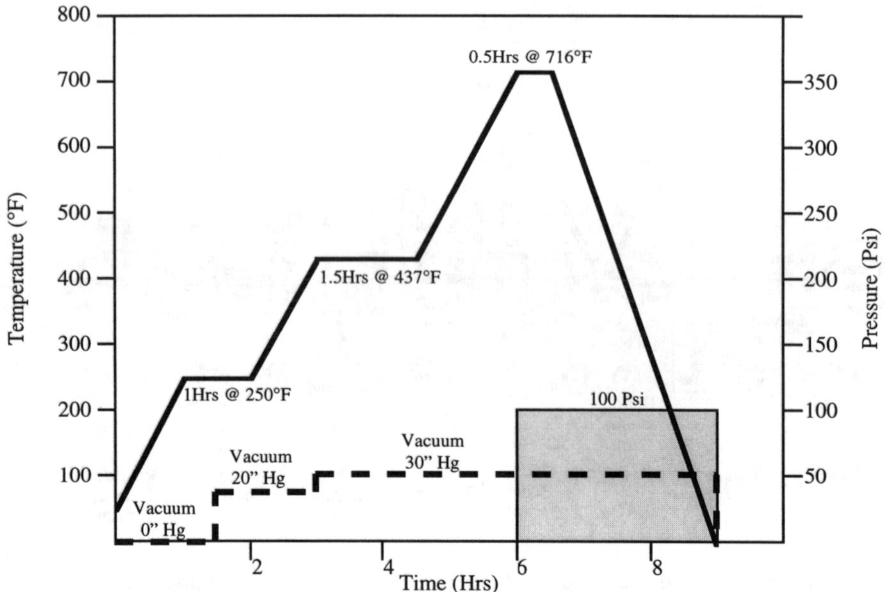

Figure 4. Cure Cycle for Polyimide Salt like Prepregs.

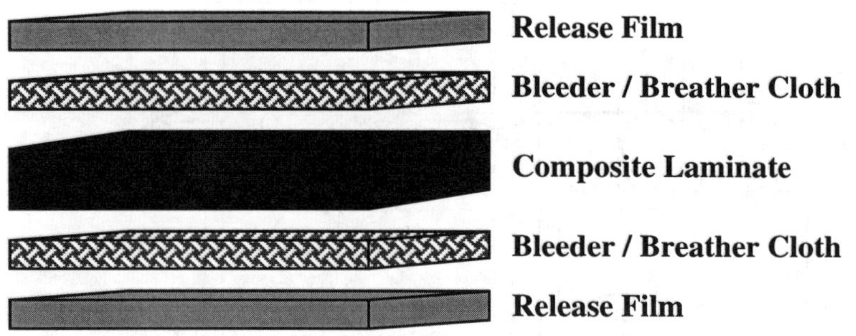

Figure 5. Composite laminate lay-up configuration.

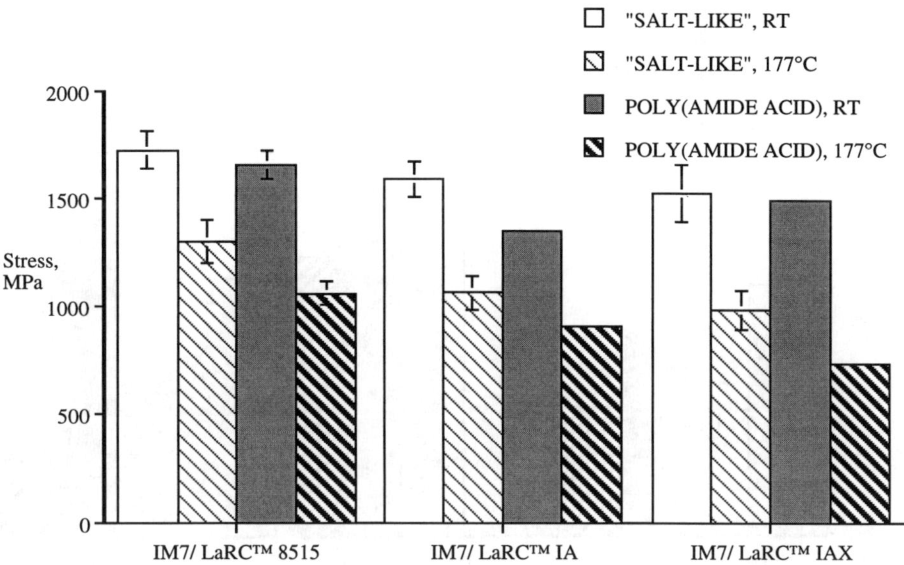

Figure 6. Zero degree flexural strength data.

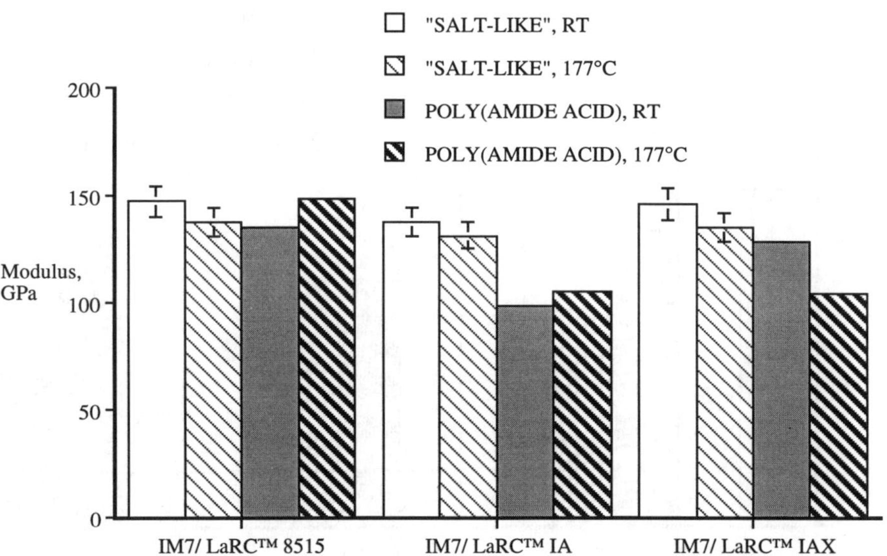

Figure 7. Zero degree flexural modulus data.

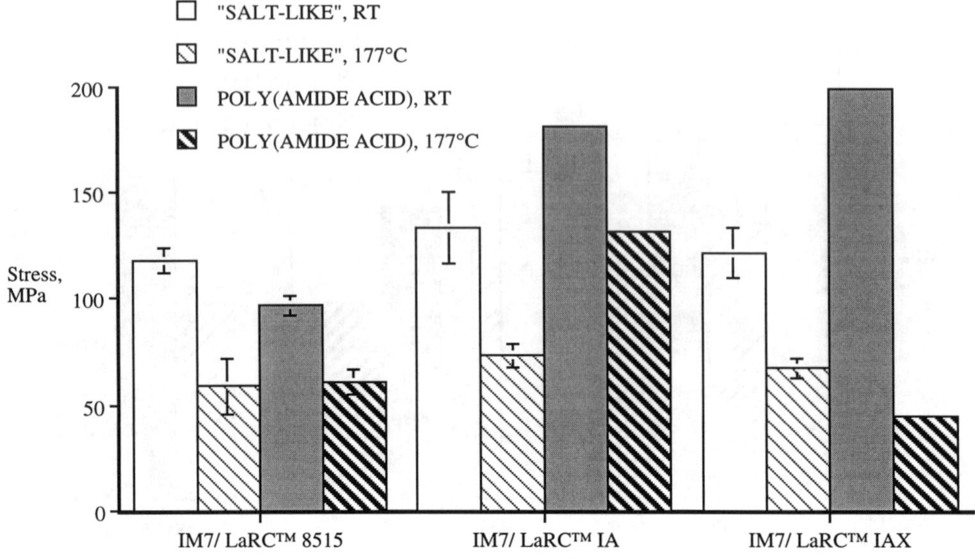

Figure 8. Ninety degree flexural strength data.

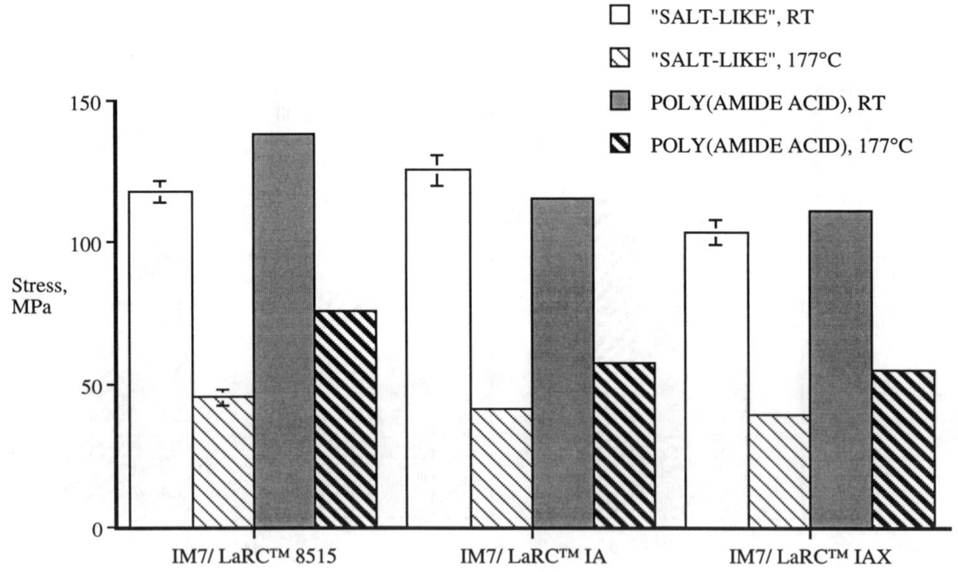

Figure 9. Short beam shear strength data.

43rd International SAMPE Symposium
May 31-June 4, 1998

THE EFFECT OF MOLECULAR WEIGHT ON THE COMPOSITE PROPERTIES OF CURED PHENYLETHYNYL TERMINATED IMIDE OLIGOMERS

J.G. Smith Jr., J.W. Connell, and P.M. Hergenrother
National Aeronautics and Space Administration
Langley Research Center
Hampton, VA 23681-0001

ABSTRACT

As part of a program to develop high temperature/high performance structural resins for aeronautical applications, imide oligomers containing terminal phenylethynyl groups with calculated number average molecular weights of 1250, 2500 and 5000 g/mol were prepared, characterized, and evaluated as adhesives and composite matrix resins. The goal of this work was to develop resin systems that are processable using conventional processing equipment into void free composites that exhibit high mechanical properties with long term high temperature durability, and are not affected by exposure to common aircraft fluids. The imide oligomers containing terminal phenylethynyl groups were fabricated into titanium adhesive specimens and IM-7 carbon fiber laminates under 0.1 - 1.4 MPa for 1 hr at 350-371°C. The lower molecular weight oligomers exhibited higher cured Tg, better processability, and better retention of mechanical properties at elevated temperature without significantly sacrificing toughness or damage tolerance than the higher molecular weight oligomer. The neat resin, adhesive and composite properties of the cured polymers will be presented.

KEY WORDS: High Temperature Polymers, Polyimides, Phenylethynyl Containing Imides, Adhesives, Carbon Fiber Composites, Oligomeric Thermosetting Imides

1. INTRODUCTION

Structural resins with an attractive combination of properties for high temperature/high performance applications are needed for advanced aircraft, in particular high speed vehicles. These advanced materials must be readily processable under pressures and temperatures compatible with conventional manufacturing equipment currently in use or under development. Minimum mechanical performance is required as dictated by the service environment that the advanced aircraft is likely to experience during its operational lifetime such as high temperature, high stress, object impact, thermal cycling, and exposure to moisture and aircraft fluids such as hydraulic fluid, deicing fluid, paint stripping solvents, alkaline cleaning solutions, and aircraft fuels.

Over the last 6 years, work in the area of phenylethynyl containing imide oligomers have yielded some promising results (1-25). Due to their oligomeric nature, these materials exhibit excellent processability during fabrication of neat resin moldings, bonded panels, and composites under pressures of 1.4 MPa or less. Upon thermal cure for ~1 h at 350-371°C, the phenylethynyl group undergoes a complex reaction involving chain extension, branching and crosslinking without the evolution of volatile by-products to afford a pseudo three-dimensional network. The reaction products are dependent on the relative concentration, location, configuration and molecular mobility of the phenylethynyl groups and the oligomer backbone. Upon curing, the glass transition temperature (Tg) of the material increases significantly relative to that of the uncured oligomer. Although the chemistry of the cured product is not well understood, the cured oligomer exhibits an excellent combination of properties that includes high Tg, high toughness, high strength, moderate modulus, and good moisture and solvent resistance.

One phenylethynyl containing imide oligomer designated PETI-5 developed at the NASA Langley Research Center has undergone extensive evaluation as an adhesive (20-2) and composite matrix resin (23-5). The material is a random copolymer prepared from 3,4'-oxydianiline, 1,3-bis(3-aminophenoxy)benzene and 3,3',4,4'-biphenyltetracarboxylic dianhydride and endcapped with 4-phenylethynylphthalic anhydride at a calculated number average molecular weight ($\overline{M}_n$) of 5000 g/mol. This material has displayed good processability and excellent mechanical properties in adhesive and composite forms; however, improved resin flow is desired in the processing of complex structural parts. In order to improve the processability of 5000 g/mol PETI-5, lower molecular weight versions (calculated $\overline{M}_n$ of 1250 and 2500 g/mol) were prepared and characterized as neat resins, adhesives, and composite matrix resins. The chemistry, physical and mechanical properties of the oligomers and the corresponding cured polymers are discussed.

2. EXPERIMENTAL

2.1 Starting Materials The following chemicals were obtained from the indicated sources and used without further purification: 3,4'-oxydianiline (3,4'-ODA, Mitsui Petrochemical Ind., Ltd., m.p. 84°C), 1,3-bis(3-aminophenoxy)benzene (APB, Mitsui Toatsu, m.p. 107-109°C), 3,3',4,4'-biphenyltetracarboxylic dianhydride (BPDA, m.p. 227°C, Allco Chemical Co.), 4-phenylethynylphthalic anhydride (PEPA, Daychem Laboratories, Inc., m.p. 152°C) and N-methyl-2-pyrrolidinone (NMP, Fluka Chemical Co.).

2.1.1 Synthesis of Phenylethynyl Terminated Imide (PETI) Oligomers
Phenylethynyl terminated imide (PETI) oligomers with a calculated $\overline{M}_n$ of 1250, 2500 and 5000 g/mole were prepared by reacting BPDA with 3,4'-ODA and APB and endcapping with PEPA. Initially, the PETI oligomers were prepared as the amide acid by first dissolving the appropriate quantities of diamines (3,4'-ODA and APB) in NMP at room temperature under nitrogen. The diamine solution was subsequently cooled via an ice water bath and the dianhydride (BPDA) and endcapper (PEPA) were added in one portion as a slurry in NMP. The solids concentration was subsequently adjusted to ~35% (w/w) with additional NMP. The reactions were allowed to stir for ~24 h at ambient temperature under nitrogen and an aliquot was subsequently removed to determine inherent viscosity and for gel permeation chromatographic analysis to assess molecular weight and molecular weight distribution. This solution was subsequently used to prepare thin films, adhesive tape or IM-7 prepreg. Imide oligomer was prepared from the amide acid solution by subsequently fitting the flask with a Dean Stark trap and condenser, adding toluene, and refluxing the solution overnight. The imide oligomers precipitated from solution during the imidization process. The powders were isolated by adding the reaction mixture to water, washing in warm water and methanol. The yellow powders were dried to constant weight with yields >95%.

2.1.2 Films NMP solutions (30-35% w/w solids) of the phenylethynyl terminated amide acid oligomers were centrifuged, the decantate doctored onto clean, dry plate glass and dried to a tack-free form in a low humidity chamber. The films on glass were imidized and cured by heating at 100, 225 and 350°C for 1 hr each in flowing air. Thin film tensile properties were determined according to ASTM D882 using four specimens per test condition.

2.1.3 Molded Specimens Powdered imide oligomer of the 2500 g/mol PETI-5 was compression molded in a 3.2 cm^2 stainless steel mold under 0.3 MPa by heating to 350 or 371°C for 1 hr. Miniature compact tension specimens (1.6 cm x 1.6 cm x 0.95 cm thick) were machined from the moldings and subsequently tested to determine fracture toughness (K_{Ic}, critical stress intensity factor) according to ASTM E399 using four specimens per test condition. G_{Ic}, (critical strain energy release rate) was calculated using the mathematical relationship $G_{Ic}= (K_{Ic})^2/E$, where E is the modulus of the material.

2.1.4 Adhesive Specimens Adhesive supported film was prepared by multiple coating of 112-E glass with an A-1100 finish with a NMP solution of the amide acid oligomer (20% solids) and stage dried in a forced air oven to 225°C after each coat. After sufficient thickness had been achieved, the film was dried to 250°C for 0.5 hr. The final volatile content of the tapes ranged from ~2-4%. Standard lap shear adhesive specimens (bond area 2.54 cm wide x 1.27 cm overlap) using titanium (Ti, 6Al-4V) adherends with a PASA Jell 107 (Products Research and Chemical Corp., Semco Div.) surface treatment were fabricated in a press at 350°C under various pressures for 1 hr. Tensile shear adhesive strengths were determined according to ASTM D1002 using four specimens per test condition.

2.1.5 Composite Specimens 35% solids solutions of the amide acid oligomers in NMP ($\overline{M}_n$ = 1250 and 2500 g/mole, inherent viscosity of 0.15-0.20 dL/g) with Brookfield viscosities of 2689 and 12,800 cps, respectively, at 25°C were used to coat unsized IM-7 carbon/graphite fiber (12k tow) on a multi-purpose prepregging machine. The unidirectional tapes (21.6 cm wide) exhibited resin contents ranging from 30-36% and volatile contents of ~15-19% and fiber aerial weights of ~145g/m^2. Laminates were fabricated in a vacuum press under 76.2 cm mercury during the entire process cycle. The laminates were fabricated by heating to 250°C over ~1 hr period and holding at 250°C for 1 hr. Pressure (0.3-0.7 MPa) was subsequently applied and the temperature ramped up over a 0.5 hr period to 371°C and held for 1 hr. The laminates were cooled under pressure to about 100°C and the pressure subsequently released. The composite panels were ultrasonically scanned (C-scanned), cut into specimens and tested for mechanical properties according to ASTM procedures.

2.1.6 Other Characterization Inherent viscosities (η_{inh}) were obtained on 0.5% (w/v) solutions of the amide acids in NMP at 25°C. Differential scanning calorimetry (DSC) was conducted on a Shimadzu DSC-50 thermal analyzer at a heating rate of 20°C/min with the T_g taken at the inflection point of the ΔT versus temperature curve. Dynamic thermogravimetric analyses (TGA) were performed on a Seiko 200/220 instrument on cured polymer powder samples at a heating rate of 2.5°C/min in air at a flow rate of 15 cm^3/min. Brookfield viscosity was obtained on a Brookfield LVT Synchro-Lectric viscometer at 25°C. Rheological measurements were conducted on a Rheometrics System 4 rheometer. Sample specimen disks, 2.54 cm in diameter and 1.5 mm thick, were prepared by press molding imide powder at room temperature. The compacted resin disk was subsequently loaded in the rheometer fixture with 2.54 cm diameter parallel plates. The top plate was oscillated at a fixed strain of 5% and a fixed angular frequency of 10 rad/sec while the lower plate was attached to a transducer which recorded the resultant torque. Storage (G') and loss (G") moduli as a function of time (t) were measured at several temperatures. Gel permeation chromatography (GPC) was performed on a Waters 150C system equipped with a model 150R differential viscosity detector and a differential refractive index detector. GPC analyses were performed on dilute solutions of the amide acids in freshly distilled NMP containing 0.02M phosphorus pentoxide. The analyses were performed using a two column bank consisting of a linear

Waters Styragel HT 6E column covering the molecular weight range of 10^3 to 10^7 g/mole in series with a Styragel HT 3 column covering the molecular weight range of 10^2 to 10^4 g/mole. A universal calibration curve was generated with Polymer Laboratories narrow molecular weight distribution polystyrene standards having molecular weights ranging from 500 to 2.75 x 10^6 g/mole.

3. RESULTS AND DISCUSSION

3.1 Synthesis of Amide Acid and Imide Oligomers of PETI-5 of Different Number Average Molecular Weights An oligomer designated PETI-5, which was developed at NASA Langley Research Center, was prepared from 3,4'-ODA, APB and BPDA at a calculated $\overline{M}_n$ of 5000 g/mol and endcapped with PEPA. Good processability was evident in the fabrication of moldings, adhesives and laminates; however, during the fabrication of complex composite structures, small areas of the part with tight bends exhibited some porosity, indicating the need for a resin with more flow. In order to improve the processability of PETI-5 (calculated $\overline{M}_n$ of 5000 g/mol), lower $\overline{M}_n$ versions (calculated $\overline{M}_n$ of 1250 and 2500 g/mol) were investigated.

The PETI-5 oligomers were prepared through the amide acid route in NMP with subsequent cyclization to the corresponding imide oligomer (Fig. 1). Initially, the diamines were dissolved in NMP and the solution cooled to ~15°C. BPDA and PEPA were then added in one portion as a slurry in NMP to the stirred solution and the solids content adjusted to ~35% with NMP. The stirred solution was then allowed to warm to ambient temperature under nitrogen. From these solutions thin films, supported adhesive film, and unidirectional IM7 carbon fiber tape were prepared. An aliquot from each of these solutions was removed to assess the molecular weight and distribution. Imide powders were prepared by cyclodehydration of the precursor amide acid oligomers by azeotropic distillation in the presence of toluene for ~24 hrs. As the imide oligomers formed, they precipitated from solution.

Oligomer characterization is presented in Table 1. For comparative purposes the data for 5000 g/mol PETI-5 is included. The initial Tgs of the oligomers were 170-210°C with melting transitions (Tms) ranging from 320-357°C. After melting, the samples failed to recrystallize upon annealing. After a thermal cure for 1 hr at 371°C, no Tms were evident. Powdered samples that were cured in a sealed DSC pan at 371°C for 1h, exhibited increases in Tg of ~60-118°C and followed the trend 1250 g/mol>2500g/mol>5000 g/mol. This was as expected since the 1250 and 2500 g/mol versions contained 64 and 34 mole % PEPA, respectively, as compared to the 5000 g/mol version which contained 18 mole % PEPA. As observed for other PETIs, the exothermic onset and peak due to the cure reaction of the phenylethynyl groups occurred at ~390°C and ~450°C, respectively. The temperature of 5% weight loss as measured by dynamic TGA in air on imide powder cured for 1 hr at 371°C ranged from 489-503°C and was comparable to that of other phenylethynyl containing imide oligomers.

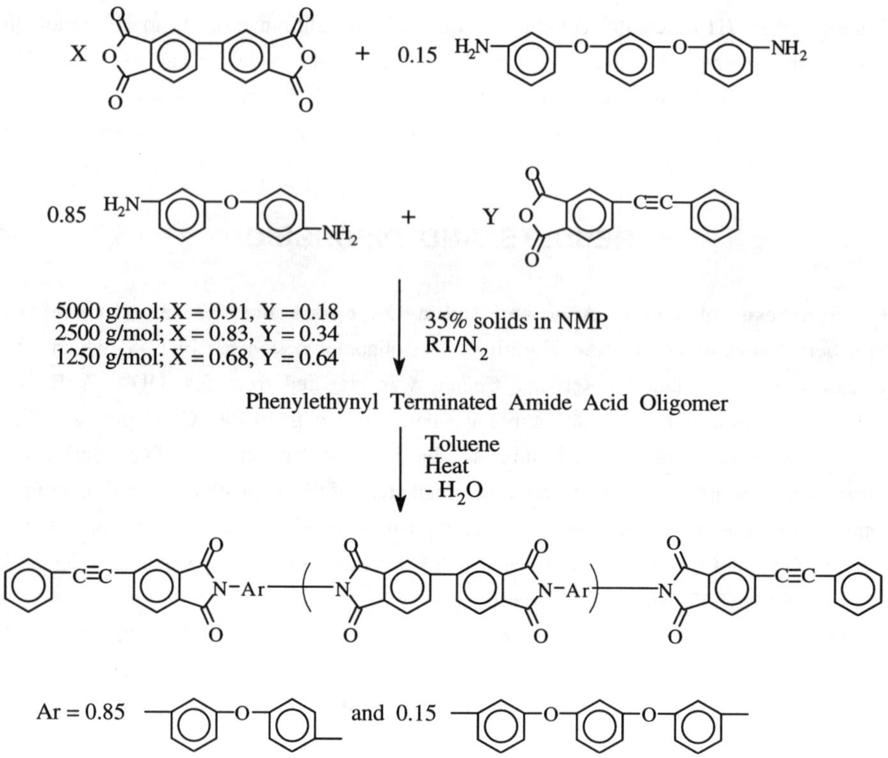

Figure 1. Preparation of PETI-5 oligomers at calculated $\overline{M}_n$ of 1250, 2500, and 5000 g/mol.

Table 1. Physical Properties

Calculated $\overline{M}_n$ (g/mole)	η_{inh}, dL/g[1]	T_g (T_m), °C[2]		TGA, 5% Wt. Loss, °C[4]
		Initial	Cured[3]	
1250	0.15	170 (320)	288	489
2500	0.20	210 (330)	277	497
5000	0.27	210 (357)	270	503

1. Determined on 0.5% (w/v) NMP solutions of the amide acid at 25°C.

2. Determined on powdered samples by DSC at a heating rate of 20°C/min.

3. Determined on powdered samples cured in a sealed DSC pan for 1 h at 371°C.

4. Determined on powdered samples in air at a heating rate of 2.5°C/min.

GPC analyses were performed on as-prepared amide acid solutions of the different $\overline{M}_n$ versions of PETI-5 to assess molecular weight and molecular weight distribution with the data presented in Table 2. In general, the $\overline{M}_n$ determined by GPC were greater than the calculated $\overline{M}_n$. Several shoulder peaks at the low molecular weight end of the molecular weight distribution curve were obtained for the 1250 g/mol PETI-5. A more Gaussian type of distribution was observed for the 2500 g/mol PETI-5 with a slight shoulder evident at the low molecular weight end.

Table 2. GPC Analysis of PETAA-5 Oligomers

Calculated $\overline{M}_n$ (g/mole)	$\overline{M}_n$, g/mole	$\overline{M}_w$, g/mole	$\overline{M}_z$, g/mole	Polydispersity
1250	2050	3022	4307	1.47
2500	3308	6022	11670	1.82
5000	7914	13865	24260	1.75

3.1.1 Films Unoriented thin films were cast from NMP solutions of the amide acid oligomers and cured in flowing air to 350°C. Thin film tensile properties are presented in Table 3. Tensile strength and modulus at room temperature of the 2500 g/mol PETI-5 were ~15% greater than those obtained for the 5000 g/mol version. When tested at 177°C, the strengths and moduli for both $\overline{M}_n$ versions were comparable. The elongation to break of both $\overline{M}_n$ versions was high and indicative of a linear thermoplastic material rather than a thermoset suggesting that the thermal cure reaction of the phenylethynyl group leads to the formation of a high degree of chain extension. A film of the 1250 g/mol PETI-5 had variable thickness due to excessive resin flow during the oven cure; however, the film was flexible and creasable indicating reasonable toughness.

Table 3. Unoriented Thin Film Tensile Properties

Calculated $\overline{M}_n$ (g/mole)	Test Temp., °C	Str., MPa	Mod., GPa	Elong. @ Break, %
2500	23	151.7	3.5	14
	177	76.6	2.2	43
5000	23	129.6	3.1	32
	177	84.1	2.3	83

3.1.2 Neat Resin Moldings Fracture toughness was determined on cured moldings of the 2500 g/mol version of PETI-5 at room temperature. The moldings were prepared from solution imidized powder in a stainless steel mold under 0.3 MPa by heating to 350 or 371°C for 1 hr. The moldings were well consolidated and exhibited a small amount of flash. The fracture toughness (K_{Ic}, critical stress intensity factor) values for the 2500 g/mol and 5000 g/mol PETI-5 cured at 350°C for 1 h were identical (Table 4). This was unexpected since typically lowering the molecular weight results in a more brittle material due to the increased crosslink density. The G_{Ic}s were different due to the difference in the tensile moduli of the two materials. The higher cure temperature (371°C) for the 2500 g/mol PETI-5 resulted in a minimal decrease in the K_{Ic}. Flash from both of the 2500 g/mol PETI-5 moldings exhibited Tgs of 279°C and no melting transitions.

Fabrication conditions were not developed to obtain good quality compact tension specimens of the 1250 g/mol PETI-5. This oligomer exhibited high resin flow. A 2.54 cm x 0.32 cm disk of the 1250 g/mol PETI-5 was tough and was fabricated by B-staging the powder to 300°C for 0.17 hr with a subsequent cure for 0.5 hr at 345°C. When these conditions were used to fabricate a compact tension specimen, the molding had voids presumably due to air. The high toughness exhibited by both $\overline{M}_n$ versions is characteristic of a thermoplastic material and further supports that the cure of the terminal phenylethynyl groups leads to a high degree of chain extension.

Table 4. Neat Resin Fracture Toughness

Calculated $\overline{M}_n$ (g/mole)	K_{Ic}, MPa·m$^{1/2}$	G_{Ic}, J/m^2
2500[1]	3.9	4261
5000[1]	3.9	4295
2500[2]	3.7	3878

1. Molding cured for 1 hr at 350°C.
2. Molding cured for 1 hr at 371°C.

3.1.3 Adhesives Supported adhesive films of 1250 and 2500 g/mol PETI-5 were prepared by multiple coating of 112 E-glass and stage-drying up to 225°C in air after each coat. After sufficient thickness had been achieved, the films were dried to 250°C for 0.5 hr. The final volatile content of the tapes were ~1% and ~4% for the 1250 and 2500 g/mol PETI-5, respectively. The high volatile content of the 2500 g/mol PETI-5 was planned in

Table 5. Preliminary Adhesive Properties

Calculated $\overline{M}_n$ (g/mole)	Bonding Conditions	Test Temp., °C	Tensile Shear Str., MPa
1250	0.03 MPa @ 350°C for 1 hr	23	21.3
		177	27.0
		200	25.9
2500	0.1 MPa @ 350°C for 1 hr	23	36.6
		177	28.6
		200	27.0
5000	0.7 MPa @ 350°C for 1 hr	23	49.0
		177	29.7

order to minimize cracking and flaking of the tape due to the $\overline{M}_n$ of the oligomer. The tape was of excellent quality and did not exhibit any cracking or flaking. A good quality tape of the 1250 g/mol PETI-5 was obtained, but it flaked upon cutting. Specimens were fabricated using titanium adherends with a PASA Jell 107 surface treatment and primed with the amide acid of the 5000 g/mol PETI-5 by heating to 350°C for 1 hr. The tensile shear strengths are presented in Table 5. For comparative purposes, tensile shear properties of 5000 g/mol PETI-5 are included.

The 2500 g/mol PETI-5 specimens exhibited lower tensile shear strengths at room temperature than the 5000 g/mol version. Comparable strengths were obtained at 177°C for both $\overline{M}_n$ versions. The failures were cohesive for both systems at both test conditions.
The 1250 g/mol PETI-5 exhibited lower tensile shear strengths at room temperature than the 2500 and 5000 g/mol PETI-5. The tensile shear strengths at 177°C were comparable to the other $\overline{M}_n$ versions. The failure mode was predominantly adhesive with increasing mixed failures at elevated temperature.

3.1.4 Rheology   Dynamic rheological properties, G' (t) and G" (t), were measured on powder molded discs of the 1250 and 2500 g/mol PETI-5. The test chamber of the rheometer was at room temperature prior to loading the specimen and a temperature profile simulating a composite fabrication cycle was followed. The specimen was heated from 23 to 250°C at a heating rate of 4°C/min and held for 1 hr. The sample was subsequently heated to 371°C at the same heating rate and held for 0.5 hr. The results are shown in Table 6. As expected decreasing the $\overline{M}_n$ lowered the minimum melt viscosity. All of these materials exhibited their respective minimum melt viscosities at temperatures where the phenylethynyl groups react. Thus, the minimum melt viscosities for all of these materials are not stable at these temperatures.

Table 6. Melt Rheology

Calculated $\overline{M}_n$ (g/mole)	Minimum Melt Viscosity, Poise	Temp., °C
1250	50	335
2500	900	335
5000	10,000	371

3.1.5 Composites   IM-7 carbon fiber composites of 1250 and 2500 g/mol PETI-5 were fabricated from solution coated unidirectional tape in a vacuum press under 0.3 and 0.7 MPa, respectively, with a final cure at 371°C for 1 hr. Acceptable tack of the 1250 and 2500 g/mol PETI-5 was obtained at a lower volatile content as compared to that for the 5000 g/mol PETI-5. The processing cycle for these laminates is presented in Figure 2 and was originally developed for the 5000 g/mol PETI-5. Other than a reduction of the applied pressure due to the better flow of the 1250 and 2500 g/mol versions, no optimization work was performed. The laminates generally exhibited excellent C-scans indicating good consolidation with little or no void content. The composite properties are presented in Table 7. For comparative purposes, IM-7 composite properties of the 5000 g/mol PETI-5 are also presented.

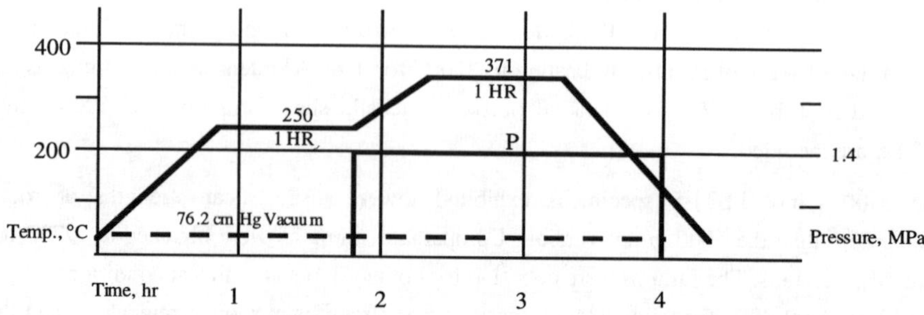

Figure 2. Laminate Processing Cycle for PETI-5 oligomers at calculated $\overline{M}_n$ of 1250, 2500, and 5000 g/mol.

The compression strengths after impact (CAI) were obtained on quasi-isotropic laminates which were impacted at a stress of 30.5 J prior to testing. The open hole compression (OHC) strengths were obtained on laminates with 58% of the plies in the zero orientation (lay-up 58/34/8). The 2500 and 5000 g/mol PETI-5 exhibited comparable OHC (RT, dry), CAI strength and modulus, and microstrain. Better retention of the OHC (177°C, dry) properties was exhibited by the 2500 g/mol PETI-5 as compared to the 5000 g/mol version. The 1250 g/mol PETI-5 exhibited comparable OHC properties to the 2500 and 5000 g/mol PETI-5,

Table 7. IM-7/PETI-5 Laminate Properties

Property	1250 g/mol	2500 g/mol[1]	5000 g/mol[1]
OHC Str., MPa			
RT (dry)	431.6	458.6	450.3
177°C (dry)	366.2	395.2	342.7
177°C (wet)	368.5	344.1	--------
CAI Str., MPa (25/50/25)	244.6	334.5	331.0
CAI Mod., GPa (25/50/25)	55.8	57.9	55.9
Microstrain, μin/in	4377	5908	5986
Thermal Cycling Microcracks/in. after 200 cycles[2]	0	0	0

1. Normalized to 62% fiber volume.

2. After 200 thermal cycles from -55 to 177°C with a 1 hr hold at each temperature and a heating rate of 8.3°C/min.

however, the CAI strength and microstrain were reduced. The 1250 g/mol PETI-5 laminates had low resin contents (26-28%) due to excessive resin flow during fabrication. Consequently, the composite properties in Table 7 were not normalized. Thus it may not be prudent to compare the laminate properties of the 1250 g/mol version with the others.

Specimens of the 1250 and 2500 g/mol PETI-5 were tested for microcrack resistance. The tests were performed on crossply laminates which were thermally cycled from -55 to 177°C at a heating rate of 8.3°C/min with a 1 hr hold at each temperature. The specimens were examined after 100 and 200 cycles under a microscope for microcracks. Neither sample exhibited any microcracks.

4. SUMMARY

PETI-5 was prepared at calculated $\overline{M}_n$s of 1250 and 2500 g/mol and characterized as unoriented thin films, neat resins, adhesives, and composite matrix resins and compared to the 5000 g/mol PETI-5. The 1250 g/mol PETI-5 exhibited excessive resin flow in film, molding, adhesive, and composite forms. Similar mechanical properties and better processability were obtained for the 2500 g/mol version as compared to the 5000 g/mol PETI-5. Thus, the 2500 g/mol PETI-5 has the potential to offer processing improvements in the fabrication of complex composite structures without sacrificing the excellent properties of the 5000 g/mol version.

5. ACKNOWLEDGMENT

The authors are grateful to Drs. Emilie J. Siochi and Tan H. Hou, Lockheed Engineering and Sciences Company, for the gel permeation chromatographic and rheological work, respectively, and to Monica Rommel and Dan Reynolds, Northrop Grumman Corporation, for some of the laminate testing.

The use of trade names of manufacturers does not constitute an official endorsement of such products or manufacturers, either expressed or implied, by the National Aeronautics and Space Administration.

6. REFERENCES

1. F. W. Harris, S. M. Padaki and S. Vavaprath, <u>Polym. Prepr.</u>, <u>21</u>(1), 3(1980).

2. F. W. Harris, A. Pamidimukkala, R. Gupta, S. Das, T. Wu and G. Mock, <u>Ibid</u>, <u>24</u>(2), 324 (1983).

3. F. W. Harris, K. Sridhar and S. Das, <u>Ibid</u>, <u>25</u>(1), 110 (1984).

4. F. W. Harris, A. Pamidimukkala, R. Gupta, S. Das, T. Wu and G. Mock, <u>J. Macromol. Sci.-Chem. A</u>, <u>24</u>,(8/9), 1117 (1984).

5. S. Hino, S. Sato and O. Suzski, <u>Jpn. Kokai Tokyo Koho JP</u>, <u>63</u>, (196), 564 (1988). <u>Chem. Abstr.</u>, <u>110</u>, 115573w (1989). U. S. Patent # 5,066,771 (1991) to Agency of Industrial Science and Technology, Japan.

6. C. W. Paul, R. A. Schultz and S. P. Fenelli, in "Advances in Polyimide Science and Technology", C. Feger, M. M. Khoyasteh and M. S. Htoo Eds., Technomic, Lancaster, PA 1993, pp 220.

7. R. G. Bryant, B. J. Jensen and P. M. Hergenrother, <u>Polym. Prepr.</u>, <u>34</u>(1), 566 (1993).

8. B. J. Jensen, P. M. Hergenrother and G. Nwokogu, <u>Polymer</u>, <u>34</u>(3), 630 (1993).

9. G. W. Meyer, S. Jayaraman and J. E. McGrath, <u>Polym. Prepr.</u>, <u>34</u>(2), 540(1993).

10. S. J. Havens, R. G. Bryant, B. J. Jensen and P. M. Hergenrother, <u>Ibid</u>, <u>35</u>(1), 553(1994).

11. P. M. Hergenrother and J. G. Smith, Jr., <u>Ibid</u>, <u>35</u>(1), 353(1994). <u>Polymer</u>, <u>35</u>(22), 4857 (1994).

12. G. W. Meyer, T. E. Glass, H. J. Grubbs and J. E. McGrath, <u>Ibid</u>, <u>35</u>(1), 549 (1994).

13. J. A. Johnston, F. M. Li, F. W. Harris and T. Takekoshi, <u>Polymer</u>, <u>35</u>(22), 4865 (1994).

14. T. Takekoshi and J. M. Terry, <u>Ibid</u>, 4874 (1994).

15. J. W. Connell, J. G. Smith, Jr., R. J. Cano and P. M. Hergenrother, <u>Sci. Adv. Mat. Proc. Eng. Ser.</u>, <u>41</u>, 1102 (1996). <u>High Perform. Polym.</u>, <u>9</u>, 309 (1997).

16. J.A. Hinkley and B.J. Jensen, <u>High Perform. Polym.</u>, <u>8</u>, 599 (1996).

17. B. Tan, V. Vasudevan, Y.J. Lee, S. Gadner, R.M. Davis, T. Bullions, A.C. Loos, H. Parvatareddy, D.A. Dillard, J.E. McGrath and J. Cella, <u>J. Polym. Sci.: Pt. A: Polym. Chem.</u>, <u>35</u>, 2943 (1997).

18. J. G. Smith, Jr., J. W. Connell and P. M. Hergenrother, <u>Polymer</u>, <u>38</u>(18), 4657 (1997).

19. J. W. Connell, J. G. Smith, Jr. and P. M. Hergenrother, <u>Intl. SAMPE Tech. Conf. Series</u>, <u>29</u>, 317 (1997).

20. R. G. Bryant, B. J. Jensen and P. M. Hergenrother, <u>Sci. Adv. Mat. Proc. Eng. Ser.</u>, <u>39</u>, 273 (1994) (closed papers volume).

21. B. J. Jensen, R. G. Bryant, J. G. Smith, Jr,. and P. M. Hergenrother, <u>J. Adhesion</u>, <u>54</u>, 57 (1995).

22. R. J. Cano and B. J. Jensen, <u>J. Adhesion</u>, <u>60</u>, 113 (1997).

23. T. Hou, B. J. Jensen and P. M. Hergenrother, <u>Composite Materials</u>, <u>30</u>(1), 109 (1996).

24. P.M. Hergenrother and M. Rommel, <u>Sci. Adv. Mat. Proc. Eng. Series.</u>, <u>41</u>, 1061 (1996).

25. M. Rommel, L. Konopka and P.M. Hergenrother, <u>Intl. SAMPE Tech. Conf. Series</u>, <u>28</u>, 14 (1996).

7. Biographies

Joseph G. Smith, Jr. is a polymer scientist in the Composites and Polymers Branch of the Materials Division at NASA Langley Research Center (LaRC). He received a B.S. degree from High Point College in 1985 and a Ph.D. from Virginia Commonwealth University in 1990. Prior to joining NASA LaRC in September 1994, he held postdoctoral research positions with the University of Akron and Virginia Commonwealth University. His work at NASA has focused on the development of high performance polymers for aerospace applications.

John W. Connell is a senior polymer scientist in the Composites and Polymers Branch of the Materials Division at NASA Langley Research Center (LaRC). He received B.S. and Ph. D. degrees from Virginia Commonwealth University in 1982 and 1986, respectively. Prior to joining NASA LaRC in January 1988, he was a research associate at Virginia Commonwealth University. Since coming to NASA, his work has focused on the development of high performance polymers for aerospace applications.

Paul M. Hergenrother is a senior polymer scientist in the Composites and Polymers Branch of the Materials Division at NASA Langley Research Center (LaRC). He received a B.S. degree from Geneva College and took graduate work at the University of Pittsburgh and Carnegie-Mellon University. Prior to coming to NASA he held various research and management positions at the Koppers Co., The Boeing Co., Whittaker Corp. and Virginia Polytechnic Institute and State University. His work at NASA has focused on the development of high performance polymers for aerospace applications.

DURABILITY CHARACTERIZATION OF BISMALEIMIDE AND POLYIMIDE-CARBON FIBER COMPOSITES

Roger J. Morgan, E. Eugene Shin, Jason E. Lincoln and Jiming Zhou
Advanced Materials Engineering Experiment Station
Michigan State University
Midland MI

Lawrence T. Drzal, Mark S. Wilenski and Andre Lee
Composite Materials and Structures Center
Michigan State University
East Lansing MI

David Curliss
U.S. Air Force Wright Laboratory
Dayton OH

ABSTRACT

Potential critical physical and chemical aging mechanisms in terms of damage initiation and propagation of bismaleimide, BMI and Polyimide, PI, -carbon fiber composites in future aerospace stress-time-temperature-moisture-chemical service environments will be reported on the molecular, microscopic and macroscopic structural levels in order to develop structural-performance phase diagrams for mechanics model-structural design analyses and associated materials structural optimization at all dimensional levels.

For BMI-C fiber composites, based on previous studies on the structure-property relations of BMI's and the fiber-matrix interfacial integrity, new fiber surface treatments will be discussed and optimal composite cure-conditions will be identified in order to inhibit composite damage initiation thresholds. The chemical characterization of the high temperature aging ($\geq 200°C$) of BMI's as a result of dehydration-induced ether crosslink formation and subsequent decomposition of these crosslinks causing T_g increases and mechanical property deteriorations will be described.

For PI-C fiber composites the critical hygrothermal induced degradation mechanisms are discussed in terms of:

(i) PI depolymerization characterization and kinetics and associated mechanical and thermal property deterioration;

(ii) Moisture evolution induced PI matrix and/or interfacial blistering upon rapid heating resulting from physically entrapped moisture and PI repolymerization.

KEYWORDS: Durability, Bismaleimides, Polyimides

1. INTRODUCTION

High temperature polymer matrix-carbon fiber composites are and will be used for aerospace structural applications for a whole range of components that will be exposed to prolonged, extreme service environment conditions. These complex service environment conditions of stress, time, temperature, moisture, chemical and gaseous environments require a thorough understanding of the physical, chemical and mechanical phenomena that control the most probable critical failure path of the composite component. Such an understanding of the critical fundamental aging mechanisms is necessary for credible long-term composite performance predictions based on experimentally observed shorter time service-environment induced composite performance deterioration mechanisms. In addition, this understanding generates meaningful information for mechanics modeling-structural design analyses and associated materials structural optimization at all dimensional levels.

The two types of high temperature polymer matrix-carbon fiber composites utilized in aerospace applications are (i) crosslinked bismaleimide, BMI, thermoset and (ii) predominantly thermoplastic polyimide, PI, -carbon, C, fiber composites.

We have in our previous studies followed a systematic durability methodology to evaluate the long term performance of these high temperature composites. This methodology involved initially studying the effects of individual or combined core test service environments upon composite performance in order to identify the most likely synergistic service environments and controlling physical, chemical and mechanical parameters of the ultimate critical failure path. This methodology leads to (i) lifetime predictions and (ii) materials optimization based on identified mechanisms rather than empiricism.

Based on previous studies (1-13), the most likely potential, critical aging mechanisms that control damage initiation of BMI-and PI-C fiber composites have been identified. The damage initiation mechanisms on the molecular or microscopic level may be different than those that control the macroscopic damage propagation mechanisms, such as delamination, that lead to the ultimate performance deterioration of the composite component. These critical aging mechanisms involve (i) further cure of BMI-carbon fiber composites with associated T_g increases, mechanical property decreases and enhanced microcrack development in service environment exposure conditions, and (ii) hygrothermal induced thermal and mechanical property deterioration as a result of physical and chemical structural changes in BMI- and PI-carbon fiber composites.

In this paper we report the following areas:
For BMI-C fiber composites:
> (i) The chemical characterization of the high temperature aging ($\geq 200°C$) of BMI matrices as a result of dehydration-induced ether crosslink formation and subsequent decomposition of these crosslinks causing T_g increases and mechanical property deteriorations will be described.
> (ii) The poor fiber-matrix interfacial integrity, optimal composite cure conditions and potential new fiber surface treatments will be discussed in order to enhance composite damage initiation thresholds.

For PI-C fiber composites the critical hygrothermal induced degradation mechanisms are discussed in terms of:
> (i) PI depolymerization characterization and kinetics and associated mechanical and thermal property deterioration;
> (ii) Moisture evolution induced PI matrix and/or interfacial blistering upon rapid heating resulting from physically entrapped moisture and PI repolymerization.

2. MATERIALS

The BMI resin system studied was based on $4,4'$-bismaleimidodiphenyl methane (BMPM)/$0,0'$-diallyl alcohol of bisphenol A (DABPA), Matrimid 5292, Ciba-Geigy. In the 100-200°C range the BMPM and DABPA monomers react to form the "ene" molecule (Figure (1)). The PI resin systems studied were:

     (i) Avimid K3B, DuPont, formed from an aromatic diethyl ester of the pyromellitic diacid and an aromatic diamine in M-methyl-2-pyrrolidone, NMP, and

     (ii) AFR700B is based on a fluorinated PI oligomer with norbornene and primary amine end cap groups, provided by U.S. Air Force Wright Laboratory (Figure (2)).

3. RESULTS AND DISCUSSION

3.1 Structure-Property Relations and BMI Composite Matrices

In the case of BMI-carbon fiber composites, the BMI cure reactions are incomplete after standard composite fabrication and post-cure procedures and further cure can occur in service environment conditions resulting in T_g increases and associated composite mechanical property deteriorations. Loseth and Rothschild (14) have observed real time T_g increases for commercial BMI-carbon fiber composites (BASF 5260) of up to 70°C after isothermal exposure in the 150-200°C temperature for up to ~ 10^4 hours. Li (15) reported a 50% decrease in impact energy after thermal aging of BMI-carbon fiber composites for 1,000 hours at 190°C. Pederson et al. (16) report an increase surface ply crack density after isothermal 177°C air exposure for 750 hours for [0/90] laminates. Also Burchan et al. (17) report transverse microcrack density increases for 16 ply quasi-isotropic BMI-carbon fiber composites aged isothermally in the 150-204°C range for 16,000 hours as a function of time, temperature and ply depth.

In our previous studies (3, 11) we conducted systematic Fourier transform infrared spectroscopy, FTIR, and differential scanning calorimetry, DSC, studies of the cure reactions of the BMPM/DABPA BMI resin system as a function of chemical composition and temperature-time cure conditions and together with literature data (18-20) it was concluded:

     (i) The principal cure reactions via the -C=C- bonds of the "ene" molecule (Figure (1)) are complete after 250°C for 3 hours; however,

     (ii) the ether crosslink reaction via dehydration of the hydroxyl groups of the "ene" molecule is only 50% complete after cure at 250°C for 9 hours.

In further systematic studies of the "cure" reactions of BMPM-DABPA BMI resins at temperatures in the 250-300°C temperature range as a function of temperature and time upon the mechanical, thermal and physical properties of BMI's and their C fiber composites our principal findings are:

     (i) Based on FTIR studies ether crosslinks decrease in concentration in the 240-300°C temperature range as a function of cure time (Figure (3)).

     (ii) The dehydration induced ether crosslink formation and subsequent rearrangement plays a critical role in the chemical and physical network structural changes of BMPM-DABPA BMI resins at high service environment temperatures and long times that can significantly modify the BMI resin thermal and mechanical response and composite resin shrinkage-induced microcrack characteristics. For example, we have recorded a ~0.2% density decrease; ~ 0.3 wt% loss; 10% modulus and a 25-40% ductility loss in the 25-177°C temperature range for these BMI resins, as a function of initial monomer compositions, that is associated with ~ 100°C increase in T_g after further cure in the 250-300°C temperature range (Figure (4)). These property changes all occur after all the double C=C bonds of the ene molecule have been consumed. The chemical and physical mechanisms associated with these BMI resin property modifications must involve thermal dissociation of the ether linkages followed by formation of less flexible crosslinks, such as -C-C- linkages, and a possible increase in overall crosslink density.

3.2 Fiber-Matrix Interface Characterization and Modification

During the fabrication and post-cure of BMI-carbon fiber composites significant residual stresses can develop within the composite because of (i) thermal expansion mismatch between the fibers
and matrix and (ii) BMI resin shrinkage during cure. Post-curing BMI-carbon fiber composites in the 250-300°C range results in matrix microcracking in 0°/90° composite laminates based on BMPM-DABPA (1:1 molar ratio) BMI resins (2,3).

Microcracking in cross-ply laminates has been studied extensively, and a variety of analyses have been used including finite element (21), analytical (22), experimental (23), and Monte-Carlo simulation (24). It has been suggested that the stress state in these composites results from the thermal expansion/resin shrinkage mismatch between the 0° and 90° plies. A shear stress transfers the load from one layer into the next, causing a combination of shear stresses at the ply interface and tensile stresses at the ply mid-plane (25) as we have directly observed by the ESEM (Environmental Scanning Electron Microscopy) (9, 26). In addition to this macroscopic stress, a microscopic stress is present due to the resin shrinkage and mismatch between the radial thermal expansion of fibers and resin.

We have used (9, 26) both finite element analysis and a closed form analytical solution to model the combination of macroscopic and microscopic stresses which result from thermal loading. In addition to the modeling, [0° /90°]$_s$ BMI/IM7 composites have been analyzed using ESEM. It was found that all laminates were microcracked regardless of cure/post-cure cycle. As expected, crack widths grow with decreased fiber volume fraction, and increase with cure advancement. Observation in the ESEM revealed poor adhesion between the IM7 C-fiber and the BMI resin for both cure induced and mechanically induced cracks. No cure induced microcracks were found in the polyimide (K3B)-IM7 C fiber composite, but poor adhesion was observed after mechanical loading.

The adhesion in the BMI/IM7 system has been measured using a variety of techniques (26). Direct evidence of poor adhesion was obtained by ESEM imaging of Mode I crack opening failure surfaces. The revealed fibers were completely devoid of resin which is seen in adhesive failures. Images of both thermally and mechanically induced cracks also show a completely adhesive failure. Four point flex testing of composites revealed very poor interlaminar shear strength as evidenced by the mid-plane shear failure of flexure samples with a 28:1 span to depth ratio. IFSS values measured using the Interfacial Testing System (ITS) were less than 10% of the values for standard epoxy/carbon fiber systems.

All evidence, to date, indicates the C fiber-matrix interface in BMI and PI composites exhibits poor mechanical integrity and is one of the weak links together with resin matrix toughness that is responsible for early microscopic damage initiation and propagation. This poor fiber-matrix interfacial integrity is more serious in BMI-C fiber composites as a result of the imposition of cure induced matrix shrinkage stresses.

There has been considerable effort to improve the fiber-matrix interface mechanical integrity for BMI- and PI-C fiber composites using a range of approaches that fall into four major categories:

(i) modification of fiber surface energy;
(ii) increase in fiber surface functional group concentration;
(iii) resin toughening;
(iv) additions of fiber sizing. (Reference (26) summarizes these literature studies.)

However, these studies to date have not resulted in significantly improving the mechanically weak interface deficiency in BMI-C fiber and to a lesser extent PI-C fiber composites. There are a number of reasons for this lack of significant improvement:

(i) The wide temperature range between composite fabrication conditions and lower extremes of service environment use temperatures produces high fiber-matrix interfacial stresses;

(ii) Functionally active C fiber surface groups, such as - COOH and -OH groups are lost from the fiber surface in the 275°C temperature region;
(iii) Lack of thermal stability of many of the thermoplastic sizings during the relatively long time - high temperature composite fabrication conditions and/or lack of sufficient toughness over wide service environment temperature ranges;
(iv) Loss of fiber strength as a result of fiber surface treatments.

3.3 Hygrothermal Induced Degradation Mechanisms
3.3.1 Background

The most critical hygrothermal induced degradation mechanisms of PI-C fiber composites as related to future aerospace service environments are (i) composite blistering upon rapid heating as a result of moisture evolution from physically entrapped moisture and, also, repolymerization of hydrolytically degraded PI that chemically produces water molecules, and (ii) direct mechanical and thermal property deterioration caused by hydrolytic chemical depolymerization (Figure (5)).

Cornelia (27) has reported significant decreases in PI-carbon fiber composite dry T_g's of >50°C after exposure to high temperature hygrothermal environments. Thorp and Crasto (28) also report smaller permanent dry T_g decreases. These observed "dry" T_g decreases can be caused by physically entrapped water molecules in glassy-state that are only released at high temperatures causing plasticization and lowering of the T_g (12, 13). With increasing temperatures new H-bonding sites within the polymer become accessible to water molecules as the physical structure of the polymer glass expands. Upon cooling, water molecules could be trapped in these new sites and are only released upon heating to a temperature that the site initially became accessible.

Polyimides are susceptible to hydrolytic polymer chain scission as a result of (i) scission of any inherent amide group defects present as a result of non-ring closure to the imide ring and (ii) imide ring opening to amide formation followed by scission of the amide group. Hydrolytic attack of the imide ring reverses the polymerization reaction, Figure (6), resulting in the formation of the polyamic acid, followed by chain scission and associated molecular weight and strength decreases resulting ultimately in regeneration of the monomers.

The hydrolytic degradation of amide linkages in polymers, particularly at high temperatures in the presence of moisture is well documented (10, 29, 30). The amide linkage is severed by this degradation mechanism, creating new polymer chain ends in the form of additional defects in the polymer, lower molecular weight and lower strengths. However, there is sparce direct data on the rates of hydrolytic attack of the imide ring. Opening of the imide ring by acidic H^+ protons or alkaline OH^- groups and resultant hydrolytic chain scission, lowering of molecular weight and associated polymer strength loss occurs in aggressive humid-acidic environments. For example, the Navy has experienced severe in service corrosion problems for their bismaleimide-carbon fiber aircraft components as a result of protonic galvanic attack of the maleimide ring (31). From the viewpoint of imide ring opening there have been a number of literature reports of acid and alkaline catalytic-induced hydrolytic degradation of PI's (32-34) that result in reported decreases in viscosity and by association molecular weight and strength. Certainly, molecular weight decreases are further accelerated by acid-catalyzed oxidative degradation of PI's. Because of the uncertainties of the amount of inherent amide link impurities in the PI's used in the literature studies of PI hydrolytic degradation, it is not possible to ascertain from these studies directly the rates of hydrolytic imide ring opening, with the exception of a model compound study conducted by Nechaev, et. al (35). The latter workers studied the kinetics of acidic catalyzed hydrolysis of phthalanilic acid and phthalanil model compounds and their data indicates the imide hydrolysis is 550X slower than the amide hydrolysis in concentrated sulfuric acid.

When physically trapped and/or chemically evolved moisture vapor pressure exceeds the local polymer matrix yield stress during dynamic hygrothermal exposure matrix cavitation will occur (36). Such matrix cavitation and associated delamination depends on a series of variables such as (i) previous humidity-time-temperature exposure and associated moisture concentration profiles; (ii) component thickness; and (iii) rapid heat-time service environment exposures, such as thermal spikes. A thermoset is more resistant to cavitation and associated macroscopic blistering than a thermoplastic because thermoset cavitation requires rupture of covalent crosslinked molecular segments.

3.3.2 Accelerated Hygrothermal Exposure

In order to ascertain the effects of high temperature hygrothermal environmental exposure upon:
  (i) physically trapped moisture;
  (ii) chemical hydrolytic degradation;
  (iii) residual mechanical, physical and thermal properties;
  and   (iv) blistering onset temperature;

K3B and AFR700B PI resins were moisture saturated in a pressure bomb at 160°C, 110 psi for 1000 hours and then isothermally dried as a function of time and temperature. (D_2O rather H_2O saturation was carried out in these studies in order to conduct ^2H NMR studies of D_2O molecular mobility in the PI's.)

The principal findings from this study are:
  (i) It is extremely difficult to remove all the moisture under vacuum from PI's even after drying above 100°C for over 500 hours (Figure (7)). The moisture desorption is directly related to the difference between the isothermal drying temperature and the dry T_g for both K3B and AFR700B PI's. These findings indicate that higher T_g PI's retain higher moisture contents for longer times and at higher temperatures, thus making them more susceptible to hydrolytic depolymerization.
  (ii) For K3B PI a 1wt% moisture concentration causes a ~25°C T_g decrease which is consistent with T_g plasticization decreases for PI's as a family (Figure (8)). In addition, upon drying the initial T_g is retrieved. For AFR700B PI, however, the T_g decrease for 1 wt% moisture is considerably greater namely ~75°C, and further T_g decreases do not occur in the 1-5 wt% moisture concentration range and also, the original dry T_g is not retrieved. These latter observations suggest that chemical hydrolytic degradation of the AFR700B has occurred in addition to physical moisture-induced plasticization.
  (iii) The extremely intense secondary glass transition T_{gg} exhibited by AFR700B at ~150°C is plasticized by absorbed moisture indicating the mobile structure is capable of H-bonding (Figure (8)). Russell and Kardos (37) report the appearance of this transition only occurs after imidization is complete after curing >300 C and, as such, they associate this peak with norbornene crosslink structures. Curliss (38) has reported such crosslinks are susceptible to hydrolysis at temperatures down to 100°C. Most recently, Thorpe and Crasto (39) find this secondary glass transition decreases in intensity upon accelerated hygrothermal exposures in the 150-200°C range.
  (iv) Accelerated hygrothermal exposure causes dramatic decreases in the flexural "dried" residual mechanical properties of AFR700B PI's with a 70% loss of strength and 85% loss of strain to failure. Such mechanical property deterioration we associate with hydrolytic chain scission of the norbornene based crosslinks. In the case of K3B PI's, smaller decreases occur in mechanical properties (5% loss of strength and 20% loss of strain to failure) for similar hygrothermal accelerated exposures but at lower drying temperatures.

3.3.3 Imide Ring Hydrolysis Rates at High Temperatures

A study has been initiated to ascertain the kinetics of imide ring hydrolysis in order to characterize hygrothermal-chemical induced damage threshold limitations for PI's at high temperatures. Based on

(i) statiscally meaningful hydrolytic induced strength loss studies caused by amide linkage scission in Kevlar 49 fibers (i.e. 80% strength loss/year at 100°C) (10) and (ii) the 550X slower hydrolytic degradation rate of the imide ring relative to an amide linkage (35) calculations indicate imide groups would depolymerize to 100% loss of PI strength in 8 days at 250°C in the presence of moisture. Initial experiments of exposure of K3B PI powder to moisture of 250°C for 8 days in a pressure bomb causes (i) complete disappearance of the IR bands at 1779 cm^{-1} and 1720 cm^{-1} associated with the imide rings; (ii) appearance of 3 endotherms at 207°C, 294°C and 337°C associated with the melting points of phthalic acid, pyromellitic acid and sublimation of terephthalic acid respectively; and the disappearance of a T_g in DSC plots all indicate significant depolymerization to the monomer has occurred.

ACKNOWLEDGMENTS

The authors thank Air Force Office of Scientific Research (AFOSR), sponsored by Dr. Charles Lee, and State of Michigan Research Excellence Funds for financial support and encouragement of this work.

REFERENCES

1. R. J. Morgan, E. E. Shin, C. Dunn, R. J. Jurek and A. Jurek, Proc. of 39th SAMPE Conf., 39, 1564, (1994).
2. H. H. Man and R. J. Morgan, Proc. of 7th ASM/ESD Advanced Composites Conf., Detroit, MI, 555, (1991).
3. R. J. Morgan, R. J. Jurek, A. Yen and T. Donnellan, Polymer, 34, 835, (1993).
4. E. E. Shin, R. J. Morgan, Proc. of ANTEC 93 on Plastic Engineering, SPE, II, 1357, (1993).
5. E. E. Shin, Q. Zheng and R. J. Morgan, Proc. of SEM 50th Anniversary Spring Conf. on Experimental Mechanics, SEM Inc., 366, (1993).
6. E. E. Shin, R. J. Jurek, L.T. Drzal, R. J. Morgan, J. K. Choi and A. Lee, Proc. of ASME Meeting, San Francisco, 183, (1995).
7. E. E. Shin, C. Dunn, E. Fouch, R. J. Morgan, M. Wilenski and L. T. Drzal, Proc. of ASME Meeting, San Francisco, 191, (1995).
8. R. J. Morgan, E. E. Shin, J. E. Lincoln, J. Choi and A. Lee, Proc. of 28th National Tech. Conf., Seattle, WA, November, 1996, pp.. 213-224.
9. E. E. Shin, R. J. Morgan, M. Wilenski, J. Zhou, J. E. Lincoln and L.T. Drzal, Proc. of 28th National Tech. Conf., Seattle, WA, November, 1996, pp. 225-235.
10. R. J. Morgan, Thermal Characterization of Composites in Thermal Analysis of Polymers, 2nd Edition, Ed. E. Turi, Academic Press, Chap. 9, (1997), pp. 2091-2261.
11. R. J. Morgan, E. E. Shin, B. Rozenberg, and A. Jurek, Polymer, 38, 639, (1997).
12. E.E. Shin, R.J. Morgan, J. Zhou, J. Lincoln, R. Jurek and D.B. Curliss, Proc. of American Soc. for Composites, 12th Technical Conf., Technomic Co., Dearborn, MI, October 1997, pp. 1113-1122.
13. R.J. Morgan, E.E. Shin, J. Lincoln, J. Zhou, L.T. Drzal, A. Lee and D.B. Curliss, Proc. of High Temple Workshop XVIII, Hilton Head, SC, January, 1998, Paper Q.
14. C. Loechett and R. Rothschild, Boeing Commercial Airplane Co., Seattle, WA, (1992), (Private Communication).
15. L. Li, Polymer Plast. Technol. Eng., 291, 549, (1990).

16. C.L. Pederson, J.W. Gillespie, R. L. McCullough, R. J. Rothschild and S.L. Stanek, Polymer Composites, 16, 154, (1995).
17. L.J. Burchan, R.F. Eduljee and J. W. Gillespie, Polymer Composites, 16, 507, (1995).
18. M. Chaudhari, T. Galvin and J. King, SAMPE Journal, July/August, 17, (1985).

19. B.J. Lee, M.A. Chaudhari and V. Blyakham, <u>Polymer News</u>, <u>13</u>, 297, (1988).

20. S. Zahir, M.A. Chaudhari and J. King, <u>Makromol. Chem. Macromol. Symp.</u>, <u>25</u>, 141, (1989).

21. D. F. Adams, <u>Journal of Reinforced Plastics and Composites</u>, **6**, pp.. 66-68, (1987).

22. Y. M. Han, H. T. Hahn and R.B. Croman, <u>Composites Science and Technology</u>, **31**, pp.. 165-177 (1988).

23. C. Wood and W. Bradley, "A New Technique to Study the Interfacial Strength and Transverse Cracking Scenario in Composite Materials," in STP 1290, "Fiber, Matrix and Interface Properties," August (1996).

24. A.S.D. Wang, P. C. Chou and S. C. Lei, <u>Journal of Composite Materials</u>, **18**, pp. 239-254, (1984).

25. N. Laws and G. J. Dvorak, <u>Journal of Composite Materials</u>, **22**, pp.. 900-916, (1988).

26. M. Wilenski, "The Improvement of the Hygrothermal and Mechanical Properties of Bismaleimide and K3B/IM7 Carbon Fiber Composites through A Systematic Study of the Interphase," Ph.D. Thesis, Michigan State University, (1997).

27. D. Cornelia, Proc. of 39th SAMPE Conf., <u>39</u>, 917 (1994).

28. K.E.G. Thorp and A. S. Crasto, Proc. of Am. Soc. of Composites 10th Tech. Conf., 601 (1995).

29. R. Deiasi, J. Mater. Sci. <u>10</u>, 1951-1958 (1975).

30. C. J. Wolf and R. S. Solomon, Proc. of National SAMPE 15th Technical Conf., Cincinnati, OH, pp. 504-510 (1983).

31. R. E. Allred, T. M. Donnellan, R. Cochran and K. Miller, "The Effect of Phenolic Finishes on the Galvanic Corrosion Resistance of Carbon/BMI Composites," Proc. of High Temple Workshop XVII, Paper R, February, 1997, Monterey, CA.

32. A. N. Krasovskii, T. A. Redrova, K. Kalnins, Yu.N. Sazonov, Vysokomol. Soedin. Ser. B, <u>24(2)</u>, 890 (1982).

33. C. E. Sroog, A. L. Endrey, S. V. Abramo, C. E. Berr, W. M. Edwards and K. L. Olivier, J. Polym. Sci., <u>A-3</u>, 1373 (1965).

34. "Polyimides Thermally Stable Polymers," Eds. M.I. Bessonov, M.M. Koton, V.V. Kudryavtsev and L.A. Lauis, Consultants Bureau, New York, 1987, pp. 154.

35. P. P. Nechaev, Yu. V.Moiseev, YA.S.Vygodskii and G.E. Zaikov, Int'l. J. of Chem. Kinetics, <u>VI</u>, 245 (1974).

36. Q. Zheng and R.J. Morgan, J. Composite Materials, <u>27</u>, 1465 (1993).

37. J.D. Russell and J.L. Kardos, Polymer Composites, <u>18</u>, 595 (1997).

38. D.B. Curliss, Proc. of High Temple Workshop XVI, Perdido Beach, AL, January 1996, Paper H.

39. K.E.G. Thorp and A. Crasto, Proc. of High Temple Workshop XVIII, Hilton Head, SC, January 1998, Paper W.

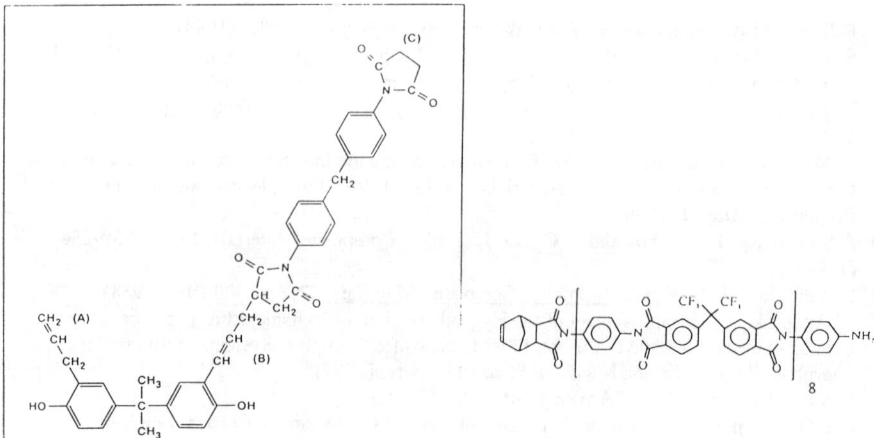

Figure 1. The chemical structure of the BMPM/DABPA "ene" adduct prepolymer. (A), (B) and (C) are double bonds capable of polymerization.

Figure 2. AFR700B PI polymeric precursor.

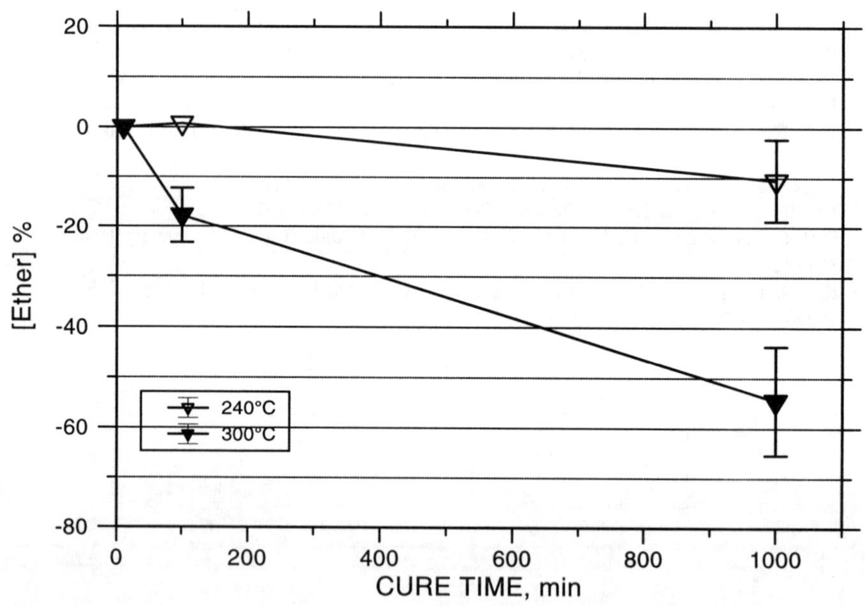

Figure 3. Infrared ether band at 1183cm^{-1} as function of cure conditions.

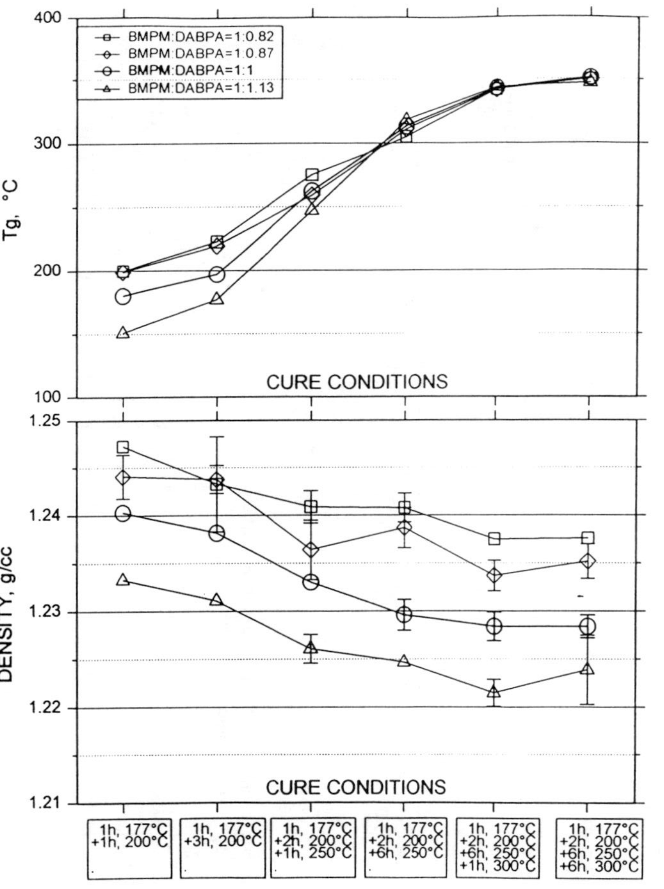

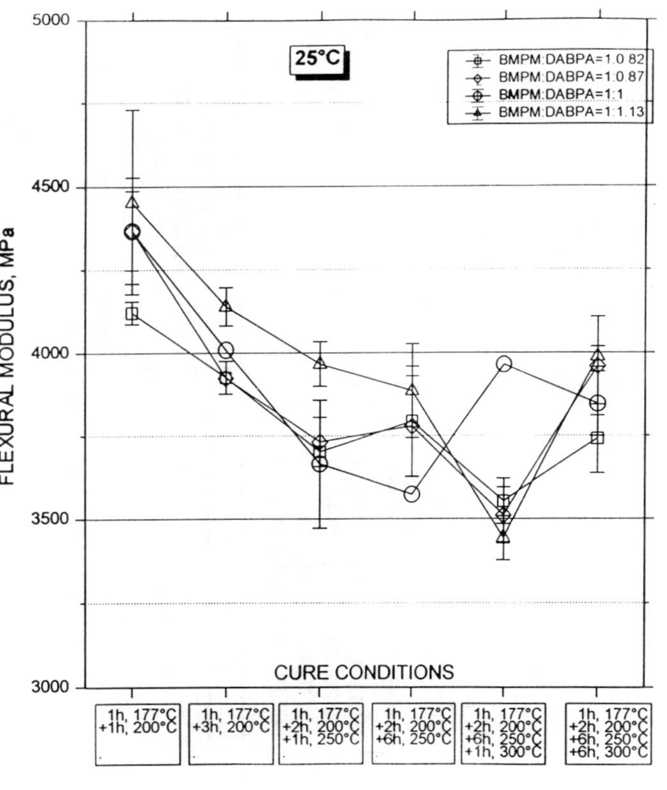

Figure 4. T_g and ambient density and flexural modulus of **BMPM-DABPA BMI's** as a function of stoichiometry and cure conditions.

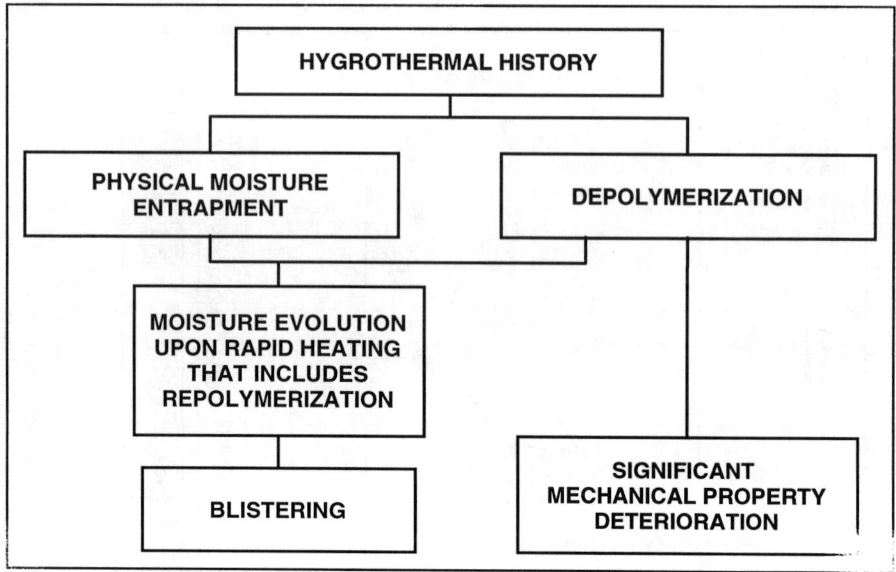

Figure 5. Critical hygrothermal induced degradation mechanisms.

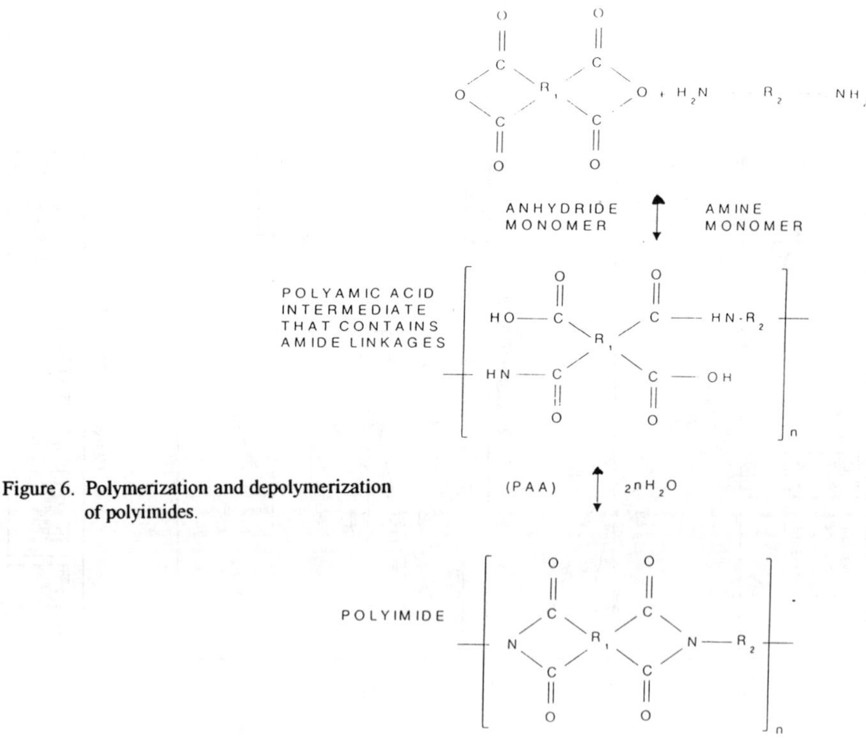

Figure 6. Polymerization and depolymerization of polyimides.

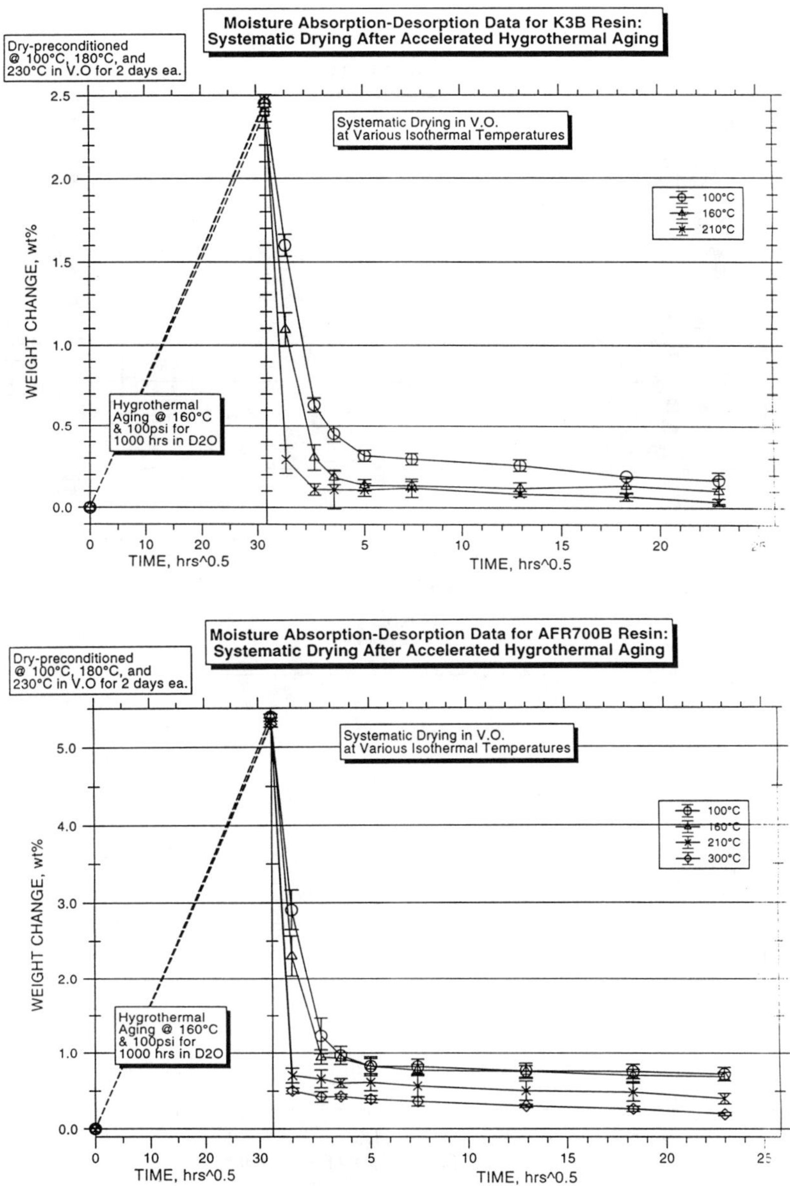

Figure 7. Isothermal moisture desorption plots under vacuum
for (a) K3B and (b) AFR700B PI's, after accelerated
high temperature hygrothermal exposure.

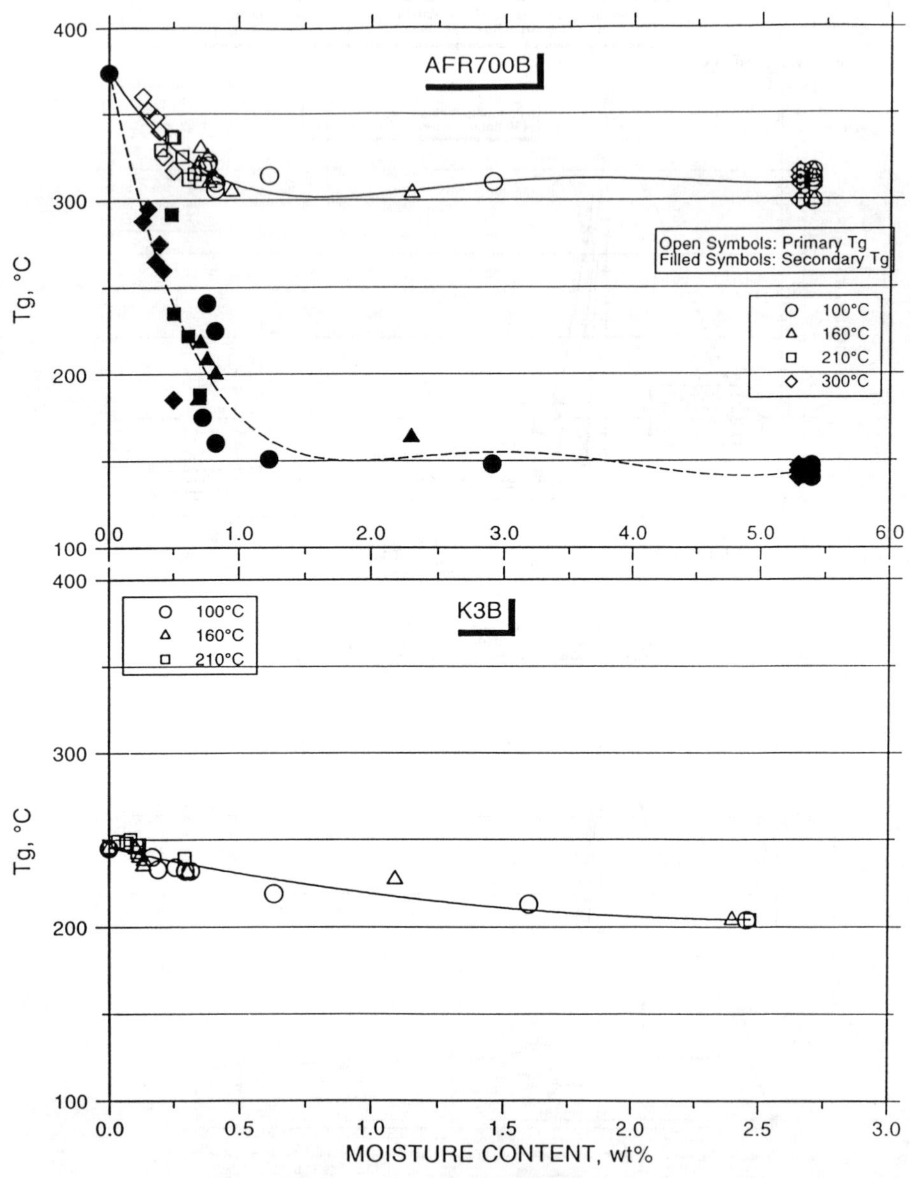

Figure 8. T_g versus moisture content for (a) AFR700B and (b) K3B PI's after accelerated hygrothermal aging followed by isothermal vacuum drying.

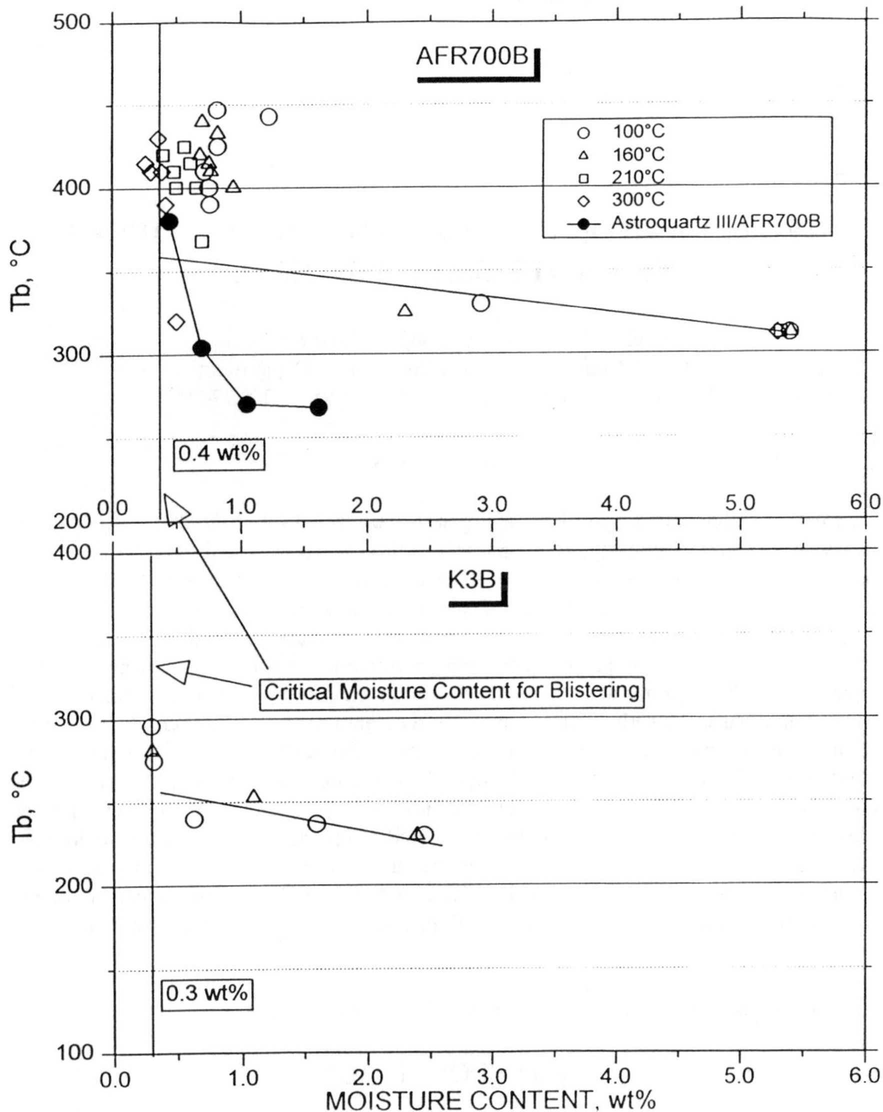

Figure 9. Blister initiation temperature, Tb, versus moisture content for (a) AFR700B and (b) K3B PI's after hygrothermal aging followed by isothermal vacuum drying.

HYDROLYTIC EFFECTS ON THE PROPERTIES OF GRAPHITE/BMI COMPOSITES

Paul O. Biney, Yang Zhong, and Jianren Zhou
FAST Center/Mechanical Engineering Department
Prairie View A&M University, Prairie View, TX 77446

ABSTRACT

Thermal and accelerated hydrolytic aging were performed on IM7/5250-4 composites. Resin loss, thermal stability, dynamic mechanical behaviors, and interlaminar shear properties of the composite were tested to investigate the synergistic effects of moisture and temperature on the physical, thermal, and mechanical properties of these composites. It was found that moisture, together with elevated temperature, induced conspicuous resin loss of IM7/5250-4 composites and had significant deteriorating effects on their interlaminar shear properties. The deterioration caused by combined moisture and temperature was more serious than that due to temperature alone. Furthermore, the drops in mechanical properties were more evident at the early stage of the hydrolytic conditioning, indicated that most of the hydrolysis reactions took place in this stage. The competing effects of postcuring and hydrolysis, together with moisture-induced plasticization and cativation of the composite, led to broader glass transition regions for the hydrolytic samples. The effects of temperature and moisture on the microstructures of the composites were examined and analyzed in terms of changes of molecular weight, crosslinking and molecular weight distribution, plasticization, and localized cavitation.

KEY WORDS: Polymer composites, Hydrolytic degradation, Bismaleimides

1. INTRODUCTION

Composites based on graphite and bismaleimides (BMI) are being increasingly used for various structural components on aircraft due to their excellent performance and ease of fabrication. These components are likely to be subjected to temperature as high as 230°C for prolonged periods. Besides, the components may encounter various weather conditions, such as rain or snow, during their service. BMIs contain various polarized groups which are prone to absorption of moisture from the environment. Moisture can cause volumetric expansion and hence plasticization of the polymer [1] and at elevated temperature, the trapped moisture can accelerate the thermal decomposition of BMI resins and may also lead to blistering in the polymer [2-4]. For composites, moisture may also cause debonding at the fiber/matrix interface [1]. These effects will deteriorate the service performance and long-term durability of the composites and may result in a premature failure of the whole component. Therefore, it is important to characterize the hydrolytic properties and to seek understanding of the hydrolytic degradation mechanisms of the composites.

In this paper, thermal and accelerated hydrolytic aging were performed on IM7/5250-4 composites. Resin loss, thermal stability, dynamic mechanical behaviors, and interlaminar shear properties of the composite were measured in an effort to investigate the synergistic effects of moisture and temperature on the physical, thermal, and mechanical properties of these composites.

2. EXPERIMENTAL

2.1 Materials and Processing Unidirectional IM7/5250-4 prepregs were manufactured by Cytec Engineering Materials Inc. Laminates with lay-up configurations of $[0]_{24}$ and $[0/-45/45/90]_{2s}$ were vacuum bag molded at 190°C for 6 hours and then postcured at 220°C for another 6 hours.

2.2 Environmental Conditioning The IM7/5250-4 laminates were cut into specimens with dimensions of either 1.4 x 0.5 x 0.085" or 0.85 x 0.5 x 0.12". Thermal aging and accelerated hydrolytic conditioning were performed on these specimens. The temperature and duration of the conditioning are listed in Table 1. In the hydrolytic process, the specimens were first saturated in humidity chamber (98%RH/23°C) and then put into sealed stainless-steel container which had been filled with sufficient amount of distilled water to ensure that the specimens were exposed to saturated steam at the desired conditioning temperature. The samples were placed on top of a packing to avoid being immersed in water. A convection oven was used for the thermal aging and the specimens were exposed to the air during the process.

Table 1 Thermal and Accelerated Hydrolytic Conditioning Parameters

Aging processes	Temperature (°C)	Duration (hr)
Accelerated hydrolytic (in a sealed container)	200	48, 96, 144, and 192.
Thermal aging (in the air)	200	96, 192, and 384.

2.3 Characterization and Testing Fiber content measurement, short beam shear (SBS) test, dynamic mechanical analysis (DMA), thermogravimetric analysis (TGA), and scanning electronic microscopy (SEM) were conducted on specimens that had been exposed to the aging conditions in Table 1. The as-manufactured specimens were used as baseline samples. After hydrolytic conditioning, the specimens were first vacuum dried before being used for the tests.

2.3.1 Fiber content measurement Fiber contents of the specimens with and without conditioning were measured by matrix digestion and bulky densities of the composites were measured by liquid displacement technique.

2.3.2 Short beam shear test Interlaminar shear strength of the composite was determined by SBS test using Instron testing machine in accordance with ASTM D2344. Specimens for the SBS test had dimensions of 0.85" in length, 0.5" in width, and 0.12" in thickness.

2.3.3 Dynamic mechanical analysis The single-cantilever clamping method was used to measure the dynamic mechanical properties of the composite using TA Instruments DMA2980. The frequency was 1 Hz and the heating ramp rate was 5°C/min.

2.3.4 Thermogravimetry analysis The thermal stability of the composite was determined using TGA2050 of TA Instruments with a heating ramp rate of 20°C/min. Samples used for TGA were small pieces of composite peeled from fractured specimens.

2.3.5 Microstructure observation An Amray 1610 SEM was used to observe the temperature and hydrolysis-induced surface damages developed during the aging processes.

3. RESULTS AND DISCUSSIONS

3.1 Thermal and Hydrolytic Effects The most obvious physical change after the thermal and accelerated hydrolytic exposures was that the composite surface lost its gloss and became degraded, especially in the latter situation. SEM observation showed that there were surface cracks or voids on the aged specimens (Fig. 1) while no surface defects were observed on the baseline samples. The thermally aged samples exhibited scattered voids about 5μ in diameter whereas hydrolytically aged samples showed irregular-shaped cracks with much lager sizes. Also, the number of surface defects in the former situation was less than in the latter with same duration. These suggest that there was BMI erosion during the aging processes and the erosion was more serious with the existence of water.

Resin content and bulk density before and after the thermal and accelerated hydrolytic conditioning are listed in Table 2. It can be seen that resin content decreased in the aged samples, which indicated that resin loss took place and the loss was more significant in the hydrolytic processes. It is also noticed that the hydrolytic conditioned composites showed evident lower densities than that of the baseline sample, which indicated a much higher void content in the hydrolytic conditioned composites. However, the bulk densities of the thermally aged composites were higher than that of the baseline sample although surface voids developed during the aging process.

Thermal properties characterized by the onset decomposition temperature and the maximum decomposition rate temperature of TGA spectra are listed in Table 2. Although the maximum decomposition rate temperatures were similar for composites with and without conditioning, the onset decomposition temperature after hydrolytic exposures did show a conspicuous drop compared with that of the baseline sample. This is believed to be due to the moisture-induced molecular weight drop and crosslink breakage in the composites exposed to moisture and elevated temperature.

Interlaminar shear strength of the composites before and after the thermal and accelerated hydrolytic exposures are also listed in Table 2. The data showed that hydrolytic exposure had a significant deteriorating effect on the interlaminar shear properties. The interlaminar shear strength of composite after 48 hour in the hydrolytic environment dropped about 40% while it remained the same in thermal aging process up to about 200 hours and decreased about 16% after 384 hours of thermal aging. The strong deterioration effect of hydrolytic exposure on the interlaminar shear properties is owed to the fact that interlaminar shear properties of composites are sensitive to the delamination and fiber/matrix adhesion and these two factors are especially easy to be damaged by moisture-induced cavitation and fiber/matrix debonding [1, 3]. Also, the samples after

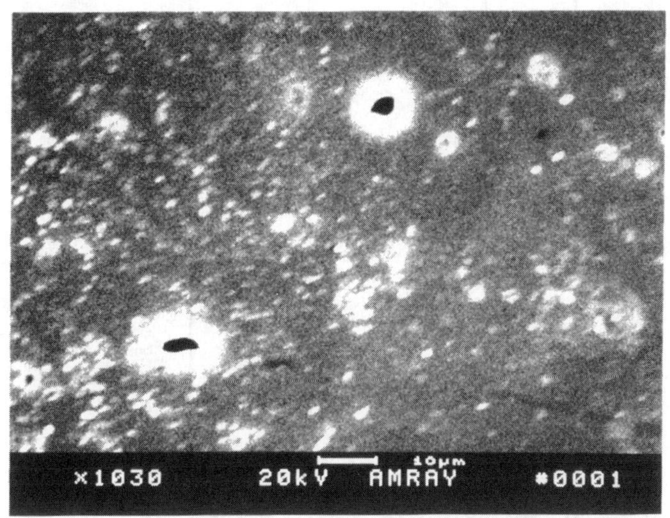

(a)

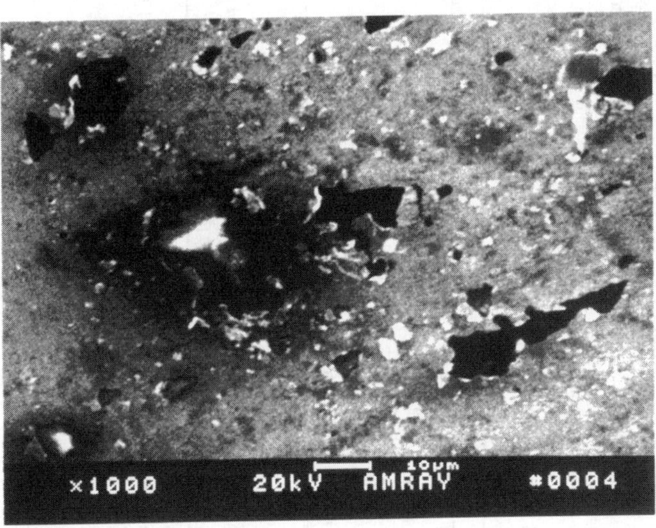

(b)

Fig. 1 Micrographs of surface cracks of IM7/5250-4 composite after (a) 192 hr thermal exposure and (b) 192 hr drolytic exposure

Table 2 Physical, Thermal, and Mechanical Properties of IM7/5250-4 Composites
Before and After Thermal and Accelerated Hydrolytic Conditioning

Environmental conditioning	Interlaminar shear strength (MPa)	Thermal stability		Resin content and composite density	
		Decomposition onset temperature** (°C)	Max decomposition rate temperature*** (°C)	Resin content (wt%)	Composite density (g/cm^3)
Baseline	75.2 (6.34)	426.9	446.6	32.1	1.60
Accelerated hydrolytic exposure at 200°C for 48 hr	46.2 (3.57*)	418.7	443.7	25.8	1.56
96 hr	52.4 (3.38)	415.0	446.4	25.7	1.56
144 hr	51.7 (2.29)	409.9	447.4	26.2	1.59
192 hr	57.9 (5.13)	412.2	449.5	26.7	1.58
Thermal exposure at 200°C for 96 hr	78.6 (3.30)	428.7	445.6	31.6	1.62
192 hr	78.6 (9.34)	427.0	442.4	29.3	1.62
384 hr	63.4 (8.92)	427.5	448.0	29.2	1.63

*   Standard deviations.

**   The decomposition onset temperature was defined as the cross-section of the two tangents in the leading edge of the transition.

*** The peak temperature of the weight loss time derivative.

hydrolytic conditioning showed obvious plastic deformation before the interlaminar shear failure while the baseline and the thermally aged samples exhibited pure interlaminar shear failure.

It is noticed from the data in Table 2 that the drop in the properties during hydrolytic processes was considerably evident at the initial stage of the exposure, especially during the first 48 hours, and after that, the drop became fairly moderate and the properties were even gradually increased. In order to have a direct observation of this tendency, the interlaminar shear strength and resin content were plotted versus duration of the thermal aging and accelerated hygrothermal aging, as shown in Fig. 2. The quick drop of properties indicated that most of the hydrolysis reactions took place in the early stage of the hydrolytic process. The increase in the properties afterwards was owed to postcuring of the BMI resin at such temperature. It can also be known that the synergetic effect of moisture and temperature are much more serious than that of elevated temperature alone.

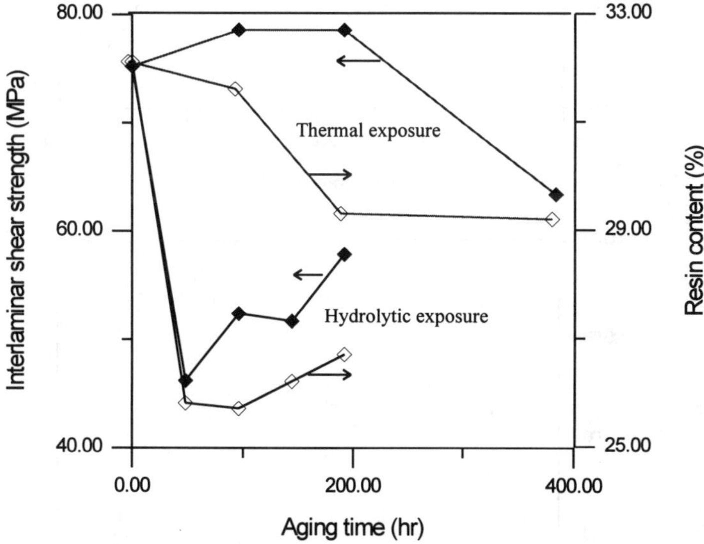

Fig. 2 Interlaminar shear strength and resin content of the composite versus aging time

Dynamic mechanical analysis can also reveal much information about structural changes in polymers. The DMA spectra of storage modulus and loss modulus versus temperature for the thermally aged samples and the hydrolytic samples are shown in Fig. 3 and 4. The increment of the storage modulus of the baseline sample after the glass transition and at temperature above 300°C indicated a postcuring of the resin. Thermally aged samples also showed a step in storage modulus at temperature above 300°C and the step became more obscure with the addition of the thermal duration, indicating that posting also occurred in the thermal aging process.

The DMA spectra of the hydrolytic samples were quite different from those of the baseline and the thermally aged samples. There was no loss modulus peak for the hydrolytic samples. Instead, their loss modulus exhibited a broad transition region together with an incomplete small peak at temperature above 300°C. It is, therefore, difficult to determine the glass transition temperature for the hydrolytic samples. The

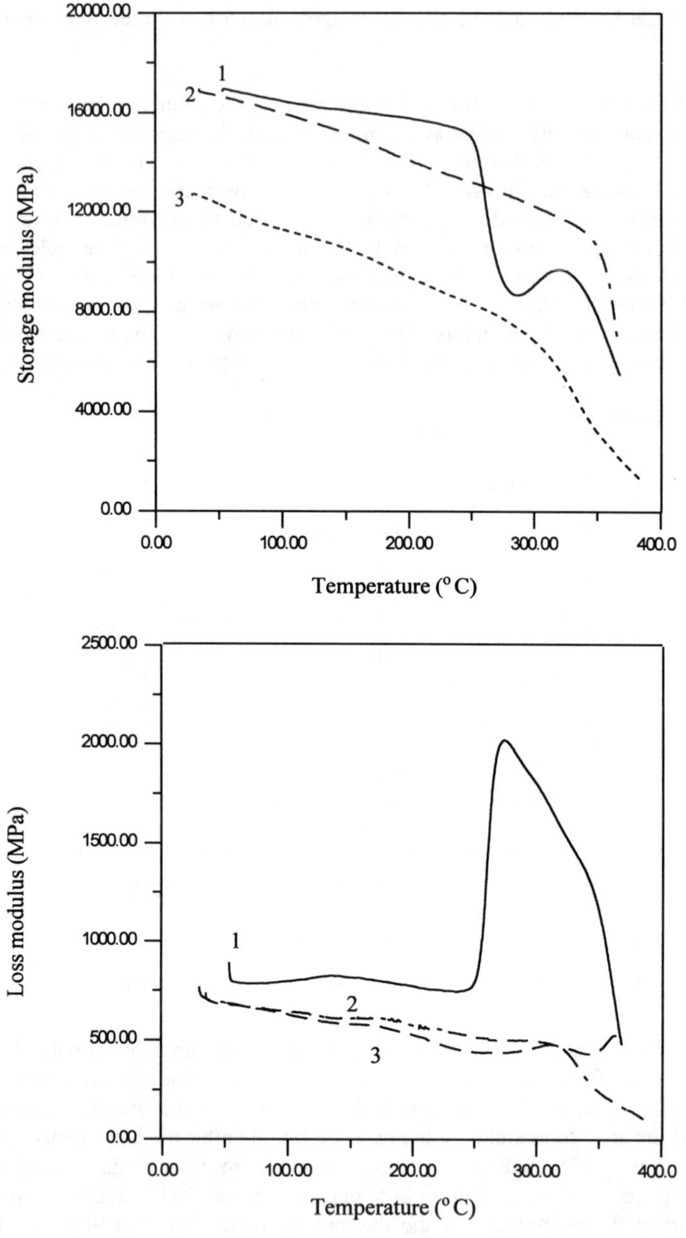

Fig. 3 DMA spectra of IM7/5250-4 composites with various hydrolytic durations.
1: baseline; 2: 96 hr hydrolytic exposure; 3: 192 hr hydrolytic exposure.

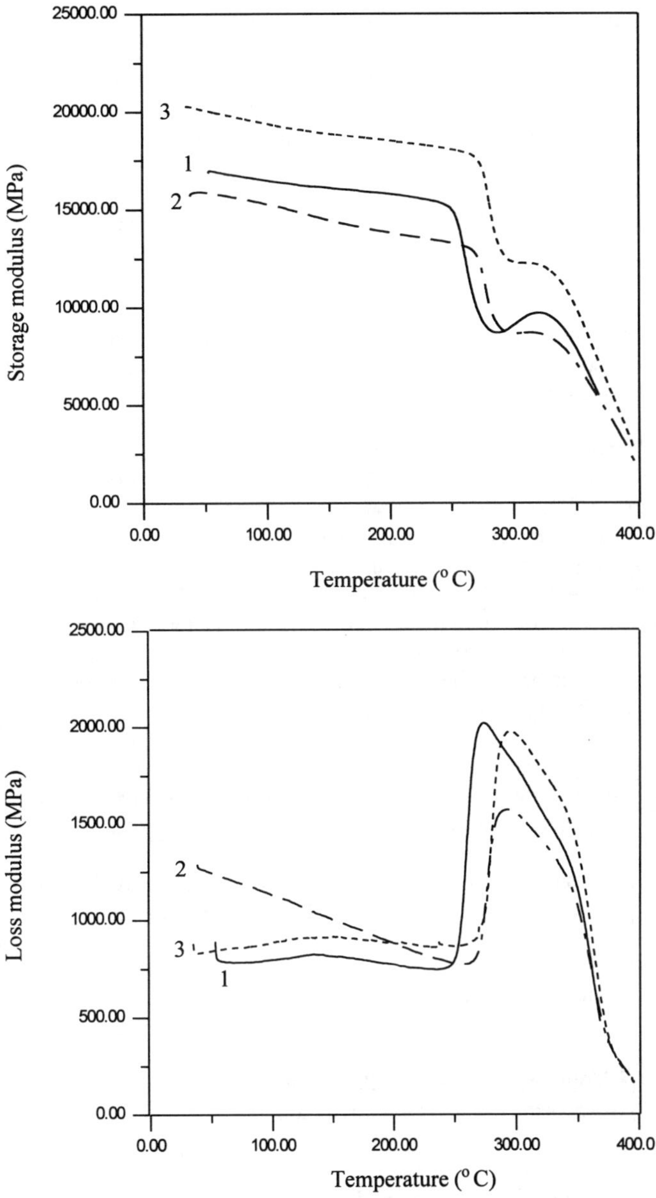

Fig. 4 DMA spectra of IM7/5250-4 composites with various thermal durations.

1: baseline; 2: 96 hr thermal exposure; 3: 192 hr thermal exposure.

storage modulus spectra showed a faster drop of the modulus with temperature compared to the baseline and the thermally aged samples between 100°C and 300°C. However, the transition temperatures were higher than that of the baseline sample. This phenomenon can be explained by the fact that hydrolysis causes molecular scission while postcuring increases the molecular weight and crosslinking. These competing factors result in a broader molecular weight distribution, which leads to a broad glass transition. Also, the hydrolysis effects are more predominant than the postcuring effect in the hydrolytic process. Only a small amount of molecules may experience the increment of molecular weight and crosslinking, so that the high temperature transition peak of the loss modulus is very small. On the other hand, moisture, together with an elevated temperature, also has plasticization and cavitation effect on composites. This increases the damping of the material and hence an increase of the loss modulus.

In addition, the DMA spectra showed that the storage modulus increased after thermal aging while it decreased after hydrolytic aging, indicating that thermal aging makes the composite become hard while hydrolysis makes the composite soft.

3.2 Analysis of Thermal and Hydrolytic Processes The microprocesses experienced by composites based on BMI during thermal and hygrothermal aging and their effects on the molecular structures of the composite are summarized in Fig. 5. It is evident that the major effects of temperature and moisture on the microstructures of composites are changes in the molecular weight, crosslinking, and molecular weight distribution, plasticization, and cavitation of the polymer matrix as well as debonding on the fiber/matrix interfaces. The combined effect of these factors determine the final performance and durability of the composite. Quantitative characterization of the contributions of these factors to the structures and properties of composites and the effective time scales of each factor need further study.

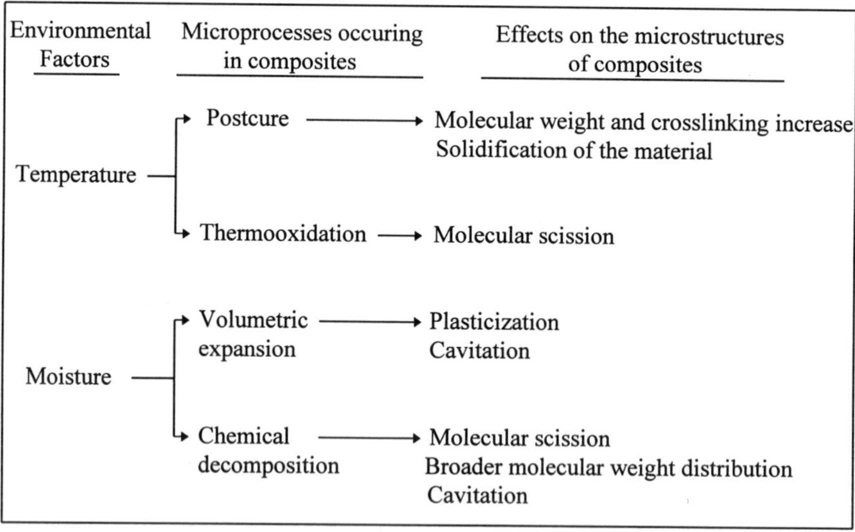

Fig. 5 Diagram of temperature and moisture effects on the structures of composites

4. CONCLUSIONS

It can be concluded that moisture, together with elevated temperature, induced conspicuous resin loss of IM7/5250-4 composites and had a significant deteriorating effects on their interlaminar shear properties. The deterioration by the combined actions was more serious than those caused by temperature alone. Furthermore, the drops in mechanical properties were more evident at the early stage of the hydrolytic conditioning, indicating that most of the hydrolysis reactions took place in this stage. The competing effects of postcuring and hydrolysis, together with moisture-induced plasticization and cavitation of the composite, led to broader glass transition regions for the hydrolytic samples. The effects of temperature and moisture on the microstructures of composites were summarized in terms of changes in molecular weight, crosslinking and molecular weight distribution, plasticization, and localized cavitation.

5. REFERENCES

1. K. Mallick, Fiber-Reinforced Composites, Marcel Dekker Inc., New York, 1993, pp. 303-318.
2. M. K. Ghosh and K. L. Mittal, Polyimides, Marcel Dekker Inc., New York, 1996, pp. 343-365.
3. R. J. Morgan, et.al., International SAMPE Technical Conference, 28, 213 (1996).
4. B. P. Rice, International SAMPE Technical Conference, 28, 778 (1996).

6. ACKNOWLEDGMENT

This research was supported by the Air Force Office of Scientific Research (AFOSR) under the grant number F49620-95-1-0512.

The IM7/5250-4 prepregs and the pipe bomb used in this study were kindly provided by University of Dayton Research Institute, Dayton, Ohio.

43rd International SAMPE Symposium
May 31-June 4, 1998

FIBER FINISHES FOR IMPROVING GALVANIC RESISTANCE OF IMIDE-BASED COMPOSITES

Ronald E. Allred, Richard E. Jensen, and Brent W. Gordon
Adherent Technologies, Inc.
Albuquerque, NM 87123

Thomas A. Donnellan and Theotis Williams, Jr.
Northrop Grumman Corporation
Bethpage, NY 11714

and

Roland C. Cochran and Kevin W. Miller
Naval Air Warfare Center
Patuxent River, MD 20670

ABSTRACT

Galvanic corrosion resistant carbon/bismaleimide (BMI) composites have been demonstrated through the use of reactive finishes to form coatings that isolate the carbon fibers from the BMI matrix. A family of novel reactive coupling agents was formulated into phenolic-based finishes that react with carbon fibers and the matrix resin during processing to form chemical bonds at the interface and in the interphase. Each fiber is coated uniformly with the phenolic finish before the BMI prepregging process. The finish serves to isolate the carbon fibers from the BMI matrix and interrupts the galvanic cell. Both static measurements on galvanic corrosion couples and dynamic electrochemical impedance spectroscopy measurements show that the finishes dramatically improve composite corrosion resistance. Flexural and short beam shear strengths show that BMI matrix composites with finished fibers are equivalent to those with unsized fibers. The mechanical data is supported by voltage contrast x-ray photoelectron spectroscopy (VC-XPS) measurements and SEM failure surface analyses which show that interfacial adhesion is superior in the finished composites. This approach has the potential to substantially improve galvanic corrosion resistance of the interface in BMI matrix composites by preventing the interfacial debonding thought to be responsible for the observed reaction acceleration. The reactive finish approach provides a cost-efficient means of improving galvanic corrosion resistance in composites with any BMI formulation.

KEY WORDS: Corrosion-Resistant, Bismaleimides, Carbon Fibers

1. INTRODUCTION

Carbon/bismaleimide (BMI) composites are the best material candidates for a number of aerospace applications that require thermal stabilities in the 120 to 175°C temperature range. There has been considerable effort expended in the synthesis, formulation, development, and qualification of BMI composites over the last fifteen years. The technology has developed to the point that state-of-the-art BMI resins even have toughnesses comparable to those of lower temperature epoxy systems. One problem that has limited the use of BMI composites is the system susceptibility to galvanic degradation.

Carbon-reinforced imide composites degrade when they are electrically connected to certain metals in the presence of saltwater. Most workers agree that the mechanism responsible for the observed effect involves hydroxyl attack on the imide ring [1-4]. Other possibilities include direct reduction of the imide ring [5,6], reactions related to the Hoffman Degradation [7] mechanism used in production of amines from amides, and polarization effects at the fiber surface [8]. Although the specifics of the mechanism are not fully known, it is clear that a requirement for the observed degradation to occur is that the BMI composite be the cathode in a galvanic cell. Any interruption of the cell circuit will prevent the BMI degradation and also the associated galvanic metal corrosion.

There is some disagreement on the effectiveness (or even the need) for galvanic protection in BMI-aluminum couples. Work performed at General Dynamics and BASF Corp. [9] indicated that composite bearing strength of galvanically coupled materials was not affected by a 2000-hour exposure to salt spray even when no surface coatings were applied to either member of the couple. A series of studies performed at the Naval Air Warfare Center (NAWC) [2,10,11] demonstrated that the type of protection scheme used in carbon/BMI-aluminum couples was critical and further that the degree of protection, as defined by visual inspection after exposure and by bearing strength measurements, was directly related to the coating quality of the specific systems used. When the coating remained tightly bonded to the surface of the composite, the amount of degradation observed in the experiment was significantly reduced relative to uncoated or "poorly" coated samples. Another interesting finding in the NAWC work was that the presence of sealant in the hole around the fastener did not improve resistance to the degradation process as anticipated, but instead resulted in a greater reduction in composite strength compared to unsealed samples. The proposed explanation was that the sealant allowed the electrolyte to be concentrated in the hole.

The effects observed in the isolation scheme study can be attributed to difficulties associated with adhesion of coatings to the BMI composites. It has been found that conventional epoxy composite-compatible paints and coatings do not adhere well to BMI composites especially after environmental conditioning. Also, it should be noted that the carbon exposed by cutting through a composite will be extremely inert and difficult to bond to.

Applying a well-bonded, nonimide uniform coating around each fiber in the composite presents the opportunity for a more dependable long-term solution to this problem. This approach functions by electrical isolation of the carbon fibers and interruption of the galvanic cell. All of the proposed degradation mechanisms should be prevented by such an action. From a reliability standpoint, the development of an improved fiber finish system that could be applied to carbon fibers and used with BMI resin systems already in existence would be the best solution to the galvanic degradation problem. In this work, we distinguish a finish as an adhesion-promoting chemistry applied to a fiber surface from a sizing, which is applied as a handling aid. Similar finishes have been shown to greatly improve the thermo-oxidative stability in polyimide matrix composites [13-16].

The approach taken was to formulate a finish consisting of a phenolic resin, a coupling agent
that is reactive with the carbon fiber and phenolic resin, surfactants, and solvents. Mixtures of
some of these ingredients are often necessary for solubility and processing reasons. The finish
is applied to carbon fiber tows using a dip process followed by thermal activation of the
coupling agent. The finish must be uniform and thick enough to serve as a corrosion barrier
while not compromising overall composite performance. The reactive finish concept is shown
schematically in Figure 1 where the finish chemically bonds to the fiber surface and to the
polymer matrix to provide a well-bonded, durable interphase to isolate the BMI matrix from
the carbon fiber.

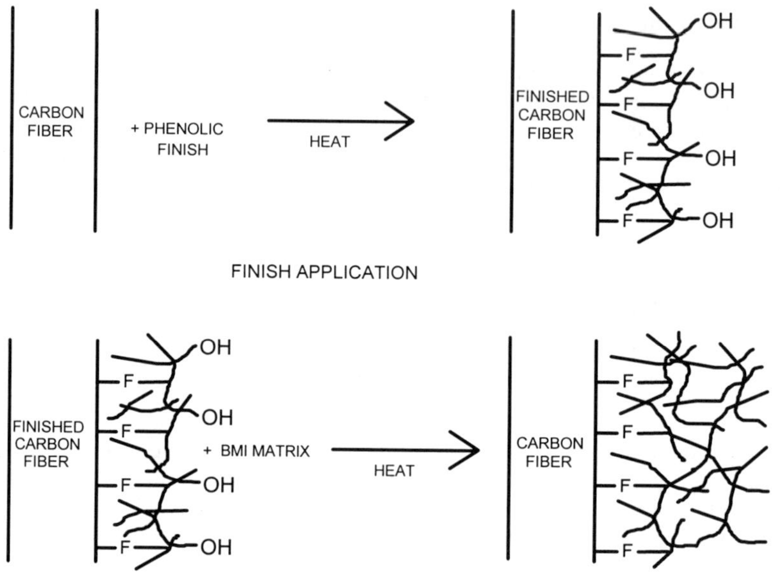

Figure 1. Reactive finish interfacial bonding concept

Results show that this approach provides well-bonded phenolic coatings to the individual carbon fibers in a 12K tow that subsequently provide a well-bonded interface to a BMI matrix in
the composite. Galvanic corrosion tests verify that the finished composites have significantly
greater corrosion resistance than control composites fabricated with unsized carbon fibers.

2. EXPERIMENTAL PROCEDURES

2.1 Materials 12K IM7 unsized carbon fibers were obtained from Hercules, Inc. (now
Hexcel, Magna, UT). Phenolic (Georgia Pacific 5236) and BMI (Ciba-Geigy Matrimid 5292)
resins were purchased from Applied Poleramic, Inc. (Benecia, CA). All other chemicals were
used as-received from Aldrich Chemical Co. (Milwaukee, WI).

2.2 Coupling Agent Synthesis and Characterization The first step in the program was to
synthesize additional quantities of reactive coupling agents using previously developed
techniques. Figure 2 depicts the generic structure of the proprietary coupling agents. The
synthesis was carried out with controlled laboratory conditions for three variations of the
structure shown in Figure 2. Verification of product purity was performed using a Mattson
Cygnus 100 FTIR spectrophotometer.

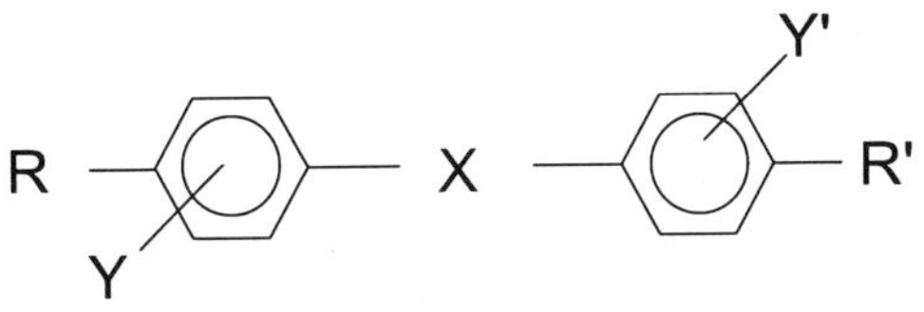

where: X = linking group, Y, Y' = ring substituents, and R, R' = functional groups

Figure 2. Schematic chemical structure of reactive coupling agents

2.3 Finish Formulation Development Finishes incorporating three versions of the reactive coupling agent structure shown in Figure 2 were prepared. The finishes were designated as PH9301F, PH9304F, and PH9305F. Finish formulations containing 1% phenolic were found to provide a uniform coating 0.75 to 1.0µ thick. Higher phenolic concentrations produced tows too stiff for subsequent processing. Bonding of the finish to the carbon fibers before prepregging is necessary so that the finish cannot be removed in the solvent of a prepreg impregnation bath or diffuse into the imide matrix during a hot melt prepreg process. Partial curing of the phenolic also improves the handling characteristics of the finished fiber tows for use in processes such as weaving, braiding, or filament winding. Finish application and bonding was accomplished in a dip process followed by thermal curing at 200°C as discussed in detail in Ref. 13. After the finishing process, the fiber tows were supple and appeared compatible with prepregging processes. Finish uniformity was determined by examination with a scanning electron microscope (SEM). Bonding was examined by soaking finished fibers in acetone for 2 hours.

Thermal properties of the finish formulations were determined using a Perkin Elmer Series 7 differential scanning calorimeter (DSC) and evaporating small droplets (≈20 mg) of the finish in the DSC pan. DSC measurements were made at a heating rate of 10°C/min.

2.4 Prepreg and Composite Fabrication The first composites produced in the program were 12K tow rods with a BMI matrix. The rods were produced in a modified version of the finishing line. A resin impregnation bath was added in place of the finish solution bath followed by a wire drawing die to form the impregnated tow into a rod and remove excess BMI resin. The BMI resin bath contained 60% solids in methylene chloride. After being pulled through the wire drawing die, the impregnated rod was drawn into the tube furnaces and held at 200°C for 30 min. The cured rods were then removed from the furnaces and cut into 30-cm lengths before being postcured in air for 1 hour at 180°C followed by 2 hours at 200°C and, finally, 6 hours at 250°C. The resultant rods were very smooth with no visible voids on their surfaces. Polished cross sections also showed no visible voids when viewed with an optical microscope.

Prepregs were manufactured by drum winding the fiber dry followed by resin impregnation using a known amount of BMI resin solids as detailed in Ref. 13. After drying, the prepreg was very stiff and boardy. Prepreg weights were recorded to determine weight percent resin. Variations in fiber and resin content in the prepregs were within ±2% from one prepreg to the next, demonstrating good reproducibility in prepreg manufacturing. Four types of prepreg were fabricated for this study: a control fiber using unsized IM7 12K and that same fiber with 3 finish types.

Unidirectional composites were fabricated from 8-ply stacks of staged prepreg using a vacuum bag hot press molding technique in hard tooling. The cure cycle consisted of ramping to 274°C at 2.8°C/min with pressure (1.4 MPa) application at 200°C. After holding 1 hour at

274°C the laminate was cooled at a rate of 2.8°C/min. The molded composite panels had a final 0.20-cm (0.080-in.) thickness.

Laminate quality was confirmed with metallographic analysis on polished cross sections, ultrasonic C-scan inspection, optical microscopy, and thermal analysis. Only materials with less than 2% voids and representative glass transition temperatures (T_gs) were selected for mechanical testing. Ultrasonic nondestructive inspection was performed on the composite panels using an Ultra Image IV system. T_gs were determined with a Rheometrics RMS 800 using a torsion/rectangular configuration at a heating rate of 5°C/min.

2.5 Mechanical Testing Flexure testing on 0°, 90°, and short-beam shear (SBS) specimens at room temperature and at 177°C was conducted in accordance with ASTM D-790 and ASTM D-2344 on a 98 KN capacity MTS Model 810 servohydraulic test frame and an upgraded 98 KN Instron Model TTD-L screw-driven test frame. A 10:1 span-to-depth ratio was used for both flexure tests and a 4:1 ratio was used for the SBS tests. Testing was performed at a rate of 0.02 cm/min.

Compression tests were conducted in accordance with ASTM D695. The initial loading rate was 0.02 cm/min to a strain of 16% followed by a rate of 0.1 cm/min until specimen failure. All elevated temperature testing was performed using an Instron Model 3111 temperature chamber at 177°C. Thermocouples were placed in contact with the test specimens in the chamber and monitored. Specimens were allowed to equilibrate for 2 minutes before the tests were run. Moisture saturation was performed at 94°C in distilled water for a period of 42 days. The average weight gain due to moisture absorption for the control and PH9304F finished specimens was 2.4 and 1.6%, respectively.

2.6 Spectroscopic Characterization Additional characterization was conducted on the 12K rod test specimens using voltage contrast X-ray photoelectron spectroscopy (VC-XPS) after the technique given in Refs. 17-19. The specimens were split to expose a fresh transverse surface and the ratio of carbon in the fiber (C_f) to carbon in the matrix (C_m) measured on a Fisons Escalab MKII instrument with 20 eV Mg α X-rays and electrons on a 5 x 2 mm spot size. The electrons selectively charge the nonconductive resin matrix portion of the surface and separate their emitted photoelectron signal from that of the conductive fibers.

2.7 Galvanic Corrosion Testing

Galvanic couple corrosion experiments were performed with BMI matrix composite rods (1 12K tow) joined to aluminum in saltwater. Rods of each type (unsized or one of 3 finishes) were coupled to a large L-shaped piece of 5052-H39 aluminum foil with a corrosion-resistant coating. The rods were connected to the foil with jumper wire. The assembly was then immersed in 3.5 wt.% saltwater approximately halfway up the rods to form the corrosion couple. A photograph of a typical sample is shown in Figure 3. The bottom of a plastic cup held in place with parafilm was used to support the composite rods. Cotton swabs were used to further support the rods where they penetrated the plastic cup lid.

Figure 3. Galvanic corrosion sample setup

2.8 Electrochemical Impedance Spectroscopy (EIS) EIS on BMI-matrix composites was run on 12K composite rods and bulk unidirectional composites following the methods given in Ref. 3. The cut ends of the composite rods were sealed with Microstop electroplater's lacquer to allow only the sides of the rods contact with the solution. Tests were run in 3.5% saltwater with a standard three electrode cell (nichrome mesh counter electrode, calomel reference electrode [sce], and the sample). A 10-mV signal was then applied over the frequency range between 10 kHz and 100 mHz, and the phase shift monitored to get the baseline data. A capacitance was then calculated that corresponds to the exposed carbon area in the composite.

Cathodic disbondment was then artificially induced by potentiostatically polarizing samples at -1.4 V_{sce}. The degree of disbondment was periodically measured using EIS. EIS was conducted by applying a 10 mV sinusoidal voltage wave at frequencies ranging from 65 kHz to 100 mHz at a sampling rate of 10 points per decade. This test was repeated several times over the exposure period.

Carbon/BMI composite samples modified with different finish compositions were tested for susceptibility to cathodic disbondment during cathodic polarization in aerated 3.5% NaCl solution. Cathodic disbondment occurs due to degradation of the polymer matrix phase that takes place when local alkalinity develops at exposed carbon fibers during the test. Two test configurations were examined in this survey. In the first configuration, composite samples were potted in epoxy end-on and polished through to a 600 grit finish using SiC papers. In this arrangement, fiber ends were exposed for testing without exposing fiber lengths. In the second configuration, surfaces of composite panels were exposed to the test solution directly. In this arrangement, fiber lengths were exposed without exposing fiber ends. Electrical connection was made to all of the fibers by abrading the opposite end of the samples and affixing a Ni-Cr wire using silver epoxy.

3. RESULTS AND DISCUSSION

3.1 Finish Characterization Properties of the reactive coupling agents used in the phenolic finishes have been previously reported in Refs. 13-15. The coupling agents display a range of melting points, reaction onset temperatures, maximum reaction temperatures, and solubilities. That variability allows finishes to be tailored to the desired processing and cure cycle for the composite.

DSC thermal characterization of the standard phenolic finish compositions containing the reactive coupling agents and neat phenolic was conducted. The phenolic resin displayed a series of melting endotherms between 75 and 90°C, a large endotherm near 100°C from water elimination, and a curing exotherm at higher temperatures. The low-temperature endotherms in the phenolic are due to melting of the base catalysts, paraformaldehyde and hexamethylene tetraamine, their decomposition products, and, perhaps, some crystallites in the phenol from hydroquinone [20].

The finishes do not show the sharp melting endotherms present in the neat phenolic between 75 and 90°C. Apparently, the crystallites dissolved in the acetone and did not reform upon drying. They do show the large water elimination endotherm followed by a strong exotherm beginning near 160°C ending below 230°C. That exotherm is followed by another exothermic region that would be expected from phenolic curing.

A summary of the finish DSC data is given in Table I. These results show that the coupling agent reaction is complete before initiation of the phenolic curing reaction. This allows the bonding reaction to be controlled independently from the polymer cure, which gives increased control to the overall process.

Table I. DSC Characterization Results at 10°C/min for Reactive Finishes

Finish	T_m (°C)	ΔH_{melt}[1] (J/g)	T_d onset (°C)	T_d max (°C)	T_d end (°C)	ΔH_d (J/g)
Phenolic, neat	69-95[2]	262	155	195	272	-178
PH9301F	70-90[2]	147	173	196	227	-136
PH9304F	111-132[2]	53	154	183	202	-95
PH9305F	80-131[2]	5	156	178	200	-149

1. Includes water evolution.
2. Series of peaks.

Finished fibers were characterized by SEM for uniformity and thickness of finish and bonding to the carbon fibers. Bonding was examined by soaking the finished fibers in acetone for 2 hours before the SEM analysis. The surface appearance of unsized IM7 fibers is smooth and featureless. SEM analysis of finished fibers before and after acetone wash showed that the finish coating is generally uniform on the fibers and that little of the finish was removed by the acetone wash. SEM observations of finished fibers bent over a 1-mm diameter radius showed no evidence of cracking at coupling agent concentrations up to 1%. This shows that the coupling agent is not embrittling the finish, which could form an undesirable interphase structure.

3.2 Screening Tests on Composite Rods Three types of screening tests to measure adhesion and corrosion resistance were run on the 12K tow composite rods before examination of composite plates: VC-XPS with SEM characterization, static corrosion of galvanic couples, and EIS.

3.2.1 VC-XPS Fiber-matrix adhesion was examined using VC-XPS analysis on the BMI matrix rods. Integration of the carbon 1s (C_{1s}) peaks for fiber and matrix and ratioing those results gives an approximate amount of interface-to-matrix cohesive failure in the sample. Spectra from both sides of the specimen were averaged to yield the results given in Table II.

Table II. VC- XPS Results on BMI-Matrix Rods

Finish	Carbon Fiber/Carbon Matrix
Unsized	1.45 ± 0.20
PH9301F	0.40 ± 0.15
PH9304F	0.25 ± 0.10
PH9305F	0.90 ± 0.15

The VC-XPS results given in Table II show that the finishes are providing better interfacial adhesion than the unsized control. The lower the carbon fiber contribution to the fracture area, the lower the amount of interface failure present in the sample. These results with the acetone soak results show that the desired carbon-coupling agent bond is occurring. That bond is the difficult link to achieve when tailoring interfacial adhesion. Adhesion controlled by the interphase region can be manipulated by finish composition providing the bond to the carbon surface is attained. SEM characterization of failure surfaces on specimens from the VC-XPS measurements showed that all the specimens displayed a mixed mode (interfacial-matrix cohesive) failure appearance except the PH9304F finished material, which had a high degree of cohesive failure. These results follow the basic trend of the VC-XPS data given in Table II and reinforce those results.

3.2.2 Static Galvanic Corrosion The second screening tests examined galvanic corrosion susceptibility using the set up shown in Figure 3. After 2 days of immersion, little aluminum corrosion product was seen with the PH9301F and PH9304F finishes. A small amount of corrosion product was seen with the controls and PH9305F finish. Al(OH)$_3$ is the expected corrosion product.

The galvanic corrosion test was continued for 14 days total exposure. This amount of exposure is severe enough to screen the finish performance. Upon removal from the saltwater, the composite samples were washed with distilled water 3 times, patted dry, and weighed. The PH9301F finished samples showed an average weight loss, while the PH9304F finished samples showed a small weight gain and the PH9305F-finished and control samples showed a larger average weight gain. Weight gains should be due to moisture absorption and weight losses to degradation of the BMI matrix. It is likely that the samples showing weight gains have absorbed a larger amount of water, which may indicate additional matrix and interface degradation creating paths for the ingress of water.

The specimens from the galvanic couple corrosion experiments were characterized using SEM. The surface appearance of the corroded specimens was examined slightly above and below and at the water line. In general, more degradation of the BMI was observed at and above the water line on all specimens, but trends between sample types were the same for each location.

The appearance of the unsized control specimens above the water line after 14 days exposure is shown in Figure 4. A low magnification view of the control specimen surface in Figure 4a shows a variation between exposed largely clean fibers and areas where the BMI matrix is degrading. A higher magnification view of the degrading resin shown in Figure 4b reveals a rough texture for the remaining resin. Only control specimens displayed this type of texture.

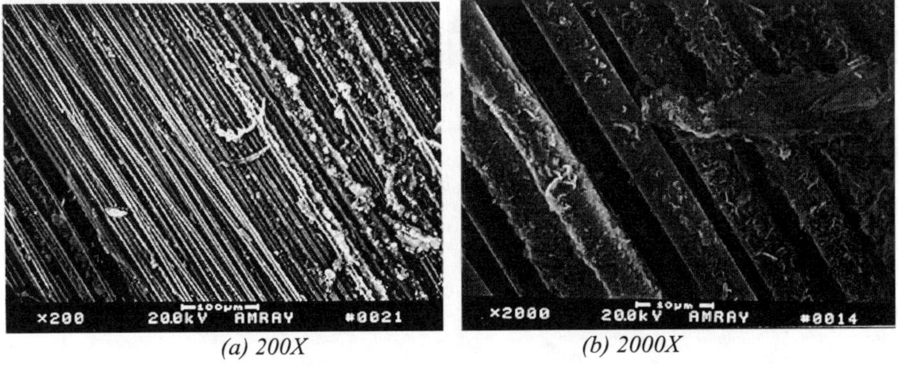

(a) 200X   *(b) 2000X*

Figure 4. Surface appearance of unsized IM7/BMI control corrosion specimens

The finished specimens showed varying degrees of resin degradation, all less than the unsized controls as shown in Figure 5. The finished samples have most of the BMI resin intact. Higher magnification views show the beginning of resin degradation exhibited in cracking and delamination.

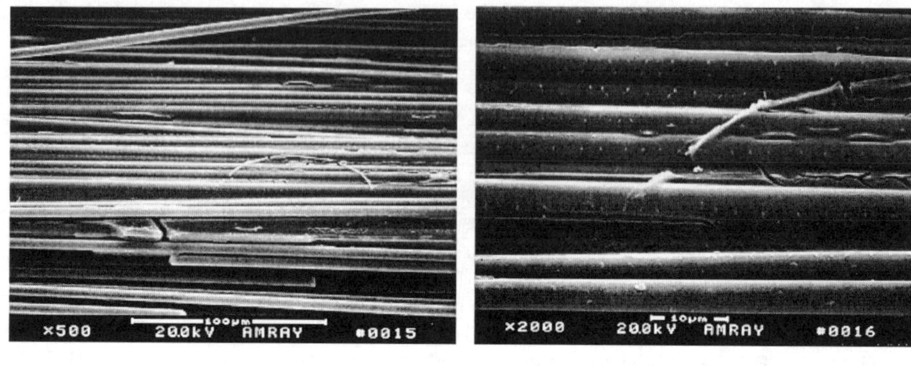

(a) 500X   (b) 2000X

Figure 5. Composite surface appearance with PH9304F finish

Examination of the composite surface structure at and below the water line in the galvanic cell shows the unsized material to be severely degraded displaying a unique texture as shown in Figure 6.

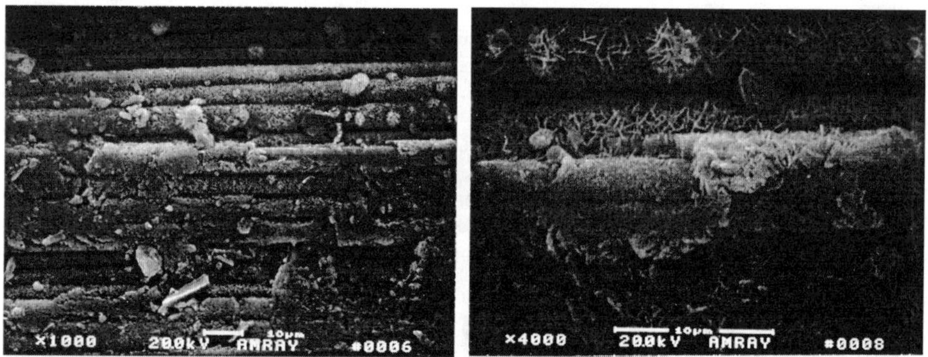

Figure 6. Surface appearance of unsized composite below water line

In contrast, the surface of the finished materials below the water line show only some cracking and beginning of delamination as shown in Figure 7.

138

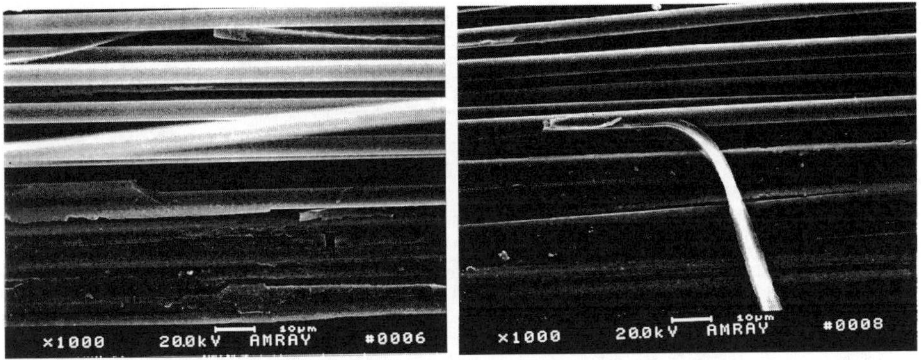

Figure 7. Surface appearance of PH9304F finished composite below water line

Additional insight into the protection from BMI galvanic corrosion afforded by the reactive finishes can be gained by examination of the extent of corrosion of the aluminum anodes shown in Figure 8. Uncoupled aluminum in the salt solution showed no evidence of attack. The aluminum electrode coupled to the unsized composite shows a substantial amount of attack and has decoupled from the upright leg of the L-shaped anode (Figure 8a). In contrast, the anodes of the finished specimens show only a small amount of aluminum corrosion and no evidence of attack at the bend leading to the upright portion (Figure 8b). These observations follow the trends seen in weight changes and the VC-XPS results for adhesion given in Table II.

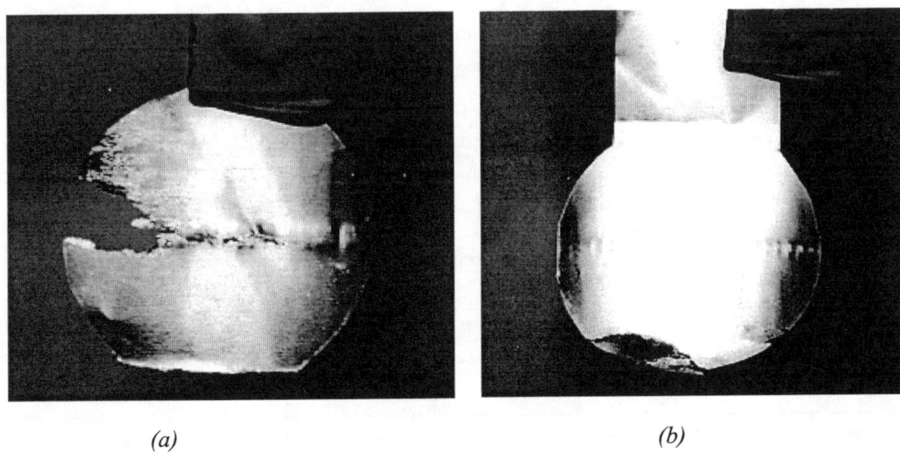

(a)   (b)

*Figure 8. Appearance of aluminum anode coupled to (a) unsized composite, and (b)
PH9304F finished composite*

3.2.3 EIS EIS experiments were run on duplicate sample rods fabricated for the static corrosion experiments. Following the baseline measurements, a capacitance was calculated that corresponds to the area of exposed carbon in the composite. The debonded area is calculated by dividing the measured capacitance by 20 $\mu F/cm^2$, which is an accepted value for the interfacial capacitance for carbon/graphite in aqueous solutions. The samples were then subjected to cathodic polarization for 1 hour at -1.4V and the impedance rerun. Results are given in Table III.

Table III. Calculated Debonded Area from EIS Measurements

Finish	Initial Debond Area (cm^2)	Final Debond Area (cm^2)
PH9301F	0.01	0.70
PH9304F	0.02	0.74
PH9305F	0.02	0.90
Unsized Control	0.147	1.05

The initial complex impedance corrosion data given in Table III show that the finished composites performed better than the controls. Taken together, the corrosion and adhesion data on the 12K rods suggest that all the finishes provide improved adhesion and galvanic corrosion protection for BMI-matrix composites.

3.3 Tests on Bulk Composites

3.3.1 EIS EIS was run on samples machined from the 8-ply unidirectional composite plates. The EIS data were analyzed to extract capacitance values, which were used to determine the relative amount of disbondment produced by cathodic polarization. EIS data were fitted to an equivalent circuit model using a complex non-linear least squares fitting routine as was done on the composite rod screening tests.

The calculated debonded area from the EIS measurements before and after cathodic polarization for the composite samples exposed along the fiber lengths is given in Table IV for the longitudinal specimens.

Table IV. Calculated Debonded Area for Longitudinal EIS Measurements

Fiber Finish	Exposure Time (hours)	Area at t = 0 h (cm^2)	Area at t = exposure time (cm^2)
Unsized	9.6	6.5	450
PH9301F	9.3	7.0	45
PH9304F	10.4	4.0	2.2
PH9305F	9.0	6.0	12.5

The calculated debond area for the end-on EIS measurements is given in Table V. These data corroborate those shown in Table IV showing that the effect of the finishes on cathodic disbondment operates independent of orientation and that the effect is significant enough to be observed in these aggressive types of tests. The EIS results provide dramatic evidence of the improvement in resistance to galvanic corrosion provided by the reactive finishes.

Table V. Calculated Debonded Area for End-On EIS Measurements

Fiber Finish	Exposure Time (hours)	Area at t = 0 h (cm^2)	Area at t = exposure time (cm^2)
Unsized	9.5	7.4	201
PH9301F	4.8	3.5	2.2
PH9301F-1	4.7	2.6	2.0
PH9304F	9.9	4.3	2.2
PH9305	9.3	2.1	47

3.3.2 Composite Mechanical Properties Specimens from the 8-ply composite panels were tested for longitudinal and transverse flexural strength and SBS strength. A concern with the approach being taken to isolate the BMI with a thick finish coating is that a second weak interface can be created at the finish/BMI interface. That would not be evident from any of the screening tests discussed above including the VC-XPS if the finish is well bonded to the carbon fiber. Results from mechanical property tests for unsized and PH9304F finished composites are given in Table VI.

Table VI. Mechanical Properties of BMI Matrix Composites

Fiber Finish	Test Condition (°C)	Fiber Orientation (degrees)	Flexural Strength (MPa)	SBS Strength (MPa)	Compressive Strength (MPa)
Unsized Control	23	0	997.7 ± 49.6	79.3 ± 4.8	888.0 ± 113.1
		90	57.9 ± 3.5	--	--
	177	0	760.5 ± 47.6	--	--
		90	53.8 ± 9.7	--	--
	177/wet	0	564.7 ± 35.9	--	--
		90	26.9 ± 11.0	--	--
PH9304F	23	0	1043.9 ± 28.3	76.5 ± 4.8	1310.7 ± 121.4
		90	42.1 ± 10.3	--	--
	177	0	772.2 ± 28.3	--	1041.1 ± 46.9
		90	48.3 ± 5.5	--	--
	177/wet	0	508.1 ± 16.6	29.0 ± 3.5	--
		90	11.0 ± 2.1	--	--

Examination of the results given in Table VIII shows that the finished material in general has similar properties to the unfinished control. The finished panel has higher compressive properties but shows more sensitivity to hot-wet conditions in this limited test matrix.

141

Additional work is required to determine the nature of the hot-wet response and potential means of improving that response in the finished material.

When optimized, these finishes should add only $1-3 per pound to the cost of BMI matrix composites. As such, this approach should provide a cost-effective means of reducing the susceptibility of carbon/BMI composites to galvanic corrosion.

4. SUMMARY AND CONCLUSIONS

This study was undertaken to improve the galvanic corrosion resistance of carbon/BMI composites through the use of a phenolic finish. In this process, each fiber is coated uniformly with the phenolic finish and bonded to the finish with reactive coupling agents before the BMI prepregging process. The finish serves to isolate the carbon fibers from the BMI matrix and interrupts the galvanic cell. Results show that this approach is a viable means of improving galvanic corrosion resistance of carbon/BMI composites and that composite mechanical properties are not compromised by the finish. Both static galvanic couples and dynamic electrochemical impedance spectroscopy measurements show that all the finishes examined dramatically improve composite corrosion resistance. Longitudinal and transverse flexure and SBS strength measurements show that BMI matrix composites with finished fibers are equivalent to those with unsized fibers. The mechanical data is supported by VC-XPS measurements and SEM failure surface analysis that show that interfacial adhesion is superior in the finished composites. This approach will provide a cost-efficient means of improving galvanic corrosion resistance in composites with any BMI formulation.

5. ACKNOWLEDGMENTS

The authors would like to express their gratitude to Dr. Rob Sorenson and Dr. Rudy Buckheit at Sandia National Laboratories for conducting the EIS measurements and Dr. Jim Miller at Amoco Performance Products for conducting the VC-XPS measurements.

6. REFERENCES

1. M. C. Faudree, "Relationship of Graphite/Polyimide Composites to Galvanic Processes," *Proc. 36th Intl. SAMPE Symposium*, April 1991, Society for the Advancement of Material and Process Engineering, Covina, CA, pp. 1288-1301.
2. R. C. Cochran, T. M. Donnellan, and R. E. Trabocco, "Environmental Degradation of High Temperature Composites," Proc. 73rd AGARD Structures and Materials Panel, San Diego, CA, Oct 1991, Report AGARD-R-785, April 1992, North Atlantic Treaty Organization, Neuilly sur Seine, France.
3. F. D. Wall, S. R. Taylor, and G. L. Cahen, "The Simulation and Detection of Electrochemical Damage in BMI/Graphite Fiber Composites Using Electrochemical Impedance Spectroscopy," ASTM-STP-1174, *High Temperature and Environmental Effects on Polymeric Composites*, C. E. Harris and T. S. Gates, eds., American Society for Testing and Materials, Philadelphia, PA, 1993, pp. 95-113.
4. M. L. Rommel, A. S. Postyn, and T. A. Dyer, "Accelerating Factors in Galvanically Induced Polyimide Degradation," *SAMPE J.*, *29*, 2, Mar.-Apr. 1993, Society for the Advancement of Material and Process Engineering, Covina, CA, pp. 19-24
5. D.W. Leedy and D.L. Muck, "Cathodic Reduction of Phthalimide Systems in Nonaqueous Solutions," *J. of the American Chemical Society*, 93, 17, 1971, pp. 4264-4270.
6. Z. Horii, C. Iwata, and Y. Tamura, "Reduction of Phthalimides with Sodium Borohydride," *J. of Organic Chemistry*, 26, 1961, pp. 2273-2276.
7. R. T. Morrison and R. N. Boyd, *Organic Chemistry*, Allyn and Bacon Inc., Boston, MA, 1973, pp. 741 and 888.

8. R. Taylor, AEROMAT 91, American Society for Materials, Materials Park, OH, May 1991.

9. J. Boyd, S. Speak, and P. Sheahen, "Galvanic Corrosion Effects on Carbon Fiber Composites: Results from Accelerated Tests," *Proc. 37th Intl. SAMPE Symp*osium, March 1992, Society for the Advancement of Material and Process Engineering, Covina, CA, pp. 1184-1198.

10. R. C. Cochran, R. E. Trabocco, J. Boodey, J. Thompson, and T. M. Donnellan, "Degradation of Imide Based Composites," *Proc. 36th Intl. SAMPE Symposium*, April 1991, Society for the Advancement of Material and Process Engineering, Covina, CA, pp. 1273-1287.

11. R. C. Cochran, T. M. Donnellan, and R. E. Trabocco, "Degradation of Imide Based Composites," *Proc. First Intl. Symp. on Environmental Effects on Advanced Materials (ADVMAT/91)*, San Diego, CA, June 1991, National Association of Corrosion Engineers, Houston TX, 1992.

12. S. Wang and A. Garton, *Proc. American Chemical Society*, Polymeric Materials Science and Engineering Division, *62*, 1990, pp. 900-902.

13. R. E. Allred and J. K. Sutter, "Fiber Finish for Improving Thermo-Oxidative Stability of Polyimide Matrix Composites," *Proc. 42nd Intl. SAMPE Symp. and Exhib.*, Anaheim, CA, May 1997, pp. 1291-1305.

14. R. E. Allred, T. M. Donnellan, and J. K. Sutter, "Thermo-Oxidative Resistant Finishes for High Temperature Polymer-Matrix Composites," *Proc. HITEMP Review 1994*, NASA Conference Publication 10146, October 1994.

15. R. E. Allred, L. A. Harrah, T. M. Donnellan, R. Meilunas, J. K. Sutter, and D. T. Jayne, "Thermo-Oxidative Stability of PMR-II-50 Matrix Composites with Reactive Finishes," *Proc. HITEMP Review 1995*, NASA Conference Publication 10178, October 1995.

16. C. H. Hooker, R. K. Eby, R. E. Allred, and M. A. Meador, *"Thermogravimetric Investigation of Reactive Polyimide Sizings on Carbon Fibers," Proc. HITEMP Review 1997*, NASA Conference Publication 10192, May 1997, Paper 10.

17. J. D. Miller, W. C. Harris, and G. W. Zajac, "Composite Interface Analysis Using Voltage Contrast XPS," *Surface and Interface Analysis*, Vol. 20, 1993, pp. 977-983.

18. J. D. Miller, R. A. Gray, D. Ward, J. B. Barr, and W. C. Harris, "Relating Thermo-Mechanical Performance to Carbon Fiber/Matrix Adhesion in PMR-15 Laminates," *Proc. High Temple Workshop XIII,* 1993.

19. J. Miller, G. Zajac, and T. Nguyen, "Interlaboratory Study of Fiber/Matrix Adhesion Using Voltage Contrast XPS," *Fiber, Matrix and Interface Properties*; Second Volume, ASTM STP 1290, C. J. Spragg and L. T. Drzal, eds., Am. Chem. Soc. for Testing and Materials, 1996, pp. 92-102, Paper K.

20. M.P. Stevens, *Polymer Chemistry An Introduction,* Addison-Wesley Publishing Company, Reading, MA, 1975.

Fabrication of Structural Components from Solution Coated Polyimide Prepreg[1]

Roberto J. Cano
Harry L. Belvin
Erik S. Weiser
David M. McGowen

NASA Langley Research Center
Hampton, VA 23681-0001

ABSTRACT

Recent advancements in high temperature materials at NASA Langley Research Center have led to the development of new polyimide resin systems with very attractive properties. The majority of the work done with these resin systems has concentrated on determining engineering mechanical properties from flat composite coupons. Since future advanced civilian aerospace components will require fabrication and testing of built-up structure from these high temperature composite materials, their processablity into structural components needed to be demonstrated.

The NASA Langley-developed polyimide matrix resin, LaRC™-PETI-5, was scaled to large quantities and solution prepregged onto Hercules IM7 carbon fiber by Fiberite Corporation. Using this prepreg which contained 20-22 percent N-methylpyrrolidinone solvent, techniques were developed to fabricated fully consolidated stiffener reinforced flat panels. The techniques included the use of movable metal tooling alone or in combination with trapped fluoroelastomer rubber. Methods were developed to remove all solvent prior to consolidation and to apply pressure uniformly to all sections of the part. Single stringer panels were made and machined into crippling and stiffener pull-off specimens. Panels with multiple stringers were also fabricated. Overall, the structural elements fabricated were of high quality and exhibited excellent mechanical properties.

KEY WORDS: Composites, Polyimides, Prepreg, Processing, and Structure

---

[1] This paper is declared a work of the U.S. Government and is not subject to copyright protection in the United States.

1. INTRODUCTION

Future civilian aerospace components may require the use of advanced resin systems that can withstand exposure to high temperatures for extended periods over the lifetime of the part. One such candidate material, a phenylethynyl-terminated polyimide given the designation LARC[TM]-PETI-5, has been developed at the NASA Langley Research Center (LaRC).[1,2] Recent work has shown the advantages of similar phenylethynyl-terminated polyimides as films, moldings, adhesives, and composite matrix resins.[3-10] Phenylethynyl-terminated oligomers provide greater processing windows than materials which either have no molecular weight control or incorporate simple ethynyl endcaps. Once the solvent is removed in a B-stage operation, these low molecular weight, low melt viscosity oligomers are relatively easy to fabricate. They thermally cure without the evolution of volatile by-products and provide an excellent means of producing polymers with high glass transition temperatures, excellent solvent resistance and high mechanical properties.

Initial work on laboratory synthesized LARC[TM]-PETI-5 was performed utilizing prepreg solution coated at LaRC on its Multipurpose Prepregging Unit.[11] The resin was then scaled to large quantities by IMITEC, Inc., Schenectady, NY, and solution prepregged onto Hercules IM7 carbon fiber by Fiberite Corporation. The majority of the earlier work done with this prepreg concentrated on determining engineering mechanical properties from flat composite coupons.[12] Since future advanced civilian aerospace components will require fabrication of built-up structure, the processablity of IM7/LaRC[TM]-PETI-5 into aircraft structural components, not just flat panels, needed to be demonstrated.

Using commercial IM7/ LARC[TM]-PETI-5 prepreg which contained 20-22 percent N-methylpyrrolidinone solvent, techniques were developed to fabricated fully consolidated stiffener reinforced flat panels. The techniques included the use of movable metal tooling and combinations of movable tooling with trapped fluoroelastomer rubber. Methods were developed to remove all solvent prior to consolidation and to apply pressure uniformly to all sections of the part. Single stringer panels were made and machined into crippling and stiffener pull-off specimens. Panels with multiple stringers were also fabricated.

2. MATERIALS[2]

Unidirectional Hercules IM7 carbon fiber/LaRC[TM]-PETI-5 prepreg from Fiberite Corporation was used exclusively in this work. The 61-cm-wide prepreg material as received from Fiberite contained 35±3 percent by weight solids, 20-22 percent by weight solvent, and a fiber areal weight of 145±5 g/m^2.

---

[2] The use of trademarks or names of manufacturers in this report is for accurate reporting and does not constitute an official endorsement, either expressed or implied, of such products or manufacturers by NASA.

Mosites 2902 cured high temperature flouroelastomer, 75 durometer with a light fabric finish, was used for all processes involving rubber expansion. The rubber was obtained in 91-cm-wide rolls, 3.2-mm-thick from Mosites Rubber Company, Inc., Fort Worth, Texas.

3. EXPERIMENTAL

3.1 Processing   The stringer panel design and lay-up is shown in Figure 1. The one- or two-bay design is a modification of an original 3-bay panel design which called for fiber in both unitape and 5-harness-satin cloth forms. Since IM7/LaRC™-PETI-5 cloth was not available, only unidirectional prepreg could be utilized. Therefore, the modified lay-up (Figure 1) was designed to have approximately the same dimensions and modulus as the original tape/cloth design.

Initial work focused on fabricating single stringer crippling specimens for mechanical testing. Panels were made such that stiffener pull-off specimens could be machined from the panel ends. Successful consolidation required debulking of the stringer plies on a steel tool in order to compact and dry the prepreg. Initial work utilizing steel tooling (Figure 2) led to the following procedure for the fabrication of single stringer crippling specimens.

1. Cut stringer plies according to the data in Table 1 which allows for ply drop-offs.
2. Tack the stringer plies together along the edge that will form the outer edge of the stringer top flange (Figures 1 and 2). A total of ten plies are required to form one half of a stringer. Tack the plies into books of 10 plies each.
3. Coat stringer tooling with release agent. (Several coats should be applied.)
4. Heat stringer tool halves and a metal plate to about 177°C.
5. Place a layer of bleeder cloth and a book of plies on the hot metal plate. Press and smooth the plies onto the hot plate with gloved hands until plies soften.
6. Align the edge of the ply book that will form the outer edge of the top flange with the edge of the metal plate. Place the stringer tool onto the ply book, aligning the lip edge of the tool with the edge of the ply book and the metal plate edge.
7. Form the ply book around the tool by slowly rolling the tool so that plies are pressed between the tool and the metal plate. (Note: Face sheet caul plates are not attached to tooling at this point. Refer to Figure 2)
8. Cover the formed plies with a C-stringer tool, clamp the lay-up onto the table and allow it to cool. (Figure 3)
9. Repeat steps 4 through 8 for each stringer half.
10. Vacuum bag and debulk stringer halves while they are still on tooling in an air oven at 177°C for one hour. Cool under vacuum.
11. Lay-up face sheet and cap strip.
12. Vacuum bag and debulk face sheet and cap strip utilizing a top caul plate.
13. Utilizing stringer halves that are still on tooling, vacuum bag stringer halves with cap strip to form the stringer. Place a fillet of prepreg material between stringer halves and cap strip.
14. Debulk under vacuum for 1 hr. at 177°C. Cool under vacuum.
15. Attach face sheet caul plates to tooling.
16. Bag debulked stringer and face sheet, adding bottom fillet between face sheet and stringer. Utilize a bottom caul plate for the face sheet.
17. Cure stringer panel in autoclave utilizing PETI-5 cure cycle. (Figure 4)

In order to expand this technique to multiple stinger panels, a method was needed to help consolidate the web. Tooling modifications were performed to incorporated Mosites rubber.

The modified tooling arrangement is shown in Figure 5. Several single stringer panels were successfully fabricated to verify the process. This technique was then used to fabricate a two stringer panel with the tooling illustrated in Figure 6. Dual stringer panels were 35.5-cm wide by 61-cm long with two stingers identical in lay-up to the single stringer panels with spacing of 17.8-cm.

3.2 Mechanical Testing   Crippling tests were performed at LaRC in a 5.34×10^5-N hydraulic test machine at a displacement rate of 0.13 cm/min. Specimens were potted, their ends ground flat and parallel, strain gauged, and tested to failure at both room temperature (RT) and 177°C. The specimen and test configuration are shown in Figure 7.

Stinger pull-off specimens were tested in a 2.45×10^5-N electronic servo-hydraulic test machine. The specimen and test configuration are shown in Figure 8. Three different boundary conditions were evaluated. The boundary conditions were varied by altering the distance from the stringer to the edge of the face sheet support (3.61, 4.88, and 6.15 cm). Specimens were tested at a rate of 0.025, 0.038, and 0.051 cm/min, respectively.

4. RESULTS AND DISCUSSION

4.1 Panel Quality   Nine single stringer crippling/pull-off specimens were fabricated. Ultrasonic inspections of each panel indicated good quality. Specimens were also sectioned and viewed under a microscope. Typical photomicrographs of a consolidated specimen are shown in Figures 9 and 10. The panels appeared to be well consolidated and did not contain significant void formation. Fiber volume fractions of 58% for the caps and webs and 54% for the face as well as void volume fractions of <1.2 % were obtained by acid digestion.

Final consolidated panel per-ply-thicknesses for 7 single stringer panels and one 2-stringer panel are shown in Table 2. Observed values for the face and cap were lower than the projected values of ~0.140 mm/ply (5.5 mil/ply). The final thicknesses of the webs were close to the anticipated thickness and averaged 0.142 mm/ply (5.6 mil/ply) for all parts. The low cap and face thicknesses may be a result of excess flow caused by the 1.4-MPa process pressure and the weight of the steel tooling.

Detailed ply thicknesses for two selected single stringer panels after drying and after consolidation are presented in Table 3. It is interesting that the 275°C/ 2hr B-stage drying led to near net per ply thicknesses for the cap and face. Additional squeeze out during final consolidation at 371°C resulted in caps and faces which were thinner than the web sections. By slightly altering the tooling process for the cap and face or, alternatively, by simply not debulking these parts but allowing the standard cure cycle to dry and consolidate those areas, this problem could be corrected, especially since these sections are essentially flat. To demonstrate the latter approach, panels AU2730 and AU2748 was debulked similarly except for one modification. The web halves of panel AU2748 were debulked with the web face down. The weight of the steel tool combined with the vacuum resulted in a dried per ply thickness of 0.152-mm and 0.141-mm for each versus 0.197-mm and 0.214-mm for panel AU2730 which was debulked with the web vertical. Thus, altering the debulking technique, web per-ply-thickness could be controlled. In both cases, however, the final consolidated thickness of each web was virtually identical (Tables 2 and 3).

The data in Table 2 shows that utilization of steel tooling modified with Mosites rubber, as shown in Figure 5, did not significantly alter the consolidation thicknesses obtained with the steel tooling alone. This would be expected for the fabrication of a single stringer panel where autoclave pressure is sufficient to easily consolidate one web regardless of which tooling was used. The difficulty arises when multiple stringers are involved. Utilization of rubber expansion for fabrication of a single stringer panel verified that the process was valid and could be applied to the fabrication of multiple stringer panels.

A dual stringer panel was successfully manufactured utilizing the tooling illustrated in Figure 6. Like the single stringer panels, photomicrographs of both webs of the final consolidated panel indicated good consolidation. The per ply thicknesses of the caps, webs, and skin of the dual stringer panel were similar to those obtained for the single stringer panels (Table 2).

4.2 Mechanical Properties   Although the data showed some scatter, nominal maximum crippling loads of 1.69×10^5 N (37,970 lbs) [range: 2.10×10^5 to 1.29×10^5 N (47,000 to 29,100 lbs)] at room temperature and 1.22×10^5 N (27,320 lbs) at 177°C were obtained. These results are slightly better than data obtained for similar panels made from IM7/Avimid™-K3B, a polyimide resin system available from DuPont. The IM7/K3B panels were manufacture by McDonnell Douglas and utilized both unitape and fabric. These panels had average maximum crippling loads of 1.54×10^5 N (34,600 lbs) at RT and 1.08×10^5 N (24,250 lbs) at 177°C. A typical crippling failure for an IM7/LaRC™-PETI-5 panel is shown in Figure 7. A typical load/strain plot for a RT crippling test is presented in Figure 11. The plot shows a change in buckling mode at approximately 9.7×10^4 N (22,000 lbs) after the panel initially buckled at 7.6×10^4 N (17,000 lbs).

Stringer pull-off data is presented in Table 4. As expected, the failure load for stringer pull-off decreased with increasing distance from the edge support (boundary condition). The IM7/PETI-5 panels afforded higher pull-off strengths than the IM7/K3B panels consistently over all three boundary conditions.

5. CONCLUSIONS

Stainless steel movable tooling was designed and verified for making high quality fully consolidated single stringer flat panels from commercially available LaRC™-PETI-5/ IM7 poly(amide acid) prepreg. Stainless steel movable metal tooling in combination with trapped flouroelastomer rubber was also designed and verified for making high quality single and multiple stringer panels from LaRC™-PETI-5/ IM7 poly(amide acid) prepreg. Procedures were developed to debulk and effectively remove the 20-22% N-methylpyrrolidinone solvent from the prepreg during the process cycle. Methods were also developed to apply pressure uniformly to all sections of the part, especially transverse to the web.

The single stinger panels afforded very good resistance to stringer pull-off and crippling loads. Average maximum crippling loads of 1.69×10^5 N (37,970 lbs) at RT and 1.22×10^5 N (27,320 lbs) at 177°C were obtained.

6. REFERENCES

1. P. M. Hergenrother, R. G. Bryant, B. J. Jensen, J. G. Smith, Jr. and S. P. Wilkinson, Soc.. Adv. Matl. Proc. Eng. Series, 39(1), 961 (1994).

2. R. G. Bryant, B. J. Jensen, J. G. Smith, Jr. and P. M. Hergenrother, Ibid.,(Closed Papers), 39, 273(1994).

3. C. W. Paul R. A. Schultz and S. P. Fenelli, U. S. Pat 5,138,028 (1992) [to National Starch and Chemical Co.].

4. C. W. Paul, R. A. Schultz and S. P. Fenelli in Advances in Polyimide Science and Technology, C. Feger, M. M. Khojasteh and M. S. Htoo, ed. Technomic Pub. Co., N.Y., 1993, p 220.

5. R. G. Bryant, B. J. Jensen and P. M. Hergenrother, Polym. Prepr., 34(1), 566 (1993).

6. B. J. Jensen, R. G. Bryant and S. P. Wilkinson, ibid., 35(1), 539 (1994).

7. S. J. Havens, R. G. Bryant, B. J. Jensen, and P. M. Hergenrother, ibid., 35(1), 553 (1994).

8. J. G. Smith, Jr. and P. M. Hergenrother, ibid., 35(1), 353 (1994).

9. J. G. Smith, Jr. and P. M. Hergenrother, Polymer, 35(22), 4857 (1994).

10. P. M. Hergenrother, R. G. Bryant, B. J. Jensen, and S. J. Havens, J. of Polymer Science: Polymer Chemistry Edition, 32 , 3061 (1994).

11. R. J. Cano, N. J. Johnston, and J. M. Marchello., Solution Prepreg Quality Control, Soc.. Adv. Matl. Proc. Eng. Series, 40, 583, (1995).

12. R. J. Cano, M. Rommel, J. A. Hinkley, E. D. Estes, Fiber Study Involving a Polyimide Matrix, Soc.. Adv. Matl. Proc. Eng. Series, 41, 1047, (1996)..

Table 1. Stringer panel ply orientations and dimensions.

| Stringer Web Halves* | | | Face Sheet** | | Cap Strip*** | |
Ply	Orientation	Width, cm (in.)	Ply	Orientation	Ply	Orientation
1	+45	8.86(3.49)	1	+45	1	0
2	-45	8.89 (3.50)	2	-45	2	0
3	0	7.62 (3.00)	3	0	3	0
4	90	7.87 (3.10)	4	90	4	90
5	0	8.18 (3.22)	5	90	5	0
6	90	8.43 (3.32)	6	0	6	90
7	45	8.64 (3.40)	7	+45	7	0
8	-45	8.92 (3.51)	8	-45	8	0
9	+45	9.19 (3.62)	9	-45	9	0
10	-45	9.19 (3.62)	10	+45	10	90
			11	0	11	0
			12	90	12	90
			13	90	13	0
			14	0	14	0
			15	-45	15	0
			16	+45	16	-45
					17	+45
					18	-45
					19	+45
					20	90
					21	0
					22	90
					23	0
					24	-45
					25	+45

*Length: 45.7 cm (18 in.) for crippling
 specimens, all plies
**Dimensions: 19.1 cm (7.5 in.)
 by 45.7 cm (18 in.)
***Dimensions: 4.4 cm (1.75 in.)
 by 45.7 cm (18 in.)

Table 2. Measured dimensions of final consolidated stringer panels.

Panel Number	Face Nominal Thickness, mm (in.)	Face Ave. Per Ply Thickness, mm (in.)	Cap Nominal Thickness, mm (in.)	Cap Ave. Per Ply Thickness, mm in.	Web Nominal Thickness, mm (in.)	Web Ave. Per Ply Thickness, mm (in.)
AU2428	2.04978 (0.0807)	0.128016 (0.00504)	4.0005 (0.1575)	0.114046 (0.00449)	2.9845 (0.1175)	0.149098 (0.00587)
AU2494	2.12344 (0.0836)	0.132842 (0.00523)	3.2131 (0.1265)	0.091694 (0.00361)	2.88798 (0.1137)	0.144272 (0.00568)
AU2512	1.75006 (0.0689)	0.10922 (0.00430)	3.6195 (0.1425)	0.10414 (0.00410)	3.05562 (0.1203)	0.152908 (0.00602)
AU2524	1.78816 (0.0704)	0.11176 (0.00440)	3.683 (0.1450)	0.10414 (0.00410)	3.05562 (0.1203)	0.152908 (0.00602)
AU2730	1.774952 (0.06988)	0.110998 (0.00437)	3.10896 (0.1224)	0.0889 (0.00350	2.917952 (0.11488)	0.145796 (0.00574)
AU2733	1.9939 (0.07850)	0.12446 (0.00490)	3.8354 (0.1510)	0.109474 (0.00431)	3.0607 (0.12050)	0.153162 (0.00603)
AU2748	1.9939 (0.0785)	0.124714 (0.00491)	N/A N/A	N/A N/A	2.84226 (0.1119)	0.14224 (0.00560)
AU2786 (2 stingers)	1.79832 (0.0708) N/A N/A	0.112522 (0.00443) N/A N/A	3.1242 (0.123) 3.3782 (0.133)	0.089154 (0.00351) 0.09652 (0.00380)	2.7305 (0.1075) 2.8194 (0.111)	0.136652 (0.00538) 0.14097 (0.00555)

Table 3. Dried and final consolidated thicknesses of single stinger panels.

	Dried* Thickness, mm/ply (mil/ply)	Final Consolidated Thickness, mm/ply (mil/ply)
Panel AU2730		
FACE	0.140 (5.50)	0.111 (4.37)
CAP STRIP	0.140 (5.50)	0.089 (3.50)
WEB HALF 1		)
web	0.197 (7.75)	0.146 (5.74)
bottom	0.132 (5.20)	***
top	0.162 (6.38)	***
WEB HALF 2		
web	0.214 (8.44)	0.146 (5.74)
bottom	0.142 (5.59)	***
top	0.215 (8.45)	***
Panel AU2748		
FACE	0.146 (5.75)	0.125 (4.91)
CAP STRIP	0.142 (5.6)	N/A
WEB HALF 1		
web	0.152 (5.98)	0.142 (5.60)
bottom	0.160 (6.29)	***
top	5.1	***
WEB HALF 2		
web	0.141 (5.54)	0.142 (5.60)
bottom	0.156 (6.14))	***
top	0.129 (5.08)	***

*Held under full vacuum at 350°F for 1 hour.
Note: Web half's top section is incorporated into final CAP.

Table 4. Stiffener pull-off data.

	IM7/LARC™-PETI-5		IM7/Avimide™-K3B	
Boundary Condition*, cm (in.)	Failure Load, N (lbs)	Stroke Rate. cm/min (in/min)	Failure Load, N (lbs)	Stroke Rate. cm/min (in/min)
3.61 (1.42)	2835.6 (637.5)	0.025 (0.01)	2336.5 (525.3)	0.051 (0.02)
4.88 (1.92)	2318.3 (521.2)	0.038 (0.015)	2166.6 (487.1)	0.051 (0.02)
6.15 (2.42)	2082.1 (468.1)	0.051 (0.02)	1942.4 (436.7)	0.051 (0.02)

*Distance from stringer to edge support.

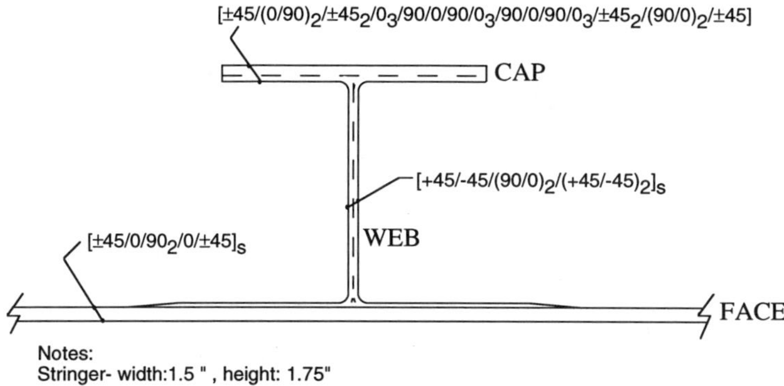

Figure 1. Stringer lay-up.

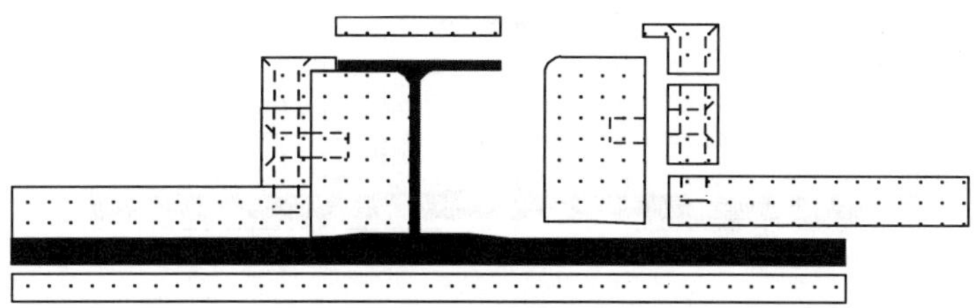

Figure 2. Single stringer panel tooling.

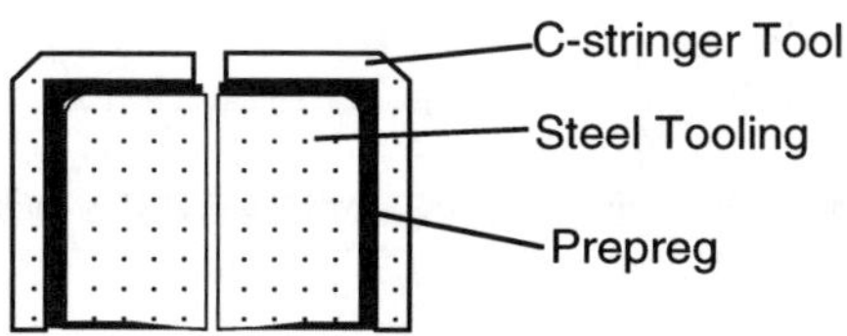

Figure 3. Schematic of debulking technique of webs.

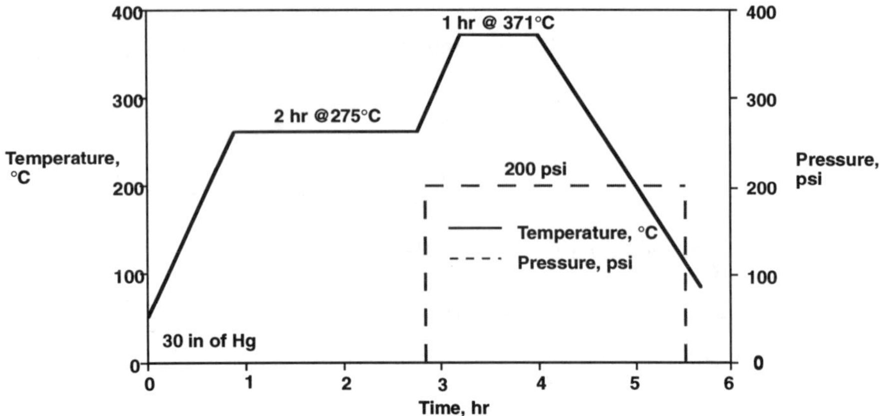

Figure 4. LaRC™-PETI-5 Cure cycle.

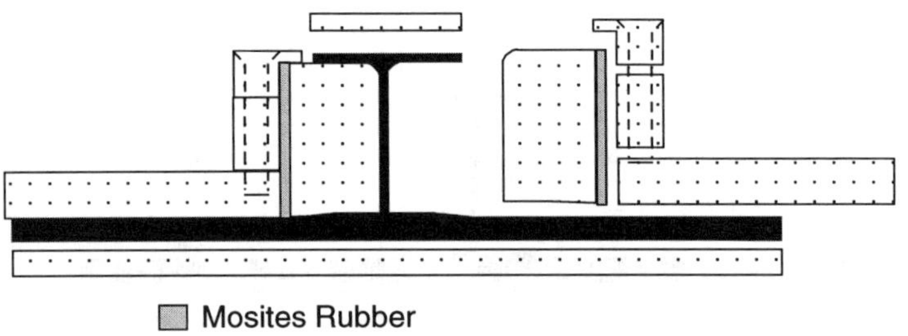

Figure 5. Modifiied single stringer tooling utilizing rubber expansion.

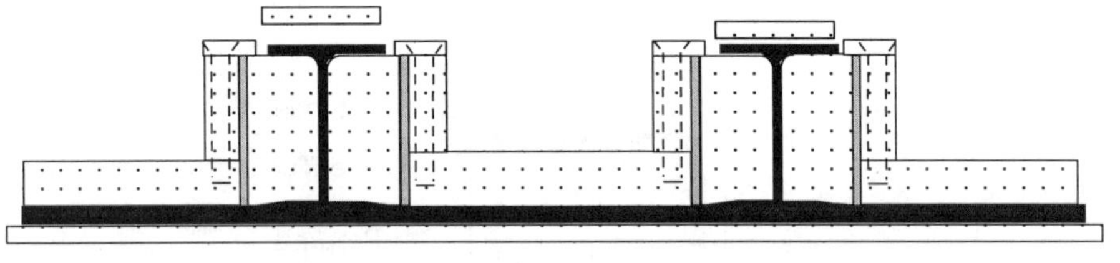

Figure 6. Dual stringer panel tooling.

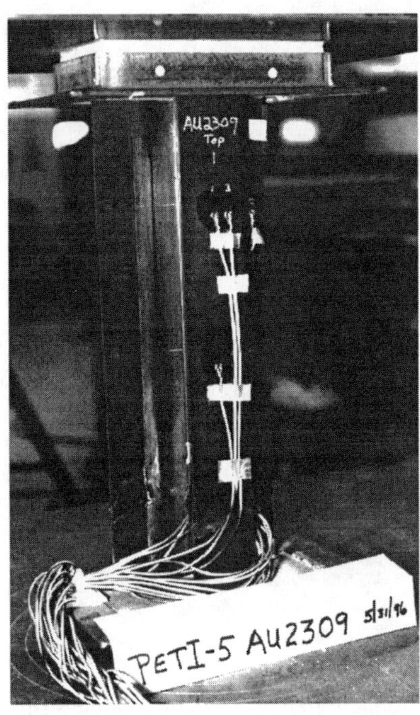

Figure 7. Crippling specimen and test configuration. Note the failure along lower left portion of specimen.

Figure 8. Stiffener pull-off specimen and test configuration.

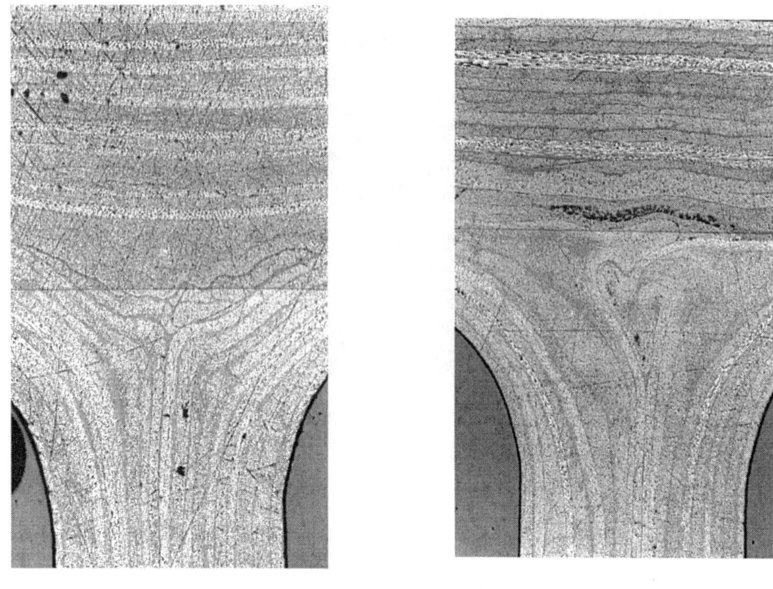

Cap   **Face**

Figure 9. Photomicrographs of a typical single stringer panel (Panel AU2730).

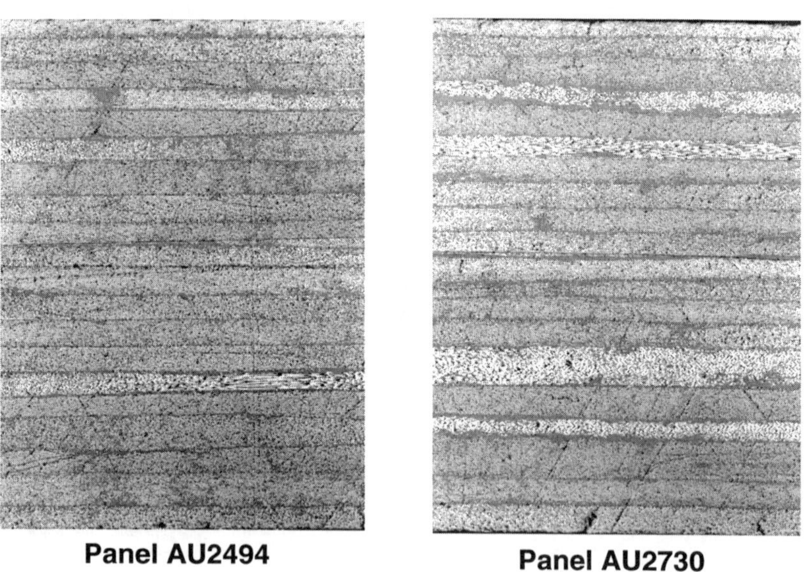

Panel AU2494   **Panel AU2730**

Figure 10. Typical photomicrographs of web.

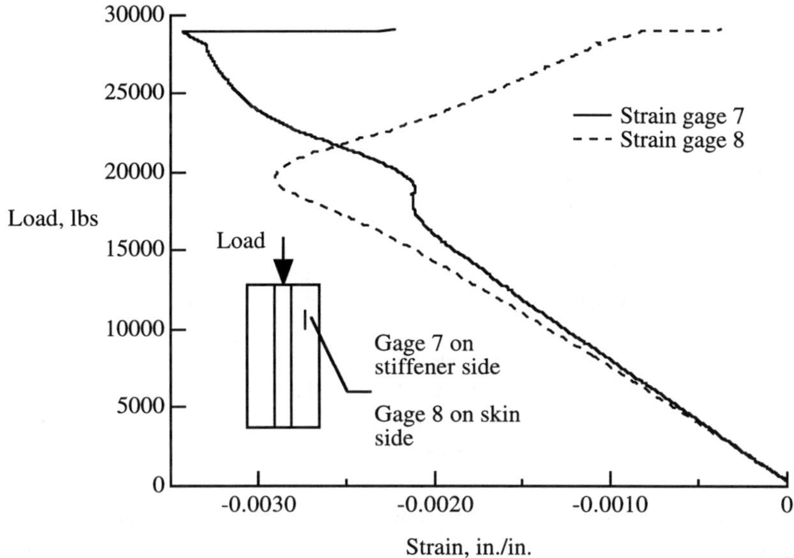

Figure 11. Load versus strain for room temperature crippling test. (Panel AU2024)

EFFECTS OF TRACE METALS ON THE THERMO-OXIDATIVE STABILITY OF 6F POLYIMIDES

James K. Sutter*, Ronnie Ortiz and Preston W. Greene
NASA Lewis Research Center, Cleveland, OH 44135

J. Marcus Jobe
Miami University, Oxford, OH 45056

ABSTRACT

Polyimides containing [Hexafluoroisopropylidene-2,2-bis(phthalic acid anhydride)] or 6F dianhydride have many potential applications in high temperature oxidative environments such as aircraft engines or electronic circuitry. In many of these applications, polyimide resins or their composites are in direct contact with metals. This study investigates the effects of trace metals in or on the surface of PMR-II-50; an addition polyimide developed at NASA Lewis Research Center.[1] Specifically, we focused on (1) determining what, if any, effect on percent weight loss of PMR-II-50 after 956 hours at 316 °C could be attributed to the type of 6F dianhydride (polymer or electronic grade), processing time at 371 °C and the presence of Kapton®, as a barrier film, to prevent or minimize diffusion of metal from the tool steel into the polyimide; (2) determining what, if any, effect on iron content in PMR-II-50 could be attributed to type of 6F dianhydride, processing time at 371 °C and the presence of Kapton® and (3) discovering any possible relationship between percent weight loss and iron content. Resin type significantly affected TOS; while processing time and using Kapton® had no significant effect on PMR-II-50 resin TOS .

Key Words: Polyimide, 6F Dianhydride and thermo-oxidative stability

INTRODUCTION

The presence of trace metals in polymer composites has been a concern since earlier studies on carbon and glass fibers identified alkali metals as an accelerant for polymer composite degradation.[2] Centers and Click[3] clearly identified the effects of metals and their oxides (Fe, Cr, Ni and Cu) on the thermal stability of polyimide adhesives although metal concentration was high, not at trace levels. They showed that effect of metals varied from retarding to degrading polyimide thermo-oxidative stability depending on the metallic additive. More recent studies focused on the effect of metals (Ag, Cu, Ni, Cr) in or on polyimides used in electronic applications.[4]

6F Dianhydride is manufactured by Hoechst Celanese and sold in two grades: electronic and polymer grade. This dianhydride is a vital monomer in polyimides requiring thermo-oxidative stabilities at temperatures >300 °C. 6F dianhydride polymer grade contains trace metals (>20ppm) such as iron (Fe); whereas, 6Fdianhydride electronic grade has smaller amounts of trace metals, typically <10 ppm. As part of this study, we examined the effects of both grades of 6F dianhydride in a thermoset polyimide, PMR-II-50, containing para-phenylenediamine and nadic acid ester as shown in Figure 1. Several other polyimides also contain 6F dianhydride such as Avimid-N®, AFR-700 and VCAP-75 and although expensive, all of these resins perform well in oxidative environments; where other non-6F polyimides degrade rapidly.

Earlier studies[5] showed that compression molding 6F dianhydride/para-phenylenediamine polyimides requires high processing temperatures, >370 °C, to form fully cured dense polyimide resin plaques. More recently, Sutter et al.[6] showed that molding and then aging three types of 6F polyimides at these temperatures in air causes the surface of the polyimides to appear a dull reddish-brown. All three 6F polyimides used in this study: AFR700B, PMR-II-50 and VCAP-75 were formulated with electronic grade 6F dianhydride only! Surface analytical techniques have identified iron as one of the prominent metals on these 6F polyimide surfaces. In order to prevent or minimize the transfer of metal from the tool steel to the polyimide when compression molding, Kapton® film was placed between the polyimide molding powder and tool steel to provide a barrier between the polyimide and the tool steel. Polyimides molded with Kapton® film did not have the reddish-brown color on their surface after aging in air at 371 °C. While this study concluded that significant differences in thermo-oxidative stability were attributed to differences in the polyimides chemical composition; the effect of using Kapton®, as a barrier for metal transport from the tool steel into the polyimide, on polyimide TOS was not significant.

In this study, a statistical design of experiments was performed to determine if there are differences in thermo-oxidative stability between PMR-II-50 polyimides containing both grades of 6F dianhydride: electronic and polymer grade. In order to determine the effects of trace metals present in the dianhydride versus metals that could diffuse into the surface of the polyimide during compression molding; Kapton® film was used as a barrier film for half of the PMR-II-50 polyimides produced for the thermo-oxidative aging test. Another factor studied was the time spent at the maximum processing temperature (371 °C). Polyimides were cured at 371 °C for either 15 minutes or two hours. If metals diffusing from the tool steel onto the polyimide have an effect on polyimide TOS, then polyimides processed for two hours may have more metal on their surface than those polyimides processed for 15 minutes and this effect may be measured by a change in the weight loss when isothermally aged at 316 °C.

EXPERIMENTAL

PMR-II-50 Molding Powder:
Monomers used in this study were electronic and polymer grades of [Hexafluoroisopropylidene-2,2-bis(phthalic acid anhydride)] or 6F dianhydride purchased from Hoechst Celanese Corp., paraphenylene diamine (PPDA) and nadic acid methyl ester (NE) from Chriskev Chem. Corp. . Both PPDA and NE were used as received;

however, 6F dianhydride (6FDA) was vacuum dried for 24 h at 125 °C before esterification with methanol. PMR-II-50 was formulated as previously reported[7] with an n value of 9. The monomer/methanol solution was gently refluxed on a hot plate in an open beaker until most methanol had evaporated. The remaining gum was place in a air circulating oven and B-staged at 210 °C for 2 hours and then 1.5 h at 288 °C. The orange-brown molding powder was ground to a fine powder using a glass mortar and pestle. FT-IR (Nicolet 510P) of the PMR-II-50 molding powder in KBr pellets confirmed complete imidization from imide carbonyl absorbtions at 1726 and 1786 cm^{-1}.

PMR-II-50 Resin Processing:
A D2 hardened cylindrical steel mold was used for processing B-staged PMR-II-50 molding powder. The mold was treated with high temperature mold release (Mono-coat® E63FF from Chem-Trend Inc.) before adding the molding powder. For the samples requiring Kapton® as a barrier between the molding powder and the tool steel, the Kapton (0.005cm thick) was also treated with mold release. Mold temperatures are recorded using a thermal couple placed in the tool near the molding powder. PMR-II-50 resin processing schedule: (1) heat press to 371 °C; (2) place charged mold in press and apply only contact pressure, then wrap mold in fiberglass insulation; (3) apply 2000 psi to mold when mold temperature is 316 °C; (4) hold mold pressure for either 15 min or 2 h, once the mold temperature is 371 °C; (5) cured PMR-II-50 is demolded when mold temperature is 212 °C. Cured PMR-II-50 resins were postcured at 371 °C for 16 h in an air circulating oven. Glass transition temperatures for all samples were measured using a TA Instruments 2940 thermal mechanical analyzer and were 355 ±5 °C.

Elemental Analysis:
All PMR-II-50 resins, monomers, Kapton® and release agent were analyzed for trace metals using ICP at Adirondack Labs in Albany, NY. Only PMR-II-50 resins and 6F dianhydride had metal contents that were detectable (>2ppm). Iron (Fe) was the most prominent metal detected confirming results from earlier surface analyses.[6]

PMR-II-50 Resin Aging:
Resin samples were uniformly cut from the PMR-II-50 cylindrical wafers using a water jet with no abrasive and were 0.5 cm square and 0.4 cm thick. PMR-II-50 resin samples were dried at 125 °C/60 cm vacuum before isothermal aging. Samples were randomly placed in three oven locations and further randomized within a location. An air circulating Blue-M oven operating at 316 ±1 °C was monitored with 9 thermal couples throughout the aging time (956 h) according to ASTM E 145-68.

RESULTS AND DISCUSSION

This study was designed and analyzed using formal statistical methodologies. The study focused on (1) determining what, if any, effect on percent weight loss of PMR-II-50 after 956 hours at 316 °C could be attributed to the type of 6F dianhydride (polymer or electronic grade), processing time at 371 °C and the presence of Kapton® as a barrier film (2) determining what, if any, effect on iron content in PMR-II-50 could be attributed to type of 6F dianhydride, processing time at 371 °C and the presence of Kapton®; and (3) discovering any possible relationship between percent weight loss and iron content.

A total of 24 PMR-II-50 samples were used in the weight loss study. All combinations of two Kapton® levels (with/without), two time levels (15 minutes, 2 hours) and two 6F dianhydride levels (electronic and polymer grades) were considered. Three sets of eight samples (each combination of the three factors, $2^3 = 8$, were represented in a given set) were randomly allocated to three locations in the oven, one set per location. The eight samples assigned to a given location were randomly assigned a slot in that location. The 24 samples were then exposed to 316 °C for 956 hours. A diagram for the test matrix is shown in Figure 2. The model for iron content in PMR-II-50 resins examined samples from the same 24 weight loss specimens; however, the effect of oven location was not included in this model.

A linear model consistent with the designed experiment and identifiable causes of % weight loss (Kapton® level, time level, dianhydride level, oven location) was constructed and is shown in Figure 3. This model permitted a statistically valid, credible analysis of the data. A least squares approach was implemented to estimate our model. The assumption of a common variance and approximate normal distribution were not seriously violated by applying normal probability plotting (Figure 4) and residual analysis techniques (Figure 5). Using statistical hypothesis testing (error probability 0.05), there was strong evidence to conclude the effect of Kapton® on percent weight loss for PMR-II-50 made with electronic grade 6F dianhydride was different from the Kapton® effect on percent weight loss for PMR-II-50 made with polymer grade 6F dianhydride. The following confidence intervals estimate the average percent weight loss for PMR-II-50 samples made with Kapton® minus those made without Kapton® (both averaged over the factor: processing time) for electronic grade (-0.438,0.375) and polymer grade (0.140, 0.953) 6F dianhydrides, respectively.

The percent weight losses shown in Figure 6 can be averaged over both processing time levels (15 min and 2 h). Note: There was no strong evidence of Kapton® x processing time or 6F dianhydride type x processing time interaction existed; i.e., the conclusion regarding a Kapton® effect are the same for both levels of processing time and the conclusions regarding a 6F dianhydride type effect are the same for both levels of processing time. Thus, it appears the average % weight loss of PMR-II-50 made with a polymer grade 6F dianhydride and Kapton® exceeds the average % weight loss of the same polymer grade 6F dianhydride type made without Kapton® from 0.14 to 0.95%. PMR-II-50 made with an electronic grade 6F dianhydride and Kapton® showed no important difference in average % weight loss from PMR-II-50 polymer resin made with an electronic grade 6F dianhydride without Kapton®. There is at least 90% confidence both conclusions are correct using Bonferroni inequality.[8] Figure 7 contains the analysis of variance (ANOVA) table and expected mean squares derived using the rules outlined by Neter et al.[8]

Evaluating the p-value for the dianhydride entry (0.000) in the ANOVA table in Figure 7 is very small suggesting there is in fact an important difference in percent weight loss from polymer versus electronic grade dianhydride. Since there is no interaction between the mean percent weight losses for polymer and electronic 6F dianhydride, the difference in overall average percent weight loss for polymer grade minus the overall average percent weight loss for electronic grade (4.839-3.843) is a legitimate comparison of the difference in percent weight loss from the two different dianhydrides. Thus, it appears there is an important difference in average percent weight loss between polymer and electronic grade dianhydrides (the average percent weight loss for polymer exceeds that for electronic by about .996 percent). A 95% interval for this difference becomes (.70862, 1.28338).

Samples from the 24 PMR-II-50 resin disks were also used to evaluate the effects of 6F dianhydride type, processing time at 371 °C and Kapton® on iron (Fe) content. All combinations of Kapton® levels (with, without), two processing time levels at 371 °C (15 minutes, 2 hours) and two 6F dianhydride levels (electronic, polymer) were considered. Thus, three samples were obtained from each combination of the three factors and the respective iron (Fe) contents (ppm) were obtained using ICP spectroscopy.

A linear model consistent with the designed experiment and the identifiable causes of iron (Fe) content (Kapton® level, time level, dianhydride level) is shown in Figure 8. This model also permitted a statistically valid, credible analysis of the data A least squares approach was implemented to estimate our model. However, applying normal probability plotting and residual analysis techniques, the assumption of a common variance was violated, as shown in Figures 9 and 10. Therefore, the data for iron content [Fe] was transformed to normalize this data set in Figure 11. A natural logarithm, ln [Fe] using the same linear model (except now the response variable, y, was ln [Fe]) permits the application of the least squares methodology to estimate the model. Once the ln transform was applied, normal probability plotting and residual analysis techniques revealed no important violation of the normality assumption or common variance assumption (Figures 12 and 13).

The analysis of variance table and expected mean squares derived using the rules outlined by Neter, Wasserman, and Kutner[8] for the analysis of ln of the iron content is shown in Figure 14. The only important effect on ln Fe content detected using formal statistical hypothesis testing was processing time (error probability less than 0.01). A 95% interval estimate of the average ln Fe content of PMR-II-50 processed for two hours at 371 °C minus the average ln Fe content of PMR-II-50 processed for 15 minutes at 371 °C was (-1.564, -0.385). This includes the ratio of the median iron content (ppm) from two hours processing at 371 °C to the median iron content (ppm) for 15 minutes processing at 371 °C with 95% confidence.

In conclusion, these analyses were averaged over both Kapton® and dianhydride levels because no important interaction was detected for these two factors. Thus, the effect of processing time at 371 °C was the same at all levels of Kapton® and 6F dianhydride. This experiment does not suggest % weight loss is related to iron content. No processing time (at 371 °C) effect was detected on % weight loss yet there was an important difference in median iron content. Iron contents for samples processed at 371 °C for 15 minutes were higher than the samples processed for 2 hours at 371 °C. No Kapton® or 6F dianhydride effect on iron content was detected; however, an important Kapton® effect on % weight loss was detected but only with polymer grade PMR-II-50. Another important conclusion was shown in Figure 7. The differences in weight loss for PMR-II-50 resins formulated with polymer grade 6F dianhydride were significantly larger than resins made from electronic grade 6F dianhydride. This conclusion was significant for both levels of Kapton® and processing time. It is expected that this difference in weight loss would increase as aging time increases. However, comparing the models and results for weight loss and iron content, there is no coorelation between iron content and %weight loss.

Finally, Deming[10-12] and more recently Hahn and Meeker[13] have detailed the important distinction between an enumerative vs. analytic study. Our study is considered analytic. The results of this study require the important assumption that variability of samples from future batches of PMR-II-50 is the same as the variability within the batch used in this experiment. Note Hahn, et al.[13], "analytic studies require the critical added assumption that the process about which one wishes to make inferences is statistically identical to that from which the sample was selected."

References

1) Vannucci, R. D. and Cifani, D.; NASA TM 100923, *700ºF Properties of Autoclave Cured PMR-II Composites*, 20th Int. SAMPE Conference, Sept. 1988

2) a) Delmonte, J., *"Technology of Carbon & Graphite Fiber Composites."* 1981 (Van Nostrand Reinhold) New York, NY
    b) Gibbs, H. H., Werdt, R. C., and Wilson, F. C., *"Carbon Fiber Structure & Stability Studies,"* 33rd Annual Technical Conference, SPI RP/C Institute, Section 24F, Washington, DC; Feb 1978

3) Centers, P. W. and Click, W. E., *"Effects of Selected   Metals and Metallic Oxides upon the Thermal Stability of Polyimide Adhesives"* AFML-TR-70-269, Sept 1970

4) a) Hahn, P. O., Rubloff, G. W., Bartha, J. W., Legoves, R., Tromp, R. and Ho, P.; Mat. Res. Soc. Symp. Proc., Vol. 40, 1985, pp. 251-263
    b) Ho, P. S., Hahn, P. O., Bartha, J. W., Legoues, F., Silverman, B. D.; J. Vac. Sci. Technol. A3 (3) May/June 1985, pp.739-745
    c) Rancourt, J. D., Porta, G. M., Moyer, E. S., Madeleine, D. G. and Taylor, L. T.; J. Mater. Res. 3 (5), Sep/Oct 1988, pp. 996-1001
    d) Green, P. F. & Berger, L. L.; Thin Solid Films 224 (1993) pp. 209-216.

5) a) Gibbs, H. H., 20th Nat. SAMPE Symposium - San Diego, CA, April 1975, pp. 212-225
    b) Gibbs, H. H., 7th Nat. SAMPE Tech. Conf., Albuquerque, NM, Oct. 1975, pp. 244-256

6) Sutter, J. K., Jobe, J. M., Jayne, D. T., Crane, E. A., and Harding, D. R., NASA CP-10146, Vol. 1, pp. 15-1 to 15-13, 1994

7) Serafini, T.T., Vannucci, R.D., and Alston, W.B.,: *Second Generation PMR Polyimides,* NASA TM X-71894, April 1976.

8) Neter, J., Wasserman, W., and Kutner, M. *Applied Linear Statistical Models,* 3rd Edition. Irwin, Homewood, IL,   1990, pp. 834, 922

9) Sutter, J. K., Jobe, J. M., Crane, E. A., J. Appl. Polym. Sci. , Vol. 57, 1995, pp. 1491-1499

10) Deming, W. E., J. Amer. Stat. Assoc., Vol. 48, 1953, pp. 244-255

11) Deming, W. E.,The American Statistician, Vol. 29, 1975, pp. 146-152

12) Deming, W. E., *"On The Use of Judgment Samples."* Reports of Statistical Applications, Japanese Union of Scientists and Engineers, Vol. 23, 1976, pp. 25-31

13) Hahn, G. J. and Meeker, W. Q., The American Statistician, Vol. 47, 1993, pp. 1-11

Synthesis of PMR II-50

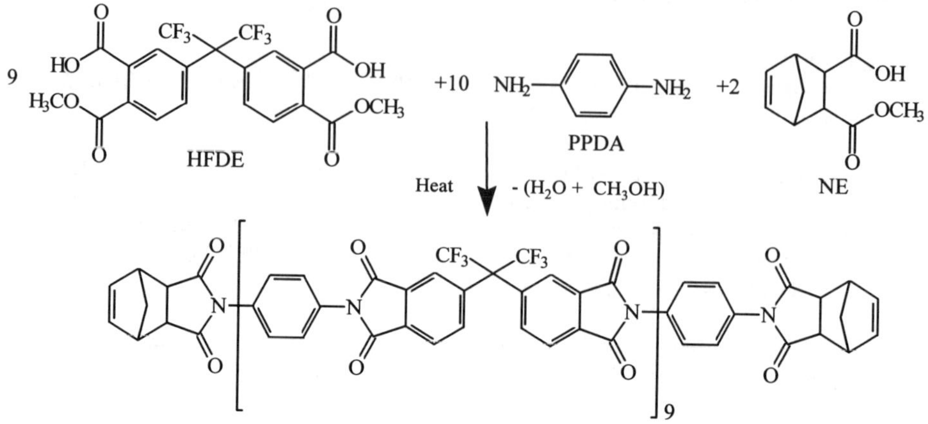

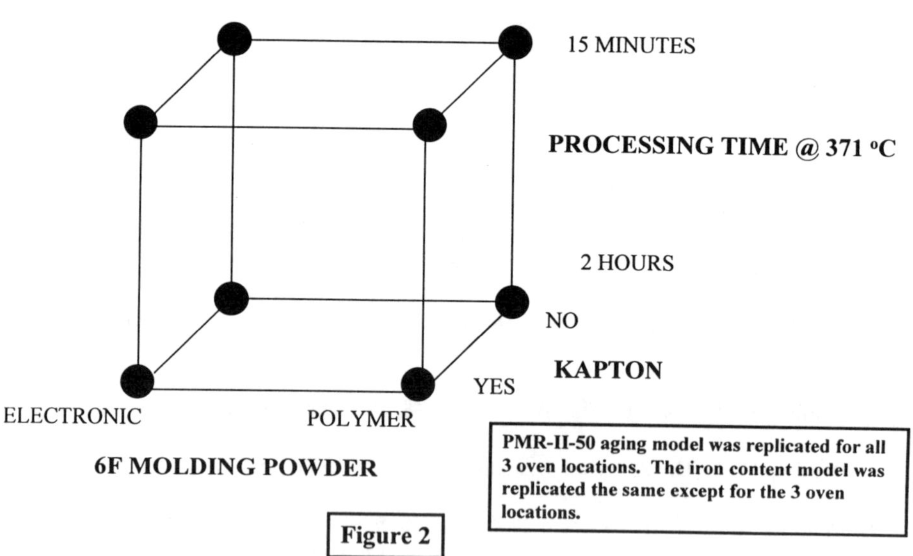

- PPDA less toxic than MDA
- Fluorinated diester leads to greater thermal stability
- Amount of nadic endcap is reduced

Figure 1

PMR-II-50 RESIN AGING MODEL

PMR-II-50 aging model was replicated for all 3 oven locations. The iron content model was replicated the same except for the 3 oven locations.

Figure 2

A LEAST SQUARES FIT OF THE MODEL:

$$y_{ijk\ell} = \mu + \alpha_i + \beta_j + \gamma_k + \alpha\beta_{ij} + \alpha\gamma_{ik} + \beta\gamma_{jk} + \alpha\beta\gamma_{ijk} + \delta_\ell + \epsilon_{ijk\ell}$$

$y_{ijk\ell}$ = % weight loss of a sample in the ℓth location from the ith kapton level, j^{th} time level and k^{th} dianhydride type exposed to 316 °C (600 °F) for 956 hours

μ = overall constant; α_i = fixed kapton effect; β_j = fixed time effect; γ_k = fixed dianhydride effect

$\alpha\beta_{ij}$ = fixed kapton by time interaction effect; $\alpha\gamma_{ik}$ = fixed kapton by dianhydride interaction effect

$\beta\gamma_{jk}$ = fixed time by dianhydride interaction; $\alpha\beta\gamma_{ijk}$ = fixed kapton by time by dianhydride interaction effect

δ_ℓ = fixed location effect; $\epsilon_{ijk\ell}$ = random error term

where:
i = 1,2	kapton level (1 = without, 2 = with)
j = 1, 2	time level (1 = 15 minutes, 2 = 2 hours)
k = 1, 2	dianhydride level (1 = polymer grade, 2 = electronic grade)
ℓ = 1, 2, 3	location (1, 2, 3)
ϵijkℓ	is a normal random variable with zero mean and variance σ^2

Figure 3

Normal Probability Plot: PMR-II-50 Resin % Weight Loss

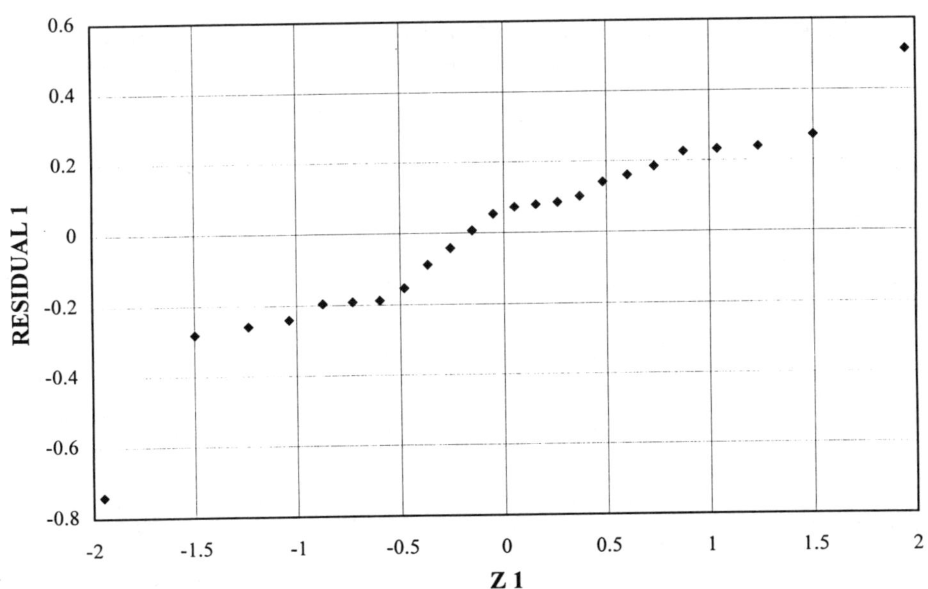

Figure 4

165

Residual Plot: PMR-II-50 Resin % Weight Loss

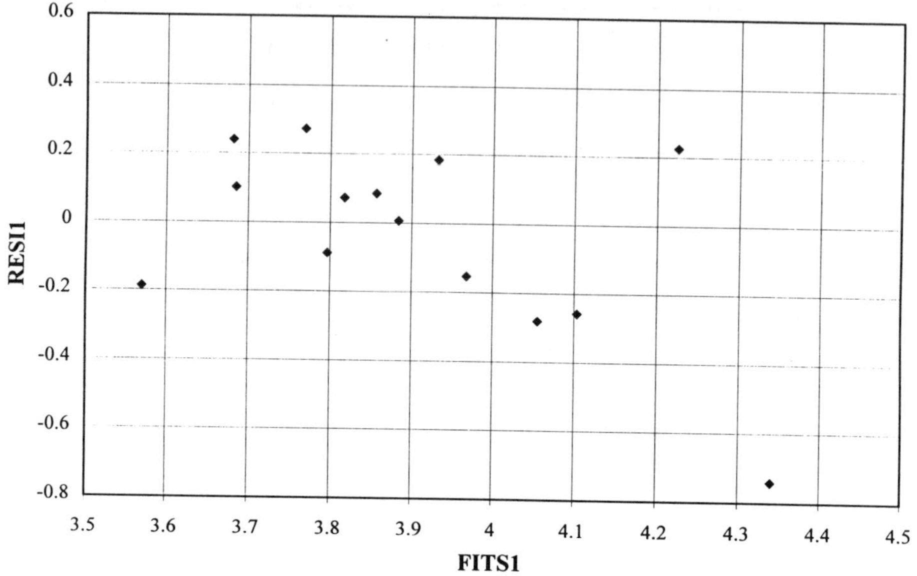

Figure 5

316°C %Weight Loss: Table of Means After 956 Hours

Polymer Grade 6F Dianhydride

Processing Time	No Kapton®	Kapton®	*Means*
15 minutes	4.358	5.104	*4.731*
2 hours	4.711	5.119	*4.945*
Means	*4.565*	*5.112*	***4.839***

Electronic Grade 6F Dianhydride

Processing Time	No Kapton®	Kapton®	*Means*
15 minutes	3.815	3.951	*3.883*
2 hours	3.903	3.704	*3.804*
Means	*3.859*	*3.828*	***3.843***

n = 3 for each cell; n = 6 for each margin; n = 12 for each Dianhydride type

Figure 6

Figure 7: Analysis of Variance for Percent Weight Loss

Source	DF	SS	MS	F	Approx P	Expected MS
Kapton	1	0.3978	0.3978	3.69	0.075	$\sigma^2 + 12\sum_i \alpha_i^2$
time	1	0.0269	0.0269	0.25	0.625	$\sigma^2 + 12\sum_j \beta_j^2$
Dianhydride	1	5.9382	5.9382	55.15	0.000	$\sigma^2 + 12\sum_k \gamma_k^2$
Kapton × time	1	0.2013	0.2013	1.87	0.193	$\sigma^2 + 6\sum_{ij} \alpha\beta_j^2$
Kapton × Dianhydride	1	0.5011	0.5011	4.65	0.049	$\sigma^2 + 6\sum_{ik} \alpha\gamma_k^2$
time × Dianhydride	1	0.1294	0.1294	1.20	0.292	$\sigma^2 + 6\sum_{jk} \beta\gamma_k^2$
Kapton×time×Dianhydride	1	0.0015	0.0015	0.01	0.907	$\sigma^2 + 3\sum_{ijk} \beta\gamma_{jk}^2$
Location	2	0.3319	0.1660	1.54	0.248	$\sigma^2 + 8\sum_\ell \delta_\ell^2$
Error	14	1.5075	0.1077			σ^2
Total	23	9.0357				

Linear Model for Iron Content

A Least Squares Fit of the Model

$$y_{ijk\ell} = \mu + \alpha_i + \beta_j + \gamma_k + \alpha\beta_{ij} + \alpha\gamma_{ik} + \beta\gamma_{jk} + \alpha\beta\gamma_{ijk} + \epsilon_{ijk\ell}$$

$y_{ijk\ell} =$ iron content (ppm) of the ℓth sample from the ith kapton level, j^{th} time level and k^{th} dianhydride type.

$\mu =$ overall constant; $\alpha_i =$ fixed kapton effect; $\beta_j =$ fixed time effect; $\gamma_k =$ fixed dianhydride effect

$\alpha\beta_{ij} =$ fixed kapton by time interaction effect; $\alpha\gamma_{ik} =$ fixed kapton by dianhydride interaction effect

$\beta\gamma_{jk} =$ fixed time by dianhydride interaction effect; $\alpha\beta\gamma_{ijk} =$ fixed kapton by time by dianhydride interaction effect

$\epsilon_{ijk\ell} =$ random error term where i, j, k are the same as in the % weight loss study

$\ell = 1, 2, 3$ replication of a sample for a particular kapton, time, dianhydride combination

Also, transforming the $y_{ijk\ell}$ with the natural logarithm, a least squares fit of the following model :
$$\ln y_{ijk\ell} = \mu + \alpha_i + \beta_j + \gamma_k + \alpha\beta_{ij} + \alpha\gamma_{ik} + \beta\gamma_{jk} + \alpha\beta\gamma_{ijk} + \epsilon_{ijk\ell}$$

Figure 8

Normal Probability Plot (No Transform): Iron Content

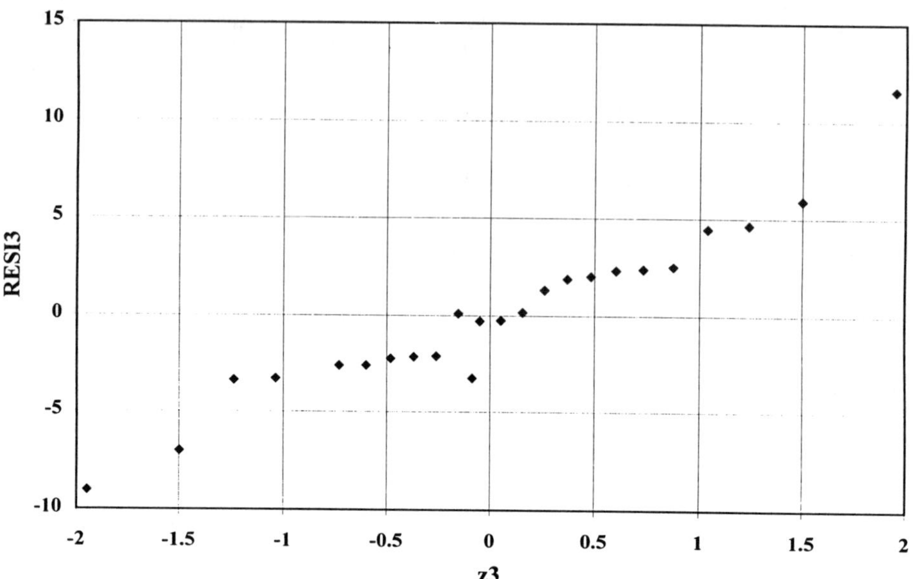

Figure 9

Residual Plot (No Transform): Iron Content

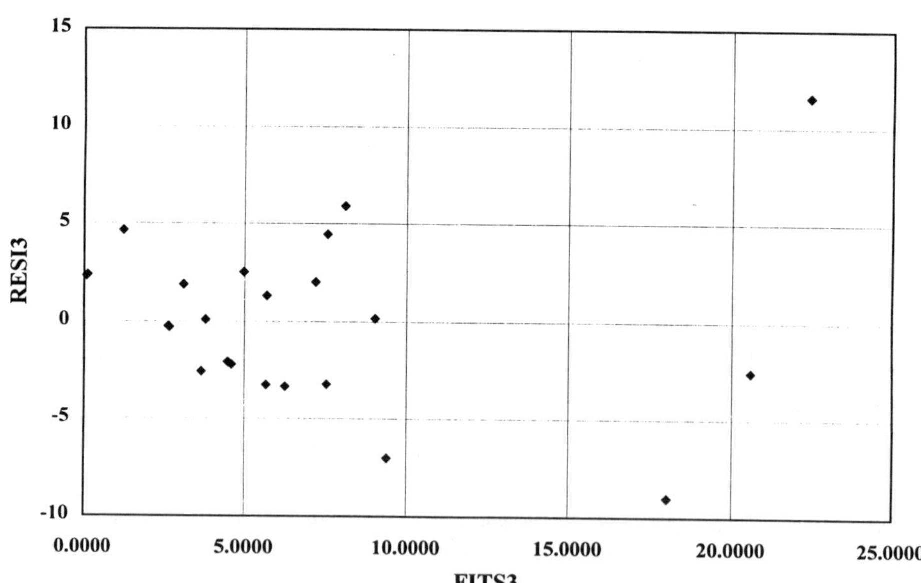

Figure 10

Natural Log of Iron Content[a] (ppm): Table of Means[b]

Polymer Grade 6F Dianhydride

Processing Time	No Kapton®	Kapton®	*Means*
15 minutes	1.8 (7.3)	1.7 (6.9)	*1.8 (7.1)*
2 hours	0.9 (2.4)	1.2 (3.6)	*1.0 (3.0)*
Means	*1.3 (4.9)*	*1.5 (5.2)*	*1.4 (5.1)*

Electronic Grade 6F Dianhydride

Processing Time	No Kapton®	Kapton®	*Means*
15 minutes	2.9 (20.3)	1.7 (5.4)	*2.3 (12.9)*
2 hours	1.3 (6.0)	0.9 (2.4)	*1.1 (4.2)*
Means	*2.1 (13.2)*	*1.3 (3.9)*	*1.7 (8.5)*

a) Values in ( ) are untransformed average iron contents measured in ppm

b) n = 3 for each cell; n = 6 for each margin; n = 12 for each dianhydride type

Figure 11

Normal Probability Plot: ln Fe Content

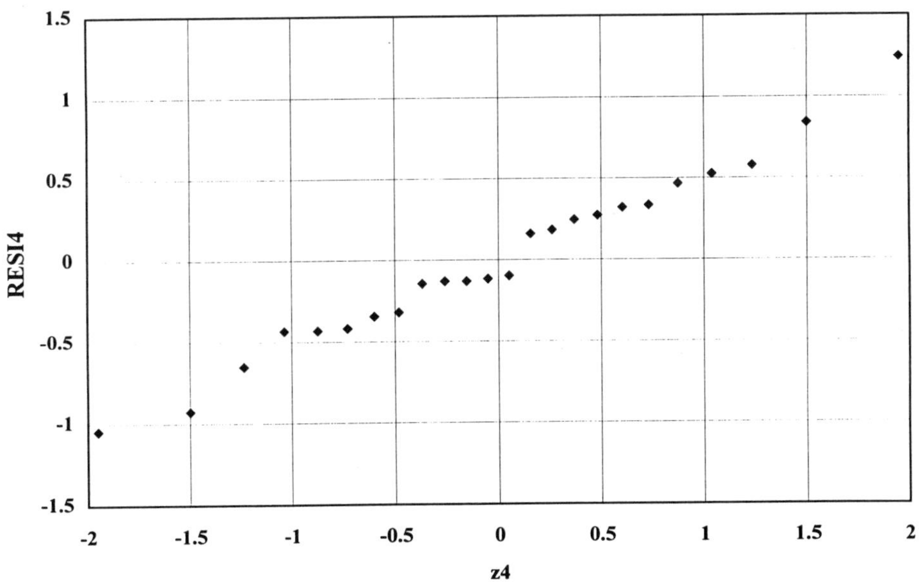

Figure 12

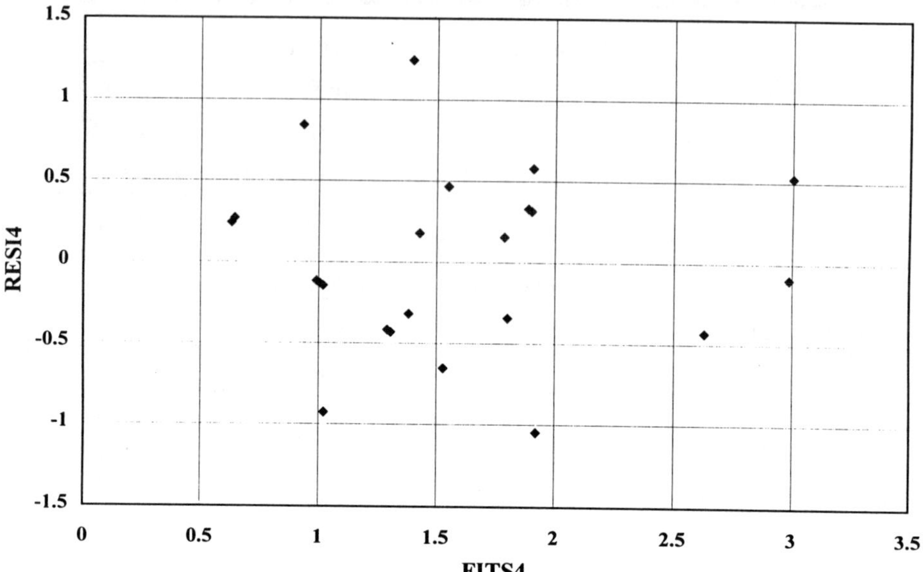

Residual Plot: ln Fe Content

Figure 13

CURE, ADHESION AND COMPOSITE APPLICATIONS OF SUBSTITUTED PHENYLETHYNYL TERMINATED POLY(ARYLENE ETHER SULFONE)S

A. Ayambem[#], S. J. Mecham[#], Y. Sun[#], T. A. Bullions[*], A. C. Loos[*] and J. E. McGrath[#]

[#]Department of Chemistry , [*]Department of Engineering Science and Mechanics
NSF Science & Technology Center: High Performance Polymeric Adhesives and Composites
Virginia Polytechnic Institute and State University
Blacksburg, VA   24061-0344

ABSTRACT

A series of poly (arylene ether sulfone) thermosets have been synthesized via oligomers endcapped with various phenylethynyl-moieties (each differing from the other systematically in the electron-withdrawing abilities of their attached substituents). The nature of the substituent was found to at once affect the electron density around the ethynyl bond as well as the terminal phenyl-ring carbon atoms. On the average, glass transition temperatures (Tg) and the modulus above the Tg of the resulting thermosets after curing were found to increase with increasing electron-withdrawing ability of the substituent on the endcapper. Upon evaluating the adhesive strengths of these materials by Ti-Ti single lap shear tests, adhesion values comparable to those of polyimides, as high as 5329 psi ($\pm$ 218), were recorded. Again, the values obtained increased with the electron-withdrawing ability of the starting oligomer. Preliminary application of these materials as a composite matrix for carbon fiber showed good consolidation and representative property values will be discussed at the meeting.

KEY WORDS: Phenylethynyl endcapped polysulfone, Adhesion, Resin Composite matrix.

1. INTRODUCTION

Poly (arylene ether sulfones) are normally linear high performance engineering polymers which display a variety of desirable properties including thermal and dimensional stability, toughness, retention of modulus at high temperatures and, radiation resistance. These combination of properties have made them applicable in many areas such as battery cases, appliances, and electrical conductors. In spite of these desirable properties however, they have been limited in their applicability in certain areas because of their relatively poor solvent resistance. For instance, the fact that they undergo crazing and cracking in the presence of such organic liquids as jet fuel, hydraulic fluids (tricresyl phosphate), and de-icing fluids (ethylene glycol) [1-6] make them unsuitable for use as structural adhesives and composite matrices in the aerospace industry. Reactive poly (arylene ether sulfone) oligomers such as (I) have been synthesized with phenylethynyl end-groups acting as sites for network formation to circumvent this drawback. [7]

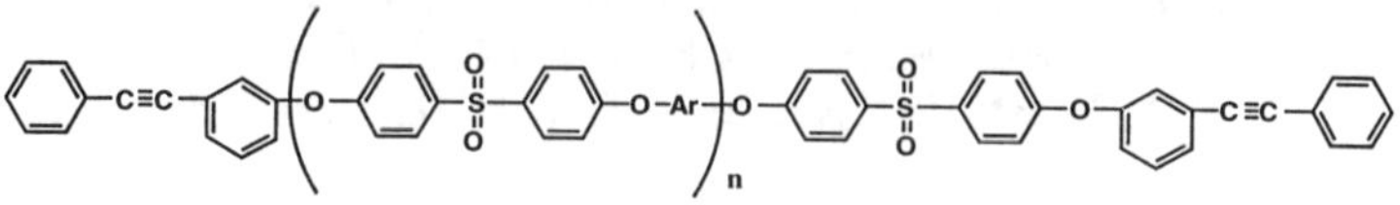

(I)

The resulting thermosets exhibited high solvent resistance in addition to many of the desirable properties of the corresponding high molecular weight thermoplastic material. The choice of the phenylethynyl moiety as endcapper was partially defined by its ability to undergo curing without the evolution of volatiles, thereby eliminating voids, which reduce adhesive and other mechanical properties. In addition, the phenylethynyl group has been shown to undergo curing at very high temperatures (above 350°C) [7-9] and, since the glass transition temperatures of the oligomers is generally around 200°C, there is a desirably large processing window for these materials which is critical for structural adhesives and polymer matrix composites.

Previous studies of cured phenylethynyl-terminated poly (arylene ether sulfone) oligomers showed glass transition temperatures which were only about 15°C above those of the high molecular weight thermoplastics. [10-11] This was in contrast to phenylethynyl-terminated polyimides which displayed glass transition temperatures that were as high as 47°C above that of control high molecular weight thermoplastics. [12] These differences could either be a function of the type of polymer, or of the electronic environment around the phenylethynyl group or, a contributive effect from both factors. Although the exact mechanism is not known, the curing reaction is generally believed to proceed by a combination of oligomeric chain extension, branching, and crosslinking. Different electronic environments around the reactive phenylethynyl endgroup (e.g electron donating or withdrawing) may therefore influence the curing reaction towards a particular direction, facilitating one process over the other. This could invariably affect the molecular weight between crosslinks (M_c) of the resulting thermoset and by extension, the glass transition temperature and certain mechanical properties of the resulting material. Previous studies examining the effect of the nature of the substituent on the curing characteristics of the phenylethynyl group [13] revealed that electron-withdrawing substituents not only reduced the cure temperatures, but also accelerated the curing process as well. The importance of the glass transition temperature lies in the fact that, to be applicable for some high-performance applications e.g as aerospace structural adhesives, the material would have to be durable at the regular operating temperatures which for supersonic aircraft can be between 177-204°C for 60,000 hours. [14] It has been proposed that for a material to withstand these conditions, it would have to have a minimum Tg of about 250°C. The literature values for linear high molecular weight bisphenol-A based polysulfone thermoplastic is 190°C while that for the biphenol-based thermoplastic is 230°C. [15-16]

The present paper describes research which has further investigated the role of the electronic environment of the phenylethynyl endgroup of poly (arylene ether sulfone) oligomers on the thermal transitions, adhesion properties and suitability composite matrix applications.

2. EXPERIMENTAL

2.1 Materials: The bisphenol-A was obtained in monomer grade purity from Dow Chemical and dried at 80°C overnight under vacuum before use. The activated halide, 4,4'-Dichlorodiphenylsulphone (DCDPS), was obtained from Amoco Chemical and was dried under vacuum at 80°C overnight before use. The biphenol was procured from Aldrich Chemicals and was also dried at 80°C under vacuum overnight prior to use. 3-aminophenol was obtained from Aldrich Chemicals and sublimed before use. Potassium Carbonate was purchased from Fisher Scientific and dried at 130°C under vacuum before use. The Phenylethynyl phthalic anhydride was synthesized by the reaction of 4-Bromo-phthalic anhydride with phenylacetylene. [16] N-Methyl-pyrollidone (NMP) and o-Dichlorobenzene (o-DCB) were purchased from Fisher Scientific and made anhydrous by drying over calcium hydride and distilling under vacuum before use. Toluene was also purchased from Fisher Scientific and was used as received. 3-Phenylethynyl phenol (PEP) was synthesized via the coupling of 3-iodo phenylacetate and phenyl acetylene. the acetate group was subsequently cleaved with base to afford the 3-phenylethynyl phenol. [17] Pure samples of 4-Fluoro-4'-Phenylethynyl-4-Benzophenone (FPEB) were kindly donated by Dr. John W. Connell (NASA Langley) and were used after drying under vacuum at 80°C overnight. 4-Fluoro-4'-Phenylethynyldiphenylsulfone (PEFDPS) was synthesized by first reacting 4-fluorobenzenesulfonylchloride with bromobenzene giving 4-Bromo-4'-Fluorodiphenylsulfone which was then reacted with phenylacetylene to give the desired product. [17] 3-Phenylethynylphthalimidophenol (PEPIP) was synthesized by the reaction of 3-phenylethynylphthalic anhydride with 4-aminophenol. [17]

2.2 Characterization: The oligomers were characterized by Nuclear Magnetic Resonance (NMR), Thermal Gravimetric Analysis (TGA), Differential Scanning Calorimetry (DSC), Gel Permeation Chromatography (GPC), Infrared Spectrometry, and lap shear tests to measure adhesive properties of the thermosets. The number average molecular weight $<Mn>$ were also obtained by end-group analysis through quantitative ^{1}H and ^{13}C-NMR. Quantitative ^{13}C-NMR spectra were run for 120 minutes with a spectral width of 25,000 Hz and a relaxation delay of 18.8 seconds, measured on a 400 MHz Varian Unity instrument. Elemental Analyses were determined by Galbraith Laboratories, Inc. A Perkin-Elmer TGA-7 was employed to determine weight loss behaviour in air. A Perkin-Elmer DSC-7 was used to identify the thermal transitions of the oligomers and cured network materials. DMA analysis were carried out by three point bending on a Perkin Elmer DMA 7e instrument at a frequency of 1 Hz and a scanning rate of 3°C/min from -150°C to 0°C, and at 5°C/min from 0°C to 350°C. Gel fractions were obtained from soxhlet extraction of the cured sample for 96 hours with chloroform. The extracted sample was then dried under vacuum at 215 - 220°C for 24 hours. The Lap shear adhesion samples were prepared by producing an E-glass cloth impregnated preform containing 85% (wt.%) polymer which was then cured at 75 psi in a hot press at 370°C for one hour between two chromic acid treated Ti (6/4) adherends. An Instron 1213 was used to test the samples following the ASTM-D1002 method with a crosshead speed of 0.05 in/min. Rheology measurements were carried out on a RMS-800 using parallel plate geometry. Discs were pressed from the uncured oligomer and trimmed to fit the 25 mm diameter of the parallel plates. Once loaded and heated to an initial temperature of 260°C, the plates were closed to a gap of 1 mm and the excess material scraped from the sides of the plates. Storage and loss moduli as well as complex viscosity were measured during a 3°C/min ramp from 260 to 390°C in a nitrogen environment. The properties were measured in the dynamic mode with an oscillation frequency of 1 rad/s. Composites were made using unsized 18-ply 3 x 3" AS-4-3K plain weave carbon fibers, each ply separated from the next by a measured quantity of uncured oligomer. Consolidation and curing was carried out between platens of computer interfaced hotpress with pressure fluctuating between 1.2 to 1.4 MPa. The mold temperature was ramped from room temperature to 200°C at 5°C/min followed by a ramp at 3°C/min to 370°C. The mold was then held at 370°C for 120 minutes before cooling to room temperature at 4°C/min. The composite panel was imaged ultrasonically via a pulse-echo arrangement using a Sonix 10 MHz transducer with a focal length of 38.1 mm.

2.3 Polymer Synthesis: A representative synthesis was conducted as follows; Into a four-necked, 250 mL round-bottomed flask fitted with a reverse Dean-Stark trap, a mechanical stirrer, a nitrogen inlet, and a thermocouple port was added 7.040 g (0.031 moles) of bisphenol-A, 8.000 g (0.028 moles) of DCDPS, and 4.908 g (0.036 moles) of K_2CO_3. NMP (55 mL) and toluene (30 mL) were then added and the reaction flask immersed in an oil bath. The reaction temperature was then raised to 145°C where it was maintained with stirring for 6 hours then raised to 160°C for 12 hours. Next, 2.004 g (0.006 moles) of the endcapper, PEFDPS, was added along with toluene (30 mL). The reaction was continued for 3 hours at 145°C, and for another 6 hours at 160°C. Work-up involved quickly filtering the salt by-product, then precipitating the hot filterate into methanol The resulting precipitate was then dried under vacuum at room temperature, then at elevated temperature.

3. RESULTS AND DISCUSSION

A series of 5,000 g/mole bisphenol-A and biphenol based oligomers of poly (arylene ether sulfone) endcapped with four different phenylethynyl moieties which contained various electron-withdrawing substituents have been synthesized by the reaction of a bisphenol with 4,4'-dichlorodiphenyl sulfone according to Scheme 1. In general the polymerization reactions followed the one-pot procedure outlined in the experimental section, but the synthetic approach for the 4-PEPIP endcapped polysulfone had to be somewhat modified. ^{13}C-NMR analysis of the product of a reaction by the one-pot approach using 4-PEPIP as endcapper did not show any ethynyl carbon peaks. An investigatory reaction between 4-PEPIP and potassium carbonate alone under identical reaction conditions gave primarily phenylethynylphthalic anhydride, indicating that the phthalimide ring undergoes base hydrolysis under those reaction conditions. To circumvent this, a two-step approach was employed by reacting the bisphenol, DCDPS and 3-aminophenol in one step, followed by the reaction of the amine-terminated product with a slight excess of phenylethynyl phthalic anhydride (in NMP/o-DCB as imidization solvent). The endcap thus became 3-PEPIP rather than 4-PEPIP, if 4-aminophenol had been used in this modified reaction.

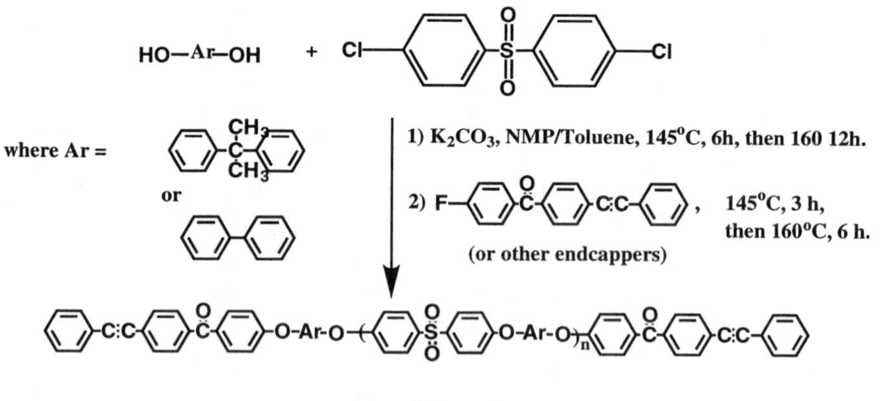

Scheme 1.

The functionally different endcappers employed were PEP, FPEB, PEFDPS, and PEPIP with the phenylethynyl phenol moiety being used as the reference endcapper.

The resonance positions of the ethynyl carbons in the ^{13}C-NMR gave good indication as to the nature of their electronic environment (Figure 1). Thus, the trend from PEP to FPEB and PEFDPS down to PEPIP was that the ethynyl carbon atom (C_β) closest to the substituent group (and which is therefore expected to be most influenced by it) progressively shows a higher chemical shift, i.e, comes to resonance at increasingly downfield positions. This observation was considered to be a good measure of the electron-withdrawing abilities of each substituent group, increasing in the order; ether< carbonyl< sulfone< imide.

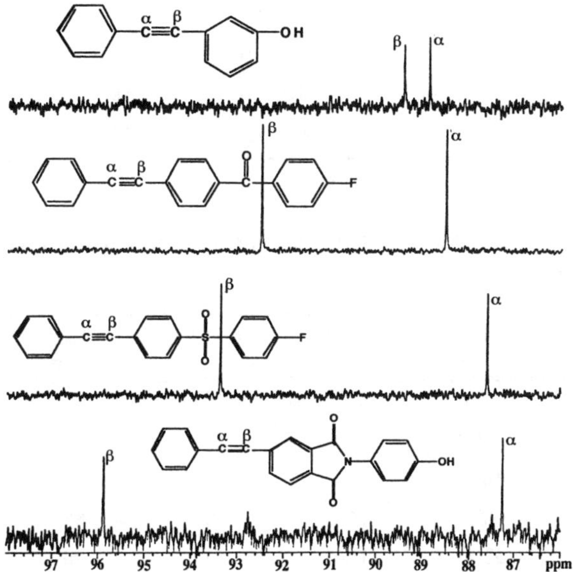

Figure 1. ^{13}C-NMR spectra of ethynyl region of the endcappers. From top to bottom, 3-PEP, FPEB, PEFDPS, AND PEPIP.

3.1 Molecular Weight And Thermal Analysis: Molecular weight determinations were carried out by GPC and by quantitative ^{13}C NMR analysis with results shown in Table 1. In general, although the molecular weights computed by end-group analysis were higher than those obtained from GPC measurements, the numbers were all within the error levels of the two measuring techniques.

Table 2 shows that the Tg's of the oligomers all fell within the same range. This is expected since any variation can be attributed to the slight differences in molecular weights. The higher Tg's of the biphenol-based oligomers would be, a consequence of the more rigid chains resulting in greater order in the material. It is also observed that the gel fractions and Tg values of the cured thermosets resulting from the 3-PEPIP terminated polysulfones appeared to be anomalous when the general trend is considered. It is possible that the phenylethynyl-imide endcap has a slightly greater preference for the chain-extension cure mechanism when compared with the other endcappers. This would result in lower gel fractions and also, lower Tgs.

Table 1. Molecular weight Characterization for phenylethynyl poly(arylene ether sulfone) oligomers (target = 5,000 g/mol)

Oligomer/Endcap	M_n^1 (GPC) g/mol	M_w (GPC) g/mol	PDI	M_n ^{13}C - NMR
Bis A/ 3-PEP	4500	9800	2.18	5500
Bis A/ FPEB	4600	8300	1.84	6200
Bis A/ PEFDPS	4000	7000	1.76	5500
Bis A/ PEPIP	6000	9900	1.74	6000
BP/ 3-PEP	5300	9300	1.75	5300
BP/ FPEB	6700	11,100	1.65	7900
BP/ PEFDPS	6800	10,700	1.58	7700
BP/ PEPIP	6000	9200	1.52	5000

[1] Run in NMP at 60°C.

Table 2. Thermal analysis and gel fractions for phenylethynyl poly(arylene ether sulfone) oligomers

Oligomer	M_n^1 (GPC) g/mol	Oligomers T_g (°C)	5% wt loss, °C (TGA)[3]	%gel fraction[4]	T_g (°C)[5] cured
Bis A/ 3-PEP	4500	153	491	91	198
Bis A/ FPEB	4600	166	514	97	206
Bis A/ PEFDPS	4000	162	498	97	215
Bis A/ PEPI-3-P	6000	165	504	85	200
BP/ 3-PEP	5300	184	549	90	234
BP/ FPEB	6700	20	542	90	240
BP/ PEFDPS	6800	206	525	91	243
BP/ PEPI-3-P	6000	191	521	88	235

[1] in NMP at 60°C, Target Mn= 5,000g/mole

[2] second heat, run at 10°C/min after heating to 280°C, then quenching.

[3] run at 10°C/min.

[4] extracted with $CHCl_3$ for 96h and dried at 215-220°C for 24h.

[5] second heat, run at 10°C/min after heating to 400°C, then quenching.

TGA results (Figure 2) showed the expected greater thermo-oxidative stability of the biphenol-based systems relative to the isopropylidene containing Bisphenol-A systems.

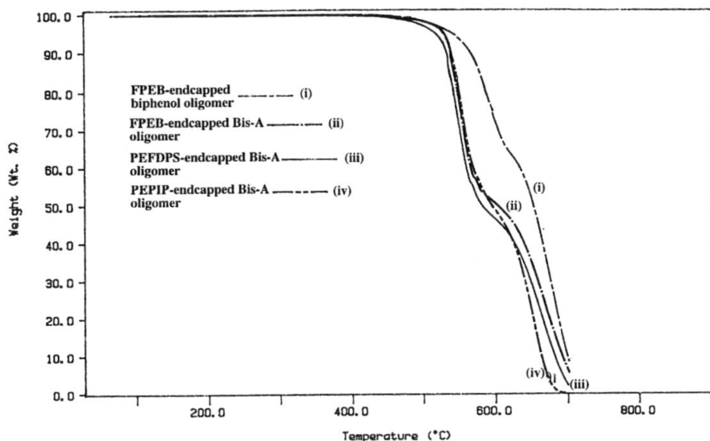

Figure 2.   TGA thermogram of phenylethynyl terminated poly(aryelene ether sulfone)

Curing was conducted in an aluminum mould under nitrogen atmosphere at 370°C for 2 hours. The cured materials all formed creasable films and gel fractions were highest for the bis-A-based thermosets. The highest Tg obtained for the cured networks was 243°C (for the biphenol-based PEFDPS endcapped polymer), nearly as high as the 250°C target. One observes that the Mn obtained by GPC in that case is 6800 g/mole. It is possible that if the target 5,000 g/mole oligomer Mn were obtained, the Tg could approach, or even surpass the goal of 250°C, because of the higher network densities. From the tabulated results, the pattern that develops is that the Tg of the thermoset appears to increase in the order of electron-withdrawing ability of the substituent in the end-group, the only anomaly being the imide endcapped polymers.

The single lap shear adhesion to titanium results are presented in Table 3. for the phenylethynyl terminated polysulfone thermosets. The sample for the bisphenol A based 3-phenylethynylphenol themoset was prepared using a Pasa Jel surface treatment which has been

Table 3.   Single Lap Shear Results for Cured Oligomers with Different Phenylethynyl Endcappers

Bisphenol	Endcapper	Adhesive Strength* (psi)
Bisphenol A	PEP	3850 ± 470**
	FPEB	4680 ± 260
	PEFDPS	4890 ± 520
	PEPIP	5090 ± 510
4,4'-Biphenol	PEP	4090 ± 250
	FPEB	4890 ± 580
	PEFDPS	5330 ± 220
	PEPIP	5230 ± 310

* average of 4 samples, cured at 370°C/120 min., chromic acid anodized, cohesive failure.
** avgerage of 4 samples, cured at 370°C/90 min., pasa jel treatment, cohesive failure.

shown to provide consistently lower adhesion strengths than the chromic acid anodized surface treated samples. The relatively lower adhesion strengths of the phenylethynylphenol based materials is pronounced. A consistent difference is present in the carbonyl and sulfone based materials as well and the imide based material is about the same as the sulfone even though the gel fractions and T_gs are lower. Failure of the adhesive bonds were cohesive and the DMA results for the Bis A-based thermosets shown in Figure 3 do correlate well with the observed trend in adhesion strengths.

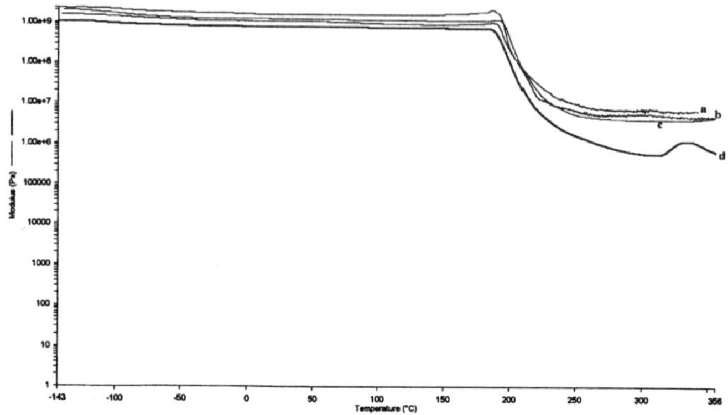

Figure 3. DMA thermogram of Bis-A based poly(arylene ether sulfone) thermosets; a) PEFDPS terminated, b) FPEB terminated, c) PEPIP terminated, d) PEP terminated.

The modulus above T_g (which is proportionally related to the crosslink density of the material [18-19]) for each of the thermosets did follow the order of adhesion strength. Thus, the 3-PEP endcapped material which had the lowest adhesion strength had the lowest modulus above the T_g. These lap shear strengths are much higher than any previously determined for the phenylethynyl polysulfone materials[10] and with values in the range of 5,000 psi, they compare very well with phenylethynyl terminated imide materials. [12] This indicates that an increased reactivity and crosslink density could have a large effect on the adhesion strength to metal adherends.

3.2 Rheology Studies And Composite Application: The rheological results giving complex viscosity, storage and loss moduli as a function of temperature of the ~7,000 g/mol biphenol-based PEFDPS phenylethynyl-terminated polysulfone are shown in Figure 4. The thermogram shows that the complex viscosity attains a minimum value of 13.718 Pa.s at about 358°C and is not much higher at 370°C. A 3x3, 18-ply composite panel was made by hot-pressing using this particular oligomer as resin. The selection of this oligomer was informed by the very good adhesion and dynamic mechanical results shown in Tables 2 and 3 and in Figure 3. In spite of the relatively high complex viscosity (a value of about 5 Pa.s is generally ideal for hot-press composite manufacture), c-scan results showed that the general characteristics of the panel were uniform across the entire surface. The consolidation was also very good.

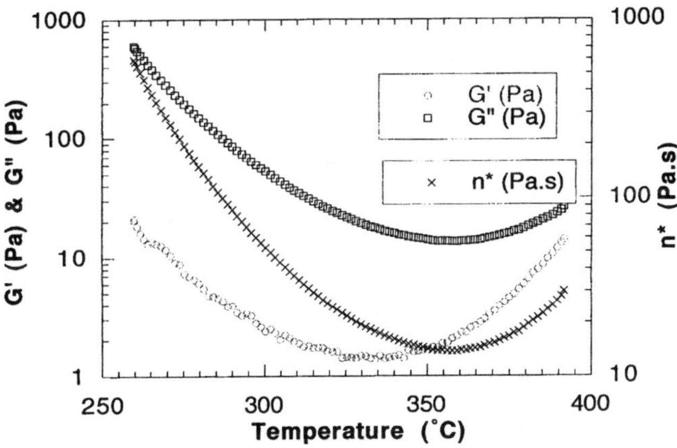

Figure 4. Rheological thermogram for PEFDPS phenyethynyl-terminated polysulfone

4. SUMMARY AND FUTURE STUDIES

Poly (arylene ether sulfones) terminated with various phenylethynyl groups, differing in the electron-withdrawing abilities of the substituents on the end-group have been synthesized and characteized. It has been established that the nature of the phenylethynyl endgroup, specifically, the electron-withdrawing ability of an attached substituent plays a major part in the glass transition temperatures, modulus above Tg, and adhesion strengths of the resulting thermosets. Application of these oligomers as resins for composite applications shows promise and mechanical evaluations of these composites are underway. Continuing efforts are geared towards employing these oligomers as resins in the powder-coating approach to composite manufacture.

5. ACKNOWLEDGMENTS

The authors would like to thank the Adhesive and Sealant Council Education Foundation and the NSF Science and Technology Center: High Performance Polymeric Adhesives and Composites (DMR-9120004) for the financial support of this work.

6. REFERENCES

1. R. N.Johnson, and A. G. Farnham, <u>J. Polym. Sci. A-1</u>, 5, 2415 (1967).
2. J. E. McGrath, A. G. Farnham, and L. M. Robeson, <u>Polym. Prep.</u>, 16(1), 476-482 (1975);
3. L. M. Robeson, A. G. Farnham and J. E. McGrath, in D. J. Meier, Ed., <u>Midland Macromolecular Institute Monographs</u>, 4, Gordon and Breach, 1978, 405-526.
4. R. Viswanathan, B. C. Johnson and J. E. McGrath, <u>Polymer</u>, 25, 1827 (1984).

5. J. L. Hedrick, D. K. Mohanty, B. C. Johnson, R. Viswanathan, J. A. Hinkley, and J. E. McGrath, J. Poly. Sci.: Polym. Chem. Ed., 23, 287 (1986).

6. P. M. Hergenrother, Angew. Chem. Int. Ed. Engl., 29, 1262 (1990).

7. S. Mecham, V. Vasudevan, S. Liu, M. B. Bobbitt, S. Srinivasan, A. C. Loos and J. E. McGrath, Polym. Mater. Sci. Engin. Proceed., 74, 61 (1996).

8. G. W. Meyer, B. Tan and J. E. McGrath, High Perf. Polym., 6, 423 (1994).

9. P. M. Hergenrother and J. G. Smith Jr. Polymer, 35(22), 4857 (1994).

10. S. J. Mecham, M. M. Bobbitt, M. E. Pallack, V. Vasudevan, and J. E. McGrath, Adh. Soc. Proc., 211 (1997).

11. S. J. Mecham, M. M. Bobbitt, M. E. Pallack, V. Vasudevan, and J. E. McGrath, Adh. Soc. Proc., 33 (1997).

12. B. Tan, C. N. Tchatchoua, V. Vasudevan, L. Dong, and J. E. McGrath, Polym. Mater. Sci. Engin. Proc., 74, 59 (1996).

13. J. A. Johnson, F. M. Li, F. W. Harris, and T. Takekoshi, Polymer, 35(22), 4865 (1994).

14. P. Hergenrother, Trends Polym. Sci., 4 (4), 104 (1996).

15. R. J. Cotter, Engineering Plastics: A Handbook of Polyarylethers:, Gordon and Breach Publishers, 1995.

16. J. L. Hedrick, D. K. Mohanty, B. C. Johnson, R. Viswanathan, J. A. Hinkley, and J. E. McGrath, J. Polym. Sci.: Polym. Chem. Ed., 23, 287 (1986).

17. S. Mecham, PhD Dissertation, VPI & State University, 1997.

18. G. Levita, S. De Petris, A. Marchetti, A. Lazzeri, J. Mater. Sci., 26, 2348 (1991).

19. E. Urbaczewski-Espuche, J. Galy, J. Gerard, J. Pascault, and H. Sautereau, Polym. Eng. Sci., 31(22), 1572 (1991).

43rd International SAMPE Symposium
May 31-June 4, 1998

SHORT CARBON FIBER-REINFORCED PMR-15
POLYIMIDE COMPOSITES

Shi Y. Yang*, Sheng Q. Gao, Jia Z. Li
Institute of Chemistry, Chinese Academy of
Sciences, Beijing 100080, China

ABSTRACT

Short carbon fiber-reinforced polyimide composites were prepared by wet impregnating chopped short carbon fibers with PMR-15 polyimide matrix resin solution followed by evaporating the solvent with strong stirring. The processing properties of the composites molding powders were investigated. The mechanical properties and thermo-oxidative stability of the composite laminates fabricated were also studied. Experimental results demonstrate that the short fiber filled molding powders exhibit good melt resin flowing characteristics and can be proceed by hot press molding technique to produce small-size, complicated composite structures which are difficult to be produced using long fiber reinforced composite prepregs by autoclave technique. Some small components, bearing hoses, were fabricated and tested for the potential applications in aeropropulsion systems.

KEY WORDS: Polyimides   Carbon Fibers   Composites   Hot Press   Laminates

1. INTRODUCTION

Thermosetting PMR polyimides which were synthesized using the process known as <u>in-situ</u> Polymerization of Monomer Reactants (PMR) are easier to process high quality composite structures than the corresponding thermoplastic polyimides(1-4). Carbon fiber-reinforced polyimide composites were usually produced by impregnating reinforced carbon fibers with PMR matrix resin followed by polymerization through crosslinking of the nadic end-capped groups and cyclodehydration to produce imide rings in polymer chains. The superior processability combined with great mechanical properties and excellent retention of mechanical properties at elevated temperature make them very attractive thermo-oxidative stable materials used at temperature as high as 320-400 ^{o}C for the application in advanced aeropropulsion systems(5-8). However, PMR polyimide composites has some notable shortcomings which severally limit their applications. One of them is the infeasibility to fabricate small-size, complicated structures due to the inadequate melt flowing properties of the prepregs. Thick, cylindrical structures are also difficulty to process due to the poor thermal conductivity of the molten fluid. To overcome these drawbacks, chopped short carbon fibers has been investigated in this laboratory as the reinforcing materials in replacing of the corresponding long carbon fibers in PMR polyimide composites, in which the short fibers were expected to be flowing easily with the molten matrix resins in the thermal curing process so that hot press molding technique may be suitable for the fabrications of some small and complicated parts.

The purpose of this study was to investigate the formulations of PMR-15 polyimide composites reinforced with short carbin fibers and to determine the effect of the formulations on processing characteristic, mechanical properties and thermo-oxidative stability of the composites.

2. EXPERIMENTAL

2.1 PMR-15 Neat Resin Chemistry   PMR-15 matrix resin solution was prepared using PMR process at room temperature by mixing the monomer solutions in anhydrous ethyl alcohol to form a 50 % solid solutions (Figure 1). The monomers employed are monoethyl ester of 5-norbornene-2,3-dicarboxylic acid(NE), 4,4'-methylenediamine (MDA), and diethyl ester of 3,3',4,4'-benzophenonetetracarboxylic acid(BTDE) which was prepared by refluxing a suspension of the corresponding dianhydride in anhydrous ethyl alcohol until all solids had been dissolved and then continuing for an additional two hours. The mole ratios of reactants NE:BTDE:MDA for PMR matrix resin is 2.000:2.087:3.087 which yields theoretically an oligomer with a calculated molecular weight of 1500.

Neat resin molding powder was prepared by placing matrix resin solution (100 g) into an air circulating oven set at 120 ^{o}C until almost all solvent had been evaporated and then dried at 220 ^{o}C at vacuum for 2 hrs to complete the imidization of the end-capped prepolymer.

2.2 Preparation of Composite Molding Powders   7.5 g of chopped T300 carbon fiber (length: 5-10 mm, diameter of 7-8 μm) was added to 85 g of 50 wt. % PMR matrix resin solution. Then the mixture was stirred mechanically for 2 hrs at room temperature in nitrogen. The solvent (ethyl alcohol) was distilled with strong stirring to produce a viscous materials in which about 15% of ethyl alcohol remained. The viscous materials was then dried in an air circulating oven at 100 ^{o}C for 4 hrs, 140 ^{o}C for 2 hrs, 200 ^{o}C for 4 hrs and 220 ^{o}C for 2 hrs, successively, to yield an imidized prepolymer molding powder with 15 wt. % of T300 fibers (Abbr. KH304-15).

KH304-7.5 and KH304-30 molding powders which contain 7.5 wt. % and 30 wt. % of T300 carbon fibers, respectively, were also prepared in the same procedure.

2.3 Melt Flowing Index(MFI) Melt flowing index of the composite molding powders were measured with XRT400 melt resin flowing index instrument in accordance with GB3682-83/ ASTM D1238-73, in which the diameter of mould exit in the sample tube was ∅1.18 mm and the loading was 7.056 kg.

2.4 Composite Laminate Fabrication The imidized prepolymer molding fine powder (50g) was placed into a 10 cm diameter metal die at room temperature. Then the die was placed into a press preheated at 220 ^{o}C. When the die temperature reached 280 ^{o}C, a pressure of 3-4 M Pa was applied; After 1 min., the pressure was released to ensure the trace of low molecular weight molecules being escaped. The pressure was applied again after 10 seconds. This pressure releasing/applying cycles repeated three times. Then the die temperature increased at a rate of 3 ^{o}C /min. to 320 ^{o}C under a pressure of 3-4 M Pa. After curing of 2 hrs at 320 ^{o}C , the die was allowed to cool under pressure to 200 ^{o}C, then the pressure was released. The composite laminate was removed from the die at room temperature.

2.5 Laminate Evaluation Prior to testing, all laminates were inspected for porosity using either ultrasonic C-scan inspection or photomicrographs of the cross sections. Flexural tests were performed on 0.600 cm wide specimens in accordance with GB1449-87 at span to depth ratios of 15-16 at rate of 1.0 mm/min.. Tensile strength tests were performed on 0.600 cm wide specimens in accordance with GB1447-83 at rate of 5.0 mm/min.. Impact strength with un-notched specimens were carried out on Izod instrument in accordance with GB1451-83.

2.6 Isothermal Aging of Composite Laminates The laminate samples (0.6 x 0.4 x 6.0 cm^3) were placed in a circulating air oven at 320 ^{o}C with an inlet air flow of 100 ml/min. All samples were weighed after aging for scheduled time intervals.

3. RESULTS AND DISCUSSIONS

3.1 Preparation of the Composite Molding Powders Figure 1 shows the chemistry for synthesis of PMR-15 polyimides. The solvent used in this study was a mixture of ethyl alcohol with some additives instead of methyl alcohol for the health and environmental concerns arising from the use of methyl alcohol. Ethyl alcohol as the solvent of PMR-15 usually causes phase separation of the resin solution, especially in winter season, due to its inadequate dissolubility. This problem can be resolved by adding some additives in ethyl alcohol. The key feature for preparing short carbon fiber-reinforced PMR polyimide composites was to impregnate matrix resin on the fiber surface uniformly. Hence, fiber aggregation or cluster due to the different gravities between carbon fibers and matrix resin solution should be avoided in the impregnation process. The method employed in this study was to immerse the chopped short fiber into the resin solution and then to remove the solvent by evaporation with mechanical stirring. A homogeneous viscous material is obtained after removing of 80-85% of the solvent, in which no phase separation occur. A series of composite molding powders were prepared from PMR-15 resin and T300 fibers. Figure 2 shows the SEM photographs of the molding powders (KH304-15). It can be seen that the single short carbon fibers were coated completely with the matrix resin and no fiber clusters or aggregation was observed.

3.2 Melt Flowing Characteristic of the Molding Powders PMR polyimide neat resins possess good melt flowing properties which ensure them outstanding processing characteristics(1). The molding powders comprised of short carbon fiber and PMR polyimide resin also exhibit good melt flowing characteristic. Figure 3 shows the effect of temperature on MFI of the molding powder (KH304-15) in the range of 270-295 ^{o}C. The experimental results demonstrate that the melt flowing behaviors was influenced intensively by temperature. First, the rate of MFI increases when temperature increases. For instance, it needs 7 minutes for the molding resin to reach a maximum MFI value at 270 ^{o}C while the time at 295 ^{o}C is only 1 minute. The higher the temperature employed, the shorter the time to reach the maximum MFI. Second, the summit width of the curve which corresponds with the lifetime of melting resin with good flowing characteristic reduced as temperature increased. MFI were determined by two factors: the melting of the matrix resin and the crosslinking of the endcap groups. MFI increases with the resin melting and decreases with the progress of the crosslinking. Third, the maximum MFI which corresponds to the lowest viscosity of the molten resin increased from 0.7 at 270 ^{o}C to an almost constant values of 3.3-3.5 above 280 ^{o}C (Figure 4), indicating that the molding powders melted completely at temperature of above 280 $^{\circ}$C. Hence, the suggested temperature used in the composite processing is in the range of 280-290 $^{\circ}$C where the MFI of the resin is high and the lifetime of molten resin with high MFI values is long enough so that it is easy to control the processing.

3.3 Composite Laminate Processing. Two factors which must be considered in the laminate processing are temperature and pressure. As suggested above, the temperature in the composite laminates fabrication was started at 280-290 $^{\circ}$C and completed at 320 $^{\circ}$C. The temperature lower than 280 $^{\circ}$C can not produce a molten resin with high flowing character to fill the whole die, while the temperature above 290 $^{\circ}$C will accelerate the crosslinking of the end-capped groups so that a proper pressure is difficulty to be applied. The factor that when and how much a pressure being applied in the molten fluid is also important in the processing. Inadequate pressure application may cause either the spillover of the molten resin from the die or the composite laminate produced with high void contents. The pressure employed in the processing of KH304-15 is 3-4 MPa in the method as described in Experimental section.

3.4 Laminate Studies Ultrasonic C-scanning of the composite laminates revealed that all the laminates reinforced with T300 fiber from 7.5-30 wt. % were of adequate quality and showed least amount of void formation (< 2%). Table 1 lists the mechanical properties of the three laminates: KH304-7.5, KH304-15 and KH304-30 along with PMR-15. KH304-15 laminate exhibits the balanced mechanical properties: flexural strength: 130 MPa; flexural modulus: 7.7 GPa; tensile strength: 98 MPa, tensile modulus: 3.9 GPa. DMA experiment indicates that KH304-15 laminate has a glass transition temperature (Tg) of 355 $^{\circ}$C. The density of the laminate is 1.55 g/cm^3.

Figure 5 shows the 320 $^{\circ}$C thermo-oxidative stability(TOS) in flowing air (100 ml/min.) of the three composite laminates along with the chopped neat T300 fiber and PMR-15 neat resin. It can be seen that the TOS of the laminates increases with T300 loadings. The weight losses of the laminates with 15 and 30% of T300 fibers are 4.3 and 3.8 %, respectively, after 240 hrs air exposure at 320 $^{\circ}$C, while that of neat T300 fiber and neat PMR resin are 6.5 and 6.0, respectively, under the same conditions. The results that the composites exhibit better TOS than that of either the fiber or the resin may be explained by the synergistic effect of matrix resin with reinforced materials in composite. A new region was formed in the interface region between polymer matrix resin and reinforced fibers in the composite, which impacts the thermo-oxidative stability of the composite. When the strength of fiber-resin bonds in the interface region is strong, air is difficulty to permeate into this region, hence the composite can withstand a long period of oxidative degradation. This is in accordance with the result reported by Bowles and Nowak(9), in which PMR-15/Celion 6K composite losses weight at a lower rate than either the matrix resin or the reinforced fibers.

Figure 6 is the SEM photograph of cross section of the laminate (KH304-15), in which the fibers were surrounded with a continuous matrix resin phase. The length to diameter(l/d) ratios of the short fiber in composite laminate was measured in the range of 30-35, which exceed the critical l/d ratio value of 20 for carbon fibers being as reinforced material in composites that ensures the effective transference of stress from matrix resin into carbon fibers.

3.5 Composite Structures for Potential Applications   As mentioned in Introduction section, the purpose of this study was to develop a short carbon fiber-reinforced PMR polyimide composite materials, of which the composite molding powders possesses good melt flowing characteristic to produce some small-size, complicated components by hot press molding technique. Figure 7 shows the photograph of a bearing hose with diameter of 9.0 mm and length of 14.0 mm. The parts produced by hot press molding has smooth surface and good mechanical properties, which can be used at temperature of 320 °C for more than hundred hours servicing in air environments. Various composite structures have also been fabricated in this laboratory using the short fiber-reinforced PMR polyimide composites investigated for the potential applications of aeropropulsion systems.

4. Conclusion

Uniform PMR polyimide composite molding powders, prepared by impregnating the chopped carbon fiber with PMR matrix resin solution and followed by removing the solvent with strong mechanical stirring, possess good melt flowing characteristics and processing properties. Composite laminates fabricated by hot press molding technique exhibit outstanding mechanical properties and good thermo-oxidative stabilities. Some small-size, complicated composite components with good quality also be fabricated using this technique, which can be used as thermo-oxidative stable parts at temperature as high as 320 °C for the potential aircraft engine applications .

5. REFERENCES

1. T.T. Serafini, P. Delvigs and G.R. Lightsey, <u>J. Appl. Polym. Sci.</u>, <u>16</u>, 905(1972).
2. R.H. Pater, <u>SAMPE J.</u>, <u>30</u>(5), 29(1994).
3. D. Garcia and T.T. Serafini, <u>J. Polym. Sci., Phys. Ed.</u>, <u>25</u>, 2275(1987)
4. J.C. Johnston, M.A.B. Meador and W.B. Alston, <u>J. Polym. Sci., Polym. Chem. Ed.</u>, <u>25</u>, 2175(1987)
5. K.J. Bowles, D. Jayne, T.A. Leonhardt, <u>J. Adv. Mater.</u>, <u>October</u>, 23(1993)
6. R.D. Vannucci, <u>SAMPE Q.</u>, <u>19</u>(1), 31(1987)
7. J.A. Hinkley and J.B. Nelson, <u>J. Adv. Mater.</u>, <u>April</u>, 45(1994)
8. P.M. Hergenrother and N.J. Johnson, <u>Resin for Aerospace</u>, C. A. May, Ed. ACS Symposium Series, <u>132</u>, 1(1980)
9. K.J. Bowles and G. Nowak, <u>J. Compos.. Mater.</u>, <u>22</u>, 966(1988)

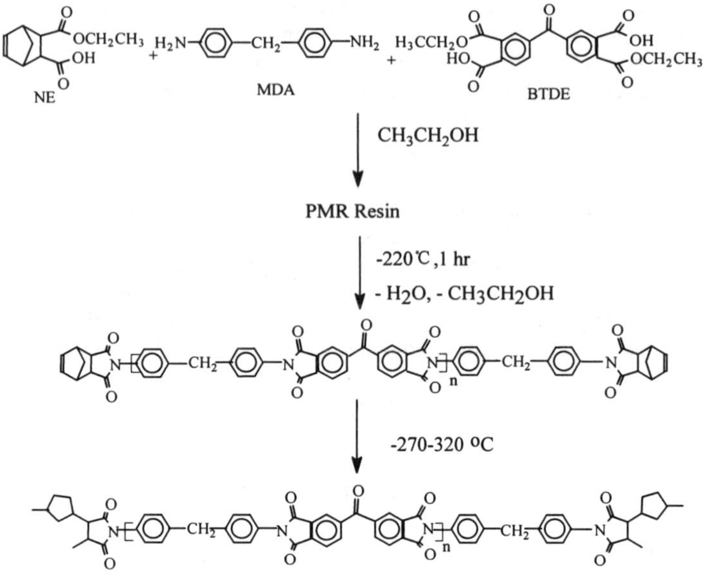

Figure 1 The chemistry of PMR-15 polyimide

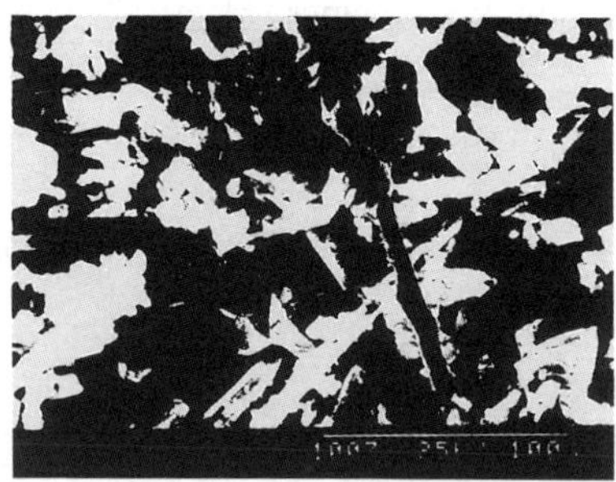

Figure 2   SEM photograph of the molding powder
prepared from PMR-15 and short T300
fiber (KH304-15)

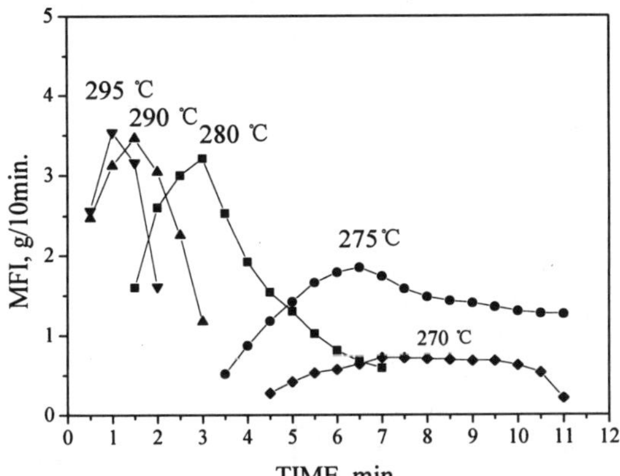

Figure 3　Melt flowing index of molding powder
prepared from PMR-15 and 15 wt. % of
short carbon fiber(KH304-15)

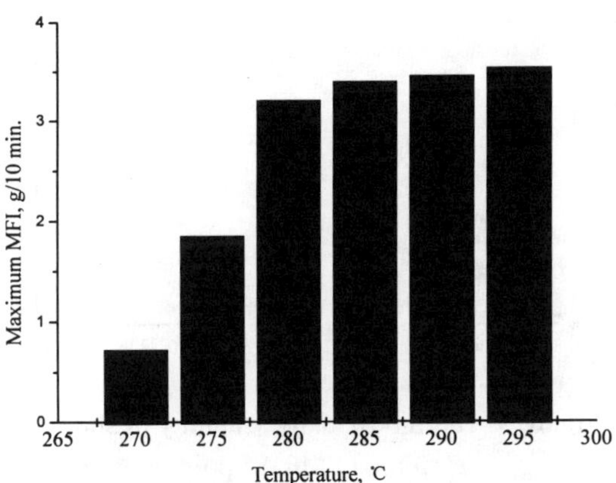

Figure 4　Effect of temperature on maximum
MFI of molding powder prepared
from PMR-15 and 15 wt.% of short
carbon fiber(KH304-15)

Table 1 The physical properties of short carbon fiber-reinforced
PMR-15 composites

	PMR-15	KH304-7.5	KH304-15	KH304-30
Decomposition temperature, ^{o}C	>400	491	497	490
Decomposition temperature at 5 wt.% of weight loss	466	490	495	505
Flexural Strength, MPa	-	106	130	117
Flexural Modulus, GPa	-	5.1	7.7	8.1
Tensile Strength, MPa	39	73	98	54
Tensile Modulus, GPa	3.9	2.4	3.9	3.3
Elongation at Breakage	-	3.4	2.7	2.3
Impact Strength, KJ/m^2	2	11	11	4.5
Weight loss in air, % 110 hrs/1 atm	4.1	3.6	2.0	1.9
250 hrs/1 atm	5.5	5.3	4.5	3.8
CTE, x $10^{-6}/^{o}C$ (20-250 ^{o}C)	43-46	46-54	52-58	48-53
Density, g/cm^3	1.33	1.45	1.55	1.76
Tg, ^{o}C	284	321	328(TMA) 355(DMA)	332

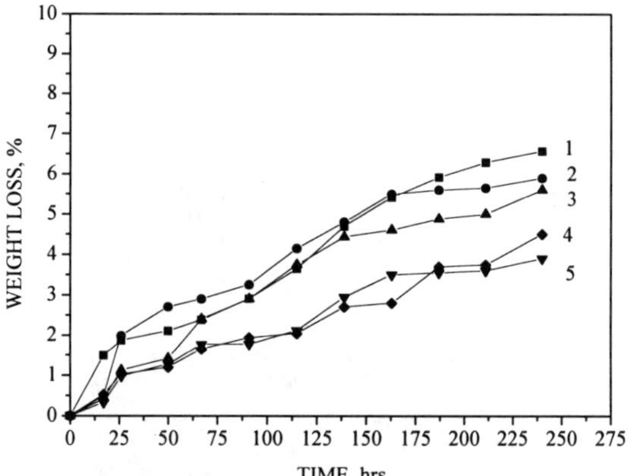

Figure 5 Thermo-oxidative stability of short fiber-
reinforced PMR-15 polyimide composite
laminates during 320℃ air exposure

1: T300 fiber; 2: PMR-15; 3: KH304-7.5
4: KH304-15;   5: KH304-30

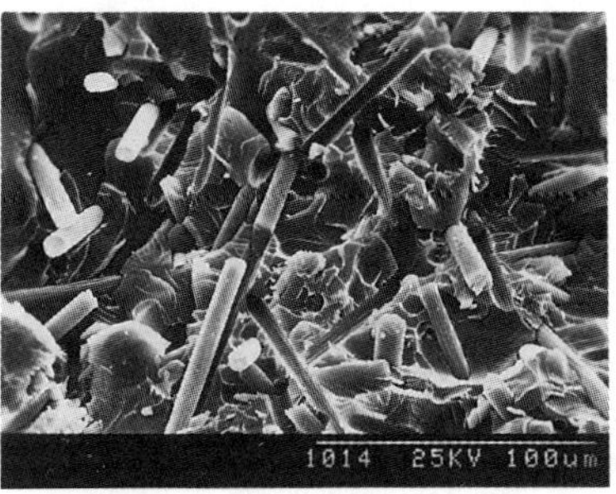

Figure 6   SEM photogragh of the cross section of
the composite laminate (KH304-15)

Figure 7   Photograph of the bearing hose

USE OF ULTRASONIC TAPE LAMINATION FOR IN-PROCESS DEBULKING OF THICK COMPOSITE STRUCTURES

Margaret E. Roylance, Foster-Miller, Waltham, MA
David K. Roylance, Department of Materials Science and Engineering, MIT, Cambridge, MA
Kathryn L. Nesmith, NAWCAD, Patuxent River, MD

ABSTRACT

Foster-Miller has developed an advanced processing technique for organic matrix composites called Ultrasonic Tape Lamination (UTL). This technique employs ultrasonic energy to insonify composite tape (either thermoplastic or B-staged thermoset) to induce controlled and virtually instantaneous viscoelastic and frictional heating in the composite. UTL can be used for in-process debulking and controlled staging of thick filament wound or fiber placed composite structures. During UTL the ultrasonic horn functions as a localized, efficient, and quick response heat source, ideally suited for a feedback control system. The horn heats only the area with which it is in direct contact, and ultrasonic loading can enhance fiber nesting to optimize debulking. With the use of UTL on-the-fly debulking during filament winding or tape placement, a net shape part can be realized without time consuming and expensive repetitive debulking steps. Problems associated with inadequate debulking, including fiber waviness (or marcelling), fiber shifting, resin migration, disbonds and voids, can be eliminated. Higher quality thick composite parts can be achieved with minor process modifications, resulting in a superior and less costly part.

KEYWORDS: Composites, Curing , Modeling

1. INTRODUCTION

The use of ultrasonic welding of metals and unreinforced polymeric materials is an important industrial process, and has been used successfully for many years. The use of ultrasound for the consolidation of polymer matrix composites containing more than 35-40% by volume of reinforcing fiber was viewed as impossible (1) until Foster-Miller developed innovative techniques for ultrasonic tape lamination (UTL) of these materials. Working in cooperation with NAWCAD, Foster-Miller has demonstrated the use of UTL for both consolidation of thermoplastic matrix composite materials and debulking of B-staged thermoset prepreg. The theoretical similarity between the two processes has been confirmed based on experiment and analysis. In both thermoplastic consolidation and thermoset debulking without staging, the ultrasonic loading must generate sufficient heat to inducing flow and provide sufficient support for consolidation. One of the singular advantages of UTL technology is that heating and cooling occur very rapidly and in a more controlled manner than during a hot gas process. Due to this inherent controllability of the process, UTL debulking of B-staged prepreg might also be used to stage the material in a controlled way. In addition, ultrasonic loading can enhance fiber nesting to optimize consolidation.

Figure 1 shows the Foster-Miller UTL head mounted on a filament winder. The horn configuration and stress applied to the composite material is critical to achieving full consolidation without inducing damage in the material. The data in Figure 2 help to clarify the importance of applied stress state to the process. Figure 2 shows the bulk (K) and shear (G) relaxation moduli for a polymeric material (in this case polyisobutylene) on a time scale reflecting the glass transition. The intensity of the relaxation and hence the amount of heat dissipated is much greater in shear or distortional loading than in bulk or dilatational loading. When the relative magnitude of the shear component of loading compared to the hydrostatic pressure is optimized, the material is heated viscoelastically with minimum fiber disruption. This can be accomplished without the use of so-called energy directors, or raised areas which must be formed on the surface of the material.

2. NATURE OF THE UTL PROCESS

Controlled use of UTL for debulking and staging of a reactive material requires a thorough understanding of the UTL process. Careful observation and monitoring of the process was undertaken to clarify the nature of the UTL process and to provide a basis for the simplifying assumptions required for numerical modeling.

One very important question which must be answered in order to develop a quantitative model of heat generation and consolidation in these materials is the relative importance of frictional and viscoelastic effects. These materials are relatively highly damping and significant viscoelastic heating will certainly occur under ultrasonic loading. The extent to which frictional heating occurs as well will change the magnitude and location of heat generation. This is due to the fact that frictional heating should generate high temperatures at the interfaces between horn and tape and between tape and part, whereas viscoelastic heating will occur throughout the volume of the affected material. In an effort to clarify the relative importance of frictional and viscoelastic heating, a study was performed of the effect of static pressure on the heat generation in AS4/ PEEK tape.

Figure 1. Ultrasonic Tape Lamination during filament winding

191

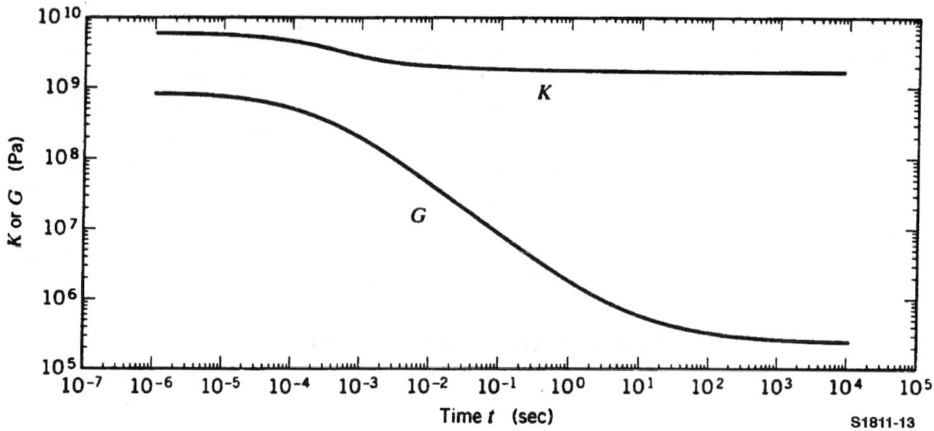

Figure 2. Bulk (K) and shear (G) relaxation moduli of polyisobutylene (from reference 2)

To clarify the relative importance of frictional and viscoelastic heating, the effects of pressure and amplitude on ULT-induced heating were measured. A thermocouple was embedded in the center of a thermoset composite substrate four inches in length. The temperature was measured during UTL of four plies of AS4/PEEK tape. The advance rate of the ultrasonic horn was constant at 2.54cm (1 inch)/minute and an ultrasonic frequency of 40 kHz was used for all the measurements. Figure 3 shows the effect of static pressure and amplitude on the measured heating of the AS4/PEEK tape during insonification.

The data in Figure 3 shows that increasing the static pressure has no observable effect on the heat generation in the material during UTL. If frictional heating were dominating the process, increasing the static pressure should increase the rate and magnitude of heat generation. This result suggests that the viscoelastic heating is dominating the UTL heat generation. It also indicates that the static pressure is sufficient to maintain contact throughout the ultrasonic loading cycle, or the viscoelastic heating would also be a function of static pressure level.

This conclusion is in agreement with the results of Tolunay et al. (3) that, for soft polymers, the interface did not have a significant effect on the amount of heat dissipated during ultrasonic welding of unreinforced materials. They observed that in this case the heating occurs over the whole volume. Further, they observed that intensive heating of the material began only after a certain temperature was reached. This temperature most probably corresponds to the glass transition temperature, since as Tg is approached, the level of viscoelastic energy dissipation as measured by loss modulus increases markedly. As the material continues to heat above Tg, the loss modulus drops again, and Tolunay et al. observed that the heating rate also generally drops until the temperature remains constant. The very rapid heating which occurs at surface asperities should soften the material in these regions quickly and allow for good uniform contact between the horn and the surface of the material.

Based on these results, our model focuses on viscoelastic energy dissipation as a volumetric heat source rather than frictional heating at the interfaces.

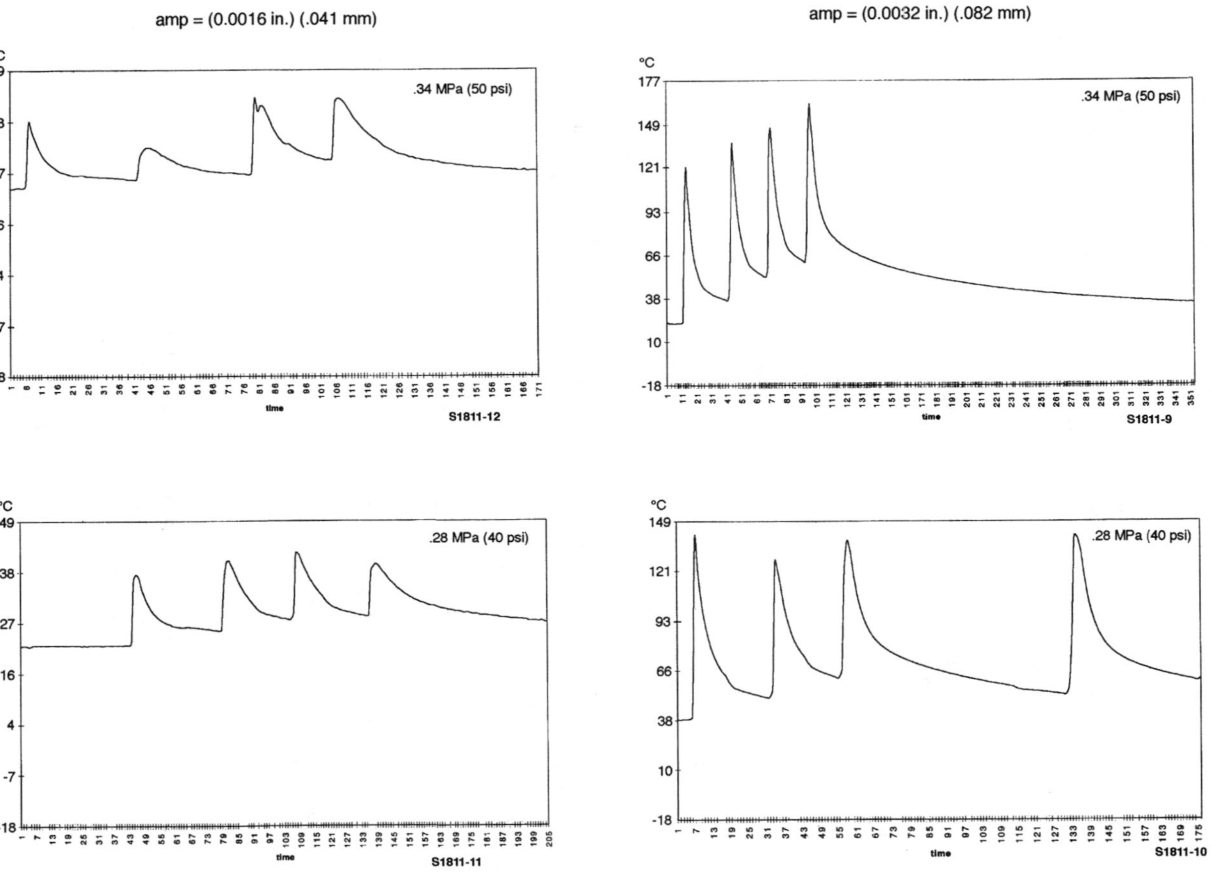

Figure 3. Effect of static pressure and amplitude on heating during insonification

3. FINITE ELEMENT MODELING OF UTL

3.1 Computational Model   We have chosen to employ a finite element analysis (FEA) developed by Roylance et al. (4) to model nonisothermal reactive processing operations, and which Benetar et al. (5) have used to predict the behavior of AS4/PEEK during attempted ultrasonic welding using energy directors. One advantage of this model is that it includes the effect of chemical reaction, which will allow us to use it to predict the extent of cure of reactive species in the B-staged prepreg. There are many experimental parameters in ultrasonic heating, to include horn angle, contact pressure, frequency and amplitude, presence of attenuating layers, and others. This makes optimization by purely experimental trials difficult and time-consuming, and computer modeling is useful in this regard.

The FEA code models the equations governing the nonisothermal flow of a reactive fluids, listed in standard texts on transport phenomena and polymer processing (e.g. References 6,7). These are the familiar conservation equations for transport of momentum, energy, and species:

$$\rho\left[\frac{\partial u}{\partial t} + u\nabla u\right] = -\nabla p + \nabla(\eta\nabla u)$$

$$\rho c\left[\frac{\partial T}{\partial t} + u\nabla T\right] = Q + \nabla(k\nabla T)$$

$$\left[\frac{\partial C}{\partial t} + u\nabla C\right] = R + \nabla(D\nabla C)$$

Here u, T, and C are fluid velocity (a vector), temperature, and concentration of reactive species; these are the principal variables in our formulation. Other parameters are density (ρ), pressure (p), viscosity (η), specific heat (c), thermal conductivity (k), and species diffusivity (D). The ∇ operator is defined as:

$$\nabla = \frac{\partial}{\partial x}, \frac{\partial}{\partial y}$$

Q and R are generation terms for heat and chemical species respectively, while the pressure gradient ∇p plays an analogous role for momentum generation. The heat generation arises from viscous dissipation and from reaction heating:

$$Q = \tau : \dot{\gamma} + R(\Delta H)$$

where τ and $\dot{\gamma}$ are the deviatoric components of stress and strain rate, R is the rate of chemical reaction, and ΔH is the heat of reaction. R in turn is given by a kinetic chemical equation; in our model we have implemented an m-th order Arrhenius expression:

$$R = k_0 \exp\left(\frac{-E^*}{R_g T}\right) C^m$$

where k_o is a preexponential constant, E^* is an activation energy, $R_g = 8.31$ J/mol-K is the Gas Constant, and m is the reaction order. The material-dependent parameters in this expression (k_o, E^*, and m) must be determined by experimentation which is able to monitor the reaction as a function of time and temperature. An example of reaction kinetics modeling suitable for this purpose has been presented by Roylance (3,4).

In treating ultrasonic curing, we take the velocities u in the governing equations to be displacements, and suppress the advective flow terms (e.g. $u\nabla u$). In this approach, the dilatational and deviatoric components of stress are considered separately, which leads to convenience in handling incompressible materials such as rubber. It also simplifies the computation of internal dissipative heating, which arises principally from the deviatoric components of stress. Once the deviatoric stresses and strains τ and γ have been computed, the dissipated heat per unit time Q is computed as

$$Q = \omega f \tau \gamma$$

where ω is the cyclic frequency and f is the fraction of strain energy dissipated per cycle. The dissipation factor f is known from dynamic mechanical testing; it is 2π times the tangent of the phase angle between cyclic stress and strain measured in these tests.

The analysis operates in iterative fashion, first solving the force equilibrium equations to determine the distribution of displacement, strain and stress generated in the body by the application of the imposed displacements at the surface. This heat dissipation associated with the viscous loss is used in the heat transfer equation, which is solved for an updated estimate of the temperature in a second iteration. In a third iteration the temperatures obtained in the previous step are used to update the rate of the chemical curing reaction. The equations are strongly coupled, with properties such as the shear modulus and chemical reaction rate depending on temperature, and the temperature depending on both viscous and reaction heating. The computer iterates repeatedly until convergence is reached. The code can also use successive iterations as time steps, so that transient situations can be modeled as well.

3.2 Results of modeling Figure 4 illustrates a typical modeling result. Figure 4a shows a stress distribution (here the vertical stress σ_y) for the case of a horn applying a vertical displacement to a portion of the upper surface of a layered specimen of cured composite and unreinforced polymer. Chemical reaction is not included in this simulation. The stresses are concentrated near the point of application, so the heat generation rate is highest here. However, heat will also be carried to other regions of the specimen by conduction. The temperature field arising from this oscillatory stress field is shown in Figure 4b. These stress and temperature contours were generated from normalized numerical results.

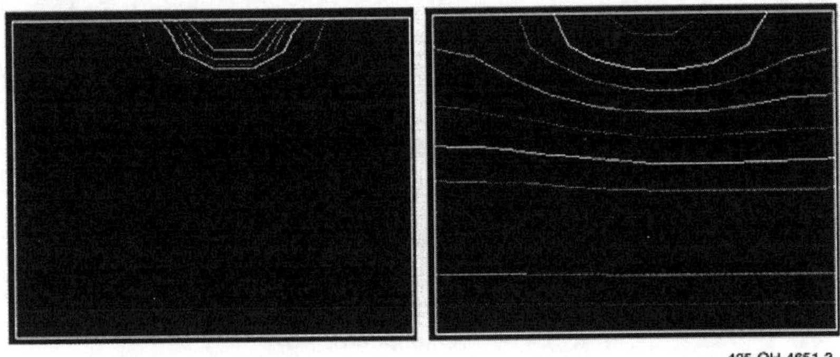

a) Stress distribution   b) Temperature distribution

Figure 4. Typical FEM results

One way of viewing the effect of ultrasonic heating is to plot the variation of temperature along a vertical line directly beneath the horn as a function of time. In Figure 5 below, the code was set to simulate a multilayer laminate consisting of unreinforced polymer with fiber reinforced composite layers on the top and bottom. A silicone rubber attenuation layer separates the upper face sheet from the ultrasonic horn. The lower composite sheet extends from $y = 0$ to $y = 0.00075$ m (it is convenient to operate the code in SI units), the polymer extends from $y = 0.00075$ to 0.00204 m, the upper face sheet extends from $y = 0.00204$ to 0.00279 m, and the silicone sheet extends from $y = 0.00279$ to 0.00359 m.

3.3 Model development   To predict consolidation of composite materials such as AS4/PEEK or B-staged thermosets, the local compressive stress and temperature of the material can be predicted. The stress and temperature required for consolidation can be experimentally determined and the model can be used to identify the ultrasonic parameters which produce these conditions. In order to extend this analysis to staging of prepreg, kinetic parameters appropriate to the epoxy curing reaction must be introduced, and the effects of the curing exotherm on the material temperature calculated. As described above, the model currently has the capability for such analysis, but the kinetic parameters for epoxy reactions have not yet been incorporated.

The extent of consolidation or debulking achieved by UTL in AS4/PEEK has been determined by density and microscopy measurements. Experimental results have shown that good consolidation of AS4/PEEK can be achieved with UTL. Apparent interlaminar shear strength of this material by short beam shear was 82.7 Mpa (12 KSI). This value is within the normal range for autoclave consolidated AS4/PEEK. Results such as these should be correlated with the ultrasonic parameters and the FEA simulations to verify the ability of the FEA model to predict consolidation.

It is clear from the governing equations that accurate modeling of UTL debulking and staging of thermoset composites requires accurate information concerning several materials parameters. These include modulus and log decrement as a function of temperature and frequency, as well as kinetic parameters for one or more epoxy curing reactions.

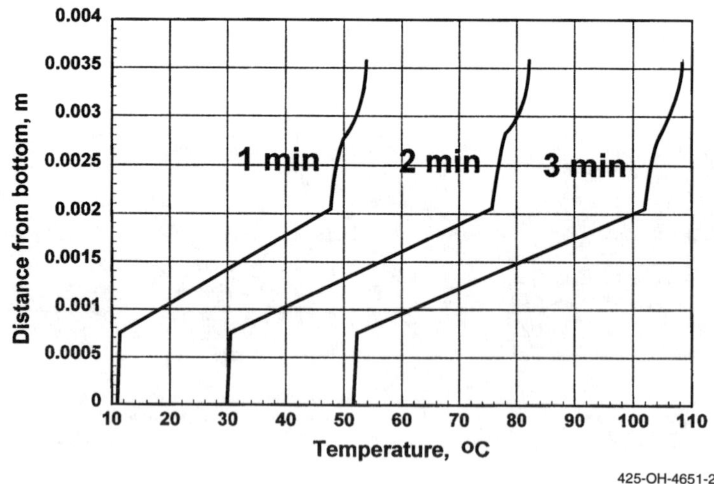

Figure 5. Predicted FEM temperature profiles

Some kinetic data for important epoxy and polyimide curing reactions are available in the literature (8,9). Some other information, such as full characterization of the effect of degree of cure on the dynamic mechanical properties of the curing material must be obtained experimentally. Torsion braid analysis (TBA) can be used to measure rigidity and loss tangent of B-staged prepreg resins during cure. An example of TBA of Cytek Fiberite 977-3 during both a slow heating cycle and staged cure at 125°C is shown in Figure 6. A series of TBA runs of this kind, coordinated with differential scanning calorimetry (DSC) to determine residual exotherm and extent of cure, would be required to characterize the dynamic mechanical properties which would control material behavior during UTL staging.

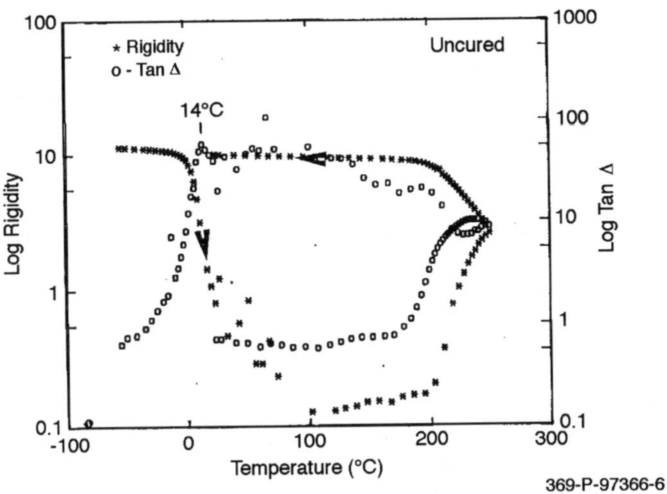

a) TBA results on B-staged IM7/977-3 prepreg at 2.5°C/min

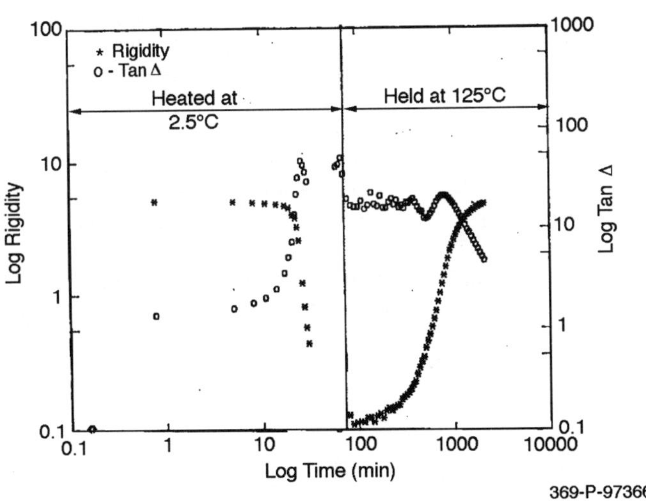

b) TBA of staging of B-staged 977-3 prepreg at 125°C

Figure 6. TBA of 977-3

197

4. CONCLUSIONS AND RECOMMENDATIONS

The overall goal of this program was to establish UTL as a feasible approach for in-process debulking and staging of composite structures, and to identify measurable and controllable parameters that can be used for automated control of UTL during automated fiber placement or filament winding. This required development of a more thorough understanding of the nature of the process as it applies to both thermoplastic and B-staged thermoset matrix composites. Quantification of viscoelastic heating and extent of cure of the thermoset matrices were identified as central to the controlled use of UTL.

In pursuit of this goal, a modeling approach has been identified and implemented which is capable of incorporating both viscoelastic heating and chemical reaction, and is thus generally applicable to both thermoplastic and thermoset matrix composites. Future work should seek to extend the FEA to include epoxy curing reactions, and to verify the analysis during model experiments monitored using techniques such as thermometry, dielectrometery and fiberoptic FTIR.

The verified FEA model could then be used to assist in optimization of in-process debulking and staging during automated fiber placement. The model would be used to predict ultrasonic parameters such as amplitude, frequency and horn pressure which will produce consolidation. It could also predict the extent of cure under these conditions, and how far the curing reaction could be advanced during UTL debulking.

5. REFERENCES

1. <u>Ultrasonic Plastics Assembly</u>, Branson Ultrasonic Corporation, 1979
2. <u>Mechanics of Materials</u>, David Roylance, John Wiley & Sons, 1996
3. Tolunay, M.N., Dawson, P.R. and Wang, K.K., "Heating and Bonding Mechanisms in Ultrasonic Welding of Thermoplastics", Polymer Engineering and Science, Vol. 23, No. 13, 1983, pages 726-733
4. Roylance, D., L. Chiao, and P. McElroy, "Versatile Finite Element Code for Polymer Composite Processing," *Computer Applications in Applied Polymer Science, II: Automation, Modeling, and Simulation,* ACS Symposium Series, No. 404, pp. 270 - 280, 1989.
5. Benetar, A. and Gutowski, T.G, "Ultrasonic Welding of Thermoplastic Composites", SPE ANTEC, 1989, pages 507-510
6. Bird, R.B.; Stewart. W.E.; Lightfoot, E.N. *Transport Phenomena,* John Wiley & Sons, New York, 1960.
7. Middleman, S. *Fundamentals of Polymer Processing,* McGraw-Hill Co., New York, 1977.
8. Hagnauer, G.L. and Pearce, P.J., "Size Exclusion Chromatography Analysis of Epoxy Resin Cure Kinetics", ACS Symposium Series, No. 245, ACS, 1984
9. Kenny, J.M. and Trivisano, A., "Isothermal and Dynamic Reaction Kinetics of High Performance Epoxy Matrices", Polymer Engineering and Science, Volume 31, No. 19, 1991, Pages 1426-1433

ON CURE INDUCED STRESSES IN LOW CURE TEMPERATURE THERMOSET POLYMER COMPOSITES

MOHAMED S. GENIDY AND
MADHU S. MADHUKAR
The University of Tennessee
Department of Mechanical and Aerospace
Engineering and Engineering Science
Knoxville, TN 37996-2210

JOHN D. RUSSELL
Air Force Research Laboratory
Materials and Manufacturing Directorate
Wright-Patterson AFB OH 45433-7750

ABSTRACT

Low temperature thermal curing resins (such as LTM-45 ELD and X343-59) are being considered by the aerospace industry for developing prototype parts. By curing at low temperatures, the parts can be cured outside of the autoclave. Therefore, expensive autoclave hardened tooling is not required. Tooling materials such as wood and foam can be used instead. This will dramatically reduce the cost of producing prototype parts. The effect of the standard cure cycles on the cure induced stresses in composites containing such resins is not characterized yet. Most of the cure induced stresses are developed after the polymer reaches the gel point. Any subsequent polymer volume change may produce significant fiber stresses. Stress relaxation acts to reduce stresses developed during cure. Relaxation time increases significantly after the polymer is gelled and thus stress relaxation becomes less effective to reduce cure induced fiber stresses. A closed loop feedback control system has been developed to monitor the cure induced fiber stresses and to determine optimum cure cycle to minimize these stresses. In this system, a polymer is cured along a given length of the pretensioned fiber. The cure induced stresses are determined by measuring the fiber tension change. This method was applied to LTM-45 ELD and X343-59 resins during standard two stages cure cycles and during a typical autoclave cure cycle used for 3501-6 epoxy. The cure induced fiber stresses were monitored during these cycles. In addition, the feedback control was used to determine optimized cure cycles to minimize these stresses.

1- BACKGROUND

Since the early 1990's, the Air Force has been studying technologies to lower the costs of advanced composite airframe structures. In these days of producing fewer aircraft in a production run, and of developing more technology demonstrator vehicles, the cost of tooling becomes an increasing percentage of the cost of the parts, since the tooling costs cannot be spread out over many parts. One method to reduce tooling costs is to cure parts outside of the autoclave. This would allow the use of very inexpensive tooling materials: wood and foam. During the Air Force program "Low Cost Composite Processing," McDonnell Douglas (now Boeing St. Louis) worked with resin suppliers to develop resins to cure outside of the autoclave around 60°C. These parts could then undergo a freestanding oven post cure to complete the cure. Boeing St. Louis has continued this work under the DARPA program "Affordable Tooling." The two resins focused on in this program are X343-59 from Cytec Fiberite and LTM-45-ELD from Advanced Composites Group. One issue not addressed with this type of cure cycle is its effect on residual stresses compared to conventional autoclave cure cycles. Therefore, the objective of this study is to use a unique single fiber stress test to study the buildup of residual stresses in these low temperature curing resins under a variety of cure cycles.

Residual stresses in polymers and polymer composites has been the subject of many previous studies (e.g. 1-16). These stresses are induced by volumetric changes occurring during the cure process as well as during cooldown. Cure induced stresses in resins were shown to depend on the composition of the resin and may constitute a big fraction of final residual stresses (1). In polymer composites, the existence of the fibers with high volume fraction appears to reduce the effect of stress relaxation even at temperatures higher than glass transition temperature (2). Models were developed to evaluate residual stresses developed during cure and cooldown (e.g. 4-9). These models assumed matrix behavior similar to that of the resin alone thus overestimating the effect of stress relaxation. Experimental work shows that controlling cure cycles may significantly reduce residual stresses, as measured by curvature of unsymmetric laminates (10,11).

A test setup was developed to monitor stresses developed during cure of plain resins, i.e. without fibers, (1). A new method was also developed to monitor stresses developed in fibers during cure of a single fiber composite (12,13). This method was modified with a closed loop feedback control to yield optimized cycles where cure induced stresses are kept to a minimum (13,14). In this paper, this method was used to study stresses development in low temperature thermal curing resins and to study the possibility of optimization the cure cycle to yield minimum residual stresses.

Two types of volumetric changes occur during cure of thermoset composites; first of which is the expansion during heating and the other is the shrinkage due to the chemical reaction (crosslinking shrinkage) and during cooling down (thermal shrinkage). The volume changes occur during different stages of the cure cycle. Typically, volume changes prior to the polymer gelation contribute little to residual stresses since the melted polymer acts like a fluid with relatively low viscosity before gelation. Gelation is usually defined by the point at which infinite network of molecules is developed inside the polymer (gel point). Beyond polymer gelation, volumetric changes cause corresponding stresses in the fibers. Through the rest of this paper, the term *effective* volume will be used to refer to volumetric changes beyond polymer gelation. These *effective* volume changes may be due to polymer expansion or shrinkage. Thus, *effective* expansion occurs during any heating beyond the polymer gelation, and *effective* shrinkage results from polymer crosslinking beyond gelation (*effective* chemical shrinkage) and during cooldown (*effective* thermal shrinkage). *Effective* shrinkage will produce compressive stresses in the fibers inside the composite where as *effective* expansion will produce tensile stresses. Since there are no external loads, compressive stresses in the fibers are accompanied by tensile stresses in the resin which may cause matrix cracking. In typical cure cycles, not much heating, if any, is applied after polymer gelation and thus *effective* thermal expansion is typically small. The sum of *effective* chemical shrinkage and *effective* thermal shrinkage is always larger than *effective* thermal expansion. Thus, *effective* shrinkage is the governing element in producing residual stresses in composites.

Since *effective* expansion produces tensile stresses, it will counteract a fraction of the compressive stresses due to shrinkage, and thus it reduces final residual stresses. These tensile stresses resemble fiber prestress which was shown to reduce residual stresses (15). Therefore, it is desirable to increase the *effective* thermal expansion. *Effective* thermal expansion is not a constant for a given resin. It depends, among other factors, on the applied cure cycle (16). For example, consider the cure cycle in which the temperature is increased rapidly to the maximum cure temperature and then held constant until the cure is completed. In such a cycle, there will be small, if any, *effective* thermal expansion since most or all the thermal expansion will occur early before gelation. Residual stresses in such a cycle were studied using the single fiber composite test (13). Generally, high residual stresses are expected in such a case. On the other hand, an example for a cycle in which *effective* thermal expansion is relatively large is the two stages cure cycles where a low temperature is applied for long time during the first stage to allow the resin to reach gelation. Then, the part is post cured by increasing the temperature to the maximum cure temperature. This heating after the polymer reaches gel point produces *effective* expansion and thus tensile stresses. This type of cycle is expected to reduce residual stresses as discussed above. Application of such cycles to low temperature thermal curing resins will be discussed in this paper.

It is possible to combine the two stages of the cycle discussed above in one stage. In such one stage cycle, low temperature is first applied for long time to allow the resin to reach gelation and then the temperature is increased directly to the maximum cure temperature. This type of cycle was used in a previous study and was shown to reduce the residual stresses as measured by the curvature of unsymmetric laminates (11). Another approach to increase *effective* thermal expansion is to use more than one temperature hold. In such cycles, the temperature is increased to a certain value, held for sometime, and then increased to a higher value and so on. This approach was applied using three-dwell cure cycle and was shown to reduce the final residual stresses (10).

Stress relaxation always works to reduce stresses, compressive or tensile, that develop in a polymer. The relaxation times increase with the advancement of polymer cure. Depending on the relaxation time and the relative magnitude of *effective* shrinkage and *effective* expansion, a composite may be in a state of either compressive or tensile stresses before cooldown begins. The compressive state of stress in fiber before cooldown is not desirable because it will be added to the compressive stresses produced during cooldown. On the other hand, the tensile stress present in the fiber before cooldown will alleviate part of the cooldown compressive stresses. Therefore, to minimize the final residual stress in composite, as much thermal expansion as possible should be applied as late as possible in the cure cycle. In this situation, the tensile stresses generated in the fiber due to thermal expansion will not be significantly relaxed. Cure cycles can be modified not only to increase the *effective* thermal expansion but also to avoid its decay by stress relaxation as will be shown later in Section 3.

2- EXPERIMENTAL METHODS

2.1. Materials: The resins were provided by Boeing, St. Louis under the DARPA program "Affordable Tooling." The resins used were LTM-45 ELD from Advanced Composites Group, and X343-59 from Cytec Fiberite. The fibers used were AS4 graphite from Hexcel.

2.2. Single Fiber Test Setup: Stresses developed during given cure cycles were monitored using the single fiber composite setup. In this test, resin is cured around a given length of a pretensioned fiber. The fiber is fixed at one end and attached to a load cell at the other end to monitor changes in the fiber tension due to volumetric changes in the resin during cure. The resin shrinkage will increase the tension in the fiber segment outside the resin. Under this shrinkage, compressive stresses will be developed in the fiber running inside the resin. Resin expansion will decrease the fiber tension outside the resin and develop tensile stresses in the fiber running inside.

In this paper, test results will be shown as the load cell reading, i.e. fiber tension outside the resin. A comparison between fiber tension data and stresses in fibers embedded in the resin shows that any increase in the fiber tension more than its initial value (as a result of the *effective* shrinkage) corresponds to the application of compressive stresses on the fibers inside the composite. On the other hand, any drop in the fiber tension below its initial value (as a result of the *effective* expansion) corresponds to the application of tensile stresses on the fibers inside the composite. Through the rest of this paper, this correspondence will be used to relate changes in the fiber tension to the stresses developed in the fibers inside the composite. Under no external loads, the matrix in the vicinity of the fibers experiences tensile or compressive stresses when fibers experience compressive or tensile stresses, respectively.

A feedback control system was applied to obtain optimum cure cycles with minimum residual stresses. Details of the test setup and feedback control can be found in previous studies (12-14). The basic idea in optimization is to use, whenever possible, a combination of stress relaxation and thermal expansion to compensate the effect of crosslinking shrinkage at any moment in the cure cycle.

2.3. Volumetric Dilatometry: The volumetric dilatometer used was the GNOMIX, Inc. PVT Apparatus. The details of the test setup can be found in reference 17.

2.4. Thermomechanical Analysis (TMA): TMA was used to determine the glass transition temperature (Tg) of the samples. The TMA used was a DuPont 943 Thermomechanical Analyzer. Samples were heated from ambient to 300°C at 10°C/min in a nitrogen atmosphere with a 1 g load.

3- RESULTS AND DISCUSSION

3.1. Standard Cure Cycles: The standard low temperature cure cycle for X343-59 resin consists of: 1.) heating from 30°C to 66°C at 1.5C/min; 2.) holding at 66°C for 14 h; 3.) cooling to 30°C at 2.5°C/min; 4.) heating to 177°C at 1.5°C/min; 5.) holding for 2 h; and 6.) cooling to 30°C at 2.5° C/min. The standard low temperature cure cycle for LTM-45 ELD consists of: 1.) heating from 30°C to 60°C at 0.5°C/min; 2.) holding at 60°C for 16 h; 3.) cooling to 30°C at 1.5°C/min; 4.) heating to 177°C at 0.25°C/min; 5.) holding for 2 h; and 6.) cooling to 30°C at 1.5° C/min. In a different set of experiments, these two resins were also cured using an autoclave cure cycle developed for 3501-6 epoxy which consists of: 1.) heating from room temperature to 116°C in 30 minutes; 2.) holding at 116°C for 60 minutes; 3.) heating from 116°C to 177°C in 25 minutes; 4.) holding at 177°C for 240 minutes; and 5.) cooling to room temperature in 50 minutes.

Figure 1 shows the volume change data of X343-59 during a low temperature cure cycle. It can be seen that the resin undergoes cure shrinkage during the 66°C hold. Figure 2 shows the variation in the fiber tension during the same cycle. As can be seen, small increase in the fiber tension occurs during the 66°C hold. As the temperature decreases from 66°C to 30°C, another increase in the fiber tension occurs due to thermal contraction. As the resin is heated to post cure temperature, 177°C, the fiber tension decreases as a result of the thermal expansion. It can be seen that at the end of the heating ramp to 177°C, the tension drops below the initial tension value. Some increase in the fiber tension can be seen during the 177°C hold time. Finally, fiber tension increases during cooldown.

As mentioned in Section 1, stress relaxation causes decay in stresses developed in the composite. Figure 2 shows that fiber tension drops beyond its initial value during heating to post cure temperature. In this case, stress relaxation works to cancel this drop causing fiber tension to increase. Therefore, it is expected that the increase in the fiber tension during the 177°C hold is due to a combination of chemical shrinkage and stress relaxation.

During the above cycle (Figure 2), curing at a relatively low temperature for a long time increases the viscosity of the resin. This causes the thermal expansion occurring during heating to post cure temperature, 177°C, to be *effective* expansion. This *effective* expansion causes a drop in the fiber tension below its initial value. As can be seen, the drop in the fiber tension is larger than the increase in the fiber tension during the 177°C hold. The final value of fiber tension before cooldown is still smaller than the initial fiber tension. This indicates net tensile stresses in the fiber before cooldown according to the correspondence shown in Section 1. These tensile stresses will cancel part of the compressive stresses developed during the cooldown and thus reduce residual stresses.

Figure 3 shows the volume change data of X343-59 using an autoclave cure cycle developed for 3501-6 epoxy. A large amount of shrinkage occurs during the first temperature hold, while some additional shrinkage occurs during the second hold. Figure 4 shows the corresponding fiber tension variation. The shrinkage during the first temperature hold causes a significant increase in the fiber tension. The thermal expansion during the second temperature ramp decreases this tension. A slight increase in fiber tension can be seen during the second hold. Finally, cooldown shrinkage increases fiber tension. Figure 4 shows that the final value of fiber tension prior to cooldown is more than its initial value. In this typical autoclave cycle, heating to the relatively high first temperature hold in short time just melts the resin and causes thermal expansion but does not develop tensile stresses in the fiber. This causes the chemical shrinkage to dominate the volume change and its effect will finally be added to cooldown shrinkage causing more residual stresses.

Figure 2 shows that when the low temperature cure cycle is applied, tensile stresses are developed in the fiber during cure. These tensile stresses will offset some of the compressive stresses that occur during cooldown resulting in less final residual stresses. On the other hand, when typical autoclave cure cycle is applied, compressive stresses are developed during cure. These compressive stresses combined with stresses developed during cooldown produce higher residual stresses. A comparison between the two cycles suggests that the autoclave type cure cycle will result in more residual stresses than the low temperature cure cycle.

Figure 5 shows the volume change data during low temperature cure cycle of LTM-45 ELD, and Figure 6 shows the corresponding fiber tension profile. It can be seen that, compared to X343-59 resin (Figure 2), more stresses are developed during the 60°C hold. There is not much change in the volume during the 177°C hold. This suggests that the cure was completed during the slow heating ramp from 30°C to the post cure temperature. Fiber tension data agrees with the volume changes data showing only a slight increase (which may be due to stress relaxation). Similar to X343-59, *effective* thermal expansion developed during heating to post cure temperature lowered the fiber tension below the initial value until the end of the cycle, i.e. fiber is under tensile stresses prior to cooldown.

Figure 7 shows the volume change data of the LTM-45 ELD under the autoclave type cycle. The behavior of LTM-45 ELD under this cycle is similar to X343-59 under the same cycle, where most of the cure occurs during the first temperature hold. Figure 8 shows the corresponding fiber tension profile. As can be seen, compressive stresses are developed in the composite prior to cooldown. Hence, similar to X343-59, low temperature cure cycle is expected to produce less residual stresses than the autoclave type.

3.2. Optimized Cure Cycles A feedback control for the single fiber stress test was shown in previous studies (13,14). This feedback control was used in this study to minimize cure induced stresses in single fiber composites with X343-59 and LTM-45 ELD resins and AS4 fibers.   The key idea is to use thermal expansion, by raising temperature, to cancel any detected change in the fiber tension caused by *effective* cure shrinkage if it was not relaxed in a certain time interval. Increasing time interval before adjusting the temperature to compensate for the cure shrinkage allows more stress relaxation. However, increasing the time intervals results in longer cure cycles with slower heating rates. Slowing the heating rate may cause vitrification where the polymer transforms to glassy state with infinite viscosity. Vitrification slows the chemical reaction significantly, and as a result, the feedback control will finally stop raising the temperature since there is no shrinkage to cancel. The feedback control has been shown to produce cure cycles with almost zero cure induced stresses, i.e. almost flat fiber tension curve were obtained (13,14).

A major assumption in determining the required thermal expansion to cancel unrelaxed stresses is that the applied temperature increase will not change the polymer reaction rate greatly. In some cases where chemical reaction is very sensitive to the increase in temperature, the temperature increment applied by the feedback control may greatly increase the rate of the reaction causing more chemical shrinkage. Again, the feedback control will apply more temperature to cancel the increase in the chemical shrinkage. The new temperature increment will again increase the reaction rate and so on. In such cases, equilibrium at each point may not be possible. In this situation, cancellation of stresses may be achieved by allowing more stress relaxation rather than heating. However, increasing time intervals to allow more stress relaxation will extend the cycle duration and may lead to vitrification preventing completion of the cure as discussed above. The X343-59 and LTM-45 ELD resins appear to be sensitive to this phenomenon.

For X343-59 resin, Figure 2, most of stresses were developed during the post cure, so that feedback control was enabled only during this part of the cure cycle. The objective of the feedback control here is to apply the heat incrementally in such a way that the fiber tension remains constant at a certain value until the resin is cured. This kind of incremental heating, if exist, prevents *effective* cure shrinkage from developing compressive stresses in the fibers during cure. Also, applying the temperature incrementally prevent large decay of tensile stresses by the stress relaxation. Preventing stress relaxation of tensile stresses, induced by *effective* thermal expansion due to heating, should finally reduce residual stresses as discussed in Section 1.

The long hold at low temperature was applied as is. Then the temperature was increased linearly to 130°C. At this point, feedback control was enabled. This temperature was high enough to start the feedback control and reach the 177°C limit in reasonable time. Figure 9 shows the obtained cure cycle and the corresponding fiber tension. As can be seen, controlling the temperature through the feedback control does not cancel shrinkage stresses completely and the curve is not flat. As discussed above, the increase in the temperature appears to also increase the chemical shrinkage through enhancing the reaction rate. This will result in net increase in the fiber tension instead of canceling it. As the reaction advances, its rate appears to decrease allowing the thermal expansion to start canceling the chemical shrinkage. However, the 177°C limit is reached before the tension returns to the initial value. The completion of the cure was checked by measuring Tg. The obtained cycle gives 171°C versus 174°C for the standard low temperature cure cycle. The difference appears to be statistically insignificant since the highest temperature 177°C is applied in both cases for sufficient time.

Figure 10 shows the case where feedback control was enabled at arbitrarily chosen lower temperature (110°C). As can be seen, better control over the reaction was possible resulting in an almost flat curve. So starting at low temperature gives the advantage of slower reaction rate but the reaction may stop before the cure is completed because of vitrification. It should be noted that starting at higher temperature is expected to increase the reaction rate and complete it in shorter time but offers less control as was seen when the feedback control was enabled at 130°C.

Other optimization runs were performed to find autoclave type cure cycles with minimum residual stresses. Figures 11 and 12 show the obtained fiber tension and resulting volume change, respectively. As can be seen again that at one stage in the cure cycle there is a region in which the feedback control does not cancel stresses because the chemical reaction rate also increases greatly as the temperature increases. However, the final value of the fiber tension is almost identical to initial value. This indicates that there are no cure induced stresses developed during such a cycle in the composite. Such an optimized cycle will produce less residual stresses than typical autoclave cure cycle shown in Figure 4 in which final value of the fiber tension is more than the initial tension value. The increase in the fiber tension above the initial value indicates compressive cure stresses in the composite. These compressive stresses will be added to the compressive stresses developed during cooldown producing more residual stresses. Glass transition temperature for such a cycle is 176°C which means that the reaction is complete

Using the optimized autoclave type cycle (Figure 11), the resin can be cured in a shorter time than for the optimized low temperature cure cycle, Figure 9. However, a comparison between the two cycles shows that the optimized low temperature cure cycle has the advantage of developing tensile stresses in the composite during cure versus almost zero stresses in the optimized autoclave type cycle. This tensile stresses will decrease the final residual stresses in the composite.

The feedback control was also used to obtain an optimized low temperature cure cycle for LTM-45 ELD resin with AS4 fibers. Similar to the case with X343-59, during the feedback control, the temperature increase also increases the reaction rate greatly preventing canceling chemical shrinkage stresses completely. After some trials, the cycle shown in

Figure 13 appears to decrease stresses during the first stage of the cure cycle. The post cure was used as is since there is no significant variation in fiber tension. Figure 14 shows the corresponding volume change data.

The optimized cure cycle shown in Figure 13 produces almost zero stresses during the first part of the low temperature cure cycle. This may be helpful to maintain dimensional stability of the parts during this stage. Also, when this optimized cycle is compared with typical low temperature cure cycle (Figure 6), it can be seen that the typical cycle results in compressive stresses at the end of the first stage of the cycle (as indicated by the increase in the fiber tension prior to cooldown from 60°C to 30°C). These compressive stresses will reduce the tensile stresses that will be developed during the heating ramp to post cure temperature. As discussed earlier, tensile stresses help to reduce residual. The optimized cycle has the advantage of almost zero stresses developed during the first stage. Thus, it keeps more tensile stresses in the composite and hence less final residual stresses.

4- CONCLUSIONS

A new test method was used to study residual stresses in low temperature curing resins for out-of-autoclave curing. The standard low temperature cure cycles consists of curing at low temperature for long time followed by post cure at higher temperature to complete the cure. The curing at low temperature for long time seems to increase total *effective* thermal expansion, which counteracts the effect of chemical and cooldown shrinkage. Shorter autoclave type cure cycles were also studied. Results indicate that higher residual stresses are expected under this type of cycles than for the low temperature cure cycles.

A feedback control system was applied to decrease residual stresses developed in these materials. Results indicate that direct cancellation of unrelaxed stresses using temperature increase is not possible during all the time intervals of the cycle. Better understanding of the reaction mechanism is needed to modify the feedback control to be applied completely to this class of polymers. However, using the current system it was possible to obtain cycles with less final residual stresses than for the standard low temperature cycles. The obtained cycles satisfy the cure requirement without increasing the duration of the cure cycles.

5- ACKNOWLEDGMENTS

This research work is supported by United States Air Force, Air Force Research Laboratory, Materials and Manufacturing Directorate under contract F49620-96-1-0085. The authors would like to thank Chad Davis, undergraduate research assistant for his effort in this work. Also, the authors would like to thank Dave Furdek and Holly Berendzen from Boeing St. Louis for contributing the resins for this work.

6- REFERENCES

1. H. Dannenberg, <u>SPE Journal</u>, July (1965).
2. A.S. Crasto and R.Y. Kim, <u>J. of Reinforced Plastics and Composites</u>, <u>12</u>, (1993).
3. S.R. White and H.T. Hahn, <u>Polymer Engineering and Science</u>, <u>30</u> (22), (1990).
4. S.R. White and H.T. Hahn, <u>J. of Composite Materials</u>, <u>26</u> (16), (1992).
5. D. Adolf and J.E. Martin, <u>J. of Composite Materials</u>, <u>30</u> (1), (1996).
6. T.M. Wang et al, <u>J. of Composite Materials</u>, <u>26</u> (6), (1992).
7. Y. Weitsman, <u>J. of Applied Mechanics</u>, <u>46</u>, (1979).
8. Y. Weitsman, <u>J. of Applied Mechanics</u>, <u>47</u>, (1980).
9. T.A. Bogetti and J.W. Gillespie, <u>J. of Composite Materials</u>, <u>26</u> (5), (1992).
10. S.R. White and H.T. Hahn, <u>J. of Composite Materials</u>, <u>27</u> (14), (1993).
11. G. Springer et al, <u>J. of Composite Materials</u>, <u>29</u> (10), (1995).
12. M.S. Madhukar et al, <u>Proceedings of the Tenth International Conference on Composite Materials</u>, Whistler, British Columbia, Canada, Volume III, A. Poursartip and K.N. Street, 1995, p. 157-164.

13. M.S. Genidy, M.S. Madhukar, and J.D. Russell, <u>Proceedings of the Eleventh International Conference on Composite Materials</u>, Gold Coast, Queensland, Australia, Volume IV, 1997.

14. M.S. Genidy, M.S. Madhukar, and J.D. Russell, <u>Proceedings of 29th International SAMPE Conference</u>, Orlando, Florida, <u>29</u>, 1997.

15. M.E. Tuttle, et al, <u>J. of Composite Materials</u>, <u>30</u> (4), (1996).

16. J.D. Russell and A. Lee, <u>Proceedings of 29th International SAMPE Conference</u>, Orlando, Florida, <u>29</u>, 1997.

17. J.D. Russell, <u>Analysis of Viscoelastic Properties of High-Temperature Polymer Using a Thermodynamic Equation of State</u>, M.S. Thesis, 1991, University of Dayton, Ohio, USA.

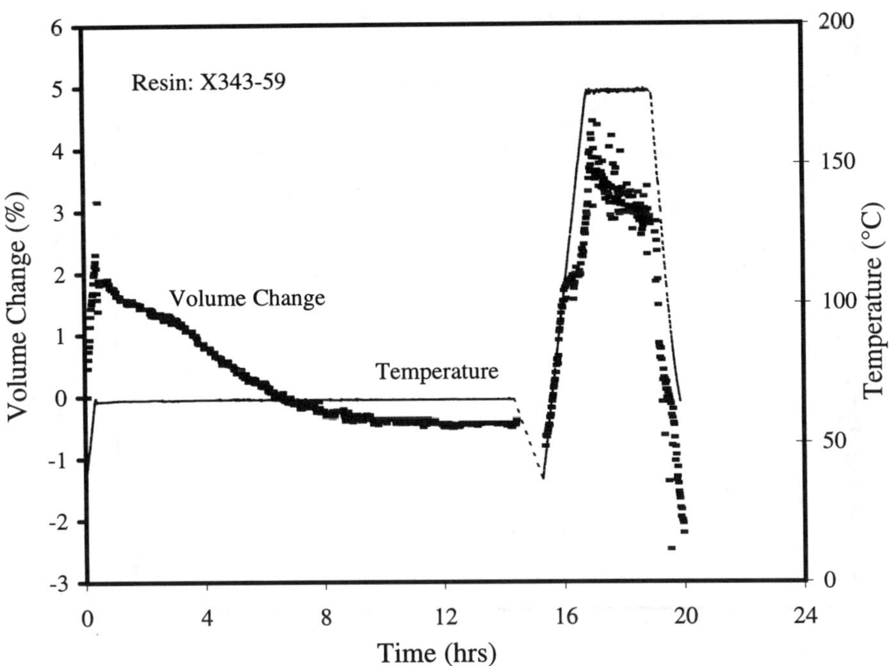

Figure 1. Volume change of X343-59 during a low temperature cure cycle.

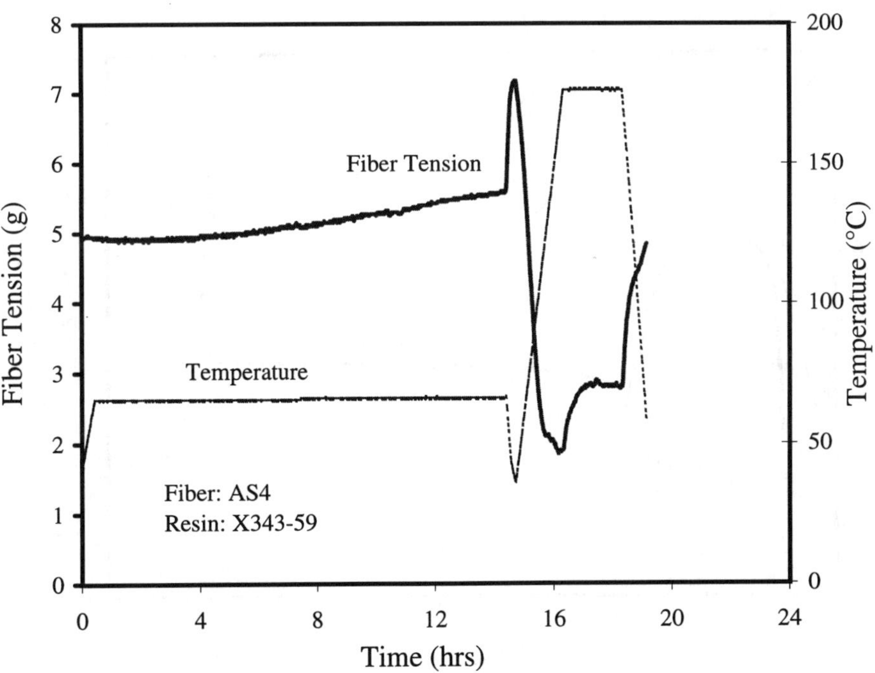

Figure 2. Fiber tension of X343-59 during a low temperature cure cycle.

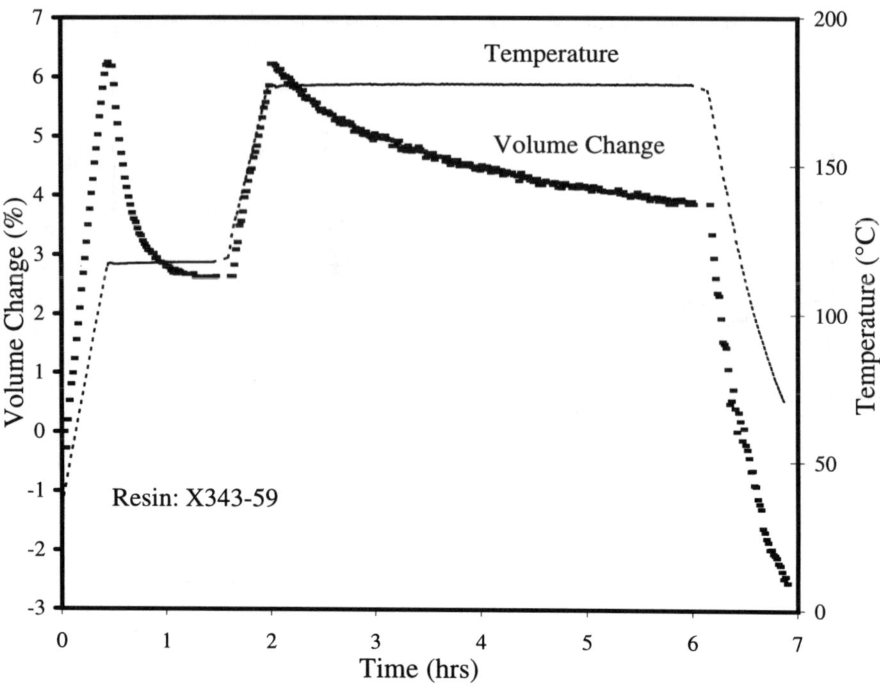

Figure 3. Volume change of X343-59 during an autoclave type cure cycle.

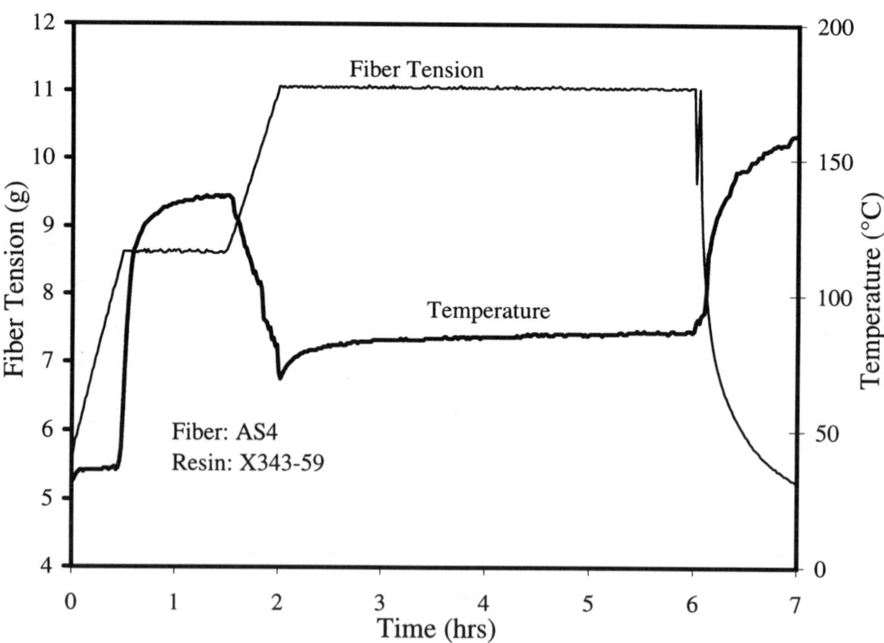

Figure 4. Fiber tension of X343-59 during an autoclave type cure cycle.

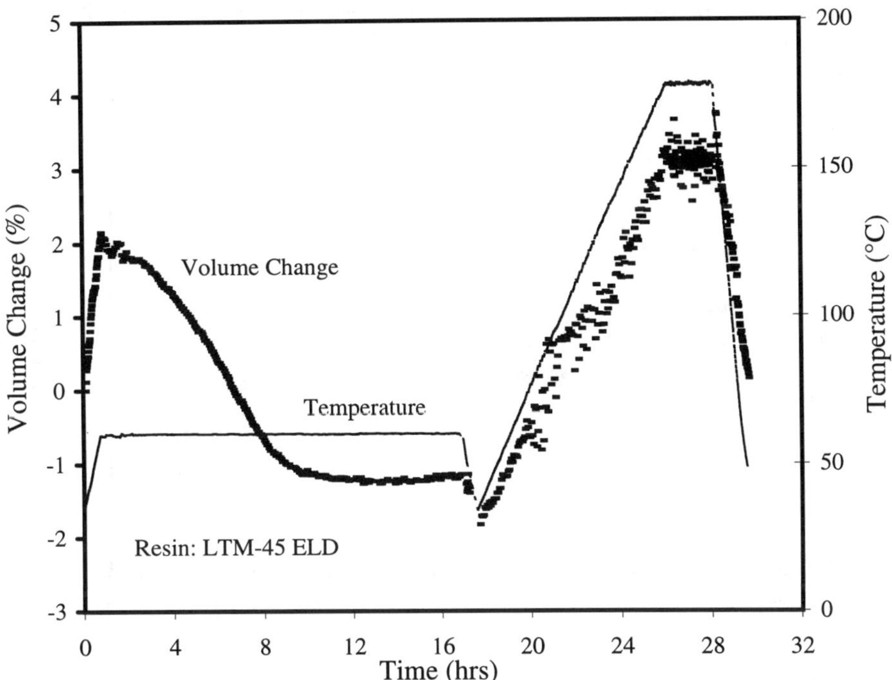

Figure 5. Volume change of LTM-45 ELD during a low temperature cure cycle.

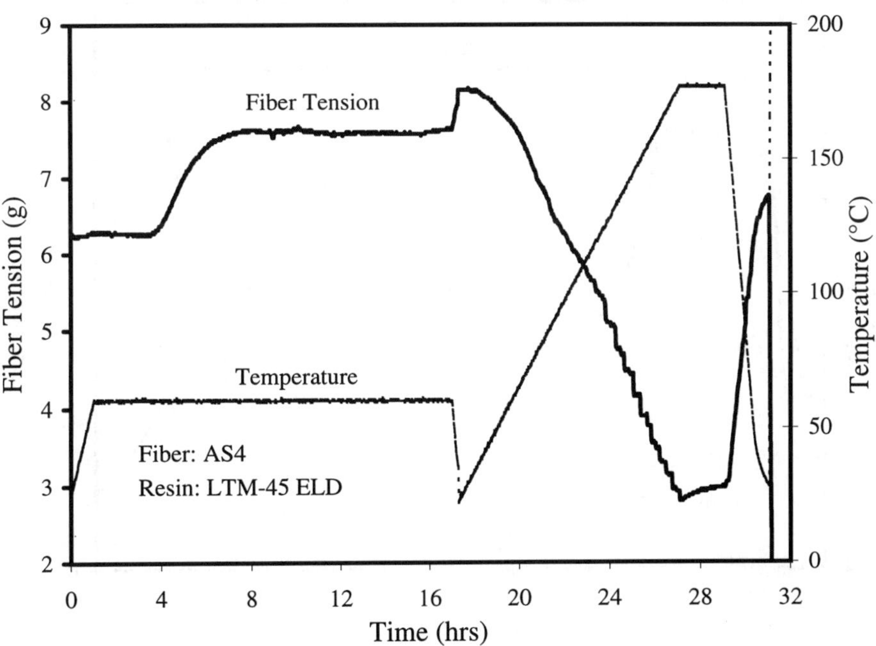

Figure 6- Fiber tension of LTM-45 ELD during a low temperature cure cycle.

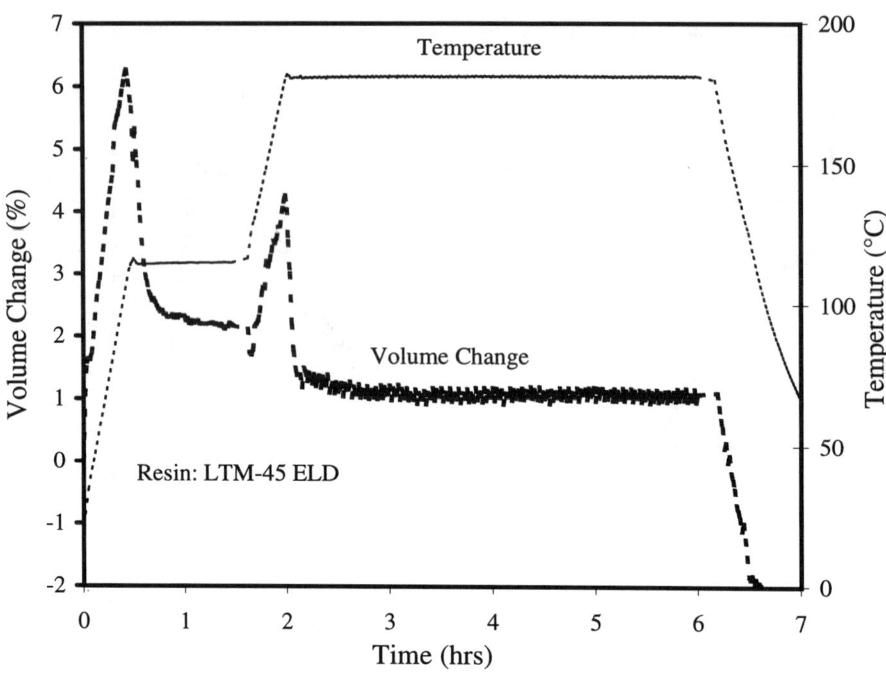

Figure 7. Volume change of LTM-45 ELD during an autoclave type cure cycle.

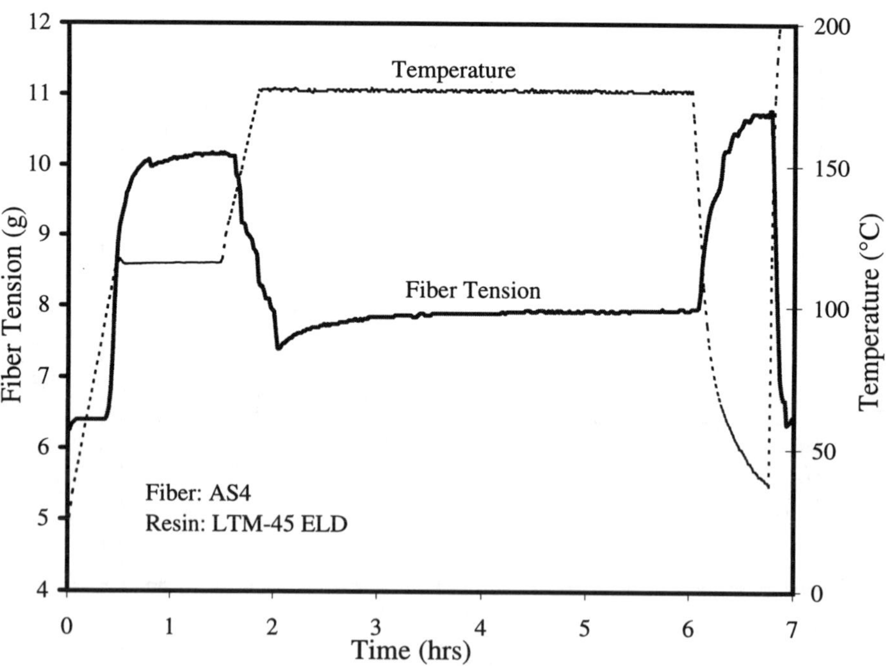

Figure 8. Fiber tension of LTM-45 ELD during an autoclave type cure cycle.

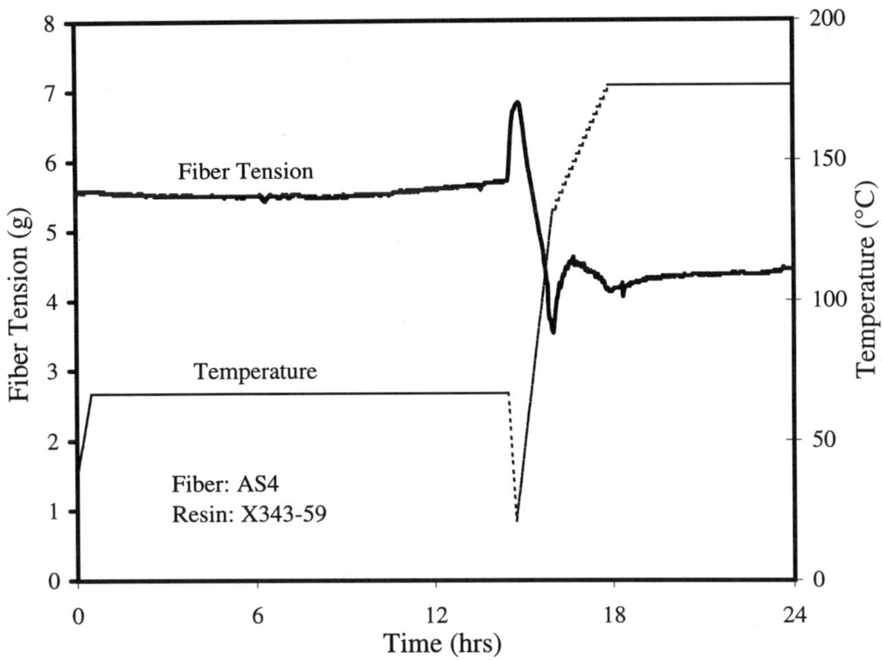

Figure 9. Fiber tension of X343-59 during a feedback control low temperaure cure cycle.

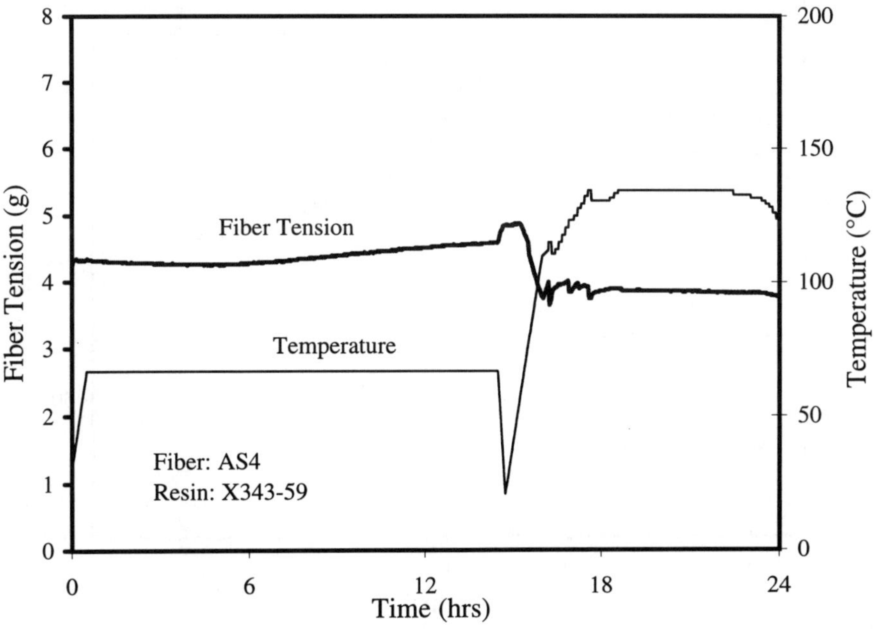

Figure 10. Fiber tension of X343-59 during a feedback control low temperature cure cycle.

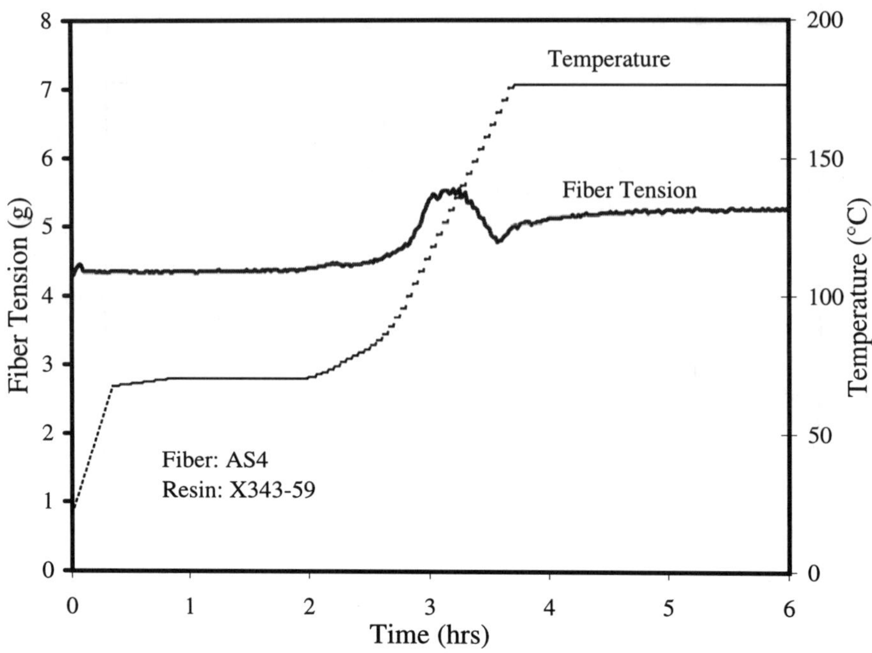

Figure 11. Fiber tension of X343-59 during a feedback control autoclave type cure cycle.

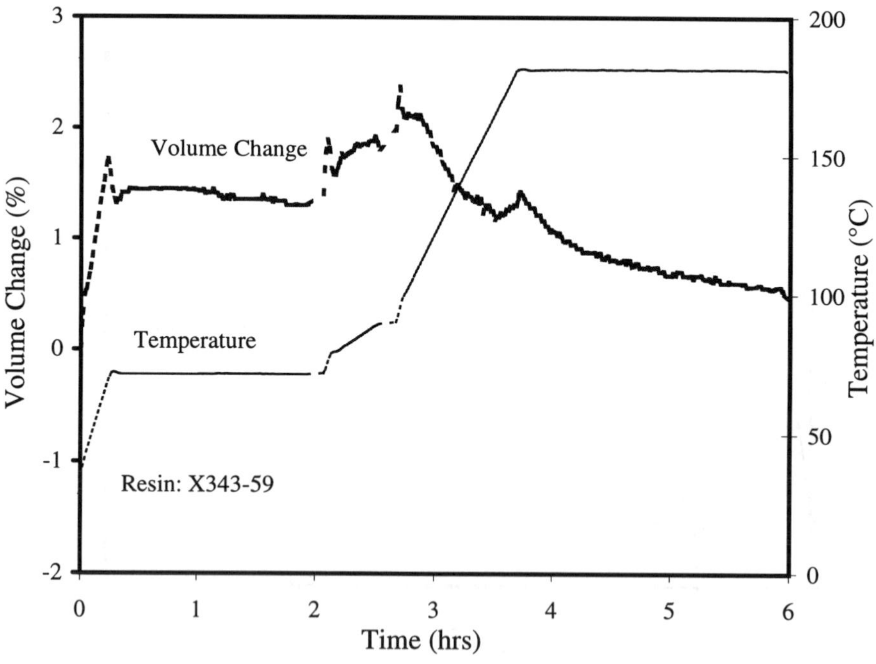

Figure 12. Volume change of X343-59 during a feedback control autoclave type cure cycle.

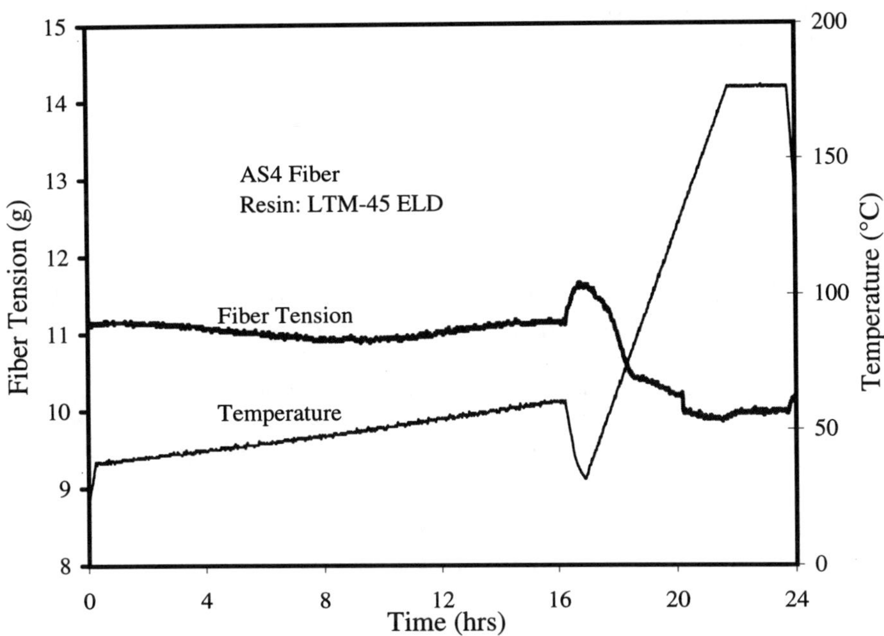

Figure 13. Fiber tension of LTM-45 ELD during a feedback control low temperature cure cycle.

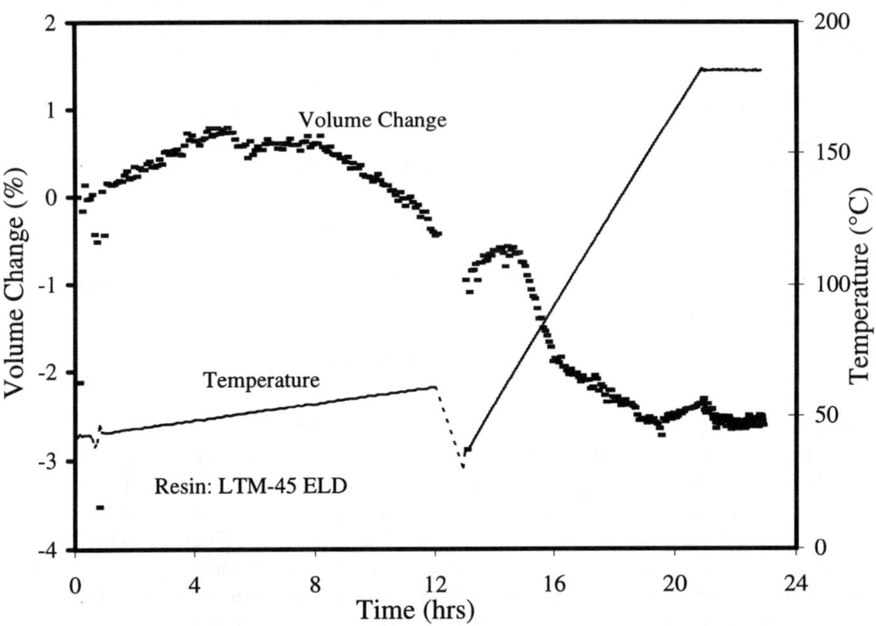

Figure 14. Volume change of LTM-45 ELD during a feedback control low temperature cure cycle.

ADAPTIVE TEMPERATURE CONTROL FOR THE THERMOPLASTIC TOW-PLACEMENT PROCESS

Dirk Heider[1], Raymond M. Foulk[2] and J.W. Gillespie, Jr.[3]
Center for Composite Materials
1) Department of Electrical and Computer Engineering
2) Department of Mechanical Engineering
3) Material Science Program
University of Delaware
Newark, DE 19716

ABSTRACT

A model-based optimization for the thermoplastic tow-placement system has been developed at the Center for Composite Materials in recent years. It predicts optimal process inputs (force, speed, etc.) on-line during the manufacturing process. This study investigates a Cerebellar Model Arithmetic Computer (CMAC) for adaptive temperature control. The CMAC is trained on- and off-line with feed-back from an AGEMA thermal camera and thereby learns the heat transfer behavior of the system. The desired heat profile under the tow-placement head predicted from the optimization is compared with the actual CMAC model output. A supplementary search finds the best suitable process inputs by minimizing the difference between desired and actual output.

KEY WORDS: Thermoplastic Tow-Placement, CMAC, Adaptive Temperature Control

1. INTRODUCTION

Thermoplastic composite tow-placement is a non-autoclave process with the potential to reduce part cost and increase product quality due to its on-line consolidation capabilities. This highly dynamic process requires an advanced control system to fully exploit these benefits. On-line model-based optimization techniques to locate optimum process inputs for a desired degree of quality have been developed for control of the robotic workcell at UD-CCM [1,2]. Some process inputs (head velocity, roller forces) can be readily controlled using the main robot controller. However, the heat transfer from the heat sources moving over the substrate is a major factor in the final part quality and is difficult to control due to its non-linear behavior. Perturbations in deposition velocity can also significantly effect system response. Thus, an intelligent and adaptive controller for the thermoplastic tow-placement technique is needed.

The system described in this study utilizes an adaptive model-based controller which predicts the heat transfer into the composite substrate. First, a Cerebellar Model Arithmetic Computer (CMAC) is trained off-line with an existing first principal heat transfer model and subsequently on-line with feedback from an AGEMA thermal camera. During the adaptation phase the CMAC can automatically adjust to existing process variations and eventually predicts accurately the thermal response. Finally, optimum set points for the heat source are found using a numerical optimization scheme.

2. TOW-PLACEMENT SYSTEM

The tow-placement system at the Center for Composite Materials (CCM) consists of an ABB robotic workcell, an AGEMA thermal camera, a supervisory SUN workstation and a tow-placement head. The head consists of two nitrogen gas torches [3], two rollers used for tow placement and consolidation, and a cut- and refeed mechanism (Figure 1). Initially, a ¼ inch wide thermoplastic prepreg (PEEK) is laid down on the substrate. Placement of subsequent tows next to previously bonded tows builds up a composite layer. A final part consists of multiple layers in different fiber directions. Hot nitrogen gas is directed onto the composite and heats up the substrate. Applying elevated temperatures and forces consolidate the material and bond the composite layers together [4-7]. Therefore, it is important to maintain the heat profile under the rollers to the required set points in order to achieve the desired final part quality.

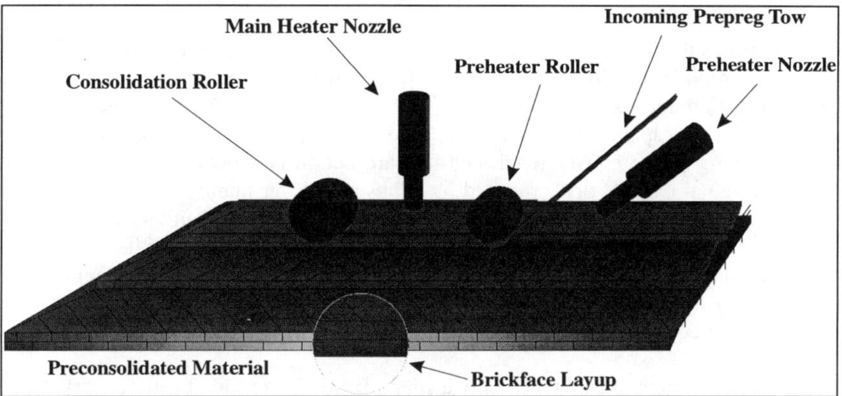

Figure 1: Schematic of the Tow Placement Head.

The distance of the torches to the substrate, head velocity, layer number, and roller and tool temperatures govern the amount of heat transfer into the material. Variations in the nitrogen gas temperature and flow rate as well as the surface quality of the substrate can greatly influence the heat profile under the torches. Typically, the heat profile has two maxima near the preheater and main torch (Figure 2). The profile shape at steady state is approximately constant throughout the process due to the fixed head design. Thus, it is sufficient to control the maximum temperature under each torch in order obtain the desired heat profile.

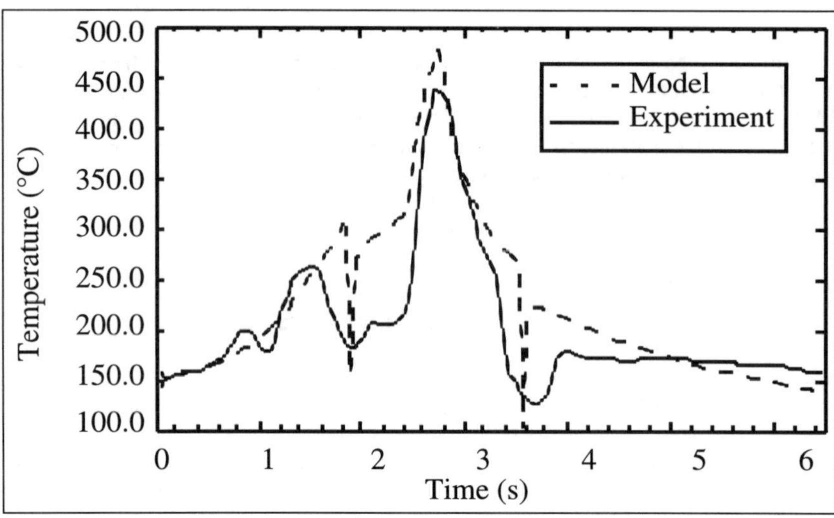

Figure 2: Center line temperature data

3. CEREBELLAR MODEL ARTICULATION CONTROLLER (CMAC)

This section explores the real-time learning capability of a CMAC which was first applied to the control of a robotic manipulator by Albus [8]. Since then it has been utilized in classification, function approximation, modeling and control [9,10]. The rapid learning rate and local interpolation capability make it a superb on-line modeling tool compared to the slow off-line training of the conventional feed-forward neural networks. There exist many variants of the original CMAC first proposed by Albus that differ mainly in how and where the network is locally trained (organization of the receptive fields, sensitivity function) as well as proposed hybrid architectures. In this study, the original CMAC architecture which provide adequate accuracy, fast training and low computation costs, is used to model on-line the heat transfer of the tow-placement system. This adaptive simulation is utilized for an intelligent predictive controller.

The CMAC is a neural network in which only a small, fixed size of the global network space is needed to calculate any particular output. This area is determined by the given input which maps into the associated CMAC space. Similar inputs stimulate similar CMAC regions and therefore have similar output values. As a result, during training local interpolation is obtained and the numbers of training cycles for specific inputs are orders of magnitude smaller than with feed-forward neural networks.

3.1. CMAC Theory In order to train a CMAC, the input vector has to be mapped into the CMAC hyperspace to obtain the activated receptive cells. The CMAC is binary if the receptive cells can only be activated or deactivated. First each input component i_n of the input vector i with n dimension has to be encoded into the normalized CMAC space c_n.

$$c_n = \frac{i_n - i_{n,\min}}{i_{n,\max} - i_{n,\min}} \qquad\qquad i_n \in \left[i_{n,\min}, i_{n,\max} \right[\qquad (1)$$

where $i_{n,\min}$ and $i_{n,\max}$ are the lower and upper bounds of the input space. The index of the activated receptive field is calculated by

$$j_n = \frac{c_n}{\varepsilon_n} + 1 \qquad (2)$$

ε_n is the maximum resolution desired by the binary CMAC. Larger ε_n gives higher resolution but lower local interpolation and longer training times. At this point, the system is a normalized look-up-table (LUT), where the network only knows trained receptive fields without any neighboring relationships. Using a layered method, each layer is divided into receptive cells of size $\rho * \varepsilon_n$ (ρ is the number of layers), and the layers are shifted against each other. Different shift arrangement can be considered but usually the shift is parallel to the hyperdiagonals of the input space. In this case, one input vector falls into exact ρ overlapping hypercubes and in every layer one cell is activated. The index of the receptive cells in each dimension and layer can be obtained by

$$r_{l,n} = floor\left(\frac{j_n + l}{\rho} \right) + 1 \quad \text{with } l=0,1,2,3,\ldots, \rho\text{-1} \qquad (3)$$

Figure 3 shows an example of a binary CMAC with three layers. Field number [6,11], which maps the normalized input range [0.333-0.4, 0.666-0.733], is activated. Therefore, the receptive cells [2,4] in layer 1, [3,4] in layer 2, and [3,5] in layer 3 are activated and have to be updated. Figure 3 highlights the interpolation range of one training point that affects not only points in the receptive cells but also neighboring data depending on their distance to the receptive field.

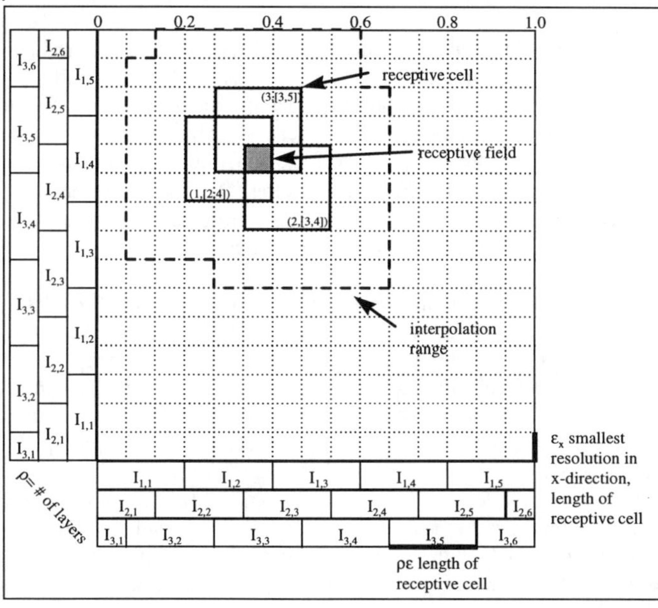

Figure 3: A binary two-dimensional with three layers (r=3) and a resolution of 0.066 (e=1/15)

Finally, all activated hypercubes have to be updated similar to the delta rule used in backpropagation of feed-forward neural networks. The advantage of CMAC is that only a fixed number ρ of receptive cells has to be updated and therefore training times decrease. The output O_{tot} for a specific input can be calculated as follows:

$$O_{tot} = \frac{\sum_{l=1}^{\rho} w_{l,act}}{\sum_{l=1}^{\rho} a_l} \tag{4}$$

where $w_{l,act}$ is the value of the activated binary receptive cell in the specific layer, and a_l indicates whether the cell is already trained.

$$a_l := \begin{cases} 1 & if \quad weight\ w_{l,act}\ is\ trained \\ 0 & else \end{cases} \tag{5}$$

During training, every activated cell has be checked for $a_l=1$. If the cell is not trained, ($a_l=0$) $w_{l,act}$ is set to the current output value. The delta rule is applied to all activated cells and the following equation for the new output $w_{l,act}^{new}$ is obtained:

$$w_{l,act}^{new} = \mu * p^* + (1 - \mu) * w_{l,act} \tag{6}$$

p^* is the target value, and μ is the learning factor of the receptive cell. If the learning rate μ goes to 1, the target value is directly stored in the receptive cell. This is not always desired, especially when the data is not free of noise. Here the receptive cells can act as a memory block from previous updates, and new noisy data can be averaged over several training cycles.

CMAC Heat Transfer Training   Initially, the CMAC space is not trained and all outputs are set to zero. A first principal heat transfer model is used to pre-train the CMAC close to the actual system response. This method provides the model with a complete description of the input-output relationships. Inputs are the head velocity, the main torch distance and the pre-heater distance, which are the primary factors in the heat transfer at a constant torch temperature of 900°C. Currently, the CMAC outputs are the maximum temperatures below the two torches. Although the whole 2-D temperature profile could be trained, the current control approach requires only the two peak temperatures.

Figure 4 shows the off-line training results after 100, 1000, 10000 cycles. The maximum temperature under the main torch is plotted versus pre-heater and main torch distance at a constant speed of 35mm/s. Here, a CMAC with 10 layers and a receptive cell width of 10% of the input range is chosen, thus providing the CMAC with a maximum resolution of 1%. After 100 cycles, only a few points are learned, the rest is kept as its initial value of zero. Further improvement are seen at 1000 cycles, and the complete input space is learned after 10,000 training patterns. Training of 10,000 pattern requires approximately 10 seconds. Therefore one training cycle takes about 1ms. This is fast enough compared to the time needed to acquire temperature values from the AGEMA camera during on-line training. After training of the CMAC with the existing first principal simulation is completed, a near match to the heat response is obtained and further on-line adaptation can be started.

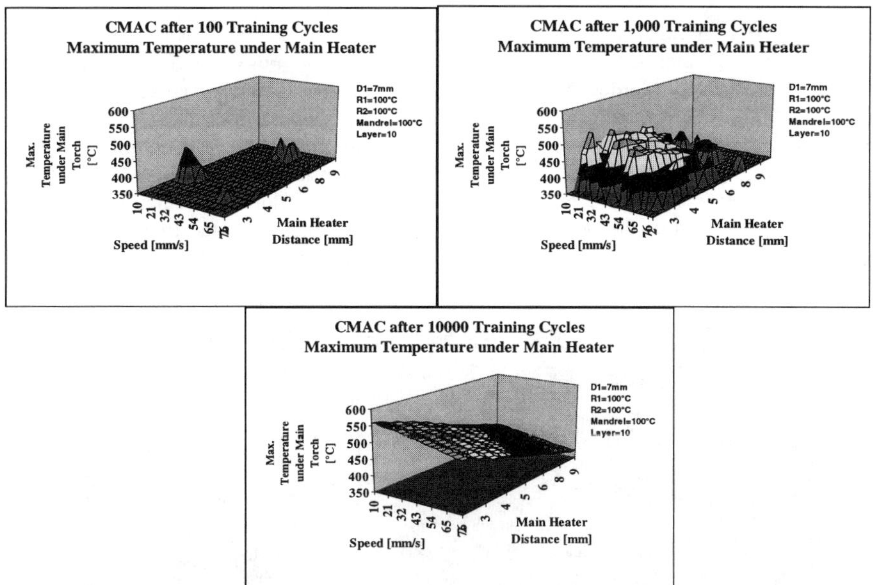

Figure 4: CMAC maximum temperature under main torch with off-line trained pattern.
Binary CMAC with 10 layers and a receptive cell width of 10%
ATP set points: Speed=35mm/s, Roller1=Roller2=Mandrel=100°C,Layer=10

4. TEMPERATURE CONTROL

4.1. Temperature Data Acquisition   There are several steps involved in transforming image files at the AGEMA thermal camera system into the CMAC temperature inputs at the SUN supervisory computer. First, the images are captured and stored on the AGEMA hard drive. Then, the image files must be transferred to the Sun system through an Ethernet connection. Finally, image processing takes place at the Sun in order to extract the maximum surface temperatures to be used as inputs to the CMAC (Figure 5).

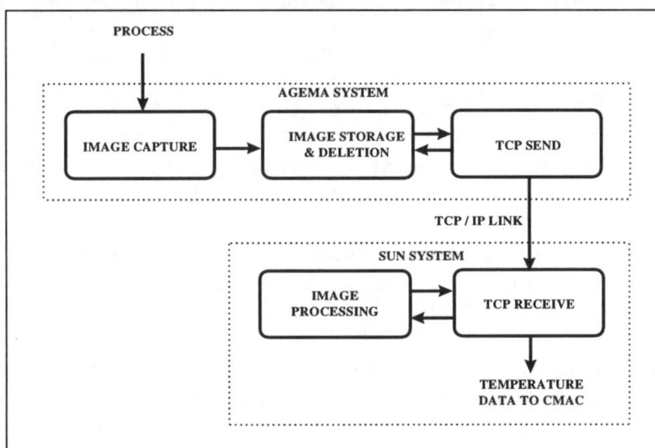

Figure 5: Image acquisition, transfer and processing schematic

The Agema system is capable of writing sequential image files from the IR camera to its hard drive at regular time intervals. A TCP/IP application sends the files to the SUN computer and deletes them afterwards in order to free disk space. Image files are received at the SUN via a Labview Virtual Instrument (Figure 6). Then, received images are passed to a Code Interface Node (CIN) where the actual image processing to find the maximum temperatures under the two torches takes place. An image file consists of a sequential series of hexadecimal values, each abstractly representing the temperature at that pixel. Address calculation is used to convert the data into a two dimensional picture. Then the pixel value must be converted into temperature data. The transfer function depends upon the temperature range setting of the IR camera.

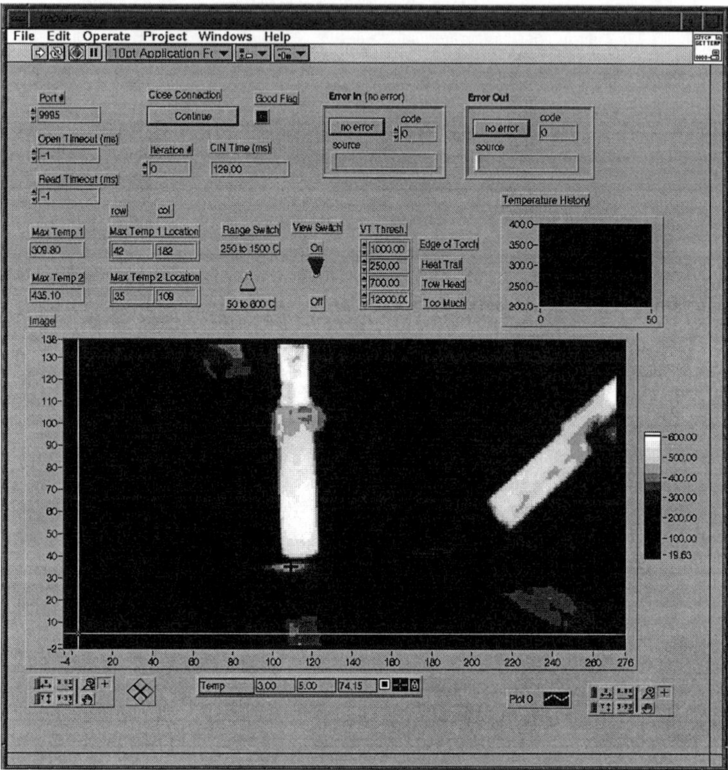

Figure 6: Transferred image on the SUN supervisory computer in a virtual instrument. Marked are the found location for the maximum temperatures under the torches.

The hottest and most easily recognizable objects in an image are the torches. For this reason, the main torch is found first in order to establish a frame of reference. Next, the search proceeds vertically along the side of the torch until the heat trail on the surface of the panel is found. From that point, a relative small area under the main torch is scanned for the maximum temperature. Care must be taken not to include the tip of the torch within this area, otherwise it may be selected as opposed to the actual surface temperature. The relative locations of the two maximum surface temperatures stay the same because of the fixed head design. Therefore, the area scanned for the pre-heater temperature is found with respect to the maximum main torch location. Finally, the pixel values at the two points are converted into real temperatures and passed back to the CMAC.

4.2. Controller Implementation   Figure 7 shows a normal heat transfer response for a head velocity of 60mm/s at 5mm pre-heater and main torch distance.   The average temperature value is about 440°C. Deviation of approximately 20°C occur at these settings. A conventional robust PID controller in this noisy environment would be limited in how fast it reacts to desired changes in the temperature profile due to different process set points. Faster response times would lead to an unstable controller which reacts more on the noise than on the desired steady state conditions. Therefore an intelligent control system has to be developed to maintain desired temperature profiles at steady state, but react fast enough when set point changes are desired.

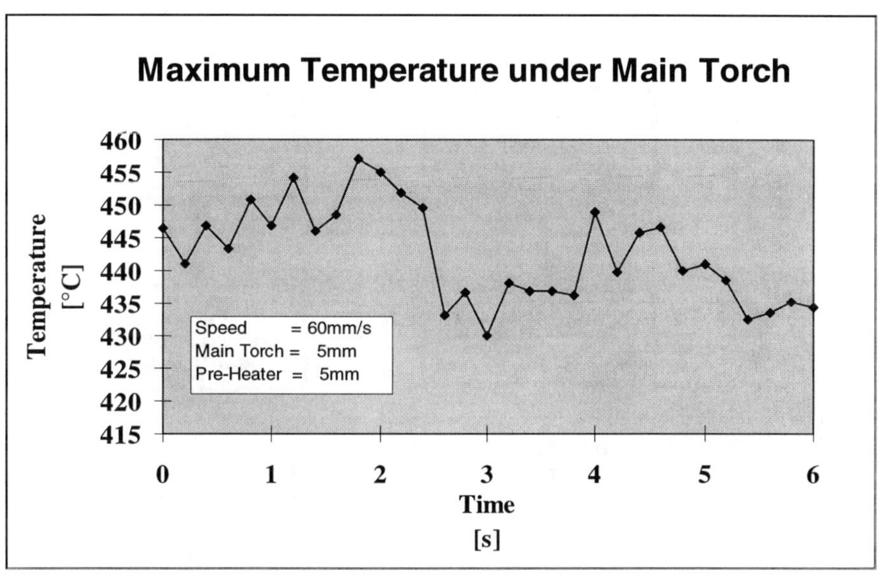

Figure 7: Steady State Temperature Noise.

The model-based predictive controller utilized in this study is shown in Figure 8. A desired temperature profile in combination with a set point for the head velocity is calculated by a neural network based optimization. The maximum temperatures under the two torches are extracted and fed into the numerical optimization. The optimization finds a close match between desired and CMAC predicted temperature at the given head speed. This is achieved by changing the input values (distance of main torch and pre-heater) to the CMAC and comparing the output values to the desired temperatures. This optimization has no direct feedback from the AGEMA camera and therefore is robust against process noise. On the other hand, desired process changes of the temperature profile directly relate to set point changes of the torch distances. After the optimization is completed, the set points have to be transferred to a stepper motor controller, which adjust the distances of the torches according to the set points. In addition, the current CMAC simulation can be trained on-line with the sensor information available from the AGEMA thermal camera. This feedback changes the CMAC model and subsequently changes the set points to the stepper motors.

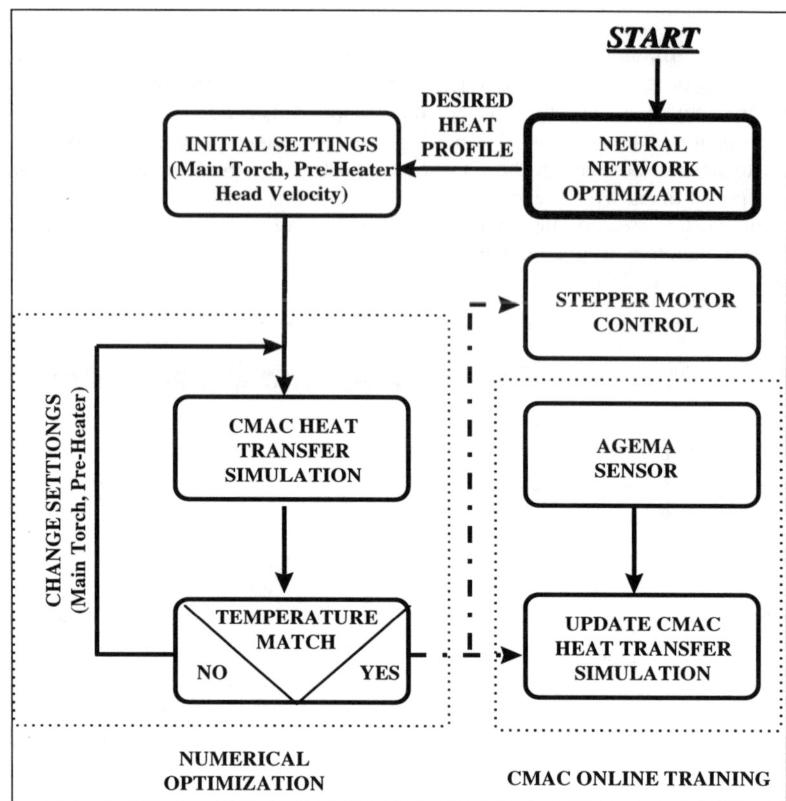

Figure 8: Flow Diagram of Temperature Control.

The performance of the model based controller was verified by tow-placing several layers of PEEK material at a constant rate of 35mm/s. Different set points for main torch and pre-heater temperature were used to evaluate steady state conditions. The CMAC subsequently learned the actual heat transfer into the substrate. Then, the set points were changed to evaluate the robustness and response time of the controller. Figure 9 shows the instantaneous temperature transition during set point change. A pre-heater distance of 2mm heats up the substrate to 250°C , 4.4mm main torch distance corresponds to 440°C. After 25 seconds the main torch temperature set point was changed to 500°C resulting into a new torch height of 2mm. The next available temperature value shows a direct increase in temperature to 510°C, close to the set point and in range of steady state noise. 5 seconds later the pre-heater temperature set point is reduced to 200°C. This results into a pre-heater distance of 8mm and temperatures of 205°C.

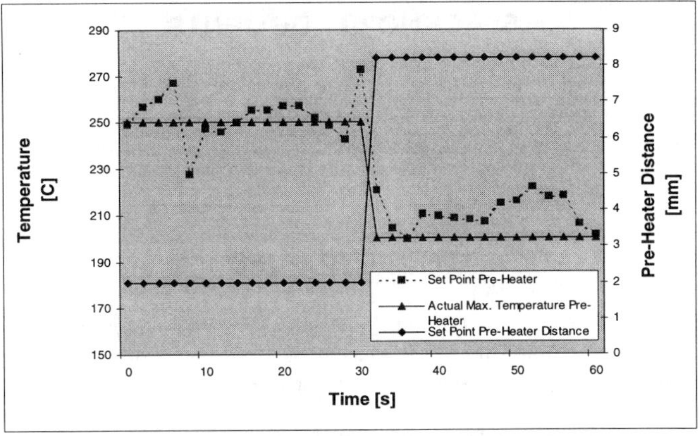

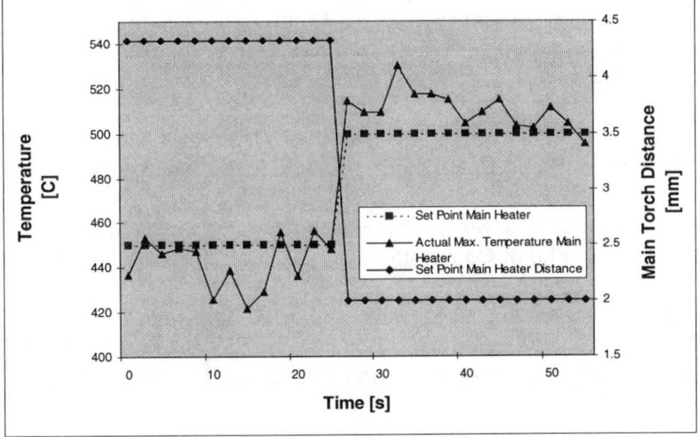

Figure 9: Controller response for different set points.
  a) Pre-heater at 2mm and 8mm; 200°C and 250°C
  b) Main torch at 2mm and 3.5mm; 440°C and 500°C

5. CONCLUSION

An adaptive predictive controller is presented which can control desired maximum temperatures for the tow-placement system. It utilizes a Cerebellar Arithmetic Model Computer (CMAC) to predict and learn the heat response of two hot nitrogen gas torches. It is used in combination with an AGEMA thermal camera and on-line optimization to adjust the stepper motor set points accordingly to a desired temperature profile. The developed approach is applicable to many other thermal processes where process simulations exist and conventional control techniques fail or are not sufficient. Future work will implement a similar control strategy on processes such as resistance and induction welding.

6. ACKNOWLEDGMENTS

This work is funded by the Department of the Army, Army Research Office under the RAPTECH-ACM Program (Grant No. DAAH04-94-G-0285). The authors would like to thank John Tierney for his work with the tow-placement model and Rod Don for his technical support.

7. REFERENCES

1. Heider, D., Piovoso, M.J., and Gillespie J.W. Jr., "Intelligent Control of the Thermoplastic Composite Tow-Placement Process," submitted to Journal of Thermoplastic Composite Materials, August 1997.

2. Heider, D., Don R.C., and Gillespie J.W. Jr., "A Neural Network Approach to Robotic Thermoplastic Tow Placement Process Control," Proceedings of the 11th Technical Conference on Composites, American Society for Composites, Technomic Publishing Company, Lancaster, PA, pp. 758-767, 1996.

3. Inventors: J. W. Gillespie, Jr., R. C. Don, S. T. Holmes, I. Howie, Adjustable Hot Gas Torch Nozzle for Rapid Heating Control; U.S. Patent No. 5,626,472: May 6, 1997.

4. Mantell, S. C., and G. S. Springer., "Manufacturing Process Models for Thermoplastic Composites," Journal of Composite Materials, Vol.26(16), pp. 2348-2377, 1992.

5. Ersoy, N. B., O. Vardar, B. K. Fink, and J. W. Gillespie Jr., "Effect of Processing Variables on Consolidation and Bonding in the Thermoplastic Fiber Placement Process," Center for Composite Materials Technical Report 95-35, University of Delaware, 1995.

6. Tierney, J. J, R. F. Eduljee, and J. W. Gillespie, Jr.," Material Response During Robotic Tow Placement of Thermoplastic Composites," Proceedings of the 11th Annual Advanced Composites Conference, pp. 315-329, 1995.

7. Tierney, J.J., Heider, D., Gillespie, J. W. Jr., "Welding of Thermoplastic Composites Using The Automated Tow Placement Process: Modeling and Control," Proceedings ANTEC 97, Toronto, Ontario, pp. 1165-1170, 1997.

8. Albus, J.S., "A New Approach To Manipulator Control: The Cerebellar Model Articulation Controller (CMAC)," Transactions of the ASME, Journal of Dynamical Systems, Measurement, and Control, pp.220-227, 1975.

9. Damarla, T. R., Zhao, W. N., "A Learning Algorithm For A CMAC-Based System And Its Application For Classification Of Ultrasonic Signals," Ultrasonics, Vol. 32, No 2, pp. 91-97, 1994.

10. Nelson, J., Kraft, L.G., "Using CMAC Neural Networks And Optimal Control," IEEE International Conference of Neural Networks, pp.2386-2390, 1995.

OPTIMIZATION OF TECHNOLOGICAL PROCESS OF SHAPEFORMING COMPOSITE AND LIGHT ALLOYS SHELLS

Bratukhin A.G., Bogolyubov V.S., Sirotkin 0.S., Lvov G.I.
ONC "Composite", 103051, Petrovka st., 24, Moscow, Russia; Kharkov State Polytechnical University, 310002, Frunze st., 21, Kharkov, The Ukraine.

Methods of Investigation of Technological Processes of Isotropic Shells Formation

The urge for developing parts with large geometric parameters in different areas of technics and providing minimum of mass, cost and controlled strength and rigidity characteristics results in wide application of thin-walled constructions. Thus, in construction building, ship-building, automobile industry, aviation industry and other fields the most characteristic constructional decisions are cylindric and dually convex non-closed shells in combination with supporting thin-walled elements of different contours. Shells, applied in aviation industry, are contour parts of complex form, made mostly of aluminium and also steel, titanium and metal-polymeric sheets.

In aviation industry the problem of manufacturing contour parts of controlled quality is rather decisive. Thus, for example, production of a kit of large-sized shells of contour parts for an airframe of a middle class passenger plane occupies 18% of labour expenditure of all preparation-stamping operations.

It should be also noted, that technological properties of materials, of which shells are made, are determined mostly by their mechanical characteristics, which, in their turn, set the sequence of operations, technological regimes of the process, straining efforts, parameters of main elements of equipment, accuracy and quality of the product.

Optimization of the processes of shapeforming contour parts is the complex scientific matter, which may be handled by solution of the problems of plasticity of shapeformed shells with account of their interaction with shapeforming tooling.

In theoretical analysis of the processes of shapeforming thin—walled contour parts, complex problems of mechanics of solid strained body, specific for investigation of all the processes of metals processing by pressure, arise. At the stage of statement of the task of process analysis it is necessary to choose correctly main outer effects on the processed blank, to admit justified assumptions of the character of interaction between the blank and tools, to define boundary conditions, corresponding to majour specific features of blank fixation.

As a rule, the processes of shapeforming constructions shell elements are accompanied by unloadings, repeated cyclic and alternate in sign loads. These phenomena require involving plasticity theories, reflecting peculiarities of material straining process under complex loading programs.

Shapeforming processes are accompanied by long travels and strains, that's why the problems are of geometrically non-linear character; in stating these problems initial and final shapes of the strained body are to be distinguished or Ailer concept of continuous medium movement is to be used.

Application of the methods of solid strained body mechanics for analysis of processes of formation started practically simultaneously with the development of plasticity theory. However, the mentioned difficulties of technological problems caused the need for introduction of assumptions and hypotheses, significantly simplifying problems statement and solution. To the number of such assumptions belong suggestions on kinematics of straining; hypotheses of flat cross-sections, neglections of certain components of travels vector; a priory statement of tensor components of stresses and strains.

For technological problems of shell theory wide use of momentless theory and models of an ideal rigidly plastic body is mostly typical. Suggestions of this type are often quite justified and the received solutions serve as theoretical basis for calculation of processes of sheet stamping. However, lack of strict evaluation of errors, connected with different assumptions, requires certain precaution in usage of such results. Moreover, practice advances such problems for the researchers, for which wide-spread assumptions are unacceptable (for example, on momentless character of strained state), and also puts forward such new problems, which cannot be solved in principle within the framework of traditional approaches.

On the other hand, wide introduction into practice of powerful means of computation permits to state and solve problems of shapeforming thin shells on more precise theoretical basis.

Statement of the Inverse Problem of Elastic-Plastic Shapeforming of Thin-Walled Parts.

For theoretical analysis of the process of straining sheet blank during shapeforming of shells by stretching-over in punches we distinguish three main stages of this process.

In the initial state, the medium surface S_0 of the shell, possessing the form of a thin plate or a gentle cylindric shell (Fig. 1), is calculated by parametric equations:

$$x_0 = x_0\,(\alpha,\beta); \quad y_0 = y_0(\alpha,\beta); \quad z_0 = z_0(\alpha,\beta) \tag{1}$$

where x_0, y_0, z_0 - Cartesian coordinates of blank medium surface points, α,β curvilinear Lagrange coordinates of these points.

In the end of the shapeforming process the blank acquires the shape, determined by the punch working surface. Shell medium surface in strained state S is set by an inexplicit equation:

$$f(x,y,z) = 0 \tag{2}$$

The determination of outer impact parameters, implementing this shapeforming (punch pressure on a blank, efforts and moments in holders), and also calculation of components of the stressed-strained state of the shell, constitute the subject matter of the special inverse problem. This problem is analyzed under the following assumptions.

Friction between the blank and the punch is not taken into account. Relative elongations and shifts are considered small, and the changes of shell thickness may be ignored. Non-uniformity of stresses and strains, spread within thickness, is of considerable value, that causes the application of momentary theory of shells.

Normal travels may considerably exceed the shell thickness. It is assumed, that shapeforming process is of monotonous character and the loading differs insignificantly from the simple one, that permits the application of deformational theory of plasticity. During straining a surface point acquires travel $\mathbf{v}$, projections of which on the axis of rectilinear coordinates system $\mathbf{OxOy}$, correspondingly, are $\mathbf{u}(\alpha,\beta)$, $\mathbf{v}(\alpha,\beta)$, $\mathbf{w}(\alpha,\beta)$. Then to achieve in the process of formation the change of medium surface in the shape (2), rectilinear coordinates of travel vector must meet the ratio:

$$f(x_0+u,\ y_0+v,\ z_0+w) = 0 \tag{3}$$

Resolving system of equations of elastic-plastic straining of thin gentle shells must include the dependence (3), for the statement of other equations of this system it is necessary to use static-geometrical ratios of the theory of thin gentle shells and the law of plastic body state.

In the general form the resolving system is as follows:

$$f(x_0+u,\ y_0+v,\ z_0+w)=0 \tag{4}$$

$$L_1(u,v,w)=x;\quad L_2(u,v,w)=y;\quad L_3(u,v,w)=z$$

The second, third and fourth equations of the system (4) represent equations of equilibrium in travels. Here L_1, L_2, L_3 are non-linear differential operators; x, y, z are projections of surface load on the axis of curvilinear coordinate system. The system (4) has six unknown functions; three components of the travel vector and three constituents of outer loading. For the problem to have an unambiguous solution it is necessary to estimate additional relationships of geometric or force character.

The first possibility suggests that conditions of the absence of tangentials to the medium shell surface, as components of outer loading, are observed:

$$x=y=0 \tag{5}$$

This accounts for the fact, that shapeforming under effect of normal pressure is the specific feature of the process of shell production by stretching-over on a punch. Within the framework of the assumptions, made with respect to straining values and angles of turn, the normal component of travel vector **w** may considerably exceed the tangential components **u** and **v**. Then the shape of the new surface **S** will be determined by the function **w**. Let the normal travel **w** be the set coordinate function:

$$w=w(\alpha,\beta) \tag{6}$$

Then the second and the third equations of the system (4) with account of the equation (5) will comprise together with boundary conditions with respect to tangential efforts (or travels) the problem for calculation of travels **u** and **w**.

$$L_1{}^*(u,v) = \phi_1(w) \tag{7}$$
$$L_2{}^*(u,v) = \phi_2(w)$$

After the solution of the boundary problem for the equations (7) from the fourth equation of the system (4) normal pressure $Z=P_3$ can be calculated. Bending moment and cutting forces at shell boundaries can not be set arbitrarily. They are determined by controlled travel **w**. This condition brings about one significant consequence: in the general case, shapeforming, determined by the equation (4), can not be carried out by effect of only normal pressure. For such shapeforming certain moments and cutting efforts are to be imposed at shell edges.

In the cases, when some blank edges are free in shapeforming process, areas of breakage of controlled geometry occur around them.

Results of Elastic-Plastic Problems Solution

Calculations of the processes of elastic-plastic straining are carried out in accordance with the analyzed above method for different geometric parameters of shells. Results of these calculations are presented in the form of diagrams of the spread of normal stresses σ_x, σ_y, in shell on symmetry lines and pressure fields, induced by the shell on the punch.

In the dwelled upon examples, the tensile effort **T** is chosen sufficient for plastic strains to take place at all shell points.

Diagram of straining sheet blank material has been approximated by the equations:

$$\varepsilon = \frac{1}{E}\,\sigma, \quad \text{when } \sigma \leq \sigma_T; \qquad \varepsilon = \frac{1}{E}\,\sigma + K\,(\sigma - \sigma_T)^n, \quad \text{when } \sigma < \sigma_T$$

in which the following parameters have been assumed: $E = 5*10^6$ Pa, $h = 2$ cm, $\sigma_T = 10^8$ Pa, $K = 10^{-14}$ Pa^{-1}.

At Fig. 2-4 there are results for a square blank of the sizes 2a=120 cm, 2b=120 cm, thickness of 1 cm, which is stretched over a punch with radii of curvature R_1=1000 cm, R_2 =200 cm wider constant tensile effort T_x =3*10^6 N/m.

Conclusion

Analysis of the received results permits to come to some certain conclusions:

- the field of normal pressure, induced on the punch in the process of stretching-over sheet parts is characterized by significant non-uniformity. Quality evaluation of the pressure field changes to a large extent with the change of curvature radii R_1, R_2 and the ratio between blank sizes;

- the developed method of calculation of the process of shapeforming of shell parts by stretching-over allows during designing of punches to determine the loads on them and to reveal the most stressed areas. Due to the considerable dependence of the pressure field on a large number of geometric parameters the most expedient form of usage of the experimental results is the computer analysis in accordance to developed programs at the stage of punch design;

- occurrence of negative pressures at some areas of the punch surface proves the loss of stability of the process and appearance of corrugation on the blank. Under increase of tensile efforts such areas vanish. The latter phenomenon discovers the prospects of developing such stretching-over process regimes, which exclude corrugation of blanks.

- optimization of regimes of technological process of shapeforming is accomplished effectively by the method of sequential coordinate descent within the space of variable parameters.

References

1. Bratukhin AG., Bogolyubov V.S. Composite Manufacturing Technology. Chapman & Hall, London, 1995.
2. Burlakov A.V., Lvov G.I. On One Class of Inverse Problems of Elastic-Plastic Shapeforming of Shells. Proceeding of RASS. Mecbanics of a Solid Body, No. 5, 1980.
3. Bogolyubov V.5., Lvov G.I., Kostromitskaya O.L. Inverse Problems of Shapeforming of Three-Layered Shells. Proceedings of RASS. Mechnics of a Solid Body. No. 5, 1989.
4. Bogolyubov V.S, Lvov G.I. Inverse Problems of Shapeformation Processes of Shells, ICIAM 95, Hamburg

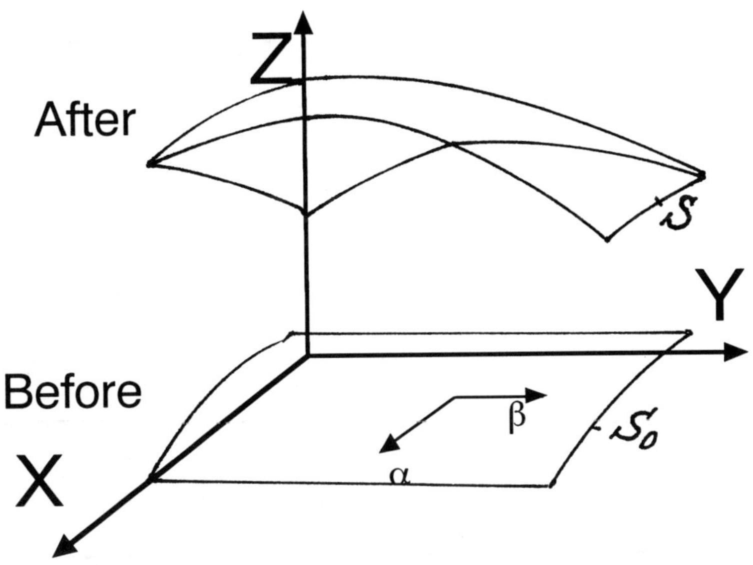

Figure 1 Medium surface of a blank before and after stretching-over

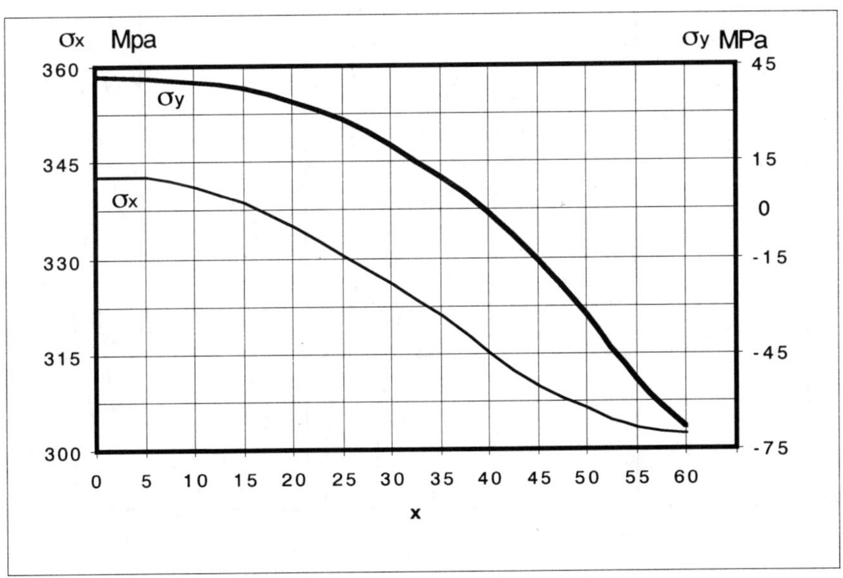

Figure 2 Spread of normal stresses along axis X

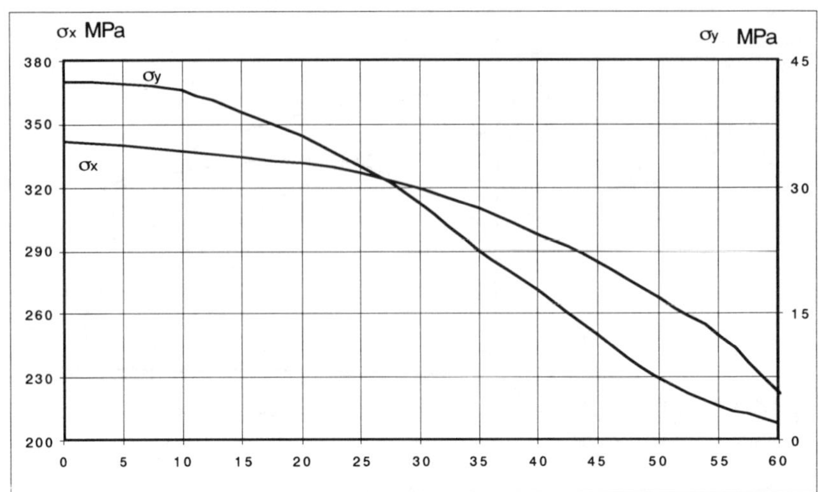

Figure 3 Spread of normal stresses along axis Y

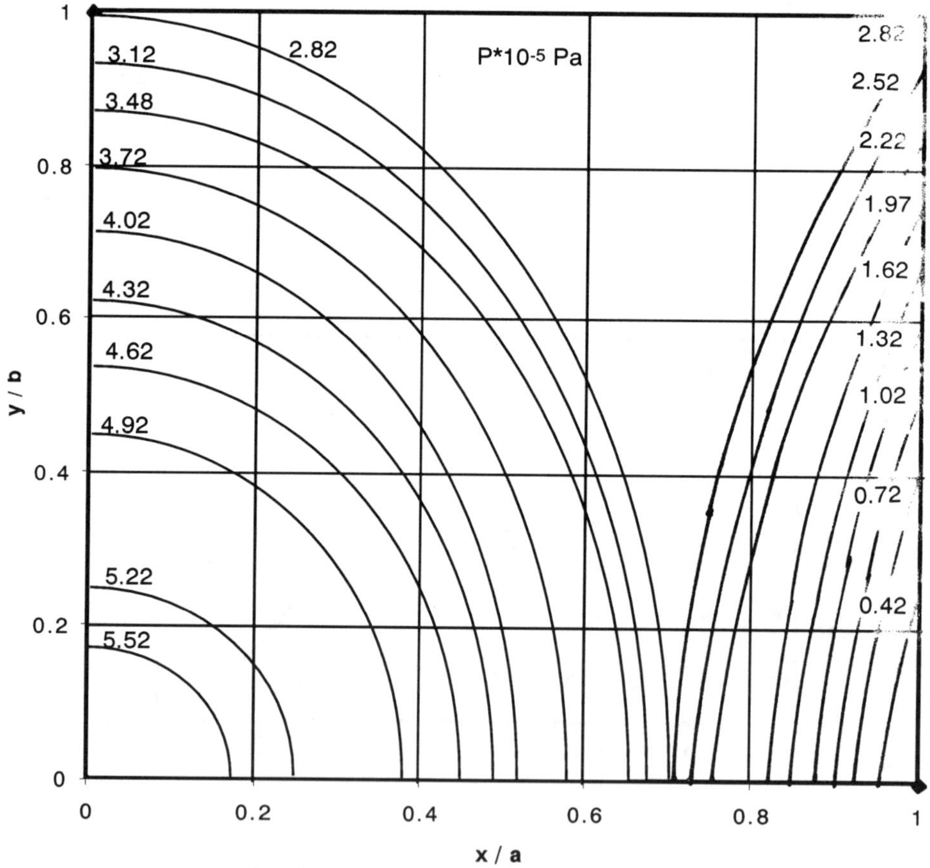

Figure 4 Curves of equal pressure

RESIN TRANSFER MOLDING PROCESS MONITORING AND CONTROL

C. William Lee and Brian P. Rice
The University of Dayton

Matthew B. Buczek and David C. Mason
Aerospace Service and Controls, Inc.

ABSTRACT

Resin Transfer Molding (RTM) offers the opportunity for significant cost savings in the fabrication of polymer matrix composite aerospace structures. However, the realization of this cost savings has been restricted by quality problems, such as voids, poor fiber volume control, etc., that are inherent to the RTM process. Similar quality issues in autoclave processes have benefited from the use of advanced process control systems such as the Air Force developed "Qualitative Process Automation" (QPA) and Aerospace Service and Control's "Composite Processing Control" (CPC) systems. These advanced systems utilize sensor feedback and a knowledge base to make control decisions and manage the process in real-time with the goal of improved end-product quality. An intelligent RTM control system based on the QPA approach has been developed. This paper presents the feasibility demonstration of its implementation on the CPC control system. A dc-resistance sensor grid was used to monitor the flow front in real-time from discrete measurements of sensor wet-out time. The CPC system was able to predict the flow front progression and alter the shape of the flow front either by zoned temperature control or by controlling resin feed rates through multiple injection ports. The same sensor grid was also capable of monitoring the cure state of the resin.

KEY WORDS: Composites, Cure Monitoring, Equipment and Machinery, Process Control, Resin Transfer Molding

1. INTRODUCTION

Resin Transfer Molding is a process where resin is injected into a closed cavity mold that is filled with a fiber reinforcement (referred to as a preform). Historically, RTM processes were the domain of manufacturers that specialized in low fiber volume (< 45%) and low performance resin systems. The aerospace industry considers RTM a process that can significantly reduce the cost of composite structures. However, aerospace structures need to have a fiber volume greater than 50% and high performance resins must be used. Both these two requirements create problems for the RTM process. Voids, lack of fiber wet-out and dry spots are some of the more common problems that are encountered. Although higher performance components have been fabricated, these processes have used very high pressures which require expensive equipment and tooling.

Considerable effort is being put into understanding the transport phenomenon of resin flow through a high fiber volume preform (>50%) and the factors associated with the successful injection of resin. Most studies center around modeling of the flow of resin through the fiber preform. The results are used to select locations of injection ports, vent ports and secondary injection/vacuum ports. Although the modeling of flow has been successful for well defined conditions, it is difficult to model the local variations in permeability, temperature, and resin state that can give rise to quality problems. The limitations of the models typically lie in the assumptions of constant permeability of the preform and isothermal condition with constant viscosity of the resin. In reality, permeability is not soley determined by the fiber volume of the preform but is also affected to a significantly extent by fiber orientation and the geometric shape of the preform. Preforms of the type used in aerospace structures are complex and frequently involve non-uniform permeability. It is very difficult and expensive to determine the permeability distribution throughout a preform. It is even more difficult, and impractical, to ensure that every preform has the same permeability distribution. For many processes, the resin viscosity can vary not only due to temperature variations, but also as a result of curing of the resin. Therefore, while modeling gives a first approximation of the mold filling process that provides insights to mold design and selection of port locations, it does not directly lend itself to solving the product quality problems.

The QPA approach of self-directed control has been implemented on the CPC system for resin transfer molding. This system utilizes a grid of discrete dc-resistance sensors to detect the resin flow front. A simple approach to predict the flow front progression in real-time is used by the system to compare to a specified desired shape of the flow front. The controller then controls either the temperatures of the various zones in the mold or the resin feed rates through multiple injection ports to alter the shape of the flow front. The results of these two demonstrations are reported in this paper although the system is capable of concurrent control of temperature, feed rates through multiple ports, vacuum application, and feed line pressure control. The dc-resistance sensor grid is also capable of monitoring the cure state of the resin.

2. EXPERIMENTAL

An experimental RTM system, designed with the capability to both monitor and control the resin flow front in real-time, was built with the following features: 1) a screw driven piston pump which can control resin flow based on volumetric flow rate and/or pressure, 2) multi-port gating control, 3) multi-zone temperature control from resin pot to the part, with 18 individual temperature zones on the tool arranged in a 6×3 grid, 3) multiple discrete dc-resistance sensors which are capable of detecting sensor wet-out by resin and monitoring resin viscosity.

2.1 Tooling The design of the tooling allowed for an extended flow path length (time) in which various control actions could be tested. It was decided to use a uniform thickness for which additional preform plies could be placed in arbitrary locations to create regions of lower permeability. It was further decided to have 18 individual heat zones and five inlet/outlet ports which could be independently controlled to affect the flow front. The tooling had overall dimensions of 61 cm × 30.5 cm (24"×12"). A 6×3 grid of heater pads, with dc-resistance sensors located at the centers of the pads, were placed as shown in Figure 1. The tooling consisted of a top plate, a bottom plate, and a clamping frame. The top tool plate was made from a 2.5-cm thick clear Lexan (polycarbonate) plate. Clear Lexan was used so that the flow front monitoring and control capabilities could be visually validated. The top tool had 18 individual point sensors, simply brass screws, centered over the heater zones. The top tool also had multiple 1/4" NPT inlets used as resin injection and vacuum ports. The top tool was fabricated to be placed over a 1.3-cm thick aluminum lower plate, which held the eighteen 7.6 cm × 7.6 cm (3"×3") heater pads and acted as the ground for the point sensors. Eighteen thermocouples, for temperature feedback control, were placed between the heaters and the lower plate. Two 1/4" steel clamping frame plates held the tooling together through the use of twelve 3/8" bolts.

2.2 Flow Front Detection Very simple electrodes fabricated from thermocouple wire can be used for dc-resistance measurement which provides a very low cost method for detecting bleeder wet-out or mold filling. As the resin completes the circuit between the two electrodes, a large decrease in resistance is sensed. Such dc-resistance sensors can be plugged into existing thermocouple ports; so no special wiring of equipment is required. In addition, most control systems are readily capable of voltage measurement, which is the signal that one measures with the use of a reference resistor. In many cases, useful information of resin cure state can also be obtained by simple dc-resistance measurement instead of using more costly dielectric sensor and instrumentation. The measured dc-resistance is effectively the same quantity as what is frequently referred to as the dielectric "ion viscosity" [1]. Figure 2 demonstrates this for the cure of a graphite/epoxy composite.

A sensor system based on dc-resistance can consist of two primary forms. One is the SMART (Sensors Mounted As Roving Threads) Weave system developed by Walsh [2], which consists of a grid of orthogonal conductive strands separated by a non-conducting medium such as a glass preform. When a conductive medium (liquid resin) wets out two strands that form a node, the voltage drops and the presence of the medium is noted. The advantage of this approach is that m+n channels are required to monitor m×n points. The disadvantages of this technique include low sensor contact area (a problem if the resin has low conductivity), inability to monitor viscosity, the need to lay-up a sensor grid for each part, and false readings caused by electrical shorts. The second arrangement can consist of discrete sensors that are tool mounted. This arrangement requires the use of one data acquisition channel for each sensor. However, the discrete dc-resistance sensors offer advantages which include the ability to monitor both wet-out state and resin state (viscosity), as well as the ability to use permanent tool mounted sensors in a production environment.

2.3 Flow Front Control While the use of sensors to monitor the flow front has been used to verify flow models and aid in tool and preform design, little has been reported on the application of this data toward on-line process control. Because slight variations in preforms can lead to local variations in permeability, it is probable that the resin flow front will vary somewhat in a production process. This variability can lead to dry spots and thus poor quality parts. Once a tool and preform have been "optimized" there are only two process variables which can be altered to affect the resin flow front. Darcy's Law describing flow through a porous medium states:

$$V = -\frac{k}{\mu}\,\nabla P \qquad\qquad\qquad (1)$$

where V is the flow velocity, μ is the resin viscosity, k is the preform permeability and ∇P is the pressure gradient. Equation (1) shows that the flow front velocity can be controlled by altering either the resin viscosity or the pressure gradient. For the cases presented in this paper, the pressure gradient developed as a consequence of the flow driven by a positive displacement pump and the resin viscosity was altered by altering the local temperature on the tool through individual control of the eighteen heater pads. Figure 3 illustrates that a temperature increase from 50°C to 100°C will result in an order of magnitude decrease in the viscosity of the resin used in this study.

2.4 Material System 3M's PR520 epoxy resin system was chosen for this study because it is a premixed resin system which has a very long out-time at room temperature [3]. This system and PR500, a similar system, are currently being used for RTM processing in aerospace applications. The viscosity and dc-resistance of the resin were measured simultaneously in a Rheometrics Dynamic Spectrometer with parallel plate fixtures. Using this technique the aluminum parallel plates acted as the sensing electrodes without compromising the viscosity measurement. The data shown in Figure 4, indicate that the resin's dc-resistance is nearly directly proportional to its viscosity.

The preforms used in this study were constructed from a plain weave glass cloth and were approximately 4 mm thick. Local variations in preform permeability were created by placing additional strips of cloth, typically 7.6 cm (3") wide, at selected locations in the preform.

3. RESULTS AND DISCUSSION

A 6×3 dc-resistance sensor grid, as shown in Figure 1, was used to obtain sensor wet-out times. We will first show that the method used for post-run processing of sensor wet-out time data allows accurate reconstruction of flow front progression. This method was used to determine the flow front progression for all experiments. We will then show two different methods for estimating the flow front locations in real-time for control purposes. The results of flow front control experiments by zoned temperature control of the mold and by feed rate control will then be presented.

3.1 Flow Front Reconstruction One can construct three imaginary paths for resin flow from the three injection ports at one end of the mold (1" below station 1) towards the exit at the other end of the mold. The sensor grid monitors the flow front along these three imaginary paths at six stations. As the flow front reaches a sensor, the resistance between the sensor electrodes decreases. The time of sensor "wet-out" is recorded. At the end of a run these 18 sensor wet-out times can be used to construct constant-time curves through interpolation. A constant-time curve over the grid represents the flow front at that time. Figure 5 shows these constant-time curves at 2-minute intervals superimposed on video clips of the flow fronts at these times. It can be seen that the reconstructed flow fronts are accurate representations of the actual flow fronts; even with such a small number of sensors and sensor data.

3.2 Real Time Estimation of Flow Front It is necessary to know the flow front location in order to perform feedback control on the flow front progression. With the 6×3 sensor grid arrangement, there are at most 18 updates of the flow front location during the course of filling the mold. For the current system running at a feed rate of about 20 cc/min, the update rate is about one per minute. Since typically only one sensor gets wetted out at a time, it is necessary to infer the shape of the flow front from past data. We tested two methods of estimating the flow front: one method updates flow front information only upon new sensor wet-out while the second method continuously updates flow front information at 5-second intervals. It should be noted that, in either case, no reliable information is available until at least one of the three station-1 sensors is wetted out.

3.2.1 Update on New Sensor Wet-Out Only When a sensor at station j on path i is wetted out, the time is recorded and the flow front velocity along this particular path is calculated by the sensor location from the inlet ($L_{j_{max}, i}$) and the wet-out time ($W_{j_{max}, i}$). The sub-subscript *max* indicates that these are the most recent values. This velocity is used for predicting the flow front location (F_i) on this particular path until the next sensor wet-out on this path occurs:

$$F_i = V_i \cdot t = \left[\frac{L_{j_{max}, i}}{W_{j_{max}, i}} \right] \cdot t \tag{2}$$

This is a very simple approach that works well for constant injection rate. A comparison of the resulting estimated flow fronts to the reconstructed actual flow fronts is shown in Figure 6. Note that the estimated flow fronts are not equally spaced in time but depends on the successive sensor wet-outs. An additional condition

$$F_i \le L_{j_{max}+1, i} \quad \text{(the predicted front must not exceed the} \tag{3}$$

next unwetted sensor location)

was added to the algorithm to correct for the over-prediction of flow front (in Figure 6, where predicted flow fronts cross).

3.2.2 Continuous Update at Constant Time Intervals In this method, the prediction uses the most current estimate of the flow front velocity:

$$F_i = V_i \cdot \left(t - W_{j_{max}, i} \right) + L_{j_{max}, i} = \left[\frac{L_{j_{max}, i} - L_{j_{max}-1, i}}{W_{j_{max}, i} - W_{j_{max}-1, i}} \right] \cdot \left(t - W_{j_{max}, i} \right) + L_{j_{max}, i} \quad \text{and} \tag{4}$$

$$F_i \le L_{j_{max}+1, i} \tag{5}$$

where t is the time since the start of the run. A comparison of the resulting estimated flow fronts to the reconstructed actual flow fronts is shown in Figure 7. The estimated flow fronts are calculated at 5-second intervals and plotted in Figure 7 at 2-minute intervals. Note that the first estimated flow front curve shown is that at 4-minutes which is in error compared to the reconstructed actual fronts. At 4 minutes, none of the sensors at station 2 had been wetted out and the lack of information resulted in the large error in prediction.

3.3 Flow Front Control The results of flow front control experiments by zoned temperature control of the mold and by feed rate control are presented in this section. The control objective for both experiments was to achieve a flat flow front.

3.3.1 Flow Front Control by Altering Resin Viscosity Through Temperature Control
According to Darcy's law for flow through a porous medium, the fluid velocity is inversely proportional to the viscosity of the fluid. Since the resin viscosity can decrease by an order of magnitude or more, it is possible to alter the flow front during mold filling by altering the resin viscosity through temperature control. To demonstrate the feasibility of this approach, we tested to see whether the controller could detect the lagging flow front along a low

permeability path and compensate for it by heating that part of the resin to bring the flow front up to meet the control objective of a flat flow front. A lower permeability region in the mold cavity was created by adding extra layers of preform cloth to path 3. The controller compared the estimated flow front and heated the zones where the flow front was behind the average value through proportional control of zone temperatures. Only one injection port at one inch before station 1 of the center path was used.

An open-loop experiment at a uniform temperature of 32°C shows that the lower permeability along path 3 resulted in lower flow velocity along that path and skewed flow fronts as shown in Figure 8(a). When proportional control was applied, heaters along path 3 were set by the controller to the maximum permitted temperature (79°C) during most of the run in order to increase the flow front velocity along path 3. The resulting flow fronts, as shown in Figure 8(b), although not quite flat, were more symmetrical and had lower curvature than those of the open-loop run in Figure 8(a). It should be pointed out that although the center path temperature was set to 32°C, it reached about 54°C because of the lack of active cooling.

3.3.2 *Flow Front Control by Varying Feed Rates Through Multiple Injection Ports*

Flow front control by controlling injection rate was tested by using three injection ports, one for each flow path at one inch before station 1. For a limited length along a flow path, increasing feed rate to one path can cause the flow front of that path to lead. Paths 1 and 3 have pneumatically controlled on-off valves and path 2 has a manual valve. The flow rates through the on-off valves on paths 1 and 3 are controlled by valve-opening time (30% controller output causes the valve to stay open 30% of the time) such that at open-loop conditions, the flow rates through the three valves are the same.

The control strategy was to manipulate flow rates (by increasing valve opening time) to path 1 and/or path 3 depending on whether the flow fronts on these paths were lagging or leading the flow front of the center path. This was done because there was no control for the center path injection valve. The control objective was to achieve a flat flow front with proportional and integral gains applied to the lead or lag distance between the flow front locations on the paths. The flow front locations were estimated and control actions were taken at 5-second intervals.

For the case shown in Figure 9, nine layers of preform cloth were used. The mold was kept at a constant temperature of 57°C. It can be seen that at about 5 minutes the controller called for a decrease in feed rate to path 1 to below the baseline value of 33% because the flow front of path 1 was leading. Flow front of path 3 started to lead at about 10 minutes which caused the feed rate to path 1 to increase. The feed rate to path 1 was kept high throughout the run because of a slight lead in path 3.

4. SUMMARY & CONCLUSIONS

A 6×3 dc-resistance sensor grid has been successfully used for flow front monitoring in resin transfer molding process. A simple 2-dimensional interpolation of sensor wet-out times allows reconstruction of flow front progression accurately. Algorithms have been developed for the control software to predict flow front locations in real-time for control purposes.

Control of flow front progression has been demonstrated with proportional temperature control and proportional-integral feed rate control through multiple injections ports.

Temperature control did alter the flow front but not to the extent desired. This was because it was not possible to maintain large temperature differentials over such a small tool area without active cooling. Greater control of flow front was achieved by controlling feed rates through multiple injection ports.

Using multiple injection ports along each path with feed rate control may improve the system performance. Such arrangement will require proper gating of injection ports. The current control system can be adapted to accomplish this easily.

We have demonstrated that self-directed control of flow front in resin transfer molding process is feasible. The methodology developed under this project is generally applicable to mold filling processes.

5. REFERENCES

1. Lee, C.W., Rice, B. P. and Buczek, M. B., "Direct Current Resistance Based Resin State Sensors", Proc. of the 41st International SAMPE Symposium, 1996.

2. Walsh, S., "Artificial Intelligence: Its Application to Composite Processing", Proc. of the 35th International SAMPE Symposium, April 2-5, 1990, pp. 1280-1291.

3. PR520 Epoxy Resin Introductory Data Sheet, 3M Aerospace Industrial Tape and Specialties Division, St. Paul, MN, November 1994.

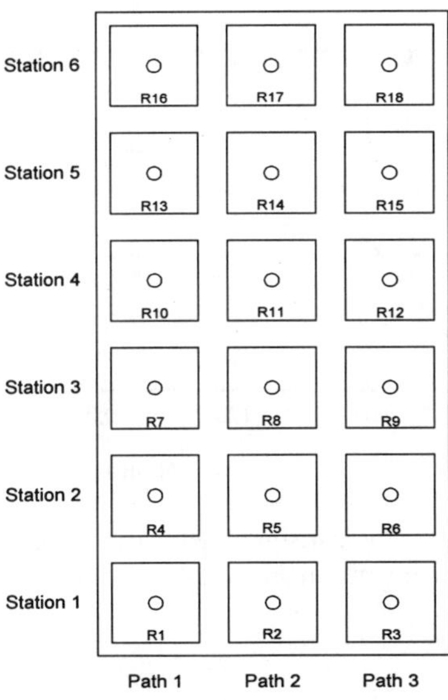

Figure 1: Sensor grid, heater pads, and flow path designations for tooling.

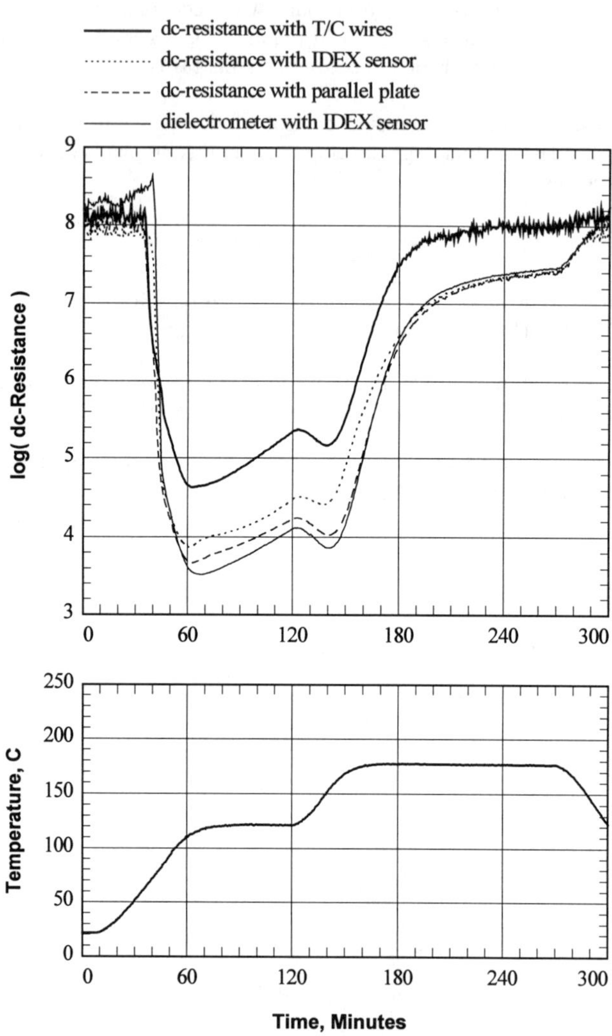

Figure 2: Comparison of dc-resistance and dielectric measurements on an 3501-6 epoxy composite [1].

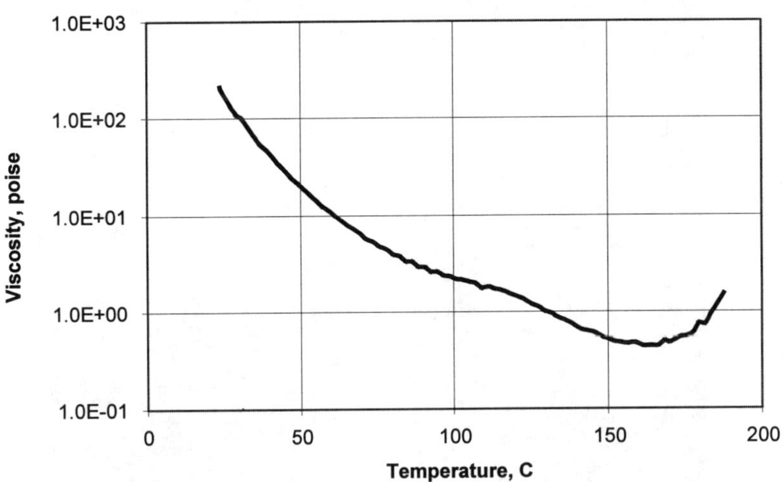

Figure 3: Viscosity profile of PR520 during a heating rate of 2°C/min.

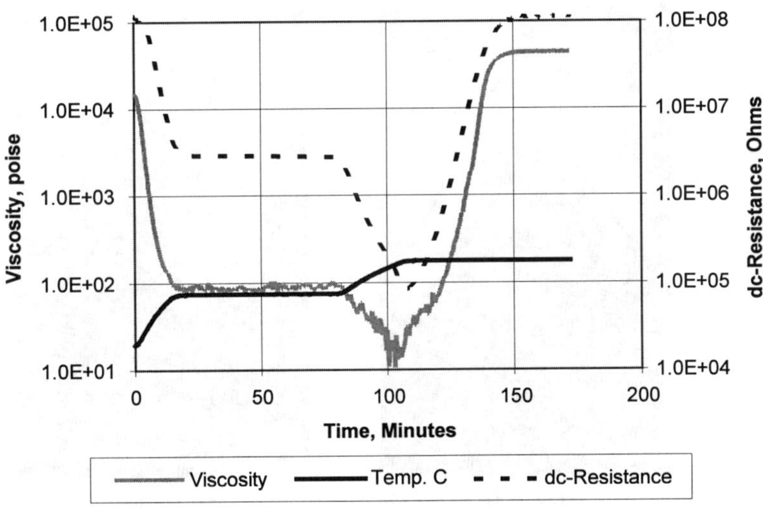

Figure 4: Simultaneous viscosity/dc-resistance measurement of PR520
using parallel-plate method.

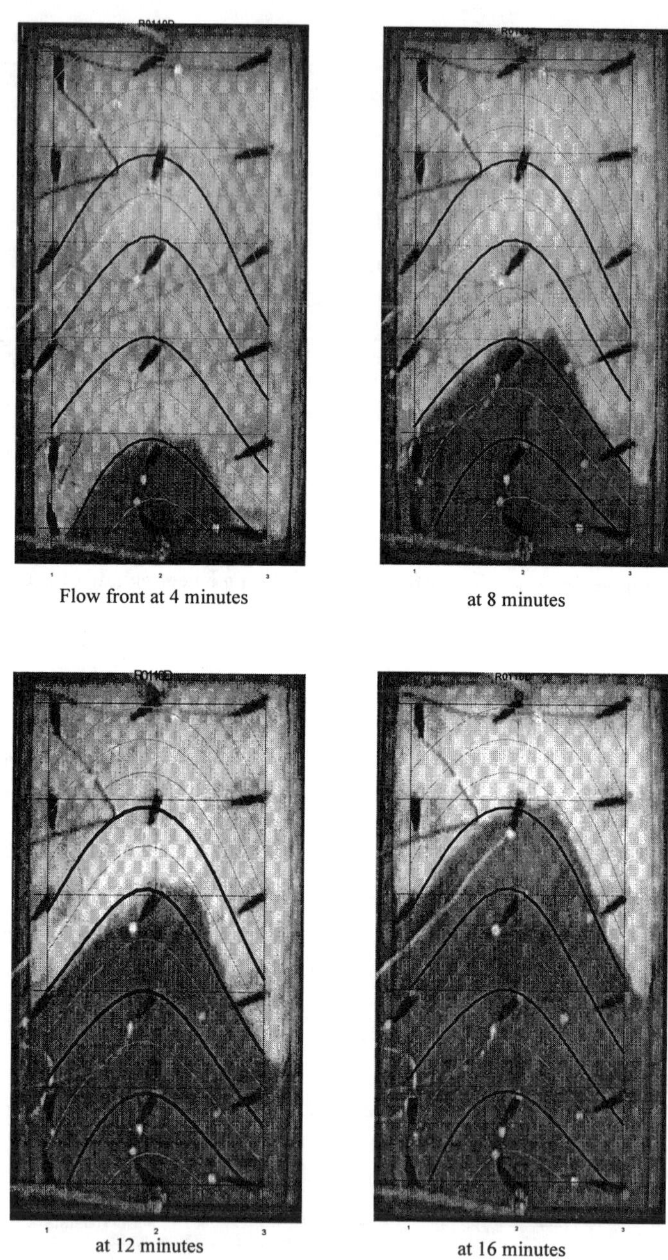

Flow front at 4 minutes   at 8 minutes

at 12 minutes   at 16 minutes

Figure 5: Comparison of actual and reconstructed flow front progression.

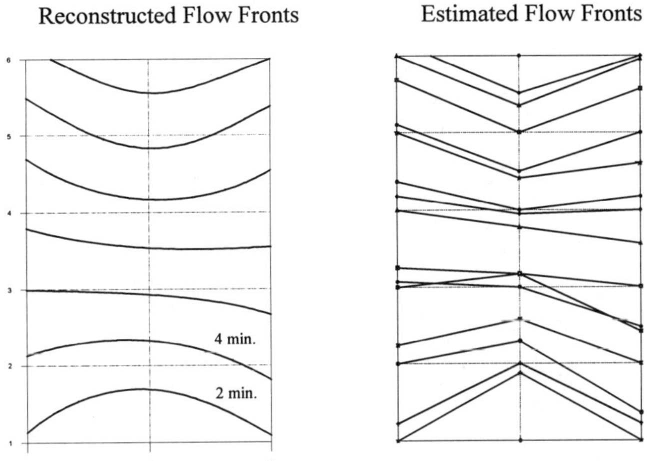

Figure 6: Estimated flow fronts at times of new sensor wet-out.

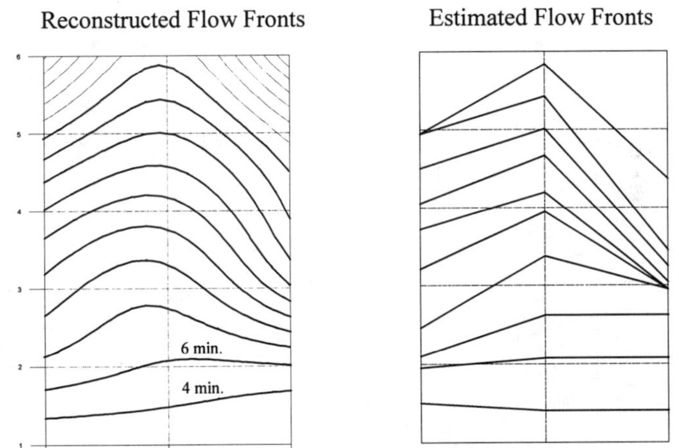

Figure 7: Estimated flow fronts based on most current flow front velocity, continuously updated at 5-second intervals. Plotted at 2-minute intervals.

241

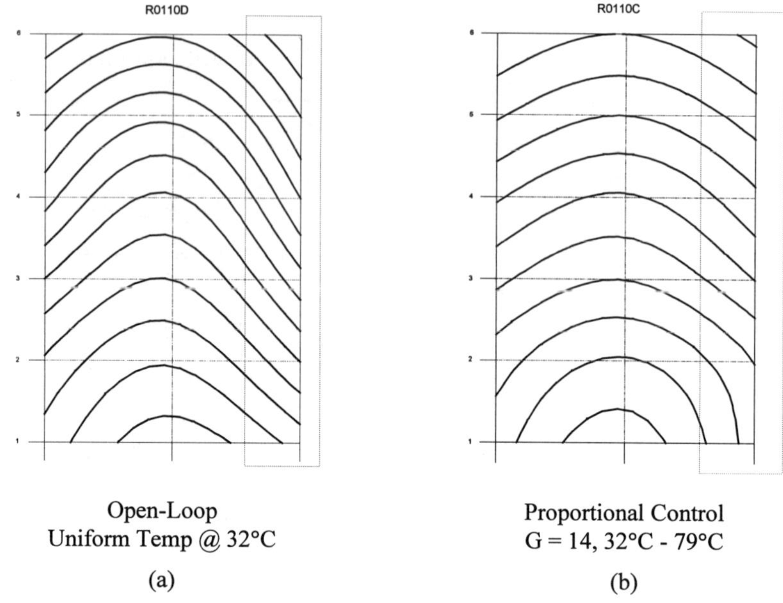

Figure 8: Proportional temperature control, update on new sensor wet-out only.
Nine layers of cloth preform with three additional layers for path 3.

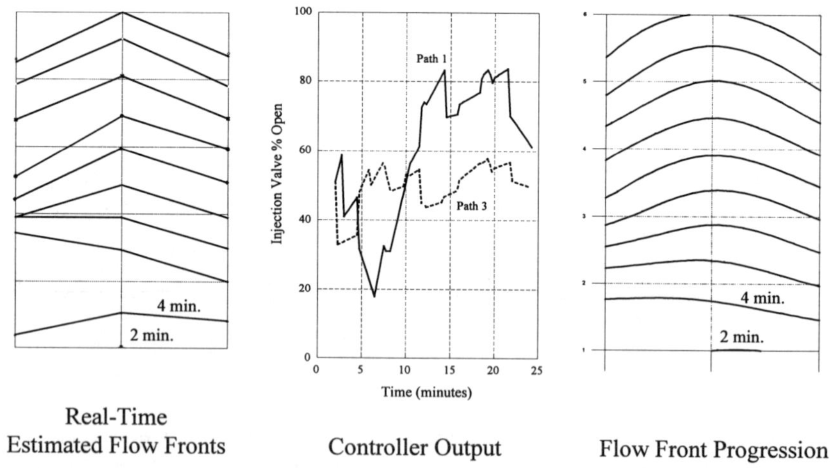

Figure 9: Control of flow front through PI control of feed rates to paths 1 and 3.

ACCELERATED CURING OF COMPOSITES USING SUPPLEMENTAL INTERNAL RESISTIVE HEATING

B. Ramakrishnan, L. Zhu, L. VanDerShuur, and R. Pitchumani*
Composites Processing Laboratory
Department of Mechanical Engineering
University of Connecticut, U-139
Storrs, CT 06269-3139

ABSTRACT

Curing of a fiber-resin mixture is often the critical and productivity-controlling step in the fabrication of thermosetting-matrix composites. Significant reduction in the overall processing times can be realized through shortening the cure cycle time, while ensuring a complete and uniform cure in the product. Toward this objective, this paper presents the use of conductive carbon mats embedded inside the composite as a means of providing additional heat generation in the form of resistance heating from within the composite during the cure process. The supplemental heating provides temperature and cure uniformity through the cross section as well as speeds up the cure process. In the context of application to resin transfer molding, systematic theoretical and experimental studies on the effects of the power supplied to, and relative placement of the resistive heating elements within the preform are presented with respect to the cure time and the product quality.

KEY WORDS: Composites processing, Resistive heating, Resin transfer molding, Thick section composites

1. INTRODUCTION

Fabrication of reinforced thermosetting composites is accomplished using techniques such as pultrusion, autoclave curing, and resin transfer molding, all of which share the common and critical step of cure. During the cure process the resin saturated preform is exposed to a prescribed temperature schedule—referred to as a cure cycle—which initiates and sustains a cross-linking polymerization reaction of the resin and transforms the fiber-resin mixture into a structurally hard composite product. In typical processing, the cure times could be on the order of hours, depending

---

*Author to whom correspondence should be addressed

upon the part thickness. An important objective of commercial composites fabrication is that of reducing the manufacturing time (and cost). Significant reduction in the overall processing times can be realized through a shortened cure cycle time, while ensuring a complete and uniform cure of the product.

During the cure process, the outer layers of the laminate, which are subject to external heating, cure far more rapidly than the inner layers, which are heated primarily by conduction from the outer layers. The differential curing rate is especially significant in thick sections and leads to structurally poor products than those intended by design. Toward addressing the above problem, alternative curing strategies using microwave energy have been explored (1,2,3). However, since microwave energy attenuates with thickness, the problem of differential curing persists, although alleviated slightly compared with conventional curing. As a result, the approach adopted in practice is the use of cure cycles of smaller magnitudes of temperatures and longer cycle durations, which has left the objective of rapid and affordable composite processing yet to be realized.

The shortcomings of the existing cure strategies in regard to the differential curing and the long cure cycle times may be avoided if the composite laminate being cured were heated from within the cross section in conjunction with peripheral heating. One approach to realizing this is the use of conductive fibers, such as carbon, as heating elements embedded within the laminate. The passage of electric current through the conductive fibers causes resistive internal heating which provides for a uniform curing through the thickness. Physically, this approach reduces to dividing the thick composite section into a number of thinner sections, each of which is cured uniformly by the resistance heating. Moreover, since the total energy input to curing the composite is increased, the cure cycle times will be reduced. The conductive fibers are left embedded in the composite since they are structural reinforcements themselves.

Initial studies on demonstrating the concept feasibility have been reported (4,5). For a viable practical implementation, however, a detailed understanding of the process is required. Toward this end, it is the intent of this study to (a) conduct a systematic experimental investigation of the effects of the number of resistive heating elements, their placement, and the power supplied to them, on the fabrication time and parameters influencing the product quality, and (b) develop a theoretical model describing the process, and to validate it with the experimental data. The experimental investigation is presented in the context of the resin transfer molding process, although the underlying methodology and results are generically applicable to other cure processes as well. The productivity enhancements afforded by the resistive heating approach are elucidated through comparison with the case of conventional curing. A processing window, identified based on quality constraints, is presented which provides the feasible range of power input as a function of the number of carbon mat layers.

The organization of the paper is as follows: Section 2 provides a brief description of the resistive heating experimental setup and Sections 3 and 4 present the thermochemical model used for a theoretical simulation of the configurations studied experimentally. The results of the study are presented in Section 5.

2. EXPERIMENTAL STUDIES

Experimental studies were conducted toward (a) investigating the effects of the three parameters—the number and the location of the resistive carbon mats embedded inside the composite, and the power supplied—on the cure time and the parameters influencing the quality of the composite, and (b) obtaining a processing window, and identifying the optimal processing conditions. A lab-scale resin transfer molding (RTM) setup, shown schematically in Fig. 1, was used for conducting the experiments. The resin transfer molding process has three unique stages involved, starting

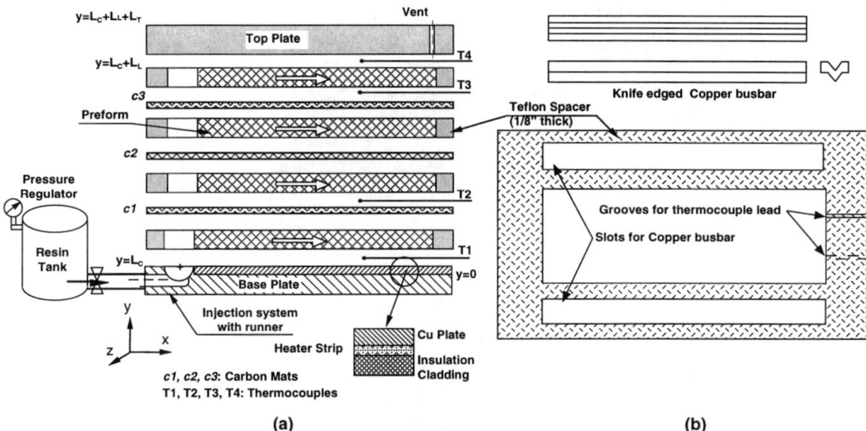

Figure 1: (a) Schematic of experimental setup used for the resistive heating studies. (b) Illustration of the Teflon spacer and Copper busbar

with the stacking up of several layers of tailored fiber mats into an enclosed mold. The second step includes the injection of a catalyzed resin mixture into the mold and its uniform permeation into the porous medium formed by the stacked fiber bed (known as a preform). Finally the whole system is heated, expediting the curing of the resin.

The main components include a pressure tank that houses the catalyzed resin and a mold in which the preform is placed. The mold consists of a sandwich construction of a bottom base plate, a stack of picture-frame spacers which determine the laminate thickness, and a top plate, all made using Teflon. Teflon was selected based on its (a) ease of machinability, (b) very low thermal conductivity, and (c) inertness towards the epoxy resin system, which facilitates demolding. The base plate is comprised of a resin inlet port which is a quarter inch diameter tapped hole that discharges into a concave runner along the width of the mold, as illustrated in Fig. 1(a). The runner ensures a uniform, essentially one-dimensional permeation of the preform along the mold length. As shown in Fig. 1(a), the Teflon mold additionally incorporates a cavity which houses a strip heater, for curing the resin. The heater cavity is closed by a copper plate that houses two the thermocouples and ensures uniform surface heating of the composite. The copper plate was surface ground to a smooth finish in order to minimize the surface interaction with the resin flow during the process, and was mounted flush with the surface of the Teflon base plate.

The experiments were conducted for a mold cavity (composite) thickness of 0.5 in., which was achieved through four spacers of 0.125 in. thickness each. Note that since the top Teflon plate acts as an insulating surface, the arrangement studied also simulates the case of symmetric curing of 1 in. thick composite. The use of four spacers provided for the incorporation of up to three resistive heating patches through the thickness of the preform. Slots were cut along the sides of the spacer frames, as illustrated in Fig. 1(b), in order to accommodate knife edged copper bus bars which not only secured the carbon mats in place and ensured firm contact throughout their length, but also provided for uniform distribution of the power supplied from the source. The power source supplied DC power with an adjustable voltage of 0 to 20 volts. Since the carbon mats are connected in a parallel arrangement to the power source, the total current gets distributed equally among them. In order to measure temperature variation through the thickness, each of the spacers also had slots cut through them on the side opposite the injection port to house the thermocouple leads. The top plate, made of 0.5 in. thick Teflon, included two air vents drilled to be in-line with the corners of the mold cavity furthest from the injection face, as shown in Fig.

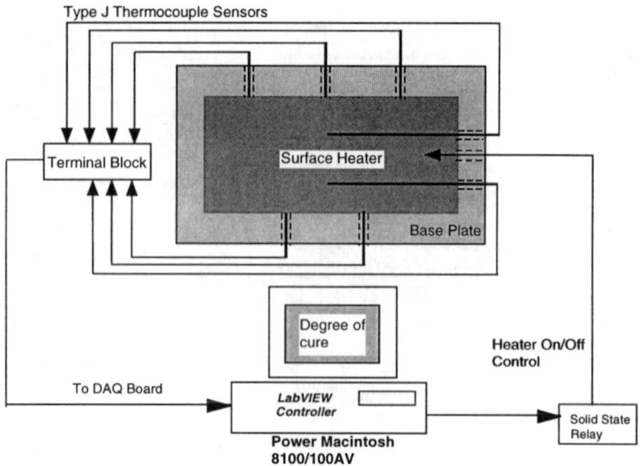

Figure 2: Schematic illustration of the temperature and cure data acquisition setup

$1(a)$. The overall sandwich setup was secured by four snap clamps.

Control of the surface heater mounted on the base plate of the mold was accomplished with a solid state relay which was regulated by a Proportional Integral Derivative (PID) control scheme implemented in LabVIEW (6). Temperature signals measured on the base plate and through the thickness of the mold by the thermocouples were acquired via a terminal block to an NB-MIO-16L-9 National Instruments data acquisition (DAQ) board as seen in Fig. 2. The locations of the thermocouples inside the mold are also illustrated in Fig. 2. The heating control was achieved by suitably adjusting the "on" (closed) and the "off" (open) duration of the solid state relay. Accordingly, the 0–100% output signal from the PID controller was scaled to control the length of time that the system heater remained at full power. The temperature through the thickness of the mold was recorded at the positions $T1$, $T2$, $T3$, and $T4$, as shown in Fig. 1, every 2 seconds. The temperature readings were used to compute the degree of cure in the composite, through its thickness, using the empirical kinetics model for the resin system, which was developed as explained in the next section. Once the degree of cure in the composite reached 97%, the system automatically shut off the power supplied to the heater and the resistive carbon mats. The time required for the composite to reach a cure of 97% throughout the cross section, was considered to be the cure time.

Continuous glass strand mat, M8610, supplied by the Owens-Corning Fiberglas Corporation was used as the reinforcement material in all the experiments. Carbon mats, G-104, supplied by Textile Technologies Industries were used as the resistive heating elements. The heating studies were conducted using epoxy resin, Epon-815, catalyzed with Epicure 3274, both supplied by the Shell Chemical Company. The resin selection for the study was based on its low viscosity which permitted quick fill times using relatively low injection pressures. The resin cure characteristics were obtained using differential scanning calorimetry explained in detail in the following section.

3. RESIN CURE KINETICS

The information on the resin kinetics is required for (a) monitoring the process as explained in the previous section, and (b) describing the exothermic source term in the thermal model presented in the next section. Since the kinetic parameters of the resin-catalyst mixture, Epon 815/Epicure 3274, was not available in the open literature, nor from the manufacturer, the cure kinetics was

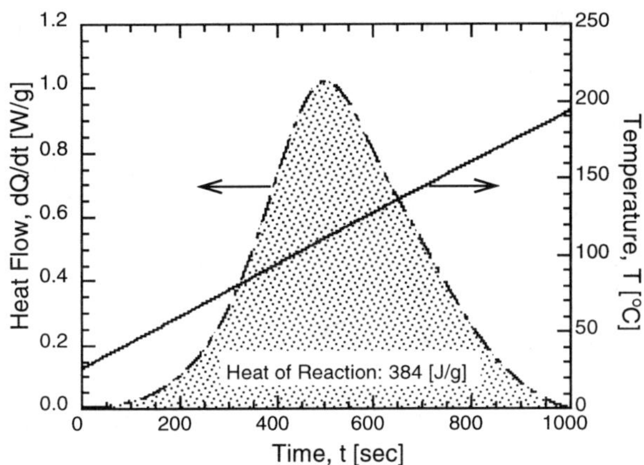

Figure 3: Estimation of the heat of reaction

characterized using differential scanning calorimetry (DSC). The goal of the characterization studies was to obtain an empirical expression for the cure rate of the form:

$$d\alpha/dt = (K_1 + K_2\alpha^m)(1 - \alpha)^n \tag{1}$$

which is shown in the literature to describe the experimentally-observed rate of cure reasonably well for epoxy resins (7). In the above equation, α denotes a fractional degree of cure, m and n are constants, and K_1 and K_2 are the rate constants that vary with the temperature following the Arrhenius relationship:

$$K_1 = k_{10}e^{-E_1/RT}; \; K_2 = k_{20}e^{-E_2/RT} \tag{2}$$

in which k_{10} and k_{20} are frequency factors, E_1 and E_2 are activation energies, and R is the universal gas constant.

First, the heat evolved during the cure reaction was measured by completing the reaction non-isothermally in which a predetermined amount of resin was heated in a differential scanning calorimeter from room temperature at the rate of 10 $°C/min$ until there was no more heat generated by reaction (which was reached at around 200 $°C$). Fig. 3 shows the rate of heat generated, dQ/dt, as a function of time, t, referred to as a thermogram for the non-isothermal DSC run. The total heat evolved due to the reaction, H_R, was obtained as the area under the thermogram by integrating the measured heat flow with respect to time.

Isothermal scans on the resin system were conducted to obtain dQ/dt versus time, t, for various temperatures. Since the total amount of heat generated by the cure reaction at any time, $Q(t)$, is directly proportional to the degree of cure, α, of the sample at that particular instant, the rate of cure, $d\alpha/dt$, can be defined as

$$\frac{d\alpha}{dt} = \frac{1}{H_R}\frac{dQ(t)}{dt} \tag{3}$$

Using the above relationship, the isothermal thermograms were cast as plots of $d\alpha/dt$ versus time, t, and the degree of cure, α, was obtained by integrating these plots with respect to time. Figure 4 shows the cure rate, $d\alpha/dt$, as a function of the degree of cure, α, at a selected few of the various isothermal conditions studied. The empirical parameters in the kinetics equation [Eq. (2)] were obtained as follows: First, the data for each temperature was fit to the right hand side of Eq. (1) to determine K_1, K_2, m and n. The natural logarithm of K_1 and K_2 were then plotted

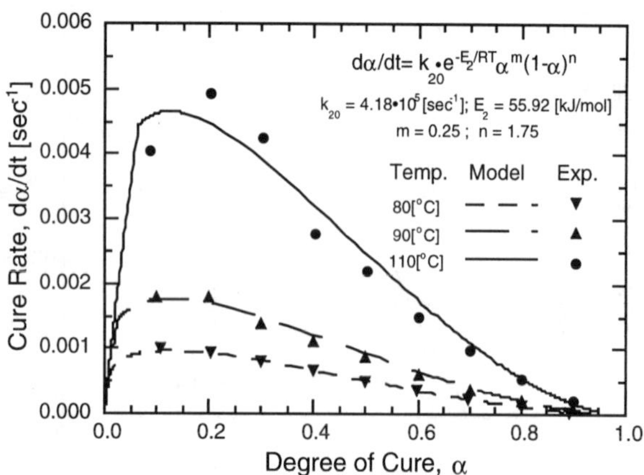

Figure 4: Model parameters for the kinetics of the resin system obtained from curve fitting the experimentally-measured variation of cure rate with respect to degree of cure

against $1/T\ [K^{-1}]$ to obtain k_{10}, k_{20}, E_1, and E_2. Further details on the procedure may be found in (7). For the Epon 815/Epicure 3274 resin system, the value of K_1 was determined to be zero, while the value of k_{20}, E_2, m and n are summarized in Fig. 4.

4. PROCESS MODEL

A numerical thermochemical model was developed to describe the cure process with the supplemental internal resistive heating. The process phenomena involved are (1) the heat transfer associated with the heating of the fiber-resin mixture, including the effects of the heat of the exothermic cure reaction and the resistive heating at the graphite mats, and (2) the kinetics of the cure reaction, described by the Arrhenius-type empirical relationship in Eq. (1).

The thermal model consists of solving the energy equation in Cartesian coordinates for the temperature distribution through the mold cross section. Assuming the heating to be uniform across the width and the length of the mold (x and z directions in Fig. 1), and the thickness of the mold to be small in comparison to the mold length and width, the energy transfer is predominantly one-directional through the mold thickness (y direction in Fig. 1). The governing equation, therefore, for the one-dimensional heat transfer in the three material domain (formed by the copper base plate, the laminate, and the top Teflon cover), accounting for the internal heat generation due to the exothermic cure reaction and due to the resistive heating of the embedded carbon mat in the composite, may be written as:

$$\frac{\partial\left[(\rho CT)_C\right]}{\partial t} = \frac{\partial}{\partial y}\left(k_C\frac{\partial T_C}{\partial y}\right) ; \ 0 \le y \le L_C \ \text{(Copper base plate)}$$

$$\frac{\partial\left[(\rho CT)_L\right]}{\partial t} = \frac{\partial}{\partial y}\left(k_L\frac{\partial T_L}{\partial y}\right) + C_{AO}H_R(1-v_f)\frac{d\alpha}{dt} + \psi\left(\frac{I}{A}\right)^2 ; \ L_C < y \le L_C + L_L \ \text{(Laminate)}$$

$$\frac{\partial\left[(\rho CT)_T\right]}{\partial t} = \frac{\partial}{\partial y}\left(k_T\frac{\partial T_T}{\partial y}\right) ; \ L_C + L_L < y \le L_C + L_L + L_T \ \text{(Teflon top plate)} \quad (4)$$

where ρC is the volumetric specific heat, T is the temperature, t and y are the time and location in the mold thickness direction, respectively, k is the thermal conductivity, C_{AO} is the initial concentration of the resin, A is the cross sectional area of the resin saturated carbon patch perpendicular to the direction of the current flow, and v_f is the fiber volume fraction. The subscripts C, L and T refer to copper base plate, laminate and the Teflon top cover, respectively. Additionally, ψ is the resistivity of the carbon mat and I is the current passing through it. The thermal conductivity of the carbon mat was obtained by using the model proposed by (8), from which the electrical resistivity, ψ, was evaluated by applying Lorenz's law relating thermal and electrical conductivities. Note that the heat generation term due to the resistive heating of the carbon mat is specific to the location where it is being placed, and is zero at all other points in the laminate cross section. Further, $d\alpha/dt$ denotes the rate of the cure reaction, which together with the heat of reaction, H_R, determines the heat release rate during the cure process. The expression for the reaction rate is given by the kinetics model, Eq. (1).

The governing equation, Eq. (4), is subjected to the following initial conditions in the three regions, namely the copper plate, the laminate and the Teflon cover:

$$T_L(y,0) = T_C(y,0) = T_T(y,0) = T_o; \alpha(y,0) = 0 \text{ in the laminate} \tag{5}$$

where T_o is room temperature. The boundary conditions associated with the governing equations, Eq. (4), are the prescribed cure temperature cycle at the bottom of the copper base ($y = 0$ in Fig. 1) and convective heat loss at the top of the Teflon cover ($y = L_C + L_L + L_T$ in Fig. 1). In addition, the temperature and the heat fluxes must be continuous at the laminate-copper interface ($y = L_C$) and the laminate-Teflon cover interface ($y = L_C + L_L$). These conditions can be represented as follows:

$$T = T_{cure}(t); \; y = 0$$

$$-k_T \frac{\partial T_T}{\partial y} = h(T_T - T_o); \; y = L_C + L_L + L_T$$

$$T_C = T_L; k_C \frac{\partial T_C}{\partial y} = k_L \frac{\partial T_L}{\partial y}; \; y = L_C$$

$$T_L = T_T; k_L \frac{\partial T_L}{\partial y} = k_T \frac{\partial T_T}{\partial y}; \; y = L_C + L_L \tag{6}$$

where h is the heat transfer coefficient and all other terms in the above equations are as defined previously. The heat transfer coefficient was obtained using the correlation for natural convection over a flat plate configuration. The thermochemical equations listed in Eq. (4) were solved for the temperature and cure profiles in the composite using a finite difference method as explained in ref. (9).

5. RESULTS AND DISCUSSION

The resistive heating experiments were conducted by systematically varying the number of carbon mats, their location in the preform and the power supplied to them. The objective of the studies was to understand the effects of these variables on the composite fabrication time and the parameters influencing final product quality. The number of carbon mats were varied from one to three, and for each case, two power inputs were considered, resulting in six different configurations. Table 1 details the six different cases along with the nomenclature used to identify the different configurations and also presents the locations of the carbon mats in each case.

The three parameters considered toward determining the quality of the composite were the maximum temperature, the maximum temperature difference in the cross section, and the maximum temperature gradient in the composite during the cure process. Increasing the cure temperature leads to faster cure times; however, very high temperatures could degrade the resin matrix. The

Table 1: Process parameters used in the studies.

Number of carbon mats	Power supplied [Watts]	Location from the mold base [inches]	Nomenclature
0	0	0	base
1	9	3/8	1cp1
	24	3/8	1cp2
2	9	3/8, 1/4	2cp1
	24	3/8, 1/4	2cp2
3	9	3/8, 1/4, 1/8	3cp1
	24	3/8, 1/4, 1/8	3cp2

temperature difference is a measure of the temperature homogeneity which determines the cure and property homogeneity within the composite. Faster curing rates can be achieved by increasing the temperature gradients but this might lead to large thermal stresses in the composite. In addition to the quality-related parameters, the cure time was investigated as a productivity measure.

The effects of the various supplemental heating configurations on the temperature distribution within the composite are shown in Fig. 5. Only four cases, the base case, $1cp1$, $2cp2$, and $3cp1$ (Table 1), are included in Fig. 5 for brevity. A common cure cycle was used for all the cases studied. The cure cycle comprised of raising the temperature of the mold base, initially at room temperature, to $130\,^{\circ}C$ within about 300 seconds and maintaining at $130\,^{\circ}C$ until the composite was cured. The discrete markers in Fig. 5 represent the temperatures recorded from the experiments while the continuous lines represent the model-simulated values for the temperature distribution within the composite. The model predictions follow the experimental temperature profiles closely for all cases, throughout the cure reaction. The model predicts the magnitude and the location of the exotherm within the composite accurately, to within the experimental error.

Figure 6 shows the effects of the various heating configurations on the resulting values of the quality-influencing parameters namely, the maximum temperature in the laminate, T_{max}, the maximum temperature difference across the section, ΔT_{max}, and the maximum temperature gradient experienced by the material during the cure process, $\dot{T}_{max}$, and the cure time, t_{cure}. Figure $6(a)$ presents the experimental results, while the data in Fig. $6(b)$ correspond to the numerical simulation of the different configurations. Further, the results are expressed in Figs. $6(a)$ and $6(b)$ as a percentage change with respect to the base case in which resistive heating was not included. Positive values on the plots denote an increase relative to the base case, while negative values represent a decrease. Overall, the predictions of the process simulations are seen to match well with the experimental trends on each of the parameters, which are described below.

With the addition of the resistive heat from the carbon mats, the non-uniformity in temperature due to peripheral heating is reduced making the temperature more homogeneous throughout the composite. As seen in Fig. $6(b)$, with the exception of the $1cp1$ case, the temperature difference within the composite is lower for all the other cases with respect to the base case. The abnormality of the temperature difference for the $1cp1$ case being larger than the base case is due to experimental error. Note that the $1cp2$ case, as seen in Fig. 6, has the lowest temperature difference throughout the composite thickness than most of the other cases.

Of the six cases studied, the maximum temperature gradient is most severe for the $1cp2$ case. The temperature gradient was found to be principally a function of the rate at which the carbon mat heated up, which is in turn a function of the resistance of the carbon mat and the magnitude of the current passing through it. As the number of carbon mats increases, the total current

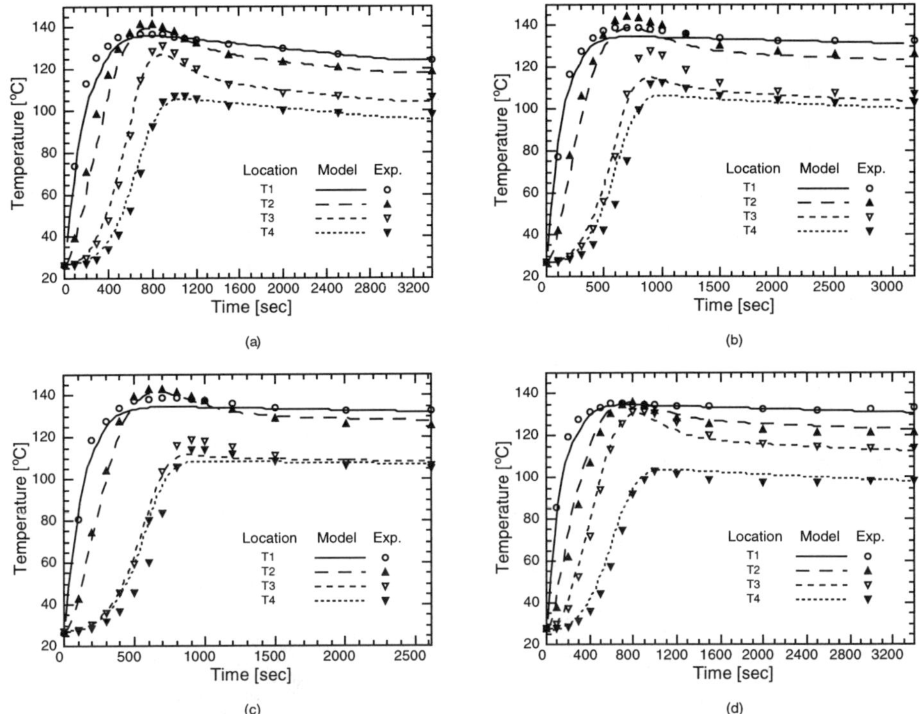

Figure 5: Variation of experimental and model predicted temperatures through the thickness of the composite for base heating only (a), and the supplemental heating configurations, $1cp1$ (b), $2cp2$ (c), and $3cp1$ (d).

supplied is divided among them, which leads to lower power dissipation locally. The case of $1cp2$ corresponds to the supplied power being dissipated entirely in one location, resulting in the highest temperature gradient.

It is also noted from Fig. 6 that, as expected, the cure time is reduced with the addition of the resistive heating elements. For a given number of carbon mats, the cure time decreases with increasing power supplied. Furthermore, the case of a single carbon mat with a power input of $24W$ (the $1cp2$ case) is seen to yield the least processing time. This configuration, therefore, represents the optimal arrangement in the absence of quality constraints.

For practical manufacturing, however, constraints on the maximum permissible temperature, T_{max}, the allowable maximum difference in temperature within the composite, ΔT_{max}, and the maximum rate of increase in temperature, $\dot{T}_{max}$, must be considered in conjunction with minimizing cure time. A processing window was developed towards determining the permissible range of power that satisfies the above-mentioned constraints. For a fixed number of carbon mats, a linear variation was assumed for each of the parameters—T_{max}, ΔT_{max}, $\dot{T}_{max}$, and t_{cure}— between the two power inputs studied (9 W and 24 W). For chosen values of the constraints, upper and lower bounds on the power were then determined by superimposing the inequality constraint specifications on the assumed linear variations. This yielded three permissible ranges of power for resistive heating using 1, 2 and 3 carbon mats.

Figure 7 shows the processing window in terms of the power input as a function of the number of carbon mats used. The shaded bands in the plot indicate the range of acceptable power which

251

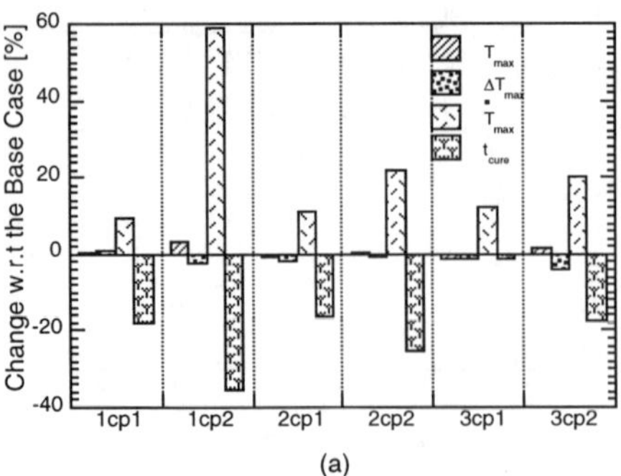
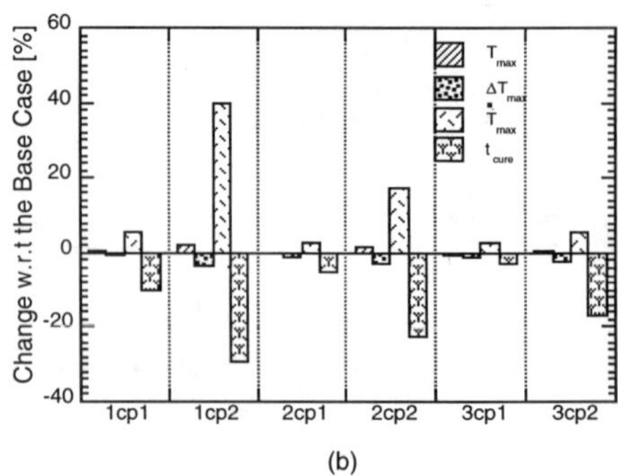

(a)   (b)

Figure 6: Comparison of (a) experimentally established and (b) model-predicted values of the parameters T_{max}, ΔT_{max}, $\dot{T}_{max}$, and t_{cure}, normalized with respect to the base case, for the different cases listed in Table 1

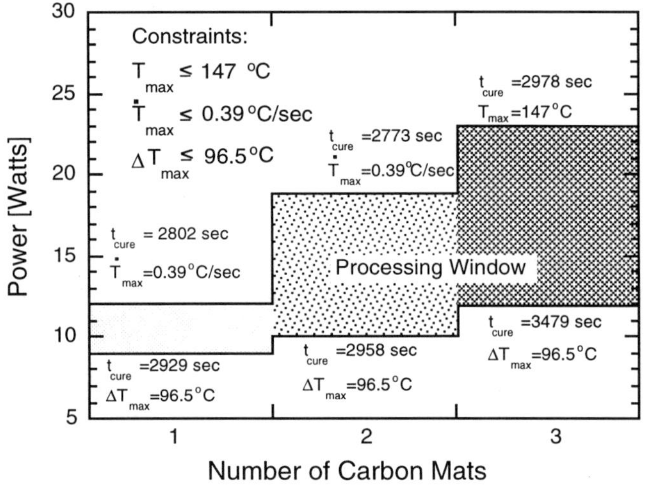

Figure 7: Processing window obtained from the experimental study

satisfies the constraints listed in Fig. 7. Also identified in the Figure are the active constraints corresponding to the upper and the lower bounds. The lower bands are seen to be governed by the permissible temperature difference within the composite (96.5 $^{\circ}C$). The upper bounds, however, are determined by the constraints on the maximum allowable temperature gradient (0.39 $^{\circ}C/sec$) and the maximum temperature within the laminate (147 $^{\circ}C$). As mentioned previously, increasing the number of resistive carbon mats embedded within the composite leads to a reduced local power dissipation within each mat. This explains the increasing trend of the upper and the lower bounds with increasing number of carbon mats, although the shift is less pronounced for the lower bound.

It is clear from Fig. 7 that the upper bound corresponds to the maximum permissible power, and therefore, to the least cure time, subject to the constraints, for a given number of carbon mats. In particular, the use of 2 carbon layers, at its corresponding upper bound on the power (about 19W) is seen to be the optimum configuration which minimizes the cure time as well as satisfies the imposed constraints. Additional constraints, such as process capability, may be added toward obtaining a more refined process window. Furthermore, process simulations conducted over a wide range of parameters may be used to obtain a better estimate of the processing window and the optimum parameters, as presented in ref. (10).

6. CONCLUSIONS

An alternative approach towards improving composite processing in terms of both quality and fabrication time by using the resistive heating of the carbon mats embedded inside the mold was presented. Both experiments as well as numerical model simulation were conducted for a range of combinations of power and carbon mat locations within the mold, and were used to describe the process parameters. The results from the numerical simulations showed a similar trend as that obtained from the experiments. A processing window was developed toward identifying the range of acceptable power that can be supplied for a given placement of the carbon mats, without violating the quality constraints.

7. ACKNOWLEDGMENTS

The authors gratefully acknowledge the financial support for the project provided by the National Science Foundation through Grant No. DMI-9522801. The Owens/Corning Fiberglas corporation and the Shell Chemical Company are acknowledged for their gracious donation of the materials used in the study.

8. REFERENCES

1. S. Carrozino, et. al., Polymer Engineering and Science, 30, (6), 366 (1990).

2. W. I. Lee and G. S. Springer, Journal of Composite Materials, 18, 357 (1984).

3. W. I. Lee and G. S. Springer, Journal of Composite Materials, 18, 387 (1984).

4. D. Butler and R. S. Engel, ICCM Proceedings, 10,(3), 269 (1994).

5. E. Sancaktar, M. Weijian, and S. W. Yugartis, Journal of Mechanical Design, 115, 53 (1993).

6. L. VanDerSchuur and R. Pitchumani, Virtual Instrumentation in Education Proceedings, 1, 10 (1997).

7. C. D. Han, D. S. Lee, and H. B. Chin, Polymer Science and Engineering, 26, (6), 393 (1986).

8. Q. Ning and T. Chou, Journal of Composite Materials, 29, 2280 (1995).

9. N. Rai and R. Pitchumani, Polymer Composites, 18, (4), 566 (1997).

10. L. Zhu and R. Pitchumani, 13th Annual Technical Conference of the American Society for Composites (1998).

43rd International SAMPE Symposium
May 31-June 4, 1998

ON-LINE PROCESS MONITORING AND ANALYSIS OF LARGE THICK-SECTION COMPOSITE PARTS UTILIZING SMARTWEAVE IN-SITU SENSING TECHNOLOGY

J. E. Bradley, J. Diaz-Perez, Dr. J. W. Gillespie Jr.,
Center for Composite Materials, University of Delaware, Newark, Delaware
Dr. B. K. Fink, U. S. Army Research Laboratory,
Aberdeen Proving Ground, Maryland

ABSTRACT

Large thick composite parts have traditionally been difficult to monitor in situ during resin infusion and cure operations due to the cost, complexity, and/or reliability of the sensing technology.   Additionally, one of the more promising low-cost sensing options, resin ionic conductance monitoring, has previously been incompatible with parts fabricated with conductive fibers due to the conductive fibers providing unwanted low impedance paths.

This paper describes how resin flow and cure status of large thick parts including electrically-conductive preforms can now be reliably and inexpensively monitored, in real time, during resin infusion and cure using zonal resin conductance measurement techniques.   Utilizing SMARTweave online sensing technology and ionic conductivity algorithms, the Army Research Laboratory (ARL) and the University of Delaware Center for Composite Materials (CCM) have developed a robust resin infusion composites manufacturing process monitoring system capable of real-time monitoring, analysis, and display of thermoset resin flow and cure state.   The system is presently capable of processing 4096 individual sensors that can be deployed in a complex multi-planar configuration.   Developed for operation in industrial environments, signal processing techniques allow unambiguous acquisition and interpretation of zonal data independent of resin system or fiber. Virtual instrumentation software graphically displays (2-D and 3-D) real-time resin flow progress during injection on a multi-color grid corresponding to the zonal matrix. Incorporating a theoretical model for ionic conductivity and experimentally-determined cure kinetics for various thermoset resin systems, the software also displays resin flow front arrival times, gelation times, and projected demold times.   Part-specific setup routines give the operator flexibility in selecting individual zonal information, and system software has been optimized to allow maximum scan speed for each I/O configuration. The new monitoring system has been successfully utilized in the manufacture of a 3' X 20' X 1" thick carbon/epoxy part.

KEY WORDS: SMARTweave, On-line Sensing, Large Thick Composite Parts

1. INTRODUCTION

VARTM (<u>V</u>acuum <u>A</u>ssisted <u>R</u>esin <u>T</u>ransfer <u>M</u>olding) is a cost-effective method for fabricating large, moderately intricate, fiber-reinforced polymer composite parts. Multiple fiber layers are placed into the tool and/or covered with a layer of bagging film to form an airtight seal. Drawn by the vacuum, resin is then allowed to flow into, and fill, the fiber preform from an injection port.

SMARTweave sensing technology, which measures changes in the conductivity of uncured resin between excitation and sense leads, is ideally suited to VARTM applications. Leads can be easily introduced between layers of fiber material. Resin cure temperatures are typically low enough to preclude damage to the leads and interface hardware.

Figure 1. Operator control station

Figure 2. Large part during resin infusion
(Sensing leads/cables at right)

Over the past several years, the research has been conducted to incorporate SMARTweave sensing technology into the VARTM manufacturing process. Prior to the system described herein, VARTM applications using SMARTweave sensing technology were limited to 64 nodes (8 sense X 8 excitation) due to signal degradation and transient noise. This noise, comprised of both cyclic and induced components, adversely affected SMARTweave's robustness, limiting its utilization in a manufacturing environment. Additionally, SMARTweave sensing technology couldn't be utilized in applications involving parts fabricated from conductive fibers, as the fiber material would produce high-resistance shorts between the excitation and sense leads. This paper summarizes recent advances to address these limitations.

1.1 Development Overview

Performance advancements of this resin flow/cure monitoring system over previous iterations include:
- Enhanced process data software filters incorporating refinements to the equivalent circuit model of the (1) system hardware interface and the (2) sensing grid internal to the composite part.
- Real-time interactive graphic displays, capable of being accessed and manipulated by an operator, detailing resin flow, nodal resin arrival times, and nodal projected resin de-mold times.
- Improved system sensitivity and response.
- Reduced cyclic and transient noise.
- Reduced signal variability between individual nodes (excitation-sense junctions).
- Broadened thermoset resin system monitoring.
- Ability to monitor parts fabricated with conductive preforms.
- Multi-planar through-thickness sensor grid configurations for 3-D flow and cure monitoring.

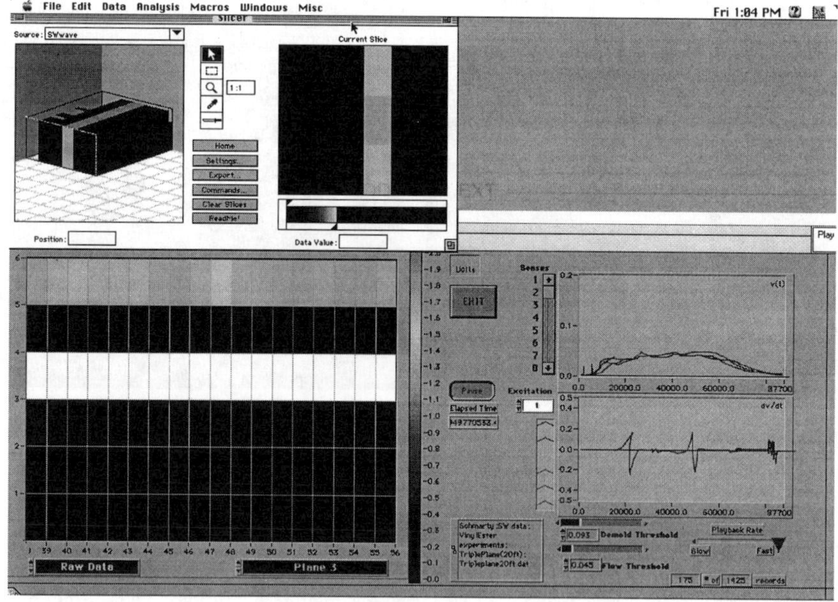

Figure 3. GUI Displaying Real-time 2-D and 3-D Resin Flow into Part

1.2 System Configuration

1.2.1 System Software

The system software utilizes National Instruments' LabVIEW virtual instrumentation software to control the sequential application of excitation voltages to individual excitation leads and to monitor, display and archive sense voltages resulting from dc current flow through the uncured resin between the excitation lead and sense leads aligned in a row orthogonally to it. During injection, resin flow progress is displayed graphically in real time on a multi-color grid corresponding to the nodes formed by the excitation-sense matrix. As a node becomes saturated with resin, the corresponding increase in conductivity is displayed as a change in color. In this manner, resin flow fronts can be easily followed by the operator. If selected by the operator, resin flow arrival times, gelation times, and projected demold times can also be displayed. In addition to the graphical display, a table of individual nodal sense voltages appears on the front panel. This information, in conjunction with the graphic displays (2-D and 3-D), provides detailed information about the resin cure status of each node.

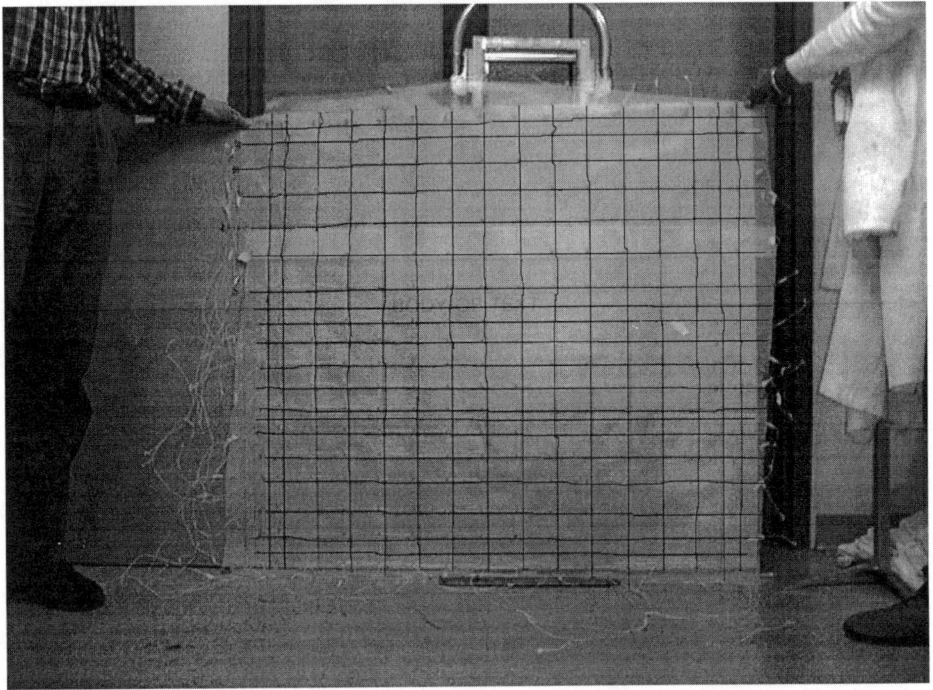

Figure 4. Pre-fabricated excitation/sense grid layer (40" X 40"). Note asymmetric
spacing conforming to part geometry

Designed to monitor complex multi-planar nodal configurations, part-specific setup routines give the operator great flexibility in selecting individual excitation and sense IO and the software has been optimized to allow maximum scan speed for each IO configuration. Since each resin system has specific conductivity characteristics, which requires a unique value for its sense resistor, the setup software provides individual initial set point adjustment for the excitation and sense scan rates based on the nodal RC time constant. A playback mode, available when the system is off line, allows the operator to review either the graphical run data (resin flow, resin arrival time, gelation time, demold time, or biaxial (voltage, time) charts of any combination of nodes.

1.2.2 System Hardware

The system hardware includes a Macintosh-resident processor and National Instruments PCI-MIO-16E Multifunction I/O card, a National Instruments SCXI chassis with two 32-channel analog output modules, and two 32-channel analog input modules which receive and multiplex sense voltage inputs from a proprietary Analog I/O Interface Chassis. Digital outputs from the MIO board are sent to the two SCXI analog output modules that, in turn, energize/deenergize excitation relays located in the Analog I/O Interface Chassis.

The Analog I/O Interface Chassis is capable of supplying 64 independent excitation voltage outputs to the in-part excitation/sense grid and receiving 64 unconditioned sense voltage inputs in response. Resin system-specific sense resistor packs allow optimization of cure voltage strength and system scan rates. Independent excitation voltage relays, with Form C contacts, minimize sense voltage signal degradation due to parallel resistance paths.

The excitation/sense grid consists of layers of rows of parallel wires. Each successive layer is separated from its neighbors by one or more layers of fabric. Layers alternate between excitation and sense function with excitation wires perpendicular to the sense wires. Each layer of excitation wire rows, coupled with a corresponding layer of sense wire rows, forms an orthogonal grid and is referred to as a sensor plane. Within each plane, the fiber material-filled gaps between the excitation and sense wires form high resistance junctions, that when saturated with resin couple with sense resistors located in the Analog I/O Interface Chassis to produce a voltage divider network for the applied excitation voltage (V_{ex}). Ignoring parallel resistance effects of neighboring junctions, the sense voltage (V_s) is defined by the equation:

$$V_s = (V_{ex} - V_s) * R_s / R_{junction}$$

$R_{junction}$ is equal to the equivalent resistance of the corresponding resin-filled fiber layer in the gap between the excitation and sense leads. The excitation/sense grid is electrically connected to the Analog I/O Interface Chassis via individually shielded cable bundles, eight cables to bundle, or in the case of large, complex parts, through part-specific wiring buses.

2. SYSTEM DEVELOPMENT

2.1 Thermoset Resin Characterization

The voltage divider equation described in the previous section was used to determine the excitation to sense line junction resistance of various Vinyl Ester and Epoxy resin systems. Using a test cell similar to that developed by England (1), resin resistance for a 25 mil gap between excitation and sense leads was found to vary between 100 MΩ and 1000 MΩ, depending on the state of the resin and, in the case of VE, percentages of Trigonox Accelerator, CoNap Initiator, and 2,4-P Inhibitor. Contact surface area and adjacent resin volume are major components in the resistance equation. An important result of England's work was the discovery and documentation of change in the sense signal voltage with time that could be directly correlated to the onset and end of gelation (Figure 5). The dV/dt_{min} point was found to accurately pinpoint the onset of gelation, with $d\alpha/dt_{max}$ corresponding to when the cure reaction becomes a pure diffusion limited process and ionic conduction ceases. System cure analysis and prediction software utilizes this dV/dt information to predict gelation for vinyl ester. Demold times for VE resin systems are defined as the time for the signal to drop within the noise background. Epoxy and phenolic resin systems exhibit no pronounced dV/dt_{min} or $d\alpha/dt_{max}$ but DSC experiments have confirmed that gelation for these resins also occur between 60% and 40% of the initial sense voltage signal, respectively. Currently, these values are used as the onset and end of gelation respectively.

2.2 Software Development

LabVIEW virtual instrumentation software by National Instruments is inherently multi-tasking with parallel subroutines in the logic flow sequence executed simultaneously. While desirable in many applications, this "lack of structure" presented a problem for flow monitoring. Quadrupling in complexity from the previous iteration, interaction between the over 100 sub-VI's requires a precise sequence of events (a sub-VI is analogous to a stand-alone subroutine in a traditional command line entry programming languages such as C). This is accomplished by dividing the flow monitoring, data analysis, storage, transfer, and visualization operations into eight discrete operations (Figure 6), ensuring that data is available when needed. This segmentation of the process steps has another benefit, its inherent modularity allows efficient enhancement and modification of system capabilities.

Guided by an on-screen display and menu, the operator manually enters pertinent plane, excitation array, and sensor array information into memory. If a part incorporates multiple sensing planes, an ancillary OEM three-dimensional display program is automatically activated to display operator-requested resin flow parameters in a real-time 3-D display appearing above the main 2-D display. Operator-selected cross-sectional "slices" for any 2-dimensional plane are also available.

The 2-D display also has several display options, including a 2-D representation of a multi-planar configuration (Figs. 1 & 3), and visualization of any single plane. At any time during the resin infusion and curing events, the operator can change the 2-D display to view resin flow, cure, gelation, or demold information for any or all planes. System software continuously calculates a dV/dt for each sensor voltage, and utilizes this information to determine projected resin gelation and demold times, based on information entered by the operator during initial configuration.

Additional parameters entered at this time include delay times between excitation scans, delay times between scans of individual sensors along a common excitation wire, sensor scan frequency, sensor data write frequency, and chart/plot range settings. Before transferring this information into memory, a pop-up menu asks the operator if resin injection will occur perpendicular to the excitation grid longitudinal axis.

If the operator enters YES, only the excitation wire closest to the injection port in each plane is initially energized. As resin infusion progresses, and resin is detected by a sensor associated with this excitation wire, the next adjacent excitation wire will be added to the scan cycle. This will continue in succession until all excitation wires are included in the scan cycle. By selectively adding excitations in this manner, system response is optimized through the elimination of irrelevant sensors.

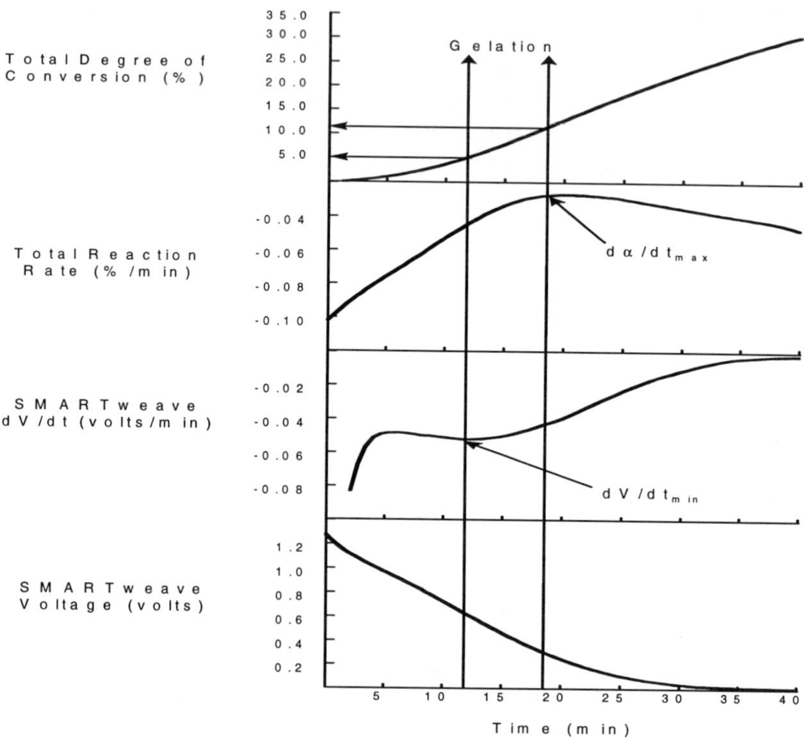

Figure 5.   SMARTweave voltage & dV/dt vs. cure/reaction rate dynamics

During initial setup, the operator is prompted to enter pertinent log information. This information includes excitation voltage, sense resistor value, resin system and fiber details, number and composition of ply layers, resin additives, cure temperatures, and initial resin resistivity. Initial resin resistivity is a derived parameter and is entered by performing a single node test on a small sample of the resin prior to injection. When requested by the operator, the excitation associated with the test node is energized, and using a voltage divider algorithm $(R_{resin} = (R_{sense} / V_{sense}) * (V_{excitation} - V_{sense}))$, the resistance of the sample is calculated and entered into memory.

2.3 Hardware Development

2.3.1 Analog IO Interface Circuit Characterization

Embedded between the reinforced fiber layers, the excitation and sense wires form an orthogonal grid as shown in Figs. 4 & 7. When injected into the fibers, the resin produces a high resistance path, R_{jxy}, between overlapping excitation and sense wires. While high resistance paths are also formed between the excitation line under excitation and the neighboring layer of sense wires above it, the scanning software designates excitation-sense pairs, and sense inputs from the above sense wire layer are not monitored. Additionally, distances between neighboring sensor planes (each plane is formed by an excitation-sense layer pair) is much greater than the distance between the excitation and sense layers forming the plane, resulting in a much higher resistance path. Figure 7 is a simple schematic of the

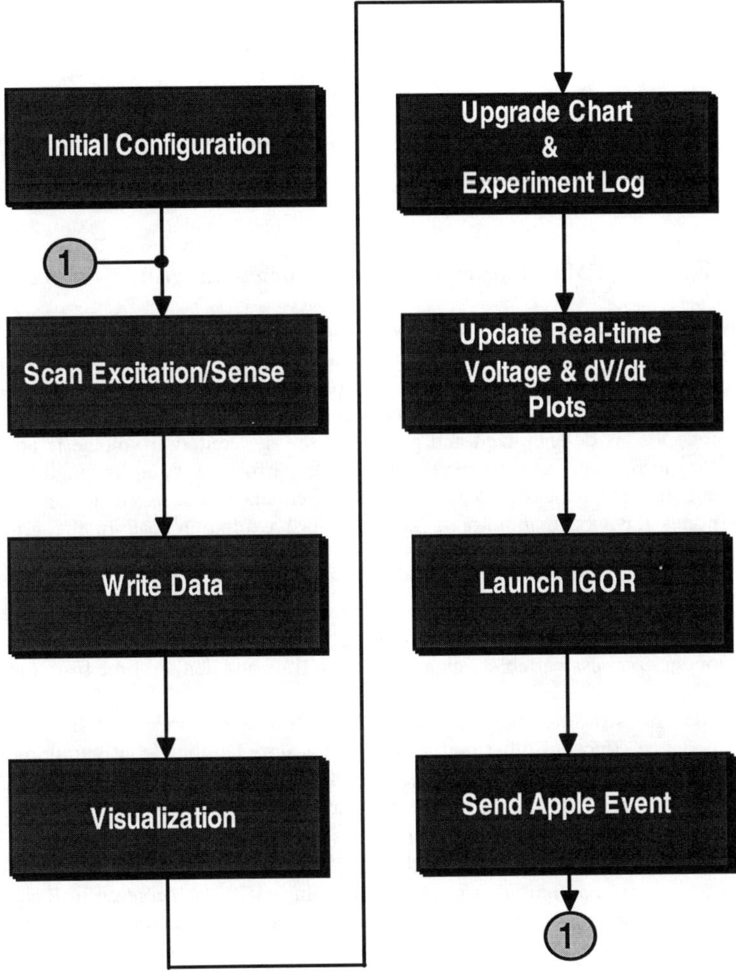

Figure 6.   LabVIEW VI software slow sequence

resistance network formed by a excitation-sense layer pair (plane), with 3 excitation wires and 3 sense wires. To monitor the sense voltage at each excitation-sense node, the system applies a

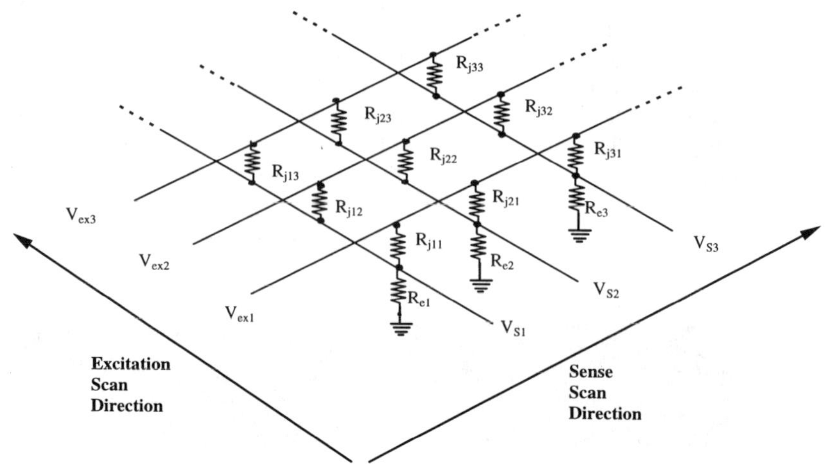

Figure 7.   SMARTweave in-part excitation/sense grid

dc excitation voltage to the first excitation wire (henceforth referred to as a lead) and monitors each analog sense, V_{sxy}, associated with that excitation lead (in this example, R_{j11} - R_{s11} , R_{j21} - R_{s21} , R_{j31} - R_{s31}) in succession. The other excitation leads, V_{ex2} , and V_{ex3} , are grounded while V_{ex1} is energized to minimize noise due to induced EMF. After all nodes along its length are monitored, V_{ex1} is de-energized and grounded, the dc excitation voltage is applied to the next excitation lead, V_{ex2} , and the sequential scan is performed along its length. V_{ex2} is then grounded and the next excitation lead in the scan sequence is scanned in the same manner. When all nodes have been monitored along the last excitation lead in the grid, the scan sequence is repeated.

V_s varies directly with R_j, which in turn varies directly with the resin cure state. System software continuously monitors V_s to determine and display resin cure properties, and then derives secondary process variables, such as gelation time and demold time from them.

Equation 1 describes the resistance network of an ideal single node grid and does take into account impedance effects of other nodes in a typical more complex configuration such as that illustrated in Figure 8. It also ignores the effects of the input filter circuitry of the SCXI analog input modules. The latter was determined to have minimal impact at the scan frequencies utilized, with input impedances of the AI modules exceeding 1 gigohm (10^3 MΩ). Increasing the number of excitation and sense lines does have an impact on V_s. Adding excitation and sense lines to the grid configuration modifies the value of R_s. A decrease in R_s relative to R_j results in a corresponding decrease in V_s. If the software fails to recognize, and compensate for, this drop-off in signal amplitude, resin cure status will be misinterpreted.

Increasing the number of sense lines has no effect on the amplitude of V_s. Each additional node does increase the current load on the excitation voltage power supply, but since current through each node is on the order of a few picoamps (10^{-9} amps), the practical effect is minimal.

Increasing the number of excitation lines has a significant effect on V_s (Equation 4). This is a result of the system scan algorithm which applies a ground to all unenergized excitation leads in order to reduce induced EMF from external sources (Figure 6). It can be seen from Equation 4 that, as j (number of excitation lines) approaches the ratio of R_j/R_s (typically 40/1) the amplitude of V_S becomes increasingly smaller.   Figure 7 illustrates the degree of signal reduction versus number of excitation lines for various resin resistivities.

$$V_s = V_{ex}/(1 + R_j/R_s) \tag{1}$$

$$R_{sequivalent} = (R_s)(R_{junction(i-1)})/(R_s + R_{junction(i-1)}) \tag{2}$$

$$R_{junction(i-1)} = R_{junction}/(i-1) \tag{3}$$

i = No. of Excitation-Sense junctions along a common sense line that are wetted w/ resin.

Since $V_s = V_{ex}/(1 + R_j/R_s)$, V_S for a sense line with multiple excitation lines that have excitation to sense junctions, R_j wetted with resin will be equal to $V_{ex}/(1 + R_j/((R_s)(R_{junction(i-1)})/(R_s + R_{junction(i-1)})))$.

$$R_{sequivalent} = (R_s)(R_j/n)/(R_s + (R_j/n)) = R_sR_j/(n\,R_s + R_j) = R_j/(n + R_j/R_s) \tag{4}$$

n = number of wetted junctions (j) along a common sense line minus the junction under test. (n = j - 1)

$$\mathbf{V_{sactual}} = V_{ex}/(1 + R_j/R_{sequivalent}) = V_{ex}/(1 + R_j/R_j/(n + R_j/R_s)) = V_{ex}/(1 + n + R_j/R_s)$$
$$= \mathbf{V_{ex}/(j + R_j/R_s)} \tag{5}$$

When there is only one excitation line, j = 1, and Equation 4 becomes identical to Equation 1.

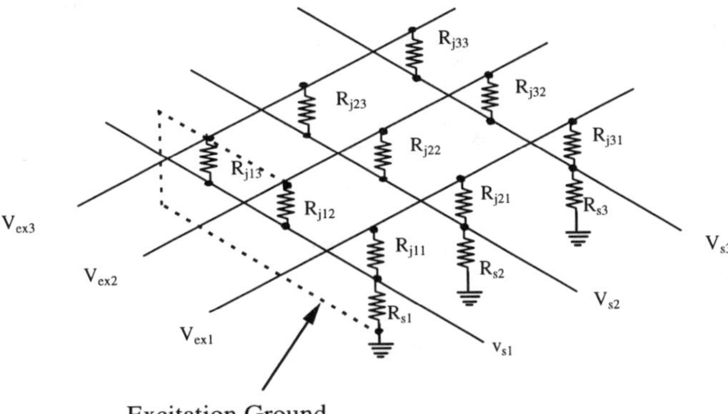

Figure 8.   Special case - multiple excitations along single sense line

2.3.2 Multi-planar Excitation/Sense Grid Development

Early multi-planar excitation/sense grid designs sought to minimize grid complexity and system overhead by "ganging", or physically connecting, excitation lines from different levels (planes) that were located in a common z-directional plane. This introduced a non-uniformity in the sense signal strengths. Sense nodes associated with ganged excitation lines exhibited higher sense voltages than their neighbors. This anomaly was further exacerbated in parts fabricated from conductive fiber. Even with special excitation and sense wires developed to allow utilization of conductive fiber (Section 3.3.6), signal strengths increased dramatically when excitation line ganging was employed. Sense signal voltages were highest for intermediate planes. In large parts with more than three planes, sense inputs were up to an order of magnitude greater than anticipated, and demonstrated a high variability. Figure 10, depicting two ganged excitation leads, illustrates why this occurred. They are actually in a common z plane. For normal system scan parameters, the junction resistance, R_{jij}, of the resin matrix-fiber layer junction is actually a series-parallel network (Frame 2). The resistance of the E-glass excitation and sense wire shrouds is essentially infinity, and the resistance of the fabric media is either infinity for non-conductive fibers, or in the case of conductive fiber, less than 1 KΩ regardless of distance from the sense node. Non-conductive fiber media, Frame 4a, provide a complex resistive component, $R_{Matrix3}/X$ in which X represents a multiplier (X > 1) corresponding to the relative distances of excitation wires Ex1 and Ex2 from the sense wire.

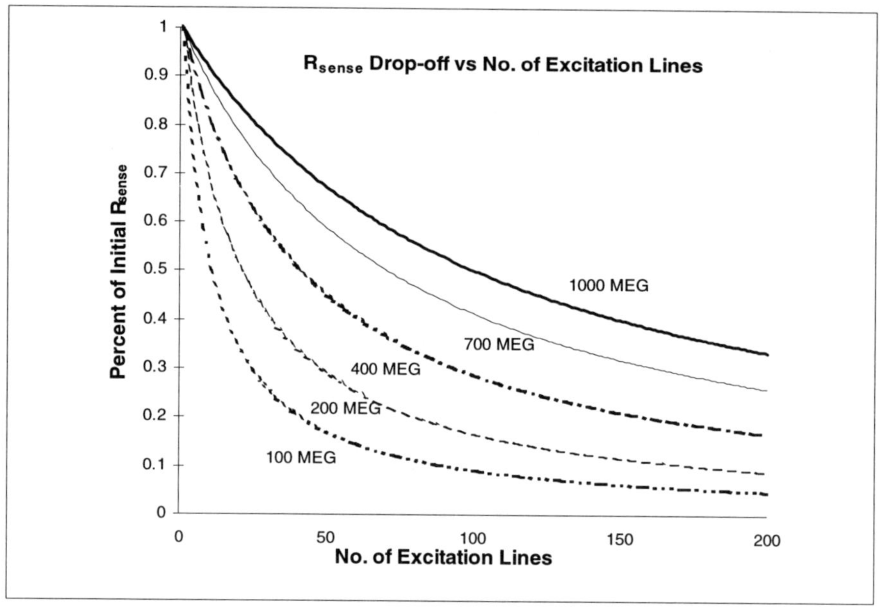

Figure 9. No. of excitation lines vs vsense for various resin resistivities

Conductive fiber media, Frame 4b, have no $R_{Matrix3}$ component as the resistance of the fiber media (< 1 KΩ) is negligible compared to the 100 MΩ + resistance of the resin matrix.

To confirm circuit response predictions, a multi-planar part was fabricated as shown in Figures 11 and 12. The part had three planes with four sense lines in each plane. Seven excitation lines were utilized (Figure 11), with excitations 2, 3, and 4 ganged to provide an excitation voltage to Planes 1 and 2. Excitation 5 similarly was ganged to provide an excitation voltage to all three planes and excitations 1, 6, and 7 were unganged and served as a control for the ganged excitations.

The part was fabricated from woven S-glass fabric layers. Two layers of the fabric material separate the excitation and sense wires, and four layers of fabric separate the individual planes. Utilizing VARTM vacuum-assisted resin infusion methodology, 411-C50 Dow Derakane vinyl ester resin was infused into the fiber media, and voltage levels at the 48 sensing nodes were recorded after total infusion of the part. Time of total infusion was approximately 5 minutes. Resin cure time was designed to be in excess of three hours to eliminate sense voltage drop-off due to loss of charge mobility as the resin approached gelation.

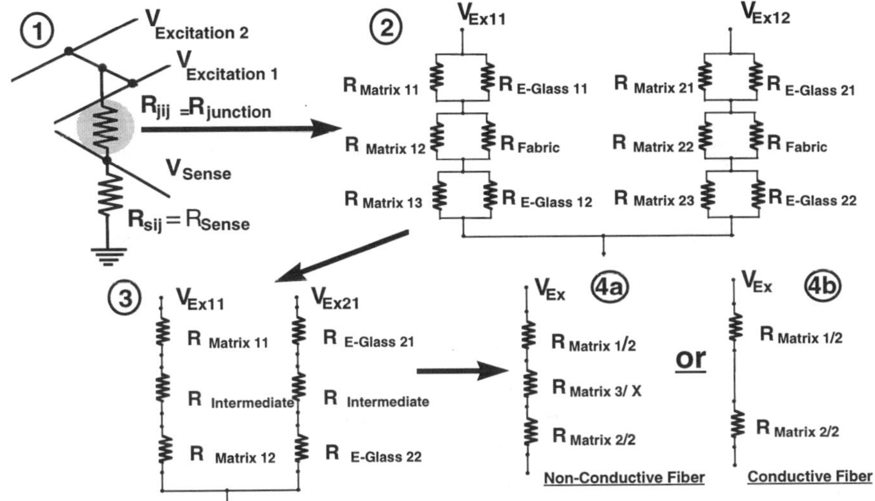

Figure 10.   Analysis of sensor node with two ganged excitation inputs

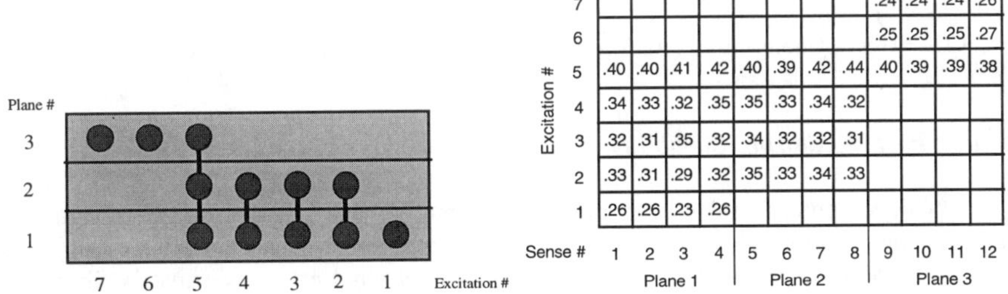

Excitation #	Sense # 1	2	3	4	5	6	7	8	9	10	11	12
7									.24	.24	.24	.26
6									.25	.25	.25	.27
5	.40	.40	.41	.42	.40	.39	.42	.44	.40	.39	.39	.38
4	.34	.33	.32	.35	.35	.33	.34	.32				
3	.32	.31	.35	.32	.34	.32	.32	.31				
2	.33	.31	.29	.32	.35	.33	.34	.33				
1	.26	.26	.23	.26								

Plane 1 = Sense # 1–4 | Plane 2 = Sense # 5–8 | Plane 3 = Sense # 9–12

Figure 11.   Cross-section of multi-planar part with ganged excitations

Figure 12.   Multi-planar part w/ ganged excitations (nodal voltages displayed)

Figure 13 graphically presents the tabular data in Figure 12. The data was extended for single, double-ganged, and tri-ganged excitations (Excitations 1, 6, 7 and Excitations 2, 3, 4 respectively) by calculating the average sense voltage for each excitation line and inserting it into the table to replace blank cells. Averages for the three ganging scenarios were calculated and included in the chart. Ganging excitation lines does dramatically increase sense voltage for nodes associated with the affected excitations. For this part, sense voltages increased by a factor of 1.3 and 1.6 respectively for double- and triple-gang configurations. As shown schematically in Figure 10, this effect is even more pronounced in parts fabricated from conductive fiber as there is no $R_{Matrix3}$ component. Non-uniform curing in the area surrounding the sense node will cause large variations in the $R_{Matrix3}$ component, producing dynamic non-uniformity in the associated nodal sense voltage. This non-uniformity is random, and cannot be anticipated, and corrected, by software algorithms. Therefore, ganging of excitation lines must be avoided

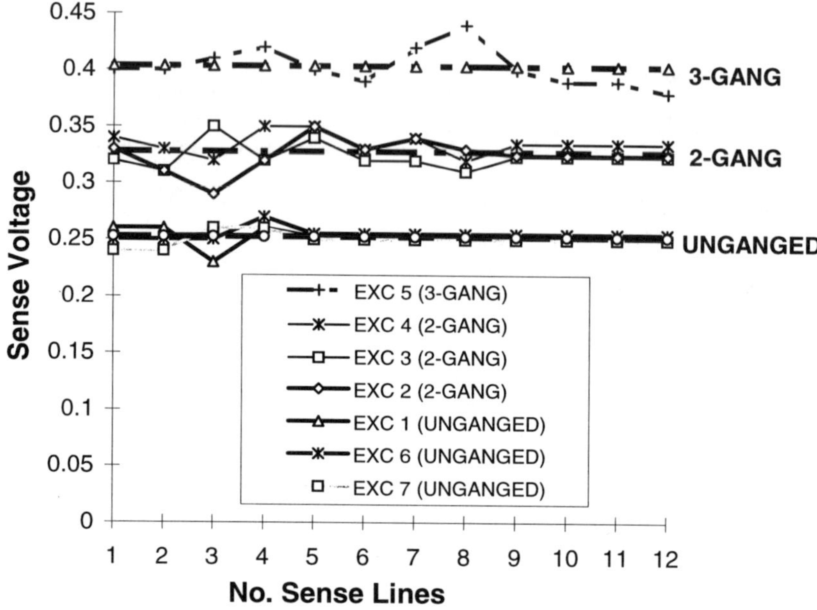

Figure 13.   Effect of ganging excitation lines on sense voltage signal strength

2.3.3 Large Part Monitoring Capability

An important development goal was to be able to monitor large parts. Employing signal to noise enhancement techniques and multi-planar configuration design analysis described in 3.3.2, this goal has been realized.) Two large parts, one a vinyl ester resin matrix/fiberglass preform composite, and the other an epoxy resin matrix/graphite preform composite, were successfully monitored from resin infusion through resin gelation. Each part, measuring 3' X 22', was divided into 336 individual sensing zones with three sensing layers, or planes. A highly desirable outgrowth of the performance enhancements is the ability to monitor through large thick parts with external grids, eliminating the need for compromising part integrity by embedding wires in the resin matrix. The modular, highly-adaptable excitation-sense grid structure can accommodate any level of complexity currently capable of being incorporated into part design.

2.3.4 Conductive Fiber Compatibility

Conductive fibers present a special technological challenge in that their relatively low electrical impedance introduces an unwanted low resistance path in the Excitation-Sense Grid circuitry. The path, between the excitation lead and the sense lead of each node in the grid, effectively allows the entire excitation voltage to be applied to the sense line even in the absence of resin.

The node, essentially "shorted out", is then completely unresponsive to variations in resin cure properties. Standard wire braiding techniques incorporating a woven fiberglass sheath failed to prevent carbon fiber shards from penetrating both the excitation and sense lead fiberglass sheathes and contacting both lead wires simultaneously. The resultant low impedance path, now in parallel with the resin junction resistance, compromises the affected sensing node and its immediate neighbors along the excitation wire.

These carbon fiber shards, released from carbon sheeting, are extremely brittle and our analysis has shown that they are unable to bend more than 45° before breaking. Utilizing this fact, wire braiding methodologies have been developed that would require a carbon fiber shard to bend at an angle of more than 60° in order to fully penetrate the fiberglass sheath and contact the wire. Subsequent evaluation of over a thousand sensing nodes embedded in carbon fiber material has failed to produce a single sensor failure due to carbon impregnation.

3. CONCLUSIONS

SMARTweave-based sensing technology, coupled with VARTM manufacturing methodologies can now cost-effectively produce large net-shape FRP composite parts utilizing a wide variety of resin systems and fiber materials. Signal-to-Noise Ratios, now exceeding 100/1, allow unambiguous acquisition and interpretation of zonal data, during the manufacture of large parts. Sensing methodologies have also been modified to allow utilization of SMARTweave in the manufacture of parts comprised of conductive fibers such as carbon, greatly expanding the manufacturing applications for this technology.

It was demonstrated that this system, utilized with the VARTM manufacturing process, is capable of monitoring, displaying, and archiving real-time multi-planar zonal flow and cure information for large parts with up to 4096 sensing nodes (64 sense X 64 excitation). Circuit analysis indicates that the present system, with hardware expansion, should be capable of monitoring over 400,000 sensing nodes without the use of a dopant to boost resin conductivity for low ionically conductive resin systems such as vinyl ester.

4. REFERENCES

(1) Direct Current Sensing of Viscosity and Degree of Cure of Vinyl-Ester Resins Masters Thesis, Material Science Program, University of Delaware, August 1997. K. M. England

(2) "Advances in Resin Transfer Molding Flow Monitoring Using SMARTweave Sensors," Proceedings of the ASME Materials Division, vol. 69-2, pp. 999-1015, 1995 Fink, B. K., S. M. Walsh, D. C. DeSchepper, R. L. McCullough, R. C. Don, B. J. Waibel, and J. W. Gillespie Jr

(3) In-Situ Sensing of Viscosity by Direct Current Measurements   England, Kenric M.; Gillespie, John, W. Jr ; Fink, Bruce K. Author Affiliation: Univ of Delaware Proceedings of the 1996 ASME International Mechanical Engg Congress & Exhibition Nov 17-22 1996 Atlanta, GA

LARGE SCALE IMPLEMENTATION OF FLOW AND CURE SENSING IN A THERMOSET RESIN INFUSED COMPOSITE STRUCTURE

Roderic Don, Karl Bernetich, and Dr. John W. Gillespie, Jr.
University of Delaware's Center for Composite Materials

Dr. Bruce K. Fink
U.S. Army Research Lab

Michael Louderback
Northrop Grumman Corporation

ABSTRACT

A highly successful demonstration of resin flow and cure sensing using the U.S. Army Research Lab's patented SMARTweave imbedded sensor technology was done on an Advanced Technology Transit Bus (ATTB) subcomponent at Northrop Grumman's El Segundo plant. The part was fabricated using single-sided tooling, and consisted of a multi-layered woven preform, sandwiched around a foam core, which was infused with resin in a modified VARTM-type process. The flow of resin into portions of the part not visible to the eye was readily detected. This experiment was the largest appplication to date of the SMARTweave technology, and was performed by researchers from the University of Delaware's Center for Composite Materials.

KEY WORDS: Flow Sensing, Cure Monitoring, Composites Fabrication

1. INTRODUCTION

A multi-use, distributed grid-type sensor system termed SMARTweave was patented by the U.S. Army (1), and utilizes a grid of conductive filaments placed on or in the reinforcement layer(s) of a composite Resin Transfer Molding (RTM) or Vaccum-Assisted Resin Infusion (VARI) preform. The SMARTweave sensor grid can be used to detect the presence of resin (and thereby determine the flow front) during the infusion process. There is sufficient sensitivity in this scheme to determine the extent of wet-out at any region after the flow front has passed. The obvious application of this sensing mode is to ensure complete infusion of the preform, or identify problem areas. On a longer time scale, SMARTweave can be used to monitor the extent of cure of the resin by detecting the change in ionic conductivity due to cure. On a much longer time scale, it is proposed that the embedded grid could be used for lifetime health monitoring of a composite laminate by detecting breaks in the sensor wires due to fatigue or through-hole damage.

Collaborative work between the U.S. Army Research Laboratory Materials Directorate (ARL-MD) and the University of Delaware's Center for Composite Materials (UD-CCM) has greatly advanced the state-of-the-art in SMARTweave flow and cure monitoring through the development of an improved sensing network scheme and better data acquisition techniques to significantly simplify the sensor network placement procedures and to greatly enhance the resolution of the system, allowing accurate detection of flow and cure progress with many RTM-grade resins (2-4).

A schematic of the SMARTweave sensing setup is shown in Figure 1. The number of sensor nodes is limited only by the capability of the data acquisition hardware. SMARTweave has been implemented successfully in a number of part geometries, including several Composite Armored Vehicle (CAV-ATD) components at United Defense (UDLP).

2. APPROACH

In this trial, the SMARTweave sensing grid was imbedded in a U-shaped subcomponent (see Figure 2) of the ATTB bus at Northrop Grumman's El Segundo plant. The intent was to monitor flow during the resin infusion process and subsequent cure at ambient and elevated temperature. The grid was placed in a region not visible to the eye (as the surface layers are, on the vacuum bag side of the part), adjacent to the tool surface in a region with a foam core. Figure 2 depicts the layout of the sensor wires in the cored portion of the subcomponent. Each crossing point of the sensor wires is a sensor "node;" in this case, there were 128 total nodes distributed on 12.7 cm centers from the 24 wires put in place.

Uncoated 28 gage soft copper wire was used as the sensing elements throughout these trials. Several initial experiments were performed at UD-CCM using the identical fabric and resin to demonstrate feasibility of using SMARTweave with this material system. The portable SMARTweave system built under contract for ARL-MD was used in all cases (see Figure 3).

3. EXPERIMENTAL SETUP AND RESULTS

3.1 Single Cell Experiment   An initial "single cell" (i.e., single SMARTweave node) experiment at UD-CCM demonstrated that the vinyl ester resin used throughout this study, Dow XUS 71928.03, was sufficiently conductive to allow use of the normal sensing system without preamplifiers or other means of boosting the signal. The catalysts were (by weight) 1.5% Trigonox and 0.2% CoNap. A reading of 0.020 V was obtained in the standard cell used at UD-CCM. This measured voltage, although small, was sufficiently above the noise level to allow rapid measurements with the multiplexed data acquisition system. A viscometer reading was taken from the same batch of resin immediately after mixing. The measured viscosity was 0.23 Pa at 24°C.

3.2 Flat Panel Experiments   Two flat 61 cm by 61 cm panels were infused at UD-CCM to further demonstrate feasibility of using this SMARTweave setup with the vinyl ester resin, foam core, and glass cloth supplied by Northrop Grumman.

The first panel consisted of two plies of E-glass fabric that were bagged in a conventional manner for VARTM-type infusion. The panel was electrically isolated from the metallic tool used for these experiments by a sheet of nylon bagging film. The first (longitudinal) set of wires was placed on top of the first ply of glass cloth. The second ply was placed on top of that assembly and the other set of wires (in the transverse direction) were placed on top of the second ply, giving an effective spacing through the thickness of 1.27 mm for the sensor nodes at the intersections. A grid of seven wires in each direction was used, giving 49 sensing nodes spaced on roughly 10.2 cm centers in each direction in the plane. The resin could be seen plainly as it infused the part. Visual verification of the sensed flow front as indicated in real time by the SMARtweave software showed a very good correlation. Figure 4 shows a typical output displayed by the software.

The second panel consisted of four total E-glass reinforcement plies, with two on either side of a foam core. The SMARTweave sensing layer was placed in the tool side of the sandwich. The infusion of this panel was unremarkable, with visual verification of the flow front in the bottom (tool side) plies when it reached the edges of the panel giving good correlation with the real time SMARTweave data display.

3.3 ATTB Subcomponent Instrumentation and Infusion   A U-shaped subcomponent of the ATTB structure was chosen for this trial. Figure 5 shows the layout of the sensing grid in the bottom portion of the subcomponent. In this case, it was desired to sense flow on the tool side of the part, under the foam core.

The first, longitudinal course (the "Excitation" leads in Figure 5) of 28 gage bare copper wires was placed directly against the tool, which had been previously coated with release agent. Small strips of Kapton tape where used to secure these wires at either end, and to insulate the wires from each other where they were run out of the bag. The first ply of glass fabric was draped over the tool, and the lateral wires (the "Sense" leads) were stretched across the part and secured at either end in the bagging ("Tacky") tape, with approximately 25 cm of wire extending beyond the edge of the part for hookup of the SMARTweave system leads. The remainder of the part including the foam core was then laid up and bagged for VARTM-type infusion. Figure 2 shows the part just before infusion. The sensor leads can be seen clearly in the foreground.

The SMARTweave system was connected and tested with vacuum applied to the bag. No shorts or false signals were detected, and resin infusion was initiated. Data collection was started upon the release of resin into the feeder tube for the part. At the highest scan rate, data was collected approximately once every 12 seconds for all 128 nodes. This data was written to disk for later replay. An intensity plot was also available for real-time display of the sensed flow. Photographs were taken of the part during infusion at regular times. The voltage readings across the grid at these same times were exported and plotted in intensity graphs. Figures 6 through 10 show intensity plots of the sensor voltage readings, which are interpolated for better interpretation of the flow front. On the same plots, a contour has been overlaid at a level of 0.01 Volts, which was used here as the threshold for determining full wet-out. The progression of the flow front can be seen easily in the plots.

Data was captured for approximately 80 minutes, at which point the instrumented area was determined to be fully wet out, and the SMARTweave system was put off-line. However, the sensor system was left connected and a further reading was taken the next morning, at some 15 hours after the start of infusion. There was a significant drop detected in the measured voltage, indicating partial cure of the resin at room temperature. The measurement is shown in Figure 11. Comparison with Figure 10 shows that the readings in a significant portion of the instrumented area had dropped below the threshold of 0.010 V.

The part was then cured at elevated temperature. It was not possible to actually monitor cure during the process due to the heating of the part in an autoclave, where it was not feasible to run the sensor leads out to the electronics. The sensor system was reconnected shortly after the part was removed from the autoclave, and the voltage readings were at the noise threshold everywhere in the grid. This is an indication that the state of cure had progressed beyond the gel point of the resin, at which point the ionic conductivity (as measured by the direct current SMARTweave system) goes to zero.

4. CONCLUSIONS AND RECOMMENDATIONS

Based on the outcome of this first trial of embedded sensors in the large ATTB subcomponent, we conclude that it would be feasible and extremely beneficial to apply the SMARTweave flow sensing technology to the VARTM process for full-scale ATTB component fabrication. Placement of the SMARTweave sensing grid in known problem areas (regions prone to poor wet-out, and not directly visible) would be most advantageous for process control and quality verification. It should be noted that there is no limitation in terms of the number of different areas that can be instrumented, given the proper equipment and channel count. Indeed, SMARTweave sensor grids have been placed at different depths through the thickness of thick-section parts to get 3-dimensional flow front determination.

Although there are some practical issues to be addressed related to bringing the sensor wires out of the mold, SMARTweave should also be considered for closed-mold processes such as RTM.

5. ACKNOWLEDGMENTS

The authors would like to thank Mr. William Ballata and Dr. Sean Walsh of the Army Research Laboratory for the use of their portable SMARTweave equipment for this study.

6. REFERENCES

1. US Patent No. 5,210,499 (May 11, 1993) S.M. Walsh (to the U.S. Army).

2. B. K. Fink, et al., "Identification of SMARTweave Compatible RTM-Grade Resins," status report for Chief, Composites Development Branch, Polymers and Mechanics Division, ARL, 1995

3. K.M. England , B.K. Fink, and J.W. Gillespie, Jr., "In-Situ Sensing of Viscosity by Direct Current Measurements," Processing and Manufacturing of Advanced Materials and Structures, T. J. Moon, et al., eds., ASME International Mechanical Engineering Congress and Exposition, Atlanta, GA, November 17-22, 1996

4. J.E. Bradley and J.W. Gillespie, Jr., "New SMARTweave-Based Monitoring System for the Manufacture of Large Composite Parts Utilizing Electrically Conductive Fibers," to be published in Proceedings of the 43rd SAMPE Symposium, Anaheim, CA, May 31—June 4, 1998

7. FIGURES

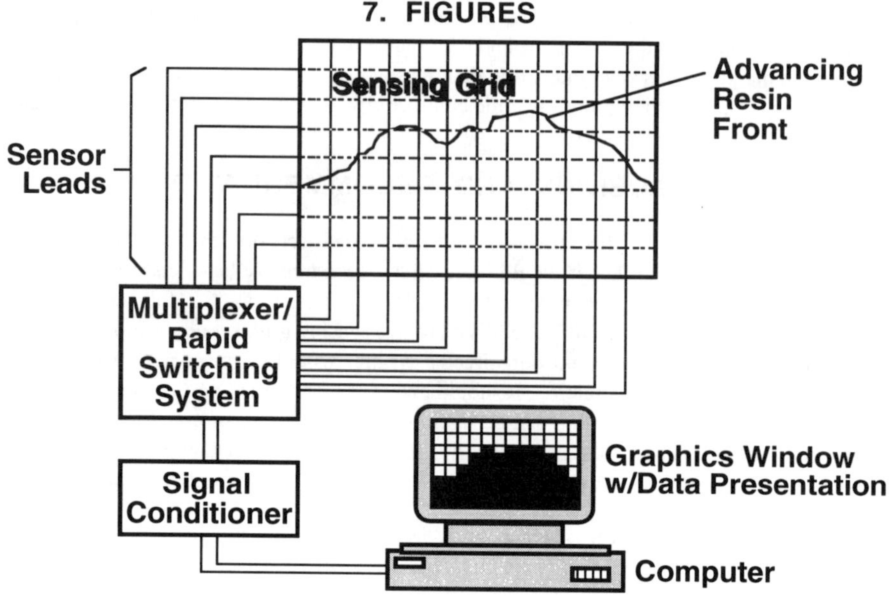

Figure 1: Schematic of the SMARTweave sensor approach

Figure 2: View of the ATTB subcomponent prior to infusion at Northrop Grumman. Note sensing wire leads (24 total connections) at the lower left.

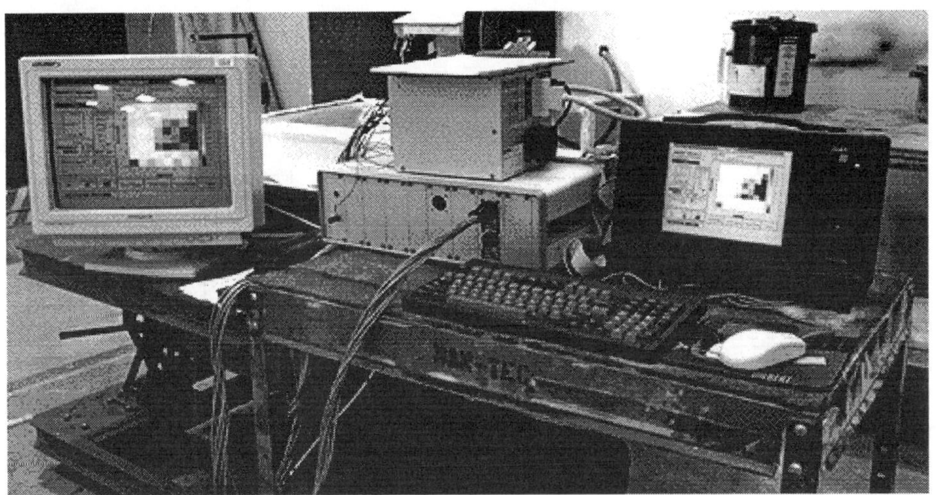

Figure 3: Portable SMARTweave equipment in place at Northrop Grumman

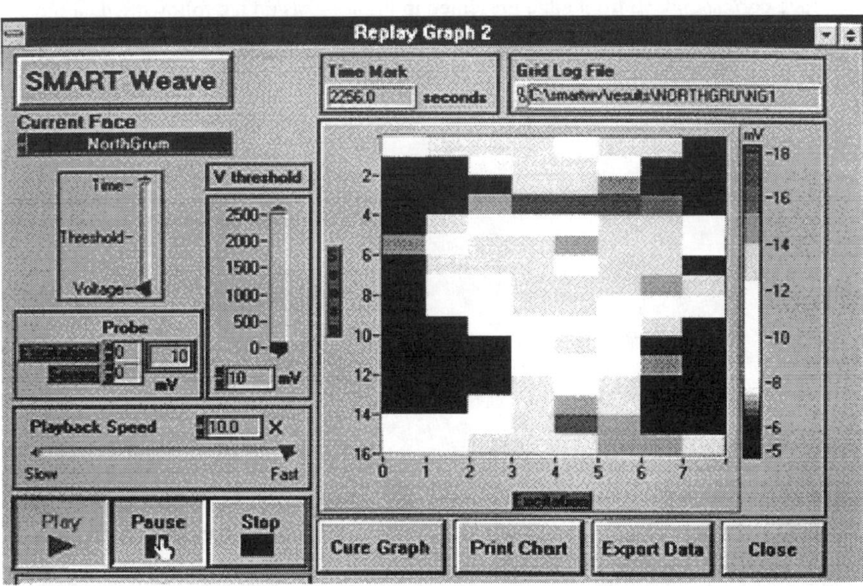

Figure 4: Screen shot of the SMARTweave computer display showing replay of the data captured during infusion of the subcomponent (paused here at about 37 minutes into the process).

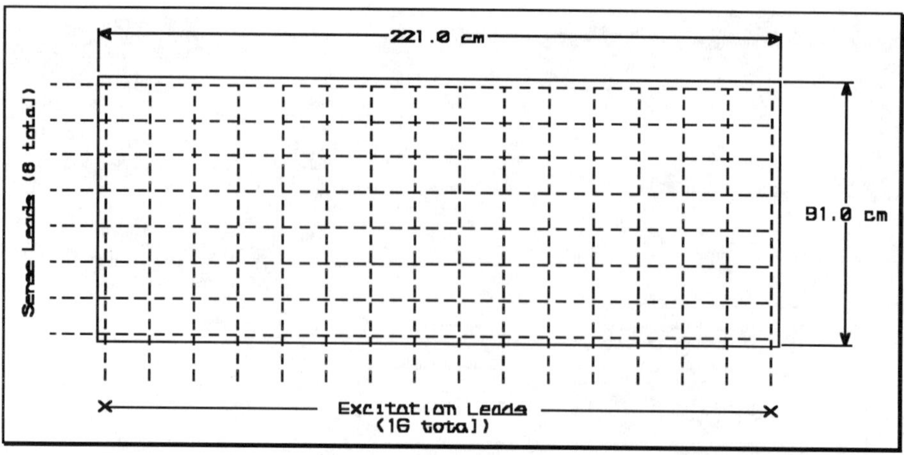

Figure 5: Layout of the SMARTweave sensor wires (dashed lines) placed below the foam core (tool side) of subcomponent. The actual instrumented area measures 89 cm by 190 cm. Lower edge corresponds to front edge presented in Figure 2, as do the following data plots.

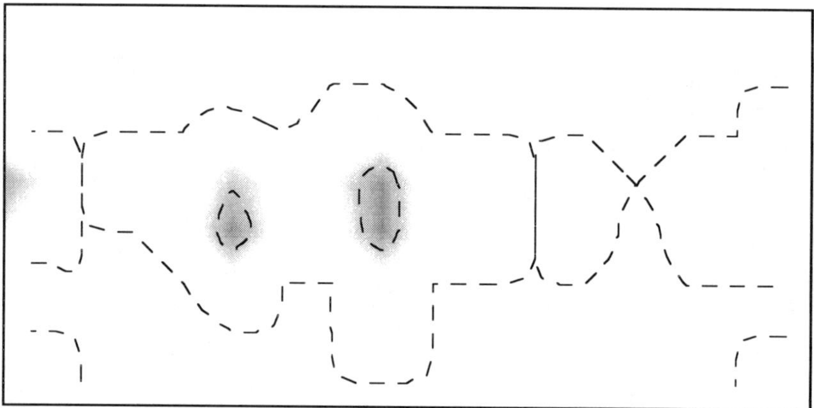

Figure 6: Intensity plot of sensed voltage at 10 minutes after start of infusion. Darker grey corresponds to higher voltage. Dashed contours indicate the extent of regions where partial wet-out is just beginning to occur. In this and subsequent plots, darker areas indicate higher voltage

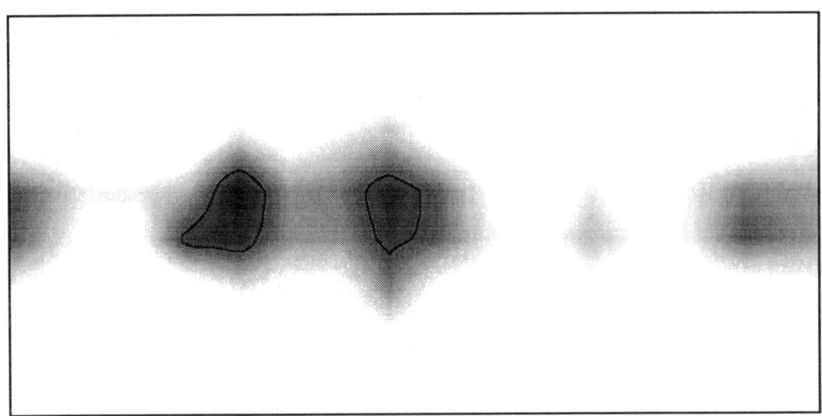

Figure 7: Sensed voltage at 15 minutes after start of infusion. Solid contour indicates the assumed threshold of 0.01 Volts for full wet-out in this and following plots.

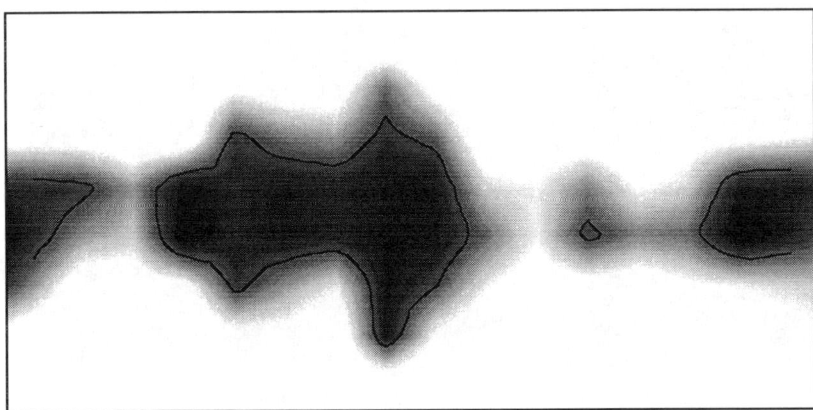

Figure 7: Sensed voltage at 21 minutes after start of infusion

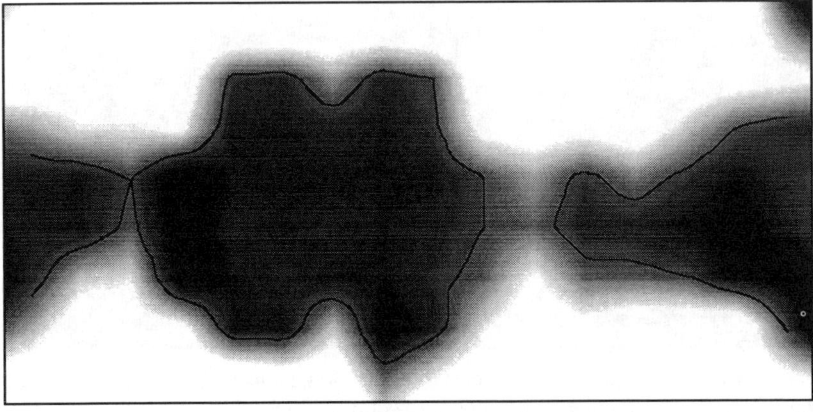

Figure 8: Sensed voltage at 30 minutes after start of infusion

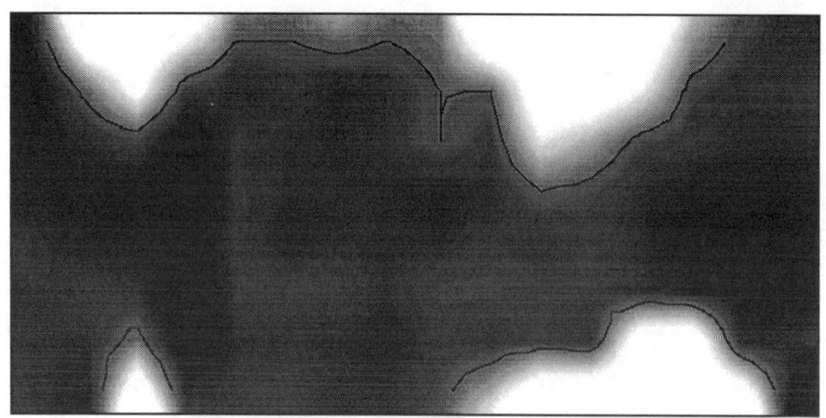

Figure 9: Sensed voltage at 45 minutes after start of infusion

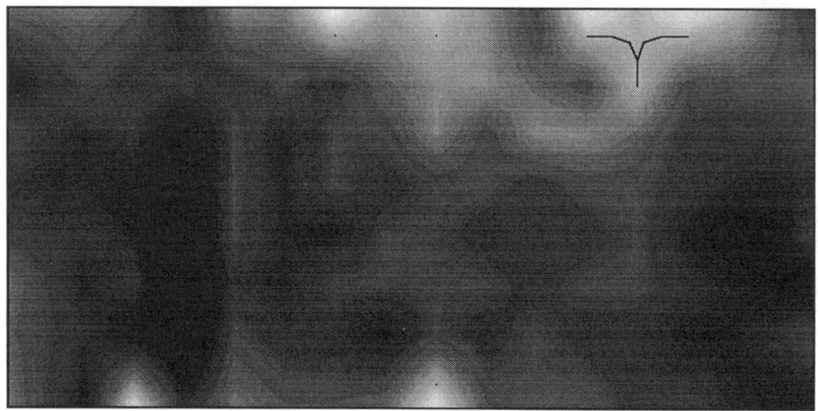

Figure 10: Sensed voltage at 79 minutes after start of infusion

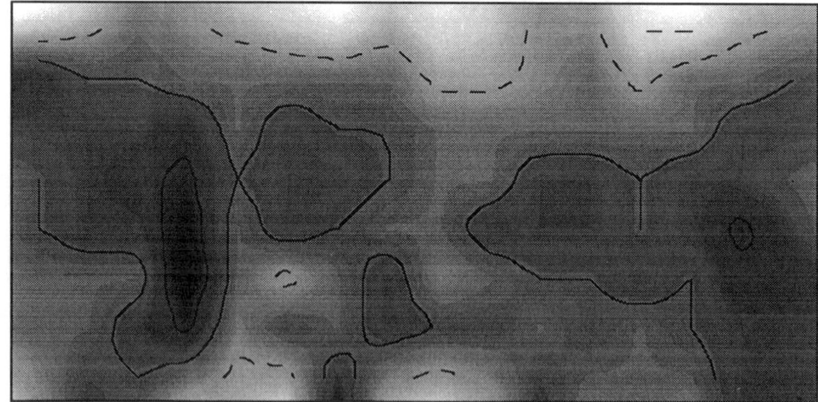

Figure 11: Sensed voltage at 15 hours after start of infusion, indicating drop in voltage due to partial curing at room temperature. Outermost solid contour indicates threshold used in previous plots to determine wet-out.

43rd International SAMPE Symposium
May 31-June 4, 1998

IN-SITU CURE MONITORING OF EPOXY RESIN USING FIBER OPTIC RAMAN SPECTROSCOPY (FORS)

Hans Rose

AvPro, Inc.

Norman, OK 73070

KEYWORDS: Cure Monitoring, Raman Spectroscopy, Autoclave

1. Abstract

Fiber Optic Raman Spectrometry (FORS) was used to monitor resin cure inside an autoclave during processing. Spectra from the cure are used to calculate a degree of cure (REI) of the resin using peak ratioing. This degree of cure (REI) is compared to a degree of cure calculated from the time-temperature history of the resin. The temperature based degree of cure is determined by Differential Scanning Calorimetry (DSC) measurements made on resin from the same material batch.

2. Introduction

AvPro, Inc., with the support of Southwest Research Institute (SwRI), has developed a cure monitoring system for obtaining in-situ Raman spectra of material curing in an autoclave. This system is in response to a SBIR Proposal solicited by the Composites Support Laboratory at Tinker Air Force Base (OC-ALC) [1]. The scope of the proposal was the development of a monitoring and control system that used Remote Raman Spectroscopy to verify material state estimates *in-situ*. This monitoring and control system is interfaced with autoclave control software (AvPro CSS200) that manages autoclave operation as well as records and manipulates Raman spectral data. The developed system has a Raman Spectrometer that uses optical fibers for gathering the spectra. The fiber optics pass into the autoclave and under a vacuum bag to monitor the cure of resin during the run.

Spectroscopy can be used for identification of specific components of resins [2]. As the resin cures, the amounts and concentrations of these components change. By tracking the spectrum of the resin as it cures, changes in its composition can be monitored. Chemical moieties in the resin that are not consumed in the reaction provide useful reference peaks in the spectrum to compare against the peaks of species that are consumed during reaction [3].

Raman spectroscopy can be used to take spectra *in-situ* much more readily than infrared (IR) spectroscopy because Raman uses visible light. Most glass is transparent in the visible region while it absorbs strongly in the infrared region of the electromagnetic spectrum [4]. This means that Raman can use long (> 10 meter) fiber optics whereas IR cannot. These fiber optics can be placed into hostile environments and transmit light to the spectrographic equipment [3,4,5].

In Raman spectrometry, UV or visible light is used to excite vibrational modes of the material under test. The intensity and frequency distribution of the scattered light is then measured [6]. A laser is used as the light source because it provides an intense monochromatic light [7]. The energy difference between the scattered photons and the incident photons is called the Raman shift. The Raman shift indicates how much energy is absorbed to excite the vibrational mode of the molecule.

3. Raman Degree of Cure

In this study, the change in the Raman spectrum of an epoxy resin with an amine-curing agent was tracked during the cure. Figure 1 shows an epoxy-amine reaction [8]. In step 1, an epoxy ring opens and substitutes one of the hydrogens of the amine. In step 2, the other hydrogen is substituted by another epoxy. Therefore, as the reaction progresses, the amount of epoxy in the mixture decreases. Figure 2 shows an epoxy in the resin used in this study. Figure 3 shows the curing agent used, diaminodiphenylsulfone (DDS) [2]. Both the epoxy and amine in this study have two reactive sites per molecule. Figure 1 shows only one active site. The sulfone group (O=S=O) in the center of the DDS does not react while the epoxies are consumed. Therefore, the sulfone peak in the Raman spectrum is convenient to ratio against the epoxy peak, for this resin system.

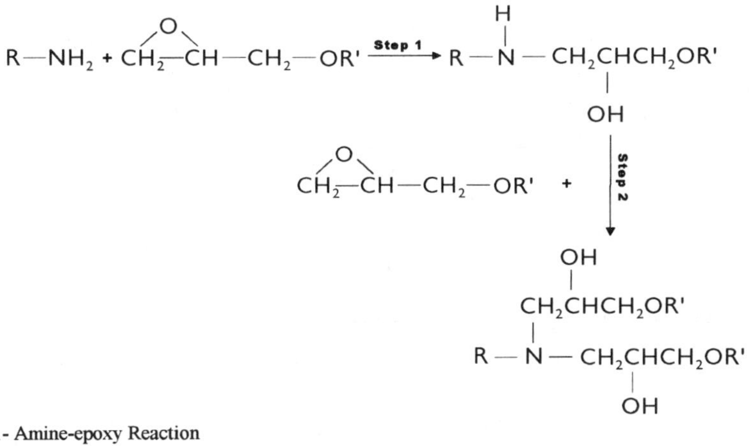

Figure 1- Amine-epoxy Reaction

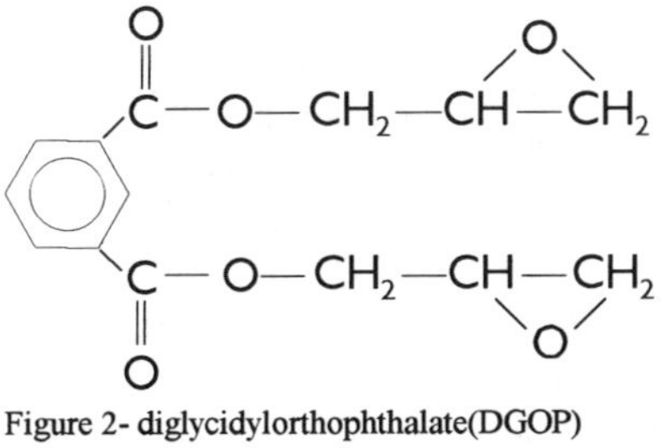

Figure 2- diglycidylorthophthalate(DGOP)

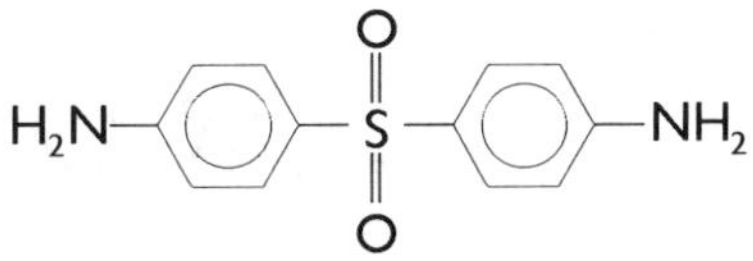

Figure 3- diaminodiphenylsulfone(DDS)

Figure 4 shows the Raman spectrum of the resin used in this study, and figure 5 is a blow up of the region from 1000cm^{-1} to 1300cm^{-1}. The spectra in figure 5 had the baseline subtracted to make the peaks easier to see. It is apparent that the epoxy peak is decreasing as the time increases while the sulfone peak remains fairly constant. The sulfone stretching peak is in the range 1165-1120cm^{-1} [9], and the epoxy ring breathing peak is in the 1280-1230cm^{-1} range [10]. This is the basis for the FORS Epoxy Index (REI). It is calculated using the following formula:

$$REI(t) = \left[\frac{\left.\frac{area\ of\ epoxy\ peak}{area\ of\ sulfone\ peak}\right|_{time=t}}{\left.\frac{area\ of\ epoxy\ peak}{area\ of\ sulfone\ peak}\right|_{time=0}} \right] \quad \text{(Eq. 1)}$$

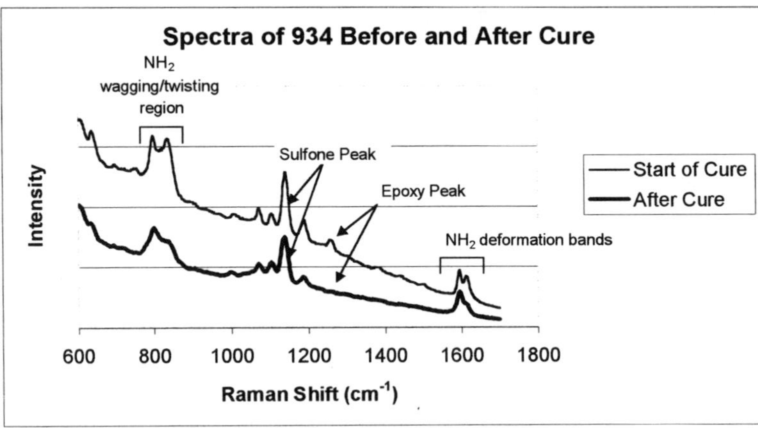

Figure 4 – Raman spectrum of epoxy-amine resin

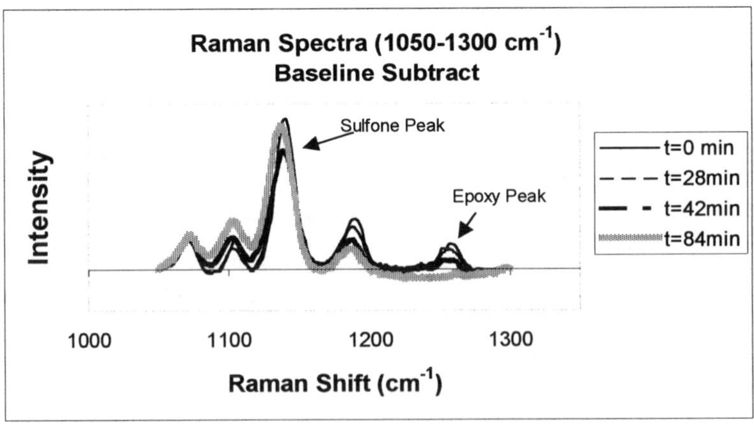

Figure 5 – Baseline subtracted Raman spectrum of epoxy-amine resin

4. DSC Degree of Cure

Model degree of cure is used to compare the ratios obtained from the Raman spectra during cure. The forms of the equations for the degree of cure models are taken from references 11 and 12. These models are based on DSC measurements of the heat of reaction of the resin. The first model is from reference 11. It will be called IsoModel on graphs. It uses the following equations:

$$\alpha_{iso} = \int_0^t \frac{H_T}{H_U}\frac{d\beta}{dt}\,dt \qquad\text{(Eq. 2)}$$

$$\frac{d\beta}{dt}=(K_1 + K_2\beta^m)(1-\beta)^n \qquad\text{(Eq. 3)}$$

$$K_{1,2} = A_{1,2}\exp\left(\frac{-\Delta E_{1,2}}{RT}\right) \qquad\text{(Eq. 4)}$$

$$\frac{d\beta}{dt} \equiv \frac{1}{H_T}\left(\frac{dQ}{dt}\right)_T \qquad\text{(Eq. 5)}$$

Where:

α_{iso}	- Degree of cure
H_T	- Total Heat/unit mass evolved at temperature T
H_U	- Total Heat/unit mass evolved when taken to full cure
$(dQ/dt)_T$	- Instantaneous rate of heat generation
t	- time
β	- Isothermal degree of cure
R	- Universal Gas Constant
T	- absolute temperature
$\Delta E_{1,2}$	- activation energies
$A_{1,2}$	- Pre-exponential factors

H_T, H_U, m, n, $\Delta E_{1,2}$, and $A_{1,2}$ are determined using the method described in reference 11. Briefly: Dynamic DSC scans are made to find H_U. Isothermal scans are run at several temperatures until $(dQ/dt)_T$ approaches zero to determine H_T for each of the temperatures. Equation (5) and its time integral are used to make a plot of $d\beta/dt$ versus β for each temperature. For each of these graphs, curve fitting is used with equation (3) to determine K_1, K_2, m, n. The parameters m and n are assumed constant with temperature. The natural log of K_1 and K_2 are plotted versus 1/T. A line fit is then used to determine $\Delta E_{1,2}$ and $A_{1,2}$. The degree of cure goes from zero (raw) to one (fully cured).

The second degree of cure calculation is from reference 12. It will be called AbModel on graphs. The equations used in this model are:

$$\alpha_{ab} \equiv \frac{Q_t}{Q_{ult}} \qquad\qquad\qquad \text{(Eq. 6)}$$

$$\frac{d\alpha_{ab}}{dt} = Z \exp\left(\frac{-\Delta E_a}{RT}\right)(1 - \alpha_{ab})^n \qquad\qquad \text{(Eq. 7)}$$

Where:

α_{ab}	- Degree of cure
Q_t	- Heat/unit mass evolved at time t (area under DSC curve)
Q_{ult}	- Total Heat/unit mass evolved when taken to full cure
t	- time
R	- Universal Gas Constant
T	- absolute temperature
ΔE_a	- activation energy
Z	- Pre-exponential factor

The parameters n, ΔE_a, and Z are determined using curve fitting to DSC data as described in reference 9. Briefly: DSC scans are made using a constant heat rate. Equation (6) is used to determine $\alpha_{ab}(t)$. With constant heat rate, $\alpha_{ab}(t)$ and $d\alpha_{ab}/dt$ can be changed to a function of T. Equation (7) is linearized using logarithms. The parameters are then found using a least squares regression. The degree of cure goes from zero (raw) to one (fully cured).

5. Experiment

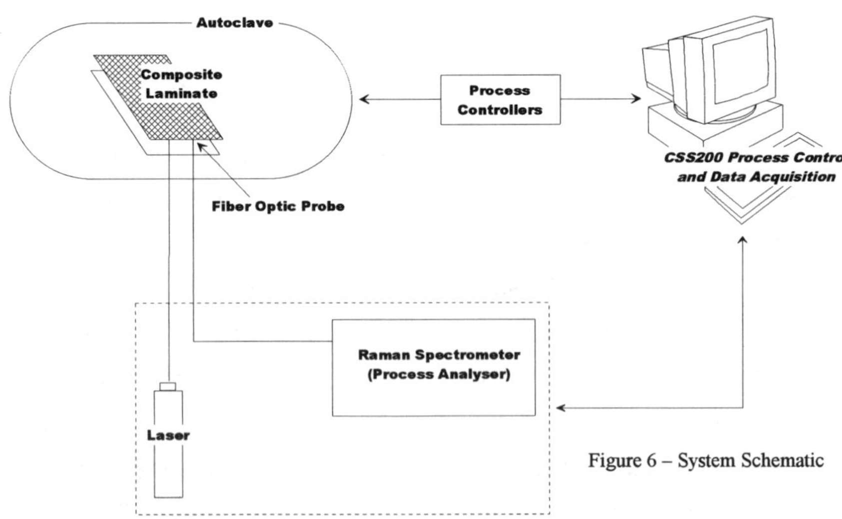

Figure 6 – System Schematic

Raman spectra were gathered from a neat resin sample as it cured. Temperature was controlled using PID time-proportioning industrial controllers and electric heaters with computer software providing the temperature setpoints. Temperature was measured using type J thermocouples placed within an inch of the sample.

The device used to take spectra was a Kaiser HoloProbe 785 System which has a SDL-8530: 785nm, 300mW, continuous wave, wavelength-stabilized laser. SMA connectors were used for making fiber optic connections. The Kaiser system uses a holographic grating system to spread the light onto a 1024 X 256 pixel charge couple device (CCD) detector cooled to -40°C by thermoelectric effect. The spectrometer has a 785nm holographic notch filter before the grating. Exposure of the CCD was set using a computer controlled shutter system. A "dark spectrum" was subtracted from the exposure to account for thermal noise in the CCD. A 600μm-diameter fiber optic was used to deliver laser light, and a separate 600μm fiber collected the scattered light for the spectrometer. The fiber ends were immersed directly in the resin.

The neat resin sample (Fiberite 934) and the fiber ends were held in a small aluminum crucible (d=5mm, h=2.5mm) wrapped in aluminum foil (Fig. 7).

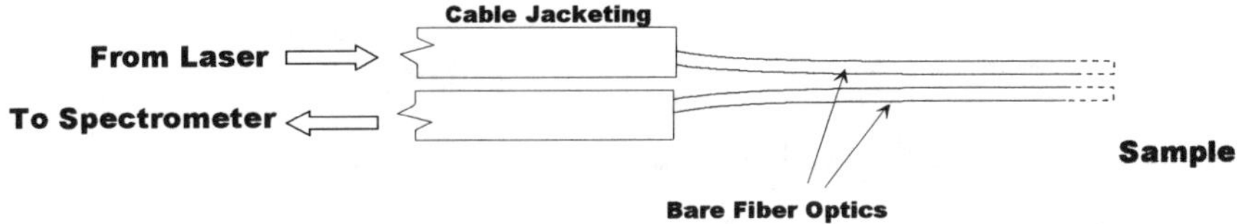

Figure 7 – Sample end of fiber optics

At the start of each run, the exposure time was adjusted to get clear epoxy and sulfone peaks without saturating the CCD. In order to increase the signal-to-noise ratio, the measurement was repeated several times and averaged. Total measurement time/spectrum was approximately one minute.

In addition, the IsoModel and AbModel numbers described in section 4 were calculated using the time-temperature data from the cure runs. For plotting with the REI, the AbModel and IsoModel have been subtracted from one. This is because degree of cure, as defined by AbModel or IsoModel, increases as the amount of reactants decreases, while REI indicates the amount of reactant (epoxy) remaining.

DSC measurements of the material were made using a Rheometrics STA-625.

6. Discussion

Figure 5 shows the portion of the spectrum used for making REI calculations.
The following graphs (Figures 8 & 9) show the Raman Epoxy Index (REI) and the DSC estimated degree of cure for different runs. It can be seen that the REI and DSC degree of cure indicate the reaction progressing at nearly the same rate. The REI actually shows the reaction starting slightly earlier than the DSC models do. This may be due to differences in staging between the time when resin was measured with DSC and when the cure run was performed. FORS allows measurement of the material in process.

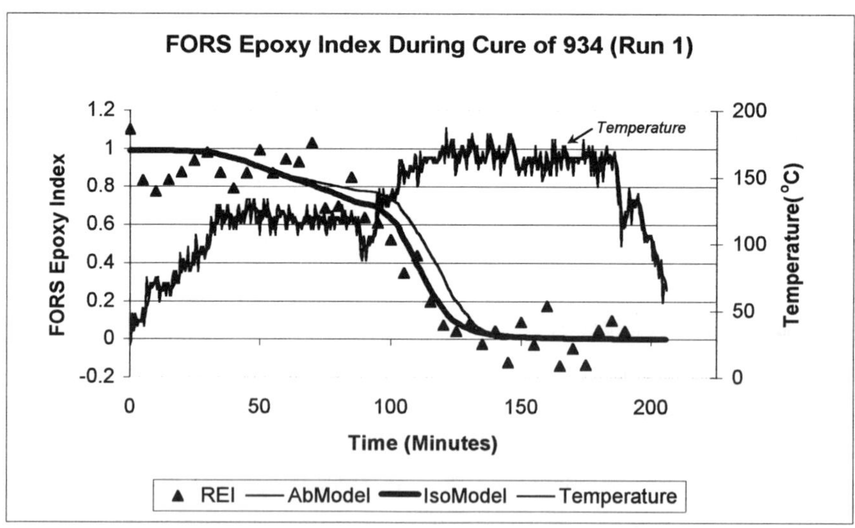

Figure 8 – Results of Run 1

284

Neat resin was used in this study because the reinforcing fibers interfere with the Raman signal. Carbon fibers tend to absorb the scattered light. Glass fiber fluorescence obscures the signal of interest [3]. AvPro and Southwest Research Institute have started to design a sensor head that will allow only the resin to reach the end of the fiber optics.

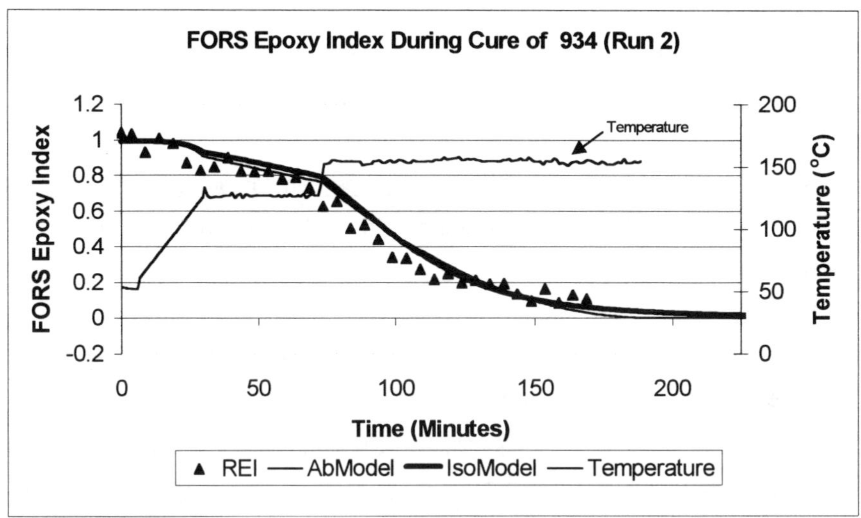

Figure 9 – Results of Run 2

7. Conclusion

FORS can be used to monitor the cure of epoxy-amine resin systems inside an autoclave during the cure process. A simple ratio of two peaks of the spectra provide a degree of cure number whose values agree with calculations for time temperature based models of the cure. There is still more work to be done with the sensor to provide a means of sampling only the resin since reinforcement cause considerable fluoresce which obscure the spectrum of interest.

8. References

1. SBIR Proposal AF94-229, 1994.
2. "Physical-Chemical Characterization Techniques, Epoxy Adhesive and Prepreg Resin Systems", Aerospace Recommended Practice ARP 1610A, Society of Automotive Engineers, Warrendale, PA, 1993.

3. J. Maguire, M. Miller, A. Orvis, <u>Laser Fiber Optic Sensors for Composite Cure</u>, Letter Report, Southwest Research Institute, San Antonio, TX, 1997.

4. J. Ferraro, K. Nakamoto, <u>Introductory Raman Spectroscopy</u>, Academic Press, San Diego, CA, 1994, pp.116-126.

5. U. S. Patent 5,262,644 (Nov. 16, 1993) J. Maguire (Southwest Research Institute).

6. N. Coltup, L. Daly, S. Wilberly, <u>Infrared and Raman Spectroscopy,</u> Academic Press Inc., New York, NY 1964, p.29.

7. Ref 4 p. 2.

8. R. Roberts, "Cure Quality Control", <u>Engineered Materials Handbook, Vol. 1: Composites</u>, ASM International, Metals Park, OH, 1987, pp.745-760.

9. Ref 6 p.308.

10. Ref 6 p.273.

11. P. Cirscioli, G. Springer, <u>Smart Autoclave Cure of Composites</u>, Technomic Publishing Co., Lancaster, PA, 1990, pp.119-122.

12. S.C. Mantell, P.R. Cirscioli, G. Almen, "Cure Kinetics and Rheology Models for ICI Fiberite 977-3 and 977-2 Thermosetting Resins", <u>J. of Reinforced Plastics and Composites, Vol. 14, No.8,</u>1995, pp.847-865.

9. Biography

Hans Rose is chief engineer at AvPro. He is an MS Candidate in Mechanical Engineering at the University of Oklahoma and received his BS in Physics from the University of California, San Diego. He is involved in extensive materials testing and research using Rheometry, Ultrasonic, DSC and other techniques to establish improved materials and process specifications for composites. He continues to be involved in installation and implementation of advanced sensor, control systems, and processing equipment for composites.

43rd International SAMPE Symposium
May 31-June 4, 1998

OPTIMIZATION OF ARCHERY BOW LIMB DESIGN

Franklin D. Meatto

and

Edward D. Pilpel

Gordon Plastics Incorporated

Montrose, Colorado 81401

1.0 ABSTRACT

KEY WORDS: Archery Bow Limb, Precured High Performance Composites, Laminate

The current focus of the archery industry towards lighter weight and higher performance products has created a demand for an expanded understanding and analysis of Archery Bow Limb Design. Advancement of these products has been fueled by higher performing composite products. Today's archery market requirements have expanded beyond strength and stiffness criteria. Speed and fatigue life have become important factors in limb design. Precured High Performance Composite materials that incorporate E-Glass, S-Glass and carbon fibers have been key in improving bow limb performance and structural durability. With the advancement of product designs and manufacturing methods, Precured High Performance Composites have satisfied the increased requirements of the materials used in bow limb construction in concert with product cost targets.

2.0 INTRODUCTION

The evolution of archery product design has been accelerated in the past several years as a result of lighter product weight and performance. In the arena of archery bow design particular focus has been directed towards the compound bow cams, riser or grip handle and the limbs.

The archery bow presents some very challenging engineering design problems and as a result has heightened the search for higher performance materials. In the quest for higher performance products it is common place to increase design stresses and approach material ultimate properties as determined in the lab. Suppliers and converters have responded by continually improving composite material properties. These improved composite materials in fact can benefit other industries. An example of this is the material handling vibrating machinery industry, where high performance composites provide longer spring fatigue life and provide long term constant spring stiffness durability compared to standard spring steel. In the past composite technology used in the archery industry has been transferred to the auto industry in the form of suspension components.

This paper will focus on the recent developments and implementation the industry is now involved with in improving overall archery limb performance.

3.0 BACKGROUND

Several factors have been instrumental in the advancement of limb performance for compound bows:

- Increase the speed of the arrow based on compound mechanical cam or limb design with improved product durability.

- The introduction of improved composite materials to the industry.

- The expanded use of narrower and quad limb designs that reduce overall limb weight.

- The basic design guideline that a lighter stiffer limb results in a faster shooting limb.

- Ultimate strength, durability or fatigue life of the limb are generally related to the outer fiber stress and shear capabilities of the design.

Compound bow limb construction has evolved from composite skins and wood to all composite construction over the last 45 years. The current constructions utilized in today's limbs are primarily "E" Glass FRP with an increased use of carbon and to an even greater extent, S-Glass. The common manufacturing methods are laminating of precured materials, molding of prepregs and compression molding. Prepreg construction has traditionally produced higher cost, heavier, and lower performing limbs, but in general provided a very durable product if properly designed. The compression molding process has typically been used for high volume limbs and was very competitive cost wise but limited in performance. Tooling is very expensive, design flexibility is limited and material properties generally do not keep pace with prepregs and precured laminates. It is interesting to note that this technology after being developed in the archery industry was embraced by the auto industry and is the basis for manufacturing composite mono-leaf springs. Precured materials have evolved in terms of replacing wood as a core material and further advancement was made through higher performance composites in the outer layers of the structure to carry outer fiber stress loads.

4.0 PRECURED COMPOSITE MATERIALS

As discussed earlier there has been an evolution of composite material input components, particularly in the fiber reinforcement and the actual converted material.

The primary materials used for the high performance requirements are primarily E-Glass, S-Glass and carbon fibers. The E-Glass fibers by far being the most commonly employed. These fibers are most commonly combined with epoxy based matrices and occasionally vinylesters.

It is important to note that the bow limb is essentially a cantilever spring subjected to extremely large deflections. This area will be described in greater detail in the next section. For this reason E-Glass is an excellent material for limbs. S-Glass with its inherent high strain ability, even greater than E-Glass, provides high strength, elongation and fatigue resistance at the outer most fibers of the cantilever beam. Carbon fiber does not have the elongation capabilities of the glass products, but it does have the high stiffness and low weight which can reduce the overall thickness of a beam and reduce the maximum outer fiber stress for a given stiffness

The three precured composite materials used have nominal properties shown in TABLE 4.1

	E-GLASS CORE	E-GLASS LAMINATE	S-GLASS LAMINATE	CARBON LAMINATE
FIBER CONTENT BY % WEIGHT	67	70	70	65
FLEXURAL STRENGTH MPa/psi	931 135,000	1551 225,000	1793 260,000	1896 275,000
MODULUS OF ELASTICITY GPa/psi	37.9 5,500,000	41.4 6,000,000	51.7 7,500,000	124.1 18,000,000
TENSILE STRENGTH MPa/psi	896 130,000	1207 175,000	1896 275,000	2275 330,000
COMPRESSION STRENGTH MPa/psi	620 90,000	655 95,000	N/A	N/A

Typically, in addition to matrix type, (normally epoxy), the mechanical properties of the composite materials are effected by the fiber bundle size. The E-glass core material is typically produced using 225 to 250 yield (yards per pound) while the laminates are produced with materials from 450 to 1200 yield. Higher yield materials in general result in an increase in mechanical properties.

The costs vs. performance and durability benefits for the composite materials given in TABLE 4.1 are part of the design process in terms of determining an optimum design for a given bow model.

FIGURE 4.1 Photograph of typical unidirectional precured "E" FRP laminates and bar stock

5.0 PRECURED COMPOSITES AND LIMB DESIGN

The primary components of proper limb design begin with overall stiffness, outer fiber stress, shear strength, fatigue resistance, limb mounting constraints and desired geometry. Typically a design starts with a desired stiffness, maximum deflection and general geometric boundaries in terms of length and width. Also included are mounting methods and an idea of the general shape.

Generally the first task is to generate an initial design based on the required stiffness and a given material, for example an all E-glass limb. When the stiffness has been achieved the outer fiber stress, (at maximum deflection full draw), is evaluated and analyzed based on the initial geometrical shape. The optimum stress design is a constant stress along the length of the cantilever beam. The design stress is generally based on the maximum compression stress and fatigue life requirements. Also considered are mounting methods at the fixed end and at the opposite end of the cantilever beam where the cams are located. It is also important fiber. The capabilities of this design are in the range of 60,000psi outer fiber stress at a maximum deflection of 50,000 cycles.

5.2 To increase the outer fiber stress capability of the limb a continuous E-Glass laminate is laminated to the contoured/compression side of the limb. (See FIGURE 5.2b). Although additional cost is added to the product, several factors are introduced that improve the structure strength by now having continuous glass fibers on both the tensile and compression side of the cantilever beam. This construction of counter balancing the lower compression strength side of the composite beam with a higher modulus and strength laminate results in a more balanced outer fiber strength design. This method offsets the neutral axis so the design is more in balance by adjusting the thickness of the compression side laminate.

5.3 To increase outer fiber stress capabilities even further a laminate of S-Glass can be laminated to the compression side and adjusted in thickness to further move the neutral axis and balancing the compression and tensile outer fiber stress even closer . (See FIGURE 5.2c).

5.4 To demonstrate the further enhancements available by laminating a combination of laminates (FIGURE 5.2d). The outer fiber stress can be reduced by incorporating inner carbon laminates that provide high stiffness. This reduces the overall section height but maintains required stiffness. Carbon laminates in combination with S-Glass on the compression side and E-Glass or S-Glass on the tensile side will result in a high performance and durable limb with light weight.

5.5 Figure 5.2e utilizes the construction of part 5.4 and laminates the components in a shaped mold to achieve the more traditional Recurve configuration.

5.6 Figure 5.2f shows the most recent design configuration on the market, where a quad-limb configuration is used which reduces the overall weight of the design. In place of the traditional mono-limb are two narrow individual limbs. In order to maintain a given stiffness these limbs are typically thicker and stressed higher, but result in an overall lighter bow. This is commonly known as a QUAD type limb design.

to assess the shear at the fixed end and shear stress at the mounting point of the cam. See FIGURE 5.1

The outer fiber stress can be evaluated in several ways based on classical mechanics of a beam in cantilever. In the simplest form a method using the radius of curvature can be employed. Also finite element methods particularly those with composite elements are commonly used. Calculations when multi-layers are involved become more complex.

Prediction methods for stiffness and strength can be successfully developed based on material properties and correlated to lab determinations of springrate and outer fiber stress. The ability to develop a constant stress beam becomes a more complex calculation and can be achieved by trial and error or through developed iteration methods by adjusting beam thickness.

Shear stress considerations can generally be simplified by calculating the shear stress developed at the fixed end based on classical material mechanics utilizing the maximum moment and the crossectional area at the clamping point.

FIGURE 5.1 Photograph of riser, limb and cam with mounting hardware.

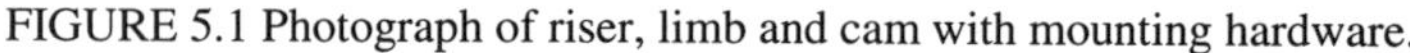

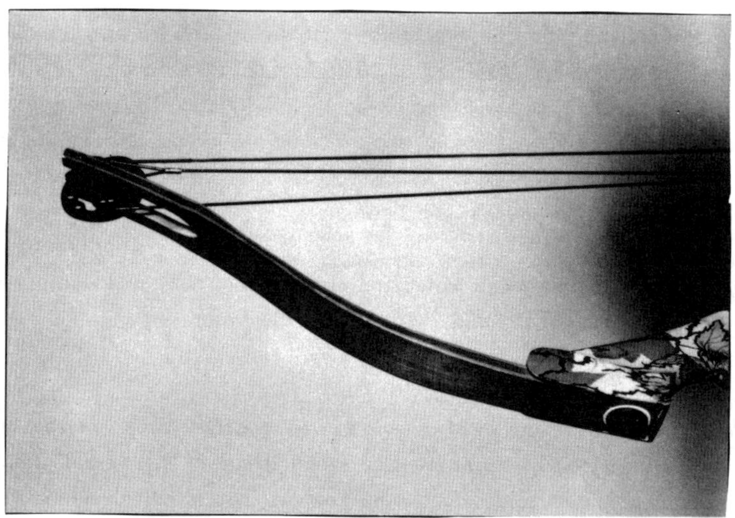

At this point in the design process a number of product decisions have to be made regarding the requirements of the limb in terms of performance, durability and the resulting cost. In a precured composite design the basic construction options are as follows:

5.1 A flat glass limb consisting of unidirectional fibers that is contoured in thickness to achieve a stress contour along the length. (See FIGURE 5.2a) This construction is the most simplistic in design and the lowest in cost.. The contoured side is always loaded in compression. This design has a limitation in outer fiber stress due to the cut

FIGURE 5.2a basic contoured limb

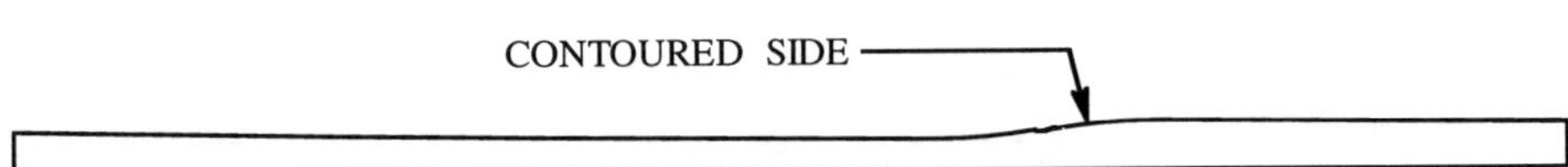

FIGURE 5.2b basic contoured limb with E-glass laminate cap

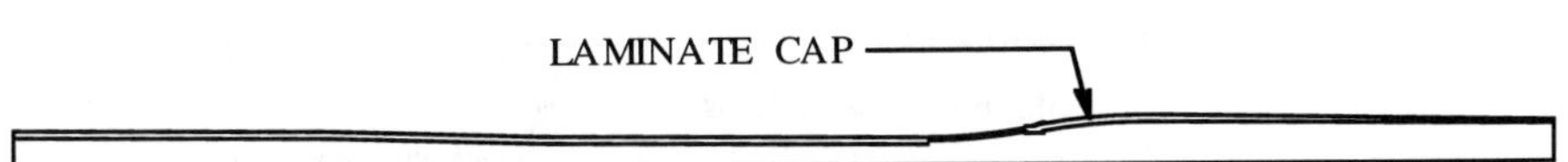

FIGURE 5.2c basic contoured limb with S-glass laminate cap

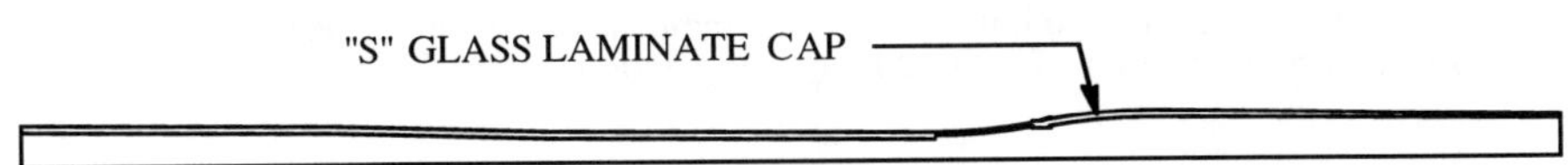

FIGURE 5.2d contoured limb with S-glass laminate cap on both sides and inner carbon laminate

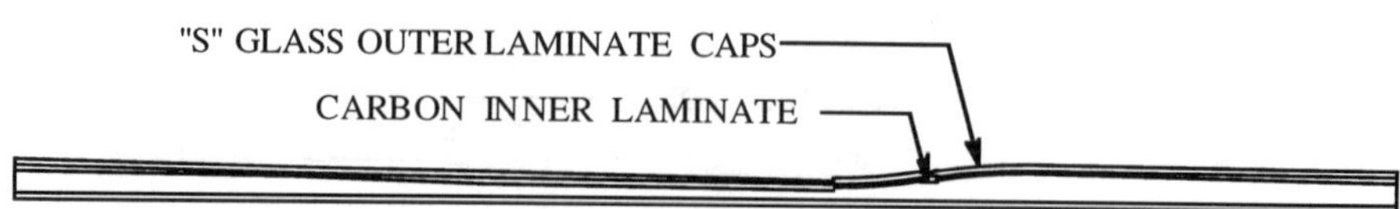

FIGURE 5.2e contoured limb with S-glass laminate cap on both sides and inner carbon laminate in a recurved limb shape.

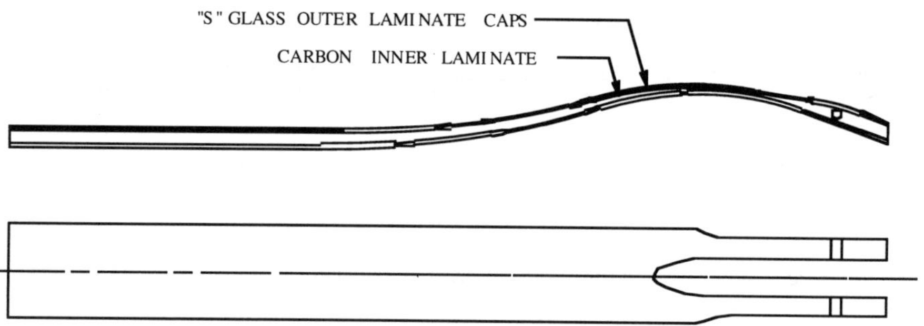

FIGURE 5.2f basic contoured limb with S-glass laminate cap on both sides in a recurved limb shape in combination with a Quad limb configuration.

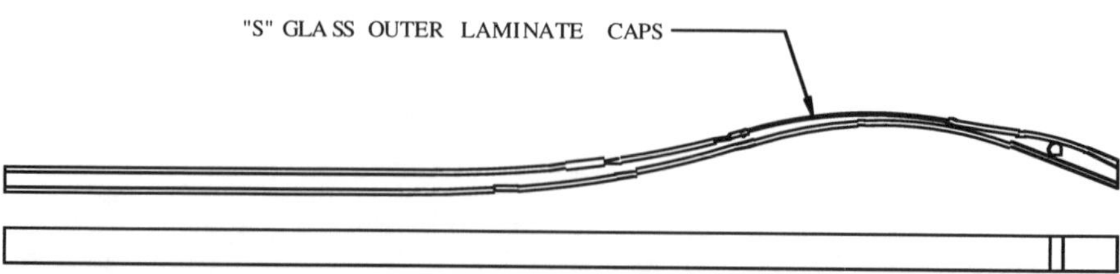

FIGURE 5.3 shows samples of typical limb designs using all High Performance Precured Composites.

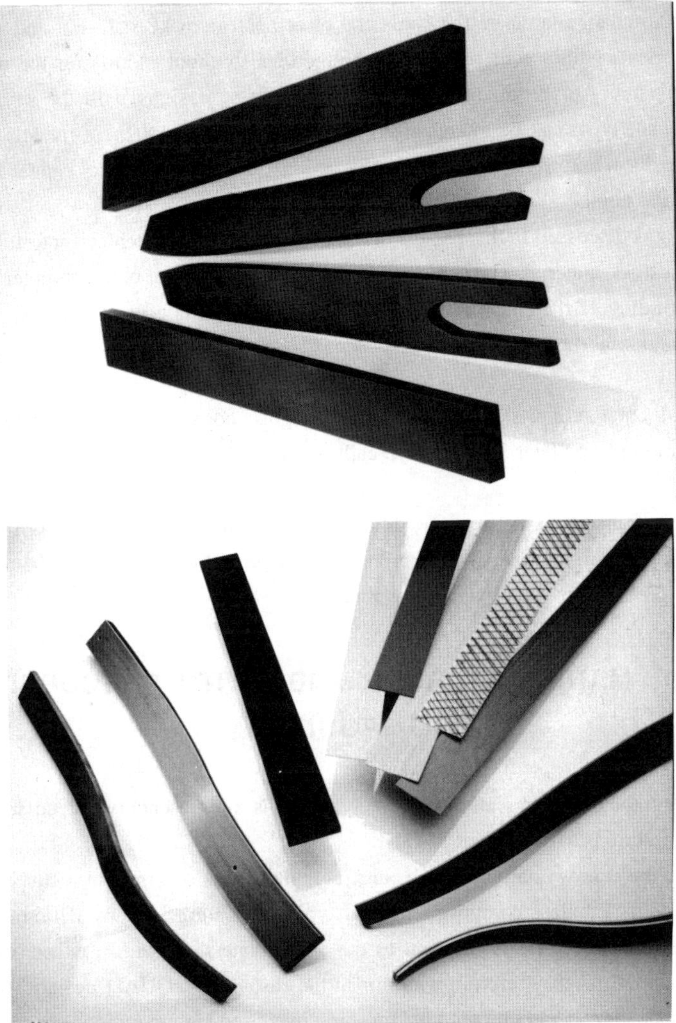

After the appropriate design considerations are optimized, the next step is to construct prototypes using an aluminum mold and verify performance and durability characteristics.

6.0 PROTOTYPE FABRICATION AND PRODUCT TESTING

Utilization of Precured High Performance Composites in limb design provides a low cost final evaluation of the product. Typically an aluminum mold cavity is produced and is large enough to produce 2 limbs per molding. In addition the mold is machined to the specifications of the final production tool. Heaters are located and are the same as will be used in the production tool. A simulation of the production tool is complete and generally in the cost range of $1500 per design. The versatility of this essentially two dimensional mold is the ability to change laminate thickness and placement to make stiffness and outer fiber stress adjustments within certain geometric boundaries, without modifying the mold. This offers the capability of producing varied overall stiffness designs with the same tool. In addition laminate types can be interchanged to produce different stiffness models within the same geometric window; a significant production advantage from a capitalization and flexibility stand point.

With all designs discussed in this paper a stress contour must be machined into the precured "E" glass bar stock material. The contouring operation is performed on a computer controlled machine. Grinding of the shape is typically done with a diamond grit tool. Again from a prototype standpoint the costs of generating shaped parts is minimal.

Designs using outer skin or skins bonded to an "E" glass core, utilize epoxy adhesives with room temperature Lap Shear Strengths typically above 27.6 Mpa (4000 psi).

Product performance and particularly durability requirements can vary from design to design, however upon completion of the prototype stage full confidence can be achieved in what will be produced in the production tooling.

7.0 MANUFACTURING LIMBS WITH PRECURED COMPOSITES

As discussed in the previous section the prototype tool is a representative precursor to actual production conditions and resultant product.

Production of the simply contoured limb construction consists of grinding a curvature into a plate of precured "E" glass composite material typically 12mm thick by 230mm wide and 650mm in length. The plate is then cut to the appropriate limb width, either for a single mono-limb approximately 40mm in width or in the case of the QUAD design 16mm. The limbs are planar shaped and prepared for finishing in subsequent operations.

Cure cycles for limbs that are laminated with epoxy adhesives range between 15 to 20 minutes depending on the number of layers and curvature complexity. Molds are configured so that multiple plates can be laminated during one pressing cycle. The 230mm wide by 650mm long molded plates are later saw cut yielding approximately 5 mono-limbs or approximately 11 QUAD type limbs.

8.0 QUALITY CONTROL OF THE PRODUCT

A key advantage to utilizing a precured product is the predictability and consistency of in use service. All composite materials used in these products are produced on a continuous production line where fiber content, strength and stiffness are monitored continually. For archery and other products such as vibratory springs it is important to be assured of consistent strength and stiffness. It is an advantage to know that key material properties are in spec along with geometric shapes before the cost of molding is added.

9.0 SUMMARY

The production of Archery Limbs utilizing precured composite components must meet rigorous design and flexibility requirements, reduced product costs, low tooling capitalization and short lead times to produce a successful product. Precured High Performance Composite materials has contributed to the advancement of Archery Bow Limb performance and durability.

10.0 REFERENCES

1. US. Pat. 5,087,503 (Feb. 11, 1992) Franklin D. Meatto (to Pacific Coast Composites Inc.)
2. US. Pat. 5,194,111 (Mar. 16, 1993) Franklin D. Meatto (to Pacific Coast Composites Inc.)

NB-304-1, Highly Toughened 250F Curing Epoxy System.

Frederick F. Saremi
Newport Adhesives and Composites
Irvine, California
(Mitsubishi Rayon Corporation)

ABSTRACT:

The performance characteristics of NB-304-1, a highly toughened system is discussed and some of the mechanical and thermal properties is compared to Newport standard NB-301. It is shown that NCT-304-1Compression After Impact (CAI) is 44.5 Ksi, and GIIC value of 7.96 in-lb/in, with Tg of 136C. Also some of the potential applications of this product in the sporting goods industry will be presented.

Key Words: Toughened Epoxy, Graphite, Prepreg, GIC, GIIC, CAI, Sporting Goods.

I. BACKGROUND:

In recent years the use of composites have increased in many areas from aerospace to recreational (1). Newport Adhesives and Composites has been one of the leaders in developing a resin system primarily designed to meet the performance requirements of sporting /recreational areas, and specifically for the golf shaft application (2-3). NCT-301controlled flow resin system has been accepted as a standard in the composites industry for this particular application. This resin system is a controlled flow, 250F curing epoxy, with some toughness applications.
There are some instances were NB-301 toughness is not adequate. To meet this demand 304-1 resin system was developed. In the past, increased toughness has been achieved at the expense of loss in modulus and glass transition temperature Tg (4). A novel approach of toughening mechanism was used in obtaining a high degree of toughness in 304-1, with little sacrifice in modulus or Tg.

II. MATERIALS:

Newport Adhesive and Composites has established the following nomenclature for identifying each product. Each resin system is identified by a set of numbers. NB prefix is used to designate the product as a film or as a coated system on a non-unidirectional reinforcement. The NCT simply designates that the reinforcement is a unidirectional tape.
301 is a 250F curing controlled flow epoxy system, and toughened 304-1 epoxy resin system, is also 250F curing, with controlled flow property. In this paper both resin systems will be mentioned as a fiber reinforced prepreg (NB-301, NB-304-1), or Uni-Tape reinforced prepreg (NCT-301, NCT-304-1).

The Graphite fiber used was from Grafil, Inc. a wholly owned subsidiary of Mitsubishi Rayon Corporation (MRC), 12K type 34-700, with 34 Msi tensile modulus, 650 Ksi tensile strength, 1.8 g/cc density, 1.9% elongation, and 6.9μ filament diameter.

III. EXPERIMENTAL:

In this section the 304-1 resin system will be compared to 301. The reactivity of both systems are identical in term of resin gel (See Table 1). The minimum complex melt viscosity of 304-1 is higher than 301, as determined by Rheometrics RDS-200 [(5mm disc diameter, 0.5 mm gap, at 1 Radian/ second, and ramp 2K (Kelvin) per minute], See Figure 1, and 2.

Mechanical properties of unidirectional graphite (NCT-301) is shown in Table 2, and NCT-304-1 is in Table 3. Grafil 12K, 34-700 fiber was used for both resins, and 275F curing results are compared. Flexure Strengths of NCT-304-1 are slightly lower, but the flexure modulus values are comparable. The compression modulus of NCT-304-1 is very close to NCT-301, whereas the compression strength values are lower. The tensile strength and modulus of both systems are close.

In Table 4, the neat resin toughness properties are compared. It is evident that 304-1 is considerably tougher than 301, GIC of 2.74 (in-lb/in2) compared to 1.50 in-lb/in2. The flexure strength is lower 21.9 Ksi versus 23.5 Ksi, but the flexure modulus values are very close 0.46 Msi, and 0.47 Msi respectively.

The 304-1 neat resin toughness translates well into a composite laminate, as seen in Table 5. Hexcel's standard AS4C unidirectional Graphite tape was used as reinforcement Both systems were cured in an autoclave, vented at 15 psi, for 1.5 hours at 275F and 30 psi.
The Compression After Impact (CAI) laminates were laid-up with 24 plies (+45/0/-45/90)$_{3s}$, the impact level was 1500 in-lb/in. with steel based jig. NCT-304-1, CAI value is 43.5% higher than NCT-301, 44.5 Ksi, and 31 Ksi respectively. Also, NCT-304-1 damaged area of the laminate is nearly half of NCT-301 laminate (Table 5).

The laminates for the Mode I and mode II interlaminar fracture testing were laid up using 20 unidirectional plies and they were cured as above. The mode I was measured using the double cantilever beam (DCB) method, The fracture specimen was pulled apart in tension at 2.54 cm/min. The crack extension was marked when a 6.35cm displacement was reached. The GIC values of NCT-304-1 and NCT-301 are similar, 2.8 in-lb/in, and 2.6 in-lb/in.

The mode II interlaminar fracture toughness was determined by the end notch flexure (ENF) test method. All specimens were pre-cracked, with the crack front of 2.54cm from the stationary post. The specimens were tested on a three point bending apparatus, and were loaded at 0.254 cm/min displacement rate. The GIIC values of NCT-304-1 (7.96 in-lb/in) is 2.4 times higher than NCT-301 (3.3 in-lb/in).

The Glass Transition Temperature (Tg) of each system was determined by different methods.
A Rheometrics RDA-700 in DMA mode was used to determine Tg. The composite laminate specimens were 2 mm x 5mm, with fixture distance of 42 mm, at 6.3 radian/ sec, and 0.15-2% strain. (See figures 3, and 4). The Tg values of NCT-301 by G' (Shear Loss Modulus) is 139.2 C, and 137.2 C for NCT-304-1. Also, the Tg with Maximum Loss Tangent (Tan δ, max) for NCT-301 is 153.9 C, and 143.7 C.

TA instruments, TMA-2100 (Thermo Mechanical Analyzer) was used to determine the Tg , at 10C/min, 5grams weight, and under Nitrogen purge. The Tg values are 136.8C, and 134.5C for NCT-301 and NCT-304-1 respectively.

It is evident from the above data that the Tg values of 304-1 are very close to 301.

IV. DISCUSSIONS:

The mechanical properties of NB-304-1, reinforced with woven graphite 3K-70P (T-250, 3K yarn, 7-mil plane weave, with 196 GSM), was determined. At two cure cycle temperatures of 250 F ,and 275 F for 1 hour. As can be seen in Table 7, the mechanical properties are good up to 160F under wet conditions (24 hour water boil). Which seem to indicate that it can be successfully used in most out door sporting goods equipment.

In addition, as the mechanical properties of unidirectional tape, in Table 8 indicate, 50K fiber results can be used where lower costs are an issue.

Also, other versions of NB-304-1 are available, such as a fast cure system, which can be cured as low as 185F for 3.3hours, with 21 days out time, this system is called NB-304-5.

The mechanical properties of NB-304-5 on 3K70P woven graphite are shown in Table 9, with press cure cycle of 20 minutes at 250F.

V. CONCLUSION:

It is shown that NB-304-1 is a highly toughened system, and this toughness is achieved with very slight sacrifice in Tg, and compares well to NB-301 which is well established in the sporting goods industry. The high degree of toughness of NB-304-1 will allow it to be used where is impact resistance is needed, along with good retention of modulus. In Golf shafts or fishing rods this can translate into the same tip strength as the standard NCT-301, with much higher impact resistance.

Also, in many of the sporting goods where damage tolerance is crucial, such as bicycles, hockey sticks, baseball bats, and wind surfing hulls, NCT-304-1is an ideal material. And when a high through-put of manufacturing is needed, fast cure NCT-304-5 can be utilized.

VI. REFERENCES

1. Steve Davis, "Racket Science Applied to Golf", SAMPE, 42nd International Symposium, Vol.42, 1997, P.860

2. Nobuhiro Takada, " Progress in Sport Goods and Technologies for Applying Advanced Composite Materials to Sport Goods", SAMPE, 41nd International Symposium, Vol.41, P.393.

3. Layton M. Bennett, " An Expert System for the Design and Selection of Planning Wind-surf Hulls." SAMPE, 41nd International Symposium, Vol.41, P.370.

4. K.J. Ahn, J.C. Seferis, T.Pelton, and M. Wilhelm, Polymer Composites, 13 (3),197 (1992)

FIGURE 1
NB-301-1

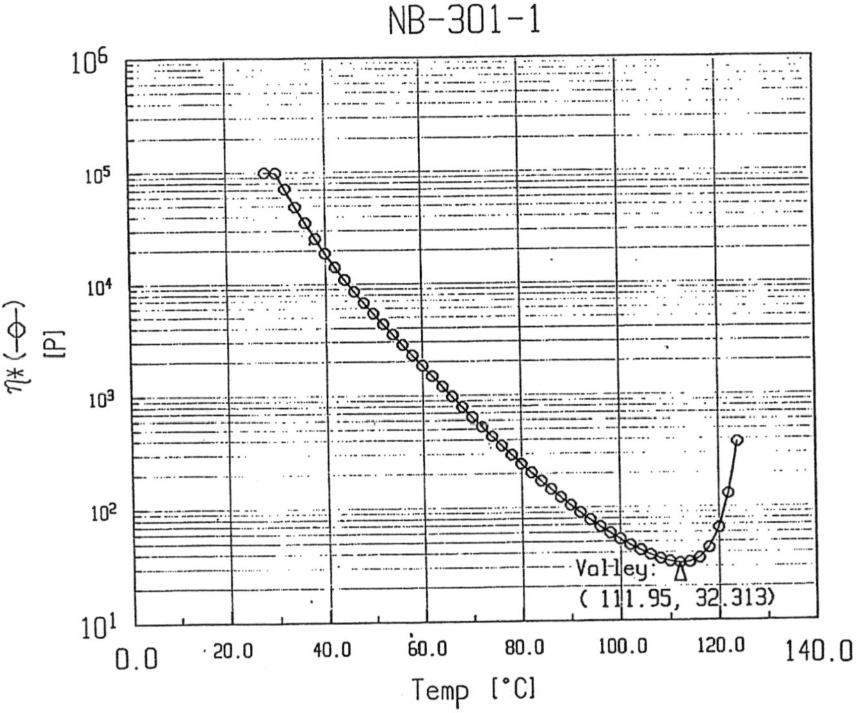

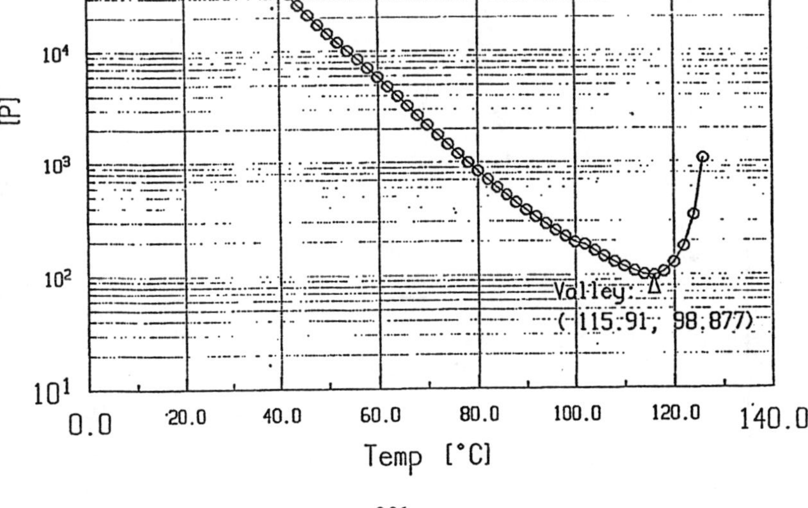

FIGURE 2
NB-304-1

FIGURE 3

NB-301-1

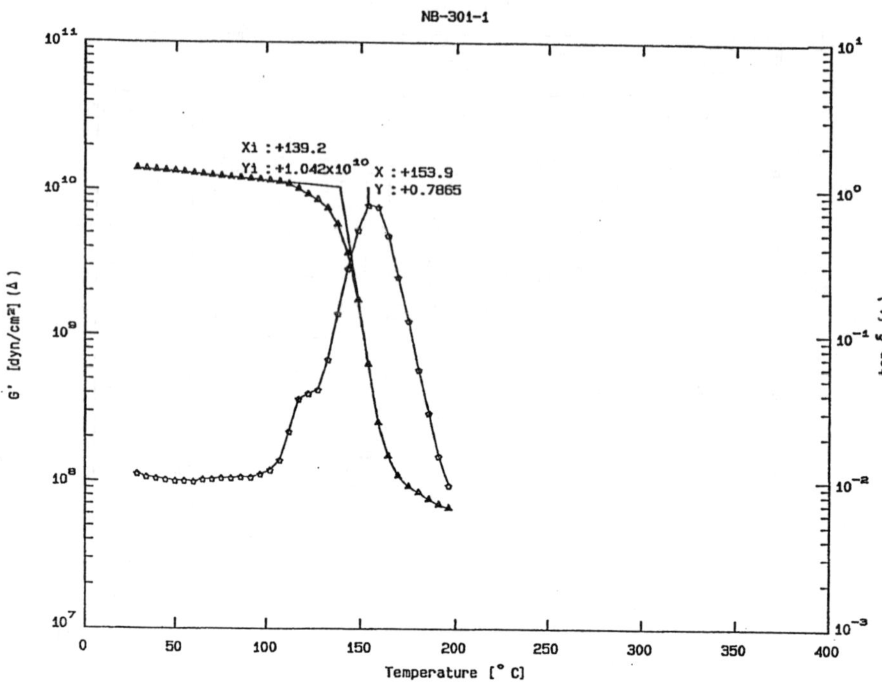

FIGURE 4

NB-304-1

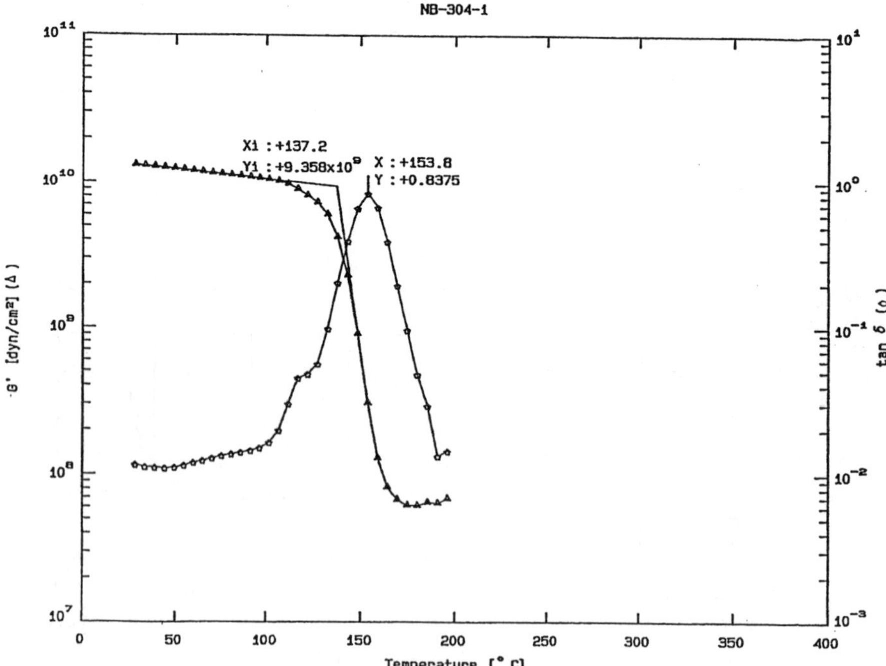

302

FIGURE 5
TMA
Sample: 304-1 GRAPHITE LAMINATE
Size: 5.3400 mm
Method: TMA 25/10/325°C
Comment: 10°C/min. NITROGEN PURGE, 5 gram load
C.T.E. from: 150-190°C
190.00°C
α=115.μm/m°C
C.T.E. from: 50-115°C
115.00°C
α=45.0μm/m°C
150.00°C
Tg
134.46°C
50.00°C
17000
8000
-1000
25
75
125
175
225
2100
Temperature (°C)
TMA
Dimension Change (µm/m)

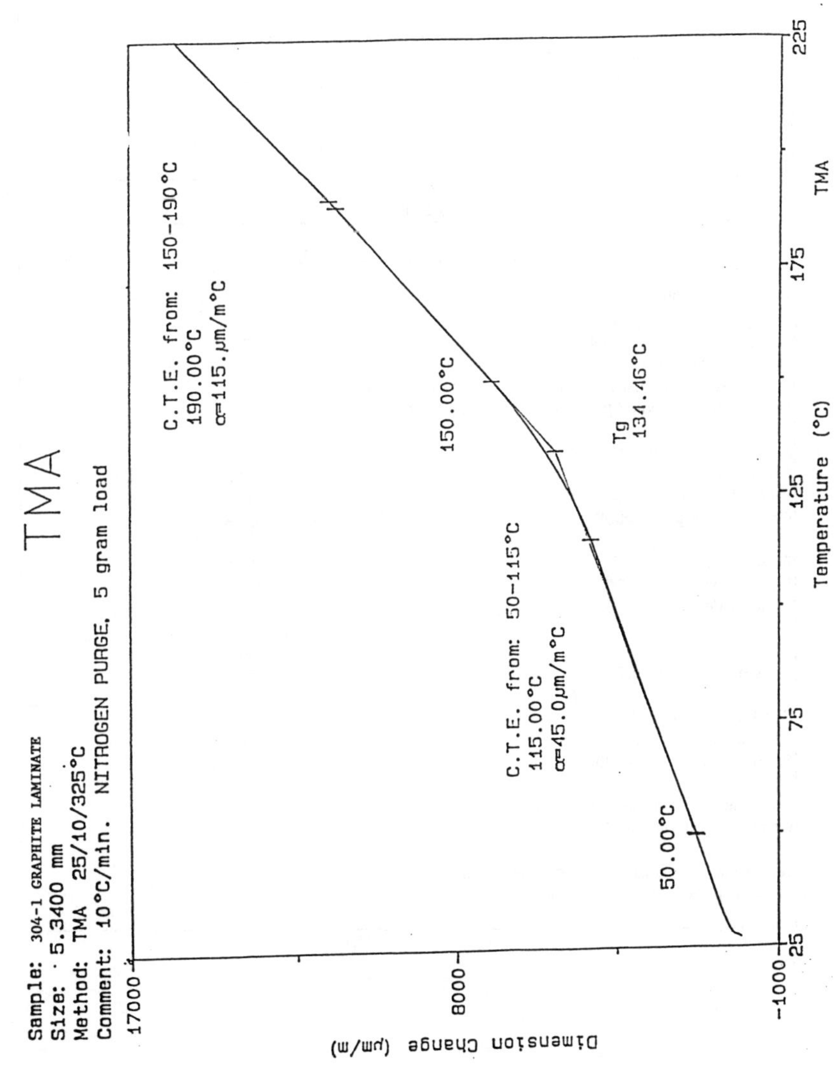

303

TABLE 1

Resin property comparison of 301 vs. 304-1

	301	304-1
Gel @275F	5-7 minutes	5-7 minutes
Melt (complex) minimum viscosity	32 Poise	99 Poise
Temperature, at min viscosity, C (F)	111.95 C (233.51)	115.91 C (240.64)
	(Figure 1)	(Figure 2)

Rheometrics RDS-200, disc diameter 5 mm, gap=0.5 mm, Ram 2K/min, at 1 rad/sec

TABLE 2

Mechanical Properties of NCT-301

Graphite: 34-700 (34 Msi Modulus, 650 Ksi tensile strength, 1.8 g/cc density, 1.9% elongation, 6.9 um filament diameter)
34-700, 12K, is from Grafil Division of Mitsubishi Rayon Corporation (MRC)
Resin Content : 35+/-2%
Fiber Areal Weight (FAW) = 120 GSM (Grams per Square Meters)
Cure Cycle: Preheated press, 275F, 50 psi, 1 hour.

All Uni-Tape properties in 0-direction

Flexural Strength @ RT.	227 Ksi
Flexural Strength @ 160F	207 Ksi
Flexural Strength @ 200F	162 Ksi
Flexural Modulus @ RT.	18.1 Msi
Flexural Modulus @ 160F	16.5 Msi
Flexural Modulus @ 200F	12.9 Msi
Tensile Strength @ RT.	290 Ksi
Tensile Modulus @ RT.	18.5 Msi
Short Beam Shear @ RT.	14.3 Ksi
Short Beam Shear @ 160F	13.0 Ksi
Short Beam Shear @ 200F	10.2 Ksi
Compressive Strength @ RT.	170 Ksi
Compressive Modulus @ RT.	19.3 Msi

Test Methods: Flexure ASTMD-790, Compression ASTMD-695, Tensile ASTMD-3039
All results Normalized to 60% Fiber Volume.

TABLE 3

Mechanical properties of NCT-304-1 as a function of cure cycle.

Resin: NCT-304-1

Graphite: 34-700 (34 Msi Modulus, 650 Ksi tensile strength, 1.8 g/cc density, 1.9% elongation, 6.9 um (micro meter) filament diameter)

34-700, 12K, is from Grafil Division of Mitsubishi Rayon Corporation (MRC)

Resin Content: 42+/-2%

Prepreg Gel Time @ 275F = 5 minutes

Lay-up:12-plies for tensile specimens tabbed with glass bonded with film adhesive.

24-Plies for Flex,Comp,SBS specimens.

Cure: Preheated Press	250F/50psi/ 1hr.	275F/50psi/ 1hr.	300F/72psi/ 15mins.
All Uni-Tape properties in 0-direction.			
Flexural Strength @ RT.	210.6	211.5	232.0
Flexural Strength @ 160F	192.0	193.1	211.5
Flexural Strength @ 200F	150.3	150.9	172.0
Flexural Modulus @ RT.	17.8	18.0	18.8
Flexural Modulus @160F	15.1	16.5	17.1
Flexural Modulus @ 200F	12.5	12.8	13.5
Compression Strength @ RT.	128.3	135.2	137.5
Compression Strength @ 160F	98.4	108.1	110.6
Compression Strength @ 200F	78.7	87.8	97.4
Compression Modulus @ RT.	17.5	18.4	18.7
Compression Modulus @ 160F	15.1	16.2	16.4
Compression Modulus @ 200F	13.9	15.1	15.7
Tensile Strength @ RT.	268.3	288.6	
Tensile Strength @ 160F	255.4	276.6	
Tensile Modulus @ RT.	18.6	18.7	
Tensile Modulus @ 160F	17.4	17.8	
Poisson Ratio(Tensile ASTMD-3039) @ RT.		0.303	

Test Methods: Flexure ASTMD-790, Compression ASTMD-695, Tensile ASTMD-3039

All results Normalized to 60% Fiber Volume.

TABLE 4

Neat Resin Toughness of NB-301 vs. NB-304-1

Cured: Resin casting degassed at 190F(88C) into a 2mmx3mm thick plate,
then cured in Autoclave, 30 psi, 275F (135C), 1.5 hours.

	NB-301	NB-304-1
Flexure Strength, Mpa (Ksi)	163 (23.5)	147 (21.9)
Flexure Modulus, Gpa, (Msi)	3.3 (0.47)	3.16 (0.46)
GIC, KJ/m2 (in-lb/in2), resin	0.265 (1.50)	0.481 (2.74)

Flexure: Three point bending
GIC: Compact Tension of 3 mm thick resin plate, assumed resin Poisson ratio of 0.38

TABLE 5

CompositeToughness of NCT-301 vs. NCT-304-1

Fiber: From Hexcel AS4C (33 Msi Modulus, 580 Ksi Strength)
Cure: Autoclave, 30 psi, 275F (135C), 1.5 hours.

	NCT-301	NCT-304-1
Prepreg Resin Content	37%	34%
CAI (Compression After Impact),Mpa(Ksi)	215 (31)	307 (44.5)
Damaged area, mm2	999	492
GIC, KJ/m2 (in-lb/in2), laminate	0.457 (2.6)	0.497 (2.8)
GIIC, KJ/m2 (in-lb/in2), laminate	0.580 (3.3)	1.399 (7.96)

CAI: 24-plies laid-up (+45/0/-45/90)3S, Impact level = 1500 in-lb/in.
CIC, CIIC: 20 plies laid-up unidirectional.
GIC: Double cantilever beam test
GIIC: Edge notch flexure test.

TABLE 6

Glass Transition Temperature (Tg) of NCT-301 vs. NCT-304-1

	NCT-301	NCT-304-1	
Thermo Mechanical Analyzer (TMA)	136.8C	134.5 C	(Fig. 5)
DMA mode, G' (Shear Storage Modulus)	139.2 C (Fig.3)	137.2 C	(Fig.4)
DMA mode, Loss Tangent, Max	153.9 C (Fig.3)	143.7 C	(Fig.4)

Cure Cycle of Graphite laminate: Autoclave at 30 psi, 275F (135C), 1.5 hours.
TMA: TA-Instruments-2100, 10C/min, 5 gram, Nitrogen purge.
Rheometrics RDA-700, DMA mode: Speciemen 2 mm x 5 mm, fixure distance = 42 mm,
6.3 rad/sec rate, 0.15-2% strain.

TABLE 7

Mechanical Properties of NB-304-1 on Woven Graphite

Resin : NB-304-1
Fabric: Woven Graphite 3K70P (T-250, 3K yarn, 7-mil plane weave) 195 GSM (Grams per Square Meter)
RC = 43+/-3%
Flow (4x4"x4",275F,25psi,15min)2-plies 7781 Bleeder = 5.2%
Vols. (275F,20 mins) = 0.04%

Lay-up: 14-plies Warp face up, parallel.

Cure: Preheated Press	250F/50psl/ 1hr.			275F /50 psi / 1 hr.		
	RT	160F	200F	RT.	160F	200F
Flex Ult.(Dry), Ksi	108.2	100.2	86.5	115.00	99.05	81.50
Flex Mod.(Dry), Msi	7.27	7.02	6.47	7.37	7.10	6.94
Flex Ult.(Wet), Ksi	100.3	59.2	32.6	97.60	43.00	36.40
Flex Mod.(wet), Ksi	7.03	6.07	3.87	7.32	5.03	4.20
Compression Ult.(Dry), Ksi	69.2	54.4	49.6	68.30	56.30	53.96
Compression Mod.(Dry), Msi	7.9	7.43	7.00	8.00	7.12	7.11
Compression Ult.(Wet), Ksi	55.1	29.7	18.9	45.20	37.80	14.50
Compression Mod.(Wet), Msi	7.2	7.00	6.92	7.30	6.93	6.00
Tensile Ult. (Dry), Ksi	83.8	74.2	75.5	85.40	78.30	58.80
Tensile Mod. (Dry), Msi	9.17	8.56	8.64	9.79	9.24	9.00
Tensile Ult. (Wet), Ksi	78.1	70.8	65.8	78.06	68.50	56.50
Tensile Mod. (Wet), Msi	8.93	8.57	6.93	9.15	8.23	7.44
Short Beam Shear (Dry), Ksi	9.03	6.98	6.19	9.33	6.67	5.93
Short Beam Shear (Wet), Ksi	6.38	3.98	2.65	6.68	3.29	2.93
Poisson Ratio(Tens.ASTMD-638)@ RT.				0.049		
Tensile Ult. (Dry, @ RT.), Ksi				87.5		
Tensile Ult. Strain(Dry, @ RT.),u in/in				9831		
Tensile Mod. (Dry, @ RT.), Msi				8.88		

Test Methods: Flexure ASTMD-790, Compression ASTMD-695, Tensile ASTMD-638

All results Normalized to 60% Fiber Volume.

Wet Condition: 24 hour water boil

TABLE 8

<u>Mechanical property comparison of Unitape 12K vs. 50K laminates</u>

Resin: **NCT-304-1**	34-700.12K	50K
Fiber Tensile Modulus, Msi	34	33
Fiber Tensile Strength, Ksi	650	550
Fiber Areal Weight (FAW), GSM (Grams/m2)	120	120
Resin Content	42+/-2%	42+/-3%
All in Zero direction.		
Flexural Strength, @ RT. Ksi	278	265.3
Flexural Modulus, @ RT., Msi	18.6	14.9
Compression Strength, @ RT., Ksi	136.9	102.4
Compression Modulus, @ RT., Msi	18.6	13.5
Tensile Strength, @ RT., Ksi	223	228.1
Tensile Modulus, @ RT., Msi	17.4	15.8
Short Beam Shear, @ RT., Ksi	13.4	11.5

34-700, 12K, is from Grafil Division of Mitsubishi Rayon Corporation (MRC)
50K fiber from Akzo, Fortafil 3(C)
Cure: Per-heated press 150 C (302 F), 72.6 Psi, 15 minutes.
All normalized to 60% Fiber Volume, except SBS.
Test Methods: Flexure ASTMD-790, Compression ASTMD-695, Tensile ASTMD-3039, Short Beam Shear ASTMD-2344.
8-plies for all the panels, except for compression 16-plies.

TABLE 9

<u>Mechanical properties of NB-304-5 on Woven Graphite</u>

Resin: NB-304-5 (Fast cure version of NB-304-1)
(Prepreg out time at room temperature 21 days)
Fabric: Woven Graphite 3K70P (T-250, 3K yarn, 7-mil plane weave) 195 GSM (Grams per Square Meter)

Resin Gel Time @275F	5 minutes
Prepreg RC%	43+/-3%
Prepreg Gel @275F	4 minutes
Flow(4x4"4" 275F,20mins,25si,15mins) 2-plies7781	10.20%
Vol.(20 mins,275F)	0.43%

Lay-up:12 plies,warp up, parallel.
Cure: Preheated press: 250F,50 psi,20 mins,

	Normalized to 60% FV
Compression, Strength,DRT,Ksi	67.3
Compression, Modulus,DRT,Msi	7.66
Compression, Strength,180F,Ksi	36.78
Compression, Modulus,DRT,Msi	7.3
Flexure,Strength. DRT,Ksi	111
Flexure,Modulus. DRT,Msi	7.02
Flexure Strength, 180F,Ksi	60.8
Flexure Modulus,180F,Msi	5.4
Short Beam Shear, DRT, Ksi	9.66
Short Beam Shear, 180F, Ksi	4.6*

Test Methods: Flexure ASTMD-790, Compression ASTMD-695, Tensile ASTMD-638, Short Beam Shear ASTMD-2344
* Not normalized

Model Epoxy/Carbon Fiber Prepreg Systems for Sporting Goods Applications

Brian S. Hayes, Fulcanelli S. Chavez, and James C. Seferis[*]
University of Washington
Department of Chemical Engineering
Polymeric Composites Laboratory
Seattle, Washington 98195

and

Edward Anderson and Jayant Angal
Zeon Chemicals Incorporated
Louisville, Kentucky 40233

ABSTRACT

A particulate elastomer was modified by changing its processing conditions and the effects were investigated in model epoxy based sporting goods prepreg systems. As a result of using the new experimental particles, the mode II fracture toughness increased by 250% when compared to the control. These carboxyl functional elastomers were found to remain in the composite interlayers. Through this type of modification, the highly stressed interlaminar region of the composite was toughened without the question of phase separation or loss of prepreg handling characteristics.

KEY WORDS: Prepreg, Epoxy Modifiers, Interlaminar Fracture Toughness

1. INTRODUCTION

Due to the increasing demand of high performance prepregs for sporting goods applications, research continues on modifiers for epoxy-based matrices. In selection of the specific resin formulation for a sporting goods application many factors must be considered. These factors not only include final thermal and mechanical properties, but also surface appearance, cure cycle flexibility, ease of lay-up, and cost. A large influence on these factors is not just the base resin components selected, but the modifiers used to increase specific properties. The most commonly used modifiers for epoxy-based resins are thermoplastics and elastomers. These materials may be used for such applications as flow control or increasing fracture toughness. Depending on the physical nature of the modifier, it may need to be dissolved in the base epoxy to form an initial homogeneous mixture. This may be accomplished by long dissolution times at relatively high temperatures or with the addition of a suitable solvent. Consequently, it may remain in the continuous phase or phase separate upon cure which is a function of material compatibility, quantity, and cure cycle.(1, 2) These variables may be eliminated by the selection of materials that are initially and remain as distinct particles. If the particles are large enough, they will remain on the prepreg surface during prepreg processing and therefore in the interlayer after cure.(3) This method of toughness modification typically yields the highest improvements in damage tolerance since ply delamination has been found to be the most life limiting failure mechanism of composites.(4, 5)

To increase the interfacial adhesion between the host epoxy matrix and the particle, the modifier materials are sometimes functionalized. A technique of pre-reacting the functional sites on the particles with the epoxy functionality is usually performed so that the maximum possible reactive sites are consumed. This is usually accomplished at elevated temperatures using a suitable catalyst while mixing. Once adducted, the final curing mechanisms remain the same.

Previously, a new family of particulate cross-linked carboxyl functional elastomers were incorporated into a base epoxy resin and the performance evaluated in unidirectional prepregs.(6, 7) Significant increases in the fracture toughness were found in the model sporting goods prepreg systems with no reduction in prepreg handling characteristics. In this study, a particulate elastomer previously investigated, was modified by changing its processing conditions and the effects evaluated in model epoxy-based sporting good prepreg systems.(7)

2. EXPERIMENTAL

2.1 Resin The base model epoxy resin formulation components and approximate composition including equivalent weights are shown in (7). The difunctional epoxy resins used were a combination of Epon® 828 and Epon® 836, both based on diglycidyl of bisphenol-A, from Shell Chemical Co. The curing agents used were dicyandiamide, Amicure® CG 1400 from Pacific Anchor Chemical Co., and diuron, from Aldrich Chemical Co. The elastomeric particle tougheners DuoMod® DP 5045, and experimental LWS "less water sensitive" DuoMod® PE 1623 and PE 1625, from Zeon Chemicals Incorporated, had a particle size between 10 and 90 μm. Glass transition temperatures for the elastomers were approximately -6°C. All of the particles were cross-linked carboxyl functional elastomers based on butadiene/acrylonitrile. Differences in the particles were a result of processing conditions. The particles were pre-reacted with the base epoxy resin at 140°C for 40 minutes using triphenyl phosphine.

2.2 Prepreg The prepreg developed in this study was a unidirectional prepreg consisting of high modulus carbon fibers (Toray T800HB 12K epoxy-sized fibers) and the previously described epoxy matrix. The carbon fibers were impregnated with the epoxy resin using a commercial scale hot-melt prepreg machine.(8) The nominal prepreg resin content was set at 32 ± 2 wt.% and the nominal prepreg grade was set at 190 g/m^2 for all experiments. Resin filming was performed at 60°C and the impregnation temperature was 82.2°C. The impregnation pressure was set to 345 kPa on a 7.62 cm diameter roller and the line speed was 0.91 m/min for all experiments.

Prepreg samples were cut into 5.1 cm by 5.1 cm squares and characterized by resin content, prepreg thickness, and tack. The prepreg thickness was determined by placing a square of prepreg between two square steel tabs and compressing the sample in a mechanical testing machine until a force of 2.67 N was reached and then measuring the final displacement. Three samples were tested for each prepreg thickness value.

Quantitative tack values of the prepregs were determined from a compression/tension experiment developed by Seferis and co-workers.(9-11) To perform the tack test, five plies of 5.1 cm by 5.1 cm prepreg were bonded between two metal tabs, and subjected to a compression to tension cycle. The prepreg stack was compressed at a displacement rate of 0.0254 cm/min to 268 N, held at 268 N compression for 30 seconds, and then pulled apart in tension at a constant displacement rate of 0.034 cm/min. From the analysis of the stress/strain data collected during the tack test, both compressive energies and toughness factors were obtained to quantitatively define the tack of the prepregs. Three tack tests were performed for each tack value. During the testing, the room temperature varied from 21 - 22°C and the relative humidity was between 34 - 37%.

2.3 Laminate Laminates were made in an autoclave from the model prepreg for mode I and mode II interlaminar fracture testing.(4) The laminates were 16 plies thick, 33 cm long, and had a 5.08 cm FEP crack starter in the midplane of the sample. The cure cycle consisted of heating at 2.8°C/min to 127°C, holding for 90 minutes, and cooling at 2.8°C/min to 27°C. The laminates were manufactured using a total consolidation pressure of 207 kPa and the vacuum bag was vented to atmosphere when the autoclave pressure reached 103.5 kPa. Each prepreg ply was pre-compacted under vacuum pressure for two minutes during the hand lay-up before another ply was positioned.

After the laminates were cured, they were cut into 1.27 cm wide samples for testing. Mode I interlaminar fracture toughness was measured using the double cantilever beam (DCB) method.(4) Each specimen was precracked in the mechanical testing apparatus to provide a sharp crack tip before testing. The fracture specimen was pulled apart in tension at 2.54 cm/min until a displacement of 6.35 cm was reached at which point the crack extension was marked. Four samples, each providing one G_{IC} value, were tested and averaged for each reported G_{IC} value. Standard deviations were calculated and are shown as error bars.

Mode II interlaminar fracture toughness was measured using the end notch flexure (ENF) test.(4) The same laminate preparation was used for the DCB specimens. A three point bending apparatus with stationary posts set 10.16 cm apart was used to create shear fracture of the specimen down the midplane. The crack front was set 2.54 cm from the stationary post and the loading point was set two inches from the post. All specimens were precracked in the mechanical testing apparatus to provide a sharp crack tip before testing. A displacement rate of 0.254 cm/min was used to load the specimen in flexure until the crack propagated. The crack front was then located with an optical microscope fixture and moved back to 2.54 cm from the stationary post. This was repeated until the sample was cracked down its entire length. Six G_{IIC} values were obtained for each specimen and averaged for the reported G_{IIC} value. Standard deviations were calculated and are shown as error bars.

The fracture surfaces were polished and optical photomicrographs were taken at 200× magnification.

Dynamic mechanical analysis (DMA) experiments on cured nine ply laminates were performed with a TA Instruments 983 DMA interfaced to a Thermal Analyst 2100 controller. A heating rate of 5°C/min to 300°C with a frequency of 1Hz and an oscillation amplitude of 0.20 mm in nitrogen was utilized.

3. RESULTS AND DISCUSSION

All of the DuoMod particles investigated were dispersed in the base epoxy without the use of a solvent. The particles dispersed evenly in the base epoxy with standard mixing when the temperature was 140°C. Each resin developed consisted of 11. 1 wt.% particles. It should be noted that although a higher concentration of particles can be added to the base resin, this may not be beneficial to the handling properties, tack and drape, including the transverse integrity of the prepreg and possible streaking. The base resin characteristics are described previously in (7).

Prepreg was made with 32±2 wt.% resin which resulted in 4 wt.% DuoMod particles in the prepreg. The addition of the DuoMod particles increased the viscosity of resin but not to a great extent when compared to other conventional toughness modifiers. Although the viscosity increased, the prepreg processing parameters were not optimized to provide a consistent prepreg thickness. Therefore, the prepreg thickness was found to increase from 0.28 mm for the control to 0.36 mm when the particles were added. This was due to the increase in viscosity causing slightly less impregnation and resulting in particles over 20 mm being sieved out of the fiber bed.(5) It must be stated that the prepreg thickness could have been reduced by changing the prepreg processing conditions. Impregnation of the prepreg was quantitatively defined through the compressive energy of the prepreg.(8) The compressive energy increased approximately 70% with the addition of the particles. This was because the resin was not fully impregnated into the fiber bed as was also seen from the thickness of the prepregs. The compressive energy of the prepreg containing DuoMod particles increased because it was harder to force the particles on the surface into the fiber bed.

The toughness factor, which is the tensile energy up to maximum stress after compression was also measured during the tack test. The toughness factor of the model prepregs increased when the DuoMod particles were added. The low glass transition temperature of the elastomers increased the strain energy to failure about 100% when the particles were added. Both adhesive and cohesive failure occurred in the prepreg stack. As a result of the particles, the quantitative tack of the prepregs increased but not excessively.

Mode I interlaminar fracture toughness was calculated as G_{IC}, the critical strain energy release rate, which was defined by the area method. The G_{IC} values are shown in Figure 1 for the

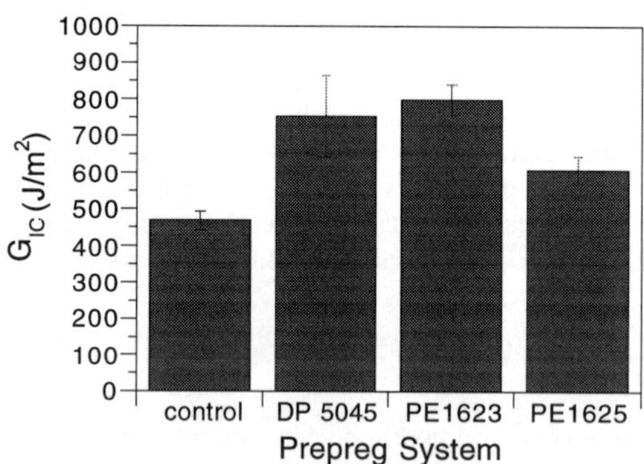

Figure 1.   Mode I interlaminar fracture toughness measured as G_{IC}.

model prepreg laminates. The G_{IC} values increased with the addition of the elastomer particles. Between each prepreg ply, particles were found in the interlayer creating a stratified composite structure. This tougher elastomer modified interface increased the fracture toughness when compared to the control, however, large differences between the particulate modified systems were not observed in mode I. Photomicrographs of the fracture surfaces of the control and DuoMod DP 5045 laminates are shown in Figure 2a and b, respectively. As shown in Figure 2a, the fracture path propagated smoothly throughout the

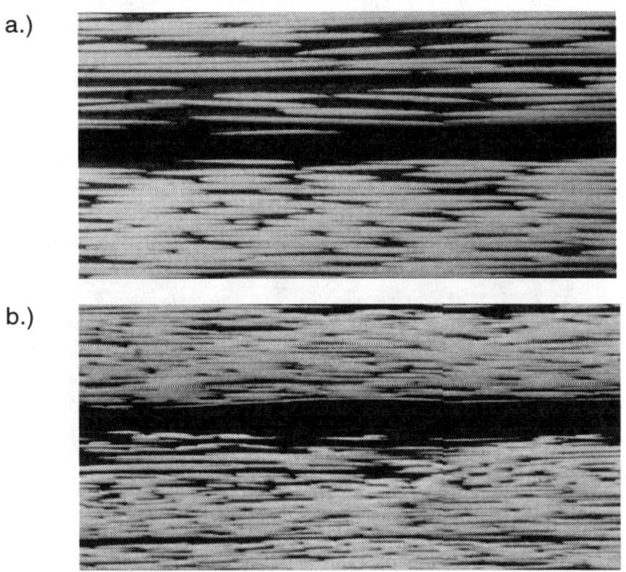

Figure 2.   Optical photomicrograph of a mode I fracture path of a.) control laminate and b.) DuoMod® DP 5045 laminate

midplane. In contrast, Figure 2b shows cavitation where the crack proceeded, due to particles deforming in the interlayer. This behavior was found in all particulate modified prepreg systems.

Mode II interlaminar fracture toughness was calculated as G_{IIC}, the critical strain energy release rate. This mode of fracture toughness is very important for prepreg systems because it has been found to directly correlate with compression after impact (CAI).(5, 12-14) As is shown in Figure 3, the G_{IIC} values increased with the addition of the particulate elastomers.

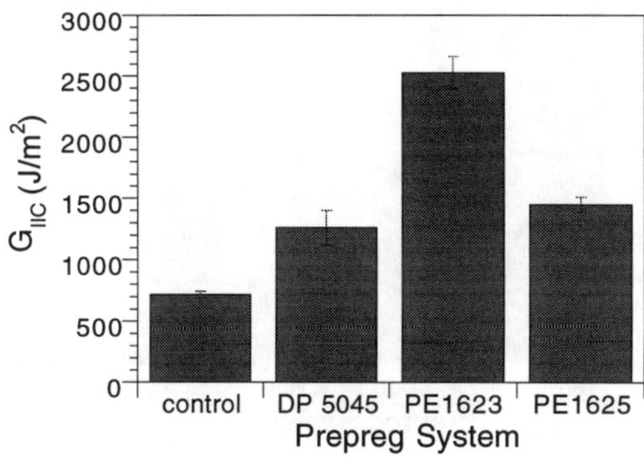

Figure 3.   Mode II interlaminar fracture toughness measured as G_{IIC}.

313

Of these particle modified prepreg based laminates, PE 1623 had a significantly higher fracture toughness in mode II. The increase in mode II fracture toughness was 250% greater than the control, and 100% greater than the other modified prepregs. The change in fracture toughness observed between the different particle modified prepregs can be explained by the average particle sizes and different surface treatments. In general, the average size of the PE 1623 particles were smaller causing the propagating crack to contact more particles. Also, a larger distribution of particle sizes was observed for the PE 1623 and PE 1625 than DP 5045, possibly causing more prepreg surface heterogeneity. In addition, surface treatments of the particles were significantly different causing different interfacial bonding between the particle and continuous phase. Figure 4, demonstrates the tortuous crack path in the midplane of the laminate containing PE 1623.

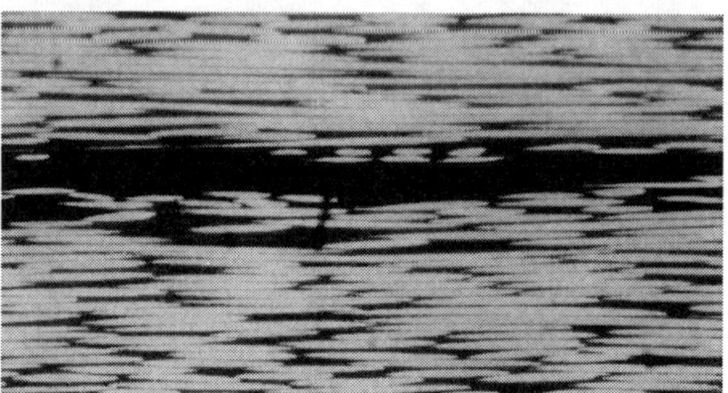

Figure 4.   Optical photomicrograph of a mode II fracture path of PE 1623 modified laminate.

Through the use of dynamic mechanical analysis (DMA) the glass transition temperature (Tg) was found using the peak of the loss modulus (E"). The glass transition temperature of all specimens are summarized in Figure 5. As shown in the figure, the glass transition temperature decreased up to 15°C with the addition of the particles. This is small reduction when considering the prepreg application and the large increases in fracture toughness. Differences between the glass transition temperatures of DP 5045 and PE 1623 laminates could be due to a difference in particle size (surface area per volume) and the compatibility between the particle matrix interface.

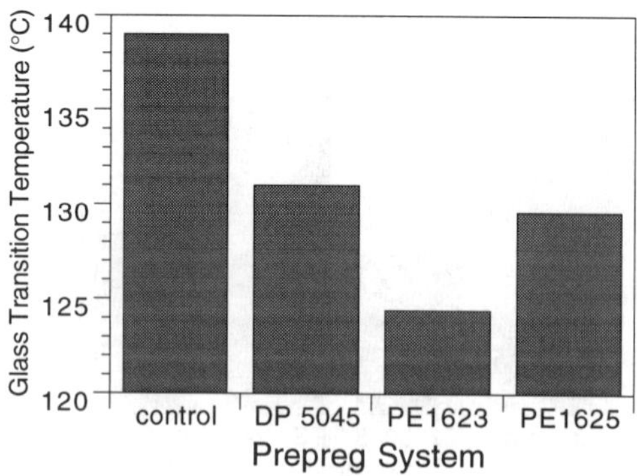

Figure 5.   Glass transition temperatures of model prepreg based laminates.

4. CONCLUSIONS

The development of particulate modifiers based on the same chemistry were processed differently to examine the effects on mechanical properties when incorporated into epoxy based prepreg systems. All particulate elastomers provided a significant increase in both mode I and mode II fracture toughness of unidirectional prepreg based laminates. As a result of the processing changes, mode II fracture toughness was found to increase by 100%, which resulted in a 250% increase when compared to the control. The ease of addition of the particles into epoxy resins without the use of solvents makes them very useful for prepreg systems where long out times without loss of tack are essential. Since these cross-linked elastomers are initially and remain as distinct particles, the morphology is independent of cure temperature and ramp rate. While the focus of this study was on modifying matrices for sporting goods prepreg systems, it should be emphasized that the particles investigated can be effectively used for increasing the performance of all prepreg systems, including but not limited to aerospace applications. Collectively, the elastomeric particles analyzed in this study provided significant increases in cured part performance when incorporated into epoxy based prepreg systems which is necessary for the next generation solvent free prepreg systems.

5. ACKNOWLEDGMENTS

The authors express their appreciation to Zeon Chemical Incorporated for project support at the PCL University of Washington.

6. REFERENCES

1. C. B. Bucknall and I. K. Partridge, <u>Polymer,</u> <u>24</u> , 639 (1983).

2. L. T. Manzione, J. K. Gillham, and C. A. McPherson, <u>J. Appl. Polym. Sci.,</u> <u>26</u> , 889 (1981).

3. M. Hoisington, and J. C. Seferis, <u>Proceedings of the 6th Technical Conference of the American Society for Composites,</u> (1991)

4. N. J. Pagano, ed. *Interlaminar Response of Composite Materials.* Composite Materials Series, ed. R. B. Pipes. Vol. 4, Elsevier, New York (1989).

5. M. A. Hoisington, <u>Process/Property Interrelations of Layered Structured Composites,</u> University of Washington, Doctorate Thesis (1992).

6. B. S. Hayes, J. C. Seferis, E. Anderson, and J. Angal, <u>J. Adv. Mater.,</u> <u>July</u> (1997).

7. B. S. Hayes, F. S. Chavez, J. C. Seferis, E. Anderson, and J. Angal, <u>29th International SAMPE Technical Conference,</u> (1997)

8. J. W. Putnam, B. S. Hayes, and J. C. Seferis, <u>J. Adv. Mater.,</u> <u>27</u> , 47 (1995).

9. K. J. Ahn, J. C. Seferis, T. Pelton, and M. Whilhelm, <u>Polym. Comp.,</u> <u>13</u> (3), 197 (1992).

10. K. J. Ahn, J. C. Seferis, T. Pelton, and M. Wilhelm, <u>SAMPE Qtr.,</u> <u>23</u> (2), 54 (1992).

11.   J. W. Putnam, J. C. Seferis, T. Pelton, and M. Wilhelm, <u>Sci. Eng. Comp. Mat.</u>, <u>4</u> (3), 143 (1995).

12.   K. Kageyama, and I. Kimpara, <u>Materials Science and Engineering</u>, <u>A</u> (143), 167 (1991).

13.   B. J. Briscoe, and D. R. Williams, <u>Composite Science and Technology</u>, <u>46</u> (3), 277 (1993).

14.   L. D. Bravenec, A. G. Filippov, and K. C. Dewhirst, <u>34th International SAMPE Symposium</u>, (May 8-11), 714 (1989).

ADVANCING THE USE OF COMPOSITES IN THE OILFIELD

Brian E. Spencer, Ph.D.
Spencer Composites Corporation
3220 Superior Street; P.O. Box 4377
Lincoln, Nebraska 68504-0377

ABSTRACT

With the need to drill deeper holes and reach father from existing platforms, designers for oilfield components are turning to composites to provide the answers for difficult applications. This paper discusses the advantages of composites as applied to oilfield applications. Specific examples are used to illustrate the significant advantages of composites.

KEY WORDS: Filament Winding, Advanced Composites, Oilfield Application

1. INTRODUCTION

Composites in general and filament winding in particular offer great benefits for oil and gas exploration and production. Advances in the last decade of the quality and capabilities of resin, fiber, and fabrication equipment and processes allow for more conversions from metal to composite. As greater numbers of engineers and designers become comfortable with the use of composites, increasing components in the oilfield are being fabricated out of composites rather than metal. In some cases composites are the only answer. Deeper holes and extended reach drilling and production are a natural for composites because of their high specific strength and stiffness. Newer resins, which are easily processed, can operate at 260° C and above. New technology and techniques in logging requires inductively transparent drill string components and casing. Composites can be designed, that are not only inductively transparent, but also provide the mechanical properties that allow direct replacement with steel while meeting API (American

Petroleum Institute) requirements. NIST (National Institute of Science and Technology) is sponsoring several joint industry projects looking at downhole tubulars, drill pipe, coiled tubing, and risers.

Figure 1 shows a typical drill string. Composites arc being evaluated for the fabrication of drill pipe, drill collars, coiled tubing, risers, and components of mud motors and downhole pumps. Filament winding and derivative techniques provide the best method for making the majority of these components.

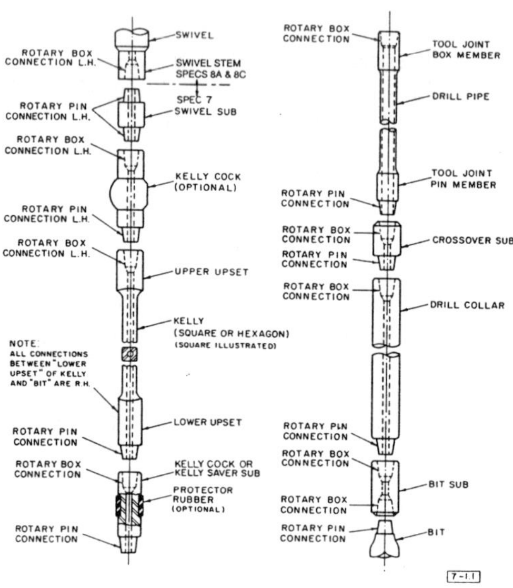

Figure 1. Typical Drill String Assembly

2. REQUIREMENTS AND ADVANTAGES OF COMPOSITES - PROBLEMS TO SOLVE

The downhole environment in the oilfield is very harsh. Composites selected for downhole use will be exposed to elevated temperature and the chemicals used to service these wells. Additionally there will be exposure to the materials in the natural formation and the abrasive and corrosive action experienced while drilling and producing the product. Wellborn fluids can have a wide range of properties including being caustic, acidic, or corrosive. Oilfield equipment is governed by API Standards, which set requirements so that products used in drilling and production can survive in the severe downhole environment. Any product offered for sale must be tested to the applicable API Standard.

Strength of Composites
Comparison with Metals (60% Fiber Volume)

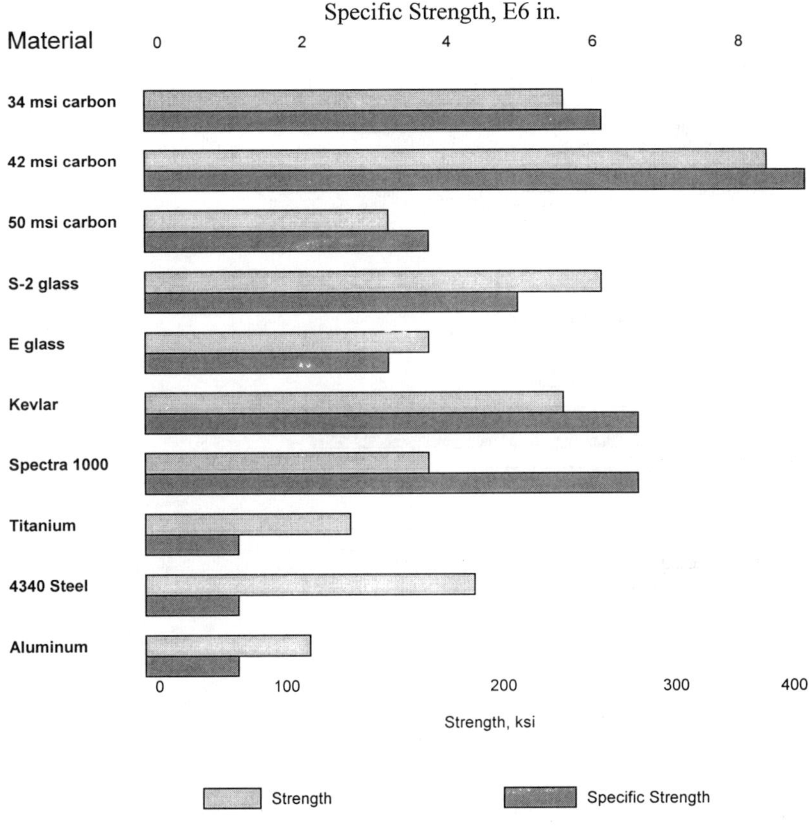

Figure 2. Strength of Composites

Stiffness of Composites
Comparison with Metals (60% Fiber Volume)

Specific Strength, E6 in.

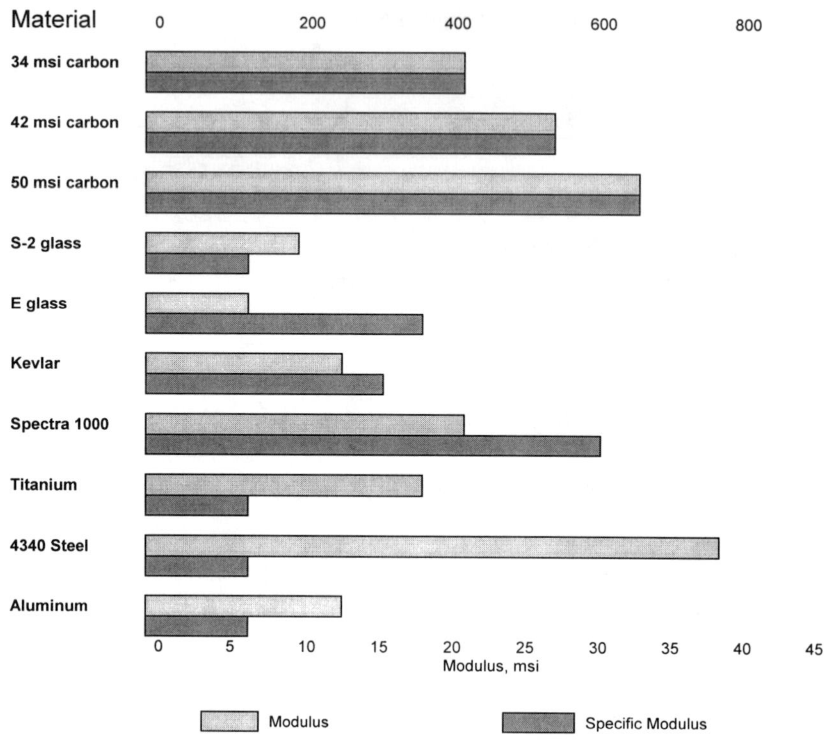

Figure 3. Stiffness of Composites

Fatigue of Composites
Comparison with Metals (60% Fiber Volume)

Tension - tension ratio at 10 to the 7th cycles

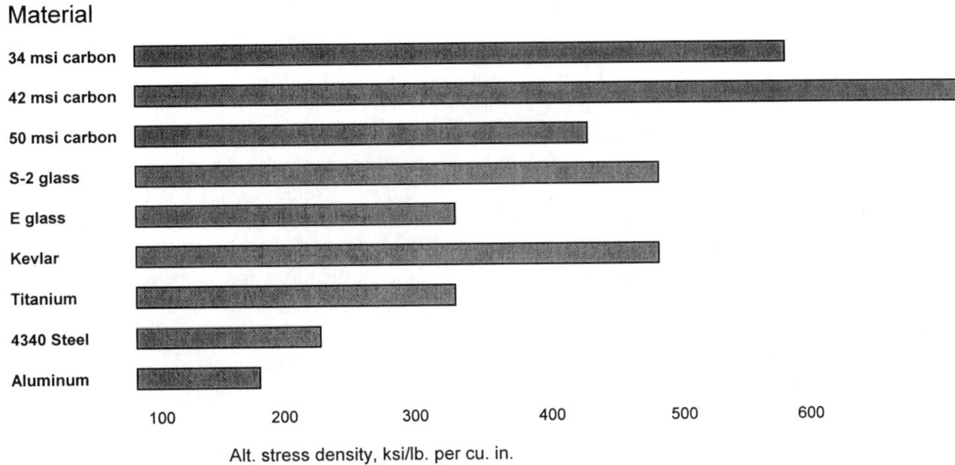

Figure 4. Fatigue of Composites

Once a bore hole is fully excavated, operators often continue to acquire formation data from the bore hole over the life of its production. In order to maintain stability in the bore hole, it is often necessary for the bore hole to be lined with a casing, normally a metal casing, cemented in place. As explained previously metal will prevent or severely attenuate the operation of sensing equipment. Accordingly a composite drill collar has been developed for LWT Instruments, that is transparent to the signal generated and received by the sensing equipment. The composite collar, because of signal attenuation, allows the sensing equipment to be run down the drill string and to take measurements without removing the string. The composite drill collar has been tested up to and has passed the mechanical tests required by the applicable API Specification. Composite drill collars are required to be inductively transparent for use with induction tools.

Stiffness is also a concern for the drill collar because the directional control of the drill string is dependent upon stiffness. Testing is underway with composite drill collars to determine the required stiffness. Composite drill collars allow logging while drilling or tripping since the logging tool can be lowered into position within the drill collar. The conventional method is to remove the drill string before lowering the instrument into position. Figure 5 shows a composite drill collar and Figure 6 shows the collar at a drilling site in Canada. The drill collar shown has a protective layer of ceramic material to counteract the effects of abrasion. The metal ends have standard pin and box type connections. The drill collar was made using the filament winding process with the metal ends being filament wound into the main composite body. Additional information on end fittings and attachment can be found in Reference (1).

There are two significant issues that need to be addressed and solved before composites will see a widespread use in components fabricated for the oilfield. These two issues are abrasion and the interface from the composite to metal. Other concerns about composite survivability, such as temperature and chemical exposure, have been addressed by the successful use of composite components in many other applications.

Composites can expect harsh handling. One of the biggest problems to overcome is abrasion, both downhole and during joint makeup and handling. Exterior coatings or barriers and internal liners have been proposed and developed to counteract the effects of abrasion but as yet are unproved due to insufficient use. Composite components, with a specifically formulated proprietary exterior coating for use downhole have been fabricated by Spencer Composites Corporation, Lincoln, Nebraska, for LWT Instruments in Calgary, Alberta, Canada. The composite component has shown promising results, and testing is in progress, but designs to address the abrasion issue are not proven. Interior liners have been successfully used for some applications for many years. Further testing is required and solutions to the abrasion issues will be found.

The concern for designing metal to composite attachments is complicated by the need for a minimal restriction on the bore while staying within a specific outside diameter. The downhole tubulars not only transmit or react axial loads but also torque loads and internal pressures. The downhole environment is severe in terms of shock loads and cyclic loads that can lead to fatigue concerns for metals. Figures 2, 3, and 4 show the advantages composites have over metals when compared on specific basis for strength, stiffness, and fatigue.

The following are applications of how composites are being used in the oilfield. These applications are used as examples of how composites are answering questions concerning their viability.

3. COMPOSITE DRILL COLLAR

In the process of excavating a bore hole it is important to acquire information concerning its formation through the use of inspection instruments. These instruments use sensing technologies and devices such as spectral gamma ray, neutron emission and detection, radio frequency tools, nuclear magnetic resonance, acoustic imagery, acoustic density, and other related devices. These measurement technologies require sophisticated devices or procedures to obtain high quality data about a formation. The level of sophistication is a direct result of the severity of the downhole operating environment. The marriage of the measurement equipment with the standard drilling equipment is limited in both the quality and type of data that can be obtained from a bore hole. In practice, drill pipe and collars are steel and accordingly limit the ability of logging or measurement tools to acquire a broad range of information because of the attenuation of the signal while passing through the steel structure. The current practice is to remove the drill pipe, drill collars, and associated equipment form the bore hole and then insert the sensing equipment.

Figure 5. Composite Drill Collar 170 mm x 8 m

Figure 6. Composite Drill Collar at Canadian Drill Site

4. COMPOSITE DRILLPIPE

It is estimated that the current limits for the drilling reach of steel drill pipe is ten kilometers. This could be increased to 15 kilometers with the use of composite drill pipe. This extended reach is significant for many offshore platforms. The extended reach using composite drill pipe is made possible by the composites higher specific strength and stiffness than steel. Steel, due to its density, is limited in its string length and the distance it can be pushed horizontally before buckling. The composite drill pipe now being developed uses metal end connections to allow the use of oilfield standard connection methods much like that explained for the drill collar in the previous paragraphs. Another advantage of composite drill pipe is that it can be fabricated to be inductively transparent which allows for the use of induction tools with little signal attenuation from the pipe. This can dramatically reduce the rig time lost while logging runs are made. The composite drill pipe is made using the filament winding process much like composite drill collars explained above. Additional information about composite drill pipe can be found in References (2) and (3).

5. COMPOSITE STATORS FOR PROGRESSING CAVITY PUMPS AND MOTORS

The Progressing Cavity (PC) pump and motor are both being evaluated for downhole use to power the drill and also pump fluids produced from the well bore. The PC concept uses a helicoidal rotor and complimentary helicoidal stator that engage each other along a sealing line to create cavities which progress axially as the rotor is rotated relative to the stator. The current technology calls for the injection of an elastomer into a steel tube with the elastomer formed to the required internal shape. This is a straightforward process and has been in use to make stators for many years. A new process for manufacturing stators using composite materials has been developed by Spencer Composites Corporation for Exoko Composite Products, Tulsa, Oklahoma. This process uses only a thin layer of elastomer with the remaining volume filled with composite. The reinforcing fiber used can be either fiberglass or carbon fiber, depending upon the stiffness and strength requirements. Figure 7 shows a schematic drawing of a PC pump. Composite stators have been designed and built that show significantly improved performance over current stator technology. Recent test data of a PC pump with a composite stator was compared to a standard PC pump. The performance improvement was as much as 400% at high speeds. Composite stators also have a higher-pressure output per stage, longer run dry life, and elimination of hystrisis problems associated with the elastomer in existing stators. Figure 8 shows a composite stator. The design uses a constant thickness elastomeric liner backed by filament wound composite. In most cases the liner thickness is less than 2.5 mm. Since the amount of elastomer is dramatically reduced the dimensional repeatability from stator to stator is more accurately controlled than with existing technology.

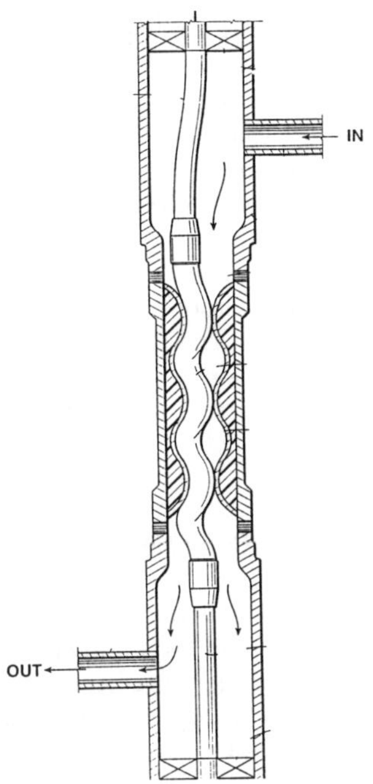

Figure 7. Schematic of a PC Pump

Figure 8. Cross Section on a Composite Stator

6. COILED TUBING

Coiled tubing is used to service wells and is seeing some service in drilling and production. Steel coiled tubing is restricted by its fatigue life. It can only be coiled and uncoiled a limited number of times. Composite tubing can be fabricated to have sufficient strength and additionally a low bending stiffness to eliminate the concern of coiling and uncoiling. Two problems to be solved in making coiled tubing are developing a liner for the tubing that can operate at 177° C and a simple yet effective metal end connection to the composite. The liner is required to provide the tubing with pressure integrity. Tubing can be built using existing liner material that will allow 100° C usage. Researchers are currently working to perfect the designs and manufacturing techniques to manufacture continuous lengths of the composite coiled tubing.

7. RIBBON ROD

Continuous composite Ribbon Rod has been developed to replace steel sucker rod. Ribbon Rod offers significant opportunities to improve the performance of beam pumping lift units. These performance improvements allow oil producers to continue using economical beam lift equipment for higher volume applications, extending upward toward other pumping technology applications. Ribbon Rod can increase beam pumping lift capacity, improve efficiency, and reduce sucker rod failures. The advantages of Ribbon Rod are summarized below:

- Higher production capability with existing installations
- Reduction of rod string weight up to 50%
- Reduction in flow restrictions
- Increased length of pump stroke
- Produces wells considered above capability of beam pumping
- Conforms to well deviations with less tubing wear
- Impervious to corrosion

Ribbon Rod is manufactured using the pultrusion process. The fiber used is 235 GPa carbon fiber with a vinyl ester resin. The standard spool length is 850 meters with the spool diameter being three meters.

8. COMPOSITE CORE SAMPLE HOLDERS

Composite core sample holders are pressure vessels used in the laboratory for examining core samples. Spencer Composites Corporation has an ongoing program of development for Temco, Inc., of Tulsa, Oklahoma. Composite core sample holders permit examination of geological core samples at high temperature and pressure. This thick-wall design allows working pressures to 138 MPa and temperatures to 225° C, while providing low X-ray attenuation. Metal vessels cannot operate in this combined pressure and temperature regime because the wall thickness required drastically attenuates the X-rays. The metal end fittings in the composite core sample holder, which have internal threads for attaching end closures, are filament wound in place. This concept utilizes the advantages of both composites and metal. The metal end fittings provide the attachment threads and the composite provides the low x-ray attenuation. A core sample holder, 2.5 meters long is shown in Figure 9. Reference (4) provides additional information on the core sample holder.

Figure 9. Composite Core Sample Holder

9. SUMMARY AND CONCLUSIONS

The use of composites and the filament winding process as a manufacturing method of choice is gradually increasing in the commercial sector. With improvements in fabrication equipment and techniques, materials, and education of the engineering community, commercial applications for composites and the filament winding process will increase.

Composites offer significant advantages over metals and other standard materials for many applications in oil exploration and production as demonstrated by the examples presented. Filament winding is the process of choice where the design requires oriented fibers. The applications for filament wound composites are only limited by the imagination.

10. REFERENCES

1.   B. Spencer, "Composite End Fitting Designs", <u>Proceedings of the Ninth International Conference on Offshore Mechanics & Arctic Engineering</u>, Vol. 3, Part A, American Society of Mechanical Engineers, New York, 1990, pp. 37-39.

2.   Editor, <u>High-Performance Composites</u>, "Update: Offshore Drilling Pipe", Jan/Feb. pp. 13, (1997).

3.   G. Hareland and W. Lyons, New Mexico Institute of Mining and Technology, D. Baldwin and G. Briggs, Lincoln Composites, R. Bratil, Saga Petroleum A/S, <u>Offshore Technology Conference</u>, "Extended Reach Composite Materials Drillpipe", (1997).

4.   C. Murray, <u>Design News</u>, May 23, "Composite Core-Sample Holder Improves X-Ray Sensitivity", (1988).

11. BIOGRAPHY

Brian E. Spencer, Ph.D., is the President and CEO of Spencer Composites Corporation, Lincoln, NE. He holds a Masters in Industrial Engineering from the University of California at Berkeley, a Masters in Mechanical Engineering from the University of California at Davis, and a Ph.D. in Engineering Mechanics from the University of Nebraska at Lincoln. He is a Registered Professional Engineer in California and Nebraska. Dr. Spencer is the 1998 recipient of the Society of Manufacturing Engineers J.H. "Jud" Hall Manufacturing Award, which celebrates outstanding advances in composites manufacturing. He has over 18 years of experience in composites. Dr. Spencer regularly gives presentations, tutorials and seminars throughout the United States, and in Canada, Australia, Belgium, and South America. He has published over 25 papers and contributed to several books on composites. He holds several patents related to filament wound composites in a wide variety of areas.

COMMERCIAL/INDUSTRIAL APPLICATIONS: FILAMENT WOUND COMPOSITE STRUCTURES

Randall J. Philpot and Douglas G. Olsen
Advanced Composites, Inc.
1154 South 300 West
Salt Lake City, Utah 84101

ABSTRACT

Fiber reinforced composite structures have found uses in an ever growing range of commercial/industrial markets. Many industrial markets are highly specialized and can justify the use of higher cost, high performance fiber reinforced composites versus lower priced metals or more conventional materials. Filament winding has proven to be a highly automated, reproducible process for manufacturing high performance composite structures. Advances in equipment, materials, design and testing capabilities, as well as performance have increased the market potential for filament wound composite structures. Some of the industrial applications in which Advanced Composites has been involved include: drive and power transmission shafts, radomes/antenna components, pressure bottles and tubing, electrical insulating structures, oil field tubing and sampling devices, mining tools, cryogenic devices and high temperature parts. The paper will discuss some of the design and manufacturing aspects of filament wound composite structures for several of these industrial applications.

KEY WORDS: Filament Winding, Composites, Design

1. BACKGROUND

1.1 Composite Design In designing filament wound composite structures for commercial/industrial applications a number of design parameters need to be considered. The following is a list of a number of the parameters that have been addressed in various filament wound structures designed and manufactured for commercial/industrial applications (1,2).

    1). Mechanical Properties
        a) Strength (Tensile, Flexural, Compressive, Shear)
        b) Modulus (Tensile, Flexural)
        c) Elongation

d) Impact Resistance/Toughness

2). Density/Weight
3). Fatigue Resistance
4). Operating Temperature (high and low)
   a) Low Thermal Expansion
   b) Thermal Conductivity (high and low)
5). Corrosion Resistance
6). Pressure
7). Electrical (Conductivity, Insulating, Transparency)

The various design requirements for industrial applications typically have required addressing at least two or more of the above listed design parameters. Some of the design parameters encountered and the means in which they were addressed will be covered in a later section for specific industrial applications.

1.2 Filament Winding Although filament winding had its origins in the aerospace/defense industry in pressure vessel type applications, it has evolved to prove useful in a wide range of applications including a number of industrial applications.

Some features of filament winding which prove useful in the automated manufacture of sturctures for industrial applications include the following (3):

1). Use of continuous filaments.
2). Ability to wet wind. Use of a resin bath to achieve fiber wet-out and control fiber to resin ratio. Fiber tension also acts as control for final fiber to resin ratio in the part.
3). Control of fiber tension which acts to consolidate the part during the winding process so that subsequent debulking steps are not necessary.
4). Reduced material costs compared to other processes. The ability to purchase fiber, resin and hardener directly from the manufacturer minimizes material costs since their are no intermediate processing steps (examples: pre-preg., custom resin formulators, weavers, braiders, etc.). Depending on the process and therefore, the starting material, raw material costs can be increased signicantly over those encountered with filament winding. It has been reported that degradation of the fibers can occur in these intermediate operations as well.
5). Flexibility in varying fiber orientation.
6). Close control over fiber orientation. Since the placement of fibers on the mandrel are controlled by a machine, operator variability is greatly reduced.
7). Hybridization of fibers. Different fiber types can be comingled within individual layers and/or changed from layer to layer to optimize part properties and/or reduce costs. One example would be the use of E-glass hoops on the outside of a predominantly graphite part for consolidation and grinding stock if subsequent grinding steps are required.
8). In wet winding, the flexibility to vary resin formulations for individual parts or limited production runs.

9). Use of pattern generation software in conjunction with analysis software to optimize part design, production rates and material costs.
10). Relatively high production rates especially for larger diameter, thicker walled parts compared to other processes.
11). With the use of a special delivery system, the ability to filament wind with various thermoplastics and the ability to achieve good part consolidation and consistency.
12). The ability to produce pre-preg. with the desired fiber orientation or pre-forms that can be subsequently molded to non-traditional or more complex shapes than can be achieved by winding on a mandrel.

2. COMMERCIAL/INDUSTRIAL APPLICATIONS

As a custom manufacturer of composite structures, contacts are made in a number of different industries in which many similar issues need to be addressed. An application exists or has just been created in which a structure is required that will meet certain design criteria. More often than not, especially in commercial applications, one important criteria that is usually brought up is cost. With cost in mind, a solution to the customer's particular application using composites can hopefully be arrived at.

If conventional materials such as metals (primarily steel or aluminum) are not well suited for the application or cannot meet the specified design criteria, then alternatives are typically sought. As an alternative to metals, there are plastics and/or thermosets, both with or without fiber reinforcing. If properties of common engineering thermoplastics are inadequate to meet the design criteria then materials with fiber reinforcing are typically investigated next. Due to the large number of processes available for producing fiber reinforced structures another determination that needs to be made is whether or not the structure requires continuous fiber reinforcing or whether a structure with primarily discontinuous or short length fibers can be produced. As this evaluation continues, the number of options available for having a structure produced that will meet the design criteria is narrowed down. Some of the commercial applications to be discussed in this paper lend themselves well to filament winding. Some could be produced by other manufacturing methods but, when all factors are considered, filament winding is often the most desirable method. Having been involved in a number of development programs, it becomes more clear on which applications are well suited for filament winding and which ones are not. Since material properties are often times a major key in achieving design criteria the manufacturing process sometimes becomes secondary.

2.1 Shafting, Converting Industry Light weight, stiff shafting with high flexural strength and excellent fatigue properties is becoming increasingly important in the converting industry. This is especially true as throughput rates are being increased to meet production demands and reduce costs. As throughput rates of the equipment increase, the rotating speed and loading that shafting used in core shafts and idler rolls is subjected to increases. Figure 1 illustrates some shafting that was designed and manufactured for these applications.

FIGURE 1. Core Shafting, Converting Industry

2.2 Antennas/Radomes TACAN (Tactical Air Navigation) antennas are used across the country for navigation of commercial aircraft. The use of composite structures in these type applications are common. Internal rotating members need to be light weight, strong, non-conductive and transparent to RF. These antennas are housed within a protective covering or radome. The radome for a particular TACAN antenna application is illustrated in Figure 2. The radome was approximately 3 ft. (1 m) in diameter and 10 ft. (3 m) tall. An access port, bearing mount for a rotating assembly and mounting interfaces with a base were incorporated into the design.

FIGURE 2. Radome, TACAN Antenna

2.3 Inner Cylinder, Collider Detector As part of a detection device for sub-atomic particles, a thin wall structural tube was required that would be transparent to the detector

and be able to support an axial compressive operating load of 29,400 lbs. (130,830 N).
The approximate tube dimensions were 32 in. (81 cm.) diameter X 120 in. (3 m) LG X
.090 in. (2.3 mm) wall thickness. A picture of the manufactured part is included in Figure
3. The inner cylinder which was subjected to a proof load of 45,000 lbs. (200,250 N)
could not have an axial deflection of more than .040 in. (1.0 mm). Ideally, with the given
part geometry, the design loads can be achieved with a minimum safety factor of 5.
Certain design aspects will be discussed further in the "DESIGN" section of the paper.

FIGURE 3. Inner Cylinder, Collider Detector

2.4 Cryogenic Valve Composites can also exhibit excellent low thermal conductivity (4).
An insulating and structural housing was required for a cryogenic valve. One end of the
housing was at the temperature of the cryogen and the other end at ambient temperature.
There was also a length limitation. So a thin wall metal housing with a number of
convolutions was overwrapped with S2 glass-epoxy composite to provide structural
integrity to the valve, as well as thermal insulation. This concept is illustrated in Figure 4.

FIGURE 4. Cryogenic Valve

2.5 Oil Field Tubing and Structures Some oil deposits in the western United States
have a high paraffin content. This material can cause problems when trying to pump
material from the well. The paraffin can form on the walls of the tubing and greatly
restrict flow and in some instances plug flow completely. Several companies have looked
at novel approaches to provide a heat source to melt the paraffin and increase the flow of
oil from the well. Composites have played a roll in several of these applications where a
non-conductive material with good mechanical properties and temperature stability at
elevated temperatures was required.

3. DESIGN

3.1 Shafting, Converting Industry As mentioned in the "APPLICATIONS" section,
the web speeds in the converting industry are ever increasing which requires stiffer,
stronger and ideally lighter weight shafting for core and roll applications. Shafting is
produced with a range of composite moduli depending on the specific application and
resulting loading conditions. Table 1 provides a range of moduli that have been produced.
Also included in the table is the densities of the different materials and a ratio of stiffness
to density.

Table 1. Modulus, Density and Ratio of Stiffness to Density, Core Shafting

Description	Modulus		Density	Stiffness to Density Ratio
	Msi	(GPa)	gms/cc	GPa/gms/cc
Aluminum	10	(69)	2.5	27.6
Std. Mod. Graphite	16	(110)	1.6	68.8
Int. Mod. Graphite	23	(159)	1.6	99.4
High Mod. Graphite	30	(207)	1.6	129.4
Steel	30	(207)	7.8	26.5

From the table it is evident that aluminum is limited in applications where higher
performance is required due to its relatively low modulus. Even though the aluminum has
a lower density than steel, the stiffness to density ratio (specific stiffness) is actually similar
to steel. Aluminum tends to find more applications for idler rolls versus core shafting
since the loading requirements tend to be lower and there is more flexibility in sizing of the
rolls. The diameter of the roll can be increased in some instances which increases the EI
value and results in an overall stiffer structure.

Shafting out of steel displays high modulus but, due to the high density it makes for a
heavy shaft. Due to the relatively low density of the graphite composite, the stiffness to
weight or density is considerably better than steel even for the standard modulus graphite
composite. Applications where flexibility exists in terms of cross sectional thickness,
allow for stiffer, lighter weight shafts to be produced compared to steel even using
standard modulus graphite. Using standard modulus graphite where possible tends to be
the most cost effective option compared to using the higher modulus graphites.

However, in applications where the diameters and cross sectional thickness is tightly constrained and a stiffness that comes closer to or matches steel is required, then using the higher modulus graphite fibers becomes necessary. Where performance is the driving force, the use of these more expensive fibers is often times justified.

Two material properties which were omitted from the table which are important design parameters are material strengths and resistance to fatigue. The graphite-epoxy composites listed in the table exhibit both superior tensile and flexural strengths and fatigue life than steel or aluminum.

In many core shaft applications the graphite composite is used in conjunction with aluminum or steel to reduce overall weight of the part while maintaining or improving the mechanical properties. Core shafting requiring slotting and holes to accommodate air actuated expansion capabilities often incorporate an external metal sleeve bonded to the composite to help minimize damage or degradation of the composite properties in these operations.

3.2 Radome, TACAN Antenna A schematic of the radome is illustrated in Figure 5. Some of the design challenges that were encountered in designing and manufacturing of this part were as follows:

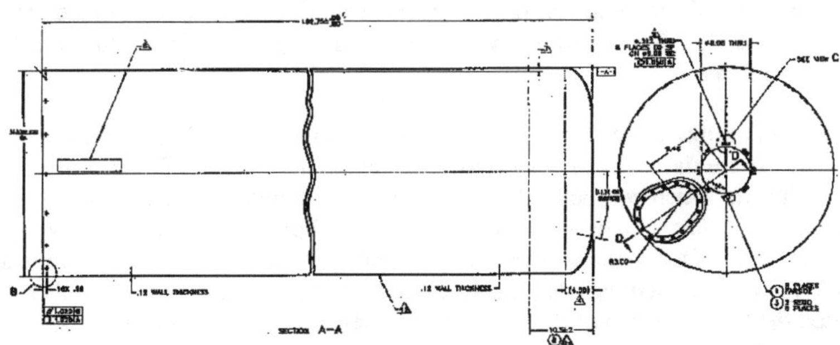

FIGURE 5. Radome Schematic

1). A dome or closed end portion that was an integral part of the radome with several cut-outs that was structural enough to prevent "oil-canning" when the antenna was being lifted by the center portion of the dome.
2). A smooth inside diameter that was uniform with precise dimensions to provide minimum drag on the mechanically scanned antenna which rotates at a speed of 900 rpm (over 120 surface ft./sec. (36.6 m/sec.)).
3). An opening at the center of the dome that served as a precision interface with a bearing housing which located and centered the central array of the antenna.
4). An access port located near the dome/cylinder transition area that was large enough for maintenance on the antenna with a flat hatch cover for easy alignment and sealing.

5). Structural strength and stiffness of the cylindrical portion to withstand severe wind loading with minimum deflection and no damage while being RF transparent.
6). Smooth durable outside finish to help minimize build-up of foreign material that may interfere with the operation of the antenna.

Utilizing the filament winding process and winding on a precision ground aluminum mandrel resolved items 1, 2, and 5. The fact that filament winding tends to lay down a constant volume of material created a natural build-up of wall thickness from the cylindrical portion towards the pole piece of the dome. The natural increase in wall thickness near the pole piece or center of the dome was sufficient that during machining operations there was ample material to machine the desired diameter flat for subsequent mounting of the bearing assembly.

The material used for winding of the radome was E-glass-epoxy. The winding pattern was predominantly +/- 23 degree helical sandwiched between layers of hoop. The outside hoop winding provided good consolidation of the helical layers which produced a composite with relatively low void content and optimum fiber to resin ratio. Excess hoop was wound on the outside for grinding stock to provide a smooth outside finish for subsequent painting. Deflection of the radome during wind load testing was a minimum and no "oil canning" or distortion of the dome occurred during lift tests, even with cut-outs at the pole and for the access port. Finished wall thickness on the cylinder portion was approximately 1/8 in. (3.2 mm) which increased to 3/8 in. (9.5 mm) towards the center of the dome.

Locating and routing out of the access port was accomplished with fixturing that referenced off the flat that was machined in the dome. Molds were fabricated and a collar cast with inserts which was bonded over the access port cut-out near the dome/cylinder transition. A flat hatch cover was then fabricated and holes drilled to match the inserts cast in the collar. Due to the high velocities of the rotating antenna, the cavity created by the molded collar caused turbulence that resulted in drag for the rotating members and increased the power to spin of the antenna. To alleviate this problem a molded plug that closely matched the shape of the cavity was fabricated and bonded to the bottom side of the hatch cover. With the plug/hatch cover assembly in place the turbulence was eliminated and the power to spin returned to acceptable levels.

A durable urethane coating was applied to the exterior surface of the radome to provide a smooth, weatherable surface. A hole pattern was placed around the open end of the radome for anchoring to a pedestal that served as the base to the antenna.

3.3 Inner Cylinder, Collider Detector Initially it was decided that filament winding would be the most cost effective way to produce a relatively thin wall graphite structure that would support a 45,000 lb. (200,160 N) axial compressive proof load with an axial deflection within the design parameters of no greater than .040 in. (1.0 mm). Initially, a design analysis was conducted to verify which proposed lay-up configurations were optimum, and whether or not the resulting safety factor exceeded the design load by a sufficient margin.

3.3.1 FEA Analysis The cylinder was modeled using the "COSMOS" FEA code with Shell4L elements as shown in the buckling model illustrated in Figure 6. The Shell4L element is a 4-node multi-layer anisotropic quadrilateral shell element with membrane and bending capabilities for the analysis of three-dimensional structural and thermal models. A nonlinear analysis was used to find the buckling values by plotting node displacement against time. A constant load rate was used to load the cylinder through time. Initially the model was tested by examining four cases from a paper by Geier et al (5) titled: "Buckling Tests with Axially Compressed Unstiffened Cylindrical Shells Made from CFRP". The analysis results correlated very well with the test results presented in the paper.

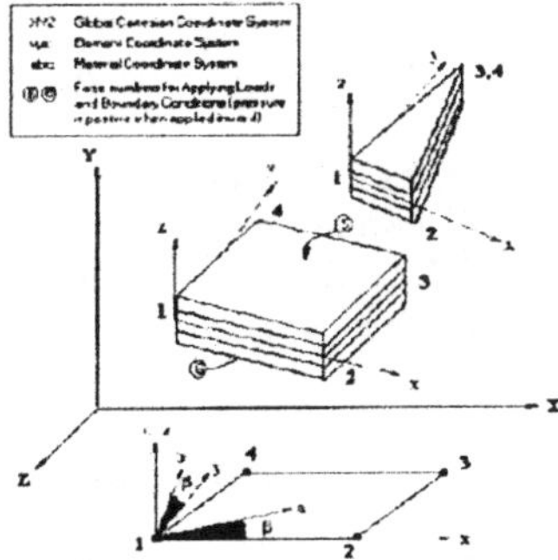

FIGURE 6. Schematic, Buckling Model

The lamination orientation proposed for the inner cylinder was then tested using the model. Analysis was done using both static and nonlinear analysis methods to examine the characteristics of the inner cylinder structure. The analysis showed that large factors of safety existed throughout the part with the exception of the ends of the part bonded to the aluminum rings.

An axial compressive load of 300,000 lbs. (1,334 kN) was applied in the model. Figure 7 illustrates the deformation that occurs in the aluminum ring portion and inner cylinder just past the aluminum rings. No significant deformation occurs in any other part of the cylinder. The projected operating load is actually 29,400 lbs. (130,830 N). The proof or test load of 45,000 lbs. (200,160 N) is approximately 1.5 times the operating load and well within the determined safety factor. The analysis showed a design safety factor of 5 against buckling which was determined to be more than sufficient. The analysis shows that the only portion of the inner cylinder where this type of safety factor becomes marginal is near the ends. A determination was made that the design produced a suitable structure with sufficient safety factor for all loading conditions anticipated.

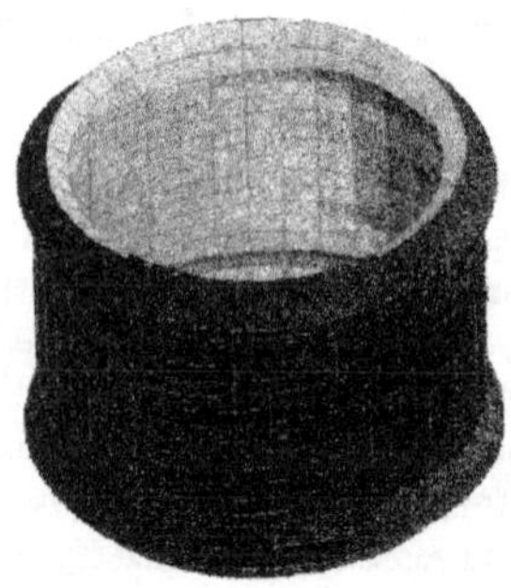

FIGURE 7. Shell Deformation, 300,000 lb. (1,334 kN) Axial Compressive Load

4. CONCLUSIONS

Filament wound composite structures have been successfully designed and manufactured for a wide range of commercial/industrial applications where design parameters were greatly varied. Some of the significant design criteria that was achievable for the varied applications with filament wound composite structures discussed in the paper included the following:

4.1 Core Shafting and Rolls for the Converting Industry

1). High stiffness to density ratio along with superior strength and fatigue properties coupled with an automated, reproducible process that was economical.

4.2 Radome

1). The ability to produce a filament wound composite structure approximately 3 ft. (1 m) in diameter by 10 ft. (3 m) long with one domed/closed end with sufficient structural integrity to allow the unit to be lifted from the dome end without "oil canning" or damage to the radome. The design criteria was accomplished with material being removed for two openings, one at the pole piece and one at the outer portions of the dome where it meets the cylinder.

2). The cylinder portion was designed to take severe wind loading conditions with minimum deflection and no structural degradation. Also, a smooth durable external surface to minimize any debris build-ups with excellent weatherability.

3). The composite was transparent to RF and did not interfere with the signal from the antenna.

4). Flat access port large enough for easy access for servicing while maintaining structural integrity.

4.3 Inner Cylinder, Collider Detector

1). Successfully design and manufacture a filament wound graphite tube with aluminum end rings bonded in place that had an axial compressive operating load of 29,400 lbs. (130,830 N) with a safety factor of 5. This needed to be accomplished while maintaining a wall thickness of approximately .090 in. (2.3 mm) for a 32 in. (81 cm) diameter X 120 in. (3 m) LG part to allow for detection of sub-atomic particles.

2). Maintain minimum axial deflection of .040 in. (1 mm) under proof load of 45,000 lbs. (130,830 N).

Filament winding in conjunction with design considerations and proper selection of materials allowed for the successful manufacturing of the above described structures to meet or exceed the design criteria. Also, many of the parts described in the paper have been subsequently produced on a production basis

5. REFERENCES

1. S.T. Peters, W.D. Humphrey and R.F. Foral, Filament Winding Composite Structure Fabrication, SAMPE, Covina, 1991.

2. S.R. Swanson, Introduction to Design and Analysis with Advanced Composite Materials, Prentice Hall, Upper Saddle River, 1997.

3. R.J. Philpot, D.K. Buckmiller, SAMPE International Symposium, 34(2), pp. 2217-2228, (1989).

4. J.A. Lee and D.L. Mykkanen, "Advanced Materials Research Status and Requirements", Huntsville, AL, 1986.

5. B. Geier, H. Klein, R. Zimmermann, "Buckling Tests with Axially Compressed Unstifened Cylindrical Shells Made from CFRP", ESA, April 1996.

6. BIOGRAPHIES

Randall J. Philpot received his B.S. Degree in Metallurgical Engineering from the University of Utah in 1977. He worked as a plant metallurgist for ASARCO and as a process engineer at the Kennecott Research Center prior to attending graduate school in Polymer Science. He received an M.S. degree in Materials Science and Engineering from the University of Utah in 1985. He is presently the President and CEO of Advanced Composites, Inc. which is involved in the design and manufacturing of filament wound composite structures. He has authored papers and received several patents on inventions related to filament wound composites.

Douglas G. Olsen received his B.S. Degree in Materials Science and Engineering from the University of Utah in 1992. He worked as a Lab Assistant at Terra Tek in Fracture Mechanics. Doug is currently a Project Engineer at Advanced Composites and is also working on his M.S. degree in Mechanical Engineering. Doug has received several patents on inventions related to filament wound composites.

High Strength End Fitting for Tubular Composites

Daniel J. Wolterman,
SpyroTech Corporation,
Lincoln, Nebraska

Abstract

Filament wound composite tubes have proven to be very viable structures for many application. However, as with nearly all composite structures, there are attachments used to introduce loads to the structure. Typically, with tubulars the attachments are at the ends of the tube. The more producible methods of attachment use bonded attachments or a series of bolts. More complex methods use interlocking geometric features between the end fitting and the composite tube. With all these methods, the structural loads are transferred over localized areas and the overall strength is limited to the strength of those areas.

This paper presents a method of attaching a metal end fitting to a composite tubular structure such that the entire length of the interface is uniformly loaded. The method discussed is applicable to both axial load (tension and compression) as well as torsional load transfer. Load paths can be isolated to provide full axial capability after torsional failure and the reverse. Data is presented to show the high strength of the fitting attachment method as well as the fail-safe isolation of load paths. Additionally, this method can be used for either low cost post wind assembly or integral winding.

KEY WORDS: Composite Tubular, End fitting, Axial Strength, Torsional Strength, Filament Winding, Joining,

1. Introduction

Filament wound composite tubes have proven to be very viable structures for many application. However, as with nearly all composite structures, there are attachments used to introduce loads to the structure. Typically, with tubulars these attachments are made at the ends of the tube. The typical construction, as shown in figure 1, has a composite body with a metal end fitting at each end to interface with other system components. While the composite material and the metal end fitting are very strong, the weak link in this structural chain is the connection between the composite tubing and the metal end fitting.

The more producible and economical methods of joining the two bodies include gluing (bonded joint) or screwing (bolts and pins) [1]. These methods have worked very well for light and moderately loaded structures that are driven more by stiffness and weight than overall strength, such as rollers for paper processing industry. However, is difficult to take full advantage of the increased strength of the composite materials because the methods of attaching the end fittings cannot meet the overall strength.

More advanced methods to increase strength involve interlocking geometry [2]. However, they come with an increase in production complexity and the potential for increased part to part performance variation.

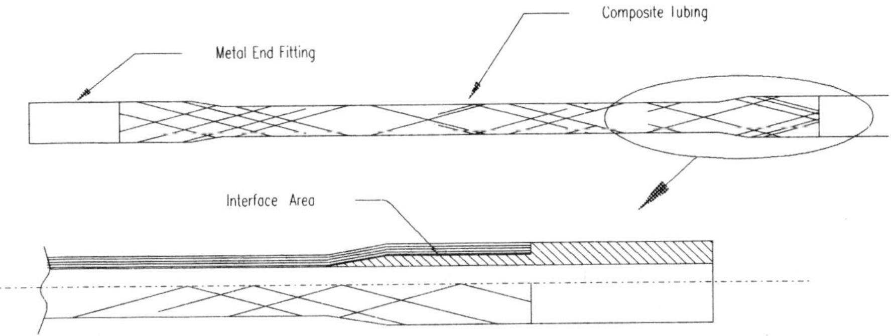

Figure 1. Typical Composite Tubular Structure with Metal End Fitting

While the strength of a composite can be five times greater that most steels, the elastic modulus of the composite tubular is generally on the order of 30 to 50% the stiffness of a steel end fitting. In order to use the full benefit of the composite, the strain field of the composite should be around 1% or slightly greater. Very few metal materials can operate in that strain range without plastic deformation and therefore limited cyclic strength. With the glued or screwed joints, the displacement field of the composite is matched to that of the metal end fitting. By constraining the displacement field and thereby the strain field, the axial load is forced out of the composite and into the steel fitting very quickly. Figure 2 shows the shear stress distribution of a unidirectional carbon composite laminate bond [3].

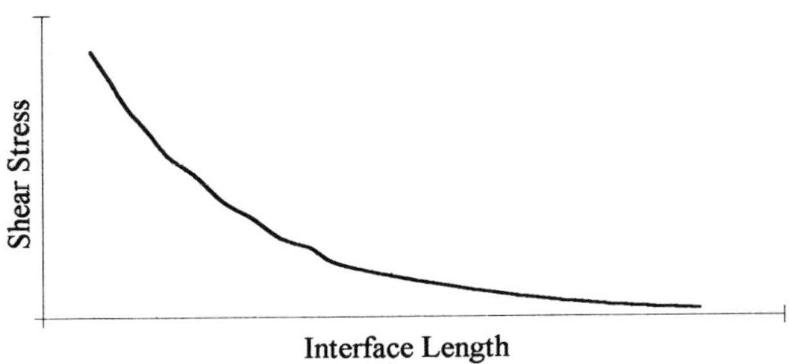

Figure 2. Stress distribution in a bonded joint. [3]

The shear stress distribution shows a very large peak at the leading edge of the interface. Increasing the length of the interface has the effect of adding length to the lower stressed center area of the joint. Failure of the bonded joint occurs at the leading edge of the bondline. Figure 3 shows experimental data for bond strength with increasing bond length [3]. Increasing the length of the interface results in very minimal increase is overall strength of the joint and therefore has very little effect on the overall strength of the bond.

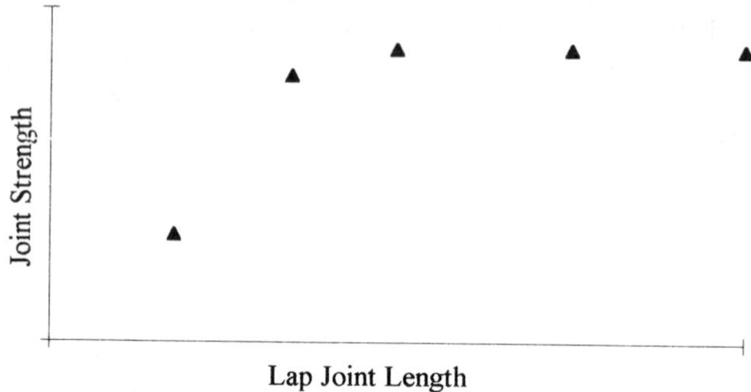

Lap Joint Length

Figure 3. Effect of added Length on bond strength. [3]

Based on the phenomenon shown above, a bonded joint can be very effect for a limited strength requirement, but will not support higher loads. A similar situation occurs with bolted joint. Compounding these problems, are also very high stresses localized around the bolts. The physics of coupling these two material results in a physical limit in strength due the limited amount of effect bond length that can be used to transfer loads between the components.

2. Design Objective

In order to increase the overall strength, the critical feature to pursue is the distribution of shear stress between the metal end fitting and the composite tubing. The strongest joint would be one where the entire interface were loaded uniformly. This would also allow the addition of length to increase the strength. Figure 4 shows the desired shear stress distribution and figures 5 and 6 show the resultant axial forces and displacements.

Interface Length

Figure 4. Desired Shear Stress distribution

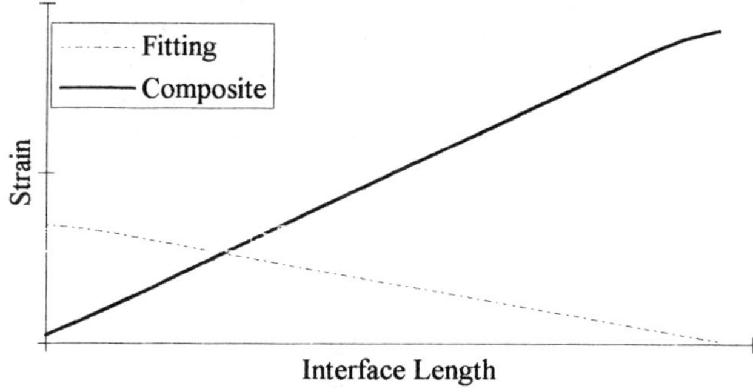

Figure 5. Strain distribution from Uniform Shear Stress distribution

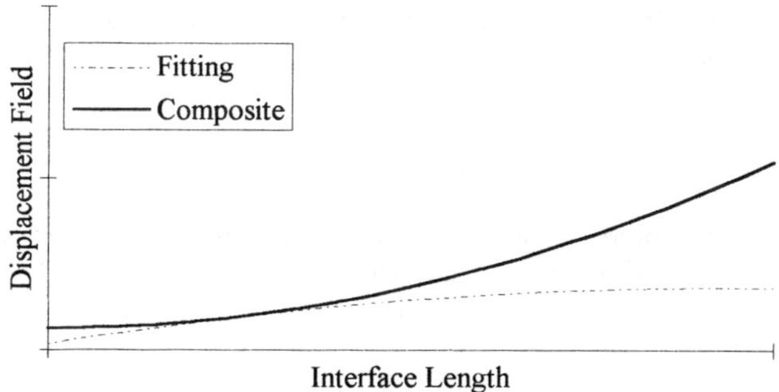

Figure 6. Axial displacement field from Uniform Shear Stress distribution

In order to achieve the desired uniform shear stress distribution, the two bodies (end fitting and composite tube) must be allowed some level of relative displacement. The use of interlocking geometric features is used to allow the relative displacement. By using a series of interlocking grooves [patent pending]. The displacement fields are no longer locked together and can be used to control the load transfer between members.

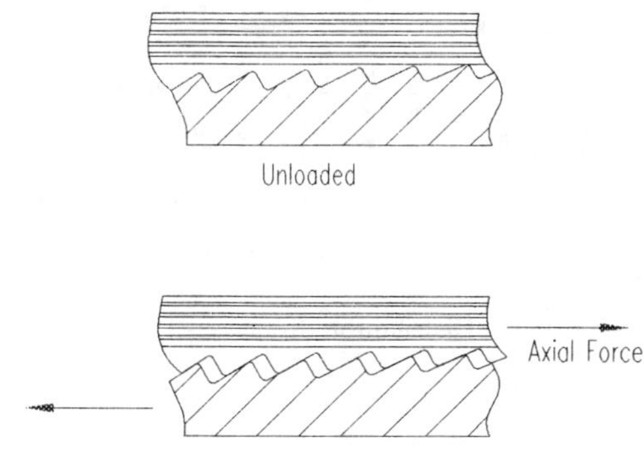

Figure 7. Composite to metal interface (free state and loaded)

As the composite tubing is pulled away from the end fitting, the grooved interfaces force the composite to displace radially, shown in figure 7. This radial displacement then generates significant interface pressure between the respect bodies. The axial loads are transferred from the composite tubing into the metal end fitting as the axial component of both the interface pressure and friction. The amount of load transferred at each groove then becomes a function of the material properties of the composite tubing, metal end fitting and the geometry of the interlocking grooves.

3. Fabrication

With all good ideas, the proof of the pudding is in the tasting. A test article was produced to evaluate the producibility of the design. The end fittings were machined on a typical CNC lathe. The composite tube was filament wound over the fitting. The grooves were formed by a combination of minimal layup and filament winding circumferential fibers into each groove. After the grooves were formed, the remainder of the structure was wound using traditional filament winding methodology. The resulting composite showed that the grooves were very well formed and the producibility of both the end fitting and the final structure was very good.

4. Testing

The two critical features to be demonstrated with the fitting technique were the uniformity of strain distribution and the overall strength.

4.1 Strain Distribution

Figure 8 shows the strain -vs- load data for a series of gages placed along the length of the interface. Figure 9 shows the same data plotted as a function of axial location.

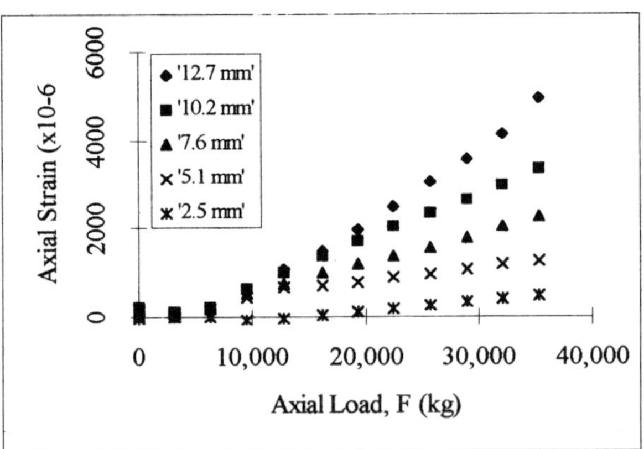

Figure 8. Axial Strain Distribution Strain -vs- Load

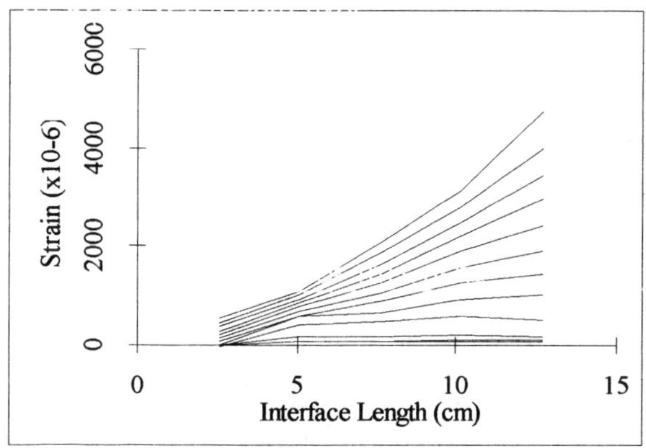

Figure 9. Axial Strain along Interface.

Comparing the strain -v- length profile from figure 9 with the theoretical goal shown in figure 5, shows very reasonable agreement. The goal of achieving uniform shear stress distribution has been demonstrated.

4.2 Ultimate Strength

With the multiple interlocking grooves, there were different possible mechanisms for failure of the fitting interface. The least desirable of these is to have a shear failure of the material, the most desirable is to have the composite fail in hoop burst from radial dilation. Figure 10 shows the load - displacement data for tensile testing.

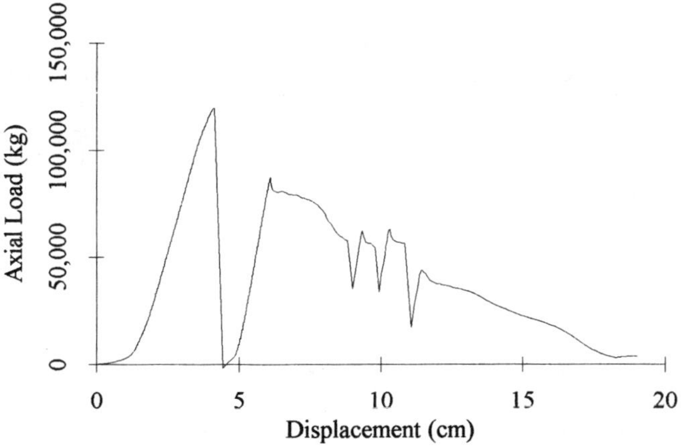

Figure 10. Axial test performance

Under tensile loading, the first mode of failure is for the composite body to dilate sufficiently to hop over one set of grooves. The composite then seats into the next grooves. Because this action is elastic, the composite seats into the next adjacent groove and still has tensile strength capability. With additional tension, the next failure mode if for the body to hoop into the next set of grooves. This process is repeated until ultimately the composite has dilated sufficiently to result in a hoop strength failure.

5. Discussion

Thus far this discussion has been focused on tensile strength of an integrally wound fitting. However, this method of attachment can be applied to torsional couplings. Additionally, the assembly of the structure can be performed after winding.

5.1 Application to Torsion
In the case of torsional loading, the discussion above remains completely valid with the exception being that the grooves would be oriented axially. The groove features with a torsional application are not as obviously machinable as the circumferential grooves. They can be machined using a CNC with X, Y, Z and A rotations. For higher production rates, the use of powder metal technology, to form the fitting would produce these features reliably and inexpensively.

5.2 Applicable to post wind assembly
Because the shear stress distribution is controlled, there is no peak shear stress in the interface. Therefore, the requirement for continuous fiber reinforcement can be reevaluated. In many cases, by increasing the interface length and lowering the shear strength a larger selection of materials can be used. This would allow the fitting to be installed using a number of structural adhesives to be used to fill the gap between the composite shell and the metal fitting. The benefits of this method would be realized in higher rate production environment. The composite tubular would be filament wound without end fittings, in it most efficient manner. The end fitting would be fabricated and whatever plating operations were required could be performed. Then the end fittings would be assembled to finish the structure. The only structural bond that would be required is between the adhesive and the composite tubular.

6. Conclusion

The end fitting discussed has proven to eliminate the effect length limit imposed by the concentration of shear stress that is typical of a bonded attachment. The strain distribution has been controlled and the resultant strength is very high. Using this methodology, the full strength of the composite material can be realized on a filament wound shaft with metal end fittings.

7.0 References

1. McConnel, Vicki P., *Composite Part Joining: Glue It or Screw It?*, Composites Technology, September/October 1997

2. Baldwin, Don, *Composite Riser Development*, OTC 84-31,OTC May 1997 Proceeding

3. John, Sabu J. *Predicting the strength of bonded carbon fibre/epoxy composite joints.* Composites Bonding ASTM STP 1227, October 1993.

43rd International SAMPE Symposium
May 31-June 4, 1998

INVESTIGATIONS ON THE REPAIR OF FIBER REINFORCED THERMOSETS

Dipl.-Ing. M. Münker, Prof. Dr.-Ing. W. Michaeli
Institute of Plastics Processing (IKV), Aachen, Germany
Dipl.-Ing. K.-U. Moll, Prof. Dr.-Ing. B. Wulfhorst
Institute of Textile Technology (ITA), Aachen
Dr.-Ing. A. Nick, Prof. Dr. rer. nat. F.-J. Wortmann, Prof. Dr. rer.nat. H. Höcker,
German Wool Research Institute (DWI), Aachen

ABSTRACT

At the Sonderforschungsbereich 332 "Production Techniques for Nonmetallic Composite Parts" different repair methods for glass and carbon fiber reinforced thermosets are examined in order to be applied in practise. Therefore test specimen are damaged in a well defined way and repaired with different textiles and under different conditions. The repair efficiency is tested with short term tensile, 3-point bending and impact tests afterwards. By using different repair textiles an optimum textile style was found. Investigations on the influence of humidity and temperature during the curing process and the use of resins with different coefficients of thermal expansion showed a different repair efficiency in static and dynamic tests. As damaged locations are glued with new laminates, the use of different adhesives for the repair of carbon fiber reinforced thermoset specimen by glueing has been examined as well. The incipient crack strength determined in the peel test increases with increasing number of polar groups at the adhesive surface. This effect becomes even more evident after chemical treatment of the specimen.

KEY WORDS: Repair, Mechanical Properties, Surface Preparation

1. INTRODUCTION

Fiber reinforced plastics (composites) are well established in high grade in the aerospace industry. Their importance in other branches is also increasing. For the production of long-fiber reinforced composites mainly thermoset matrix materials are used. Compared to thermoplastic matrix materials they behave extremely brittle. Already at low loads in the non-reinforced directions damages occur. Examples for this kind of stress are interlaminar shear stresses and impacts (1).

The substitution of damaged composite parts might require a lot of work for demounting the damaged part and integrating the new part and hence might be cost intensive. Also the new parts are not always as fast available as necessary.

In this paper different investigations on the repair of fiber reinforced thermosets are described. An important role plays the bonding between the damaged part and the repairing textile. The influences of different adhesives and several surface modifications are shown. The repair methods with different repair textiles are evaluated and the reconstruction methods are described. Also different curing conditions and results with resins with different coefficients of thermal expansion are presented.

2. DAMAGES AND REPAIR METHODS

2.1 Damages Surface damages of composites are detected visually. They might cause secondary damages by mechanical loads of the reinforcing fibers or by penetrating media. These damages should therefore be repaired in a short period of time after their detection.

The fracture of the composite occurs due to a complete debonding of the single layers or by locally bordered delaminations. A delamination is a debonding of composite layers. Delaminations especially reduce Young's modulus of the composite because of the non-existing load transfer between the single layers.

If glass fiber reinforced plastics are in use, delaminations can be detected by differences in brightness. This is not possible in case of carbon fiber reinforced plastics, where a test of the natural frequency or an ultrasonic inspection of the part has to be done.

The repair of delaminations is possible. Usually the method of resin injection is used. Channels are drilled to the debonded area. Through this channels a resin is injected. Afterwards the composite part is compressed and the matrix material cured. A complete bonding of the single layers takes place. Furthermore a combined fibre-matrix-fracture can occur. In most cases this damage is detectable visually. A further control of the region around the damage is necessary as there might be some visually non detectable damages as well (2-9).

2.2 Repair Methods In the past fibre-matrix-damages have been repaired by the application of metal or composite reproducers. These reproducers were fixed with screws or rivets. This method only covers the damaged area. Usually the strength of composites which are repaired by this way is low. Sometimes it is not higher as the strength of the damaged part without repair. The failure of the repaired composite occurs in the area of the screws or rivets.

For optical and functional reasons (aerodynamic functional surfaces) the reconstruction of the surface is necessary. Therefore the development and the evaluation of new repair methods is needed (1,3).

2.1.1 Reconstruction Method for the Repair of Composites The reconstruction method is a composite compatible repair method. A composite is used as repair material which is placed into the damaged area. Therefore a reconstruction of the surface of the damaged part is possible. The reconstruction method is used recently in the outer layers of gliders and boats. Now investigations are done by aircraft suppliers and airliners to make use of this method (3,8,10).

For the preparation of the reconstruction method the damaged area is removed. Afterwards reinforcing material is laminated into this areas and cured.

Two standardized applications of the reconstruction method in the described experiments are used. For the examination of different matrices and different curing processes the following scheme is applied:

1. Manufacturing of composites by a standardized hand lay-up and compression process with a glass woven fabric.
2. Cutting the plates in two parts and milling step-likely on the one side with a scarf joint ratio (SJR) of 1 : 25.
3. At this scarf joint, consisting of four steps, the plates are completed with a new laminate as described in point 1.
5. Consolidation of the matrix material with different curing processes.
4. Sawing out test specimen sized to 30 x 40 mm^2 prependicular to the scarf joint.

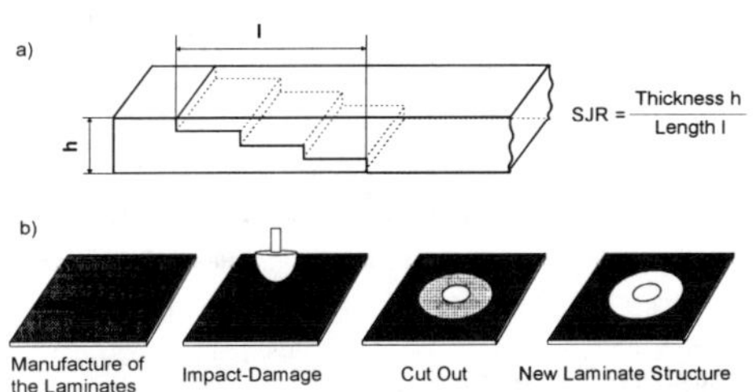

Fig 1: Application of Reconstruction Method

The step-likely scarf joint is shown in Figure 1a.

The tests for varying different repair textiles needs other test specimen. To minimize the influence of the different fiber orientations the processing is as follows:

1. Manufacturing of composites by a standardized method, cutting the composites into parts.
2. Defined damage of composites by penetration; the analysis of the necessary energy allows an evaluation of the similarity of the test items.
3. Inspection of the damaged area (brightness differences or ultrasonic investigations).
4. Milling out the damaged area with a defined shape.
5. Wet laminating of the reinforcing repair textiles by using the same matrix material and afterwards consolidation of the matrix material.

This process is presented in Figure 1b. The mechanical properties of all repaired parts are tested in tensile tests, bending tests, dynamic tensile and bending tests. To evaluate the success of the repair-procedure the repair-efficiency is introduced. It is a ratio of a mechanical property of the repaired composite to the non-damaged composite:

$$repair\text{-}efficiency = \frac{mechanical\ property\ of\ repaired\ composite}{mechanical\ property\ of\ non\text{-}damaged\ composite} * 100$$

First-of-all the influence of different humidities during the curing process is presented, also the influence of the use of resins with different coefficients of thermal expansion.

3. VARIATION OF MATRICES AND CURING PROCESSES

The test specimen for this experiments, which are carried out at the IKV, consists of 18 layers of a glass woven fabric. The 18 repair glass woven fabrics are divided to the four steps like 5/4/4/5 textiles per step. The reinforcing textiles are oriented in the same directions as those of the composite to be repaired (11,12).

3.1 Matrices with Different Coefficients of Thermal Expansion For these tests two different resin systems with a significant difference in their coefficients of thermal expansion (2 : 1) are used. The results of the tensile tests for the repaired specimen at two different temperatures are presented in Figure 2a. There is no significant influence of both parameters, temperature and

Using the same resin system the tensile strength decreases 6 % by a rise of 40 K of the testing temperature for the repaired specimen. The observed behavior corresponds to the course of the tensile strength for higher temperatures. In the contrary the tensile strength increases with increasing test temperatures using different matrices. This effect is based on the post curing of

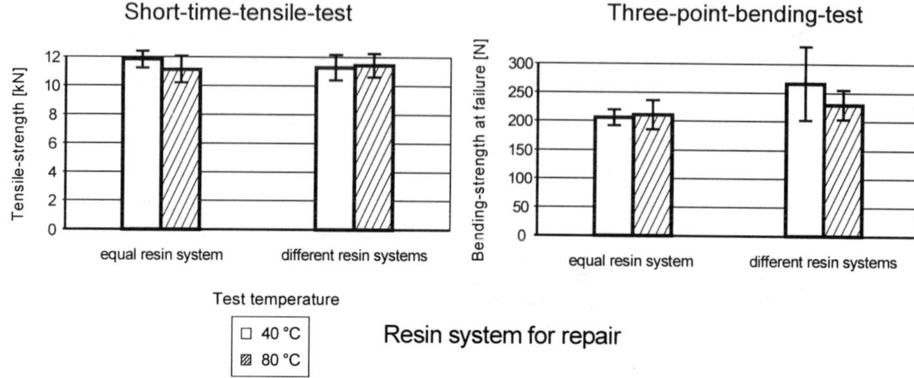

Fig. 2: Mechanical Properties of Repair Composites Using Different Test Temperatures and Resin Systems

the test specimen by conditioning the test samples at 80 °C for 16 h. That result is supported by DSC-analysis of the different cured resins (11).

The three point bending tests present a similar result. An influence of test temperature and matrix system cannot be observed, too. The standard deviation of the mechanical properties are higher for the specimens manufactured with two resin systems (Fig. 2b). The repair efficiency is approximately 82 %. The elevated repair-efficiencies obtained during the bending test are caused by an improved load transmission into the repaired area compared to the tensile tests.

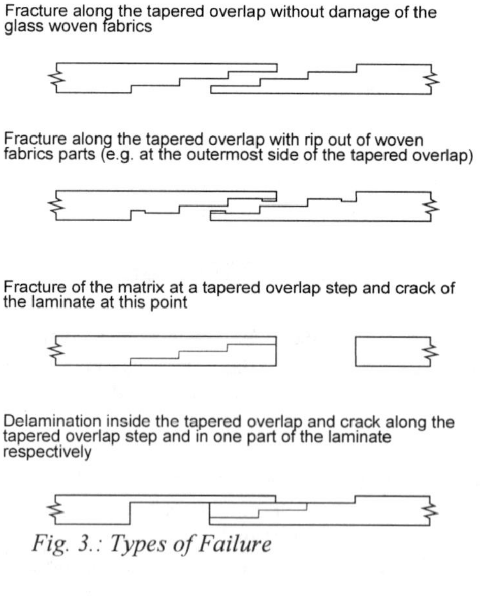

Fracture along the tapered overlap without damage of the glass woven fabrics

Fracture along the tapered overlap with rip out of woven fabrics parts (e.g. at the outermost side of the tapered overlap)

Fracture of the matrix at a tapered overlap step and crack of the laminate at this point

Delamination inside the tapered overlap and crack along the tapered overlap step and in one part of the laminate respectively

Fig. 3.: Types of Failure

The mechanical properties of the repaired parts are not influenced by the testing temperature and the resin system but the kind of failure during the tensile tests changes. Generally four different kinds of failure are observed (Fig. 3). The two failures presented above are delaminations of the whole repaired laminate. Partly some woven fabric parts rip out, too. The third and the fourth failure are fractures of the repaired area. Often this fracture starts at the outer steps. Due to the repair method this place is a notch in the composite structure where the fracture starts. During the bending tests with equal resin systems the relation of delamination failure to fracture is 1 : 1. By using different resin systems only delaminations can be observed. This indicates that the two different resin system used as matrix decreases the bonding more than one used resin system.

3.2. Different Curing Conditions For these tests a cold-curing resin system is used for the damaged and repaired composite. The repair composite is cured at different humidities (50 % and 95 % relative humidity of air) and different temperatures (25 °C and 80 °C) in an air-conditioned oven.

Figure 4 shows the results of the tension test and the three-point-bending test. A decrease of the strength due to the humidity is observed at both temperatures. There are more influences at a curing temperature of 80 °C (10,6 % tensile stress at break and 12,5 % flexural bending strength) than at 25 °C (2,1 % respectively 9,3 % decrease corresponding to the tension tests respectively the three-point-bending-tests). The standard deviation of the bending properties at "25 °C and 50 %" is too large to make a representative statement (12).

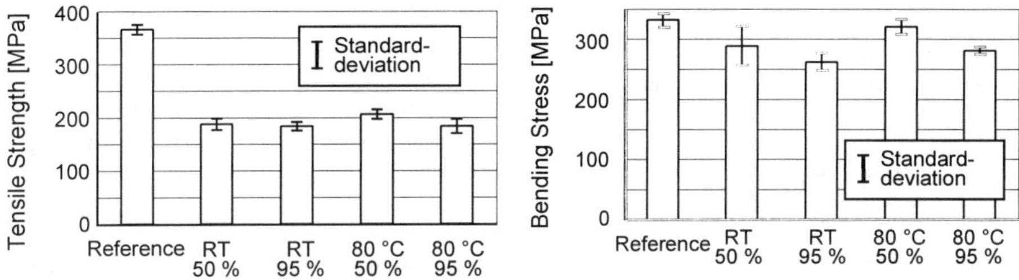

Fig. 4: Mechanical Properties of Repair Composites with Different Final Curing Conditions

A higher repair-efficiency at the bending loads of approximately 95 % is observed, too (Fig. 4). The repair efficiency at the tensile loads is approximately 50 %. The devated tensile and bending strengths of the test samples cured at 80 °C are caused by a longer curing time of the resin.

Summing up, the influences of humidity during curing is higher than the influence of different resin systems. Both effects are not very significant. Other irregularities, caused by the way of making the test stripes, and non avoidable notch-effects are responsible for a wide scatter within the several test columns.

4. COMBINATION OF DIFFERENT REINFORCING REPAIR TEXTILES

4.1 Repair Textiles For these examinations, carried out at the ITA, the test items have been manufactured of 18 layers of a glass woven fabric or 2 layers of a warp knit multiaxial textile (WIMAG) manufactured from glass. The test items were orthotropic. They had a thickness of 2,8 mm, a width of 150 mm, and a length of 300 mm. They have been repaired alternately with the same reinforcing textiles (Fig. 5). The reinforcing textiles are oriented in the same directions as those of the composite to be repaired (13,14).

Earlier investigations show that the repair-efficiency strongly depends on the shape of the prepared repair area. For eliminating these influences and to do the investigations with high practical relevance the scarf joint ratio was set to 1 : 25 like during the experiments described above. Thereby the prepared repair area is limited in a way that the investigations are not only relevant for test specimens but also for real parts.

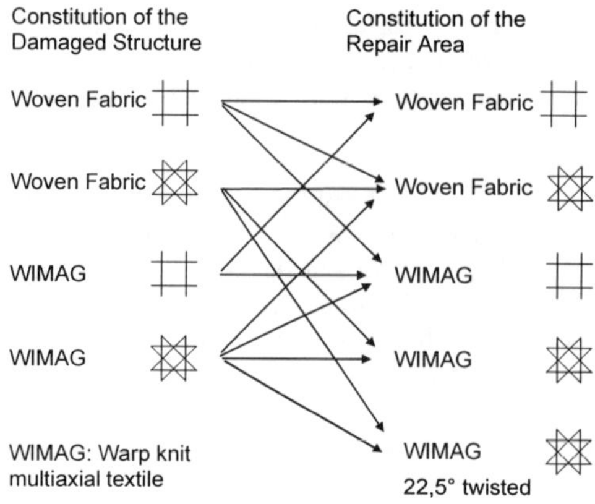

Fig. 5: Repair with Different Textiles

4.2 Results of Repair Investigations

The surface of the repaired specimen was reconstructed. The analysis of the repaired composites showed no residual damages. Delaminations or other damages were not detectable during ultrasonic and brightness investigations. These structures have a similar weight as the non-damaged specimens. This is especially for lightweight-constructions an important factor (10,13,14).

The specimen were tested after repair by short-time tensile test and three-point - bending test. Five specimen were tested by means of each test method. The repair efficiencies were calculated for Young's modulus, strength and elongation. Figure 6 shows the repair-efficiencies related to the strength of the specimen.

Fig. 6 shows that the repair efficiencies are clearly different depending on the load (tension, bending). The bending tests lead to higher repair-efficiencies than the tensile tests. The specimen manufactured from the woven fabric and repaired with the WIMAG achieve a repair-efficiency of 97 %. In this case a reconstruction of the mechanical properties is succeeded. The repair of the same specimen with the woven fabric leads to a repair efficiency of 90 %. In most cases this will fulfill the requirements.

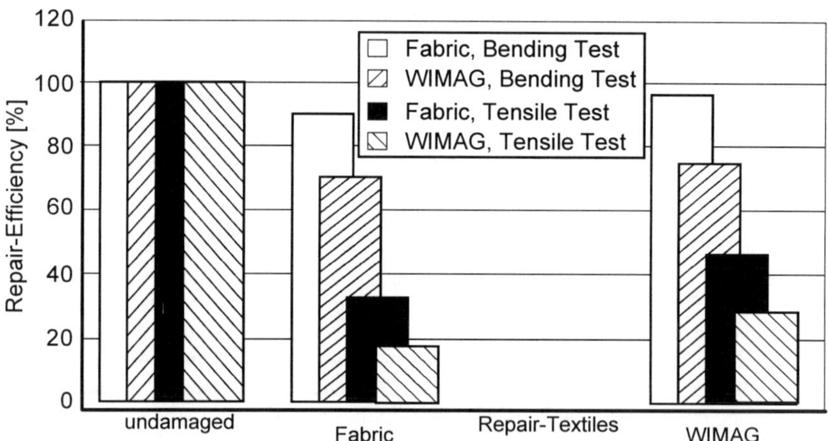

Fig. 6: Short-Term-Tensile and Bending Tests of Orthotropic Structures

The repair-efficiencies of the specimens made from WIMAG are nearly independent of the reinforcing textile used for the repair. They are in average at 75 %. The reduced repair efficiencies of the WIMAG specimen are caused by the higher Young's modulus of the non-damaged specimen compared to these made of the woven fabric. The higher modulus results from the non-crimped architecture of the reinforcing fibers. Therefore it is more difficult to reproduce the higher modulus.

Similar relations are obtained by the tensile tests. The repaired composites made from the woven fabric achieve higher repair-efficiencies than the WIMAG specimen. The fabric specimen which are repaired with the WIMAG obtain repair-efficiencies of 50 %, those which are repaired with the same fabric reach repair-efficiencies of 40 %. The results of the WIMAG-specimens are lower not depending on the repair material.

All examinations, varying the curing process, using different matrices, and combining different repair textiles, show a similar behavior. For a single type of loading repair efficiencies were measured which are suitable for practical use. For obtaining these results also in tensile loads further optimizations of the reconstruction method are necessary.

The bonding between the composite and the repair structure has an important effect on the repair-efficiency, too. That's why examinations of the bonding of composites are made at the Sonderforschungsbereich 332 "Production Techniques for Nonmetallic Composite Parts".

5. EXAMINATION OF BONDING OF CARBON FIBER REINFORCED THERMOSTES

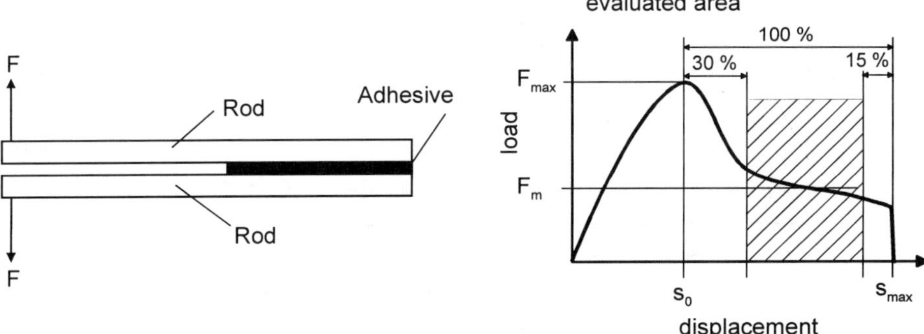

Fig. 7: Experimental Set up of the Peel Test

The following examinations are carried out at the DWI. The peel test according to DIN 53282 (15) is a suitable test method to compare different adhesives and surface preparations. The test described in (15) is normally used for films. A modification is necessary, too (Fig.7) (16-18).

The peel tests were carried out in a temperature range from -55 °C up to 81 °C. The test evaluation of DIN 53282 is modified in the following way: The maximum load during the peel test is defined as peel strength F_{max}, the peel length at these point is s_0. The average peel force F_m is defined as the arithmetic mean of the measured force from s_0 up to the test end without the first 30 % and the last 15 % of this part of the peel length s (Fig. 7).

5.1 Materials and Surface Preparations As test material a carbon fiber reinforced epoxy resin is used. The test specimen were cut out in 100 x 15 x 1 mm^3 parts as they are shown in Figure 7. The rods are bonded at a length of app. 50 mm, the thickness of the adhesive is fixed with spacers of 0,1 mm. The applications of load is realised with rings. As adhesives four different resin-catalyst systems are used with a composition as follows:

Type A: cold curing epoxy/polyamine-system
Type B: heat curing epoxy resin with hardener and accelerator
Type C: cold curing polyurethane/isocyanate-system
Type D: cold curing epoxy/polyamine-system with another composition than in type A

Every bonding needs an intensive cleaning of the specimen surface. The surface preparation is done in a combined mechanical and chemical way. Two different cleaning agents are used.

On the one hand the specimen were ground with a grit of P120 and cleaned with a methyl ethyl ketone (MEK-treatment). On the other hand additionally the specimen were stored in a 65 % nitric acid (HNO_3-treatment) for five minutes. All test samples were purged with water after the surface preparation and dried for 20 h in a drying plant at room temperature.

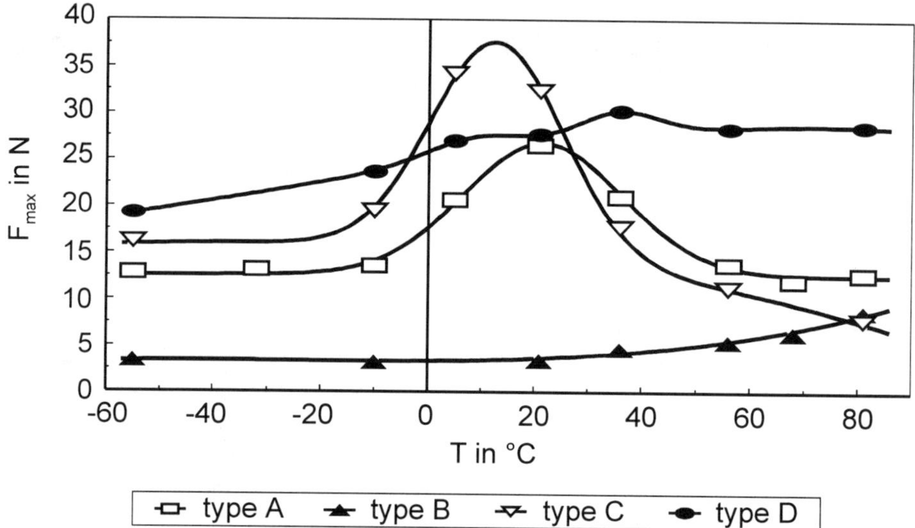

Fig. 8: Average Values of the Standardized Peel Strength F_{max} of the Adhesive (MEK-treatment)

5.2 Results of the Peel Test Five tests are carried out for each test conditions. The peel strength is determined as it is described above. The peel strength is standardized from the real sample width of an ideal width of 10 mm. They are presented in Figure 8 for the MEK-treatment (18).

The different adhesives show different levels of strength and a different behavior depending on the temperature. The Type A and C have a significant strength maximum at room temperature. However, Type D shows another behavior though its composition is similar to Type A: No significant maximum at room temperature is observed. The peel strength is also retained at high temperatures. Therefore this adhesive is suitable for a large range of temperatures.

Type B shows increasing peel strength with increasing temperature on a very low level. This is caused by an imperfect curing of the adhesive. The heat resistance of the epoxy resin is poor, so that the curing process can not be realized. Also the viscosity of this system is very low at high temperatures so that the resin creeps out of the rods.

Concerning the average peel force F_m a similar behavior was observed. Only Type D shows an increasing average peel force F_m. At 81 °C the specimen break. Therefore no values could be measured at this temperature.

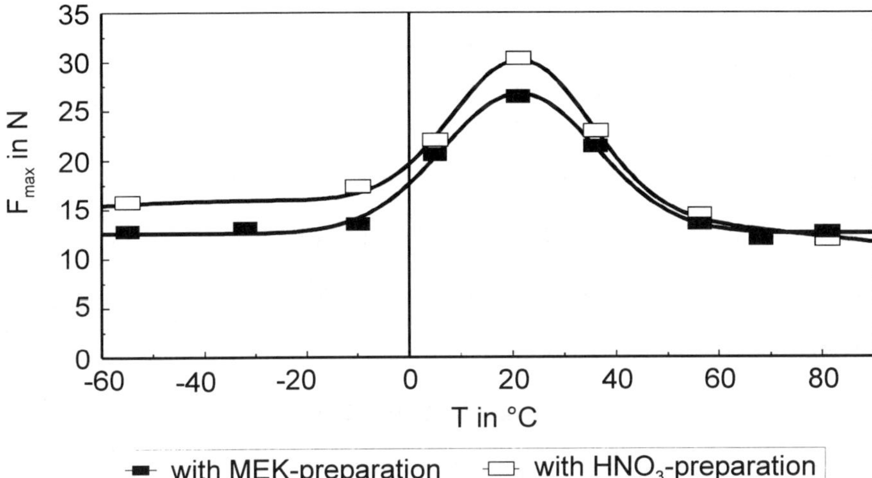

Fig. 9: Comparison of the Peel Strength of the MEK and HNO₃ surface preparation (adhesive: type A)

Additional peel tests are carried out for the adhesive Type A with the HNO_3- treatment. The results are presented in Figure 9. They are compared with the results of the MEK-treatment. The chemical preparation causes increased peel strength up to 28 %. This effect cannot be observed at higher temperatures. The principle shape of the curves with a maximum at room temperature is not changed due to the other surface preparation.

5.3 Further investigations The results of the peel tests were confirmed by further investigations (18). Thermomechanical analysis show a softening point for the Types A and C at 20 °C and for Type D at 80 °C. At the softening points stress peaks are decreased because of the ductile behavior of the adhesives. A better bonding is reached.

A determination of the chemical elements and their bondings is possible with x-ray photoelectron spectroscopy (XPS). By measuring the fractions of hydrophilic and polar groups, especially of the C-O-groups, an estimation of the adhesion loads gets possible (19,20).

It is observed, that the percentage of the polar groups is correlated with the peel strength at room temperature in Figure 8. For example Type C shows the highest peel strength and also the biggest parts of polar groups at the surface.

6. CONCLUSION

For the elimination of damages of composite parts special repair methods are necessary. The repair might obtain advantages regarding the ecological and economical aspects and the saved time, compared to demounting a damaged part and mounting a new one.

The elimination of fiber-matrix-damages is possible by the reconstruction method. For special loads (bending test) repair-efficiencies of app. 95 % can be reached. The use of different resin systems had no significant effect on the repair-efficiency. Curing processes with different humidities show an influence of the tensile and bending strength of repaired composites. The use of different repair textiles has shown that it is not necessary to use exactly the same reinforcing textile as used in the damaged composite.

The investigations of the surface preparation show an influence on the peel strength. A better bonding of the repair structures can be reached with a HNO_3-preparation and with cold-curing epoxy or polyurethane systems as adhesives. The milled-out repair area is like a big notch. At the border of the notch multiaxial tensions will occur. Further investigations will show wether semi-isotropic repair structures are able to accept this loads in a better way especially in case of tensile loads. More investigations about a mechanical surface preparation are also planned.

7. ACKNOWLEDGEMENTS

We like to thank Deutsche Forschungsgemeinschaft (DFG) for financial support. The investigations are integrated in the Sonderforschungsbereich 332 (SFB 332) at the University of Technology, Aachen (RWTH Aachen), Germany.

8. REFERENCES

1. R.B. Heslehurst, <u>Sampe J.</u>, <u>Vol. 33</u>, No.6. (1997), p. 16 - 21
2. H.W. Bergmann, <u>Konstruktionsgrundlagen für Faserverbundbauteile</u>, Springer-Verlag, Berlin, Heidelberg, New York, 1992
3. A.A. Baker, Repair Techniques for Composite Structures, in: D.H. Middelton, ed.<u>Composite Materials in Aircraft Structures</u>, Longman Scientific and Technical, Harlow, 1990
4. K. Rüsenberg, <u>Der Plastverarbeiter</u> 27 (1976), p. 477-481
5. S.D. Labor, <u>16th National SAMPE Technical Conference</u>, Albuquerque NM, 1984, p. 119-128
6. R.M. Stone, <u>12th National SAMPE Technical Conference</u>, Seattle, 1980, p. 688-701
7. E. Heitz, <u>Der Plastverarbeiter 28</u>, 1977, p. 469 - 476
8. K.B. Armstrong, <u>Proceedings of the Institute of Mechanic Engineers, 203</u>, 1987, 92, p. 105-112
9. R. Jones, R.S.Callivan, K.C. Aggarwal, <u>Engineering Fracture Mechanics, 17</u>, 1983, H. 1, p. 37 - 46
10. B. Wulfhorst, K.-U. Moll, W. Michaeli, K. Kocker, "Berwertung verschiedener Verfahren zur Reparatur von Faserverbundwerkstoffen" in: VDI-Gesellschaft Werkstofftechnik, ed,. <u>Effizienzsteigerung durch innovative Werkstofftechnik</u>, 1995, Düsseldorf, p. 515 -523
11. M. Bittner, "Untersuchung des Versagensverhaltens reparierter FVK mit verschiedenen duroplastischen Matrizes im Laminatneuaufbau", thesis at IKV, Aachen, 1997
12. F. Lindner, "Untersuchung des Versagensverhaltens reparierter FVK mit duromerer Matrix bei verschiedenen Feuchtegehalten der Umgebung", thesis at IKV, Aachen, 1996
13. B. Wulfhorst et al., Aufbau, Prüfung und Aufmachung der Harze und Fasern, Arbeits- und Ergebnisbericht der Deutschen Forschungsgemeinschaft zum SFB 332, Teilprojekt 1, RWTH Aachen, Aachen, 1995
14. B. Wulfhorst, K.-U. Moll, "Analyse der Materialeigenschaften von Fasern und Matrices für Laminate aus FVK, Industrielle Anwendung der Faserverbundtechnik II, Symposium, 4. and 5. October 1995, RWTH Aachen, Aachen, 1995
15. DIN 53282, September 1979
16. M. Rasche, <u>Adhäsion, 10</u>, 1986, p. 18
17. D. Knoll, <u>Adhäsion, 5</u>, 1990, p. 36 - 39

18. A. Nick, "Untersuchungen zum Kleben und zu den viskoelastischen Eigenschaften von Verbundkunststoffen aus kohlenstoffaserverstärktem Epoxidharz", dissertation, RWTH Aachen, 1997

19. R. Kaufmann, H. Höcker, <u>GIT Fachz.Lab.</u>, <u>4</u>, 1994, p. 348 - 352

20. H. Brockmann, <u>6. Int. Symp. SWISS BONDING</u>, Basel, 1992, p. 63-70

9. BIOGRAPHY

Prof. Dr. rer. nat. Hartwig Höcker, born 1937, is head of the German Wool Research Institute (DWI) at the Aachen University of Technology, Germany.

Prof. Dr.-Ing. Walter Michaeli, born 1946, is head of the Institute of Plastics Processing (IKV) at the Aachen University of Technology, Germany.

Prof. Dr.-Ing. Burkhard Wulfhorst, born 1936, is head of the Institute of Textile Technology (ITA) at the Aachen University of Technology, Germany.

Prof. Dr. rer. nat. F.-J. Wortmann, born 1952 is head of department of polymer physics at the German Wool Research Institute.

Dr.-Ing. Albrecht Nick, born 1965, studied mechanical engineering at Aachen University of Technology. Since 1994 he was scientific assistant and achieve 1997 his Ph.D. degree at the DWI at Aachen University of Technology.

Dipl.-Ing. Klaus-Uwe Moll, born 1968, studied mechanical engineering at Aachen University of Technology since 1993 he has been scientific assistant to achieve his Ph.D. degree at the ITA at Aachen University of Technology.

Dipl.-Ing. Michael Münker, born 1968, studied mechanical engineering at Aachen University of Technology since 1995 he has been scientific assistant to achieve his Ph.D. degree at the IKV at Aachen University of Technology.

ENVIRONMENTAL EFFECTS ON THE MECHANICAL PROPERTIES OF NOTCHED E-GLASS/VINYL ESTER COMPOSITE MATERIALS

Stephanie E. Buck, David W. Lischer, and Sia Nemat-Nasser
Center of Excellence for Advanced Materials, University of California, San Diego,
La Jolla, CA 92093-0416

ABSTRACT

E-glass/vinyl ester composite specimens have been tested with and without notches to determine the combined effects of load, temperature, and moisture on the mechanical properties of this class of composites. Many of the results for biaxial and quasi-isotropic unnotched samples are presented elsewhere (1,2). Notches have been created in biaxial samples to study the changes that might be brought about by seemingly minor damage during the construction of civil engineering structures. Preliminary tests have demonstrated the expected drop in strength and life due to the presence of the notch. Crack growth from the notch tip under axial tension and delamination in the vicinity of the notch are found to vary according to the specific environmental conditions used in the study.

KEY WORDS: Environmental Durability, Mechanical Properties, Composites

1. INTRODUCTION

Composite materials have been considered for structural applications. In particular, polymeric composites with thermosetting resins may be used for such applications as civil engineering applications, due to the improvement in specific strength and stiffness they provide over other materials. A variety of materials may be used for both the resin and the reinforcement for these applications. Vinyl ester resin provides good matrix properties as well as good moisture resistance and E-glass fibers provide reasonable strength and stiffness while keeping overall costs down. A variety of methods may be used to manufacture these composites. Processing the composites by SCRIMP (Seemann Composite Resin Infusion Molding Process) keeps costs lower than the use of such processes as RTM (Resin Transfer Molding) while providing a relatively fast and reproducible set of materials. A variety of external factors may affect the performance of a composite in a given application. Moisture, elevated temperature, and a sustained load, particularly in combination, may degrade the mechanical properties of a material. The presence of a defect such as a notch, which may be created during the manufacture or use of a composite part, may exacerbate this degradation.

Numerous researchers (3-14) have described experiments and models used to assess the performance of notched composite specimens, but only limited studies have been performed to observe environmental effects in the presence of a notch (15). This paper describes preliminary studies on the effects of the environment in combination with quasi-static axial tension on the performance of a polymeric composite material.

2. EXPERIMENTS

2.1 Materials   The composite specimens consisted of a thermosetting resin, Derakane 411-350 vinyl ester, reinforced with E-glass fibers. The layup was $[0/90]_{2s}$, with each layer of the woven material also containing a random mat. The composites were manufactured by SCRIMP, a processing method resulting in panels which have one face, the tool surface, with a polished appearance, while the other side, the vacuum bag surface, follows the wave of the fibers. The nominal thickness of the material used was 0.61 centimeter. Both sides of the gauge section were left unmachined. The material had a fiber volume fraction of approximately fifty-two percent.

2.2 Apparatus   Samples were cut in a dogbone shape using a water jet. The notches were produced by a file and a sharpened saw blade. The nominal notch length was 1.5 millimeters, leading to a sample width/notch length ratio of 9.5. The nominal radius of curvature for the notches was 0.04 millimeters. Figure 1 shows the sample design. A two-part epoxy was applied to the vacuum bag surface of the tab regions of each sample. This area was then milled parallel to the other sample face. The tab regions of the tool surface of each sample were sanded to roughen the surface to help reduce slipping of the samples in the grips of the conditioning load frames and the test apparatus. Samples were weighed and measured before and after conditioning. They were post-cured at 120°C for two and one half hours prior to conditioning.

The environmental conditioning was performed in tanks which were specifically designed for this purpose. Each tank holds tap water (replaced as it evaporates with deionized water to maintain constant mineral content) at elevated temperatures and submerged samples which are loaded in specially designed pneumatic load frames. The load frames are able to provide up to 3000 pounds of force. The samples were conditioned at 21°C, 49°C, 57°C, and 66°C for at least 210 hours.

Some samples were observed microscopically after conditioning; others were tested to failure. A scanning electron microscope was used to observe microcracking, fiber breakages, and other damage in the composites. The elastic moduli and the load and strain to failure were measured using a servohydraulic MTS machine at strain rates of 10^{-5} s^{-1} for the elastic tests and $6 \bullet 10^{-4}$ s^{-1} for the failure tests. Samples were loaded by a triangular waveform generated with a WaveTek device. Loading in the tests to failure was discontinued manually when the load had dropped to approximately half its peak value. Strain was measured with an extensometer for the elastic tests and with the LVDT for the tests to failure. All data were collected with a LabView data acquisition system. Samples were typically videotaped during testing.

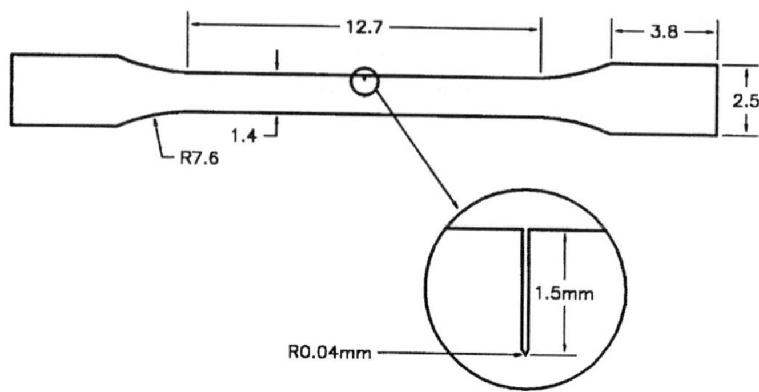

Figure 1. Sample design. Dimensions are in centimeters unless otherwise noted.

3. RESULTS AND DISCUSSION

Post-conditioned samples typically had a crack which extended from the pre-existing notch tip in a direction transverse to the sustained load. In some instances, there was evidence of delaminations which grew from either side of this crack in the direction of the applied load. With few exceptions, the notched samples aged up to 620 hours had strengths 15 ± 2% below the values for the unnotched samples tested under the same conditions. A complete set of data will be presented at a later date. Failure of the notched samples occurred in the notch region and involved the formation of transverse matrix cracks in the vicinity of the notch and the extension of the transverse crack at the notch tip, followed by longitudinal delamination of inner layers, leading to the catastrophic failure.

The presence of a notch exacerbates the degradation associated with the environmental conditioning. For unnotched specimens, the water absorption at the lowest temperature used, 21°C, was 0.04 percent by weight, which remained constant between 140 hours and 620 hours of submersion, while that for the highest temperature studied, 66°C, was nearly 0.07 percent by weight, and remained constant after 140 hours of submersion. The combination of moisture and elevated temperature was previously found (1) to diminish the mechanical properties of the material studied here. The addition of the notch has been found in these studies to further reduce the mechanical performance of the material. In addition to the possibility that the presence of a notch allows additional moisture absorption simply due to the increase in the surface area of the sample, the localized damage created by the presence of the notch reduces the capability of the fibers to perform their function of slowing, stopping, or changing the direction of crack growth, in addition to greatly reducing the strength of the composite in the neighborhood of the notch. These findings are significant due to the inevitable formation of notches and other defects during the construction and use of composite components in civil engineering applications.

Figures 2-5 show photographs and scanning electron micrographs for two samples after conditioning. The macroscopic images of the samples, both aged for 620 hours, one at 21°C and the other at 57°C, show damage in the vicinity of the notch. Both samples show a crack initiated at the notch tip and longitudinal delaminations of interior layers emanating from the crack at the notch tip. Both the internal delaminations and the superficial damage are more extensive in the sample aged at the higher temperature. The micrographs for these samples show fiber damage along the notch wall for both cases. The sample conditioned at the lower temperature shows a matrix crack near the left side of the image. The sample aged at the higher temperature shows not only matrix damage, but also several fiber breakages in the vicinity of the crack extending from the notch tip. Similar comparisons were found for samples conditioned at a given temperature for different lengths of time. These results may indicate some change in the earliest damage mechanism from matrix microcracking in unnotched specimens to a combination of matrix microcracking and fiber breakages for the notched material.

The results indicated a general trend of greater damage with longer conditioning times and higher conditioning temperatures, but the trends were not monotonic. Deviations from the overall trend may be the result of microscopic irregularities introduced during the manufacturing process or to the lack of the exact reproduction of the notch shape, although the notch tip radii were measured to be similar in each case. Many further studies are in progress, particularly for longer conditioning times, to clarify the trends. However, the indication that the failure strengths so far obtained have relatively little scatter in them does point to notch sensitivity of the material (12), necessitating further studies to pinpoint the failure mechanisms.

4. CONCLUSIONS

1.   The presence of an edge notch in the composite material studied here causes a significant reduction in tensile strength when compared to unnotched samples subjected to the same conditioning times and temperatures.

2.   The presence of an edge notch may change the damage mechanisms which lead to failure of a composite under the conditions studied. Microscopic observations show that the damage in the vicinity of the notch is increased by conditioning time and temperature.

5. ACKNOWLEDGMENTS

This research was supported by the Army Research Office under grant DAAH04-95-7-0369 to the University of California, San Diego. Thanks are also due to DuPont, Seemann Composites, Dow, Knytex, and Hanna Resins for supplying the materials, to the Center of Excellence for Advanced Materials for providing the experimental facilities, and to the Institute for Mechanics and Materials for supplementing the research.

6. REFERENCES

1. S. E. Buck, D. W. Lischer, and S. Nemat-Nasser, <u>SAMPE Int'l Symposium</u>, <u>42</u> (1997).
2. S. E. Buck, D. W. Lischer, and S. Nemat-Nasser, <u>J. Comp. Mat.</u>, in press (1998).
3. K. E. Perry, Jr., and J. McKelvie, <u>Experimental Mech.</u>, <u>36</u>, 55 (1996).
4. Y. Zhao, S.-S. Pang, and C. Yang, <u>Trans. ASME</u>, <u>118</u>, 542 (1996).
5. A. Laksimi, X. L. Gong, and M. L. Benzeggagh, <u>Comp. Sci. Tech.</u>, <u>52</u>, 85 (1994).
6. J. Xiao and C. Bathias, <u>Comp. Sci. Tech.</u>, <u>52</u>, 99 (1994).
7. J. Karger-Kocsis, T. Harmia, and T. Czigány, <u>Comp. Sci. Tech.</u>, <u>54</u>, 287 (1995).
8. L. K. Jain and Y.-W. Mai, <u>Comp. Sci. Tech.</u>, <u>55</u>, 241 (1995).
9. S. L. Bazhenov, <u>Comp.</u>, <u>26</u>, 125 (1995).
10. A. Afaghi-Khatibi, L. Ye, and Y.-W. Mai, <u>J. Comp. Mat.</u>, <u>30</u>, 333 (1996).
11. R. Marissen, H. R. Brouwer, and J. Linsen, <u>J. Comp. Mat.</u>, <u>29</u>, 1544 (1995).
12. V. M. Karbhari and D. J. Wilkins, <u>Polym.-Plast. Tech. Eng.</u>, <u>31</u> (1&2), 103 (1992).
13. J. Awerbuch and M. S. Madhukar, <u>J. Reinf. Plast. Comp.</u>, <u>4</u>, 3 (1985).
14. S. Weihe, M. König, and B. Kröplin, <u>Comput. Mat. Sci.</u>, <u>3</u>, 254 (1994).
15. P. Lee-Sullivan and G. Lu, <u>Wear</u>, <u>176</u>, 81 (1994).

7. BIOGRAPHIES

Stephanie E. Buck received her B. S. in Engineering and Applied Science from the California Institute of Technology and her M. S. in Engineering Science from Dartmouth College. She is currently a Graduate Student and Research Assistant in Materials Science at the University of California, San Diego, where she is studying the effects of the environment on composite properties.

David W. Lischer is a Development Engineer for the Applied Mechanics and Engineering Sciences Department at the University of California, San Diego. He received a B. S. in Applied Mechanics in 1972 and a M. S. in Engineering Physics in 1975 from UCSD. His current fields of study with the Center of Excellence for Advanced Materials include durability of composites, dynamic fracture toughness of ductile materials, and optical techniques for the measurement of material properties.

Sia Nemat-Nasser, John Dove Isaacs Chair in Natural Philosophy and Professor of AMES at UCSD, holds a Ph. D. from UC Berkeley. He is currently Director of the Center of Excellence for Advanced Materials as well as Director of the Institute for Mechanics and Materials. Research interests include micromechanical and constitutive modeling of nonlinear response and failure modes of materials; analytic, computational, and experimental mechanics and mechanics of materials, especially those of ceramics, ceramic composites, advanced metallic and polymeric composites, high strength alloys and superalloys, as well as rocks and geomaterials. He has authored, co-authored, or edited over nineteen books and proceedings, has organized over thirty-five scientific workshops and meetings, and has published over 300 scientific articles.

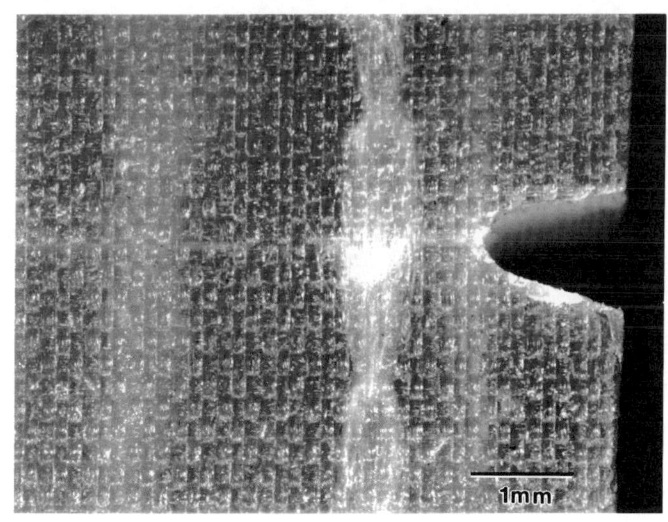

Figure 2. Sample aged for 620 hours at 21˚C.

Figure 3. Sample aged for 620 hours at 57˚C.

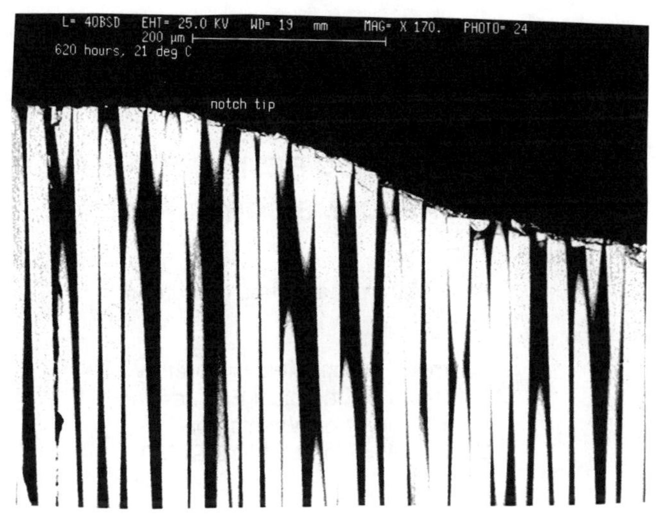

Figure 4. Sample aged for 620 hours at 21˚C.

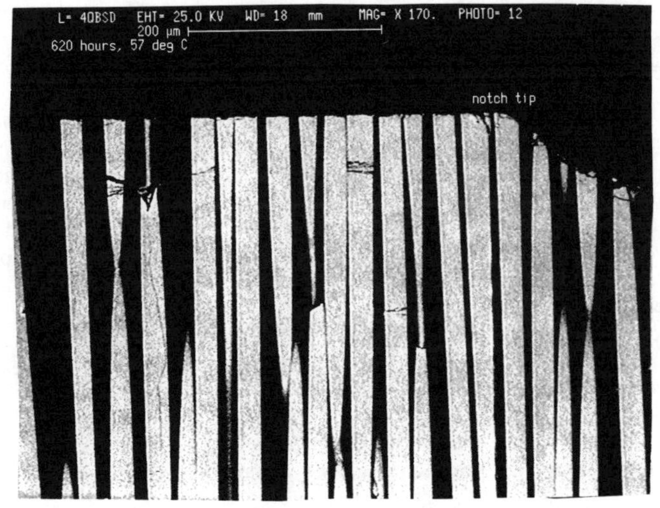

Figure 5. Sample aged for 620 hours at 57˚C.

EFFECT OF PREPREG RESIN COMPOSITION
ON HONEYCOMB CORE CRUSH

Cary J. Martin and James C. Seferis*

Polymeric Composites Laboratory
Department of Chemical Engineering
University of Washington
Seattle, Washington 98195-1750

ABSTRACT

Past research has shown that the level of core crush experienced in the autoclave manufacture of honeycomb cored composite structures varies widely between prepreg systems. This differing manufacturing behavior has been attributed to differences in prepreg frictional resistance at processing conditions. At similar levels of impregnation, elastomer modified epoxy-based prepreg systems show consistently higher levels of core crush than standard high flow prepregs. Therefore, the role of resin composition in honeycomb core crush was investigated to determine the feasibility of developing crush resistant elastomer modified epoxy-based prepregs. This was accomplished through development of model prepregs with varying elastomer concentration. The kinetics of the reactive carboxyl-epoxide adduction were also investigated. The core crush tendencies of the model prepregs were evaluated through the measurement of prepreg frictional resistance. It was found that prepreg frictional resistance decreased with increasing elastomer concentration. These differences were related to differences in the viscosity and elasticity of the impregnated resins as determined through oscillatory rheology. It is anticipated that these differences would result in an increased incidence of core crush.

KEY WORDS: Prepreg, Epoxy Resins, Honeycomb Core

1. INTRODUCTION

One problem associated with the autoclave manufacture of prepreg based honeycomb sandwich structures is core crush. Historically, the origins of this defect were not well understood and its occurrence was often unanticipated. In commercial production, several techniques such as internal septumization and tool modification through grit strips are used to mechanically restrain the core from collapsing inward. These approaches, while reducing the incidence of core crush, have produced no fundamental understanding of the underlying process or mechanism.

Research efforts to understand core crush have utilized the force balance approach, first conceived by Brayden(1). In this technique, essentially a mechanics free-body analysis, the forces acting on a honeycomb structure are conceptually isolated and individually examined. The driving force for core collapse has been identified as a pressure differential between the autoclave and core, while the resisting forces are internal core pressure, the lateral compressive strength of the skin/core combination, and the frictional resistance of the prepreg plies.

In a separate work, Seferis and co-workers investigated the effects of five processing variables on core crush. They found that prepreg resin type and vacuum level used during lay-up and staging were the major influences on core crush(2, 3). Internal pressure measurements confirmed that core crush could be related to the in-situ and post-cure internal core pressure. Panels that utilized high vacuum levels during lay-up showed greater crush than panels produced with a reduced vacuum level. In addition, they found that a highly rubberized, controlled-flow epoxy prepreg showed consistently greater core crush than a standard epoxy prepreg, indicating that core crush is a material dependent phenomenon.

While the investigations of internal pressure established that it indeed influenced core crush and that by controlling internal pressure during cure, core crush could be controlled or eliminated, they failed to elucidate the nature of its material dependence. Prepreg permeability and intralaminar frictional resistance are two possible differences. Differing levels of permeability would effect changes in internal core pressure and change the crush driving force. However, these studies controlled the driving force directly and core crush differences were still observed.

The second material dependent factor, prepreg frictional resistance, was initially discounted as insignificant. When core crush occurs, the prepreg layers are observed to move inward with the core. Therefore, slippage must be initiated either between the prepreg plies or at the prepreg/tooling interface. It has recently been shown that significant differences exist in the frictional resistance of commercial woven carbon fiber prepregs which correspond to the observed differences in core crush behavior(4). In an investigation of the core crush behavior of a unidirectional prepreg, these differences in frictional resistance and hence core crush behavior were found to be dependent on the impregnation parameters used during the prepreg manufacture(5). These investigations have identified frictional resistance as an important prepreg process parameter and established a technique by which prepreg materials can be screened to determine their tendency to cause core crush during manufacture.

While these studies have identified the material dependent origins of core crush, so far they have not identified a set of requirements which could be used to develop a crush resistant prepreg material. Of the systems examined to date, a controlled flow elastomer modified epoxy prepreg showed the highest levels of core crush and lowest frictional resistance. A standard low viscosity prepreg system and a high viscosity thermoplastic modified epoxy system showed increased frictional resistance and decreased core crush. The use of carboxyl functionalized elastomer modification of epoxy resins to increase matrix toughness is widely acknowledged is well established(6-8). However, the effect of these modifications on the handling and manufacturing properties of these materials is only now being investigated.

As the use of elastomer modification is widespread, there is a need to understand its effect on honeycomb core crush. While the past research has shown that an elastomer modified prepreg showed high levels of core crush, it has not been shown that this behavior is necessarily the result of rubber incorporation. Therefore, in this study the effect of elastomer concentration on core crush behavior will be investigated. This will be done by first developing experimental prepregs with different concentrations of a solid elastomer. Next, the crushing tendency of these materials will be evaluated by measuring the prepreg/tooling frictional resistance as a function of temperature.

2. EXPERIMENTAL

A requirement of this investigation was the development of a elastomer modified resin system which could be impregnated via hot melt impregnation without the use of solvents. To investigate the role of elastomer concentration on core crush, the experimental work was divided into four sections – resin mixing, rheological evaluation of the resins, production of the prepreg materials, and evaluation of the core crush characteristics of the trial materials. Experimental details of these operations are described below.

2.1 Resin preparation   Epoxy based experimental resins were produced with elastomer concentrations varying from 2 to 10% of the total mass. Hycar 1472, a carboxyl-modified solid acrylonitrile-butadiene elastomer (CMBN) with randomly distributed carboxyl groups was used to increase resin viscosity and elasticity. Available in a crumb form, the solid particles were washed with methanol to remove a protective talc prior to dissolution in methyl ethyl ketone. The base resin consisted of a 50:50 mass blend of diglycidylether of bisphenol-A (DGEBA) [Epon 826, Shell Chemical] and triglycidyl p-aminophenol (TGPAP) [Epon HPT 1076, Shell Chemical]. The primary curing agent, diaminodiphenyl sulfone (DDS) [Ciba Geigy HT976], was used in a 60% stoichiometric ratio. Dicyanamide (DICY) [Amicure CG2100, Pacific Anchor Chemical] at a level of 1 phr was used as a cocuring agent. Chromium(III) 2-ethylhexanoate (also referred to as chromium octoate) [Shepherd Chemical Company] was used as an epoxy/carboxy esterification catalyst. Available as a solid paste, the chromium compound was dissolved in a 70 weight percent MEK solution to aid in resin mixing.

Resin solutions were prepared by first combining the epoxies at 60°C under high shear mixing to form a base solution. The rubber solution and curing agents were then added while mixing at room temperature. After mixing for ten minutes, the esterification catalyst was added using a microsyringe. After complete mixing, excess solvent was removed at 40°C under full vacuum. Resins were stored at 0°C until the prepregs were manufactured.

2.2 Rheological evaluation   Resin viscosity and elasticity were evaluated using a TA Instruments CSL-100 controlled stress rheometer operated in an oscillatory mode. Parallel platens with a diameter of 4 cm and a gap of 0.5 mm were used. An oscillating stress of 50 μN·m was applied at a frequency of 1 Hz. A heating rate of 5°C·min^{-1} was used for dynamic experiments.

2.3 Prepreg processing   The prepregs were produced used a laboratory scale prepregger described elsewhere(9). The fiber reinforcement used a 3K-70 plain weave carbon fiber fabric (Toho) cut into rolls 10.2 cm in width. Resin films of thickness 0.38 mm and width 50.8 mm were cast onto a release film and applied to the fiber bed from both sides. Impregnation was provided in a heated impregnation zone with three sets of force controlled rollers. Force on the rollers was 22, 68, and 178 N for the first, second, and third sets. Impregnation temperatures were 35, 55, and 85 °C for the 2, 6, and 10% rubber resins, respectively. These temperature were chosen so that the viscosity of each resin would be approximately 2 Pa·s during impregnation. After impregnation, the top backing paper was removed. The prepregs were then cut into 21.6 mm lengths and staged at 135°C for ten minutes in an air circulating oven to remove any residual solvent and promote the epoxy-carboxy esterification, increasing resin viscosity and elasticity. The dry, unimpregnated regions were trimmed from the sides and the prepregs were stored at –10°C. Resin content was determined by recording the mass of each prepreg and subtracting the known areal weight of the fabric (193 g·m^{-2}).

2.4 Core crush   The core crush tendencies of the trial prepregs were evaluated through the measurement of prepreg frictional resistance using the apparatus and procedure described elsewhere(4). Prepreg samples were 21.6 by 5.1 cm. The frictional resistance was evaluated at conditions simulating those experienced during processing. To minimize specimen contamination, measurements were performed with the surface covered by the backing paper against the tooling. Normal consolidation pressure was 310 kPa. Measurements were conducted at five temperatures from 37.8 to 93.3°C at increments of 13.9°. Three individual measurements were performed for each prepreg at each temperature. Data were reported as the maximum force or force at a displacement of 1.9 mm.

3. RESULTS AND DISCUSSION

3.1 Development of resin system   The kinetics of the non-catalyzed epoxy/carboxy esterification reaction were studied using a 1:6 mass blend of CMBN/DGEBA. This mixture was used as a model blend to eliminate the effect of curing agents. Reaction progress was monitored using oscillatory rheology. The viscosity of this blend as a function of temperature is shown in Figure 1. This reaction was observed by an increase in viscosity between 150 and 160°C. Above this temperature only a slight additional increase was observed. The isothermal reaction is investigated in Figure 2, which shows the effect of isothermal hold time on viscosity at four temperatures from 130 to 160°C. At the highest temperature, the reaction was complete within ten minutes. As hold temperature was decreased, reaction onset and completion were delayed. At 150°C, no further increases in viscosity were observed after 15 minutes while nearly 30 minutes was required at 140°C. These results were consistent with those reported by other researchers and indicated the need for an esterification catalyst(10). These adduction reactions are often performed using compounds such as triphenyl phosphine, as well as others(11).

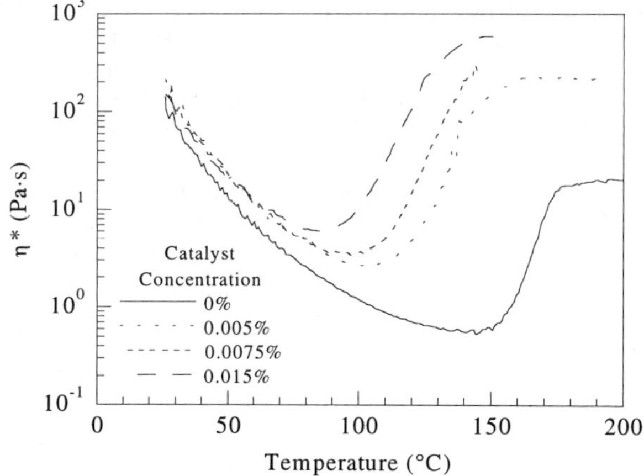

Figure 1 Dynamic conversion of CMBN/DGEBA blends

One class of compounds which has shown a wide utility for the catalysis of a number of epoxide reactions are coordinated transition metal catalysts, particularly those containing chromium in the $^+3$ oxidation state(12-19) Notable for this application is the work of Browning, who describes the development of a low flow prepreg system designed for honeycomb structures. In this work, chromium octoate was used to react the epoxy and carboxy groups without promoting epoxy homopolymerization. This reaction was described as having a reaction temperature which could be varied by varying catalyst concentration.

As this catalyst seemed ideal, the effect of catalyst concentration on the viscosity and elasticity of the previously described CMBN/DGEBA blends was investigated. These results are shown at low concentrations in Figure 1. A catalytic effect was indeed observed. At high concentrations, a semisolid mass could be produced at room temperature. At much lower concentrations, minimum viscosity was found to occur between 85 and 105°C. At a heating rate of 5°C·min^{-1}, this reaction continued for approximately 40°C. For all concentration levels used, the maximum viscosity occurred before 150°. Above this temperature viscosity was seen to decrease.

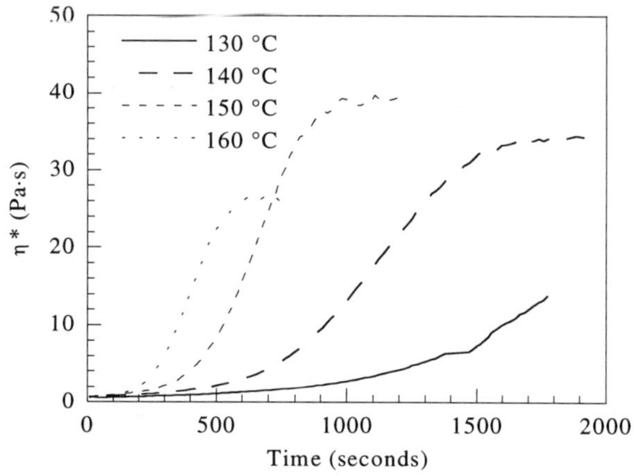

Figure 2 Isothermal conversion of CMBN/DGEBA blend

Based upon these findings, a catalyst level of 0.01% of the total resin mass was used for the trial prepregs. CMBN concentration was set at 2, 6, and 10% of the total resin mass. To produce a resin with an initially low viscosity, the base resin consisted of a 1:1 mass blend of a difunctional and trifunctional epoxide (DGEBA and TGPAP). The viscosity of these as-impregnated resins is shown in Figure 3 while the viscosity of the resins after staging is shown in Figure 4. As was expected, the viscosity of the staged resins was greater than that of the as-impregnated resins. The largest differences were observed for the resin containing 10% CMBN. For example, the viscosity of this resin at 40°C increased from 20 to 260 Pa·s while the phase lag between the applied stress and material response, a direct measure of specimen elasticity, decreased from 82 to 32°. The phase lag of the staged resins is shown in Figure 5. For the 6 and 10% CMBN prepregs, resin elasticity increased with increasing temperature. Similar levels of impregnation were observed for all three materials, with a large amount of resin present on the prepreg surface.

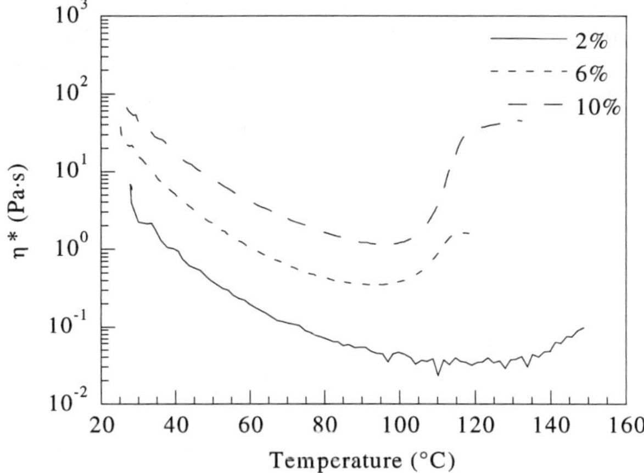

Figure 3 Viscosity of as-impregnated resins as a function of temperature (heating rate $5°C·min^{-1}$)

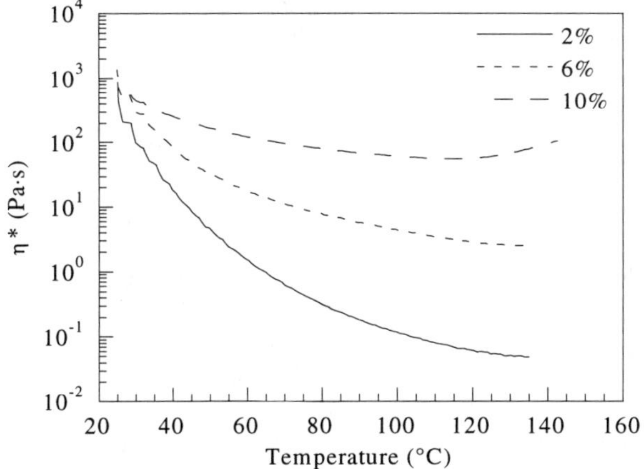

Figure 4 Viscosity of staged resin as a function of temperature (heating rate $5°C·min^{-1}$)

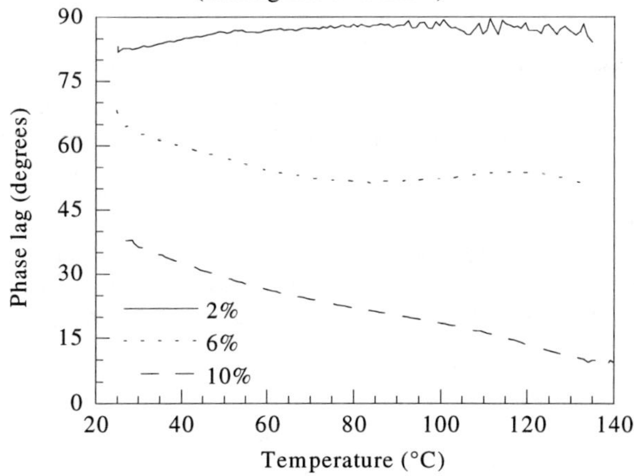

Figure 5 Phase lag of staged resin at a heating rate of $5°C·min^{-1}$

3.2 Core crush Finally, the core crushing characteristics of the experimental prepregs were evaluated through measurement of their frictional resistance. The frictional resistance of these materials is shown in Figure 6 as a function of temperature. An analysis of variance (ANOVA) showed that frictional resistance was effected by measurement temperature and rubber concentration as well as the temperature-concentration interaction, as determined by significance at the 1% level. The overall frictional resistance of the experimental prepregs is shown in Figure 7.

Of the three materials, the prepreg containing 2% CMBN had the highest overall resistance. As CMBN content was increased a decrease in frictional resistance was observed. The resistance of the 2% material was essentially unaffected by temperature. For the 6% prepreg, an increase was observed with increasing temperature while the frictional resistance of the 10% prepreg decreased with increasing temperature. For the 10% material, a minimum resistance of 75 N was observed at 79°C. The minimum resistance of the 6% prepreg was 103 N at 39°C.

In past work, it was found that minimum frictional resistance with respect to temperature in the range of 50 to 100°C correlated with the level of core crush experienced in a series of honeycomb test panels. Therefore, it is expected that the 10% CMBN prepreg would experience the highest degree of core crush while the 2% prepreg would experience the least amount of core crush. These results indicate that the level of core crush a prepreg system will experience will increase with increasing amount of solid elastomer.

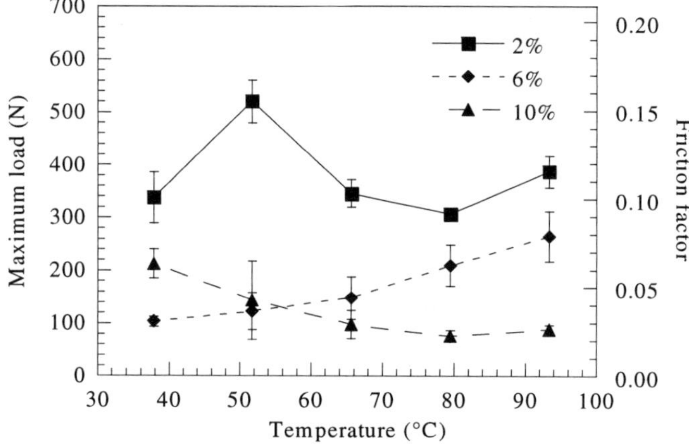

Figure 6 Frictional resistance of experimental prepregs

These differences in frictional resistance and core crush behavior can be understood through a consideration of the rheological properties of the as-impregnated resin. After the measurement of frictional resistance, the prepreg samples were examined via optical microscopy to determine the degree to which resin distribution was changed by the application of heat and pressure. While the surface resin distribution was initially the same, differences were observed after testing. It was found that for the 6% material, the high point of the weave architecture, where the fiber tows cross, contained little resin. In contrast, the 10% prepreg contained a large amount of resin in this area after testing. This is shown in Figure 8.

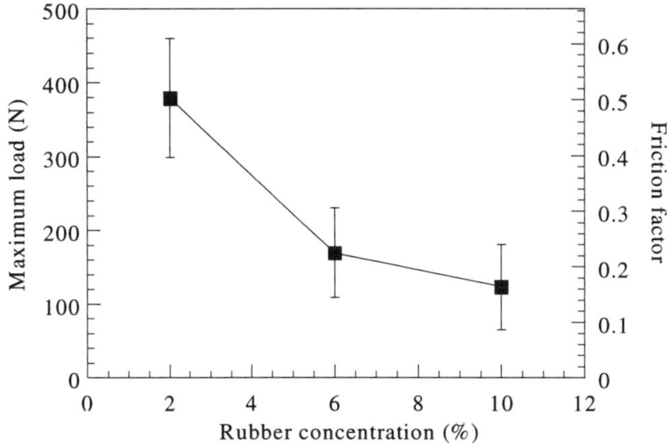

Figure 7 Mean frictional resistance of experimental prepregs

This suggests that as temperature is increased, the resin in the 6% material is forced into the interstitial areas, allowing greater fiber-tooling contact and increasing the frictional resistance. This was confirmed by the frictional behavior of the 2% modified prepreg, which had a low viscosity and a high frictional resistance. In contrast, the resin containing 10% CMBN had the highest viscosity and lowest phase lag, preventing the resin from flowing out of the fiber-tooling interface. Therefore, this resin acts as a lubricant, resulting in a decrease in frictional resistance with increasing temperature. This behavior is similar to that previously observed for the commercial controlled flow system.

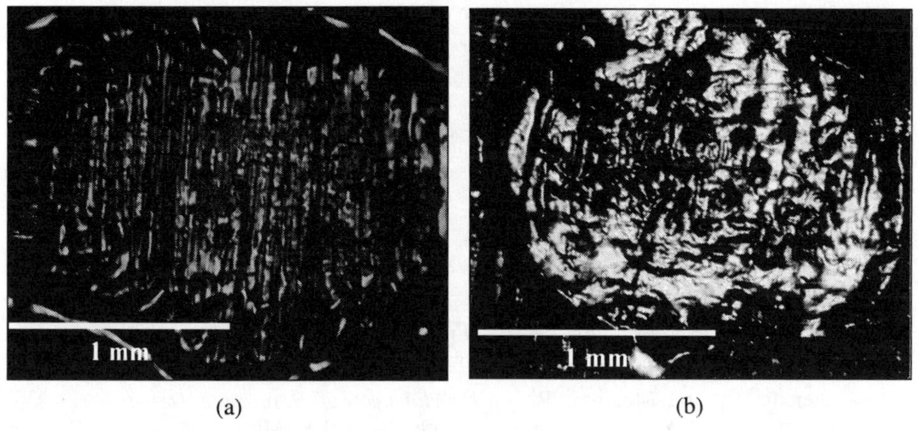

(a)   (b)

Figure 8 Micrographs of prepreg surfaces with (a) 6% and (b)
10% CMBN after frictional measurement at 79°C

4. CONCLUSIONS

Through an examination of the frictional resistance of experimental epoxy/carbon fiber prepregs with varying concentration a CMBN elastomer, it was found that an increase in CMBN concentration resulted in a decrease in frictional resistance. This decrease in frictional resistance is expected to correspond to an increase in the core crush experienced in the autoclave manufacture of honeycomb sandwich structures. It was found that as CMBN concentration increased, resin viscosity increased and the phase lag or degree of elasticity increased. Furthermore, the prepregs with high levels of CMBN exhibited a decrease in frictional resistance with increasing temperature, while the prepregs with lower concentrations showed an either an increase or no temperature effect.

These results suggest that for poorly impregnated woven fabric prepregs, an increase in core crush will be observed with an increase in elastomer concentration. It raises the possibility that crush-resistant elastomer modified prepregs can be produced by changing the degree of impregnation, allowing greater contact between the fibers and tooling, thereby increasing the frictional resistance and decreasing the incidence and severity of honeycomb core crush.

5. ACKNOWLEDGEMENTS

The authors would like to acknowledge the support of the Boeing Company through the Polymeric Composites Laboratory consortium at the University of Washington.

6. REFERENCES

1. Brayden, T. H. and Darrow, D. C., *Effect of Cure Cycle Parameters on 350°F Cocured Epoxy Honeycomb Core Panels*, 34th International SAMPE Symposium, 861 (1989).

2. Tulleau, T., Processing Studies in Advanced Composites, Masters of Science in Engineering, University of Washington (1992).

3. Renn, D. J., Tulleau, T., Seferis, J. C., Curran, R. N. and Ahn, K. J., *Composite Honeycomb Core Crush in Relation to Internal Pressure Measurement*, J. Adv. Mater., **27**, 31-40 (1995).

4. Martin, C. J., Seferis, J. C. and Wilhelm, M. A., *Frictional Resistance of Thermoset Prepregs and its Influence on Honeycomb Composite Processing*, Composites Part A, **27**, 943-951 (1996).

5. Martin, C. J., Putnam, J. W., Hayes, B. S., Seferis, J. C., Turner, M. J. and Green, G. E., *Effect of Impregnation Conditions on Prepreg Properties and Honeycomb Core Crush*, Polym. Comp., **18**, 90 (1997).

6. Ting, R. Y. and Moulton, R. J., *Fracture Properties of Elastomer-Toughened Epoxies*, 12th National SAMPE Technical Conference, 265-274 (1980).

7. Diamant, J. and Moulton, R. J., *Development of Resins for Damage Tolerant Composites - A Systematic Approach*, 29th National SAMPE Symposium, 422-436 (1984).

8. Jordan, W. M., Bradley, W. L. and Moulton, R. J., *Relating Resin Mechanical Properties to Composite Delamination Fracture Toughness*, J. Comp. Mater., **23**, 923-943 (1989).

9. Lee, W. J., Seferis, J. C. and Bonner, D. C., *Prepreg Processing Science*, SAMPE Qtr., **January** 58-68 (1986).

10. Lee, B. L., Lizak, C. M., Riew, C. K. and Moulton, R. J., *Rubber Toughening of Tetrafunctional Epoxy Resin*, 12th National SAMPE Technical Conference, 1116-1126 (1980).

11.   Drake, R. and Siebert, A., *Elastomer-Modified Epoxy Resins for Structural Applications*, <u>SAMPE Qtr.</u>, **July** (1975).

12.   U.S. Pat. 3,635,869 (January 18, 1972) Steele, R. B., Katzakian, Jr., A., Scigliano, J. J. and Burry, J. W. (to the Aerojet-General Corporation)

13.   U.S. Pat. 3,855,176 (December 17, 1974) Skidmore, P. H. (to the United States of America)

14.   U.S. Pat. 3,977,996 (August 31, 1976) Katzakian, Jr., A., Weyland, H. H., Steele, R. B. and Gold, M. H. (to the Aerojet-General Corporation)

15.   U.S. Pat. 3,978,026 (August 31, 1976) Katzakian, Jr., A., Weyland, H., H., Steele, R. B. and Gold, M. H. (to the Aerojet-General Corporation)

16.   U.S. Pat. 4,016,022 (April 5, 1977) Browning, C. E. and Reinhart, J., Theodore J. (to the United States of America)

17.   U.S. Pat. 3,962,182 (July 12, 1974) Steele, R. B., Katzakian, Jr., A., Scigliano, J. J. and Hamel, E. E. (to the Aerojet-General Corporation)

18.   U.S. Pat. 4,131,715 (December 26, 1978) Frankel, L. S. (to the Rohm and Haas Company)

19.   Smith, J. D. B., *Novel Epoxy-Anhydride VPI Resins Containing Chromium Latent Accelerators*, <u>18th Electrical Electronics Insulation Conference</u>, 262-268 (1987).

DAMAGE TOLERANCE EVALUATION OF COMPOSITE HONEYCOMB STRUCTURES

Ric Abbott, Snr Principal Engineer
Raytheon Aircraft Company

ABSTRACT

The properties of honeycomb composites enable the design of aerospace structures which have excellent potential for reduced manufacturing cost based on lower part count and elimination of joints and subassembly operations. The damage tolerance properties of these structures must be taken carefully into account in order to supply products which are robust, safe and long-lasting. This paper gives the background leading to the latest aerospace composite honeycomb manufacturing techniques and provides examples of (commercial aircraft) damage tolerance evaluation applied to those structures. A brief description of in-service experience of durability and repairability are also included.

KEY WORDS: Damage Tolerance, Honeycomb, Composites.

1. INTRODUCTION

The first airplanes were built of wood, fabric, and resin. In a way, today's composite airplanes are returning to those basics, except now, the fibers are carbon and Kevlar and these are set in high temperature curing epoxy resins. Also, extensive use is being made of honeycomb sandwich structures. In earlier times sandwich cores would be of balsa or plywood, whereas now the cores are of aramid, fiberglass, and even carbon fiber. The benefits of modern composite sandwich construction are obvious: low weight, high bending stiffness, and the ability to fabricate very large structures with compound curvature. These may be cured in a single piece, eliminating parts, joints, sub-assemblies, and associated inspection costs. The civil airplane certification of such structures involves all the strength, stiffness, and damage tolerance evaluations normally applied to metallic structures; however, in damage tolerance evaluation of composite honeycomb structures, although the same *principles* apply as those for metallic structures, the application of these principles must take into account the particular properties of composite honeycomb structures.

It is the purpose of this paper to introduce to the reader the latest composite honeycomb manufacturing practice and to give examples from a typical damage tolerance evaluation program such as will be required to certify commercial airplane composite honeycomb primary structures.

2. MODERN HONEYCOMB STRUCTURES

Composite honeycomb structures have been used extensively on commercial airplanes for fairings, control surfaces, radomes, and leading and trailing edge structures. In recent years, in pursuit of lower weight *and* lower cost, composite honeycomb has found its way into aircraft primary structures. The Beech Starship, the first all carbon fiber airplane to be FAA certified, was designed to take advantage of the benefits of sandwich construction with an essentially all carbon fiber honeycomb airframe. In fact, the main fuel tank and the control surfaces were the only parts not of honeycomb construction. However, the cost of the airframe was excessive because of the labor intensive nature of the hand lay-up methods and multiple sub-assemblies employed at the time.

Times have changed, and a new business jet, the Raytheon Premier I, is under development. This employs carbon fiber composites with updated manufacturing methods in construction of the fuselage, horizontal tail, vertical tail, flaps, ailerons, and spoilers, see figure 1. The fuselage is the major component to be of carbon fiber honeycomb construction. As part of the plan for greater efficiency, the fuselage shells are fabricated on special tooling and utilize the Cincinnati-Milacron *Viper* Automated Fiber Placement (AFP) system. The tooling and the system allow fabrication of the fuselage shell in just two pieces--the forward fuselage (pressure cabin and nose section) and the aft fuselage (the non-pressurized section supporting the tail and engines and containing most of the of the airplane systems as well as the main baggage compartment).

The AFP system is controlled by *Acraplace* software which translates the CATIA solid models representing the fuselage composite design into precise three dimensional instructions. This enables the machine to place narrow tapes of carbon fiber prepreg in exact position on a male tool to achieve maximum strength with minimum weight. The machine builds up carbon fiber plies for the inner and outer face sheets in between which is the honeycomb core. The final ply on the outer surface is a hybrid carbon fiber/aluminum woven fabric for lightning strike protection which is, at present, manually placed.

Figure 1
Major Composite Structures on the Raytheon Premier I

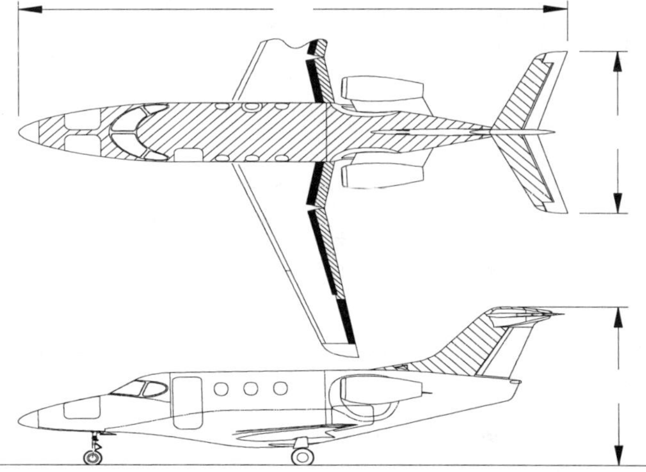

This type of construction exhibits high resistance to bending and buckling and therefore does away with the maze of internal frames, stringers, longerons, rivets, doublers and sub-assemblies seen in conventional thin skin assembly. It is in fact a true monocoque structure with the required strength and stiffness built into the shell itself.

When each fuselage section is finished on the AFP system, it is transferred to a different tool for autoclave cure. In the case of the forward fuselage the result is a pressure cabin formed in a single piece--this includes integral window, windshield, and door frames, with no joints, rivets or leak paths. This cabin weighs about 20% less than the equivalent structure in a metallic business jet. See figure 2.

Figure 2
Raytheon Premier I Forward Fuselage

3. DAMAGE TOLERANCE EVALUATION

3.1   Regulatory Basis   Damage tolerance evaluation has been the norm for Transport Category Airplane structures (metal or composite) certified under Part 25 of the Federal Aviation Regulations since the late 1970's. But the Starship was the first airplane to be certified to damage tolerance requirements under Part 23 Small Airplane regulations. Raytheon engineers worked in cooperation with FAA specialists to establish Special Conditions for Fatigue and Damage Tolerance Evaluation which were first published for application specifically to the Starship in 1986. These conditions have since been codified into the main body of Part 23, Federal Aviation Regulations.

The intent of damage tolerance evaluation is the same regardless of the size of the airplane, even though the regulations may contain different wording. In general terms, the intent is to ensure long term safety based on published inspection procedures considering manufacturing quality intrinsic to the processes used and recognizing that certain damage may occur during service.

3.2 Typical Damage Scenarios and Related Requirements Three different damage scenarios will normally be considered:

3.2.1 Scenario 1, Initial Quality This covers items intrinsic to the manufacturing process and the inspection standards: Scenario number 1 represents the as-delivered state and therefore the structure must be capable of meeting all requirements in terms of strength, stiffness, safety, and longevity.

3.2.2 Scenario 2, Damage During Assembly or Service Damage from scenario number 2 must exhibit predictable growth, or no growth, during a period of in-service loading (usually expressed in number of inspection intervals) and must be detectable by the specified in-service inspection methods. Also, the residual strength of the structure with such damage must always be at least equal to the applicable residual strength requirements.

3.2.3 Scenario 3, Damage from Discrete Sources Damage resulting from scenario number 3 will be obvious to the crew during a flight (or be detected during a preflight inspection) and therefore a specific set of residual strength criteria apply which are concerned with safely completing a single flight.

3.3 Damage Source and Modes Up this point no details of damage mode, damage magnitude, or structural response have been discussed. The author believes that it simplifies the evaluation to first recognize the generic scenarios and potential damage sources, and from those identify the possible damage modes and the desired structural response.

From the above definitions it is not too difficult to build up a matrix such as the one shown in Table 1.

Table 1
Example Matrix: Damage Source and Potential Modes

SCENARIO 1		SCENARIO 2		SCENARIO 3	
Source	Damage Modes	Source	Damage Modes	Source	Damage Modes
Manufacturing process	Small imperfections within the inspection sensitivity and acceptance criteria: - Porosity - Voids - Disbonds	Tools Baggage Hail Gravel	Resin cracking Delamination Core crush Puncture	Severe lightning	Plies burned Puncture
		Lightning	Resin burn Delamination Loose rivets	Bird strike	Delamination Core crush Puncture
		Water intrusion	Core cell damage	Rotor burst	Puncture Severed elements
		Cyclic loading	Delamination growth Disbond growth	Engine fire	Resin burn Delamination
		Bleed air	Resin burn	Ground equipment Hangar doors	Puncture

The damage modes from scenario 1 are typically not a significant problem from the load capability point of view. However, the potential modes from intrinsic manufacturing quality must be identified and controlled by the manufacturing specifications and acceptance criteria. Given this, it is usually easy to demonstrate that these small imperfections will not grow under cyclic loads typical of commercial airplane service.

Scenario 3 imparted damage is at the opposite end of the scale: these modes of damage are easily detectable and will need attention before further flight (except maybe for an authorized ferry flight to a repair facility). So inspection and longevity under cyclic loads are non-issues.

The scenario which creates the most need for investigation is scenario 2, and a typical test program is described in the following section.

3.4 Element Testing To evaluate composite honeycomb structural performance under the various damage modes, element testing is usually performed. It is *possible* to conduct these evaluations on the full scale test articles, but this is a risky approach and the results will be too late to guide the design to a minimum weight and cost configuration.

3.4.1 Static Tests Testing to validate tolerance to the damage modes in scenario 2 will include impact tests without puncture, puncture tests of detectable size and larger, water ingression tests with freeze/thaw cycles, and lighting strike tests. Strength testing will be performed for the failure modes shown to be critical based on the internal loads analysis, typically a finite element analysis.

The static strength portion of the element test matrix is shown in table 2:

Table 2
Element Test Matrix--Static Loading

TEST TYPE/ DAMAGE MODE	TENSION (Fuselage Top)		COMPRESSION (Fuselage Bottom)	SHEAR (Fuselage Side)
	Hoop	Longitudinal		
Virgin	3	3	12	3
Impact	3	3	3	3
Detectable Puncture	3	3	3	3
Large Puncture	3	3	3	3

NOTE: Numbers in cells indicate number of replicates.

A larger number of undamaged specimens may be tested at a selected loading in order to validate the laminate analyses by comparison of mean and B-Basis test results to analytical predictions. This may also be desirable in order to establish that undue variability is not introduced by a particular manufacturing process.

3.4.2 Cyclic Tests The test matrix for cyclic loading follows the same pattern, except now loading at multiple stress levels is desirable to establish the sensitivity of flaw growth to cyclic stress level. Again an increased number of specimens may be tested at a selected condition to identify variability, in this case as it affects flaw growth. Generally, cyclic testing of undamaged composite panels is not of great interest. Also, composite honeycomb panels are less sensitive to flaw growth under tension loading. This insensitivity can be demonstrated by testing at a single constant amplitude of 67 percent of the maximum stress test result from similar specimens under static tension loading. See table 3. In addition to constant amplitude stress level testing, spectrum loads representing lifetime varying amplitude loads should be tested as there is, today, no industry-wide acceptance of analytical methods predicting flaw growth rates under lifetime variable amplitude loading.

Table 3
Element Test Matrix - Cyclic Loading

	TENSION			COMPRESSION				SHEAR		
STRESS LEVEL	1	2	3	1	2	3	Spectrum Loading	1	2	3
Impact	3			3	3	3	3	3	3	3
Detectable Puncture	3			3	12	3	3	3	3	3

NOTE: Numbers in cells indicate number of replicates.

3.4.3 Pressure Cabin Shell Residual Strength Honeycomb construction has a particular advantage in maintaining residual strength after incurring large size damage from sources such as those described scenario 3. This is due to the honeycomb shell stiffness imparting great resistance to crack bulging which in thin skin structures is a source of high crack extension forces. Tests to validate residual strength in the presence of large puncture damage are usually conducted on cylinder wall samples loaded to simulate internal pressure or a combination of pressure and shear.

3.5 Test Results Selected examples of element test results in typical presentation formats are shown in the following figures. The results shown were obtained from samples representing fuselage shell construction on a business jet. However, scale matters, and different results might be obtained from tests on samples representing large transport airplanes because of different face sheet thicknesses and core densities required to carry the basic pressure and bending loads.

Figure 3
Tension - Hoop

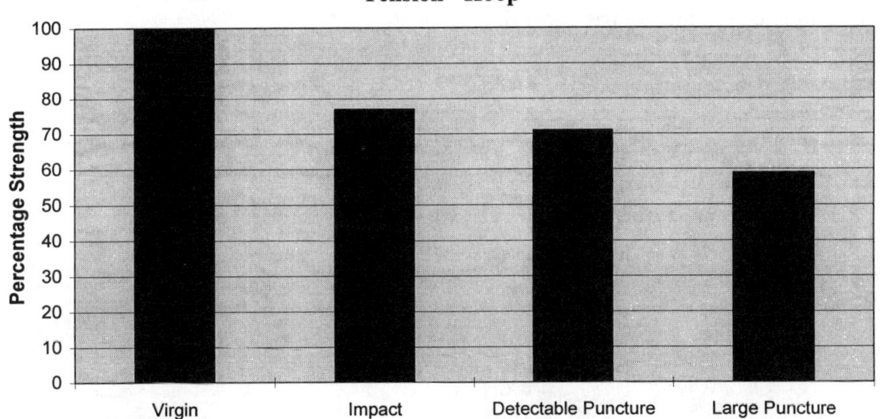

3.5.1 Tension From figure 3, hoop tension loading from internal pressure, it's clear that designing for damage tolerance need not impose a serious weight penalty. Ultimate design pressure must be carried with the undamaged panel and this means that just a little additional material will enable the panel to meet the required residual strength load with large puncture damage. This is because residual strength required for the pressure case is about 60 percent of the ultimate pressure. In the case of longitudinal tension loading from fuselage bending, figure 4, the residual strength requirement is limit load, ie., about 67 percent of the ultimate pressure, and again to carry that load with a large puncture a small amount of material must be added. In both cases a more robust structure would result if ultimate loads were to be carried with impact damaged panels. And this may be required by the regulations unless impact damage is readily detectable.

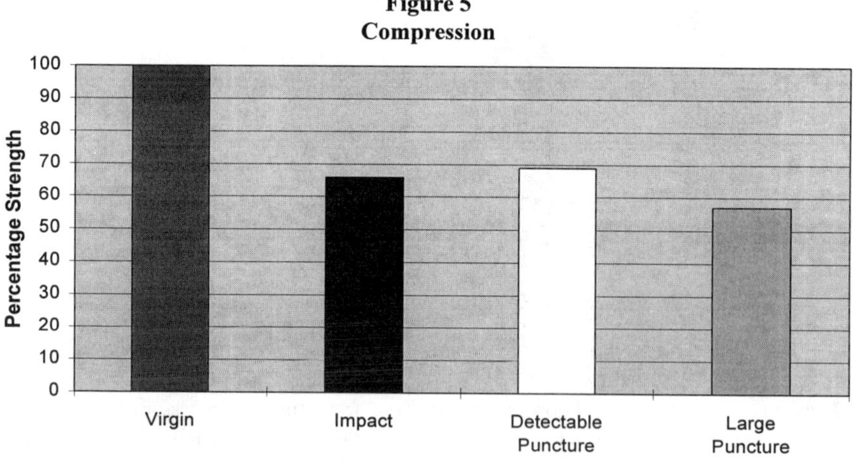

Figure 4
Tension - Longitudinal

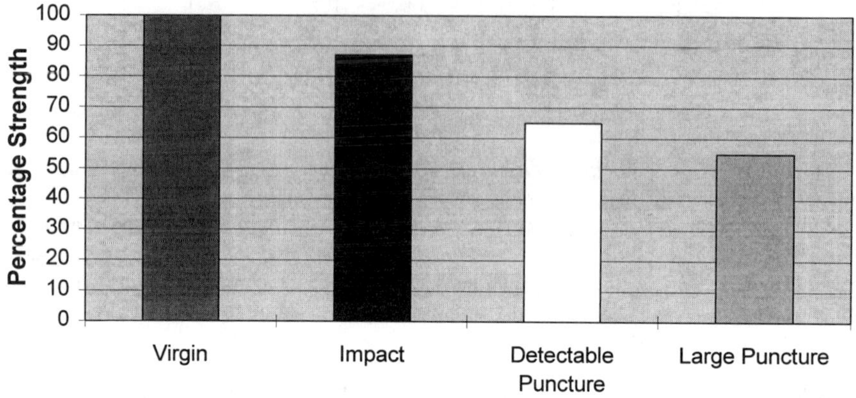

3.5.2 Compression As shown in figure 5, a similar situation exists in the case of compression loading from fuselage bending. Designing to the residual strength requirement with large puncture damage would imply using approximately 85 percent of the allowable virgin strength; but if impact damage is to be good for ultimate load then only 65 percent of the virgin strength will be used. This may not be a serious penalty as the maximum compression loads occur in the fuselage bottom from down-bending load cases and the lower fuselage is usually reinforced by cargo or passenger floor structure.

Figure 5
Compression

3.5.3 Shear Maximum shear loading occurs along the side of the fuselage. Again, designing to carry limit load (67 percent of ultimate load) with large puncture damage is a slight weight penalty, approximately 82 percent of the maximum virgin strength can be used. But in this case, 82 percent of virgin strength will accommodate impact damage at ultimate load.

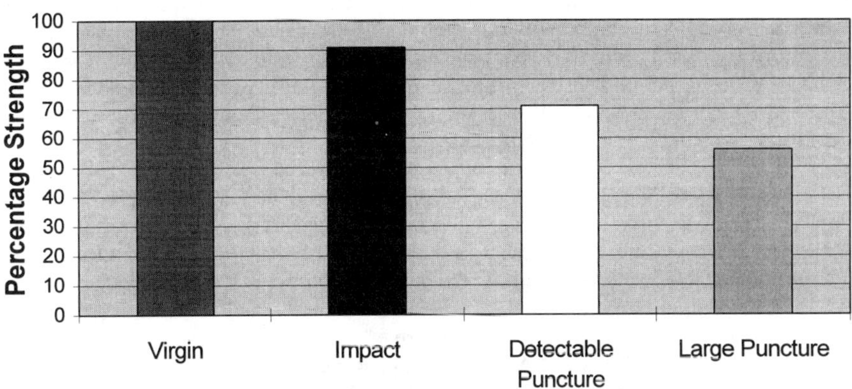

Figure 6
Shear

3.5.4 Pressure Cylinder Results Figure 7 presents an assemblage of tests results illustrating the trend from no damage to massive damage. Massive damage being the type of wall puncture that could only occur from a serious collision with ground support equipment such as steps, generator carts, refueling equipment, baggage handling equipment, and so on. As mentioned previously, this type of damage should be obvious and should be detected before flight, but, just in case... tests were conducted to determine cylinder wall residual strength with massive damage. The trend revealed has proven typical of test results from big and small airplane honeycomb pressure containment structure in that with larger and larger damage a residual strength threshold becomes apparent. This confirms the benefit of honeycomb structure in resisting skin bulging and uncontained damage growth.

Figure 7
Cylinder Wall Test Results
Tension - Hoop

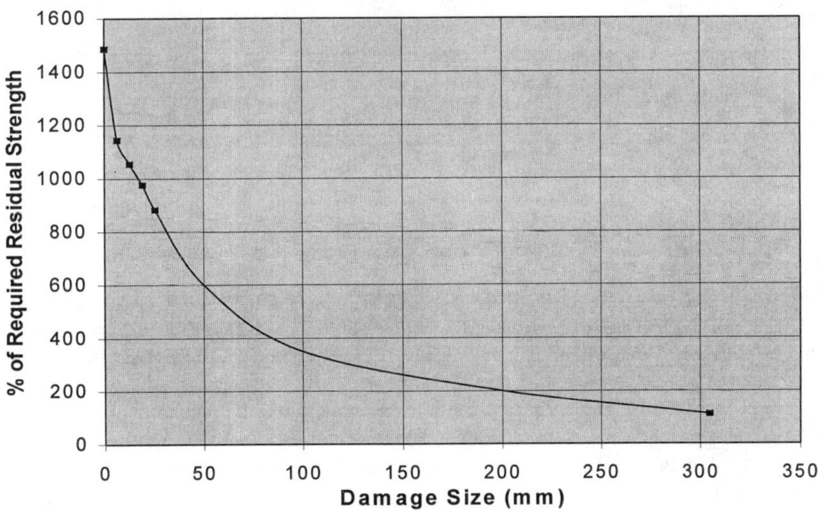

383

3.5.5 Cyclic Test Results An example is presented in figure 8 of typical constant amplitude test results for large puncture damage under constant amplitude compression loading. The most useful presentation for these data is as shown, ie., a best-fit straight line. This type of plot can be used to assess important damage tolerance characteristics of the materials tested. The significance of scatter in flaw growth life can be assessed by plotting a B-Basis stress-life line assumed to be parallel to the mean life plot. The flaw growth threshold may then be determined by extrapolation of the B-Basis life line to ten million cycles (100 million in the case of rotating equipment with high cycle loading).

A prime use of these data is to structure the full scale cyclic test to run in an economical, yet rational, manner. All stress cycles below the flaw growth threshold be may eliminated from the full scale test spectrum. Also, scatter in flaw growth life may be accounted for by a factor increasing the applied loads and therefore reducing the number of test lifetimes.

Figure 8
Constant Amplitude Cyclic Test Results -
Large Puncture Damage Under Compression Loading

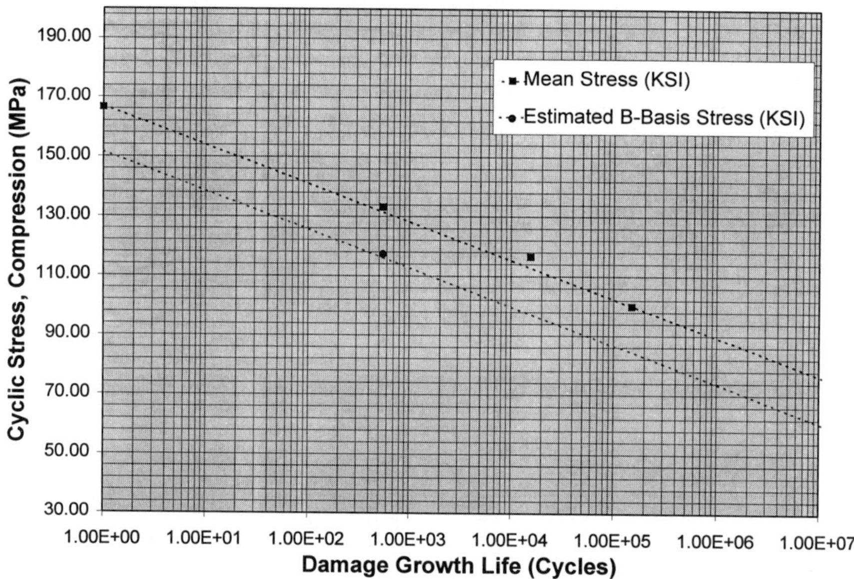

3.6 Full Scale Tests
*** 3.6.1 Full Scale Cyclic Tests*** These days, certification of major load carrying structure requires that components such as wings, fuselages, and tail structures are tested through a sequence of loads representing at least two lifetimes of expected missions. Each lifetime consists of thousands of load cycles, including wing lift, fuselage reactions, tail loads, pressure cycles, and landing loads. During these tests damage will be mechanically inflicted in the structure to simulate in-service damage. These situations include lighting strike, hail damage, runway damage, and tool impacts. These damage modes will be tested through as much as one lifetime of fatigue testing to prove that the structure is, in fact, damage tolerant in the full size articles, ie., that damage will not grow in an unpredictable manner and will always be detected by the inspections specified.

Larger damage may be inflicted later in the full scale cyclic testing to simulate impacts with ground service equipment, impacts with hangar doors and other aircraft (hangar rash) and poor maintenance practices; all bad things that occasionally happen in the loading, handling, and maintenance of commercial airplanes. The larger damage modes should be detected before the next flight and so the demonstration for these modes may consist of a relatively few flights of cyclic loading and inclusion in the residual strength tests.

3.6.2 Full Scale Residual Strength Tests After completion of the lifetimes of cyclic testing, the major components of wing, fuselage and tail will be subjected to load tests to verify that, in spite of all the load cycles and inflicted damage, the remaining structure will still carry the required residual strength loads (flight loads and or pressure loads expected to be encountered during the service life of the aircraft, except for the larger damage modes which are associated with specific residual strength criteria).

3.7 Continued Airworthiness Inspections Based on interpretation of the test results obtained, inspection procedures, threshold time, and frequency of inspections will be established and published in the airplane manuals. A factor is usually applied so that allowance is made for the damage to exist over several inspection intervals, depending on the criticality of the structure. A further factor may be needed to account for scatter revealed in the flaw growth test results.

4. SERVICE EXPERIENCE

The service experience with composite structures in business aviation has been excellent. Starships have been flying since the late 80's and no problems with major structure have been encountered. Composite stabilizer structures have been in use on model 1900 commuters flying typically 2500 hours per year since the mid 80's; again no problems related to the composites have been reported.

Safety in the event of an emergency landing has also proven to be outstanding. A nose landing gear collapsed during a landing of one of the Starship test airplanes; the airplane was flown home and was back in service in 10 days. The repairs were made by procuring blank parts from the factory, cutting out the needed replacement sections, and splicing these into place by bonding and fastening.

An even more spectacular event in occurred Denmark in February, 1994. Starship number 35 ran off the runway into a snow bank at approximately 130 mph after an aborted take off. Crew and passengers were shaken but otherwise unhurt. No fuel was spilled, no seats came loose, no windshield or window glass was broken, or even cracked, and the cabin was undistorted enabling the cabin door to open normally. The crew and passengers unbuckled their seat belts and walked away. The right hand main gear collapsed, the other main gear and the nose gear were sheared off (not torn out, but the aluminum forgings severed) from the force of hitting the snow bank at high speed. The right hand wing tip was dragged along the ground and as a result suffered damage to the flaps, tip sail, and rudder. The nose section was damaged by the nose gear being severed and forced upward into the structure. The cabin underbelly was crushed through skidding along without the landing gear but the damage was localized to the area between the seats.

A team was sent to survey the damage and list the replacement parts needed. Later the airplane was repaired on-site by a crew of five technicians plus one engineer, one inspector, and one service manager. Some parts with localized damage were repaired using techniques published in the Starship Structural Repair Manual which allows damage to be repaired on-site by trained service staff. For more extensive damage, blank parts were delivered from the factory and were used as stock from which to cut replacement panels which were then bonded and/or fastened into place. Of course, aircraft systems such as landing gear, hydraulics, antennae, etc., were simply replaced with factory spare parts. The repairs were finished and the airplane rolled out for flight test in July of 1994. This was ahead of schedule and under budget, much to the surprise of the insurance company and the Danish aviation authorities who were both convinced that a metal airplane would have suffered much greater damage and would have been totaled by such an incident.

5. CONCLUSIONS

Modern manufacturing methods enable the fabrication of composite honeycomb primary load carrying structures for commercial aircraft use which are low cost as well as low weight. These structures require a damage tolerance evaluation for certification to Part 23 or Part 25 of the FAA regulations. A rational damage scenario and supporting element test program will considerably assist the damage tolerance evaluation.

Composite honeycomb structures can be designed to tolerate large puncture damage with little weight penalty. It may not be required (by regulation) to design to carry ultimate loads after impact damage (it depends on the inspections specified). However, a more-economical-to-operate product will probably result when composite honeycomb structures are designed to carry ultimate load with impact damage, in fact, the FAR 23 regulations are quite specific in this area and composite structures on the Premier I are designed for ultimate load with impact damage.

Composite honeycomb structures are relatively insensitive to cyclic loading, and a flaw growth threshold may be defined from the test results. Scatter in flaw growth life should be examined in order to establish a relationship between test lifetimes and service lifetimes for full scale cyclic testing. This will also enable the full scale test to be conducted more economically than on equivalent metal structures.

With the combination of careful analysis, rational testing, and advanced manufacturing techniques we can expect to see further applications of composite honeycomb primary structures.

INVESTIGATION OF THE THERMAL AGING BEHAVIOR OF CYANATE ESTER/CARBON FIBER COMPOSITE

Kimo Chung and James C. Seferis
Polymeric Composites Laboratory, Department of Chemical Engineering
University of Washington, Seattle, Washington

ABSTRACT

A fundamental problem of advanced airplane composite systems is the lack of understanding of the aging process and how it affects the material properties associated with degradation. A cyanate ester resin/carbon fiber composite system was utilized to analyze long-term properties in an accelerated aging environment. The change in viscoelastic properties of cured laminates in the range of the glass transition temperature was studied. Dynamic mechanical properties were evaluated by Dynamic Mechanical Analysis (DMA) through time-temperature equivalence. Consequently, a modeling methodology has been developed which provides an understanding of the aging process of fiber reinforced composites quantitatively in isothermal environments. Collectively, this work was focused on providing an understanding of long-term properties of polymers and composites in relation to their accelerated aging behavior.

KEY WORDS: Composite Systems, Novel Cyanate Ester, Viscoelastic, Time-Temperature Equivalence, Dynamic Mechanical Analysis, Modeling Methodology

1. INTRODUCTION

Polymeric composites are becoming more important for advanced aerospace applications primarily due to their high strength-to-weight ratio. The most common materials currently used in aircraft composite structures are epoxies because of their wide range of properties and ease of processing. As supersonic commercial aircraft is being considered, high temperature polymeric composites with matrices such as bismaleimides and cyante esters are under consideration. Cyanate ester resins are regarded as advanced engineering materials because they have high performance properties such as toughness, Tg, thermal properties, superior adhesion, and low dielectric properties as cured thermosets. The resins have already been used in the electronic industry, and also considered as a optimum candidate for radome structures in the aircraft industry.(1) A fundamental problem that still remains, however, is that the long-term properties of composites are often inferior to metals such as aluminum, titanium. Accordingly, in the aircraft industry there is a tremendous need for reliable prediction of the maximum lifetime for polymeric composites which have been exposed to extreme environmental conditions such as hygrothermal cycling, load cycling, ultraviolet radiation, jet fuel, hydraulic fluid, and paint stripper during the aircraft service.(2)

When the matrix candidates are selected for a specific application, evaluation and validation processes should be performed to predict long-term properties with respect to temperature and moisture since they can have detrimental effects on polymer composites (3). In this study, cyanate ester resin/carbon fiber prepreg was developed and used to investigate long-term property change after exposure to temperature near the glass transition. Master curves focused to the glass transition temperature of unaged material were obtained for both unaged and aged specimens, and the modeling methodology was applied to quantitatively evaluate the degree of aging on the specimens in the temperature environment. Collectively, in this study, the concept of time-temperature equivalence was reviewed and applied to aging of composites.

2. BACKGROUND / THEORY

Since the relaxation time of polymers can be significantly long, it may be a time consuming process to investigate the long-term properties of viscoelastic materials. However, it can be more readily measured if a variety of temperatures with a fixed time scale are used, which is the basis for time-temperature superposition.(4,5) With this theory, the viscoelastic relaxation phenomenon can be studied systematically by dynamic mechanical experiments that vary the time or frequency of testing. The master curve can be accomplished at specific temperature for the study on the properties of the material. The shift factor is thought to be proportional to the time required for relaxation of applied stress, therefore, it is expressed as,

$$a_T = \frac{\tau(T)}{\tau(T_g)}$$

[1]

where a_T : Shift factor

    $\tau(T)$: Relaxation time at temperature T

    $\tau(T_g)$: Relaxation time at temperature T_g

The shift factor may be unity if T_g is considered as the reference temperature, T_0. William-Landel-Ferry theorized that most polymers have the same master curve and the shift factor can be described as a function of temperature as below.(4)

$$\text{Log}[a_T] = \frac{-C_1(T - T_0)}{C_2(T - T_0)}$$

[2]

where C_1, C_2 are constants, and T_g is taken as a reference temperature. In addition, the shift factor can be related to an activation energy through kinetic theory and described by using an Arrhenius type equation.

$$\text{Log}[a_T] = \frac{E}{2.303R}\left[\frac{1}{T} - \frac{1}{T_0}\right]$$

[3]

As a general rule, the WLF equation commonly fits the experimental results above T_g whereas the Arrhenius type equation fits below the T_g, and the activation energy for the WLF relationship can be described by C_1, C_2 at the reference temperature as follows.(4)

$$E = 2.303RT_0^2 C_1 / C_2$$

[4]

In this study, the standard linear solid (SLS) model is used based on the mechanical model having one spring and two dashpots. The relation between stress and strain can be described as shown in the differential equation. (5)

$$\sigma + \tau\left(\frac{d\sigma}{dt}\right) = G_r\varepsilon + G_u\tau\left(\frac{d\varepsilon}{dt}\right)$$

[5]

where G_r : Relaxed modulus

    G_u : Unrelaxed modulus

τ : Relaxation time ($\tau = \mu/(G_u\text{-}G_r)$)

If an oscillatory mechanical condition is applied to the equation, the complex modulus (G^*), composed of the storage modulus (G') and loss modulus (G''), is obtained and it provides an approximate representation of the observed behavior of polymers in their viscoelastic range.

$$G^* = G' + iG'' = G_u - \frac{G_u - G_r}{1 + i\tau\omega} \qquad [6]$$

Dillman and Seferis suggested a modified model of the standard linear solid model using an α parameter to consider asymmetric relaxation time distribution and a β parameter to consider symmetric relaxation time distribution.(6,7)

$$G^* = G' + iG'' = G_u - \frac{G_u - G_r}{\left[1 + (i\tau\omega)^\beta\right]^\alpha} \qquad [7]$$

where α is a shape parameter that represents asymmetric relaxation time distribution, and β represents the symmetric relaxation time distribution ranging from 0 to 1, respectively. Then Equation [7] can be arranged to represent storage modulus (G') and loss modulus (G'') as follows.

$$G' = G_u - \frac{(G_u - G_r)\cos(\alpha\theta)}{\left[1 + 2(\tau\omega)^\beta \cos(\beta\pi/2) + (\tau\omega)^{2\beta}\right]^{\alpha/2}} \qquad [8]$$

$$G'' = \frac{(G_u - G_r)\sin(\alpha\theta)}{\left[1 + 2(\tau\omega)^\beta \cos(\beta\pi/2) + (\tau\omega)^{2\beta}\right]^{\alpha/2}} \qquad [9]$$

$$\text{with} \quad \tan\theta = \frac{(\tau\omega)^\beta \sin(\beta\pi/2)}{1 + (\tau\omega)^\beta \cos(\beta\pi/2)} \qquad [10]$$

3. EXPERIMENTAL

The carbon fiber reinforced material investigated in this study included a cyanate ester resin obtained from Ciba Geigy and a carbon fiber (T800) from Toray. Ten plies of prepreg were consolidated for each laminate, cured in an autoclave at 177°C for 1 hour, and postcured at 210°C for 1 hour and 250°C for 2 hours according to the material specifications.(8) The heating/cooling rate was constant at 2.77°C/min. A consolidation pressure of 100 Psi (690 KPa) was utilized for laminate curing.

The laminates were cut into a group of specimens measuring approximately 16.05 x 9.5 x 1.95 mm. A group of specimens was put into a Lindberg tube furnace and thermally aged for 10, 20 and 30 hours, respectively, at 275°C with an air flow rate of 250 ml/min. Dynamic mechanical experiments were carried out with a TA Instruments DMA 983 apparatus coupled to a TA Instruments 2100 Thermal Analyzer for data acquisition and reduction. Seven oscillation frequencies were used including 0.01, 0.04, 0.1, 0.3, 0.5, 1, 3 Hz. Temperature was increased in 5°C/min steps from 100°C to 350°C. The oscillation amplitude was set to 0.15 mm and nitrogen gas was used with a flow rate of 300 ml/min for the thermal aging study to prevent oxidation during the DMA experiments.

4. RESULTS AND DISCUSSION

4.1 Material Evaluation   Figure 1 shows the cross-section of a cyanate ester resin/carbon fiber cured laminate taken by optical microscopy.   It was found from the figure that the interface between resin and fiber was uniform and the void contents was less than 2 %. Figure 2 shows the storage modulus and Phase lag as a function of temperature at 1 Hz.   The storage modulus was found to be constant up to 150 °C and decrease slightly up to 220 °C. The glass transition of an unaged specimen was observed at 275°C from the tan-δ peak at 1 Hz.   As aging proceeded, the glass transition of the material as determined from the peak of tan-δ was found to decrease dramatically, from 275 °C to 190 °C within 30 hours, implying that the relaxation time was decreased by the aging process.(7)   The decrease in glass transition temperature may attributed to the degradation of epoxy sizing on the surface of carbon fibers during the aging process.   The other reason may be related to the bulk matrix degradation that causes the breakage of chain bonds reducing crosslinking density in the matrix and giving the lower glass transition to the material, or degradation of the materials unreacted in the matrix during the aging process.   More work is necessary to clarify the phenomena.

4.2 Master curve construction   Figure 3 shows the storage modulus (G') of unaged specimen measured using DMA.   The reference temperature was chosen for the modulus master curve construction, and horizontal shifts were applied on a logarithmic time scale, so that the modulus curves from other temperatures join as smoothly to the curve at the reference temperature.   The glass transition temperature, 275°C, was chosen as a reference temperature for the unaged specimen from the tan-δ peak at 1 Hz for this study.   The shift factor was evaluated by the ratio of relaxation times at the temperature considered and at the reference temperature using Equation [3].   After performing the horizontal shift for the glass transition modulus in Figure 3, the activation energy for the glass transition was evaluated at 275°C. The same steps were applied for the aged specimen for 20 hours to evaluate activation energies and construct the master curves.   Figure 4 shows the master curves of both unaged and aged specimens drawn with above procedures.   Comparing the two master curves, it was found that the configuration was changed due to the thermal aging effect as well as the changes glass transition temperature of the material.   It may be analyzed by viscoelastic modeling of these master curves.

4.3 Modeling of Master Curves   The relaxation time was assumed to be a function of temperature and calculated to evaluate the effect of thermal aging by rearranging Equation [1] as below.(9)

$$\tau = \tau_0 a_T = \tau_0 \exp\left[\frac{E}{2.303R}\left(\frac{1}{T} - \frac{1}{T_0}\right)\right] \qquad [11]$$

The a_T and the temperature dependence of τ were determined from the master curve construction using Equations [3] and [11].   Equation [8] may be rearranged using an h parameter which is a function of $\tau\omega$, α and β for the modeling giving

$$G' = G_u - (G_u - G_r)h(\tau\omega, \alpha, \beta) \qquad [12]$$

$$h(\tau\omega, \alpha, \beta) = \frac{\cos(\alpha\theta)}{\left[1 + 2(\tau\omega)^\beta \cos(\beta\pi/2) + (\tau\omega)^{2\beta}\right]^{\alpha/2}} \qquad [13]$$

It can be seen that G' has a linear relationship with h($\tau\omega,\alpha,\beta$) if h is determined by proper selection of the parameters, $\tau\omega$, α and β.   A trial and error method was used to evaluate the parameters.   Figure 5 shows the procedure to evaluate the parameters, τ_0, α and β, for an unaged specimen using the linear relationship between G' and h.   The pre-exponential factor, τ_0, has a characteristic for the master curve to shift in parallel to log ($\tau\omega$) axis and it could also be evaluated from the relationship. G_u and G_r can be evaluated from the line in Figure 5 by using Equation [12] after the parameters τ_0, α and β has been set. The unrelaxed modulus, G_u, represents the modulus at high frequency or low temperature, where the material is in the glassy state, while the relaxed modulus, G_r, represents the modulus of the material at low frequency or high temperature, where the specimen is in the rubbery state. They may be assumed to be constant throughout the relaxation for such a reversible process as glass transition. As can be seen in Figure 4, however, G_r was found to continuously decrease over the experimental temperature range in this study, therefore G_r was assumed to be a function of frequency for the modeling purpose. Same procedure was performed to evaluate the parameters for aged specimen. Figure 6 shows the comparison of the model with the master curves for both the unaged and aged specimens as a function of frequency (1/f). The model description is in good agreement with the data in the transition region of the storage modulus. Table 1 shows the values of all parameters evaluated in this study. It was found that value of α, which represents asymmetric relaxation time distribution, increased, and β, which represents symmetric relaxation time distribution, decreased as the aging proceeded. It can be concluded that the aging process affects molecular structure of the matrix resin. In the result, the asymmetric form of the master curves was the results from an increase in α and decrease in β decreases as the aging proceeds. From the results, it was proved that the change in the values of the model parameters could express the degree of thermal aging of the specimens quantitatively used.

5. CONCLUSIONS

Cyanate ester/carbon fiber cured laminates were prepared and their aging behavior in an air atmosphere was investigated.

Optical microscopy revealed that the interface between resin and fiber was uniform and void content in the resin matrix was less than 2 volume % in the cured specimens. Master curves were constructed to compare the aging effect of the specimens. As the aging proceeded, the glass transition temperature was found to decrease dramatically, implying that the relaxation time decreased by the aging process. The model used to fit the data was in good agreement and described the aging effect successfully in a quantitative manner.

6. ACKNOWLEDGMENT

The authors would like to express their appreciation to the U.S. Air Force Office of Scientific Research, who provided financial assistance for this work through AFOSR Grant Number F49620-97-1-0163.

7. REFERENCES

1.   Ising, S.J., Z.F. Crawley, and G.L. Merriman, 5th Int'l SAMPE Elec. Conference, Jun. 18-20, 286, (1991)

2.   Apicella, A., L. Nicolais, G. Astarita, and E. Drioli, Polymer, 20,1143 (1979)

3.   Keenan, J.D., J.C. Seferis, and J.T. Quinlivan, J. Appl. Polym. Sci., 24, 2375 (1979)

4.   Progelhof, R.C. and J.L. Throne, Polymers,Engineering Principle Hanser, 1993, pp.250,257

5.   Ward, I.M., Mechanical Properties of Solid Polymers, John Wiley & Sons, 1983, pp.79,107

6. Dillman, S.H. and J.C. Seferis, <u>J. Macromol. Sci.-Chem.</u>, 227 (1983)

7. Nam, J. D., <u>Polymer Matrix Degradation</u>, Doctoral Dissertation, University of Washington, Seattle, Washington, 1991

8. <u>AroCy B-30 Cyanate Ester Semisolid Resin</u>, Technical Brochure, Ciba

9. Nam, J.D. and J.C. Seferis, "Viscoelastic Characterization of Phenolic Resin/Carbon Fiber Composite Degradation Process", <u>Macromolecues</u>, In Preparation, 1998

Table I Summary of Model Parameters

Parameters	0 Hr	20 Hrs
α	0.12	0.41
β	0.37	0.25
τ_0 (sec)	0.12	7×10^{-7}

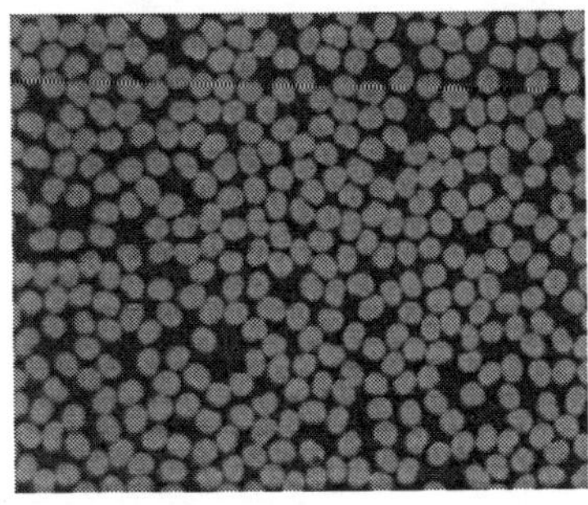

Figure 1 Cross-section of cured laminate (x60)

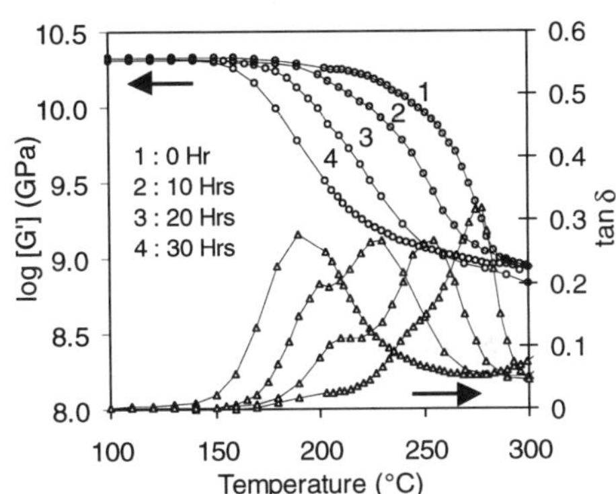

Figure 2 Comparison of storage modulus and tan δ as a function of thermal aging at 1 Hz.

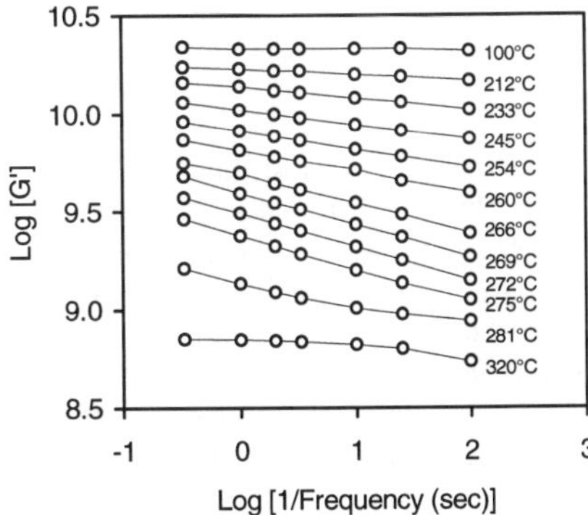

Figure 3 Storage modulus as a function of time and frequency applied to unaged specimen

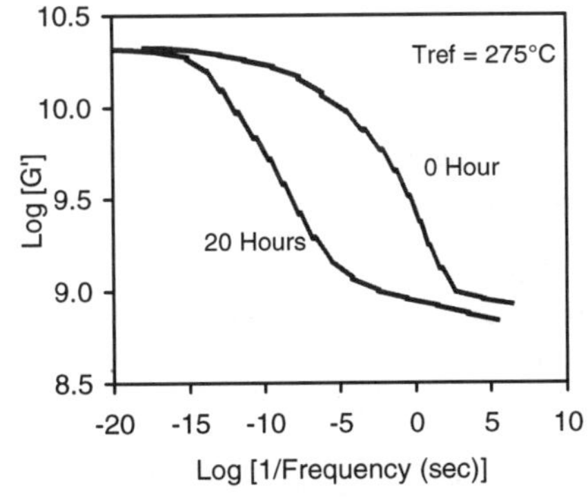

Figure 4 Comparison of mater curves for both unaged and aged specimens

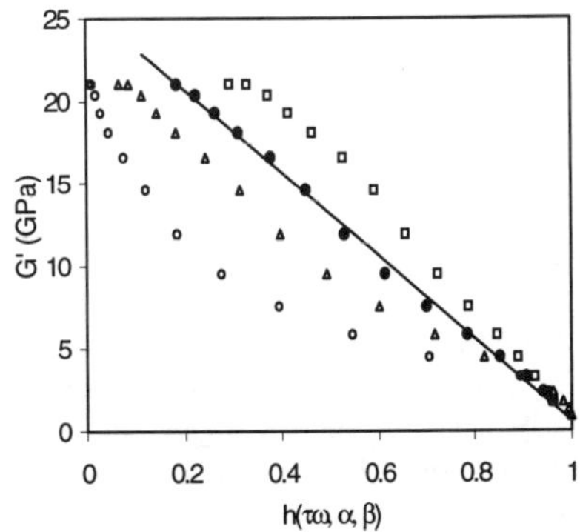

Figure 5 G' as a function of h to evaluate Model parameters

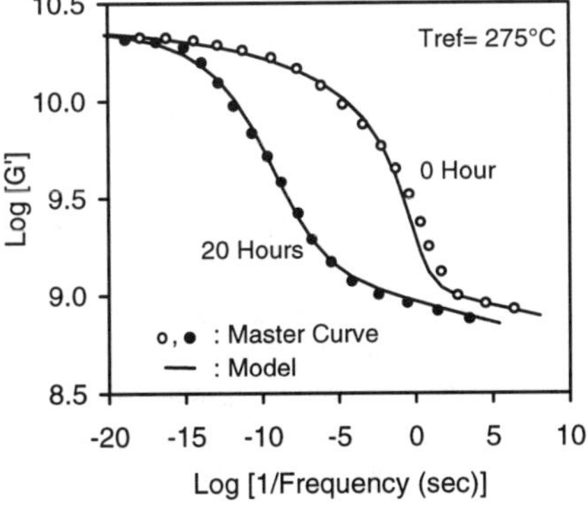

Figure 6 comparison of model with master curves

RAIN IMPACT DEGRADATION PHENOMENA OF CORE IN SANDWICH STRUCTURE

Ron Banuk
Northrop Grumman Corporation
One Hornet Way
El Segundo, CA 90245

ABSTRACT

Using the rotating arm facilities at the Naval Air Warfare Center (NAWC) in Patuxent River, Maryland, the database presented in paper "Rain Erosion/Impact Durability of Core for Sandwich Structure" [1] was expanded to include a Nomex honeycomb of two densities and seven varieties of closed-cell poly (vinyl ester) PVC and poly (ether imide) PEI foam cores. Impact durability curves are then plotted in the form of speed versus time in a 25.4-mm/hr (one-inch/hr) rainfield. This format is shown to be a paradigm for point design rain impact qualification. The rather surprising degradation of sandwich core (Nomex honeycomb, PVC foam, and PEI foam) as a function of time in a rainfield is quantified in terms of progressive modulus reduction, progressive strength reduction, and progressive increase in permanent set. An attempt to simulate the degradation phenomenon is done with a laboratory compression-compression fatigue test. A failure criterion for comp-comp fatigue in cellular materials is discussed. New data on tension after comp-comp fatigue for pristine and once-overloaded materials is presented. A correlation is drawn between the post ultimate compressive strength, F_{C-PU}, and the apparent strength endurance limit of certain core materials subjected to rain impact.

KEY WORDS: Foam, Honeycomb, Rain Erosion

1. INTRODUCTION

This paper concerns itself with the effect of rain impact (not rain erosion) on sandwich structure. Once the impact properties of a given core are quantified for a standard facesheet, the durability of that core can be compared to other similarly measured cores and generalized to other external loads. These external loads need not be created solely by a rainfield, but could include light hail, heel loads on walkways, and abusive impact on containers.

Rain and sand erosion are largely surface-effect phenomena. Rain erosion is greatly influenced by the elastomeric coating on the skin and to some extent by the skin thickness itself, if the core is of sufficient density to withstand the rain impact [1,2]. Rain erosion is best measured at the USAF WL/UDRI rotating arm facility in Dayton, Ohio.

Rain impact is largely a core-effect phenomenon. It is not influenced by the elastomeric coating, but is influenced by the thickness, stiffness, and strength of the skin [1]. To avoid the concurrent influence of these three variables, all test data presented in this paper have been normalized to 1.6-mm (0.063-inch) thick 7075-T6 aluminum skins. Rain impact is best measured at the U.S. Navy's NAWC rotating arm facility at Patuxent River Base in Lexington Park, Maryland [3]. Key aspects of the Naval rotating arm are described in Figure 1.

Even though both the Naval and Air Force facilities advertise one-inch-per-hour (25.4-mm/hr) rain rates with 2-mm diameter drops, coupons that fail by impact will last only 10% as long in the Naval chamber because of less air entrainment, larger drops, more drops on target, and less drop breakup. The accelerated environment of the Naval chamber provides a quick way to compare the impact durability of one core (honeycomb or foam) to another.

The data presented in this paper can be added to that presented in the seminal paper on this subject: "Rain Erosion/Impact Durability of Core for Sandwich Structure" [1].

2. PARADIGM FOR RAIN IMPACT DURABILITY

A leading edge or radome point design requires a minimum time at a certain speed in a one-inch-per-hour rainfield with the operative skin in place. The impact durability of the core can be more easily compared against other candidates if the skin is standardized and time to core failure is plotted against speed through the rainfield. The paradigm shown in Figure 2 was determined in a previous work and includes supersonic speeds. Because time was at a premium in the present series of tests and because the strobe was not functioning, only the subsonic leg of the curves was tested and plotted.

3. RAIN IMPACT DURABILITY OF CORE MATERIALS

The material properties of the two eighth-inch-celled (3-mm) Nomex honeycombs, two PEI closed-cell foams, and five PVC closed-cell foams are shown in Table 1. The manufacturer's properties are compared against test values from actual rain impact coupons. Recommended coupon dimensions and three off-axis holding fixtures are shown in Figure 3.

Test results are shown in Figure 4. The higher the line, the more durable the core. The 805-km/hr (500-mph) qualification speed is drawn as a point of reference. Even though the Navy would like to see a 10-minute life at this speed, the spec calls out 5 minutes. If a range of densities for any one core had been tested, say 50, 75, 100, 200, and 300 kg/cu m, than the minimum density required to meet 1, 5, and 10 minutes at 500 mph could have been plotted for each material. Since that was not done, what can be said is that the minimum density to meet 500 mph for 5 minutes is Baltek's B-3.75 PVC and for 10 minutes it is either Baltek's B-6.25 or Divinycell's R100 PVC.

In summary, the PVC foams appear the most durable. Fire retardant additives to PVC C71.75 ET degrade the properties of its B-3.75 baseline. The macrocellular PEI foams (R82) are almost as durable as the experimental microcellular PEI foam made by 3M and tested in a previous paper [1] and shown in Figure 2. Finally, while HRH 10-1/8-4 Nomex honeycomb definitely does not meet the Naval criterion for durability, the 8-pcf (132 kg/cu m) variety does. If thinner glass or quartz skins are used, however, the sandwich assembly would have to be requalified for that point design.

All failure points in Figure 4 were determined by squeezing coupons manually while under visual examination and making a subjective evaluation of failure. But what do these failures look like, and how do foam failures compare to honeycomb failures?

4. FAILURE CRITERIA FROM VISUAL INSPECTION

Figures 5, 6, and 7 are photographs of three core materials (Nomex honeycomb, PVC foam, and PEI foam) shown pristine and after two stages of use. From these observations, Table 2 can be constructed to verbally describe the stages of failure for honeycomb and foam. Notice that very light foam and honeycomb will fail differently than their denser counterparts.

The question that can now be asked is "Is there an associated physical property degradation corresponding to the visual indications?".

5. PHYSICAL PROPERTY DEGRADATION

The following results are rather surprising. Prior to testing cylindrical test coupons exposed to varying amounts of rain impact at a given speed, I hypothesized that there would be an incubation period during which no degradation would occur. Results were contrary. Figure 8 shows that Nomex honeycomb loses its compressive modulus in a linear fashion beginning with the first drop. Figures 9 and 10 show that foam cores have a precipitous wear-in period with PEI being more declivitous than PVC. Unlike honeycomb, the foams have what appears to be a modulus endurance limit. Superposed on the modulus degradation curves are visual indications of wear and visually determined periods of useful life. Figure 11 shows how the compressive modulus was determined for a pristine and a fatigued coupon.

Figures 12, 13, and 14 show what appear to be compressive strength endurance limits. The endurance limit is roughly equivalent to the post-ultimate compressive strength, F_{C-PU}, as determined by static test. Figure 15 shows the difference between the ultimate compressive strength, F_{CU}, a one-time event and the post-ultimate compressive strength, F_{C-PU} (a repeatable property). The tensile strength after F_{C-PU} (F_{TAC-N} or $F_{TAC-PUe}$) is an indication of the toughness of the core and was presented in a previous paper [1].

The process of core deformation under rain impact takes the following sequence. The core material will deform during repeated high-speed impact, recover a portion of this deformation elastically when the cyclic loading is terminated, recover most of its creep within 5 minutes, and be left with some permanent set after being allowed to stand overnight. Only Nomex honeycomb showed an incubation period for this phenomenon. See Figure 16. For all three materials, the permanent set curves were roughly the inverse of the modulus degradation curves.

Can these degradation phenomena mentioned above be simulated by constant amplitude compression-compression fatigue in the laboratory?

6. LABORATORY TEST SIMULATION

If pristine closed-cell PVC foam (R75 Klegecell) is subjected to constant amplitude comp-comp fatigue at some percentage of F_{CU}, then the curve shown in Figure 17 will result. If, however, the coupon is not pristine, but has been subjected to an initial overload such that F_{C-PU} is reached, then the second cycle will be F_{C-PU} and it is hypothesized that the subsequent curve will fall under that of the coupon not subjected to an initial overload. See the dashed curve in Figure 17. Notice that the comp-comp curve for the pristine material has no endurance limit or dual-slope tendencies. The same material when subjected to tension-tension fatigue also shows no endurance limit, in fact, it displays a negative dual-slope curve! See Figure 18. (It is not clear at this time, however, if the degradation is due to adhesive or foam cellular breakdown.)

It was pointed out in a previous paper [1] that the durability and damage tolerance of PVC foam can be attributed to it being able to retain 100% of its tensile strength, $F_{TAC-PUe}$, even after a single compressive overstrain of up to 1000%. See Figure 19. Tensile strength after comp-comp fatigue, F_{TAC-N}, however, is a good indication of what is left in tension and shear after constant amplitude loading and should be a part of every comp-comp test. See the two paired curves in Figure 20 for paradigm format. Typical stress-strain curves for tension-after-compression, F_{TAC-N}, tests of varying cycles are shown in Figure 21.

What constitutes foam compressive failure and what happens after failure? The comp-comp fatigue curves I have generated were done with a deflection control of about one half millimeter. The residual tensile strength, F_{TAC-N}, is an inverse function of how much creep is allowed. The two curves plotted together as shown in Figure 20 are self correcting. If too much creep is allowed then the residual tensile strength, F_{TAC-N}, will be low. If too little, then F_{TAC-N} will be higher, but the comp-comp- fatigue curve will be lower. There is an optimum creep allowance for each material--a value that I have not found. Figure 22 shows the trend in creep after the initial 0.75 mm indication. For the foam under study, about two thirds of the life of the foam was spent at the time of the initial indication. This value was confirmed by continuing comp-comp testing until the time between 0.75 indications was reduced to only a few cycles of load.

Finally a PVC foam under constant amplitude comp-comp fatigue will show modulus and strength reductions as well as creep and permanent set accretion similar to that seen during whirling arm rain impact testing. See Figures 23, 24, and 25. Figure 25 shows the difference between creep and permanent set measurements for a closed-cell PVC foam.

7. CONCLUSIONS

1. Sandwich core degrades in the presence of repeated skin impact with no incubation period. This is true even if the core shows no visible evidence of wear to the eye or evidence of permanent set as is true in some honeycombs.

2. The degradation of the physical properties of a core material can be simulated by compression-compression fatigue at 80% F_{CU}.

3. Paradigms for the presentation of core material fatigue data are as follows:
   A) F_{C-PU}: See Table 1 and Figure 15. This static property is an indication of the compressive strength endurance limit of core material subjected to the NAWC whirling arm rain impact test. It is the post-ultimate compressive strength of the core.

B) $F_{TAC-PUe}$: See Figure 19. This static property is an indication of the toughness of a core material after being subjected to overstrains of 1000% or more. It is the post-ultimate tension after compression strength as a function of overstrain.

C) F_{TAC-N}: See Figure 20. This static property when used in conjunction with the comp-comp fatigue curve, both legitimizes the comp-comp curve and shows the residual strength of correlative properties. It is the tension-after-compression strength of a pristine coupon subjected to comp-comp fatigue for a certain number of cycles (N) at some stress (%F_{CU}).

D) Comp-comp fatigue of a core that has been subjected to an initial overload beyond F_{C-PU} will take the hypothesized shape shown in Figure 17. This is a measure of the durability of a core after a severe impact.

E) Time vs. speed plot of core durability for NAWC rain impact: See Figure 2. Core material should be characterized across the entire flight regime both as a means of comparison and to establish the database for a Miner's Rule evaluation were the impacting spectrum actually known.

4. Modulus (E), Strength (F), and Permanent Set (δ) degradation: See Figures 8-16. The residual modulus, strength, and permanent set of a core material have been shown to vary with the number of cycles of comp-comp fatigue. These characteristics should be published as part of the basic material properties. A loose correlation has been shown between constant amplitude comp-comp fatigue cycling at 80% F_{CU} and rain impact loading at NAWC.

9. REFERENCES

[1] Banuk, Ron, Northrop Grumman Corporation, "Rain Erosion/Impact Durability of Core for Sandwich Structure", <u>SAMPE International Symposium</u>, <u>42</u>, 341, (1997)

[2] Schmitt, G.F., Jr., AFML, "The Subsonic Rain Erosion Response of Composite and Honeycomb Structures", <u>SAMPE Journal</u>, 15, <u>(5)</u>, 6, (1979)

[3] Hickman, John H. and Gilligo, Tom A., Code 5021 NAWC, "Test Report: Rain Erosion and Impact Evaluation of State of the Art Aircraft Materials", 6 April 1992

P13-10

Table 1. Material Properties of Core Materials

ID	Mfg	Chemical	Name	Designation	Density Test kg/cu m	F_{CU} Test/Mfg Mpa	$F_{C\text{-}PU}$ Test $\%F_{CU}$	E Test/Mfg Mpa
#1	Hexcel	Phenolic	Nomex	HRH 10-1/8-8	132.1	13.1/12.6	16	393/538
#2	Hexcel	Phenolic	Nomex	HRH 10-1/8-4	67.4	4.0/4.0	36	193/193
#3	Baltek	PVC	Air Lite	B-6.25	110.5	2.3/1.8	93	90/110
#4	Baltek	PVC	Air Lite	B-3.75	61.7	0.9/0.8	96	28/55
#5	Baltek	PVC	Herex	C71.75 ET	70.7	1.6/1.6	84	69/145
#6	Baltek	PEI	Airex	R82.80	77.5	1.0/0.8	92	41/48
#7	Baltek	PEI	Airex	R82.110	135.1	1.5/1.7	92	69/90
#8	Diab	PVC	Klegecell	R75	61.9	1.0/1.2	80	48/83
#9	Diab	PVC	Klegecell	R100	112.0	2.1/1.7	81	110/124
---	BASF	EPP	Neopolen P	85 g/l	84.6	0.4/0.6	100	16/---

1) Nomex Honeycomb: Orthogonal Weave Aramid, Hexagonal Cell, TP Adhesive Bond, Phenolic Dip
2) Foams: Closed Cell Macrocellular
3) $F_{C\text{-}PU}$: Post -Ultimate Compressive Strength (an indication of the rain impact comp strength endurance limit)
4) Coupons: All tests done on rain impact coupons. (D=25.4 mm, t_C=10 mm)
5) Crosshead Speed: 2.54-mm per Minute
6) E (Modulus) determined via crosshead deflection. [Author recommends deflectometer measurements.]
7) Test results are the average of three specimens.

P13-12

Table 2. Visual Indications of Progressive Sandwich Core Damage When Subjected to Rain Impact in the NAWC Whirling Arm

Light Honeycomb Nomex HRH 10-1/8-4	Heavy Honeycomb Nomex HRH 10-1/8-8	Light Foam PVC Kleg R75	Heavy Foam PVC Kleg R100
1) OK	1) OK	1) OK	1) OK
2) Resin Flecks	2) Resin Flecks	2) Softening	2) Permanent Set
3) Incipient Buckling	3) Incipient Cracking	3) Skin-to-Foam Delam or Mid-Plane Fracture	3) Softening
4) Incipient Cracking	4) Incipient Buckling	4) Permanent Set	4) Skin-to-Foam Delam or Mid-Plane Fracture
5) 360° Cracking	5) 360° Cracking		
6) Crush	6) Crush		

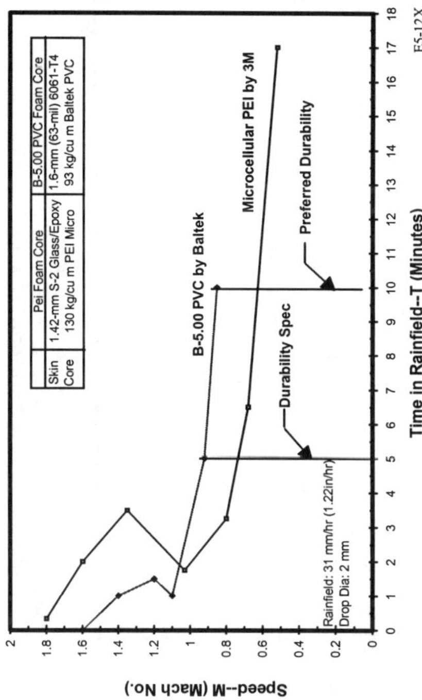

Figure 2. Paradigm for the Impact Durability of Foam at NAWC

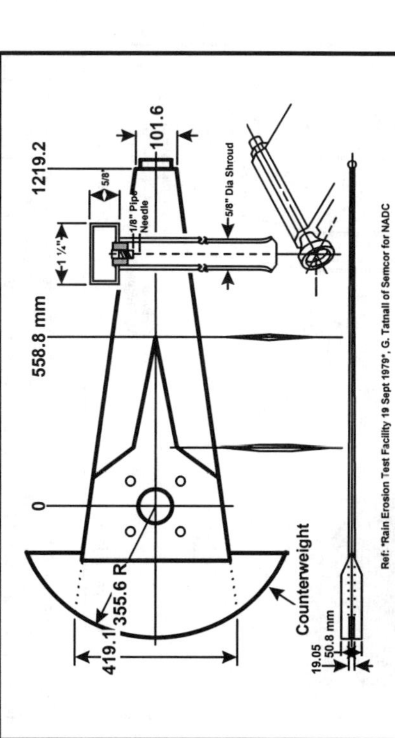

Figure 1. NAWC Rotating Arm Blade Geometry

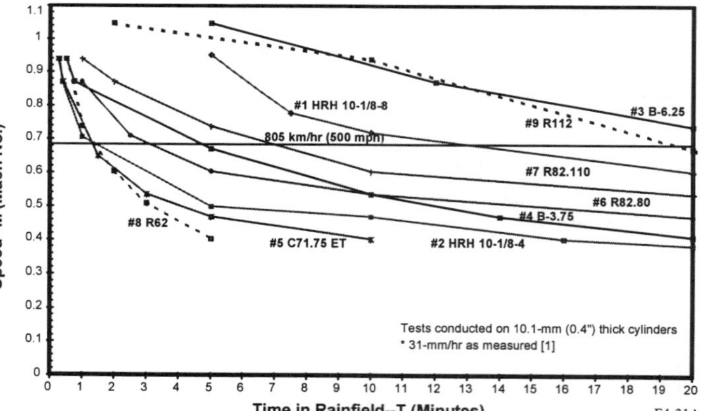

Figure 3. Standard NAWC titanium specimen holder and an exploded view of a sandwich coupon. This holder exposes 22.22-mm diameter and accepts a maximum thickness of 12.7 mm (.50 in.) which includes a 1.60-mm (.063 in.) spacer

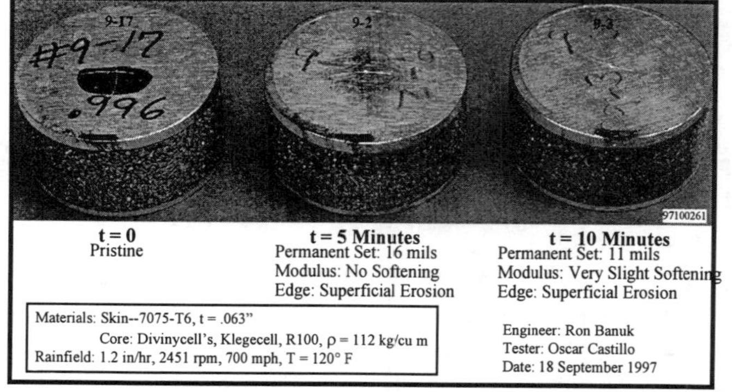

Figure 4. Visual indications of progressive sandwich core damage when subjected to rain impact in the NAWC whirling arm

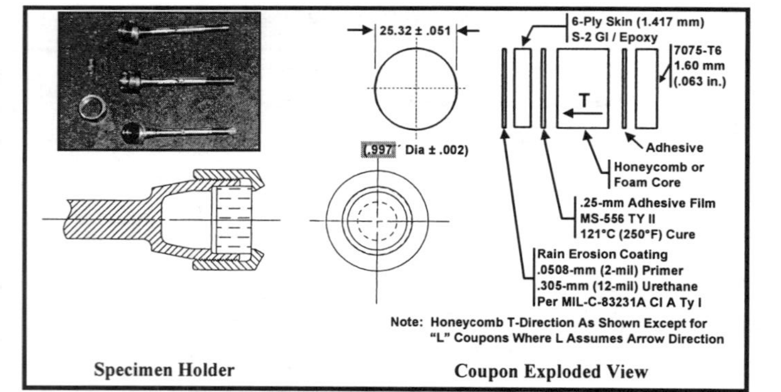

t = 0	t = 5 Minutes	t = 14 Minutes
Pristine	Front Face: Incipient Cracking	360° Cracking
	Back Face: Incipient Buckling	Front Face: Crush

Materials: Skin--7075-T6, t = .063"
Core: Hexcel's Nomex HRH-10-1/8-8, ρ = 132 kg/cu m
Rainfield: 1.2 in/hr, 2451 rpm, 700 mph, T = 120° F

Engineer: Ron Banuk
Tester: Oscar Castillo
Date: 18 September 1997

PZ2-1X

Figure 5. Stages of damage for Nomex honeycomb core. The pristine coupon is shown to the left. The center coupon has been whirled in the NAWC chamber at mach 0.94 for 5 minutes and that on the right for 14 minutes. Observed damage is recorded under the pictures.

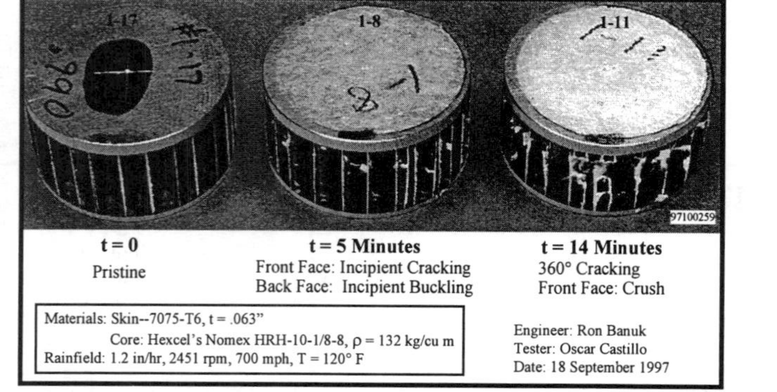

t = 0	t = 5 Minutes	t = 10 Minutes
Pristine	Permanent Set: 16 mils	Permanent Set: 11 mils
	Modulus: No Softening	Modulus: Very Slight Softening
	Edge: Superficial Erosion	Edge: Superficial Erosion

Materials: Skin--7075-T6, t = .063"
Core: Divinycell's, Klegecell, R100, ρ = 112 kg/cu m
Rainfield: 1.2 in/hr, 2451 rpm, 700 mph, T = 120° F

Engineer: Ron Banuk
Tester: Oscar Castillo
Date: 18 September 1997

PZ2-3X

Figure 6. Stages of damage for PVC foam. The pristine coupon is shown to the left. The center coupon has been whirled in the NAWC chamber at mach 0.94 for 5 minutes and that on the right for 10 minutes. Observed damage is recorded under the pictures.

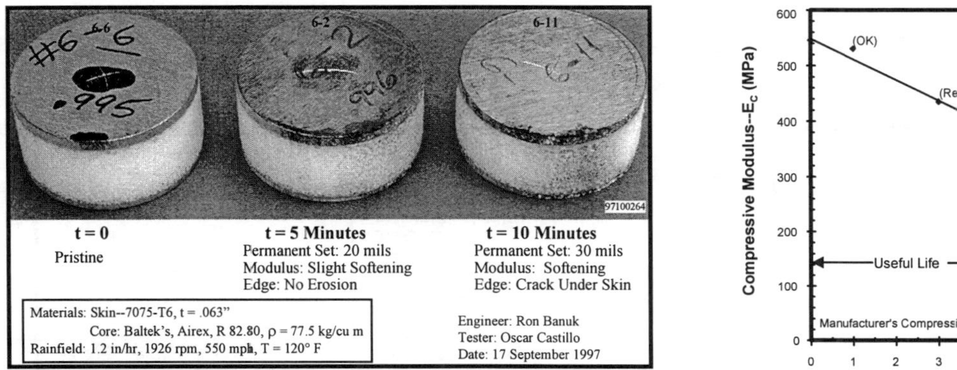

Figure 7. Stages of damage for PEI foam core. The pristine coupon is shown to the left. The center coupon has been whirled in the NAWC chamber at mach 0.74 For 5 minutes and that on the right for 10 minutes.   Observed damage is recorded under the pictures.

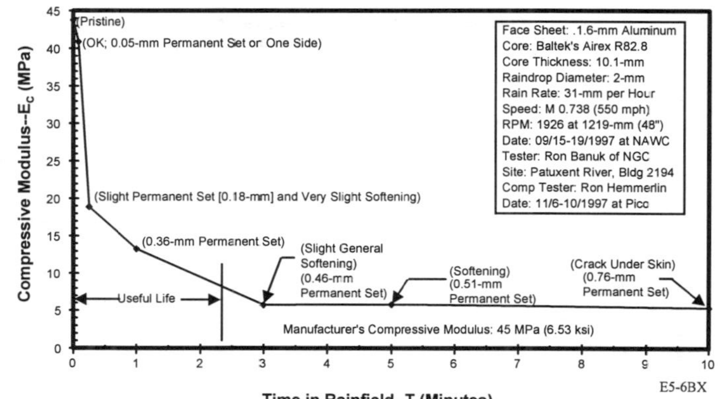

Figure 4. Compressive modulus degradation for Nomex honeycomb core in a rainfield

Figure 9. Compressive modulus degradation of PVC foam in a rainfield

Figure 10. Compressive modulus degradation of PEI foam in rainfield

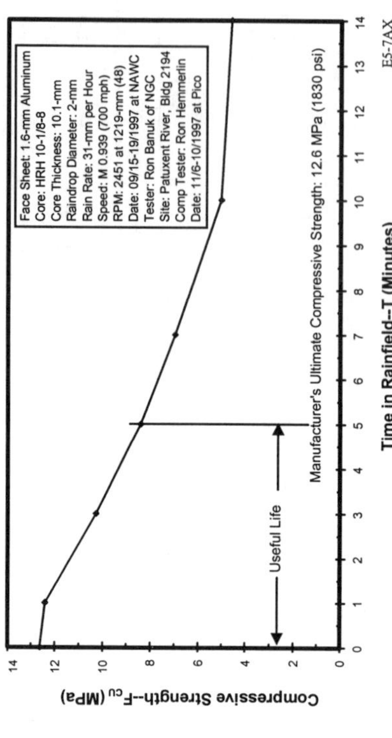

Figure 11. Foam Compressive Modulus Measurement for Two Conditions

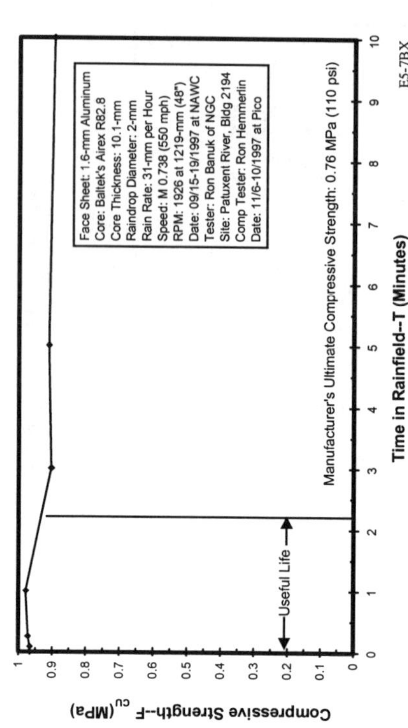

Figure 12. Compressive strength degradation for Nomex honeycomb core in a rainfield

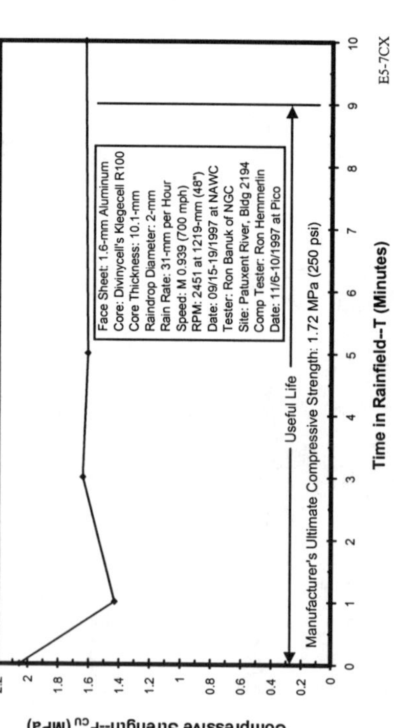

Figure 13. Compressive strength degradation of PVC foam in a rainfield

Figure 14. Compressive strength degradation for PEI foam in a rainfield

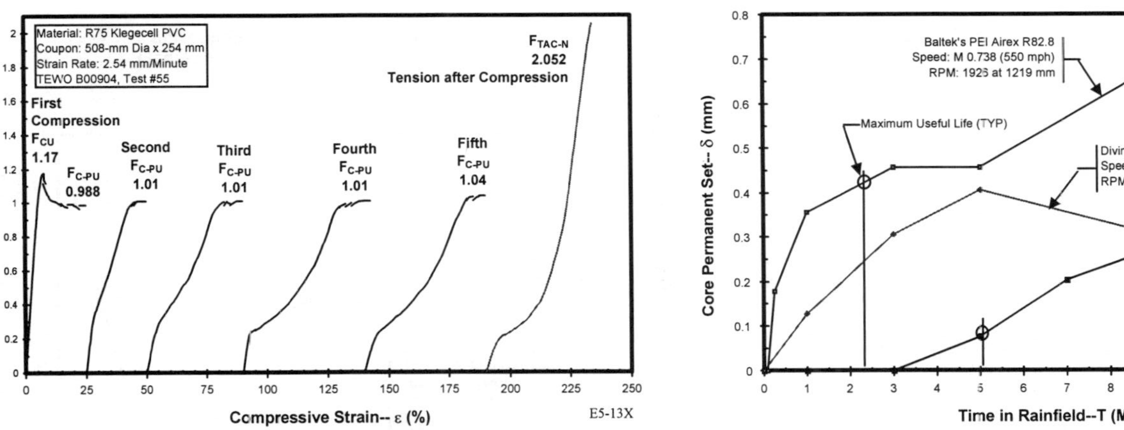

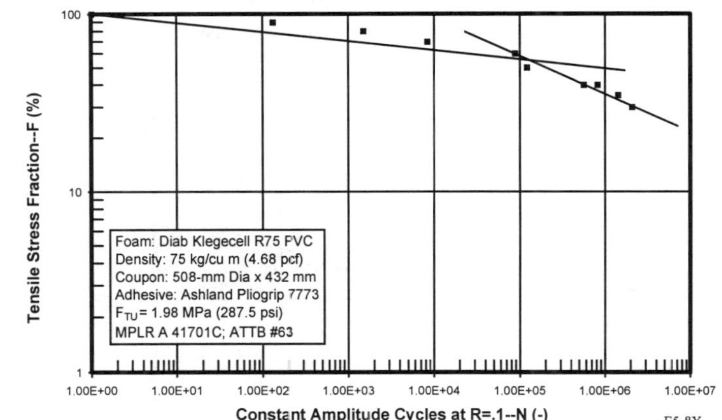

Figure 15. Tension after multiple post-ultimate compressions (F_{TAC-N})

Figure 16. Core material permanent set in a rainfield

Figure 17. Compression-compression fatigue for PVC foam

Figure 18. Tension-tension fatigue for PVC foam

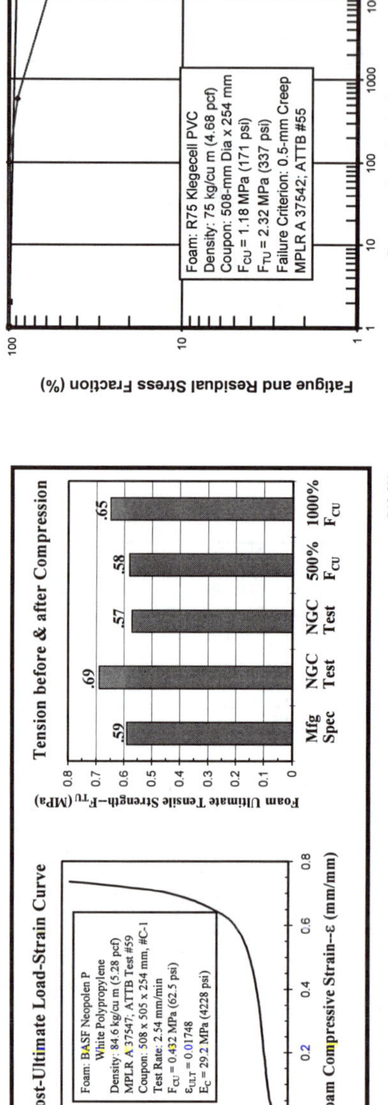

Figure 19. Polypropylene post-ultimate compression followed by tension-after-compression strength

Figure 20. Paradigm for tension after compression-compression fatigue data presentation

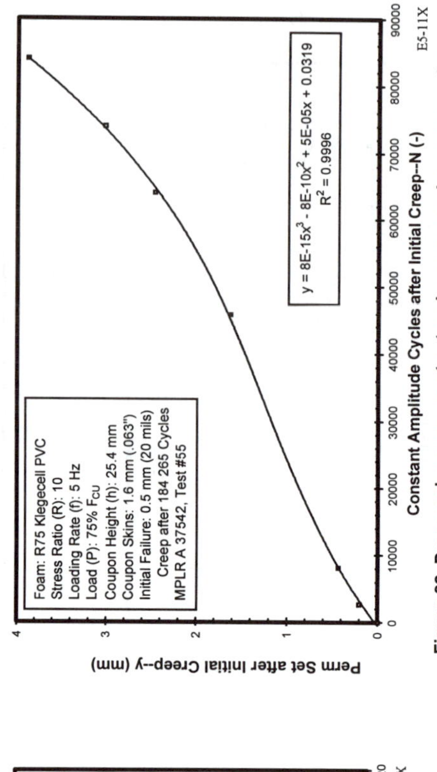

Figure 22. Progressive permanent set under compressive-compressive fatigue at 75% F_{CU} after an Initial Incubational Creep of 0.5 MM

Figure 21. Tension after varying amounts of compression-compression fatigue at 80% F_{CU}

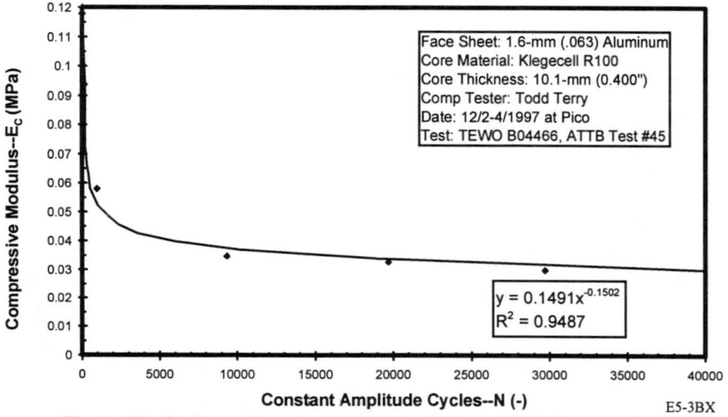

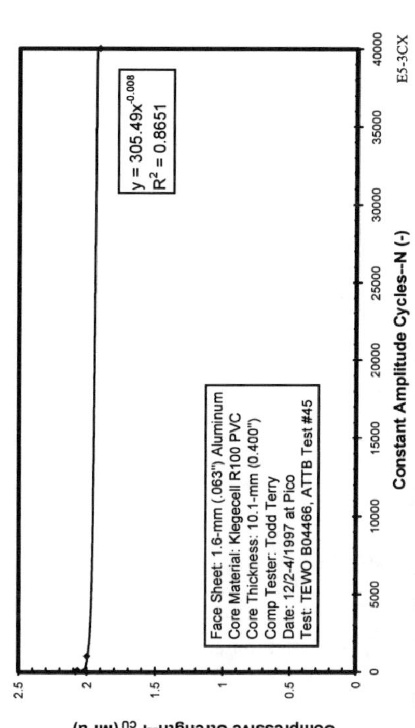

Figure 23. Compressive modulus degradation of PVC foam under constant amplitude compressive-compressive loading at 80% F_{CU}

Figure 24. Compressive strength degradation of PVC foam under constant amplitude compressive-compressive loading at 80% F_{CU}

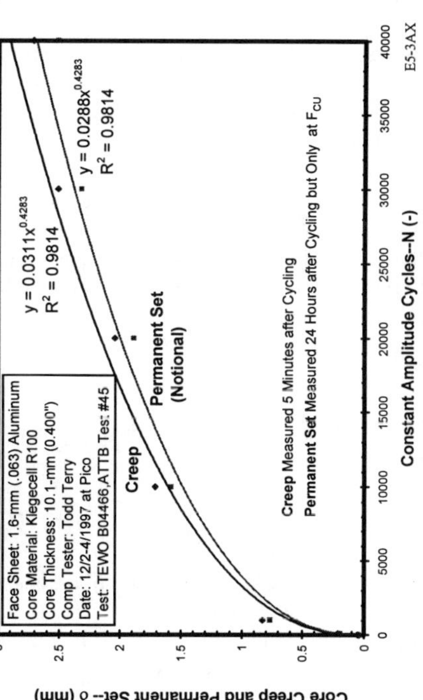

Figure 25. Creep and notional permanent set for comp-comp fatigue at 80% F_{CU}

DESIGN OF DURABLE, REPAIRABLE AND MAINTAINABLE AIRCRAFT COMPOSITES

William Cole (contact)
Commercial Aircraft Composite Repair Committee
Design Task Group
Warrendale, PA 15096

ABSTRACT

The early successes of composite materials in the commercial aviation industry led to their use in more complex and structurally significant components. With this came damage phenomena that was new and complex when compared to traditional aluminum structures. This in turn caused airlines to adopt new and more complicated maintenance procedures to preserve operational efficiencies.

In 1991, the Commercial Aircraft Composite Repair Committee (CACRC) was created by the Air Transport Association (ATA), the International Air Transport Association (IATA), and the Society of Automotive Engineers (SAE) to establish an international forum to address and standardize the many facets of operating commercial aircraft with composite components. The Design Task Group of the CACRC, composed of a diverse group of manufacturers and operators, was tasked with creating awareness and expanding knowledge in the design community, based upon operator experience, so that the durability, reparability, and maintainability of aircraft composites could be improved.

This effort has resulted in a design document containing potential problem-solving recommendations for in-service concerns of various composite components. The document includes various component designs and considers operational environments of composite materials; alternates that can be implemented to improve existing designs; and case studies focusing on recurrent situations, which include damage and repair descriptions in addition to economic factors.

KEY WORDS: aircraft, composites, repair

1. INTRODUCTION

Composite materials, mostly glass fiber, began to replace aluminum secondary components on commercial aircraft in the early 1960's. The introduction of advanced fibers, such as carbon/graphite and aramid, offered increased strength, reduced weight, and improved corrosion and fatigue resistance over aluminum. The airframe industry introduced advanced composites cautiously to assure the capabilities of these new material systems.

The early successes of these first components led to the use of composites for more complex components, such as flight controls and engine nacelles. The increased specific stiffness and strength of composites over aluminum, coupled with weight-driven requirements, led to the design of lighter thin-skinned sandwich structures. These sandwich constructions proved to be less damage resistant than previous aluminum designs. The damage phenomena, such as delamination and microcracking, were new and complex in comparison to traditional aluminum structures. This complicated damage assessment, inspection, cleaning, painting, and repair. These problems, not foreseen at the design stage, made in-service problems a frequent occurrence. This situation was further aggravated by the lack of operator technician training for composite damage evaluation and repair.

Aircraft operators raised their concerns to manufacturers worldwide stating that the damage occurring to composite structures was inconsistent with their abilities to maintain them. In 1991, an international committee, the Commercial Aircraft Composite Repair Committee (CACRC), was formed to address these concerns. The CACRC consists of experienced personnel from aircraft manufacturers, airlines, composite repair facilities, material manufacturers, and regulatory agencies. The Design Task Group, one of several task groups created within the CACRC to work on specific problems, was challenged with compiling information relative to composite component damage so that the shortcomings of present designs can be improved and avoided in future designs.

To achieve this, the Task Group conducted an international survey of airlines and aircraft manufacturers who were asked to compile the damage data that had occurred to composite components. This information was the basis for creating the <u>Design of Durable, Repairable, and Maintainable Aircraft Composites</u>[1] which is a document highlighting many historical in-service concerns with certain composite aircraft component and sub-component designs. The design guide attempts to identify problems that have occurred and recommend potential solutions. The intent is not to criticize any design, but instead create awareness and expand knowledge in the design community based upon operator experience, which can lead to improved designs having reduced repair and maintenance costs.

This paper is intended to create awareness and introduce this design guide so that the information contained in it can be utilized more extensively. Each chapter of the design guide was created to communicate the environment in which composite materials operate, how existing designs have performed, and how future designs might be improved. These chapters include Design Considerations, highlighting construction types and damage sources; Database of Current Performance and Alternative Considerations, showing problematic designs and possible alternatives; and Detailed Design Case Studies, specifying the successes or shortcomings of entire composite components.

2. DESIGN CONSIDERATIONS

The design guide demonstrates that durability, reparability, and maintainability are essential elements in the effective use of composite components. These elements are determined to a significant extent during the design phase. Consequently, repair designs should be developed along with the initial component design because of the similarity of the parameters considered. These include material selection, ply orientation, joining methods, fastener selection, and accessibility.

In evaluating a particular design, the overall effect on performance, weight, reliability, manufacturing cost, durability, spares provisioning, maintainability, and reparability must be thoroughly addressed and weighed. All of these elements will have a significant effect on the life cycle cost of the component, which is a process that quantifies not only the manufacturer's cost but also the total cost of operating the unit over its estimated lifetime. Designers should perform trade studies that estimate these total costs and recognize that their choice will be a compromise to obtain an optimal balance.

The evaluation of damage to determine if a composite design has reached this balance depends upon the type of construction chosen. Sandwich and stiffened skin structures are the two most common construction methods; the design benefits and detriments of each are discussed. Face sheet material selection, type of stiffening, and the process by which they are joined are design choices that significantly influence the success of both of these construction methods.

The sources of damage of a component are a contributing factor that will determine how it will perform throughout its life. The types of damage, its frequency, and its severity are in many cases dependent upon the location of the composite component on the aircraft. The design guide gives a thorough evaluation of these parameters as they would apply to mechanical impact, lightning strike, overheat, erosion, manufacturing deficiencies, fluid ingression, chemical contamination, corrosion, operational stresses, and disassembly. In addition to an aircraft illustration with potential damage locations shaded, the damage sensitivity and list of damage effects are given. An example of damage considerations is presented in Figure 1.

2.3.4   Erosion

<u>Damage Sensitivity.</u> Due to forward facing steps, even though they may be very small, unprotected leading edges will erode. After erosion has begun, subsequent peel back of individual composite plies occurs rapidly, which significantly increases the extent of damage.

- Radomes.
- Engine inlet cowls, fan blades, fan OGVs.
- Leading edge fairings.
- Environmental control system plenum chamber and ducting.

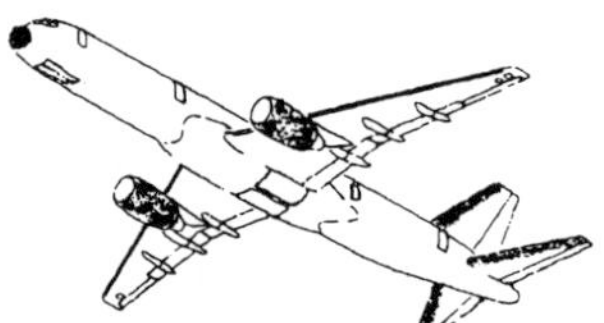

Figure 1. Example of Sources of Damage.

3. DATABASE OF CURRENT PERFORMANCE AND ALTERNATIVE CONSIDERATIONS

This chapter of the design guide contains a database that the composite design engineer can use to acquire historical performance information of some composite components and sub-components. This database was developed using in-service airline experience to evaluate specific designs. These evaluations are arranged using the ATA code format so that those familiar with using commercial aviation manuals can more easily extract information.

For each airline experience, there are two pages of information. The existing design is described and illustrated in terms of its materials, construction, and configuration. Potentially affected areas of a typical transport aircraft are indicated by a shaded figure.

The problem statement appears at the bottom of the first page. It highlights the source and/or type or damage followed by a more detailed description of the problem cause and effect. Finally the severity of this problem is rated based upon the industry-wide data compiled.

This severity rating index is a subjective ranking given to each specific concern. Some of the parameters affecting this rating include:

- Spares availability

- Shop capacity

- Aircraft schedules

- Material availability

- Problem frequency

- Economics

- Safety

Alternate considerations are illustrated on the second page of the evaluation. These considerations include a description of recommended changes in design, materials, processing, or construction. These alternate considerations provide a recommendation to the composite design engineer for possible future design improvements. Figure 2 provides an example of how the information in this section of the design guide is presented.

- Stiffened panel
- Bonded stiffeners, as shown below in a generic detail

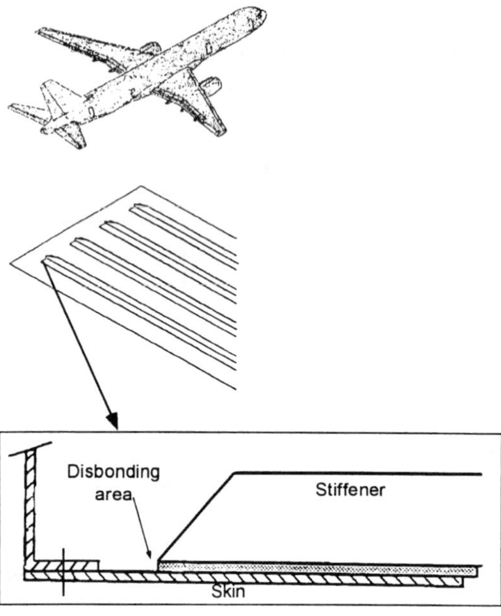

Problem

Peel force/stiffener runout

Peel forces between the stiffener and skin at the termination of the stiffener initiate disbonding and delamination that can spread along the length of the stiffener. This is difficult to repair if disbond is only partial and if the skins are too thin to allow the use of countersunk attachments.

Severity

High

Figure 2A. Example of Current Component Performance and Alternative Considerations: Existing Design.

<table>
<tr><td>Alternate
Consideration</td><td>Distribute peel stress into the stiffener
Consider all potential sources of peel loads at the stiffener ends. Consider the use of "peel stop" fasteners to arrest disbonds and delaminations. Build up stiffener flange(s) instead of the skin and run the flange all the way to the panel edge. Also consider providing a smooth transition for the stiffener load into the skin.</td></tr>
</table>

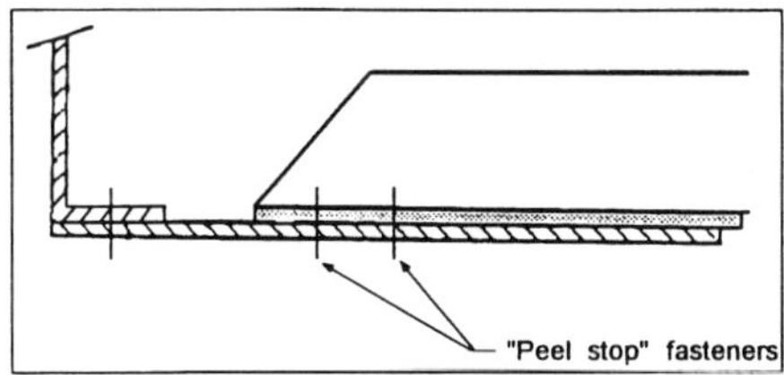

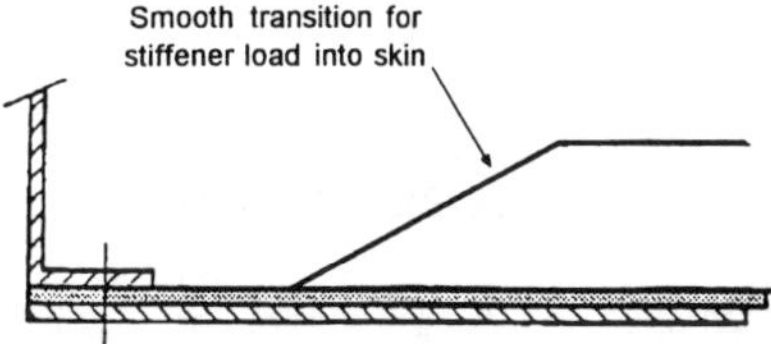

Figure 2B. Example of Current Component Performance and Alternative Considerations: Alternate Consideration.

4. DESIGN CASE STUDIES

The final chapter of the design guide provides detailed case studies of specific problems or successes of complete composite components. They provide specific information in areas of component function, performance, chronic damage, repair description, and economic consideration. The purpose of these case studies is to offer the designer a better perspective by presenting a complete discussion of the efforts needed to maintain a specific design for a particular function.

For those case studies in which damage has been a problem, the damage sources include disbonds, lightning strikes, overheats, erosion, and fluid ingress. All designs, specific damage, or a combination of both are illustrated by way of drawings or photographs.

Each design case study is organized and presented as follows:

a. Design and Component Function

b. Damage Description:
   - Source
   - Type
   - Size
   - Frequency (in terms of fleet size and components per aircraft)

c. Repair Description (if case study includes damage):
  - Documentation
  - Tooling
  - Repair material
  - Repair accomplishment
  - Concerns

d. Performance in Operation (if case study is for design commendation):

e. Economic Considerations:
  - Material
  - Labor hours
  - Tooling
  - Delay and cancellation
  - Spares and leasing

f. Summary: Design concerns or commendations

5. CONCLUSION

Due to the significant performance and design benefits offered by composite materials, they have become an integral aspect of commercial aircraft structures. As operator experience has demonstrated, all of these structures will experience damage. If composite structures include durability, reparability, and maintainability as design requirements, the reliability, and therefore the cost, will significantly improve. <u>Design of Durable, Repairable, and Maintainable Aircraft Composites</u> was created to promote awareness and expand knowledge in the design community, based upon operator experience, in the hope that improved designs having reduced repair and maintenance costs will result.

6. REFERENCES

1. SAE, Commercial Aircraft Composite Repair Committee, <u>Design of Durable, Repairable and Maintainable Aircraft Composites</u>, Document Ref. AE-27, Warrendale PA, 1997.

ACTIVE AND PASSIVE DAMPING CONTROL OF INTELLIGENT STRUCTURE VIBRATION

Khaled A. Alsweify, Faysal A. Kolkailah, Ph.D., P.E
California Polytechnic State University
San Luis Obispo, CA 93407
and
Said H. Farghaly, Ph.D.
Associate Dean,
Faculty of Engineering (Mataria), Helwan University, Cairo, Egypt

ABSTRACT

This paper presents an overview of the development for studying intelligent structure, which is the concept of a higher form of materials, and structures by providing the necessary life function of sensing, actuating, control, and intelligence to those structures. Such actuating, sensing, and signal processing elements are incorporated into a structure for the purpose of influencing its states or its characteristics, by they mechanical, thermal, optical, chemical, electrical, or magnetic properties. From this overview, the application of intelligent structure technologies will have a tremendous impact in reshaping the technology and economic bases of the international business environment during the next two decades.

KEY WORDS: active and passive damping, intelligent structure, vibration, controls.

1. INTRODUCTION

This article presents an overview of the development of intelligent structures. Intelligent structures are those, which incorporate actuators and sensors that, are highly integrated into the structure and have structural functionality, as well as highly integrated control logic, signal conditioning and power amplification electronics. Such actuating, sensing, and signal processing elements are incorporated into a structure for the purpose of influencing its states or its characteristics, by they mechanical, thermal, optical, chemical, electrical, or magnetic properties. For example, a mechanically intelligent structure is capable of altering both its mechanically states (its position or velocity) or its mechanical characteristics (its stiffness or damping), piezoceramic transducers (PZT) worked will as sensors, and appeared their sensitivity to a particular resonant frequency was a function of position with respect to corresponding node line(s). To define the intelligent structures there are many definition, one of this definitions, that those structures which have actuators distributed throughout are defined as actuated, or structural element that contains an integral closed loop system consisting of sensor, actuator and control logic with adjustable control parameters.

Some Typical applications of smart or intelligent structure technologies are;
- Automotive and transportation industries (engine mounting, dampers).

- Aerospace industry (smart wings to detect damage, large space structures).
- Defense industry (submarines to reduce drag and noise, stealth technologies, structures supporting weapons and antenna).
- Biomedical devices (artificial hands and joints).
- Robotics and industrial machinery (active balancing, dynamically tuneable robot arms).
- Sporting goods (tennis rackets, fishing poles).

Any vibratory system, in general, includes a means to store potential energy (PE) (spring or elastic behavior), a means of storing kinetic energy (KE) (mass or inertia) and a means (damper) by which energy is gradually dissipated (dampers). The vibrations in a system involve the transfer of its PE to KE and vice versa. most of the engineering structures, in general, are flexible in nature concerning the design involved and in selecting the geometrical and physical properties of the structures. Thus the amount of PE and KE stored in the structure may vary across the cross-section of the structure. Such structures are termed as distributed parameter systems (DPS). They have infinite number of degrees of freedom (continuos system). Their dynamics can be described by partial differential equations (PDE). Typical example (other than the once mentioned before) are, transverse vibrations of strings, beams, plates, etc., rods under axial vibration, shafts in torsion or bars in bending.

2. VIBRATIONAL ANALYSIS

2.1. Vibrational analysis and control of cantilever beams The cantilever beam is a very delicate structure and required much time and energy to construct. some parts of it are very sensitive to precision and some are more forgiving. Therefore, the cantilever beam is chosen as a model it typified some common problems associated with large space structures. The problem of vibrational analysis of cantilever beams has been studied by many authors, hence, only a few related and selected references are presented. Resonance frequencies, structural damping, responses, impedance, and force transmissibility properties are presented by *Y. P. Lu, H. C. Nelson and A. J. Roscoe* (1). The subject of vibration characteristics and the related force transmissibility properties of composite structures are addressed for a composite beam made of *Hercules AS4/3501-6* graphite/epoxy with a layered structure sequence of (0, 0, 30, -30). The vibration properties of individual composites are evaluated. The criterion defined for performance comparison between composite materials and conventional materials is also discussed. In their study, it is observed that the impedance of the composite beam are lower compared with the steel beam, because the system has lighter weight, and the resonance frequencies are higher, due to the ratios of stiffness and mass are higher. A method for predicting the dynamic properties of laminated composite beams is developed by *R. G. Ni and R. D. Adams* (2). A comprehensive mathematical technique is developed for predicting dynamic modulus and damping of laminated composite beams, based on Adams and Bacon's technique. The damping prediction is improved by taking bending-twisting coupling into account. The good agreement is obtained between the theoretical and experimental values of four types of laminate and these values show that the mathematical technique is satisfactory developed for predicting the flexural modulus and damping values of symmetric, laminated composite beams. A design and analysis of an active damper for a thin cantilever beam, describes the apparatus and procedures are used in the preliminary testing of the damper by *T. Bailey and J. E. Hubbard Jr.* (3). Using distributed- parameter actuator and distributed- parameter control theory, some of the tradeoffs, such as truncation of the model can avoid. The distributed- parameter actuator is a piezoelectric polymer, poly (vinylidene fluoride). A control law is developed using *Lyapunov's second method* for distributed- parameter systems. Since no modes are truncated in the analysis, this control law will theoretically control all of the modes of vibration, and it can be avoid any structural problems with uncontrolled modes. Preliminary testing of the active damper is done using only the first mode of vibration of a small cantilever beam because the angular velocity of the tip of the beam is not available. Three controllers are developed to test the damper on the first mode. Testing is performed using two of these controllers, the constant-gain and the constant-amplitude controllers. The constant-gain controller is a linear controller and provides approximately double the baseline damping. The constant-amplitude controller is nonlinear controller and provides nonlinear damping- double the baseline loss factor for large vibrations increasing by a factor of 40 to at least 0.040 for small vibrations. These results are achieved with a simple damper configuration, simple control algorithms, and moderate voltage levels compare to the breakdown voltage of the poly (vinylidene fluoride). A theory is developed by *S. Hanagud, M. W. Obel and M. Meyyappa* (4) to quantitatively identify changes in a damping matrix of a structural dynamic system when electronic damping is applied to the system. The specific problem considered is the quantitative identification of electronic damping, contributed by Piezoceramic materials bonded to structure. Electronic damping experiments are conducted on a cantilever beam under excitation conditions. Piezoceramic transducers are used as both sensors and drivers with a velocity feedback. The mass, stiffness and damping matrices of a beam before and after application of the electronic Damping are identified by a parameter identification technique that is capable of considering general linear viscous damping matrices. Total effects of the baseline or inherent damping and the electronic damping contributed by piezoceramic transducers are identified by computing the structural dynamic system matrices M, C, K of a selected system. The identified system matrices is compared with the corresponding matrices of the same structural dynamic system without electronic damping. The difference between the two system matrices are then related to combined effects of piezoceramic transducer properties, gain, phase shift, selective filtering parameters and electron-mechanical coupling. A techniques for modeling induced strain actuation of a beam like components of intelligent structures are developed by *E.F.Crawley and E.H.Anderson* (5). Two analytical models and one numerical model describing the detailed mechanics of induced strain actuators bonded to one dimensional structure are presented .The models illustrate the extension, bending, and localized shearing deformations induced. The specific characteristics of one type of induced strain actuator, Piezoceremic materials, are discussed, and implications for practical use of Piezoceramic actuator are outlined. A series of experiments is conducted in order to verify the accuracy of the surface-bonded bending models of induced strain actuation. The major objective of these tests is to provide comparative strain data for evaluation of the models. In addition, the response of the beam is calculated to verify the basic static and dynamic models. It is also expected that the influence of the nonidealities in piezoelectric actuation strain and the finite bond stiffness will be observed. The analytic model for structures with distributed piezoelectric actuators is verified experimentally for both surface bonded and embedded actuators by *E. F. Crawley and J. De Luis* (6). The results from tests performed on dynamic beams are in good agreement with the analytically predicted responses. A technique for selecting the location of a surface bonded or embedded piezoelectric actuator based on the characteristics of the strain distribution on a structure is developed. This technique indicates that segmented actuators are always more effective than a continuous actuator since the output of each actuator can be individually controlled. In addition, a systematic method based on maximizing the piezoelectric- mechanical effectiveness of the actuator and the structure is used to select the piezoelectric actuator material. The development of on efficient finite model for dynamic analysis of laminated beams treated by a constrained viscoelastic layer is described by *C.T.Sun, B.V.Sankar and V.S.Rao* (7). The finite element model is designed so as to represent the viscoelastic core shear accurately . The laminated beam and constraining layer are modeled by using the offset beam element, which takes shear deformation into account. The viscoelastic core is modeled by using plane finite element which are compatible with the beam elements. System damping and tip displacement are computer and compared. With those measured experimentally by using the impulse -frequency response technique. Results show that dynamic response is improved by using of such damping treatments. On The basis of the numerical and experimental results presented in their work, it is observed that the finite element model which is presented has an accurate and efficient representation of a sandwich beam and that viscoelastic surface layer treatments can be used to significantly improve the dynamic response of structures. Increases in overall system damping and large reductions in response amplitude are achieved by using damping treatment. Results also show, for each mode of vibration, that there exists a length, location and a thickness of the damping tape, for a given thickness of the constraining layer, for which the overall system damping is maximized. The effects of dynamic coupling between a structure (cantilevered beam) and on electrical network through the Piezoelectric effect are modeled by *N. W. Hagood, W. H. Chung and A. Y. Flotow* (8). The coupled equations of motion of an arbitrary elastic structure with piezoelectric elements and passive electronics are derived. The equations are applied to the case of a cantilever beam with surface mounted piezoceramics and indirect voltage and current drive. The theoretical derivations are validated experimentally on an actively controlled cantilevered beam test article with indirect voltage drive. The general models are specialized to the case of a cantilevered beam to demonstrate the applications of the model to a simple structure. The predicted dynamics of the beam using this model are found to be in close agreement with experiments conducted to test the model's accuracy. These non-collocated control experiments are verified the predicted effects of the electrical circuit on the beam dynamics in both open loop and closed loop configuration. A new type of structural damping mechanism is presented by *N. W. Hagood and A.Van flotow* (9), which based on Piezoelectric material shunted by passive electrical circuits. A model for general shunting of these materials subject to arbitrary elastic boundary conditions is developed to determine the 6x6 material compliance matrix when the material is shunted . Resonant circuit shunting of Piezoelectric is modeled and shown to exhibit behavior very similar to the well-known mechanical tuned vibration absorber. Experiments are conducted on a cantilevered beam, which validated the shunted piezoelectric models . the models developed are able to predict the influence of the shunted Piezoelectric on the cantilevered beam damping in both the resistive and resonant shunting cases. In both cases, the models also correctly predicted the optimal tuning parameters and effect of variations away from the optimal parameters. Great benefits for base system energy dissipation is attained by shunting the electrodes of the piezoelectric material with appropriate passive circuits. How far use can be made of Piezoelectric elements, integrated as sensors and actuators in a closed control loop, to improve the modal damping behavior of elastomechanical systems (beams) is shown by *R.Freymann and E.Stumper* (10). Focused is pointed on some special characteristics of piezoelectric elements, such as their dynamic behavior, their energy transfer capabilities when being used as actuators and their geometric filtering possibilities when being used as sensors. It is shown that piezoceramic actuators, due to their linear behavior, when fed with voltages, which are not opposed, to their direction of polarization, are easy to be handled in analytical and experimental investigations. The experimental part of the work is concentrated on the active damping augmentation in the low frequency range of a flexible beam shaped structural system with a total length of four meters. The active control of flexural waves on a slender beam for a secondary source array which comprises either point forces or pairs of moments generated by piezoelectric actuators is discussed by M. J. Brennan, S. J. Elliott and R. J. Pinnington (11). A model of the secondary source array is first developed and coupled into the beam dynamics by using the wave approach. The behavior of the beam when three active control strategies are applied is investigated. One of these is a wave suppression strategy and the other two are based on consideration of power, specifically that of maximizing the power absorbed by a secondary force array, and that of minimizing the total power supplied to the beam by the primary and secondary forces. The best overall control strategy for global control on a finite beam is shown to be that of minimizing the total power supplied to the beam.

2.2. Passive and Active Control of Vibrations Most of the causes for failure/damage of structures, machines and reasons for excessive noise exterior to the machinery are due to the vibrating members of the machine elements. Thus, it is often necessary to kill these vibrations, if possible, at the root itself. That is, vibrations are due to be suppressed in each vibrating member itself. The subject of vibration control in flexible structures has attached wide interest over the past several decades. One of the major driving force behind such interest is the real possibility of large flexible space structures with stringent mission requirements. The question arises whether the techniques presently available would be sufficient to handle such stringent requirements. The main strategy behind the control of vibrations is to introduce damping and stiffness into the structures so as to counteract the vibration induced exceptions. There are two main techniques that are being followed conventionally in vibration control efforts. They are designated as passive and active controls. In order to control the vibrations in these structures passively, additional materials are added on an *ad hoc* basis to introduce mass to the structure and damped the vibrations. By adding mass to the existing structures, the natural frequencies are shifted away from the driving frequencies of excitation. The advantage of this type of control is that there are no moving mechanical parts involved and hence, little or no maintenance is required. However, the method has several disadvantages. The additional mass introduced may increase the over all mass of the structure which may violate the design constraints imposed on the overall weight. Quit often, it may not be possible to use this method over a broad range of operating conditions. Moreover, this method would prove to be costly to implement. Engine mounts and vehicle suspension systems are some of the examples in which passive control methods are used traditionally. Hence researchers have investigated the use of external devices that can absorb vibration energy from the structure. Such a mechanism is termed active control. Relevant schemes generally dissipate the vibration energy by opposing the vibration with use of external and discrete actuators to impart a motion or force to the structure to counteract the driving dynamics of the vibrations. These actuators and the associated sensors are called "active elements". The advent of inexpensive microprocessors has opened the door to use these active elements with feedback and automatic control feasibilities towards active control of vibrations. The use of such active elements in dissipating the energy from the structure can be equated to introducing stiffness and damping into the structure. Thus, the purpose of the active control is similar to that of passive control, but performed actively, thereby allowing certain flexibility to the designer. For example, active control allows robust mechanisms for controlling the vibrations. Some article of this technique is discussed in this section and the focus is made mainly on the control of vibrations specific to flexible structures. The development of the support electronics for these sensors and actuators is focused by *D. K. Linder and W. A. Tabisz* (12). The power electronics for driving an actuator comprised of piezoelectric material is considered. The miniaturization of power conversion/amplifier electronics for these actuators suitable for these smart structure applications is discussed. The power converters to minimize thermal dissipation and the benefits of the decentralized power converters for the power bus distribution systems are also pointed out. The technology requirements, approaches to meet the requirements using adaptive structures, and the recent jet propulsion laboratory research results in adaptive structures are described by *B. K. Wada and J. A. Garba* (13). They summarizes the research in adaptive structures at jet propulsion laboratory and the potential of active structural systems in meeting some requirements for future missions are presented. The development and experimental verification of the induced strain actuation of plate components of an intelligent structure is presented by *E. F. Crawley and K. B. Lazarus* (14). Equations relating the actuation strains created by induced strain actuators, to the strains induced in the actuator/substrate system are derived for isotropic and antisotropic plates. Simple test articles are used to verify the accuracy of the basic induced strain actuator/substrate system models, and cantilever plate test articles   consisting of a single structural plate with piezoceramic actuators symmetrically bonded to both surfaces are built and tested to verify the ability of the models to predict the strain induced in systems with extensive stiffness coupling and complicated boundary conditions. The ability of the models to predict the induced strains for more representative systems is shown, and that relatively large deflections can be obtained through induced strain actuation. substantial agreement between the experimentally measured and analytically predicted deformations is found, verifying the models developed, and demonstrating that induced strain actuation is an effective means of controlling plate deformations. A transient dynamic finite element analysis for studying the response of laminated composite plates due to transverse foreign object impact is presented by *Hsi-yung T Wu and Fu-Kuo Chang* (15). These analysis to evaluate the distributions of all the stresses and strain through laminate thickness during transverse impact, and to study the effect of laminate thickness and ply orientation on these distributions is developed. The analysis of stresses and strain in the laminates is based on 3-dimensional linear elasticity with the consideration that, in each layer, the materials are homogeneous and orthotropic. The effect of the initial velocity of the impactor on the contact force distribution, the displacements, and the transient stress and strain histories inside the plate during impact are studied analytically and numerically. An excellent agreement is found between two studies. A description of the dimensional elasticity equations and the associated finite element model for natural vibrations of laminated rectangular plates is presented by *J. N. Reddy and T. Kuppusamy* (16). The work is motivated by a lack of three-dimensional finite element results for natural frequencies of relatively thick laminated composite plates, and therefore a three-dimensional finite element is used to predict natural frequencies accurately. The numerical results for natural frequencies are compared with those obtained by a shear deformable plate theory. These results indicate that, although the element based on the three-dimensional elasticity predicts frequencies more accurately than the element based on the first-order shear deformation theory, the computational expense precludes the use of such three-dimensional elements in problems that require a large number of elements. A number of cross-ply and angle-ply rectangular plates are analyzed for natural frequencies. For relatively thick plates, the plate element predicts frequencies higher than those predicated by the 3-D element. The effect of fiber orientation and boundary conditions on the vibration behavior of stepped orthotropic square plates by using finite element method is studied by *S. K. Mathotra, N. Ganesan and M. A. Veluswami* (17). The results are compared with those of constant thickness plates. From these results it is shown that, the main characteristics of these structures (beam, plates) are that they have infinite natural frequencies with different extents of damping pertinent to each one of them. The level of damping in these structures depend on the type of material used and the environmental conditions, while the frequencies depend upon the physical properties and the geometry of the structure itself. The specific damping capacities, mode shapes, and natural frequencies of various free-free carbon and glass fiber-reinforced plastics plates are predicted and measured by *D. X. Lin, R. G. Ni and R. D. Adams* (18). The objective of this investigation is to predict the natural frequency and specific damping capacity of anisotropic laminated composite plates in various modes of vibration by using, and extending a finite element method and damped element model for predicting the damping of narrow, angle-ply laminated plates. These finite element method included transverse shear deformation and rotary inertia to predict the natural modes of free-free carbon and glass fiber-reinforced plastics plates. To investigate the accuracy of the theoretical prediction a comparison is made with the experimental results from various laminated carbon and glass fiber-reinforced plastics and graphite and glass fiber-reinforced plastics rectangular plates in the free-free condition for the six modes of vibration. The results, therefore, suggest that the finite element technique using the damped element model is a good purpose tool for the analysis of structures fabricated from composite materials.

The first of two companion papers on active control of plate vibrations are presented by *L. A. Walker and P. P. Yaneske* (19, 20). In the first of these companion paper a description is given of the essential features of an active system for deriving specific control of plate motions in the interests of damping the vibrations of a lightweight acoustical barrier, or reinforcing those of an enclosure. A signal control unit is analyzed for the feedback of energy via a corrective point force derived from local sensed motion of the plate. Conditions for the maintenance of stability over a wide frequency range depend on the restriction of phase variation between those plate points on the closed loop at which the motion sensor and control force are applied. In the second paper analysis shows that the characteristics of multiple and signal systems are essentially similar, and is concerned with four basic arrangements of motion sensors and force actuators designed to improve efficiency by utilizing the ability of a multiple system to act selectively on specific modes. Particularized optimum forward gains is applied and a gain stability depends on preservation of like transfer symmetry between sensing and forcing points. Also, this paper extends the analysis to consider possible advantages from the employment of an array of signal transducers on the plate surface. A mathematical model of the mechanical system, the sources of perturbing vibrations, the control system, and the different absorption criteria to be compared are defined by *E. Luzzato and M. Jean* (21). The problems are set in infinite dimension spaces, and the approximate problems are derived in finite dimension spaces. Two methods of resolution are proposed and the solutions are compared to solve problems of damping of vibrations in some given domain of a continuous viscoelastic structure, which is the aim of this work. Numerical results simulation the response of square plate submitted to flexural vibrations are presented for three different absorption criteria, thus permitting a quick evaluation of the comparatives of the chosen criteria. The active control of disturbances propagating along a wave-guide is considered with particular reference to the control of flexural vibrations in thin beams by *B. R. Mace* (22). The control system consists of a number of point sensors and point exciters distributed in some manner along the wave-guide. Active control of flexural waves in beams is then considered and two possible active control system are examined. The first consists of measurements of displacement and rotation at a point force and point moment while the second comprises displacement measurements at two points and the application of two point forces. In both cases the ideal active controller is derived and some aspects of the systems performances examined. In this approach to active vibration control the overall size of the structure being controlled is irrelevant, because it may be very large and hence a very large number of modes may be effected, and the control system itself may however be compact. A new technique for vibration suppression in large space structures in laboratory experiments on a thin cantilever beam is investigated by *J. L. Fanson and T. K. Caughey* (23). This technique, called positive position feedback, makes use of generalized displacement measurements to accomplish vibration suppression. The experiments control the first six bending modes of a cantilever beam and make use of piezoelectric materials for actuators and sensors, simulating a piezoelectric active-member. An analytical and experimental investigation of the possibility of actively blocking the propagation of bending waves along a uniform beam is described by *D. J. Pines and A. H. Von Flotow* (24). The investigation has the form of a case study; a thin brass plate-beam is used, it is excited with a short duration impulse, and the resulting disturbance spreads dispersively as it travels along the beam. A short portion of the beam is used as an active block. Strain gauge sensors used to drive thin piezoceramic bending moment sensors are used to drive thin piezoceramic bending moment actuators through a dynamic compensatory. The analytical work described includes the nominal design and its performance, the performance degradation due to modeling errors, and the performance degradation due to approximate implementation of the dynamic compensatory.

How the frequency considerations influence the structural design, their sensitivity, and constraint approximations, and how different type of optimization algorithms are used in solving the frequency problem are addressed in the review by *R. Grandhi* (25). Several survey papers exist in the field of structural optimization covering optimization algorithms, constraint approximations, sensitivity analysis, shape optimization, etc., are presented.

2.3. Active materials for sensing and actuation (Piezoelectric Materials) As explained in the previous section, active control which is a robust and efficient way of controlling vibrations in structures, requires active element to carry out its objective. In effect, the complete system (including the sensors and actuators) should be capable of changing the physical properties namely, stiffness and damping. Though a variety of devices can be considered as sensors and actuators, most of them may not be suitable for controlling the vibrations efficiently. There are, however, certain materials, which have the unique ability in changing their physical properties through application of external electrical stimuli. Such materials are termed as " intelligent/smart " materials. Some of the examples are piezoelectric (PZ) materials, electroheological (ER) fluids and shape memory alloys (SMA). First, if a stress is applied to certain crystalline materials they develop an electric moment whose magnitude is proportional to the applied stress. This is Known as the direct piezoelectric effect. Such crystals are known as piezoelectric materials. Conversely when an electric potential is applied to these crystals, a mechanical stress is in induced due to the reverse piezoelectric effect. Examples of the materials processing these properties are *Quartz, Polyvinyldene fluoride* (PVDF), *Lead Zirconate Titanate* (PZT) and *Barium Titanate* Commercially available (PVDFs) are often film type of various dimensions, while (PZTs) are available as thin wafers. A simple example for (PZT) actuator-driven one-degree of freedom spring-damper-mass system is used by *C. Liang, F. Sun and C. A. Rogers* (26). They illustrated the methodology used to determine the dynamic characteristics of (PZT) actuators and the structural response based on the concept of structural impedance, because the dynamic excitation provided by a (PZT) actuator depends on its dynamic characteristics and structural mechanical impedance. The work presented that the conventional method of using the statically determined induced moment or the force of (PZT) actuators as forcing functions is not correct. An approach to determining the structural impedance corresponding to two types of actuator loading is provided. Experimental and analytical studies to compare the performance of *PolyVinylidene Fluoride* (PVDF) films and (PZT) for modal analysis are conducted by *S. V. Hanagud, S. G. Savanur, G. L. Negeshbabu, C. C. Won and A. V. Srinivasan* (27). Possible effects of the geometry of the transducer on the actuation Forces are also investigated. They get a conclusions that the piezoceramic actuation indicates more flexibility and low applied voltage in comparison to (PVDF) films. The spatial distribution (PZT) of small dimensions will also make mode estimates easier. Also, (PZT) actuation with transducers located at selected subdomains provides a superior actuation with less noise and potential flexibility in the identification of modes. Experimental and theoretical results for the deformation modes of a polymer piezoelectric bimorph structure are presented by *M. Cundari and B. Abedien* (28). A relation which describes the deflection from rest axis of the beam in terms of piezoelectricity generated moment, damping factor, cantilever shape function, and time is developed using virtual work methods and it is confirmed for an oscillating three layer beam subjected to uniformly distributed piezoelectric moments and structural damping. Experiments are conducted to verify the theory and they find a discrepancy between they prediction and the actual deflections at frequencies higher or lower than the resonant frequencies. Experimentally, the derived relations is shown to be accurate in predicting tip-deflection near the mechanical resonance of the system but inaccuracies exit at other frequencies in the tested spectrum. More importantly, the alignment of theory and prediction is quite good at the resonant condition, this is where the maximum deflection for a given input signal occurs and is of most practical interest. Several models of the mechanical interaction between piezoelectric actuators and structures to which they are bonded are presented in a lot of works. Some of these different models of the strain field to developed their equivalent forcing functions are used by *C. M. Lapeter and H. H. Cundney* (29). These forcing functions are used together with finite element models to predict the forced response of beams and plates. The force response shape of these structures is simulated for a frequency signal sent to the piezoelectric actuator. This simulation is repeated at several frequencies, both on resonance and off resonance. The response of the actual beams and plates modeled in the simulation is measured using a scanning laser vibrometer. It is shown that, this instrument allowed the forced response of the beams and plates to be measured at a spatial resolution higher than the finite element model, and allowed a detailed comparison of the predicted and actual forced response shapes is presented. A coupled electro-mechanical analysis of piezoelectric ceramic (PZT) actuators integrated in mechanical systems to determine the actuators power consumption and energy transfer in the electro-mechanical systems is presented by *C. Liang, F. Sun and C. A. Rogers* (30). In these work, the power consumed by the (PZT) actuators consists of two parts: the energy used to drive the system which is dissipated in terms of heat as a result of the structural damping, and energy dissipated by the (PZT) actuators themselves because of their dielectric loss and internal damping. The coupled analysis presented a simple model for using, a (PZT) actuator-driven one-degree-freedom spring-mass-damper system, to illustrate the methodology used to determine the actuator power consumption energy flow in the couple electro-mechanical systems. The theoretical model is validated by experiment. It is presented that, this model can be applied easily to continuous mechanical system provided that the structural impedance corresponding to the actuator load is determined.

2.4. Smart Structures Smart/intelligent structures contain cohesively sensors, actuators and control mechanisms. Thus, smart structures have built-in capabilities to actively change the geometry or the physical properties of a structure. Some of the active elements used to build smart structures are discussed as before in section 2.3. Analytic and experimental development of piezoelectric actuators as elements of intelligent structures are presented *by E. F. Crawley and J. De Luis* (6). These structures has highly distributed actuators, sensors, and processing networks. Static and dynamic analytic models are derived for segmented piezoelectric actuators that are either bonded to an elastic structure are embedded in a laminated composite. A scaling analysis is performed to demonstrate that the effectiveness of piezoelectric actuators is independent of the size of the structure and to evaluate various piezoelectric materials based on their effectiveness in transmitting strain to the substructure. By other meaning, the scaling analysis also provided a systematic method based on maximizing the piezoelectric-mechanical effectiveness of the actuator and the substructure that can be used to select the piezoelectric actuator material. Three test specimens of cantilever beams are constructed, an aluminum beam with surface-bonded actuators, a glass/epoxy beam with embedded actuators. The response of the specimens is compared well with those predicted by the analytic models. The flexural static models for both surface-bonded and embedded actuators are integrated into dynamic models for a cantilevered *Bernolli-Eular* beam. The forcing applied by the actuators is shown to consists of a passive stiffness component due to the presence of the actuator and a voltage-dependent component that relates to the ability of the piezoelectric actuator to impart a force on the structure. Also, static tests determined how the elastic properties of the composite are effected by the presence of an embedded actuator are carried out on glass/epoxy laminates, where, the ultimate strength of the beam is reduced by approximately 20%, while not significantly affecting the global elastic modulus of the specimen. Incorporating the piezoelectric effect into classical laminated plate theory, distributed sensors and actuators capable of sensing controlling the modal vibration of a one-dimensional cantilever plate are derived theoretically by *C. K. Lee, W. W. Chaing and T. C. O'Sullivan* (31). It is shown that critical damping can be achieved using such a sensor/actuator combination for a one-dimensional cantilever plate. In these work, the sensor signal was proportional to the time derivative of the modal coordinate, which was easily used for velocity feedback control. Also, by eliminating electromagnetic interference and ground loop noise, critical damping is experimentally demonstrated for the first mode of a one-dimensional cantilever plate using (PVF) as the sensor/actuator material. An analytical investigation to study the mechanical and electrical response to laminated composite plate containing distributed piezoelectric ceramics acting as actuators and sensors is performed by *S. K. Ha and F. K. Chang* (32). A finite element analysis is developed for simulating the response of the structures subjected to mechanical or electric loading. The development of the analysis is based on the theory of elasticity for antisotropic and inhomogeneous materials integrated with electric theory for piezoelectric. An eight-node three-dimensional composite brick element is adopted for the analysis, and three incompatible modes are considered to account for the bending response of the composites due to deformations of the piezoelectric. The analytical, numerical calculations are performed to simulate the experimental results obtained from the literature. The behavior of two dimensional patches of piezoelectric material bonded to the surface of elastic distributed structures and used as vibration actuators is analytically investigated by *E. K. Dimitriadis, C. R. Fuller and C. A. Rogers* (33). A static analysis used to estimate the loads is induced by the piezoelectric actuator to the supporting elastic structure. They then apply to develop on approximate dynamic for the vibration response of simply supported elastic rectangular geometry. The results demonstrate that modes can be selectively excited and that the geometry of the actuator shape markedly affects the distribution of the response among modes. In there study, It is appeared that it is possible to tailor the shape of the actuator to either excite or suppress particular modes leading to improved control behavior. A finite element formulation for vibration control of a laminated plate with piezoelectric sensors/actuators is presented by *W-S. Hwag and H. C. Park* (34). The total charge developed on the sensor layer is calculated from the direct piezoelectric equation. The piezoelectric sensor is distributed, but is also integrated since the output voltage is dependent on the integrated strain rates over the sensor area. Also, the piezoelectric actuator induces the control moments at the ends of the actuator. Therefore, It is shown that the number, size, and location of the sensors/actuators are very important in the control system design. Classical laminate theory with the induced strain actuation and *Hamilton's* principal are used to formulate the equations of motion. By selective assembling of the element matrices for each electrode, responses with various sensor/ actuator can be investigated. The static of piezoelectric bimorph beam are calculated. Finally, for a laminated plate under the negative velocity feedback control, the direct time responses are calculated and the damped frequencies and modal damping ratios are derived by modal state space analysis. An overview and assessment of the technology leading to the development of intelligent structures are presented by *E. F. Crawley* (35). An overview of the critical component technologies necessary to infuse distributed control functionality into structures is presented, followed by examples of synthesis into contemporary applications. The paper concludes with a summary of the art and research needs and a vision of the developments of the future. Smart structures as a technology for next generation aircraft is discussed by *W. Schmidt and C. Boller* (36). Actual research work is concentrated on active vibration control for a cantilever beam made of carbon fiber reinforced polymer (CFRP) and is equipped with an accelerometer (sensor) and an actuator mode of piezoelectric ceramics. Sensor signals are fed into the signal generator which generate the signal for the actuator. The test objective is to look at the actuator performance. The results shows that damping is significantly increased when using the piezoelectric actuator with control. Also, recent developments in piezoelectric and electrostrictive sensors and actuators for smart structures are discussed by *L. E. Cross* (37). A recent advances in the fabrication of lightly perfect thin films of (PZT) type compositions on silicon, by both vapor and liquid phase techniques provides a new high force electromechanical power source for microelectromechanical systems are presented. The current status of the evaluation of electromechanical response in (PZT) on silicon is briefly reviewed. The optimal location of actuators for active damping of vibration is discussed by *J. Holnicki-Szulc, F. Lopez-Almansa and J. Rodellar* (38). A computationally efficient procedure for selection of actuators is described in (28). This method formulated in modal coordinates and is based on the so-called progressive collapse analogy of an adjoin structure made of a hypothetical elastic-brittle material and having the same shape as the eigen modes of the structure to be controlled. The computer simulation of progressive collapse for the adjoin structure provides with the location of the most sensitive spots, the best for placement of actuators. Analytical techniques, with experimental results for using piezoelectric actuators as elements of intelligent structures are presented by *E. F. Crawely, J. de Luis, N. W. Hagood, and E. H. Anderson* (39). Mechanical models for the interaction of piezoelectric with one- and two-dimensional structures are derived. In addition, a model coupling the electrical and mechanical properties of the piezoelectric is presented. This model is used to derive optimal resistive and tuned electrical circuits to maximize energy dissipation in the structure. These results are implemented on a prototype space-truss. Finally, a technique for embedding piezoelectrics inside laminated composite structure is presented. A technique, which allows a single piece of piezoelectric material to concurrently sense and actuate in a closed loop system is developed by *J. J. Dosch. D, J. Inman, and E. Garcia* (40). The motivation behind this technique is that a self-sensing actuator will be truly collocated and has applications in active and intelligent structures, such as vibration suppression. A theoretical basis for the self-sensing actuator is given in terms of the electromechanical constitutive equations for a piezoelectric material. In a practical implementation of the self-sensing actuator an electrical bridge circuit is used to measure strain. The bridge circuit is capable of measuring either strain or time rate of strain in the actuator. The usefulness of the proposed device is experimentally verified by actively damping the vibration in a cantilever beam. A single piezoelectric element is bonded to the base of the beam functioned both as a distributed moment actuator and strain sensor. Also by using a feedback control law, the first mode of vibration is suppressed and by using a positive position feedback law the first two modes of vibration are suppressed. A set of piezopolymer devices based on a composite laminate theory for piezoelectric polymer materials is developed by *C. K. Lee and F. C. Moon* (41). By using different combinations of ply angles and electrode patterns, a Piezopolymer/metal shim plate structure is built that exhibited both bending and torsion deformation under an applied electric field. A set of torsion-beam sensor structures is also built that could distinguish between bending and torsion or between different vibration modes. The experimental results agreed quite closely with the theoretical predictions. A techniques for modeling induced strain actuation of beam-like components of intelligent structures are developed by *E. F. Crawley and E. H. Anderson* (42). Two analytical models and one numerical model describing the detailed mechanics of induced strain actuators bonded to and embedded in one-dimensional structures are presented. The models illustrate the extension, bending, and localized shearing deformations induced. The range of parameters for which the simpler analytic models are valid is also establishes. The specific characteristics of one type of induced strain actuator, piezoceramic materials, are discussed and experimental results are used to validate the beam actuation models presented. A study on the vibration control of elastic or flexible robot structures is presented by *H. S. Tzou and G. C. Wan* (43, 44) in two parts. Effect of distributed passive (part I) and active (part II) actuators on elastic robot structures are studied. The proposed distributed passive viscoelastic actuator (part I) is a layer (or layers) of viscoelastic polymer directly attached to the flexible robot element, the oscillation of which is to be controlled. The passive actuator is activated by the oscillation of the robot structure and it automatically dissipates vibration energy and constrains the undesirable motion to eliminate the disturbance and to maintain a precise robot trajectory. Experimental results and finite element simulations of the electromechanical sensor/actuator are presented and discussed.

4. CONCLUSIONS AND RECOMMENDATIONS

All structures must have sufficient damping to keep vibration and acoustic response within acceptable limits. The current interest in large and flexible space structures provides new motivation and requirements for structural damping enhancement for vibration control. Vibration and noise can be reduced by a number of means. These can be broadly divided into two groups active and passive control methods. Active control can be used to suppress the vibration (or noise) by the use of certain active elements such as actuators, which alter the dynamic (or acoustic) response of the structure. It requires the use of special purpose hardware, and real-time control algorithms. In passive control, energy dissipation can be achieved by external add-on damping devices such as isolators or constrained viscoelastic layers. Recent developments in understanding the dynamic properties of viscoelastic materials have resulted in many innovative means to enhance the inherent damping in structures. Damping is now recognized as a design parameter in all stages of design and development of structures subjected to dynamic loading. In this paper the previous and current research efforts for viscoelastic passive damping or active damping to control the vibroacoustic response of structures (beams, plates) are presented. The effect of both passive and active damping to control the dynamic behavior of laminated composite structures (especially plates) should be considered for future work. The fiber orientation used in some of the previous research was unidirectional and cross-ply. There is many other fiber orientation that are used like quasi-isotropic and angled plies. The effects of passive damping on laminates with these different orientation should be studied.

5. REFERENCES

1. Y. P. Lu, H. C. Neilson, and A. J. Roscoe, <u>Composite Materials</u>, <u>27</u>[16], 1598-1605, [1993].
2. R. G. Ni, and R. D. Adams, <u>Composite Materials</u>, <u>18</u>, 104-121, [1984].
3. T. Bailey, and E. Hubbard Jr., <u>Guidance, Control, and Dynamics</u>, 8[5], [1985].
4. S. Hanagud, M. W. Obal, and M. Meyyappa, <u>AIAA/ASME/ASCE/AHS Structures, Structural dynamics and Materials conference 26th</u>, Orlando, Florida,443-452, 15-17 April, [1985].
5. E. F. Crawley, and E. H. Anderson, <u>AIAA/ASME/AHS/ASC Structures, Structural Dynamics and Materials Conference 30th</u>, Mobile, Ala., [1989].
6. E. F. Crawley, and J. de Luis, <u>AIAA/ASME/ASCE/AHS Structures, Structural Dynamics and Materials Conference 27th</u>, San Antonio, Texas, 116-124, 19-21 May, [1987].
7. C. T. Sun, B. V. Sankar, and V. S. Rao, <u>Sound and Vibration</u>, <u>139</u>[2], 277-287, [1990].
8. N. W. Hagood, W. H. Chung, and A. von Flotow, <u>AIAA/ASME/AHS/ASC Structures, Structural Dynamics and Materials Conference 31st</u>, 2242-2256, [1990].
9. N. W. Hagood and A. von Flotow, <u>Sound and Vibration</u>, <u>146</u>[2], 243-268, [1991].
10. R. Freymann, and E. Stumper, <u>AGARD structures and materials panel meeting 75th</u>, Lindau, Germany, 30.1-30.12, 5-7 Oct., [1992].
11. M. J. Brennan, S. J. Elliott, and R. J. Pinnington, <u>Sound and Vibration</u>, <u>186</u>[4], 657-688, [1995].
12. D. K. Linder and W. A. Tabisz, <u>North American Conference on Smart Structures and Materials</u>, 1-14, Jan31-Feb4, [1993].

Materials, 1-14, Jan31-Feb4, [1993].

13. B. K. Wada, and J. A. Garba, <u>AGARD structures and materials panel meeting 75th</u>, Lindau, Germany, 28.1-28.13, 5-7 Oct., [1992].

14. E. F. Crawley, and K. B. Lazarus, <u>AIAA/ASME/AHS/ASC Structures, Structural Dynamics and Materials Conference 30th</u>, 1451-1461, Mobile, Ala., [1989].

15. H.-Yung, T-Wu, and F.-Kuo Chang, <u>computers and structures</u>, 31[3], 453-466, [1989].

16. J. N. Reddy, and T. Kuppusamy, <u>Sound and Vibration</u>, 94[1], 63-69, [1984].

17. S. K. Malhotra, N. Ganesan, and M. A. Veluswami, <u>Computer and Structures</u>, 31[5], 657-661, [1989].

18. D. X. Lin, R. G. Ni, and R. D. Adams, <u>Sound and Vibration</u>, 139[2], 277-287, [1990].

19. L. A. Walker, and P. P. Yaneske, <u>Sound and Vibration</u>, 46[2], 157-176, [1976].

20. L. A. Walker, and P. P. Yaneske, <u>Sound and Vibration</u>, 46[2], 177-193, [1976].

21. E. Luzzato, and M. Jean, <u>Sound and Vibration</u>, 86[4], 455-473, [1983].

22. B. R. Mace, <u>Sound and Vibration</u>, 114[2], 253-270, [1987].

23. J. L. Fason, and T. K. Canghey, <u>AIAA/ASME/AHS/ASC Structures, Structural Dynamics and Materials Conference 28th</u>, 588-598, [1987].

24. D. J. Pines and A. H. von Flotow, <u>Sound and Vibration</u>, 142[3], 391-412, [1990].

25. R. Grandhi, AIAA Journal, 31[12], 2296-2303, [1993].

26. C. Liang, F. Sun, and C. A. Rogers, <u>Proceedings, Smart Structures and Materials</u>, 1-4 Feb., [1993].

27. S. V. Hanagud, S. G. Savanur, G. L. Nageshbabu, and C. C. Won, <u>ASME, Smart Structures and Materials J.</u>, 123, 19-24, [1991].

28. M. Cundari, and B. Abedian, <u>ASME, Smart Structures and Materials J.</u>, 123, 25-31, [1991].

29. C. M. Lapeter, and H. H. Cudney, <u>ASME, Smart Structures and Materials J.</u>, 123, 139-143, [1991].

30. C. Liang, F. Sun, and C. A. Rogers, <u>Proceedings, Smart Structures and Materials</u>, 1-4 Feb., [1993].

31. C.-K. Lee, W.-W. Chiang, and T. C. O'sullivan, <u>AIAA/ASME/AHS/ASC Structures, Structural Dynamics and Materials Conference 30th</u>, San Antonio, Texas, 1451-1461, [1989].

32. S. K. Ha, and F. K. Chang, <u>AIAA/ASME/AHS/ASC Structures, Structural Dynamics and Materials Conference 31st</u>, 2323-2330, [1990].

33. E. K. Dimitriadis, C. R. Fuller, and C. A. Rogers, <u>ASME, Smart Structures and Materials J.</u>, 113, 100-107, [1991].

34. W.-S. Hwang, and H. C. Park, <u>AIAA J.</u>, 31[5], 930-937, [1993].

35. E. F. Crawley, <u>AIAA J.</u>, 32[8], 1689-1699, [1994].

36. W. Schmidt, and C. Boller, <u>AGARD Structures and Materials Panel Meeting</u>, Lindau, Germany, 1.1-1.14, 5-7 Oct., [1992].

37. L. E. Gross, <u>AGARD Structures and Materials Panel Meeting</u>, Lindau, Germany, 16.1-16.33, 5-7 Oct., [1992].

38. J. H-Szulc, F.L-Almansa, and J. Rodellar, AIAA Journal, 31[7], 1274-1279, [1993].

39. E. F. Crawley, J. de Luis, N. W. Hagood, and E. H. Anderson, <u>American Control Conference 7th</u>, Atlanta, 1890-1896, [1988].

40. J. J. Dosch, D. J. Inman, and E. Garcia, <u>Intelligent Material System and Structures</u>, 3, 166-185, [1992].

41. C. K. Lee, and F. C. Moon, <u>Acoustic society of America,</u>, 85[6], 2432-2439. [1989].

42. E. F. Crawley, and E. H. Anderson, <u>Intelligent Material System and Structures</u>, 1, 4-25, [1990].

43. H. S. Tzou, and G. C. Wan, <u>Computer and Structures</u>, 35[6], 669-677,[1990].

44. H. S. Tzou, C. I. Tseng, and G. C. Wan, <u>Computer and Structures</u>, 35[6], 679-687,[1990].

DYNAMIC RESPONSE OF CARBON/EPOXY CANTILEVER PLATES WITH PASSIVE DAMPING

Khaled A. Alsweify, Cecilia Booker, Eltahry I. Elghandour,
Faysal A. Kolkailah, Ph.D.,P.E
California Polytechnic State University
San Luis Obispo, CA 93407
and
Said H. Farghaly, Ph.D.
Associate Dean,
Faculty of Engineering (Mataria), Helwan University, Cairo, Egypt.

ABSTRACT

This paper presents an experimental investigation of the free vibration of rectangular composite plates with passive damping. The type of passive damping used in this study is an interlaminar viscoelastic damping layer, in which the vibrational energy is dissipated through shear deformation. The damping material employed is a 3M material (SJ-2015 ISD112) with peak damping properties in the ambient temperature range. A number of 100 mm x 200 mm graphite (IM7)/epoxy cantilever plates with and without damping were tested. The lay-up sequence for the undamped plates is $[0/90/0/90]_s$, and for the damped plates, a viscoelastic damping layer was added. Fabrication and test procedures are described. The natural frequencies, amplitude and damping ratios were determined. The variation of damping ratio with damping layer thickness is discussed, and a comparison between the time and frequency response for plates with and without damping are presented. The results show that by adding a viscoelastic damping layer to the composite structural system, the amplitude and natural frequencies decrease.

KEY WORDS: Damping, Passive, structure, Vibration

1. INTRODUCT1ON

Conventional structural designs are often unacceptable in coping with modern problems of structural resonance caused by the complex nature of the dynamic environments. Current interest in large flexible space structures provides new motivation and requirements of structural damping enhancement for vibration control. The objective of enhancing damping in structural elements is to control the response of the elements in order to prevent catastrophic failure due to excessive deformation. Advanced composite materials are used in a variety of components for aerospace applications involving dynamic loading conditions. Structural damping needs to be improved to reduce fatigue failure, noise and unwanted vibrations caused by dynamic loading. This may be accomplished by optimizing the internal structural configuration, or by using some type of vibration control technique. Vibration control techniques may be classified into active sensors and actuators. Passive control enhances the damping of a structure by using materials with better internal damping.

In this study, passive damping is used to enhance the damping of a carbon/epoxy cantilever plate under free vibration by adding a viscoelastic damping layer to the middle surface of the plate. Cantilevered flat plates are convenient simulations for structural components, such as wings, horizontal and vertical stabilizers, and compressor fan blades for aircraft engines.

A number of papers have appeared which explore the effects of interlaminar damping of beam and plate structures. Ditaranto and Mcgraw [1] investigated the dissipation of vibratory energy in sandwich plates with a viscoelastic core. Only transverse inertial effects were included in the analysis. The solution for the damping was given for simply supported edges and a relation between modal frequency and loss factor was obtained. Khatua and Cheung [2] used a finite element technique for the study of elastic multi-layer beams and plates. Orthotropy was included in the analysis, but rotatory and translatory inertia effects were neglected. Rao and Nakra [3] analyzed the flexural vibration of unsymmetrical sandwiched beams and plates with viscoelastic cores. Longitudinal translatory and rotatory, and transverse inertial effects were included. Muckhopadhyay and Kingsbury [4] studied the effect of facing anisotropy on the damped response to harmonic transverse dynamic loads of a rectangular sandwich plate with a viscoelastic core. The work pointed out that when a composite structure undergoes flexural deformation, the anisotropic facings will deform under normal strain and undergo shear deformation in their own plane. Barrett [5] developed a comprehensive model to predict the damping of composite laminated plates with a viscoelastic layer. The effect of stress coupling and compliant layering was examined. Saravanos and Pereira [6] developed a discrete layer laminate theory for composite laminates with damping layers by incorporating a piecewise continuous displacement field through the thickness. Nadella and Rao [7] applied the modal strain energy method proposed by Johnson and Kienholz [8] to estimate the modal parameters of the multi-damping layer anisotropic laminated composite beams. Gerst, Rao and He [9] presented experimental results for composite beams with single and double damping layers, and demonstrated that cocuring is an effective way of fabricating highly damped composite structural components.

2. EXPEREMENTAL PROCEDURE

Three different types of flat composite cantilever plates were fabricated and tested to determine the natural frequencies, damping ratios and modal amplitudes associated with each resonant peak. The materials used are listed in Table 1.

Material	Specifications
Carbon/epoxy prepreg Bryte Technologies, Inc.	fM7 carbon fiber, 977-2 epoxy resin 0.1625mm thick
3M Viscoelastic Damping Layer	SJ2015 Type 1205, SJ2015 Type 1210 Pe operating range 32-140'F
Piezoelectric Ceramics Morgan Matrix Inc.	Lead zirconate titanate 0.275 mm thick

Table 1. Material Specifications

The first type of the plates fabricated consisted of 8 layers of prepreg in a cross-ply sequence, $[0/90/0/90]_s$. The prepreg was layed-up in 200 mm x 225 mm sheets, cured in a hot press using the cycle presented in Figure 1, and finally the specimens were cut into 100 mm x 225 mm laminates.

The other two types of plates are the same size and have the same sequence as the first type with an added 3M viscoelastic damping layer. The damping layer thickness for the second plate is 0.125 mm and for the third 0.25 mm. These plates were fabricated by laying up 4 layers of prepreg in a cross-ply sequence, [0/90/0/90], then cured using the same cycle presented in Figure 1. The laminates were cut into 100 mm x 225 mm plates and an adhesive damping layer of the same size was sandwiched between two 4-layer laminates. A 8.649 KPa compressive force was applied to the sandwiched plate for 24 hours at 90° F in order to obtain a good bond between the adhesive and laminates.

Piezoelectric ceramic sensors were bonded to the plates with a conductive epoxy at a distance of 22 mm from the clamped end. The conductive epoxy allows the surface of the plate to be used as a common negative pole for the sensors.

Two 25 mm x 100 mm x 6.25 mm aluminum bars were bolted to one edge of the plate and then mounted on an electromagnetic shaker. Dynamic testing was performed at room temperature. Time response and frequency response data was acquired using National Instruments AT-NUO-16F-5 Data Acquisition (DAQ) card in conjunction with National Instruments LabViEW for Windows Network Analyzer Virtual Instrument (VI).

3. RESULTS AND DISCUSSION

Figure (2) presents the undamped and the damped composite plates frequencies. The response of the damped composite plate with double thickness (3M) material exhibits the expected low damping as seen in the same figure. Measurements for the undamped case should be used only as a base line. The decrease in frequency and the increase in damping can be seen in the damped composite plate with single and double thickness (3M) material damping layer. Figure (2) also shows the comparison between the one layer and two layers damping material combinations. It can be seen that the addition of another layer of damping causes more effect in the higher modes. Because of the shear deformation of the viscoelastic layer these variation causes mechanical energy to be dissipated into heat. As seen in figure (3) the addition of double thickness layer of viscoelastic material further reduce the amplitude of vibration and he damping ratio in all the fundamental bending modes frequencies. The increase in the damping ratio is due to additional shear deformation in the viscoelastic layer of the composite plate. Table (2) shows a comparison for the natural frequency between the experimental and numerical values for the cantilever composite plates without damping layer. The numerical analysis values are obtained by using the finite element program "COSMOS/M". One can see that the agreement between theory and experiment for the natural frequencies is very good.

Lay-up	Mode 1 (Hz)		Mode 2 (Hz)		Mode 3 (Hz)	
	Num.	Exp.	Num.	Exp.	Num.	Exp.
$[0/90/0/90]_s$	44	41	277	259	775	712
$[0/90/0/90/3M]_s$	---	33	---	---	---	---
$[0/90/0/90/2(3M)]_s$	---	25	---	---	---	---

Table .2 Comparison of numerical and experimental for the natural frequency of laminated composite plates with fixed-free end conditions

4. CONCLUSION

The objective of this research has been to verify that the addition of viscoelastic materials into carbon epoxy composite plates is physically possible to suppress the vibration and increase the damping in the structure. This was accomplished by fabricating plates of carbon epoxy composite with and without damping material. The plates were tested to determine the damping ratios and natural frequencies. The addition of one viscoelastic material layer gives an excellent increase in the damping ratio with a corresponding decrease in the dynamic stiffness of the structure. The addition of two layers of viscoelastic material provides a little increase in the damping ratio compare to the one layer with a further decrease in the dynamic stiffness of the structure, and with a little decrease in the structure's natural frequencies. This study has shown that the amount of damping is strongly dependent on natural frequency and the mode shape of the structure. Finally, very good agreement between numerical and experimental was found for the cantilever composite plate without the damping viscoelastic material.

5. REFERENCES

1. Ditaranto, R. A., and J. R. McGraw, Jr., Journal of Engineering for Industry, 91, 1090 (1969).
2. Khatua, T. P., and Y. K. Cheung, International Journal of Numerical Methods for Engineers. 6(1), 11 (1973).
3. Rao, Y. V. K- Sadasiva, and B. C. Nalua, Journal of sound and Vibration 34(3), 309 (1974).
4. Mukhopadhyay, A- K., and H. B. Kingsbury, Journal of sound and Vibration 47(3), 347 (1976).
5. Barrett, D. J., Journal of sound and Vibration 154(3), 453 (1992).
6. Saravanos, D. A., and J. NL Pereira, AIAA Journal 30(12), 2906 (1992).
7. Nadella, Satish, and Mohan D. Rao, Proceedings of the 13th international modal analysis conference, Nashville, Tennessee VI, 1995, PP. 233-239.
8. Johnson, Conor D., and David A. Kienhoiz, AIAA Journal 20(9), 1284 (1981).
9. Gerst, D., M. D. Rao, and S. He, Proceedings of the Damping of multiphase Inorganic Materials Symposium, Chicago, Illinois, 2-5 November 1992, PP. 85-93.

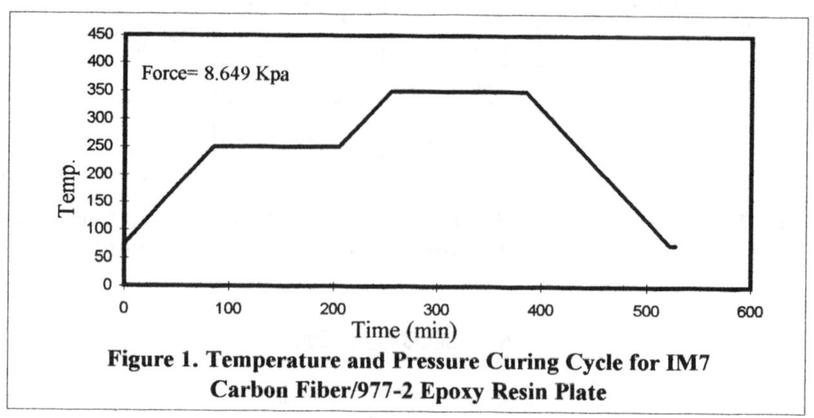

Figure 1. Temperature and Pressure Curing Cycle for IM7 Carbon Fiber/977-2 Epoxy Resin Plate

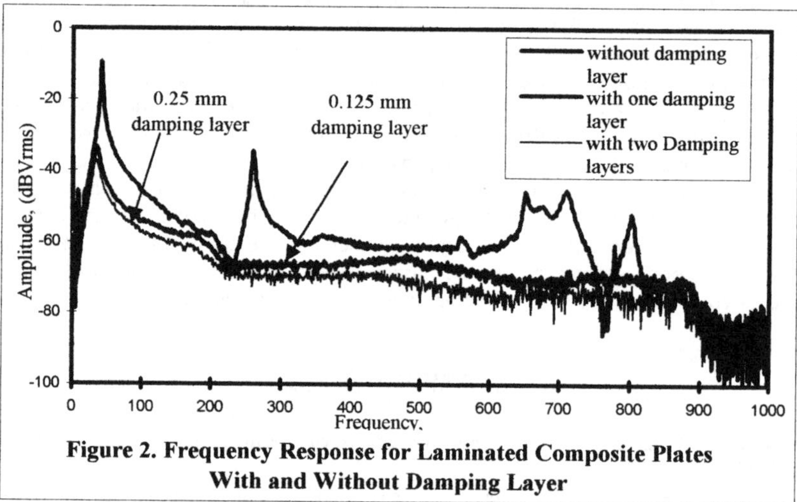

Figure 2. Frequency Response for Laminated Composite Plates With and Without Damping Layer

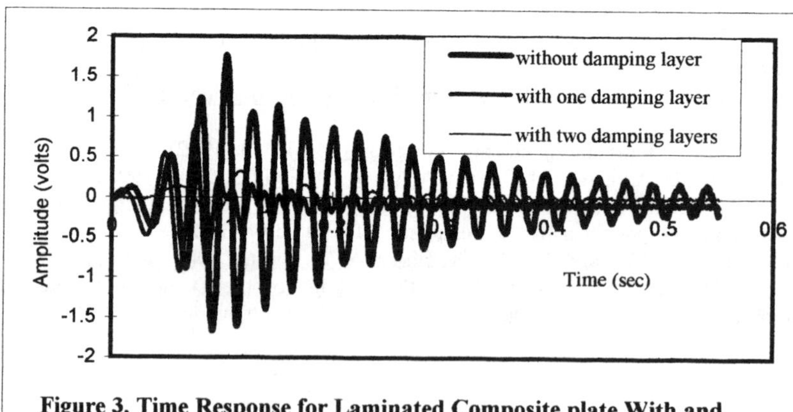

Figure 3. Time Response for Laminated Composite plate With and Without Damping Layers

VARIABLE COMPLIANCE
VIA
MAGNETO-RHEOLOGICAL MATERIALS

A. J. Millar Henrie
Boeing Commercial Airplane Group
Everett, WA 98203
L. L Howell
Mechanical Engineering Department
Brigham Young University
Provo, UT 84602

ABSTRACT

This paper presents the behavior of compliant, magneto-rheological segments and the change in rigidity that occurs when they are placed in a magnetic field.   Segments were fabricated out of flexible tubing filled with magneto-rheological material. Force/deflection patterns of the segments were obtained where the magnetic field strength, segment length, tubing size and volume fraction of iron filings in the magneto-rheological material were varied. A finite-element analysis of these force deflection patterns provided an equivalent modulus of elasticity E_C for the segments. E_C varied from 1.38 Mpa (200 psi) to 6.90 Mpa (1000 psi) when the magnetic field was increased. This resulted in a five fold increase in the rigidity of the segments. From E_C the modulus of elasticity of the magneto-rheological material, E_{MR}, was determined. The relationship between increase in E_{MR} and increase in magnetic field strength, H_o, was assumed to be linear in the range of magnetic field strengths used. Statistical analysis showed this relationship E_{MR}/H_o, or κ, depends upon the volume fraction or iron filings, the length of the segment, the surface area contact of the magneto-rheological material with the tubing and the volume of magneto-rheological material in the segment. An equation is provided that predicts the increase of rigidity for various segments.

KEY WORDS: Smart Materials, Magneto-Rheological, Compliant Mechanisms

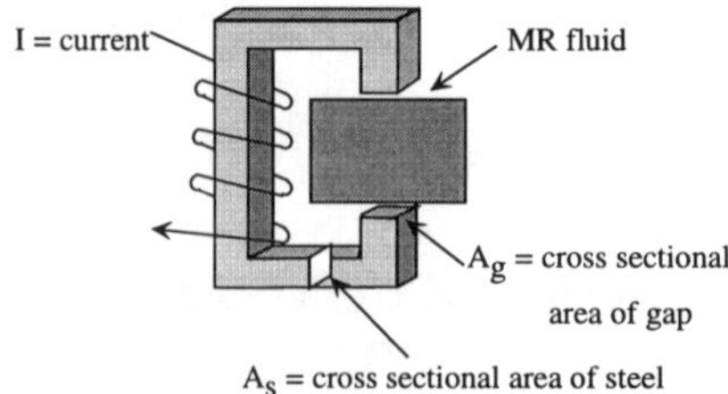

Figure 1: Magnetic circuit and MR material.

1.   INTRODUCTION

Magneto-Rheological (MR) materials change from a liquid state to a solid state when placed in a magnetic field. This can cause a change in rigidity of compliant segments made from this material. Cantilevered compliant segments constructed of flexible tubing filled with magneto-rheological material were subjected to end loads in a magnetic field. A model to predict the change in apparent modulus of elasticity, E_{MR}, as a function of magnetic field strength H_o was developed. This relationship allows predictions of force/deflection behavior of compliant mechanisms made with such segments to be made. Once this relationship is known, compliant magneto-rheological segments may be implemented into compliant mechanism designs. Mechanisms constructed with these segments would have behavior that is dependent upon the applied magnetic field. Segments would vary from the extremes of a flexible segment with no resistance to deflection, to a rigid segment that allows no deflection. This would result in mechanisms whose behavior can be changed as a function of time.

2.   MAGNETO-RHEOLOGICAL MATERIALS

A magneto-rheological material is a material that changes its internal structure, and as a consequence its rheological properties, when placed in a magnetic field. Changes include varying viscosity, elasticity, plasticity, optical, electrical, acoustic and volumetric and mechanical properties (4,10). MR fluids typically consist of micron-size, magnetically polarizable particles suspended in a fluid, most commonly mineral oil or silicone oil. When a magnetic field is applied across this mixture the particles align forming chains that restrict fluid flow (11) (Figure 1). Yield stresses in excess of 80 kPa are easily obtained (9). When the magnetic field is removed, the material immediately regains its fluid properties.

There is no current universally accepted theory for the causes of the MR effect, however it is generally agreed that the change in properties is due to the realignment of particles in the fluid to form fibrils, or long strands of suspended particles, that resist shear (Figure 2). The observed alignment of particles is believed to be related to the displacements and torque produced in the medium by the field and the translational motion and relocation of particles

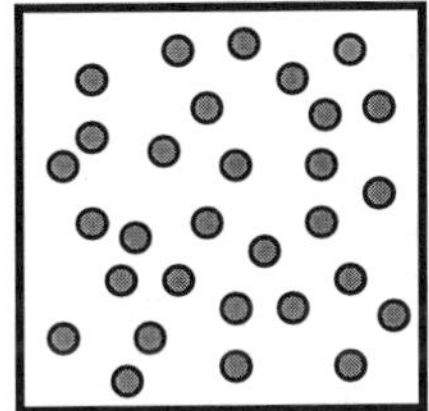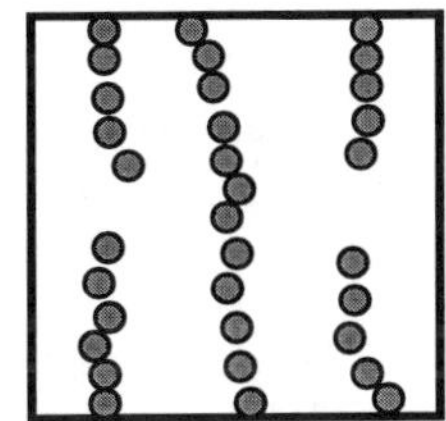

Figure 2a: Suspended iron particles before a magnetic field is applied, b: aligned iron particles after a magnetic field is applied.

to positions having minimum potential energy (4,10). A Bingham plastic model is often considered sufficient to model MR devices for design purposes.

3. MODELING

Modeling compliant cantilever beams, constructed as shown in Figure 3 with a PVC shell and a magneto-rheological filling, posses several unique difficulties when experiencing large deflections in a magnetic field. Various models are used to explain the behavior of these segments. Explanations of those models and their assumptions follow.

4.1 The Effect of Volume Fraction   The average magnetic induction B_o is measured in the air gap of the magnet. To find the applied magnetic field H_o within the MR fluid the effective permeability of the MR suspension μ_{eff} must be determined where μ_o is the normal permeability assumed here to be 1 and the relationship between them is :

$$H_o = \frac{B_o}{\mu_o \mu_{eff}} \tag{1}$$

Tang and Conrad (9) determined that the effective dielectric constant for heterogeneous dielectrics could be modeled by assuming that the particles form many simple cylindrical aggregates of a simple cubic lattice. At low magnetic fields the relationship can be assumed to be linear and a function of volume fraction (Figure 4).

4.2. Torsion Due to a Magnetic Field When a magnet is placed in a uniform, external magnetic field B_o with σ, the dipole moment of the material, making an angle θ with B_o a torque τ will act on the magnet to restore it to its lowest energy level which is

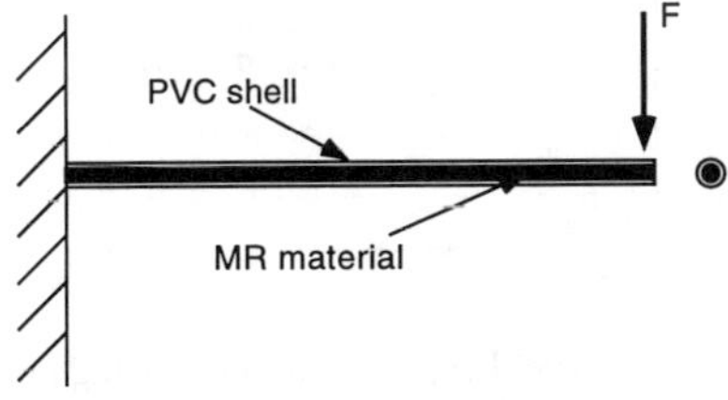

Figure 3: End force F, applied to a cantilever MR segment.

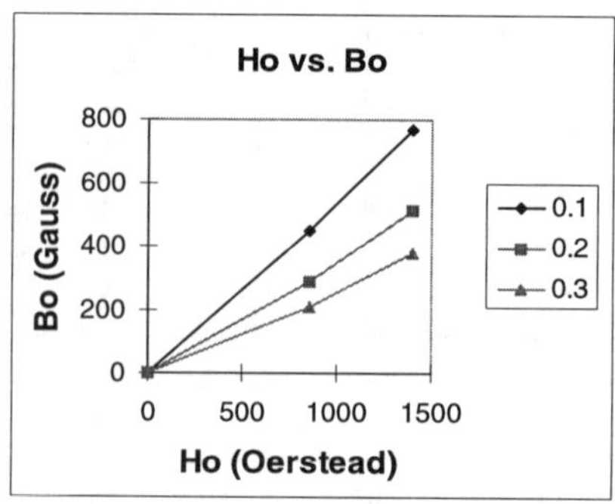

Figure 4: Ho vs. Bo as volume fraction changes.

parallel with the magnetic field (Figure 5). This relationship is defined as:

$$\tau = \sigma \times B_0 \tag{2}$$

The magnetic dipole moment of MR materials σ has not yet been determined but is assumed to be a function of volume fraction of iron particles and volume of MR material

$$\sigma = c\sigma_o(\varsigma V) \tag{3}$$

where σ_o is the magnetic dipole of the iron filings, ς is the volume fraction of iron in the MR, V is the volume and c is an unknown constant. The torque that will act on the magnet is then found by:

$$\tau = \sigma B \sin(\theta) \tag{4}$$

The angle of deflection θ, of the individual fibrils formed in the MR material is not known. It was originally assumed that the lowest energy arrangement of the fibrils would be continuous strings from end to end of the beam aligned in the field (Figure 6a). It was assumed that the fibrils would maintain this relationship when rotated through the field. The angle of deflection of the long fibrils would then increase along the length of the beam and also the beam should become more stiff as the angle increased.

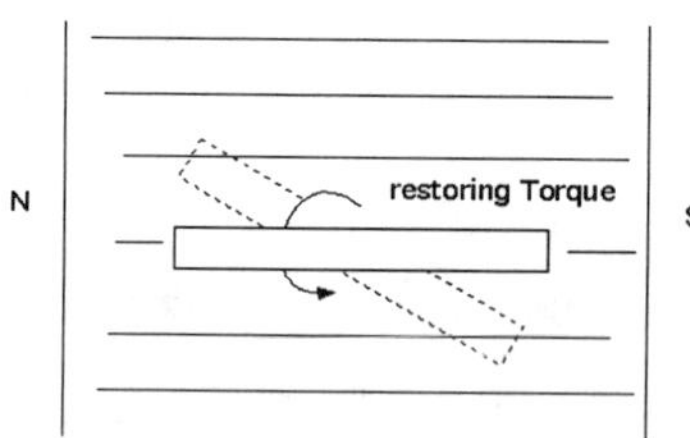

Figure 5: A magnet in a magnetic field experiences a restoring torque when it is displaced from its lowest energy position which is paralell to the magnetic field.

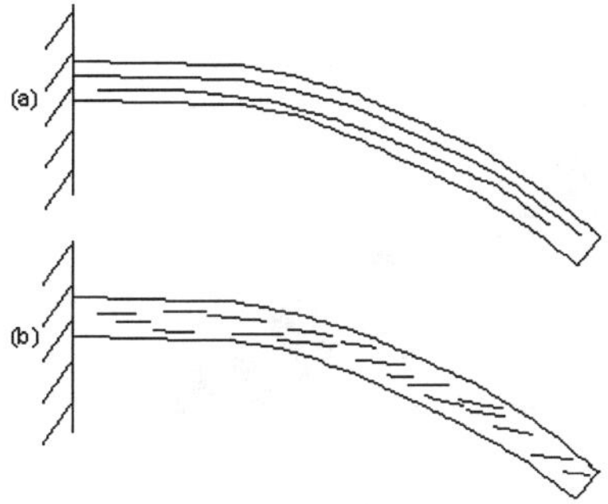

Figure 6: (a) fibrils formed the entire length of the beam, (b) short fibrils aligning with the magnetic field.

Statistical analysis showed that the angle of deflection was not an important factor in the change of stiffness. A second theory is that smaller fibrils align with the field as the beam is bent, experiencing only small angles of rotation (Figure 6b). Individual fibrils break off to maintain the lower energy position of remaining parallel with the magnetic field. The increase in stiffness then becomes largely due to the interaction of fibrils.

It was also shown that the length of the beam is very important in determining the change in stiffness as a magnetic field is applied. Longer beams experience a much greater change in stiffness than do short beams. For two similar compliant MR beams that differ only in length (Figure 7) and assuming similar angles of deflection for individual small fibrils in the MR fluid, then:

$$T_a = c\sigma_o\varsigma\frac{\pi}{4}d_i L_a B\sin\theta \tag{5}$$

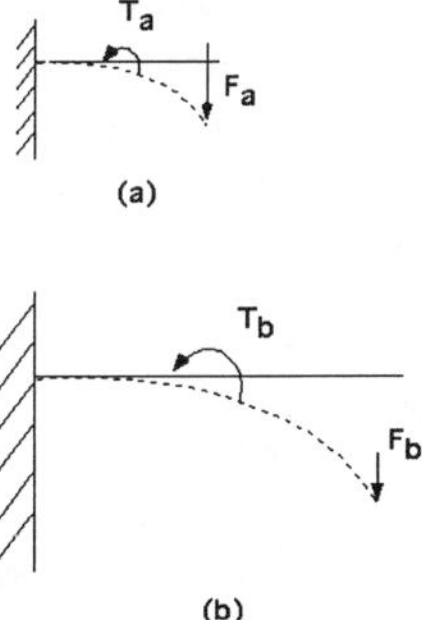

Figure 7: Two fixed-free segments deflected to the same angle that differ only in length. Tb > Ta which results in beam b changing stiffness in a magnetic field much more than beam a.

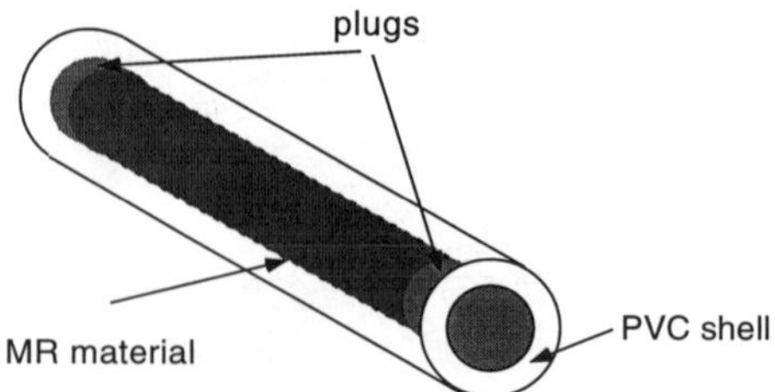

Figure 8: Magneto-rheological compliant segment made of PVC tubing sealed at both ends with the MR fluid in the center.

$$T_b = c\sigma_o\varsigma\frac{\pi}{4}d_i L_b B\sin\theta \tag{6}$$

where $L_b > L_a$ and therefore $T_b > T_a$. This relationship is also followed for increasing volume and volume fraction. This results in the change in stiffness of the longer beam, or the beam with the larger volume fraction or larger volume of MR material, to be much greater. The interaction of these elements will be covered in greater detail later.

4.   EXPERIMENTATION

Experiments measured the deflections of cantilevered MR beams due to applied forces at the ends of the beams. Parameters varied included volume fraction of iron filings ς, magnetic field H_o, inner diameter of PVC tubing d_i, length of beam L, and angle of deflection θ. These force/deflection relationships were then used to determine the change in rigidity of MR materials when a magnetic field is applied. This allows for predictions of forces necessary to obtain desired deflections of MR beams.

5.1 Experiment Set-Up   The compliant magneto-rheological segments consisted of a PVC tubing shell filled with MR material sealed at both ends (Figure 8). The MR material used was prepared by manually stirring iron filings (between 100 - 200 um) into mineral oil. Three different volume fractions of iron filings were used ς = 0.1, 0.2 and 0.3. Figure 9 shows the experimental set up which consists of (a) an electromagnet, (b) a cantilevered force/deflection measuring system, and (c) the compliant segment.

Forces were applied by hanging masses from different lengths along the beam. Force/deflection analysis of the mechanisms were done at three levels of magnetic field (0, 800 and 1400 Gauss) three beam lengths (7.62, 6.35 and 5.08 cm, 3.0, 2.5 and 2.0 inches) and three tubing sizes 0.159x0.0794, 0.318x0.0794 and 0.476x0.0794 cm, (1/32x 3/16, 1/32x2/16 and 1/32x1/16 inches).   The cantilever beams were loaded and allowed to equilibrate for one minute.

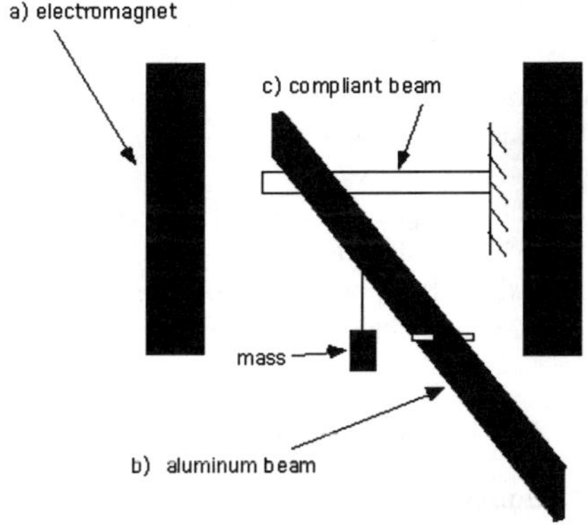

Figure 9: Experiment set up, a) electromagnet, b) aluminum beam for measuring force and deflection and c) compliant MR beam.

5.   COMPOSITE SEGMENTS

Compliant MR segments are constructed of two materials that act together to determine the rigidity of the composite segment. The composite rigidity $(EI)_c$ is assumed to be (4):

$$(EI)_c = (EI)_{pvc} + (EI)_{MR} \tag{7}$$

where $(EI)_c$ is the rigidity of the entire segment, $(EI)_{PVC}$ is the rigidity of the PVC shell and $(EI)_{MR}$ is the rigidity of the magneto-rheological material. This assumption enables the behavior of the MR material to be separated from the behavior of the composite beam. The moments of inertia of the composite beam, PVC shell and magneto-rheological material are assumed to be respectively:

$$I_c = \frac{\pi}{64} d_o^4 \tag{8}$$

$$I_{pvc} = \frac{\pi}{64}\left(d_o^4 - d_i^4\right) \tag{9}$$

$$I_{MR} = \frac{\pi}{64} d_i^4 \tag{10}$$

where d_o and d_i are shown in Figure 10.

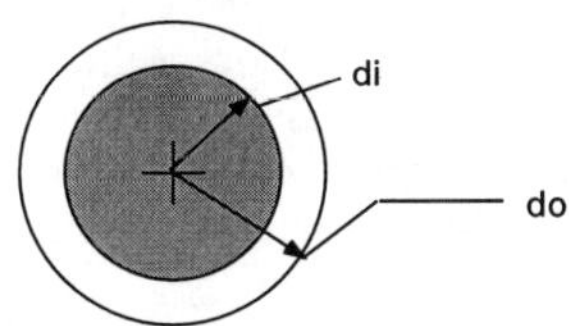

Figure 10: Cross-sectional view of a compliant MR segment.

6.   SURFACE AREA CONTACT

Contact between the magneto-rheological material and the shell affects MR behavior. Surface roughness at this contact has been shown to make a large difference in the shear stress of MR fluids at low magnetic fields (8). This has much to do with the formation of the fibrils along the wall and the strength of the bond between the MR fluid and the surface it is contacting. The larger the ratio of surface area to volume the more bonds are able to form between the PVC wall and the fluid per volume. These bonds keep a rigid connection between the two materials. If this bond begins to shear the assumption of

$$(EI)_c = (EI)_{pvc} + (EI)_{MR} \qquad (11)$$

fails to hold. It is this assumption that enables the behavior of the MR material to be separated from the behavior of the composite beam. Also the assumption that the MR fluid is behaving as a Bingham plastic before it begins to flow becomes invalid and the material will creep significantly with time. It is important that this bond remains secure enough to prevent yielding of the material and shearing along the PVC wall. This can be accomplished in two ways.

First, a maximum Surface Area/Volume ratio (A/V) of the MR fluid can be provided. All geometry used in this research were hollow tubes. A more successful pattern may be a star or doughnut shape to increase the (A/V) ratio. Using cross sectional areas that are too small increases the chance of bubbles forming in the segment because of surface tension, especially if larger volume fractions are used because of their high viscosity. This proved to be a problem in segments with an inner diameter of 0.159 cm (1/16 inch). Bubbles provide inhomogenities in the segment and may introduce weak spots. As mentioned, the surface roughness of the interface appears important at relatively low magnetic fields where this research was focused. Increasing the surface roughness along the interface may also increase fibril yield stress.

7.   FINITE-ELEMENT ANALYSIS

Once data were obtained to determine a force/deflection relationship for a segment, a commercial finite-element code capable of non-linear analysis was used to determine the composite rigidity $(EI)_c$ and rigidity of the PVC tubing $(EI)_{PVC}$. Two-dimensional elastic beam elements were used to construct the segment. The segments were meshed into twenty elements each and the deflections obtained from the experiments were specified for each data point. The density of the segment was also specified and a gravity load applied. Geometry of the segments was specified by assuming that the moments of inertia of the composite beam and PVC tubing were respectively shown in equations (9) and (10). This assumes that at zero magnetic field the MR material supplies no rigidity to the beam.

Non-linear analysis was necessary because of the large deflections of the beams. An initial modulus of elasticity of the beam was assumed to be $E_c' = 1,380$ kPa (200 psi). The resultant force F_R necessary to obtain the deflection specified was then determined by finite element analysis. This force was compared with the actual force applied to the segment F_A, and the ratio was used to determine the true modulus of elasticity of the composite beam E_c.

$$E_c = \frac{F_A}{F_R} E'_c \tag{12}$$

An iterative process was needed to determine E_c. For data with a magnetic field of $H_o = 0$, it was assumed that the MR material supplied no rigidity to the beam and

$$E_c = E_{PVC} \tag{13}$$

This allowed for both the rigidity due to the composite beam and the PVC shell to be determined. Using equations (8) and (11) the equivalent modulus of elasticity of the MR material E_{MR} can be determined by

$$E_{MR} = \frac{(EI)_c - (EI)_{PVC}}{I_{MR}} \tag{14}$$

which then allows the change in rigidity to be determined as a function of magnetic The relationship of E_{MR}/H_o is assumed to be a linear relationship at these magnetic fields as is shown in Figure 11 for one beam.

8. BOX-BEHNKEN STATISTICAL ANALYSIS

A Box-Behnken analysis takes into consideration four variables at three levels each. Twenty-seven experiments were ran in three blocks with one variable set repeated in each block to determine variation between blocks. The levels for each variable were as follows: volume fraction ς : 0.1, 0.2 and 0.3; beam length L:, 7.62, 6.35 and 5.08 cm, (3.0, 2.5 and 2.0 inches) (actual construction produced variations on these lengths); surface area/volume was determined by the inner diameter of the tubing at 0.159x0.0794, 0.318x0.0794 and 0.476x0.0794 cm, (1/16x1/16, 2/16x1/16 and 3/16x1/16 of an inch); angle of deflection, θ: 20, 30 and 40 degrees (the nature of the experiments produced some variation in angle of deflection). Larger angles were initially used but the segments began to creep at a high rate, showing that the material had exceeded its maximum stress and was experiencing visco-

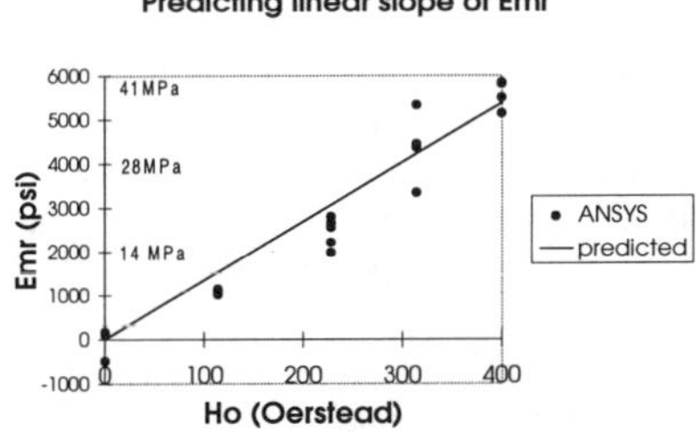

Figure 11: The relationship E_{MR}/H_o is assumed to be linear at these magnetic fields as shown for beam #22.

Term	Coef	Stdev	t-ratio	p
Constant	8.62	2.6818	3.216	0.006
vf	-44.11	13.9498	-3.162	0.006
AL/V	-0.28	0.0843	-3.284	0.005
vf*vf	29.75	28.5735	1.041	0.314
AL/V*AL/V	0.00	0.0008	2.992	0.009
vf*AL/V	0.98	0.1439	6.832	0.000

s = 0.6320   R-sq = 96.1%   R-sq(adj) = 94.9%

elastic creep as identified by the Bingham-plastic model. The response variable was κ, the slope of the line showing the relationship between E_{MR} and H_o.

Several response surface regression analyses were performed which identified significant variables, helped define non-dimensionalised parameters and provided a mathematical model for κ.

9. RESPONSE SURFACE REGRESSION

The angle of deflection, θ, was found to be insignificant in determining E_{MR} and it was dropped from the analyses. Two non-dimensionalized parameters were defined from the experimental variables, volume fraction ς, and AL/V where A is the surface contact of the MR material, L is the length of the beam and V is the volume of material in the beam. Only tubing of diameters 0.318 and 0.476 cm (2/16 and 3/16 inch) were considered because air bubbles had formed in the smaller tubing resulting in erratic results. The statistical model is shown in Table 1.

This model has an R squared of 96.1% which means that 96.1% of the data are accounted for by this model. The standard deviation 0.6320. The mathematical model for determining the slope of the linear relationship between E_{MR} and the magnetic field H_o, or κ is:

$$\kappa = 8.62 - 44.11\varsigma - 0.28\frac{AL}{V} + 29.75\varsigma^2 + 0.001\left(\frac{AL}{V}\right)^2 + 0.98\varsigma\frac{AL}{V} \tag{15}$$

10. REPEATABILITY OF MEASUREMENTS

Experiments on one segment were repeated once in each block for a total of three replicates. This shows any variation between blocks or scatter in the data. The force/deflection pattern of this segments is shown in Table 2. The average deflections and standard deviations of the deflections are also shown in the table. This is a maximum of a 7% change in deflection showing that the repeatability of measurements is satisfactory.

A comparison of the force predicted and the experimental force necessary for the deflection obtained, F_A-F_R provides the residuals of equation (15) for this segment. Error due to the experimental set-up is +/- 0.0133 N (+/- 0.003 lbf).

Table 2 Repeatability of Measurements

actual force N	actual force (lbf)	Ho O	300	525
0.058	0.013	0.951		
0.093	0.021		0.924	
0.111	0.025			0.872
0.058	0.013	0.938		
0.093	0.021		0.938	
0.111	0.025			0.859
0.058	0.013	0.964		
0.093	0.021		0.938	
0.111	0.025			0.807
avg (in)		0.951	0.933	0.846
stdev (in)		0.013	0.008	0.034

11. RESULTS

The results of the response surface regression show strong correlation between length, surface area contact and magnetic field strength in determining the stiffness of compliant MR fixed-pinned segments. A mathematical model was developed to predict the slope of a linear relationship between E_{MR} and H_o. This relationship is given in equation (15). The standard deviation of κ was 0.632 and predictions of E_{MR} can now be made within +/- one standard deviation. Once E_{MR} is determined, E_c can be calculated by rearranging equation (15) as:

$$E_c = \frac{E_{PVC} I_{PVC} + E_{MR} I_{MR}}{I_c} \qquad (16)$$

E_c, along with the geometry and desired deflection of each segment, can be input into a finite element code and the resultant force necessary to obtain that deflection can be calculated. Graphs that show the force/deflection pattern of the beam at different magnetic fields can then be constructed.   The predicted and actual force/deflection pattern for one beam is shown in Figure 12.

Once the force necessary for the deflection specified has been determined for several points, a force/deflection curve for the segment can be constructed at several levels of magnetic field.   These predicted curves are shown in Figure 12 compared with measured force/deflection curves of the same segment, beam #22. Major trends are followed but the force is underpredicted. The modulus of elasticity of the PVC was underestimated in this case by about 690 kPa (100 psi) and could have greatly contributed to this error where rigidity can be changed as a function of time.

12. CONCLUSIONS

Compliant magneto-rheological filled segments can be constructed where the rigidity of the segment can be changed as a function of magnetic field strength. Factors that largely affect the change in rigidity of the segment include: the magnetic field strength inside the segment H_o determined by the magnetic induction B_o and volume fraction of iron filings in the magneto-rheological material, the volume fraction of iron filings in the MR material and the parameter AL/V (contact area of the fluid A, length of the beam L and volume of MR material). Rigidity of the segment increases with increasing magnetic field increasing volume fraction and increasing AL/V.

Assumptions were made that for the range of magnetic fields H_o studied here, the relationship between the change in rigidity and the change in magnetic field can be modeled as linear. This slope, κ was empirically estimated to be:

$$\kappa = 8.62 - 44.11\varsigma - 0.28\frac{AL}{V} + 29.75\varsigma^2 + 0.001\left(\frac{AL}{V}\right)^2 + 0.98\varsigma\frac{AL}{V} \tag{17}$$

With a predicted modulus of elasticity E_{MR}, force/deflection patterns of segments can be predicted using a finite element program. This will allow segments to be used in design applications.

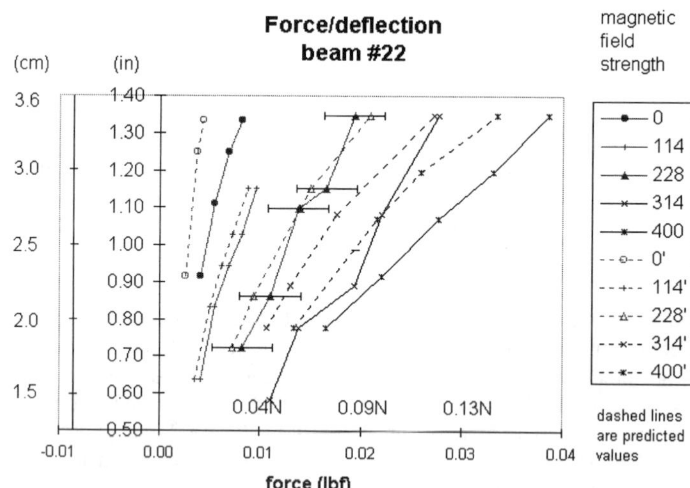

Figure 12: Force/Deflection pattern at different magnetic field strengths.

REFERENCES

1. S. Choi, Y. Park, and M. Suh, <u>AIAA J.</u>, <u>32</u> (2), 438 (1994).
2. J.P. Coulter, K.D. Weiss and J.D. Carlson, <u>Proceedings of the Conference on Recent Advances in Adaptive and Sensory Materials and Their Applications</u>, Part II: Applications, pp. 507-523 (1992).
3. D. Haase, <u>The Physics Teacher</u>, <u>31</u>, 218 (1993)
4. H.C. Hibbler, <u>Engineering Mechanics, Statics</u>, Macmillan Publishing Co., New York, 1992.
5. L.W. Hunter, et. al, <u>Proceedings of the Conference on Recent Advances in Adaptive and Sensory Materials and Their Applications</u>, 483 (1992).
6. D.J. Klingenberg, <u>Scientific American</u>, Oct.,112 (1993)
7. S. Morishita, and T. Ura, <u>Proceedings of the Conference on Recent Advances in Adaptive and Sensory Materials and Their Applications</u>, 537 (1992).
8. X. Tang and H. Conrad, <u>J. Rheo.</u>, <u>40</u> (6), 1167 (1996)
9. K.D. Weiss, J.D. Carlson and D.A. Nixon, <u>Journal of Intelligent Material Systems and Structures</u>, <u>5</u>, 772 (1994).
10. K.D. Weiss, J.P. Coulter and J.D. Carlson, <u>Proceedings of the Conference on Recent Advances in Adaptive and Sensory Materials and Their Applications</u>, 605 (1992).
11. W.M. Winslow, <u>Journal of Applied Physics</u>, <u>20</u>, 1137 (1949).

BIOGRAPHY

Larry L. Howell has been an Assistant Professor in the Mechanical Engineering Department at Brigham Young University since 1994. He received a B.S. from Brigham Young University, a M.S. and Ph.D. from Purdue University, and is a licensed professional engineer. His research has centered in the areas of compliant mechanisms and MEMS (microelectromechanical systems) and is currently funded by the National Science Foundation under a NSF CAREER Award. Industrial experience includes structural design on the YF-22 at General Dynamics, finite element analysis for Engineering Methods, Inc., and consulting for various companies in the U.S. and Canada

Alisa J. Millar Henrie is currently a design engineer for Boeing's Commercial Airplane Group in Everett, Washington. She graduated from Brigham Young University in 1997 with a master's degree in mechanical engineering. Her master's thesis research was with magneto-rheological materials and compliant mechanisms.

STRUCTURAL HEALTH MONITORING OF FILAMENT WOUND COMPOSITE PRESSURE VESSELS WITH EMBEDDED OPTICAL FIBER SENSORS

Richard Foedinger and David Rea
Technology Development Associates, Inc.
Wayne, Pennsylvania
Dr. James Sirkis, Dr. Richard Wagreich and John Troll
University of Maryland Smart Materials and Structures Research Center
College Park, Maryland
Robert Grande and Craig Davis
Pressure Technology Incorporated
Hanover, Maryland
Terry L. Vandiver
U.S. Army Aviation and Missile Command
Redstone Arsenal, Alabama

ABSTRACT

This paper presents the results of an on-going Phase II Small Business Innovation Research (SBIR) program to develop an improved technique for health monitoring of filament wound composite pressure vessels using embedded fiber optic sensor (FOS) arrays. Photomicrograph inspections and finite element analyses have been conducted to evaluate the influence of optical fiber diameter, coating and orientation on the local stress/strain distribution and overall structural integrity of the composite structure. An optical fiber pay-out system was developed for automatically embedding optical fibers into a filament wound composite structure without damaging the fibers or compromising typical manufacturing speeds. Bragg grating and In-Line Fiber Etalon (ILFE) optical fiber sensor arrays have been produced for embedding in standard ASTM 14.6 cm. (5.75 inch) diameter, filament wound carbon/epoxy composite pressure vessels. Initial demonstration of embedded sensor arrays for measuring thermal and mechanical strain was performed using an S-glass/aluminum pressure vessel. Further demonstration experiments with standard carbon/epoxy pressure vessels and multiple Bragg grating and ILFE sensor arrays are in progress.

KEY WORDS: Composite Structures, Fiber Optic Sensors, Filament Winding

1. INTRODUCTION AND BACKGROUND

Because of their small size, low weight, immunity to electrical interference, corrosion resistance and compatibility with composite materials and process conditions, fiber optic sensors offer significant potential for health monitoring of filament wound composite pressure vessels. Fiber optic sensor (FOS) arrays embedded in a composite structure can be used to monitor internal strain and temperature during curing, detect damage and interrogate the structural integrity of a pressure vessel throughout its service life. The successful implementation of embedded FOS arrays for filament wound pressure vessel applications will depend on several key issues, including: (1) characterization of the effects of the embedded FOS arrays on composite structural integrity and the influence of expected manufacturing and operating conditions on the optical performance of the sensor, (2) the development of automated manufacturing methods to embed the FOS arrays without damaging the sensors or compromising typical manufacturing speeds, (3) the design and fabrication of robust sensor arrays which can survive the manufacturing process conditions and expected operational loading environments, and (4) demonstration of improved ingress/egress techniques which protect the optical fiber leads and provide an interface for connecting to the external monitoring system. Many of these issues are being addressed as part of a Phase II Small Business Innovation Research (SBIR) Program sponsored by the U.S. Army Aviation and Missile Command (AMCOM) and conducted by Technology Development Associates, Inc. (TDA), Pressure Technology Incorporated (PTI) and the University of Maryland Smart Materials and Structures Research Center (UM SMSRC).

A significant objective of this research is to demonstrate the performance of multiple FOS arrays embedded in filament wound carbon/epoxy composite pressure vessels for health monitoring of composite rocket motor cases for small diameter tactical missile applications. The baseline composite structure identified for this research, illustrated in Figure 1, is a small diameter (7.5 to 23 cm), thin wall (0.13 to 0.19 cm) carbon/epoxy pressure vessel with integrally wound forward and aft domes. The burst pressure for this design is 51.7 MPa (7500 psi). Operating and storage temperature requirements range from -4 to 82 °C. The general requirements identified for the sensor/instrumentation include small fiber diameter (< 125 μm), transverse strain sensitivity less than 3%, maximum strain capability of 21,000 μ in/in, maximum pressure of 68.9 MPa (10,000 psi) temperature capability of -4 to 149 °C (expected maximum cure temperature of composite material), response time compatible with motorcase dynamic loading rates and multiplexing capability.

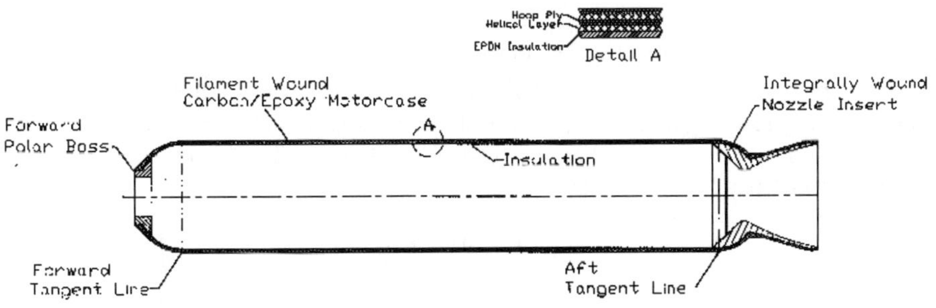

Figure 1. Representative geometry of filament wound composite motor case.

The composite material system selected for this research is a Thiokol Corporation Resin (TCR) carbon/epoxy prepreg. This material is a high performance carbon/epoxy prepreg that can be stored at ambient temperature for up to one year. Compared with typical wet winding material systems, it offers improved control of resin content and flow during cure. These processing advantages were considered to be especially important for embedded fiber optic sensor applications since there is better control of resin content and flow around the optical fiber, allowing improved characterization of the optical fiber/composite interaction mechanics and reduced fiber slippage resulting in better control of fiber orientation. Both T700 and T1000 carbon/epoxy TCR prepreg materials have been evaluated for FOS embedding application, with the lower cost T700 carbon fiber having more commercial pressure vessel application potential and the higher strength T1000 carbon fiber having more specific applications for high performance rocket motor cases.

An important goal of this research program is to develop and demonstrate improved optical fiber sensors and sensor arrays which meet the distributed strain and temperature sensing requirements and which can survive the filament winding, curing and operational loading environments. Two different fiber optic sensor types are being considered: (1) Bragg grating sensors and (2) In-Line Fiber Etalon (ILFE) sensors. Bragg grating sensors, Figure 2, have attracted considerable interest for embedded sensor applications because of their intrinsic nature and unparalleled optical multiplexing potential. In this type of sensor, optical Bragg gratings are written into the core of a conventional optical fiber. Different methods of producing such gratings have been developed, all of which take advantage of the photorefractive effect. One method involves holographically writing gratings into a Ge-doped fiber by side exposure to ultraviolet (UV) interference patterns. The Bragg grating sensor is illuminated using a broadband light source. The narrow wavelength component reflected by the sensor is determined by the Bragg wavelength $\lambda_B = 2n\Lambda$, where n is the mean index of refraction of the fiber core and Λ is the period of the index modulation of the core induced by the UV exposure. Changes in strain to which the optical fiber is subjected result in a shift in this wavelength leading to a wavelength-encoded optical measurement.

For surface mounted FBG sensors in which there is no transverse strain sensitivity effect, the measured Bragg grating wavelength shift is generally related to axial strain in the optical fiber core by the following relationship:

$$\frac{\Delta\lambda}{\lambda} = \varepsilon \left[1 - 0.5\, n^2 \left\{ P_{12} - \nu (P_{11} + P_{12}) \right\} \right] + \xi\, \Delta T, \qquad [1]$$

where $\Delta\lambda/\lambda$ is the normalized wavelength shift, ε is the axial strain in the optical fiber core, P_{ii} are strain-optic constants, ν is Poisson's ratio of the fiber, n is the refractive index of the fiber that the propagating mode sees, and ξ is the thermo-optic coefficient of the fiber. For a typical low-birefringent optical fiber, n = 1.45, $P_{11}=0.113$, $P_{12}=0.252$ and $\nu = 0.19$, so that Equation 1 simplifies to:

$$\frac{\Delta\lambda}{\lambda} = 0.79\, \varepsilon + \xi\, \Delta T \qquad [2]$$

For the case where Bragg grating sensors are embedded into a composite structure such as a pressure vessel, the sensor can experience significant transverse loading effects, depending on the fiber orientation and load conditions, so that the wavelength-strain relationship in Equation [2] can not generally be used in all cases. The results of hydrostatic pressure tests of filament wound composite pressure vessels with a single embedded Bragg grating sensor, previously conducted as part of the prior Phase I investigation, showed that for a non-parallel optical fiber orientation, bifurcation in the reflected Bragg grating occurs due to transverse strain sensitivity, causing non-trivial errors (up to 12%) in the measured strain data. An important objective of the photomicrograph inspections, finite element analyses and pressure experiments conducted as part of this Phase II research is to expose and characterize such transverse strain sensitivity errors.

In comparison with conventional Bragg grating sensors, In-line Fiber Etalon (ILFE) sensors have zero transverse strain sensitivity and the intrinsic thermal response is orders of magnitude smaller. This type of sensor, illustrated in Figure 3, is fabricated by fusing a small section of hollow core fiber between two ordinary single mode fibers. The hollow core fiber is specifically manufactured to have the same outer diameter as the lead-in/out fibers to provide a well-defined gage length and a robust, smooth fusion weld at the air/glass interfaces. The sensor cavity length can range from 20 μm to 1000 μm depending on the strain gradients involved and the characteristics of the optical source. The operation of ILFE sensors is based on the 4 % Fresnel reflection caused by the refractive index mismatch at the two air/glass interfaces. These interfaces form a low finesse Fabry-Perot interferometer whose optical phase is dependent on the change in cavity length. Fabrication and testing of ILFE sensors to date has demonstrated that improved quality control of the fusion splice is critical to achieving the desired failure strain such that no damage occurs during embedding of the sensors.

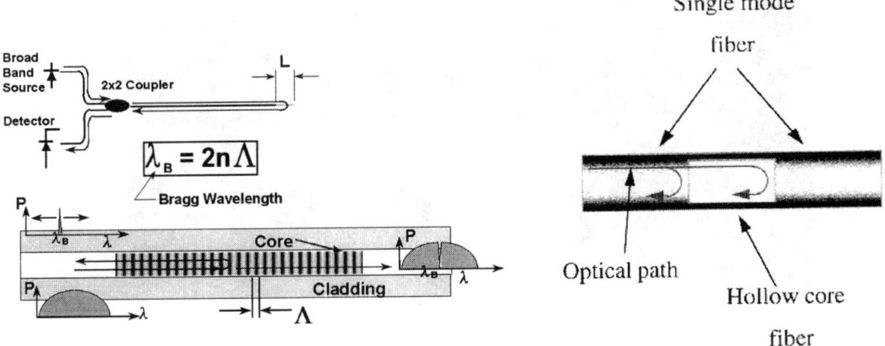

Figure 2. Fiber Bragg Grating (FBG) Sensor.   Figure 3. In-Line Fiber Etalon Sensor.

2.0 OPTICAL FIBER PAY-OUT SYSTEM

In order to efficiently integrate the FOS arrays into the filament winding process, a controlled tension pay-out system was developed. The system was designed to be compatible with typical filament winding speeds without causing damage to the optical fibers. As illustrated in Figure 4, the optical fiber is fed directly from the spool through a specially designed tensioner and winding eye. The optical fiber passes through a thin Teflon tube

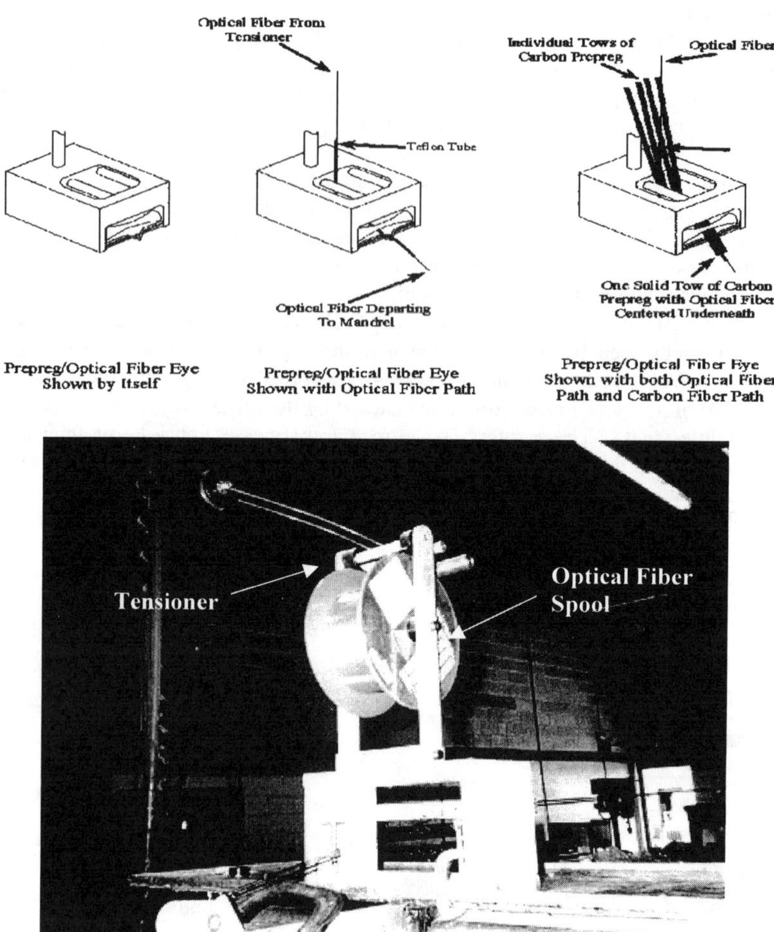

Figure 4. Automated optical fiber pay-out system

inserted in the winding eye such that the optical fiber is centered beneath the carbon prepreg tows as it exits the winding eye. In this manner, the optical fiber is automatically located below the carbon prepreg ply as it is wound onto the mandrel.

The optical fiber pay-out system was demonstrated during filament winding of 7.62 cm (3 inch) diameter cylinders with 125 μm diameter optical fibers embedded along each helical and hoop ply. These cylinders were sectioned to provide specimens for photomicrograph inspection of the embedded optical fibers and surrounding composite plies. No damage to the optical fibers was observed during filament winding of the cylinders. Further demonstration of the optical fiber pay-out system is being performed for a standard ASTM pressure vessel geometry.

3. MICROGRAPH INSPECTION OF EMBEDDED OPTICAL FIBERS

FOS arrays embedded in composite structures introduce perturbations in the composite microarchitecture, resulting in local stress and strain concentrations which can negatively influence the overall structural integrity of the composite pressure vessel. As previously observed, the effect of the embedded optical fiber on composite structural integrity is determined by the mechanical interaction between the fiber and host composite material, which is influenced by several factors, including the diameter of the optical fiber, coating material and thickness, location of the optical fiber in the composite lay-up and orientation of the embedded optical fiber relative to the adjacent composite plies. In addition to influencing composite stress/strain distribution, the presence of voids or resin pockets resulting from embedded optical fibers can also affect the interpretation of the embedded sensor data. The presence of voids can cause local strain gradients which can smear the optical spectrum reflected from a Bragg grating sensor, for example. Stress concentrations can also produce birefringence, which makes sensor data interpretation difficult.

In order to examine the effects of such factors on the composite microarchitecture, three 7.62 cm. (3 inch) diameter carbon/epoxy cylinders were filament wound to provide specimens for microscopic inspection of the embedded optical fibers. Filament winding was performed using an Entec 418HC three-axis, two-spindle, computer controlled winding machine. The first two cylinders were wound with T700SC-12K-50C/UF3325-10 prepreg and the third cylinder was wound with T1000/UF3325 carbon/epoxy prepreg. The lay-up for the cylinders was [$\pm$ 30, 90_2, $\pm$ 30, 90_2]. A Corning 125 μm diameter, single mode fiber with an acrylate coating was embedded along each helical and hoop ply,. In addition, an acrylate coated and a polyimide coated fiber were manually placed longitudinally into one of the cylinders between the inner helical layer and first hoop ply.

Several photomicrograph specimens were prepared from the filament wound cylinder sections. CCD images and Polaroid™ pictures were obtained at approximately 50X magnification. General observations from the photomicrograph inspections suggest that the local microarchitecture depends primarily on the orientation of the optical fiber relative to the reinforcing fibers (i.e., parallel or perpendicular).

Figure 5 shows a representative comparison of photomicrographs of optical fibers embedded parallel to and perpendicular to the carbon reinforcing fibers for a T700 cylinder. For the optical fibers embedded parallel to the adjacent composite plies, only small voids or resin pockets, approximately 10 to 70 μm in diameter, were observed in at least one quadrant adjacent to the embedded optical fiber. From the photomicrographs examined, it appeared that the location of the voids/resin pockets was related to the filament winding direction of the carbon/epoxy prepreg tow. For the optical fibers embedded perpendicular to the adjacent composite ply, a much larger lenticular resin pocket is formed around the embedded optical fiber.

The influence of the optical fiber coating was also considered as part of the photomicrograph inspection. Lower cost, standard acrylate coated optical fibers and more expensive specialty polyimide coated fibers were examined. An important result of the photomicrograph inspections was that the acrylate coatings appeared to survive the embedding, gelling and curing process without any degradation. This was considered to be significant since the cure temperature of 132 °C (270 °F) approached the theoretical breakdown temperature of the acrylate coating.

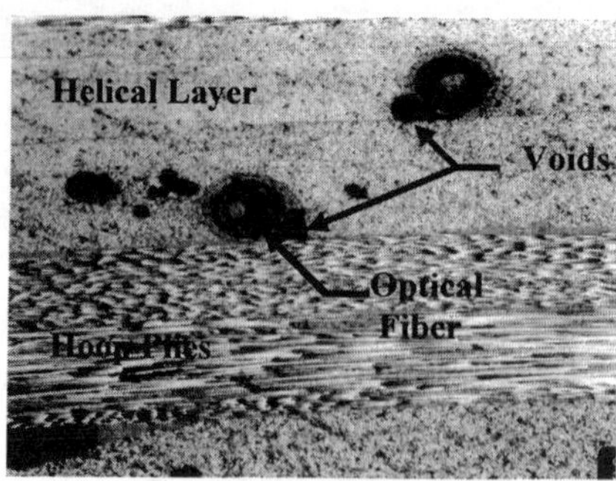

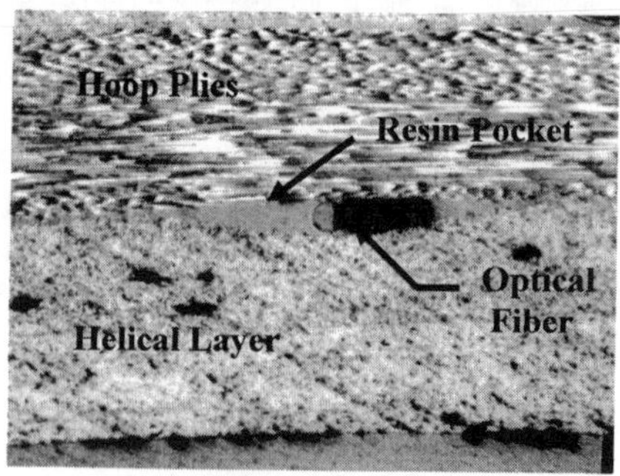

<table>
<tr><td align="center">(a) Parallel Orientation,
Acrylate Coating</td><td align="center">(b) Perpendicular Orientation,
Polyimide Coating</td></tr>
</table>

Figure 5. Comparison of photomicrographs of optical fibers embedded in filament wound cylinders at different orientations relative to adjacent composite plies.

4. ANALYSIS OF FIBER OPTIC/COMPOSITE INTERACTION

Analytical models are being developed to investigate the effects of the embedded sensors on composite stress/strain distribution and evaluate the effects of transverse loading on the response of the embedded sensor. Resin pocket prediction models are being used to define the geometry of the embedded sensor, resin pocket and distorted composite laminate for detailed three-dimensional finite element models. The photomicrographs taken from the filament wound cylinder sections were compared with the resin pocket prediction models prior to developing the finite element models. A representative correlation between the resin pocket prediction model and photomicrographs is shown in Figure 6. The photomicrograph shown is of a 125μm diameter polyimide coated optical fiber embedded in a T1000G/UF3325 carbon/epoxy composite with a lay-up of [±30/90$_2$/±30/90/OF/90] in which the optical fiber is embedded perpendicular to the outer hoop plies. In this case, very good agreement between the resin pocket model and the photomicrograph was observed, with the model only slightly over-predicting the length of the resin pocket. The displacement functions also agreed very well with the laminate distortion observed from the photomicrograph. In general, results obtained for the photomicrographs inspected to date show good agreement for optical fibers embedded perpendicular to the adjacent plies, but over-predict the size of the void/resin zone for parallel or near-parallel optical fiber orientations.

Figure 7 illustrates an ANSYS® 3-D finite element model of a section of a composite laminate with an optical fiber embedded in the axial direction (0°) adjacent to outer hoop plies. The resin pocket geometry was taken from the photomicrograph for this configuration. The results of the analysis conducted to date have shown relatively good agreement for the predicted composite axial strains compared with strain gage data obtained from prior Phase I experiments. However, the predicted strains in the embedded optical fiber are considerably lower than the measured strains in the embedded Bragg grating sensor. The results of the parametric studies performed to date have demonstrated that the size of the resin pocket and the orientation of the embedded optical fiber relative to the adjacent composite plies have a significant influence on the stress distribution in the composite. The location of the optical fiber within the stacking sequence also influenced the stress distribution. The maximum stresses were predicted for orientations where the optical fiber was oriented perpendicular to the adjacent composite plies, as expected.

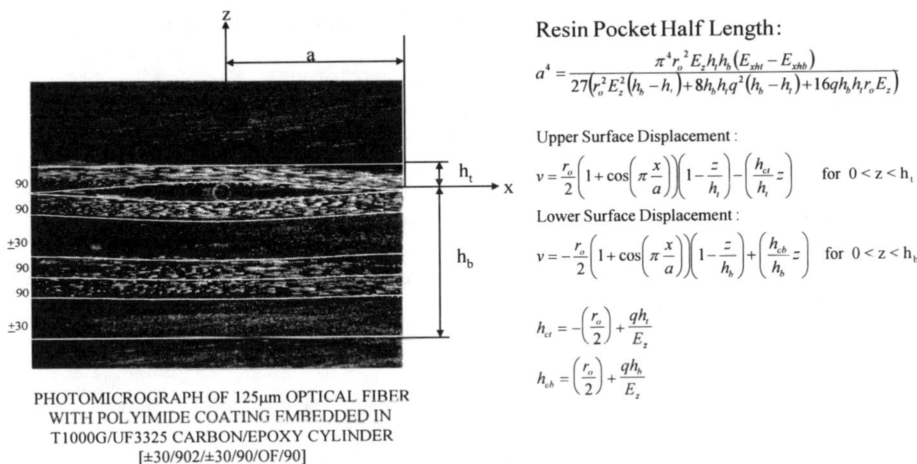

$$a^4 = \frac{\pi^4 r_o^2 E_z h_t h_b \left(E_{xht} - E_{xhb}\right)}{27\left(r_o^2 E_z^2 \left(h_b - h_t\right) + 8h_b h_t q^2 \left(h_b - h_t\right) + 16 q h_b h_t r_o E_z\right)}$$

$$v = \frac{r_o}{2}\left(1 + \cos\left(\pi \frac{x}{a}\right)\right)\left(1 - \frac{z}{h_t}\right) - \left(\frac{h_{ct}}{h_t} z\right)$$

$$v = -\frac{r_o}{2}\left(1 + \cos\left(\pi \frac{x}{a}\right)\right)\left(1 - \frac{z}{h_b}\right) + \left(\frac{h_{cb}}{h_b} z\right)$$

$$h_{ct} = -\left(\frac{r_o}{2}\right) + \frac{q h_t}{E_z}$$

$$h_{cb} = \left(\frac{r_o}{2}\right) + \frac{q h_b}{E_z}$$

Figure 6. Comparison of resin pocket prediction model with photomicrograph results.

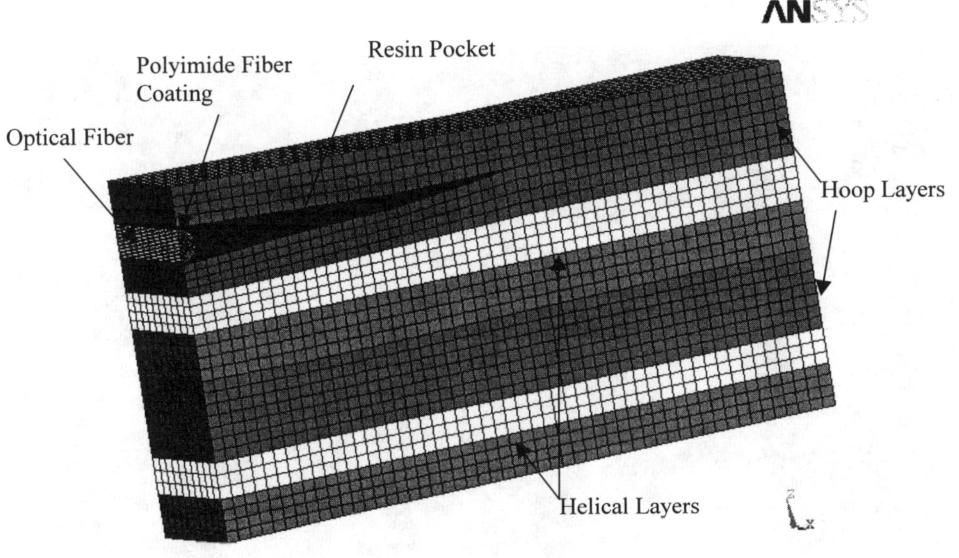

Figure 7. Three-dimensional finite element model of embedded optical fiber in composite laminate.

5. PRESSURE VESSEL EXPERIMENTS

In order to evaluate the performance of embedded FOS arrays and provide data for use in interpreting thermal and mechanical strain measurements, initial Bragg grating and ILFE sensor arrays were produced and embedded into a 10.16 cm. (4 inch) diameter, 23 cm. (9 inch) long, filament wound S-glass/aluminum composite pressure vessel. Two optical fiber arrays with two Bragg gratings in each array were manufactured. The two gratings in each array were separated by approximately 30.5 cm (12 inches) of optical fiber, with roughly 3.05 meters of lead-in and lead-out fiber. Two additional fibers with one ILFE sensor in each were also manufactured for embedding in the filament wound pressure vessel. The ILFE sensors were located towards one end of each of the fibers which were approximately 2.45 meters long.

The four Bragg gratings and two ILFE sensors were embedded into the filament wound pressure vessel. In order to protect the optical fiber from potential stress damage at the points of entry into and exit from the composite, or ingress/egress points, the fibers were inserted into Teflon tubing and fixed in position with silicon rubber adhesive prior to filament winding. The ingress/egress points were located at the base of the neck of the pressure vessel. The excess non-embedded fiber from each sensor was coiled and taped to a small card near the neck of the bottle to prevent accidental entanglement during the filament winding process. A photograph of the mechanical arrangement of the filament winding process for embedding the optical fiber sensors is shown in Figure 8.

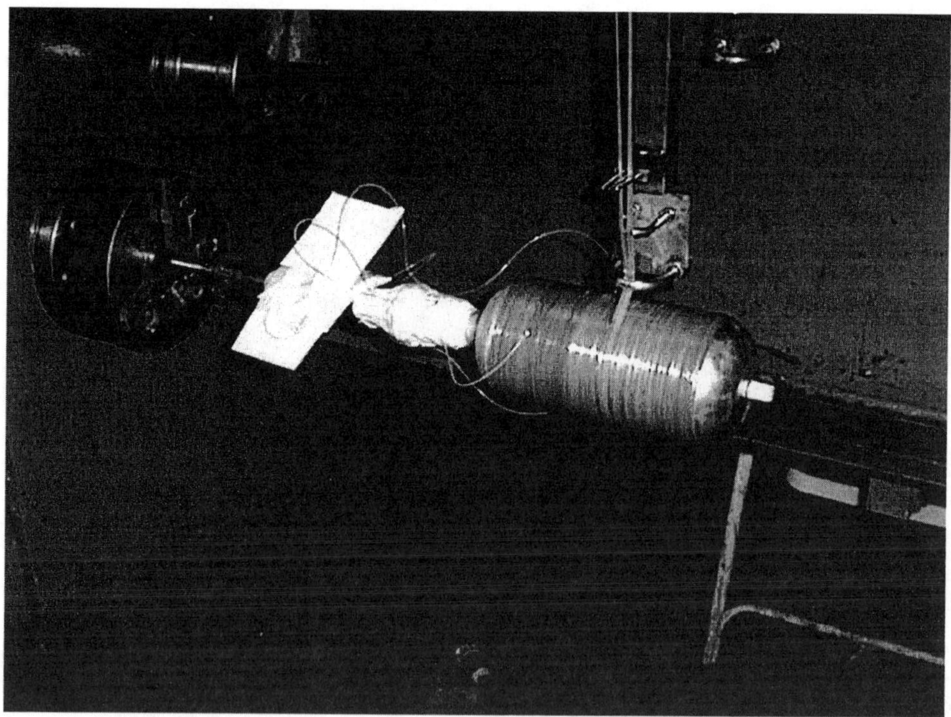

Figure 8. Embedding optical fibers during winding of S-glass/aluminum pressure vessel.

During the embedding process, the sensors were carefully aligned along the direction of the filament winding in order to reduce the effects of transverse strain. Two Bragg gratings and one ILFE were embedded in the hoop windings and the same number were embedded in helical windings. The location of the Bragg grating sensors was not precisely marked; however, the distance of the sensors from the neck of the bottle and the orientation of the sensors was recorded. The approximate location and orientation of the sensors in the pressure vessel is depicted on the outline diagram of the pressure vessel shown in Figure 9. The transition regions close to the cylinder domes is noted in this figure in order to better explain the different strain effects observed by the sensors during the pressurization test. In this diagram, all sensors are depicted on the front face of the vessel to clarify the distance between the sensor and the vessel neck. In fact, some of the sensors were actually embedded at locations around the axis of the vessel.

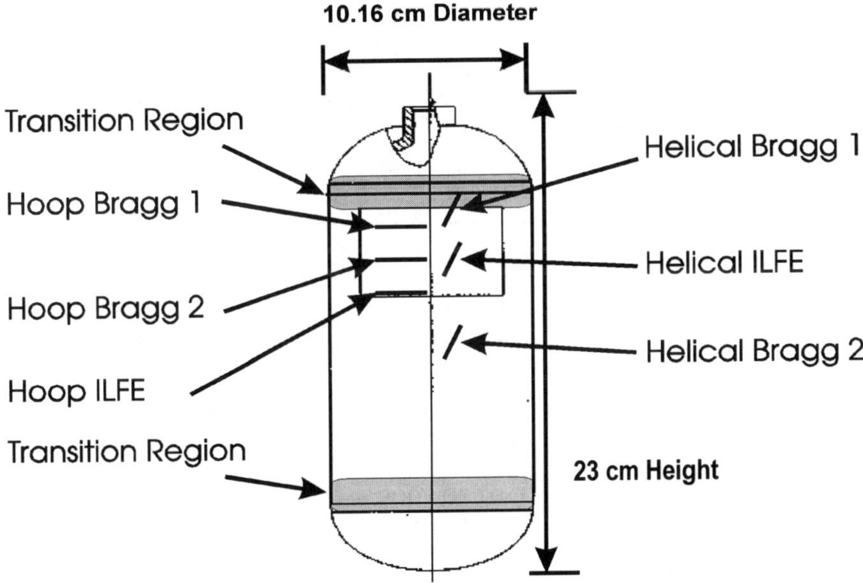

Figure 9. Approximate location of four Bragg grating sensors and two ILFE sensors embedded in filament wound S-glass/aluminum pressure vessel.

5.1 Cure Monitoring Initial calibration of the Bragg grating sensors was done prior to the gel process as well as immediately following the gel process prior to the curing process. After the gel process was complete it became evident that the second array of fiber Bragg gratings was damaged within the pressure vessel in two locations. This was determined by the unusually low level of light that was reflected back from both Bragg gratings in this second fiber array. After various interrogation procedures, it was determined that the fiber was cracked on both sides of the sensor array within the vessel. Typical reflection spectrum monitoring of the Bragg gratings did not give sufficient light to accurately measure changes in the grating wavelength during the tests. In order to improve the signal to noise ratio, the data acquisition process was performed by observing the transmitted light through the two Bragg grating array fibers on an optical spectrum analyzer (OSA). Intensity dips were observed from the transmitted light due to the Bragg grating reflection spectrum. The ILFE sensor response could not be ascertained before or after the gel process.

Prior to curing, the pressure vessel was allowed to gel at a temperature of 104 °C (220 °F) while continually rotating. After the gel process was complete, the spectral dips due to the Bragg sensors were monitored and recorded throughout the cure process. This process lasted approximately two hours and involved raising the temperature of the vessel to approximately 132 °C (270 °F) within the first hour and maintaining this temperature level for the second hour. The temperature of the vessel was monitored during the cure process using a vessel-mounted thermocouple.

During the cure process, the Bragg grating wavelengths changed due to both temperature changes in the sensor itself and due to the thermal expansion of the pressure vessel. This change in the Bragg wavelength is a known function of both temperature and strain, as previously discussed. By using the monitored temperature measurements of the vessel, the temperature effects on the gratings were subtracted out from the results. In this way, the strain on the gratings due to the expansion of the pressure vessel was determined. The strain experienced by each of the Bragg gratings in the vessel, as a function of the vessel temperature during the cure process, is shown in Figure 10.

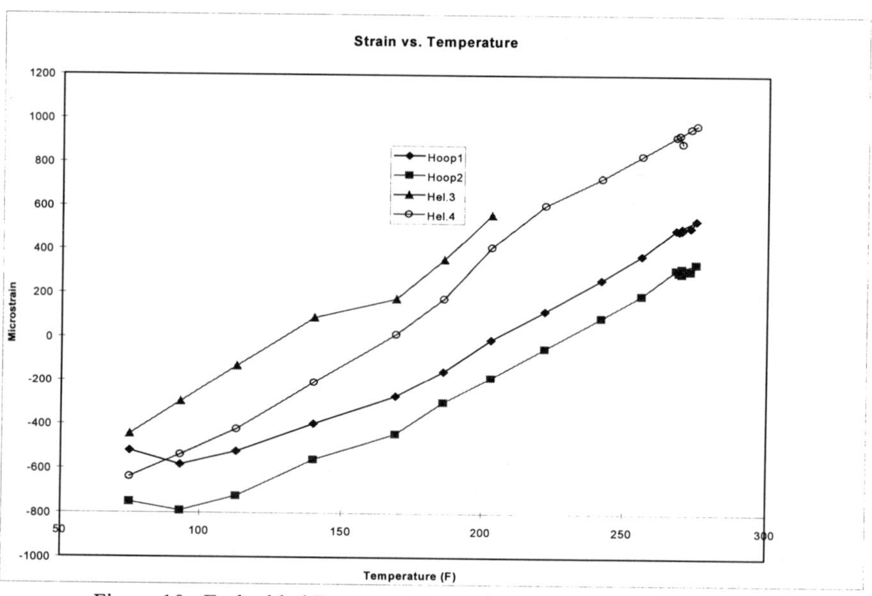

Figure 10. Embedded Bragg grating strain vs. temperature during curing.

The initial negative residual strains experienced by the gratings were most likely due to the cross-linking process during the gel cycle. The bonding between the filament windings and the optical fibers occurred at an elevated temperature when the vessel has undergone thermal expansion. After the gel process, the vessel was brought back to ambient room temperature and the thermal contraction led to negative residual strain on the sensors. The variation in the initial residual strains may be due to non-uniform cross-linking during the gel cycle. According to theoretical models of the curing process, the strain change on the sensors as a function of temperature was similar to expected values.

454

5.2 Pressurization Experiment The filament wound S-glass/aluminum pressure vessel with embedded sensors was subjected to hydrostatic pressure testing. The vessel containing the embedded sensors was positioned within the pressure chamber as shown in Figure 11. The embedded optical fibers ingress and egress from the neck of the vessel were fed out of the pressure chamber to the interrogation equipment, which consisted of a light source and an OSA.

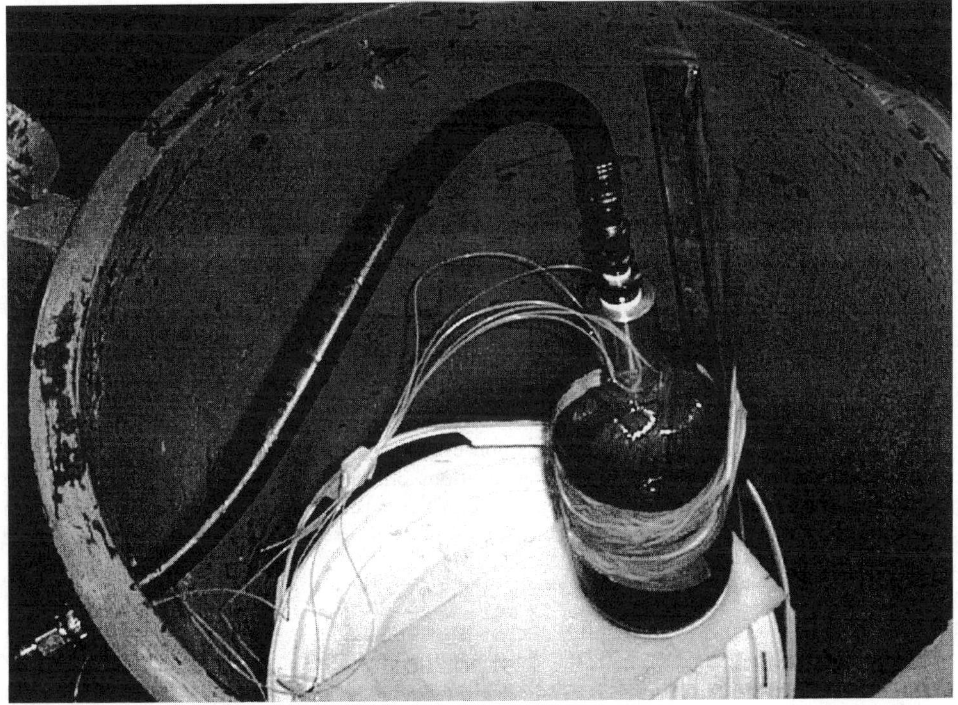

Figure 11. Filament wound S-glass/aluminum pressure vessel in test chamber.

Two pressure tests were performed. During the first test, the pressure within the vessel was raised to 37.9 MPa (5500 psi) in 3.45 MPa (500 psi) intervals, which is above the autofrettage pressure of the cylinder. The pressure was then lowered directly back to 0 MPa. This was done in order to observe the effects of the yielding of the aluminum shell on the filament windings. Second, the pressure within the vessel was raised to 20.7 MPa (3000 psi) in 6.89 MPa (1000 psi) intervals and lowered again to insure the strain on the vessel remained linear as a function of pressure over the entire observation region. The transmission dip of the four gratings was again monitored as a function of vessel pressure using the OSA. The strain on the Bragg grating sensors was determined by the change in the observed wavelength dip. The strain determined from the four embedded Bragg gratings as a function of the vessel pressure for the first and second tests are shown in Figures 12 and 13.

455

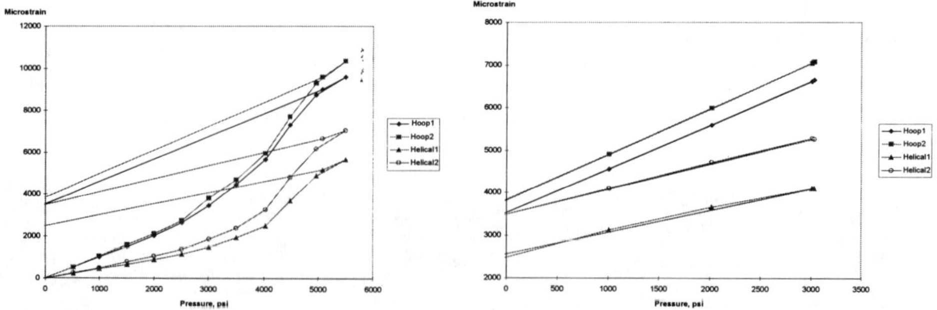

Figure 20. Embedded Bragg grating strain vs. pressure for S-glass/aluminum pressure vessel.

The observed sensor strain was similar to expected values from theoretical models. During the first pressurization of the vessel, the strain was a linear function of pressure up to the autofrettage pressure of the cylinder when the aluminum shell began to yield. After this pressurization point, and in all subsequent measurements, the strain was seen to be a linear function of pressure as expected. Since the vessel was designed to burst in the hoop layers, the hoop windings should experience more strain than the helical windings. This was observed in both pressurization tests.

The variation in the sensor response to pressure was most likely the result of differential strains measured at varying locations on the pressure vessel. Sensors in close proximity to the transition zone on the dome of the pressure vessel experience higher strain than the sensors located in the central region of the bottle. This accounts for the higher strains measured by Hoop Grating 1 and Helical Grating 1.

No spreading of the optical spectrum of any Bragg gratings was observed during either the temperature tests or the pressurization tests. This indicates that the sensors experienced sufficiently little transverse strain when embedded in the direction of the filament windings.

Because no response was obtained from the embedded ILFE sensors, it was suspected that they might have broken during embedding. It also appeared that there was a break in one of the Bragg grating arrays, because the signal decreased as the pressure increased (the gratings could only be viewed through transmission, and after video averaging). To inspect the embedded fibers for breaks and serious loss, intense red light from a Helium-Neon (HeNe) laser was coupled into the fibers. In this kind of fiber, the propogation of light at 633 nm is multimode, and is therefore more lossy than light propagating in the fundamental mode. Thus the light leaking out at bends, breaks, terminations, and areas of high transverse strain can be seen. Preliminary results suggest that in each of the embedded sensors, there was serious loss in the portion of the optical fiber embedded in the dome region. It's not known whether this loss comes from the embedding procedure, the curing, or the autofrettage that the sensors experience. Further monitoring of the gelling and curing stages using the HeNe laser may help to reveal the cause of this loss.

6. SUMMARY AND CONCLUSIONS

The research effort described in this paper has demonstrated that fiber optic sensor arrays can provide a useful tool for structural health monitoring of filament wound composite pressure vessels. An automated manufacturing technique was developed to embed optical fiber sensors in a filament wound composite pressure vessel without damaging the optical fibers or compromising typical manufacturing speeds. Photomicrograph inspections have provided further insight into the influence of optical fiber diameter, coating material and fiber orientation on the local composite architecture. Optical fibers embedded parallel to adjacent reinforcing fibers produce only small voids or resin-rich zones while optical fibers embedded perpendicular to adjacent composite plies produce much larger resin pockets which may result in higher stress concentrations. For embedded Bragg grating sensors, perpendicular optical fiber orientations also result in transverse strain sensitivity errors, making sensor data interpretation more difficult. A significant result of the photomicrograph inspections is that the less expensive, commercially available acrylate coatings survived composite cure temperatures up to 132 °C (270 °F). Analytical resin pocket prediction models agreed fairly well with the photomicrographs for perpendicular fiber orientations, but over-predicted the size of the resin pocket for near-parallel optical fiber orientations. The results obtained to date from the 3-D finite element analyses show good agreement for the composite axial strain, but under-predicted the optical fiber strain. Further model development and validation with pressure test data is in progress.

The initial cure monitoring and pressure experiments with the S-glass/aluminum pressure vessel generally showed good agreement with theoretical models. Strain data from the Bragg grating sensors was consistent with expected trends. Problems experienced with the ILFE sensors demonstrated the importance of improved quality control in fusion splicing the hollow core fiber. Further experimentation with standard ASTM carbon/epoxy pressure vessels and multiple embedded Bragg grating and ILFE sensor arrays is in progress.

7. ACKNOWLEDGEMENTS

This research was conducted under U.S. Army Aviation and Missile Command Contract Number DAAH01-97-C-R165. The authors gratefully acknowledge the support of individuals at Technology Development Associates, Pressure Technology Inc. and the University of Maryland Smart Materials and Structures Research Center who supported the analyses, fabrication, experimentation, and data evaluation activities.

VALIDATION OF BENDING STIFFNESS VALUES DERIVED FROM LAMB WAVE MEASUREMENTS OF HEAT-DEGRADED LAMINATES

Jeffrey R. Kollgaard
Boeing Commercial Airplane Group

Wei Huang
Digital Wave Corporation

ABSTRACT

Composite laminates, because of their excellent engineering properties, are finding their way into critical applications on aircraft. In some of these applications the potential for heat damage is a concern. Short term heat exposures from accidents such as fires, or long-term heat exposures from high speed flight, can cause material degradation which is not detectable with conventional nondestructive test techniques.

Currently, destructive methods are necessary to identify the presence and degree of degradation. These methods are not suitable for in-service aircraft. A method is needed which yields data on mechanical properties nondestructively. While researchers have made encouraging progress toward this objective, their techniques have not been embodied in a commercial instrument.

In this report, Boeing uses a prototype lamb wave scanner manufactured by Digital Wave Corporation to extract bending stiffness properties from degraded laminates. The scanner operates by indexing over the part surface, obtaining phase velocity readings of the A_0 lamb wave mode over a pre-selected range of frequencies. The bending stiffness D_{11} and the extensional transverse shear stiffness A_{55} are determined from the measured phase velocity dispersion curve.

The D_{11} values measured by the scanner were validated by mechanical testing per ASTM D790.

KEY WORDS: Composite Structures, Mechanical Properties, Non-Destructive Evaluation, Ultrasonic Lamb Waves

1. INTRODUCTION

Composite laminates are finding their way into increasingly critical applications on aircraft. Examples include the laminate empennage structure and floor beams on commercial airliners, the composite wings on some military aircraft, the composite outer wing box on one commuter airplane (1), and the composite fuselage on at least one business jet. The increasing demands on composite structure create an increasing need for in-service, non-destructive characterization of its properties. In some cases, degradation of mechanical properties can occur which can not be detected by either visual or conventional nondestructive techniques. This is particularly true in cases of fire damage, or in situations of long-term heat exposure such as those envisioned in future high speed aircraft. In one case involving a military jet with a composite wing, the aircraft was grounded for an extended period because no nondestructive methods existed to determine residual strength near areas visibly fire-damaged.

The most common method for measuring the stiffness of a composite laminate is to mechanically load representative specimens and measure strain versus applied stress. This can be both time and cost prohibitive, and can not be applied *in situ*. Recently, advances in high frequency ultrasonic velocity measurement techniques have demonstrated the ability to recover the stiffnesses in composite laminates. To name just a few: a leaky Lamb wave technique was developed by Mal and Bar-Cohen (2), and numerous bulk wave methods were proposed by Every and Sachse (3), Rokhlin and Wang (4), and Hosten et al (5). However most of the methods utilize mode conversions at a liquid-solid interface. In other words, the measurements need to be performed by immersing the composite laminate plate inside water.

It was recently shown by Gorman (6) and Elliot (7) that by placing a pair of wide band transducers normally on the same side of a thin plate one could effectively generate and receive the lowest order antisymmetric lamb mode (A_0 mode) at lower frequencies. At higher frequencies the lowest order symmetric plate mode (S_0 mode) starts to dominate, while the higher order A and S modes exist only at even higher frequencies where the ultrasonic wavelength is about the same as the plate thickness. Huang et al (8) utilized the method of placing dry-coupled pulser/receiver on the same side of the plate to generate and receive the A_0 mode at low frequencies, and applied it to measure stiffnesses and to inspect defects in laminated composite plates. Such an experimental setup does not require immersion of the specimen nor does it require access to both sides of the laminate.

In the area of nondestructive measurement of heat degradation, efforts accelerated after the prolonged grounding of an A-6 aircraft, following a hydraulic fire, in 1991. Mehrkam, Armstrong-Carroll, and Cochran (9) investigated a number of nondestructive techniques during the assessment of this damage and reported limited success. In subsequent papers (10, 11) they concluded that DRIFT (Diffuse Reflectance Infrared Technique) and ultrasonics were the most promising of those studied; however, the conventional ultrasonic techniques employed did not detect damage short of physical delamination. Lamb wave methods were not used.

Further studies by the above authors (12) showed that flexural strength of high-temperature composites decreased significantly with exposure to thermal spikes (flame impingement). Since the fundamental modes of lamb waves are directly influenced by the flexural stiffness of composite materials, this would seem to suggest that their use in these situations might be warranted. However, the sensitivity must be such that property changes are detected before the appearance of visible damage or physical delamination. Seale and Smith (13), using the fundamental symmetric Lamb wave mode (S_0 mode) and samples exposed to temperatures from 550 °F to 1000 °F for periods of 3 to 10 minutes, showed that this sensitivity can be achieved. Their study did not include mechanical testing of the heat-degraded samples. Similar lamb wave methods were also suggested by Frame et al (14).

The present study uses a prototype lamb wave scanner, the first of its kind, to examine heat-degraded laminates. The scanner, now marketed as the F-Scan™ and described in reference (8), collects phase velocity data and uses it to determine the bending stiffness D_{11} and extensional transverse shear stiffness A_{55} in real-time. The scanner was used to extract D_{11} bending stiffness values from a series of heat-degraded laminates. The values were then correlated directly with the results of mechanical tests.

2. MATERIALS

The graphite epoxy material used for the study was Toray P2302-19, a carbon prepreg used on the 777 empennage. All of the specimens were uniaxial 12-ply tape panels. Panel thickness was 2.35 mm (0.0925 inches). Specimens 1 to 12 were cut from one large panel. Specimens 19-24 were cut from a second large panel. Both parent panels were cured at 121 °C (250 °F) per Boeing manufacturing specifications on a flat tool. Specimens 13-18 were 20-ply quasi-isotropic panels. While lamb wave results were consistent on these panels, they could not be mechanically tested in time to include them in this study.

3. PROCEDURES

The cured panels were square in shape and 305 mm (12 inches) on a side. Each was divided by a template into 16 coupon-sized areas. Eight of the areas were oriented parallel to the fiber direction and the remaining eight were oriented perpendicular to the fiber direction. See Figure 1.

The prototype F-Scan™ system was employed at three locations on each coupon to measure the phase velocities of the A_0 mode in a selected frequency range. The F-Scan™ system consists of a arbitrary function generator board, a low frequency pulse amplifier, transmitting and receiving sensors, a 2-D scanning bridge, an analog signal conditioning module, a high speed analog-to-digital converter, a CPU, and data capture and analysis software. A functional block diagram of the F-Scan™ system is shown in Figure 2. A photograph of the prototype system appears in Figure 3.

Figure 1. Test Panels Before and After Heat Exposure

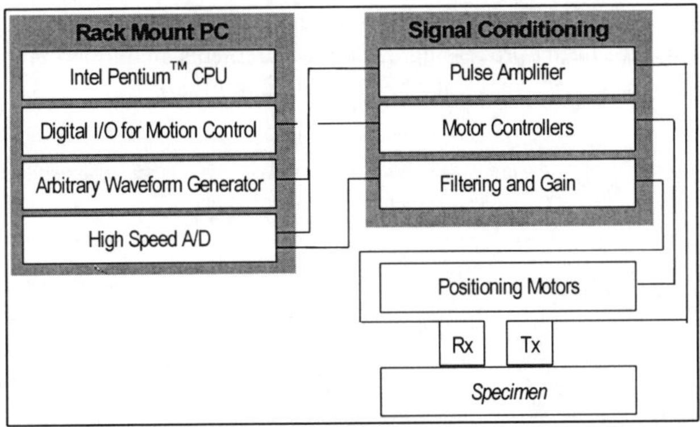

Figure 2 Functional Block Diagram of the F-Scan™ System.

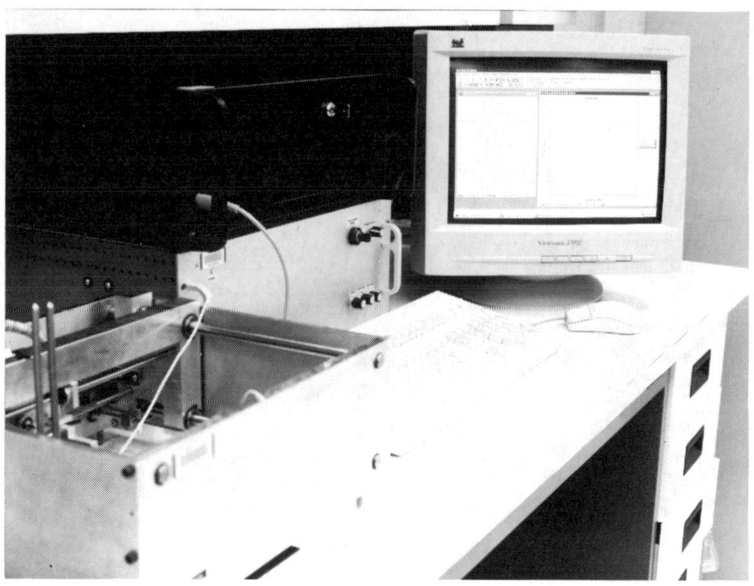

Figure 3 . Photograph of Prototype F-Scan™

The scanner employs one dry-coupled, normal-incidence transmit transducer (Tx) and one similar receive transducer (Rx). In addition to the x and y scanning capability of the bridge, the receiver Rx can move in the Δx-direction independent of the pulser Tx. There is also a z-axis motor to press the transducers down to the plate surface providing the necessary contact pressure. The measurable frequency range is between 20 KHz to 1 MHz. At each phase velocity measurement spot the scanner steps through a pre-selected range of frequencies. During processing these measurements are plotted in dispersion curve form as phase velocity vs. frequency, as shown in Figure 4.

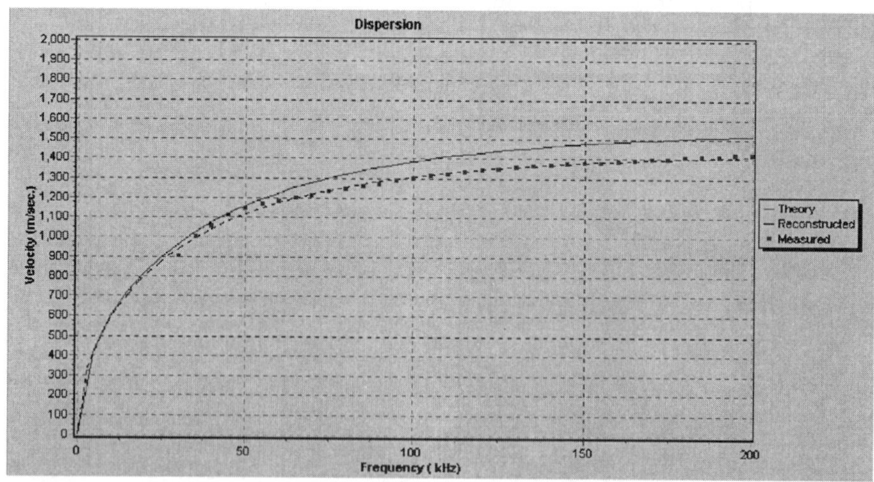

Figure 4. Example of Plotted Phase Velocity Measurements

The points shown in Figure 4 are the phase velocities measured experimentally. The two curved lines are the computed theoretical (upper) and reconstructed (lower) dispersion curves. To calculate the theoretical curve, the user inputs the known lamina properties and laminate lay-up into the software. The software then uses the classical laminated plate equations to compute the A, B, D stiffness matrices as follows:

$$\left(A_{ij}, B_{ij}, D_{ij} \right) = \int_{-h/2}^{h/2} \left(Q_{ij} \right)_k \left(1, z, z^2 \right) \; dz, \quad i,j = 1,2,6, \tag{1}$$

$$A_{ij} = \int_{-h/2}^{h/2} \left(\kappa_i \kappa_j Q_{ij} \right)_k \; dz, \quad i,j = 4,5. \tag{2}$$

where $(Q_{ij})_k$ are the reduced stiffnesses for the kth lamina in the laminate, A_{ij}, B_{ij}, and D_{ij} are the laminate extensional, coupling and bending stiffnesses, and h is the plate thickness. κ_4 and κ_5 in Equation [2] are the shear correction factors which can be defined as the ratio of the Rayleigh surface wave velocity over the out-of-plane shear wave velocity in the x and y directions respectively, see reference (8). Finally the phase velocity for A_0 mode propagating along the 0-direction in an orthotropic laminate (B_{ij} =0, $A_{16} = A_{26} = A_{45}$ =0, D_{16} =D_{26} =0) is determined by the solution to the wave number k in the following equation (reference 16)

$$(D_{11}k^2 + A_{55} - I\omega^2)(A_{55}k^2 - \rho h \omega^2) - A_{55}^2 k^2 = 0, \tag{3}$$

where $k=\omega/V$, ω is the circular frequency and V is the phase velocity. $I=\rho h^3/12$ is the area moment of the plate.

The reconstructed dispersion curve is calculated using Equation [3] with the D_{11} and A_{55} values determined by minimizing the sum of squares of the deviation between the experimental and calculated velocities considering D_{11} and A_{55} as variables in a 2-dimensional space (reference 8):

$$\min_{(D_{11},A_{55})} \sum_{i=1}^{m} (V_i^e - V_i^c)^2, \tag{4}$$

where m is the number of velocity measurements in different frequencies, V^e is the experimental phase velocity and V^c is the velocity calculated using Equation [3] for a given set of D_{11} and A_{55} values with user-input laminate density and thickness. In Figure 4, one sees that the reconstructed dispersion curve matches closely with the experimental points.

After taking these measurements, the scanner lifts from the part and moves to a new transmit location. There were three locations per coupon and sixteen coupons per panel.

Following the initial measurements, the panels were soaked in an oven for various intervals. Two of the panels (panel 1 and panel 7) were left undegraded as "control" panels. The resulting schedule of exposures appears in Table 1 below:

Table 1. Heat Exposure of Test Specimens

Panel	Exposure Time (minutes)	Temperature	Result
1	0	(Control)	No change
2	10	246 °C (475 °F)	Warpage
3	20	246 °C	Warpage
4	30	246 °C	Warpage
5	60	246 °C	Delamination
6	90	246 °C	Delamination
7	0	(Control)	No change
8	10	274 °C (525 °F)	Delamination
9	15	274 °C	Delamination
10	20	274 °C	Delamination
11	25	274 °C	Delamination
12	30	274 °C	Delamination
19	30	246 °C (475 °F)	Warpage, discoloration, blisters near edges
20	40	246 °C	Warpage, discoloration, blisters near edges
21	35	246 °C	Warpage, discoloration, blisters near edges
22	33	246 °C	Warpage, discoloration, blisters near edges
23	32	246 °C	Warpage, discoloration, blisters near edges
24	31	246 °C	Warpage, discoloration, blisters near edges

Those panels which had been exposed to heat were again tested with the scanner in the same programmed locations as before. Once these results were obtained, the panels were cut into coupons and mechanically tested to the requirements of ASTM D790, Test Method II (four point loading), Table 2, L/d ratio of 40 to 1 (see Reference 17).

Following mechanical testing, beam theory was used to calculate the D_{11} stiffness coefficient from the coupon spring rate. The equation used is as follows:

$$D_{11} = \frac{\left[\frac{P}{y}\right] \cdot (a) \cdot (L)^2 \cdot \left[3 - 4\left(\frac{a}{L}\right)^2\right]}{2 \cdot 24 \cdot b} \qquad [5]$$

where:

D_{11} = bending stiffness in Newton meters

P = load in Newtons

y = deflection in meters

a = distance from loading nose to nearest support in meters

L = span between supports in meters

b = width of beam in meters

These D_{11} stiffness coefficients are compared with their nondestructively derived counterparts in the section of this paper that follows. Similarly, one can obtain the D_{22} values by taking ultrasonic and mechanical measurements in the 90-degree direction.

4. RESULTS

The results of the ultrasonically-measured and mechanically-measured tests are shown in Table 2 below:

Table 2. D_{11} and D_{22} Stiffness Values Obtained through Mechanical and Nondestructive Tests

	Mechanical D_{11} (N*m)	NDE D_{11} (N*m)	Mechanical D_{22} (N*m)	NDE D_{22} (N*m)
Average for Panel 1 (control panel)	138.48	149.30	7.85	8.03
Standard Deviation	1.26	16.87	0.18	0.77
Average for Panel 7(control panel)	135.24	156.81	8.56	9.72
Standard Deviation	0.50	7.76	0.30	1.40
Ave. for Panel 2 (10 minutes at 475 F)	137.14	161.23	8.19	7.95
Standard Deviation:	1.15	9.15	0.11	0.83
Ave. for Panel 3 (20 minutes at 475 F)	135.50	142.30	8.18	7.72
Standard Deviation:	2.08	20.03	0.11	0.69
Ave. for Panel 4 (30 minutes at 475 F)	137.46	148.64	8.08	8.34
Standard Deviation	0.80	22.06	0.16	0.65
Ave. for Panel 19 (30 minutes at 475 F)	172.13	169.91	8.70	9.63
Standard Deviation	8.47	32.22	0.40	0.31
Ave. for Panel 20 (40 minutes at 475 F)	142.74	145.12	8.39	9.83
Standard Deviation	2.15	7.27	0.41	0.85
Ave. for Panel 21 (35 minutes at 475 F)	167.32	172.94	8.93	9.45
Standard Deviation	19.84	27.70	0.77	0.88
Ave. for Panel 22 (33 minutes at 475 F)	144.53	163.62	8.49	9.95
Standard Deviation	2.36	11.62	0.04	0.69
Ave. for Panel 23 (32 minutes at 475 F)	145.44	168.45	8.53	10.67
Standard Deviation	2.37	12.63	0.06	0.42
Ave. for Panel 24 (31 minutes at 475 F)	144.95	163.47	8.76	10.16
Standard Deviation	1.97	21.03	0.27	0.22

Note that the averaged D_{11} and D_{22} values in Table 2 were obtained from the 8 coupons in the same panel. One sees that overall the D_{11} and D_{22} values measured ultrasonically are slightly higher and have greater variations than those measured mechanically. The former may be due to an insufficiently large specimen length/depth ratio for such an anisotropic panel with significant shear deformations.

Table 2 indicates that the stiffnesses of the test specimens did not change significantly with heat exposure as long as there was no gross delamination present. There were some stiffness differences among different panels with similar exposure treatments as shown by the D_{11} and D_{22} values of control panels #1 and #7, and panels #4 and #19. Differences such as these could mask any discernible effect that the heat exposure had on the panels.

Nondestructive measurements were also performed on panels 19-24 in their undegraded condition. Before and after comparisons of these panels appear in Table 3. In the 0 degree (fiber dominated) direction, NDE results indicate an increase in stiffness after heat exposure, while in the 90 degree (matrix dominated) direction there is an equally distinct decrease. However, the standard deviation of the ultrasonic measurements is too great to make a definite conclusion about property changes. The causes for the relatively large scattering in ultrasonic measurements will be discussed in the next section.

Note that the stiffness values before thermal degradation could not be verified by mechanical testing using the same coupons since they were needed later for use in heat exposure tests. One may also note that the D_{22} values in panels 19-24 are greater than the two control panels #1 and #7. This may be due to a difference in the curing process since panels 19-24 were cut from a different composite plate as described in Section 2.

Table 3. D_{11} and D_{22} Values Measured Ultrasonically Before and After Heat Exposure

	Before Exposure D_{11} (N*m)	After Exposure D_{11} (N*m)	Before Exposure D_{22} (N*m)	After Exposure D_{22} (N*m)
Ave. for Panel 19	150.96	169.91	11.00	9.63
Standard Deviation	32.14	32.22	0.39	0.31
Ave. for Panel 20	132.93	145.12	10.62	9.83
Standard Deviation	30.59	7.27	1.11	0.85
Ave. for Panel 21	125.33	172.94	10.89	9.45
Standard Deviation	27.7	27.70	0.43	0.88
Ave. for Panel 22	123.37	163.62	11.07	9.95
Standard Deviation	25.81	11.62	0.65	0.69
Ave. for Panel 23	137.85	168.45	10.88	10.67
Standard Deviation	30.19	12.63	0.62	0.42
Ave. for Panel 24	149.42	163.47	11.26	10.16
Standard Deviation	28.39	21.03	0.49	0.22

5. DISCUSSION AND CONCLUSIONS

Results of the study indicate two things:

1.   In the test specimens, which might represent structure exposed to a nearby fire for a period of time, the bending stiffness properties of the material did not change significantly with increasing duration of heat exposure. The properties changed only when gross delamination was imminent.

2.   Lamb wave measurements to nondestructively extract bending stiffness properties were, at the time taken and with the equipment used, less precise than those from destructive tests. However, they appear to offer great potential. Nondestructive measurements are ideal for use in continuously monitoring the thermal degradation of the same panel over a long period of time. This would eliminate the variations among different panels observed during this study.

Although the prototype F-Scan™ system had more scatter in the results than the mechanical data, part of the problem involved the characteristics of the prototype. Improvements in hardware, software, and sensors have been incorporated in the commercial version of the scanner. These were not available at the time of testing and it is expected that they would have reduced measurement variation. Among the factors contributing to the data scatter were the following:

1.   The fixture for moving, placing, and indexing the transducers was subject to some free play which may have affected the Δx displacements used to calculate the phase velocity.

2.   Some of the panels, especially those exposed for longer durations, had delaminations present near the edges as seen in Figure 1. These delaminations affected both the mechanical and nondestructive test results, contributing to data scatter. The nondestructive measurements were affected more severely, since they are more localized measurements. An effort was made to remove visually-delaminated test coupons from the study, but some delamination may have escaped detection and caused outlying data points.

Ultrasonic test methods of the type used in this study offer certain advantages. Unlike spectroscopic techniques such as DRIFT (Diffuse Reflectance Infrared Technique) and LIF (Laser-Induced Fluorescence, reference 18), flexural wave techniques give a direct readout of stiffness properties which can be considered an absolute measure. A correlation of a material "signature" to external calibration standards or test data is not required. The technique is also dry-coupled, with both transducers interrogating the part from one side. This is advantageous for field applications or inspection of large parts. However, further refinements of the equipment would be necessary to produce a field-portable system with broad inspection capabilities. Currently the system is best suited to production applications.

Property changes caused by thermal spikes were not significant enough to be detected by this technique. More profound changes caused by long-term heat exposure (19) may lend themselves to ultrasonic detection. There are on-going studies at Boeing using both the mechanical and ultrasonic tools to monitor stiffness changes as a function of long-term heat exposure. Preliminary results showed that some reduction in the stiffness values occurred. This was verified with both mechanical and ultrasonic measurements and may become the subject of a future paper.

Evaluation of process variables such as porosity may be another future application, as well as evaluation of repaired laminates. Lamb waves have also been used (reference 20) to detect transverse cracking in laminates. Finally, local time-of-flight/amplitude mapping, a current function of the F-Scan™, may prove useful in detecting defects in layered materials which are difficult to test with through-transmission ultrasonics (reference 8).

Recently, using a much-refined production F-Scan™ system, D_{11}, D_{22}, A_{44}, and A_{55} stiffness values for a 16 ply, uni-directional AS4/3501 graphite/epoxy plate were measured to be within 4% of the material manufacturer's published values. Furthermore, a statistical study carried out by the authors showed that the repeatability of the velocity measurement using the current fixturing is within plus or minus 1%. The new version of the F-Scan™ may have accuracy high enough to be comparable with mechanical testing results.

6. REFERENCES

1. J. E. Raasch, "Regional Aircraft Fly Farther, Faster", <u>High Performance Composites</u>, January/February, 19 (1998)

2. A. K. Mal and Y. Bar-Cohen, "Ultrasonic Characterization of Composite Laminates", <u>Wave Propagation in Structural Composites</u>, ASME AMD Vol. 90, A.K. Mal and T.C.T. Ting, Ed., pp. 1-16, 1988.

3. A. G. Every and W. Sachse, "Determination of the Elastic Constants of Anisotropic Solids from Acoustic Wave Group Velocity Measurements," <u>Phys. Rev.</u> B42, pp. 8196-8205, 1990.

4. S. I. Rokhlin and W. Wang, "Double Through-Transmission Bulk Wave Method for Ultrasonic Phase Velocity Measurement and Determination of Elastic Constants of Composite Materials", <u>J. Acoust. Soc. Am.</u> 91, pp. 3303-3312, 1990.

5. M. Deschamps Hosten and B. R. Tittmann, "Inhomogeneous Wave Generation and Propagation in Lossy Anisotropic Solids. Application to the Characterization of Viscoelastic Composite Materials", <u>J. Acoust. Soc. Am.</u> 82, pp. 1763-1770, 1987.

6. M. R. Gorman, "Plate Wave Acoustic Emission", <u>J. Acoust. Soc. Am.</u> 90, pp. 358-364, 1991.

7. B. B. Elliott, "Flexural Wave Propagation in Anisotropic Laminates and Inversion Algorithms to Recover Elastic Constants Using Phase Velocity Measurements", <u>M.S. Thesis</u>, Naval Postgraduate School, 1992.

8. W. Huang, S. M. Ziola, J. F. Dorighi, and M. R. Gorman, "Stiffness Measurement and Defect Detection in Laminated Composites by Dry-Coupled Plate Waves", <u>Proceedings of SPIE</u>, Vol. 3396, edited by R. H. Bossi and D. M. Pepper, 1998.

9. P. A. Mehrkam, E. Armstrong-Carroll, and R. Cochran, "Fire Damage Assessment of A-6 Composite Wing BUNO 152951, Final Report", Naval Air Warfare Center, Warminster, PA, Report Number NAWCADWAR-93058-60.

10. P. A. Mehrkam, E. Armstrong-Carroll, and R. Cochran, "Heat Damage Evaluation of Painted Graphite/Epoxy Composites", <u>Conference on Characterization and NDE of Heat Damage in Graphite Epoxy Composites</u>, Orlando, Florida, 1993, pp. 113 - 123.

11.   P. A. Mehrkam and E. Armstrong-Carroll, "Detection of Composite Heat Damage by DRIFT Spectroscopy", 38th International SAMPE Symposium, May 10-13, 1993, pp. 217-225.

12.   P. A. Mehrkam, E. Armstrong-Carroll, and R. Cochran, "Evaluation of the Effect of Heat Damage on Bismaleimide Composites", 26th International SAMPE Technical Conference, October 17-20, 1994, pp. 354-364.

13.   M. D, Seale and B. T. Smith, "Lamb Wave Propagation in Thermally Damaged Composites", Review of Progress in Quantitative Nondestructive Evaluation, Vol 15, 1996, pp.261-266.

14.   B. J. Frame, et al, " Composite Heat Damage, Part I. Mechanical Testing of IM6/3501-06 Laminates, Part II. Nondestructive Evaluation Studies of IM6/3501-06 Laminates", ORNL/ATD-33, 1990, Oakridge National Laboratory, Oak Ridge, TN.

15.   J. E. Shigley and C. R. Mischke, Mechanical Engineering Design, 5th Ed., McGraw-Hill, New York, 1989, Table A-9, p. 739.

16.   B. Tang, E. G. Henneke II, and R. C. Stiffler, "Low Frequency Flexural Wave Propagation in Laminated Composite Plates," Proc. of Acousto-Ultrasonics; Theory and Application, edited by J. C. Duke, Jr., (Plenum Press, New York, 1988), pp. 45-65.

17.   "Standard Test Methods for Flexural Properties of Unreinforced and Reinforced Plastics and Electric Insulating Materials", ASTM Standard D790-95a, American Society for Testing and Materials, 1996.

18.   W. G. Fisher, et al., "Laser Induced Fluorescence Imaging of Thermal Damage in Polymer Matrix Composites", Materials Evaluation, 55 (6), pp. 726-729.

19.   G. A. Matzkanin, "Nondestructive Characterization of Heat Damage in Graphite/Epoxy Composites: A State-of-the-Art Report", Proceedings of the Conference on Characterization and NDE of Heat Damage in Graphite Epoxy Composites, Orlando, Florida, 1993. Nondestructive Testing Information Analysis Center Publication, p. 24.

20.   V. K. Kinra and V. Dayal, "Nondestructive Evaluation of Composite Material using Ultrasound", Acousto-Ultrasonics: Theory and Application, Plenum Press, New York, 1988, pp. 143-150.

A NON-DESTRUCTIVE TEST METHOD FOR THE DETERMINATION OF PERCENT RESIN CONTENT, FIBER AREAL WEIGHT AND PERCENT FIBER VOLUME OF COMPOSITE MATERIALS

Kathy M. Kelly, Hexcel Satellite Products
and
Dr. Peter R. Ciriscioli
Tempe, Arizona 85284

ABSTRACT

Historically, prepreg percent resin content and fiber aerial weight, as well as cured laminate percent fiber volume have been determined by ultrasonic wash and acid digestion, both destructive test methods. There are several problems associated with destructive methods including lack of repeatability, use of solvents, long test times, and the loss of fibers during the wash phase which results in erroneous values.

A non-destructive procedure has been developed which is based on sample area, conservation of fiber mass and known fiber mass variation. Property values for the nondestructive and destructive methods were analyzed using standard statistical techniques. Results of the analysis indicate that the nondestructive method for composite material resin and fiber related physical properties has less error, hence greater precision. The nondestructive method is more repeatable, does not use solvents, saves significant cost and time and is more precise than the destructive method.

KEY WORDS: Non-Destructive Evaluation, Fiber Areal Weight (FAW), Fiber Volume

1. INTRODUCTION

Most material properties of interest in composite materials are dependent upon the relative amounts of resin (matrix) and fiber (reinforcement). Data normalization is a common technique involving post-test data manipulation that attempts to eliminate unrealistic artificial variation caused by local changes in fiber volume. There are many methods which provide a means for determination of fiber volume. These methods fall into one of two main categories: destructive or non-destructive.

Destructive procedures using heat and acid have been the methods in vogue (1,2). There are many problems associated with destructive methods including lack of repeatability, long test times, large variation in data values, and the use of non-environmentally friendly solvents. Non-destructive procedures exist based on laminate thickness (3, 4). However, they are not commonly used.   Another non-destructive technique based on the conservation of fiber mass using sample area has largely been ignored by the composites community. By knowing the uncured material's fiber areal weight, cured sample surface area and number of plies, the fiber mass component for fiber volume can easily be determined. Fiber areal weight can be calculated by simply multiplying the number of fiber tows by the fiber's weight per unit length. These two techniques are certainly not new, but much more accurate, easier and faster than the destructive counterparts.

Due to increased business pressures to reduce cost and time associated with testing and environmental impact, a study was undertaken to identify the individual error components as well as the overall error in the destructive and non-destructive methods for fiber related property determination.

2.0 EXPERIMENTAL

2.1 Statistical Approach   Intuitively, the percent fiber volume in a laminate is dependent upon the amount of fiber, or fiber areal weight in a sample. As with laminate evaluation, destructive and non-destructive methods exist for fiber areal weight determination, with the destructive approach via ultrasonic solvent wash the most common. The obtained property values irrespective of the method is the goal of the testing. The question is, how does one know which value is the most accurate?

By using the statistical technique known as propagation of error, also known as stacking of tolerances or quadratic sum, the amount of error in a value derived from a given test method can determine which method yields the most representative value.   The general formula is

$$\sigma^2 = \sum_{i=1}^{N} (\partial f/\partial x_i)^2 \, \sigma^2 \, x_i \qquad\qquad [1]$$

where the variance or uncertainty of the property of interest is equal to the sum of the product of each individual variable squared multiplied by the variance of each component. Each individual components contribution is determined by the use of partial derivatives. The larger the individual coefficient, the larger the contribution of that term to the overall error of the system. This formula allows for the calculation of the overall error of a system by the quadratic sum of the individual variance components. The formulas for fiber areal weight, percent resin and fiber volume are all first order or linear equations, which allows for use of the quadratic sum (5). It is also important to point out that Fiber Volume is linear in the range of interest, which is 45-65 % (6).

When we know the functional relationship connecting the final result with the variable components, (i.e., the formula) we can study how the variation is transmitted to the final result. In particular, we can decide which components are the major cause of variability by evaluation of the resultant error coefficients. This allows for process focus on the important variable(s).   The larger the coefficient, the larger the weight, hence the larger the contribution of a particular variable to the overall error of the system.   Error represents the upper and lower bounds on the system, thus uncertainty is never any worse than this, therefore preventing an over estimation of the error (7). The derived error range represent the worst case scenario for each method under evaluation.

The choice of data used in the evaluation is critical. It is necessary that the data be uncorrelated, therefore independent (8). Each data point must be independent of all the others, chosen at random, and exhibit linear relationships. This is the case for the properties under study. From the independence property, the variance of the property under evaluation equals the sum of the variances of the individual components (9), which is exactly what is specified by the quadratic sum formula [1].

Figure 1 illustrates the error sources in the entire process.

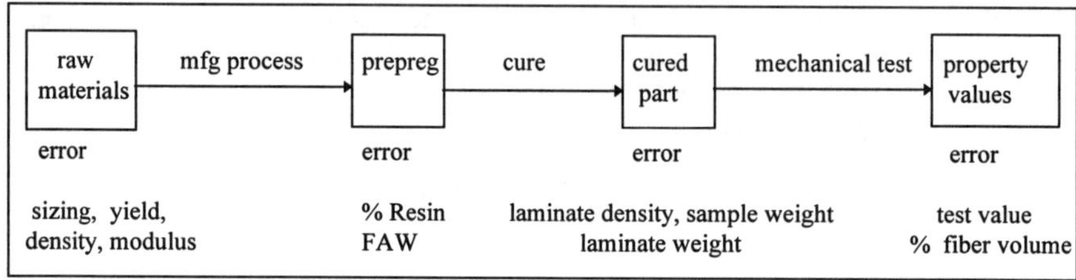

Figure 1: Schematic showing that uncertainty can be estimated for the error sources at each step.

2.2 Material   The data evaluated represents independent and uncorrelated data for prepreg properties as well as laminate properties. For the prepreg, one roll per lot was used as an individual data point. One laminate per lot was evaluated for percent fiber volume and tensile modulus.   The materials consisted of unidirectional tape and woven fabrics made with high modulus pitch and PAN fibers employing various epoxy and cyanate resin systems.   All measurements were made using calibrated scales, micrometers, and templates.

2.2.1 Prepreg   The basic calculations for percent resin, and fiber areal weight are:

Fiber weight, less sizing (non-destructive only) $((F_{ys}) - (F_{ys} * sizing))$ [2]
FAW :   non-destructive $(F_y * tows* k_w)$ [3]
           ultrasonic wash $(W_f * k_{ss})$ [4]
Percent Resin: non-destructive $[1- ((F_y * tows * k_w)/(sw *k_{ss})) *100$ [5]
           ultrasonic wash $[1 - ((W_f)/(sw))] * 100$ [6]

The main difference in the non-destructive and destructive methods is the use of fiber yield and tows per width versus ultrasonic solvent extraction for fiber areal weight. Tows are considered known with no uncertainty, hence it is treated as a constant. The amount of sizing is accounted for by Equation [2] for the non-destructive method since sizing becomes part of the resin when the ultrasonic wash is used.

2.2.2 Laminate   The calculated fiber volume is the reinforcing fiber's contribution to the total volume. Although void content does affect the laminate fiber volume, it is not a factor in the calculation since it contributes to the total volume in the same way as the resin or any other non-reinforced component. Since the calculation requires the number of plies, the method is applicable to forms with distinct plies or accurate knowledge of the number of plies contained in the specimen. Also, since the method is dependent upon surface area, an accurate method of determining surface area is critical. In the case of flat laminates, this can be determined by direct measurement using a calibrated micrometer.

The formulas for percent fiber volume for the destructive method based on acid digestion and the non-destructive methods based on sample area and thickness are:

Acid digestion	$((W_f * \rho_e) / (W_1 * \rho_f)) *100$	[7]
Thickness	$((FAW) / (k_t * CPT * \rho_f))$	[8]
Area	$((FAW * Plys * Area* \rho_e)/(1550* W_1 * \rho_f)) * 100$	[9]

Acid digestion calculates fiber volume based on the mass of the fibers remaining after chemically treating the cured material at an elevated temperature. It is well known that there is a relationship between laminate (or specimen) thickness and fiber volume for given values of fiber areal weight and fiber density (10). The other non-destructive method, which uses the sample area, is based on the conservation of fiber mass throughout the prepreg and subsequent laminate (part) manufacturing process. This method involves measuring the laminate or specimen surface area and calculating the fiber volume based on the measured area, the number of plies in the laminate, and previously determined values of fiber aerial weight, laminate weight, laminate density and fiber density. Since the number of plies in a laminate are known with absolute certainty, it is treated as a constant. The FAW, sample area, and plies, when divided by the conversion factor represents the fiber content (fiber weight) of the sample under study.

2.2.3 Tensile Modulus   The translation of tensile modulus to laminate or part tensile modulus is highly dependent upon percent fiber volume.

Specimens	$((L_E * normalization\ factor) / (FV))$	[10]

3.0 RESULTS

3.1   Analysis Techniques   Statistical analysis was performed using standard techniques. Variance components for each of the variables under evaluation were based on the standard deviations of the measurement samples. All variables except fiber density, yield, sizing and tensile modulus were based on direct measurements. Fiber properties values were taken from supplier certifications which routinely supply the mean, standard deviation and sample size for values reported.   In the case of multiple fiber lots, the values having the largest standard deviation was used. This was done to make the estimates as conservative as possible.   Hypothesis testing procedures using t-tests compared the resultant data population means and variances. In all cases the null hypothesis was the means of the populations under evaluation were equal, and the alternative was that the means were unequal. The analysis is illustrated by use of graphs based on pooled standard deviations of the data by a technique known as Analysis of Variance (ANOVA). Please refer to any standard statistics reference for more detail on the subject. The actual analysis for the data in this paper was performed using MINITAB Release 9 statistical software.

3.1.1 Prepreg   Table 1 summarizes the data generated for the non-destructive property estimation and standard ultrasonic wash technique.

Fiber yield is the largest contributor to error for the FAW determination by the nondestructive method, while the dry fiber weight is the largest contributor for the wash

Property	FAW		% Resin	
Method Error Sources	<u>Non-destructive</u> fiber yield	<u>Wash</u> dry fiber	<u>Non-destructive</u> fiber yield sample weight	<u>Wash</u> sample weight dry fiber weight
+/- Error	0.49 g/m2	1.00 g/m2	0.64%	0.85%
Sample Size	132	132	132	132
ANOVA	< Wash	> non-destructive	< Wash	> non-destructive

Table 1. Prepreg Summary
Notes: error sources are listed in decreasing order of influence

method. For Percent Resin determination, fiber yield again is the largest contributor for the non-destructive method with the sample weight contributing the largest amount of error for the ultrasonic wash, and the dry fiber weight also being a significant error source. The Analysis of Variance graphical representation clearly chows two distinct curves, each representing a data set (population) obtained as a result of applying a particular test method. The more distinct the curves, the greater the differences in the populations. The curve to the left in Figures 2 and 3 represent the error obtained for fiber areal weight and prepreg percent resin content respectively by the non-destructive approach. In other words, the error for the non-destructive method is significantly lower than the error for the destructive methods.

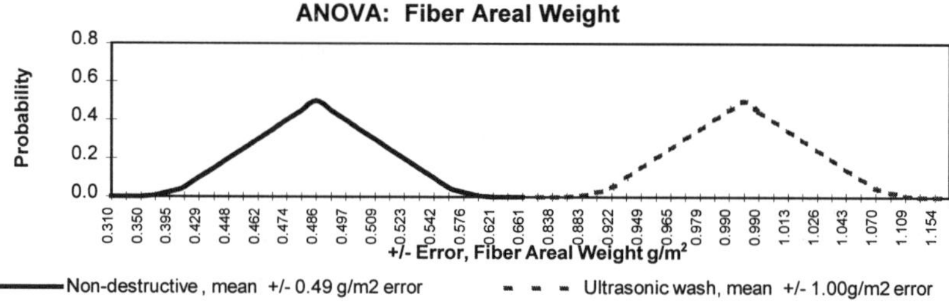

Figure 2. ANOVA: Fiber Aerial Weight

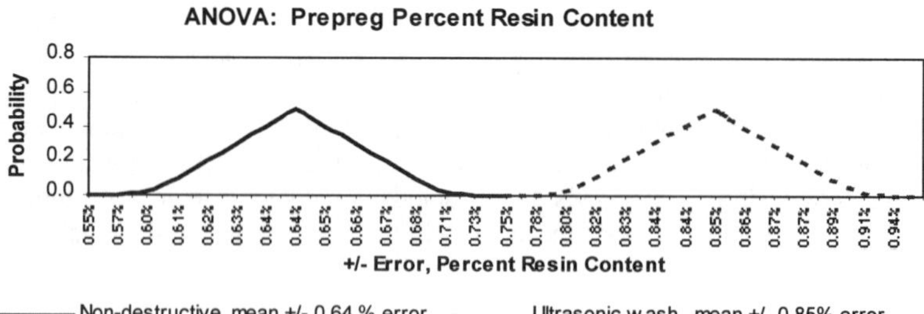

Figure 3. ANOVA: Prepreg Percent Resin Content

3.1.2 Laminates

3.1.2.1 % Fiber Volume: Table 2 summarizes the results for percent fiber volume for the two non-destructive methods and acid digestion methods.

Property	Percent Fiber Volume		
Method	<u>Area</u>	<u>Thickness</u>	<u>Acid Digestion</u>
Error Sources	area laminate weight laminate density fiber density fiber weight	fiber density cured ply thickness fiber areal weight	fiber weight laminate weight laminate density fiber density
+/- Error	0.59%	0.74%	0.80%
Sample Size	69	69	69
ANOVA	< Thickness < Acid Digestion	> Area = Acid Digestion	> Thickness > Acid Digestion

Table 2. Percent Fiber Volume Summary
Notes: error sources are listed in decreasing order of influence

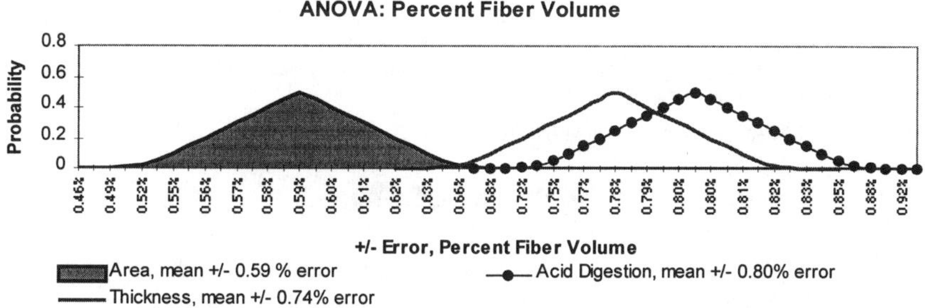

Figure 4. ANOVA: Percent Fiber Volume

The significant contributors to the error term are sample area, fiber density and fiber weight for percent fiber volume determination by area, thickness and acid digestion respectively. It must be noted that the cured ply thickness (thickness based method) and laminate weight (acid digestion) also contribute significantly to the overall error as well. The ANOVA results are illustrated in Figure 4. The ANOVA shows two distinct populations for percent fiber volume error determined by area and acid digestion. The area based method is clearly superior to the other two methods when evaluated in terms of test uncertainty. The error associated with the thickness based method is marginally less than acid digestion, but greater than the area method.

3.1.2.2 Tensile Modulus   Table 3 summarizes the results for normalized Tensile Modulus error based on the three percent fiber volume determination methods. In all cases, percent fiber volume is the largest contributor to the error associated with normalized tensile modulus values. The ANOVA results are depicted in Figure 5. The ANOVA shows the area method yielding normalized tensile modulus variation less than either the thickness based or acid digestion normalized values. Thickness based is marginally better than the acid digestion derived fiber volume normalization values.

Property	Normalized Tensile Modulus		
Method	<u>Area</u>	<u>Thickness</u>	<u>Acid Digestion</u>
Error Sources	% fiber volume	% fiber volume	% fiber volume
	test value	test value	test value
+/- Error	0.52 Msi	0.68 Msi	0.69 Msi
Sample Size	61	61	61
ANOVA	<= Thickness	= Acid Digestion	>= Thickness
	<= Acid Digestion	<= Acid Digestion	>= Acid Digestion

Table 3. Laminate Tensile Modulus
Notes: error sources are listed in decreasing order of influence

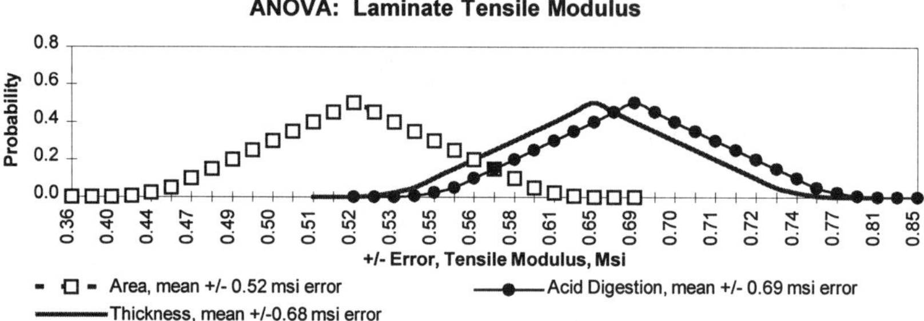

Figure 5. ANOVA: Laminate Tensile Modulus

4.0 DISCUSSION

Test methods are not perfect. They all have their advantages and disadvantages. The objective of this study was to determine which test method gives the property values of interest with the least amount of associated error. The advantages and disadvantages will be discussed below.

4.1 Prepreg
4.1.1 Non-Destructive Method The major advantages are greater precision, less time required to determine property values and the elimination of hazardous chemicals when compared to the ultrasonic wash method.

There are a few disadvantages which can be overcome with careful attention to the manufacturing process. It is obvious that fiber areal weight is grossly dependent upon fiber yield, or weight per unit length. Fiber yields with percent coefficient of variations of greater than 1% will affect the resultant fiber areal weight and fiber volume precision no matter which test method is used. Where possible, fiber yields should be determined per creel set-up for manufacturing runs rather than using manufacturer certification data.

4.1.2 Destructive Method The only advantage is in the absence of reliable information concerning the prepreg material, percent resin content and FAW can be somewhat approximated. Many times material will be housed in a freezer and necessary certification and testing information is lost or difficult to obtain. The disadvantages are the advantages of the non-destructive methods. Fiber weight is the largest contributor to error primarily due to the difficulty in recovering the fiber during the wash phase. Fibers can also be lost during drying if not properly contained.

4.2. Cured Laminates/Parts In general, significant fiber wash is always a potential problem and is not method specific. This error source can be greatly reduced if samples are taken at least one inch from a laminate or part edge.

4.2.1 Non-Destructive Methods The major advantages for both the area and thickness based methods are greater precision, shorter test times, elimination of chemical and solvents hence a safer, less expensive method. When using the area method, which yields less overall error, care must be taken to insure that the surface area of the sample under study is determined with great accuracy, since this is the major contributor to percent fiber volume error. Where possible, irregular shapes are to be avoided. The thickness approach has its drawbacks. Surface irregularities, which are a characteristics of fabrics, or laminates processed with peel-ply, can give erroneous thickness measurements. Care must be taken in the choice of measurement device end-types. Thickness measurements will vary depending upon the end-types used, and can be accentuated with textured surfaces.

4.2.2 Destructive Methods The only advantage to this approach is, in the absence of reliable information concerning FAW and ply count, percent fiber volume can be approximated. The disadvantages are the advantages of the non-destructive methods. An additional disadvantage is the potential for fiber degradation during the actual acid digestion process. While a correction factor can be estimated, it is a separate step in the procedure and adds even more uncertainty to the obtained value. Fiber weight is the largest contributor to error primarily due to the difficulty in recovering the fiber during the wash phase. Fibers can also be lost during drying if not properly contained.

4.2.3 Void Measurement This is a derived measure irrespective of the test method employed, and very difficult to measure with any degree of certainty. By using laminate density as a guide, a range can be established that the density needs to fall within in order to rule out excessive voids. This approach was investigated by Simon and Strunk in 1987 (11). Ultrasonic scans give a qualitative indication of porosity. Image analysis is also another method which can be used to quantify void volume. When using resin systems with low volatiles, voids should not be an issue if the laminates/parts are properly debulked, bagged, consolidated and cured. Interestingly, the calculation of void content is based on a system of no voids, which is why negative void volumes is not uncommon (see Section 4.2.4). The precision of percent void property estimation has been reported to be +/- 0.5% (12). For a discussion on the effect of voids on mechanical properties, the reader is directed to Reference 13.

4.2.4 Percent Resin Content of Laminates This is another derived property and assumes no voids in the calculation (matrix weight = sample weight - fiber weight). It also employs resin density in the overall calculation. This is a property that is typically only measured during a qualification and/or scale phase of process development. Neat resin casting is an expensive and time consuming to fabricate and typically only a nominal value is used for calculations once a resin system has reached the manufacturing stage.

4.2.5 Hybrid Laminates While the calculation is a little more tedious using the area method, this should pose no problem as long as the number of plies, fiber areal weight and fiber density is known for each component material.   In all cases, the fiber density values can be averaged based on the weight of each fiber lot used in a particular lot.

5.0 CONCLUSIONS

When a method for determining any property, in this case, percent resin content, FAW or percent fiber volume is used which gives a value with the least amount of error (highest precision), the resultant values will provide property values with higher precision. Based on the results of this study, the non-destructive methods for percent resin content and fiber aerial weight for prepreg, and percent fiber volume of cured laminates result in values with a much greater precision, hence less uncertainty than their destructive testing counterparts. In regards specifically to laminate percent fiber volume, the method based on sample surface area and conservation of fiber mass is superior to the acid digestion and non-destructive thickness based methods.   The method reduces test time, is repeatable, uses no environmentally damaging solvents, while providing values with less uncertainty.

6.0 ACKNOWLEDGMENTS

Special thanks to Dr. James E. Ashton, S. Charles Maurice and Nancy Chow.

7.0 KEY TO TERMS

Term	Description	Comments
AFY	actual fiber yield	sizing weight removed
sizing	%sizing on fiber	supplier certifications
%Resin	prepreg matrix content	
FV	% Fiber Volume of the laminate	
F_{ys}	Fiber Yield, g/m, w/sizing	supplier certification
F_y	Fiber Yield, g/m, no sizing	
tows	# of tows per width	treated as a constant
k_w	conversion factor width, inches to meters	1 meter = 39.37 inches
W_f	fiber weight, grams	
1550	conversion for in^2 to $meters^2$	treated as a constant
ss	size of original prepreg sample, in	treated as a constant
k_{ss}	conversion factor, sample size	treated as a constant
sw	original sample weight, grams	
Plies	# of plies in laminate	considered a constant
k_t	conversion factor, CPT	0.254, to account for units
ρ_f	fiber density, g/cm^3	from supplier certification
ρ_e	laminate density, g/cm^3	
FAW	fiber aerial weight, g/m^2	
CPT	cured ply thickness, mils	
AREA	area of cure laminate sample, $inches^2$	
L_E	laminate tensile modulus, msi	
F_E	fiber tensile modulus, msi	supplier certification
norm factor	factor used to normalize test values	treated as a constant

8.0 REFERENCES

1. ASTM Test Method D 3171M, _Annual Book of ASTM Standards,_ Vol. 15.03, American Society for Testing and Materials,   West Conshohocken, PA.
2. SACMA Recommended Method (SRM) 23R-94, _Suppliers of Advanced Composite Materials Association,_ Arlington, VA.
3. R.C. Kausen, _SAMPE International Symposium,_ 20, 95-107, (1975).
4. SACMA Recommended Method (SRM) 10R-94, _Suppliers of Advanced Composite Materials Association,_ Arlington, VA.
5. G.E.P. Box, W.G. Hunter, and J.S. Hunter, _Statistics for Experimenters,_ John Wiley & Sons, New York, 1978, pp. 563-564.
6. MIL-HDBK-17-1E, Normalization Methodology, U.S. Army Materials Laboratory, Materials Directorate, Aberdeen Proving Ground, MD, p. 2-75.
7. J.R. Taylor, _An Introduction to Error Analysis: The Study of Uncertainties in Physical Measurements,_ University Books, Mill Valley, CA, 1982, p. 63.
8. Box, et.al., pp 563-564.
9. N.D.Cox, _How to Perform Statistical Tolerance Analysis_ (11), American Society for Quality Control, Statistics Division, Milwaukee, WI, 1986, p. 2.
10. MIL-HDBK-17-1E, p. 2-75.
11. S. Simon and L. Strunk, _SAMPE International Symposium,_ 32, (1986), 116-122.
12. S.R. Ghiorse, _MTL TR 91-13,_ U.S.Army Materials Technology Laboratory, Watertown, MA (April 1991), p. 15.
13. S.R. Ghiorse, _SAMPE Quarterly,_ 24 (2), 54-59, (1993).

9.0 BIOGRAPHY

Ms. Kelly obtained her Master's of Science Degree in Industrial Engineering from Arizona State University and has over 10 years experience in advanced materials testing and manufacturing. She is a Certified Quality Engineer and is currently Quality Manager at Hexcel's Satellite Products Division (formerly Fiberite Tempe Space Products) in Chandler, AZ.

Dr. Ciriscioli holds a PhD. in Mechanical Engineering from Stanford University and has over 20 years of materials and composites experience. His experience includes positions as Director of Research for Fiberite, a Group Engineer for Lockheed Missiles and Space, and a Senior Engineer at LearFan. Dr. Ciriscioli is the author of over 20 publications in the areas of mechanical properties, processing and environmental stability of composites, chemical and refinery plant corrosion, patent law and ceramic fiber composites, the SECURE and SIMULATOR computer codes (currently distributed by Stanford University) and the book _Smart Autoclave Cure of Composites._

An Adaptation of Torsion Braid Analysis for Characterization of Viscoelastic Changes During the Cure of Thermally Polymerized Resin Systems Using Dynamic Mechanical Analysis

Elizabeth J. Hutton[1], Paul F. Reboa, Randall Willard, James A. Harvey[2]

Hewlett-Packard Company
1000 N.E. Circle Blvd, Corvallis, OR 97330

ABSTRACT

A brief overview of dynamic mechanical analysis theory and history is given, followed by a discussion of the challenges of obtaining mechanical measurements from substances which begin as viscous liquids then cure to form stiff solids during the measurement. A technique to overcome this problem is described, and its utility is illustrated with examples from two different adhesive systems; one thermosetting and the other thermoplastic.

1. INTRODUCTION

Dynamic mechanical analysis (DMA) measures the response of a material to a sinusoidal or some other type of periodic stress. Since the stress and strain are not in phase, two quantities can be determined: a modulus and a phase angle or a damping term. Dynamic mechanical analyses give more information about a material than static mechanical tests, although theoretically the other types of mechanical tests can give the same type of information. Dynamic tests over temperature and frequency ranges are sensitive to the chemical and physical structure of the material in question, highlighting chemical and physical changes which occur.

The variety of information that one can obtain from dynamic mechanical analysis of polymers includes glass transition temperature, secondary transitions, morphology of crystalline polymers, aging effects, and cure profiles. These quantities in turn can be useful in obtaining solutions to practical problems, as well as in exploring the fundamental behavior of resin systems.

It is convenient for the end user of a thermally polymerized resin system if the material can be applied, without mixing, in a liquid form, to efficiently wet the surfaces of the parts to be joined or the surface to be coated. It is also desirable to understand the mechanical changes these systems undergo as they polymerize. Unfortunately, the

KEYWORDS: Dynamic Mechanical Analysis, Epoxy, Polyamic Acid, Curing

---

[1] To whom correspondence should be addressed
[2] Also, Adjunct Professor, Oregon Graduate Institute of Science and Technology, Portland OR 97291-1000

instruments best used to examine the viscoelastic properties of liquids, rheometers, cannot continue to measure these systems once they solidify. Likewise, the instruments best used to examine the viscoelastic properties of solids, DMAs, require that a sample be physically clamped, an impossibility with a liquid.

The shortcomings of the two instruments force the use of a less conventional approach to characterize a thermally polymerized resin. A glass cloth/resin composite holds the liquid in place, allowing it to act as a coherent system. As polymerization proceeds, the composite sample stiffens, eventually allowing the resin to be measured without significant interference from the mechanical properties of the glass cloth.

2. THEORICAL ASPECTS

The results from a dynamic mechanical analysis are usually expressed as complex moduli or compliances. The following notation will be illustrated in terms of shear modulus, G, but an exactly analogous notation holds for Young's modulus, E. The complex moduli are defined by

$$G^* = G' + iG''$$

where G^* is the complex shear modulus, G' the real portion of the modulus, and G'' the imaginary portion of the modulus, with $i = \sqrt{-1}$. G' is called the storage modulus and G'', the loss modulus. Loss modulus is an energy dissipation term. The angle that reflects the time lag between the applied stress and strain is δ, and it defined by a ratio called loss tangent or dissipation factor:

$$\tan \delta = \frac{G''}{G'}$$

Tan δ, a damping quantity, is a measure of the ratio of energy dissipated as heat to the maximum energy stored in the material during one cycle of oscillation. Tan $\delta = 1$ occurs at the crossover point of G'' and G'. If this point is present, it is indicative of the gel point. The loss modulus, G'', is directly proportional to the heat, H, dissipated per cycle:

$$H = \pi G'' \gamma_0^2$$

where γ_0 is the maximum value of the shear strain during a cycle. Other dynamic mechanical terms expressed by complex notation include the complex compliance J^* and the complex viscosity η^*.

$$J^* = J' - iJ''$$

$$\eta^* = \eta' - i\eta''$$

Some of the interrelations between the complex quantities are

$$G' = \omega\eta''$$

and

$$G'' = \omega\eta'$$

where ω is the frequency of the oscillations in radians per second.

3. PRACTICAL ASPECTS

If one examines the response of the different types of polymeric materials to various forms of stimuli, one will see that dynamic mechanical analysis offers a versatile technique for the study of materials. To be specific, let us look at the uncured thermosetting resin formulations.

An adhesive formulation contains many ingredients depending upon its polymer type and what additives are needed to meet HP product and process requirements. A typical epoxy adhesive formulation contains at least an epoxy resin and a co-reactant (hardener). Accelerators, fillers, viscosity modifiers, and other additives may be present.

During a typical cure, the formulations will start out at a relative medium range viscosity. As temperature is applied to the system, some components melt and other ingredients thin, thus an overall reduction in viscosity has occurred. As the temperature is increased and the onset of the cure exotherm is reached, the formulations start to react and thus the viscosity increases. Any loss of solvent will also increase the viscosity. The viscosity continues to increase until the system has reached a state of total achievable cure. At this point the viscosity does not increase further, giving the appearance that the system has reached total cure. By monitoring viscosity one can use the fingerprint to monitor the quality and consistency of the uncured resin formulation, the curing process, and product performance. Dynamic mechanical analysis is ideal for this type of characterization. All modern dynamic mechanical instruments have the computing capability to calculate the dynamic viscosity from the moduli data and display it as a function of temperature, simplifying data analysis.

4. ANALYSIS OF ONE-COMPONENT EXPOXY

4.1 Experimental Techniques

The analytical procedure is relatively simple. It consists of coating the uncured resin formulation on glass cloth. In essence, one is preparing a composite prepreg specimen. The size of the specimen and the number of plies used to make the specimen depends upon the instrument used and the data obtained from a preliminary analysis. For the TA Instruments 983 system we have found that a 3" x 0.5", 5 ply specimen is suitable for the analysis of one-component epoxies. We wrap the ends of the specimen with aluminum foil to prevent bonding the specimen to the clamps during analysis. We use a minimal clamping force, as the sample cannot retain its shape under load. When excessive clamping force is used, resin is squeezed from the cloth in the clamp region. Our analyses were all performed using cantilever clamps to produce simple bending motion.

Because sample and clamp masses can produce thermal gradients, a temperature ramp rate of no more than 3° Centigrade per minute is recommended for most DMA analysis. We use a rate of 2° Centigrade per minute for this analysis. When using a fixed oscillation frequency, the sample should not pass through resonance at any time during the experiment. We have had good success with a frequency of one cycle per second. In selecting the oscillation amplitude, the resin system and sample dimensions are important factors, as is the sensitivity of the instrument. The amplitude must be sufficient to create a large signal, but not so large as to excessively disturb the sample. In the TA 983, we used an amplitude of 2% of the sample length. The temperature range should be selected such that the analysis includes the cure process, but not decomposition of the resin. Finally, it should be pointed out that it is a good analytical practice to cool down the specimen to ambient after it has reached its maximum temperature and reanalyze the specimen to insure total achievable curing and to confirm other observations.

4.2 Interpretation of Data

In order to obtain accurate DMA measurements, it is necessary to prepare samples of known and consistent dimension, preferably perfect cylinders or rectangles. The epoxy-impregnated glass cloth only approximates this condition because it is too soft to hold its shape until the analysis is well underway. Moreover, the measurements are taken on the combined resin-cloth system, never on the resin alone. For these reasons, the actual magnitudes of the measured values should be regarded with reasonable skepticism, and reproducibility must be established before the experimental results can be of practical value. The variations of any given quantity with respect to temperature do not suffer from these drawbacks, and can be used with more confidence. For example, if a minimum in viscosity appears in the data, it is possible to determine the temperature at which this occurs, but not the magnitude of minimum viscosity itself. If reproducibility could be established, it would be possible to compare minimum viscosities between formulations, but the differences would not be truly quantifiable.

These constraints on data interpretation may sound excessive, but a great deal of useful information is contained in the data. The examples which follow concern an epoxy formulation which is applied such that it is not fully supported during cure. If it develops unexpected mechanical properties at any temperature from ambient to the gel point, it may change its shape and location unpredictably. In this event, the adhesive cannot fulfill its design functions.

In Figure 1, the loss modulus, E" is detectable throughout the analysis. The storage modulus, E', is initially below the threshold of detection for the instrument, and does not appear until the reaction has begun to build the polymer network, around 68° Centigrade. During the interval in which E' does not appear, tan delta is unmeasurably large. The next feature of interest is the apparent gel point at 89° Centigrade. The glass cloth itself had been found not to contribute to the signal under the analysis conditions, so it is reasonable to suppose that this really is the gel point. The features at around 145° Centigrade are not fully understood. The development of the polymer network is clearly seen in the continuous increase of E' from its first appearance to the end of the analysis.

Figure 1. DMA thermogram of supported resin

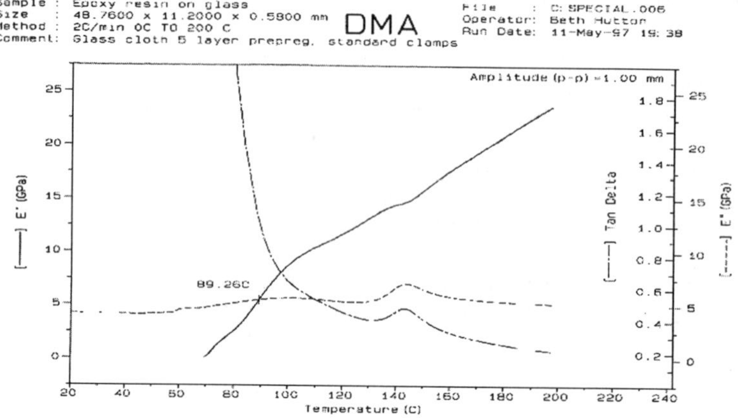

In Figure 2, the loss modulus and its derivative with respect to temperature are plotted. We have found derivative signals useful when analyzing a signal's change with respect to temperature.

Figure 2. DMA thermogram of supported epoxy resin

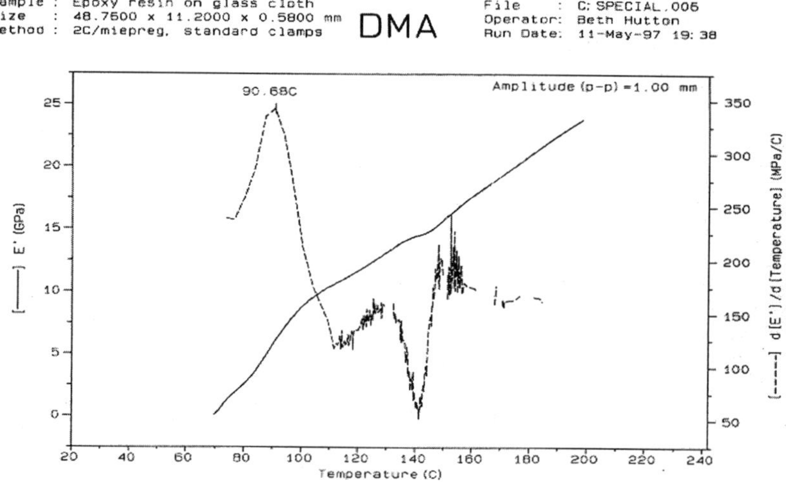

4.3 Application of the Data

We became interested in the mechanical properties of the adhesive shown in Figures 1 and 2 because it would sometimes move during processing, resulting in misshapen parts. This behavior was associated with specific batches of adhesives, but none of our supplier's tests indicated any differences between the better and worse batches. Our investigations also failed to shed any light on the matter, until we obtained data using the glass cloth technique, when a consistent pattern finally emerged.   In the poorly performing batches of adhesive, DMA revealed that the early portion of the cure process was shifted to higher temperatures as compared with the better lots. This is seen in the data as the temperatures at which E' appears, the first inflection point in the E' curve, and the apparent gel point. We were able to determine the extent to which the early cure kinetics could vary in temperature and still allow the adhesive to function well within the available range of processing parameters. This makes intuitive sense, as a higher temperature cure would leave the partially cured adhesive susceptible to extraneous forces for a longer period of time. The bottom line, though, is that we can now predict the production line performance of this material before we use it, saving us money through reduced line scrap.

5. THE THERMAL ANALYSIS OF FIBERGLASS CLOTH IMPREGNATED WITH POLYAMIC ACID

5.1 Experimental Techniques

Woven fiberglass cloth was found to be a good "carrier" for determining the curing profile by DMTA of a proprietary polyimide solution. The fiberglass cloth is either soaked or impregnated with the precursor polyamic acid solution, dried at 100°C to remove most of the solvent and to stiffen the composite prepreg specimen. Normally, three layers of fiberglass cloth were processed together and found to have adequate modulus. The composites were then either tested in tensile or bending mode.

Procedure: Samples were prepared by cutting and piecing together three fiberglass strips, about 1" x 3". The ends were wrapped together with aluminum foil, then the strips were soaked in a Petri dish with our polyamic acid solution for 5 minutes, making sure to leave the ends dried. The strips were lifted out of the Petri dish, allowed to drip for about a minute, then transferred to an oven (BlueM) where one end of the aluminum foil ends was held with tension clips and the other end of the clips was mounted on the upper rack of the oven. Similarly, the other other end of the strip was clamped with the clip and mounted on the lower rack of the oven, so that the entire clip was held in a vertical direction with just a small amount of tension. Drying was done at 100°C to drive off the solvent for 30 minutes. The dried samples were stiff and suitable for mounting on the DMTA after cutting to the proper size with shears.

The DMTA is a Rheometrics Mark III instrument. In tensile mode, the sample size was approximately 12-12.5 mm wide, 8-11 mm long, and 0.1 mm thick. We were not interested in absolute modulus numbers, simply in the phase transitions, especially curing. The tensile measurements were done at 1 Hz, 16 μm peak-to-peak strain (0.046%), and with a static force of 2-8 N. The temperature scan rate was 3 degrees/minute, usually from room temperature to 400°C. We found it necessary to reclamp the sample at 40 cNm at 100°C, because of the large softening that occurred. Also, severe shrinkage occurs with this polyimide, limiting the usefulness of the tensile data except to note where a change in the slope of E' occurs.

In bending mode, our conditions were 1 Hz, 16 μm peak-to-peak strain, and in single cantilever mode. We used the medium frame for an 8 mm length, and torqued at 40 cNm. The bending mode was friendlier to these samples and no special requirements or changes were needed during the runs.

5.2 Experimental Results for Tensile Analysis

The data shown below pertains to a proprietary polyimide, with and without a photoresist. The curing profile was necessary in order to optimize the baking conditions. However, this polymer is provided as the polyamic acid solution and after application, is cured to the polyimide. Also of interest, was the effect of the photoresist on the curing.

Application of the polyamic acid / photoresist solution to the fiberglass cloth allowed us to follow curing with our DMTA and correlate it with FT-IR data taken on thin films. Figure 3 shows the curing profile of this polyimide/photoresist in tensile mode. The Young's modulus decreases rapidly as the temperature is increased then flattens out at 125°C. It remaining relatively flat until about 375°C, where it increases until 450°C. The tan δ curve shows a sharp increase up to 125°C, then slowly declines up to 400°C, followed by a slight increase to 425°C. The absolute values of E' are not useful, since we cannot isolate the contribution of the glass cloth from the overall modulus. Therefore, only the changes in slope of E' and tan δ are important and not the absolute amount.

Figure 3. Tensile curing of the polyamic acid with photoresist. Curing is seen at about 125°C.

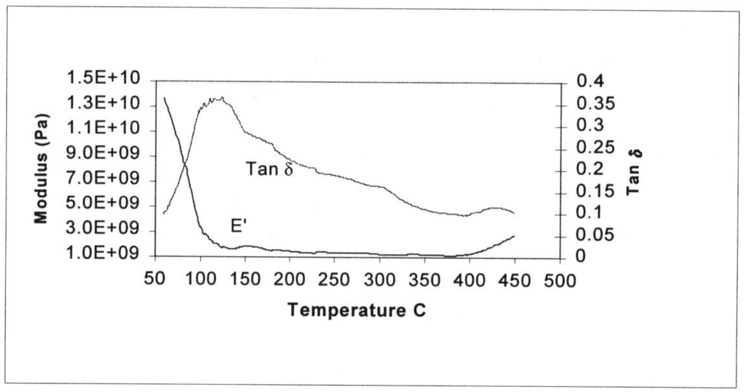

In Figure 3, the initial decrease in E' is attributed to softening of the polyamic acid. By FT-IR, it is known that this polyamic acid converts to the polyimide between 120°C to 150°C, completing the condensation reaction at 150°C. We interpret the levelling out of the modulus as the onset of curing, since an increase in the modulus due to curing would offset the softening of the amic acid. It should be noted that due to severe shrinkage the samples had to be re-tightened after heating to 100°C.

To improve upon the above data, the sample in Fig. 3 was heated to 100°C, clamped and the tension load adjusted. Data collection began at 100°C. The result is shown in Figure 4. Under these conditions, softening can be compensated and it is now clear that curing is indeed occurring by the large increase in E' and concurrent tan δ decrease. E' continues to increase until 200°C, then declines slightly. Previous work on cured films indicates this is the glass transition region. Again, as in Figure 3, an increase in E' is seen at 375°C. The reason for this increase is unknown.

5.3 Experimental Results for Bending Mode Analysis

Based on the above results, we decided to try out the same procedure but in bending mode, specifically cantilever bending. Results are shown in Figure 5 below. Similarly to Figure 3, the flattening of the E' curve indicates curing is occurring as expected. Unlike Fig. 3, tan δ does not continue to decrease; instead it too flattens out until around 200°C where it decreases. This change corresponds to the glass transition. Normally, the glass transition is observed by a decrease in E'. However, the woven fiberglass support exerts its own effect on E'. The polyamic acid converts in situ to the polyimide while absorbed on the glass fibers – it is in essence an adhesive to the fiberglass fibers (alternatively, the fibers can be viewed as a huge filler loading). Therefore, the stiffness of the fiberglass does not decrease even though the polyimide is undergoing the glass transition because the fibers restrict the polymer's translational capability.

Figure 4 illustrates the effect on curing of the polyimide alone, without the photoresist. The polyimide by itself is now seen to undergo curing at a slightly elevated temperature, beginning at 150°C instead of 125°C. This may be reasonable due to the possible formation of a salt complex with the photoresist and polyamic acid. DSC data also correlates with this data; the polyimide with photoresist cures at a slightly lower temperature than the polyimide by itself.

Figure 4. Bending data of same polyimide as in Figure 3, but without photoresist. Curing is now slightly shifted to 150°C. It is not known whether the second peak in the tan δ curve is due to a shifted Tg.

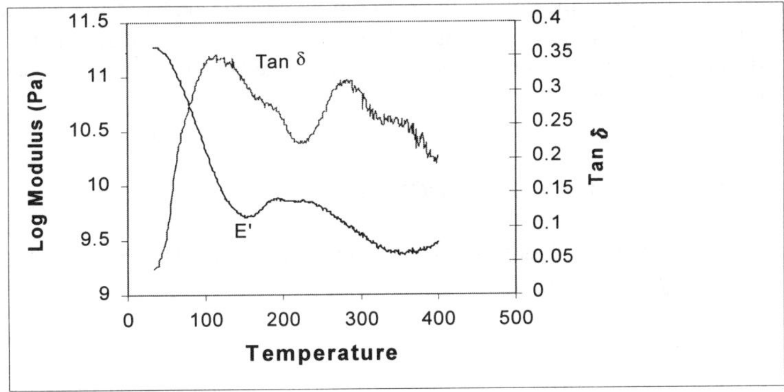

Also worth noting is the second peak in the tan δ curve at about 300°C. This peak is very reproducible in all the experiments run with the polyimide by itself. The peak is reminiscent of a glass transition. However, we have not investigated this point in detail so its origin is unknown for now. It is interesting to speculate that the photoresist may be plasticizing the polyimide to such a large extent.

Figure 5. Bending data of polyimide of Figure 3. Re-clamping at 100°C was not needed in bending mode (single cantilever). The flattening of E' corresponds to curing. Decrease in tan δ at around 200°C is the Tg.

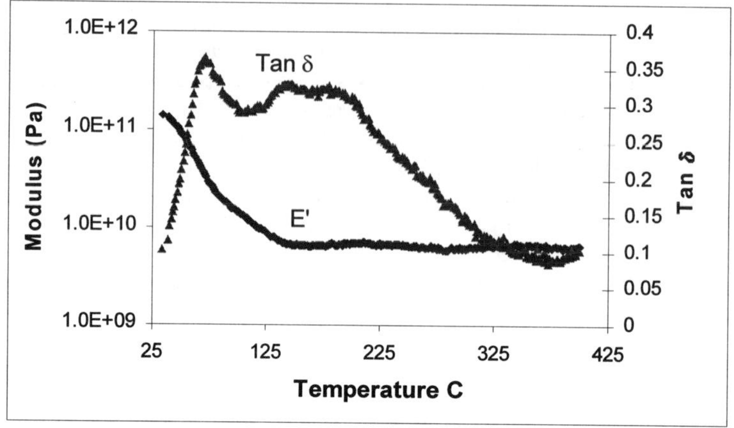

6. CONCLUSIONS

The dynamic mechanical characterization of thermally polymerized materials is possible through the application of the materials to a glass cloth substrate. This rapid and reliable method can be applied in both R&D and manufacturing settings. It can complement other characterization techniques to provide the researcher thorough understanding of a material's static and dynamic properties. This understanding can allow a researcher to fine tune material to the processing needs, or enable awareness of material quality issues before significant yield loss occurs.

7. ACKNOWLEDGEMENTS

The authors would like to thank Trent Hering, Vladek Kasperchik, David Neel, and Jones Oliver for their insight and assistance.

NDE DATA FUSION

Richard H. Bossi and James M. Nelson
Boeing Information, Space & Defense Systems Group
Seattle, WA 98124-2499

ABSTRACT

The manufacture of modern complex aerospace systems is trending toward large unitized structure, utilizing precise process steps and requiring sophisticated nondestructive evaluation (NDE) of the final product. The quality assessment of the product is moving from a uniform defect size specification to zoned criteria and fitness for service determination. An integrated approach to handling nondestructive evaluation and process control measurements is proving to be a useful and logical approach for the review, evaluation and disposition of processed articles. An NDE data fusion work station allows process and/or NDE measurements taken by various techniques or at different times and locations to be reviewed in a single workspace, increasing confidence in accept or reject decisions. An essential element of proper data fusion is the registration of the data to part fixed coordinates, i.e. mapping the data to a geometry model of the part. Using finite element definition of the geometry allows the NDE information to be readily available for other uses such as structural analysis. The finite element data model is the fundamental building block enabling NDE data fusion. By employing a visual programming methodology with an object oriented tool set in the design, the NDE Data Fusion workstation can be used not only for well developed applications but allows rapid creation of new data handling and comparison schemes.

KEY WORDS: Nondestructive Evaluation (NDE), data fusion, ultrasonics, computed tomography, eddy current

1. INTRODUCTION

The term "data fusion" refers to the process of integrating data from diverse process measurements and nondestructive observations of an object into a consistent description of the condition of the object. [1] The objective of data fusion methods is to co-register, collocate, and combine multiple data sets into useful forms. Combining carefully selected "modes" or alternative views of process and NDE data leads to reduced ambiguity, but increases inspection and evaluation labor unless convenient software tools for data fusion are available. Combining data from different process steps and NDE tests require systematic engineering methodologies which are often outside the norms of conventional (i.e. single mode) data analysis. In particular, precise cross-registration techniques and techniques for normalizing, cross-calibrating, and re-sampling are necessary. Of these, systematic registration of data with respect to the evaluated part is the fundamental requirement.

An approach to managing process monitoring and NDE data from multiple sources is shown in figure 1. [2] The approach displaces the familiar "image" data model in favor of an "object" or finite element data model. In the object model not only the value of data is described, but also its geometry. This approach has been implemented in INDERS (Integrated Nondestructive Evaluation Data Reduction System), a collection of software programs for integrating data from computed tomography (CT), ultrasonic (UT), eddy current (ET), and other techniques. [3] Essentially the data acquired from process monitoring or NDE operations are expressed in a part fixed coordinate system using techniques commonly used in finite element analysis. Features of interest can be calculated by appropriate combinations of inputs from the various sources which are transformable in a common coordinate system. The results can be mapped back onto any of the original test system representations or presented in standard computer graphics, visualization or data base representations. This approach, embedded in the INDERS software package, has won acceptance on various Boeing military and commercial programs, and has been used by these programs to support numerous studies, investigations, and qualification activities.

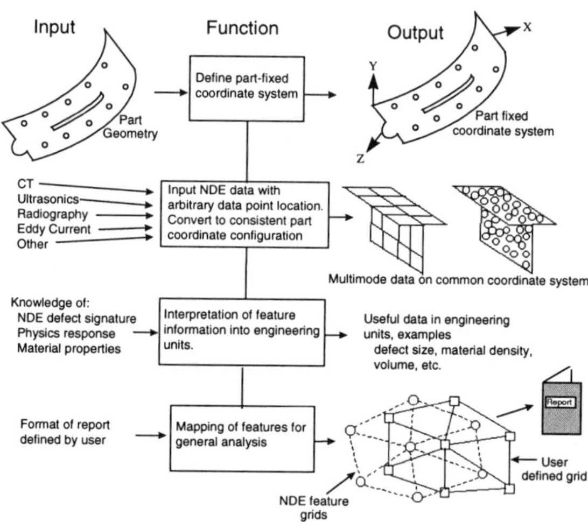

Figure 1 Managing process monitoring and NDE data with INDERS.

2. DATA FUSION WORKSTATION

Practical process and NDE data fusion requirements encompass great variety. Variations occur in data types, formats, interpretation, dimensionality, operator sophistication, and throughput requirements. Each of these parameters affects the design and/or selection of data management tools and the potential for justifiable return on investment. To be of greatest benefit, a data fusion workstation must be developed which is suitable for locating in engineering, manufacturing, operations, and maintenance facilities. The workstation hardware must be compatible with the facility requirements of the operations environment, including space available, power, and cleanliness. In designing the user interfaces, the workstation and associated software must suit a variety of skill levels corresponding to the personnel who are to use the workstation. In this way, the workstation can be seen as an asset and not a hindrance. For routine NDE personnel use, simplicity and robustness are required. For certain engineering and/or investigative uses, access to complex tools and a programming environment may be required. The user interfaces must be intuitive to learn, providing tools that solve the real problems conveniently. Visualization tools are used to enhance intuitive comprehension of complex data relationships. Often-repeated mechanical steps may be automated, such as locating test and geometry data, inputting, and saving data, and compiling reports. Finally, the workstation must be capable of performing its functions in a timely fashion, consistent with engineering manufacturing, operations and maintenance needs.

The key emphasis is on analyzing and presenting the data in a format most beneficial to the user. There are two major problems associated with the effective use of NDE data. The first is the lack of a systematic approach to data acquisition, interpretation, and analysis, and the second is the high cost of human interaction and judgment on the very high volume of data that emerging and future NDE techniques will generate. To remedy the lack of systematic approaches to NDE data interpretation and analysis, Boeing has developed a philosophy for NDE data interpretation and a software framework (INDERS, discussed in the previous section) for NDE data fusion. To remedy the high cost of human interaction with the large volumes of NDE data produced by advanced techniques, packaged software end items must be assembled for specific applications.

Most newer NDE hardware is "fully functional" digital-based equipment, which means that ready access to measurement data is available to the workstation. Those systems often create digital images or data files which require interpretation, normalization, and registration software to be written to convert them to the workstation standard. Process data is often tabular information. The INDERS based data fusion workstation handles the information from tabulations and images in a manner analogous to a spreadsheet. A spreadsheet program simultaneously satisfies needs ranging from solving quick arithmetic problems to complex financial computation to database operations to scientific simulation. It does this successfully, because it provides (1) a way of viewing and operating on data that is easily understood, (2) a powerful set of simply-accessed tools, and (3) a means for building higher-level solutions whose inner details can be ignored by solution users. Like a spreadsheet, simple tools can be used for ad hoc reading, interactive viewing, and processing and storing data from a large number of modes. These tools can be used interactively for purposes of diagnostic exploration or technique development, or they can be combined in permanently-storable solutions, to be later recalled and reused as though they were part of the original system functionality.

The features described above are implemented by layering INDERS on a powerful commercial product: AVS Express™, a product of Advanced Visualization Systems, Inc. AVS Express™ is an object oriented scientific visualization tool that enables data visualization applications to be programmed graphically. [4] Each AVS Express™ module performs some specific task (such as an image processing or visualization operation). Data flows between modules over connecting paths drawn in a click-and-drag style by the programmer/user. The network, its interactive controls, and its displayed images are presented in windows. Once a network solution has been created, it and others can be arranged in menus, and the underlying network can be hidden, making a turnkey application.

3. EXAMPLES

3.1 Rocket Nozzle Evaluation Carbon-carbon rocket motor nozzles are very high value items because of their cost to manufacture, their long schedule time, and the extremely high value of their payload. Multiple mode NDE is applied to rocket motor nozzle exit cones to assess both the "as manufactured" quality and flight worthiness after storage and handling. Multimodal methods used include ultrasonics (UT), eddy current (ET) and X-ray computed tomography (CT). Each inspection modality (e.g. CT or UT) can indicate the presence of an anomaly without necessarily distinguishing the material defect. For example, CT images of a dry ply and a delamination on two different parts are essentially indistinguishable. Conversely, different NDE modalities may respond to an anomaly to differing degrees. This concept of using several modalities to distinguish an anomaly is generalized into a multi-modal parameter representation of anomaly types illustrated in figure 2. [5]

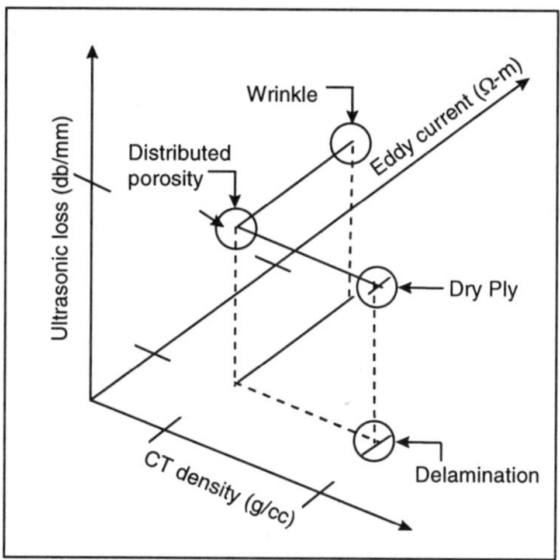

Figure 2 NDE parameter representations of anomalies.

In order to correctly identify manufacturing defects and damage, it is critical that the data from each mode be viewed in precise co-registration. Figure 3 shows how data from ultrasonic, computed tomography, and eddy current can be presented together in part fixed coordinates. The rocket nozzle exit cone shape is shown as a surface with features from each methodology displayed over the critical (i.e. highly sressed) forward region of the nozzle exit cone as shown. The quantitive parameters evaluated here are eddy current impedance, minimum CT density through the wall, CT density averaged through the wall, and ultrasonic amplitude. Low density CT indications can be due to wrinkling, "dry ply" conditions or delaminations. The "dry ply" condition can have sufficient across-ply tensile and inter laminar shear strengths to support the mission loading. A delamination, however, would result in mission failure if misdiagnosed as a dry ply. Only by considering complementary features of the ultrasonic data and the CT data can the two conditions be distinguished. In this example, the eddy current scan allows a correct interpretation of an otherwise benign low density indication in the CT scan as a wrinkle whose presence would almost certainly lead to flight failure.

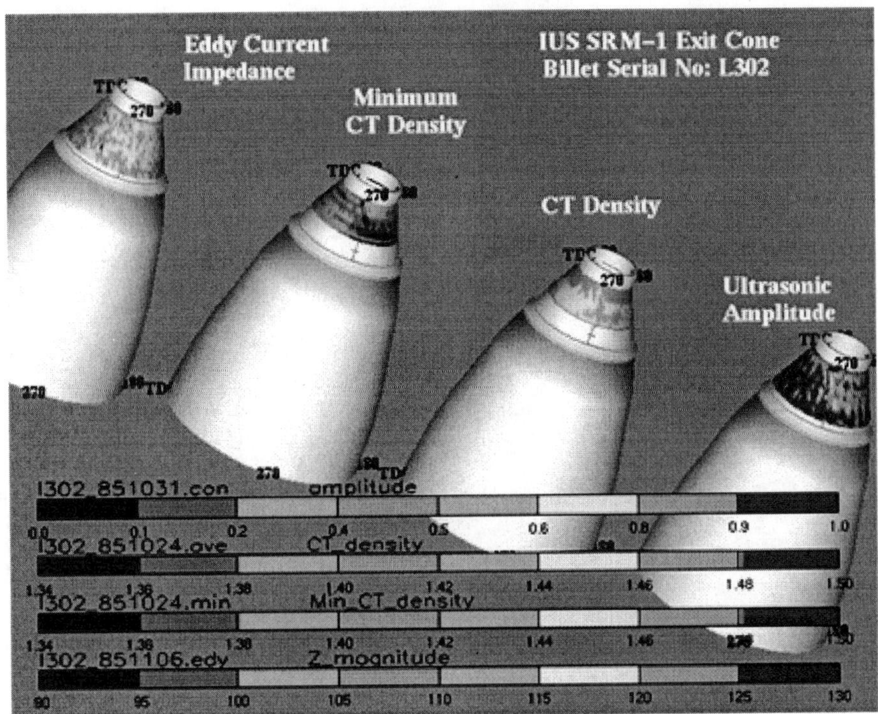

Figure 3 NDE multimode data fusion for a carbon-carbon rocket nozzle exit cone.

3.2 Radome Paint Thickness The performance of certain radomes containing sensitive instruments such as radar systems is critically dependent on its paint and paint thickness distribution. Inspection of the radome for paint thickness is an essential NDE activity. The measurement of the thickness is obtained ultrasonically. The raw data, however, is processed in the data fusion workstation by taking the 25 MHz pulse echo ultrasonic measurement and performing an autocorrelation of the Hilbert transform based analytic signal. This measurement is displayed on the geometry model of the radome as shown in Figure 4. Not shown in this figure, but added to this model the reviewing engineering staff, is the corresponding radar range data for the radome performance. Using these measurements and the display capability in a single workstation allows improved interpretation and disposition of radome manufacturing issues.

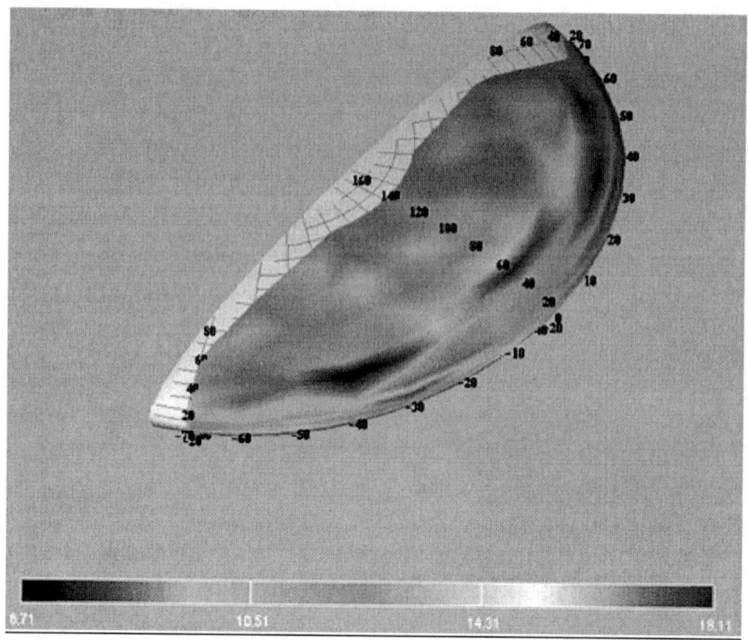

Figure 4 Radome ultrasonic thickness measurement.

3.3 Composite Structure Another example of data fusion for a complex process is the integration of process monitor data and NDE measurements for thermoplastic induction welding. This process uses a metallic susceptor between thermoplastic surfaces as shown in figure 5. The susceptor is inductively heated causing the surfaces to fuse. Time, temperature, and pressure are the critical parameters that must be sufficient for welding but not over processed to cause voiding in the skin or cap composite structure. During welding the time and temperature are monitored at specific locations in the weld zone using thermocouples and may be adjusted by the induction coil controls. The time at temperature for the weld interfaces has been determined to be the most important measurement for the process control. The pressure to the tooling is a fixed value assigned for specific welding conditions.

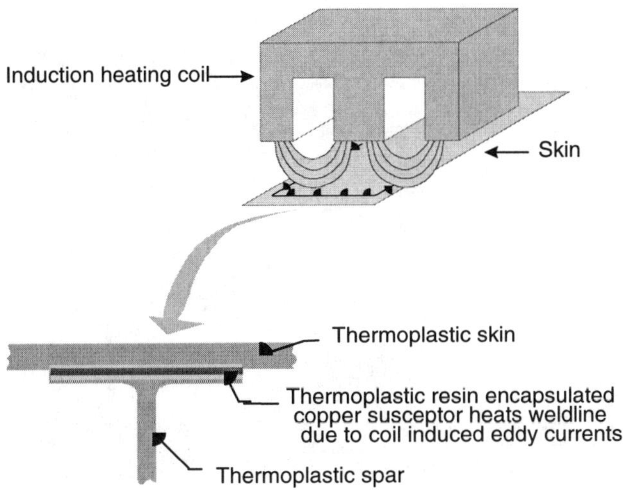

Figure 5 Induction welding process.

The temperature data is obtained from thermocouples placed at various locations along the weld and also at either the spar surface or between ply layers around the susceptor. The equivalent time (in minutes) to a process target temperature is calculated for each thermocouple and is plotted as a function of position along the weld.

Ultrasonic inspection of the weldline can be performed at various stages of the welding process. During development, UT scans have been taken at various times in the process to understand the state of the welding and its progress. The most reliable inspections, however, are after the welding is completed and are taken with a full waveform digitizing scanning system. This full waveform data is processed to compare signal strengths from the weldline and the back of the spar cap signal for an ultrasonic pulse echo inspection taken from the skin surface. Figure 6 is an illustration of combining test data from both the final ultrasonic inspection and the processing time at temperature. This data fusion workstation display shows the geometric configuration of a welded skin to a T spar. The geometry data includes chalkmarks for position information. The data fusion workstation display in this case shows a three dimensional view of the weld. The time at temperature data are shown at the top of the figure 6 image display by colored squares at the position of thermocouples in the test object. The depth locations in the object are very close, but can be shown by manipulation of the image display and also by the use of increasing square size with depth. The ultrasonic models shown in the lower portion of figure 6 are processed image data that are overlayed on to the geometry model of the weld. The bottom model is the ratio of the weldline echo to the front face surface echo. Low intensity (dark blue) level is the most desirable. The middle model is a "weld factor" calculation which is a point by point calculation of the ratio of the difference of the back of spar cap signal to the weldline signal to the sum of the back of spar cap signal and the weldline signal. This value is positive for a good weld and negative for a poor weld. This "weld factor" model is processed data which essentially fuses ultrasonic amplitude data information from different depths in the part. Because there is no back of spar cap signal over the web of the T spar, the "weld factor" along this longitudinal region is not a valid representation.

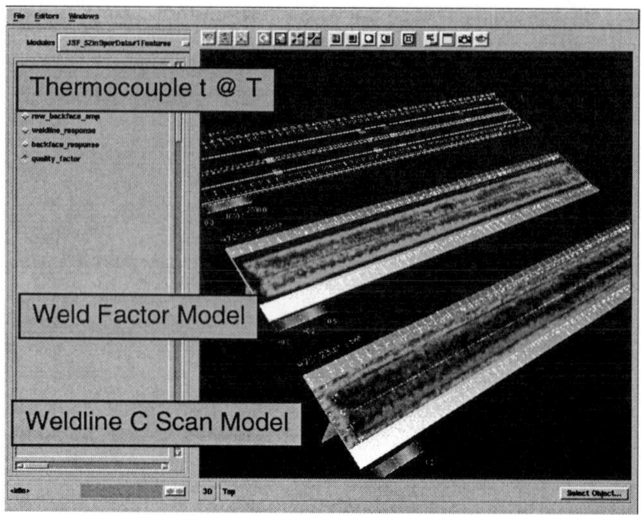

Figure 6 Induction welding data fusion models of process data and NDE measurements.

Using the data fusion workstation display, the process engineer is able to note that the time at temperature values are decreasing from right to left in the data. The weldline echo model also shows that in the lower time at temperature region, (on the right) there is an indication that begins to appear over the web. This greater ultrasonic reflection over the web relative to other areas is not reflected in the "weld factor" model because the "weld factor" calculation over the web is invalid. This increased signal over the web has been noted in other welds and appears when the welding parameters are near the process bounds. Additional data processing can be performed in the data fusion workstation. For example, histograms may be taken over regions of the weld factor model for quantitative analysis. Studies have shown that when the peak of the histogram is above zero a good weld is indicated.

4. CONCLUSION

The integrated approach implemented in the NDE Data Fusion Workstation to handling nondestructive evaluation and process control measurements is proving to be a useful and logical approach for the review, evaluation and disposition of processed articles, particularly large unitized structures. The fundamental requirement for NDE data fusion is the co-registration of data obtained from various techniques or at different times and locations in part fixed coordinates. This allows mapping the data to a geometry model of the part and allowing that data to be reviewed in a single workspace, increasing confidence in accept or reject decisions. Finite element definition of the geometry data from INDERS data processing is the fundamental building block enabling NDE data fusion and allows the NDE information to be readily available for other uses such as structural analysis. The implementation of a visual programming methodology with an object oriented tool set in the NDE Data Fusion workstation allows rapid creation of new data handling and comparison schemes. The ability to visualize both process information and reduced inspection data on a part fixed coordinate system is critical to the interpretation of the quality of modern structural products and greatly aids the development of product acceptance criteria.

5. REFERENCES

1. R. Bossi and J. Nelson, "Data fusion for process monitoring and NDE," in Nondestructive Evaluation for Process Control in Manufacturing, Richard Bossi, Tom Moran Editors, Proc. SPIE 2948 pp. 62-71 (1996).

2. James Nelson, David Cruikshank and Stuart Galt, "A Flexible Methodology for Analysis of Multiple Modality NDE Data," Review of Progress in Quantitative NDE, Vol 8A pp. 819-826 1989.

3. Integrated Nondestructive Evaluation Data Reduction System (INDERS) for Nozzle Components Software, Boeing Document, D180-31159-1.

4. D. Argyle and J. Youngberg, "Software Design Document for a Nondestructive Evaluation Data Fusion Workstation," The Boeing Company, D658-10665-1, 1997.

5. G. E. Georgeson, B. M. Lempriere and J. E. Shrader, "Integrated Nondestructive Evaluation Data Reduction System (INDERS)," JANNAF, Silver Springs, MD, October 1989.

ACKNOWLEDGEMENT

This work is sponsored in part by the Air Force Research Laboratory, NDE Branch under contract F33615-95-C-5234. Laura Mann is the AFRL program manager.

ELEVATED TEMPERATURE MEASUREMENTS OF ELASTIC CONSTANTS IN POLYMER COMPOSITES

Donald E. Yuhas and Bruce Isaacson
Industrial Measurement Systems Inc.
245 W Roosevelt Rd.
West Chicago IL 60185

ABSTRACT

This work describes an ultrasonic technique which has been used successfully to measure the elastic constants in polymer composites over the temperature range from 23 °C to 325 °C. Our method exploits the fundamental relationship between sound speed in specific propagation directions and the material elastic constants. Results obtained at room temperature are found to compare favorably with those obtained by conventional, static, mechanical testing methods. Elevated temperature measurements of the shear and longitudinal modulus on continuous fiber, graphite/epoxy composites and short fiber composites are reported. The moduli generally are found to decrease with increasing temperature. The shear moduli are more sensitive to temperature changes than the longitudinal moduli. Those properties which are fiber dominated exhibit less temperature dependence than the matrix dominated properties. Unusual behavior was observed for the major axis modulus of the graphite/epoxy. Values at 325 °C are 8% higher than those at room temperature.

KEY WORDS: Mechanical Properties, Testing Equipment, Polymer Composites

1. INTRODUCTION

Knowledge of the relationships between mechanical properties of polymer composites and composite structures is important in the design as well as the processing of materials for optimal performance. Increasingly, computer simulation and modeling are used as product development tools to speed component design and materials selection. As input to the structural models, accurate material property data e.g. density, thermal expansion, and elastic moduli are needed. Unfortunately, methods to measure these properties over the range of temperatures encountered in use and processing have not kept pace with the simulation tools development. Specifically, the Young's modulus, shear modulus and Poisson's ratio, which are essential inputs to any structural composite finite element analysis, are rarely measured. In order to accurately estimate the loads and/or vibration modes, it is desirable to have these parameters measured over the operational temperature range of the structure.

Unidirectional composites are the basic structural elements from which many advanced aerospace composite structures are fabricated. Quantitative measurements of the elastic properties of these are critical to the design and modeling of these structures. In particular, the shear properties are difficult to measure by conventional test methods since they often require extensive fixturing and specialized sample preparation. In this paper we exploit the fundamental relationship between ultrasonic propagation speeds and the elastic constants in order to determine all relevant mechanical parameters. Ultrasonic methods have been used for the past forty years to determine the elastic constants in anisotropic polymer composites which possess transverse isotropy (1-5). We have extended the measurement capabilities to temperatures as high as 325 °C.

2. EXPERIMENTAL METHODS

2.1 Ultrasonic Measurement Technique   The method used in this study is an ultrasonic through transmission technique for measuring the velocity of sound in material. Ultrasonic testing methods are directly analogous to the familiar testing methods of radar and sonar. The basic concept of ultrasonic testing is illustrated in the top of the schematic diagram shown in Figure 1. A short burst of high frequency sound is generated at the transmitting transducer and propagates through the sample to the receiving transducer. Measurements are made of the transit time and the signal amplitude. If the thickness is known, the ultrasonic velocity can be determined from the transit time measurement. By measuring velocities of shear and longitudinal wave modes for different sample orientations, the material elastic constants can be determined.

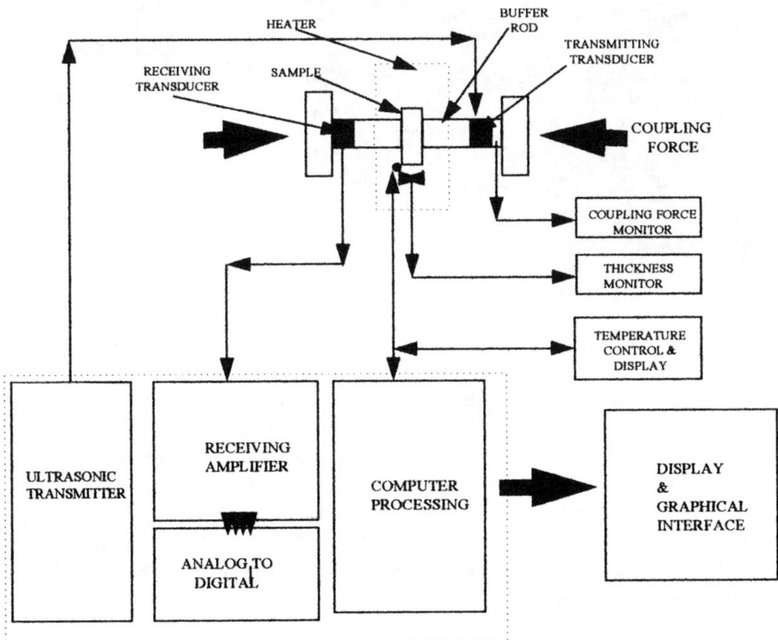

Figure 1   Schematic Showing the Basic Components of the ETEK and the Concept for Through-Transmission Ultrasonic Measurements

The instrumentation used in this study is the **ETEK I** system which is depicted in Figure 2 (6). For elevated temperature measurements, the samples are mounted in a holding fixture which is compatible with the small sample heater. With reference to Figure 1, the measurements are made in the following way. The ultrasonic transducers, which are attached to fused quartz buffer rods, are then coupled to the sample by inserting the buffers through the appropriate ports in the sample heater. The buffer rods provide the ideal combination of good ultrasonic coupling while thermally isolating the hot sample from the transducers. The transducer/buffer/sample combination is held in contact by a spring loaded test fixture which applies a uniform coupling force. The coupling force is monitored continuously throughout the testing cycle. Coupling force can range from 1 Kg to 250 Kg. The sample thickness change due to thermal expansion is monitored throughout the temperature cycle.

The transducers are conventional piezoelectric elements. Several different pairs are needed to generate and detect both shear and longitudinal modes. Both wave modes are necessary in order to extract the elastic constants. For high temperature measurements a silicon-based liquid is used which will maintain contact for temperatures up to 325 °C, which is the maximum operating temperature for the system used in this study.

Figure 2   Elevated Temperature Elastic Konstant (ETEK) System

The sample heater consists of a solid copper block, approximately 75 cm by 50 cm by 30 cm with embedded resistive heating elements. A small cavity is machined in the block to accept the sample holder. The sample holder is also fabricated from copper and is design so that the samples (typically 20 mm by 15 mm by 8 mm) can be attached mechanical while providing access ports for coupling of the buffer rods. The sample holder is also made from copper and functions both to mechanically hold the sample and promote good thermal contact with the heater. Heat is transferred to the sample primarily by direct conduction through the copper. Temperature is measured by a thermocouple located directly adjacent to the sample holder.

A short burst of ultrasound is transmitted from the transmitter/buffer combination through the sample and is received by the receiver/buffer combination. The system is capable of operating up to frequencies as high as 30 MHz. The received radio frequency signal is captured by a high speed analog to digital converter (50 MHz) and displayed on the monitor. By measuring the time difference in the delay with and without the sample present the sound velocity can be measured. By propagating difference wave types through different sample directions, the various elastic constants in anisotropic materials can be measured. The system is equipped with echo tracking algorithms as well as automatic gain control so that once the temperature cycle is initiated the transit time is measured automatically.

The time required for the sample to reach thermal equilibrium with the heater is dependent upon the sample size and its thermal properties. The temperature controller is designed so that user defined heating cycles are possible. For the work reported here, stepwise heating cycles were used where the temperature controller was programmed to move to a specific set point or series of set points and maintain this temperature for a specific time. For these studies the maximum temperature was 325 °C and the temperature steps were 28 °C. The hold time or soak time ranged from 5 to 10 minutes per point which was sufficient to allow the system to attain thermal equilibrium. Transit time measurements were made at the end of the soak time interval.

In addition to the transit time measurement, which is used to calculate the ultrasonic velocity, the signal amplitude is monitored. From this parameter, the signal loss is measured. The signal loss measures the variation of the attenuation within the sample. It is also influenced by the efficiency of coupling at the buffer rod sample interface which may be temperature dependent. For longitudinal waves, the transmission across the interface is quite stable. For example, the losses attributable to coupling are typically less than 3 decibels from ambient to 325 °C. This is not the case for the shear wave mode. In this case, the signal loss due to coupling at the interface may be substantial greater than 30 decibels. Therefore, whereas the monitored signal loss for longitudinal waves may unambiguously be ascribed to the sample properties, the increased signal loss in shear wave experiments must be treated with caution.

2.2 Sample Preparation Two very different composite types were used in this study, one continuous fiber advanced composite and one chopped fiber composite. Both material types possessed anisotropic propagation characteristics and have transverse isotropy.   The sample preparation is not extensive, however because the sample orientation relative to the direction of ultrasonic propagation is important, care must be exercised to document this orientation. Best results are obtained when the samples are flat and have parallel sides. Standard machining tolerances of +/- 25 microns are adequate. It is possible to make all of the required measurements from a single 20 mm. by 15 mm. by 8 mm.

The advanced composite is a unidirectional graphite/epoxy sample (AS4 graphite fibers in an epoxy matrix). Only one fiber volume fraction ,60 percent, was available for these studies and this sample had a density of 1.59 gm/cc The original sample plate was of 20 cm by 20 cm with a thickness of 0.8 cm. From this plate, small segments 20 mm by 15 mm by 8 mm were cut. The fibers are oriented along the 20 mm. dimension of this sample, the "unique axis". The choice of sample size is primarily governed by compatibility with the sample heater and associated holding fixture.

The second composite investigated in this study was an automotive friction material. This material is more common, although less well known, than the advanced composite. The material consists of a variety of chopped metal, ceramic , and polymer fibers as well as a host of particulate mineral fillers bound in a phenolic resin. The exact composition is proprietary.   Our friction material sample belongs to a class of materials referred to as a semi-metallic and has a density of 3.13 gm/cc. The resin content is typically less than 20 % by weight and the structure may have as much as 10% porosity.

The friction materials used in this study were taken from intact automotive brake pads. Samples were cut to dimensions 20 mm. by 15 mm. by 8 mm.. In this case, the 8 mm dimension , which is the thickness dimension of the original friction pad, is the unique axis. These materials are designed primarily for their friction performance and not their modulus or strength. However, the vibration induced noise generated by these materials is dependent upon there elastic properties, thus, these properties are of considerable commercial importance.

2.3   Ultrasonic Velocity/Engineering Constant Relations   For materials with transverse isotropy, it is necessary to make five independent measurements. For this work, we will define a standard Cartesian coordinate system. The "unique" axis of the sample is always oriented parallel to the "3" direction.   The plane of isotropy is then defined by the "1-2" coordinate axes.   Two longitudinal modes, one along the unique axis, V_{33}, and one perpendicular to the "unique" axis, $V_{22}=V_{11}$, are measured   Two shear wave modes, one along the "unique" axis, polarized in the plane of symmetry, $V_{31}=V_{32}$ and one propagating in the symmetry plane with polarization also in the plane. $V_{21}=V_{12}$.   Finally, a quasi- shear mode or quasi- longitudinal mode which propagates at 45 degrees relative to the unique axis is needed to complete the measurement suite. This is called the "V_{45}"

Once these five measurements are made, the elements of the elastic stiffness matrix are determined and then by a matrix inversion the engineering constants are calculated. The conversion algorithms have been presented elsewhere and will not be reproduced here. The reader is referred to excellent articles by Ledbetter and Buchur (1,2).

3. EXPERIMENTAL RESULTS

3.1 Ambient Temperature Data   Generally, prior to measuring the elevated temperature properties of composites,   all five independent elastic constants are measured   at ambient temperature. This is most easily done by making use of the **ETEK** instrumentation and directly coupling the samples to the transducers without using the buffer rods. The buffer rods, which are necessary for the elevated temperature work, add additional interfaces and shift the echo position in time and thus unnecessarily complicate the measurement process.

Figure 3 and Figure 4   show the results obtained on our unidirectional graphite/epoxy composite sample. In Figure 3, we present the ultrasonic velocity data. In Figure 4, we show the calculated engineering constants. These results were obtained at a frequency of 2.25 MHz. Only one sample region was analyzed so it is difficult to estimated the variability attributable to sample non-uniformity. For the velocity measurements, three repeat measurements were made for each mode. The maximum error from the mean values plotted in Figure 4   do not exceed   0.5%. The nomenclature used to describe the various mode is relatively standard. All modes are designated in tensor notation, V_{ij}, where the first index indicates the axis along which the sound is propagation and the second index indicates the polarization (direction of wave particle motion). The only exception to this rule is the V_{45} mode which is the quasi-shear mode which propagates at 45 degrees relative to the unique axis. As one might anticipate, the velocity is highest along the "3" which is parallel to the orientation of the high modulus graphite fibers.

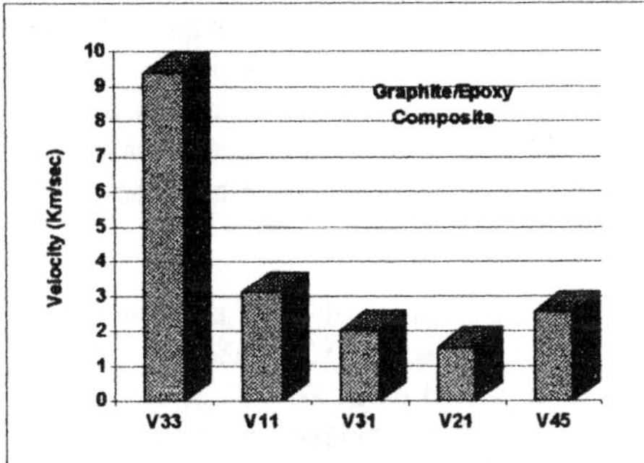

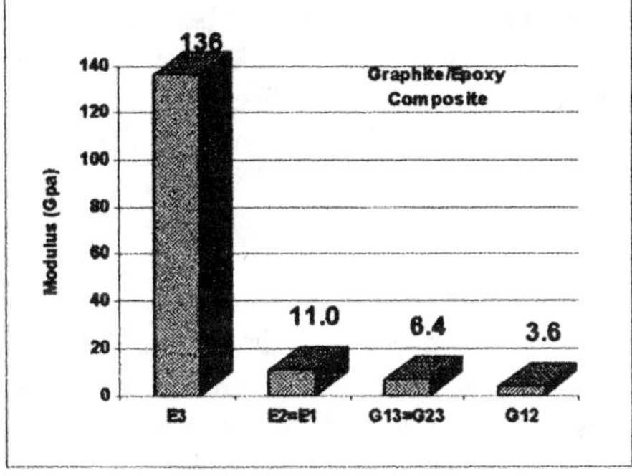

Figure 3 Independent Velocity Modes in Graphite/Epoxy Composite

Figure 4 Young's Moduli and Shear Moduli Calculated from Velocity Data

Figure 4 shows the four major engineering parameters used in finite element analyses. They are the major axis (unique axis) Young's modulus, E_3, the minor axis Young's modulus, $E_2 = E_1$, the major axis shear modulus, $G_{13} = G_{32}$, and the minor axis shear modulus, $G_{12} = G_{21}$. It is of interest to compare the results of Figure 5 with those found in the literature for comparable AS4 graphite fiber/epoxy composites. In particular, we are interested in seeing how these results compare with those obtained through conventional static mechanical tests. Unfortunately , there does not appear to be many situations where the full complement of engineering properties have been measured. This is a testament to the difficulty in making these measurements using conventional static methods.

We were able to find some partially complete data on similar composite material and also the results of other ultrasonically based measurements. The comparisons are shown in Table I. There appears to be excellent agreement between our measurements of moduli and those found in the literature for the Young's moduli., both the major axis, E_3 and the minor axis, E_2. For the shear properties, G_{ij}, the agreement is within 15 %. For the Poisson's ratio, p_{ij}, there appears to be more variability. However , it should be noted that the data presented by Weeton suggests that considerable variation in conventional testing methods is found, particularly for the shear properties (9). Variations as large as 40% have been reported. Thus, we can conclude that there is agreement between our measurements, other ultrasonic methods and the available conventional static test methods.

**Table I   Comparison of Engineering Constants
For AS4 Graphite/Epoxy Composite**

	E_2 Gpa	E_3 Gpa	G_{23} Gpa	G_{12} Gpa	p_{12}	p_{31}	p_{13}
This work	10.99	136.	6.44	3.61	.52	.21	.02
Tsai & Hahn[7] (static)	8.96	138	7.1			.30	
Wu & Ho[8] (static)	10.34	138	6.55	2.60			
Wu & Ho[8] (Ultrasonic)	11.1	137.4	6.4	4.0	.48	.35	.003
Weeton[9] (Static)	11.4	136.6	5.1		.29		

The second composite material to be analyzed in this study is an automotive friction material taken from a semi-metallic brake pad. The term semi-metallic is a misnomer, these pads are polymer based composites with some metal fiber filler. In Figure 5 and Figure 6, we present the results of our ambient temperature (23 °C) analysis. The ultrasonic velocity data appears in Figure 5 while the computed moduli are shown in Figure 6. These measurements were also obtained at a frequency of 2.25 MHz.

The values plotted in Figure 5 are the average values obtained from the analysis of three different samples. As such, the variability combines both the measurement error (0.5%) and the sample variability which was +/- 2. % for this particular material. It should be noted that an additional velocity mode appears in Figure 5, V_{32}. Whereas, the symmetry of the unidirectional composite is known a priori, that of the brake pad is not. However, by exploring the magnitude of the ultrasonic propagation in different directions one can ascertain the symmetry. In Figure 5, we show both the shear wave mode propagating along the unique "3" axis with polarization parallel to the "2" and the "1" direction. The similarity of the two velocities confirms the high degree of symmetry in the plane of the pad. This justifies treating it as a material with transverse isotropy.

In Figure 6, we show the calculated moduli which can be compared to the unidirectional advanced composite. While the symmetry of this material is the same as the unidirectional composite, this material, in almost all other respects, is at the other end of the elastic spectrum. In the unidirectional, continuous fiber composite, the unique axis exhibits the highest modulus. In contrast, the unique axis in the friction material has the lowest modulus.

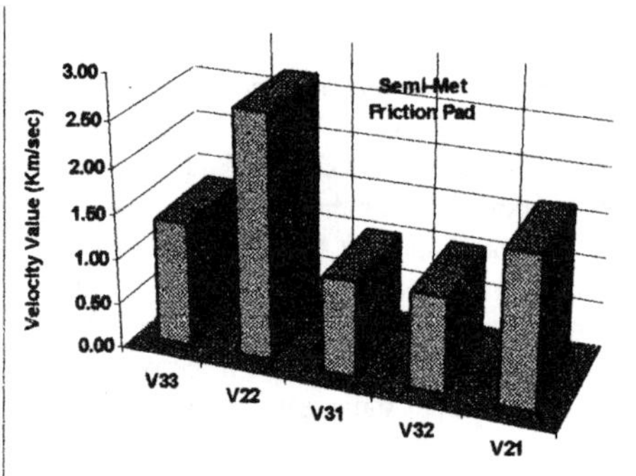

Figure 5 Velocity Modes For Friction Material

Figure 6 Calculated Moduli for Friction Material Shown in Figure 5

3.2 Elevated Temperature Data Before presenting the data on the composite properties, it is of interest to investigate the behavior of the "neat" resins. Both of the composites investigated in section 3.1 have polymer binders. For the unidirectional composite, the binder is an epoxy while that of the friction material is a phenolic. In Figures 7 and 8, the longitudinal modulus of an epoxy resin and a phenolic are measured as a function of temperature. In both cases, the resins are fully cured. The modulus data plotted in Figure 7 shows the results of two separate temperature cycles. The variability is such that the data for the first run is indistinguishable from that from the second run. The first cycle points are plotted as a squares while the repeat measurement points are plotted as circles. Thus, the data for repeat cycling of the same material is highly reproducible. The results for both resins indicate that the modulus of the binders can vary substantially with temperature.

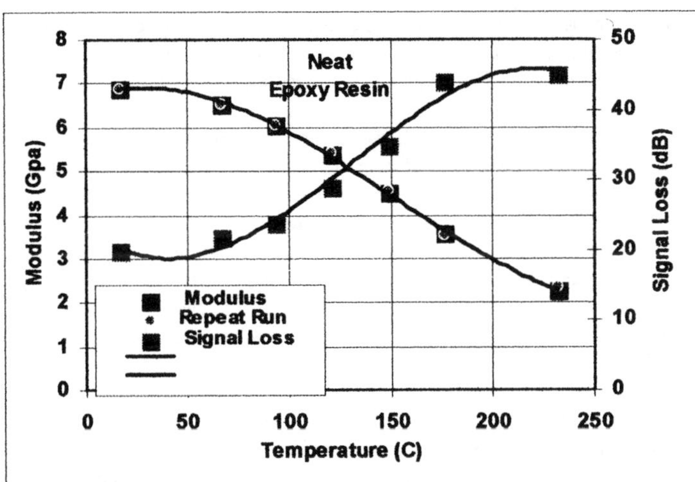

Figure 7 Longitudinal Modulus and Signal Loss on a Fully Cured Epoxy Sample

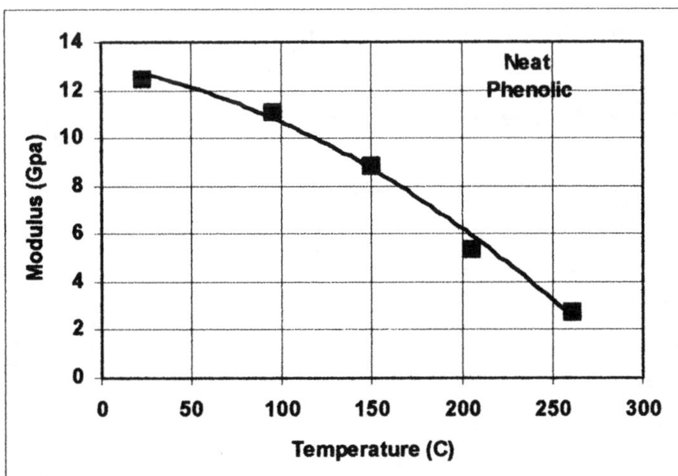

Figure 8 Longitudinal Modulus of a Fully Cured Phenolic

For the epoxy data shown in Figure 7, we have shown the signal loss data. Although this does not enter into the modulus calculations, it may be of interest. The signal loss is plotted on a logarithmic scale in decibels where 20 decibels corresponds to a decrease in signal strength by a factor of 10. No effort is made to calibrate the signal loss in this presentation. The signal loss is primarily used as a "diagnostic tool" for the measurement process. It is tracked as the samples are heated and cooled to insure that no irreversible damage was done to the material and that the ultrasonic coupling throughout the measurement process was consistent.

Using the same samples that were analyzed in Section 3.1, a series of elevated temperature experiments were conducted. This was accomplished by the methods described previously. The test samples were mounted in the fixture and the ultrasonic velocity and signal loss was monitored as a function of temperature. A stepwise heating cycle was used where the temperature controller was set to move in 28 degree Centigrade increments, up to a maximum of 325 °C. At each increment, a five minute holding time was used to make sure that the sample achieve thermal equilibrium. In some cases, the holding time was extended to 10 minutes, however, we found that the resulting data was indistinguishable from that obtained using a five minute holding period.

Results of the temperature runs for the four ultrasonic modes are plotted in Figures 9 through 12. In each case, the modulus data is plotted as closed triangles while the signal loss data is plotted as closed squares. The scaling is of course different for the two measurements. In each case, the axis on the left is the modulus axis and that on the right is the signal loss. Figures 9 and 10 show the longitudinal modulus for the major and minor axis. These moduli are determined from the measurement of longitudinal waves. For the major axis longitudinal modulus, we observe a slight increase with temperature from an initial value of 140 Gpa to a value of 151 Gpa at the highest temperature measured. In contrast, the minor axis longitudinal modulus decreases with temperature by approximately 50 percent.

The shear moduli are plotted in Figures 11 and 12. These moduli typically show a greater dependence on temperature than the longitudinal moduli. The minor axis shear modulus G_{12} shows the greatest percentage decrease. Accompanying the decrease in modulus is a substantially increased signal loss. In fact, at the highest temperatures measured, the signal loss is sufficiently large that the signal to noise for the system is approaching unity.

Surprisingly, even though the composites were heated to 325 °C and held there for up to 10 minutes there was no measurable degradation in the properties. Both the signal loss and the transit time measurements returned to there pre-thermal cycling levels. Repeat measurements on these samples produced identical modulus values within a few percent.

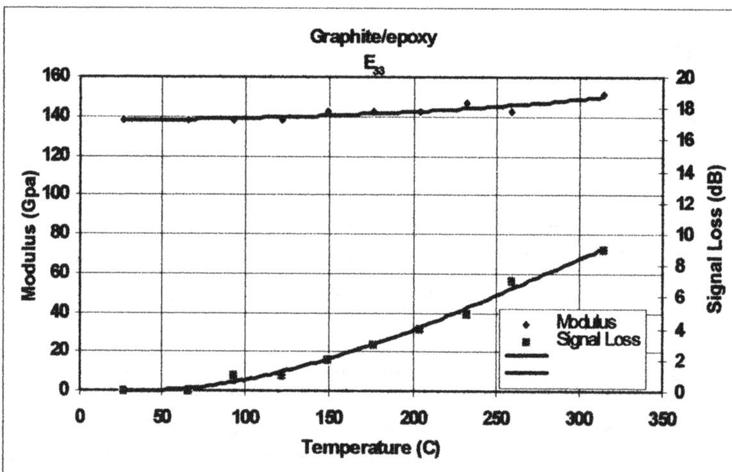

Figure 9 Major Axis Longitudinal Modulus and Signal Loss for Graphite/epoxy Composite

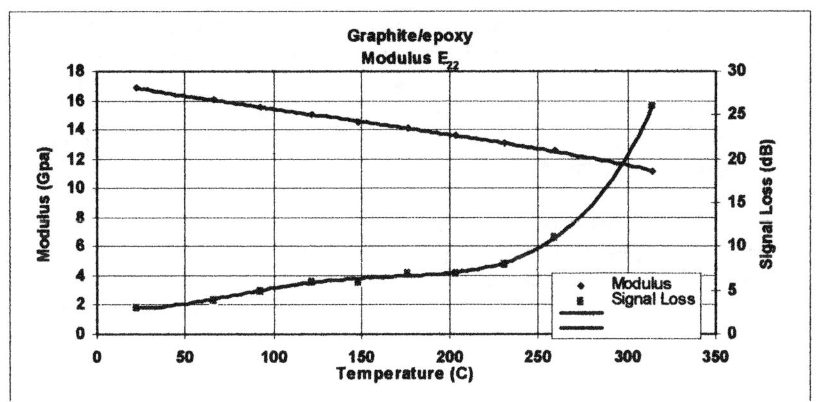

Figure 10 Minor Axis Longitudinal Modulus and Signal Loss for Graphite/epoxy Composite

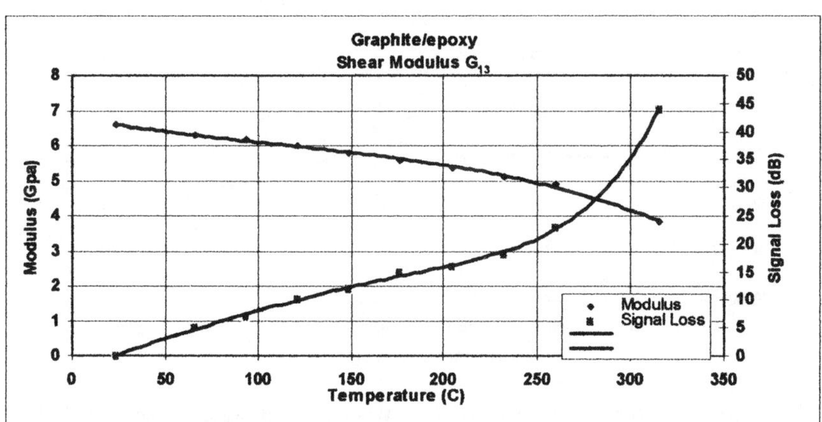

Figure 11 Major Axis Shear Modulus and Signal Loss for Graphite/epoxy Composite

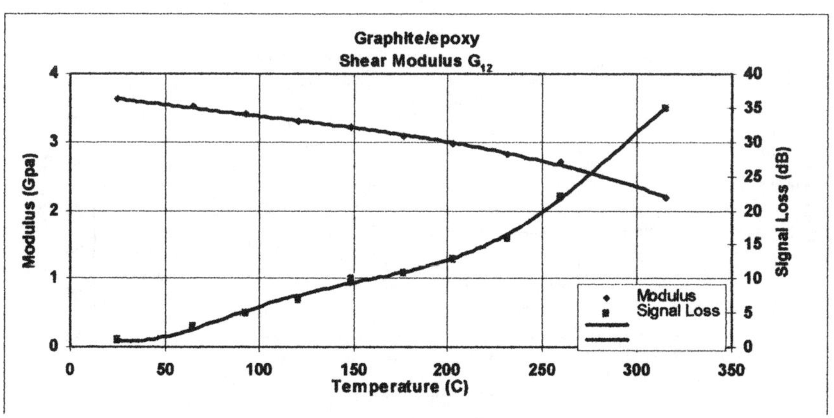

Figure 12 Minor Axis Shear Modulus and Signal Loss for Graphite/epoxy Composite

The analogous data on our friction material sample is shown in Figures 13 through 16. The longitudinal moduli are plotted in Figures 13 and 14 and the shear moduli in Figures 15 and 16. As mentioned previously, the friction material is the antithesis of a undirectional composite. The reinforcement is oriented perpendicular to the unique axis. As such, the temperature dependence of the moduli are greater along this composite's "major" axis. Perhaps, the terminology "unique" axis would be more appropriate in this case.

For all cases, the moduli in this chopped fiber composite show greater dependence on temperature. Similarly, the variation of signal loss with temperature is also greater compared to the unidirectional material. It is also interesting to note that the baseline attenuation in these materials is quite high owing to their inhomogenious structure and porosity. The unidirectional composite is essentially "lossless" at ambient temperature. The initial room temperature loss in the friction material may be as high as 40 decibels.

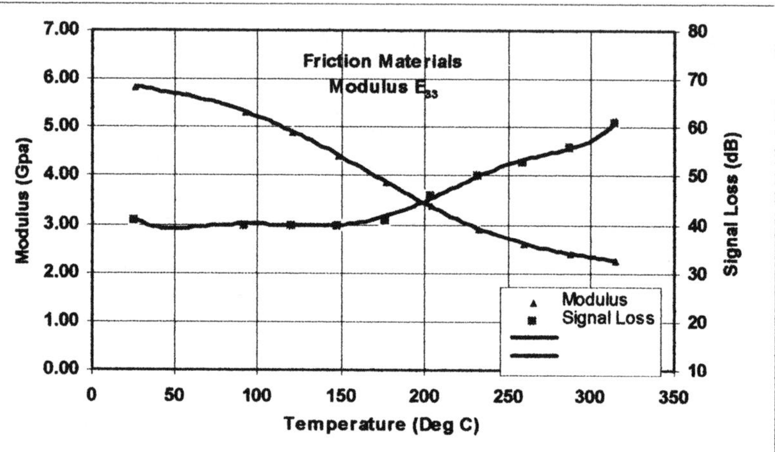

Figure 13 Major, "unique", Axis Longitudinal Modulus and Signal Loss for Friction Material Composite

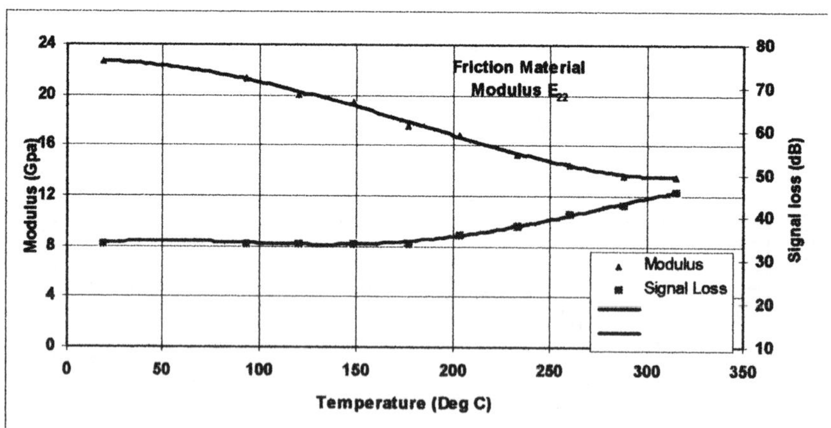

Figure 14 Minor Axis Longitudinal Modulus and Signal Loss for Friction Material Composite

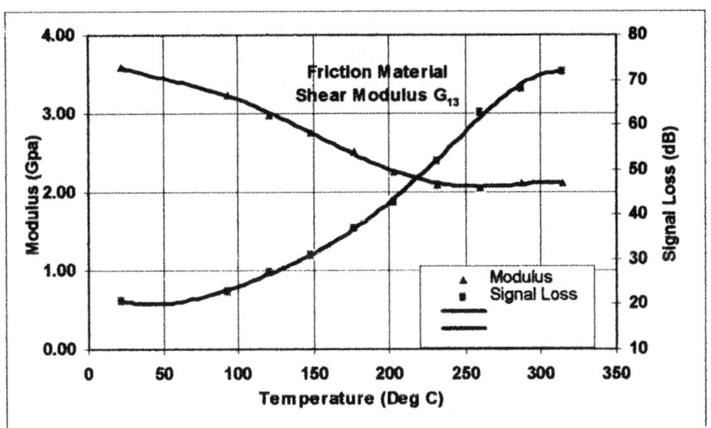

Figure 15 Major "Unique" Axis Shear Modulus and Signal Loss for Friction Material Composite

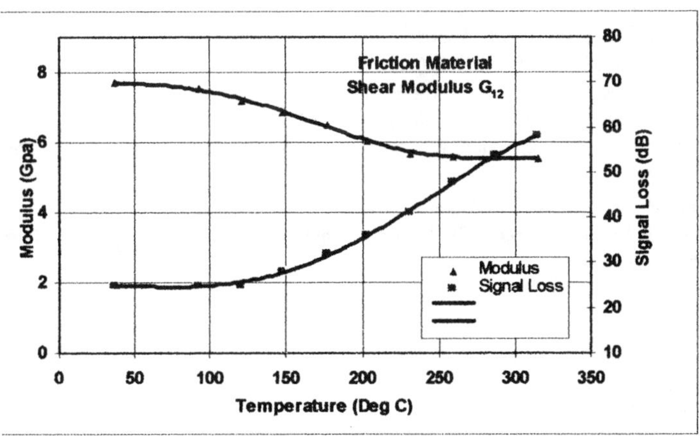

Figure 16 Minor Axis Shear Modulus and Signal Loss for Friction Material Composite

4. SUMMARY

Using ultrasonic measurement techniques, we have presented the first elevated temperature measurements of the elastic properties in transversely isotropic polymer composites. All relevant moduli for a chopped fiber composite and a high modulus continuous fiber composite were measured from 23 °C to a maximum temperature of 325 °C. For the unidirectional composite the major axis modulus increased slightly with temperature. This is unusual behavior but may be related to the unique properties if the graphite fibers which dominate the modulus in this direction. All other moduli decrease in value as the temperature rises with the shear modulus showing a stronger dependence on temperature. Longitudinal modulus measurements made on "neat" resin casting showed a modulus change of more than a factor of 3 from 23 °C to 250 °C. The composite materials exhibit substantially less temperature dependence.

We have also demonstrated the ability to use ultrasonic velocity measurements to determine the complete set of engineering constants of transversely isotropic material. For the unidirectional AS4 composite we have compared our results with those obtained from the literature. The results are in agreement with both the data obtained by conventional static test methods as well as other ultrasonic investigations.

Future directions for investigations involve a more systematic study of the temperature dependence of the "neat" resin and composites of differing fiber volume fractions. It would be of interest to compare the high temperature ultrasonic results with those calculated using available micromechanics models. Additional work is planned on extending the temperature range of the existing ultrasonic system. Lastly there is a need to gain a better understanding of the signal loss data and its potential as a useful characterization parameter.

5. REFERENCES

1. H. M. Ledbetter and D.T. Read, Jour. App. Phys, 48, 5 (1977).
2. V. Bucur Acoustics of Wood, CRC Press, Boca Raton , 1995, pp. 35-55.
3. J. E. Zimmer and J. R. Cost, Jour. Acoustic Soc. Amer., Vol. 47, #3(part2) , (1970).
4. R.E. Smith, J. Appl. Phys., Vol. 43, #6, (1972).
5. Industrial Measurement Systems Inc. Application Note #101,(1995).
6. Commercially available from Industrial Measurement Systems Inc. West Chicago IL
7. W. Tsai and H. T. Hahn, Introduction to Composite Materials, Technomic Pub. Company Inc. 1980, pp. 19-20.
8. T. T. Wu and Z. H. Ho, Exp. Mechanics, 12,(1990).
9. J. W. Weeton, D. M. Peters, and K. L. Thomas, Engineers' Guide to Composite Materials, ASM Press, pp. 6-53, (1987).

Cost-Competitive Carbon-Based Passive Thermal Management Products for Electronics Cooling

Eyan Lee, William E. Davis
Applied Material Technologies, Inc.
Santa Ana, CA 92704

ABSTRACT

Electronic components are becoming smaller in size and developing unprecedented power dissipation characteristics as processing speeds increase. The state of the art in electronics has moved from individual chips encased in heavy, large protective metal carriers with gull wing leads soldered to a printed wiring board (PWB) to the chips, minus the carrier, being wire bonded directly onto the PWB—chip-on-board—or even "flipped" over and bump soldered to the PWB without leads—flip-chip-on-board (FCOB). However, along with the higher electronics packaging density and greater levels of heat dissipation generated by smaller components comes the need to more effectively manage the increasing heat levels generated by these new electronic components.

KEY WORDS: Carbon Carbon Composites, Carbon Fibers, Ceramic Materials/Composites, Composite Components, Composites, Die Attach, Dimensional Stability, Glass Laminates, Multichip Modules, Packaging, Printed Wiring Boards, Silicon, Surface Mount Technology, Thermal Management, Thermally Conductive Materials

1. INTRODUCTION

One solution with excellent potential is passive thermal management using high conductivity carbon-based materials. Carbon fiber materials typically exhibit high thermal conductivity (up to 1100 W/m·K), low coefficient of thermal expansion (~0 ppm/°C), light weight (~0.07 lbs/in^3 vs. ~0.1 lbs/in^3 for aluminum), and high stiffness (up to ~42 msi vs. 10 msi for aluminum). All of these properties are favorable for use in upcoming electronic packaging and PWB dimensional constraint applications.

2. NEW PACKAGING TRENDS AND ISSUES

New technologies already in early stages of implementation are direct attach methods such as flip-chip-on-board and higher power generating, smaller physical "footprint" dies and chips. Direct attach methods eliminate the need for low conductivity and bulky wiring substrate layers consisting of e-glass/epoxy (FR4) or e-glass/polyimide (G10). Instead, the metal circuitry is etched, brazed, or laser-etched directly to a highly (thermally) conductive substrate--"metallization." The higher power dissipating, smaller silicon chips, then, are directly bump soldered—in the case of FCOB—to the metallized, conductive substrate. This is still under development. Currently, an "intermediate" solution being produced is a high density metal interconnect (HDMI) "decal." In this case, fine pitch, high density wiring layers are laid upon one another, separated by micro-thin dielectric layers, to form the decal. This decal is then bonded to a single substrate, usually a ceramic such as alumina, thereby eliminating the need for discrete PWB layers.

The need for incorporation of material with carbon-based composite properties is becoming evident with the advent of new chip packaging technologies. The ongoing trend is towards miniaturization of electronics. Higher levels of power dissipation, lower micron levels of chip fabrication, and closer proximity chip placement are resulting in silicon-based chips with unprecedented power densities and corresponding thermal management needs being placed closer together in order to fit more functionality into a smaller physical volume. Next generation consumer electronics, such as personal communications systems (PCS) telephones, personal digital assistants (PDAs), and laptop computers, and smaller, lighter communications satellites and aircraft electronics represent a few examples of ready applications for miniaturized electronics. However, increased power dissipation over a smaller "footprint"—area of the chip/die in contact with the substrate/PWB through which heat is conducted—leads directly to higher operating temperatures and significantly decreased corresponding mean time between failure, i.e. lower reliability. In order to counteract this effect, thermal management of the higher power density electronic components has become of paramount importance. A highly thermally conductive substrate is needed to conduct the added heat generated by the high power density components to a thermal bus.

Another reliability issue, especially in the case of direct die attach, remains for CTE matching of the substrate with the silicon-based chip or component. Silicon-based chips exhibit CTEs in the 3 to 5 ppm/°C range, whereas e-glass/epoxy and e-glass/polyimide expand at ~14 ppm/°C. This expansion differential leads to stress build-up and susceptibility to fatigue failure at solder joints leading to lower overall reliability. Direct attach techniques, such as FCOB, seek, in part, to remedy this by eliminating the discrete PWB layers and relying on a closely matched CTE substrate.

Another deciding factor in solder joint durability is PWB/substrate stiffness. Lower in-plane stiffness directly leads to higher out-of-plane dynamic deflections. This, in turn, causes increased maximum shear stresses and, ultimately, premature fatigue failure in solder joints. A secondary effect of out-of-plane deflections involves components populating closely spaced modules in a rack, i.e. a standard electronics module form factor "E" (SEM-E) rack. Here the electronics modules are attached on both sides of a heat sink/doubler substrate. Over the planar modules are assembled protective covers with just several thousands of an inch separating the inside surface of the cover from the top surfaces of the electronic components. High overall assembled module stiffness is required to keep the electronic components from striking the protective covers during dynamic deflections and becoming damaged.

The final major issue facing the successful implementation of upcoming high power density chips and direct attach packaging methods is one of cost. Lighter weight FCOB substrates, for example, are being sought in order to lower payload launch costs or to incorporate more functionality (processing power, memory, communications capability, etc.) at the same overall weight in spacecraft. For avionics applications, lighter weight wiring or FCOB substrate material will offer lower aircraft weight, which could correspond to increased range (fuel mileage) and maneuvering performance.

3. CARBON-BASED PASSIVE THERMAL MANAGEMENT SOLUTIONS

Promising passive thermal management solutions for imminent high power density, miniaturized electronics applications are taking shape in the form of carbon-based material currently under development by Applied Material Technologies, Inc. AMT is currently working on developing cost-competitive carbon-based thermal management products for the high performance electronics industry.

AMT's aluminum, and resin, infiltrated Affordable High Performance Composite (AHPC) materials offer a wiring and direct attach solution to imminent electronics miniaturization technology. These materials are presently under commercial development in AMT's "Affordable High Performance Composites" Phase II Small Business Innovative Research (SBIR) project under the direction of the Materials Directorate at Wright Patterson Air Force Base. AHPC is a low cost, ~35 msi continuous fiber carbon material which is processed under AMT specifications to deliver stiffness and thermal conductivity properties comparable to much more expensive 100+ msi carbon fiber composites. Properties of both resin and aluminum infiltrated versions have been demonstrated to show the following:

Material	Density, lbs/in^3	Modulus, msi	CTE, ppm/°C	k, W/m·K
Metal Infiltrated AHPC	0.078	42.1	0 to 3.3	351
Resin Infiltrated AHPC	0.060	39.5	-1.2 to 1.4	213
Aluminum (6061 Alloy)	0.097	10	24	170

The above table illustrates the potential for AHPC to enable forthcoming high power dissipating, close proximity, direct attach chip placement technology. Carbon-based AHPC can achieve more than double the conductivity of aluminum and CTE properties approaching those of silicon-based chips—3 to 5 ppm/°C. In addition, AHPC material has been tested to over 40 msi modulus at nearly twenty percent less weight per given volume than aluminum. AHPC is showing the high thermal conductivity, CTE characteristics, high stiffness, and light weight potential necessary for enabling direct attach FCOB chip placement methods involving high power dissipating, close proximity techniques.

For electronics applications requiring dimensional stability at a prescribed board/PWB CTE value to match expansion characteristics with a specific electronic component, AMT is currently developing a carbon-based constraining core, STABLCOR™, for use in PWBs. This work is currently being performed as part of AMT's "Lightweight, Dimensionally Stable Printed Wiring Boards" Phase II SBIR under the direction of NASA Lewis Research Center. The thickness ratio of the carbon-based core to the e-glass/polyimide or e-glass/epoxy layers determines the overall CTE of the board. In order to augment the commercialization of this product, AMT is developing an analytical tool for in-house design of STABLCOR™-cored PWBs to customer-supplied CTE requirements. A STABLCOR™ PWB can be designed as low as ~4 ppm/°C to match many surface mount technology (SMT) leadless chip carriers (LCCs) and as high as ~12 ppm/°C to match plastic integrated circuit package (IC) and encapsulant CTEs. This technology makes use, primarily, of the low CTE characteristics exhibited by carbon-based composites. AMT performed a CTE test simulation to predict STABLCOR™ PWB CTEs as part of its Phase I "Lightweight, Dimensionally Stable Printed Wiring Board" effort. The CTE predictions and correlations to testing are shown in Figure 1 and 2 with their CTE defined by (strain vs. temperature) slope values.

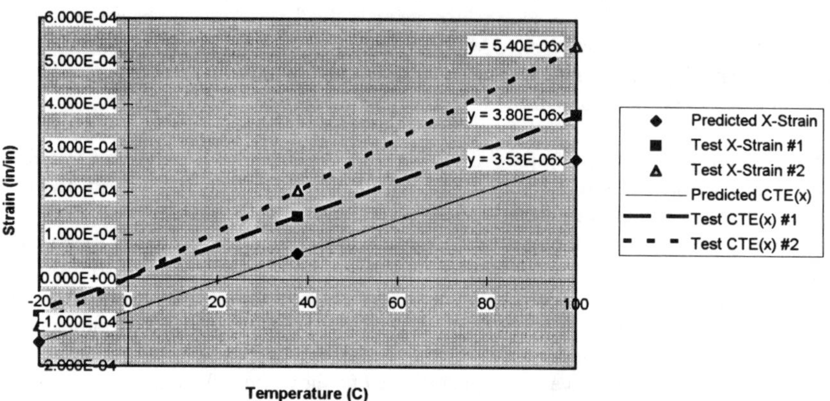

Figure 1. Predicted and Tested In-Plane (x-dir.) CTE for STABLCOR™ PWB

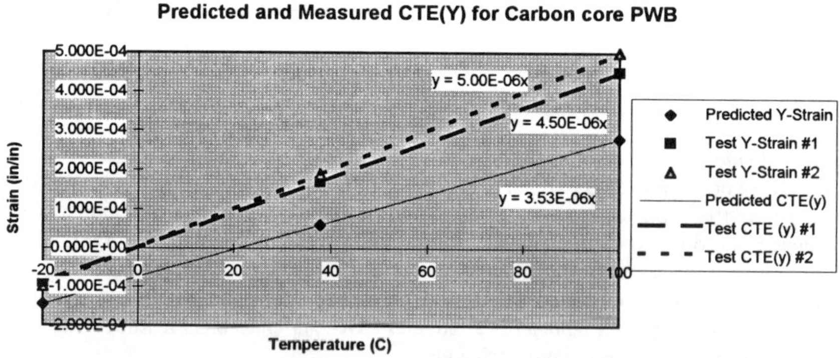

Figure 2. Predicted and Tested In-Plane (y-dir.) CTE for STABLCOR™ PWB

In addition, STABLCOR™ PWB CTEs can be tailored to match those of state-of-the-art glass-filled alumina chip carriers(~4.5 ppm/°C) all the way up to the CTEs required by plastic packages (~10-14 ppm/°C). A comparison plot of CTE (or TCE) values exhibited in testing by non-cored PWBs and PWBs cored with various materials is shown in Figure 3.

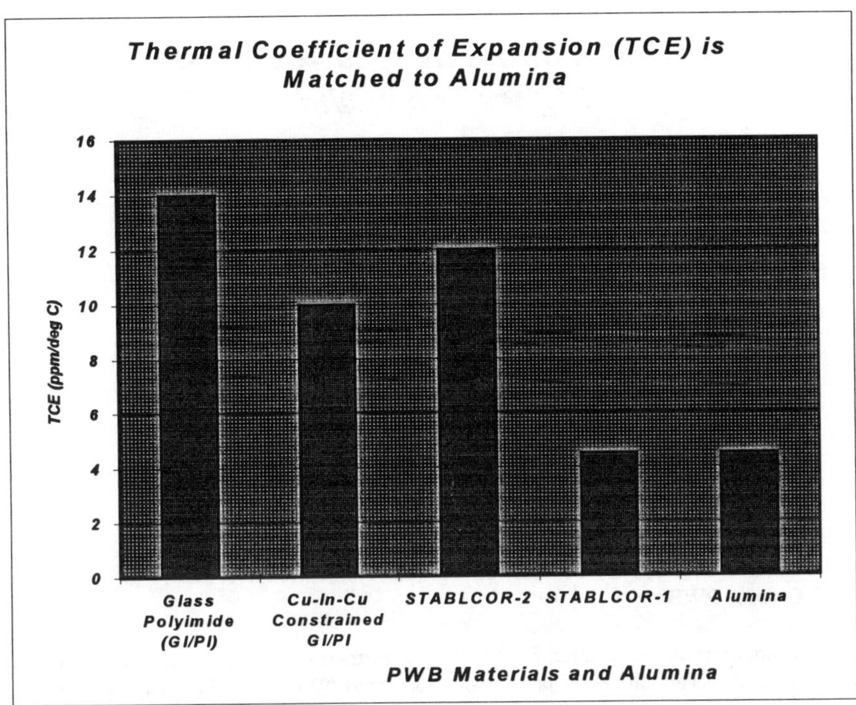

Figure 3. STABLCOR™ PWB CTE Matched to that of Alumina Chip Carriers

The table below displays STABLCOR™-cored PWB tested properties to date.

Material	Density, lbs/in^3	Modulus, msi	CTE, ppm/°C	k, W/m·K
STABLCOR™ PWB	<0.07	10	4-12	20-80
Non-cored PWB	~0.06	3.5	16	~7

In addition to the SMT CTE-matching capabilities of STABLCOR™ PWBs, compared to non-cored PWBs, STABLCOR™ offers more than double the stiffness and more than ten times the in-plane conductivity at virtually no increase in weight. Figure 4 illustrates the superior thermal conductivity characteristics of STABLCOR™ over state-of-the art non-cored and copper-invar-copper-cored PWBs.

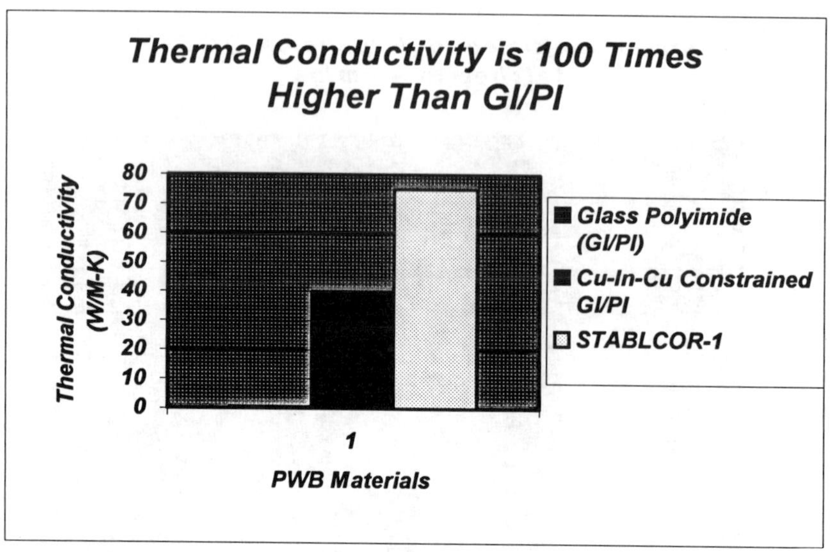

Figure 4. Thermal Conductivity Comparison Between Non-Cored, STABLCOR™, and C-I-C-cored PWBs

For reasons similar to the ones stated for AHPC, higher thermal conductivity, lower stiffness, and increased stiffness are all beneficial to SMT-populated PWBs. By eliminating gull wing leads, surface mount components allow for closer proximity electronic component placement and, therefore, increased packaging density per given PWB area. However, gull wing leads are essentially wires with "slack" soldered, on one end to the component, and, on the other end, to the PWB circuitry. And, by eliminating this, the ability of the leads to help "compensate" for differential expansion of the PWB in relation to the component is lost. Therefore, CTE-matching between the PWB and the electronic component is crucial in order to avoid solder joint failure, where the solder joint is the mechanical interface between the component and the underlying PWB. Increased thermal conductivity aids in lowering peak IC operating temperatures, and higher PWB stiffness lowers dynamic deflections and subsequent solder joint stresses. Both of these factors significantly increase electronic component MTBF. By increasing reliability and avoiding added weight, STABLCOR™ PWBs will lower costs associated with board and component replacement as well as averting increased payload launch costs in spacecraft applications.

4. CONCLUSIONS

Through its SBIR efforts AMT is developing a new and exciting line of carbon-based products for upcoming electronics packaging and direct attach trends. The unique combination of high thermal conductivity, low CTE, high stiffness, and low weight of AHPC and STABLCOR™ will play an important role in enabling the dominant, thermal management-dependent, electronics packaging processes of the imminent future.

POLYMER-BASED, LOW-COST INTEGRATED SUBSTRATES FOR POWER ELECTRONICS

W. Kowbel, X. Xia and J.C. Withers
Materials & Electrochemical Research Corporation, Tucson, AZ 85706

ABSTRACT

Rapid advances in high power electronic packaging require the development of new heat-sink and substrate materials. Advanced composites designed to provide thermal control as well as improved thermal conductivity have the potential to provide benefits in the removal of excess heat from electronic devices. Carbon-carbon (C-C) composites are under consideration for numerous electronic packaging applications. The very high cost (up to $20,000/kg) of C-C composite has hindered their wide spread commercialization. A new, polymer-based manufacturing process has been developed to produce high thermal conductivity (400 W/mK) C-C composite at greatly reduced cost (less than $100/kg). Several types of polymer slurry based low dielectric coatings were applied to the C-C composites. Processing schemes were developed to produce crack-free coatings exhibiting good thermal conductivity. Metallization of the dielectric coating was performed for the process integration with electronic devices. Thus, integrated substrates for the power electronics were fabricated without the need of conventional metal/ceramic joining and associated high stresses. The properties of this new composite material for power electronics substrates will be presented.

KEY WORDS: polymer-based, C-C, dielectric insulation, power electronics

1. INTRODUCTION

Future spacecraft will utilize electronic components with much higher power densities than present systems. The electronic components will also be designed to weigh less and cost less than existing components. The higher power densities will challenge existing heat dissipation technologies. Spacecraft, by their nature, must employ conductive and radiative cooling since forced-air convection used in most ground-based systems is not a viable heat rejection mechanism in space.

Present satellites dissipate heat using radiator panels constructed from organic matrix composites made with high conductivity pitch based fibers. These composites offer in-plane conductivities that are equivalent to aluminum and copper and exhibit densities that are significantly lower than metals. However, the use of organic composite radiators will be limited in future systems with higher power densities because of the low through-thickness conductivity of the material. The through-thickness conductivity of a composite controlled by the matrix in organic composites typically have values of about 1 W/mK.

Carbon-carbon composites offer in-plane properties that are equivalent or better than organic composites, while at the same time exhibiting much higher through-thickness thermal conductivities, typically 20 W/mK. Thus carbon-carbon composites appear to be more attractive than metals or organic composites for heat dissipation components. Carbon-carbon composites, developed at MER, are much less expensive and offer much faster fabrication times. The MER carbon-carbon composites offer low-cost because they employ a filled matrix precursor that achieves he desired composite density in the first fabrication cycle.

Electronic component reliability significantly decreases with increasing temperature. Thermal dissipation increases as power system requirements increase. Further, thermal density increases as components are made smaller to increase functional performance and to save space and weight. Chip-on-board (COB) technology provides for possible simplification of current electronic packaging [1].

Currently used printing wiring board (PWB) in space systems have two major limitations: high CTE (about 15×10^{-6} cm/cm °C) and very low thermal conductivity (about 0.2 W/mK). These two factors greatly hinder thermal management technology by imposing severe limitations on reliability.

The failure in space electronics can be caused by: (i) temperature increase on the chip, (ii) the CTE mismatch between the board and chip seeks the most conductivity path to reach the sink. The sinks to which the heat is eventually transferred are the cooling fluids and boards. The very low conductivity of the PWB greatly lowers its performance.

Currently, the typical solution to these problems is to include a metallic thermal conductor in the electronic packaging or mounting board. Aluminum is the material of choice for many commercial applications where power dissipation is not a problem and operating temperature ranges are narrow. For more demanding thermal conditions encountered in military and high power commercial electronics, copper clad sheets of molybdenum or Invar are included in the mounting boards to constrain thermal expansion and promote thermal conduction. Similarly, materials such as copper-tungsten are utilized in electronic packaging for thermal management purposes.

The CTE mismatch between the board (15×10^{-6} cm/cm °C) and the chip carrier (6.8×10^{-6} cm/cm °C) results in greatly reduce fatigue life. This problem is greatly aggravated in COB, when the silicon chip (2.6×10^{-6} cm/cm °C) is placed directly on board. Two solutions are applied in COB to enhance the fatigue life: (a) modification of board expansivity, (b) organic encapsulants. The modification of board expansivity involves the use of metal core. Severe thermal mismatch problems arise with this approach at the metal/PWB interface leading to premature failure. The second approach consists of using low stiffness materials. The draw-back of this technology is the very low thermal conductivity of the organic encapsultants and thus greatly compromised thermal performance.

2. EXPERIMENTAL

2.1 MATERIALS High thermal conductivity C-C composites were produced using a one step process describe elsewhere [2]. P-30X 4k fiber were used combined with thermoset-based carbon material. Fiber-tow process was used to make pre-preggs which were subsequently molded onto 0/90 configuration. A single heat treatment up to 2800 $^\circ$C was used. 15 cm x 15 cm x 0.05 cm plates were produced for space applications, while 5 cm x 5 cm x 0.05 cm plates were produced for COB applications. An inorganic resin infiltration was performed to improved the C-C composites mechanical properties for space applications. Several dielectric insulation-type layers were examined for COB applications; including CVD Si, Ceraset/BN and Ceraset/AlN. CVD Si layer was applied by using a plasma-enhanced CVD (PECVD) system via a decomposition of $SiCl_4$ in hydrogen. Ceraset/BN and Ceraset/AlN coatings were applied by using a spin-on-disc technique. Metallization of dielectric layers was performed by Au evaporation.

2.2 MATERIAL CHARACTERIZATION/EVALUATION

SEM was used to analyze composite cross section and coating uniformity. XRD was used to analyze CVDSi coating chemistry. Dielectric constant was measured using an impedance analyzer. Thermal diffusivity was measured using a laser flesh method while heat capacity was measured by DSC. Thermal conductivity was calculated as a product of density, heat capacity and thermal diffusivity. Flexure strength was measured using a four-point bending 40:1 span-to-depth ratio. Tensile and compressive strength was measured using a dogbone specimen.

3. RESULTS AND DISCUSSION

3.1 SPACE APPLICATIONS Increased power density and miniaturization of satellites calls for the use of new materials for space radiators. Experimental programs such as EO1 examine the use of C-C composites for space thermal management. Key characteristics enabling the use of C-C composites in these applications include low-cost, short processing time, high compressive strength and high thermal conductivity. Competing technologies include Al and graphite/cyanate-ester composites.

Figures 1a and 1 show SEM micrographs of a high thermal conductivity C-C composite. Virtually no transverse cracking is observed combined with good fiber bundle penetration.

The one-step C-C composite exhibit a high density of 1.85 g/cm^3. The novel processing and an active filler provide for the minimization of matrix cracking. Slurry formulation and fiber tow spreading was optimized to provide for good fiber bundle slurry penetration. Figure 2 shows the corresponding XRD pattern, showing a well-developed three dimensional graphite.

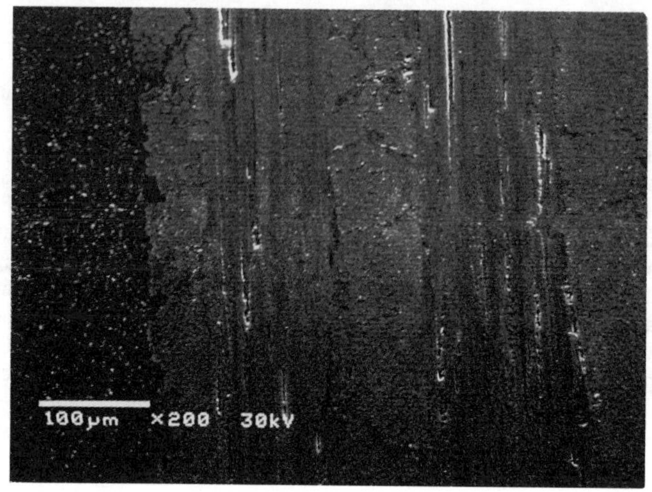

(a)

(b)

Figure 1 SEM of high conductivity C-C composite , (a) low magnification, (b) high magnification.

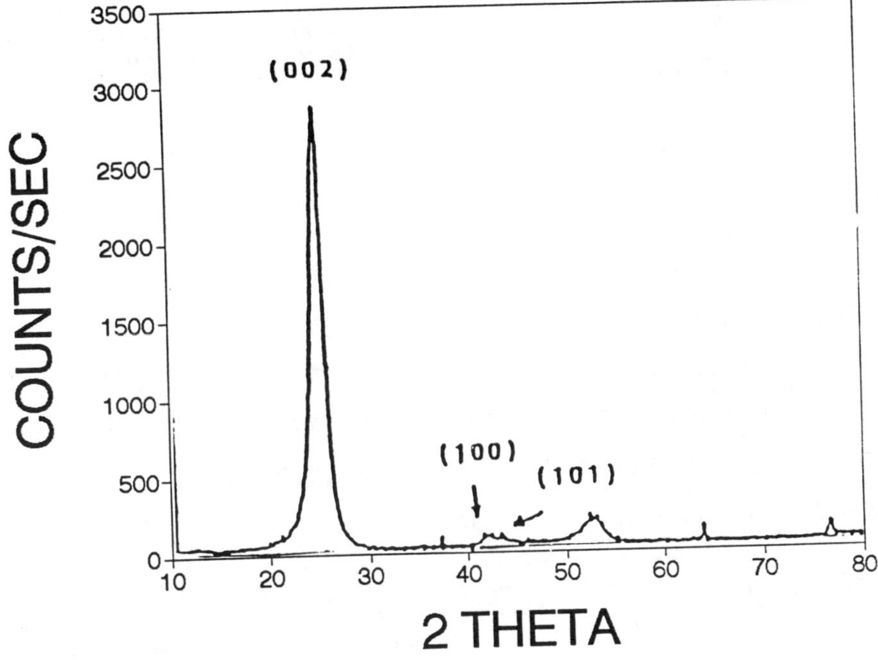

Figure 2.   XRD pattern.

Table I shows the properties of high thermal conductivity C-C composites as compared to Al and a graphite/cyanate composite. The C-C composite offers the specific thermal conductivity equivalent to that of graphite/cyanate-ester but much higher through-the-thickness thermal conductivity and compressive strength. The compressive strength and the through-the-thickness thermal conductivity are expected to play a critical role in the future satellites systems with increased power density.

Table II shows the properties of a hybrid C-C/polymer composite. The compressive strength is further improved without any detrimental effect on thermal properties

The inorganic resin used for C-C composite infiltration offers very low outgasing and excellent rheological properties at room temperature. The improved mechanical properties are attributed to the resin ability to penetrate small pores within a C-C composite. Thus, a hybrid C-C/inorganic resin composite provides good potential for future space thermal management applications.

Table I. Properties of high conductivity C-C composites

Properties	C-C, 0/90	Graphite-Poly-Cyanate	Aluminum 6061
Thermal Conductivity (in plane), (W/mK)	300	275	180
Thermal Conductivity (through-the-thickness), (W/mK)	40	0.6	180
Density, (g/cm^3)	1.85	1.65	2.71
Specific Conductivity (W-cm^3/mK-g)	162	166	66.4
Flexural Strength MPa (ksi)	450 (70)	450 (70)	180 (26)
Flexural Modulus GPa (Msi)	350 (50)	175 (25)	70 (10)
Tensile Strength MPa (ksi)	300 (45)	420 (60)	455 (65)
Tensile Modulus GPa (Msi)	280 (40)	280 (40)	77 (11)
Compressive Strength MPa (ksi)	245 (35)	126 (18)	350 (50)
Compressive Modulus GPa (Msi)	280 (40)	175 (25)	20 (3)
ILT, MPa (ksi)	6 (.9)	7 (1)	N/A
ILS, MPa (ksi)	10.5 (1.5)	10.5 (1.5)	N/A
CTE, (10^{-6}cm/cm-°C)	- 0.5	0	27

Table II. Properties of hybride composites

Properties	C-C, 0/90	C-C, 0/90 + Resin
Thermal Conductivity (in plane), (W/mK)	300	295
Thermal Conductivity (through-the-thickness), (W/mK)	40	35
Density, (g/cm^3)	1.85	1.9
Specific Conductivity (W-cm^3/mK-g)	162	155
Flexural Strength MPa (ksi)	450 (70)	560 (80)
Flexural Modulus GPa (Msi)	350 (50)	210 (30)
Tensile Strength MPa (ksi)	300 (45)	420 (60)
Tensile Modulus GPa (Msi)	280 (40)	280 (40)
Compressive Strength MPa (ksi)	245 (35)	350 (50)
Compressive Modulus GPa (Msi)	280 (40)	210 (30)
ILT, MPa (ksi)	6 (.9)	10 (1.5)
ILS, MPa (ksi)	10.5 (1.5)	14 (2)
CTE, (10^{-6}cm/cm-°C)	- 0.5	0

Table III shows the cost analysis for high thermal conductivity C-C composites. The projected cost at 4/kg/day production is about $100/kg/no profit). For a comparison, the graphite/cyanate composite cost is about $3,000/kg due to the use of very expensive $2,000/kg graphite fibers and an expensive resin. The projected cost of $100/kg would actually bring the high thermal conductivity C-C composite cost to the level comparable to the Al space-based space materials cost ($300/kg due to difficulty with machining).

Table III.   Cost analysis

	NET Cost ($/kg) Fiber tow type
Raw Materials	$59.40
Supplies	$4.40
Utilities	$1.85
Labor	$22.00
Burden	$23.14
Total (no profit)	$110.79

***   Production rate = 4 kg per day**

The one step C-C composite process offers the net-shape capability, thus resulting in further cost reduction. In summary, this novel technology has a great potential for space thermal management applications.

3.2 COB APPLICATIONS Figure 3 shows an SEM of PECVD Si metallized coating deposited on a C-C composite. The PACVD Si coating thickness was about 20 µm and the Au layer was about 0.2 µm thick. Relatively smooth surface is observed combined with no cracking. Silicon offers good strain isolation capability. Low modulus of Si and relatively low CTE provide for good thermal match with the C-C substrate. XPS confirms pure Si layer formed by the PACVD method. Silicon is compatible with IC technology thus provides a good candidate for the chip substrate. In addition, thermal oxidation of silicon (silica formation) can offer very good dielectric insulation. Thus, this technology appears to have a good potential for COB applications.

Figure 4 shows a cross section of AlN/polymer coated C-C composite after metallization. A crack-free polymer-based coating is observed. The inorganic polymer used for the slurry coating has a very low modulus (1 GPa) and low dielectric constant (2.7). Addition of AlN resulted in the increased thermal conductivity of the AlN/polymer composite (5 W/mk). The low modulus of the coating provides excellent strain isolation [3] (no cracking). The AlN/Ceraset composite had a dielectric constant around 4.5.

523

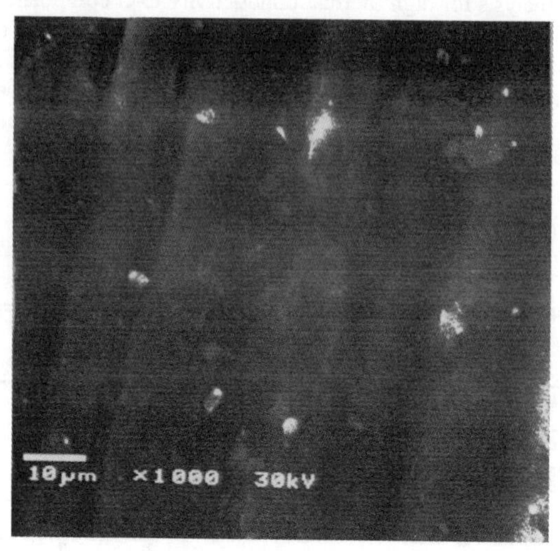

Figure 3.   SEM of PECVD Si.

Figure 4.   SEM of AlN/polymer coated C-C composite.

Figure 5 shows the corresponding surface SEM. Good planarization is observed. The AlN/polymer slurry was applied at room temperature and cured at 180 °C. The T_g of the AlN/polymer composite was found to be around 400 °C. The AlN/polymer slurry was optimized with respect to the filler content (above 60%). The increased filler content resulted in increased thermal conductivity and increases slurry viscosity. Spin-on-disc process optimization was performed to obtain high quality, 50 μm thick AlN/polymer coating. A 0.2 μm thick Au film was deposited on the AlN/polymer coating. Good adhesion between the AlN/polymer layer and Au layer was observed. The AlN/polymer coating exhibited an increased T_g over the polymer (T_g = 180 °C). Thus, the use of filled inorganic polymers such as Ceraset offers good potential for electronic applications. Thus, good dielectric isolation for COB applications was achieved.

Figure 6 shows a cross section of BN/polymer coated C-C composite after metallization. The Ceraset/BN coating is about 50 μm thick and the Au layer is about 0.2 μm thick. Good adherence of the Au layer to the Ceraset/BN coating is observed. No cracking is observed due to very low modulus of the polymer. The BN/polymer composite had a dielectric constant around 3 and its thermal conductivity was calculated to be around 10 W/mk.

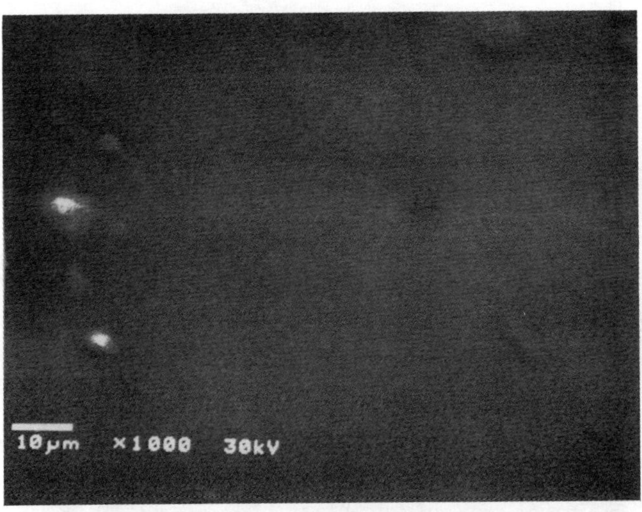

Figure 5. SEM of AlN/polymer composite coating surface.

Figure 7 shows the corresponding surface SEM of a metallized coating. Good planarization combined with no cracking is observed. The T_g of the BN/polymer composite is about 450 °C.

Thus, the BN/Ceraset composite offers excellent dielectric isolation for COB applications. The BN/Ceraset coating offers good thermal and dielectric properties well-suited for COB applications.

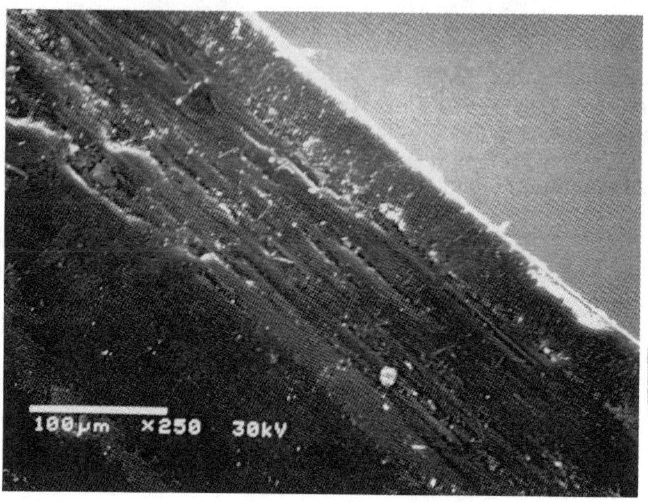

Figure 6. SEM of BN/polymer coated C-C composite.

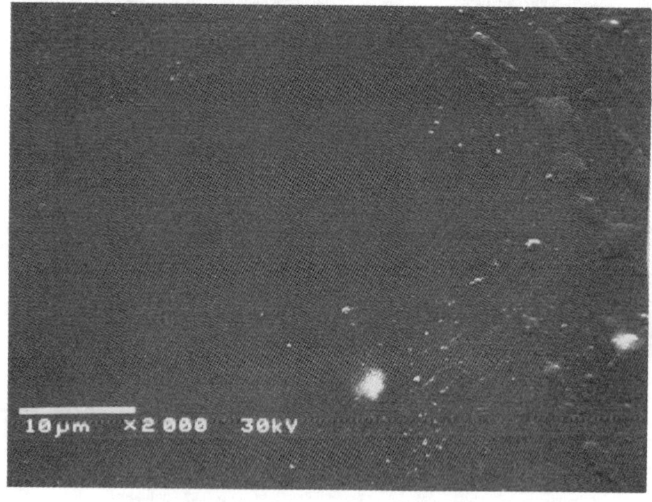

Figure 7. SEM of BN/polymer coating surface.

4. CONCLUSIONS

Low-cost, high thermal conductivity C-C composites offer potential in the use of space thermal management. Providing great cost saving of this new process can be fully implemented, C-C composites can compete with advanced metals such as Al and copper alloy in the area of commercial heat sinks.

Proper dielectric insulation combined with a sufficient thermal conductivity can provide for numerous COB applications of this new technology.

5. REFERENCES

1. D.D. Doanne, Multichip Modulus Technology and Applications, New York, 1993, p.502.
2. W. Kowbel and J.C. Withers, MER Proceeding, 445, 37(1997).
3. D.D. Doanne, Multichip Modulus Technology and Applications, New York, 1993, p.471.

POLYANILINE CONTAINING TERNARY IMMISCIBLE POLYMER BLENDS:
A MELT PROCESSABLE CONDUCTIVE SYSTEM

M. Zilberman [1], A. Siegmann [1], and M. Narkis [2]
Departments of [1] Materials and [2] Chemical Engineering,
Technion - Israel Institute of Technology, Haifa 32000, Israel.

ABSTRACT

In the present study, novel conductive PANI / thermoplastic matrices, binary and ternary blends were developed through melt blending. The binary blends' investigation focused on their morphology in light of the interactions between the components, and on the resulting electrical conductivity. Similar solubility parameters of the two components lead to fine PANI particles dispersion within the matrix, and to formation of conducting paths at low PANI contents. The morphology of a conducting network is described by a primary structure of small dispersed PANI particles, interconnected by a secondary, short range fibrillar structure. Ternary blends presently studied consist of two immiscible thermoplastic polymers, forming a co-continuous phase morphology, and PANI. The PANI preferably locates in one of the co-continuous matrix phase, affecting therefore, the blend's morphology. This concentrating effect enables relatively high electrical conductivity at a low PANI content. The blending sequence slightly affects the blends morphology and electrical conductivity.

KEY WORDS: Polyaniline, Conductive Polymer Blends.

1. INTRODUCTION

The conventional method for imparting electrical conductivity to common polymers, is by admixing conductive solid fillers, such as carbon black or metal particles. The relatively high filler content required to attain for high electrical conductivity may adversely affect the materials processability and properties. Therefore, there is an increasing interest in processible polymeric materials, using conventional melt processing techniques, whose electrical conductivity can be tailored for a given application, having attractive mechanical properties. Intrinsically conductive polymers (ICPs) are expected to yield this attractive combination of properties [1]. The presently most common ICPs are: polyaniline, polypyrrole, polythiophene, polyacetylene, poly(p-phenylene) and poly(p-phenylene sulphide). Due to the conjugated double donds in their chains, they exhibit extraordinary electronic properties, such as low ionization potential and high electron affinity, thus, can be easily reduced or oxidized . These polymers undergo "doping" upon exposure to a protonic acid and become electrically conductive [2]. Due to their chain structure and strong intermolecular interactions, undoped ICPs are insoluble in common organic solvents, and melt processing is impossible, due to their early decomposition while still in the solid state. Therefore, they are categorized as intractable, infusible, and difficult to process materials [2].

Polyaniline (PANI) is presently one of the most promising ICPs for a number of important reasons: its conductive form has an acceptable chemical stability combined with relatively high levels of electrical conductivity, that can be controlled; the monomer (aniline) is relatively inexpensive and its polymerization is a straightforward, high yield reaction [3]. PANI offers potential technological applications such as static films for transparent packaging of electronic components, electromagnetic shielding, rechargable batteries, light emitting diods, non-linear optical devices, sensors for medicine and pharmaceutics, and membranes for separation of gas mixtures [2]. The main disadvantage of conductive PANI, like other ICPs, is its limited processability. One method, still at a development stage, to process PANI, without altering the polymer's structure, is by blending with conventional polymers. These blends may combine the desired properties of the two components. i.e., high electrical conductivity of PANI together with physical and mechanical properties of the matrix polymer. The morphology of such blends has a dominant effect on their properties.

Heeger and co-workers [3-5] reported the use of functionalized protonic acids to both, dope emeraldine base (EB, the non-conducting base form of polyaniline) and, simultaneously, solubility in common organic solvents. In most recent studies of electrically conductive polyaniline containing polymer blends, blending was performed in a solution. Only a few studies concerned with blends prepared by melt processing have recently been reported [1,6-8]. Shacklette et. al. [5] reported that Versicon^{TM} (Allied-Signal Inc.), a conductive form of polyaniline, is dispersible in polar polymers such as polycaprolactone and poly(ethyleneterephthalate glycol). The percolation threshold for conductivity in these blends was observed in the range of 6-10 %v/v of Versicon. Ikkala et. al. [1] reported on the electrical and mechanical properties of conductive polymer blends, prepared by blending thermoplastic polymers with a Neste PANI Complex, (Neste Oy), using conventional melt processing techniques. According to Passiniemi et. al. [8], certain blends with polyaniline can be processed by ordinary processing methods, such as injection molding, film blowing and fiber spinning. They suggested that the key feature is the plasticizer technology, developed by Neste Oy (Finland).

In our recently reported studies, various PANI complexes were melt blended with thermoplastic polymers, and the resulting binary blends were investigated [9-11]. These studies focus on the blends' morphology, in light of the interactions between their components, and the resulting electrical conductivity. In the present study, PANI was melt blended with two immiscible thermoplastic polymers and a novel conductive ternary system was developed. The blends' complex morphology and its effect on the resulting electrical conductivity were investigated.

2. EXPERIMENTAL

2.1 Materials

Polyaniline: Versicon^{TM}, conductivity = 6 S/cm; Zipperling Kessler & Co, Germany.
Matrix Polymers:
Polystyrene (PS), Galirene HH-102-E, Carmel Olefins, Israel.
PS plasticized with di-octyl phthalate (PS/DOP), PS:DOP = 85:15.
Copolyamide 6/6.9 (CoPA), a random copolymer of 51 % [HN - $(CH_2)_5$ - CO] and 49 % [HN - $(CH_2)_6$ - NH - CO - $(CH_2)_7$ - CO], EMS, Switzerland.

2.2 Blend Preparation

Binary and ternary polymer blends, containing a single polymer or two immiscible polymers and a doped polyaniline, were prepared by melt mixing in a Brabender plastograph, at 50 rpm for 12 minutes. The blending temperature varied with the matrix polymer as follows: blends containing PS were blended at 180 ^{0}C, pasticized PS - 150 ^{0}C, CoPA - 165 ^{0}C. Flat plaques, 3 mm thick, for conductivity measurements, were prepared by compression molding.

2.3 Conductivity Measurements

Electrical conductivity measurements were performed using the "four probe technique" (ASTM D 991-89) for samples ($12 \times 1.2 \times 0.3$ cm^3) of conductivity levels $> 10^{-7}$ S/cm and the "two electrodes technique" (DIN-53596) for the less conductive blends. Samples for the latter technique were first coated with a silver paint, to reduce sample-electrode contact resistivity.

2.4 Morphological Characterization

High Resolution Scanning Electron Microscopy (HRSEM) of cryogenically fractured surfaces was performed using a Leo Gemini-982, at accelerating voltage of 1 kV. Transmission Electron Microscopy (TEM) of ultramicrotom samples was performed using a Jeol 2000 FX, at an accelerating voltage of 100 kV.

3. RESULTS

3.1 Binary Blends

The electrical conductivity as function of PANI content, in blends with PS/DOP, CoPA and PS, is presented in Fig. 1. A percolation threshold of 10 wt% PANI is observed for the PS/DOP blend series. The (PS/DOP) / PANI 85/15 blend is already conductive ($\sim 3*10^4$ S/cm), due to formation of conducting paths of PANI within the PS/DOP matrix. This blend series exhibit a conductivity "jump" of more than 9 orders of magnitude in the percolation region. Beyond percolation, the conductivity level slowly increases with the PANI content, due to generation of a conductive network of an improved quality, attaining 0.3 S/cm for the 70/30 blend. This maximal observed conductivity level is similar to that of the neat PANI after exposure to 150 ^{0}C, the melt blending and compression molding temperature. The CoPA / PANI blend series exhibits a different electrical conductivity behavior. The conductivity increase with PANI content is more gradual than that observed for the previous blend series, and the maximal conductivity level ,$7*10^4$ S/cm for the 70/30 blend, is lower. The reason for this different behavior will be discussed below. Interestingly, the PS / PANI blends are insulative even at a PANI content as high as 20 wt%.

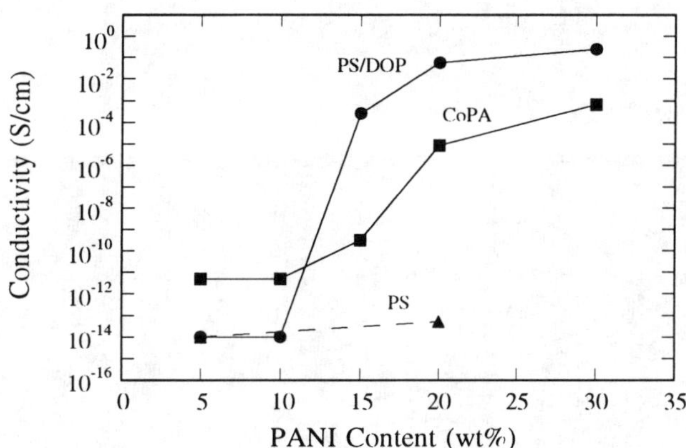

Fig. 1: Electrical conductivity vs. PANI content for various matrix polymer / PANI blend series

The blends' morphology was studied to elucidate the electrical conductivity-structure relationships. HRSEM micrographs of cryogenically fractured surfaces of PS, PS/DOP and CoPA blends with 20 wt% PANI, are presented in Fig. 2, compared with the fractured surface morphology of the neat matrix polymers. Large domains of PANI (1-20 µm) are observed within PS, indicating the absence of a continuous network of PANI and explaining the insulative behavior of the PS / PANI blends, even at 20 wt% PANI. In contrast, the blends of PS/DOP or CoPA with PANI exhibit small PANI particles within the thermoplastic matrix. The PANI particles in the PS/DOP matrix (0.1 - 0.5 mμ) are smaller than those observed in the CoPA matrix (0.1 - 2 mμ). The high conductivity at relatively low PANI contents in PS/DOP and CoPA is, thus, related to the high levels of dispersability and structuring of the PANI particles within the matrix polymer.

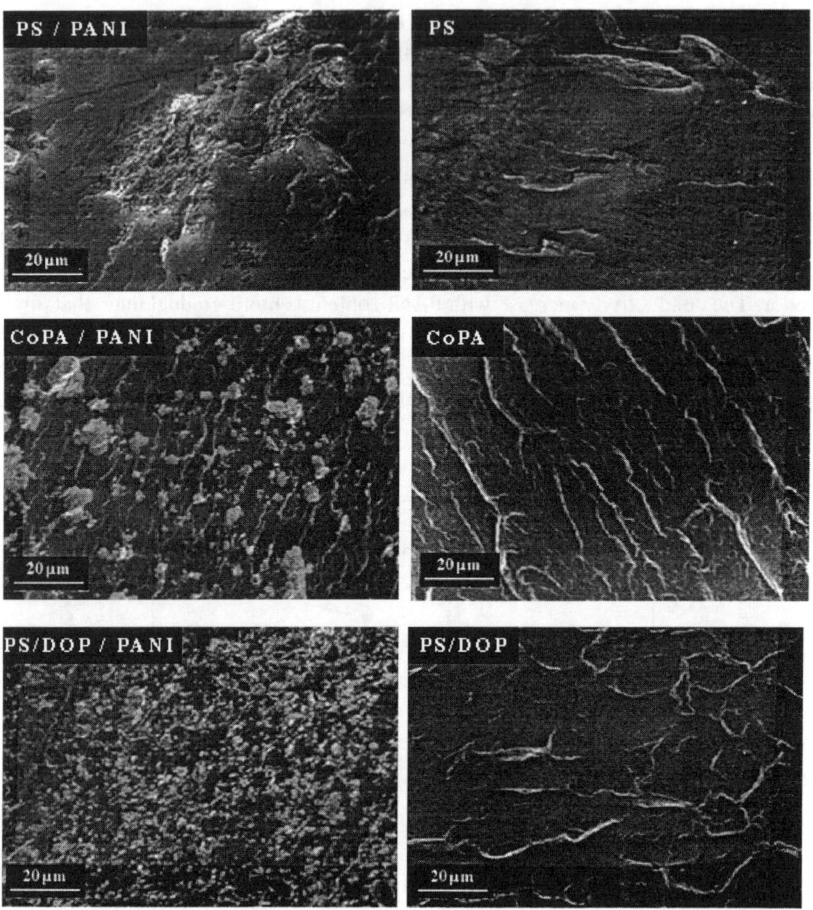

Fig. 2: HRSEM micrographs of various matrix polymer / PANI 80/20 blends, compared with their neat matrix polymers

The level of PANI particle mechanical fracturing, during melt mixing with a thermoplastic matrix, depends on the specific matrix used. In polymer blending, both, kinetic and thermodynamic considerations, determine the degree of polymer dispersion within another polymer [12]. Thus, during melt blending, the matrix polymer interactions with PANI particles influence the level of PANI fracturing and dispersion. High interaction levels enable a higher level of PANI fracturing (at similar matrix viscosity and shear level), due to better interphase shear stress transferability [9]. This should occur in systems comprising components of similar solubility parameters, i.e., low interaction parameter (interaction parameter decreases with increasing solubility). Molecular simulations were performed [13], to calculate solubility and interaction parameters for the studied polymers (Table 1). It should be noted that the calculated solubility parameters take into account weighted contributions of the various functional groups in the polymer chain, ignoring specific interactions. The solubility parameters of the PANI and CoPA are similar, whereas the solubility parameter of PS is well below that of the PANI. Accordingly, the CoPA based blends are compatibile and conductive, while the PS containing blends are non-compatibile. These considerations support the electrical conductivity results (Fig. 1) and the morphological observations (Fig. 2).

TABLE 1. CALCULATED SOLUBILITY AND INTERACTION PARAMETERS

Polymer	solubility parameter $(J/cm^3)^{0.5}$	Interaction parameter with PANI
PANI	23.9	
CoPA	24.2	0.006
PS	19.5	0.79
DOP	20.9	

In order to reduce the processing temperature of the PS based blends from 180 ^{0}C to 150 ^{0}C, the PS matrix was plasticized with 15 wt% DOP. While the PS / PANI blends are insulative, due to the poor dispersion of PANI, the (PS/DOP) / PANI blends are highly conductive (Fig.1), due to the high level of PANI dispersion (Fig. 2). DOP in the present case acts not only as a plasticizer, reducing the processing temperature. The addition of DOP, $\delta = 20.9$ $(J/cm^3)^{0.5}$, to PS, $\delta = 19.5$ $(J/cm^3)^{0.5}$, increases the latter's solubility parameter towards that of PANI, $\delta = 23.9$ $(J/cm^3)^{0.5}$. This small change, predicted by molecular simulation, is, however, insufficient to dramatically affect the blends morphology and conductivity. It is suggested that specific interactions, unaccounted for by molecular simulation, may exist in the (PS/DOP) / PANI system. The plasticizer can cause conformational changes in the PS chain, generating enhanced Van-der-Waals interactions of PS phenyl rings with PANI. Another possibility is that some DOP migrates during blending into the PANI phase, affecting its rheological behavior. The specific interactions, suggested to exist in the (PS/DOP) / PANI system, may be the reason for the better PANI dispersion and structuring and, thus, for the higher conductivitiy of this system compared to those of PS / PANI and CoPA / PANI systems. The effect of DOP content on the electrical conductivity of (PS/DOP) / PANI 80/20 blends is depicted in Fig. 3. Interestingly, the conductivity sharply increases with the increase in DOP content, up to 10 wt% DOP, and then levels off. The shape of this curve resembles that of a percolation curve; a relatively low plasticizer (DOP) content changes the insulative PS / PANI 80/20 blend to a conductive one.

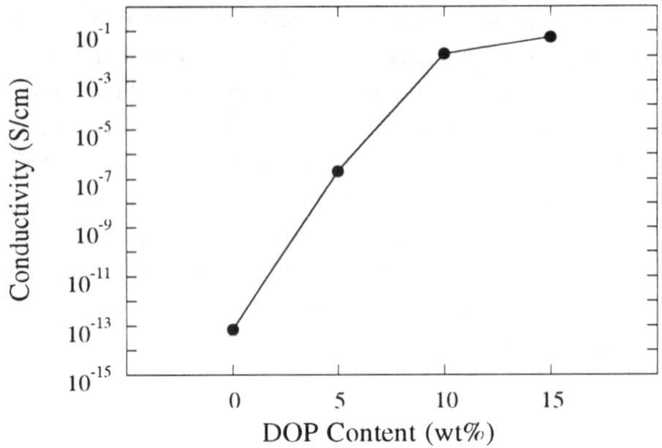

Fig. 3: Electrical conductivity vs. DOP content for (PS/DOP) / PANI 80/20 blends

HRSEM observations of the blends enable to estimate the PANI particles' size and their level of dispersion within the matrix polymer, but not the expected continuous network structure in conductive blends. Therefore, the blends' morphology was also studied by TEM. For example, a micrograph of CoPA / PANI 80/20 blend is presented in Fig. 4. This blend shows two types of regions : a region rich in PANI, in which continuous paths of PANI (in black) were developed within the CoPA matrix, and a region containing apparently discrete PANI particles. It is suggested that the discrete particles are also part of the three dimensional network, however, observed at an angle perpendicular to the network's direction.

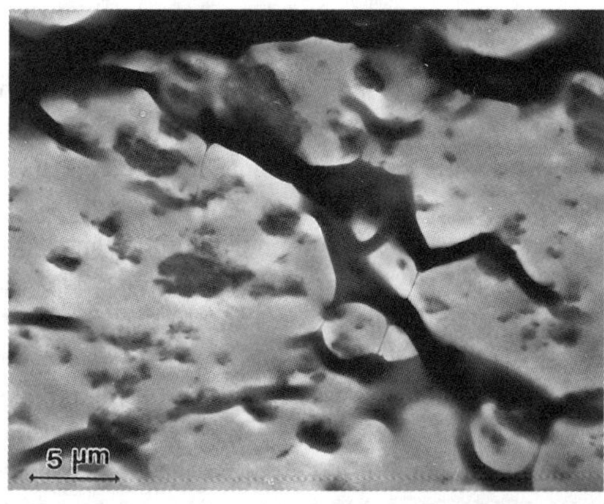

Fig. 4: A TEM micrograph of CoPA / PANI 80/20 blend

Figure 4 also indicates that the continuous network is actually composed of small particles (0.5-1 μm), consisting of even smaller particles, about 0.1 μm apart. The inter-particle gaps are, presumably, too large to enable hopping of charge carriers between particles. It has, thus, been suggested [9], that in addition to the observed (TEM) structural features, a very fine short-range PANI fibrillar-like morphology exists, presently unobservable by both, HRSEM and TEM, which interconnects the observable fine dispersed PANI particles, contributing to the continuity of the electrical paths. Since the solubility parameters of CoPA and PANI are quite similar, partial dissolution during blending of low molecular PANI fractions in the molten CoPA matrix may occur. Upon cooling, at least partial precipitation of this PANI fracture occurs, forming the secondary morphology, a short-range fibrillar-like structure. The secondary fine fibrillar structure resembles the structure obtained in some solution cast polymer / PANI blends [3].

3.2 Ternary (PS/DOP) / CoPA / PANI Blends

Further to binary blends, ternary blends of the two immiscible polymers (PS/DOP and CoPA) and PANI were investigated. The electrical conductivity as a function of CoPA content, for blends containing 10, 15 and 20 wt% PANI is presented in Fig. 5. The conductivity level of the blends containing 20 wt% PANI increases with a decrease in the CoPA content, and is higher than that of the CoPA / PANI 80/20 binary blend and lower than that of the (PS/DOP) / PANI 80/20 binary blend. HRSEM micrographs of cryogenically fractured surfaces of ternary blends containing 80 wt% matrix polymers and 20 wt% PANI are presented in Fig. 6, compared with the morphology of the corresponding matrix polymers. These micrographs show that the PANI (in white) is preferably located in the CoPA phase and its mode of dispersion resembles that observed for the CoPA / PANI binary blend (Fig. 2). The PS/DOP phase does not contain any PANI particles. Hence, the effective PANI content in the CoPA phase is higher than its nominal content in the blend. Since the two matrix polymers form, in the composition range studied, a co-continuous blend structure, and since the PANI containing phase is conductive, their ternary blends are conductive. This concentration effect is quite impressive in ternary blends containing 10 and 15 wt% PANI (Fig. 5). While the (PS/DOP) / PANI 90/10 and CoPA / PANI 90/10 binary blends are practically insulative, the ternary blends containing 10 wt% PANI are highly conductive (at least 10^{-4} S/cm), due to the above percolation PANI content in the CoPA phase. The ternary blends containing 15 wt% PANI exhibit an intermediate behavior.

The preferable location of PANI in the CoPA rater than in the PS phase should be addressed. CoPA and PANI have similar solubility parameters, that are higher than that of PS (Table 1). Addition of the DOP plasticizer to PS slightly increases its solubility parameter towards that of PANI, but as mentioned before, the (PS/DOP) / PANI system is more compatibile than the CoPA / PANI system. Therefore, it should be expected that the PANI will prefer the PS/DOP as a host matrix. The opposite, migration of PANI into the CoPA phase, is obtained probably due to an earlier migration of DOP from PS to CoPA. Hence, the less plasticized PS is not as compatible with the PANI and therefore, the PANI prefers the CoPA matrix. Moreover, a slight decrease in the solubility parameter of the CoPA, probably occurs due to the addition of DOP, enhances its compatibility with the PANI.

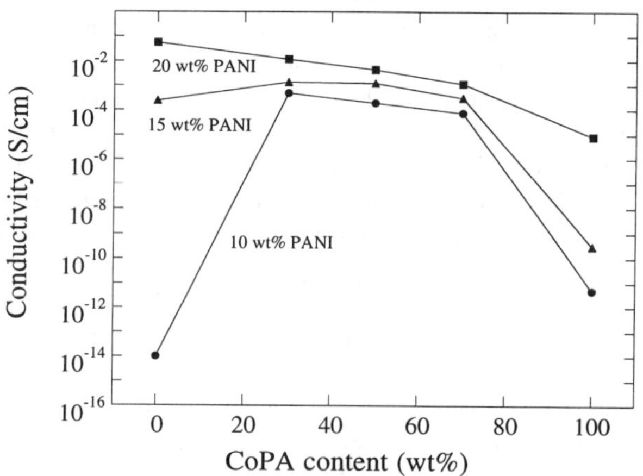

Fig. 5: Electrical conductivity vs. CoPA content for (PS/DOP) / CoPA / PANI ternary blends (CoPA content + PS/DOP content = 100% matrix polymers)

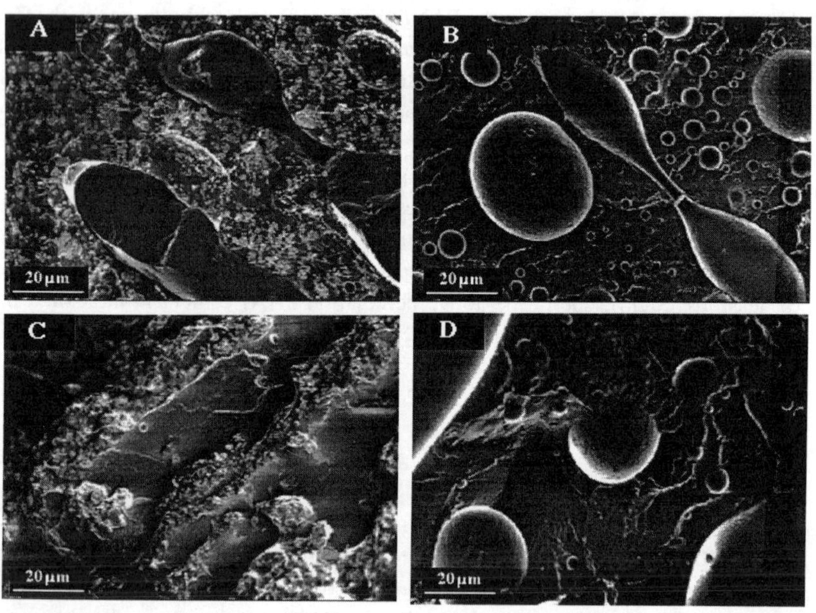

Fig. 6: HRSEM micrographs of (PS/DOP) / CoPA / PANI ternary blends containing 20 wt% PANI and their matrices: (A) (PS/DOP) / CoPA 30/70 blend, (B) (PS/DOP) / CoPA 30/70 matrix, (C) (PS/DOP) / CoPA 70/30 blend, (D) (PS/DOP) / CoPA 70/30 matrix

536

The three ternary blend series, containing 10, 15 and 20 wt% PANI, show an increase in conductivity with the decrease in CoPA content, due to the concentration effect, but the measured conductivity levels are equal or higher than the calculated ones, assuming that the PANI is only located within the CoPA (Table 2). The probably higher compatibility of CoPA/DOP with PANI, than the neat CoPA, may be the reason for a better PANI structuring and, therefore, higher measured conductivities in the CoPA/DOP containing blends. The difference between the measured and the calculated blends conductivity increases with both, the increase in CoPA content and the decrease in PANI content, hence, with the decrease in the effective PANI content. It seems, therefore, that the CoPA/DOP effect is more significant in the percolation region than in the higher PANI contents.

TABLE 2. MEASURED AND CALCULATED CONDUCTIVITY OF (PS/DOP) / COPA / PANI BLENDS

PANI Content (wt%)	CoPA content[a] (wt%)	effective PANI content[b] (wt%)	measured conductivity (S/cm)	calculated conductivity (S/cm)
10	30	27	$7*10^{-4}$	$3*10^{-5}$
	50	18	$2*10^{-4}$	$1*10^{-6}$
	70	14	$1*10^{-4}$	$1*10^{-10}$
15	30	37	$1*10^{-3}$	$1*10^{-3}$
	50	26	$1.5*10^{-3}$	$1*10^{-4}$
	70	20	$6*10^{-4}$	$1*10^{-5}$
20	30	45	$1*10^{-2}$	$1*10^{-2}$
	50	33	$7*10^{-3}$	$1*10^{-3}$
	70	26	$2*10^{-3}$	$1*10^{-3}$

[a] – CoPA content + (PS/DOP) content = 100% matrix polymers
[b] – assuming that the PANI is located only within the CoPA phase

In order to study the effect of sequence of components mixing on the blends morphology and electrical conductivity, ternary blends containing 20 wt% PANI were prepared in three diferent ways: (a) - mixing the three components together in the Brabender mixer; (b) - (polymer 1 + PANI-pTSA) + polymer 2, i.e., mixing polymer 1 with PANI-pTSA and then adding polymer 2; (c) - (polymer 2 + PANI-pTSA) + polymer 1" i.e., mixing polymer 2 with PANI-pTSA and then adding polymer 1. The conductivity results are presented in Fig. 7. Generally, the three blend series show similar conductivity behavior, i.e., the conductivity increases with the decrease in CoPA content, due to the concentrating effect of PANI in the CoPA. When PANI is first blended with PS/DOP and then the CoPA is added, the PANI migrates to the CoPA. HRSEM observations indicate that some of the PANI particles remain at the interphase between the two matrix polymers, and probably do not contribute to the blend's conductivity. As a result, The [(PS/DOP + PANI) + CoPA] blends are less conductive than the [(CoPA + PANI) + PS/DOP] blends.

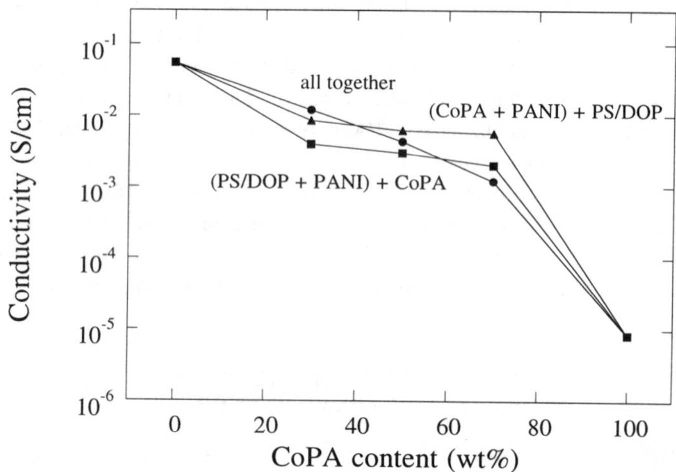

Fig. 7: The effect of mixing sequence on the electrical conductivity of (PS/DOP) / CoPA / PANI ternary blends containing 20 wt% PANI (CoPA content + PS/DOP content = 100% matrix polymers)

4. CONCLUSIONS

Binary and ternary blends of PANI with PS, DOP plasticized PS and CoPA were prepared by melt processing. The investigation focused on the blends morphology, the interaction between the components, and the resulting electrical conductivity.

Binary Blends

The level of interaction between the blend's components determines the PANI particles level of fracturing and mode of dispersion within the matrix polymer, and thus, the blend's morphology and electrical conductivity.

Similar solubility parameters of the matrix polymer and the doped PANI lead to high level fracturing of PANI and the formation of conducting paths at low PANI contents. Therefore, blends with CoPA become conductive at a low PANI content while blends with PS stay insulative in the prsently studied PANI contents. The (PS/DOP) / PANI blends are highly conductive, probably due to specific interactions between the blends' components.

The conductive network consists of two levels of structures: A primary structure of small dispersed PANI particles, and a short-range very fine fibrillar structure, interconnecting these particles.

Ternary Blends

A system consisting of PANI and two immiscible thermoplastic polymers, PS/DOP and CoPA, was investigated.

The PANI is preferably located in the CoPA phase; its content in that phase is higher than its nominal content in the blend. Therefore, these ternary blends, consisting of a matrix having a co-continuous phase structure, exhibit high electrical conductivity at low nominal PANI contents.

CoPA is the "preferred" matrix, rather than the PS/DOP phase, probably due to DOP migration from the PS to the CoPA.

Since the PANI can migrate during blending from the PS/DOP to the CoPA phase, the blend preparation sequence only slightly affects the ternary blends morphology and electrical conductivity.

5. REFERENCES

1. O.T. Ikkala, J. Laakso, K. Vakiparta, E. Virtanen, H. Ruohonen, H. Jarvinen, T.Taka, P. Passiniemi and J.E. Osterholm, Synthetic metals, 69, 97 (1995).
2. J.E. Frommer and R. R. Chance, Electrically Conductive Polymers, in Encyclopedia of Polymer Science and Engineering, H.F. Mark, Ed., John Wiley & Sons, New York, 1988, Vol 5.
3. A.J. Heeger, Synthetic metals, 57, 3471 (1993).
4. Y. Cao, P. Smith and A.J. Heeger, Synthetic metals, 57, 3514 (1993).
5. C.Y. Yang, Y. Cao, P. Smith and A.J. Heeger, Synthetic metals, 53, 293 (1993).
6. L.W. Shacklette, C.C. Han and M.H. Luly, Synthetic metals, 57, 3532 (1993).
7. S.J. Davides, T.G. Ryan, C.J. Wilde and G. Beyer, Synthetic metals, 69, 209 (1995).
8. P. Passiniemi, J. Laakso, H. Ruohnen and K. Vakiparta, Mat. Res. Soc. Symp. Proc., 413, 577 (1996).
9. M. Narkis, M. Zilberman and A. Siegmann, Polym. Adv. Tech., 8, 525 (1997).
10. M. Zilberman, A. Siegmann and M. Narkis, J. Macromol. Sci. B37, 301 (1998).
11. M. Zilberman, G.I. Titelman, A. Siegmann, Y. Haba, M. Narkis and D. Alperstein, J. Appl. Polym. Sci., 66, 243 (1997).
12. Y.J. Lee, I. Manas-Zloczower and D.L. Feke, Polym. Eng. Sci., 35, 1037 (1995).
13. D. Alperstein, M. Narkis, M. Zilberman and A. Siegmann, "Computer Aided selection of a Polymer Matrix for Polyaniline Conductive Blends", Polym. Adv. Tech., submitted.

Acknowledgement

This research was partially supported by the Israel Ministry of Science.

DESIGN GUIDELINES FOR ADVANCED
SILICON-CARBIDE ALUMINIUM COMPOSITE MATERIALS

RMK Young* and Alan Moses**
(*AEA Technology plc, Harwell, UK , ** Ametek Inc, Wallingford, CT.)

High volume fraction silicon carbide - aluminium (SiC-Al) composites are proving to be highly effective for advanced thermal management applications in electronics. This follows from property combinations which include high thermal conductivity, low thermal expansion, low density and good mechanical properties . This paper examines the properties and attributes of one class of these materials (known as HIVOL™) in which both material constituents and processing have been engineered to optimise the key properties and deliver them at an affordable price. However, substitution for existing designs is often inappropriate and guidelines should be available so that designers can successfully implement these materials to best advantage. As these are relatively new materials there is still little design information available. In the light of this, some suggestions for good design practice are offered for different thermal management contexts.

KEY WORDS: HIVOL ™ ,Silicon Carbide Aluminium, Thermal Management, Electronics Packaging, Design Guidelines

1. INTRODUCTION

High volume fraction SiC-Al - aluminium metal matrix composites (MMCs) are materials which can be processed by infiltrating porous ceramic beds with liquid aluminium. These materials have property combinations which are attractive for packaging materials for electronics thermal management Key potential benefits available with the use of this technology include:

- reduced component mass (for weight critical applications) following from low density relative to conventional expansion alloys
- improved dissipation and hence increased power densities supported resulting from high thermal conductivity
- increased reliability available with good dissipation AND good expansion matching to reduce the expansion mismatch strains experienced on operational thermal cycling
- low cost for net shaped engineered systems
- isotropic behaviour simplifies design

However, these potential benefits can only be realized if correct design practice is applied to the parts in question - net shaping in particular is only credible if the parts are designed with this in mind.

There are several application areas where these and other similar materials are being used. These areas include aerospace where lightweight thermal management is at a premium - and more recently in automotive applications in which the severe demands of power control thermal management can only be effectively met using MMC technology.

Wider uptake of these materials is being hampered by a shortage of technical information and specification data and design guidelines. Substitution for existing designs is often inappropriate and can lead to unnecessary engineering difficulties. Guidelines should be available so that designers can successfully implement these materials to best advantage. As these are relatively new materials there is still little design information available. In the light of this, some suggestions for good design practice are offered for different thermal management contexts.

2. HIVOL™ PROCESSING

HIVOL™ materials are processed using specially developed low pressure infiltration processing methods to force liquid aluminium into a porous bed of particulate SiC [1].

The technique is based on the following stages:

1. Forming a porous bed of SiC particles using a proprietary mould filling method
2. Placing the moulds into a pressure vessel
3. Preheating and evacuation of the SiC - filled moulds
4. Introduction of liquid aluminium alloy into the moulds and pressurization to force infiltration of the aluminium alloy into the porous beds.
5. Directional cooling and solidification
6. Stripping of parts from moulds and finishing (if required).

The ceramic bed itself is engineered to meet the volume fraction requirements while maintaining a structure which is sufficiently permeable to allow full melt infiltration during processing. Additional constraints on the minimum and maximum allowable particle sizes in the bed mean that the particle size distribution - and hence the bed forming process - is critical for producing material which is optimised for thermal performance.

The HIVOL™ process differs from other MMC processes in that the porous bed is formed directly in the same mould cavity used for the melt infiltration process. This avoids the requirement for a separate preform moulding stage.

With the porous SiC bed in place, the melt infiltration takes place. This melt infiltration process in HIVOL™ is based on the principle of isothermal infiltration. This is a process technique in which the thermal design is arranged so that the

timescales for infiltration and cooling can be separately controlled. Benefits available
with this process include:

- no separate preform stage is required so overall tooling costs are lower
- minimum possible process temperatures so that interfacial reactions are minimised
- controllable timescales to allow full infiltration.
- low gate velocities so that the porous beds are not damaged during infiltration
- direct net shaping is available using mould inserts so minimum finishing is required
- tailoring of expansion to other values is possible

3. MATERIAL ATTRIBUTES AND QUALITY

HIVOL™ materials have been designed to meet the thermal management requirements
of the electronics industry While many different variants are possible with appropriate
selection of volume fraction and matrix materials - most of the application
requirements can be met with the two grades which are currently commercially
available (HIVOL™ B and HIVOL™ C). Details are discussed below.

3.1 Constitution   HIVOL™ B is material produced with approximately 70% SiC
particulate and infiltrated with a matrix of an aluminium alloy containing 4% Cu. This
system can be heat treated if required to develop metallurgical hardening.

HIVOL™ C is similar except that the matrix alloy contains silicon which results in a
lower thermal expansion system.

3.2 Thermal Expansion   The thermal expansion is a key design property which is
determined by the SiC volume fraction, the matrix alloy and also the internal stress
state.

The most significant factor affecting the thermal expansion is the volume fraction of
the SiC bed..   It is important to note that while standard grades of HIVOL™ are
produced with volume fractions of 70% - the process can be controlled to produce
materials with other volume fractions within set limits. Combining this with the use of
matrix alloys with different expansion coefficients means that material with expansion
matched to other target values can be developed relatively easily.

The effects of even small volume fraction changes can be significant in determining mismatch stresses between the composite part and either ring frame materials or - for example - direct bond copper substrates which are soldered on.

Thermal expansion behaviour is usually quoted as a simple expansion coeffecient but this can be misleading owing to non-linear effects.   Over higher temperature ranges the expansion is non linear owing to the temperature dependency of the constituents and this is further compounded by the influence of the internal stress state.   This is apparent in the hysteresis between the expansion values on heating and cooling (see Figure 1).

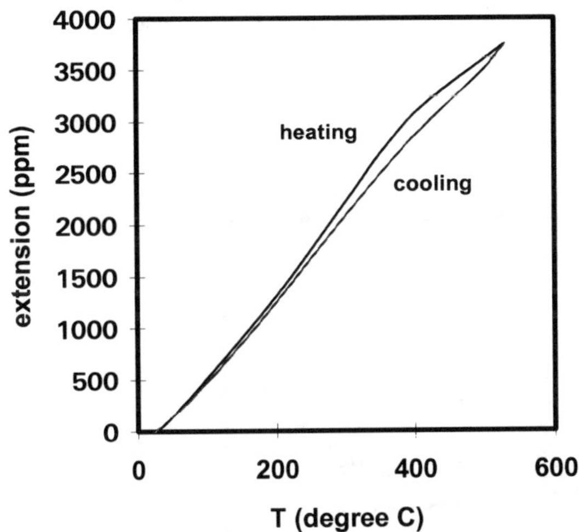

Data for the expansion behaviour of the two commercially available variants is shown in Table 1 [2].

Table 1

Temperature range	Thermal expansion coefficient 10^{-6} K^{-1}	
	HIVOL B	HIVOL C
	over 3 cycles	over 4 cycles
-60 to 0	5.6±0.2	4.7±0.2
0 to 50	6.5 ±0.1	5.8±0.1
50 to 100	7.6±0.1	6.7±0.2
100 to 150	8.6±0.2	7.5±0.2
150 to 100	7.8±0.1	6.9±0.1
100 to 50	7.1±0.1	6.5±0.0
50 to 25	6.6±0.2	5.8±0.1

3.3 Thermal Conductivity   The thermal conductivity of the material is determined by many factors which include: the intrinsic thermal conductivity of the material constituents, the volume fraction of SiC, the physical size distribution, the nature of the SiC-matrix interface (including aluminium carbide formation), and the porosity content. The constituents and processing of HIVOL™ materials have been designed to generate microstructures which return the highest possible thermal conductivity within the constraint of expansion matching to target values.

In common with many materials, the thermal conductivity decreases with increasing temperature (Figure 2). Thermal conductivity is not significantly affected by porosity [3] but this has not been studied systematically studied.   HIVOL™ material is essentially free from porosity and has very high thermal conductivity, typically 220 W/mK minimum.   The data shown was determined using laser induced transient thermography.

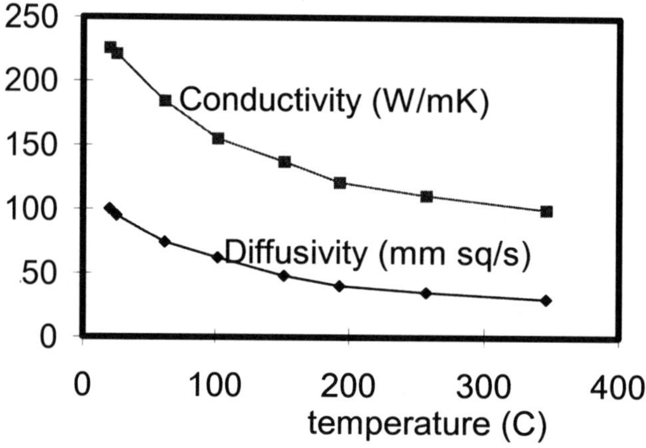

Figure 2

3.4 Elastic Properties   Determination of mechanical behaviour of the material by mechanical testing reveal that apparent modulus reduces significantly with increasing strain (shown schematically in Figure 3).   There appears to be no well defined elastic limit - possibly because the matrix is in a condition of incipient tensile yield as a result of expansion mismatch stresses generated on cooling the material during processing.[4]. Micromechanical models incorporating the effect of thermal residual stresses can account for the non linear load extension behaviour under tensile loading (shown in Figure 4).   Elastic behaviour is also adversely affected by microcracking in the SiC particles. This is not unexpected as the largest particles used are relatively coarse (80 microns diameter).   Statistical models indicate that cracks are more likely to arise in the larger particles and this is responsible for the eventual failure of material at the ultimate tensile strength.   Microcracking can be monitored through the observation of acoustic emission events during a tensile test. This demonstrates that microcracking can start well below failure stresses and that high acoustic activity correlates well with decrease in modulus [4].

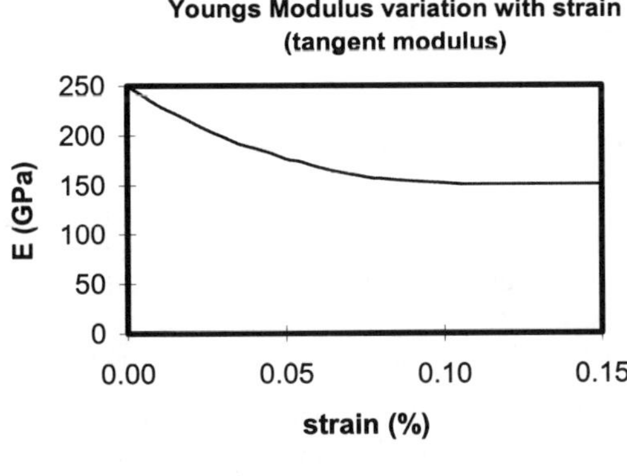

Figure 3

3.5 Mechanical Properties - strength and toughness   Failure initiation by cracking of SiC particles in a region of high stress can account for a strong volume sensitivity on apparent strength.   The effect is significant;   failure stress levels of around 600 MPa have been measured using 3 point bend testing on samples conforming to ASTM requirements (span 30 mm, width 8 mm, thickness 1.5 mm) although lower flexural strength values are specified (365, 307 MPa in HIVOL B and C respectively) owing to the statistical nature of failure. In practice strength is also found to depend on annealing or heat treatments if the matrix is metallurgically active - as is the case with HIVOL™ B which reacts well to T6 solution and ageing treatments. In common with many materials, tensile strength values are lower than flexural strength values (225 MPa) which is consisent with statistical strength models.

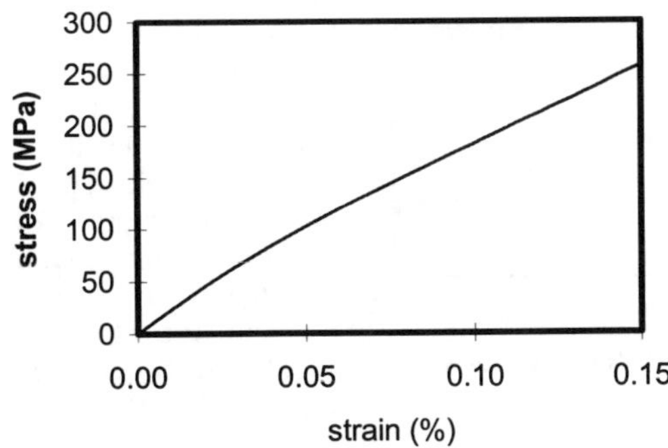

Figure 4

The availability of some plastic behaviour in the material prior to fracture - even with high SiC loadings imparts some toughening effect on the material.   Fracture energy determination on HIVOL™ B using a chevron notch technique suggest values in excess of 20 MPam$^{1/2}$ while measurements based on short compact test pieces on HIVOL™ C indicate values of around 8-9 MPam$^{1/2}$ [5].   As with the measurement of mechanical strength, and careful studies on consistent material of known temper using establish test techniques is required in order to establish meaningful limits.

3.6 Non destructive test techniques for expansion determination

At present,   HIVOL™ MMC materials are used for their functional thermal and expansion behaviour.   For reliable manufacture of consistent product it is important to have access to test techniques - preferably non-destructive - to validate quality of commercial material. Direct measurement on complete components is possible using advanced interferometry techniques but this is inappropriate for production monitoring.

As thermal expansion is controlled principally through the volume fraction of the SiC constituent techniques are required which are well adapted for this.   As the volume fraction of the SiC also controls the elastic properties and the velocity of sound in the material the measurement of these properties the measurement of these parameters is a natural correlation parameter for material quality.

The measurement of elastic modulus by mechanical testing suffers from several disadvantages but techniques based on measurement of resonant frequency can be useful providing the material is in a simple and known geometry as is often the case, and the density is known. In such cases modulus values can be extracted which can be used to determine the volume fraction - and hence thermal expansion coefficient using correlation models. This method is only capable of returning an average value of the modulus for the component being evaluated.   In more complex geometries the technique can still be a useful quality parameter even if absolute modulus values cannot be extracted without detailed analysis.

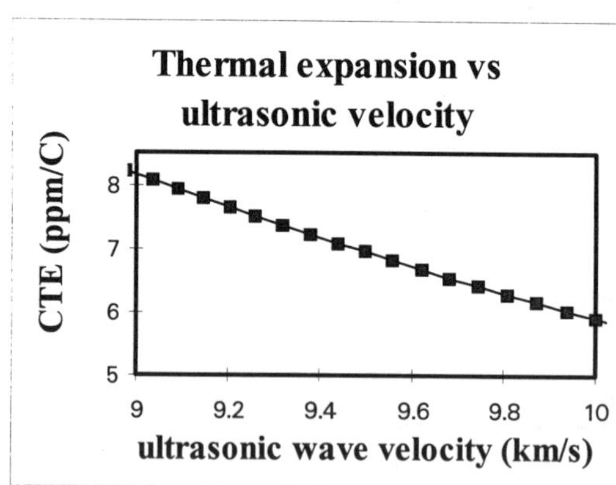

An alternative technique is based on the direct measurement of ultrasound velocity as it propagates through the material which also correlates with volume fraction of SiC. This benefits from the possibility of mapping distributions over a plate and more advanced methods allows determination of attenuation which can be used to extract information about porosity contents.

Figure 5

3.7 Non destructive test techniques for thermal conductivity Thermal conductivity of high conductivity materials is best measured indirectly by determination of the thermal diffusivity using transient thermography methods which is relatively straightforward. Specification data is obtained on small samples but validation of thermal quality on large components is of great importance for advanced thermal management applications. Conventional thermography based on Xenon flash thermography is probably inappropriate for thick section material because the energy densities are too low. However, other variants of scanning themography [6] (SLITT) which can directly map variations in thermal properties across the surface may be more appropriate than the alternative process of flash thermography owing to the higher energy densities - and hence greater penetration depth which is available.

3.8 Electrical conductivity The electrical conductivity is not a key property driver for many application areas but this is still a useful quality parameter which is rapid and easy to measure. An acceptance window on material conductivity can exclude material which is porous which has low conductivity or material or defects associated with higher local matrix contents which lead to higher conductivity.

4. DESIGN GUIDELINES

4.1 Flatstock Manufacture

2-D Flatstock Capacity limits exist which are determined by the capacity of the process equipment. Typical dimensions of large plates which have recently been manufactured include 235 by 150 mm, 190*140 mm and 140*130 mm. At the other end of the scale, 1mm heatsink cubes can be made by dicing flat plate. Flatstock can be machined to and in section thickness ranging from 0.5 mm to 20 mm. Typical "standard" gauges are 1, 1.5, 2, 3 and 5 mm thick. In volume - parts can be manufactured to net shape or with minimal machining - while for lower volume work the machining techniques described below are appropriate. Flat plates are usually finished by grinding prior to surface finishing. Surface finishes of less than 0.5 microns R_a are readily achievable on the ground plate.

Smooth perimeter forms are generally preferred - but some projecting features can be tolerated provided the overall expansion differences between the mould and the plate do not cause binding or locking of the part in the mould (see Figure 6). Good surface release coats and generous draft angles - up to 3° - are recommended to facilitate separation from the mould. In the HIVOL process the parts are usually separated from the moulds at ambient temperatures so there is potential for expansion mismatches to develop between the mould and the casting. The interaction between the mould and the part is determined by the expansion difference between the two - which is again strongly influenced by any variations in the SiC bed density. One result is that different grades of HIVOL with different expansion behaviour require different mould materials to ensure the most appropriate expansion management in casting.

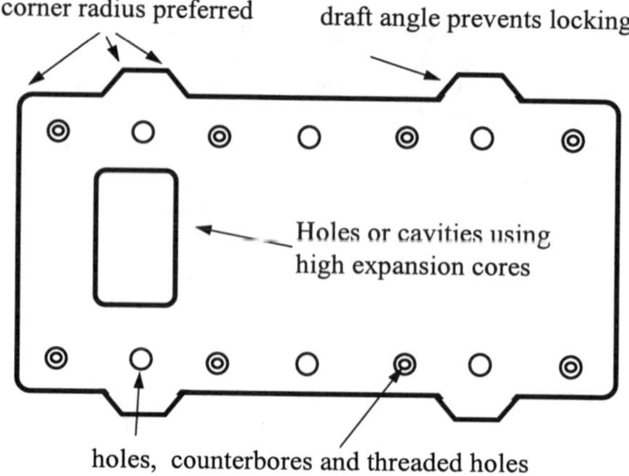

Figure 6

For lowest possible tooling costs, moulds are usually manufactured by milling - so that tight internal radii (less than 2 mm) are difficult to achieve. Moulds can also be manufactured using die sinking processes to realise more demanding designs but this is generally discouraged owing to the higher tooling cost required. The process can also be adapted to allow SiC free zones to allow holes to be cast or machined in - which is required for low cost manufacture in volume.

2½-D Flatstock Components which are basically flat but with additional relief - such as plinths or depressions - referred to as 2½ d structures - can be engineered by modifying the moulds. These are frequently required in microwave or radar applications. The provision of plinths is only really possible if these are cast in.
Recesses - usually rectangular -can also be allowed for if the mould itself is engineered with a plinth. In this case however - the expansion factors are reversed and these type of features can only easily be engineered by using moulds with expansion behaviour very close the that of the HIVOL material and if generous draft angles are provided. As before, tooling costs can be minimised if appropriate radii are allowed in internal radii for the plinths.

The potential for mould lock up is most if locking features - such as plinths or recesses - are large and widely separated so that larger absolute mismatch stresses build up.

In the finishing of 2½ d structures there is little scope to modify the surface prior to plating. In such cases , the as-cast surface finish is of great importance and so moulds may require polishing and release coats carefully controlled to maintain good finish (typically around 2 microns Ra). In the event that improved surfaces are required - vibro-rumbling or shot blasting with soft media (plastic) can be appropriate although this can increase edge radius.

3-d Net shaping Where designers specify 3-d components (boxes) - there is an economic imperitive that these be as near fully net shaped as possible as machining processes are generally expensive. For 3-d castings the same considerations apply as for 2½-d structures.   In order to create generic box structures,   deep moulds and removable cores are required.   Thin walls - to around 1 mm can be produced (subject to aspect ratio) although the preform moulding process may need adapting to ensure good cavity fill.   Matching clamshell moulds can be used to process more complex net shape structures.

Techniques for net shaping must also make provision for casting in inserts of materials which are machineable.   This is of particular importance for machining of small threaded holes which cannot readily be machined in the MMC material itself.

4.2 Machining   Machining techniques available include electrodischarge machining (EDM), laser techniques, abrasive waterjet machining and diamond milling, drilling, dicing, grinding and lapping
- EDM   This is widely used for prototype or small volume manufacture.   It is the technique most appropriate for high precision and becomes less economic for thicker grades.   High precisions are readily available.
- laser techniques   Carbon dioxide, YAG and copper vapour laser (CVL) systems have all been evaluated. Except for thin components, edge quality is poor owing to hard deposits which are difficult to remove.   CVL is still too costly for production work.
- abrasive waterjet machining is a promising technique which is compromised by precision levels which are just short of adequate.   This is expected to change as the technology continues to evolve.
- diamond machining requires careful life management of the tooling to be economic.
- dicing works well for fabrication of large numbers of small rectangular or square components.   Very high precisions are available using semiconductor dicing machines.
- lapping is appropriate for finishing smaller parts but becomes less competitive in relation to grinding for larger components.   The surface finish with lapping is rather different to that obtained with grinding.

These techniques all have their relative strengths and weakness but in all cases high tolerances add considerable cost and where possible the tolerances should be relaxed to meet the most appropriate technique.

4.3 Assemblies   Many package structures are built up by attaching ring frames containing feedthroughs to metallized MMC bases.   This approach fits well into conventional integration methods but suffers from the disadvantage that costly braze materials may need to be specified to provide for die attach processes.   Compatibility between ring frames and the MMC base becomes increasingly important with the physical scale of the ring frame. This is because any expansion mismatch builds up between the ring frame and the MMC base on cooling from the braze temperature and this leads to a structure with locked-in residual stresses which generally impairs hermetic performance or in severe cases causes shear failure of the ring frame at the solder joint or at the plating interface. This stress state also leads to bowing of the package base.   Limited bowing in which the MMC surface opposite the ringframe is generally advantageous in that it provides good thermal contact between the package and the secondary heatsink.   As the bow increases the clamping or screw down forces on the package body required to flatten it increase to levels which threaten the corners and the shear stresses between the ring frame and package base also increase. Stresses rise to a maximum at the corners and susceptibility to failure rises with the degree of mismatch and the severity of the stress intensity of the ring frame corner.

The susceptibility to damage arising from excessive corner shear stress can be decreased by following some of the design guidelines associated with adhesive joints, i.e. by providing the largest possible radius on external radii and possibly by machining a flanged ring frame, ideally with tapered sections to increase the transfer length and reduce stress concentration. These measures should reduce the shear stress to within the limits acceptable for the joint. However, the freedom to design in this way is restricted both by the cost penalty arising from the extra machining involved and also the marginal increase in the footprint of the package assembly.

In the light of this it is clear that designers and suppliers should work together to develop an integrated approach to package design with proper selection of both ring frame and the grade of MMC materials used. Ideally, designers would specify the CTE behaviour required to develop a neutral or low residual stress package which can be met using the flexibility available in MMC processing and manufacture.

4.4 MMC - Ringframe Compatibility The susceptibility to mismatch strains may be estimated or modelled using known expansion data from representative braze temperatures. In practice, relaxation processes can occur so that these calculations must be regarded as upper bounds for the mismatch stains. These do not develop in a linear manner because of the non linearity of both the MMC materials and the expansion alloys. This nonlinearity also means that the residual strains developed do not necessarily scale with the brazing temperature Table 2 provides a summary of the brazing strains developed after cooling to 25°C in some of the systems of interest. Negative values correspond to combinations for which the advantageous convex bow on the base is developed. This indicates that N48 (48 Alloy) appears to be appropriate with both HIVOL™ B and C with low braze temperatures (220 C) while titanium appears to be well adapted to HIVOL™ B at both high and low braze temperatures. In addition, titanium is a system of interest in the light of its low density and low elastic modulus which leads to proportionately lower stresses in the brazed assembly.

An important inference is that the mismatch strains - and hence the shear stresses in the brazed joint - are small and can be affected by small variations in the volume fraction of the ceramic bed. Control of the thermal expansion through process control can therefore be used to good advantage.

Table 2 Mismatch strains with candidate ringframe materials

		braze T	36 Alloy	42 Alloy	48 Alloy	F-15	Titanium
mismatch%	C	380	1.35E-03	0.069	-0.044	0.098	-0.058
mismatch%	C	220	0.066	0.069	-0.037	0.048	-0.03067
mismatch%	B	380	0.041	0.11	-0.0050	0.137	-0.019
mismatch %	B	220	0.081	0.084	-0.022	0.063	-0.016

4.5 Bow Thin plates are prone to bowing owing to the the surface stresses imparted by grinding. Thin section is naturally more susceptible to this than thicker plate since the resistance to flexure varies with the cube of the thickness. Even with thin plate (0.5

mm) flatness ratios of better than 1% can be preserved with appropriate finishing processes and annealing treatments. Controlled bows may be specified for power semiconductor applications.

4.6 Surface Engineering Most applications for HIVOL™ require a metallization for functional reasons. HIVOL™ responds well to plating with electroless nickel - although plating processes have had to be developed to obtain good strike coats owing to the different activation requirements of the SiC and aluminium constituents. For optimum solderability electrolytic nickel is preferred as this passivates over much longer timescales than electroless nickel plate. Alternative presentations include electroless nickel followed by tin/lead (also to maintain solderability) , and gold for high reliability applications. Plating has to withstand bake tests commensurate with the maximum temperatures experienced in subsequent processing.

Alternatively , the HIVOL process can be adapted to offer a surface presentation of aluminium which can be anodised or given a chromate conversion treatment.

5. SUMMARY

1. High volume fraction metal matrix composite systems have powerful property combinations which can meet the requirements of electronics thermal management.
2. Thermal expansion values can be controlled relatively. New grades offering different expansion behaviour can be manufactured relatively easily.
3. Good non-destructive test techniques are required to validate the quality of production material in terms of volume fraction of SiC and porosity contents. The known techniques need to be calibrated with composites of known volume fraction of SiC and known porosity content. The development of assay techniques for this is of great importance.
4. All material for mechanical properties determination should be screened using the developed inspection techniques before measurements are made.
5. Cost effective manufacture of MMC components requires realistic tolerancing.
6. The successful engineering of MMC components requires good communication between designers and process engineers.
7. HIVOL™ materials have all the appropriate properties to meet all of these requirements.
8. Titanium ringframes appear to be well adapted to HIVOL™ B material in terms of good expansion matching, low stress development and low weight.

REFERENCES

1. Patent Specification: Metal Matrix Composites: HIVOL
   F9411024.4 (June 1994)

2. National Physical Laboratory Report DMM 0280, April 1994

3. R. Morrell , <u>Handbook of Properties of Technical and Engineering Ceramics,</u>(Part 1, An Introduction for the Engineer and Designer) HMSO 1989 p 85

4   João Quinta da Fonseca <u>Mechanical Properties of High Volume Fraction Metal Matrix Composites,</u> School of Materials, University of Leeds, May 1997

5. Private Communication, Societe Francaise de Ceramique,   November 1997

6. Private communication, Laurie Forrest, AEA Technology February 1998

ACKNOWLEDGEMENT

The support of the Corporate Level Funding Programme of AEA Technology and the UK Department of Trade and Industry (CARAD MEA Programme) is gratefully acknowledged.

TO FUSE OR NOT TO FUSE-THAT IS THE QUESTION

Chester H. Haynoski
Reflow Fusing Company
Van Nuys, California 91406
Sue Troup-Jones
Associated Plating Company
Santa Fe Springs, California 90607

ABSTRACT

The decade of the 90's has seen great changes for the aerospace and electronics industries. As requirements grow more stringent, the components have had to change to meet the new needs. This paper will look at BRIGHT TIN, ALKALINE TIN, and FUSED TIN and how these finishes can be properly used in aerospace and electronic applications. It will review the pros and cons of all three processes and point out where each finish can be used to its best advantage. The main objective of this paper is to help educate OEMs so that engineering and purchasing can make informed, technically correct decisions when they specify a process.

KEY WORDS: Electroplating, Tin Reflow, Tin Fusing

1. INTRODUCTION

The solderability of components is an important issue for companies manufacturing aerospace and electronic packages. It is important that the best coating is chosen to insure that parts will solder and stay soldered for the expected life of the product whether that life is five minutes or fifty years.

Various finishes can be applied to components that will make those components solderable. This paper will discuss the relative merits of bright tin, alkaline tin and fused tin.

2. SOLDERABILITY

Solderability is defined as the ability to join parts together using solder. The resultant "solder joint" shall be capable of meeting defined mechanical and electrical parameters. Anything that causes a solder joint to deviate from this definition is considered a failure.

Obviously, if the joint does not stay joined, it is a failure. However, a solder joint that looks fine after assembly can also be a failure. It can fail in shear test. It can fail electrically in assembly or in the field. These failures are usually caused by stress fracturing or build up of corrosion products.

Another aspect of solderability is shelf life or "how long can we keep these parts in the stockroom and still be able to solder them?" An expectation of <u>one year of shelf life</u> that is actually only <u>three months of shelf life</u> can cause many problems in production.

Still another issue in solderability is the formation of a copper/tin intermetallic layer. Tin is strongly cathodic to copper. This means that the tin will show a marked tendency to migrate into the copper. The resultant "intermetallic" (Cu_3Sn and Cu_6Sn5) is not solderable. This intermetallic is a result of time and temperature. Long storage times favor this formation. High temperatures accelerate the growth. Best available technology says that parts plated with .00015 (150 microinches) of tin which has produced an intermetallic thickness of less than .0001 (100 microinches) will not encounter problems. Parts having an intermetallic thickness of more than .000125 (125 microinches) will almost always have major problems with solderability.[1] This intermetallic condition becomes a major aspect when determining plating thicknesses, storage time, and temperature conditions.

The last issue that can cause problems with <u>soldered</u> parts is called whiskering, purple plague, purple pest, tin pest and other colorful euphemisms. This phenomenon, though not a result of soldering, can cause soldered parts to fail electrically because of shorting. It is a condition related to choosing tin plating as a finish.

3. BRIGHT ACID TIN

3.1 Attributes and positive aspects

Bright acid tin electroplating is a process where metallic tin is deposited on a substrate from an acid electrolyte. The tin is deposited in the +2 or stannous condition. This stannous condition can be achieved from a variety of chemistries including sulfate, fluoborate, and sulfonate. The resultant coating has a bright and shiny appearance. The coating is bright because of the codeposition of various organic additives that are in bright acid tin electrolytes. These additives deposit in the interstitials of the crystal causing the lattice to line up in very straight and even patterns in relation to the next crystals. This coating has good cosmetic value, resists fingerprinting and scratching, has good corrosion resistance, and is very solderable. It is also considered inexpensive especially compared to silver and gold.

3.2 Negative Aspects

Bright acid tin electroplated tin components have a very high field failure rate.[2] One of the reasons for this appears to be that the co-deposited organics can cause dewetting, bubbling or boiling of the coating. This happens because the organics are being released during soldering. This dewetting causes an incomplete solder joint. Deposits from acid solutions form loosely adherent acicular growths rather than smooth uniform coatings.[3, 4]

Another form of failure is called "wicking." Wicking occurs when a terminal that is being soldered is heated to the melting point of the solder. The bright acid tin separates from the copper or copper plated layer leaving a broken bond in the process. Once this separation occurs it cannot be reattached in its original form. It leaves voids. Many things can happen at this point. A loose bond causes poor electrical conductivity. Capillary action can trap foreign material. In addition, the soldering acids are covered (trapped) during the cooling process. This can lead to the formation of corrosion cells.[3, 4]

The most striking example of repeat failure of electronic devices is the phenomena called whiskering. These metallic dendrites cause subsequent electrical degradation due to the formation of leakage paths across insulator surfaces. This condition (also called MMRS-migrating metal resistive shorts) can occur with almost any metal including gold, silver, lead, copper, and tin. Historically tin has been the most notorious. The presently accepted explanation for whisker extrusion is that they are a stress relief mechanism.[5, 6]

The last negative aspect in this current discussion is the shelf life of bright acid tin plated components. The expected shelf life is 3 to 6 months for a tin thickness of .000050 (50 microinches). A thickness of .0001 (100 microinches) has a projected shelf life of one year.[2] It is important to know the shelf life of any finish before leaving parts in the stock room for a long time.

Additionally, bright acid tin will oxidize rapidly if exposed to high humidity, elevated temperature, and/or any slightly corrosive atmosphere such as salt air. These oxidation products are not very solderable. Proper procedure requires the use of soldering aids such as acid flux or acid core solder. The problem with using these aids is that acid entrapment can occur during the soldering process. Cleaning after soldering is effective only on the outer surfaces.[3] The internally trapped acid sets up a corrosion cell which can ultimately cause the joint to fail literally from the inside to the outside. The point that is being made here is that as an end user, you need to know not only how long you can leave the parts in the storeroom, but also the conditions that exist in that room.

4. ALKALINE TIN

4.1 Attributes and Positive Attributes

Alkaline tin electroplating is a process where metallic tin is deposited on a substrate from an alkaline electrolyte. The tin is deposited in the +4 or stannate condition. This stannate condition can be achieved with the addition of an appropriate hydroxide such as sodium or potassium hydroxide. The resultant coating has a dull, matte appearance. This coating is very soft and easily scratched and fingerprints readily. It does have good corrosion resistance, and is very solderable. It is also considered inexpensive especially compared to silver and gold. The electrolyte that is used for reflowing is alkaline tin.

4.2 Negative Attributes

Alkaline tin electrodeposits do not exhibit as many negative aspects as bright acid tin mostly because much of the plating is reflowed which removes the problems encountered with dewetting (no organic brighteners) and wicking. However, unreflowed alkaline tin is extremely subject to whiskers. It also plates much slower than acid tin (valance +4 rather than +2).

Alkaline tin's real strength lies in its ability to reflow, fuse or flow brighten.

5. REFLOW FUSING

5.1 Definition and Methodology

Before beginning any discussion about reflow fusing, there must be an understanding about the terminology. The process has been called FUSING,

REFLOWING, REFLOW FUSING, and FLOW BRIGHTENING. For the rest of this paper, the term that will be used will be REFLOW FUSING.

Reflow fusing is a unique hot process that changes the molecular structure of the tin plating. The alkaline tin plated part is raised to the melting temperature (or FLOWING temperature) of the tin. Immediately upon reaching this flow temperature, the part is quenched in a lower temperature fluid. This process causes the "flowing" tin to set up as a smooth, very shiny, nonporous, amorphous-like, continuous coating.

5.2 Benefits of choosing alkatine tin electroplating followed by reflow fusing

Conventional plating is fine grained and stressed with microscopic porosity. Reflow fusing seals the pores thereby making the coating more resistant to corrosion. Reflow fusing also stops whisker growth because the process removes the stress inherent with crystalline structures. By changing the crystalline structure to a large grain, amorphous-like structure, most of the stress in the coating is removed. Whisker growth is the result of the coating needing to be stress relieved. An amorphous-like coating is not as stressed, so whiskers do not form. It has long been recognized that tin whisker growth in electronic equipment can be virtually eliminated by using reflow fused tin.[2]

Reflow fusing also increases electrical conductivity by eliminating a lot of crystalline resistively. The slow cooling leaves the coating in a permanent compressive state. This compressive state allows better electron flow that increases conductivity. These structural changes also make the parts more solderable by providing a smooth, clean, tin-rich surface ready for soldering.

Reflowing is also a very good quality control tool. It is a purification process (similar to the technique of using a hot pack to draw infection to the surface of the skin) that brings the impurities to the surface of the part. This "drawing out" will expose any plating defects that are inherent in poor plating practices.

To insure consistent quality for plated parts it is advisable to reference industry accepted standards that may be used as a yardstick for performance and quality control. For bright tin, matte tin, and reflow fused tin the use of ASTM-B-545-92 is recommended. The specification recommended for solderability is MIL STD 202 Method 208.

Finally, when parts are alkaline tin electroplated followed by reflow fusing, the shelf life is greatly extended.[4] Because reflow fusing causes the small grained electrodeposited coating with large grain boundaries to change to large, amorphous-like grains with very little grain boundary, the coating becomes nearly pore-free. A pore-free coating is not as subject to oxidation that is the first stage of corrosion. The very bright reflow fused coating will oxide, but very slowly.

6.0 CONCLUSION

Reliable solderability along with excellent, long-term electrical characteristics and very good field corrosion resistance is the goal of all soldered assemblies. All of these goals can be met by choosing the proper coating for the parts prior to assembly. This paper has examined the pros and cons of tin plating. It has recommended that for superior solderability, excellent electrical characteristics, and very good corrosion resistance, ALKALINE TIN ELECTROPLATING followed by REFLOW FUSING is one of the very best processes that could be chosen to produce a superior, long lasting, field reliable assembly.

7. REFERENCES

1. A. Audelo, "Failure Mechanism of Tin Plating in the Solder Industry", The American Welding Society, Second International Soldering Conference, 1973
2. W.G. Bader and R.G. Baker, "Solderability of Electrodeposited Solder and Tin Coatings After Extended Storage", Bell Telephone Laboratories, Murray Hill, NJ
3. R.E. Horn, "An Overview of Tin and Tin Alloy Plating Particularly as Applied to the Electronic Industry", CARTS, 1992 pp 71 and 74 Solderability
4. C.H. Haynoski, "Reflow Fusing for Small Electronic Parts Tin or Tin-Lead Plated to Enhance Solderability and Shelf Life", Reflow Fusing Company, Van Nuys, CA
5. J.R. Devaney, "Corrosion and Dendrite Growth and Other Metallurgical Phenomena in Microelectronic Packages", ASM's 3rd Conference on Electronic Packaging (1987)
6. S.C. Britton, "Spontaneous Growth of Whiskers on Tin Coatings: 20 Years of Observation", Tin Research Institute, 1974

AN ULTRA LOW MOISTURE POLYMER ADHESIVE FOR MICROELECTRONIC PACKAGING

My N. Nguyen, Carl S. Edwards and Irving Y. Chien
Johnson Matthey Electronics
10080 Willow Creek Road
San Diego, CA 92131

ABSTRACT

A new type of paste adhesive based on a modified olefin thermoset polymer has been developed. Due to its hydrophobic molecular structure, moisture absorption is inherently very low. For example, it absorbs less than 0.05% moisture @ 85°C/85% RH/168 hrs. Even in HAST test (150°C/85% RH/168 hrs.), moisture content is only 0.1%. Commercial epoxy adhesives absorb about 0.5-2.0% under similar test conditions. The excess moisture leads to package cracking/delamination during solder reflow process The olefinic polymer was formulated as a low stress, electrically and thermally conducting paste adhesive. Its other principal features include excellent adhesion, good processability, fast cure time and low temperature cure. Typical cure times are 5-30 minutes at temperatures of 100-150 °C. This material appears to have the flexibility to address a broad range of applications such as die attach, underfill, thermal interface materials for laminate packages, flexible circuits. It will provide significant cycle time reduction and reliability improvement for advanced device packaging.

1. INTRODUCTION

Interfacial delamination and cracking of plastic packages during PCB mounting continue to be a serious assembly problem as VLSI packages become thinner and die sizes increase. The industry standard specifications JEDEC/IPC[1] outline experimental procedures for rating the moisture sensitivity levels of plastic packaged integrated circuits. There are six levels of sensitivity ranging from moisture insensitive (Level 1) to extremely sensitive (Level 6). Levels 1 and 2 classified packages are highly desirable because they are more robust and have very little assembly risk. Unfortunately, the new higher pin-count products are most often housed in moisture sensitive packages such as PBGA, TQFP, etc. (levels 3-6) where assembly and reliability risks are of major concern.

The root cause for the "popcorning " problem is well-established and is due to the moisture absorbed by the polymer packaging materials. This moisture vaporizes at the high solder reflow temperatures (220-240°C), which induces high steam pressure throughout the package.

Figure 1 illustrates the corresponding steam pressure generated during reflow as a function of moisture concentration[2,3]. There is a critical moisture concentration above which the hygro-thermal stresses become greater than the interfacial fracture toughness of the mold compound and/or the die-attach adhesive. Delamination areas would then be initiated[4]. This threshold moisture level has been determined empirically, using moisture weight gain data for various types of packages[4,5]. It appears that if the moisture concentration is less than 0.1% by weight, the package would be free of delaminations after solder reflow.

The moisture content of commercially available epoxy based molding compounds and die attach materials after Level 1 preconditioning (1 week at 85°C/85%RH) are shown in Figure 2. At the high moisture levels of 0.3-0.5% for EMC, and 1-2% for die attach, it is clear that the current materials are not capable of producing a moisture insensitive package.

This paper describes a new type of low moisture thermoset adhesive that potentially can provide solution to "popcorning". We will examine the material structure and properties, and its performance in various microelectronic packaging applications.

2. MATERIAL CHEMISTRY

The new material is based on a modified cycloolefin thermoset (MCOT) system[6,7]. The prepolymer is a liquid which cures thermally to form MCOT polymer having the general structure shown in Figure 3.

Figure 4 compares its moisture absorption characteristics against epoxy and cyanate ester. Since MCOT polymer is essentially comprised of non polar olefins, moisture absorption is significantly lower when compared to epoxies.

3. MATERIAL CHARACTERIZATION

The MCOT adhesive consists essentially of thermally and/or electrically conducting fillers dispersed in liquid prepolymers. Upon heating, the prepolymers are combined via additional polymerization to form a thermoset network. Since there are no outgassed by-products formed by this reaction, the weight loss during cure is very low[7]. The typical properties of the two MCOT based products (JM7100 and JM7200) are given in Figure 5. Adhesive strength for some common substrate materials are shown in Figure 6.

In Figure 7, we examined the effect of moisture stresses vs. adhesion strength. Even in a very severe test, (150°C/100%RH/5atm for 168 hours in the Parr bomb), there is no degradation in adhesive strength, and moisture absorption remains low. This lack of moisture susceptability is a unique feature of the olefin based adhesive. Although its adhesive strength is less than that of epoxies, its moisture induced stresses is also much lower. If the die bond strength is sufficient to withstand the stresses induced by subsequent assembly steps (e.g. wire bonding, molding, etc.) without inducing delamination, the olefin based die attach is likely to perform well in package moisture preconditioning tests.

4. MCOT ADHESIVE APPLICATIONS

This material was developed to fill an industry need for an ultra low moisture and low stress adhesive combined. The unique combination of the above properties with low temperature curing, fast curing and long pot life allow for a broad range of applications. Table I hightlights some of the potential applications along with the desirable attributes for the given application.

4.1 MCOT Adhesive Cure Process The current adhesive cure process for polymer laminate packages is quite complicated. It usually requires bakeout of the laminate substrate to remove absorbed moisture and the adhesive must be slowly cured over several hours to ensure void-free bonding for high reliability. As a result, this entire process is not only lengthy, but is also difficult to control.

Most of the voids are caused by vaporization of surface moisture on the organic substrate during the cure process. They are created by the outgassed moisture which becomes permanently entrapped as the adhesive is setting up. Due to the low moisture and low temperature, fast cure properties of the olefin based adhesive, it is possible to eliminate the substrate bakeout step and to reduce cure times to about 5 minutes.

Figure 8 showed a typical fast cure temperature profile using heat tunnel equipment manufactured by ASM.

4.2 Die Attach Die attach performance of the MCOT adhesive has been reported in previous publications[7,8]. The material was evaluated for BGA and QFP packages, and passed all reliability tests. Its principal attributes are no die attach delamination, reduced package warpage and reduced cure cycle.

4.3 Flexible Circuit Board Flexible circuit subtrates – there are two major types of flex circuit substrates: Polyimide and Polyester. Polyimide has excellent high temperature stability, chemical resistance and electrical properties and is the dominant material for flexible circuit use and has been used extensively for high reliability applications. For all its strengths, there are some weaknesses, including high moisture aborption and high cost. On the other hand, Polyester is a low cost, moisture stable material that requires low temperature processing (<120° C). this material is popular for many heat form applications such as membrane switches. Shortcomings for polyester include, limited use with solder applications and dimensional stability. [9] Together, both materials cover a broad range of applications within the flexible circuit industry. For both substrate materials, polyolefin is a good match for flip-chip underfill and die attach applications. Here, the low stress, low moisture and the low temperature/fast cure materials provide a good match for flexible packaging systems that include not only flex circuits but also Tape BGA and Chip-scale packaging.

Figure 9 shows the typical cure time/temperature relationship for this material. Precise onset of the cure mechanism at 70° C allows for long pot life at room temperature and very rapid cure above the onset temperature. Figure 10 shows DSC curves on uncured material, and on material that has been cured at 150° C for 3 minutes and 10 minutes. There is no significant residual exotherm peak over the 70°C to 150°C temperature range, indicating the material is almost fully cured under these conditions.

Adhesion reliability performance on Kapton film is shown in Figure 11. There are no significant change in adhesion after 1000 thermal cycles, -65° C to 150°C or after HAST, 500 hours @ 130 °C/85 %RH

4.4 Thermal Interface Adhesive Among a wide variety of interface materials such as elastomer tapes and phase change materials, thermal grease is the most commonly used. Thermal greases still give the lowest thermal resistance[10, 11] perhaps due to its ability to spread in very thin layers. Typical thermal impedence values range between 0.6-1.6 $°C-cm^2/W$. The main drawback for grease is its thermal performance deteriorates significantly after thermal cycling. In effect, the grease is squeezed out, leaving behind some gaps at the interface.

The MCOT adhesive, due to low compliance property, is a good candidate for thermal interface. Preliminary evaluations employing olefin based interface materials are showing capability of <0.5 $°C-cm^2/W$. For this type of material, conductive fillers such as Boron Nitride or silver can be selected to provide thermal conductivity of more than 2W/m °C. Unlike thermal grease, thermal performance of olefin adhesive will not degrade after thermal cycling or power cycling of IC devices.

5. CONCLUSIONS

The new olefin based die attach adhesive has shown great potential for advanced IC packaging. Since its unique chemistry combines very low moisture absorption and adequate adhesive strength, as well as the ease of processing, this material is very robust and has the flexibility to address a broad range of applications. It may promise to play an important part of the total material solution for many contemporary problems in polymer based packages.

6. REFERENCES

1. JEDEC Test Method A-113, "Preconditioning of Plastic Surface Mount Devices prior to reliability testing", 1995.

2. K. Sawada, T. Nakazawa, N. Nawamura, and T. Sudo, "Package deformation and Cracking Mechanism due to Reflow Soldering", 1993 Proc. Japan Int. Electron. Manuf. Tech. Symp., p.295-298.

3. B. Bhattacharyya, W. Huffman, W. Jahsman, and B. Natarajan, "Moisture absorption and mechanical performance of Surface Mount Plastic Packages", 1988 Proc. Electron. Comp. Conf., p.49-58.

4. L. Nguyen, K. Chen, J. Schaefer, A. Kuo, and A. Slenski, "A new Criterion for Package Integrity under Solder Reflow conditions", 1995 Proc. Electron. Comp. Tech. Conf, p.478-490.

5. B. Poborets, Q. Ilyas, M. Potter, and J. Argyle, "Reliability and Moisture Sensitivity evaluation of 225-pin Overmolded BGA Package", 1995 Proc. Electron. Comp. Tech. Conf, p.434-439.

6. M. Nguyen, U.S. Patent Pending.

7. M. Nguyen, "Development of an Ultra Low Moisture Polymer Adhesive for Die Attach Applications", 1997 Proc. Int. Electron. Manuf. Tech. Symp., Austin, Texas.

8. M. Nguyen, I. Chien, "An Ultralow Moisture Polymer Adhesive for BGA Packages". Proceedings of the 1997 ISPS Conference, San Diego, CA, p.119.

9. J. Fjelstad, "An engineer's guide to flexible circuit technology", Electrochemical Publications, LTD., Bristol, England, 1997.

10. Z. Celick, D. Copeland, A. Mertal, "Thermal Enhancement and Reliability of 40 mm EPBGA Packages with Interface Materials", 1997 IEEE/CPMT, Symposium Proceedings, Austin, Texas, p. 376-385.

11. T. Dolbean and T. Tarter, "Thermal Management of Pin Grid Array Packaging for Flip Chip connected X86 Microprocessors", 1997 ISPS Proceedings, San Diego, CA, p. 279-284.

7. ACKNOWLEDGEMENTS

The authors wish to thank the staff at Johnson Matthey San Diego with particular thanks to Kevin Taylor for preparation of laboratory data.

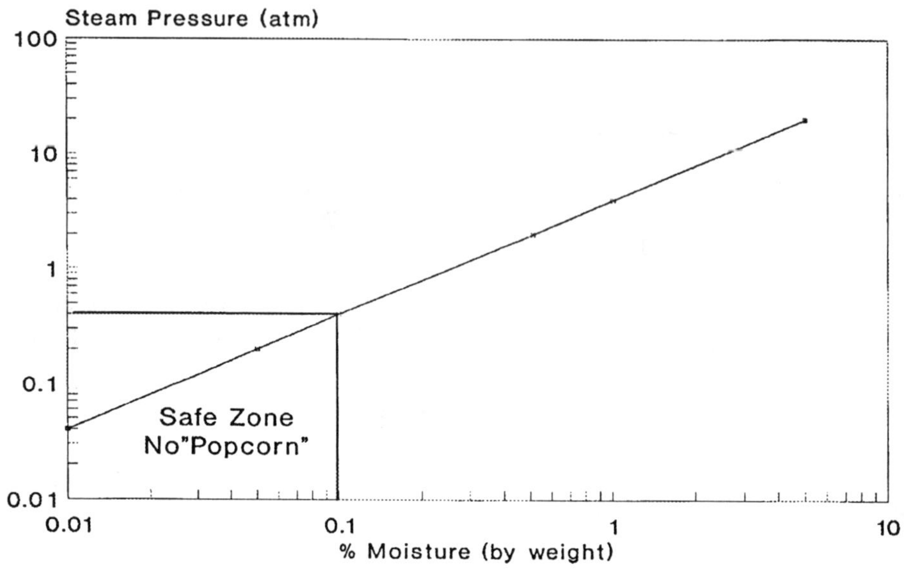

Figure 1. Effect of moisture concentration of epoxy

MATERIALS	% ABSORBED H_2O (*)
Encapsulant/EMC/Laminates	0.3 - 0.5
Die Attach	1 - 2
JEDEC Moisture Classification	3 - 4

(*) 85%/R/H/85°C/168hrs.
Excess Moisture Content → Popcorning in Polymer Packages

Figure 2. Moisture sensitivity of commercial packaging materials.

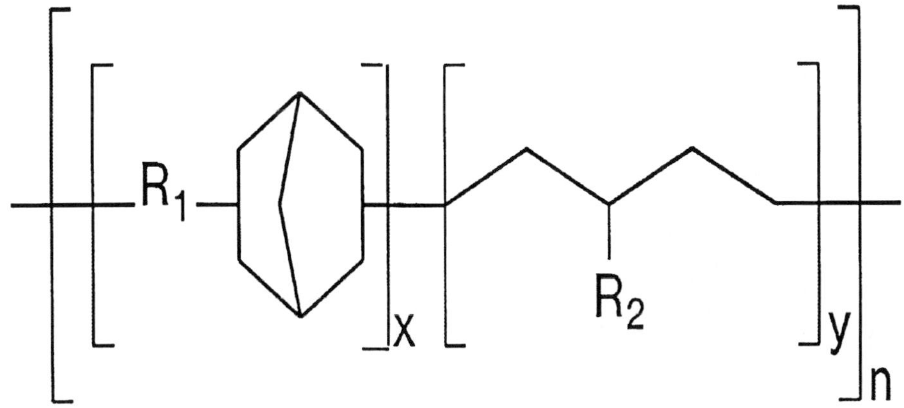

Figure 3. General chemical structure of MCOT adhesive.

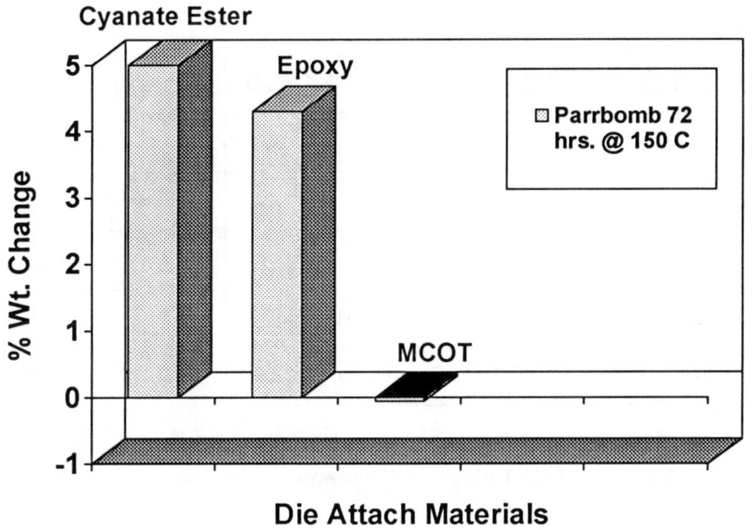

Figure 4. Moisture absorption characteriatics of MCOT adhesive.

Description	JM 7200	JM 7100
% Filler	40% BN	80% Ag
Viscosity, Pa-s	10	10
Cure Temp	100-150° C	100-150° C
Cure Time	<30 min.	<30 min.
Pot Life @ 25° C	16 hrs.	16 hrs.
Shelf Life @ -40° C	>6 months	>6 months
Extractable Ions (Na, Cl, K. F)	<10 ppm	<10 ppm
% Weight Loss (up to 300 C)	<1	<1

Cured Properties	Thermal Conducting Paste	Electrical Conducting Paste
CTE (20-200° C)	85 ppm	78 ppm
Modulus @ 25° C	1000 Mpa	1000 Mpa
Modulus @ 200° C	150 Mpa	165 Mpa
Glass Transition	65° C	65° C
Bulk Thermal Conductivity (watts/m° C)	1.0	2.3
Volume Resistivity (ohm.cm)	NA	1.5 x 10^{-4}

Figure 5. Typical properties of MCOT based adhesives (JM7100 and JM7200).

Cure Temperature (° C)	100	125	150	175
Cure Time (min.)	2	2	2	2
% Cure	44	55	65	78
% Water Absorption (*)	0.13	0.12	0.13	0.12
Adhesion Strength After Cure (**) (psi)	625	750	860	1030
Adhesion Strength After Parr Bomb (psi)	2450	2820	2650	2540

(*)   150° C / 100% RH / 168 hrs. Parr Bomb Test
(**)   Die Shear Test 80 x 80 mils Si on Copper Substrate

Figure 6. Effect of moisture stresses on adhesive strength.

Substrate Type	Adhesive Strength @ 25°C (kg)	Adhesive Strength @ 185°C (kg)
Bare Copper	28	7
Ni Plated Copper	30	9
Ag Plated Copper	30	7
Black Oxide Copper	27	8
Anodized Aluminum	28	8
Solder Mask BT	22	4
Au Plated BT	25	4
Kapton	24	5

Die size 0.2 x 0.2 sq. in.

Figure 7. Adhesive strength on various substrates.

Set Temperature (°C)	Time (sec)
80	40
85	80
100	40
125	40
150	80

Total Cure Time = 280 sec.

Figure 8. A typical fast cure temperature profile for MCOT adhesives.

Cure Temperature (^{o}C)	Min. Time in Minutes
150	5
130	10
100	30
85	60
65	120

Figure 9. Typical cure schedule for olefin adhesives.

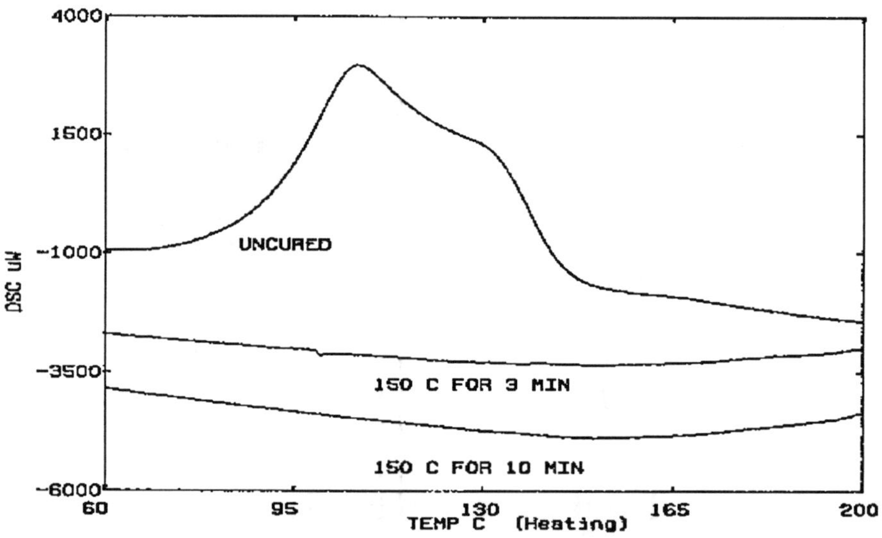

Figure 10. Cure characteristics of MCOT adhesive

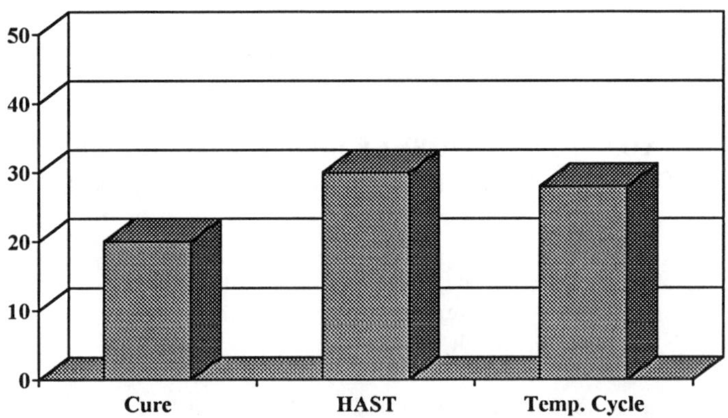

Figure 11 Effect of environmental testing on adhesion

DEGRADATION OF HUBBLE SPACE TELESCOPE METALLIZED TEFLON® FEP THERMAL CONTROL MATERIALS

Patricia A. Hansen, Jacqueline A. Townsend
NASA/Goddard Space Flight Center
Greenbelt, Maryland 20771

Yukio Yoshikawa, J. David Castro
Lockheed Martin Technical Operations
Greenbelt, Maryland 20770

Jack J. Triolo and Wanda C. Peters
Swales Aerospace
Beltsville, Maryland 20705

ABSTRACT

The mechanical and optical properties of the metallized Teflon® Fluorinated Ethylene Propylene (FEP) thermal control materials on the Hubble Space Telescope (HST) have degraded over the seven years the telescope has been in orbit. Astronaut observations and photographic documentation from the Second Servicing Mission revealed severe cracks of the multi-layer insulation (MLI) blanket outer layer in many locations around the telescope, particularly on solar facing surfaces.

Two samples, the outer Teflon® FEP MLI layer and radiator surfaces, were characterized post-mission through exhaustive mechanical, thermal, chemical, and optical testing. The observed damage to the thermal control materials, the sample retrieval and handling, and the significant changes to the radiator surfaces of HST will be discussed. Each of these issues is addressed with respect to current and future mission requirements.

KEY WORDS: Hubble Space Telescope, Multi-Layer Insulation, Teflon® Fluorinated Ethylene Propylene (FEP)

1. INTRODUCTION

The Hubble Space Telescope (HST) was launched in April 1990 and deployed at an orbital altitude of 598 km (320 nmi) and 28.5° orbit inclination. The Telescope's mission is to spend 15 years probing the farthest and faintest reaches of the cosmos. Crucial to fulfilling this promise is a series of on-orbit manned servicing missions to upgrade scientific capabilities. The First Servicing Mission (SM1) took place in December 1993, 3.6 years after deployment. The Second Servicing Mission (SM2) followed in February 1997, 6.8 years after deployment. Two subsequent servicing missions are also scheduled for late 1999 and early 2002.

Astronaut observations and photographic documentation from the SM2 revealed severe cracks of the multi-layer insulation (MLI) blanket outer layer in many locations around the telescope, particularly on solar facing surfaces. Two samples were retrieved during the SM2, the outer Teflon® FEP MLI layer and radiator surface, and were characterized post-mission through exhaustive mechanical, thermal, chemical, and optical testing. This paper details the observed damage to the thermal control materials, the sample retrieval and handling, and the significant changes to the radiator surfaces of HST. Each of these issues is addressed with respect to current and future mission requirements.

2. HST DESCRIPTION

As shown in Figure 1, the HST houses the five scientific instruments (SIs) and three Fine Guidance Sensors in the Aft Shroud (AS). The Support Systems Module (SSM) Bays, located directly above the AS, house the batteries, tape recorders, and house keeping equipment. The Forward Shell (FS) and Light Shield (LS) is the structure enclosing the primary and secondary mirror and stray light rejection baffles. The Aperture Door (AD) is closed only during servicing periods or safe hold modes to protect the primary and secondary mirrors from direct sun impingement and contamination.

The HST uses several thermal control materials to passively control temperatures on-orbit. These materials are of two primary types: radiators and MLI blankets. As shown in Figure 2, the radiator surfaces are the AS, Aft Bulkhead, AD, and SSM Bays and louvers on the SMM Bays. The radiator surfaces are 5 mil Teflon® FEP/Ag/Inconel bonded to the vehicle substrate (aluminum) with acrylic adhesive. All other surfaces are blanketed.

3. HST ON-ORBIT ENVIRONMENT

The HST thermal control system (TCS) is exposed to the low Earth orbit (LEO) environment which includes solar radiation, particle radiation and atomic oxygen. Solar exposure, including near ultraviolet radiation, vacuum ultraviolet radiation, and soft x-rays from solar flares, may cause surface damage in polymeric materials such as Teflon® FEP.

Trapped electron and proton particle radiation may cause mechanical or chemical property changes in the bulk of the polymeric material. Atomic oxygen can erode polymeric materials such as Teflon® FEP through chemical reactions with gaseous oxide products.

Table 1 summarizes the HST environmental exposures experienced by the HST samples. The fluences shown in Table 1 do not take into account scattering of atomic oxygen or solar radiation off of other surfaces on the telescope.

Table 1. HST Sample Environmental Fluences

Sample	Equivalent sun hr (ESH)	X-ray fluence (J/m^2)	Trapped electrons and proton fluence > 40 keV ($\#/cm^2$)	Plasma fluence ($\#/cm^2$)	Atomic Oxygen (atoms/cm^2)
SM2 MLI	33,638	0.5-4Å: 16 1-8Å: 252.4	electrons: 2.14×10^{13}	electrons: 4.66×10^{19}	1.64×10^{20}
SM2 CVC	19,308	0.5-4Å: 6.1 1-8Å: 96.9	protons: 1.83×10^{10}	protons: 1.63×10^{19}	

4. THERMAL SUBSYSTEM DESCRIPTION

The HST TCS is designed to control temperatures of SSM components and the structure that interfaces with the Optical Telescope Assembly (OTA) and SIs. The TCS provides thermal control for all SSM equipment during all mission phases and is passive to the maximum extent possible. The SSM uses passive thermal control design consisting of insulation, component arrangement, and mounting configurations augmented with thermostatically controlled heaters. Louver systems are used for the battery bays.

4.1 MLI Blankets MLI blankets are used on over 80 percent of the external surface area of HST. These blankets have a FOSR (Flexible Optical Solar Reflector) outer layer (5 mil aluminized Teflon® FEP) which is reflective to energy in the solar spectrum to minimize the effect of solar orientation and day/night cycling on spacecraft temperatures. The MLI blankets insulate the structure from the external thermal environment, thus conserving heater power and minimizing temperature extremes and gradients. The baseline MLI design is 15 layers of embossed Double-aluminized Kapton® (DAK) with an outer layer of 5 mil FOSR and an inner layer of 1 mil single Aluminized Kapton® (SAK). The surface properties of this outer layer control the external blanket maximum temperatures to 20°C or less. No spacers are used between the layers since the embossing pattern reduces layer-to-layer conduction and meets the Orbiter flammability requirements.

The blankets are closed out on all four sides with a taped cap section and the layers are tied together throughout the blanket using a pattern of 10 mil acrylic transfer adhesive film. Where the blankets were cut to fit around stantions, handrails, portable foot restraint sockets, etc., the blanket was closed out by taping a cap section using 10 mil acrylic transfer adhesive film. In addition, where the blankets were vented ("X" cuts), the outer layer was reinforced using aluminized Kapton® scrim tape. Velcro® was stitched to the internal layer, which was reinforced with aluminized Kapton® scrim tape. To indicate fold lines on the SSM Bay doors blankets, the blankets were stitched through all the blanket layers.

4.2 Radiators   The radiator surfaces are perforated 5 mil Silver FOSR bonded with acrylic adhesive to the (aluminum) substrate. This material was purchased in 4-inch (width) rolls with the adhesive already applied. The FOSR was applied in sections and a Teflon® wand was used to minimize air entrapment between the FOSR and the substrate. Damaged FOSR was replaced as required during the buildup of the telescope. The AD external surface is 5 mil Aluminized FOSR.

5. SM2 OBSERVATIONS

One of the objectives during the SM2 was to photo document the condition of the telescope external surfaces. During EVA (extravehicular activity) periods, photography of the telescope was performed highlighting damaged MLI areas as requested. During Crew sleep periods, the telescope was systematically photographed.

5.1 On-Orbit   During the HST SM2, damage of the MLI was observed on the +V3 side (sun side) with several large cracks in the light shield MLI outer layer (above the stowed High Gain Antenna). This large crack is shown in Figure 3. Upon further visual observations of the vehicle, additional cracks were seen on all MLI surfaces. Cracking of the MLI outer layer was seen all over the telescope, with the most damage seen on the LS, SSM Bay 8, SSM Bay 7, and SSM Bay 10. The SSM Bay 8 is shown in Figure 4; the MLI is cracked around most of the bay perimeter, and the MLI appears as though it is partially detached and lifted away from the telescope.

Prior to patching the corners of the SSM Bay 8 MLI, the astronauts performed two tests: a Velcro® cycling and Teflon® FEP bend test. The Velcro®, attaching the MLI blanket to the bay, was cycled to determine its integrity – was it still attached to the telescope substrate and did the hook still hold the pile securely? The astronauts reported that the Velcro® appeared to be securely fastened to the vehicle substrate and the hook seemed to hold the pile securely. The astronauts also bent a piece of the Teflon® FEP over on itself to determine if manipulating it during the patching process would cause significant damage. The astronauts reported that the Teflon® FEP did not crack.

Although the damage to the MLI was visually dramatic, there was no measurable thermal effect due to the cracks and the exposure of the DAK to the telescope environment. However, during the SM2 mission, a program decision was made to reconfigure flown MLI patches to patch the worst of these damaged areas.

5.2 Damage Map   Post mission analysis of the photographic documentation indicated that the MLI outer layer was cracked extensively on all MLI surfaces. The cracks were noticeably larger and more numerous on the sun side (+V3) of the telescope. The cracks in the FEP Teflon on the anti-sun side (-V3) of the telescope tended to start and end at stress concentration points. Whereas, the cracks on the sun side (+V3) of the telescope start at stress concentration points and either end where two cracks meet or in the middle of an MLI blanket.

Figure 5 shows a map of the MLI damage, using SM2 photo documentation (Reference 3). This map depicts the damage and notes where the cracks have opened on the LS and SSM bays. The location of the features on the map is approximate and shown for illustration purposes only. The most important item to note is that the damage is seen all over the vehicle and is noticeably worse on the sun side (+V3).

6. SAMPLE DESCRIPTION

6.1 MLI Sample   An MLI sample, shown in Figure 6, was taken from the upper LS crack prior to patching the area. The LS MLI cracks are shown with patches installed in Figure 7. The sample was cut right to left with a change in the initial direction of the cut as the astronauts realized they were cutting through the sample. The sample is shown flat in Figure 8; the astronaut cuts, on-orbit cracks and handling cracks are identified. The sample was handled carefully and stored in an EVA trash bag for the duration of the EVA. The trash bag was transferred to the crew cabin and the sample was placed in a reclosable polyethylene bag and stowed in a middeck locker for the duration of the mission. The sample was requested as an early destow item and was turned over to the HST project within 6 hours of the Orbiter landing.

The sample was transferred to a poylcarbonate container and transported via over night courier service to the Goddard Space Flight Center (GSFC). Upon arrival at GSFC, a small portion of the sample was photo documented. The sample was then stored in a laboratory until testing.

6.2 Radiator Sample   As part of the installation of the Near Infrared Camera Multi-Object Spectrometer, a cryogen vent line was routed through the Aft Bulkhead to expel Nitrogen boil off from the instrument. As a result, the Cryogen Vent Cover (CVC) was removed and returned to Earth at project request. The outside of the CVC (radiator surface) had been exposed to the HST ambient environment and provided a good data point for the thermal degradation of the radiator surfaces. This Aft Bulkhead (-V1) surface was exposed to direct solar incident radiation, however, it was significantly less than the MLI sample from the LS. The telescope sun angle history has been compiled and is shown in Table 1 as equivalent solar hours (ESH).

7. DISCUSSION

7.1 MLI Sample   The MLI sample was characterized through exhaustive mechanical, thermal, chemical, and optical testing. The testing revealed several changes to the mechanical, chemical and thermal properties of the MLI. Details of these tests may be found in References 4-7. The on-orbit cracks are nearly featureless indicating that they occurred very slowly under low load and in the presence of a damaging environmental factor. This crack mechanism is similar to stress-corrosion cracking in metals and slow crack growth in ceramics. The material shows significant reduction in ultimate strength and elongation, indicating through-thickness embrittlement of the Teflon® FEP. The surface of the Teflon® FEP is de-fluorinated from interaction with the orbital environment. Increases in the solar absorptance of the Teflon® FEP correlate well with the number of days spent in space, rather than with equivalent solar hours. The aluminum backing is cracked in a "mud tiling" pattern due to the difference in thermal expansion of the Teflon® FEP and the aluminum and the 40,000 thermal cycles HST experienced. It is expected that the condition of the MLI will continue to degrade, and that it will be fragile by the next servicing mission.

7.2 Radiator Sample   The radiator sample was characterized through exhaustive mechanical, thermal, chemical, and optical testing. Testing of the CVC sample showed that the solar absorptance of the bonded silver Teflon® FEP on the AS of HST has increased. Flight temperature data plots of the maximum and minimum temperatures over several years show an increase of the maximum temperatures to indicate a solar absorptance increase whereas the minimum temperatures remained constant to indicate no emittance change (or no Teflon® FEP thickness decrease). As with the MLI, the Teflon® FEP bonded to the aft shroud is embrittled, and the silver backing is extensively cracked. The application process used (the angle of application and rubbing to insure adhesion), has been shown to induce cracking in the metal layer. The acrylic adhesive can bleed through the cracks and then darken in the presence of UV and VUV, thus increasing the absorptance of the film. Post mission ground testing has shown that removing the release strip from the Teflon® FEP tape caused a significant amount of cracking of the silver even before application to the substrate.

Additional exposure of the CVC sample to ultraviolet radiation showed that the solar absorptance did not continue to degrade. However, newly bonded 5 mil Teflon® FEP, exposed at the same time to ultraviolet radiation experienced a significant increase in solar absorptance during the first 2000 hours of exposure. The newly bonded Teflon® FEP was bonded using both standard techniques and a "gentle" technique to minimize the cracking of the silver. The standard technique produced the largest increase in the solar absorptance; however, the "gentle" technique still produced a significant absorptance increase. This indicated that the techniques used to bond the Teflon® FEP to the substrate causes the silver to break in the application process and thus allows the adhesive to darken under exposure to ultraviolet radiation. Using a Teflon® wand to smooth the Teflon® FEP also caused localized absorptance increases. The expected degradation of the HST TCS (increase in solar absorptance) is given in Figure 9 with the CVC sample degradation indicated.

8. FUTURE REQUIREMENTS

To maintain the operational capability during HST's 15-year mission, the TCS must maintain the HST operational temperatures during all sun pointing attitudes. To continue to meet this requirement, candidate replacement materials have been selected to repair and/or replace degraded MLI during subsequent servicing missions. References 8-9, detail the evaluation of the candidate materials and the current HST metallized 5 mil Teflon® FEP. Additional testing of 5 mil aluminized Teflon® FEP will be on going to predict the expected condition of the Teflon® FEP during the future servicing missions.

9. ACKNOWLEDGEMENTS

As with any large test program, many people, behind the scenes, need to be acknowledged for their contribution: Diane Kolos and Robert Gorman for their excellent photo documentation, Teri Gregory and Janet Barth for their analyses. In addition, the authors would like to gratefully acknowledge the contributions of the entire HST MLI Failure Review Board.

10. REFERENCES

1. *Hubble Space Telescope Thermal Control System Description and Operating Manual*, LMSC/F420173A, 15 October 1991.
2. *HST 2ⁿᵈ Servicing Mission Media Reference Guide*, February 1997.
3. *Hubble Space Telescope SM2 Multi-Layer Insulation (MLI) Damage Assessment*, NASA Document JSC-27943, August 7, 1997.
4. J.A. Dever, K.K. de Groh, J.A. Townsend, L.L. Wang, "Mechanical Properties Degradation of Teflon® FEP Returned From the Hubble Space Telescope", NASA/TM-1998-206618.
5. J.A. Townsend, P.A. Hansen, J.A. Dever, J.J. Triolo, "Analysis of Retrieved Hubble Space Telescope Thermal Control Materials", Science of Advanced Materials and Process Engineering Series, 43, 000 (1998).
6. C. He and J.A. Townsend, "Solar absorptance of the Teflon® FEP Samples Returned from the HST Servicing Missions", Science of Advanced Materials and Process Engineering Series, 43, 000 (1998).
7. L. Wang and M. Viens, "Fractography of MLI Teflon® FEP From the HST Second Servicing Mission", Science of Advanced Materials and Process Engineering Series, 43, 000 (1998).
8. J.A. Dever, J.A. Townsend, J.R. Gaier, A.I. Jalics, "Synchrotron VUV and Soft X-Ray Radiation Effects on Aluminized Teflon® FEP", Science of Advanced Materials and Process Engineering Series, 43, 000 (1998).
9. J.A. Townsend, P.A. Hansen, J.A. Dever, J.J. Triolo, "Evaluation and Selection of Replacement Thermal Control Materials for the Hubble Space Telescope", Science of Advanced Materials and Process Engineering Series, 43, 000 (1998).

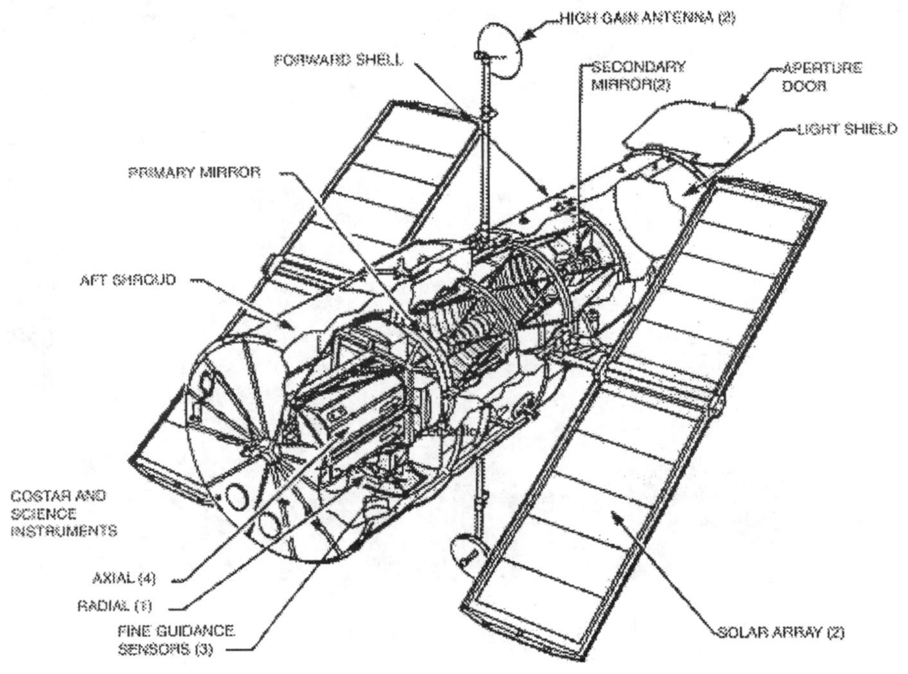

Figure 1: HST Configuration

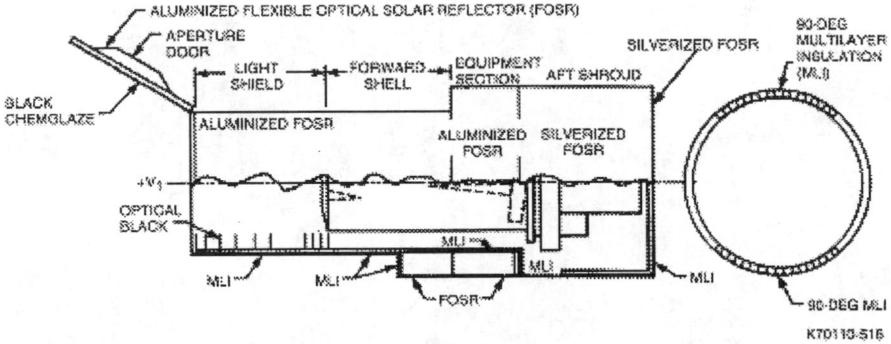

Figure 2: HST Thermal Control Subsystem Materials Location Map

Figure 3: Light Shield MLI Cracks

Figure 4: SSM Bay 8 MLI Crack

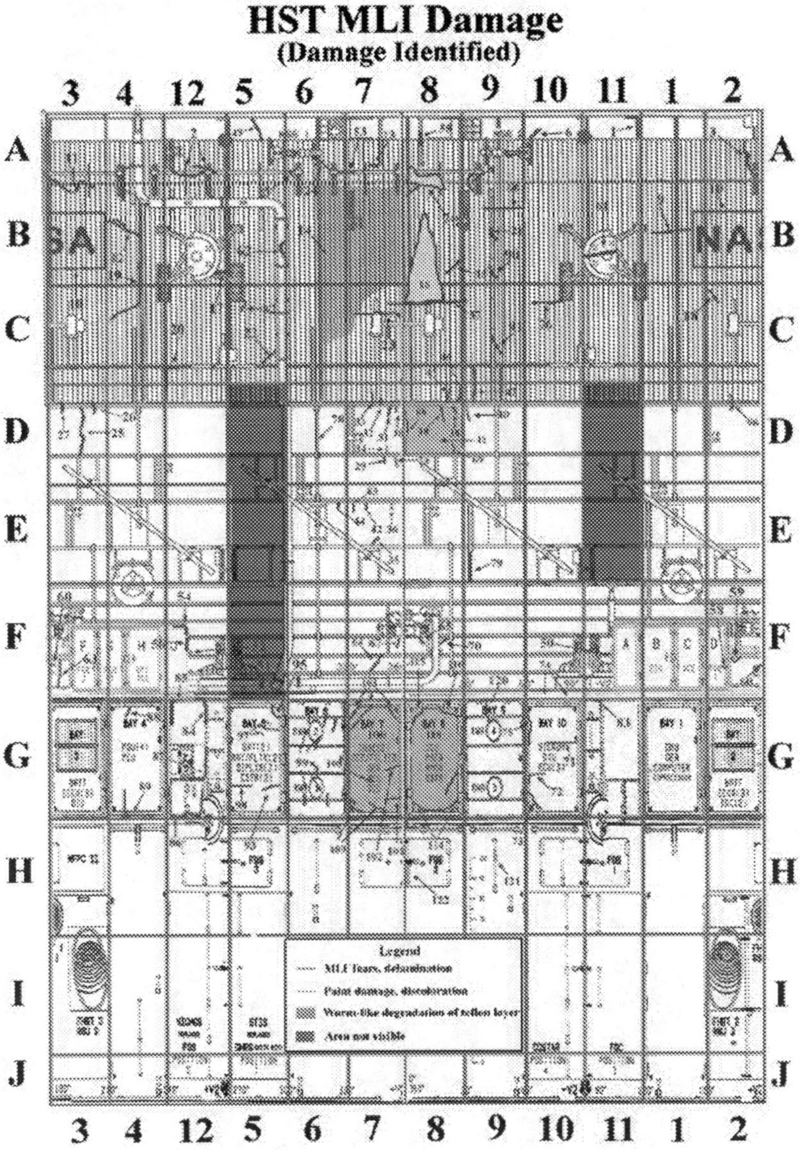

Figure 5: HST Thermal Control Subsystem Degradation Map

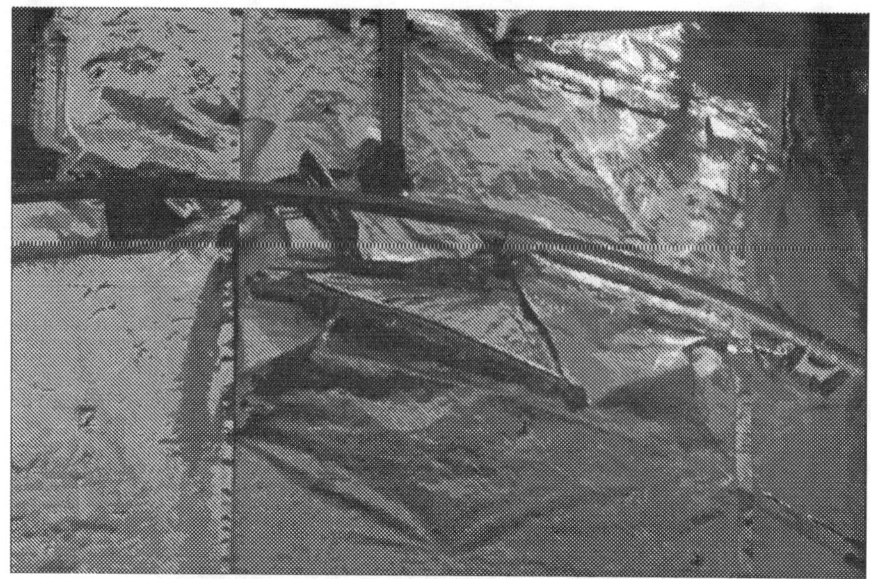

Figure 6: Upper Light Shield MLI Crack (SM2 Sample)

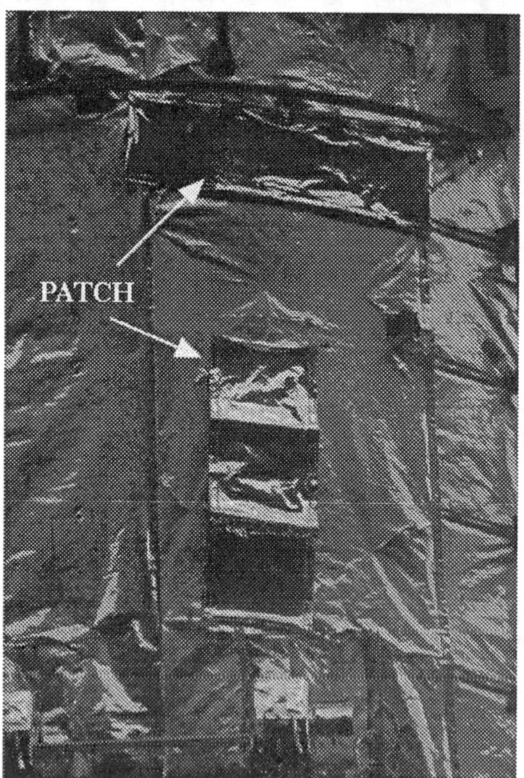

Figure 7: HST Light Shield MLI Cracks with Patches Installed

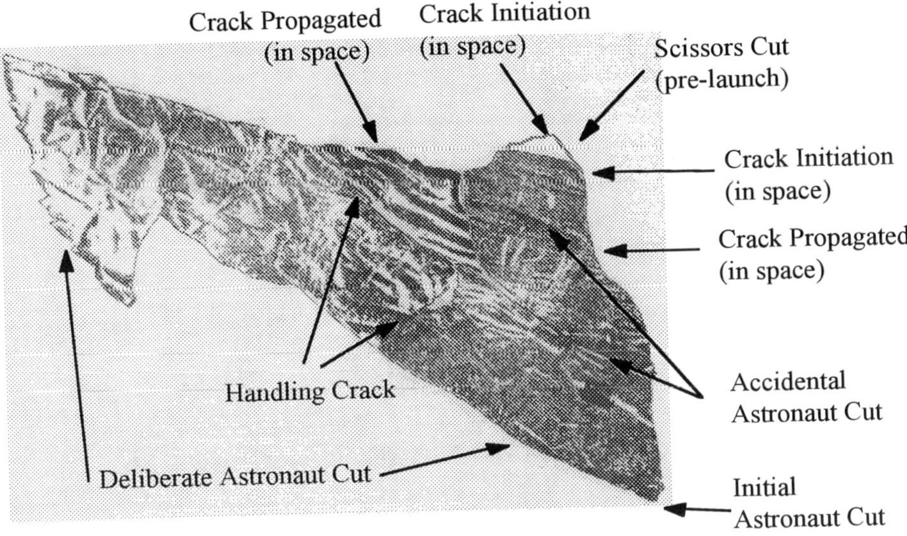

Figure 8: SM2 Teflon FEP Sample

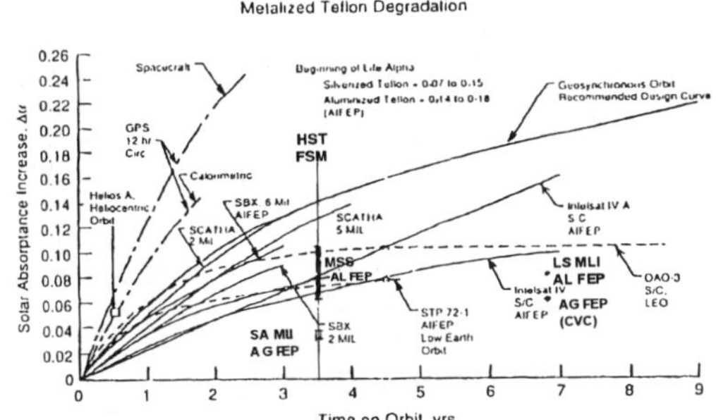

Metalized Teflon degradation (in geosynchronous orbit unless otherwise indicated)

Reference: "Satellite Thermal Control Handbook", D. G. Gilmore, Editor, Aerospace Corporation

Figure 9: HST Solar Absorbtance Measurements versus Predictions

ANALYSIS OF RETRIEVED HUBBLE SPACE TELESCOPE THERMAL CONTROL MATERIALS

Jacqueline A. Townsend, Patricia A. Hansen,
NASA Goddard Space Flight Center, Greenbelt, Maryland, 20771 USA

Joyce A. Dever
NASA Lewis Research Center, Cleveland, Ohio 44135 USA

Jack. J. Triolo
Swales Aerospace, Beltsville, Maryland 20705 USA

ABSTRACT

The mechanical and optical properties of the thermal control materials on the Hubble Space Telescope (HST) have degraded over the nearly seven years the telescope has been in orbit. Astronaut observations and photographs from the Second Servicing Mission (SM2) revealed large cracks in the metallized Teflon® FEP, the outer layer of the multi-layer insulation (MLI), in many locations around the telescope. Also, the emissivity of the bonded metallized Teflon® FEP radiator surfaces of the telescope has increased over time. Samples of the top layer of the MLI and radiator material were retrieved during SM2, and a thorough investigation into the degradation followed in order to determine the primary cause of the damage. Mapping of the cracks on HST and the ground testing showed that thermal cycling with deep-layer damage from electron and proton radiation are necessary to cause the observed embrittlement. Further, strong evidence was found indicating that chain scission (reduced molecular weight) is the dominant form of damage to the metallized Teflon® FEP.

KEY WORDS: LEO Environmental Effects, Teflon® FEP (fluorinated ethylene propylene), Hubble Space Telescope

1.   INTRODUCTION

The Hubble Space Telescope was launched into Low Earth Orbit (LEO) in April 1990 with two types of thermal control surfaces: Multi-Layer Insulation (MLI) blankets and bonded radiator surfaces (1). During the First Servicing Mission (SM1) in December 1993 MLI blankets were retrieved and were subsequently analyzed in ground-based facilities. These studies revealed that the outer layer of the MLI, aluminized Teflon® FEP (fluorinated ethylene propylene), was beginning to degrade. Close inspection of the Teflon® FEP revealed through-thickness cracks in areas with the highest solar exposure and stress concentration. Mechanical tests showed that the ultimate strength and elongation of the Teflon® FEP had reduced significantly (2). During the Second Servicing Mission (SM2) in February 1997, astronauts observed and documented severe cracking in the outer layer of the MLI blankets on both solar facing and anti-solar facing surfaces (1). During the repair process, a small specimen of the outer layer was retrieved from the Light Shield (LS) region and was returned for ground-based analysis. In addition, as part of an instrument installation, a sample of the bonded Teflon® FEP radiator surface was returned on the cryo-vent cover (CVC).

Since the damage to the outer layer was so severe at SM2, a Failure Review Board was convened to, among other tasks, determine the mechanism of the damage. This effort consisted of two major investigations. First, the specimens retrieved during the servicing missions were characterized exhaustively using mechanical and chemical analysis methods to understand the mechanism of the damage. Second, pristine samples of the material were exposed to simulated space environments in an effort to duplicate the condition of the retrieved specimens and determine what element(s) of the space environment caused the damage. Although some of the individual results are detailed elsewhere in this volume (3, 5, 14), this paper summarizes all of the results and draws overall conclusions about the failure mechanism based on those results, the observations of HST, and the calculated fluences.

2. MATERIALS

Two types of thermal control materials were investigated. The first type, the MLI blanket, was composed of several underlying layers of aluminized Kapton®, and a top (space-exposed) layer of aluminized Teflon® FEP. The top layer was 127 μm (0.005 in) Teflon® FEP with roughly 100 nm of vapor deposited aluminum (VDA) on the back (FEP/VDA). The layers of the MLI were bonded together at the edges of the blanket assembly with an acrylic adhesive. The bottom layer was attached to the spacecraft with Velcro® (1).

The HST radiator surfaces used the second type of thermal control material. The material consisted of 127 μm (0.005 in) Teflon® FEP with roughly 100 nm of vapor deposited silver (VDS) on the back (FEP/VDS). The silver side was coated with Inconel and then with an acrylic adhesive. The entire sheet was then bonded directly to the spacecraft (1).

MLI blankets were removed from the HST magnetometers during SM1. A nominal specimen, from a region with average solar exposure [11,339 equivalent solar hours (ESH)], was designated MLI SM1 for this investigation. A second MLI specimen was retrieved from the LS area during SM2 and was designated MLI SM2. A sample of the radiator material (FEP/VDS) was retrieved during SM2 on the cryo-vent cover (CVC). This specimen was designated CVC SM2. In addition to the flight samples, a control sample of the MLI was provided by Lockheed Martin Missiles and Space, and was designated "pristine".

In this paper, "MLI specimen" refers to the outer layer only (FEP/VDA or FEP/VDS), not the full MLI blanket. The specimen designations are summarized in Table 1 below.

TABLE 1: SPECIMEN NAMES AND DESCRIPTIONS

Specimen Name	Description
Pristine	MLI from Lockheed Martin (FEP/VDA layer), received 4/15/97
MLI SM1	MLI from magnetometer cover (FEP/VDA layer), retrieved at SM1
MLI SM2	MLI from light shield (FEP/VDA layer), retrieved at SM2
CVC SM2	Radiator FEP/VDS from cryo-vent cover, retrieved at SM2

3. ANALYSIS OF RETURNED SPECIMENS

The SM2 flight specimens were fully documented using macro photography, optical microscopy, and scanning electron microscopy (SEM). Then the MLI specimens from SM1 and SM2 were characterized through exhaustive mechanical, optical, and chemical testing.

3.1 Scanning Electron Microscopy (SEM) and Optical Microscopy   The first task was to document the MLI SM2 specimen and assemble the four pieces received into the single specimen that was cut in orbit. Both SEM and optical microscopy were utilized in this effort. Once the original configuration had been determined, the edges were identified as either a deliberate cut, a handling artifact, or an on-orbit fracture (see Figure 1). From this information, the fracture initiation site became apparent.

The fractures that resulted in the MLI SM2 specimen initiated at an edge of the MLI that had been cut to fit around a handrail. From small defects in this cut edge, two fractures developed and propagated in orbit almost normal to one another, resulting in a roughly triangular specimen. The VDA was completely missing from the MLI SM2 specimen in regions where the Teflon® FEP was bonded to the rest of the blanket, which included the region where the cracks initiated.

Although the blankets were relatively flat when deployed, photos of the MLI SM2 specimen in orbit showed that it was tightly curled, with the space-exposed Teflon® FEP surface as the inner surface and the VDA exposed (1). This curling indicated a volume gradient in the specimen. Based on the diameter of the curl (1.5 cm) the estimated strain difference between the outer and inner surface of the Teflon® FEP was ~1.5% (3).

SEM images of the initiation region show clear differences between the scissors cut that occurred prior to launch or on orbit, the fractures that propagated while in orbit, and cracks from subsequent handling (3; figures 3-8). The featureless nature of the orbital fracture is unique, and attempts to duplicate this smooth fracture with the SM2 specimen under bending or tensile stress resulted in fractures with more fibrous features (3; figures 1, 2, 9). The inability to duplicate the featureless fracture indicated that the fractures propagated in orbit very slowly, in the presence of relatively low stress and under the influence of radiation and other environmental factors. This type of "slow crack growth" has never been studied in Teflon® FEP (3).

Homogeneous mud-like cracking (mud-tiling) and buckling of the VDA were also apparent in the SEM and optical images. A mismatch between the coefficient of thermal expansion (CTE) of the Teflon® FEP and the VDA was most likely the cause (3, 19). Tensile cracks would develop in the aluminum from low cycle fatigue as the material was cycled above room temperature, and buckling would occur when the material was cycled below room temperature (3; figure 17).

The mud-tiling of the metal backing was apparent in all of the specimens. In the CVC specimen, handling and processing procedures such as bending and pressing the FEP/VDS while adhesive bonding it to the spacecraft surface most likely created the cracks. SEM images of the surface show recession and texturing common in polymers exposed to a sweeping ram fluence of atomic oxygen.

3.2 Mechanical Analyses Mechanical properties were studied using the following instruments and techniques: Instron Mini Tester, Confocal Microscope, bend testing, and Nano Indenter II Mechanical Properties Microprobe (MPM).

3.2.1 Instron Mini Tester The most obvious indication of degradation in the MLI specimens was found in the tensile test results. Reference 3 contains details of the testing and analysis. Table 2 (below) is a summary of the strength test results. In terms of strength, the MLI SM2 specimen was obviously most degraded. The CVC SM2 specimen was less degraded, and the strength of the MLI SM1 specimen degraded the least. This ranking was most apparent in the elongation data.

TABLE 2: SUMMARY OF STRENGTH TEST RESULTS (4)

Material	Yield Strength (MPa)	Ultimate Strength (MPa)	Elongation (%)
pristine	13.8	24.8	340
	14.3	26.5	360
	14.3	28.1	390
MLI SM1	14.3	15.4	196
	14.3	16.6	116
CVC SM2	11.0	12.1	25
	15.4	16.0	25
	N/A	11.0	15
MLI SM2	N/A	13.2	0
	N/A	2.2	0

*3.2.2   **Confocal Microscope***   Thickness measurements were made using a Confocal microscope. Several measurements were made of samples that had been potted and cross sectioned for other analyses. Table 3 (below) gives the thickness measurements of the materials.

TABLE 3: THICKNESS MEASUREMENTS

Material	Thickness (μm)	
pristine	121.4	
MLI SM1	120.9	±0.6
MLI SM2 (region 1)	111.2	±0.5
MLI SM2 (region 2)	113.8	±1.4

The MLI SM1 sample was taken from one of the two returned magnetometer covers. This thickness is considerably higher than that reported in reference 2 for a region with comparable solar hours, however, this specimen was from a different magnetometer cover.

*3.2.3   **Bend Testing***   Because of the condition of the MLI during SM2, astronauts were directed to bend the outer layer on Bay 8 to determine how fragile it was. The Bay 8 specimen, which had not curled, was bent 180 degrees so that the two VDA surfaces touched. Following this, the astronaut found no obvious damage to the material (1). The returned SM2 specimen proved much less durable.

Bend testing was performed on MLI specimens from SM1 and SM2, and on radiator specimens from SM2 (CVC SM2); detailed results are reported in reference 4. Each small specimen was bent manually to 180 degrees around successively smaller mandrels. Following each bend, the specimen was examined with an optical microscope to detect crack length and features. As expected, the pristine material showed no cracking when bent around the smallest mandrel, a strain of 15 percent (4).

Each of the two MLI SM2 samples formed a full-width crack when bent around the first or second large mandrel with the space-exposed surface in tension. Examination showed that this single, full-width crack went most of the way through the thickness of the sample, although the strain from the mandrel diameter was only 2 to 2.5 percent. SEM analysis of the fractures showed the fibrous features of a handling crack. Bending two other MLI SM2 samples around the smallest mandrel with the space-exposed surface in compression did not produce cracks, even at the resulting 15 percent strain. This implied that the space-exposed surface was more brittle than the back surface (4).

The MLI SM1 specimens and the CVC SM2 specimens cracked quite differently from the MLI SM2 specimens. Instead of a single, catastrophic crack, the specimens developed several very short, shallow cracks that eventually joined to form a long, jagged crack across the surface at much smaller mandrels (higher strain). Existing flaws from vent cuts or handling reduced the strain at which cracks first appeared. Unlike the MLI SM2 samples, these samples appeared to retain considerable fracture toughness (4).

*3.2.4   **Nano Indenter II Mechanical Properties Microprobe (MPM)***   The surface micro-hardness of MLI specimens from SM1 and SM2 and from radiator specimens from SM2 (CVC SM2) were measured by Nano Instruments using their patented Continuous Stiffness Measurement technique. The results are reported in detail in reference 4. All of the space-exposed specimens showed an increased hardness at the surface that decreased with depth. By 500 nm, the hardness of all the exposed specimens was indistinguishable from that of pristine at 500 nm. Although the SM1 materials seemed to show a trend of increasing hardness with increasing solar exposure, the SM2 materials, which had the highest solar exposure, did not follow this trend (4).

3.3   Optical Analyses   The optical properties were studied using a UV-Vis-NIR Spectrophotometer (Cary 5E, Varian) and a Laboratory Portable Spectroreflectometer (LPSR).

Significant effort was spent in determining the appropriate method for measuring the solar absorptance of the flight materials. Because of the mud tiling and the delamination of the metal coatings, traditional methods gave results that either over- or under-estimated the changes to the solar absorptance. Reference 5 details the different methods that were considered and the results of the various tests.

For the MLI SM2 specimen, most of the 0.08 solar absorptance increase of the material was attributed to increases in the solar absorptance of the Teflon® FEP, rather than to cracking in the VDA. With the VDA removed, the solar absorptance of the MLI SM2 specimen was still 0.06 higher than pristine. No clear correlation was found between solar absorptance increase and equivalent solar hours (ESH).

Literature values for solar absorptance of pristine FEP/VDS were found between 0.06 to 0.09. The increase in the solar absorptance of the CVC SM2 specimen was attributed to darkening of the acrylic adhesive that was used to bond the material to the spacecraft. During the bonding process, the material was repeatedly bent and deformed, which created the mud tiling cracks in the silver deposit. The adhesive bled through these cracks in the silver and was exposed to sunlight. Acrylic adhesives are known to darken when exposed to UV (1, 5).

3.4   Chemical Analyses   The chemical composition was studied using Time-of-Flight Secondary Ion Mass Spectrometry (TOF-SIMS), Fourier Transform infrared microscopy (μ-FTIR), Attenuated Total Reflectance infrared microscopy (ATR/FTIR), and X-ray Photoelectron Spectroscopy (XPS).

3.4.1   Time-of-Flight Secondary Ion Mass Spectrometry   Time-of-Flight Secondary Ion Mass Spectrometry (TOF-SIMS) was used to determine the ion composition of the first mono-layer (0.3 nm) of the specimens and to image ion intensities on the cross sections (6).

For pristine Teflon® FEP, the most common fragmentation point was at the bond between CF_2 molecules. Some minor contamination of the surface was found (6).

The spectra of the MLI SM2 specimen had the most evidence of chemical changes. There were many oxygen-containing species, mostly in C_x-O_y-F_z bonds. In high mass regions of the spectra, C-O-F bonds were more prevalent than C-F bonds. There was some indication of de-fluorination on the surface, but the dominance of the C-O-F bonds weakened the evidence. In the cross section, the highly oxidized ions were present to a depth of 5-10 μm. Ion concentrations in the 10-110 μm depth were similar to those of the pristine Teflon® FEP. Several silicon-containing ions were detected on the surface (6).

The most striking change detected in the CVC SM2 specimen was evidence of de-fluorination on the surface. Ions were detected at intervals that represented only an additional C atom, rather than an additional CF_2 molecule. Although oxygen was detected in a few of the low-mass fragments, the C-F bonds dominated the spectra. Unlike the MLI SM2 specimen, most ions did not contain oxygen. Analysis of the cross section showed a spectrum very similar to pristine Teflon® FEP, indicating that the de-fluorinated region was on the very surface of the specimen. Very few silicon-containing contaminants were found on this surface (6).

The spectra of the MLI SM1 specimen most closely resembled the pristine. There was some evidence of oxidation and de-fluorination, but not to the extent present in either of the other two flight specimens. Silicon-containing contaminants were detected on the surface (6).

An in-depth analysis of the contamination was not conducted for this investigation. However, earlier investigations into this type of contamination proved that spacecraft surfaces can be contaminated by silicones in the shuttle bay and solar arrays.

3.4.2   Fourier Transform Infrared Microscopy   Fourier Transform infrared microscopy (μ-FTIR) analysis was performed as described in reference 13, and the details of the analysis can be found in reference 7. The testing conducted for this effort did not confirm that this method can detect crystallinity changes. So, although this method showed no significant differences in the crystallinity of the of the pristine, MLI SM1, MLI SM2 or CVC SM2 specimens, the test was inconclusive. Also, only MLI SM1 showed a significant amount of oxidation in the first 3 to 5 μm of the material (7).

3.4.3   X-ray Photoelectron Spectroscopy   X-ray Photoelectron Spectroscopy (XPS) was performed on the MLI SM2 and CVC SM2 specimens and a pristine specimen. The analysis depth of the XPS is roughly 100 Å. In this case, a change in the ratio of carbon to fluorine (C/F) was defined as damage (8).

The C/F ratio of pristine Teflon® FEP was 8.05 with an oxygen concentration of 0.2 atom%. The CVC SM2 specimen appeared to be the most damaged with a measured C/F ratio of 6.3, and an oxygen concentration of 1.9 atom%. A typical region of the MLI SM2 specimen had a C/F ratio of 6.8 with an oxygen concentration of 0.8 atom%. A region of the MLI SM2 specimen that appeared contaminated was the least damaged with a measured C/F ratio of 7.9, an oxygen concentration of 1.5 atom%, and a trace contaminant of either silicone or hydrocarbon (8)

3.5 Molecular Structure Analyses   The molecular structure was investigated using X-ray Diffraction (XRD), density gradient column, and Solid-State Nuclear Magnetic Resonance Spectroscopy (NMR).

3.5.1   X-ray Diffraction   X-ray Diffraction (XRD) was used to detect changes in the crystallinity of the returned MLI specimens from SM1 and SM2. The details of this analysis can be found in reference 9, and the results are summarized in Table 4 (below).

The pristine specimen had a crystallinity of 28-29%. Specimens with various ESH returned during SM1 showed a crystallinity of 28-32%. These measurements were within the uncertainty of the instrument, so MLI SM1 specimens had a crystallinity that was indistinguishable from pristine. The SM2 specimens showed a significant increase, with a crystallinity of 46-47% (9).

3.5.2   Density Gradient Column   The density of the specimens was found using a density gradient column. This data was then converted to crystallinity values using a table provided by DuPont. This method was outlined in reference 10, and the results are summarized in Table 4 (below).

The calculated crystallinity of the SM1 specimens were indistinguishable from pristine material at 50%. The crystallinity of the MLI SM2 specimen was higher, at 65% (10).

Although there were differences between the absolute value of the crystallinity determined using XRD and the density method, the change in crystallinity is identical. Both methods show an increase in crystallinity of 15%. The different absolute values of the two methods was not surprising because the principles involved were so different. Based on literature data comparing XRD to various methods, a difference in absolute crystallinity of up to 14% is not uncommon (9).

TABLE 4: SUMMARY OF CRYSTALLINITY RESULTS

Sample	ESH	XRD		Density Gradient Column	
		# Tested	Crystallinity (%)	Density (g/cm^3)	Crystallinity (%)
Pristine 5 mil FEP	0			2.1400	50
Pristine FEP/VDA	0	6	28-29	2.1394	50
MLI SM1	4,477	1	30	2.1375	49
	6,324 or 9,193	1	29	2.1381	50
	9,193 or 6,324	1	32	2.1381	50
	11,339	2	29-30	2.1378	50
	16,670	1	32	2.1406	50
MLI SM2	33,638	2	46-47	2.1836	65

3.5.3   Solid-State Nuclear Magnetic Resonance Spectroscopy   Solid-State Nuclear Magnetic Resonance (NMR) was performed at the University of Akron on MLI specimens from SM1 and SM2. The results are summarized in reference 10.

NMR performed on the pristine material showed a CF_3 abundance of 7.5%. Analysis of the MLI SM1 specimen detected no significant changes in chemistry or morphology. However, the analysis of the MLI SM2 specimen showed evidence of changed morphology. The results indicated that the SM2 specimen had undergone chain scission, and that either an increased crystallinity or cross-linking also occurred. No quantitative analysis was feasible.

4. SPACE ENVIRONMENT SIMULATIONS

Several different simulations were employed individually and in combination to determine which elements of the space environment were most likely to cause the damage observed in the returned specimens. Teflon® FEP specimens were exposed to combinations of electrons, protons, and thermal cycling at NASA Goddard Space Flight Center (GSFC). Vacuum ultraviolet and soft x-ray radiation exposures were carried out at Brookhaven National Labs (BNL), National Synchrotron Light Source (14).

4.1 GSFC Radiation and Thermal Cycle Exposures   The initial purpose of the electron and proton radiation exposures carried out at GSFC was to determine the dose at which Teflon® FEP would shatter with gentle contact. Specifically, at what servicing mission would the HST MLI outer layer shatter if astronauts tried to remove it or came into contact with it. The approach was to expose specimens of the material to increasing fluences of electrons and protons and then perform tensile tests to determine the changes to the yield and ultimate strengths. When initial testing revealed little change in tensile test data at SM2 fluences, the decision was made to add thermal cycling to the test matrix. The modified test procedure and results are outlined below (11, 12).

4.1.1 Procedure   Twenty-eight tensile test specimens (ASTM D1822, Type L Die) were cut with identical orientation from a single sheet of Teflon® FEP. The GSFC Radiation Effects Task Group exposed a set of three specimens to each of the fluence of electrons and protons listed in Table 5 below. Each fluence was based on the estimated fluence at a specific HST servicing mission.

TABLE 5: FLUENCES FOR GSFC RADIATION EXPOSURES

Run	Protons (1 MeV) $\times 10^{10}$/cm^2	Electrons (0.5 MeV) $\times 10^{13}$/cm^2	Equivalent HST Mission	Fluence Years	Number of Thermal Cycles ($\pm$ 50)
1	1.956	1.949	SM2	6.8	-
2	2.771	2.740	SM3	9.6	39,712
3	3.567	4.130	SM4	13.2	56,304
4	5.861	6.040	EOL	20	77,088
5	11.72	12.08	2xEOL	40	116,800
6	29.30	30.20	5xEOL	100	-

Following the irradiation of a set of specimens, one specimen was thermal cycled. The other two were tensile tested to determine the effect of the radiation alone on tensile properties. An un-exposed control specimen was tensile tested along with each set to verify the consistence of the test procedure (11).

The temperature limits of the thermal cycling were based on the nominal limits for the MLI outer layer in orbit. Based on the thermal properties of the FEP/VDA, the MLI outer layer of solar-facing surfaces reached +50 °C when in the sun, and dropped to -100 °C when in shadow (1). Although these limits changed when the MLI SM2 specimen curled and exposed the VDA, since most of the damaged surfaces on HST did not curl, these limits were used for the experiment.

Rapid thermal cycling between -100 °C and +60 °C took place in a nitrogen atmosphere in a modified thermal cycle chamber. Liquid nitrogen vapor and a heat gun were added to the chamber to reduce the period of the cycles to 15 seconds. Temperatures were monitored with thermocouples all around the test specimen, and the cycle was driven by a thermocouple affixed with epoxy to a control specimen mounted adjacent to the test specimen (12).

Following thermal cycling, the specimens were tensile tested and changes in tensile properties were noted (11).

4.1.2 Results   Table 6 below summarizes the results of the combined testing. The loads and elongation can be calculated from the following gauge dimensions: area, 0.127 mm x 318 mm; length, 1.905 cm.

The data indicate that yield strength was unchanged by radiation and slight evidence that yield strength increased due to subsequent thermal cycling. The ultimate strength clearly was reduced by radiation, with roughly 25% loss at the 20 year HST end-of-life (EOL) fluence. Adding thermal cycles reduced the ultimate strength by another 15%. As with the returned SM2 specimens, the changes are most apparent in the elongation data. With the EOL radiation fluence, the elongation had lost 18% of the pristine value. Following thermal cycling, the elongation lost another 28%. Although these EOL fluences were significantly higher than what HST experienced at SM2, the trends in the data compared favorably.

TABLE 6: TENSILE TEST RESULTS FOLLOWING
RADIATION AND THERMAL CYCLING (11)

Run	Radiation Fluence (years)	Thermal Cycles	Yield Strength (MPa)	Ultimate Strength (MPa)	Elongation at Failure (%)
Control (10 specimens)	0	0	14.2 ± 0.2	25.1 ± 0.3	356 ± 8
1	6.8	0	14.0	23.2	345
		0	14.5	25.9	329
		0	13.9	25.4	377
2	9.6	0	13.5	20.2	314
		0	13.8	21.0	321
		39,000	14.3	17.6	284
3	13.2	0	13.8	19.9	301
		0	13.8	21.5	280
		56,804	14.4	18.0	267
4	20	0	13.8	19.5	301
		0	13.8	19.0	280
		77,088	14.3	15.4	267
5	40	0	13.8	18.2	293
		0	13.8	16.9	263
		116,800	14.9	14.3	132
6	100	0	13.5	14.3	233
		0	13.7	13.4	180

4.1 Brookhaven National Labs Exposures   Earlier investigations indicated that soft x-rays from solar flares may have caused the degradation observed on SM1 specimens (13). The National Synchrotron Light Source (NSLS) at Brookhaven National Labs (BNL) can provide very high flux radiation in the desired energy with very tight bandwidths. Using two different beamlines, Teflon® FEP specimens were exposed to radiation at several energies in the vacuum ultraviolet (VUV) and soft x-ray region of the spectrum in an attempt to confirm the earlier work and determine if there was a region of the spectrum to which Teflon® FEP was particularly sensitive. The procedure and results of these experiments are detailed in reference 14 of this volume.

The study found that exposure to synchrotron radiation of narrow energy bands between 69 eV and 1900 eV was capable of causing degradation in the mechanical properties of Teflon® FEP. However, Teflon® FEP samples exposed to synchrotron radiation of doses significantly greater than HST end-of-life (EOL) doses did not show loss of tensile strength or elongation comparable to that of the severely embrittled SM2 specimens. Based on these results, it was concluded that VUV and soft x-ray radiation alone were not sufficient to cause the severe degradation in mechanical properties observed in Teflon® FEP materials exposed to the HST environment for 6.8 years. Some evidence of wavelength-dependence in the damage was observed for samples exposed to 290 eV synchrotron radiation which is at the carbon absorption edge. These samples showed surface cracking after tensile testing.

5. DISCUSSION

5.1 Analysis of Returned Specimens   The mechanical properties of specimens that were returned from the second servicing mission were significantly degraded. Curling in the MLI SM2 specimen indicated a volume shrinkage gradient through the thickness, and bend test results confirmed that the space-exposed surface was more embrittled than the inside surface. Fractographic examination of the cracks that occurred in orbit indicated that they propagated very slowly under relatively low stress in the presence of radiation or other environmental effects. Similar featureless fracture surfaces were found in the few small cracks in the SM1 specimens as well. This "slow crack growth" phenomenon has never been studied in Teflon®.

Crack patterns in the vapor deposited metal coatings on the back of the thermal control materials resembled homogeneous mud cracks. This "mud tiling" can be caused by thermal cycling or handling. When the material was bonded, as with the radiator surfaces on HST, the adhesive bled through these cracks in the metal and darkened in the presence of ultraviolet radiation, causing increases in the solar absorptance.

The chemical analysis techniques used did not yield consistent results. TOF-SIMS data (analysis depth of 0.3 nm) indicated that the MLI SM2 specimen was the most damaged, with oxygen-containing ions dominating the mass spectra. The XPS data (analysis depth of 10 nm) indicated that the SM2 CVC specimen was most changed, with the lowest C/F ratio. Infrared microscopy (analysis depth of 3 μm) was inconclusive with respect to crystallinity, although some contamination was found on the MLI SM1 specimen. The differences may have been simply a function of the analysis depths and sensitivities of the different techniques. Limited attempts to determine the chemical composition deeper into the bulk of the material with these techniques found no changes at significant depth. Therefore, it is unlikely that composition changes (e.g. de-fluorination, oxidation) can explain the changes to the bulk properties observed in the retrieved specimens.

The reduced elongation in the SM2 specimens, as evidenced by the tensile and bend test results, demonstrated the material's loss of plastic deformation capability. Since plastic deformation is a function of chain entanglements and, thus, chain length, this indicated a reduced molecular weight in the returned Teflon® FEP. This implied that chain scission, rather than crosslinking, was the dominant damage mechanism in the SM2 materials (18). Although it was not as pronounced, similar reduction in elongation and ultimate strength occurred in the SM1 materials.

The density measurements and XRD analysis of the MLI SM2 specimen revealed a 15% increase in crystallinity. The NMR analysis confirmed that both chain scission and increased crystallinity occurred in the MLI SM2 specimen and found no change in the bulk molecular structure of the MLI SM1 specimen.

With these analytical results, the condition of the returned specimens was well documented, and the type of damage was well characterized, but the cause of the damage was still not apparent.

5.2 Space Environment Simulations   Based on the results from the synchrotron radiation exposures, it was clear that neither VUV nor soft X-ray radiation alone could cause the observed bulk damage to the HST thermal control materials. These wavelengths did reduce elongation at extremely high fluences, however, even at doses several orders of magnitude higher than experienced by HST at SM2 there were no comparable bulk property changes (14).

The documentation of the condition of the blankets in various locations around HST during the two servicing missions was revealing. At the first servicing mission, there were very few macroscopic cracks. A few were discovered near the NASA logo on the anti-solar-facing side of the spacecraft, and a few were found on close examination of the returned materials from the solar-facing side. However, in general, the outer layer of the MLI blankets appeared to be intact (1). During SM2, cracks all around HST were visible to astronauts and in photographs. The damage appeared to be worse on the solar-facing side of HST, but the MLI on the anti-solar side was still significantly damaged. Cracks were prevalent on both sides of the spacecraft; the cracks on the solar-facing side tended to be longer (1).

Note that the anti-solar-facing side of HST only received albedo ESH equivalent to roughly 10 % of the solar-facing ESH (1). This meant that at SM1 the solar-facing surfaces of HST had received five times more ESH than the anti-solar facing surfaces had received at SM2. If any component of the ESH was the dominant damaging environmental factor, the damage to the solar-facing side should have been far worse at SM1 than the anti-solar facing surfaces. It can be argued that the damage to the solar-facing materials should have been worse at SM1 than the damage to anti-solar facing surfaces at SM2. The photographic evidence of HST clearly contradicts this supposition. This, coupled with the BNL results, casts strong doubt on the idea that any component of the solar spectrum, including soft x-rays, could be solely responsible for the damage observed on HST.

Because the damage to HST did not appear to coincide with ESH, components of the space environment that are more closely homogeneous were suspect. The likely candidates were electron and proton fluences and thermal cycling. The GSFC simulations explored the combined effects of radiation and thermal cycling.

The GSFC experiments showed that electron and proton radiation alone affected the tensile properties of the Teflon® FEP. The reduced ultimate strength and elongation was apparent at fluences comparable to the HST end-of-life (20 years). Subsequent thermal cycling between -100 and +60 °C reduced these properties further. These particle radiation exposures coupled with thermal cycling produced damage that most closely resembled the HST specimens. However, the study did not duplicate the degree of damage observed on the returned SM2 specimen with SM2 doses of radiation and thermal cycling at nominal limits (-100 to +50 °C).

The MLI SM2 specimen had curled while in orbit, exposing the underlying VDA to the sun. Once the aluminum was exposed, the material cycled from -100 to +200 °C with each 90 minute orbit. Literature data for Teflon® FEP showed two second-order transitions, one at -80 °C and another at +80 °C. Cycling through the upper transition temperature could easily affect the nature of the damage. However, since most of the damaged surfaces on HST did not curl, the nominal limits were chosen for the experiment. Further tests are needed to determine the effect of the higher temperature cycling.

Some of the differences between the MLI SM1 specimen and the MLI SM2 specimen could be attributed to the higher temperature cycles of the curled SM2 specimen. Additionally, the astronaut observation following the "bend test" in orbit revealed that a damaged region that did not curl maintained more fracture toughness in orbit than the curled MLI SM2 specimen exhibited in the ground testing. That qualitative difference could be from either the different temperatures experienced by the Bay 8 specimen and the MLI SM2 specimen or from changes that occurred after the SM2 specimen was exposed to atmosphere. Further testing is required to understand the effects of atmosphere on vacuum irradiated Teflon® FEP

6.   CONCLUSIONS

Analysis of the returned specimens showed that both the MLI SM1 and MLI SM2 specimens underwent chain scission. Evidence of increased crystallinity was found only in the MLI SM2 specimen. Solar absorptance changes in the MLI SM2 specimen were attributed to changes in the Teflon® FEP and mud tiling in the VDA. Solar absorptance changes in the CVC SM2 specimen were attributed to mud tiling from handling and subsequent darkening of the acrylic adhesive.

The conclusions of the HST MLI FRB were based on the combined evidence of HST damage and data uncovered in ground-based experiments. The FRB concluded the following:

The observations of HST MLI and ground testing of pristine samples indicate that thermal cycling with deep-layer damage from electron and proton radiation are necessary to cause the observed Teflon® FEP embrittlement and the propagation of cracks along stress concentrations. Ground testing and analysis of retrieved MLI indicate that damage increases with the combined total dose of electrons, protons, UV and x-rays along with thermal cycling.

Tests continue in order to determine the effects of the higher temperature limit the MLI SM2 specimen experienced.

7. ACKNOWLEDGMENTS

The analysts in the Materials Engineering Branch of NASA, Goddard Space Flight Center who performed tests are Diane Kolos, Len Wang, Charles He, Charles Powers, Bruno Munoz, Alex Montoya, Mary Ayres-Treusdell, Tom Zuby, and Mike Viens. Other GSFC contributors include Steve Brown, Claude Smith, Bob Gorman, Wanda Peters, and Dave Hughes. Analysts at NASA, Lewis Research Center who performed tests are Bruce Banks, Don Wheeler, and Jim Gaier. The TOF-SIMS analysis was performed under contract by Mark Nicholas of Evans East. The Nano Indenter analysis was performed by Barry Lucas and Angela Gizelar of Nano Instruments, Inc. NMR analysis was performed by Matthew Espe and Daveen England of the University of Akron. The work of all these analysts and the contributions of the entire HST MLI Failure Review Board are gratefully acknowledged by the authors.

8. REFERENCES

1) P.A. Hansen, J.A. Townsend, Y. Yoshikawa, J.D. Castro, J.J. Triolo, and W.C. Peters, "Degradation of Hubble Space Telescope Metallized Teflon® FEP Thermal Control Materials", <u>Science of Advanced Materials and Process Engineering Series</u>, <u>43</u>, 000 (1998).

2) T. Zuby, K. DeGroh, and D. Smith; "Degradation of FEP Thermal Control Materials Returned from the Hubble Space Telescope," NASA Technical Memorandum 104627, December 1995.

3) L. Wang, J.A. Townsend, and M. Viens, "Fractography of MLI Teflon® FEP from the HST Second Servicing Mission", <u>Science of Advanced Materials and Process Engineering Series</u>, <u>43</u>, 000 (1998).

4) J.A. Dever, K.K. deGroh, J.A. Townsend, L.L. Wang, "Mechanical Properties Degradation of Teflon® FEP Returned from the Hubble Space Telescope", NASA/TM-1998-206618

5) C. He and J.A. Townsend, "Solar Absorptance of the Teflon® FEP Samples Returned from the HST Servicing Missions", <u>Science of Advanced Materials and Process Engineering Series</u>, <u>43</u>, 000 (1998).

6) M. Nicholas, "Problem 24-230, TOF-SIMS," Evans East Analytical Report, April 23, 1997.

7) A. Montoya, "Analysis of Multi-layer Insulation from HST Servicing Mission (#AM0497g)," NASA Internal Memorandum, April 30, 1997.

8) J. Dever, "Summary of HST MLI Materials Analysis Results", NASA Internal Memorandum, June 23, 1997.

9) C. He, "The Crystallinity of the FEP samples returned from HST First and Second Servicing Missions," Unisys Internal Memorandum No. MA-1997-047, June 26, 1997

10) J. Dever, "Solid-State Nuclear Magnetic Resonance (NMR) and Electron Spin resonance (ESR) Characterization of HST Materials", NASA Internal Memorandum, November 13, 1997.

11) M. Viens, "Tensile Strength of Irradiated and Thermal Cycled FEP/VDA," NASA Internal Memorandum, January 15, 1998

12) C. Powers, "Completion of Thermal Cycle Life Tests of HST MLI Samples," NASA Internal Memorandum, January 8, 1998

13) A. Milintchouk, M. Van Eesbeek, F. Levadou and T. Harper, "Influence of X-ray Solar Flare Radiation on Degradation of Teflon® in Space" <u>Journal of Spacecraft and Rockets,</u> <u>Vol. 34</u>, No. 4, July-August, 1997.

14) J.A. Dever, J. A. Townsend, J.R. Gaier and A.I. Jalics, "Synchrotron VUV and Soft X-Ray Radiation Effects on aluminized Teflon® FEP", <u>Science of Advanced Materials and Process Engineering Series</u>, <u>43</u>, 000 (1998).

15) J.A. Townsend, P.A. Hansen, J.A. Dever, and J.J. Triolo, "Evaluation and Selection of Replacement Thermal Control Materials for the Hubble Space Telescope", <u>Science of Advanced Materials and Process Engineering Series</u>, <u>43</u>, 000 (1998).

16) J. Dever, "Summary of HST MLI Materials Analysis Results", NASA Internal Memorandum, June 23, 1997.
17) B. Banks "Report for 'Potato Chip' Testing of the Candidate HST MLI Replacement Materials", NASA Internal Memorandum, November 21, 1997.
18) R. Thompson, *Report on GSFC HST MLI Failure Analysis*, July 3, 1997.
19) B. Banks, T. Stueber, S. Rutledge, D. Jaworske, W. Peters, "Thermal Cycling-Caused Degradation of Hubble Space Telescope Aluminized FEP Thermal Insulation," AIAA 98-0896, Presented at the 36th Aerospace Sciences Meeting & Exhibit, Reno NV, January 12-15, 1998

FIGURE 1: MLI SM2 SPECIMEN WITH CRACKS IDENTIFIED

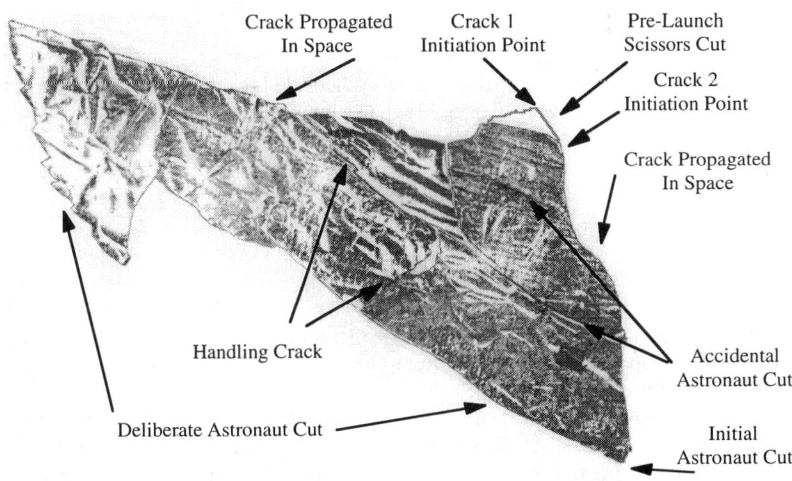

MECHANICAL PROPERTIES AND FRACTOGRAPHY OF MLI FEP FROM THE HST SECOND SERVICE MISSION

Len Wang
Unisys Federal Systems at Materials Engineering Branch, GSFC, NASA
Michael Viens and Jacqueline Townsend
Materials Engineering Branch, GSFC, NASA

ABSTRACT

Tensile tests and fractography of MLI Teflon[TM] (FEP) outer surface layer retrieved during the Hubble Space Telescope (HST) second servicing mission were performed as part of the effort to investigate the failure mechanism of the MLI observed during the second service mission. The fracture surfaces of a tensile test sample as well as a bend-cracked sample were also analyzed for comparison purpose. The surface morphology of the Teflon and aluminum was compared with that of a pristine sample. Fractography analysis shows that the samples exhibit three kinds of distinctive fracture surfaces, although the material has been severely embrittled in space. They are classified as shear fracture, fibrous fracture and smooth fracture based on the fracture surface morphology. The shear fracture is due to a cut by scissors or other cutting devices. Bending and/or tension of the sample during the handling created the fibrous fracture. The smooth fracture occurred in space. The fracture that occurred in space can not be duplicated by tension and/or bending of the embrittled samples. This indicates that cracks were probably developed and propagated very slowly in the material under the influence of the radiation and/or other environmental effects and relatively low stresses and/or stress intensity factors. This "slow crack growth" property of Teflon material has never been studied.

KEY WORDS: Fractography, FEP Materials, Hubble Space Telescope

1. INTRODUCTION

Cracking of the Teflon[TM] (FEP) outer surface layer of the multi-layer insulation (MLI) light shield (LS) blanket was first observed [1] on close examination of samples retrieved from the first Hubble Space Telescope (HST) servicing mission (SM1), after the HST had been in orbit for 3.6 years. During the HST second servicing mission (SM2), astronauts observed and photographed severe cracking of the FEP surface layer on many locations around the telescope. The second service mission was conducted 6.8 years after HST had been in orbit. A small piece of the outer surface layer of the MLI blanket material was trimmed off of the light shield and brought back for analysis. Tensile tests and fractography of MLI FEP outer surface layer retrieved during the HST second servicing mission were performed as part of the effort to investigate the failure mechanism of the FEP material.

2. MATERIALS AND EXPERIMENTAL PROCEDURES

The MLI blanket material that is used on the HST light shield is comprised of the following layers: 1) The outer surface layer is 0.005 inch thick FEP with inner side vapor deposit aluminum (VDA). 2) The inner surface layer is 0.001 inch thick Kapton with inner side VDA. 3) Between the outer and inner surface layers, there are 15 layers of 0.00033 in. thick embossed Kapton with VDA on both sides. All of the analyses were performed on the outer FEP layer where cracking was observed [2,3].

In addition, the radiator surfaces on the HST equipment bays and aft shroud have an outer layer of silver (0.005 in.) FEP that is bonded to the spacecraft with an acrylic adhesive. A sample of this bonded material was also retrieved during the second servicing mission on a cryo-vent cover (CVC) [2,3].

Three types of FEP were tensile tested: pristine, perforated (CVC sample) and non-perforated (MLI light shield) material brought back from HST during the second servicing mission. ASTM D1822 type L specimens were tested at a speed of 2 to 5 inches per minute on an Instron Mini Tester. The loading of the specimens was monitored using a full-scale sensitivity of 5 pounds.

The fractography analysis was performed using optical and scanning electron microscopy (SEM) on two pieces of the FEP outer layer sample that was retrieved during the HST second servicing mission. They are labeled as SEM1 and SEM2. The locations of the two samples are as shown in Figure 0. The fracture surfaces of tensile test samples as well as bend-cracked samples were also analyzed for comparison purpose. The surface morphology of the FEP and aluminum was compared with that of a pristine sample.

3. RESULTS AND DISCUSSION

3.1 Tensile Test and Bend-cracked Samples

The most noticeable degradation of the SM2 Light Shield sample material is probably indicated by the curling and embrittlement. The samples were curled outward indicating a volume shrinkage gradient. The samples were severely embrittled in space and lost almost all the plasticity and toughness, as shown in Table 1. Uncurling of the sample did propagate and even generate cracks. The tensile test of the samples revealed almost no elongation (Figure 1) and the samples can easily be fractured by uncurling and bending against the surface which has been exposed to the space, hereafter referred as to 'exposed surface'. Tensile test and flexure bending of the material show that cracks were always initiated from the exposed surface indicating that the exposed surface is more brittle than the inner material and there probably exists a degradation gradient along the thickness direction [4].

Although the material has been severely embrittled in space, the fracture surface of both the tensile tested and bend-cracked samples exhibit fibrous features, as shown in Figure 1 and 2. The significance of this observation is that the fractures that occurred in space are very different from those created by tensile test and flexure bending.

3.2 Sample SEM2 (SM2 Light Shield)

Figure 3 is photograph of sample SEM2. According to a survey of the cracking sites and the sampling locations, it is believed that sample SEM2 contains two crack initiation sites as indicated in Figure 3. SEM analysis shows that the cracks of this sample were most likely initiated from an edge with scissors cut, as shown in Figure 4 to 7. This edge is believed to be the original edge of the MLI blanket which indicates that the scissors cut was probably performed during the installation of the MLI in order to fit the blanket with the HST. The fracture surface of a scissors cut shows apparent shear and tear (Figure 7).

Table 1. Tensile Strength Testing on different FEP Materials

Material	Yield Load (lbs)	Ultimate Load (lbs)	Elongation at Failure (%)	Comments
pristine	1.25	2.25	340	
pristine	1.30	2.40	360	
pristine	1.30	2.55	390	
MLI LS non-perforated	N/A	1.2	0	specimen tore at grip
MLI LS non-perforated	N/A	0.2	0	specimen failed at crack
VCV perforated	1.0	1.1	25	failure initiated at hole
VCV perforated	1.4	1.45	25	
VCV Perforated	N/A	1.0	15	

Elongation calculated assuming a 0.75 inch gauge length..
Stress can be estimated assuming a 0.125-inch wide by 0.005-inch thick gauge area.

The fracture that occurred in space (Figure 8) shows an extremely smooth fracture surface which is different from the fracture surfaces created by tensile test or flexure bending (Figure 1 and 2). Figure 8 presents the typical fractograph of the fracture that occurred in space. This smooth fracture can not be duplicated by tension and/or bending of the embrittled samples. This important observation also helped us to distinguish them from the cracks which were generated during sample handling since they are similar to those created by tension and/or bending of the sample, as shown in Figure 9.

Therefore, three types of distinctive fracture surfaces were observed. They are tentatively classified as shear fracture, fibrous fracture and smooth fracture based on the fracture surface morphology. The shear fracture is due to a cut by scissors or other cutting devices. The fibrous fracture was created by bending and/or tension of the sample during the handling of the embrittled material, and the smooth fracture happened in space.

3.3 Sample SEM1
The three types of fracture surfaces were confirmed by sample SEM1. This sample does not contain any crack initiation site. The locations of the three types of fractures in sample SEM1 were indicated in Figure 10. Figures 11 to 13 display the three typical fractographs of the fractures. As a comparison, Figure 14 shows a SEM micrograph of a scissors cut edge of a pristine FEP sample. Shear and tear are apparent.

3.4 Surface Morphology
Compared to the surface of a pristine FEP surface (Figure 15), erosion of the FEP surface is apparent, as shown in Figure 16. It is consistent with atomic oxygen errosionfrom a sweeping fluence, rather than a ram-oriented fluence.

Mud like cracking and buckling of the aluminum reflecting layer is evident and homogeneous (Figure 17). They are most likely due to the thermal mismatch between the

FEP and the aluminum. Tensile cracking is due to the thermal cycling above the room temperature, and the buckling is due to thermal cycling below room temperature and volume shrinkage. Calculated low cycle fatigue life of the aluminum layer due to tensile cracks is 10,000 to 20,000 cycles based on a thermal cycling temperature range of 25^0C above room temperature.

3.5 Fracture Mechanism
Tension and/or bend of the embrittled samples can not duplicate the fracture that occurred in space. This indicates that cracks probably developed and propagated very slowly in FEP material under the influence of the radiation and/or other environmental effects and relatively low stresses and/or stress intensity factors.

Tensile stresses can be generated by at least two different mechanisms, i.e. the thermal mismatch between the MLI and the telescope and the volume shrinkage of the FEP material. The thermal mismatch between the MLI and the telescope body will cause a 3% mismatch strain in a temperature range of 200 ^{0}C. Both the FEP and the aluminum could accommodate this kind of strain by a permanent plastic deformation when the materials were new, and no substantial stresses would be developed, later. However, around areas where the MLI is bonded to the telescope body, cyclic strain/stress will persist due to the thermal mismatch.

The volume shrinkage has a gradient in the thickness direction which caused the curling of the fractured FEP film. According to a measurement of the curling diameter, which is ~0.6 in., assuming that the curled film is nearly stress free and the shrinkage strain is linear in the thickness direction, the estimated strain difference between the outer and inner surface of the FEP film is ~1.5%. This strain will generate a tensile stress of ~700 Psi on the outer surface of the FEP film in keeping the film flat (a Young's modulus of 10^5 Psi is used in the calculation). Stress of this magnitude is probably enough to create crazing and crack initiation in the degraded FEP material. This can be especially true around the scissors cut edges. The sustained tensile stresses are also responsible for the slow crack propagation under the influence of radiation and/or other environmental effects.

The initiation and propagation of cracks in FEP under relatively low stresses and/or stress intensity factors as well as the influence of radiation and/or other environmental effects have never been studied and characterized. This property resembles the stress corrosion cracking and/or corrosion fatigue in metallic materials and the slow crack growth phenomenon in glass/ceramics materials. There are well established test methods for metallic and ceramics materials which can be applied to the polymeric materials with some modifications. It is necessary to characterize the FEP and other potential polymeric materials used in long duration space flight applications in terms of slow crack growth properties.

4. CONCLUSIONS

1. Fractography analysis shows that both the SEM1 and the SEM2 samples exhibit three kinds of distinctive fracture surfaces, although the material has been severely embrittled in space. They are classified as shear fracture, fibrous fracture and smooth fracture based on the fracture surface morphology.
2. The shear fracture is due to a cut by scissors or other cutting devices. The fibrous fracture was created by bending and/or tension of the sample during handling and the smooth fracture occurred in space.
3. The fracture that occurred in space can not be duplicated by tension and/or bending of the embrittled samples. This indicates that, under the influence of the radiation and/or other environmental effects and relatively low stresses and/or stress intensity factors, cracks

were probably developed and propagated very slowly in FEP materials. This "slow crack growth" property of FEP material has never been studied.
4. According to a survey of the cracking and sampling locations, it is believed that sample SEM2 contains two crack initiation sites. SEM analysis shows that the cracks of this sample were most likely initiated from an edge with scissors cut, which probably happened during the installation of the MLI.
5. Erosion of the FEP surface is apparent. The morphology is, however, different from the generally observed surface morphology due to atomic oxygen erosion. The mechanism is unknown.
6. Mud like cracking and buckling of the aluminum reflecting layer is evident and homogeneous. They are most likely due to the thermal mismatch between the FEP and the aluminum. Calculated low cycle fatigue life of the aluminum due to tensile cracks is 10,000 to 20,000 cycles.

5. REFERENCE

1. Zuby, T. M., de Groh, K. K, Smith, D. C., "Degradation of FEP Thermal Control Materials Returned from the Hubble Space Telescope," <u>NASA Technical Memorandum</u> 104627, December 1995.

2. Hansen, P.A., Townsend, J.A., Yoshikawa, Y., Castro, D., Trolo, J.J., Peters, W., "Degradation of Hubble Space Telescope Metallized Teflon^{TM} Thermal Control Materials", Sci. Adv. Matl. Proc. Eng. Ser., 43, 000(1998).

3. Townsend, J.A., Hansen, P.A., Triolo, J.J., "Analysis of Retrieved Hubble Space Telescope Thermal Control Materials", Sci. Adv. Matl. Proc. Eng. Ser., 43, 000(1998).

4. Dever, J.A., de Groh K.K.,Townsend, J.A., Wang,L., "Mechanical Properties Degradation of Teflon^{TM} FEP Returned From the Hubble Space Telescope", NASA TM-1998-206618, AIAA-98-0895.

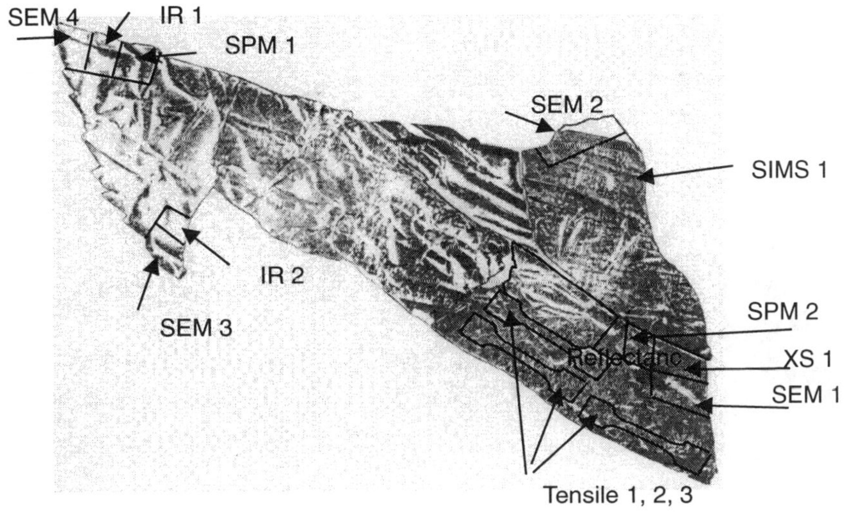

Figure 0, Location map of samples SEM1 and SEM2
on the SM2 Light Shield sample.

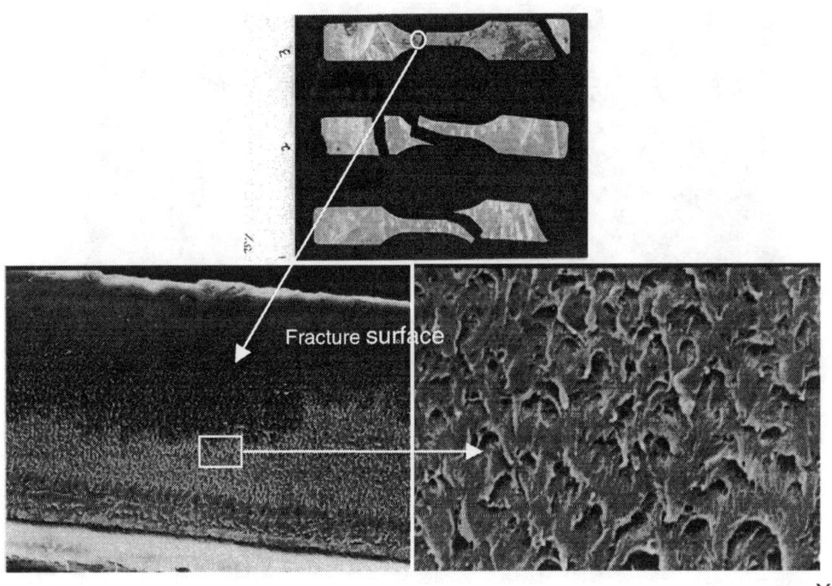

Figure 1. Tensile test samples and fracture surface.

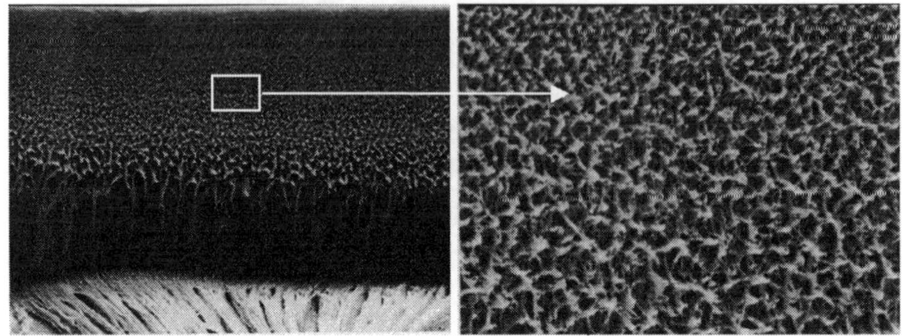

Figure 2. Fracture surface of a bend-cracked sample.   5000X

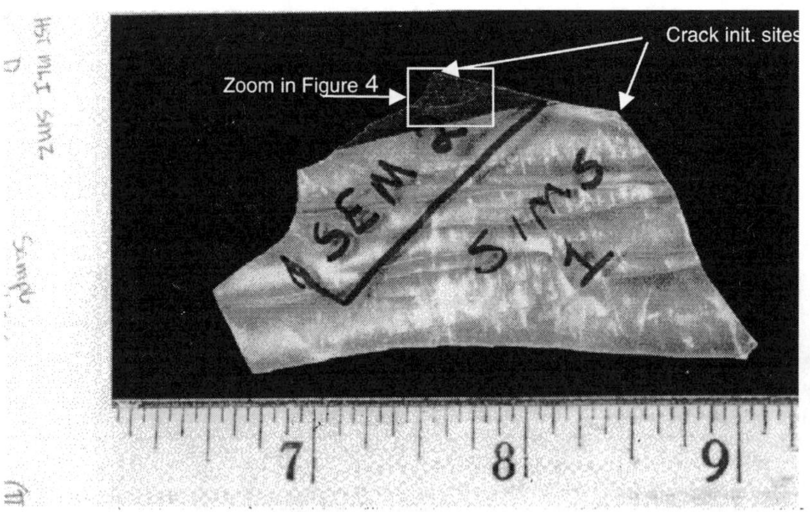

Figure 3. Sample SEM2 and crack initiation sites.

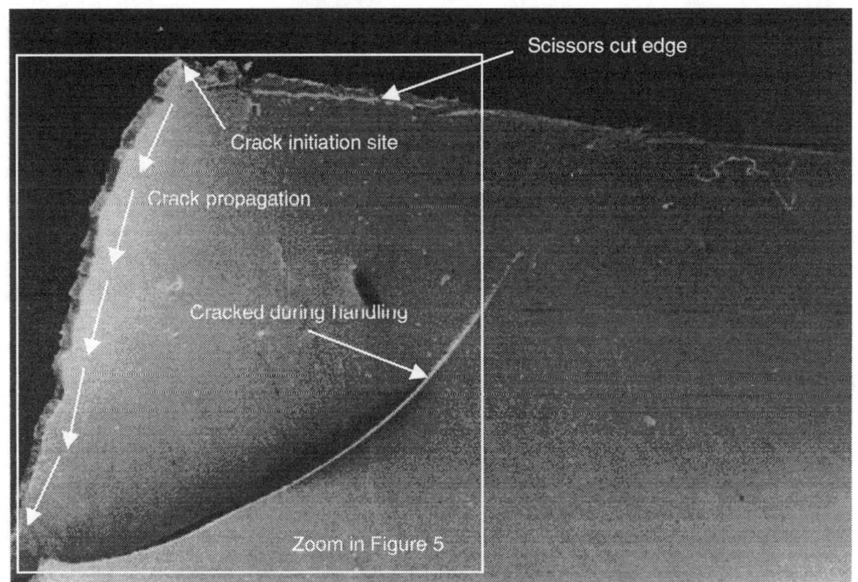

Figure 4. Crack initiation on sample SEM2.   25X

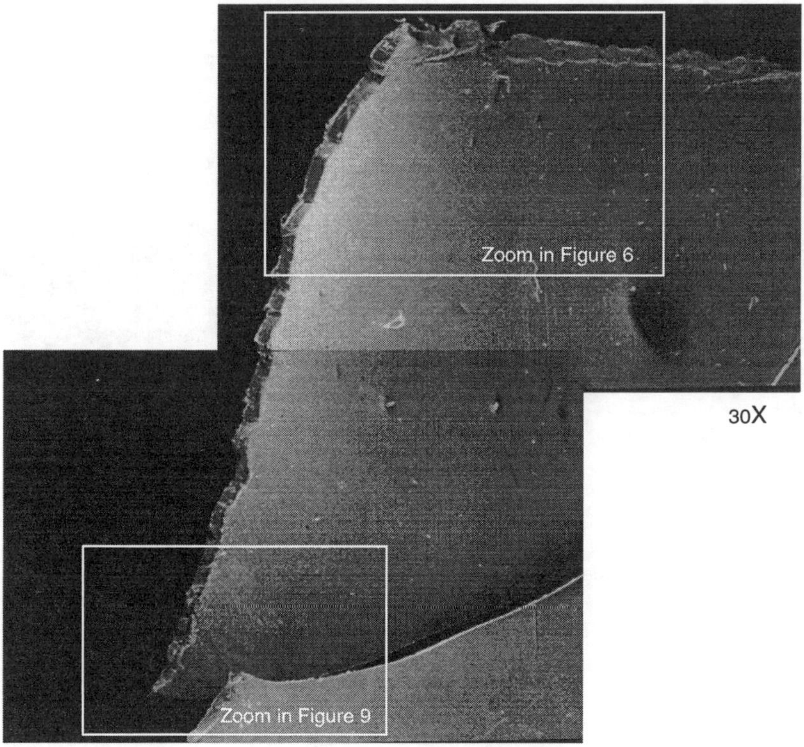

30X

Figure 5. An enlarged view of Figure 4.

601

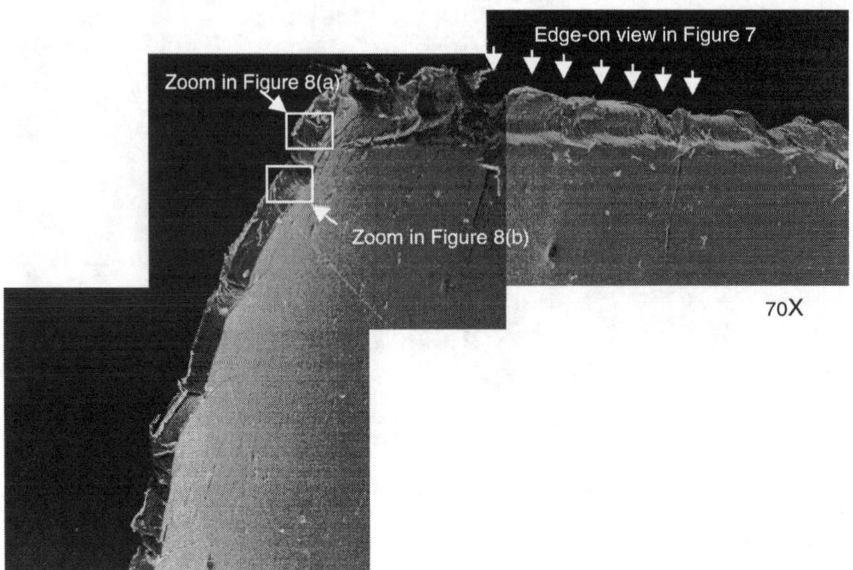

Figure 6. An enlarged view of Figure 5.

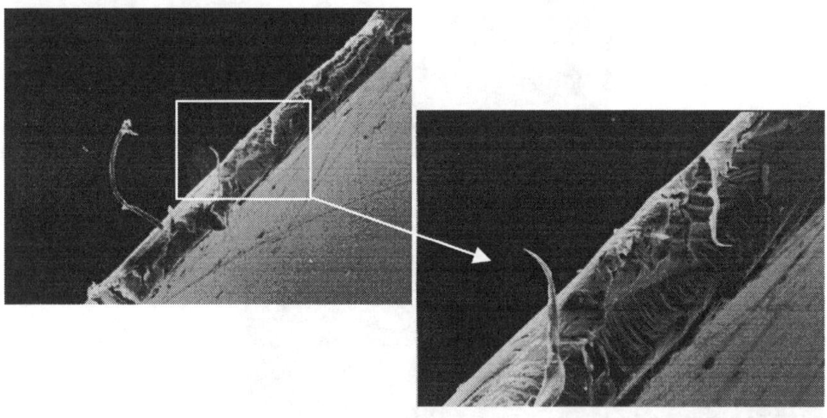

Figure 7. An edge -on view of Figure 6, scissors cut, sample SEM2.

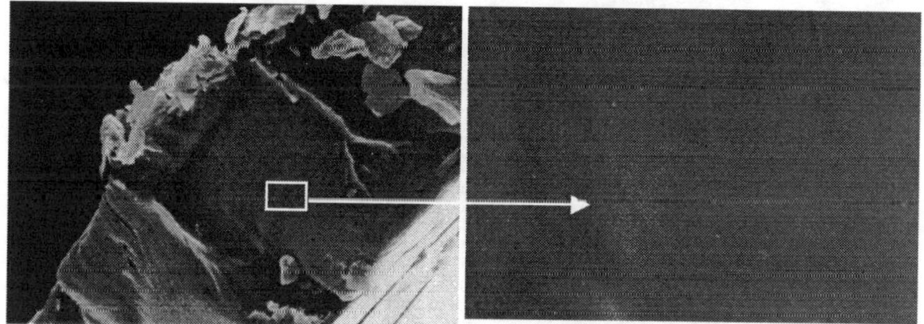

5000X

Figure 8(a) Fracture occurred in space, sample SEM2.

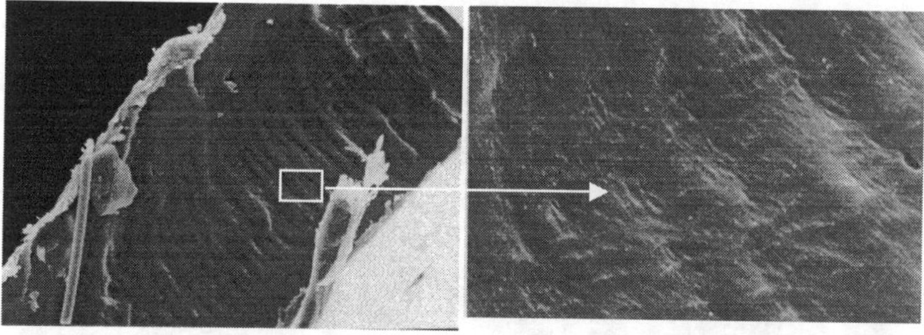

5000X

Figure 8(b) Fracture occurred in space, sample SEM2.

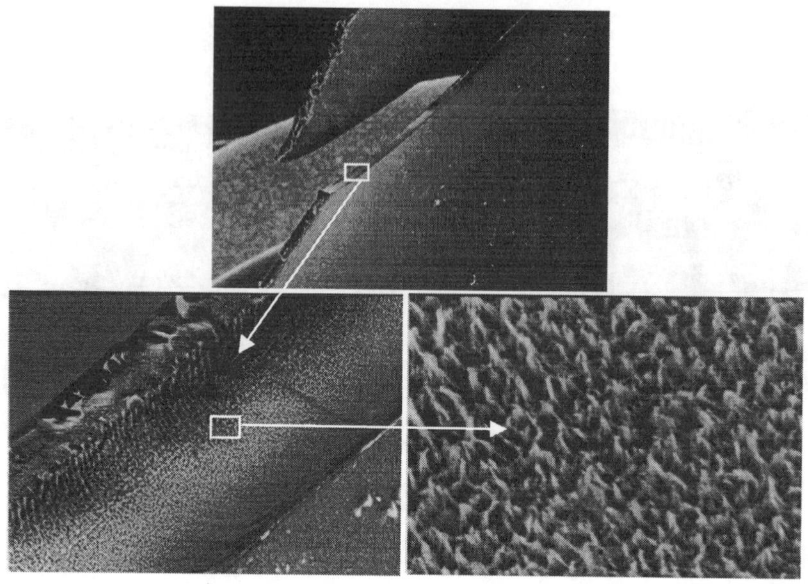

5000X

Figure 9. A crack created during the sample handling in sample SEM2.

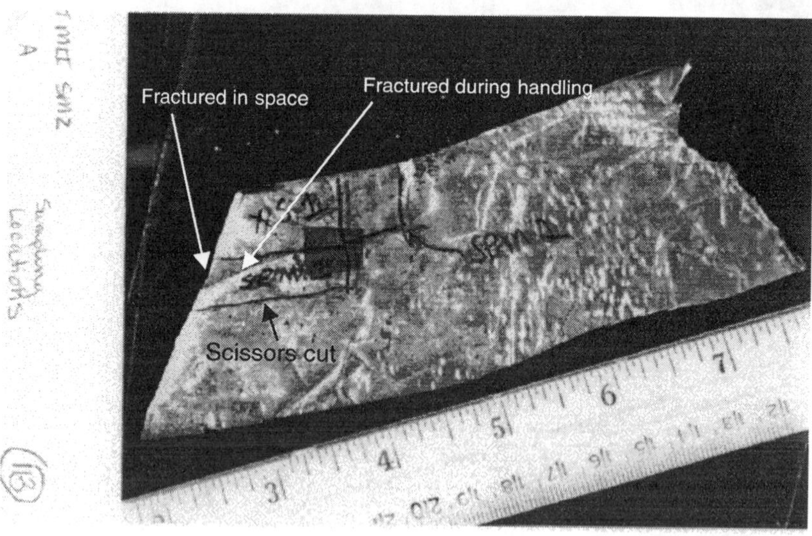

Figure 10. Sample SEM1.

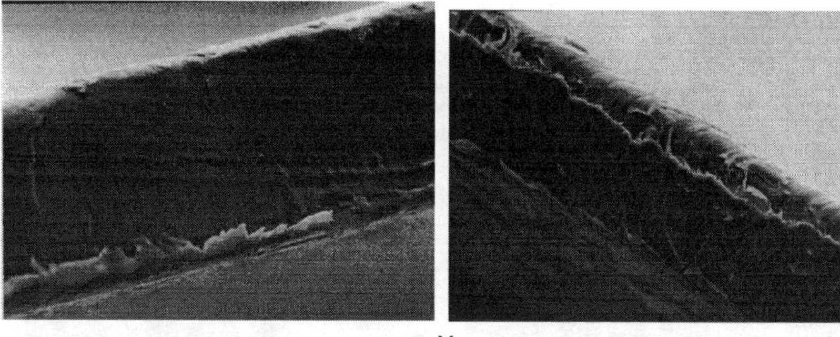

500X

360X

Figure 11. Sample SEM1,
fracture occurred in space.

Figure 12. Sample SEM1,
scissors cut edge (sampling).

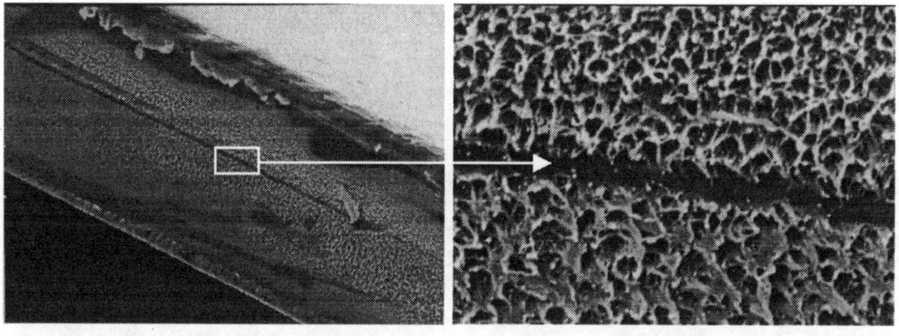

3600X

Figure 13. Fracture created during sample handling, sample SEM1.

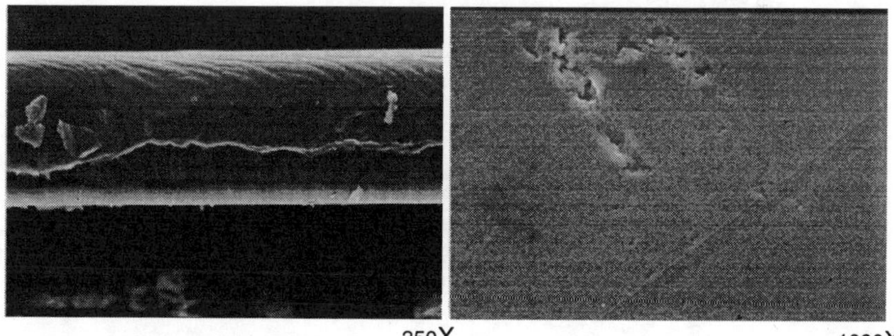

350X

1000X

Figure 14. Scissors cut edge
of a pristine FEP sample.

Figure 15. Surface morphology
of a pristine FEP sample.

605

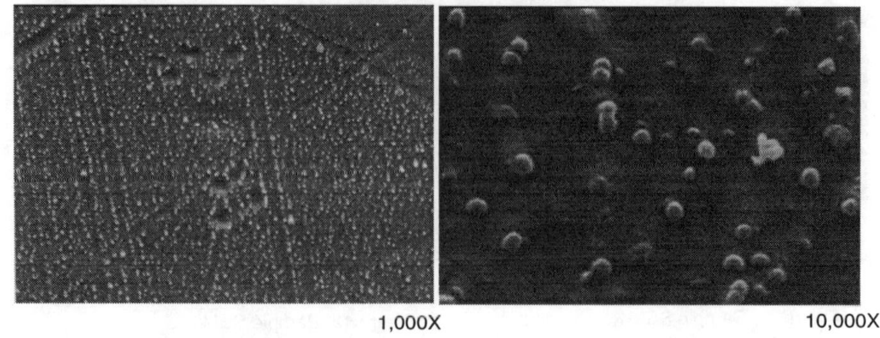

1,000X   10,000X

Figure 16. Surface morphology of the FEP, sample SEM1.

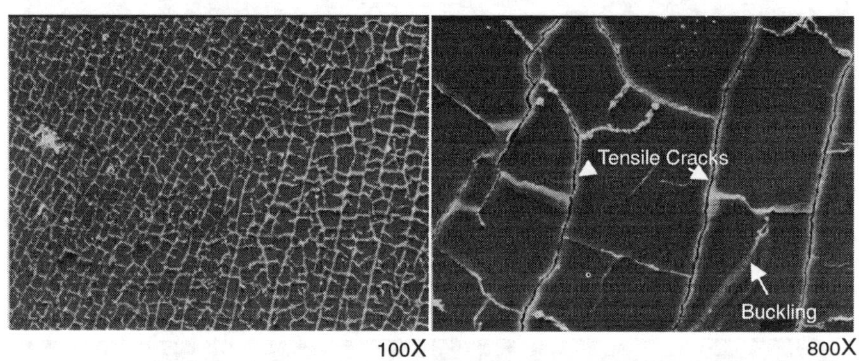

100X   800X

Figure 17. Mud like crack/buckling of the aluminum layer, sample SEM1.

SOLAR ABSORPTION OF THE FEP SAMPLES RETURNED FROM THE HST SERVICING MISSIONS

Charles He
Unisys Federal Systems at NASA Goddard Space Flight Center
Greenbelt, Maryland 20771

Jacqueline A. Townsend
Materials Engineering Branch, NASA Goddard Space Flight Center
Greenbelt, Maryland 20771

ABSTRACT

The solar absorption of several fluorinated ethylene propylene (FEP) samples that were retrieved from two Hubble Space Telescope Servicing Missions (SMs), was measured to assess the effect of the space environment on the optical properties of the material. The solar absorptance of the silver/FEP on the Cryo Vent Cover retrieved from the HST Second Servicing Mission (HST SM2) was significantly higher than that of the same type of FEP from the First Servicing Mission (HST SM1), the Long Duration Exposure Facility (LDEF) Mission and the Solar Max Mission. Crazing of the silver coating and darkening of the adhesive beneath the coating are the major causes of the absorption increase. The aluminum coating on the aluminum/FEP samples retrieved from the HST SM1 and SM2 was found to be crazed and delaminated. The Al/FEP samples from both SM1 and SM2 showed an increase in solar absorptance compared to pristine samples but there was large scatter in the measurements due to non-uniform crazing and delamination of the aluminum coating. The crazing of the aluminum coating and the degradation of the FEP are responsible for the solar absorptance increase. The solar absorption of the FEP itself (Al coating removed) increased for both SM1 and SM2 samples.

KEY WORDS: Hubble Space Telescope (HST), Fluorinated Ethylene Propylene (FEP), Solar Absorption

1. INTRODUCTION

The HST was launched on April 24, 1990. The First Servicing Mission (SM1) took place in December 1993 and the Second Servicing Mission (SM2) in February 1997. Both Al/FEP and Ag/FEP samples were brought back for analysis from both servicing missions. Zuby et al.

investigated the FEP materials from the returned HST solar array assembly from the SM1 [1]. Damage to the FEP layer of the MLI was most pronounced on the solar facing surfaces. Through-thickness cracks were observed on the FEP surface with high solar exposure before any testing of the material. Bending tests revealed that the tensile stress during the test caused cracking for the FEP with high solar exposure, but not for the FEP with low solar exposure. The crack depth (across the thickness of the material) from the bending test was also measured and the results showed that the bending test induced deeper cracks for the high solar exposure area than for the low solar exposure area. The mechanical properties such as tensile strength and elongation decreased significantly compared to pristine samples. Van Eesbeek et al. also investigated the FEP thermal control material from the SM1[2]. Infrared spectra of the FEP indicated that the crystallinity of the material increased during the flight. Electron Spectroscopy for Chemical Analysis (ESCA) found the presence of C=C bonds and defluorination on the surface layer. The loss in mechanical strength on the HST SM1 samples was in the order of 30-50% compared to the virgin FEP material, consistent with Zuby et al's results [1].

The FEP samples from the HST SM2 have been analyzed at NASA Goddard Space Flight Center. As a part of this analysis, this paper presents the results of the optical property analysis of the samples. The solar reflectance and absorptance of Ag/FEP and Al/FEP samples from HST SM1 and SM2 were measured according to ASTM Standard E903. Pristine Al/FEP was also measured for comparison.

2. EXPERIMENTS

A Cary 5E UV-Vis-NIR Spectrophotometer with an integrating sphere was used to measure the reflectance of the samples. The spectral range was 200 nm to 2500 nm. Step scan was used at each wavelength that corresponds to the Solar Spectral Irradiance Curve specified in ASTM Standard E490. The time at each step is 1 second. The solar reflectance was calculated using the weighed ordinate method specified in ASTM Standard E903. The solar absorptance was calculated by subtracting the solar reflectance from 1. The typical uncertainty for such measurements is around 0.02, according to ASTM Standard E903.

3. RESULTS AND DISCUSSION

3.1 The HST SM2 Ag/FEP on the Cryo Vent Cover

The HST Cryo Vent Cover (CVC) was retrieved during the SM2 and is illustrated in Figure 1. The silver coated FEP (Ag/FEP) material on the CVC was bonded to an aluminum plate by an acrylic adhesive. Crazing of the Ag coating and localized discoloration were observed under optical microscope. Two areas with visible brown discoloration were measured, each with a diameter of 25.4 mm and the results are shown in Figure 1. The difference in absorptance between the two areas is not statistically significant. The average absorptance is 0.125. Table 1 compares the absorptance data for the Ag/FEP material from previous flights.

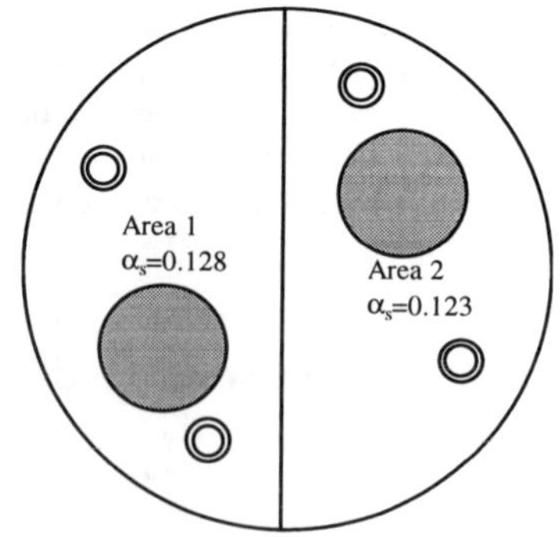

Figure 1. The locations of measurements of solar reflectance on Cryo Vent Cover returned from HST Second Servicing Mission.

Table 1. The solar reflectance and absorptance of Ag/FEP materials

	α_s, preflight	α_s, post-flight	Reference
Pristine Ag/FEP	≤ 0.09	------	[3]
HST SM2 Cryo Vent Cover (Ag/FEP)[a]	------	0.125	This work
HST SM1 SA-I Drive Arm MLI(Ag/FEP)[b]	------	0.097-0.102	[1]
LDEF TCSE Ag/FEP[c]	0.06 - 0.07	0.08 - 0.10	[3]
LDEF Ag/FEP[d]	0.061 - 0.064	0.059-0.087	[3]
LDEF Ag/FEP[e]	0.070	0.073	[3]
Solar Max (Ag,Inconel)/FEP[f]	0.05 - 0.07	0.11 max.	[3]

a. HST second servicing mission. Duration: 83 months.
b. HST First Servicing Mission. Duration: 43 months. Orbit: 614 km to 587 km, inclination 28.5°, Solar UV exposure: 4,477-16,670 ESH; AO fluence: 2.4×10^{20} atoms/cm^2 maximum.
c. LDEF TCSE: Long Duration Exposure Facility Thermal Control Surfaces Experiment. Duration in space: 69.2 months; orbit: 472 km to 333 km, 28.5° inclination; AO fluence: 8.99×10^{21} atoms/cm^2; solar UV exposure: 11,200 ESH; thermal Cycles: ~ 34,000 cycles from -29 C° to 71 C°, ± 11 C°; radiation at surface: 3.0×10^5 rads.
d. Absorptance as a function of locations. Solar exposure: 7,100 - 10,500 ESH; AO fluence: 2.31×10^5 - 8.43×10^{21} atoms/cm^2.
e. LDEF Experiment S0010 Exposure of Spacecraft Coatings, Tray B on Row 9.
f. Solar Max mission. Duration: 50 month; orbit: 575 km to 495 km, 28.5° inclination.

The measured absorptance of the HST SM2 CVC Ag/FEP material is significantly higher than the pristine sample and the Ag/FEP material returned from other flight missions. The CVC surface did not receive direct sun illumination for the most of the mission. The Ag/FEP on HST SM1 SA-I Drive Arm had an absorptance of 0.101 for a surface with the highest solar exposure (about 16,000 ESH) [1]. The significant degradation of the CVC Ag/FEP material from HST SM2 compared to the same material from SM1 could not be explained by the longer duration of SM2.

One possible explanation is that the degradation in solar absorptance of the Ag/FEP material is caused by the crazing of the Ag coatings and the subsequent discoloration of adhesive beneath the Ag layer. Optical examination of the cover revealed that there was severe crazing of the Ag coating, especially in areas that were heavily handled during the assembly process. Those areas had the most severe discoloration (became brownish). Handling of pristine FEP material can generate crazing in the Ag coating. Thermal cycling alone can also generate the crazing. It was found [4] that 40,000 cycles from 50 °C to 150 °C caused the aluminum coating of a pristine Al/FEP sample to craze and the crazing pattern was similar to the crazing pattern observed on the samples from the SM2. This crazing increased the absorptance by 0.03, or 25%. Therefore, different crazing patterns and adhesive discoloration could be one of the reasons for the disparity in solar absorptance of Ag/FEP samples between SM1 and SM2.

3.2 The HST SM1 Al/FEP and Ag/FEP samples   There are six Al/FEP samples and three Ag/FEP samples from the HST SM1. The Al/FEP samples (designated as MSS-B α-1 through α-6) were obtained from a Magnetometer cover MLI. The Ag/FEP samples (designated as SA-I α-7 through α-9) were from the Solar Array Drive Arm MLI. The description of the samples is summarized in Table 2. The detailed description of how the samples were cut from the MLIs can be found in [1]. The SA-I Ag/FEP samples have a fiber reinforced fabric layer bonded to the Ag side with an acrylic adhesive and therefore are opaque. The MSS-B Al/FEP samples have wrinkles and fold lines. Figure 2 shows the optical micrographs under transmitted light. Under high magnification, those samples show the crazing, or mud-tiling, of the aluminum coating. The mud-tiling at those wrinkles and folding lines are much more severe than at other areas. This mud-tiling phenomenon has been observed on the HST SM1 samples, the HST SM2 samples, and the thermocycled pristine Al/FEP samples. Both handling and thermal cycling could result in such mud-tiling. Those cracks have a significant effect on measured solar absorptance of the material. Besides the mud-tiling cracks, scratches and delamination of the aluminum coating are also visible on all samples without any backing layer.

The sample MSS-B α-3 was obtained from an area where the top layer of Al/FEP was bonded to the under layers of the MLI and thus is opaque. More importantly, there were discoloration spots on the sample, which affect the optical properties of the sample. The mud-tiling pattern was also observed on the SA-I Ag/FEP samples but was much less severe than those observed in MSS-B samples. The SA-I samples have a fabric backing layer and show no wrinkles or folds on the surface. Two standard coatings, one aluminum coating on a glass substrate and the other a gold coating on a glass substrate, were also tested and the results were compared with the literature value [5].

The SM1 samples are 26 mm diameter disks. Since the mud-tiling on the samples is not uniform, four measurements were made for each sample by rotating the sample to four

Table 2. Description of the HST SM1 samples.

Sample	Description
MSS-B α-1	Al/FEP, wrinkles and folding lines
MSS-B α-2	Al/FEP, wrinkles and folding lines, Al delamination spots
MSS-B α-3	Al/FEP bonded to layers of Al/Kapton, discoloration spots
MSS-B α-4	Al/FEP, wrinkles and folding lines
MSS-B α-5	Al/FEP, wrinkles and folding lines, Al delamination spots
MSS-B α-6	Al/FEP, wrinkles and folding lines
SA-I α-7	Ag/FEP bonded to a fabric backing layer, flat, no wrinkles
SA-I α-8	Ag/FEP bonded to a fabric backing layer, flat, no wrinkles
SA-I α-9	Ag/FEP bonded to a fabric backing layer, flat, no wrinkles
Pristine Al/FEP	flat, no wrinkles, and no Al delamination
Standard Al	Al coating on glass, mirror-like
Standard Au	Gold coating on glass, mirror-like

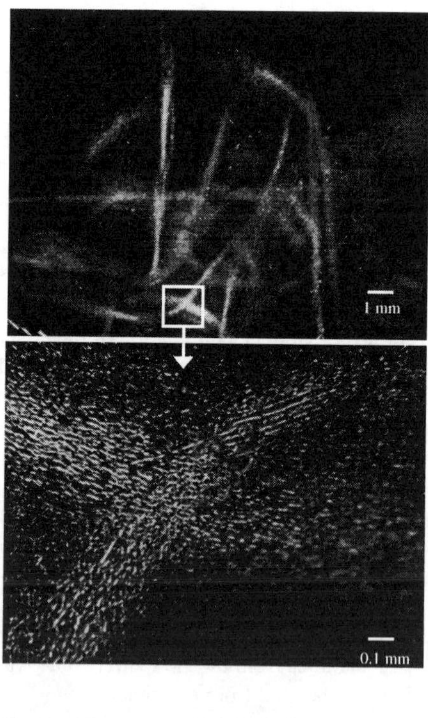

Figure 2. Optical micrograph of MSS-B α-1 under transmitted light. The aluminum coating shows mud-tiling cracks, more severe at wrinkled areas.

positions, as illustrated in Figure 3. For all the measurements, a diffuse standard with 100% reflectivity was placed behind the sample. This procedure was used to eliminate the effect of large delamination of the aluminum coating on the solar absorptance calculation.

The results are listed in Table 3. It should be noted that ASTM E903 indicates that the errors due to measurement and calculation could be ± 0.02. The absorptance for the aluminum coating is 0.071, compared with 0.08 in [5]. The absorptance for the gold coating is 0.182 compared to 0.19 in [5]. Both data are in agreement with the literature data.

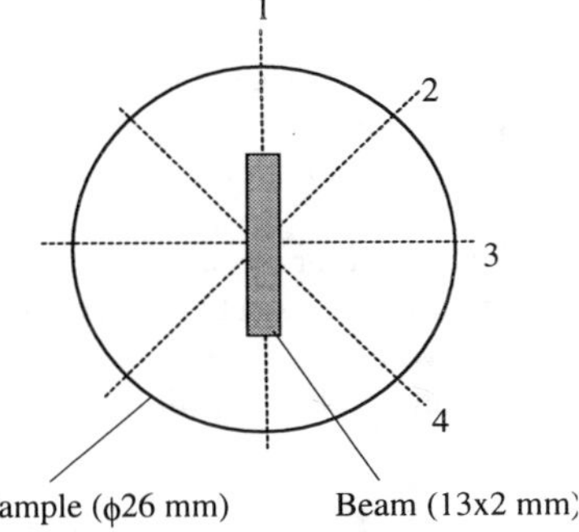

Figure 3. Positions to measure the absorptance of HST SM1 MSS-B and SA-I samples.

Six pristine samples from various locations were measured. The range of the scattering in the absorptance data is only 0.006 and the standard deviation is only 0.002 (Table 3). This scattering contains *both* sample variation and the variation due to the instrument. The scattering due to the instrument alone could be even smaller than the above value. The data from different instruments, however, could have higher variation due to the difference in the configuration of the instruments.

The solar absorptance of the magnetometer cover samples (MSS-B) show large scatter by simply changing the orientation of the sample during the measurement. Since the solar absorptance is calculated by subtracting the solar reflectance from 1, the transmittance due to the mud-tiling of the aluminum coating is counted as absorptance. The mud-tiling patterns are

Table 3. Solar absorptance of HST First Servicing Mission Magnetometer cover (Al/FEP) samples.

Sample	ESH [1]	α_s, range (min. - max.)	α_s, average	Literature[a]
Al coating	-	-	0.071	0.08 [5]
Au coating	-	-	0.182	0.19 [5]
Pristine Al/FEP	-	0.119 - 0.125	0.12 ± 0.002	0.13
MSS-B α-1	11,339[b]	0.189 - 0.250	0.23 ± 0.04	0.22
MSS-B α-2	11,339[c]	0.163 - 0.315	0.21 ± 0.06	0.19
MSS-B α-3	16,670	0.205 - 0.263	0.22 ± 0.02	0.22
MSS-B α-4	4,477	0.165 - 0.197	0.17 ± 0.01	0.17
MSS-B α-5	9,193 or 6,124	0.181 - 0.211	0.20 ± 0.01	0.18
MSS-B α-6	6,124 or 9,193	0.220 - 0.289	0.26 ± 0.03	0.22
Pristine Ag/FEP	-	-	-	<0.09
SA-I α-7	Highest	0.097 - 0.107	0.102 ± 0.005	0.101
SA-I α-8	Middle Value	0.086 - 0.089	0.087 ± 0.001	0.097
SA-I α-9	Lowest	0.096 - 0.107	0.102 ± 0.005	0.102

a. Values are from [1] except where noted.
b. the sample was removed adjacent to the 16,670 ESH surface.
c. the sample was removed adjacent to the 4,477 ESH surface.

not uniform across the samples, and therefore the solar absorptance of a sample is highly dependent on where the reflectance is measured. This is the dominant reason for such large scatter in the absorptance of the MSS-B samples. For the sample MSS-B α-3, which is opaque and thus has no transmission contribution to the absorptance, the large scatter is the result of the localized discoloration on the sample.

The previous data for the same MSS-B samples [1] fall into the range of the data from this work. The differences between the two sets of data are less than 0.02 in most cases, except for MSS-B α-6 where the difference is 0.04. The differences are caused predominantly by the non-uniformity of the mud-tiling patterns and to a lesser extent by the difference in instruments.

The solar absorptance of the six Al/FEP (MSS-B) samples ranges from 0.17 to 0.26, larger than for the pristine sample, 0.12. Part of this degradation in solar absorptance is due to the mud-tiling in the aluminum coating. The data for the sample MSS-B α-3 show that the effect of discoloration of the FEP can significantly increase the absorptance of the material (as indicated by the large scatter due to the discoloration spots).

Table 3 also lists the Equivalent Solar Hour (ESH) for each sample published in [1]. Considering the large scatter of the data for each sample, there is no clear trend between the solar absorptance and the ESH. The relationship between the absorptance and the ESH could be clearer when the absorptance data of the FEP alone (without the aluminum coating) are measured.

The Ag/FEP samples are opaque and there are no wrinkles and folds on the surface. It can be seen that the scatter in the data for the SA-I samples is much less than those for Al/FEP samples (Table 3). The data are in agreement with the data in [1].

3. 3 The HST SM2 MLI Al/FEP samples   The HST SM2 samples were retrieved from the torn top layer of the MLI blanket. The samples were embrittled, cracked, and curled in space. The handling of the sample by the astronauts during the retrieval process may have introduced additional damage to the sample. During the documentation process, the curled sample was flattened, which causes additional through-thickness cracks. The delamination and scratching of the aluminum coating were observed in some areas. Similar to the measurement of HST SM1 Al/FEP samples, a diffuse standard with 100% reflectivity was placed behind the sample to eliminate the effect of delamination and cracking. Due to the limited sample size, only two measurements were made. Table 4 lists the results for pristine, SM1, and SM2 samples.

The increase in solar absorptance is 0.05-0.11 for the SM1 samples and 0.07 for the SM2 samples. Since the SM1 and SM2 samples experienced a different space environment such as solar fluence and radiation dose, those factors should be taken into account when comparing the degradation in solar absorptance of the two samples.

Table 4. Comparison of solar absorptance of Al/FEP samples from HST SM1 and SM2

Sample	α_s	Increase in α_s
Pristine Al/FEP	0.125	-
HST SM1 MLI	0.17 - 0.26	0.05 – 0.14
HST SM2 MLI	0.196	0.07

3. 4 Solar Absorption of the FEP without the aluminum Coating   One of the causes for the increase in solar absorptance for both SM1 and SM2 samples is the cracking and delamination of the aluminum coating. Another reason could be the degradation of the FEP itself. This degradation could be detected by removing the aluminum coating and measuring the absorptance of the FEP. In order to measure the degradation of the FEP, the aluminum coating on the samples was removed using a NaOH solution. Pristine FEP and the SM2 FEP samples were immersed in a NaOH solution until the aluminum coating was etched off and then rinsed with water. The aluminum coating on the SM1 samples was removed using a cotton swab soaked with the NaOH solution. The MSS-B α-3 is a multilayer MLI blanket and the removal of the aluminum coating was not possible. Therefore, no measurement was made on the MSS-B α-3.

Table 5 lists the solar absorptance data for all the samples, including the absorptance values of the samples before the aluminum coating was removed. The solar absorptance, α_s, of pristine FEP is 0.01. The α_s for the five SM1 MSS-B samples ranges from 0.02 to 0.05. The SM2 samples have a α_s value of 0.07. Table 5 also compares the increase in α_s for the FEP samples (without aluminum coating) with those for the Al/FEP samples (with aluminum coating). The former represents the α_s degradation of the FEP itself. The later represents the α_s degradation of the aluminum coating (mud-tiling) as well as the FEP. The differences between the two are the degradation due to the mud-tiling of aluminum coating alone, assuming no other degradation mechanisms exist. For the SM1 sample, the FEP degradation is smaller than the mud-tiling degradation. This is consistent with the observation that the wrinkles and the folding lines on the SM1 samples generated severe mud-tiling of the aluminum coating. The SM2 samples have cracks and large aluminum delamination. Those cracks and delamination, however, do not contribute to the measured absorption because a highly reflective backing was used. The mud-tiling degradation is quite small for the SM2 samples. Therefore, the increase in the absorptance of the FEP accounts for the majority of the overall increase in the solar absorptance of the Al/FEP samples.

It should be noted that the SM2 samples were immersed in NaOH solution during the etching process so that the space-exposed surfaces had the opportunity to react with NaOH. Although there is no evidence that the NaOH reacts with pristine FEP, the space-exposed FEP may have radicals resulting from the space radiation that can react with NaOH. However, the SM1 samples, whose exposed surfaces did not contact the NaOH solution during the etching, showed an increase in α_s. It is reasonable to assume that the α_s for the SM2 samples should increase too. If additional SM2 samples are available, further study should be taken to confirm

Table 5.   Solar absorptance of the HST FEP materials.

Sample	# of samples	α_s of Al/FEP (as-received)	Increase in α_s of Al/FEP	α_s of FEP (Al removed)	Increase in α_s of FEP
Pristine FEP	6	0.12 ± 0.002	--	0.01 ± 0.001	--
SM1 MSS-B α-1	1	0.23	0.11	0.04	0.03
SM1 MSS-B α-2	1	0.21	0.09	0.02	0.01
SM1 MSS-B α-4	1	0.17	0.05	0.02	0.01
SM1 MSS-B α-5	1	0.20	0.08	0.04	0.03
SM1 MSS-B α-6	1	0.26	0.14	0.05	0.04
SM2	2	0.20	0.08	0.07	0.06

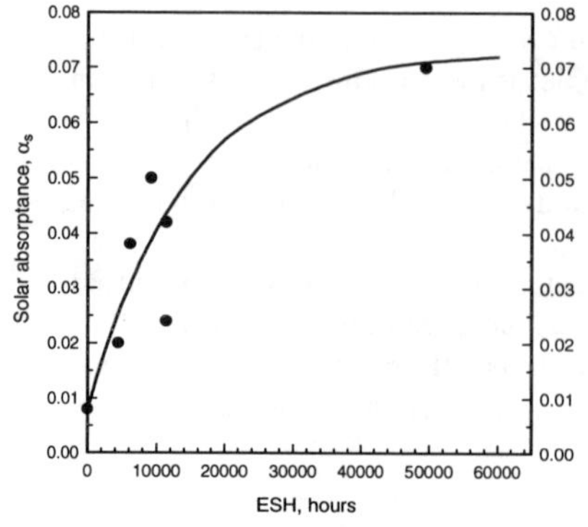

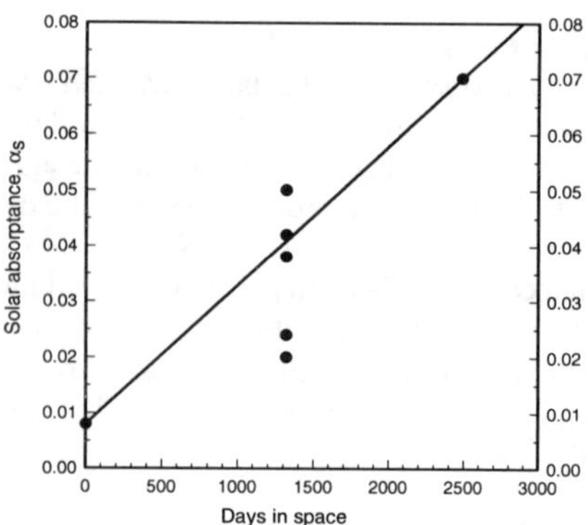

Figure 4 Increase of solar absorptance of the HST FEP material (without coating) as a function of Equivalent Solar Hours (ESH).

Figure 5 Increase of solar absorptance of the HST FEP material (without coating) as a function of time in space.

that the SM2 samples do not react with NaOH.

Figure 4 plots the solar absorptance of the HST FEP as a function of the ESH. Figure 5 plots the solar absorptance against the duration in orbit. Both figures confirm that the solar absorptance of the FEP increases with the ESH or/and duration in orbit. A correlation between the absorptance and the ESH could indicate the role of vacuum UV radiation while the relationship between the absorptance and the duration in space could show the effect of other space environment factors such as electron radiation. Since the scattering in the SM1 data is large and there is only one data from the SM2, the exact functional relationship between the α_s and the ESH or the duration in orbit remains unclear at this point. More data points with different ESH values or duration in orbit are needed to establish the functional correlation and to make predictions.

4. CONCLUSIONS

1. The average solar absorptance of the HST SM2 Cryo Vent Cover (Ag/FEP) was 0.125 (pristine Ag/FEP <0.09). This value is significantly higher than the same type of sample from the HST SM1, LDEF and Solar Max Mission. The difference could not be explained by the difference in duration alone. Crazing of the aluminum coating and discoloration of the adhesives could be the cause.
2. The solar absorptance of the Al/FEP samples from the HST SM1 Magnetometer MLI ranges from 0.17 to 0.26. The crazing and delamination of the aluminum coating played a significant role in the absorptance increase. The absorptance of the FEP also increases.
3. The solar absorptance of the HST SM2 Al/FEP sample was 0.199 (pristine Al/FEP 0.125). Most of the increase in the solar absorption can be attributed to the degradation of the FEP material. Delamination of the aluminum coating also contributed to the absorptance increase.

5. ACKNOWLEDGMENT

The authors would like to thank Mr. Bradford Parker for reviewing the manuscript and useful comments.

6. REFERENCES

1. T. M. Zuby, K.K. De Groh, D.C. Smith, "Degradation of FEP Thermal Control Materials returned from the Hubble Space Telescope," <u>Proceedings of Hubble Space Telescope Solar-Array Workshop</u>, European Space Agency, 1996, pp. 385
2. M. Van Esbeek, F. Levadou, A. Milintchouk, "Investigation on FEP from PDM and Harness from HST-SA1," in ref. 1, pp. 403
3. E.M. Silverman, "Space Environmental Effects on Spacecraft: LEO Materials Selection Guide," <u>NASA Contractor Report 4661</u>, 1995, pp. 10-130
4. J.A. Townsend, P.A. Hansen, J.A. Dever, J.J. Triolo, "Evaluation and Selection of Replacement Thermal Control Materials for the Hubble Space Telescope," in this proceedings.
5. J.H. Henninger, "Solar Absorptance and Thermal Emittance of Some Common Spacecraft Thermal-Control Coatings," <u>NASA Reference Publication 1121</u>, 1984

43rd International SAMPE Symposium
May 31-June 4, 1998

SYNCHROTRON VUV AND SOFT X-RAY RADIATION EFFECTS ON ALUMINIZED TEFLON® FEP

Joyce A. Dever
NASA Lewis Research Center
Cleveland, OH

Jacqueline A. Townsend
NASA Goddard Space Flight Center
Greenbelt, MD

James R. Gaier
NASA Lewis Research Center
Cleveland, OH

Alice I. Jalics
Cleveland State University
Cleveland, OH

ABSTRACT

Surfaces of the aluminized Teflon® FEP multi-layer thermal insulation on the Hubble Space Telescope (HST) were found to be cracked and curled in some areas at the time of the second servicing mission in February 1997, 6.8 years after HST was deployed in low Earth orbit (LEO). As part of a test program to assess environmental conditions which would produce embrittlement sufficient to cause cracking of Teflon® on HST, samples of Teflon® FEP with a backside layer of vapor deposited aluminum were exposed to vacuum ultraviolet (VUV) and soft x-ray radiation of various energies using facilities at the National Synchrotron Light Source, Brookhaven National Laboratory. Samples were exposed to synchrotron radiation of narrow energy bands centered on energies between 69 eV and 1900 eV. Samples were analyzed for ultimate tensile strength and elongation. Results will be compared to those of aluminized Teflon® FEP retrieved from HST after 3.6 years and 6.8 years on orbit and will be referenced to estimated HST mission doses of VUV and soft x-ray radiation.

KEY WORDS: Synchrotron Vacuum Ultraviolet and Soft X-Ray Radiation, Teflon® Degradation

1. INTRODUCTION

Since the Hubble Space Telescope (HST) was deployed in low Earth orbit (LEO) in April 1990, two servicing missions have been conducted to upgrade its scientific capabilities. The first servicing mission (SM1) was conducted in December 1993, 3.6 years after deployment. The second servicing mission (SM2) was conducted in February 1997, 6.8 years after deployment. The HST servicing missions provided an opportunity for on-orbit examination and retrieval of second-surface metalized Teflon® FEP (fluorinated ethylene propylene) used as the top layer of multi-layer insulation (MLI) blankets and on radiator surfaces. Minor cracking of FEP surfaces on HST was first observed upon close examination of samples with high solar exposure retrieved during SM1 (1). During SM2, astronaut observations and photographic documentation revealed cracks in the FEP layer of the MLI on both solar-facing and anti-solar facing surfaces of the telescope.

The efforts reported in this paper were conducted as part of a test program to identify the LEO environmental constituent(s) responsible for cracking and embrittlement of Teflon® FEP on HST (2). Soft x-ray radiation from solar flares has been investigated previously as a possible cause for mechanical properties degradation of Teflon® FEP (3). This paper describes an investigation of the effects of vacuum ultraviolet (VUV) and soft x-ray radiation on Teflon® FEP. Samples of aluminized Teflon® FEP were exposed to synchrotron radiation of various VUV and soft x-ray wavelengths between 18 nm (69 eV) and 0.65 nm (1900 eV), and doses and fluences were compared to those estimated for the HST mission. Synchrotron radiation exposures were conducted using the National Synchrotron Light Source (NSLS), Brookhaven National Laboratory (BNL). Tensile testing was conducted on synchrotron radiation-exposed samples to determine tensile strength and elongation.

2. X-RAY AND SOLAR EXPOSURE ENVIRONMENT ON HST

Table 1 provides information about the x-ray and solar exposure environment on HST from deployment to SM1, SM2 and end-of-life (EOL) (4-6). Radiation in the x-ray wavelength regions 0.05-0.4 nm (3-25 keV) and 0.1-0.8 nm (1.5-12 keV) is attributed primarily to solar flares (3). Based on the x-ray fluence data for these two wavelength regions given in Table 1, it is evident that the majority of the solar flare x-ray fluence is in the 0.4 - 0.8 nm wavelength range, or 1.5-3 keV. The data in Table 1 are for the incident x-ray fluence and do not necessarily indicate the radiation dose absorbed by the Teflon® material.

TABLE 1. X-RAY AND SOLAR EXPOSURE ENVIRONMENT ON HST (4-6)

	SM1 3.6 years	SM2 6.8 years	EOL 20 years
X-ray fluence, 0.05-0.4 nm or 3-25 keV (J/m^2)	14.7	16	47.15
X-ray fluence, 0.1-0.8 nm or 1.5-12 keV (J/m^2)	222.6	252.4	699.6
Solar exposure for solar-facing surfaces (equivalent sun hours)	16,670	33,638	100,000

In addition to the data shown in Table 1 for the two regions of x-ray radiation, Reference 7 provides spectral flux data for VUV to soft x-ray wavelengths. From these data, HST fluences for narrow wavelength bands centered on wavelengths between 18 nm (69 eV) and 1.77 nm (700 eV), which were included in the synchrotron exposure experiments, were calculated and are shown in Table 2. Data for the additional synchrotron exposure energies up to 1900 eV were not available from this reference.

TABLE 2. SOFT X-RAY AND VUV FLUENCES FOR NARROW ENERGY BANDS ON HST

Energy Range, Ave. (eV)	Wavelength Range, Ave. (nm)	Moderate Solar Activity Flux (photons/cm²s) (7)	Estimated Fluence on HST, Assuming Moderate Solar Activity (J/m²)		
			SM1 (3.6 yrs.)	SM2 (6.8 yrs.)	EOL (20 yrs.)
68.6-69.4, 69	17.9-18.1, 18.0	1.10E+08	515	969	2943
283-297, 290	4.18-4.38, 4.28	4.00E+06	126	238	722
490-532, 510	2.33-2.53, 2.43	2.00E+06	133	251	761
663-742, 700	1.67-1.87, 1.77	1.00E+06	65	123	373

In order to determine the radiation dose absorbed by the Teflon® layer of the MLI, it is necessary to know how radiation is absorbed as a function of energy in Teflon®

3. RADIATION ABSORPTION IN TEFLON®

Figure 1 shows the attenuation length of Teflon®, modeled as C_2F_4, as a function of energy (8). Attenuation length is defined as the depth into the material where the radiation intensity has fallen to $1/e$ (0.368) of its intensity at the surface. As shown, attenuation length is dependent on energy or wavelength. The absorption edges at approximately 290 and 700 eV are for carbon and fluorine, respectively. Crosses shown on the curve indicate the energies used for exposure of samples to synchrotron radiation, 69-1900 eV.

In order to calculate the absorbed radiation dose in Teflon®, it was necessary to use the fluence data for 0.05-0.4 nm and 0.1-0.8 nm along with the data for attenuation length. The equation for photon intensity at a specific energy as a function of depth in a material, I_E, is given as follows:

$$I_E = I_{0E} \, exp(-(\mu/\rho)x) \tag{1}$$

where I_{0E} is the incident intensity at a specific energy, μ is the extinction coefficient, ρ is the density, and x is the depth into the material. The term μ/ρ is referred to as the linear absorption

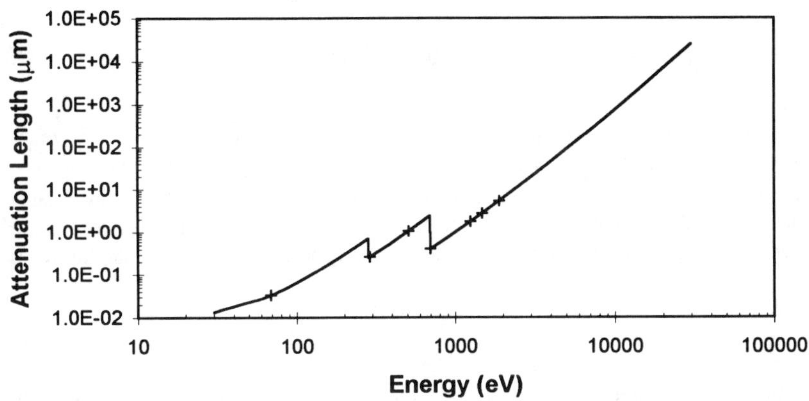

Figure 1 - Attenuation length of Teflon®, modeled as C_2F_4 (8).

coefficient. The definition of attenuation length can be stated as the value of x where

$$I_E = I_{0E}/e. \qquad [2]$$

Substituting Equation 2 into Equation 1 gives:

$$I_{0E}/e = I_{0E}\, exp(-(\mu/\rho)x). \qquad [3]$$

Solving Equation 3 for μ/ρ gives:

$$1/x = \mu/\rho \qquad [4]$$

or

$$\text{attenuation length} = \rho/\mu.$$

Figure 1 provided x vs. energy, so, $1/x$ vs. energy can be obtained from the data in Figure 1. At energies above 1.5 keV, the energies provided for the HST exposure environment, the $1/x$ vs. energy data was curve-fitted to provide the linear absorption coefficient, μ/ρ, as a function of energy, E:

$$\mu/\rho = 3.46 \times 10^9\, E^{-3.12}. \qquad [5]$$

Then, this was substituted into Equation 1 to give:

$$I_E = I_{0E}\, exp((-3.46 \times 10^9\, E^{-3.12})x). \qquad [6]$$

Equation 6 provides the intensity of radiation at depth x into the material for a specific energy, I_E. Therefore, the absorbed intensity at a specific energy, A_E, is given as

$$A_E = I_{0E} - I_E \qquad [7]$$

or

$$A_E = I_{0E} \, (1 - exp((-3.46 \times 10^9 \, E^{3.12})x)). \tag{8}$$

Because I_{0E} and I_E represent intensity, they are expressed in units of W/m^2 or J/m^{2}s. Therefore, they can be described as values of x-ray flux at a specific wavelength or energy. A_E is then the absorbed x-ray flux at a particular energy. The incident HST x-ray fluence values indicated in Table 1 are reported in units of J/m^2 over a range of energies to indicate flux integrated over a specific mission duration. This can be represented by the following equation:

$$F_0 = \int I_{0t} \, dt \tag{9}$$

where F_0 is the incident fluence over a specific mission duration given in J/m^2, I_{0t} is the incident flux at a specific time, and t is time. Multiplying both sides of Equation 8 by time, we obtain the following:

$$F_{AE} = F_{0E} \, (1 - exp((-3.46 \times 10^9 \, E^{-3.12})x)) \tag{10}$$

where F_{AE} and F_{0E} are the absorbed and incident values of fluence, respectively, at a specific energy. Because fluence values in Table 1 are for the rather broad energy/wavelength bands indicated rather than for discreet energies/wavelengths,

$$F_{0E} = \frac{F_{0,E_1-E_2}}{E_2 - E_1}. \tag{11}$$

Equation 11 was then substituted into Equation 10 and the result was numerically integrated over each energy range, 1.5-12 keV and 3-25 keV, using the HST incident fluence data from Table 1. For example, the numerical integration approximated the absorbed fluence for the 1.5-12 keV range given by the following equation:

$$F_{A,1.5\text{-}12\,keV} = \frac{F_{0,1.5-12keV}}{12 - 1.5} \int_{1.5keV}^{12keV} (1 - exp((-3.46 \times 10^9 \, E^{-3.12})x)) \, dE. \tag{12}$$

The absorbed x-ray radiation fluence was thus determined for specific depths, x, into the Teflon$^{®}$ material. Fluence of absorbed x-ray radiation was converted to dose, in units of rads, using the following equation:

$$\text{Dose, rads} = F_A(1/\rho) \, (1/x) \, (1 \text{ rad}/(0.01 \text{ J/kg})) \tag{13}$$

where absorbed fluence over a particular energy range is F_A, in units of J/m^2; density, ρ, is in units of kg/m^3; and thickness, x, is in units of m. Because the outer layer of MLI on HST is comprised of 127 µm thick Teflon$^{®}$ FEP, absorbed radiation dose was calculated for this thickness and is shown in Table 3.

TABLE 3. X-RAY RADIATION DOSE ON HST FOR 127 μm FEP

	SM1 3.6 years	SM2 6.8 years	EOL 20 years
X-ray dose, 0.05-0.4 nm or 3-25 keV (krads)	0.99	1.08	3.18
X-ray dose, 0.1-0.8 nm or 1.5-12 keV (krads)	39.2	44.4	123

4. SYNCHROTRON RADIATION EXPOSURE CONDITIONS

Samples of second-surface aluminized Teflon® FEP were exposed to synchrotron radiation of various energies between 69 eV and 1900 eV. Two different beamlines at the NSLS were used to conduct experiments. As indicated in Table 4, Beamline U16B was used to provide radiation between 69 eV and 700 eV, and Beamline X8A was used to provide radiation between 510 eV and 1900 eV. Table 4 shows the energies and wavelengths used in these experiments including the calculated bandwidth at full-width, half-maximum and the estimated spot size for the beam. Also shown are attenuation lengths for each energy, as shown in Figure 1. As indicated by the attenuation lengths, the majority of radiation from these synchrotron radiation exposures is deposited within the first few micrometers of material.

TABLE 4. SYNCHROTRON RADIATION EXPOSURE CONDITIONS

NSLS Beamline Name	Energy (eV)	Wavelength (nm)	Approx. Spot Size, h x w (mm)	Attenuation Length, μm
U16B	69 ± 0.1	18.00 ± 0.026	3 x 5	0.033
	290 ± 1.7	4.28 ± 0.025	3 x 4.5	0.262
	700 ± 10	1.77 ± 0.025	3 x 4	0.403
X8A	510 ± 2.6	2.44 ± 0.012	10 x 4	1.06
	700 ± 3.5	1.77 ± 0.0089	10 x 4	0.403
	1256 ± 6.3	0.99 ± 0.005	10 x 4.5	1.74
	1489 ± 7.4	0.83 ± 0.004	10 x 4	2.74
	1900 ± 9.5	0.65 ± 0.003	10 x 4	5.36

5. EXPERIMENTAL PROCEDURES

5.1 Samples Samples for synchrotron radiation exposure were fabricated from Teflon® FEP of 127 μm thickness with approximately 100 nm of vapor deposited aluminum. Samples were exposed to the synchrotron radiation beam such that the FEP surface faced the beam and the aluminum surface was the back surface of each sample. Samples for tensile testing were "dog bone" shaped and were die-cut with a die manufactured according to ASTM

Standard D638-95, Type V (9). For this sample size, the width of the narrow section of the dog bone was 3.18 mm, and the gauge length was 9.53 mm. The beam size for each synchrotron energy used is shown in Table 4 for comparison. For experiments at both beamlines, the beam covered the width of the narrow section of the dogbone. For experiments conducted at Beamline X8A where the beam was approximately 10 mm in height, the full gauge length of each specimen was illuminated by the beam. For samples exposed at Beamline U16B, the full gauge length was not covered by the beam. Additional samples included in the experiments which were not intended for tensile testing were nominally 3 to 5 mm maximum width and 15 mm length.

5.2 Apparatus Figure 2 shows the sample/detector holder which was able to hold up to seven pairs of samples as shown. When pairs of samples were exposed, each pair consisted of one tensile specimen and one non-tensile specimen.

Figure 2: Sample holder for synchrotron radiation exposure of FEP samples.

In most cases, the tensile specimen was centered on the synchrotron beam, and the other sample received the "tail" of the synchrotron beam, typically a small fraction of the intensity provided by the beam center. Shown at the left of the sample holder is the silicon photodiode used to measure the synchrotron beam intensity for each sample exposure. The sample/detector holder was attached to a linear positioner with a movement scale allowing positioning accurate to $\pm$ 0.01 mm. Prior to installation in the vacuum chamber, distances between the photodiode and samples were measured with digital calipers to assure accuracy in placement of the samples in the synchrotron beam. Measurements of synchrotron beam intensity were made before and after each exposure and were averaged to determine photon fluence for each sample exposed. A mask was used over the photodiode providing a measurement area 3 mm wide by 10 mm high, comparable to the gauge size of a tensile specimen.

5.3 Tensile Testing Tensile testing was conducted to determine degradation in tensile strength and elongation for synchrotron radiation-exposed FEP samples as compared to pristine material. Sample dimensions were described in Section 5.1. Samples exposed at Beamline X8A were tensile tested using an Instron 4505 load frame with Instron Series IX data acquisition software. Samples exposed at Beamline U16B were tensile tested using an automated bench-top tensile tester and data acquisition software. A strain rate of 0.212 mm/s was used for all tests. Because samples from each beamline were tested using different instruments at different times, results for each set of synchrotron-exposed samples were compared to a set of pristine samples tensile tested at the same time.

6. RESULTS AND DISCUSSION

Table 5 shows results of tensile testing of synchrotron radiation-exposed second surface aluminized FEP. For each sample the exposure energy, measured fluence, absorbed dose, and comparison to orbital doses are provided. Samples U5T and U2T were exposed to both 290 eV and 700 eV radiation, sequentially, as shown. The objective of these experiments was to determine whether exposure to radiation at both the carbon and fluorine absorption edges would produce synergistic degradation.

The HST EOL 0.05-0.4 nm (1.5-25 keV) doses at the attenuation length of the exposure energy were calculated using the data in Table 1 and equations 1-11. For 69-700 eV the narrow-band energy doses for HST EOL at the attenuation length of the exposure energy are also shown for comparison. These narrow-band energy dose values were calculated using the data in Table 2 and equations 2, 7, 9, and 13. As shown in Table 5, in most cases, the sample dose significantly exceeded the HST EOL dose.

The ultimate tensile strength (UTS) and elongation were measured for each tensile specimen, and data are provided in Table 5 with the percent changes from pristine specimens. The tensile test data for pristine specimens are given in Table 6. The tensile test data in Table 5 can be compared to the changes in tensile strength and elongation experienced for aluminized FEP materials retrieved from HST during the first and second servicing missions which are provided in Table 7 (10).

For most synchrotron radiation-exposed samples, some degradation in tensile strength and elongation was observed. However, the degradation in tensile strength and elongation experienced by the synchrotron radiation-exposed samples was not comparable to that demonstrated by samples retrieved from HST after 3.6 years and 6.8 years. The worst degradation in tensile strength was 30.3%, for sample X1T, which received 256 times the HST EOL x-ray dose or an equivalent of over 5000 years in space. This result can be compared to that of an HST specimen retrieved during SM1 after 3.6 years in space. This material showed a degradation in tensile strength of 37-42%, greater than that of the worst damaged synchrotron-exposed sample. The synchrotron radiation-exposed sample which experienced the greatest degradation in elongation was X14T which showed a degradation in elongation of 83%. This sample received an equivalent of over 1500 times the HST EOL x-ray dose or an equivalent of over 30,000 years in space. Compare this result to the complete loss of elongation for the material retrieved from HST after 6.8 years. It is obvious from these results that at the energies used for these synchrotron radiation exposures, VUV radiation and soft x-ray radiation were not sufficient to cause the degradation in mechanical properties displayed by 6.8 year HST-exposed aluminized FEP.

All synchrotron-exposed tensile specimens were examined using optical microscopy for evidence of surface embrittlement or cracking. This effect was described in Reference 3. The only specimens showing evidence of surface cracking after tensile testing were those samples exposed to 290 eV synchrotron radiation, samples U1T and U4T. Figure 3 shows a photomicrograph of sample U4T after tensile testing as compared to sample U6T, exposed to 69 eV synchrotron radiation of a higher absorbed radiation dose. Sample U6T was comparable to all other synchrotron-exposed specimens in that it showed no evidence of surface cracking. It is possible that because 290 eV is at the carbon absorption edge, more severe degradation in the radiation-absorbing layer may occur.

TABLE 5. TENSILE TESTING OF SYNCHROTRON RADIATION-EXPOSED FEP SAMPLES

Sample Label	Energy (eV)	Incident Fluence (J/m^2)	Dose at atten. length (krads)			Ultimate Tensile Strength, UTS (Mpa)	Elongation (%) at Break or at Max. Load*	% Change in UTS from pristine	% Change in % Elongation from pristine
			Sample Dose	HST EOL 0.05-0.8 nm Dose	HST EOL Narrow Energy Band Dose				
U3T	69	5304	4.73E+06	839	2.7E+06	19.48	217	0.2	-1.5
U6T	69	5076	4.52E+06			17.29	179	-11.0	-18.8
U4T	290	3136	3.52E+05	825	5.6E+04	15.08	110	-22.4	-50.1
U1T	290	3099	3.48E+05			14.76	118	-24.1	-46.5
U5T	290	1583	1.78E+05	825	5.6E+04	14.34	74	-26.2	-66.4
	700	1534	1.12E+05	816	2.1E+04				
U2T	290	1611	1.81E+05	825	5.6E+04	14.25	91.3	-26.7	-58.6
	700	1521	1.11E+05	816	2.1E+04				
X6T	510	474	1.31E+04	777	1.6E+04	22.36	649	16.0	11.1
X17T	700	973	7.10E+04	816	2.1E+04	16.18	434	-16.1	-25.7
X15T	700	3055	2.23E+05			17.60	489	-8.7	-16.3
X16T	700	10347	7.55E+05			16.97	484	-12.0	-17.2
X10T	1256	17982	3.04E+05	742	N/A	16.23	406	-15.8	-30.4
X3T	1489	161	1.73E+03	695	N/A	20.34	626	5.5	7.2
X2T	1489	8211	8.81E+04			20.56	629	6.7	7.7
X1T	1489	16123	1.73E+05			13.45	194	-30.3	-66.8
X8T	1489	21938	2.35E+05			14.36	322	-25.5	-44.8
X14T	1489	95859	1.03E+06			13.85	98	-28.2	-83.2
X4T	1900	32	1.76E+02	599	N/A	17.30	516	-10.3	-11.6
X5T	1900	3014	1.65E+04			19.32	596	0.2	2.0
X12T	1900	6632	3.64E+04			16.58	440	-14.0	-24.6

* For U samples, % strain is given at maximum load; for X samples, % strain is given at break.

TABLE 6. TENSILE TESTING OF PRISTINE ALUMINIZED FEP

Specimen Description	Ultimate Tensile Strength (MPa)	Elongation (%)*
X7T, Unexposed witness	20.48	600.3
Pristine, Avg. for 6 specimens, comparison for U samples	19.43 ± 1.9	220 ± 32
Pristine, Avg. for 3 specimens, comparison for X samples	19.28 ± 3.2	584 ± 126

* For U samples, % elongation is obtained at maximum load; for X samples, % elongation is obtained at break.

TABLE 7. TENSILE TEST DATA OF HST-EXPOSED MATERIALS (10)

Sample	Yield Strength (MPa)	Ultimate Tensile Strength, UTS (Mpa)	% Change in UTS from Pristine	Elongation (%)	% Change in % Elongation from Pristine
Pristine aluminized FEP, 3 samples	14.1±0.3	26.5±1.7		363±25	
SM1-retrieved MLI, 3.6 yrs., 11,339 ESH	14.3 14.3	15.4 16.6	-42 -37	196 116	-46 -68
SM2-retrieved MLI, 6.8 yrs., 33,638 ESH	N/A N/A	13.2 2.2	-50 -92	0 0	-100 -100

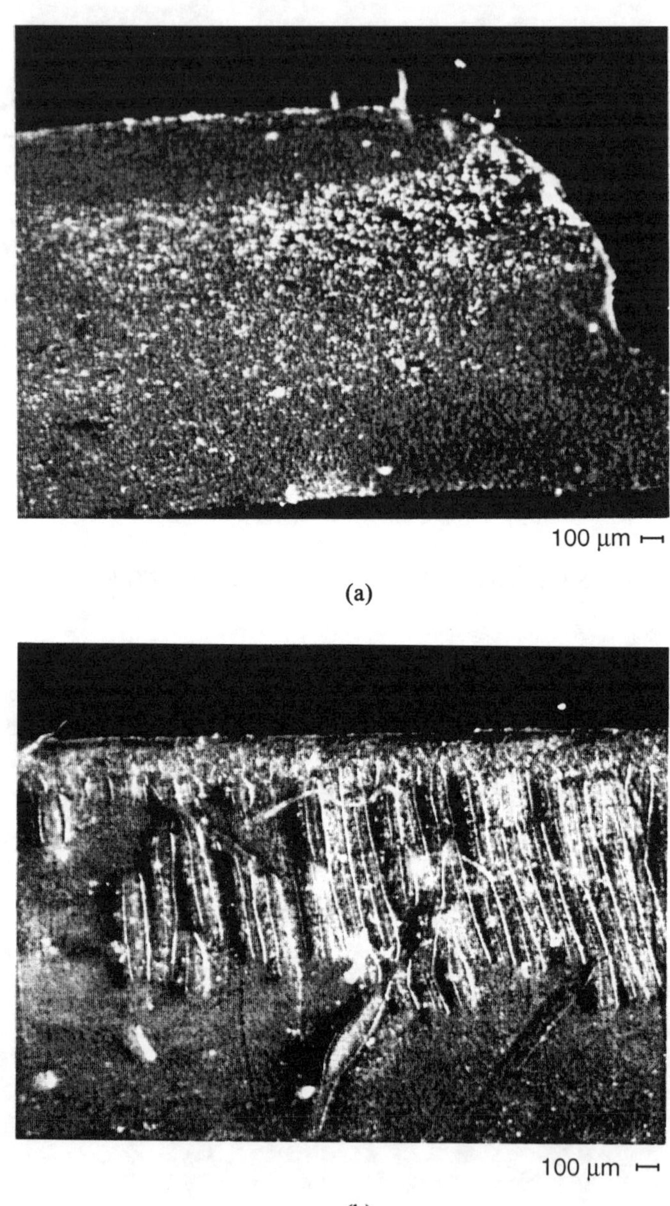

Figure 3: Photomicrographs of tensile-tested, synchrotron radiation-exposed samples (a) U6T exposed to 5076 J/m^2 at 69 eV and (b) U4T exposed to 3136 J/m^2 at 290 eV.

7. CONCLUSIONS

Exposure to synchrotron radiation in the VUV and soft x-ray range of energies is capable of causing degradation in the mechanical properties of Teflon[®] FEP. However, FEP samples exposed to synchrotron radiation of doses significantly greater than HST EOL doses did not show loss of tensile strength or elongation comparable to that of the severely embrittled aluminized FEP retrieved from HST during SM2 which experienced 6.8 years on orbit. Some evidence of wavelength-dependence of damage was observed for samples exposed to 290 eV synchrotron radiation which is at the carbon absorption edge. These samples showed surface cracking after tensile testing. Based on these results, exposure to VUV and soft x-ray radiation alone is not sufficient to cause the severe degradation in mechanical properties observed for FEP materials exposed to the HST environment.

8. ACKNOWLEDGMENTS

Research was carried out, in part, at the National Synchrotron Light Source, Brookhaven National Laboratory, which is supported by the U. S. Department of Energy, Division of Materials Sciences and Division of Chemical Sciences. The authors gratefully acknowledge the support of Michael Sagurton of SFA/Los Alamos National Laboratory, and Steven Hulbert of the NSLS, BNL during experiments conducted at the NSLS. The authors also acknowledge the following for their technical contributions to this paper: Bruce Banks of NASA Lewis Research Center; Thomas Stueber, Edward Sechkar and Mark Forkapa of NYMA, Inc.; Demetrios Papadopoulos of the University of Akron; and Elizabeth Gaier of Manchester College.

9. REFERENCES

1.   Zuby, T. M., de Groh, K. K, Smith, D. C., "Degradation of FEP Thermal Control Materials Returned from the Hubble Space Telescope," NASA Technical Memorandum 104627, December 1995.

2.   Townsend, J. A., Hansen, P. A., Dever, J. A., and Triolo, J. J., "Analysis of Retrieved Hubble Space Telescope Thermal Control Materials," <u>Sci. Adv. Matl. Proc. Eng. Ser.</u>, 43, 000 (1998).

3.   Milintchouk, A., Van Eesbeek, M., Levadou, F. and Harper, T. "Influence of X-ray Solar Flare Radiation on Degradation of Teflon[®] in Space," <u>Journal of Spacecraft and Rockets</u>, Vol. 34, No. 4, July-August 1997, p. 542-8.

4.   NASA Memorandum from J. Barth to P. Hansen, July 7, 1997.

5.   Jackson and Tull Technical Note from T. H. Gregory to P. Hansen, May 12, 1997.

6.   Jackson and Tull Technical Note from T. H. Gregory to J. Townsend, June 25, 1997.

7. Tobiska, W. K., "A Brief History of Empirical Modeling of the Solar EUV Spectral Irradiance," in <u>Proceedings of the Workshop on the Solar Electromagnetic Radiation Study for Solar Cycle 22</u>, Richard F. Donnelly, ed., July 1992, pp. 338-53.

8. Calculation done interactively using Lawrence Berkeley Laboratory Center for X-Ray Optics, Internet Website, http://www-crxo.lbl.gov.

9. ASTM D638-95, "Standard Test Method for Tensile Properties of Plastics," 1995.

10. Dever J. A., de Groh, K. K., Townsend, J. A., Wang, L. L., "Mechanical Properties Degradation of Teflon FEP Returned from the Hubble Space Telescope," AIAA-98-0895, presented at the AIAA 36th Aerospace Sciences Meeting, Reno, NV, Jan. 12-16, 1998, NASA Technical Memorandum 1998-206618.

A SENSITIVITY ANALYSIS OF MODELLING PREDICTIONS
OF THE WARPAGE OF A COMPOSITE STRUCTURE

Andrew Johnston[1], Pascal Hubert[2], Karl Nelson[3], Göran Fernlund[2] and Anoush Poursartip[2]
1-Institute for Aerospace Research, NRC, Ottawa, CANADA
2-Composites Group, Department of Metals and Materials Engineering,
The University of British Columbia,Vancouver, BC, CANADA
3-Boeing Defense & Space Group, Seattle,WA, USA

ABSTRACT

The full potential of composite materials in commercial aerospace applications can only be realised if the relatively high material costs are offset by exploiting their inherent suitability for direct manufacture into large monolithic structures, thereby significantly reducing production costs. However, the manufacture of such structures presents a number of difficult challenges, of which one is variability in process-induced deformation. This paper discusses the application of a comprehensive 2-D finite element processing code, COMPRO, to the prediction of process-induced deformation in a family of stiffened structures. Baseline predictions are compared to experimental measurements and a sensitivity analysis of a large number of parameters is then performed. Although there are many assumptions in the model and gaps in the input data, the sensitivity analysis provides insight into the relative importance of the parameters for this stiffened structure.

KEY WORDS: Composite Structures, Process Modelling, Warpage

1. INTRODUCTION

Modelling of materials processes is a well-established methodology. Over the past two decades, a number of increasingly capable computational models for the autoclave processing of thermoset matrix laminated composite materials and structures have been developed [e.g. 1-10]. Some of these models have focused largely on a single phenomenon such as heat transfer and resin reaction kinetics [e.g. 4,6,7], resin flow [e.g. 2], or stress development [e.g. 8,9] while others have been more broad in scope examining a number of important processing phenomena and their interactions [e.g. 1,3,5,10].

Process modelling is approaching the point where it might be used in support of production of reasonably complex composite structures. However, there are still significant gaps in our understanding there are large numbers of unverified assumptions in the models, and there is great difficulty and cost in fully characterising and calibrating all input parameters. These problems mean that fully quantitative and independently calibrated predictions are not available yet, but an interim useful strategy is to calibrate the model with some baseline experimental results, and then perform a sensitivity analysis of the input parameters.

1.1. COMPRO Process Modelling Software

In this paper we use COMPRO, a two-dimensional finite element process modelling software developed by the Composites Group at The University of British Columbia specifically to analyse industrial autoclave processing of composite structures of intermediate size and complexity. COMPRO incorporates analyses of a number of important processing mechanisms including component internal temperature, resin degree of cure, resin flow and the development of residual stress and deformation. This software can examine complex 2-D structures with multiple composite and non-composite materials, including the effects of process tooling and autoclave characteristics. A more detailed description is given in [10].

A modular approach is employed for the overall program structure, similar to that described by Loos and Springer [1]. The main body of the program consists of a series of 'modules', each responsible for performing a single task such as calculating resin flow (the 'flow-compaction' module) or development of internal stresses (the 'stress-deformation' module). The various modules are called as needed by a controlling routine as the solution marches forward in time. At the beginning of each time step, an 'autoclave controller' module, simulating automated autoclave control, updates all process variables including the autoclave air temperature, the autoclave pressure and the vacuum bag pressure. A central database containing a description of the modelled components (i.e. composite, tool, inserts, etc.,) is updated by each solution module as it is called.

1.2. Aim of This Paper

The purpose of this paper is to show the results of a comprehensive sensitivity study of the effect of key analysis parameters, thermophysical and mechanical properties, and boundary and initial conditions on the predicted warpage of a realistic structure. This is done in full realisation that there are still many unproven assumptions in the model and that the input data is not fully characterised. However, this type of sensitivity analysis can be used not only to gain insight into the manufacturing of this particular structure, but also can be used to guide further development, modification, and validation of the modelling work.

2. PREDICTIONS FOR A J-STIFFENED STRUCTURE

This paper builds on a previous analysis of a J-stiffened structure [11]. These structures are illustrated in Figure 1. The face sheets for all parts were made of 12-plies of Hercules AS4/8552 grade 190 prepreg tape with [45/90/-45/0/45/90]$_s$ lay-up. A single pre-cured braided J-frame made by RTM was centred on the top face sheet and bonded to it using Cytec Metlbond 1515 structural adhesive, which is also used to bond the skins to the Hexcel HRP 0.1 kg/m^3 (8 lb/ft^3) honeycomb core. No special tooling was used to prevent the frames from shifting during cure. Finite element descriptions of all parts were generated for COMPRO using the PATRAN prepocessor. An example of the FE discretization is shown in Figure 2.

The parts were assembled on a large aluminum tool with a constant radius of curvature of 3.10 m (122") in the y-direction and cured using the process cycle shown in the inset in Fig. 2. To examine part-to-part variability, three samples of all components were built. As a baseline, three samples of skins and honeycomb structures with no frames were also built.

Dividing the analysis into a series of four problems allows independent examination of warpage sources in the simpler structures where perhaps only one mechanism is dominant before tackling the most complex problem (the stiffened honeycomb structure), in which several competing mechanisms may be present.

The first of the structures examined is the unstiffened skin. For this simple structure, the dominant source of deformation was found to be the tooling constraint. As discussed in [11], the level of tool/part interaction, modelled via a thin shear layer between the part and the tool, is a significant unknown. This structure is ideal for examining tool/part interaction since this is the dominant deformation mechanism and the low thickness of the part (12 plies) and its large span (30 cm total) result in easily measured warpage.

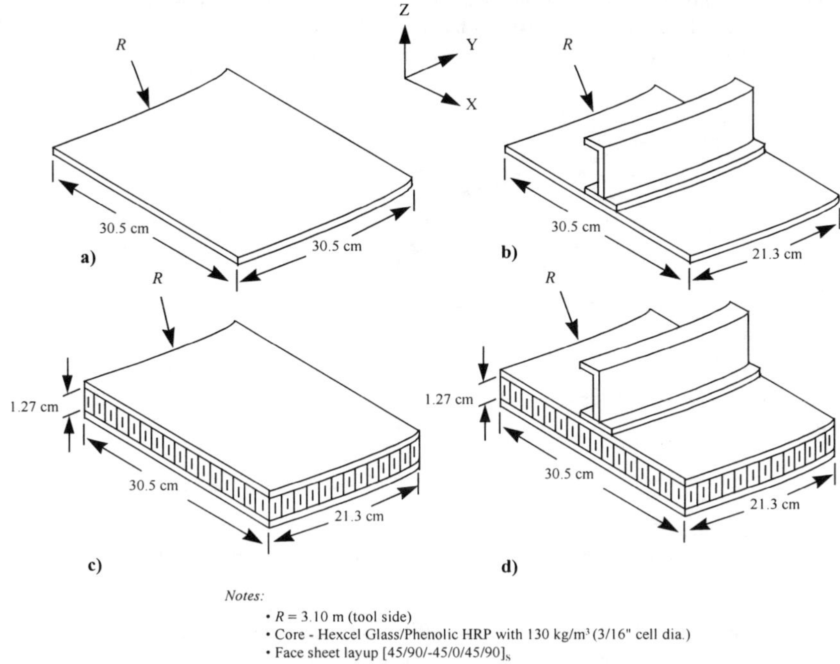

Figure 1   Structures examined in this paper: a) unstiffened skin laminate, b) stiffened skin laminate, c) unstiffened honeycomb structure, d) stiffened honeycomb structure.

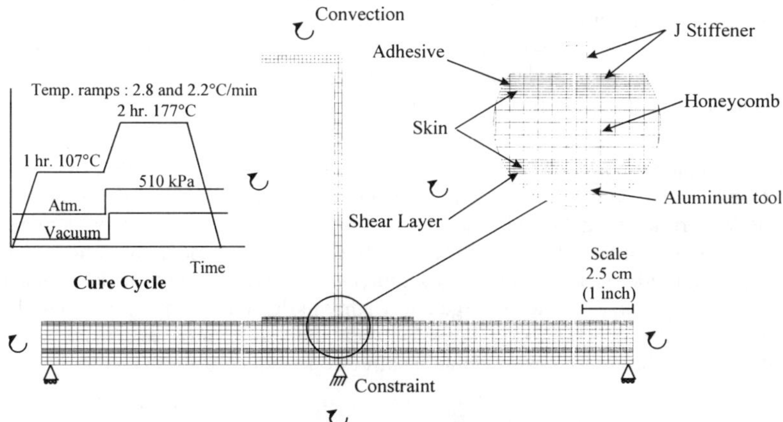

Figure 2   Finite Element Representation of Honeycomb and J-frame Structure

The effect of shear layer on the predicted unstiffened skin warpage was examined by varying its modulus over a wide range. Using a low shear layer modulus results in a nearly symmetric residual strain profile through the skin thickness and thus only a very low level of warpage.

The significant constraint of a high shear layer modulus, however, results in generation of an unsymmetric residual strain profile, and thus much higher warpage.

As shown in Figure 3, the maximum predicted deflection is found to be very highly dependent on shear layer modulus. Comparing predicted warpage to that actually measured, a shear layer shear modulus, G_{SL}, of 5×10^4 Pa has been chosen as the baseline. Figure 4 shows that this choice results in a very good fit to experimental measurements in both warpage shape and magnitude.

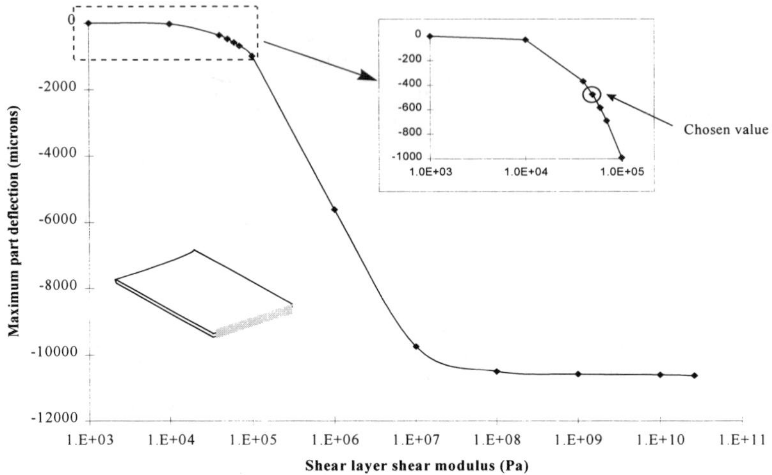

Figure 3.   Shear layer calibration using unstiffened skin part.

Two major sources of process-induced deformation are apparent for the unstiffened honeycomb structure. One is tool/part interaction, the same as for the unstiffened skin. The other is the rather large lag between the temperatures of the top and bottom skins due to the large thermal mass of the tool and the low air/tool heat transfer rates in this case. As illustrated in Figure 5, this temperature gradient results in corresponding gradients in the cure rate of the two skins and consequently in differences in the rate of resin modulus development. The combination of these hardening gradients with the rapid change in thermal and cure shrinkage strains at this point in the process results in the development of residual stress.

Figure 6 shows a comparison between measured and predicted warpage for this structure. It is difficult to say how good the agreement is in this case because of the large amount of scatter in the experimental data and since two of the specimens warped predominantly upwards and one downwards. All that can truly be concluded is that the deformation is of the same order of magnitude as measured. Two sources of potential error in this prediction should be pointed out. One is the rather large uncertainty in the honeycomb moduli employed in the prediction. Also, the adhesive layer between the honeycomb and skins was not modelled. It is unclear how predictions would be affected by either of these factors.

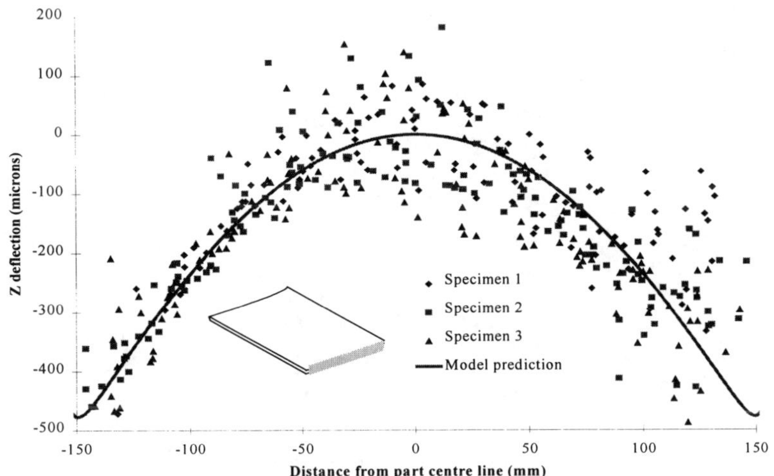

Figure 4. Warpage prediction for unstiffened skin with calibrated shear layer.

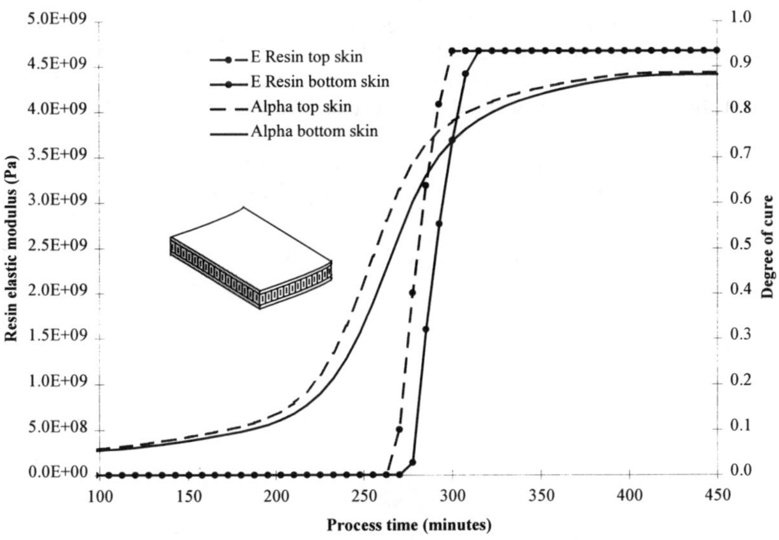

Figure 5. Predicted resin degree of cure and resin modulus in top and bottom skins of unstiffened honeycomb structure.

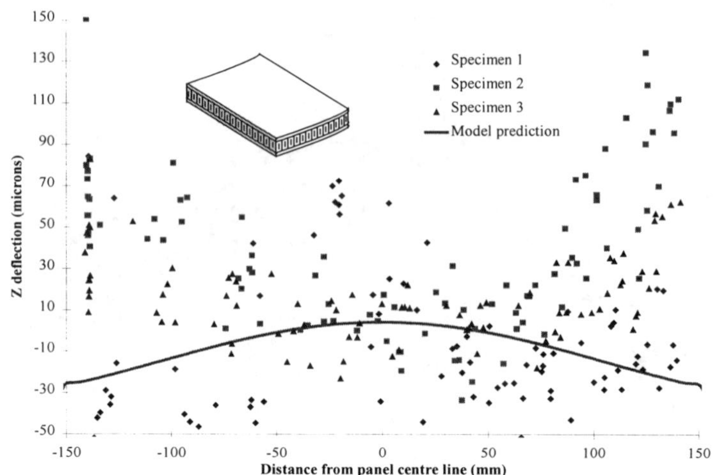

Figure 6. Model warpage prediction for unstiffened honeycomb panel.

The next part examined was the stiffened skin structure. For this structure, it was clear from the observed warpage pattern that the major source of warpage was an interaction between the skin and the J-frame. However, initial attempts to model this part resulted in very poor results, indicating warpage in the *opposite direction* to that actually measured. Closer examination of the problem revealed that an important factor had been overlooked. The J-frame includes on its underside an adhesive noodle that had not been included in the original model. Re-doing the analysis, this time incorporating a geometrically crude representation of this noodle, resulted in the predicted post-processing shape shown in Figure 7.

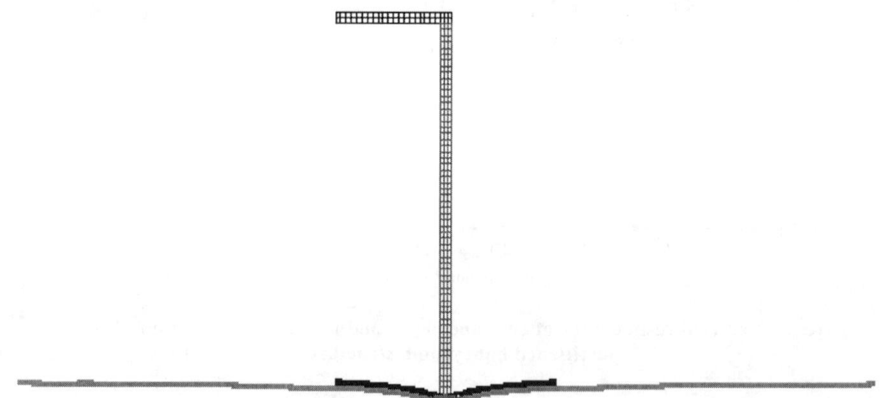

Figure 7. Predicted post-processing shape of J-stiffened skin part with modelling of adhesive noodle (displacements exaggerated by a factor of 10).

The magnitude of the predicted deflection from this new model agreed much better with experimental results. However, the bulk of the warpage was predicted to be much too localized, reflecting the crudity of the used geometric representation of the noodle. Unfortunately, model limitations on the number of elements prevented a more detailed geometric description of this region of the foot from being used. For this reason, the effect of the noodle was instead incorporated into the analysis by calibrating the *CTE* of the entire J-frame foot such that good agreement was obtained with measured total part deflection. Figure 10 shows that neither this 'distributed strain' approach nor the crude noodle model accurately predict the shape of the warpage directly beneath the foot, but act as bounds to the actual shape. A more accurate model of the noodle structure would be expected to produce a better fit in this region.

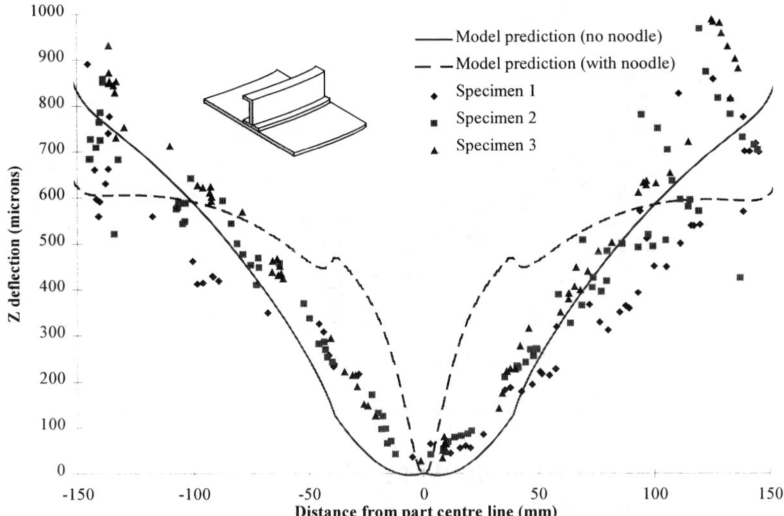

Figure 8. Model warpage prediction for J-stiffened skin structure showing best fit prediction and prediction with crude adhesive noodle model.

The final part examined, representing a synthesis of the other three, is the stiffened honeycomb structure. All of the deformation mechanisms acting in the other structures play a role in determining the final shape of this structure. Fortunately, since analyses of these deformation sources have already been performed in the simpler parts, modelling of this structure is quite straightforward. The predicted underside deformation for this case is compared to measurements in Figure 9. As shown in this figure, a somewhat lower maximum deflection is predicted than obtained from experiment (about 0.12 mm versus 0.15 mm). Away from the foot of the frame, the 'slope' of the predicted deformation is in quite good agreement to that measured, but warpage in the area directly beneath the foot is not well predicted. This points to the most likely source of the observed disagreement as being the distributed foot *CTE* representation of the adhesive noodle. However, as is discussed in the following sensitivity analysis, the predicted warpage in this case is quite sensitive to input parameters, and other factors may also be involved.

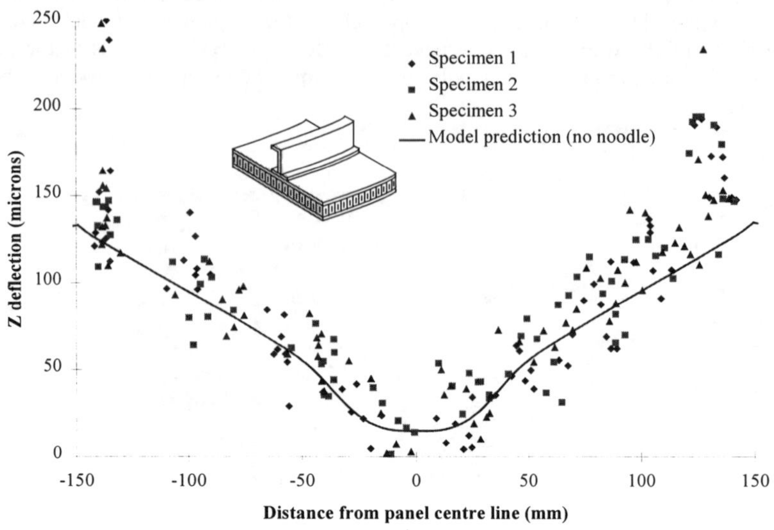

Figure 9. Model warpage prediction for J-stiffened honeycomb panel (nominal case).

3. SENSITIVITY ANALYSIS

The focus of the sensitivity analysis is the effect of variability in process parameters on process-induced deformation. In this analysis the *maximum* underside deflection of the stiffened honeycomb structure is examined. The parameters examined in this sensitivity study are outlined in Table 1, which also shows the 'high' and 'low' values employed in each case.

Sensitivity analysis predictions for a number of parameter categories are summarized graphically in Figures 10-12. Analysis parameters had little effect on predicted deflection, nor did any composite thermophysical properties with the exception of resin cure kinetics (Figure 10).

As shown in Figure 11, boundary and initial conditions (including tool material) were of intermediate importance, with all parameters except initial degree of cure having a significant influence on predicted warpage. Varying the mechanical properties of the various materials in the structure (Figure 12) had the greatest effect on deformation predictions. For the range of property values examined, predicted deflections ranged from -0.06 mm to 0.23 mm.

Table 1. Parameters examined in sensitivity analysis.

Parameter	Nominal Value	High Value	Low Value
Analysis parameters			
Maximum Overall Time Step (s)	30	100	10
Maximum Degree of Cure Step (Stress)	0.05	0.10	0.025
Maximum Percentage Change in E_{resin} (Stress)	10	20	5
Thermophysical properties			
Resin Specific heat capacity (J/kgK)	$1300 + 4.29 * T - 369 * \alpha$	$2790 - 3.80 * T$ [1]	$1005 - 369 * \alpha$ [2]
Resin Thermal conductivity (W/mK)	$0.148 + 3.43 \times 10^{-4} * T + 6.07 \times 10^{-2} * \alpha$	$0.185 + 4.293 \times 10^{-4} * T + 7.59 \times 10^{-2} * \alpha$ (+25%)	$0.155 + 6.07 \times 10^{-2} * \alpha$ [2]
Honeycomb	$0.0774 + 3.4 \times 10^{-4} * T$	$0.0968 + 4.25 \times 10^{-4} * T$ (+25%)	$0.0581 + 2.55 \times 10^{-3} * T$ (-25%)
Fibre V_f (-)	0.573	0.602 (+5%)	0.544 (-5%)
Resin heat of reaction (J/kg)	540×10^3	590×10^3 (+10%)	490×10^3 (-10%)
Mechanical properties			
Resin modulus development model	[12] Model #2, Parameters in [12] Table 6.2	[12] Model #1 [6] (see [12] Table B.5) $\alpha_{C1} = 0.608$, $\alpha_{C2} = 0.750$	
Resin modulus development: Timing (modulus development model #2)	Parameters in [12] Table 6.2	$T_{C1a}^* = -56.7$ K, $T_{C2a}^* = -23.0$ K [3]	$T_{C1a}^* = -34.7$ K, $T_{C2a} = -1.0$ K [3]
Resin modulus development: Initial modulus (modulus development model #2)	Parameters in [12] Table 6.2	$E^0 = E^\infty/10^2$	$E^0 = E^\infty/10^4$
Resin cure shrinkage: amount (shrinkage model #1)	Parameters in [12] Table 6.5, ($\alpha_{C2} = 0.67$ actually used)	$V_r^{Sx} = 0.124$ (+25%)	$V_r^{Sx} = 0.074$ (-25%)
Resin cure shrinkage: timing (shrinkage model #1)	Parameters in [12] Table 6.5	$\alpha_{C1} = 0.05$, $\alpha_{C2} = 0.77$	$\alpha_{C1} = 0.05$, $\alpha_{C2} = 0.57$
Composite CTE_3 ($\times 10^{-6}$/°C)	28.6	$27.4 + 6 \times 10^{-2} * T$ [4]	25.7 (-10%)
J-Frame foot CTE_1 ($\times 10^{-6}$/°C)	8.50 (#)	2.9	12.0
Honeycomb moduli (Pa)	$E_{11} = 43.6 \times 10^6$, $E_{33} = 113 \times 10^6$, $G_{13} = 16.6 \times 10^6$	$E_{11} = 87.6 \times 10^6$, $E_{33} = 226 \times 10^6$, $G_{13} = 33.2 \times 10^6$	$E_{11} = 21.9 \times 10^6$, $E_{33} = 56.5 \times 10^6$, $G_{13} = 8.3 \times 10^6$
Honeycomb CTE_1 ($\times 10^{-6}$/°C)	10	15	5
Layup accuracy	Perfect	-4° to +4° random angle deviation, both faces	-4° to +4° random angle deviation, both faces
Boundary and initial conditions			
Tool material	Aluminum 'equivalent'	invar 'equivalent'	
Effective heat transfer coefficients (W/m^2 K) P is gage pressure	Top Side: $4.05 + 2.41 \times 10^{-5} * P$ Bottom Side: $36.6 + 2.00 \times 10^{-4} * P$	Top Side: $6.08 + 3.62 \times 10^{-5} * P$ (+50%) Bottom Side: $54.9 + 3.00 \times 10^{-4} * P$ (+50%)	Top Side: 4.05 Bottom Side: 36.6 [4]
Initial degree of cure (-)	$\alpha_0 = 0.05$	$\alpha_0 = 0.15$	$\alpha_0 = 0.01$

Notes:

[1]   Measured variation above Tg

[2]   Room temperature value for nominal case

[3]   Shifts modulus development curve by α = +/- 0.05 at a given temperature. Estimated error is based on measurements of modulus development for other materials tested at a range of frequencies

[4]   Based on crude estimates of the variation of specimen CTE_3 with temperature from the original ply measurements (taken from a single chart since no numerical data was available).

[5]   From previous cure kinetics analysis by The Boeing Company

[6]   # - Calibrated value.

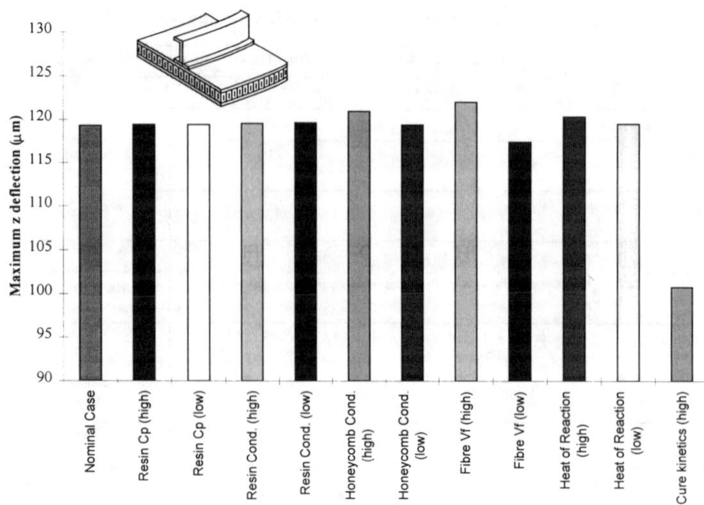

Figure 10. Predicted sensitivity of maximum deformation of stiffened honeycomb structure to variation in thermophysical properties.

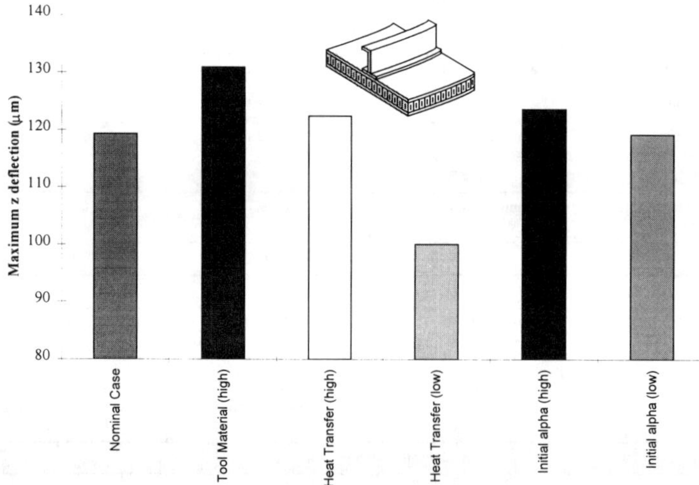

Figure 11. Predicted sensitivity of maximum deformation of stiffened honeycomb structure to variations in boundary and initial conditions.

The predicted sensitivity of structure warpage to mechanical property variations (Figure 12) was significant, with virtually all parameters examined causing at least a 10% shift in model predictions. The most significant was the J-frame foot thermal expansion coefficient, variations in which resulted in changes in model predictions of over 100%. While not insignificant, neither tool material nor initial resin modulus were predominant sources of variation in this analysis.

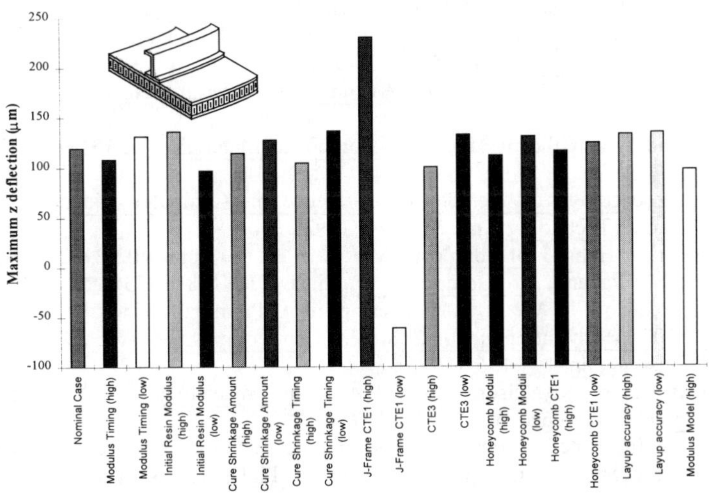

Figure 12. Predicted sensitivity of maximum deformation of stiffened honeycomb structure to variations in mechanical properties.

4. CONCLUSIONS

For the sensitivity analysis performed on the structure presented in this paper, the following conclusions can be drawn:

- It is critical to have a complete and sufficiently detailed representation of the structure, including the tool and the effect of autoclave characteristics.

- Normal variations in composite thermophysical properties had little effect on process model predictions.

- The resin cure kinetics model played an important role, primarily through the effect of cure model predictions on resin hardening and cure shrinkage behaviour rather than its effect on temperature distribution.

- Examination of very local behaviour may be required in some instances in order to get accurate predictions, for example when studying the effects of the J-frame noodle.

5. ACKNOWLEDGMENTS

This paper summarizes work performed over a number of years under funding by the Science Council of British Columbia, the Natural Sciences and Engineering Research Council of Canada, The Boeing Company, The National Aeronautics and Space Administration and Integrated Technologies Inc. We would also like to gratefully acknowledge the significant interaction and support from our colleagues at The Boeing Company and The University of British Columbia.

6. REFERENCES

1. A.C. Loos and G.S. Springer, J. of Composite Materials, 17, (2), pp. 135-169 (1983).
2. R. Davé, J.L. Kardos and M.P. Dudukovic, Proceedings of the American Society for Composites, 1st Technical Conference, Technomic Publishing, pp. 137-153 (1986).

3. A.R. Mallow, F.R. Muncaster and F.C. Campbell, <u>Proceedings of the American Society for Composites, Third Technical Conference</u>, Technomic Publishing, pp. 171-186 (1988).
4. J. Mijovic and J. Wijaya, <u>SAMPE J.</u>, <u>25</u>, (2), 35-39 (1989).
5. T.A. Bogetti and J.W. Gillespie Jr., <u>J. of Composite Materials</u>, <u>26</u>, (5), pp. 626-660 (1992).
6. T.A. Bogetti and J.W. Gillespie Jr., <u>J. of Composite Materials</u>, <u>25</u>, (3), pp. 239-273 (1991).
7. J.M. Kenny, <u>Proceedings of the Third Conference on Computer Aided Design in Composite Materials Technology</u>, pp. 530-544 (1992).
8. S.R. White and H.T. Hahn, <u>J. of Composite Materials</u>, <u>26</u>, (16), pp. 2402-2422 (1992).
9. H.T. Hahn and N.J. Pagano, <u>J. of Composite Materials</u>, <u>9</u>, 91-108 (1975).
10. P. Hubert, A. Johnston, R. Vaziri, A. Poursartip, in A. Poursartip and K.N. Street, eds., <u>Proceedings of The 10th International Conference on Composite Materials (ICCM-10)</u>, Woodhead Publishing, pp. 149-156, (1995).
11. A. Johnston, P. Hubert, K. Nelson, and A. Poursartip, <u>Proceedings of 38th International SAMPE Technical Conference</u>, pp734-744 (1996).
12. A. Johnston, "An Integrated Model of The Development of Process-Induced Deformation in Autoclave Processing of Composite Structures", Ph.D. Thesis, The University of British Columbia, April 1997

43rd International SAMPE Symposium
May 31-June 4, 1998

RESIDUAL STRESSES IN CONTINUOUS GLASS FIBER/ POLYPROPYLENE COMPOSITE THERMOFORMED PARTS

Y. Youssef and J. Denault

Industrial Materials Institute, National Research Council

75 De Mortagne, Boucherville (Québec) Canada J4B 6Y4

ABSTRACT

Thermoplastic composite matrix material undergoes successive thermal phase transformations during the thermoforming process. The processing conditions control these transformations and hence are responsible for the final crystalline morphology, the inerface quality, the mechanical properties as well as for the residual stresses developed in the composites.

The objective of this work was to develop a model in order to be able to predict the development of residual stresses during the molding process of the continuous glass fiber/polypropylene composites, PP/G. For this purposes, the classical lamination theory model was adapted in order to include temperature-dependent parameters and relaxation phenomena. Stress relaxation phenomena were taken into account by considering the viscoelastic behavior of the matrix and the transverse stress cracking in the composite plies. The model incorporates also the effect of thermal history and crystallization kinetics which are known to control the matrix microstructure and the interface quality of the composites.

KEY WORDS: Thermoplastic composites, Polypropylene, Glass fiber, Interfacial strength, Residual stress, Stress relaxation

1. INTRODUCTION

The processing of thermoplastic composite materials induces inevitably thermal residual stresses due to the mismatch in the fiber and matrix coefficients of thermal expansion. These stresses result in dimensional distortions of the molded parts and may reduce its mechanical performance. The understanding of the development of these stresses, the identification of the variables controlling them may help in reducing or controlling these stresses and may also enable to predict their effects on part warpage and mechanical performance.

Many investigations of the process-induced residual stresses topic have been carried out [1-6] but most of the reported work is related to the graphite/polyetheretherketone system. In this case, most significant portion of the residual stresses builds-up between the glass transition temperature, T_g where relaxation effects are much less important resulting in increasing residual stresses with increasing cooloing rates [3]. Polypropylene based composites are very different since their T_g is lower than the ambient temperature. Also, the fiber-matrix interaction in these composites is very particular due to the inert character of PP and brings particular behavior characteristics. The growing interest in this class of thermoplastic composites, particularly for automotive applications, justifies more research efforts in order to understand and characterize their behavior starting with the processing phase.

In previous work on thermoforming of continuous fiber polypropylene/glass composites, it has been shown that processing conditions, particularly molding temperature and cooling rate, are responsible for the final matrix morphology and hence control the mechanical performance of the composite [7]. The effect of the cooling rate was found particularly significant in the development of residual stresses [8]. It was also shown that residual stress predictive models should take into account temperature-dependent parameters and should account for stress relaxation phenomena to be more effective.

In the present work, residual stresses in cross-ply unsymmetric laminates are experimentally evaluated by the measurement of curvatures on narrow strips of two polypropylene/glass systems characterized by different fiber-matrix interface quality. Classical lamination theory model for residual stress prediction is adapted in order to include temperature-dependent parameters and relaxation mechanisms. The prediction of the model are than compared to the measurements.

2. [0$_n$/90$_n$] LAMINATE RESIDUAL STRESS MODEL

Thermal stresses rising at ply level in an unsymmetric $[0_n/90_n]$ composite laminate due to a change of temperature from T_i to T_f are given by the following equation:

$$\sigma_x^T = \int_{T_i}^{T_f} \left(\overline{Q}_{11}\alpha_x + \overline{Q}_{12}\alpha_y \right) dT$$

$$\sigma_y^T = \int_{T_i}^{T_f} \left(\overline{Q}_{12}\alpha_x + \overline{Q}_{22}\alpha_y \right) dT \tag{1}$$

where $\overline{Q}_{ij}$ and α_j are respectively the transformed stiffnesses and the coefficients of thermal expansion (CTE) of the ply. The resulting thermal forces, N_i^T, and moments, M_i^T, in a $[0_n/90_n]$ laminate of thickness h are:

$$N_x^T = \int_{-h/2}^{h/2} \int_{T_i}^{T_f} \left(\overline{Q}_{11}\alpha_x + \overline{Q}_{12}\alpha_y \right) dT\, dz$$

$$N_y^T = \int_{-h/2}^{h/2} \int_{T_i}^{T_f} \left(\overline{Q}_{12}\alpha_x + \overline{Q}_{22}\alpha_y \right) dT\, dz \tag{2}$$

and

$$M_x^T = \int\limits_{-h/2}^{h/2} \left(\int\limits_{T_i}^{T_f} \left(\overline{Q}_{11}\alpha_x + \overline{Q}_{12}\alpha_y \right) dT \right) z\, dz$$

$$M_y^T = \int\limits_{-h/2}^{h/2} \left(\int\limits_{T_i}^{T_f} \left(\overline{Q}_{12}\alpha_x + \overline{Q}_{22}\alpha_y \right) dT \right) z\, dz \tag{3}$$

During the processing of thermoplastic composite laminates in a hot press, the laminate is kept flat until the end of the process. When the pressure is released (open mold), the laminate is free to shrink and warp due to the thermal forces and moments built-up during the cooling phase. The laminate mid-plane strains and curvatures, ε_i^o and k_i, are then calculated by solving the governing equation:

$$\left\{ \begin{array}{c} \varepsilon_i^o \\ k_i \end{array} \right\} = \left[\begin{array}{c|c} A & B \\ \hline B & D \end{array} \right]^{-1} \left\{ \begin{array}{c} -N_i^T \\ -M_i^T \end{array} \right\} \tag{4}$$

where A, B and D are the stiffness and coupling matrices of the laminate. For the considered $[0_n/90_n]$ laminate, the solution of equation (4) for curvatures leads, in dimensionless form, to:

$$hk_x = -hk_y = \frac{24\, h \left(Q_{11} - Q_{22} \right) N_x^T + 96 \left(Q_{11} + Q_{22} + 2\, Q_{12} \right) M_x^T}{h^2 \left(Q_{11}^2 + 14\, Q_{11}Q_{22} + Q_{22}^2 - 16\, Q_{12}^2 \right)} \tag{5}$$

where Q_{ij} are the ply stiffnesses at the final temperature, T_f. If the material properties are considered independent of temperature and equal to those measured at T_f, equation (5) reduces to:

$$hk_x = -hk_y = -\frac{24 \left(Q_{11}Q_{22} - Q_{12}^2 \right) \left(\alpha_1 - \alpha_2 \right)}{\left(Q_{11}^2 + 14\, Q_{11}Q_{22} + Q_{22}^2 - 16\, Q_{12}^2 \right)} \left(T_f - T_i \right) \tag{6}$$

These equations constitute the basics of the residual thermal stress model based on classical laminate theory (CLT). This model predicts anticlastic saddle shapes for unsymmetric laminates regardless the dimensions of the panel. Hyer [9] investigated experimentally the shapes of unsymmetric laminates and reported that for small thickness to in-plane dimensions ratios, unsymmetric laminates can take right cylinder curvatures. Hyer [10] developed an extended laminate theory (ELT) to account for the non-linear out-of-plane displacements when the thickness is small compared to the in-plane dimensions of the laminate. For a given thickness to in-plane dimensions ratio or when curvatures are measured on narrow strips, the solutions of both CLT and ELT are very close [3]. Since the curvatures studied in the present work are measured on narrow strips, the CLT model will be used.

3. RESIDUAL STRESS MEASUREMENTS

The thermoplastic composite considered in the present study is glass fiber reinforced polypropylene. Two different unidirectional prepreg tapes were used. The first one, named PP/G, consisting in a pure PP matrix reinforced with 52 wt% of E-glass fibers, and the second, named mPP/sG, consisting in a 50/50 blend of chemically modified PP with pure PP reinforced with 60 wt% of E-glass fibers coated with a specific thermoplastic sizing. The use of these two different materials aims at highlighting the role of fiber-matrix interface quality in the residual stress build-up. PP/G prepreg has a nominal thickness of 0.4 mm and mPP/sG tape has a nominal thickness of 0.25 mm. These two materials were obtained from BAYCOMP Canada.

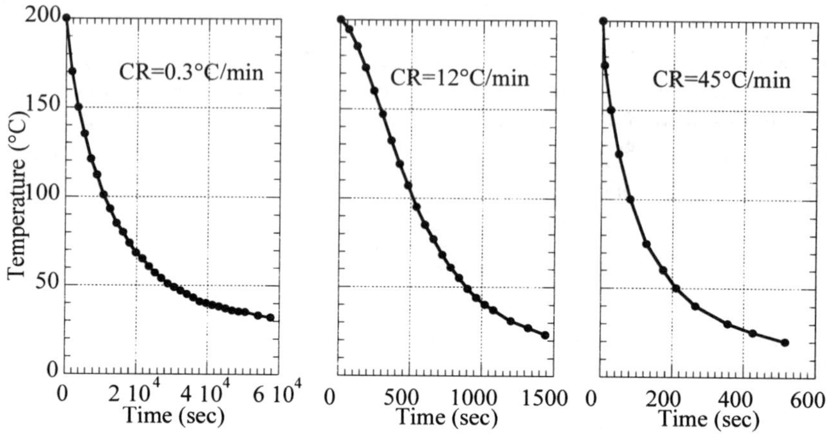

Figure 1. Temperature variation during the cooling phase of the molded panels from 200°C to room temperature

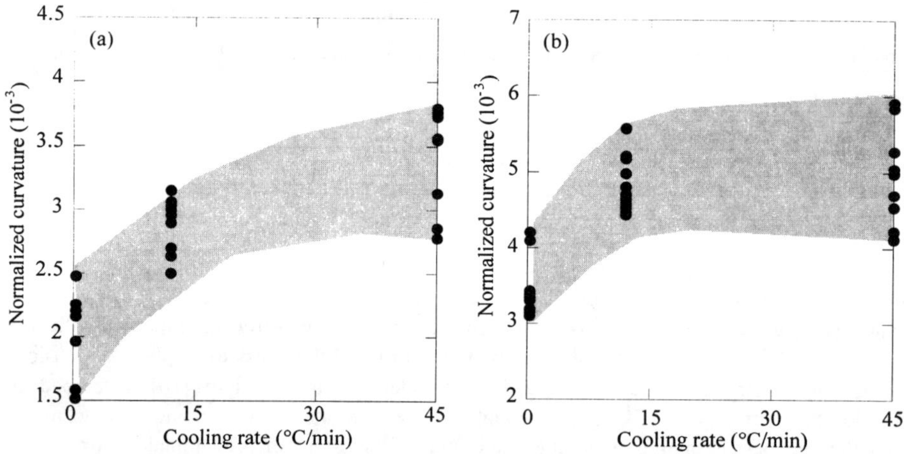

Figure 2. Normalized curvatures ($k \times h$) as a function of the cooling rate measured on cross-ply unsymmetric laminates of PP/G (a) and mPP/sG (b) composites

Square panels (150 mm×150 mm) of cross-ply unsymmetric laminates were molded in a flat matched mold. Molding temperature was 200°C, molding pressure was 0.7 MPa, residence time at molding temperature was 5 min and three cooling scenarios were applied resulting in the time-temperature curves shown on Figure 1. The average cooling rates in the temperature range of 140 to 80°C (covering the crystallization range) were evaluated respectively as CR=0.3, 12 and 45°C/min. Six [$0_4/90_4$] panels of mPP/sG and six [$0_3/90_3$] panels of PP/G were molded. All panels took a saddle-like shape when demolded at room temperature. Three to five narrow strips 15 mm×150 mm were cut from each panel and their curvatures were measured. Figure 2 shows the distribution of the measured curvatures normalized by laminate thickness ($k \times h$) for both materials as a function of the cooling rate. The scatter in the curvature measurements may be attributed to the usual error sources in composites, e.g., precision of measurements, fiber misalignment, disorientation of plies and non-symmetry of ply thickness. These results will be compared to theoretical predictions.

4. TEMPERATURE-TIME DEPENDENT BEHAVIOR

The mechanical and thermal properties of the composite systems are defined by the well-known rule of mixtures. The stiffness and the coefficient of thermal expansion of glass fibers can be reasonably assumed constant over the temperature range 23°C to 200°C. The temperature-dependent properties are those of the matrix and their variations are discussed here after. The matrix properties are sensitive to temperature and stress/strain history and hence, the composite transverse properties are also temperature-and-time dependent.

4.1 Mechanical Properties of the matrix

DMTA measurements have been carried out on both matrix materials under sinusoidal loading at a frequency of 1 Hz and R=0.1 and for temperature varying from -40°C to the complete softening of the material at around 150°C at a rate of 2°C/min. The results of these tests are shown in Figure 3. It should be noted that the DMTA does not allow to scan temperature from the melt state back to room temperature. In order to verify the behavior of the material during cooling, complementary measurements of the compression modulus (linear combination of Young's and bulk modulus) using ultrasonic waves in heating and cooling at a rate of 2°C/min have been carried out, details of this method can be found in Ref. [11]. The variation of the compression modulus (which is proportional to Young's modulus) with temperature for PP is also represented in Figure 3. The ultrasonic measurements show that the material softens while being heated until the melt temperature ($T_m \approx 165$°C). It could be observed that the cooling path differs from the heating one only between T_m and the crystallization temperature ($T_c \approx 120$°C). Since the residual stress calculations are limited to temperatures below T_c, the modulus variation during the cooling phase can be assumed identical to that measured during heating by DMTA without significant error.

The values of the DMTA measured modulus at T=23°C (1860 MPa for PP and 1980 MPa for mPP) are in the same range than the Young's modulus values of polypropylene (from 1600 to 2000 MPa) obtained by tensile tests. Quasi-static tensile tests have been conducted on unidirectional specimens loaded in the transverse direction and on cross-ply laminates loaded at 45° angle. It was found that the lamina transverse stiffnesses measured experimentally on the composites vary with the cooling rate used when processing the laminates. The slower is the cooling, the stiffer is the lamina and this was explained by the difference in the matrix microstructure [8]. However, these values are generally lower than the nominal moduli (calculated by rule of mixture) especially in the case of PP/G. The difference between the transverse stiffness measured directly on the composite and that calculated by the rule of mixture and nominal PP properties is certainly related to the fiber/matrix interface quality as well as to inter-spherulitic adhesion through the remaining amount of amorphous phase to bind crystalline units. In fact, the rule of mixture does not account for these components and assumes perfect bond between fibers and matrix. To take this parameter into account, the transverse stiffnesses of the lamina in the model evaluation will be those measured experimentally at room temperature. The variation of this properties with temperatures will be scaled to the trend measured by DMTA. Doing so, the interface contribution to the lamina stiffness will be implicitly accounted for.

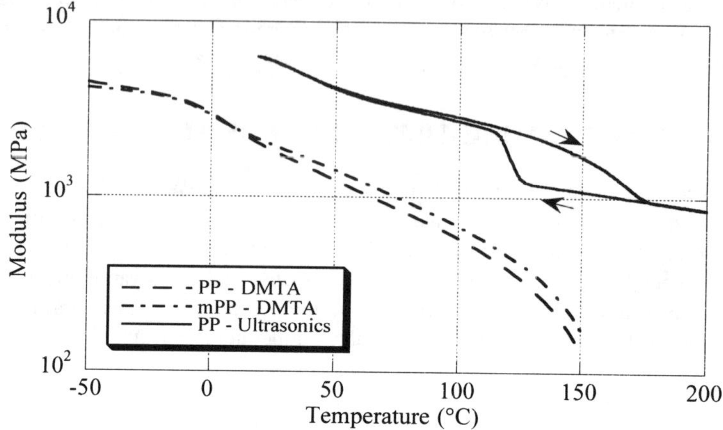

Figure 3. Modulus variation with temperature for PP and mPP matrices.

Also, from transverse tensile tests, transverse strengths of the composite systems under study have been measured. Regardless the cooling rate, mPP/sG systems showed transverse strengths around 18 MPa. PP/G systems, due to the poor quality of the fiber-matrix interface, failed at stresses as low as 4 MPa. This low transverse strength value can be of prime importance when evaluating residual stresses.

4.2 Coefficients of Thermal Expansion of the Matrix

Pressure-Volume-Temperature (pvt) tests have been conducted on mPP samples for three pressure levels and with temperature varying from 220°C to 40°C at a cooling rate of 1°C/min, a description of the technique used can be found in Ref. [12]. The variation of the specific volume at the molding pressure of 0.7 MPa and for cooling cycle from 220°C to room-temperature at 1°C/min is shown in Figure 4. The specific volume shows a marked shrinkage at 131°C corresponding to the crystallization temperature of the mPP cooled down at 1°C/min. After crystallization, the specific volume decreases almost linearly with decreasing temperature. The coefficient of thermal expansion of the tested material is extracted from this data using the following approximation [13]:

$$\alpha_m = \frac{1}{3} \frac{1}{v(T_o)} \frac{\partial v}{\partial T} \tag{7}$$

where v is the specific volume, T is temperature and T_o is a reference temperature. The calculated CTE values are reported in Figure 4 and show mainly a linear variation with temperature. Results of PP are not shown but they are almost identical except for the crystallization temperature which is 123°C instead of 131°C for the mPP.

646

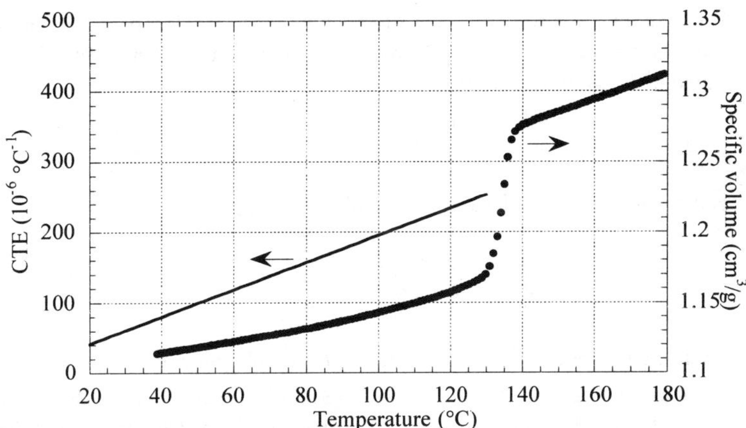

Figure 4. Variation of the measured specific volume under constant pressure of 0.7 MPa and the calculated CTE with temperature for mPP

4.3 Stress-Free Temperature

The stress-free temperature is the temperature at which thermal stresses start to build-up during the cooling phase. In the case of semi-crystalline thermoplastics, it is common to take the crystallization temperature as stress-free temperature. The crystallization onset temperature can be considered to account for the crystallization shrinkage, however Unger and Hansen has shown that the contribution of the chrystallization shrinkage to the overall thermal stress can be neglected due to the very low stiffness value (Strain precedes modulus hypothesis) [3]. Hence, for the two matrix systems used in the present study and based on results of DSC characterization presented in Ref. [8], the stress-free temperature considered for calculations is the crystallization peak temperature and will be 131, 124 and 109°C for mPP/sG at slow, moderate and fast cooling rates respectively. For the PP/G system, these temperatures will be 123, 111 and 94°C for the same cooling conditions.

4.4 Stress relaxation considerations

In order to take into account the effect of the viscoelastic nature of the PP matrix in the residual stress relaxation but without dealing with the complexity of this subject, a simple model used in the numerical calculation of residual stresses in injected parts [13] together with relaxation data PP found in litterature [14] were included in the evaluation of the residual stress model. The basic principle is to evaluate the thermal stress build-up by incrementing the temperature (and time) from the stress-free temperature to room temperature by a constant step. For each temperature step, it consists in evaluating the increase of stresses in the individual plies (Eq. 1) and add it to the cumulative thermal stress. If the cumulative transverse stress (same as stress in the matrix) overpasses the equilibrium stress of PP for that temperature, the stress is reduced by a factor that is function of temperature and the time spent at that temperature. The relation between relaxed and unrelaxed stresses is [13]:

$$\sigma_m^r = \frac{1}{1 + \Delta\xi/\tau}\sigma_m^u \tag{8}$$

where the subscript m refers to matrix, the superscripts r and u refer to relaxed and unrelaxed states, τ is the relaxation time and $\Delta\xi$ is the reduced time increment. This reduced time at time t (t being related to T) is defined by [15]:

$$\xi(t) = \int_0^t a_T\,(T(s))ds \qquad (9)$$

where a_T is the shift factor calculated using the following approximation [15]

$$log(a_T) = \frac{-8.86\,(T - T^*)}{101.6 + (T - T^*)} \qquad (10)$$

with the reference temperature, T^*, fixed to 43°C (T^* should fall in the range $T_g \pm 5$°C). With this reference temperature, eq. (10) agrees well with the experimental values published by Ariyama *et al.* [14]. A lower limit for stress relaxation is defined by the equilibrium stress which is temperature dependent [14]. These calculations are repeated for each temperature/time step.

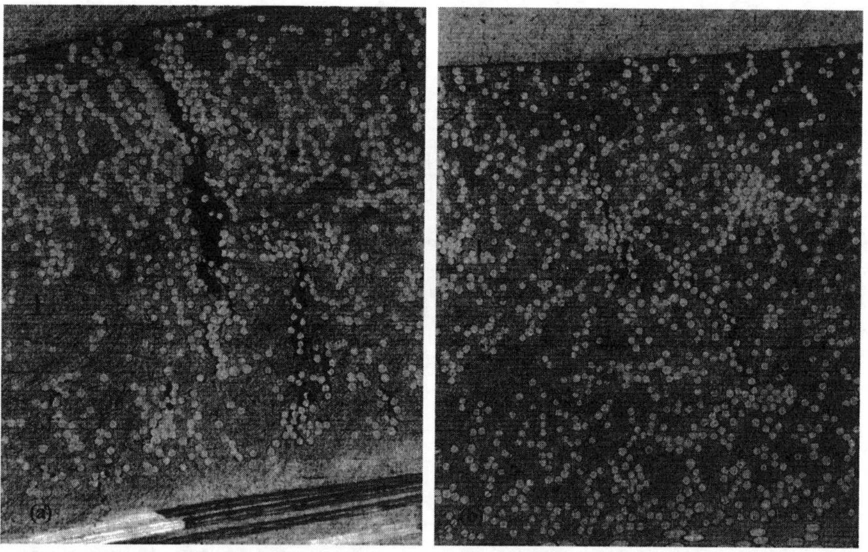

Figure 5. Optical micrographs showing complete transverse crack (a) and partially initiated crack (b) in PP/G sample

Another stress relaxation mechanism that might take place in the studied systems is the transverse cracking of the single plies. The presence of transverse cracks has been verified by optical microscopy and typical observations are shown in Figure 5. The PP/G samples molded with cooling rates of 0.3 and 12°C/min show obvious transverse cracks covering the laminate half-thickness and other initiated cracks that did not propagate to the surface of the laminate. Since the lamina transverse strength is very low, it is possible that the thermal stresses developed in the transverse direction of a ply overpass the transverse strength of the ply. In this case, transverse cracks will appear progressively relieving partially the residual stresses. A simple way to consider this effect is to set the transverse strength of the material considered as an upper limit for transverse stresses. Kim *et al.* studied this phenomena by

Finite Element Analysis and showed some correlation between curvature and crack density in unsymmetric graphite/PEEK laminates [2].

At the end of these iterative calculations, the resulting stresses are used to evaluate the resulting thermal forces and moments (Eqs. 2 and 3) which are used to evaluate the curvatures (Eq. 5).

5. MODEL VALIDATION

The model described in the previous section has been applied to evaluate the curvatures of the PP/G and mPP/sG $[0_n/90_n]$ laminates and the results are compared to the measured curvatures (Fig. 6). The predicted curvatures are marked Model 1, 2 or 3 depending on the way they have been evaluated. Model 1 does not consider any relaxation mechanism but accounts for temperature-dependence of the matrix properties (Eq. 5), Model 2 includes viscoelastic relaxation (Eq. 8) while Model 3 takes account of possible transverse failure as described previously.

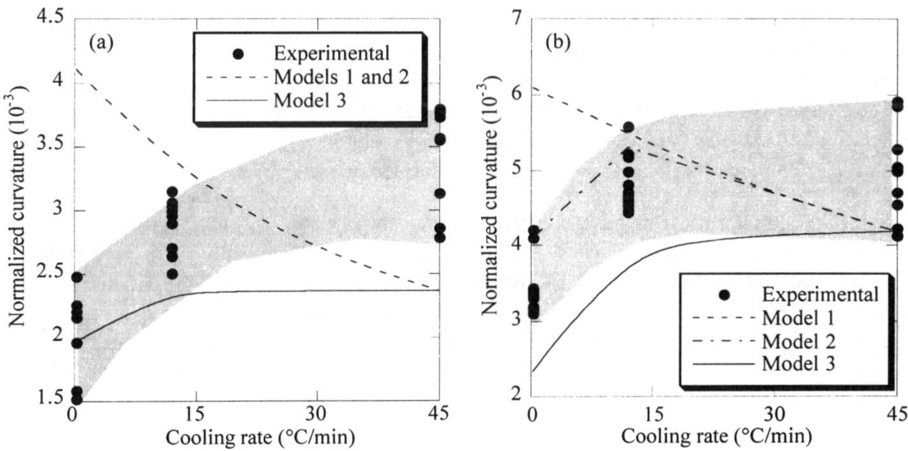

Figure 6. Comparison of measured and calculated normalized curvatures for PP/G (a) and mPP/sG (b) as a function of the cooling rate

From Figure 5, it is clear that without consideration for stress relaxation phenomena, residual stress predictions (Model 1) are ineffective and unrealistic. The predicted curvatures decrease with increasing cooling rate while they show experimentally significant decrease for the slow cooling rate. The fact that these calculations consider temperature-dependent stiffnesses and coefficients of thermal expansion does not make them effective.

Calculations considering the stress relaxation due to the viscoelastic nature of the matrix material are closer to the experimental measurements for mPP/sG where the fiber-matrix interface contributes to the stress transfer. In the case of PP/G, characterized by weak fiber-matrix interface, the stresses built-up during the cooling phase are always lower than the equilibrium stress and hence, no viscoelastic relaxation can take place (Model 1 and Model 2 results are identical).

The results of Model 3 draw a trend identical to that of experimental measurements. However, the absolute values are generally lower than the experimental ones. The shift in the calculated curvatures can be attributed to the approximate values of some variables used in the calculations and to the simplifications in the development of the model

These results demonstrate the contribution of the viscoelastic nature of the matrix and transverse cracking in stress relaxation due to processing thermal cycle. Moreover, these results demonstrate the importance of the fiber-matrix interface quality in the overall performance of the composite. In effect, a weak interface composite can not stand the process-induced residual stresses and can incorporate initially transverse cracks that may be detrimental for its in-service performance.

6. CONCLUSION

In this work, process-induced residual stresses in thermoformed polypropylene/glass composite laminates have been experimentally measured using cross-ply unsymmetric laminates on two composite systems presenting different interfacial properties. Classical lamination theory model for residual stress prediction has been adapted by taking into account the variation of the physical and mechanical properties of the matrix and by accounting for stress relaxation either by the viscoelastic behavior of the matrix or by transverse cracking in the composite plies. The model formulated this way incorporates the effect of the thermal history and processing kinetics (cooling rate) on the composite properties, particularly those controlled by the matrix microstructure and the interface quality.

The results showed that the residual stress model should consider temperature-dependent parameters and relaxation phenomena in order to be comparable to experimental measurements. The study has also shown that the residual stress build-up depends on the cooling rate since this parameter controls the matrix morphology and the interface properties. The comparison of the two composite systems has emphasized the importance of the fiber-matrix interface quality in the development of the composite stiffness and strength. Weak interface systems did not stand the process-induced residual stresses.

7. ACKNOWLEDGEMENT

The authors wish to thank P. Gagnon, P. Sammut and G. Chouinard from the Industrial Materials Institute , National Research Council for their technical support.

8. REFERENCES

1   G. Jeronimidis and A.T. Parkyn, J. Compos. Mater., 22, 401 (1988)

2   K.S. Kim, H.T. Hahn and R.B. Croman, J. Compos. Tech. Research, 11, 47, (1989)

3   W.J. Unger and J.S. Hansen, J. Compos. Mater., 27, 108, (1993)

4   J.A. Barnes and J.E. Byerly, Compos. Sc. Technol., 51, 479, (1994)

5   J.T. TZENG, J. Thermoplast. Compos. Mater., 8, 163, (1995)

6   C. Wang and C.T. Sun, J. Compos. Mater., **31**, 2230, (1997)

7   J. Denault and J. Guillemenet, International Plastics Engineering and Technology, **2**, 1 (1996)

8   Y. Youssef and J. Denault, 42nd International SAMPE Symposium and Exhibition, Anaheim , CA, 134, (1997)

9. M.W. Hyer, J. Compos. Mater., **15**, 175 (1981)

10. M.W. Hyer, J. Compos. Mater., **15**, 296 (1981)

11 A. Sahnoune, F. Massines and L. Piché, J. Polym. Sc., B, Polym. Phys., **34**, 341 (1996)

12 P. Zoller and D.J. Walsh, Standard Pressure-Volume-Temperature Data for Polymers, Technomic Publishing Co., Lancaster, 1995, pp. 1-17

13 K.K. Kabanemi and M.J. Crochet, International Polymer Processing VII, **1992**, 60, (1992)

14 T. Ariyama, Y. Mori and K. Kaneko, Polym. Eng. Sc., **37**,1, (1997)

15 S.L. Rosen, Fundamental Principles of Polymeric Materials, 2nd Ed., John Wiley & Sons, Inc., New York, 1993, pp.298-343

43rd International SAMPE Symposium
May 31-June 4, 1998

CONTROL OF WARPAGE AND RESIDUAL STRESSES DURING THE AUTOMATED TOW PLACEMENT PROCESS

John Tierney
R. F. Eduljee, J. W. Gillespie Jr.

Materials Science Program
Center for Composite Materials
University of Delaware, Newark, Delaware 19716

ABSTRACT

The automated tow placement process is being investigated as a rapid affordable non-autoclave method for the fabrication of large thermoplastic composite sections. During the process, tows are in-situ consolidated layer by layer to form the final composite laminate. Residual stresses and warpage develop internally due to temporary unsymmetric conditions, coupled with local thermal gradients, and thermal expansion mismatch between matrix and fiber.

In this study, the internal stresses and resulting warpage are modeled from a modified classical lamination theory solution. First ply tension and spring-back prediction methods are used to minimize the final curvatures using a generalized [ABD] input/output algorithm coupled with an iterative searching function.

A pretensioning loading apparatus which can independently control the first ply preload in both the x- and y- directions is used to verify this optimization model. A laser measurement unit measures the developing curvatures by scanning the surface of the panel during fabrication and after release of the mechanical preload. Experimental results to date for both unconstrained and prestressed conditions, have shown good agreement with model predictions.

KEYWORDS: Composites, Residual Stresses, Process Modeling.

1. INTRODUCTION

The thermoplastic composite tow-placement technology (ATP) is a non-autoclave consolidation process that offers the potential to significantly reduce fabrication costs for large scale single component structures. ATP involves the continuous consolidation of a thermoplastic prepreg on an already consolidated substrate material. This is achieved by the concurrent use of localized heat and pressure so that bonding is achieved at the newly formed interface and that consolidation is achieved within the material.

One of the inherent problems with the processing of composites with the ATP process is the development of internal stresses and resulting warpage which can result in parts with dimensional tolerances that do not meet design specifications. The final curvatures of a composite laminate manufactured with the ATP process are influenced during manufacture by the laminate stacking sequence, the through the thickness temperature and material property gradients as well as panel-tool interactions. Residual stresses are further magnified during part manufacture by the one sided deposition and consolidation required to achieve the desired thickness and laminate stacking sequence. These stresses then induce undesirable curvatures due to non-symmetric thermal loading coupled with the coefficient of thermal expansion mismatch between the consolidated layers. Thermal gradients also exist in the in-plane directions since the part is heated locally. These gradients can significantly effect internal stress gradients at a local level due to the stacking sequence variation.

Much work has been carried out at the Center for Composite Materials (UD-CCM) Composites manufacturing laboratory at the University of Delaware to investigate material property gradients within the bulk material and at the locally heated consolidation region. [1-4] Results from experimental investigations and model predictions have shown that large thermal gradients develop which induce significant material property gradients. The use of annealing has been shown to improve the final quality of the material through multiple passes. This non-symmetric processing history increases the through thickness internal stress profile which can lead to the development of visible warpage even in symmetric laminates. Unsymmetric laminate stacking sequences also exist during processing, which contribute significantly to curvature development at these intermediate steps.

The configuration used at the CCM is shown in Figure 1. The tow placement head is composed of a series of hot nitrogen torches and consolidation rollers and is mounted on an ABB IRB 6400 six-axis robot. The preheater torch and roller are used to develop intimate contact at the interface. The second torch and roller are used to provide supplemental through thickness heating so that good bond strength and consolidation is achieved at the interface and substrate layers. This localized traveling heat source generates internal stresses through the thickness of the laminate which can result in undesirable warpage. This investigation examines the general mechanisms responsible for internal stresses and develops methods for controlling final warpage using a model based predictive approach. [5]

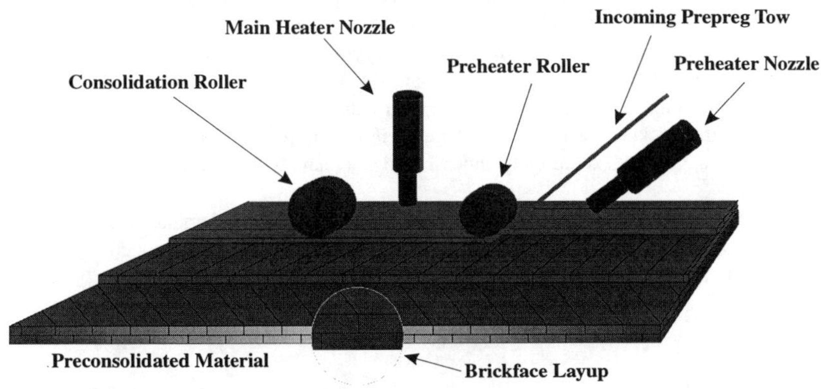

Figure 1. Schematic of the Automated Tow Placement Process.

2. Residual Stress Model

A residual stress solution, utilizing Classical Lamination Theory is used to determine the incremental residual stresses and warpage developed during the ATP process. CLT theory has been extensively used to describe the behavior of composite materials under mechanical, thermal, and hygothermal loading conditions. Many references are available where classical lamination theory is utilized to describe composite material behavior. The development of this theory is omitted here for the sake of brevity but can be found in numerous textbooks [6,7].

A number of important assumptions are made in this simplified residual stress model in order to avoid the complexities associated with a moving localized heat source on a composite laminate. The first assumption in the model is that the panel undergoes the same thermal cycle across the panel surface. Since classical lamination theory is time independent (not including the effects of viscoelastic relaxation) this assumption is valid for a constant velocity manufacturing process, since every point on the surface observes the same thermal processing cycle. So although this analysis cannot provide information about localized residual stresses during panel manufacture it can predict the overall response of the laminate based on this assumption.

A simplified thermal history is applied to this model to examine the effects of non-symmetric heating on the final warpage. A surface ply is added at an initial processing temperature T_i, which exceeds the glass transition temperature, T_g, and is cooled to ambient temperatures ($T_f = 25$ °C). The remaining substrate plies do not undergo any thermal history and remain at ambient temperatures. This step is repeated until the desired number of plies are bonded to the surface. This assumption leads to an overestimation of the thermal contribution to residual stresses for the first ply due to the fact that this temperature drop is applied over the laminate surface and it's thermal moment component is significant due to the non-symmetric condition.

Previous studies have indicated that the glass transition temperature is exceeded for several layers beneath the surface [2,4] though the area above T_g on each ply decreases significantly with increasing depth. The effect of neglecting multiple ply cooling results in an underestimation of thermal stresses based on bulk heating, but an overestimation of non-symmetric conditions on the laminate since thermal shrinkage is limited to one ply rather than a series of plies. This overestimation is more significant for thin laminates as the temperature distribution in the manufacturing process is more uniform through the thickness which is not the case in the simplified thermal model. The effect of multiple ply thermal gradients is presently under investigation and will be included in future studies on this topic.

The effective in-plane force and moment resultants of a composite laminate can be determined from CLT by summing the load resultants for each ply within the laminate through a second order tensoral transformation as follows ,

$$
\left(\Delta N_x^T, \Delta M_x^T \right) = \sum_{k=1}^{N} \left(\overline{Q_{11}} \Delta \varepsilon_x^T + \overline{Q_{12}} \Delta \varepsilon_x^T + \overline{Q_{16}} \Delta \varepsilon_{xy}^T \right) \left(\Delta z_k, \frac{\Delta z_k^2}{2} \right)
$$

$$
\left(\Delta N_y^T, \Delta M_y^T \right) = \sum_{k=1}^{N} \left(\overline{Q_{12}} \Delta \varepsilon_x^T + \overline{Q_{22}} \Delta \varepsilon_y^T + \overline{Q_{26}} \Delta \varepsilon_{xy}^T \right) \left(\Delta z_k, \frac{\Delta z_k^2}{2} \right) \tag{1}
$$

$$
\left(\Delta N_{xy}^T, \Delta M_{xy}^T \right) = \sum_{k=1}^{N} \left(\overline{Q_{16}} \Delta \varepsilon_x^T + \overline{Q_{26}} \Delta \varepsilon_x^T + \overline{Q_{66}} \Delta \varepsilon_{xy}^T \right) \left(\Delta z_k, \frac{\Delta z_k^2}{2} \right)
$$

where $\Delta \varepsilon_i^T$ represents the transformed process-induced incremental stress-free strains in each ply in the global coordinate system, N is the number of plies, and Q_{ij} represents the transformed plane stress stiffness matrix coefficients obtained from the instantaneous composite properties of each ply. The resulting plate loads are then used to compute the effective laminate response in terms of the in-plane strains, $\Delta \varepsilon^\circ$ and curvatures, $\Delta \kappa^\circ$, through the inverted [ABD] matrix as follows:

$$\begin{bmatrix} \Delta\varepsilon^0 \\ \Delta\kappa^0 \end{bmatrix} = \begin{bmatrix} A & B \\ B & D \end{bmatrix}^{-1} \begin{bmatrix} \Delta N^T \\ \Delta M^T \end{bmatrix} \qquad (2)$$

From this equation the transient process-induced stress and strain distributions through the laminate thickness can be obtained by the cumulative summation of the respective increments computed over each time step. Since AS4/PEKK material is used in the present investiagation, the strain contribution due to crystallization is not modeled as it has been shown PEKK does not crystallize under highly non-isothermal conditions [8]. In this analysis, the focus of study is the relationship between the laminate loads to the final strains or curvatures within a composite laminate. The nature of this process is such that the stacking sequence or geometry of the system continually changes with time. Thus by summing the stress and resulting curvature changes after the placement of each ply, it is possible to determine the effect of such incremental changes in geometry on the final stress and resulting warpage state of the material.

Partial heating of the laminate is incorporated into the CLT solution by including a series of empirical constraint vectors that account for the solid restraining boundary which exists around the heated region as shown in Figure 2. When this locally heated region cools it attempts to contract, but is limited by the solid boundary. The result is two competing forces, the mechanical constraint load presented by the boundary and the thermal load generated at the processing zone upon cool down. The degree or amount by which the mechanical force exceeds the thermal load is represented by the degree of constraint acting on the laminate.

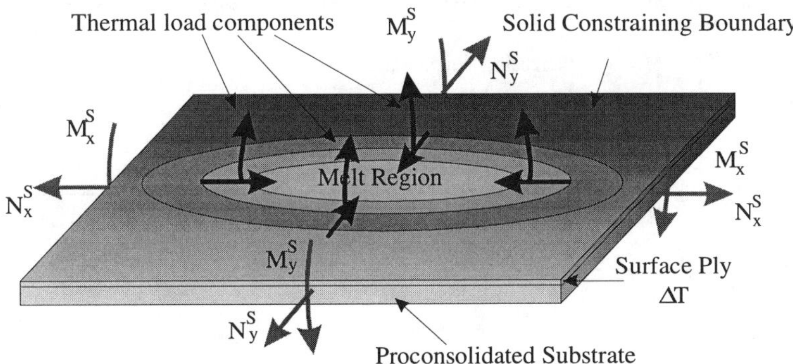

Figure 2. Schematic of partial constraint vectors used to accommodate effect of partial heating of composite substrate.

The degree of constraint is accommodated in the model by applying a set of empirical constraint vectors to the applied thermal loads and moments as follows:

$$\begin{aligned} \Delta N_i^T &= \theta_i \Delta N_i^T \\ \Delta M_i^T &= \theta_i \Delta M_i^T \end{aligned} \quad (i = x, y, xy) \qquad (3)$$

where θ varies form 0 to 1. With θ equal to zero, the panel is clamped in that direction, and vice versa. These vectors act to inhibit the growth of curvature by essentially holding or clamping the material in both the in-plane as well as out of plane directions. The inclusion of partial constraint vectors will result in a decrease in final curvatures as the thermal component of both in-plane and moment loading is reduced. The magnitude of these partial constraint vectors are found through experiments for a range of laminate structures and are presented in this study.

The effect of clamping of the panel during processing can be applied to the model by applying the condition of full constraint on the material, thus eliminating the effective in-plane strains and out of plane curvatures found in Equation (2). As a result, the internal stresses which develop during the process are related to the thermal component of in-plane strains alone. Upon release of the laminate the effective in-plane force and moment loads are found by summing the internal stresses over the cross sectional area of the laminate as follows:

$$\left(\Delta N_x^T, \Delta M_x^T\right) = -\sum_{k=1}^{N} \sigma_x^k \left(\Delta z_k, \frac{\Delta z_k^2}{2}\right)$$

$$\left(\Delta N_y^T, \Delta M_y^T\right) = -\sum_{k=1}^{N} \sigma_y^k \left(\Delta z_k, \frac{\Delta z_k^2}{2}\right) \tag{5}$$

$$\left(\Delta N_{xy}^T, \Delta M_{xy}^T\right) = -\sum_{k=1}^{N} \sigma_{xy}^k \left(\Delta z_k, \frac{\Delta z_k^2}{2}\right)$$

where σ_t^k is the final stress state at each ply within the material. The effective in-plane strains and out of plane curvatures are then found from these loads using Equation (2). Clamping of the laminate is a prerequisite condition for laminate prestressing, and is included in the warpage control procedures presented in this study.

An example of how internal stresses vary with partial constraint is shown in Figure 3 below for a fully clamped 16 ply unidirectional lamina. The fully constrained profile is characteristically tensile at the surface and compressive at the substrate due to the non symmetric thermal load applied at the surface.

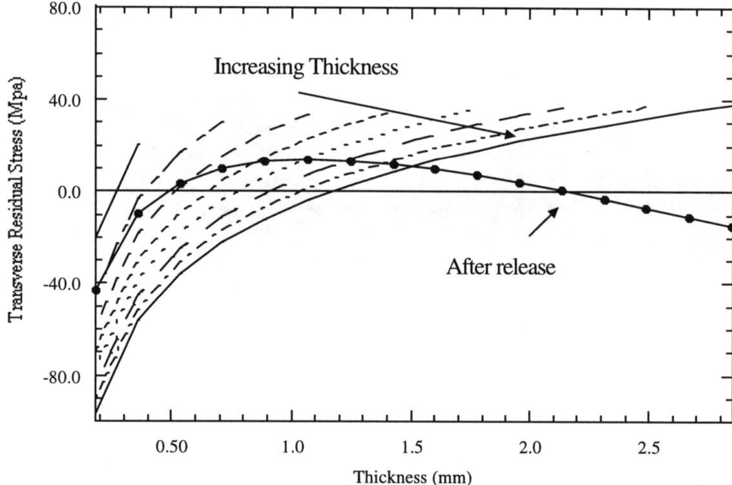

Figure 3. Plot of the transverse through the thickness internal residual stress profile with increasing thickness for a 16 ply unidirectional lamina.

The magnitudes of the stresses, both compressive and tensile, increase with subsequent placement of plies. When the panel is released, the profile is significantly altered. The substrate region remains in compression, the center of the lamina goes into tension and the surface is in compression. The difference between these two profiles, shown in Figure 4, represents the restoring clamping moment which is applied to the lamina in order to prevent out-of-plane strains. The reverse of this moment would then correspond to the actual moment required to hold the panel in place during processing.

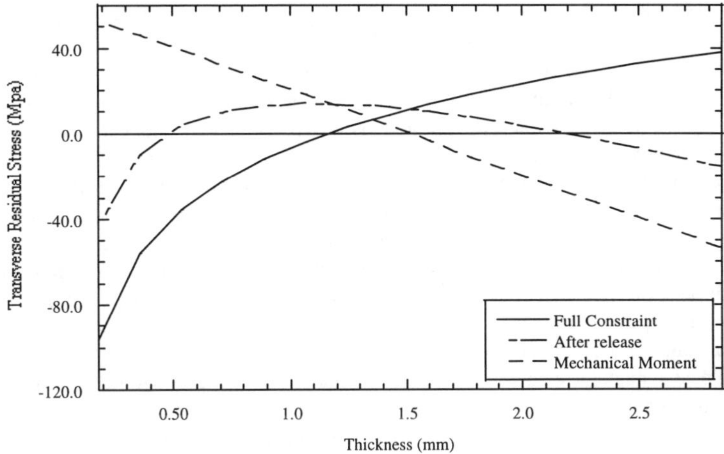

Figure 4. Internal residual stress profile for a unidirectional lamina for a completely constrained panel before and after release.

3. Methods for Controlling Internal Stresses and Warpage

Two methods have been identified for controlling the warpage induced in the ATP process. Both methods rely on a model based predictive approach which controls either the initial stress field and/or curvatures. The first method involves the use of prestressing or pretensioning the substrate plies. The release of the laminate after manufacture results in a compressive moment with acts to restore the shape to the desired curvatures. This model based prediction method uses a search algorithm to determine optimum initial in-plane loading such that release of the laminate results in negligible warpage. The important point to note is while the warpage may be reduced the internal stresses are not. This compressive restoring moment acts to significantly change the internal stress profile, since the external load is applied locally to the substrate plies. In the model the substrate loading is included as a restoring force and moment resultant in the CLT solution as follows:

$$\begin{bmatrix} \overline{N} \\ \overline{M} \end{bmatrix}_{opt} = \left\{ \begin{bmatrix} N^M_{k=1,2} \\ M^M_{k=1,2} \end{bmatrix} + \begin{bmatrix} N^T \\ M^T \end{bmatrix} \right\} \begin{bmatrix} A & B \\ B & D \end{bmatrix} \begin{bmatrix} \varepsilon^o \\ 0 \end{bmatrix} \tag{4}$$

where $N^M_{k=1}$ = the in-plane mechanical loads applied to the material at the first and second plies and $M^M_{k=1}$ are the corresponding moments generated by these in-plane loads upon release of the laminate after manufacture. Note that this method for controlling the panel curvatures is successful for laminate stacking sequences where the load can be realistically applied through the fibers, for example the warpage developed in a unidirectional laminate cannot be controlled by a single substrate layer in the same direction. The warpage for this lay-up develops in the matrix dominated transverse direction and as such can only be orthogonally loaded in the fiber direction through an additional cross ply layer.

The second method for "controlling" the final shape of composite structures produced by the ATP process involves the use of spring-back techniques, where the final warpage is included in the design methodology of the initial tool surface. Similar to the first ply tension methods, the final curvatures are first predicted using analytical or numerical studies. The tool surface is then modified to account for the incremental curvature buildup during processing. This technique has the advantage that the material is not placed under increasing stress as with the pretension method, and since the development of warpage is essentially a relaxation of stresses the final internal stress profile is in fact reduced. Also, no special mechanical devices are necessary for constraining the material. Using this technique however requires an accurate prediction for the spring-back response as large expensive molds are used and must be designed to account for the various mechanisms attributed to warpage development.

The Poisson's ratio of the composite structure can have a significant effect on the final warpage of a composite laminate. This is even more evident when the material is placed under loading as is the case with first ply tension methods. By applying tension in one direction a net compressive force is induced in the transverse direction which opposes the controlling force for that particular direction. This Poisson's interaction effect is accounted for in the model by switching the optimization search loop between both the x- and y- directions such that the final force in both directions compensates for the Poisson's interaction effect induced by the transversely applied load.

Both warpage control methods rely on the use of a generalized [ABD] searching function where the desired outputs, (in this case, being either first ply loads, or initial curvatures) are determined by carrying out the following procedure.

1. An initial value for the loads or curvatures is applied to the substrate plies.
2. The laminate is manufactured with the desired stacking sequence and then released from the tool after manufacture.
3. The initial loads (or curvatures) are then modified such that the final curvatures approach the desired values. (including the Poisson's interaction effect)
4. This process is repeated either to determine the optimal initial loading condition for the first ply tension method or the optimal initial shape for the "spring-back" prediction method.

The initial mechanical loading in the substrate plies is increasing if the curvatures are found to remain negative after release and vice versa. The empirical partial constraint vectors, which are found experimentally, are applied to both optimization models to account for the effect of partial heating on the composite laminate.

4. Experiments

A number of unconstrained and prestressed panels were manufactured to verify the results from the classical lamination theory model. The material used consisted of a 6.35mm wide AS4/PEKK composite prepreg tow. The thickness of the prepreg tape used was 1.778×10^{-4}m (0.007in.) A clamping mechanism, shown schematically in Figure 3, was designed and built to apply and control the in-plane loads during processing. A cross-type configuration was designed to apply in-plane tensile pretension in the $0°$ and $90°$ directions. The substrate material was composed of 12" wide APC2/PEEK composite prepreg tape which was clamped at the edges of a 1" thick steel table. The in-plane loads were applied to the substrate through the use of a moment arm clamping mechanism which includes a series of bolts with embedded strain gages. Cutouts are made in the substrate ply to facilitate a series of clamping bolts. Two rubber strips are placed between the substrate ply and steel clamp to provide supplemental traction. Two load bolts are then mounted at each clamp. By carrying out a number of load-strain calibrations, it is possible to determine the load-strain response for each bolt. Using simple moment calculations it is then possible to determine the load applied to the substrate material when each bolt is tightened. The lamina preload is then applied uniformly across the substrate material through independent control of each load bolt. Each bolt is tightened until the required loads are reached. The initial load was found to remain constant during processing which indicated that the substrate plies did not relax out during processing.

In the cross type configuration, two plies are used to apply the pretension load on the laminate. These plies are clamped and loaded to the desired tension levels required for curvature control. The substrate plies are then bonded together under tension. The extension strains resulting from mechanical loading on these substrate plies are not applied to the laminate structure until the panel is released. Upon release, the substrate plies attempt to contract, and as a result, induces a restoring moment which alters the final shape of the panel. The laminates examined in this study were comprised of symmetric laminates with the first two plies being part of the laminate design.

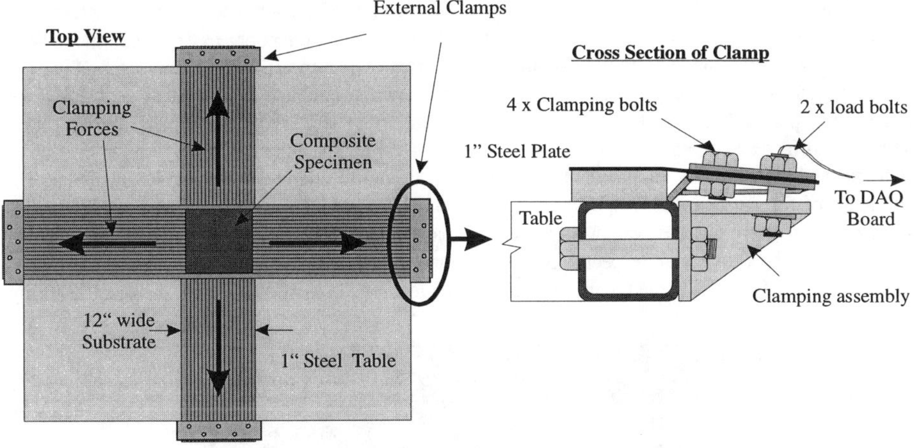

Figure 3. Clamping mechanism used to control process induced warpage.

The final panel curvatures are measured using a Keyence™ LB-11 laser displacement sensor. This laser sensor, shown schematically in Figure 4, can accurately measure the surface distance to within 0.1mm and has an operating range of 80mm (60-140mm). The sensor is mounted on the ATP head and travels across the surface in both the x- and y- directions to determine the relative surface height. The data is sent as a voltage reading to a DAQ board where a LabVIEW™ interface converts the signal using a scaling function to a three dimensional plot of surface heights. A thousand voltage readings are channeled to the LabVIEW™ interface at every point on the surface and averaged to minimize the noise level generated by external electromagnetic interference. A median filter is then used to remove any voltage spikes due to scattering of the laser.

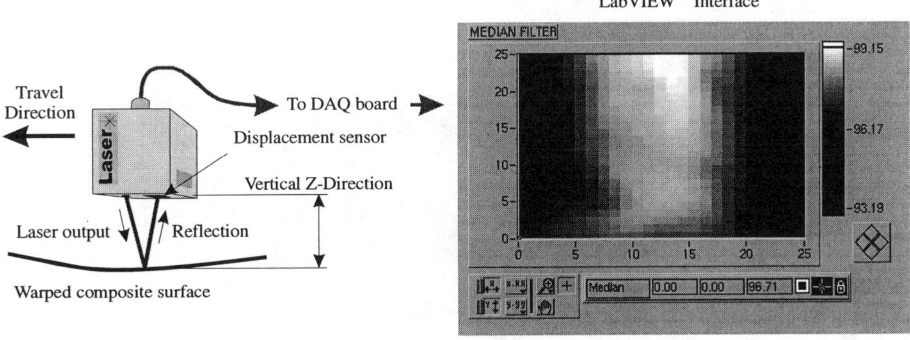

Figure 4. Schematic of laser displacement unit and corresponding LabVIEW™ output.

5. Results

The different stacking sequences examined in this study comprised of simple $[90°/0°]_{2S}$ and $[90°/0°]_{4S}$ symmetric cross ply laminates and $[0/90/45/-45]_S$ and $[0/90/45/-45]_{2S}$ quasi-isotropic laminates. A number of unsymmetric laminates were also examined to test the clamping mechanism.

Figure 5 shows the resulting warpage from an unconstrained [90°/0°]$_{2s}$ laminate after manufacture. The panel warps in the upward or negative direction in both the x- and y-directions as expected, due to both the thermal gradients and temporary unsymmetric conditions experienced during manufacture. Figure 6 is a laser displacement scan of the resulting warpage from this laminate. The derivative of this three dimensional plot is determined in both the x- and y-directions. The average of each derivative is found along the surface and averaged to minimize experimental error. The equation of the average derivative is found and the slope of this line, which represents the double derivative of surface height, is the curvature of the surface with respect to the direction of interest. This method is applied to all the specimens to determine the warpage of the panel after manufacture. Figure 7 shows the average derivative of this surface with respect to both the x- and y-directions. Note that the slope is different in both directions. This is attributed to unsymmetric conditions which exist early during the manufacturing process.

Figure 5. Unconstrained [90/0]$_{2s}$ laminate manufactured with the ATP process.

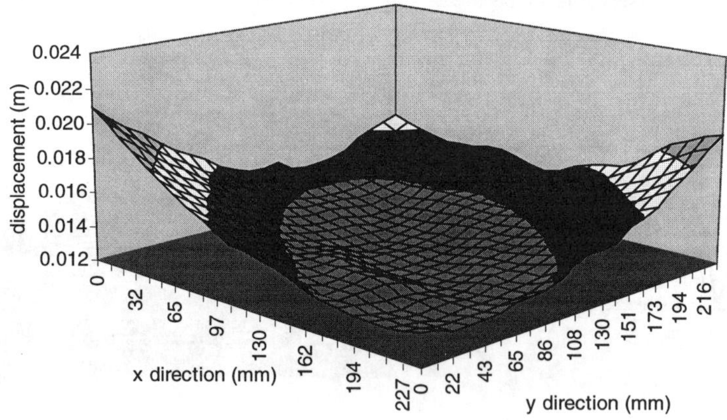

Figure 6. Three dimensional plot of the resulting for a [90/0]$_{2s}$ laminate manufactured with the ATP process.

The degree of curvature builds up quickly due to the fact that a larger percentage of the material is undergoing a thermal cycle, which limits the volume of solid boundary material. This early development of warpage was also observed from the CLT model for all laminate stacking sequences examined in this study. Figure 8 is a plot of model predictions of the developing curvatures for a

[90°/0°]₂S laminate with variation in degree of partial constraint. Note that the curvatures are unrealistically high when the degree of partial constraint is not taken into account, i.e. when $\theta = 1$. On the other hand, as the degree of constraint approaches zero the curvatures also approach zero, as the solid boundary acts to completely inhibit the growth of warpage.

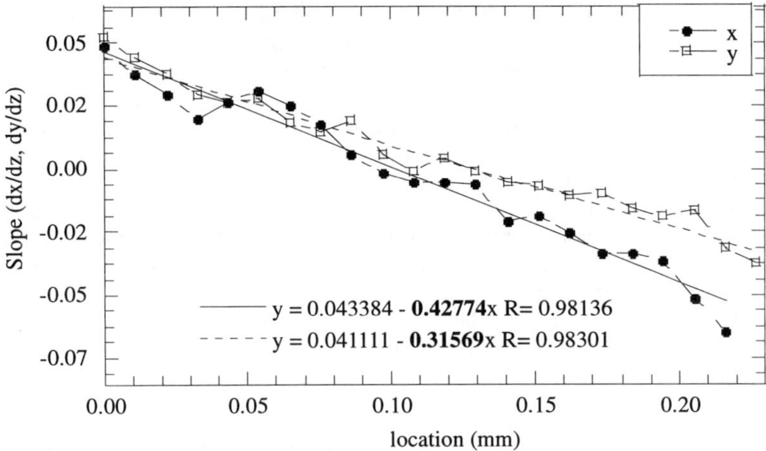

Figure 7. Plot of average derivative of a [90°/0°]₂S laminate including equation fit to line. Slope (shown in bold) corresponds to panel curvature in that direction.

The largest growth in warpage occurs in the fist few plies which suggest that this point in the process represents the critical time for which warpage control should be implemented. The deviation in curvatures in the x- and y- directions is seen to develop at this early stage in the process and remain the condition throughout the remainder of the tow placement process. This unsymmetric growth in warpage at the first few plies was also observed for all the laminate stacking sequences examined in this study. The optimization procedure takes into account this early development of unsymmetric warpage and is evident in the optimal pretension search results.

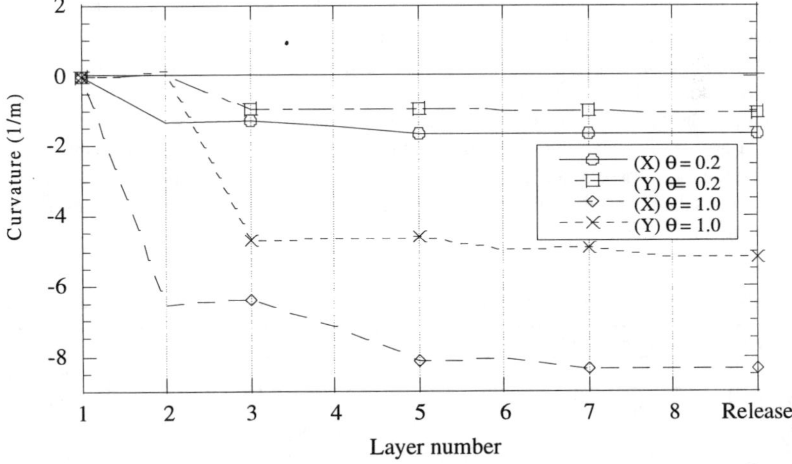

Figure 8. Plot of model predictions for developing curvatures of a [90°/0°]₂S cross-ply laminate with variation in partial constraint

The resulting curvatures from the model for completely unconstrained panels were found to be much higher than experimental results. For the case of completely constrained panels the warpage was found to be zero for all cases except unsymmetric laminates since these panels exhibit uncontrollable warpage which cannot be constrained. Experimental results were found to be much closer to fully constrained model predictions indicating that the panel itself behaves as a natural constraint against warpage. The unsymmetric panels exhibited severe warpage since non-symmetric conditions existed for most of the processing time. From the experimental results, it is possible to determine the empirical degree of constraint due to the solid constraining boundary by using a linear approximation based on the model predictions for constrained and unconstrained panels. The resulting constraints are then applied to the model for both optimization procedures to determine either optimal loading or optimal initial curvatures. Table 1 shows the resulting degree of constraint for the panels examined in this study. Note that the degree of constraint is a function of stacking sequence though the values were found to be very similar for all lay-ups. It was not possible to determine the degree of constraint for the highly unsymmetric panels since experimental curvatures fell outside the predicted range for constrained and unconstrained panels. These panels exhibited a large degree of curvature and significant geometrical nonlinear response which violate assumptions made in the classical lamination theory model. For the symmetric panels, one obtains results very similar to experimental values, indicating that the linear assumption made with the degree of partial constraints is valid in this analysis.

Table 1   Degree of partial constraint and resulting curvatures from model predictions and experimental results.

Laminate Stacking Sequence (Panel ID)	Resulting degree of constraint θ^n_x, θ^m_x	θ^n_y, θ^m_y	Model Predictions (m^{-1}) κ^i_x	κ^i_y	Experimental results (m^{-1}) κ_x	κ_y
[90/0]$_{2s}$ (1)	0.051	0.061	-0.416	-0.313	-0.427	-0.315
[90/0]$_{4s}$ (2	0.030	0.055	-0.236	-0.325	-0.242	-0.323
[0/90/45/-45]$_s$ (3)	0.038	0.074	-0.413	-0.425	-0.431	-0.387
[0/90/45/-45]$_{2s}$ (4)	0.037	0.064	-0.394	-0.416	-0.399	-0.399

Table 2 shows the optimal substrate loading conditions for the unconstrained panels examined in this study using the degree of constraints given in Table 1. Since these laminates are thin, the strain, or in-plane load, required to restore the curvature to zero is small.

Optimum initial curvatures were found to be equal in magnitude and opposite in sign to the predicted curvatures with partial constraint included. These initial curvatures are thus quite high and basically involve an inversion of the mold shape. For practical mold shape design this may not be possible, and as a result may require a combination of both pretension loading and initial curvature design so that the mold shape requires minimal change. Future studies combining both initial curvature and pretension loading are being considered as a method for optimal design of composite laminates produced using this process.

Table 2.   Resulting curvatures using optimal initial loading criterion and optimal initial curvatures.

Panel ID.	Optimal in-plane loading(N) N^m_x	N^m_y	Resulting Curvatures(m^{-1}) κ^i_x	κ^i_y
(2)	338	341	7.73×10^{-9}	-6.62×10^{-9}
(3)	335	552	-9.22×10^{-10}	-8.29×10^{-9}
(5)	223	410	-2.99×10^{-9}	1.53×10^{-9}
(6)	376	644	-2.87×10^{-9}	-1.72×10^{-4}

From these results, it is also evident that the final curvatures for thicker panels are similar in magnitude to those of thinner panels indicating that the curvatures are developed early in the process when few consolidated plies are available to constrain warpage development. The curvatures as expected, were also found to be not the same in both directions due to the intermediate non-symmetry conditions during processing. Model predictions also showed these trends for the stacking sequences examined in this study.

Figure 9 is a plot of final warpage after release for a $[90°/0°]_{2s}$ laminate with varying initial substrate tension loading. Also shown in the figure is the resulting curvatures for the unconstrained panel. The maximum load that could be applied to the substrate without damaging the strain gages was approximately 22270N (5000lb). A test was carried out to determine the maximum effect of pretension loading on the composite. It was found that severe positive curvatures could be achieved with this pretension method. Significant curvatures were also observed for panels subjected to 4500 N (1000lb). Minimum warpage was obtained when the loads were set to 330 N, as predicted by the residual stress model. It must be noted that this small load was difficult to control as the corresponding strains within the load bolts were small. Also note that the panels exhibited different curvatures in the principle coordinate directions due to non-symmetry during processing Other methods of accurately applying smaller loads are being investigated as model predictions require no more than 650N to control the warpage in the laminates examined in this study. The thickness of these laminates is small enough that a small amount of compression strain in the substrate plies can completely restore the laminate shape. It must be noted however that large pretension loads are required for significant changes in laminate curvatures.

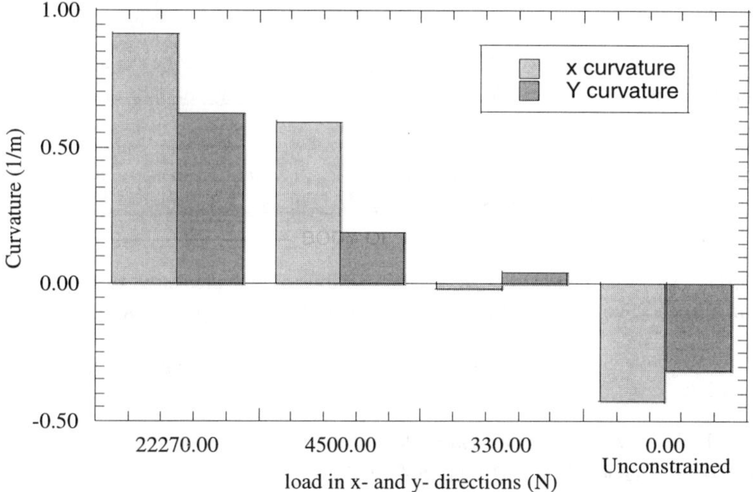

Figure 9. Experimental curvatures in the x- and y- direction with variation of pretension load for a $[90°/0°]_{2s}$ laminate. Also shown is the results for optimal initial pretension conditions.

6. Conclusions

A classical lamination theory model was developed to determine the internal stresses and resulting warpage for parts produced by the ATP process. A simplified analysis was carried out with this model to determine the final response of a composite laminate under a simple thermal load on the surface ply. The degree of partial constraint is used as an approximate means to compensate for a solid boundary which is not included in CLT theory. Experimental results for symmetric laminates gave results between the predicted clamped and unclamped predictions. By using a linear assumption, the degree of constraint in the x and y directions were determined and applied to the model. The resulting curvatures based on this degree of constraint were found to closely agree with experimental results. Two methods for controlling the final warpage were presented; substrate or first ply tension and spring-back prediction based on optimal initial curvature design. Initial results using both in-situ pretension and spring-back prediction methods show promise as a means of controlling the final shape of composite panels. Experimental studies for both optimization techniques are currently being carried out to validate model predictions.

7. ACKNOWLEDGMENTS

This work is funded by the Department of the Army, Army Research Office under the RAPTECH-ACM Program (Grant No. DAAH04-94-G-0285). The author would like to thank Raymond Foulk for his work with the ABB robotic programming, Dirk Heider with his help on the LabVIEW™ interface programming and Rod Don for his technical support.

8. REFERENCES

1. Pitchumani, R., R. C. Don, J. W. Gillespie Jr., and S. Ranganathan, "Analysis of On-Line Consolidation During the Thermoplastic Tow-Placement Process," in Heat and Mass Transfer in Composites Processing, M. K. Alam and R. Pitchamuni eds., ASME Press, 1994.

2. Tierney, J. J., Heider, D., Gillespie, J. W. Jr., "Welding of Thermoplastic Composites Using The Automated Tow Placement Process: Modeling and Control", ANTEC 97, Toronto, Ontario, pp. 1165-1170.

3. Tierney, J. J. R. F. Eduljee and J. W. Gillespie Jr. "Residual Stress and Warpage Development During the Automated Fiber Placement Process," Proceedings of the 11th ASC Technical Conference, pp.777-786, 1996.

4. Tierney, J. J, R. F. Eduljee, and J. W. Gillespie, Jr., "Material Response During Robotic Tow Placement of Thermoplastic Composites," Proceedings of the 11th Annual Advanced Composites Conference, pp. 315-329, 1995.

5. Heider, D., Don, R.C., and Gillespie, J.W. Jr., "Model-Based Predictive Control for the Tow-Placement Technique," Proceedings of QNDE, San Diego, CA, August 27-September 2, 1997.

6. Pister, K. S., Dong S. B., "Elastic Bending of Layered Plates," J. Eng. Mech. Division, ASCE, 1959, pp.1-10.

7. Reissner, E. Stavsky, Y., "Bending and Stretching of Certain Types of Heterogeneous Aeolotropic Elastic Plates," J. Appl. Mech., 1961, pp. 402-408.

8. Ferrara, J. A. Seferis, J. C. "Processing characteristics of PEKK matrices and their composites" 23rd SAMPE Technical Conference pp. 1137-1149, 1991.

THERMOSTRUCTURAL APPROACH FOR LOW COST FABRICATION PROCESSING

Christos C. Chamis[1]
NASA Lewis Research Center
Cleveland, Ohio 44135

ABSTRACT

A new approach is described which is based on adapting coupled thermal/structural computer codes for low cost fabrication processing. The coupled thermal structural behavior is simulated by using a coupled multidisciplinary computer code. Through this approach the temperature gradients during processing and the respective evolution of thermomechanical properties during cooling are simultaneously evaluated. The approach is demonstrated by applying it to ice forming and casting process to fabricate a simple and a complex component. The volume of the part and its respective mold is modeled by 3-D mixed field finite elements which accommodates solidification, heat transfer, stress/structural analysis. Results obtained show that the temperature gradients are functions of the pouring ports and built-up rapidly in the mold reaching magnitudes approaching those of the melt. The approach is amenable to formal optimization for mold material and multiple pouring ports in order to minimize the thermal gradients which reduce residual stresses and thereby increase part service life.

KEY WORDS: Ice Forming, Casting, Mold Filling.

1. INTRODUCTION

To produce better, cheaper, faster products and with early-time to market, the fabrication process must be expedited by substantial time reductions and costs. Fabrication of near-net-shape components must be accomplished to minimize costly reworking. Casting is a well known fabrication process (1) for the low cost products. One approach to further expedite the casting process is to computationally simulate it. Shrinkage and distortions are common occurrences in castings. It would seem logical that the simulation must accommodate simultaneously heat transfer and structural analysis as the solidification proceeds. Coupled thermostructural computer codes have been developed to effectively describe the complex behavior of hot engine components (2). It is reasonable to expect that those types of codes may be equally applicable to simulate the casting process of components. The application of similar codes to the fabrication process of metal matrix fiber composites (3) further reinforces that expectation. The objective of the present article is to describe results from three different

---

[1] Senior Aerospace Scientist Research and Technology Directorate. (SAMPE Fellow)

investigations to model fabrication processes by using available coupled thermostructural computational computer codes. More specifically the objective is to demonstrate what can be done by describing what has been done.   The three different investigations deal with: (1) ice formation, (2) casting forming of a simple component and (3) investment casting of a complex space shuttle main engine (SSME) component. Deterministic computer codes are used for the first two and probabilistic for the last one in order to quantify uncertainties associated with the complex SSME component.

2. FUNDAMENTAL CONSIDERATIONS/APPROACH

The application of coupled thermostructural computational simulation methods to casting is based on the following fundamental considerations:

1. The part has been designed and its mold configured.

2. The mold cavity and surrounding structure are modeled by 3-D finite elements.

3. The material for the casting is in molten form hot enough to flow easily to fill the cavity.

4. The mold is filled at known or prescribed flow rates.

5. The mold is filled from one port at the volume rate of one 3-D finite element, adjacent finite elements continue filling at that rate until all the elements in the bottom layer have been filled, then, proceeding with successive higher layer until the mold cavity is filled. The filling process assumed is illustrated schematically in Figure 1.

6. Heat-transfer analysis is performed continuously as each 3-D finite element is filled until the mold-cavity is filled, and the mold cools to ambient or other prescribed conditions.

7. Stress analysis is performed following each heat transfer analysis first to evaluate the stresses and deformations in the mold and second to evaluate the stresses in the casting as it begins to solidify and cools down to final temperature.

The above steps are fairly representative of the casting physics and permit direct simulation as the process evolves from initial pour to final cool-down temperature tracking temperature profiles, solidification fronts and developing stresses. In this approach component-shrinkages, component-distortions and other component-anomalies are quantified as well as appropriate process modifications are identified to remedy those anomalies.

A slightly modified simulation procedure but using the same coupled thermostructural computer code was applied by the author and his collaborators for ice formation, (4). It is instructive to present select results from that simulation to demonstrate the versatile applicability of the computer code.

3. SIMULATION OF ICE FORMATION

Schematics of the water region to freeze and the finite elements model are show in Figure 2. A block diagram of the coupled thermostructural analysis computer code is shown in Figure 3. Note the ice properties fare modeled by using composite mechanics. The details are described in (4). Herein we show typical results that are relevant to casting simulation - especially solidification and mechanical properties evolution.

The evolution of solidification in the volume of water in Figured 2 is shown in Figure 4: 4a - temperature profile; 4b - ice thickness formation versus time. The mechanical properties evolution as solidification proceeds are shown in Figure 5: 5a - modulus; 5b - strength. As would be expected, both properties initially increase rapidly and they level-off as the solidification process approaches completion. Note the comparisons with the limited experimental data is relatively very good. Obviously other mechanical properties as well as thermal properties can readily be predicted simultaneously with the modulus and strength (4).

The important observations from the above discussion are: (1) coupled thermostructural codes can effectively be used to simulate solidification process from a liquid state and, (2) the magnitude evolution of thermomechanical and other properties can be quantified as solidification proceeds.

4. SIMULATION OF CASTING FORMING

Casting forming with respect to mold filling and the evolution of the temperature profiles in the melt and in the mold casing are described below.

The fundamental considerations on which the simulation is based were described earlier. Herein we present and discuss some typical results from that simulation by using CSTEM in combination with MFIM. CSTEM is a coupled computer code configured to simulate thermostructural , acoustic, and electro-magnetic behavior with optimization and composite modeling capabilities (2). MFIM is a multi factor interaction model which is used to model coupled material behavior in complex environments (5).

The metal melt pour starts as is schematically depicted in Figure 1. The temperature profile in the cavity and in the mold after 8 and 18 seconds is shown in Figure 6. As can be seen at 8 seconds only one 3-D finite element was filled. At 18 seconds several elements were filled. It is also seen that the temperature diffuses throughout the mold relatively quickly. The amount of mold volume filled at 40 and 67 and respective temperature profiles are shown in Figure 7. The mold temperature rises and spreads as more volume gets filled. The amount of volume filled at 90 and 148 seconds with respective temperature profiles are shown in Figure 8. More than half the volume has been filled and the temperature in the mold approach that of the melt in the region closer to the filling port. The volume filled at 174 and 211 seconds with respective temperature profiles are shown in Figure 9. As can be seen in Figure 9b, the cavity volume is completely filled and the mold temperature approaches uniformity. At this point the pouring stops and the solidification starts.

The important observation from the afore discussion is that the thermal profiling part of the pour is properly described and the solidification part for stresses and displacements remains to be done in order to complete the simulation. Results for the solidification have not been obtained as yet.

5. SIMULATION OF SSME COMPONENT

In this section a probabilistic simulation case is described which addresses quantification of uncertainties in that fabrication process. It is the degree of that quantification which contributes directly to producing a better, cheaper, faster, product to market. It is that quantification which reduces risk to an acceptable level and permits the fabrication of cost effective products and <u>not</u> perfect products. The simulation case described is about investment casting process which is low cost manufacturing for complex shaped parts.

The simulation method is demonstrated by applying it to a component of the SSME (Space Shuttle Main Engine) HPOTP HEX (High Pressure Oxidizer Turbopump Heat Exchanger). It was selected to demonstrate the analysis technique. (6). The analyzed part is the discharge turning vane which is investment cast from Inconel 625 nickel based superalloy. The casting as well as the mold, core and surrounding insulation is modeled in detail. The effects of contact resistance between the casting and mold are included, and the heat of fusion is accounted for in the capacitance of the casting material.

The turning vane (component) and the investment casting process simulated are shown in Figure 10. Results obtained from that simulation are shown in Figure 11. In Figure 11a, the dominant variables that contributed to the reliability of the component along with mean values of the dominant variables and their scatter ranges are listed. In Figure 11b , the probabilistic evaluation of the porosity formation is plotted. Note that the cumulative distribution and probability density are shown as well as cutoffs for one standard deviation and 50 percent safety factor. In Figure 11c the probability sensitivity factors are ordered in bar chart form for three probability levels. It can be seen that the gab conductance contributes the most to probability. This means that controlling this variable during the casting process would reduce the porosity scatter most effectively and thereby result in a lower cost product. The important observation from the last sample case is that probabilistic simulation can be used to quantify uncertainties in the casting process. Probabilistic sensitivities can be used to identify dominant variables to control in order to reduce most effectively the cost of the fabricated component.

6. GENERAL REMARKS

The importance of the observations noted previously are further amplified by the following general comments:

1. The three different simulations described collectively demonstrated that computational simulation method developed for thermostructural analysis can be effectively used to simulate the entire casting process. The application is very much dependent on the innovativeness of the user for that particulate fabrication process.

2. It is not necessary to couple these methods with traditional casting simulation methods. Those traditional methods were not configured to accommodate simultaneously heat transfer, solidification and evolution of the thermomechanical properties, stresses, distortions etc., as the solidification proceeds.

3. More complete simulation of the thermodynamics of the process will require coupling of the thermostructural simulation with computational chemistry.

4. The simulation methods described are amenable to optimization for mold material and multiple pouring ports in order to minimize temperature gradients, reduce residual stresses and thereby increase component service life while reducing fabrication costs.

7. CONCLUSIONS

The significant conclusions from an investigation to use coupled thermostructural computer codes to simulate casting processes for low cost products are as follows:

1. These methods can be effectively used to simulate ice forming, melt pouring in molds, and controlling porosity in investment casting of complex components.

2. The simulation can simultaneous describe, temperature profiles, solidification fronts, and thermomechanical properties evolution.

3. The mold temperature can increase rapidly with pouring very close to the pouring port and spread throughout the mold wall as the cavity is filled.

4. Probabilistic simulations can be used to quantify the uncertainties inherent in fabrication process.

5. Probabilistic sensitivities can be used to identify and control dominant fabrication variable that contribute most to uncertainties.

6. The sample cases described collectively demonstrate that available coupled thermostructural computer codes can be used to expedite fabrication of better, cheaper, faster products to market.

8. REFERENCES

1. D. M. Stefanescu, Metals Handbook ®, Ninth Edition, Vol. 15 - Casting, ASM International, Metals Park, Ohio, 1988.

2. M. S. Hartle, by, R.L. McKnight, H. Huang and R. Holt "Coupled Structural/Thermal/Electromagnetic Analysis/Tailoring of Graded Composite Structures." Final Status Report (GE Aircraft Engines, Cincinnati, OH), NASA CR 189153, April 1992.

3. M.R. Moral, D. A. Saravanos, and C. C. Chamis: "Tailored Metal Matrix Composites for High-Temperature Performance." NASA TM 105816, March 1992.

4. M. T. Tong, S. N. Singhal, and C. C. Chamis: "Computational Simulation of the Formation and Material Behavior of Ice." NASA TM 106702, January 1994.

5. A. R. Shah, and C. C. Chamis: "Cyclic Load Effects on Long Term Behavior of Polymer Matrix Composites." NASA TM 107007, May 1995.

6. G. E. Orient, et. al. "Probabilistic Casting Porosity Prediction Methodology," Proceedings of the 38th AIAA/ASME/ASCE/AHS/ASC Structures, Structural Dynamics and Materials Conference , April 7-10, 1997, Paper No. A-97-1405.

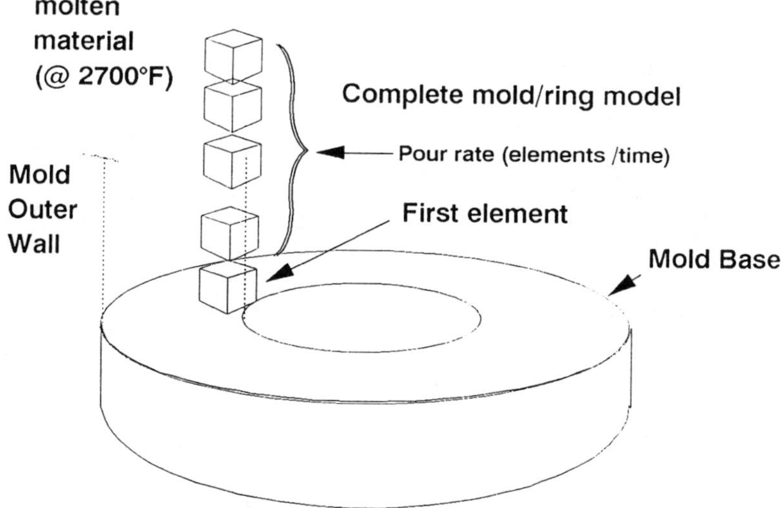

Figure 1 - Mold-filling schematic.

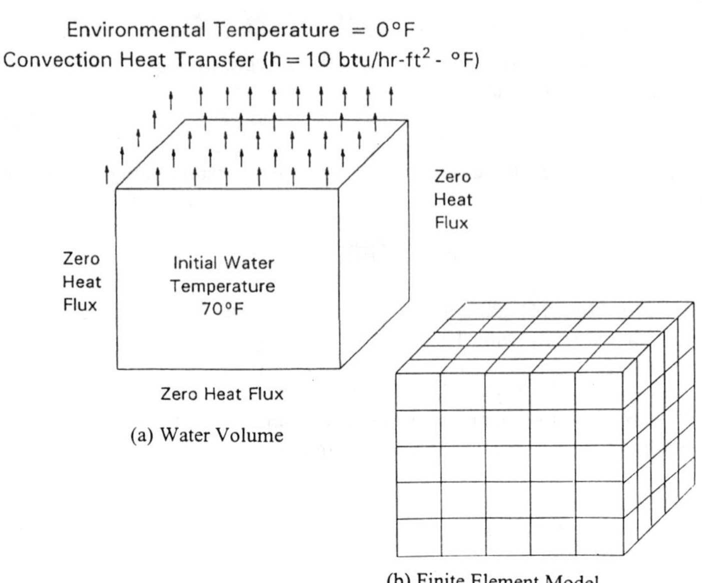

Figure 2 - Schematic of water volume and finite element model

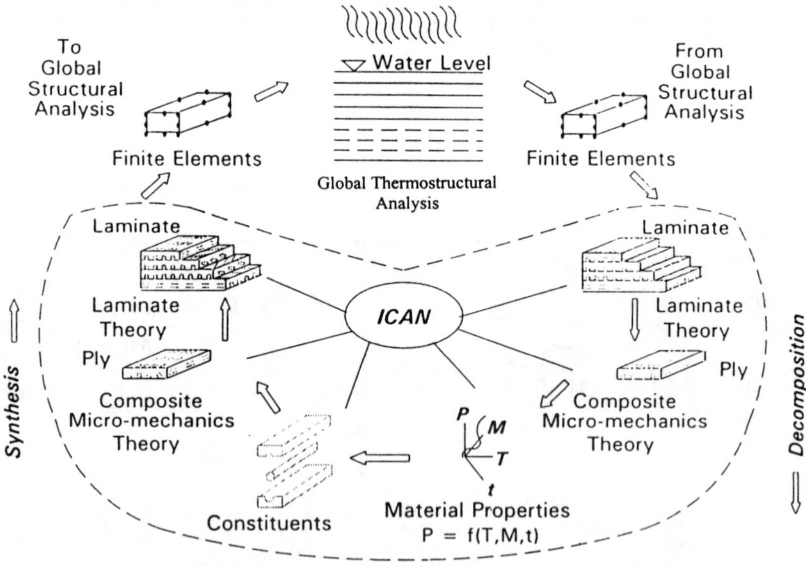

Figure 3 - Coupled thermostructural simulation of solidification process.

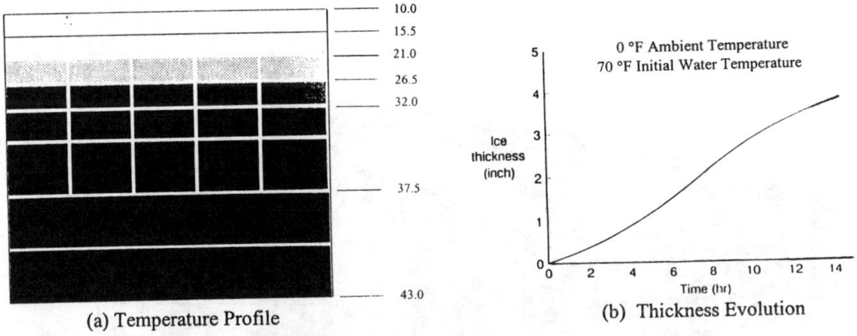

(a) Temperature Profile   (b) Thickness Evolution

Figure 4 - Evolution of Ice Solidification.

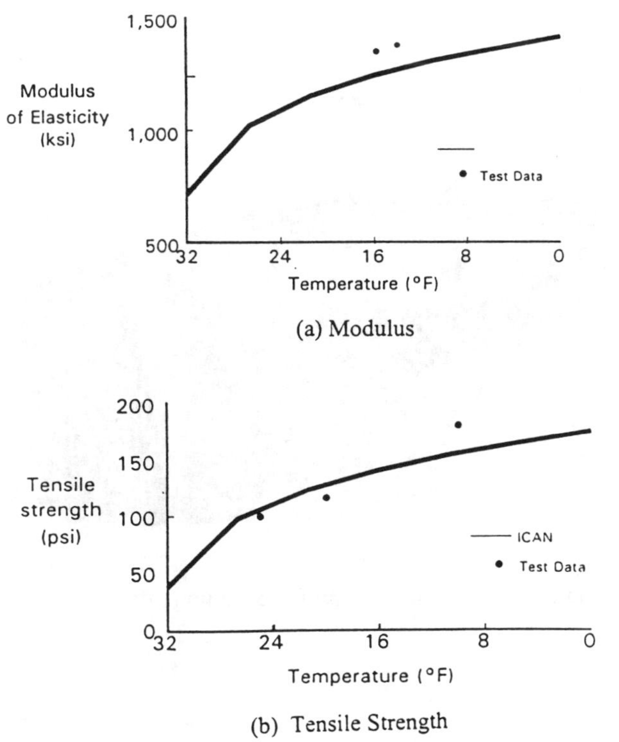

(a) Modulus

(b) Tensile Strength

Figure 5 - Magnitude of ice mechanical properties during solidification.

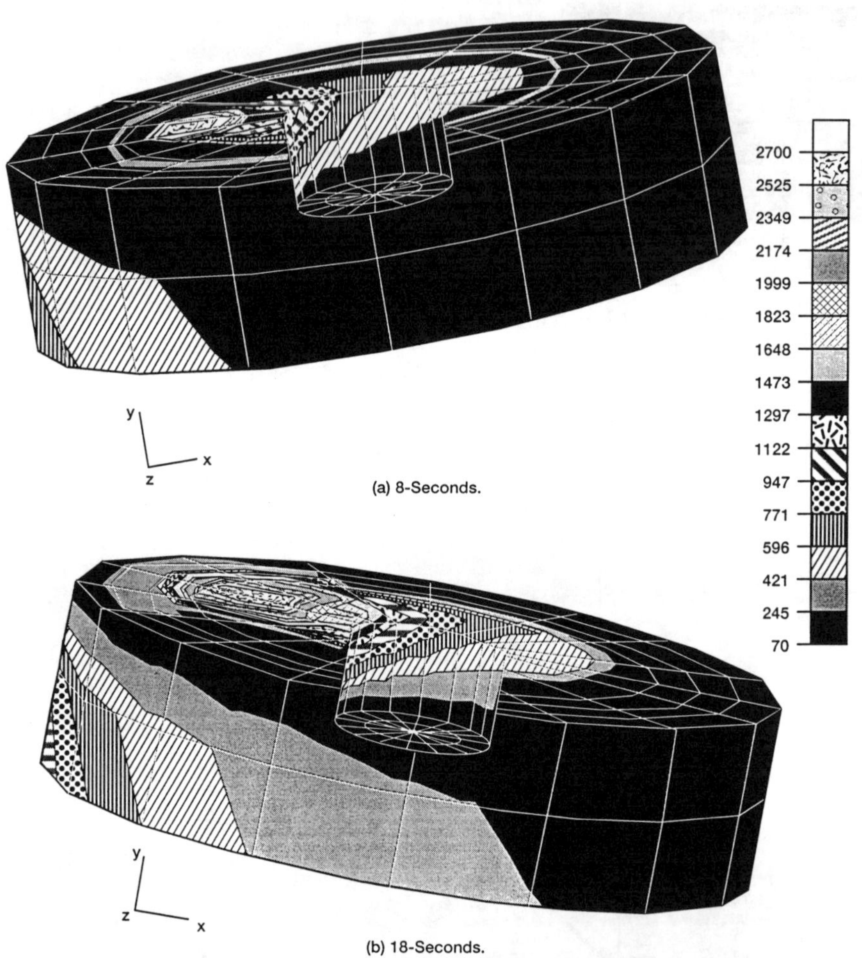

Figure 6 - Casting forming-temperature profile evolution (after 8 and 18 seconds).

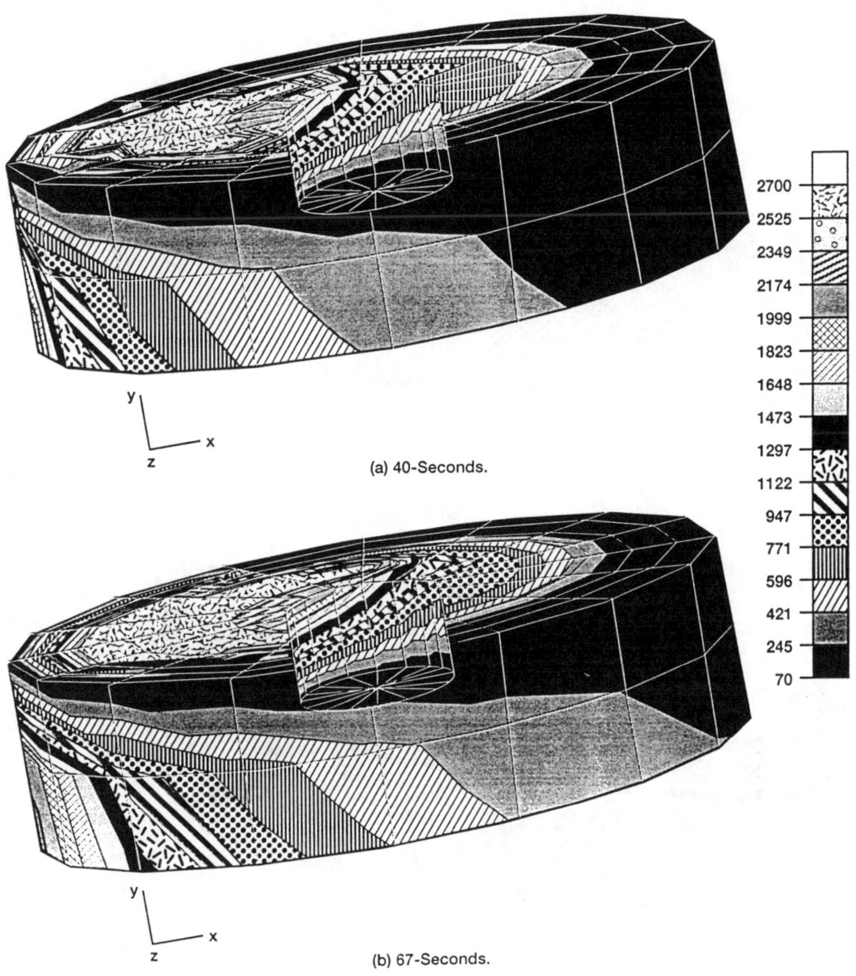

Figure 7 - Casting forming-temperature profile evolution (after 40 and 67 seconds).

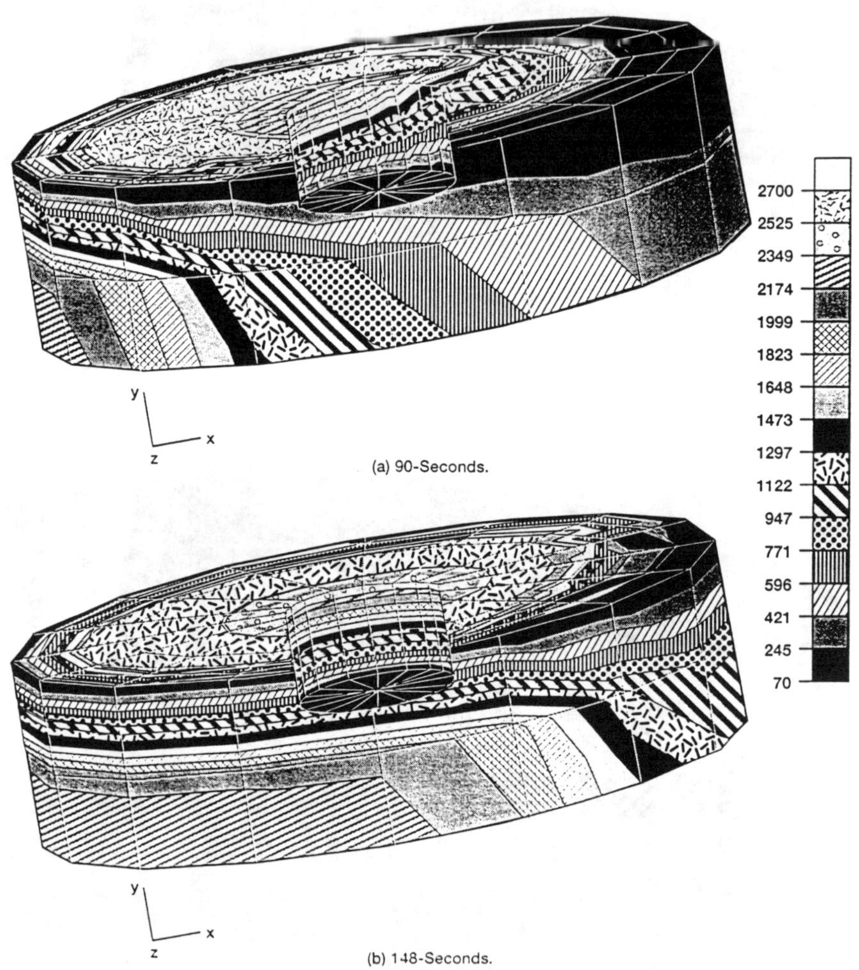

Figure 8 - Casting forming-temperature profile evolution (after 90 and 148 seconds).

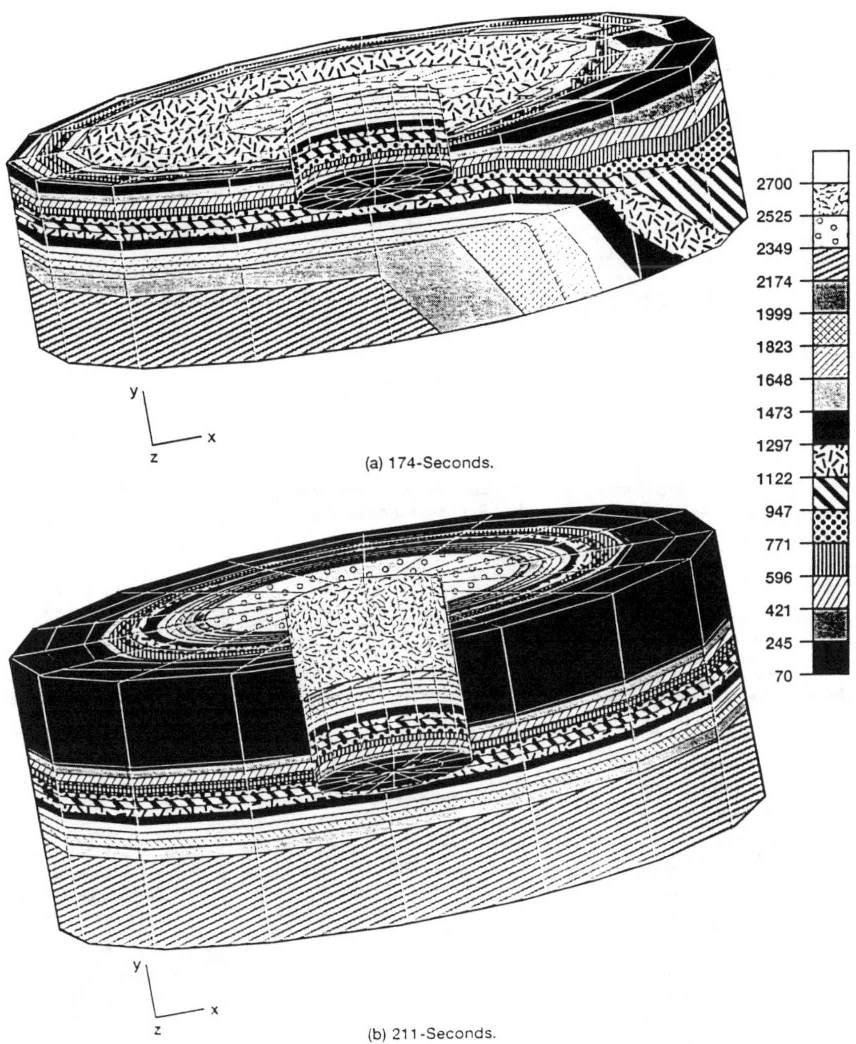

Figure 9 - Casting forming-temperature profile evolution (after 174 and 211 seconds).

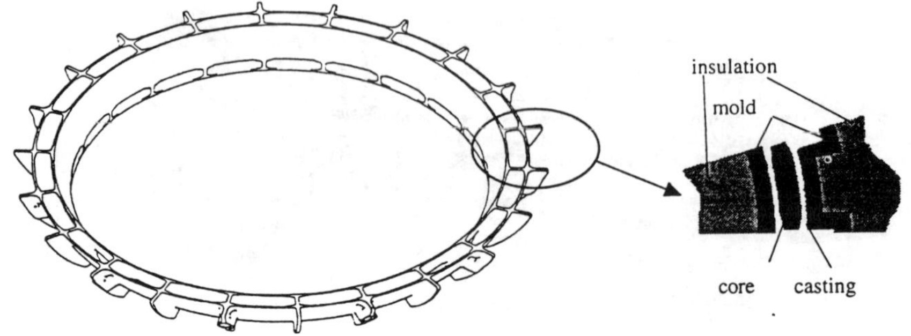

(a) Turning Vane

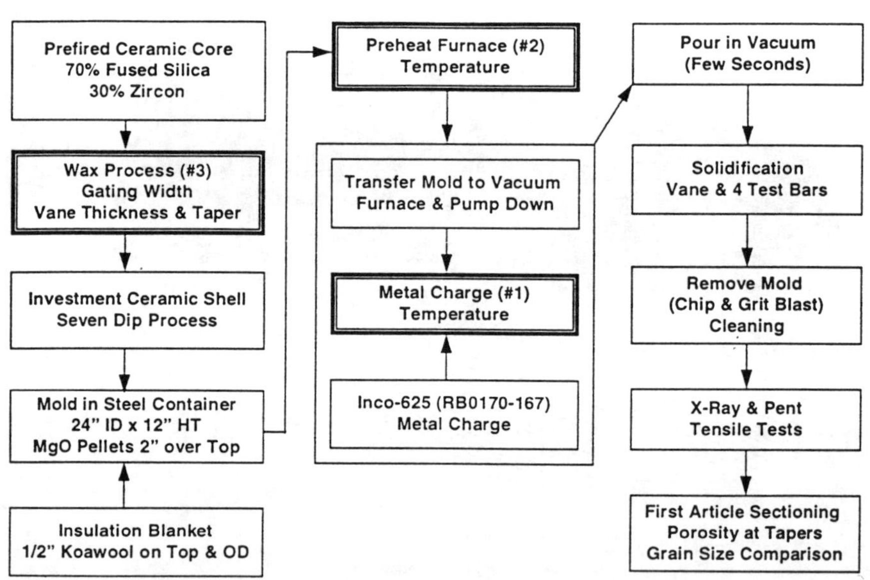

(b) Investment Casting Process

Figure 10 - Space shuttle main engine (SSME) turning vane and casting process schematics.

Variable name	Mean	Std. Dev.	C. o. V.	Meaning
TPOUR	2800	8.33	336	Pour temperature
TPREHT	1800	5.00	360	Preheat temperature
THKVAN	0.07	0.0033	21.2	Vane thickness
HCNTCT	0.0025	0.0002	12.5	Gap conductance
TCMELT	0.02087	0.00522	4	Molten metal conductivity

(a) Random Variables

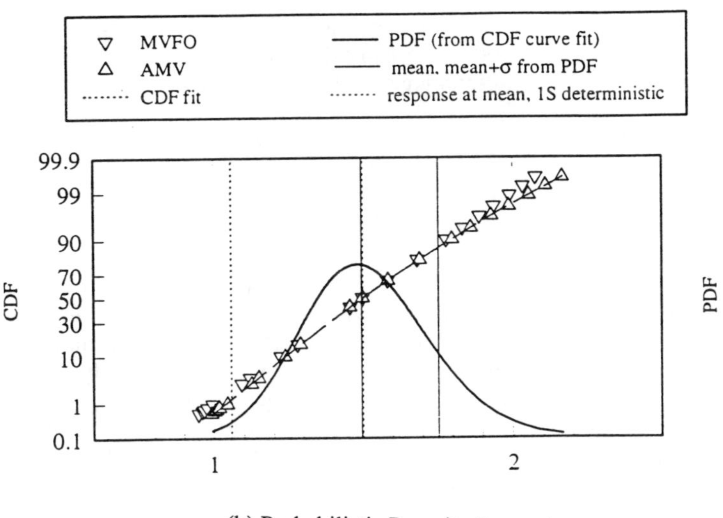

(b) Probabilistic Porosity Formation

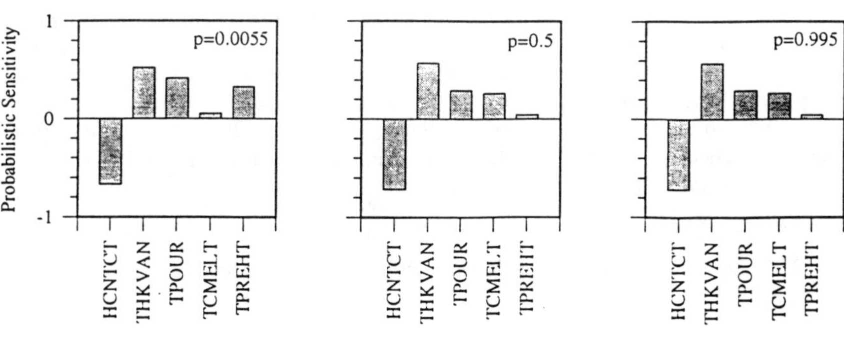

(c) Probabilistic Sensitivities

Figure 11 - Investment Casting Probabilistic Assessment of Space Shuttle
Main Engine (SSME) Turning Vane.

BEYOND RAPID PROTOTYPING - A SEAMLESS TRANSITION FROM CAD TO MANUFACTURING

Yehoram Uziel
Soligen Technologies, Inc.
Northridge, CA 91324

ABSTRACT

Successfully launching a new product depends on a fast and efficient development program coupled with quick and smooth transition to mass production. Once the product design is completed, the ability to create production tooling and debug the manufacturing processes determines the speed and the success of launching the new product. The ultimate goal of a product launching program, is to produce the production tooling fast, only once, and right on the first time. The rationale behind the use of Rapid Prototyping (RP) to produce prototype tooling is to reduce the risks associated with tooling design, and to enable an early use of prototype parts which were manufactured in a process which is as close as possible to the final production process.

The purpose of this paper is to outline a new approach for bringing new products to market, in a faster and less costly manner than currently practiced. This paper presents a case study of launching a new automotive engine and outlines a paradigm shift in the process of bringing an intake manifold to production. This paradigm shift becomes possible with the use of Direct Shell Production Casting (DSPC®), a rapid manufacturing process which makes traditional casting obsolete. DSPC bypass the need for temporary tooling, and produces the actual ceramic casting molds directly from a CAD file. Consequently, the use of DSPC yields functional cast parts and thereby eliminates tooling design and fabrication until after the cast part is functionally tested and approved for production.

KEY WORDS: Direct Shell Production Casting DSPC, Computer Aided Design CAD, Rapid Prototyping RP

1. INTRODUCTION

Successfully launching a new automotive engine depends on a fast and efficient development program coupled with quick and smooth transition to mass production. Many engine parts, especially those with complex core cavities such as engine blocks and cylinder heads must be cast in metal. Fabricating cast metal parts, especially parts that incorporate complex core cavities, remains one of the slowest, most difficult and costly steps of manufacturing. This is because of the need to produce tooling (dies, patterns and core boxes) that are then used to create the sand (or ceramic) molds for the metal casting process. Once the product design is completed, and the engine has been functionally tested, the ability to cost effectively create production tooling and debug the manufacturing processes to minimize scrap and meet target production costs, determines the speed and the success of launching the new product. Long lead times and the high costs of production tooling are the main weaknesses of the transition process from the product design phase to its mass production. To shorten time to-market and optimize profitability, the ultimate goal of a product launching program, is to be able to produce the production tooling fast, only once, and right on the first time.

The rationale behind the use of Rapid Prototyping RP to produce prototype tooling is to reduce the risks associated with tooling design, and to enable an early use of prototype parts which were manufactured in a process which is as close as possible to the final production process. Prototype tooling is less expensive than production tooling and is designed to last for limited quantities. These quantities are sufficient for the design iteration and testing process. Each time that the engineer modifies the part's design, making substantial changes to the part's geometry, or requesting a different alloy, the tooling needs to be redesigned and remade. This multi-step design iteration process is expensive and time consuming.

Once a decision on commercialization is reached, production tooling needs to be made. Creating production tooling requires a new tooling design that often causes substantial inconsistencies between the prototype and the production part. Differences in part geometry or production yield, can create a "domino effect", that can derail the next level of assembly. Production tooling is made from different materials to last longer than the prototype. It is also designed for automatic molding lines, different gating, and other casting parameters. Differences between prototype and production tooling could increase the probability of "unpleasant manufacturability surprises" when it comes to mass production. The price of these surprises is exorbitant and often risks the timely launching of a new product.

2. DIRECT SHELL PRODUCTION CASTING (DSPC®)

Unlike Rapid Prototyping, DSPC eliminates the need for temporary tooling. DSPC is a patternless casting process, that makes conventional casting techniques obsolete for creating a first article part. It uses the customer's CAD to create the actual casting molds and therefore enables the production of any cast part before tooling (prototype or production) are made or even designed.
Eliminating the need for temporary tooling expedites the design and the functional testing of the cast parts. It enables the customer to perform an indefinite number of design iterations, including testing the design with different alloys, without the costly and time consuming need to produce tooling. The use of CAD enables the ultimate configuration control assuring that all design changes are properly documented. For production quantities, production tooling (steel or aluminum patterns and core boxes) are cast as net shape tools from the same CAD file of the approved part. This production tooling is now created only once, guaranteeing a smooth and cost effective transition from the prototype stage to production. By combining DSPC and conventional casting practices, long lead times, costs and the geometrical inconsistencies between the different tooling are eliminated.

The DSPC machine is like a three dimensional printer that uses the designer's CAD to create the actual ceramic casting molds. The DSPC system operator uses the designed part's CAD file to design the ceramic casting mold by adding the gating system to the part geometry and converting the resulted file into a cavity file in CAD space. This is a one-time process which is performed off line to the DSPC machine. The computer file of the ceramic mold is used to automatically generate the actual ceramic mold in layers in the DSPC system. The DSPC fabrication process involves three steps per layer. First, the ceramic shell model is "sliced" to yield a cross-section of the ceramic mold. Second, a layer of fine powder is spread by a roller mechanism. Third, a multi-jet printhead moves across the layer, depositing binder in regions corresponding to the cross-section. The binder penetrates the pores between the powder particles and adheres the particles together into a rigid structure. Once a given layer is completed, the ceramic shell model is sectioned again at a slightly higher position, and the process is repeated until all layers are formed. The DSPC mold is cleaned of excessive powder, fired and poured with molten metal. The following illustrates the steps of the DSPC process:

Step 1: The part is designed on CAD software and exported in .STL format.

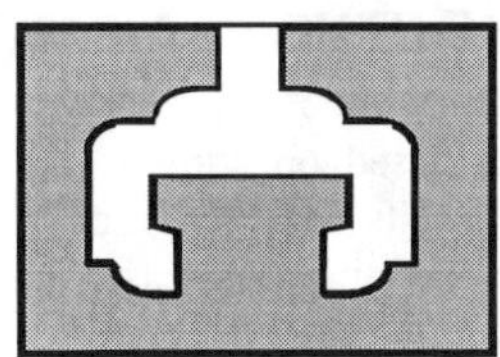

Step 2: Soligen's software designs the casting mold and "slices" it into layers.

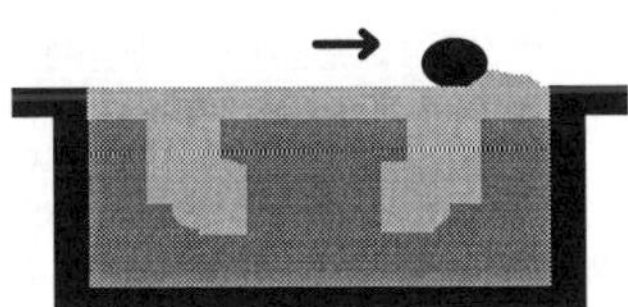

Step 3: A thin layer of powder is deposited for each layer of the casting mold or "shell."

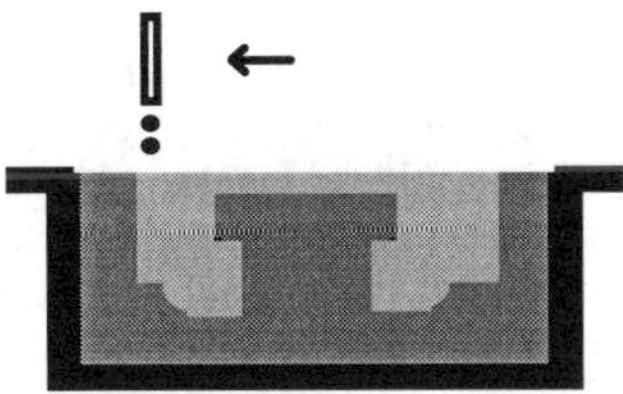

Step 4: An ink-jet printhead deposits binder that solidifies the powder into ceramic.

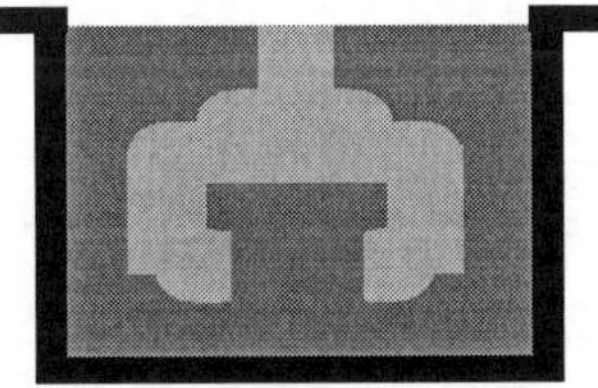

Step 5: The process is repeated, until all layers of the shell have been formed.

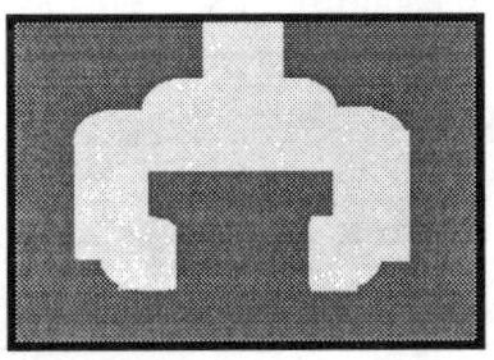

Step 6: Loose powder is removed from the completed shell.

Step 7: The shell is fired, and filled with molten metal.

Step 8: The shell is broken away from the part. The part is finished, and shipped.

A DSPC mold may contain integral ceramic core, allowing a hollow metal part to be produced. Virtually any molten metal can be cast in DSPC molds. Automotive parts have already been manufactured in aluminum, magnesium, ductile iron, stainless steel, etc.

3. A CASE STUDY - AN AUTOMOTIVE INTAKE MANIFOLD.

This case study is based on a time compressed effort to develop a new race car intake manifold. Similar results were observed in short run production programs of cylinder heads, turbochargers and transmission cases. In this specific case, by combining DSPC and conventional casting practices the customer was able, within the original budget and schedule, to test 5 different versions of the manifold design, select the optimal design and receive 150 manifolds identical to the selected version. Traditionally, or even with the help of RP this schedule of 8 weeks would have not allowed any design iteration. Moreover, had the customer been forced to change the manifold design, the cost of the change would have substantially exceeded the original budget.

The following table and graphs illustrate a comparison between traditional sand casting assisted by RP to produce temporary tooling and DSPC in producing a series of this new automotive intake manifold. In the design iteration mode, three manifolds are needed for each dynamometer testing iteration. For the purpose of conservatism, we assumed that the external shape of the manifold is changed only twice and each iteration requires only two new sets of core boxes. Had we assumed that each version of this design requires a new pattern and 4-6 core boxes (a case that represent the level of complexity of this program), the differences between DSPC and sand casting for the design verification phase would have ballooned. Another assumption was that the transition to production does not require a new set of production tooling since the production quantities (150 manifolds) are relatively small. This particular manifold is for a V-6 engine. The following table outlines the cost and the production lead time assumptions:

TABLE 1. Cost and Lead Time	DSPC	Sand Casting
Cost of patterns (one time)	no cost	$20,000
Cost of a set of core boxes	no cost	$12,000
Cost of each casting	starting $8,500 declining to $3,400	$300
Time to receive tooling	no need for tooling	12 weeks
Time to receive a new set of core boxes	no need for tooling	6 weeks
Time to cast a set of 3 manifolds	2 weeks	1 week
Transition to production tooling	at no cost	none
Price for production parts (each)	$120	$120

Tooling for conventional casting dominates the cost and lead time required to produce a first article part. With DSPC, the part's cost or delivery time are independent of the part's geometry.

The following charts show the cumulative of time and costs between traditional casting and DSPC for this automotive manifold.

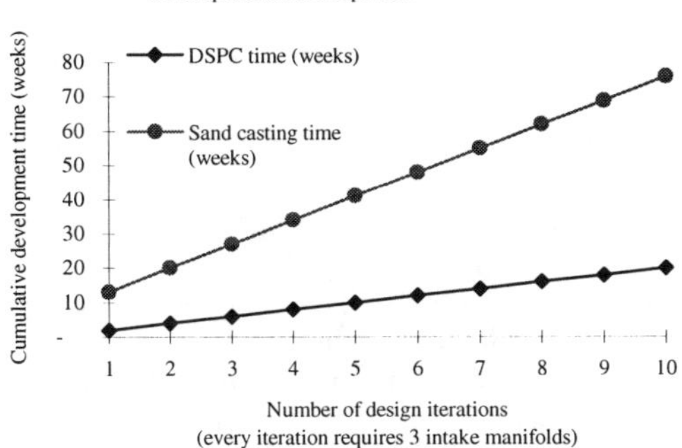

Using DSPC enables the engineer to make many more design iterations in a short period of development time. Had the program called for a change in alloy (such as from aluminum to iron), the difference in the lead time between iterations of sand casting would be drastically increased as well as the program cost. With DSPC such changes would not affect the delivery time at all. It may, however, slightly affect the cost (by the difference in cost of the alloys)

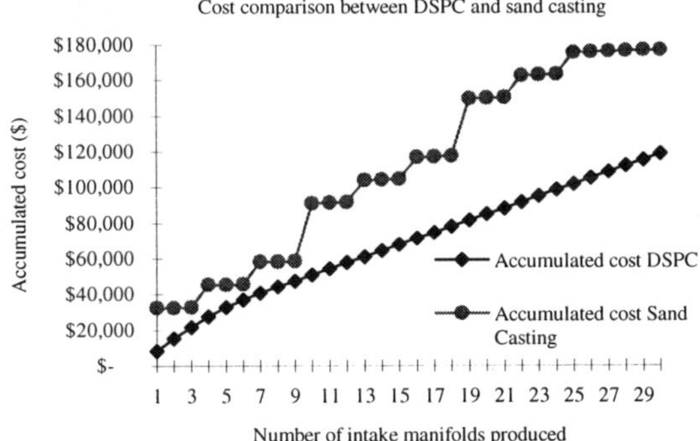

Once the design has been verified and the manifold first article batch has been successfully tested and approved, the company has quickly moved to produce the short production run. The CAD file of the approved manifold was directed and a matchplate pattern and three core boxes were electronically created. Soligen has used its DSPC technology to cast the matchplate and the core boxes. This process took 10 business days, after which these tools were used to produce 150 manifolds (another week). In parallel to the casting process, CNC code was generated to machine the cast manifold (spot face the flange and drill and tap the needed threads). The process of converting the CAD data to files for the pattern and the coreboxes was concluded at no additional cost to the customer. The transition from CAD to conventional casting has proven itself seamless to the customer. By creating the production tooling from the same CAD file of the manifold, the part was approved by the customer, eliminating interpretation errors in the design process of the production tool.

4. CONCLUSIONS

By combining DSPC with conventional casting and CNC methods, the following advantages can be realized:

- **Speed**—Substantial reduction of time to market due to the elimination the need for tooling at the design phase.
- **Cost reduction**—The ability to incorporate design changes without the cost and lead time associated with producing or modifying patterns or tooling.
- **Design flexibility and versatility** — The ability to test unlimited number of design iterations including testing a design with different alloys.
- **Seamless transition to from CAD to conventional casting**—Design and fabrication of production tools can be delayed until after the design is verified.
-

About the Author   *Yehoram Uziel* is Founder, CEO, and Chairman of Soligen Technologies, Inc. From January 1989 to January 1992, he served as Vice President of Engineering at 3D Systems, Inc. Prior to 1989, Mr. Uziel was Cofounder, Opto-Mechanical Division Head and Director of New Business Development of Israel-based Optrotech, Inc., a manufacturer of inspection equipment for the printed circuit board industry. From 1990 to 1993, he was Chairman of the Board of ConceptLand Ltd., a rapid prototyping service bureau based in Israel. Mr. Uziel received a B.Sc. in Mechanical Engineering from the Technion Institute of Technology in Israel.

PROTOTYPE TOOLING AND MANUFACTURING THROUGH LAMINATED OBJECT MANUFACTURING (LOM)

Sung S. Pak
Helisys, Inc.
24015 Garnier Street
Torrance, CA 90505

ABSTRACT

In today's competitive business environment, the ability of a manufacturing firm to prototype and qualify parts and tools quickly and economically is of paramount importance. Amongst a myriad of prototyping technologies, Laminated Object Manufacturing (LOM) has emerged as one of the leaders due to its ability to fabricate large, complex tools of high accuracy and adequate thermo-mechanical properties without the traditional programming and fixturing requirements in unprecedented time. Examples in sand casting, investment casting, blow molding, hydroforming, and injection molding are presented herein.

KEY WORDS: Rapid prototyping, tooling, composites, ceramics, ceramic matrix composites (CMC), polymer matrix composites, injection molding, hydroforming, plastics, blow molding, sand casting, investment casting

1. INTRODUCTION

Rapid prototyping (RP) is a term used to denote a class of additive fabrication processes by which complex, 3D objects are created automatically from a computer-generated database (1). With traditional model-making processes, several weeks are typically consumed before the designer receives his or her prototype. With RP, however, the designer could receive the prototype in a just few days, if not hours. The power to evaluate designs only days after their completion is allowing unprecedented savings in development time and cost. (2). However, even greater time and cost savings can be realized by extending RP to manufacturing, especially in low volume production. This paper describes how a recently developed RP process called Laminated Object Manufacturing (LOM) can be applied to existing low to medium volume manufacturing processes to achieve unprecedented savings in time and cost.

2. LAMINATED OBJECT MANUFACTURING

Laminated Object Manufacturing (LOM), as the name implies, creates complex three dimensional objects by electronically slicing the intended object into planar cross-sections, and sequentially reconstructing the object by laminating sheet materials and cutting the corresponding cross-sections with a focused laser beam. Fig. 1 illustrates the commercially reduced process. Lamination of each layer is achieved by way of a heated roller traveling across the top surface in one single reciprocal motion. The heat-sensitive adhesive under the top sheet of paper is then activated momentarily and provides the necessary tack for adhesion. The power of the laser beam is adjusted so that the depth of cut is limited to the thickness of each layer. The excess material, which supports the model during the fabrication process, is diced up to facilitate its easy removal after the entire object has been constructed.

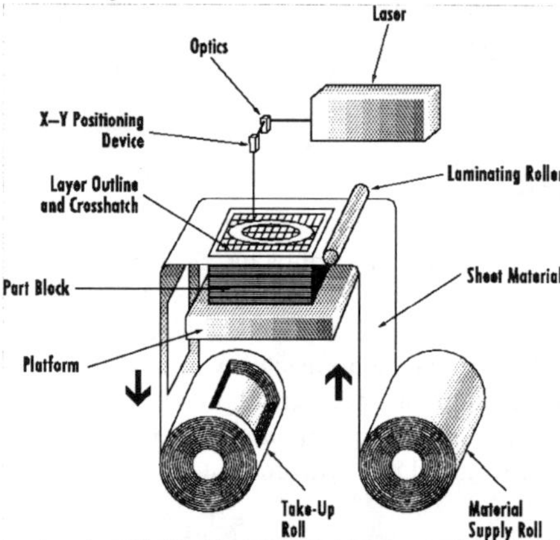

Fig. 1. Schematic of the LOM process.

The ability of the LOM process to handle virtually any sheet material allows the utilization of a wide variety of materials, including paper (3), plastics (4), polymer composite prepregs (5), metals (6), ceramics, and even ceramic matrix composites (7-8). Paper, however, has been the most popular because it is inexpensive and possesses many of the same properties of wood, the most popular pattern material. The choice of material depends on the application. For visualization and concept modeling, paper is ideal. For snap-fit and limited functional testing where flexibility and high interlaminate strength are needed, plastics are superior. For intermediate temperature and pressure tooling applications, such as injection molding, polymer composites may provide a solution. High temperature (>1000 $^\circ$ C) structural parts and tooling will necessitate ceramic materials.

3. APPLICATIONS

3.1 Sand Casting Sand casting is a metal-forming process where objects are produced by casting molten metal into sand molds. It is one of the most versatile metal-forming processes, providing unmatched breadth in size, shape, and material (9).

Sand molds have traditionally been created from meticulously hand-crafted wooden patterns by highly skilled workers. The detail and artistry required for these patterns involve several weeks of preparation. With the advent of LOM and its ability to quickly and inexpensively produce wood-like sand casting patterns, the traditional, labor intensive pattern making is being drastically refined. It should be noted, however, that instead of supplanting the traditional pattern/mold making process, LOM streamlines it because the technology is fully compatible with all the secondary techniques that are familiar to the foundryman.

When just a few castings are needed, an LOMPaper™ pattern is used to make direct impressions in sand. First, an LOM pattern with appropriate shrinkage factors and draft angles is created. It is then sealed (with epoxy or lacquer), filled, coated with a primer and sanded until an acceptable surface finish is obtained. Finished patterns that are simple in shape and contain a flat surface are mounted onto a flat, wooden board. This pattern is called a lose pattern. For more complicated patterns, a special wooden follow board is created by boring out a cavity in a wooden plank up to the depth of the parting location. The follow board thus determines the parting line. The top half of the pattern is placed into the follow board so that the bottom half is exposed above the board's surface.

The bottom half of the sand mold called the drag is formed first by placing sand in the volume defined by the bottom half of the pattern and a surrounding flask. The drag is turned over and the board is removed (Fig. 2). The top half of the sand mold called the cope is then formed by packing sand over the drag. Next, the two halves are separated and the pattern is removed. The mold is completed by creating gates and risers, which is typically accomplished by manually cutting into the mold. The mold is then ready for pouring molten metal.

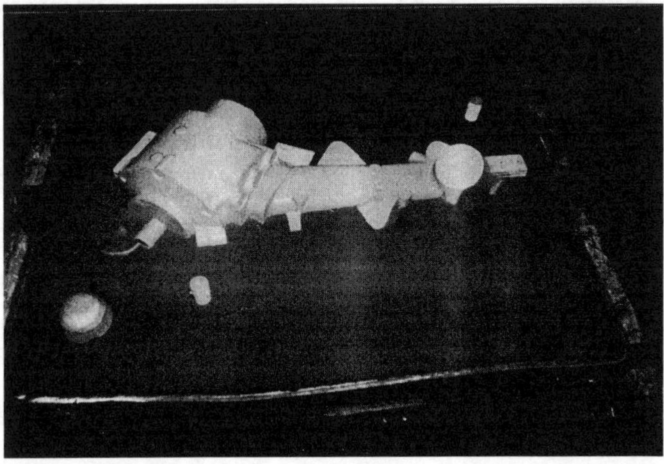

Fig. 2. A sand mold formed by pouring green sand into a volume defined by an LOM pattern, a flask, gates, and risers.

The life of an LOM pattern varies depending on the material, geometry, and the type of sand being used by the foundry. Typically, an LOM pattern can create up to 50 sand molds. Upwards of 100 replications are possible if proper care is taken (10).

If a much higher number of castings are required, a match plate pattern is produced from an LOM master. Much like the lose pattern described above, a match plate pattern is a split pattern with the cope and the drag on opposite sides of a plate that defines the parting line. It is different from a lose pattern in that it contains strategically placed gates and risers. A match plate pattern is produced by first creating a silicone rubber negative and casting two-part plastic reactants, such as epoxy or polyurethane, into the cavity under vacuum. These materials provide a superior wear and stress resistance in automated molding machines. For higher throughput, several parts can be replicated on the same match plate. A sprue, gates, runners and risers are then positioned to complete the process.

If even a higher number of castings are required, the LOM process is used to create aluminum or cast iron tooling. This can be achieved by creating a master LOM pattern and casting a match plate. Or, using the master pattern to make an EDM electrode by use of the abrading process.

3.2 Hydroforming Hydroforming is a well established sheet metal forming process where a metal sheet is placed over a tool and formed via hydraulic pressure. Aluminum, steel, copper, nickel, and sometimes titanium are formed at pressures as high as 83 MPa. In this application, high compressive and shear strengths are important for maintaining the integrity of the LOM tool against the stresses induced by the "flowing" metal sheet.

Hydroforming tools are typically made by CNC machining aluminum or steel stocks. Their cost ranges widely from a few hundred to several thousand dollars depending on the size, material, and geometry. This may not seem costly, but for small volume orders and prototyping jobs, a major portion of the procurement cost is the tooling cost. Moreover, these shops typically have to devote a large floor space for storing these tools in anticipation of winning more orders in the future. However, the order may never come. It would be of tremendous economic benefit if inexpensive, expendable hydroforming tools for limited production and prototyping runs were made available.

The shear and compressive properties of LOMPaper is sufficient for forming thin sheet metal parts, such as the aluminum door panel of a passenger car. For more demanding applications, a higher performing material must be chosen. A fiber glass/epoxy composite material is often the preferred LOM material. To form a sheet metal over the tool, a thin layer of lubricant, such as limonin oil, is applied over the surface of the tool to facilitate an easier metal "flow." Then, a metal sheet sandwiched with a sacrificial rubber layer is placed between the metal and the bladder. Next, the assembly is pressed until a predetermined level of pressure and dwell time are reached. In the case of a 0.46 mm thick electronic enclosure shown in Fig. 3, 55 MPa was sufficient to form copper with all the critical dimensions satisfied.

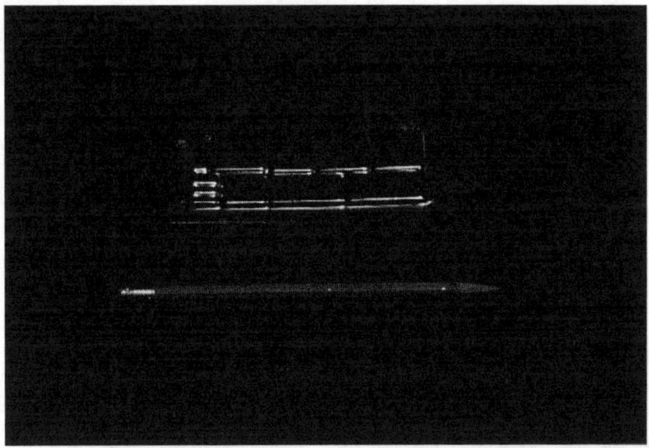

Fig. 3. Copper electronic enclosure hydroformed over an LOM composite tool.

The main mode of LOM tool failure in hydroforming is due to shear stress-derived delamination. The failure rate can be minimized by avoiding thin walls. For example, in the case study shown above, the male tools with block-like protrusions were much more durable than female tools that contained thin, free-standing ribs.

3.3 Investment Casting Investment casting is a process by which metal duplicates are created from an expendable master. The master, typically wax, is repeatedly dipped in a ceramic slurry and dried until a thick shell is created. The master is then extracted from the shell by heating the assembly in an oven beyond the melting point of the wax. The empty shell is then densified by further heating. Next, a molten metal is cast into the empty shell. The metal duplicate is obtained by breaking open the shell.

LOMPaper^{TM} masters can be used as a one-to-one replacement of wax masters (direct investment casting) or as a tool into which up to 100 wax patterns can be injected (indirect investment casting). In the case of direct investment casting, the low thermal expansion coefficient of paper and the low softening point of the adhesive allow the pattern to collapse and prevent the shell from cracking during heating.

LOM masters are different from wax and hence must be treated as such. Firstly, LOM masters have to be sealed tightly to prevent moisture ingression from the ceramic slurry. Acrylic enamel clear coat is a preferred choice because it not only seals out moisture, but leaves little or no ash residue in the shell as well. Secondly, a higher temperature must be used to extract the master. With wax, temperatures in excess of 77 $^{\circ}$C is sufficient to melt the master. It then flows out due to gravity, leaving an empty shell. LOM masters on the other hand have to be burned by heating at temperatures greater than 574 $^{\circ}$C in an oxidizing environment. More typically, heating temperatures ranging from 871 to 1094 $^{\circ}$C are used (11). The obtained shell is cooled and the remaining ash is either blown out with compressed air or flushed out with water.

Indirect investment casting is used when more than several duplicates are needed. It is much more economical to fabricate a single LOM mold, and inject wax to create hundreds of duplicates. The wax patterns can then be taken through the normal investment casting process as outlined above to replicate them into a metal of choice.

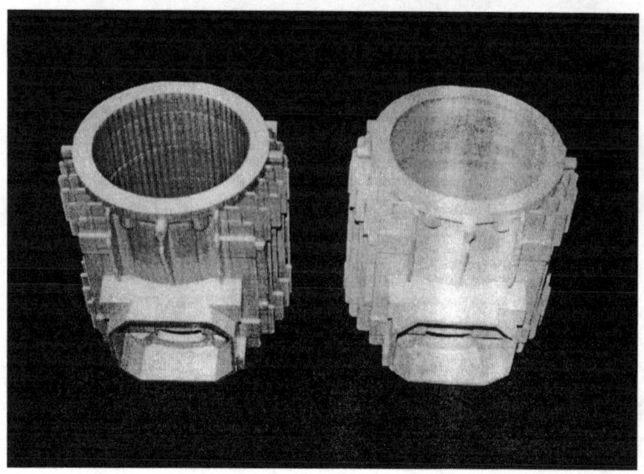

Fig. 4. An LOM paper master (left) and an investment cast aluminum replica (right).

3.4 Blow Molding Blow molding is the most popular plastic-forming process for hollow objects, such as soft drink containers and toy wheels. In this process, objects are formed by extruding a partially molten tubular preform called a parison into a mold and inflating the preform against the mold cavity with a gas pressure in the range of approximately 0.68 MPa.

Designing blow molds poses special challenges as it requires special cooling, pinch-off, and venting considerations. LOM provides an excellent platform for quick evaluation of mold designs in production settings. The thermo-mechanical properties LOMPaper^{TM} molds are sufficient for direct injection and molding of low melting thermoplastics such as LDPE and HDPE (12). For higher melting plastics, however, either the paper mold has to be coated with a thin layer of metal, or composites or higher performance materials must be used.

3.5 Injection Molding Injection molding is undoubtedly the most popular of all the thermoplastic forming processes.   Its ability to mass-produce complex shapes quickly and economically has made injection molding a multibillion dollar business. It is a challenging process for LOM as the temperature of the melt can range from 149 to 316 $^\circ$ C and the pressure from 3 to 207 MPa. For low melting plastics, LOMPaper^{TM} molds can be sprayed with a thin metal coating, such as kirksite, for injections in the upwards of a few hundred parts. For higher melting and abrasive systems such as glass-filled nylon, an epoxy/polyester fiber-based LOM material may be more appropriate. Fig.4 shows such a mold with a pair of polypropylene parts. The molten plastic temperature was approximately 216 $^\circ$ C and the injection pressure 3.4 MPa. Besides polypropylene, similar composite molds have been used successfully to inject up to 50 parts of neoprene, polyethylene, acetal, glass-filled acetal, and nylon parts.

Fig. 5. Composite injection mold and polypropylene parts.

Due to the lower thermal conductivity and heat deflection temperature, the composite mold has to be treated slightly differently from metal molds. Firstly, the mold cannot be heated and cooled as quickly, and hence a comparatively extended cycle time has to be applied. A cycle time of three to five minutes is typical with one to two minute dwell time. Secondly, longer runner lengths and larger gate openings help to reduce the temperature of the molten plastic in order to minimize overheating.

As with metal molds, the high density and machinability of the material allow cooling lines to be drilled directly into the mold with little or no leakage. This can shorten the cycle time. However, it should be noted that the cycle time for limited production runs and prototyping is not critical. Only when several hundred replicas are required does the increased cycle time begin to become a concern.

4. SUMMARY

Laminated Object Manufacturing offers a cost-effective temporary tooling solution to today's most demanding manufacturing needs. Its ability to produce complex three-dimensional masters and tools out of a diverse array of materials allows quick evaluation of designs and fabricability. The feasibility of using LOM to produce tooling for sand casting, investment casting, blow molding, hydroforming, and injection molding has been established and, in most cases, reduced to practice.

5. REFERENCES

1. M. Burns "Fabricators in Japan," Ennex Fabrication Technologies," Los Angeles, CA, pp. 4, 1996.

2. R. Chartoff, A. Lightman, and J. Schenk, <u>Proceedings of the Fifth International Conference on Rapid Prototyping</u>, University of Dayton, Dayton, 1994.

3. S.S. Pak and G. Nisnevich, "Interlaminate Strength and Processing Efficiency Improvements in laminated Object Manufacturing," <u>Proceeding of the Fifth International Conference on Rapid Prototyping</u>, R. Chartoff, A. Lightman, and J. Schenk (eds.), University of Dayton, Dayton, 1994, pp. 171-180.

4. S.S. Pak, "Fabrication of Plastic LOM Objects," <u>Proceeding of the Sixth International Conference on Rapid Prototyping</u>, R. Chartoff, and A. Lightman (eds.), University of Dayton, Dayton, 1995, pp. 35-41.

5. D. Klosterman, B. Priore, and R. Chartoff, "Laminated Object manufacturing of Polymer Matrix Composites," <u>Proceeding of the Seventh International Conference on Rapid Prototyping</u>, R. Chartoff, A. Lightman, M. Agarwala, and F. Prinz (eds.), University of Dayton, Dayton, 1997, pp. 283-292.

6. C. Chi, L. Dodin, and S. Pak, "Development and Fabrication of Metallic LOM Objects," <u>Proceeding of the Seventh International Conference on Rapid Prototyping</u>, R. Chartoff, A. Lightman, M. Agarwala, and F. Prinz (eds.), University of Dayton, Dayton, 1997, pp. 293-299.

7. G. Kalmanovich, "Curved-Layer Laminated Object Manufacturing," <u>Proceedings of Solid Freeform Fabrication Symposium</u>, D. Bourell, et.al (eds.), University of Texas, Austin, 1996, pp. 273-279.

8. D. Klosterman, et.al, "Structural Composites via Laminated Object manufacturing," <u>Proceedings of Solid Freeform Fabrication Symposium</u>, D. Bourell, et.al (eds.), University of Texas, Austin, 1996, pp. 105-115.

9. P. O'Meara, L. Wile, J. Archibald, R. Smith, and T. Piwonka, "Bonded Sand Molds," in <u>ASM International Handbook</u>, Vol. 15, Casting, ASM International, Materials Park, 1992, pp. 222-230.

10. B. Mueller, "Coating of LOM Sand Casting Patterns to Improve Their Resistance Wear," Helisys Internal Report, 1996.

11. B. Hinckley, "Direct Investment Casting of LOM Rapid Prototypes," M.S. Thesis, Brighham Young University, Provo, UT, 1995

12. L. Wu, "Feasibility Study of Rapid Prototype Molds for Blow Molding Using LOM," M.S. Thesis, Brighham Young University, Provo, UT, 1994

Direct Fabrication of Ceramics and Composites through Laminated Object Manufacturing (LOM)

**Don Klosterman, Richard Chartoff, George Graves, Nora Osborne,
Allan Lightman, Gyoowan Han, Akos Bezeredi, Stan Rodrigues**

Rapid Prototype Development Laboratory, and
Ohio Rapid Prototype Process Development Consortium
University of Dayton
Dayton, OH 45469-0130

ABSTRACT

The capability of rapid prototyping has recently been extended to the direct fabrication of structural ceramics and ceramic matrix composites (CMCs). Summarized herein is the status of an ongoing program involving a novel approach for fabricating structural composites. The Laminated Object Manufacturing (LOM) process was custom adapted for fabricating both flat layer and curved layer composite laminates. A new LOM machine was developed which uses fiber prepregs and ceramic tapes as a feedstock. The machine stacks and bonds layers of these materials and cuts them with a laser. The output is a three dimensional "green" form that is capable of being further processed to a seamless, fully dense ceramic structure using traditional techniques. The LOM process is also applicable to polymer matrix composites (PMCs). LOM is fully automatic and requires only a computer graphic file of the part to be created. Intricate tools and molds are not required anywhere in the overall process cycle.   This report reviews how the complete process was demonstrated for monolithic SiC parts and continuous fiber SiC/SiC composites. The focus of the paper is a detailed description of the various materials and processing issues involved.

KEY WORDS : Ceramic Materials/Composites, Custom Processing, Low Cost Manufacturing/Fabrication/Processing, Tooling, Rapid Prototyping, Automation

1. INTRODUCTION

1.1 Rapid Prototyping, Current Capability
Rapid prototyping (RP), also know as solid freeform fabrication (SFF) or layered manufacturing, is a relatively new area of technology involving the automated fabrication of dimensionally accurate, solid prototypes or models directly from computer graphic files. There are several commercially available technologies, all of which operate under the same principle of building an object in a layer by layer fashion. First, the layerwise geometric information is extracted from a computer graphic file that is sliced with a series of

horizontal planes. Then, the article is physically reconstructed, cross section by cross section, using wax, low strength plastic, plastic coated powder, or adhesive paper. RP is widely used in industry for fabricating conceptual models and prototypes of limited functionality.   Recently, there has been much activity in developing new materials and improving the existing processes so that fully functional articles can be constructed from engineering materials such as composites, ceramics, metals, and engineering plastics.

Laminated Object Manufacturing (LOM) is a popular, commercially available rapid prototyping process. Prototypes are constructed from adhesive coated paper or plastic, as illustrated in Figure 1. The machine consists of an elevator platform, a hot laminating roller, and a CO_2 laser. A software algorithm slices a computer graphic model into thin cross sectional layers. The model is physically recreated in a layer-by-layer cycle consisting of laminating a single sheet of material to the existing stack, laser cutting the perimeter of the part cross section, and laser-dicing or "cubing" the waste material. During the build, the waste material is left in place to serve as a support. After all layers have been completed, the part block is removed from the platform, and the excess material is removed, or "decubed", to reveal the three dimensional part. Full scale prototype parts of up to 81 cm x 56 cm x 51 cm (32 in. x 22 in. x 20 in.) are produced in 6-72 hours, depending on their size and complexity. The process is fully automatic, with the exception of decubing and post finishing. LOM parts are used as conceptual models, casting patterns, and functional prototypes for low temperature ($<50^{\circ}C$) and low strength applications.

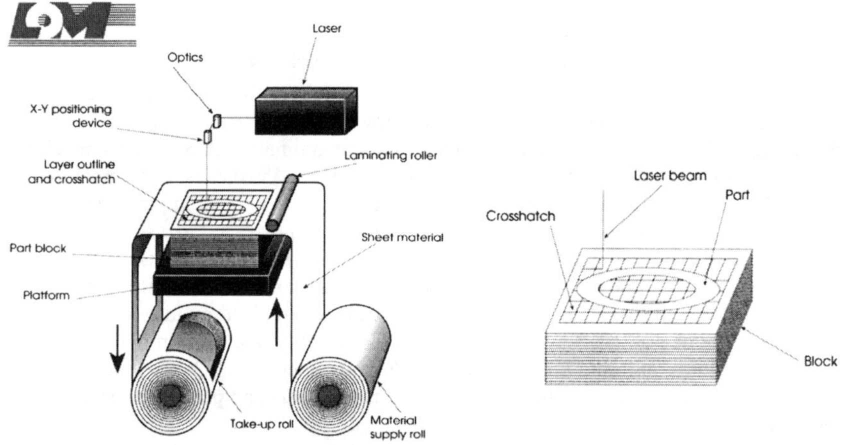

Figure 1 : Schematic of the standard LOM process (courtesy Helisys, Inc.). In this example, a hollow cylinder is being created.

1.2 Rapid Prototyping, State-of-the-Art

A new solid freeform fabrication (SFF) technology has been developed that provides industry with an enhanced capability for fabricating structural ceramic and CMC articles. LOM was custom adapted for fabricating monolithic ceramics and continuous fiber CMCs. An ambitious three-year demonstration effort was recently completed that involved materials and process research, machine design, and system software adaptation. In addition to expanding the envelope of available materials to include structural ceramics and composites, an entirely new building paradigm was implemented: the *curved* layer build style. Instead of being limited to building with flat layers, the machine is now capable of building in a curved-layer-by-curved-layer manner. The new curved layer LOM process allows continuous fiber composites to maintain their fiber continuity in the plane of curvature in order to achieve optimum mechanical performance.

The rationale for using this particular RP process for ceramic and composite fabrication is that LOM is uniquely suited for processing flat sheet materials such as tapes and prepregs. In addition, it produces geometrically complex objects and operates with a high degree of automation. Compared to other RP techniques, LOM has the largest build envelope. The ceramic LOM process holds promise for lowering the cost of fabricating monolithic and CMC parts by virtue of its automation, which reduces fabrication time and eliminates the need for tooling. Complex shaped, thin cross section parts that cannot be produced by machining also can be made by this method. The process can best be applied by industry as a product development tool for fabricating molds, tooling, or testable prototypes, or used directly for small lot production.

Described in this report are the results of the three-year technical effort. The first step was to demonstrate the feasibility of making monolithic ceramic articles using the standard, flat-layer LOM process with ceramic tapes as a feedstock. The second step involved developing the necessary technology for making curved layer monolithic ceramic parts. Finally, materials and processes were developed to fabricate curved layer, continuous fiber CMCs. Material selection was limited to one non-oxide monolithic and one composite material system. Monolithic SiC and continuous fiber SiC/SiC composites were chosen as the focus materials. Inherent to the process of making the CMCs was the production of a precursor composite structure, consisting essentially of a polymer matrix composite (PMC), which was subsequently converted to a CMC. Thus, the results demonstrate that PMC fabrication is also viable with the new LOM process.

2. MONOLITHIC CERAMICS, FLAT LAYER PROCESS

The overall process for making net shape monolithic ceramics via LOM is illustrated in Figure 2. The feedstock to the process was a 250µm-thick, flexible sheet called a ceramic tape. Ceramic tapes were fabricated from SiC powder, carbon and graphite powders, and a polymer binder system using a standard doctor blade tape casting process. The bimodal SiC powder contained large (60 µm) and small (2-3 µm) diameter particles. Figure 3 illustrates the microstructure of the tape, showing a bimodal particle size distribution, which is typical for reaction bonding powder formulations. The binder system contained a thermoplastic polymer and organic plasticizer to the extent of about 20 wt% of the total composition.

The LOM process for monolithic ceramics entails sheet feeding, solvent spraying, hot roller laminating, and laser cutting (Figure 2). Although flat layers are used, the final part shape can be quite complex. The ceramic tapes must be sprayed with a solvent immediately before the lamination step to enhance lamination efficiency and repeatability. The advantages afforded by applying a solvent are documented elsewhere [1-3]. When the LOM process is complete, parts are removed from the build platform encased in a block; the waste material is manually removed ("decubed") to isolate the part. This part is a "green" form which must undergo further processing to produce a fully dense ceramic structure.

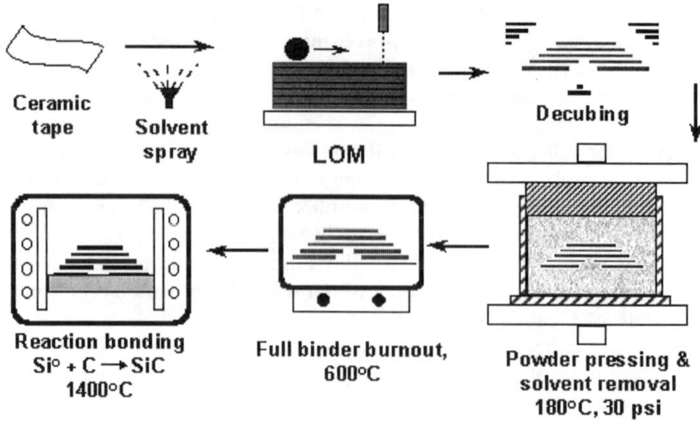

Figure 2 : Overall process for making monolithic ceramic parts with flat layer LOM.

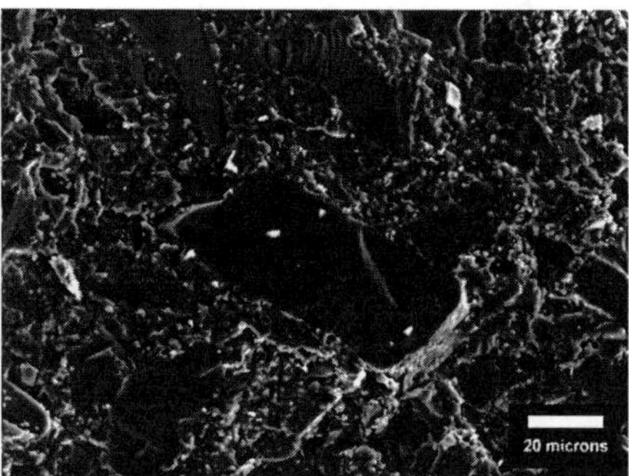

Figure 3 : Microstructure of SiC ceramic tape showing bimodal particle size distribution.

To improve layer bonding, decubed green parts from the LOM process are placed in a cylindrical chamber which is subsequently back filled with powder and fitted with a ram (see Figure 2). A solvent removal cycle is applied using a heated, programmable, uniaxial press. The powder medium may be SiC, or a foreign material such as silica if the part is able to be fitted with a porous Teflon bag. Shrinkage during this step is 5% ± 2% in the z direction. Scaling the computer graphic file in the z direction by 5% using LOM software can compensate this dimensional change.

After the press cycle, the part undergoes freestanding binder burnout in argon. If necessary, the underside of the part can be supported by a bed of powder. Alternatively, it is possible to perform the entire burnout cycle in the powder chamber during the previous step. Further work is needed to assess the merits of each case. After burnout is complete, the part is densified via reaction bonding. The reaction bonding process involves infiltration of liquid silicon into the burned-out, porous part. The silicon reacts with the *in-situ* carbon to form additional SiC. The primary advantage of reaction bonding as a densification technique is that little or no shrinkage occurs, resulting in near net shaped parts.

Various parts, given in Figure 4 and in previous reports [1-4], effectively illustrate the geometric complexity that can be produced with the ceramic LOM process. The minimum feature size that can be achieved is 250-500 μm. The overall dimensional tolerance of the LOM build process, in general, is about 250-500 μm for parts of approximately 50 cm in width. However, less is known about the dimensional changes that accrue during the subsequent post processing steps. Fortunately, most or all of the dimensional change occurs in one step, during powder bed pressing.   It is also fortunate that this step is a potentially well controlled and repeatable process, although further work is needed in this area.

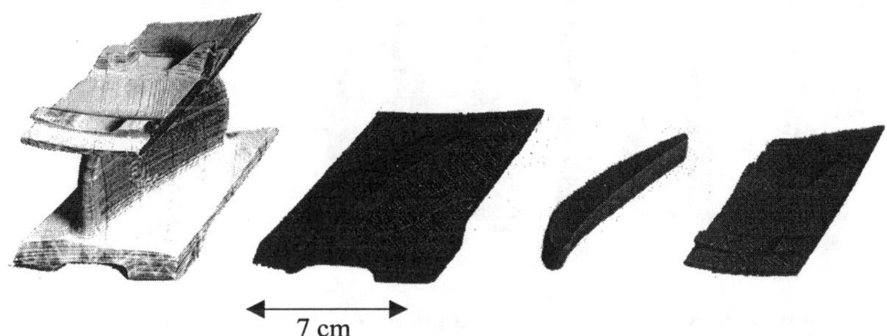

Figure 4 : Complex sections of a turbine engine blade, faithfully reproduced with SiC using the flat layer LOM process. The full model at the far left was made from standard LOM adhesive paper.

Figure 5 shows the polished cross section of reaction bonded LOM part interior illustrating two major points. First, the infiltration of silicon (white spots) during the reaction bonding process is very effective. Second, evidence of a layered structure has been largely, although not completely, removed. In addition, there is little porosity. This microstructure is comparable to reaction bonded SiC produced with traditional processes like pressing and casting.

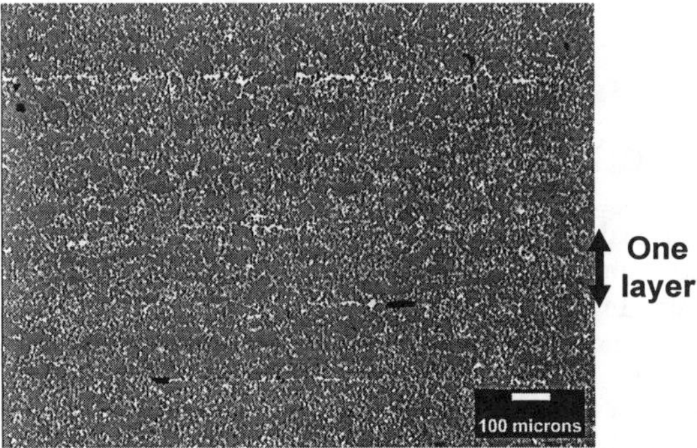

Figure 5 : Cross section of reaction bonded, monolithic SiC bar made with the ceramic LOM process.

An intensive characterization of the mechanical behavior of LOM SiC parts was made.
MOR bars (4 mm x 3 mm x 50 mm) were fabricated using the ceramic LOM process and
densified via reaction bonding. Over 50 bars were made in each of six different LOM runs.
The results of the 4-point flexure tests are given in Figure 6. The specimens were tested as
received without any polishing. The standard deviation of each of the six runs ranged from
10 to 34 MPa, a relatively narrow variation.

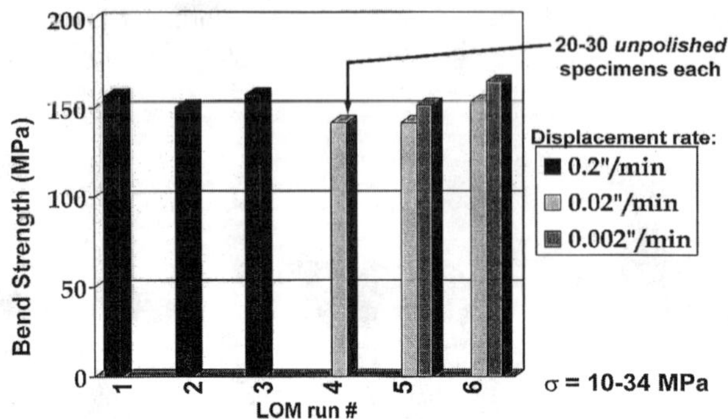

Figure 6 : Ambient, 4-point bend strength of reaction bonded LOM SiC.

It is important to make some comparison of LOM-produced SiC to that fabricated by
traditional methods. The consistency of the bend strength from LOM run to run is
characteristic of SiC as a material in general. The fact that the bend strength does not
change with strain rate indicates that this material does not exhibit slow crack growth,
another favorable characteristic of SiC in general. The magnitude of the bend strength is
difficult to compare to commercial material because these specimens were unpolished.
With proper preparation, the bend strength of LOM SiC is expected to significantly increase
from 150 MPa, quite possibly by a factor of two. The 4-point bend strength of reaction
bonded SiC cited in the literature [5, 6] ranges from 210 MPa to 350-540 MPa as the
specimen specific gravity ranges from 2.45 to 3.18 respectively. The specific gravity of
specimens in this study is approximately 2.90 to 3.0. Testing of polished LOM SiC
specimens is in progress at this time, and the bend strength is expected to fall within the
range cited in the literature.

3. CURVED LAYER MONOLITHIC CERAMICS

The flowchart for making *curved* layer monolithic ceramic parts is exactly the same as in
Figure 2 except for the way the tape layers are processed within the LOM machine. The
details of the curved layer lamination step are given in Figure 7. The first step is to
fabricate a custom mandrel for the part. This task is easily accomplished using the flat
layer LOM process to fabricate a LOM-paper mandrel. The next step is to mount the
mandrel on the LOM platform. Layers of ceramic tape are laminated over the mandrel
using a flexible thermoforming apparatus. After lamination each layer is cut with a laser
that is controlled with newly developed hardware and software. The laser beam is kept
focused on the surface of the part through simultaneous movements of the laser beam in the
X and Y direction and movement of the platform in the Z direction.

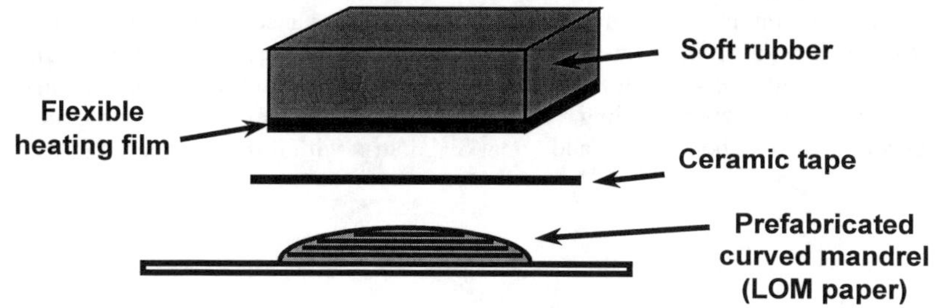

Figure 7 : Simplified illustration of curved layer LOM bonding mechanism.

Parts made with the curved layer LOM process are shown in Figure 8. The advantages of the curved layer process are numerous compared to the flat layer process (for making curved layer parts). As seen in Figure 8, the stair step effect on the primary curved surfaces is eliminated. In addition, the speed of building is greatly increased, especially for thin shell parts, since fewer layers are required. This also leads to drastically reduced waste. Finally, decubing of the primary curved surfaces is eased or eliminated.

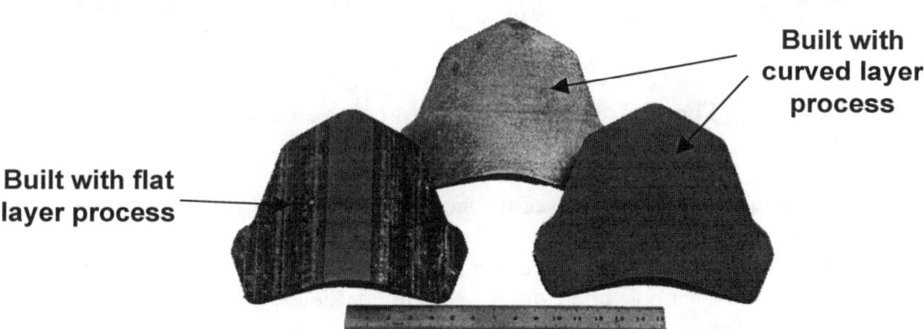

Figure 8 : Aircraft engine flame holders built with curved layer LOM. The dimensions of each are 1.5 cm x 9.5 cm x 1.0 cm with a shell thickness = 0.2 cm.

4. CURVED LAYER CERAMIC MATRIX COMPOSITES

The curved layer LOM process for SiC/SiC is described in Figure 9. The feed material, a "tape-preg", is a sheet that combines a SiC fiber prepreg and a SiC ceramic tape (described in more detail below). The tape-preg layers are laminated to a curved mandrel, but due to the adhesives in the tape-preg, no solvents are needed as in the monolithic LOM process (see Figure 2). A special laser (described later this section) is used to cut the SiC fibers with high quality. Parts are removed from the mandrel and decubed. Thermoset polymer resin contained in the fiber layers is fully cured and partially pyrolyzed in the powder pressing chamber. A pressure of 60 psi is required to avoid delamination in subsequent processing steps. Compaction during this step is $7\% \pm 2\%$ in the z direction. Again, scaling the computer graphic file in the z direction by 7% using LOM software can compensate this dimensional change. At this stage the structure is a CMC precursor composite, containing a fully cured thermoset resin that is subsequently converted to carbon by pyrolysis in an oven under an argon atmosphere. The green parts are placed in an oven freestanding or may be supported by a bed of powder. The resulting structure is finally densified through reaction bonding. Metallic silicon infiltrates the porous structure and reacts with *in-situ* carbon to form additional SiC matrix with little or no part shrinkage.

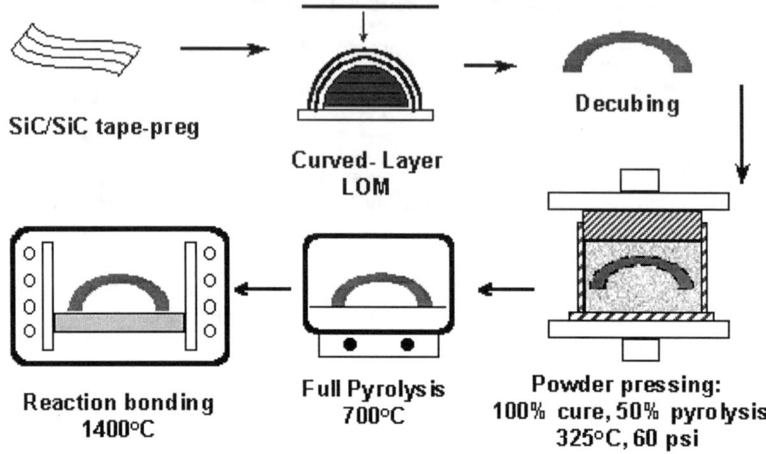

Figure 9 : Curved layer LOM process for CMCs.

The "tape-preg" was a novel approach used for the fabrication of SiC/SiC composites. The central idea is an alternating fiber prepreg layer and ceramic tape layer architecture. The advantages of the alternating layer technique are the relative ease of preparation of the preforms, elimination of fiber abrasion from the ceramic particles, moderately high fiber volume fraction in the final CMC, and low cost resin binder.

The tape-preg fabrication process is illustrated in Figure 10. First, fiber prepregs were made by winding SiC fibers (Nicalon, Dow Corning, carbon coated grade) onto a drum and brushing them with a solution of furfural-phenolic thermosetting resin (FurCarb UP440, QO Chemicals, Inc.) and methanol. The furfural resin was chosen because it produces a relatively high char yield ($\cong 50\%$) when heated to 700°C. The furfural resin serves a dual role: as a binder during the part fabrication and as a carbon source during the subsequent reaction bonding process. After drying in air overnight, the prepregs were removed from the drum, flattened into a single sheet (250 µm thickness), and placed on top of a SiC ceramic tape (250 µm thickness). This dual layer stack was vacuum bagged and B-staged in an oven at 105°C for 30 minutes. "B-staging" advances the cure state of the furfural resin in order to optimize the lamination characteristics during the LOM process as well as the compaction behavior during powder bed consolidation. The resulting one-piece sheets, referred to as tape-pregs, were stiff, nontacky, and board like, although they softened and conformed to a curved mandrel with the application of heat. Overall, the tape-preg is a robust sheet material suitable for the LOM process. The fiber content in the tape-preg is approximately 25% by weight.

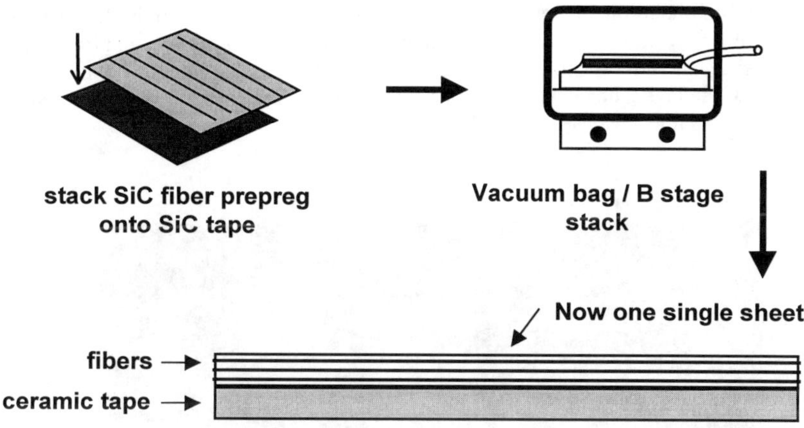

Figure 10 : Tape-preg fabrication process.

The primary advantage of the curved layer process for fabricating CMCs is that the fibers retain their continuity in the direction of curvature, thus enabling structures to achieve optimum mechanical strength and performance. In addition, all the advantages of the curved layer process for *monolithic ceramics* apply to CMCs. Reduction of waste assumes an increasingly importantly role as the expense of CMC materials is even higher than monolithic ceramics.

An aircraft engine flame holder was used to demonstrate the curved layer LOM process (see Figure 8). The SiC/SiC flame holder in the center has undergone all of the processing steps. In comparison, the SiC/SiC flame holder on the right, which has not been fully densified, serves to illustrate that there is very little or no shrinkage associated with the reaction bonding process. The monolithic flame holder (on the left) was built with the flat layer process; this piece dramatically illustrates the improved surface finish that is obtained with the curved layer process.

A major challenge facing the program from the onset was high quality laser cutting of continuous ceramic fibers. All available lasers were deemed unsuitable due to their cutting mechanism: selective burning through the target material. This cutting mechanism is unacceptable for continuous ceramic fibers because the melt temperature of the fibers is too high to be reached under reasonable laser illumination and/or the high power needed causes excessive heat damage in the cut zone. Furthermore, fibers such as SiC are thermally conductive, so the heat damage zone will extend axially. To overcome this problem, a copper vapor laser was used throughout the program. This laser operates by a completely different cutting principle: photoablation. The copper vapor laser is a high pulsed repetition frequency, medium-power, metal-vapor laser that can be tuned to operate in the green

(511nm). As can be seen in Figure 11, the laser cuts were excellent, with no fiber end damage, narrow cut channels, and minimal burn-back zone.

The microstructure of LOM SiC/SiC panels at two different stages is shown in Figure 12. The first stage is immediately after the cure process, before any substantial pyrolysis has taken place. This photomicrograph illustrates the alternating tape/fiber layer arrangement, excellent compaction, and minimal porosity. The photomicrograph on the right is of a specimen that has undergone the entire processing cycle (i.e., reaction bonding). There is quite a bit of porosity, although much of this may be attributable to particle and fiber pull-out that occurred when the specimen was polished. In any case, the flatness, continuity, and integrity of the layers have been well maintained. This microstructure is compared to commercially available CMC systems [7].

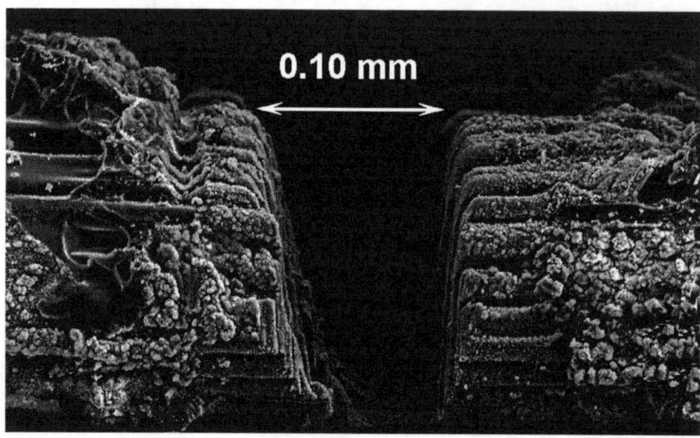

Figure 11 : SiC fiber / furfural resin prepreg cut with a copper vapor laser. The fiber diameter is approximately 15 μm. High quality laser cutting is characterized by the presence of nearly vertical walls, clean and undamaged fiber ends, and minimal resin burn back.

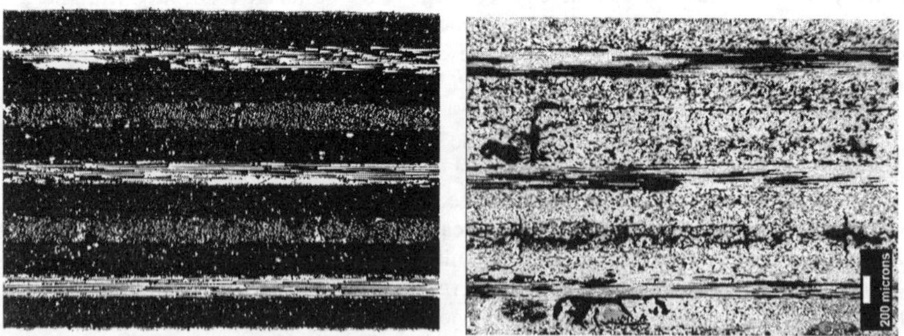

Figure 12 : Polished cross section of (left) eleven layer SiC/SiC LOM composite prior to pyrolysis, and (right) after pyrolysis and reaction bonding. From top to bottom the layers are: ceramic tape, 90° fiber layer, ceramic tape, 0° fiber layer, ceramic tape, etc.

Upon closer examination of the LOM SiC/SiC microstructure (Figure 13), it is evident that although the fibers are intact, there is fiber damage particularly at the fiber interface. This damage occurs as a result of the high temperature involved with the reaction bonding process (>1400°C). There are several potential solutions to this problem, such as developing appropriate fiber coatings and using silicon alloys that will infiltrate the composite at lower temperature. However, the physical mating of the layers is intimate and the overall infiltration efficiency is quite good relative to commercial CMC systems.

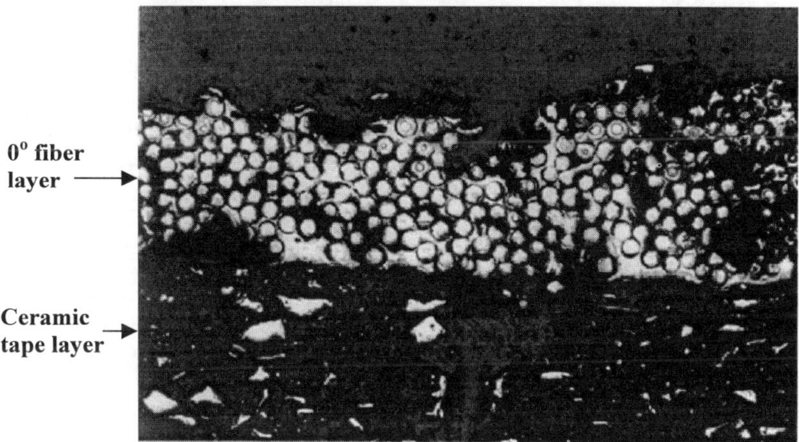

Figure 13 : Photomicrograph of reaction bonded SiC/SiC microstructure showing mating of various layers, infiltration efficiency, and fiber condition.

Final testing of polished SiC/SiC specimens in four-point bend is currently in progress. Exceedingly high performance is not expected since the fiber coating problem has not been solved. The major result to report here is that the entire process of fiber preform to near-net-shape, densified part has been fully demonstrated. Densified samples such as the flame holder and body armor are handleable, and photomicrographs illustrate that the microstructure compares quite favorably to commercial CMC systems. Further work is needed to fully characterize and assess the overall process reproducibility and accuracy.

5. CONCLUSIONS

The feasibility of using LOM for net shape, freeform fabrication of monolithic ceramics, continuous fiber CMCs, and curved layer composites has been fully demonstrated. Monolithic SiC and continuous fiber SiC/SiC were used as demonstration material systems. Commonly available preforms such as ceramic tapes and fiber prepregs are entirely suitable as feedstocks to the process. The overall process methodology involves use of a LOM machine to produce dimensionally accurate green forms directly from CAD files, followed by post processing steps to bring the part to full density. Both flat layer and curved layer laminates can be built with this technology. The curved layer paradigm is critically important for fabricating continuous fiber, curved shell composites in order to maintain fiber continuity. Monolithic SiC parts displayed microstructures and mechanical properties comparable to commercially available material made with traditional processes. Complex shapes can be produced. Curved layer SiC/SiC composites were also produced with excellent microstructure, although mechanical properties are not expected to be as high as commercial CMC systems due to ineffective fiber coatings. Inherent to the process of making the CMCs was the production of precursor composite structure, consisting essentially of a polymer matrix composite (PMC), which was subsequently converted to a CMC. Thus, the results demonstrate that PMC fabrication is also viable with the new LOM process.

Although the parameters of the various processing steps are expected to change with different materials systems, the overall LOM process is expected to be generically applicable to a wide range of high performance materials. Thus, an entirely new fabrication capability is now available for designers and manufacturers. The new LOM process holds promise for lowering the cost of fabricating monolithic ceramic, CMC, and PMC parts by virtue of its automation, which reduces fabrication time and eliminates the need for tooling. The process can best be applied by industry as a product development tool for fabricating molds, tools, or testable prototypes, or used directly for small lot production.

6. REFERENCES

1. Klosterman, D., R. Chartoff, et al., "Laminated Object Manufacturing (LOM) of Advanced Ceramics and Composites," *Solid Freeform Fabrication Symposium*, Austin, TX, August, 1997.

2. Klosterman, D. R. Chartoff, B. Priore, S. Pak, "Rapid and Affordable Manufacture of Structural Composites via Laminated Object Manufacturing (LOM)," *42nd International SAMPE Symposium & Exhibition*, Anaheim, CA, May 5-8, 1997.

3. Klosterman, D., R.Chartoff, N. Osborne, G. Graves, "Laminated Object Manufacturing, a New Process for the Direct Manufacture of Monolithic Ceramics and Continuous Fiber CMCs," *Ceramic Engineering and Science Proceedings, 21st Annual Confernce on Composites, Advanced Ceramics, Materials, and Structures - B*, the American Ceramic Society, Vol. 18, Issue 4, 1997, pp.113-120.

4. Klosterman, D.A., R.P. Chartoff, N.R. Osborne, G. Graves, and A. Lightman, "Structural Ceramic Components via Laminated Object Manufacturing," *Proceedings of the International Conference on Rapid Product Development*, Messe Stuttgart, Germany, June 10-11, 1996, pp. 247-256.

5. Leatherman, G.L., and R.N. Katz, *Superalloys, Supercomposites and Superceramics,* Academic Press, Boston, 1989, p. 673.

6. Lee, W.E., and W. M. Rainforth, *Ceramic Microstructures*, Chapman & Hall, London, 1994, p.427.

7. Chartoff, R., S. Pak, D. Klosterman, "Net Shaped Structural Ceramics and CMCs by Laminated Object Manufacturing," *DARPA Defense Sciences Manufacturing PI's Conference,* Washington DC, November 5, 1997.

7. ACKNOWLEDGMENTS

The work reported here was performed under a grant by DARPA/ONR, N00014-95-1-0059, with Dr. William Coblenz and Dr. Steven Fishman as program officers.

8. BIOGRAPHIES

Donald A. Klosterman is an Associate Research Engineer for University of Dayton's Rapid Prototype Development Laboratory (RPDL) and Center for Basic and Applied Polymer Research. He is currently involved in the area of solid freeform fabrication of polymer and ceramic matrix composites using Laminated Object Manufacturing (LOM). Dr. Klosterman has conducted research in other areas of advanced manufacturing such as fiber optic sensor development for polymer composites, modeling of polymer crystallization, and inverse gas chromatography. He holds a Ph.D. in Materials Engineering from the University of Dayton where he studied process models for composites, sensor technology, and the phenomenon of void formation in composites. Dr. Klosterman is currently Second Vice Chairman of the Midwest Chapter of SAMPE.

Richard P. Chartoff is Professor of Materials Engineering at the University of Dayton (UD) and a Senior Polymer Engineer in the University of Dayton Research Institute's (UDRI) Center for Basic and Applied Polymer Research. He teaches graduate courses in polymer science and engineering and supervises laboratory activities in the Polymer Center. Dr. Chartoff is also a co-founder of the UDRI's Rapid Prototype Development Laboratory, which is engaged in various research programs associated with materials processing in automated layered manufacturing systems. UD is one of the only Universities carrying out R&D programs in materials processing for rapid prototyping. The program at UD in this area is internationally recognized. Dr. Chartoff has numerous publications and patents and has served as a consultant for many American and European industrial organizations. He has been recognized as a research fellow by the Royal Norwegian Council for Scientific and Industrial Research and has been elected as a Fellow of the North American Thermal Analysis Society (NATAS). His activities in technical societies include service on the executive board of NATAS and the Dayton Section of the American Chemical Society as well as President of the Miami Valley Section of the Society of Plastics Engineers.

George Graves is Group Leader of the Ceramic and Glass Laboratories at the University of Dayton Research Institute. This laboratory performs a variety of materials fabrication and characterization studies. Research areas include characterization of structural ceramics, hot pressing of oxides and intermetallic materials, fracture analysis of various glass and ceramic materials, and fabrication of ceramic matrix composites and metal matrix composites.

Nora R. Osborne is an Associate Research Engineer in the Metals and Ceramics Division of UDRI. She received her undergraduate degree in Ceramic Engineering from Ohio State University, and an M. S. in Material Science from the University of Dayton. She has worked primarily with high-temperature structural ceramic materials, and sol-gel monolithic glass lenses and coatings for titanium aluminides. She is currently working with UDRI's Rapid Prototype Development Laboratory to develop Laminated Object Manufacturing technology for ceramics.

GUIDE TO CONVERTING FROM 12K TO 50K PREPREG TOW FOR FABRICATION TIME AND COST SAVINGS

D. DeWayne Howell
Storage Technology Corporation, Composites Group
2270 South 88th Street, MS-0091
Louisville, Colorado 80028

ABSTRACT

As a result of the resurgence of the consumption of 12K carbon fiber by the aerospace industry, the price of this fiber has increased and availability has decreased. In an effort to control costs and throughput of products currently in production at Storage Technology Corporation (StorageTek), we have converted much of our 12K prepreg tow consumption to 50K prepreg tow. The conversion represents a $22.05/kg ($10.00/lb.) cost savings over the 1997 year 12K carbon fiber prepreg cost. Fundamental differences between 50K and 12K such as increased filament wound layer thicknesses, lower material stiffness and increased variability of the 50K fiber made the conversion non-trivial. Other major development issues included: spool tensioning, tow twisting, catenary, splicing knots, edge splintering and dealing with greater dimensional variation of the finished parts. The advantages of lower cost and decreased production times, however, made the development effort well worth while and was a success.

Due to the recent introduction of 50K prepreg tow into the market, the positive potential for using 50K tow rather than traditional 12K tow on products produced across the composites industry are great. This paper offers a guide to the differences, advantages, challenges and application of 50K prepreg tow to volume production of filament wound products.

**Figure 1
50K Fiber Box
Beam**

KEY WORDS: Filament Winding, Prepreg Tow, 50K Fiber, 12K Fiber

1. INTRODUCTION

The increase in carbon fiber consumption around the world has led to shortages in the supply of 12K carbon tow. Possibilities exist in an alternative carbon tow product for increased availability and reduced cost, mainly, 50K carbon tow. StorageTek has converted most of its products made using 12K fiber prepreg tow to 50K prepreg tow. The conversion process required a moderate degree of effort to insure that the 50K fabricated products performed as well as the 12K products.

This paper is intended serve as a guide to those who are also interested in converting from 12K to 50K towpreg. Conversion must account for the thicker winding layers and the lower stiffness of the 50K prepreg. Much practical experience from this conversion has been gained and is disseminated in this paper. Even though there are a number of challenges associated with converting from 12K to 50K, the rewards far out weigh them.

2. FUNDAMENTAL DIFFERENCES BETWEEN 12K AND 50K

2.1 Target Markets: For years 12K and smaller filament count carbon tow markets have been primarily dedicated to industries requiring a high degree of quality control and finished part consistency, namely, the aerospace industry. In the 1990's, due to reduced price and greater availability, the trend has shifted somewhat to other markets including industrial, commercial, sporting goods and the infrastructure. However, the 1997 price increase ranging from roughly $37.48/kg ($17/lb.) to $41.89/kg ($19/lb.) for standard modulus (approximately 227.5 GPa (33 Msi)) 12K fiber can be quite prohibitive for non-aerospace applications. 50K fiber tow has also been produced for several years, but has received minimal use in industries requiring a high degree of quality control and finished part consistency. The trend in 50K tow today is towards lowering the price below the current range of $20/kg ($9/lb.) to $24.25/kg ($11/lb.) price, while simultaneously incorporating higher quality.

2.2 Filament Count/End Area: The obvious difference between 12K and 50K tow is the filament count per tow. 12K consists of 12,000 filaments of carbon per tow and, correspondingly, 50K consists of 50,000 filaments per tow. The 50K number, however, is somewhat misleading. Actually, the two major 50K fiber producers, Akzo and Zoltek, really produce 48,038 (Ref. 1, estimated from) and 45,714 (Ref. 2) filaments per tow respectively. Reported tow end area is 2.071 mm^2 (0.00321 in^2) (Ref. 3, derived from) and 1.968 mm^2 (0.00305 in^2) (Ref. 2) respectively. This is a very important factor when it comes to determining the part thickness yielded using these two, or any other, fibers.

2.3 Engineering Properties: The properties, as reported by the suppliers (Ref. 1,2,4), for 12K and 50K fibers are shown in Table 1.

Table 1
Supplier Reported Engineering Property Comparison

Properties (impreg. strand)	Zoltek 48K Panex 33	Akzo 50K Fortafil 3(C)	Amoco 12K T-300
Tensile Modulus	227.5 GPa (33 Msi)	227.5 GPa (33 Msi)	240.0 GPa (33.5 Msi)
Tensile Strength	3,619 MPa (525 ksi)	3,792 MPa (550 ksi)	3,654 MPa (530 ksi)
Density	1.77 Mg/m^3 (0.064 lb/in^3)	1.80 Mg/m^3 (0.065 lb/in^3)	1.77 Mg/m^3 (0.064 lb/in^3)

Although the data in Table 1 suggests an almost identical set of properties for 50K and 12K, the translation of such to a structure is not as identical. For example, Thiokol Corporation reported in reference 5 that the specific burst strength of Panex 33 is 85.0 MPa/kg (5.55 ksi/lb.) as compared to 91.4 MPa/kg (5.965 ksi/lb.) for T-300. If strength were directly transferable the burst strength for Panex 33 should have been 90.5 MPa/kg (5.909 ksi/lb.) Table 2 gives the percentage difference between supplier reported properties and actual test properties.

Table 2
Tested Engineering Property Comparisons

	% of T-300 12K values (supplier based data)	% of T-300 12K values (based on test Data)
Tensile Strength (Panex 33, Ref. 2)	99.1%	93.0%
Tensile Modulus (Fortafil 3C, Ref. 1)	98.5%	89.5%
Shear Modulus (Fortafil 3C)	not available	61.1%

In addition, StorageTek has measured the bending and torsional stiffnesses of Akzo's Fortafil 3C 50K versus Amoco's T-300 12K in box beams. A vibration test was used to determined the bending and torsional vibration frequencies of a 17.78 cm x 17.78 cm x 211.3 cm (7 in x 7 in x 83.2 in) beam (Figure 2.) The resulting fiber tensile modulus in translation was determined to be 204.8 GPa (29.7 Msi) as compared to 228.9 GPa (33.2 Msi) for T-300. Similarly the fiber shear modulus was determined to be 22.8 GPa (3.3 Msi) for Fortafil 3C as compared to 37.2 GPa (5.4 Msi) for T-300. Section 4.2 gives the detail on how these reduced properties for 50K were determined.

2.4 Variability: The amount of variation in fiber yield and stiffness differs when comparing 12K to 50K fiber. Table 3 gives the comparisons for fiber yield. The 12K fiber has almost three times less variation than does the 50K fiber. Fiber yield is based on one of, or a combination of tow cross sectional area and fiber density. As is shown in Section 4.3, the yield variation can be credited almost entirely to tow cross sectional area variation rather than density variation. Thus, 50K fiber will result in higher part dimensional variation by about three times.

Figure 2
Tube Vibration Test

Table 3
Fiber Yield Variation

	Average Yield	Standard Deviation	Standard Deviation Percent of Average
Amoco T-300 12K	1277 m/kg (1884 ft/lb)	12.9 m/kg (19 ft/lb)	1.00%
Akzo Fortifil 3C 50K	264 m/kg (390 ft/lb)	7.5 m/kg (11 ft/lb)	2.82%
Zoltek Panex 33	288 m/kg (425 ft/lb)	N/A	N/A

Table 4 illustrates the variation of translated, i.e. as applied to a structure, beam vibration frequencies as measured with the StorageTek beam vibration test. Note that the beam vibration equation is a function of beam stiffness, mass, length and end mass (Equation 1.)

Table 4
Vibration Frequency Variation

	Vibration Frequency (Hz.)	
	ω_n	σ
T-300 Bending	85.93	0.19
Fortifil 3C Bending	84.18	0.66
T-300 Torsion	89.21	0.27
Fortifil 3C Torsion	88.24	0.55

$$\omega_n = \sqrt{\frac{3EI}{[(\text{end mass}) + 0.23(\text{tube mass})] \, L^3}} \qquad \text{Equation 1}$$

Since the end mass and beam length are constant, and the contribution to the frequency due to the mass of the beam is small, the variation in frequency between the 12K and 50K is due primarily to beam stiffness, EI. As can be seen in Table 4, the variation in bending for 50K is 3 times that of 12K and for torsion the variation is only twice as much for 50K as compared to 12K.

2.5 Prepreg Characteristics: In addition to differences in the fibers themselves, the characteristics of the prepreg also varies. Thiokol's TCR division prepregs all of the fiber we use at StorageTek. Thiokol reports that as spooled fiber bandwidth and resin content varies as shown in Table 5. However the tow bandwidth and thickness at the part being fabricated is dependent upon ones fiber delivery system. Our system holds each 12K tow to 22.2 N (5.0 lb.) tension and each 50K tow to 44.48 N (10 lb.) tension. Each tow rolls over seven fiber redirect surfaces before being deposited onto the winding mandrel. The resulting differences in as wound bandwidth and thickness is given in Table 6.

Table 5
As Spooled Tow Bandwidth & Resin Content
of Thiokol Prepregs

	Tow Bandwidth	Resin Content by weight
T-300/UF3327	3.81 ±0.762 mm (0.15 ±0.030 in)	32.5 ±1.0%
Fortifil 3C/UF3327	8.38 ±2.540 mm (0.33 ±0.100 in)	32.5 ±1.5%

Table 6
As Wound Band Bandwidth & Thickness
of Thiokol Prepregs

	Band Bandwidth	Tow or Band Thickness
T-300/UF3327	16.76 ±2.540 mm (0.660 ±0.100 in)	0.278 ±0.102 mm (0.011 ±0.004 in)
Fortifil 3C/UF3357	11.81 ±1.905 mm (0.465 ±0.075 in)	0.635 ±0.152 mm (0.025 ±0.006 in)

3. ADVANTAGES

Three major advantages to using 50K fiber are availability, cost and product through-put.

3.1 Availability: Unlike 12K fiber , 50K fiber is readily available. 12K is the standard of the aerospace industry and is being consumed at the rate in which it can be produced. 50K fiber, on the other hand, is not being consumed as fast as it is being produced. Although the fiber industry is gearing up to produce more fiber to meet and exceed the demand, 50K is currently more readily available. In fact, both major 50K suppliers are also expanding their production of fiber. Zoltek plans to have its new 2,267,000 kg/year (5,000,000 pound/year) plant operational soon.

3.2   Lower Cost: Low cost is the most attractive feature of the 50K fiber. As stated above, the raw fiber alone represents a $15.43/kg ($7/lb) to $17.64/kg ($8/lb) cost savings. The prepreg tow cost can be as much as $22.05/kg ($10/lb) less than the 12K prepreg.

3.3 Higher Through-put: Since 50K has more filaments in a tow, winds thicker per layer, and occupies fewer tensioners on the tensioning creel, parts can be filament wound much faster than with 12K (Figure 3.) We converted from a 6 tow 12K winding band (a total of 72K filament count band) to a 2 tow 50K winding band (a total of 96K filament count band.) Considering the decreased total bandwidth and increased band thickness, we were able to reduce the actual winding time by 34%. A greater reduction in winding time could yet be realized by going to a 3 tow or greater 50K winding band. If the winding fabrication time for fabricating a 12K structure is known ($Time_{12K}$), then Equation 2 can be used to approximate the 50K winding time. This

Figure 3
Filament Winding of Beams

formulation assumes that winding machine speeds are set to the same rate for both 50K and 12K windings and that the part can be considered "thin walled".

$$\text{Time}_{50K} = \left[\frac{(\text{Thick}_{\text{Laminate 50K}}) \, (\text{Thick}_{\text{Tow 12K}}) \, (\text{BW}_{\text{Band 12K}})}{(\text{Thick}_{\text{Laminate 12K}}) \, (\text{Thick}_{\text{Tow 50K}}) \, (\text{BW}_{\text{Band 50K}})} \right] \text{Time}_{12K} \qquad \text{Equation 2}$$

Where, $\text{Thick}_{\text{Laminate}}$ is the total thickness of the laminate, $\text{Thick}_{\text{Tow}}$ is the thickness of a single tow, and BW_{Band} is the bandwidth of the multi-tow winding band.

4. DESIGN CONVERSION

4.1 Layer Design considerations: Due to the increased tow bandwidth and tow thickness, conversion from 12K product designs to 50K designs require redesigning the layup of the product. Thin laminates of less than 1.905 mm (0.075 in) may prove difficult to redesign using 50K fiber since anything less than this limits the 50K layup to only one un-lapped and un-gaped $\pm\theta$ helical layer. Gaps occur when the winding band does not lie down flush next to the neighboring winding band. Laps occur when a winding band overlaps onto another winding band, thus not lying adjacent and flush to the neighboring band. Although laps and gaps are generally avoided in 12K layups, it may be necessary to allow this in 50K layups in order to allow for flexibility in adjusting for the desired part thickness.

As an example, we converted a 9 layer, 4.267 mm (0.168 in) thick 12K laminate to a 4 layer 50K laminate. This required the winding bands of the helical layers to have a slight overlap. Another product we converted resulted in one helical layer in the laminate to be gaped by 100% between winding bands. The following conversion process steps are useful:

1. Determine the total thickness of the various angle layers in the 12K laminate. If the total thickness of any of the different angle layers is less than 1.27 mm (0.050 in) then a "gaped" winding will be necessary for that layer. If it is more than 1.27 mm (0.050 in) then the layer will have overlapping bands or can be separated into multiple layers of the same winding angle.
2. The wound thickness of a 50K layer is represented by Equation 3, and is a function of these six factors: 1) equivalent part diameter, D_i. 2) fiber tow end cross sectional area, A_{tow} 3) fiber content of the cured part, V_f 4) winding angle, θ 5) number of tows in a winding band, N_{tow} 6) number of winding circuits to complete a winding layer, N_{cir}, where one circuit is defined as one pass up and then back down the winding mandrel. Having defined these factors, the average layer thickness can be calculated using this equation or the "Winding Thickness" form in the commercially available "CompositePro" software package (ref. 6.) This calculation assumes low void content.

$$T_{\text{layer}} = \left[\sqrt{\frac{8 \, A_{\text{tow}} \, N_{\text{cir}} \, N_{\text{tow}}}{\pi \, V_f \, \text{Cos}(\theta)} + D_i^{\,2}} \; - D_i \right] / 2 \qquad \text{Equation 3}$$

The key to using this method is to realize that this is an average winding thickness, and assumes that the total number of winding circuits in a layer are equally distributed around the circumference of a part. We have had good success predicting as cured part thicknesses applying this technique. This method, however, does not account for the lower stiffness of the 50K fiber, which will now be addressed.

4.2 Adjusting for Reduced Stiffness: As stated in Section 2.3, the extensional and shear moduli for 50K fiber as translated into a structure is about 10% less than 12K for the extensional modulus and about 39% less for the shear modulus. Structural applications at StorageTek are usually stiffness critical and not strength critical; therefore, matching the original 12K structures layer angles and thicknesses alone, as shown in Section 4.1, does not yield an equivalent stiffness structure. Data from a beam vibration test was used to determine the effective fiber properties for Fortafil 3C versus T-300. The following procedure was developed for determining the effective 50K material properties using the CompositePro software (Ref. 6.)

1. Several 17.78 cm x 17.78 cm x 211.3 cm (7 in x 7 in x 83.2 in) tubular box beams were filament wound with both T300 and Fortafil 3C prepreg in a series of helical and hoop windings to a thickness of 4.191 mm (0.165 in.) Both beam types had a fiber volume of 62.5% (Ref. 7.) Each beam was frequency tested (Figure 2) to determine the bending and torsional frequencies.

2. Using T-300 fiber properties and 3501-6 type epoxy properties (Thiokol UF3327 engineering properties were not available) a micromechanics analysis was run to predict T-300/Epoxy lamina properties.

3. The layup of the 12K beams was modeled using the T-300/Epoxy properties and analytical beam bending stiffness (EI_{12K}) and torsional stiffness (GJ_{12K}) were calculated.

4. Per the discussion of section 2.4, the difference in the frequencies between the 12K and 50K beams is due to the stiffnesses only. Therefore, Equation 4 determines the ratio relationship used to calculate the expected 50K beam stiffnesses (EI_{50K} and GJ_{50K}) using the frequency test data and the predicted 12K stiffnesses. The relationship for torsional stiffness is of similar form.

$$EI_{50K} = \frac{(EI_{12K})\,(\omega_{n50K})^2}{(\omega_{n12K})^2} \qquad \text{Equation 4}$$

5. Having calculated the expected 50K stiffnesses, the layup of the 50K beams was modeled using micromechanics generated Fortafil 3C/Epoxy lamina properties. During the micromechanics calculations, the 50K fiber stiffnesses along the fiber, E_{1f}, across the fiber, E_{2f}, and in shear, G_{12f}, were iteratively modified until the iterated lamina properties for Fortafil 3C/Epoxy drove the predicted EI_{50K} and GJ_{50K} to match the expected values as calculated in Equation 4. The fiber Poisson's ratio, ν_{12f}, was held constant at 0.20 for this analysis since its effects were small. Thus, the effective 50K fiber properties were calculated and are shown along with the 12K properties in Table 7.

Table 7
Effective Fiber Properties

	12K T-300	50K Fortafil 3C
E_{1f}	228.9 GPa (33.20 Msi)	204.77 GPa (29.70 Msi)
E_{2f}	14.48 GPa (2.10 Msi)	14.89 GPa (2.16 Msi)
G_{12f}	37.23 GPa (5.40 Msi)	22.75 GPa (3.30 Msi)
ν_{12f}	0.20	0.20

6. Now that the 50K fiber and lamina properties are known, a layup can be modeled and iteratively modified to yield predicted $EI_{50K} = EI_{12K}$ and $GJ_{50K} = GJ_{12K}$.

For the 17.78 cm x 17.78 cm x 211.3 cm (7 in x 7 in x 83.2 in) tubular box beams, which we have converted to 50K, the lower fiber stiffnesses required us to increase the thickness of the beam from 4.267 mm (0.168 in) for the 12K to 4.521 mm (0.178 in) for the 50K. The stiffness match is shown in Table 4 of Section 2.4. Although the above technique was developed for box beams, the general concept should apply to other structures. Keep in mind that the 50K fiber properties calculated here are effective properties based upon a translated structural response. It is possible that different fabrication techniques, curing scenarios and structure geometries may give rise to different degrees of fiber property translation. *Of special note is that 50K fiber suppliers claim to be working on ways to increase their fiber stiffnesses to match the translated stiffness of 12K products.*

4.3 Accounting For Yield Variation: The fiber yield (length per unit weight) variation for 50K fiber is three times that of 12K fiber (Table 3,) when considering a 3 sigma range. Since fiber yield is a function of the cross sectional area and density of a fiber, it was necessary to determine which, if not both, of these parameters contribute most to the yield variation. If cross sectional area is the driving factor then laminates made with 50K fiber will show a large variation in average laminate thicknesses. If density is the driving factor then these will show a large variation in average laminate weight.

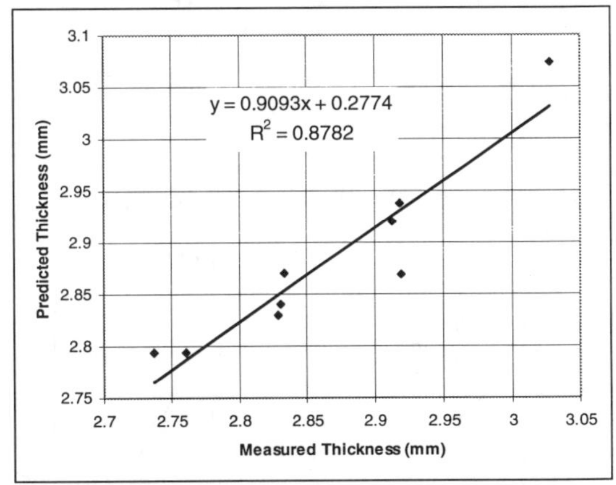

Figure 4
Predicted Vs. Measured Avg. Thickness

To assess which is the main contributing factor, ten 10.16 cm x 10.16 cm x 201.93 cm (4.0 in x 4.0 in x 79.5 in) tubular box beams were fabricated using different, but known, fiber yields for each beam. Assuming that the yield variation was due mainly to cross sectional area variation, CompositePro (Ref. 6) was used to predict the average part thickness as a function of fiber cross sectional area. The fabricated beams were then measured for average part thickness. Figure 4 illustrates that the correlation of predicted versus measured thicknesses are very good (the closer the R^2 correlation factor is to one the better the correlation). This means that the assumption that the yield variation in the fiber is due mainly to varying cross sectional area is valid.

There are a couple of ways to control the variation in average part thickness due to the yield variation, other than getting the fiber supplier to change their process. *One* method would be to sort each spool of fiber according to its yield and use these spools with one member of family of winding programs. Each member of the family of winding programs compensates for a different fiber yield to produce the desired part average thickness by winding a greater or lesser number of winding circuits in a layer. Although this method would work nicely, it presents a logistics problem coordinating which spools to use with which winding program. *Another method*, which is the one which we have chosen, is to have the prepreger, Thiokol, sort the fiber for us and deliver prepreg which falls within only one sigma of the nominal yield. Refering to Table 3, this effectively reduces the "Standard Deviation Percent of Average" to 0.94%, which is comparable to the 12K value. With this method we only need to use one winding program and does not require us to sort the fiber.

The resulting variations of the 12K versus the 50K 17.78 cm x 17.78 cm x 211.3 cm (7 in x 7 in x 83.2 in) tubular box beams are shown in Figures 5 and 6 and are based upon a minimum population of 20 beams. The data collected for the 50K and 12K beam thicknesses depicts a reasonable normal distribution of thickness for the 12K, whereas the 50K distribution is less normal. The 12K shows a standard deviation in thickness of 0.019 mm (0.0007 in) less than the

Figure 5
12K Thickness Distribution

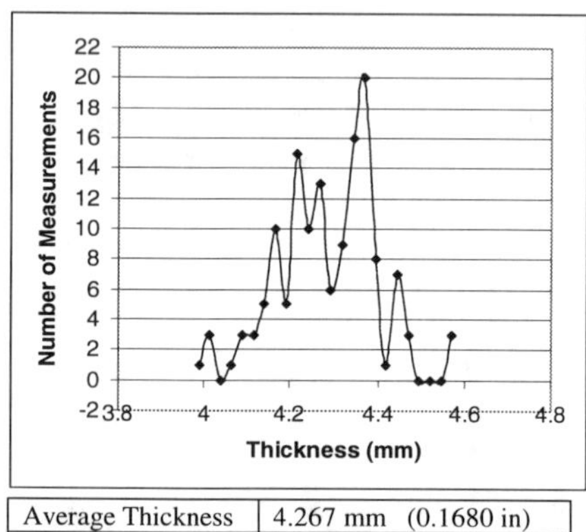

Average Thickness	4.267 mm (0.1680 in)
Standard Deviation	0.117 mm (0.0046 in)
Range	0.584 mm (0.0230 in)

Figure 6
50K Thickness Distribution

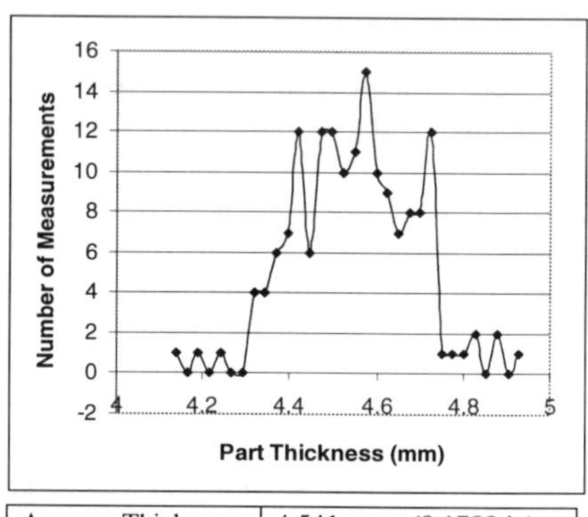

Average Thickness	4.541 mm (0.1789 in)
Standard Deviation	0.136 mm (0.0053 in)
Range	0.790 mm (0.0311 in)

50K. Since we have an equivalent fiber yield range between the 12K and the 50K, due to the sorting of the 50K fiber, one might expect for the thickness variation to also be equal. However, a close examination of the standard deviation plots in Figures 5 and 6 shows that the 50K distribution is more flat or squared off at the center than is the 12K plot, and that the range of thicknesses is 0.206 mm (0.008 in.) greater for the 50K. We believe this can be attributed to the fact that since we are using only 1σ of the yield on the 50K fiber as opposed to the full 3σ of the 12K, that we are actually using more fiber at the extremes of the 50K yield than at the extremes of the 12K yield. Figure 7 illustrates this effect. In addition, the 6 tows used in a band for the 12K has more of an averaging effect on the thickness than does the 2 tows used in a 50K band. That is, a 12K part may be made up of 6 different spools of 12K, whereas a 50K part may be made up of only 2 spools.

Figure 7
Fiber Yield Distribution Effect

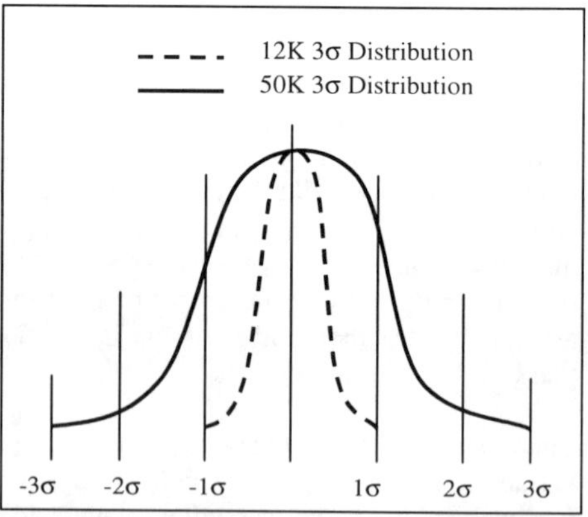

5. FILAMENT WINDING FABRICATION CONSIDERATIONS

5.1   Spool Tensioning: When winding with prepreg it is necessary to do so with tension on the spool, as seen in Figure 8. This provides the necessary peeling force to allow the prepreg to de-spool without sticking to the underlying fiber on the spool and winding up on itself. It is our experience that about 17.8 N (4 lb.) to 22.2 N (5 lb.) of tension per tow is adequate to de-spool 12K prepreg at a rate of 51.85 m/min (170 ft/min.) Comparatively, the 50K prepreg needs a higher tension to de-spool without wrapping up on itself. For 50K, we wind with about 35.6 N (8 lb.) to 44.48 N (10 lb.) tow tension (the maximum limitation of our current tensioners) at a maximum rate of 47.28 m/min (155 ft/min) At this fiber tension, higher winding rates cannot be achieved because the tow sticks to the spool and causes a "snap" of the fiber leading to a breakage. Thiokol recommends a higher tension of at least 66.72 N (15 lb.) to 88.96 N (20 lb.) per spool to prevent tow breakage. The temperature of the fiber on the spool as the fiber de-spools is very important. The fiber tensions and feed rates mentioned above are good for room temperature (approximately 21°C ±5°C (70°F ±10°F)). At temperatures lower than 16°C (60°F), the prepreg has a tendency to stick as it de-spools, which leads to fiber breakage. In general, due to these factors and fiber splintering (see Section 5.4), breakage during the winding of 50K tow is much more common than for the 12K tow. A higher fiber tension may reduce breakage.

Figure 8
Fiber Tensioning Creel

5.2   Tow Twisting: Unlike 12K prepreg tow, occasionally a 50K prepreg tow will twist during winding. Twist is defined as a tow flipping over so that the side that was against the winding mandrel is now oriented away from the mandrel. We have observed about 2 twisted tows for every 0.45 kg (1.0 lb.) of 50K prepreg wound, which amounts to only 0.2% of total part volume.

5.3   Fiber Catenary: Tow catenary is quite evident in the 50K fiber whereas the 12K fiber shows very little or no catenary. The 50K parts actually exhibit portions of tow edges lifting off of the mandrel after the beam is wound. Catenary can be seen in Figure 9 at the edges of the winding bands in the center of the figure where they are lifted off the surface. A visual comparison of the 12K vs. 50K parts after autoclave cure at 4.481 bar (65 psi) shows that the 50K fibers, although fairly straight, exhibit an undulation in the thickness direction in the regions where catenary was evident on the as wound, pre cured part. This is a factor which could reduce the compressive strength of a 50K filament wound laminate.

Figure 9
50K Fiber Catinary

5.4   Splice Knots: Splicing of spools is necessary whenever one spool runs out of fiber. Spools for 50K are currently available from Thiokol in the range of 3.63 kg (8 lb.) to 5.44 kg (12 lb.) Using two winding tows per band, it is usually necessary for us to have at least 2 splices for every 9.07 kg (20 lb.) of prepreg wound, excluding splices due to a tow breakage. Thiokol is experimenting with a 8.16 kg (18 lb.) spool to reduce the number of splices required. We splice by simply tying a square knot at the two fiber ends and then continue to wind (Figure 10.) Instead of cutting the knots out of the part, once the tow has been deposited onto the mandrel, we leave the knot in place. For our 4.521 mm (0.178 in) thick parts the knots are completely compressed into the laminate under the 4.481 bar (65 psi) autoclave pressure. It is difficult to measure even 0.0254 mm (0.001 in) of thickness deviation due to the knots. Figure 10 also shows gapping of the fiber bands. This is due to the close proximity of this knot to the ends of the beam, and is a common occurrence in the filament winding of 12K or 50K box beams.

Figure 10
50K Splice Knot

5.5   Tow Splintering: Another thing to be aware of when using 50K prepreg tow is fiber splintering. Unlike 12K prepreg, the 50K prepreg tow has a considerable number of splinters in the tow. This is not a "fuzz" on the fiber but rather actual splinters of separation from the main tow. Sometimes these splinters remain attached to the spool during de-spooling and cause the tow to split and then break. Both the 50K fiber supplier and the prepreger are working to reduce this splintering phenomenon.

5.6   Tow Separations: A phenomenon also unique to 50K prepreg tow is fiber separation. Occasionally, as the fiber de-spools, the 50K tow can be seen to separate within the tow forming 2 or 3 separate and distinct bands within a tow. 12K does not exhibit this characteristic. This may be a leading factor which produces the catenary in the 50K tow. It also may contribute to tow breakage, since one of the separated bands within the tow may stick more readily during the de-spooling process.

5.7   Delivery System: Since the bandwidth of a 50K tow is greater than the 12K tow, the delivery system for the prepreg fiber will need to be modified. This modification, however, need only be minimal. Our delivery system consists of tensioners, a tension sensing tower, redirect rollers and a final delivery head, for a total of 7 rolling surfaces of contact for each tow of fiber. The conversion to 50K required only a modification to the final delivery head. Instead of having 6 grooves in the delivery head rollers (the two closely spaced light colored roller in Figure 11,) we changed to only 2 grooves, i.e. 6 tows for a 12K band as opposed to 2 tows for a 50K band.

Figure 11
50K Delivery Head

6. CONCLUSIONS

The differences between filament winding with 12K and 50K prepreg tow are more than the obvious higher filament count of 50K. In a market where 12K carbon fiber is in limited supply, 50K is readily available. 50K offers 34% or higher reductions in filament winding time for an equivalent structure. Costs are approximately 42% less per kg for 50K prepreg compared to 12K, leading to substantial product cost savings.

The conversion to 50K prepreg tow, however, is not a trivial task. The differences in tow thickness and bandwidth between 50K and 12K necessitates the need for reprogramming filament winding machines and outfitting your winding machine with a modified final delivery head. The fact that 50K has lower stiffnesses and strength must be considered when trying to achieve a structure which matches 12K performance. The increased variation in fiber yield for 50K should also be taken into account in order to manage product dimensional variations. Additionally, consideration should be given to the required increased tow tension for proper de-spooling, larger splice knots in the parts fabricated, increased catenary in the tow, and tow splintering and separation which can lead to an increase in fiber breakages during the winding process.

All things considered, 50K fiber prepreg is a viable and recommended replacement for 12K filament wound applications. At StorageTek, we have realized substantial cost saving on our production products, and a significant decrease in production time. Current 50K suppliers are working on ways to improve their fiber quality, increase fiber stiffness and reduce the cost even further. As a bonus, many of the current 12K fiber suppliers are looking at possible production of 50K or higher filament count fiber. The future of 50K tow for filament winding of commercial, infrastructure, sporting goods and even aerospace applications looks bright!

7. REFERENCES

1. Fortafil 3(C) Continuous Carbon Fiber, Akzo, Rockwood, Tennessee, 1990.
2. PANEX 33 Continuous Carbon Fiber, Zoltek Corp., St. Charles, Missouri, 1996.
3. Yield Distribution for Fortafil 3(C), Correspondence with Akzo technical staff, December 17, 1996.
4. Typical Properties THORNEL PAN Based Fibers, AMOCO, Ridgefield, Connecticut, 1988.
5. K. Fisher, "Filament Winding With Heavy Towpreg," High Performance Composites, p. 38, May/June 1996.
6. CompositePro for Windows, Peak Composite Innovations, LLC, Littleton Colorado, 1997.
7. J. He, J. Self, "Resin Content and Resin Flow of Big Tow Prepregs and Resin Content of Z-Tubes," Storage Technology Report, November 4, 1996.

8. BIOGRAPHY

D. DeWayne Howell is a graduate of the University of Cincinnati, with a BS in Aerospace Engineering. He is a Senior Development Engineer for the Composites Group at Storage Technology and has worked in the composites industry continuously for 13 years.

CARBON NANOFIBERS FROM POLYACRYLONITRILE

AND MESOPHASE PITCH

Iksoo Chun, Darrell H. Reneker, Xioayan Fang, Hao Fong
Joe Deitzel*, Nora Beck Tan*, Kristen Kearns**
Institute of Polymer Science, The University of Akron, Akron, OH 44325-3909
* U.S. Army Research Laboratory, Weapons and Materials Research Directorate, Aberdeen
Proving Ground, MD 21005-5069
** Department of The Air Force Wright Laboratory (AFMC), Wright-Patterson AFB,
OH 45433-7750

ABSTRACT

Carbon nanofibers were produced from both polyacrylonitrile and mesophase pitch. Stabilization and carbonization processes were used to convert as-spun nanofibers to carbon fibers. The nanofibers are arbitrarily long, although a way to make nanofibers less than a millimeter long, with high aspect ratios, was found. The diameters of typical carbon nanofibers are in the range from 100 nanometer to a few microns. A log normal distribution provides a good representation of the measured distribution of diameters. The carbon nanofibers were observed by polarized optical microscopy, scanning electron microscopy, transmission electron microscopy, and wide angle x-ray diffraction. As-spun mesophase pitch fibers are transparent, red or orange in transmitted light, and birefringent. Electron diffraction patterns were obtained from individual carbon nanofibers. The interplane spacing of (002) planes of mesophase pitch based carbon nanofibers was measured using wide angle x-ray diffraction. Nanofibers provide a higher ratio of surface area to mass than carbon fibers ordinarily used in composites. Carbon nanofibers can be useful in filters, as a support for catalysts in high temperature reactions, in composites to improve mechanical properties, or for thermal management in semiconductor devices. Nanopores in carbon fibers were produced using nitrogen gas saturated with water vapor.

KEY WORDS: Carbon Nanofibers, Electrospinning, Mesophase Pitch

1. INTRODUCTION

Electrospinning is a method for making very small diameter fibers. An electric field provides the driving force. When an applied electric field overcomes the surface tension of a polymer solution or melt, an electrically charged jet is formed. The jet is pulled toward an electrically conducting collector held at an attractive potential. As the electrically charged jet is stretched by the electric field, its diameter decreases and the radial component of the electrical force increases. The single jet divides into two or more jets and each of the divided jets divides repeatedly. The "splayed" jets are collected as a form of nonwoven fabric on a metal screen, aluminum foil, woven cloth, or other surfaces.

The electrospinning process has been known for a long time. In the early 20th century, threads of molten sealing wax were produced using a high electric field (1). Formhals patented a process and an apparatus to make fine polymer fibers using high electric field in 1934 (2) and other patents followed. In 1952, Vonnegut and Neubauer (3) invented an electrostatic fountain and found when the spray of droplets was illuminated with a beam of parallel white light, colors were seen. Taylor (4) studied the disintegration of water drops in an electric field. He demonstrated that a drop elongates and a cone which has semi angles close to 49.3° forms at each end. In 1971, Baumgarten (5) produced acrylic microfibers, using electrical forces, with diameters in the range from 0.05 to 1.1 microns. He observed the relationships between fiber diameter, jet length, solution viscosity, solution feed rate, and surrounding gas. Larrondo and Manley (6-8) made continuous fibers of polyethylene and polypropylene using electric fields as the driving force. The diameters of these polyolefin fibers depended on the electric field and were around 50 microns. Reneker and Srinivasan (9) electrospun nanofibers from a liquid crystal solution of poly(p-phenylene terephthalamide) in sulfuric acid. Reneker and Doshi (10) used electrospinning to make fibers of water soluble poly(ethylene oxide) with diameters of 0.05 to 5 microns. Reneker and Chun (11) produced polyester nanofibers from solution, from the melt in air, and from the melt in vacuum. Reneker and Kim (12) made a nanofiber reinforced epoxy composite using a fabric composed of partially aligned nanofibers.

The carbon fibers, with diameters around 10 microns, now used are made from polyacrylonitrile and mesophase pitch. The acrylic polymer precursor used contains about 85 weight percent acrylonitrile (13), with acrylic acid or methacrylic acid comonomers. Melt spinning is difficult since polyacrylonitrile degrades below its melting temperature, so solution spinning is used. The as-spun polyacrylonitrile fibers are heated in air to 200 to 400 °C, under tension, to stabilize them for treatment at higher temperature. The changes in chemical structure during stabilization are described by Gupta and coworkers(14), and by Watt and Johnson (15). The effects of tension in the fibers, the heat treatment temperature and the treatment medium on the stabilization process were observed. Stabilized fibers are converted to carbon fibers by heating to 1700 °C under inert gas. In this carbonization process, all chemical groups such as HCN, NH_3, CO_2, N_2, and hydrocarbons are removed. The carbonized fibers become graphite-like. After carbonization, the fibers are heated up to 2000 to 3000 °C under tension. This process, called graphitization, makes carbon fibers with aligned graphite crystallites. Many observations (16-20) of the microstructure of stabilized and carbonized fibers based on polyacrylonitrile, using scanning electron microscopy, transmission electron microscopy, and wide angle x-ray diffraction, were reported.

Mesophase pitch is also an important precursor of carbon fibers. Fibers made from aligned molecules of liquid crystal pitch transform into carbon fibers with highly aligned graphitic planes. Mesophase pitch is insoluble, so fibers must be spun from the melt. Matsumoto (21) observed the structure and morphology of carbon fibers made from pitch and reported that the tensile strength of the carbon fibers continuously increases with the temperature of heat treatment. The pitch fibers must be stabilized by heating in air to prevent melting or fusing during the high temperature treatment which is required to obtain high strength and high modulus carbon fibers. Carbonization of the stabilized fibers is carried out at temperatures between 1000 °C to 1700 °C, depending on the desired properties of carbon fibers. Many papers on the characterization of mesophase pitch based carbon fibers with diameters of around 10 microns were published (22-26).

Carbon nanofibers have many potential applications as filters for separation of small particles from gas or liquid, supports for high temperature catalysts, and heat management materials in aircraft and semiconductor devices. Another potential application is in composites. When nanofibers are placed between crossed plies in a composite, resistance to crack growth is improved. Electrospun carbon nanofibers with high surface area, electrical conductivity, and thermal conductivity are candidates for applications in small electronic devices, rechargeable batteries (27), and supercapacitors (28-31).

2. POLYACRYLONITRILE

2.1 Carbon Nanofibers From Polyacrylonitrile

2.1.1 Experimental   Polyacrylonitrile (#18,131-5) was purchased from Aldrich Chemical Company, Inc. A ten percent by weight solution in dimethylformamide was used. The experimental set-up was similar to that described in earlier publications (32, 33). The polymer solution was held in a glass pipette. The electric field was provided by a high voltage power supply (Gamma High Voltage Research, Ormond Beach, Florida) through a metal electrode immersed in the solution. The electrically charged jet that emerged from a liquid drop at the tip of the glass pipette was attracted toward an electrically grounded carbon fabric on a rotating drum winder. The applied potential between the spinneret and the winder was 10 to 20 KV. The distance between the spinneret and the winder was 10 to 20 cm. The fibers dried in flight and collected on the carbon fabric in the form of a nonwoven fabric. As-spun fibers were stabilized at 270 °C for 15 minutes in air, and then carbonized at 800 °C for one hour under an inert atmosphere. The linear dimensions of sheet of nanofibers shrank about 10% during the heat treatment, and detached from the carbon fabric. Some of carbonized nanofibers were heated, at 650 °C or 800 °C, for 1 hour, in nitrogen gas that had flowed through a flask of boiling water.

2.2.2 Observation Using Scanning Electron Microscopy   Figure 1 shows a thin layer of electrospun polyacrylonitrile nanofibers supported on a carbon fabric which was made with 7 micron diameter fibers. The dramatic difference between the diameter of the electrospun fibers and that of the conventional carbon fibers is evident. The diameters of the electrospun fibers ranged from 100 to 500 nanometers. Polyacrylonitrile fabric sheets that were 90 cm long, 30 cm wide and up to 100 microns thick were produced. The nanofibers are partially aligned parallel to the winding direction. A cut edge of a sheet of carbonized nanofibers is

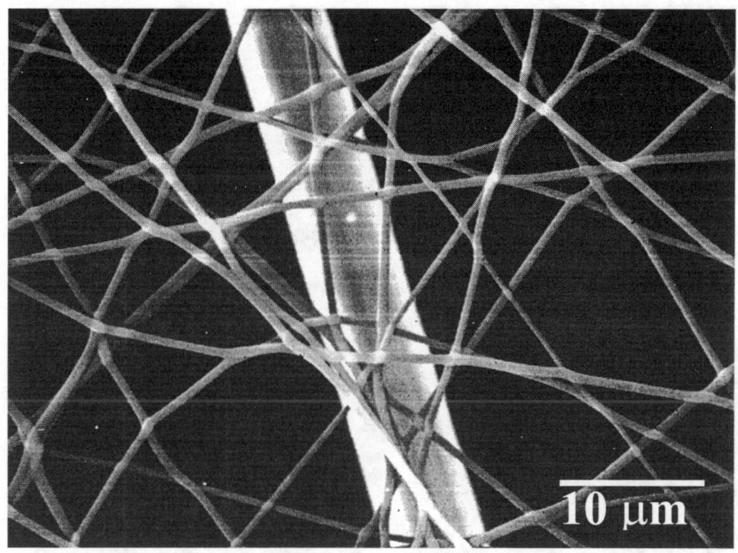

Figure 1. As-spun PAN nanofibers on a conventional carbon fiber.

shown near the left edge of figure 2. The higher magnification image in figure 3 shows the diameters of the carbonized electrospun polyacrylonitrile fibers.

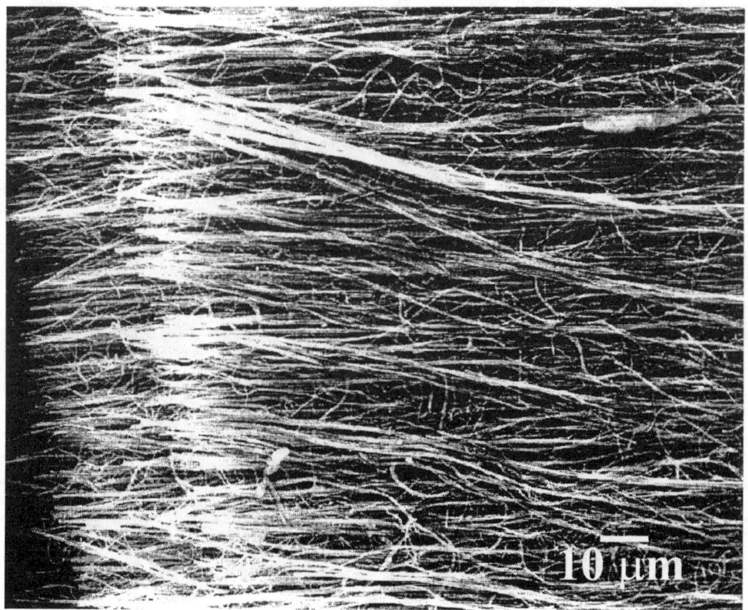

Figure 2. Carbonized PAN nanofibers.

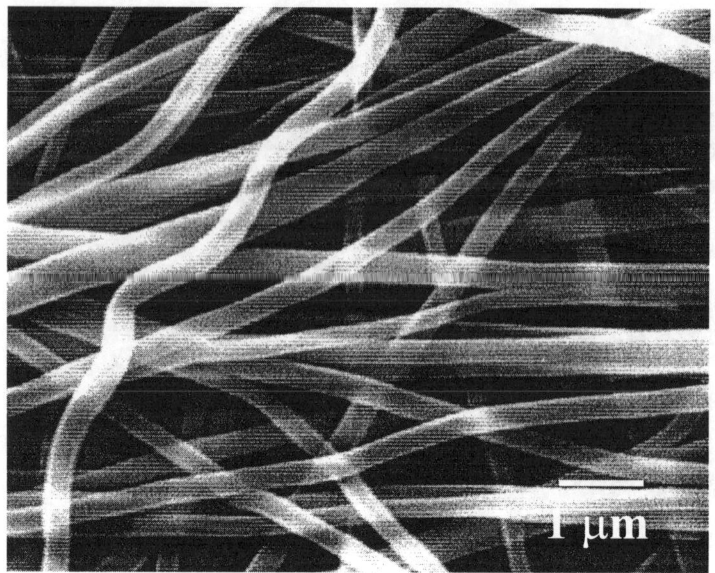

Figure 3. Carbonized PAN nanofibers.

2.2.3 Nanopores In The Surfaces Of Carbon Nanofibers   Some of the carbon fibers were treated at temperatures of 650 and 800 °C for one hour in a nitrogen stream that contained water vapor. This process has been shown to increase the chemically accessible surface area of the carbon. The scanning electron micrograph in figure 4 shows that nanopores, with

Figure 4. PAN nanofibers activated with water vapor.

diameters equal to a fraction of the fiber diameter, were created in some of the fibers. Pores of this size would not have a large affect on the amount of chemically accessible surface area but their presence indicates that the water reacted with the nanofibers.

3. MESOPHASE PITCH

3.1 Carbon Nanofibers Electrospun From Molten Mesophase Pitch   Mesophase pitch electrospun from the melt, were converted to carbon fibers using the same stabilization and carbonization procedures described above for the polyacrylonitrile nanofibers.

3.1.1 Experimental   Mesophase pitch (Mitsubishi AR made by Mitsubishi Gas Chemical Company, Inc) was supplied by the Air Force Materials Laboratory at Dayton, Ohio. The apparatus used was similar in concept to the pipette apparatus described in a previous publication (11), but modified to work at temperatures up to 450 °C. The experimental set up is shown in figure 5. The sample was placed in a 1.2 cm diameter metal tube, with a stainless steel tube of the sort commonly used in hypodermic needles as a tip. The metal tube was sealed with a flanged lid and placed in an electrically heated oven. Nitrogen gas was supplied to the top of the sealed tube to prevent oxidation and to push molten pitch, at a temperature around 350 °C, through the smaller tube. An electric field was established at the open end of the smaller tube by applying a potential of up to 30,000 volts to a metal screen placed 10 to 30 cm below the tip of the small tube.

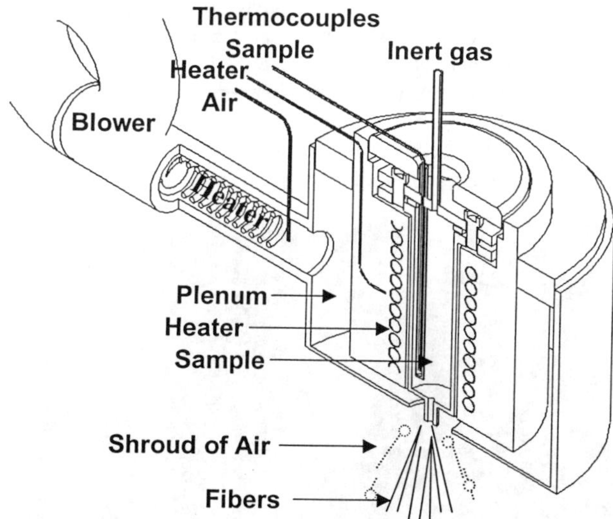

Figure 5. Apparatus for molten mesophase pitch.

Since molten pitch is a liquid (more precisely, a thermotropic liquid crystal), surface tension will cause a jet of molten pitch to tend to contract into droplets. Yet, the jet must remain soft enough to be stretched by the electrical forces after it leaves the tip. Careful control of the temperature of the air surrounding the jet is required. This was accomplished by using a heat gun to deliver air at a controlled temperature to a metal plenum around the tip end of the apparatus. The plenum had a hole that projected a cone shaped stream of hot air along the trajectory of the jet. The temperature of this moving shroud of air was adjusted so the jet was soft enough to be stretched, but was not liquid enough to collapse into droplets.

3.1.2 *Microscopic Observation Of Pitch Fibers* Figure 6 is a scanning electron micrograph of representative electrospun fibers. Figure 7 shows the fibers observed between

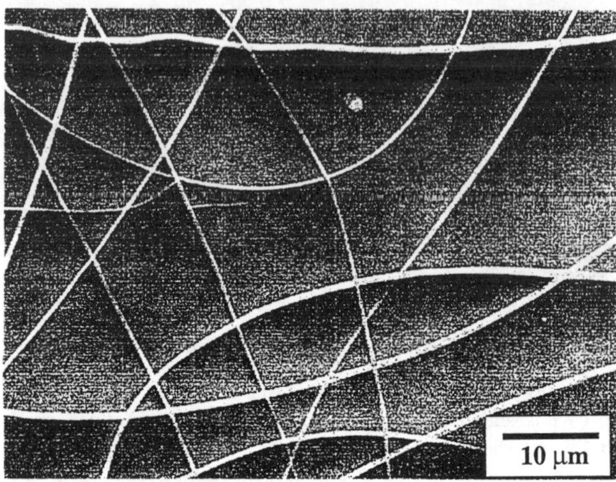

Figure 6. As-spun mesophase pitch fibers.

crossed polarizers in an optical microscope. The image of the fibers are bright everywhere except where the fibers are oriented at 45° with respect to the analyzer and the polarizer, in which case the image of the fibers is dark. This shows that the fibers are birefringent, as is consistent with the x-ray and electron diffraction patterns. The anisotropy created, during spinning is retained during the heat treatment of the mesophase pitch fibers. The short fibers in figure 7 were produced when some of the jets of molten pitch began to collapse toward droplets because the shroud of air surrounding the jet was too hot.

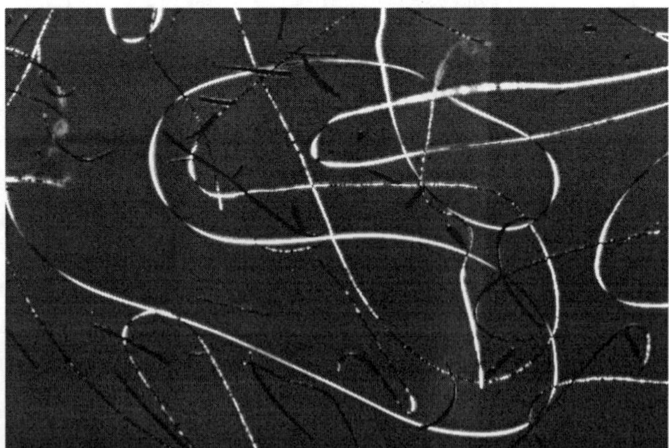

Figure 7. As-spun mesophase pitch fibers between crossed polarizers in an optical microscope.

Electron diffraction patterns were obtained from carbon nanofibers made from mesophase pitch. A fiber with a diameter of around 600 nanometers and its electron diffraction pattern are shown in figure 8. The fibers are so thin that they are easily penetrated by the electron beam of an ordinary transmission electron microscope, without thinning or coating. The electron diffraction patterns exhibited a (002) reflection on the equator, showing that the normal to the graphitic planes (002) is perpendicular to the long axis of the nanofibers, as is the case for the larger, conventionally spun carbon fibers (25).

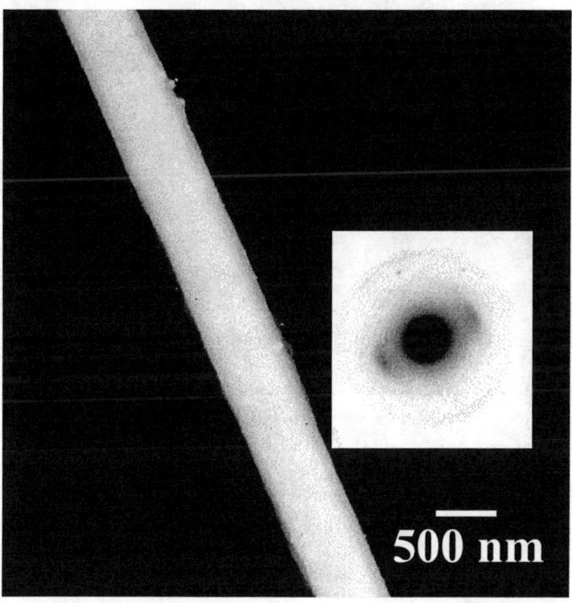

Figure 8. Electron diffraction pattern and an electron microscope image of a carbonized mesophase pitch fiber.

3.2 Carbon Nanofibers Spun From Molten Mesophase Pitch By A Related Higher Productivity Process   Mesophase pitch nanofibers are produced from a single tip at a rate of a few grams per hour in this process, which is still under development.

3.2.1 Scanning Electron Micrographs   A scanning electron micrograph of as spun mesophase pitch fibers is shown in figure 9. Long fibers with small diameters and short fibers with relatively large diameters are shown. The short fibers were produced in the same way as those shown in figure 7. This sample shows that short fibers can be produced, although conditions are generally adjusted to produce very long nanofibers. Figure 10 shows the distribution of fiber diameters in a typical sample of fibers and figure 11 shows the cumulate distribution. The range of the fiber diameters from the high productivity process is higher than for the carbon nanofibers produced by the method described at section 3.1.1. Figure 10 shows that the distribution of the diameters is described better as a log normal distribution than as a normal distribution. In this sample, the geometric mean diameter was 535 nanometers.

3.2.2 Wide Angle X-Ray Diffraction   Figure 12 shows the intensity of the (002) diffraction peak from a nonwoven sheet of carbon nanofibers made from mesophase pitch and also

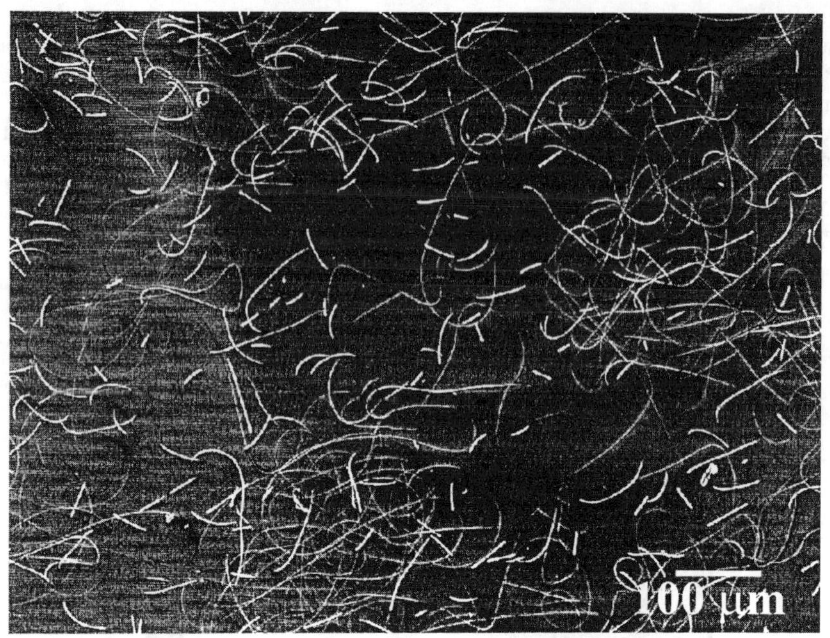

Figure 9. Short fibers of mesophase pitch.

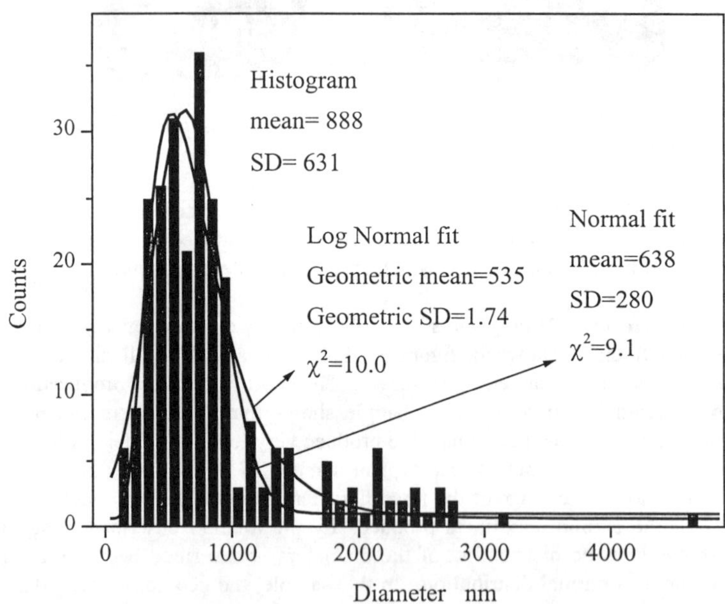

Figure 10. Distribution of diameters of as-spun mesophase pitch fibers.

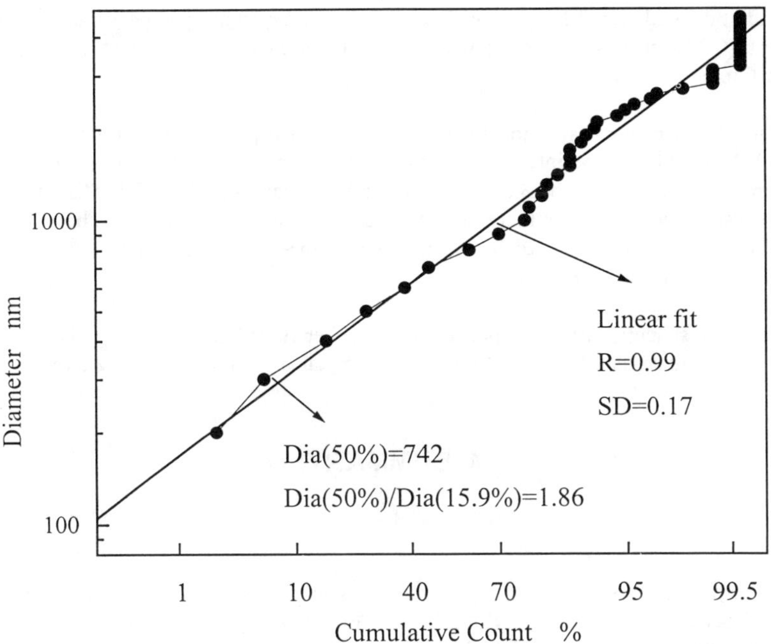

Figure 11. Cumulative log normal distribution of as-spun mesophase pitch fibers.

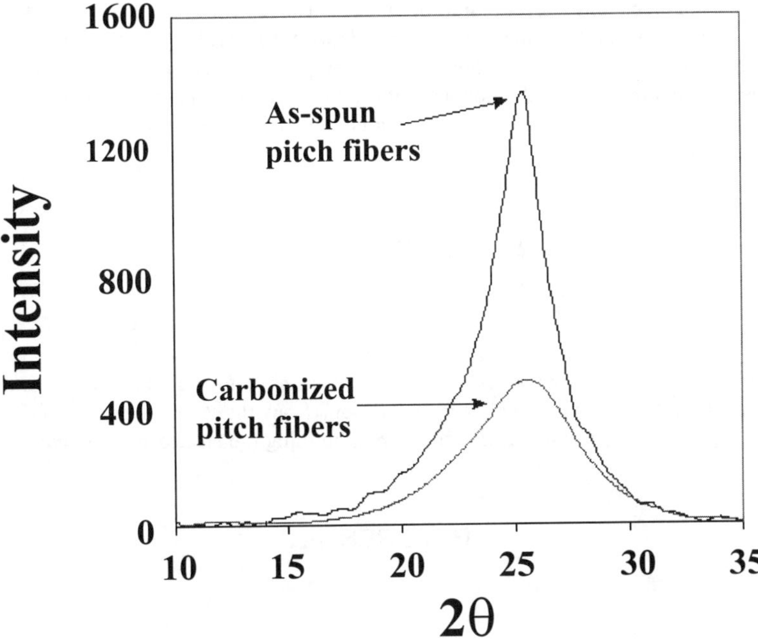

Figure 12. Wide angle x-ray diffraction peak for as-spun and carbonized mesophase pitch fibers.

from a sheet of nanofibers of as-spun pitch. A Siemens D5005 x-ray diffractometer using CuKα radiation with λ= 0.154 nm was used. The interplanar spacing of the carbonized nanofibers (0.349 nm) is consistent with values reported in the literature (26) for larger diameter fibers.

In the carbonized mesophase pitch fibers, the crystal size parameter derived from the full width at half height of these intensity curves was 1.68 nm. In the literature (26), the crystal size parameter is reported to be 3.5 to 4.5 nm for fibers carbonized at 1300 °C. It is known that the crystal size parameter increases with carbonization temperature. The small value observed here confirms that the carbonization temperature of 800 °C was too low to produce the highest possible crystalline order.

The interplanar spacing in the as-spun pitch nanofiber (0.351 nm) was slightly larger than that in the carbonized nanofiber (0.349 nm). The crystal size parameter was 2.54nm.

4. SUMMARY

Carbon nanofibers were produced using electrospinning techniques. Electrospun carbon precursor fibers, based on polyacrylonitrile and mesophase pitch, having diameters in the range from 100 nanometer to a few microns, were stabilized and carbonized. These carbon nanofibers have a very high aspect ratio. The nanofibers have a high ratio of surface area to unit mass because of their small diameters. As spun mesophase pitch nanofibers are transparent, colored, birefringent and oriented. The spacing of (002) planes between graphitic planes of carbon nanofibers agreed with reported values even through the carbonization process was carried out at lower temperatures. Nanopores were produced in carbon nanofibers made from the polyacrylonitrile by a high temperature reaction with water vapor carried in nitrogen gas, a process which is known to increase the surface area per unit mass of carbon black.

ACKNOWLEDGMENTS

This work was supported by the Air Force Materials Laboratory in Dayton, Ohio, the Army Research Laboratory in Maryland, and by a grant from the Army Research Office. The authors are grateful to Mr. Dale Ertley for help in building the electrospinning equipment.

REFERENCES

1   J. Zeleny, Physical Review, X (1), 1 (1917).
2. U.S. Pat. 1,975,504 (Oct. 2, 1934) A. Formhals.

3. B. Vonnegut and R. L. Neubauer, Journal of Colloid Science, 7, 616 (1952).
4. G. Taylor, Proceedings of the Royal Society of London Series A, 280, 383 (1964).
5. P. K. Baumgarten, Journal of Colloid and Interface Science, 36, 71, (1971).
6. L. Larrondo and R. St. J. Manley, Journal of Polymer Science, Polymer Physics Edition, 19, 909 (1981).
7. L. Larrondo and R. St. J. Manley, Journal of Polymer Science, Polymer Physics Edition, 19, 921 (1981).
8. L. Larrondo and R. St. J. Manley, Journal of Polymer Science, Polymer Physics Edition, 19, 933 (1981).
9. G. Srinivasan and D. H. Reneker, Polymer International, 36, 195 (1995).
10. J. Doshi and D. H. Reneker, Proceedings of IEEE Industry Applications Society Meeting Part 3, 3 (1993).
11. I. Chun, Fine Fibers Spun by Electrospinning process from Polymer solutions and Polymer Melts in Air and Vacuum: Characterization of Structure and Morphology on Electrospun Fibers and developing a New Process Model, Dissertation, The University of Akron, (1995).
12. J. Kim, Improved Mechanical Properties of Composites Using Ultrafine Electrospun Fibers, Dissertation, The University of Akron, (1997).
13. L. H. Peebles, Carbon Fibers - Formation, Structure, and Properties, CRC Press, Inc., Boca Raton, Ann Arbor, London, Tokyo, 1995, pp. 7.
14. A. K. Gupta, D. K. Paliwal and P. Bajat, Journal of Macromolecular Science-Reviews in Macromolecular Chemistry and Physics, C31, (1), 1 (1991).
15. W. Watt and W. Johnson, Applied Polymer Symposia (9), 215 (1969).
16. D. J. Johnson, Philosophical Transactions of Royal Society of London Series A, 294, 443 (1980).
17. T. Ko, H. Ting and C. Lin, Journal of Applied Polymer Science, 35, 863 (1988).
18. C. P. Ju, J. Don and P. Tlomak, Journal of Materials Science, 26, 6753 (1991).
19. J. Liu, P. H. Wang and R. Y. Li, Journal of Applied Polymer Science, 52, 945 (1994).
20. Y. Huang and R. J. Young, Carbon, 33, (2), 97 (1995).
21. T. Matsumoto, Pure and Applied Chemistry, 57, 1553 (1985).
22. M. Guigon and A. Oberlin, Composites Science and Technology, 25, 231 (1986).
23. S. N. Magonov, H. J. Cantow and J. B. Donnet, Polymer Bulletin, 23, 555 (1990).
24. S. Kumar D. P. Anderson and A. S. Crasto, Journal of Materials Science, 28, 423 (1993).
25. Huang and R. J. Young, Journal of Materials Science, 29, 4027 (1994).
26. I. Mochida, S. H. Yoon, N. Takano, F. Fortin, Y. Korai and K. Yokoawa, Carbon, 34, (8), 941 (1996).
27. N. Takami, A. Satoh, M. Hara and T. Ohsaki, Journal of the Electrochemical Society, 142, (8), 2564 (1995).
28. A. J. Juodilis, Electronic Design, 159 (September 30, 1982).
29. A. Yoshida, I. Tanahashi, Y. Takeuchi, and A. Nishino, IEEE Transaction on Components, Hybrids, and Manufacturing Technology, CHMT-10, (1), 100 (1987).
30. S. T. Mayer, R. W. Pekala and J. L. Kaschmitter, Journal of the Electrochemical Society, 140, (2), 446 (1993).
31. C. Arbizzani, M. Mastragostino, L. Meneghello, and R. Paraventi, Advanced Materials, 8, (4), 331 (1996).
32. J. Doshi, The Electrospinning Process and Applications of Electrospun Fibers, Dissertation, The University of Akron, (1994).
33. D. H. Reneker and Iksoo Chun, Nanotechnology, 7, 216 (1996).

HIGH TEMPERATURE STRUCTURAL FOAM[1]

Erik S. Weiser
NASA Langley Research Center
Hampton, VA 23681

Faye F. Baillif
Lockheed Martin Michoud Space Systems
New Orleans, LA 70129

Brian W. Grimsley and Joseph M. Marchello
Old Dominion University
Norfolk, VA 23529

ABSTRACT

The Aerospace Industry is experiencing growing demand for high performance polymer foam. The X-33 program needs structural foam insulation capable of retaining its strength over a wide range of environmental conditions. The High Speed Research Program has a need for low density core splice and potting materials. This paper reviews the state of the art in foam materials and describes experimental work to fabricate low density, high shear strength foam which can withstand temperatures from -220°C to 220°C.

Commercially available polymer foams exhibit a wide range of physical properties. Some with densities as low as 0.066 g/cc are capable of co-curing at temperatures as high as 182°C. Certain polyimide foams have moduli of elasticity as high as 0.19 Mpa, with tensile strengths of 3.7 MPa and compressive strengths of 3.6 Mpa, however, they cannot withstand liquid hydrogen temperatures. Other polyimide foams withstand temperatures from -250°C to 200°C, but they do not have the required structural integrity.

The research activity at NASA Langley Research Center focuses on using chemical blowing agents to produce polyimide thermoplastic foams capable of meeting the above performance requirements. The combination of blowing agents that decompose at the minimum melt viscosity temperature together with plasticizers to lower the viscosity has been used to produce foams by both extrusion and oven heating. The foams produced exhibit good environmental stability while maintaining structural properties.

Key Words: Polymer foam, low density structural foam, thermoplastic, polyimide

---

1.0 INTRODUCTION

Low density porous materials, otherwise known as foams, were first developed in the United States and Europe in the mid-to-late 1930's (1). The term polymer foams or cellular polymers refers to a two-phase gas-solid system in which the solid polymer is continuous and the gaseous cells are dispersed throughout the solid (2). Polymeric materials are foamed to meet various application needs, such as weight-reduction, insulation, buoyancy, energy dissipation, conveyance, and comfort. These polymeric foams can be produced by several different methods including extrusion, compression molding, injection molding, reaction injection molding, and solid state methods.

Rigid extruded foamed polystyrene was first produced commercially in the United States in the mid-1940's followed by polyurethanes (1). A solution of polystyrene in methyl chloride is heated above its normal boiling point while under pressure within an extruder barrel. As the material exits the barrel of the extruder, the solvent evaporates, simultaneously foaming the plastic and cooling it below its glass transition temperature (Tg). Foaming and dimensional stabilization thus are carried out in one step (3). Later, polystyrene was fabricated from beads in one of the first closed mold operations. The polystyrene was prefoamed into beads and later placed in a closed mold, where the beads were allowed to expand further and fuse together as the temperature was raised above Tg. Concurrently, phenolic and urea-formaldehyde foams were used in Europe but did not appear in the United States until the mid-1960's and the 1970's when the market for building thermal insulation was expanded by the oil energy crises of the 1970's (1).

As late as 1978, the largest single application of cellular polymers involved comfort cushioning, and the market was dominated by flexible open-cell polyurethane foam and to a lesser extent latex foam rubber. Thermal insulation was the second largest application and was clearly the largest application for rigid closed-celled materials (1). It is estimated that in 1995 close to 6 billion pounds of foamed plastics were produced and consumed in the United States alone, and it is projected that this usage will grow at 3 - 4% annual rate to about 7 billion pounds by the year 2000 (4).

In the late 1960's a new type of polymeric foam was developed by Monsanto and DuPont based on a polyimide precursor. These new polyimide foams showed improved properties in areas such as thermal stability, non-flammability, radiation resistance, improved toughness, and reduced smoke and toxic fume generation (1). Recently, the aerospace industry has had a growing need for high performance polymer foams for applications such as cryogenic insulation, flameproofing, energy absorbers, etc, and has turned to polyimide foams to meet these needs. For example, the X-33 Single Stage to Orbit (SSO) Space Transportation Vehicle requires a lightweight structural foam which retains its structural integrity at temperatures ranging from -250°C to 250°C (5). The High Speed Research Program has a need for low density core splice material with a density of 0.50 g/cc and also potting material with a density of 0.56 g/cc (6). This paper will review the state of the art in polyimide foam materials and will describe experimental work being done to fabricate low density, high shear strength, foam which can meet upcoming critical needs in aerospace projects.

2.0 FOAM CHARACTERISTICS

There are two main types of polymeric foams: thermoplastic and thermoset foams. Thermoplastic foams can be reprocessed and recycled, while thermoset foams are intractable since they are usually highly crosslinked. Within these classes, the polymeric foams are further classified as rigid, semi-rigid, semi-flexible, or flexible, depending upon their compositions, cellular morphologies, thermal characteristics such as Tg, percent crystallinity, extent of crosslinking, etc. (4)

2.1 Foam Material The cellular void space in any polymeric foam material is formed through the use of blowing agents. There are two types of blowing agents used to produce foams: chemical blowing agents and physical blowing agents (4). Chemical blowing agents give off gases during the foaming process, either due to a polymerization reaction or by thermal decomposition of a secondary agent. Physical blowing agents are simply inert gases such as nitrogen, carbon dioxide, freon, etc. The three most important characteristics of a secondary blowing agent are the decomposition temperature, the volume of gas generated per unit weight, and the nature of the decomposed residue. To produce uniform cells the blowing agent must be uniformly dispersed or dissolved, uniformly nucleated, and rapidly and smoothly decomposed over a narrow temperature range. Foam stabilization occurs through the attainment of a high viscosity or gelation of the polymer system (3). If stabilization occurs before the blowing agent has begun to decompose and release gas, large fissures or holes may form. Cells formed too soon before stabilization may collapse and give a coarse, weak, spongy material. Stabilization of thermoplastic foam occurs when the material is cooled below Tg, while in a thermoset foam, increased crosslink density or gelation will cause a foam to stabilize.

2.2 Cell Structure: Open or Closed Foams are usually classified as closed-cell or open-cell. Closed cell foams are generally rigid, while open cell foams are generally flexible (4). The presence of cells acts to substantially decrease both the density and the strength of the solid polymer. The foam cells also act as a good heat insulator by virtue of the low conductivity of the gas, usually air, contained in the system. Foamed materials have a much higher flexural modulus to density than unfoamed materials due to the cells, e.g. a beam of foam deflects less than an unfoamed beam of the same length, width, and weight under the same load. Finally, any foam will have energy-storing or -dissipating capacity operating through a greater displacement than the unfoamed material (3).

If the cells are discrete and the gas phase of each is independent of that of other cells the material is termed closed-celled (2). Closed cells result when the decomposition and gelation are carried out in a closed mold almost filled with polymer and blowing agent. The closed mold is raised in temperature and held under pressure. After the heating cycled is completed, the article is cooled until the foamed material is dimensionally stablebefore the pressure is released. The growth of a hole or cell in the molten polymer is controlled by the pressure difference between the inside and the outside of the cell, the surface tension of the molten polymer, and the radius of the cell.

$$DP = 2(g)/r \qquad (2)$$

Where DP is the difference between the pressure generated by the blowing agent inside the cell and the pressure of the fluid phase of the molten polymer. Gamma (g) is the surface tension of the molten polymer and r is the radius of the cell.

Opened-cell foams are foams which have discrete interconnecting pockets which are porous. Open cells result when a mixture of polymer and blowing agent are spread on a substrate and allowed to fuse without a second confining surface present to hinder the cell formation. In general, open cell foams have a greater effect on reducing the overall strength of the material. An open-cell foam acts as a sponge, soaking up liquid by capillary action. A closed-cell foam makes a better buoy or life jacket because the cells do not fill with liquid (3).

3.0 HIGH PERFORMANCE FOAMS[i]

Polyimide foams are used for their superior thermal and acoustic insulation properties over polystyrene, polyethylene, and other low-cost polymer foams in applications where performance is more important than cost. Commercially available high performance polyimide foams are very limited. Several companies produce polyimide foams which could be considered for use by the aerospace industry: Imi-Tech Corporation in Elk Grove Village, Illinois; Schuller International Inc. in Littleton, Colorado; and Rohm Tech Inc. in Malden, Massachusetts. Imi-Tech Corporation and Schuller International Inc. both produce a highly aromatic polyimide foam, while Rohm Tech Inc. produces a rigid foam based on polymethacrylimide. In all three cases the foams produced by these companies can only meet certain requirements demanded by the aerospace industry. None of the foams produced can meet all the requirements of future aeronautics and space programs.

3.1 Imi-Tech Corporation - *SOLIMIDE*® Imi-Tech Corporation produces polyimide foams under the tradename *Solimide*®, of which thereare four main types produced. TA301 and AC430 are low density materials, 0.0088 and 0.0054 g/cc, respectively, used primarily as light-

---

weight thermal acoustic insulation where non-flammability, low smoke generation, and weight are important factors. Both materials have very high Tg's (~275°C) but lose a significant amount of their initial tensile strength after 1000 hours at 260°C: 60-70% for TA301 and 25% for AC430. They therefore are not suitable for elevated temperature applications. A high temperature version called HT430 can withstand 260°C for 1000 hours with only a 4% tensile strength loss. This version of the *Solimide*® foam is used mainly for insulation due to its high temperature stability. Imi-Tech Corporation also offers various densified foams, from 0.032 grams per cubic centimeter to several different customized densities depending on the application, under the *Solimide*® tradename however, most of these foams are used for thermal insulation and acoustic cushioning in aircraft interiors. The Imi-Tech densified foams have found limited use due to their low strength capabilities and high material costs. Table 1 shows some selected properties for various *Solimide*® foams.

Table 1. Selected Properties for *Solimide*® Foams.

Property	Test Method	Solimide TA301	Solimide AC430	Solimide HT340	Densified Solimide
Density (g/cc)	ASTM D-3574 (A)	0.0088	0.0053	0.0064	0.0320
Thermal Conductivity (W/mK)	ASTM C-518 @ 75°F	0.042	0.050	0.046	0.032
Tensile Strength (MPa)	ASTM D-3574 (E)	0.069 ± 50%	0.048 ± 40%	0.041 ± 50%	0.317
Friability (% wt. loss)	ASTM C-421	1	0.9	-	-
Optical Smoke Density (D_m) Flaming Non-Flaming	ASTM E-662	 5 3	 3 2	 3 1	 - -
Dielectric Constant	ASTM D-150	1.015	-	1.013	1.039
Compressibility (MPa)	ASTM D-3574 (C)	0.007 ± 43%	-	-	0.041

3.2 Schuller International Inc. - *INSULMIDE*™ Schuller International has had an established core technology centered on closed cell polyurethane foam production at least since the early 1980's (1). Shuller's polyimide foam technology, however, has not become a viable entity until mid-1992 when they developed *Insulmide*™ foams. *Insulmide*™ foams are available in a variety of forms from 1.22 m. x 2.44 m. blocks as thick as 0.178 meters to sheets of any smaller size and as thin as 3.18 millimeters. *Insulmide*™ is also available as precut pipe insulation in lengths up to 1.22 meter with shiplap joints and factory-applied jackets with adhesive. The standard density of Schuller foam is 0.0088 grams per cubic centimeter. *Insulmide*™ foam could find use as a thermal and acoustic insulation in marine applications or any other application where a low density material is required. Like other polyimide based foams, *Insulmide*™ foam has improved fire resistance, is non-toxic, has a low smoke emission, is a good thermal insulator, and has superior acoustic insulation capabilities to many foams that are available. *Insulmide*™ foam is not a structural foam and therefore, does not meet the requirements demanded by the aerospace industry for applications which require good tensile and compressive strengths. Table 2, below, shows some selected properties of *Insulmide*™ foams.

Table 2. Selected Properties for *Insulmide*™ Foam.

Property	Test Method	Insulmide
Density (g/cc)	ASTM D-3574 (A)	0.0088
Thermal Conductivity (W/mK)	ASTM C-518 @ 75°F	0.042
Tensile Strength (MPa)	ASTM D-3574 (E)	0.033
Friability (% wt. loss)	ASTM C-421	1.2
Optical Smoke Density (D_m) Flaming Non-Flaming	ASTM E-662	 1.8 2.0
Compressibility (MPa)	ASTM D-3574 (C)	0.012

3.3   Rohm Tech Inc. - *ROHACELL*® Rohm Tech Inc. markets many foams under the tradename *Rohacell*®. *Rohacell*® is a polymethacrylimide rigid foam (PMI) which is manufactured by hot foaming of methacrylic acid/methacrylonitrile copolymer sheets. *Rohacell*® foams come in various grades and within each grade the foam is fabricated to different densities. The industrial grade, IG, is a closed-cell rigid foam and has a maximum operating temperature of 121°C and comes in various densities (0.032 to 0.192 g/cc) (1). IG foam is used the aircraft, marine, electronics, and radiation industries. This type of foam has also found uses in sporting goods as structural stiffeners and in freight containers. Within the IG grade there is an industrial grade with certified material property test results called AIR. The aircraft grade, WF, and heat-rated aircraft grade, WFHT, are used for construction applications. *Rohacell*®'s WF and WFHT foams are also closed-celled rigid foams with densities ranging from 0.048 to 0.304 grams per cubic centimeter and a maximum use temperature of 177°C. *Rohacell*® foams have excellent tensile and shear strengths compared to other polyimide foams. These foams are used in helicopter blades, radomes, antennae, and nose tips for aircraft. Table 3 gives some selected data on *Rohacell*® foams.

Table 3. Selected Properties for *Rohacell®* foams.

Property	Test Method	Rohacell 31	Rohacell 51	Rohacell 71	Rohacell 110	Rohacell 170
Density (g/cc)	ASTM D-1622-63	0.032	0.051	0.075	0.111	0.170
Thermal Conductivity (W/mK)	ASTM C-177-63 @ 68°F	0.031	0.029	0.030	-	-
Tensile Strength (MPa)	ASTM D-638 68	0.98	1.86	2.74	3.43	7.38
Flammability	DIN 4102	class B2	class B2	class B2	class B2	class B2
Dielectric Constant	@ 2.0 GHz	1.08	1.07	1.08	-	-
Compressive Strength (MPa)	ASTM D-1621-64	0.39	0.88	1.47	2.94	6.37
Shear Strength (MPa)	ASTM C-273-61	0.39	0.79	1.28	2.35	4.41

The foams discussed in this section represent the mainstay of high performance polyimide foams. Each material has qualities which can bring it to the forefront for the applications which fit it best. When these foams are compared to the strict requirements for a reusable launch vehicle (X-33) or commercial supersonic aircraft (HSR), however, they do not entirely meet the grade. *Insulmide™* does not have the reproducibility to make it a viable candidate. *Solimide®* foams require too much secondary manufacturing to densify the foam into a structural type material. Finally, *Rohacell®* foams do not have the flammability resistance nor the hydrophobic qualities required to withstand the harsh environmental conditions necessary in aerospace applications. For this reason, NASA has begun to look at more advanced polyimides to meet the future needs of these applications.

4.0 FOAM RESEARCH

4.1 Polyimide Characterization   Two NASA Langley polyimides, LaRC™ IA, oxydiphthalic anhydride / 3,4-oxydianiline (7), and LaRC™ IAX, oxydiphthalic anhydride with a 90:10 ratio of 3,4-oxydianiline and 1,4-phenylenediamine (8); and *Mitsui Toatsu's* PIXA with a 4% stoichiometric imbalance, were tested to identify melt rheology and crystallization kinetics. The minimum melt viscosity was found using a *Rheometrics System 5* rheometer. The viscosity of the polymer is the critical element to allow foaming to occur in an extruder. Melt rheology results from the polyimides tested indicated that the viscosities needed to be lowered. To reduce the viscosity to an acceptable level a plasticizer was added to the polymers in small quantities. Plasticizers allow the Tg of a polyimide to be lowered thus softening the material and reducing the viscosity. For these experiments, a NASA Langley plasticizer was used: LaRC™-LV232. Table 4 provides the rheological data which include the effect of a plasticizer on the viscosity of the polyimides. A *Shimadzu DSC-50* differential scanning calorimeter was used to ascertain the glass transition point of the polyimides. The results in table 4 helped to determine the temperatures necessary to allow foaming to occur in an extruder.

Table 4. Polyimide Characterization.

Polyimide	Tg (°C)	Minimum Viscosity (poise)	Minimum Viscosity Temperature (°C)
LaRC™ - IA	248.83	3855.0	400
LaRC™ - IAX	257.72	4418.6	400
Mitsui - PIXA	302.87	3985.4	400
90%-IA + 10% LV232	-	1932.4	400
90%-IAX + 10% LV232	-	2237.1	400
90%-PIXA + 10% LV232	-	1153.3	400

4.2 Chemical Blowing Agents The TGA data were obtained using a *Seiko TG/DTA 220* thermo-gravametric analyzer at 2.5°C/min. in flowing (40 ml/min.) air. Table 5 gives the results from the TGA analysis and the amount of gas generated by the blowing agent decomposition. The gas generation data was obtained using a gas/bomb collector. A measured amount of blowing agent was placed in a steel pipe attached to a sealed collector. The pipe was heated to the blowing agent's decomposition temperature and the resulting gas was collected and measured.

Table 5. Blowing Agent Characterization.

Blowing Agent	TGA		Gas Generation	
	(°C)	% wt Loss	(°C)	gas (ml/g.)
BASF Theic tris (2-hydroxethyl) isocyanurate	280	100	250-300	105
Espirit Chemicals Barium Salt of 5-phenyltetrazole	400	70	350-400	110
Polyvel Inc. 25% Barium Salt of 5-phenyltetrazole in copolymer carrier	420	80	350-450	48

The TGA data show that *BASF's Theic* blowing agent, tris (2-hydroxethyl) isocyanurate, has 100% weight loss. However, the decomposition temperature of 280°C is below the Tg of the high strength polyimides used in this project and therefore it will not be able to foam the polymers adequately. The Barium salt of 5-phenyltetrazole, by Espirit Chemicals, therefore appears to be the best blowing agent tested. Its decomposition temperature matches the minimum viscosity temperature of the polyimides and it is capable of generating the same amount of gas as the *Theic*. The Polyvel, 25% Barium Salt of 5-phenyltetrazole in copolymer carrier, has a good decomposition temperature, but it does not produce enough gas to make it a viable candidate.

4.3   Extruder Experiments   Tests were conducted to identify a chemical blowing agent capable of producing a low density foam from the LaRC™-IA, -IAX, and Mitsui PIXA polyimides. The effect of the LV232 plasticizer on foam density was also investigated.

A C.W. Brabender™ 3/4" single screw laboratory extruder was used to generate the polymer foam. This instrument was chosen for its ability to simulate small scale production while still allowing for an intimate blending of the polymeric material with fillers and blowing agents. The material is conveyed through three heat controlled zones by a rotating venting screw with a 5:1 compression ratio. The speed at which the screw rotates determines the amount of shear imparted to the material and the amount of time it is subjected to heat. The ideal screw speed was found to be 20 revolutions per minute. In the first two heat zones the material is blended and softened. The temperature in the third zone is chosen to reach the lowest viscosity of the polymer and cause decomposition of the chemical blowing agent. Here the shear imposed by the screw aids in disbursement of cells as they nucleate in the molten polymer. The foamed material exits the third zone and is forced into the die attachment where it is formed. A circular die with a 0.64 centimeter orifice was utilized to produce foam rods with diameters of 0.95 to 1.91 centimeters. The rod diameter is further influenced by the convey belt which carries the foam away from the heat of the die. The foam extrusion tests yielded reproducible foams with densities ranging from 0.14 g/cc to 0.75 g/cc (Table 6).

Table 6. Extruded foam test data.

Powder Mixtures	Foam Density (g/cc)
90% PIXA + 10% Polyvel	0.51
90% IAX + 10% Polyvel	0.53
90% IA + 10% Polyvel	0.54
80% IAX + 10% Polyvel + 10% LV232	0.44
90% PIXA + 10% Barium Salt	0.21
90% IAX + 10% Barium Salt	0.17
90% IA + 10% Barium Salt	0.17
80% IA + 10% Barium Salt + 10% LV232	0.14
80% PIXA + 10% Theic + 10% LV232	0.75
80% IAX + 10% Theic + 10% LV232	0.71

4.4   Closed Mold Experiments

4.4.1   Solution Preparation   In a 1 liter three neck round bottom flask equipped with a mechanical stirrer, N_2 inlet and drying tube was placed 3,4'-ODA (80.1g, 0.4 mol.) and 109 grams of Fisher Scientific's Tetrahydrofuran (THF) / Ethanol (70/30). The solid dissolved within one hour and a slurry of ODPA (122.8g , 0.796 mole), PA (1.1849 g, 0.008 mol) and

307 grams of THF / Ethanol (70/30) was added to create a final solution with 30% solids by weight. After an hour of stirring the mixture became warm to the touch and an orange viscous solution formed. The mixture was stirred continuously for 15 hours.

4.4.2   Pellet Preparation   The polyamide acid solution containing 30% LaRC™-IA and 70% solvent THF / Ethanol (70/30) was poured through a 0.64 centimeter adapter into a shallow water bath to form "ropes" with a diameter of 0.635 - 0.95 cm. The outside of the ropes formed a white skin which encapsulated the polyamide acid solution.   After further leaching of the THF / Ethanol solvent, the ropes were removed and air dried overnight.   The dried ropes were pelletized in a grinder. The resulting pellets contained 20 - 25% THF / Ethanol solvent by weight as found by thermal gravametric analyzer (TGA).

4.4.3 Foaming   The polyamide acid pellets containing 25% solvent by weight were placed in a circular aluminum mold with a volume of 13.5 cubic centimeters.   The base of the mold contained a sheet of kapton release film and the top surface was covered with a graphite block, to allow venting while the volume was contained.   The block was preheated in the oven to 260°C. The aluminum mold and block were placed in an oven at 260°C for 30 minutes, then removed and quenched in an ice bath. The resulting foam cylinders have densities ranging from 0.08 g/cc to 0.18 g/cc.

5.0   DISCUSSION

5.1 Extruded Foam Low density foams are possible by extrusion. Table 6 shows that by using a chemical blowing agent that has a decomposition temperature which matches the minimum viscosity point of a polymer, a low density foam can be formed by extrusion. It is clear from the data that Espirit Chemicals, Barium salt, chemical blowing agent is indeed the most favorable blowing agent among the three tested. This is because the blowing temperature closely matches the polymer's minimum melt viscosity temperature. The densities of LaRC™-IA, LaRC™-IAX, and Mitsui PIXA foams produced using the single-screw extruder and the Espirit Chemical's blowing agent all display densities below the desired 0.20 g/cc. In addition, the use of 10% LaRC™-LV232 plasticizer gave the polyimide foams with lower densities than those without the plasticizer.

The foam extrusion experiments also prove that the decomposition temperature of the BASF Theic is below the minimum melt viscosity temperature of the polyimides. Figure 1 is a photomicrograph of LaRC™-IA and the BASF Theic blowing agent, no gaseous cells formed within the polymer. The lack of cellular pockets within the polymer can be attributed to the Theic decomposing at a temperature below the minimum melt viscosity of the material. The Theic blowing agent decomposed while the polymer was in a solid state and thus did not foam the material.

Figure 2 shows a nearly uniformed closed-cell structure produced by the barium salt chemical blowing agent. Ten percent barium salt was used to produce the LaRC™-IA foam with a density of 0.17 g/cc. The photomicrograph also indicates that a better cellular size distribution is needed to enhance the quality of the foam. The cells within this foam structure have a wide variation in size, which could be due to poor distribution of polymer and blowing agent. However, it is clear that the use of Espirit Chemical's barium salt will produce a low density foam by extrusion.

The Polyvel blowing agent, 25% Barium Salt of 5-phenyltetrazole in copolymer carrier, mixed with LaRC™-IA, figure 3, resulted in a relatively high density foam as compared to the foam fabricated with 100% barium salt. This can be attributed to the low amount of gas generated during decomposition, 48 ml/g. compared to 110 ml/g. for the 100% barium salt. As seen in Table 6, the density produced with the Polyvel blowing agent was nearly double that given by the Espirit Chemicals barium blowing agent for every experiment. Figure 3 displays a distribution of cells throughout the micrograph. Once again the cell size is not uniform and thus the foam lacks homogeneity.

5.2 Close Mold Foaming Initial closed mold experiments have resulted in the fabrication of several low density foams. Densities ranging from 0.08 to 0.14 g/cc were produced. These densities are well below the target value of 0.20 g/cc and the foams appear to have a poor cellular structure (fig. 4). The poor cellular structure can be attributed to the lack of a surfactant in the polymer mix and the coarsness of the polymer powder. Surfactants are used in polymer foams to control the gaseous cell growth. The surfactant strengthens the interface between the gas formed by the chemical blowing agent and the polymer melt. Finely ground polymer powder would allow more uniform cell growth due to the increased surface area. At the rapid heating rate used, the blowing agent may explode at such a high pressure that the polymer can not hold its cellular structure and the foam walls may breakdown. Figures 4 shows a photomicrograph of LaRC™-IA with poor cellular structure due to a rapid processing cycle and large polymer particle size. Figure 5 shows LaRC™-IA foam with improved cellular structure produced using finely ground polymer precursor and a more controlled heating cycle in which the polymer was slowly ramped to its finally cure temperature of 260°C. Comparing figures 4 and 5, it should be noted that the controlled heat ramp resulted in a slightly higher density foam, due to a more stable foam process. However, a more uniform cell structure should result in better physical properties of the foam.

6.0 CONCLUSIONS AND FUTURE WORK

In this study it was shown that low density polyimide foams are possible by both extrusion and closed-mold processing. Espirit Chemicals Inc., Barium Salt, chemical blowing agent has the best decomposition properties (110 mg/cc gas evolution and 400°C decomposition temperature) to allow low density foams to be extruded. The addition of plasticizers to the polymer solution aids in the reduction of melt viscosity and allows chemical blowing agents to successfully foam high melt temperature polyimide materials. Foams produced using plasticizers had uniform cell distribution.

In the period ahead mechanical properties of extruded and closed-mold formed foams will be determined. The poor cellular uniformity in the initial foams may result in less than optimum mechanical properties. The use of surfactants and finely ground polymer precursors in later experiments should allow uniform cell size and distribution and thus increase the strength of the foam.

7.0 REFERENCES

1. NASA Contractor's report, NASA Contract No. NASW-4367, Research Triangle Institute, North Carolina, 1994, Closed Material.

2. J. L. Holmes, "Cellular Materials," Encyclopedia of Polymer Science and Technology, Vol. 3, John Wiley and Sons, Inc., New York, 1985, pp. 80-98.

3. F. Rodriguez, "Principles of Polymer Systems," Hemisphere Publishing Corporation, New York, 1989, pp. 403-416.

4. K. Khemani, "Polymeric Foams: Science and Technology," American Chemical Society, Washington D.C., 1997, pp. 1-7 & 195-205.

5. V.P. McConnell, NASA Gets Hands-On With X-33 Design, High-Performance Composites, T. J. Baucom, ed., Elseevier, New York, May, 56-58 (1997).

6. K. Fisher, RTM and Core Materials Offer Product Advances, High-Performance Composites, T. J. Baucom, ed., Elseevier, New York, Sept., 33-35 (1997).

7. T. H. Hou, N. J. Johnston, T. L. St. Clair, IM-7/LaRC™-IA Polyimide Composites, High Performance Polymers, 7 (2), 105-124 (1995).

8. T. H. Hou, N. J. Johnston, E. S. Weiser, and J. M. Marchello, Processing and Properties of I-M7 Composites Made From LaRC™-IAX Polyimide Powders, Soc. Adv. Matl. Proc. Eng. Series, 27, 135, (1995).

Polyimide Foam Micrographs

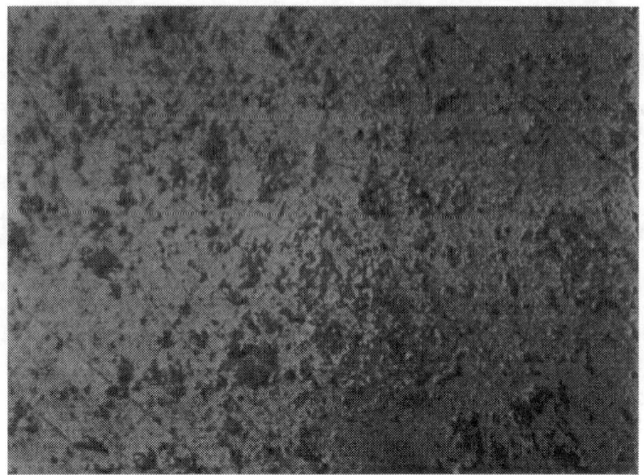

Figure 1. 90% LARC™-IA and 10% THEIC
Foam Density 0.69 g/cc

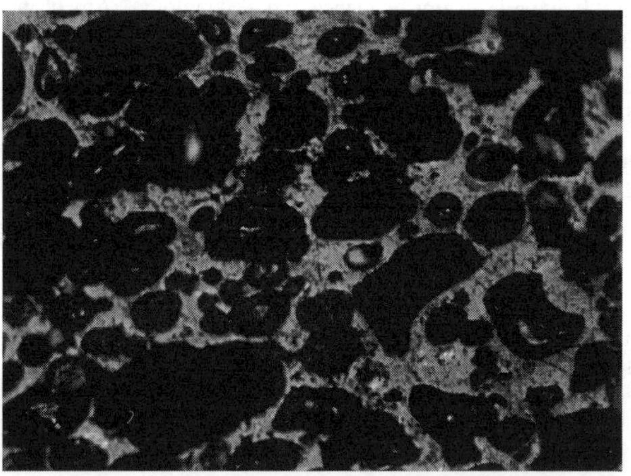

Figure 2. 90% LARC™-IA and 10% Barium Salt
Foam Density 0.17 g/cc

Polyimide Foam Micrographs

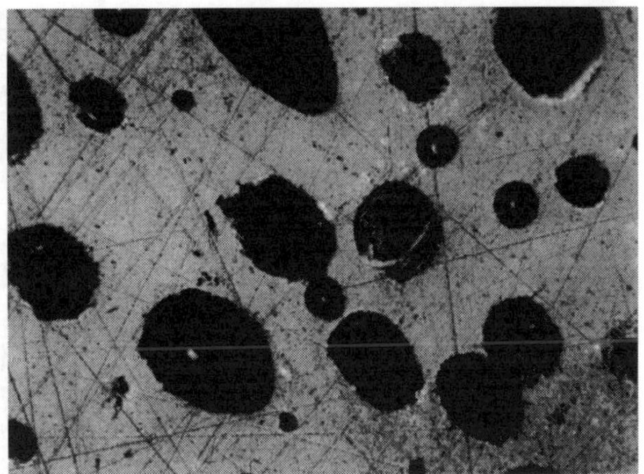

Figure 3. 90% LARC™-IA and 10% POLYVEL
Foam Density 0.54 g/cc

Figure 4. 90% LARC™-IA and 10% THF
Foam Density 0.10 g/cc

Polyimide Foam Micrographs

Figure 5. 90% LARC™-IA and 10% THF
Foam Density 0.14 g/cc

43rd International SAMPE Symposium
May 31-June 4, 1998

HIGH THERMAL CONDUCTIVITY, MESOPHASE PITCH-DERIVED CARBON FOAM

James Klett

Carbon and Insulation Materials Technology Group, Metals and Ceramics Division,
Oak Ridge National Laboratory, Oak Ridge, TN, 37831-6087

ABSTRACT

A relatively simple technique for fabricating mesophase pitch-based carbon foam has been developed at Oak Ridge National Laboratory. This technique produces graphitic foam with an open cell structure and an extremely high bulk thermal conductivity, >100 W/m·K. The cell walls have a highly aligned graphitic structure, similar to high-performance carbon fibers, exhibiting interlayer spacing (d_{002}) of 0.336 nm, coherence length ($L_{a,100}$) of 203 nm, and a stacking height ($L_{c,002}$) of 442 nm. Consequently, the foam cell walls (struts) exhibit a thermal conductivity between 700 and 1,200 W/m·K. Because of the low density (ρ) of 0.5 g/cm^3, the specific thermal conductivity of the foam is more than four times greater than that of copper.

KEY WORDS: Carbon Foam, High Thermal Conductivity, and Mesophase Pitch

1. INTRODUCTION

The extraordinary mechanical properties of carbon fiber result from the unique graphitic morphology of the spun (extruded) filaments (1). Contemporary advanced structural composites exploit these properties by creating a disconnected network of graphitic filaments held together by a matrix suitable for the application. Carbon foam derived from a pitch precursor, on the other hand, can be considered as an interconnected network of graphitic ligaments and, thus, should exhibit isotropic material properties (2, 3). The foam represents a potential reinforcing phase for structural composite materials. Because of the continuous graphitic network, the foam-reinforced composites will display higher isotropic thermal conductivities than carbon fiber reinforced composites. Furthermore, the lack of interlaminar regions, which develop in traditional prepregged carbon fiber reinforced composites, should result in enhanced mechanical properties such as shear strength and fracture toughness.

Recent research into fiber-reinforced composites has been driven by the need for increased mechanical properties such as strength, stiffness, creep resistance, and toughness in structural engineered materials. Such improvements have been achieved through the enhancement of fiber properties, fiber/matrix interfaces, and matrix materials.

In recent years the use of carbon fibers has evolved from structural reinforcement to a thermal management material, with the emphasis in applications such as high-density electronic modules, communication satellites, and automotive systems. The high cost of carbon fibers has stimulated research into both novel reinforcements and new composite processing methods. Other than low cost, the primary concerns in thermal management applications are high thermal conductivity, low weight, and low coefficient of thermal expansion (4). Such applications have focused on sandwich structures (a high thermal conductivity material encapsulating a structural core material) to provide the required mechanical properties (4). However, since structural cores are typically low-density materials, the thermal conductivity of the overall composite through the thickness is relatively low (~3-10 W/m·K for aluminum honeycomb (5, 6)). Metallic foams are being explored as a potential core material; however the thermal conductivities are still low, 5 - 50 W/m·K (6). New pitch-derived graphitic foams present a unique solution to this problem by offering high thermal conductivity with a low weight.

In order to produce high stiffness and high conductivity graphitic foams a mesophase pitch must be used as the precursor, thus assuring a graphitic structure in the struts (2, 7, 8). Typical processes utilize a blowing technique, or pressure release, to produce foam of the pitch precursor (7, 9-12). The pitch foam is stabilized by heating in air or oxygen for many hours to cross-link the structure and "set" the pitch so it does not melt during further heat treatment (13). Stabilization is a very time consuming process and can be expensive (depending on the part size and equipment required). The "stabilized" pitch is carbonized in an inert atmosphere to temperatures as high as 1100°C, and graphitized at temperatures as high as 3000°C to produce a thermally conductive graphitic structure.

A new, less time consuming process for fabricating pitch-based graphitic foams without the traditional blowing and stabilization steps has been developed. It is believed that this new foam will be less expensive and easier to fabricate than traditional foams. Therefore, it should lead to a significant reduction in the cost of carbon-based thermal management and structural materials (i.e. foam reinforced plastics and foam core composites).

2. EXPERIMENTAL

2.1 Foam Processing. The foam is produced via a proprietary method developed at Oak Ridge National Laboratory in the Carbon and Insulation Materials Technology Group. The process does not utilize a thermodynamic flash (blowing) agent to produce the foam and, most importantly, the unique method eliminates the requirement to stabilize the foamed pitch prior to carbonization (typically an oxidative stabilization step). The method is fairly versatile and can be easily adjusted to control pore/cell size and density. For this research, three mesophase pitches were used to produce graphitic foam: Mitsubishi ARA24 naphthalene-based synthetic pitch, and two propriatary mesophase pitches from Conoco Corporation labeled Conoco A and Conoco B. The properties of each pitch are listed in Table 1. All foam samples were graphitized at 5°C/min in Argon to 2800°C and soaked for 1 hour.

Samples of the graphitized ARA pitch-derived foam were vacuum impregnated with an epoxy resin [100 parts DER (Dow) 332 resin mixed with 46 parts of Jeffamine T-403 (Huntsman) hardener] to produce a foam-reinforced composite with a density of 1.26 g/cm³.

Some samples of the graphitized ARA pitch-derived foam were densified with carbon by vapor phase (methane) deposition to a final density of 1.3 g/cm^3.

Table 1. Properties of mesophase pitch used to produce graphitic foam.

Mesophase	Softening Point [°C]	Mesophase Content [%]	Carbon Yield @1000 °C, N$_2$
Mitsubishi ARA24	237	100	78
Conoco A	285	100	73
Conoco B	355	100	87

2.2 Foam Characterization. The physical characteristics of the foams, such as surface area and pore diameter, were measured on a Micromeritics Autopore II 9220 mercury porosimeter. In order to develop a fundamental understanding of the foam structure and graphitic morphology, several examination techniques were employed. Optical microscopy with cross-polarized light and a first-order red wavelength retarder were performed on a Nikon Microphot-FXA microscope. Samples were examined using a JOEL scanning electron microscope. X-ray diffraction data were taken with a Scintag PAD V powder diffractometer, using Cu Kα radiation.

Compression testing was performed to characterize the strength and stiffness of the foam and foam-reinforced composites (not all samples were fully characterized due to lack of material). Compression tests were performed on ½-in diameter x ½-in thick samples using ASTM standard C695-91 on an Instron test rig with a 1000 lb. load cell.

The thermal conductivity of the foam was measured with two different techniques. First, a xenon flash diffusivity technique was performed on samples ½-in. diameter by ½-in. thick on a custom built machine in the High Temperature Materials Laboratory at Oak Ridge National Laboratory. Also, thermal conductivity was measured on ½-in. x ½-in. by 1-mm thick samples with a thermal interface tester using the thermal gradient method at a major computer chip manufacturer.

3. RESULTS AND DISCUSSION

3.1 Processing All three mesophase pitches produced acceptable foams, although the two with low melting points produced foams with a density gradient. The Mitsubishi ARA 24 and the Conoco-A pitches produced foams with the lower 50% of the foam exhibiting a density of about 0.55 g/cm^3 and the upper 50% exhibiting a density of about 0.35 g/cm^3. Figure 1 is typical foam produced from Mitsubishi ARA 24 pitch prior to carbonization. The lower density material was weaker than the dense material. This became evident as the samples were physically handled. The low-density portion would crumble, while the dense portion could be handled without damage. Henceforth, only the dense portion of these foams will be discussed.

The Conoco B pitch produced foam with no density gradient. It is believed that this is because of the higher melting point of the pitch and a narrower temperature bandwidth at which decomposition gases evolve. A more viscous fluid combined with a shorter foaming period prevents settling and coalescing of the bubbles, virtually eliminating the gradient.

Figure 1.   Foam produced from ARA24 mesophase pitch.

3.2 Characterization Table 2 presents the physical characteristics of the foams after graphitization. The foams produced from the two pitches with low melting points had similar densities, but significantly different mean pore diameters and specific surface areas. On the other hand, the foam produced from the high melting point mesophase pitch (Conoco B) had both a lower density and mean pore diameter with a significantly higher specific surface area. Figure 2 is a SEM micrograph of the Mitsubishi foam illustrating that the foam exhibits a predominately open cell structure.   The struts of the foam appear completely different from vitreous glassy carbon foams produced commercially; the struts being significantly thicker and exhibiting a spherical structure. The layering effect visible in the struts shows that the material is graphitic

Figures 3 (a) and (b) are optical micrographs under cross-polarized light. It is clear that from the micrograph of the struts (Figure 3(a)), that the graphene layer planes are highly oriented parallel to the surface of the bubbles. As the foam is produced, the shear stresses from the expansion of the bubbles cause the liquid mesophase crystals to align parallel to the surface of the bubbles. During carbonization and graphitization, the resultant aligned carbon forms the highly ordered graphitic structures. The lack of an oxidative stabilization allows the crystals to form very large graphitic crystals. Figure 3(b) is a high magnification picture of the junctions of the struts, illustrating that the region is highly graphitic, but there is more fold-sharpening and crystal misalignment in these regions. These regions will serve to weaken the foam structurally and reduce the overall thermal conductivity.

Table 2. Properties of graphitic foams produced from different precursors.

Foam	Density [g/cm³]	Mean Pore Diameter [μm]	Specific Surface Area m²/g
Mitsubishi ARA24	0.54	93	4.0
Conoco A	0.52	114	2.0
Conoco B	0.46	73	7.2

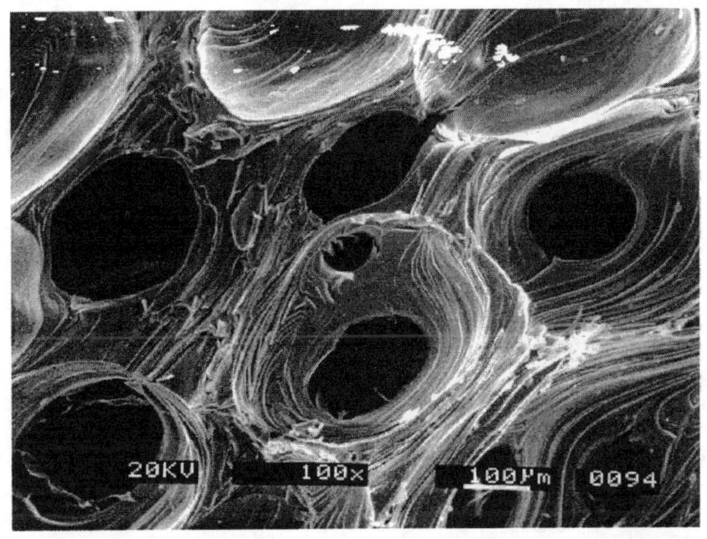

Figure 2.   Scanning electron micrograph of ARA pitch-derived foam graphitized at 2800°C.

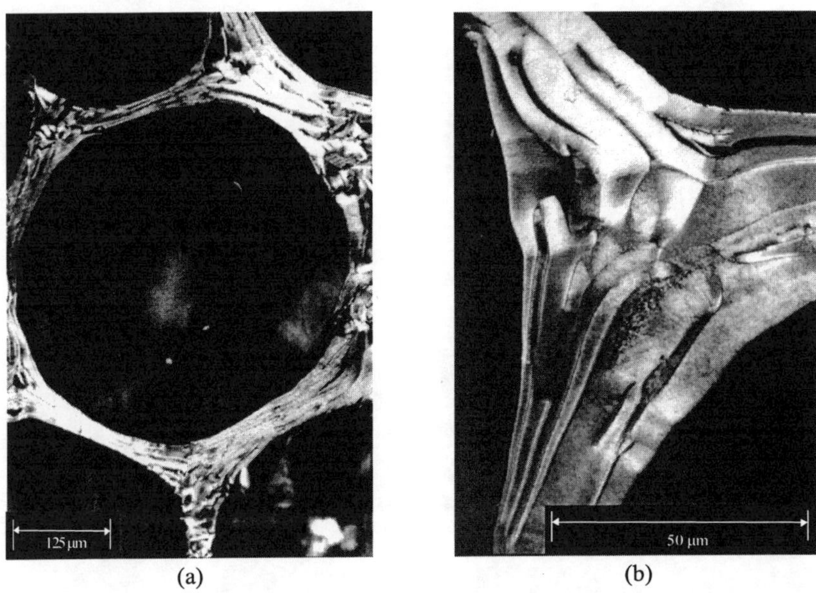

(a)   (b)

Figure 3.   Optical micrograph of ARA pitch-derived foam graphitized at 2800°C illustrating the struts (a) at 40x and the junctions between the struts (b) at 100X.

3.3 X-ray Analysis Lattice parameters were determined from the indexed diffraction peak positions. The X-ray method for crystallite size determination has been extensively reviewed elsewhere (14). The 002 and 100 diffraction peak breadths were analyzed using the Scherrer equation to determine the crystallite dimensions in the a- and c- directions.

$$t = \frac{0.9\lambda}{B\cos(2\theta)},$$

where t is the crystallite size, λ is the X ray wavelength, B is the breadth of the diffraction peak (full width half maximum (FWHM) minus the instrumental breadth), and 2θ is the diffraction angle.

As shown in Figure 4, the 002 peak (which is characteristic of interlayer spacing), was very narrow and asymmetric, indicative of highly ordered graphite. The interlayer spacing calculated with the Scherrer method was 0.3362 nm, significantly closer to pure graphite (0.3354 nm) than most high performance pitch derived carbon fibers (15). Table 3 is a comparison of heat treatment temperatures and X-Ray diffraction results of the graphite foam and various carbon fibers (15). The foam has the lowest d-spacing and the highest degree of graphitization. The crystallite size in the c-direction was calculated from these data to be 442 nm, and the 100 peak (or $10\overline{1}0$ in hexagonal nomenclature) was used to calculate the crystallite size in the a-direction of 203 nm. These crystallite sizes are larger than typical high thermal conductivity carbon fibers (15, 16), and therefore, the foam material should perform similarly to high order pyrolytic carbon and high thermal conductivity carbon fibers such as K1100 and vapor grown carbon fibers (VGCF).

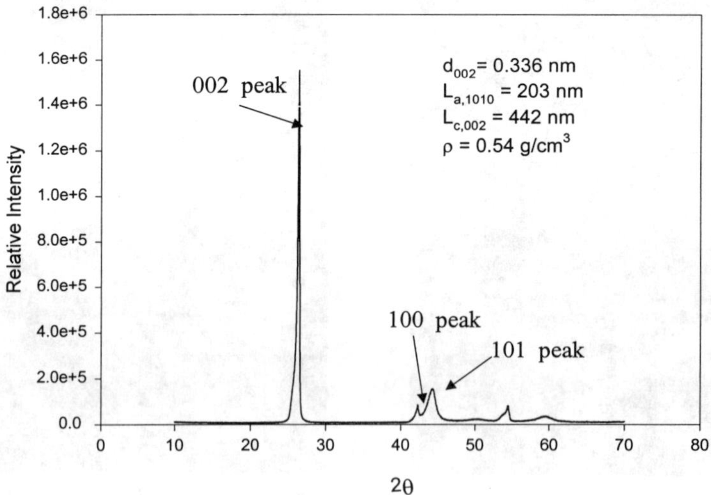

Figure 4.   X-ray analysis of graphtized foam (2800°C).

Table 3. Comparison of X-Ray diffraction results and the degree of graphitization of various carbon fibers and the graphite foam.

Material	Heat Treatment	d-spacing	g_p*
	[°C]	[nm]	[%]
PAN fiber [15]	2500	.342	23
P-120 (pitch) [15]	--	.3392	56
Floating catalyst VGCF[15]	--	.3385	64
Fixed catalyst VGCF[15]	--	.3449	--
Fixed catalyst VGCF[15]	2200	.342	23
Fixed catalyst VGCF[15]	2800	.3366	86
Graphitized Foam	**2800**	**.3362**	**91**

*g_p= degree of graphitization and is defined as (0.3440-d-spacing)/(0.3440-0.3354) where 0.3440 and 0.3354 are the d-spacing of turbostratic graphite and single crystal (perfect) graphite (15).

3.4 Mechanical Properties The mechanical properties of the foam and foam-based composites are presented in Table 4. The compressive strength of the foam was rather low (3.4 MPa) compared to carbon fibers. However, it compares well with aluminum and Kevlar® honeycombs. When the samples were impregnated with the epoxy resin the compressive strength increased by an order of magnitude to 34.3 MPa, and the flexural strength, 19.5 MPa, approached that of commercial thermal management panels. Although similar compressive strengths, 31.6 MPa, and flexural strengths, 19.4 MPa, were achieved when the foam was densified with CVD carbon, the mode of failure was different, as shown in Figure 4. While both the raw graphitic foam and the resin filled foam exhibited a high work of fracture, the CVD/foam material had a more brittle failure.

Table 4. Mechanical properties of foam and other thermal management panels.

Material	Specific Gravity	Flexural		Compressive	
		Strength	Modulus	Strength	Modulus
		MPa	GPa	MPa	GPa
ARA Foam	*0.54*			*3.4*	*.180*
ARA Foam /Epoxy	*1.26*	*19.5*	*--*	*34.3*	*.560*
ARA Foam /Carbon CVI	*1.3*	*19.4*	*2.3*	*31.6*	*.850*
EWC – 300X[17] K1100 (4K PW)/ERL 1939-3 resin	1.72	29.5[†]	13.1[†]	18.5	--
Aluminum Honeycomb[5] (CRIII 5052) ⅛ -in. cell size, 1 mil wall	0.07	--	--	3.7	1.030
Aluminum Honeycomb[5] (CRIII 5052) ⅛ -in. cell size, 3 mil wall	0.19	--	--	18.6	1.030
Kevlar Honeycomb[5] (HRH®-49) ¼" cell size	0.03	--	--	0.90	.172
Aluminum Foam[18]	0.5	--	--	~1.0	~1.0

[†]Tensile properties.

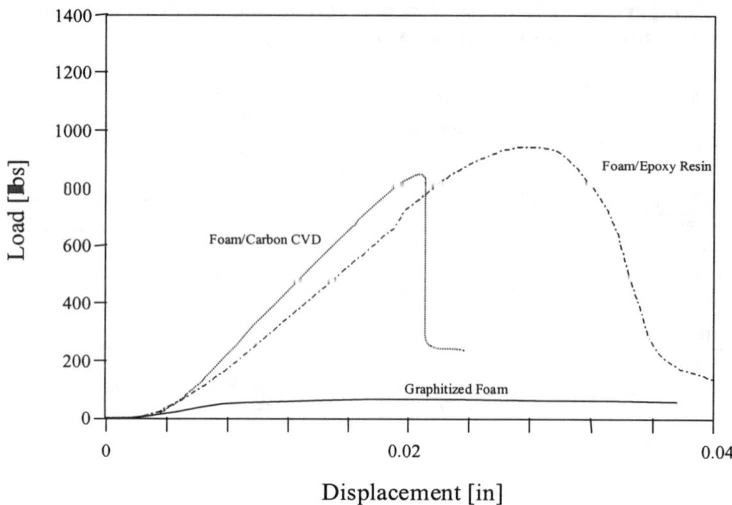

Figure 4.   Compression tests of ARA 24 Pitch-derived carbon foam and foam derived composites.

3.5 Thermal Diffusivity The validity of the flash diffusivity method and whether the open porosity would permit penetration of the heat pulse into the sample had to be established. Deep penetration of the pulse in samples typically causes a change in the characteristic heat pulse on the back face of the sample. Thus, errors in the reported diffusivity can be as high as 20% (19). However, the rather large struts and small openings of the foam limits the depth of penetration to about two pore diameters (140-220 μm), or less than 2% penetration. Therefore, it was believed that this technique would yield a fairly accurate value for the thermal conductivity. This was confirmed by testing samples with both the flash diffusivity method (20) and the thermal gradient method (21). The measured conductivities varied by less than 5%, verifying the flash method as a viable method to measure these foams. If the pore structure changes significantly, the flash method will likely yield inaccurate results.

The thermal diffusivity of the foam was very high as shown in Table 5. The thermal conductivity of the graphitized ARA 24 foam was as high as 106 W/m·K. This is remarkable for a material with such a low density, 0.54 g/cm^3. The foam exhibits thermal conductivies comparable to the in-plane thermal conductivity of some other thermal management materials and significantly higher than in the out-of-plane directions of the other thermal management materials. Although several of the other thermal management materials have higher in-plane thermal conductivities, their densities are much greater than the foam, i.e., the specific thermal conductivity of the foam is significantly greater than all the available thermal management panels. In fact, the specific thermal conductivity is more than four times greater than copper, the preferred material for heat sinks.

It is clear that for thermal management, where weight is a concern or where un-steady state conditions occur often, the graphitic foam is superior to most other available materials. The advantage of isotropic thermal and mechanical properties should allow for novel designs that are more flexible and more efficient.

Table 5. Thermal properties compared to composite thermal management panels

	Specific Gravity	Thermal Conductivity		Specific Thermal Conductivity*	
		In-plane	Out-of-plane	In-plane	Out-of-plane
		[W/m·K]	[W/m·K]	[W/m·K]	[W/m·K]
ARA Foam	*0.54*	*106*	*106*	*198*	*198*
Other Thermal Management Materials					
EWC – 300X K1100 (4K PW)/Cyanate Matrix [17]	1.72	29.5	13.1	17	7
Copper [17]	8.9	400	400	45	45
Copper(10%)-Tungsten [17]	6.5	180	180	27	27
Copper(70%)- EWC-300X (K1100) [17]	6.5	240	100	36	15
K321 (3K phenolic densified) [22]	1.6	54.8	8.3	34	5
K321 (2K AR pitch densified) [22]	1.88	233.5	20.4	123	11
Aluminum Honeycomb [5] (CRIII 5052) $\frac{1}{8}$ -in. cell size, 1 mil wall	0.07	--	~5	--	71
Aluminum Honeycomb [5] (CRIII 5052) $\frac{1}{8}$ -in. cell size, 3 mil wall	0.19	--	~10	--	142
Aluminum Foam [18]	~0.5	12	12	24	24

*defined as thermal conductivity divided by specific gravity

3.6 Applications

3.6.1 Internal Combustion Engines A piston for internal combustion engines made from a foam-reinforced aluminum (rather than the more conventional aluminum alloy) will have higher creep resistance and lighter weight, thus improving efficiency and reducing emissions. Also, foam-reinforced plastics could be utilized as a piston or engine block material. Although some plastics can withstand the temperature of an internal combustion engine (not generally higher than 300°C), the low thermal conductivity of the plastic prevents heat removal during the cycles. Therefore, the system overheats and the plastic melts or decomposes. However, a carbon foam-reinforced plastic will have a thermal conductivity similar to aluminum pistons (which typically don't experience temperatures higher than 300°C), and therefore should remove heat at a similar rate to aluminum pistons. A cyanate ester/foam piston will save as much as 40% the weight of the pistons, reducing the slung weight of the engine. This will increase power output and improve efficiency. It is even conceivable that the entire engine block could be made from a foam-reinforced polymer.

3.6.2 Heat Exchangers The combination of open porosity and large specific surface achieved with the foam allows for the improvement of heat exchangers. If the shell side of the exchanger is filled with the high thermal conductivity foam, there will be effectively several orders of magnitude increase in surface area to transfer heat to the working fluid (preferably a gas due to the pressure drop). Heat will be rapidly transferred from (or to) the shell side fluid through the foam, and then to the heat exchanger tubing and the tube side fluid. Such an increase in surface area will allow for a reduction in size of the heat exchanger, offsetting the increase in pressure drop through the foam. A reduction in size of heat exchanger will reduce weight and improve efficiency in many applications, such as automobiles and aircraft.

4. CONCLUSIONS

The manufacture and properties of a high thermal conductivity foam have been reported here. The existence of very sharp 002 and 100 peaks confirms that the graphite crystals are very large and are highly graphitic (nearly 91%).   Under cross-polarized light, very large monochromatic regions in the struts of the foam are visible, suggesting that these struts will behave like high thermal conductivity carbon fibers, such as K1100 and VGCF. In fact, the d-spacing and crystallite sizes were better than VGCF, which have a thermal conductivity as high as 1950 W/m·K. These properties, combined with the continuous graphite network throughout the foam, result in an isotropic thermal conductivity greater than 100 W/m·K and a specific conductivity over 4 times that of copper, an industry standard for thermal management. While the mechanical properties were similar to honeycomb structures, the isotropic thermal conductivity of a foam-core composite will provide far superior thermal management characteristics. This should lead to more efficient thermal management materials.

The densified foam test data indicate that the foam can be utilized as a replacement for carbon fiber in some applications, thereby reducing costs. This should allow carbon reinforced plastics, carbon, ceramics, and metals to enter markets not usually considered.

Although the data and discussion presented in this paper illustrate the potential of this material to be an enabling technology for many applications, significantly more work is needed. The ability to add chopped fibers and particulates to control pore size, thermal conductivity, and mechanical properties must be evaluated. A full characterization of the kinetics of the foaming reaction must be undertaken in order to allow optimization of the process. The effects of different pitches on the foam characteristics need to be studied. The ability to produce a gradient in the material might be of interest in some applications. Finally, the effects of pore size on the mechanical and thermal properties should be evaluated.

5. REFERENCES

1. D. D. Edie, Carbon Fibers, Filaments, and Composites, Edited by J. L. Figueiredo, et. al., Kluwer Academic Publishers, pp. 43-72 (1990).
2. W. Hagar and Max L. Lake, Mat. Res. Soc. Symp., 270, 29-34 (1992).
3. Hagar, W. and Max L. Lake, Mat. Res. Soc. Symp., 270, 41-46 (1992).
4. Shih, Wei, "Development of Carbon-Carbon Composites for Electronic Thermal Management Applications," IDA Workshop, May 3–5, 1994, supported by AF Wright Laboratory under Contract Number F33615-93-C-2363 and AR Phillips Laboratory under Contract Number F29601-93-C-0165.
5. Hexcel Product Data Sheet (1997).
6. Gibson, L. J., Mat. Sci. and Eng., A110, 1-36 (1989).
7. Hagar, W. and Max L. Lake, Mat. Res. Soc. Symp., 270, 35-40 (1992).
8. White, J. L., and P.M. Sheaffer, Carbon, 27 (5), 697-707 (1989).
9. Gibson, L.J., and M. F. Ashby, Cellular Solids: Structures & Properties, Pergamon Press, New York (1988).
10. Knippenberg, W. F., and B. Lersmacher, Phillips Tech. Rev., 36 (4), 93-103 (1976).

11. Aubert, J. H., <u>Mat. Res. Soc. Symp.</u>, <u>207</u>, 117-127 (1990).

12. Cowlard, F. C., and J. C. Lewis, <u>J. of Mat. Sci.</u>, 2, 507-512 (1967).

13. Kearns, Kris, "Graphitic Carbon Foam Processing," 21st Annual Conference on Ceramic, Metal, and Carbon Composites, Materials, and Structures, Janyary 26-31, Cocoa Beach, Florida, pp. 835-847 (1997).

14. H.P. Klug and L.E. Alexander <u>X-Ray Diffraction Procedures for Polycrystalline and Amorphous Materials</u>, 2nd edn., J. Wiley, New York, USA, (1974).

15. Lake, M., *Mat. Tech.*, <u>11</u> (4), 131-144 (1996).

16. Jones, S. P., C.C. Fain, and D. D. Edie, <u>Carbon</u>, <u>35</u> (10), 1533-1543 (1997).

17. Amoco Product Literature (1997).

18. Steiner, K., J. Banhard, J. Baumister, and M. Weber, "Production and Properties of Ultra-Lightweight Aluminum Foams for Industrial Applications," Proceeding from the 4th International Conference on Composites Engineering, Edited by David Hui, July 6-12, pp. 943-944 (1997).

19. Inoue, K., <u>High Temp. Tech.</u>, <u>8</u> (1), 21-26 (1990).

20. Cowan, R. D., <u>J. of App. Phys.</u>, <u>34</u> (4), 926-927 (1962).

21. Kobayashi, K., <u>J. Japan Welding Soc.</u>, <u>561</u>, 38-42 (1987).

22. Olhlorst, C. W., W. L. Vaughn, P. O. Ransone, and H-T Tsou, "Thermal Conductivity Database of Various Structural Carbon-Carbon Composite Materials," NASA Technical Memorandum 4787, November 1997.

ACKNOWLEDGEMENTS

The author wishes to thank Ernie Romine of Conoco corporation for supplying the proprietary pitches for this research, Mike Rodgers of ORNL for performing mercury porosimetry, Barbra Bennet of WVU and Andy Payzant of ORNL for performing the X-Ray analysis of the foams, Marie Williams of ORNL for preparing optical microscopy samples, Mike Wood of AlliedSignal for performing carbon CVI densification of the foam, and Lynn Klett of ORNL for filling the foams with epoxy and thermoplastic resins.

Research sponsored by the Laboratory Directed Research and Development Program of Oak Ridge National Laboratory, managed by Lockheed Martin Energy Research Corp. for the U. S. Department of Energy under Contract No. DE-AC05-96OR22464.

MICROCELLULAR PITCH-BASED CARBON FOAMS
BLOWN WITH HELIUM GAS

Heather J. Anderson[1], David P. Anderson[2], and Kristen M. Kearns[1]

[1]Materials and Manufacturing Directorate, US Air Force, 2941 P St. Ste. 1,
Wright Patterson Air Force Base, Ohio 45433-7750

[2]University of Dayton Research Institute, 300 College Park Avenue,
Dayton, Ohio 45469-0168

ABSTRACT

Microcellular graphite foams were produced with graphitic alignment along the struts from anisotropic pitch. The foams were created by first pressurizing the pitch with helium gas. The pitch temperature was raised above its softening point. At temperature, more helium was added to the system and held for a length of time. The system was finally vented to the atmosphere, generating the porous material (foam). The pores created in the foam, form a three dimensional, interconnected network of struts. The graphite planes are aligned along the struts by the expansion of the cells in the foam, similar to the spinning mechanism for fibers. The foams were carbonized in a similar manner as pitch based fibers. Open-celled graphitic foams possess excellent specific mechanical properties. Current work including foam morphology, and microscopy will be discussed.

KEY WORDS: Carbon foam, Microcellular structural foam, and Foam processing.

1. INTRODUCTION

Structural composites consist of individual fibers bound by a matrix. The matrix transfers the load to the aligned fibers and contributes some impact strength. The fibers are melt spun from a pitch precursor, which give the fibers strength and stiffness, translating those properties along the fiber direction in the composite. If a microcellular foam were created, containing a network of interconnected struts that simulate graphite fibers, a truly unique material would be at hand. A composite made with the foam would have properties similar to that of a carbon fiber composite material with the advantages of an interconnected reinforcement. Foams with a density of approximately 0.1 Mg/m^3 have been predicted to have a modulus as high as 2000 MPa [1].

Carbon foams have been blown with a variety of other gases; nitrogen [2] and carbon dioxide [3], producing different structures and properties. Helium gas was examined here to determine the advantages it gave over other blowing gases. Comparison of foams blown with the various gases used in our laboratory has recently been completed [4]. Microcellular, open-cell foams are produced from an anisotropic pitch with graphitic planes aligned along the struts. These blown foams undergo the same processing steps as do pitch fibers, which include stabilization, carbonization and graphitization.

2. EXPERIMENTAL

2.1 Materials The foam precursor is an anisotropic pitch, produced by Mitsubishi Gas Chemical Co., and is denoted by the trade name AR Resin. A catalytic polymerization of naphthalene is used to create a 100% anisotropic pitch with a softening point of 239°C, which is supplied in pellet form. The glass transition temperature of AR Resin resides in the range of 230°-260°C.

2.2 Processing The pitch pellets were jet milled into 1-3 μm particles and then pressed in a mold to create a disk. The pitch disks were then placed into a Parr® pressure vessel, and purged with helium gas. The reactor was then pressurized with helium (heating pressure) and heated to the desired final temperature, referred to as the saturation temperature. Once the disks reached the saturation temperature, more gas was added to the system, thus increasing the pressure (saturation pressure). Once the system reached the desired saturation pressure, it was held for a predetermined length of time (saturation time). The gas was then rapidly vented to the atmosphere. The foamed disks were removed and placed in a forced air oven, where they were oxygen stabilized. Then they were carbonized in a nitrogen environment to 900°C at 1°C/min.

2.3 Design of Experiment A 2^4 factorial design of experiment matrix was employed to determine how the morphology is effected by each processing variable. Conditions for the test matrix were based upon earlier work completed using nitrogen gas to blow foams [2].

Table 1. Processing matrix for blowing helium foams.

Sample ID	Heating Pressure	Saturation Temperature	Saturation Pressure	Saturation Time
1	+	+	+	+
2	-	-	+	+
3	-	+	-	+
4	+	-	-	+
5	-	+	+	-
6	+	-	+	-
7	+	+	-	-
8	-	-	-	-

2.4 Microscopy Samples for each of the foams were vacuum infiltrated with a fluorescently-dyed epoxy potting resin and then polished by ordinary resinographic techniques. The fluorescent resin allows the distinction between the resin and carbon material. It also tells whether the foam structure contains open or closed cells. If the resin penetrates through the pores then the cellular structure is open. Conversely, if it does not penetrate then the structure is found to be closed. The polished plugs were examined with a Nikon Optiphot FXL microscope in bright-field, cross polarized light, and fluorescent illumination. Images were obtained using a CCD digital camera. Image-1 image processing software was used to measure the bubble sizes and porosity in the foams.

SEM's were obtained digitally using an R. J. Lee Instruments Personal Scanning Electron Microscope (PSEM) operating at 10 kV. The PSEM was used to evaluate the microstructure of the foam samples and to support the macroscopic observations.

2.5 Density Measurements   Weights and dimensions of the foam samples were taken between each processing step for physical evaluation and to give a non-destructive indication of the effects of each processing step. The volume expansion was calculated from the physical dimensions before and after foaming. Density measurements were made on stabilized pitch foam specimens using the Archimedes displacement method with water (ASTM C 373-72 for porous ceramics). This method gives apparent porosity, apparent bulk density, and material density.

3. RESULTS AND DISCUSSION

The processing conditions were altered during the blowing part of the process to determine what effect they had on the foam morphology. The heating pressure, saturation temperature, saturation pressure, and saturation times were held at two different levels according to the 2^4 factorial matrix. An example of the sensor outputs used to monitor the processing conditions is shown in **Figure 1**.

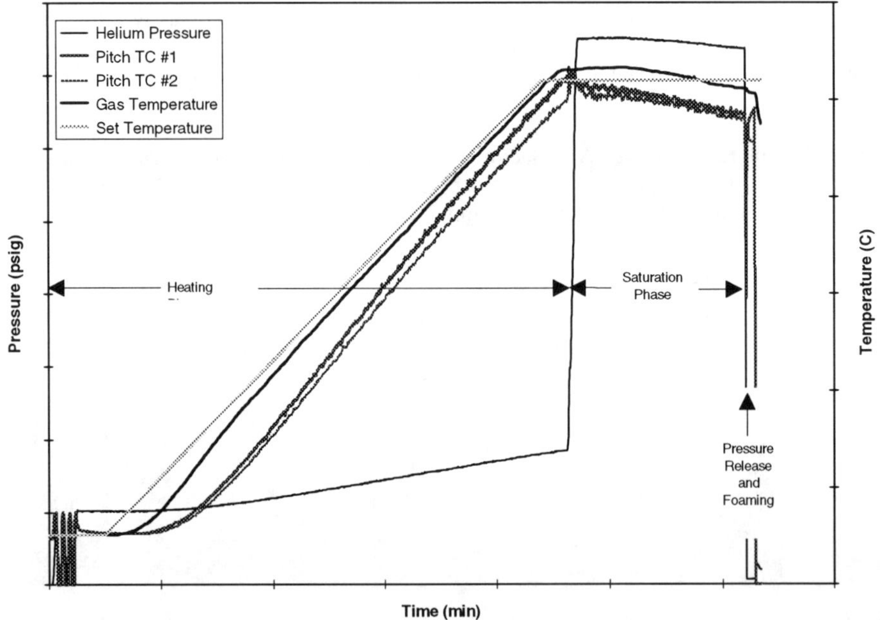

Figure 1. Example of foam processing sensor output.

Bloating occurred in all the samples during carbonization and thus were not processed any further, nor were mechanical tests conducted. However, on further examination the bloating was found to be confined to the unconstrained surface only. Because of the bloating more detailed microscopic examination of both stabilized and carbonized foams was conducted.

The microscopy and density measurements were used to determine the amount of open and closed cells. When the foam is filled with fluorescent resin this proves that it is indeed open celled. If it were closed celled, the resin would not have penetrated through the foam structure.

Figure 2 shows a stabilized foam with a distinct open cell microstructure; which is the desired structure. In this study all foams produced at the lower temperature were closed cell. The foam sample containing the smallest pores appeared to have cells that collapsed (**Figure 3**). In addition one of the higher temperature foam samples contained closed cells.

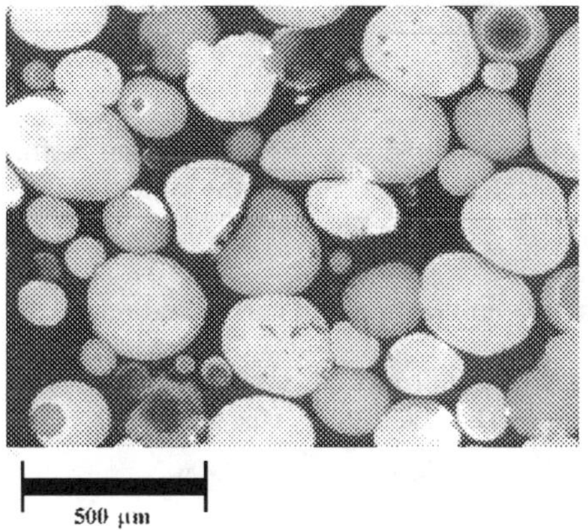

Figure 2. Fluorescent micrograph of stabilized foam showing open cellular structure.

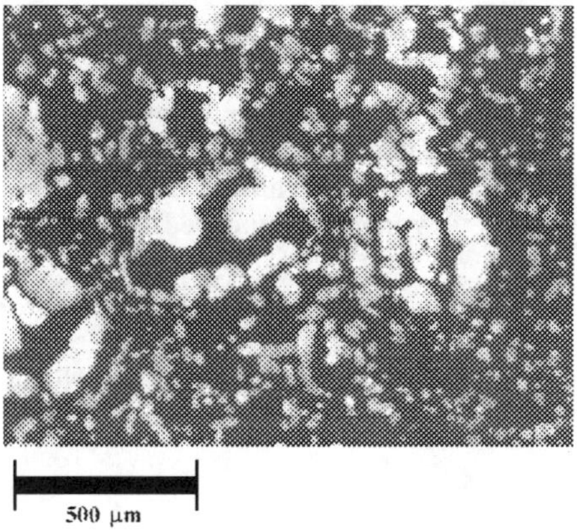

Figure 3. Fluorescent micrograph of stabilized foam showing collapsed cellular structure. Note foam was re-infiltrated with fluorescent resin to give contrast.

All the samples bloated during carbonization; a typical bloated structure is shown in **Figure 4**. Both the SEM micrographs and optical analysis of the blown foams showed that the microstructure contained both open and closed cells that varied in size and uniformity. After carbonization, many of the blown foams containing closed cells collapsed or coalesced.

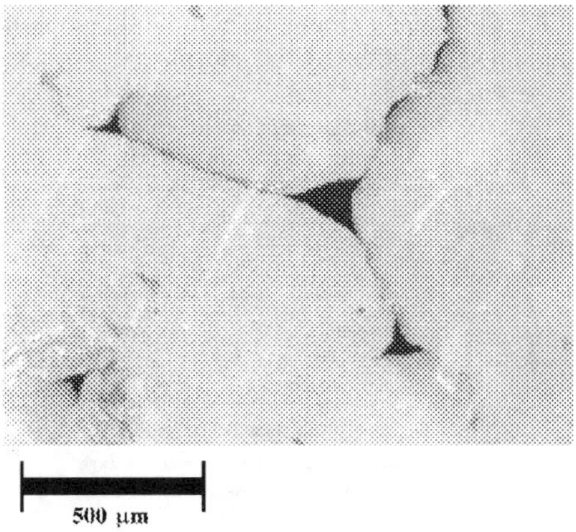

The cell sizes were measured from optical microscopy using the observed cell diameter method described in ASTM D 2856-70. The stabilized data is shown in **Figure 5**. The porosity data shown in **Figure 6** shows the same trends for each set of processing conditions. **Table 2** shows the average cell diameters change the most between stabilization and carbonization for samples which were blown with the lower saturation pressure. This highlights that the cells in some of the stabilized foams are coalescing, collapsing or bloating during carbonization.

Figure 4. Fluorescent micrograph of bloated foam sample (after carbonization) showing large cells.

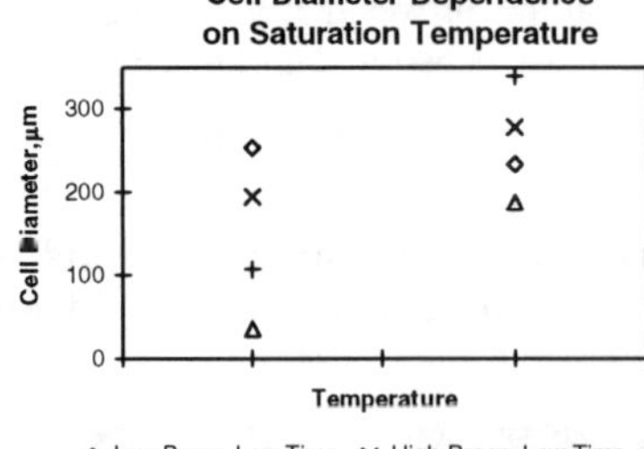

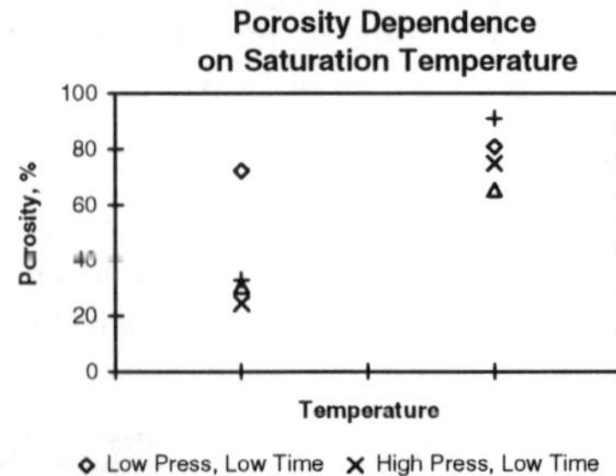

Figure 5. Average cell diameter calculated using optical microscopy and fluorescent resin. Heating pressure of the system is compared.

Figure 6. Porosity calculated using optical microscopy with fluorescent resin. Heating pressure of the system is compared.

Table 2. Foam average cell diameters after stabilization and carbonization.

Sample ID	Saturation Temperature	Saturation Temperature	Stabilized Cell Diameter (μm)	Carbonized Cell Diameter (μm)
1	+	+	440	480
2	-	-	35	39
3	+	+	188	434
4	-	-	107	217
5	+	+	234	257
6	-	-	194	165
7	+	+	278	201
8	-	-	254	*

*   Carbonized foam was completely bloated with cells too large to measure.

The processing conditions were correlated with volumetric expansion, density, cell diameter, and porosity. As previously reported in nitrogen blown foams [2] and carbon dioxide blown foams [3], the saturation temperature and heating pressure effected the morphology of the

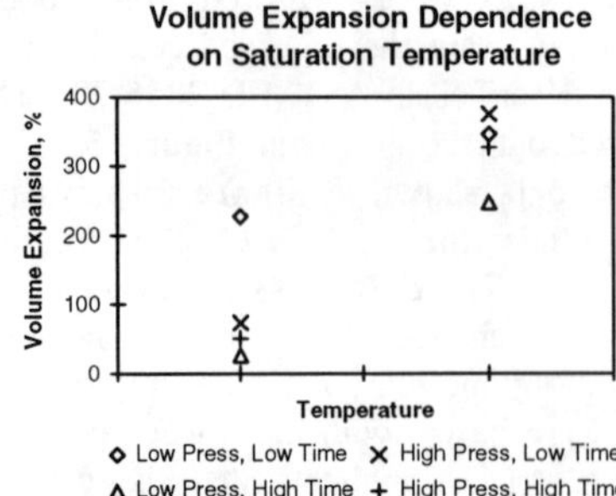

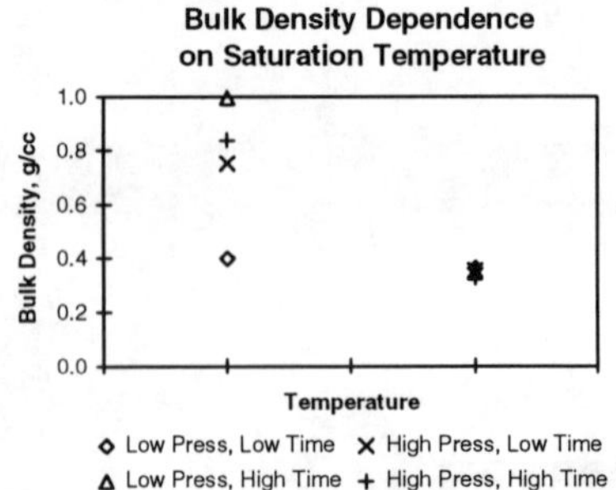

Figure 7. The volumetric expansion was measured after foam stabilization. Heating pressure of the system is compared.

Figure 8. Density calculated using the Archimedes method. Heating pressure of the system is compared.

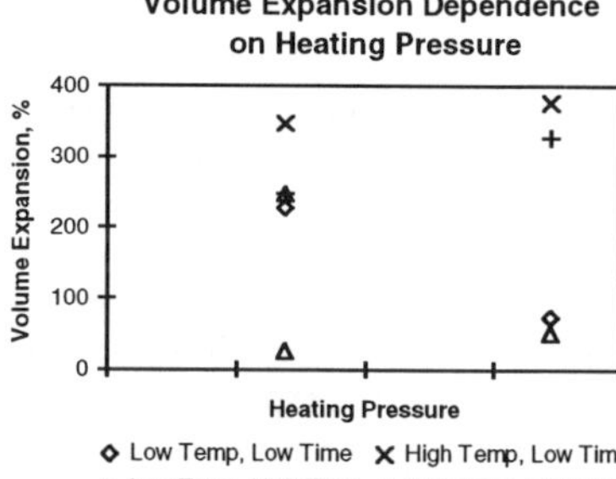

Figure 9. Typical plot of another processing condition showing no correlation.

foam. However, in the helium blown foams only the saturation temperature had an effect (**Figure 7 - Figure 9**). The foams processed at the higher saturation temperature increased in volume by 250-450% and those processed at the lower saturation temperature increased only 20-250% by volume as seen in **Figure 7**. Bulk density at the lower saturation temperature varied from 0.4-1.0 Mg/m^3 while, at the higher saturation temperature the densities all clustered near 0.4 Mg/m^3 (**Figure 8**). The density measurements give unusually low material density values when the foam is closed celled. As can be seen in **Figure 9**, the variation of volume expansion with heating pressure showed no correlation. This doesn't follow the results observed in the nitrogen and carbon dioxide blown foams.

Figure 10 is an SEM of a foam that was processed at the low saturation temperature, low heating and low saturation pressures, and short saturation time. The size of the cells were approximately 100 µm and were uniformly distributed. However they were primarily closed celled. The foam in **Figure 11** was blown at the high saturation temperature, high heating and high saturation pressures, and long saturation time. These foams contained open cells, which were not uniform throughout and were much larger.

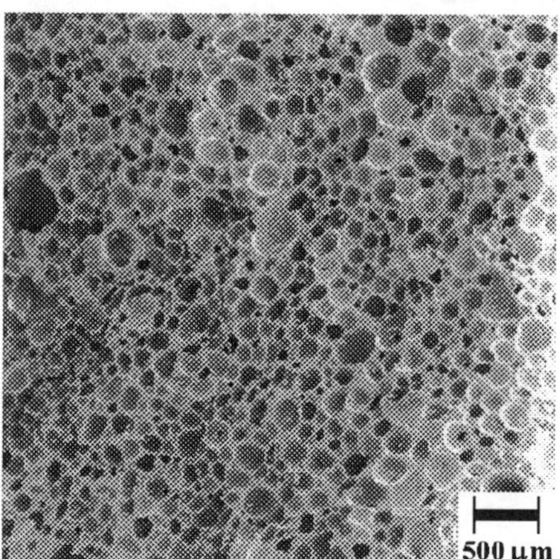

Figure 10. SEM of uniform celled foam as stabilized.

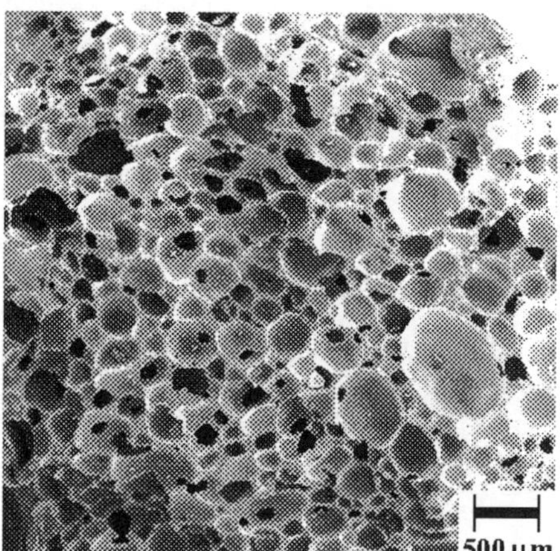

Figure 11. SEM of non uniform celled foam as stabilized.

4. CONCLUSIONS

Increasing the material temperature above its softening point, increases the gas mobility, allowing it to penetrate the material. A sudden release of pressure causes the gas to expand the pitch and form bubbles or pores. The growth of the bubbles depends on the pitch viscosity and most likely the nucleation density (not yet examined). The viscosity depends on the temperature of the material.

The foams blown at the lower saturation temperature contained smaller pores. This is a result of the higher viscosity, which limited the volume expansion of the material. These foam samples primarily contained closed cells after foaming. When they were carbonized, the cells had either coalesced or collapsed. The viscosity also limits the gas from blowing out the cell walls preventing an open cellular structure. At the higher temperature, the lower pitch viscosity allows the gas to expand at a much faster rate which enables the cells to grow more freely. This produces larger cells and giving the foam a higher volume expansion. The foams processed at the higher temperature also had lower densities.

The foams did not stabilize sufficiently to prevent bloating during the carbonization cycle, although the samples should have stabilized based on their times in the forced air oven. Most of the foams produced, contained a closed cell morphology which partially explain the bloating. The fact that all the samples bloated shows that the oxygen did not penetrate throughout the material sufficiently to stabilize it. Additional work needs to be done to examine why the pitch did not stabilize.

5. ACKNOWLEDGEMENTS

This work was partially supported by the U. S. Air Force under Contract number F33615-95-D-5029.

6. REFERENCES

1. R.B. Hall, and J.W. Hager, <u>21st Biennial Conference on Carbon Extended Abstracts</u>, 1993, pp. 100.

2. K. M. Kearns and D. P. Anderson, <u>Proceedings of the ICCM-11</u>, pp. 825 (1997).

3. H. J. Anderson, D. P. Anderson, and K. M. Kearns, <u>23rd Biennial Conference on Carbon Extended Abstracts</u>, <u>II</u>. pp. 4 (1997).

4. K.M. Kearns, "Structural Graphitic Carbon Foams", <u>Proceedings of MRS</u> Spring 1998 (in press).

THE CELLULAR STRUCTURE OF NET-SHAPED PITCH-BASED MICROCELLULAR CARBON-FOAMS

David P. Anderson[1], Kristen M. Kearns[2], Cindy Tucci[3], and Gerry Mestemaker[3]

1 - University of Dayton Research Institute, 300 College Park, Dayton, OH 45469-0168
2 - Materials and Manufacturing Directorate, US Air Force, 2941 P St., Ste. 1,
Wright-Patterson Air Force Base, Ohio 45433-7750 USA
3 - Brookville High School, 106 S. Hill St., Brookville, OH 45309-1499

ABSTRACT

Net-shaped microcellular carbon-foams were made from an anisotropic synthetic naphthalene-based pitch. The techniques used to produce net shapes included free expansion of a shaped preform, shaping a newly blown foam, and foaming into a mold. These foams were all blown using nitrogen gas under previously identified foaming conditions [1]. Quantification of bubble sizes, uniformity and porosity was performed using optical microscopy and image analysis. These foam structural parameters were measured as a function of location within the foams for each of the shaping methods. Porosity did not vary significantly among the processing variables or location. The bubble sizes and uniformity did vary in the foams, particularly with each foam having some large bubbles on its bottom surface. Density and porosity were also measured using water displacement with similar results. Free expansion of a shaped preform produced the most variability, and foaming into a mold produced the least variability and best net-shape.

KEY WORDS: Foams, Cellular Structure, Net-Shape Processing

1. INTRODUCTION

Graphitic foams have the potential for use as core replacements between composite sheets or used directly as a composite reinforcement. Model graphitic foams were predicted [2] to have a compression modulus of approximately 2 GPa with a density of about 0.1 Mg/m^3. Not many other foams or core materials have a density and compression modulus near these values. Structural composites are based on disconnected carbon fibers integrated into some type of matrix. The strength and stiffness of commercial carbon fibers are due to the graphitic morphology that originates from the melt spinning of the precursor pitch. If an interconnected network of struts could be produced which possessed a similar morphology to the carbon fiber, a new generation of composite reinforcement could emerge. Foams are an example of such materials. An open cell structure is the desired architecture for resin infiltration as a composite preform and drainage as a core material.

Microcellular, open-cell foams can be produced from anisotropic pitch with graphitic planes aligned along the struts [1]. The process sequence includes blowing, stabilizing, carbonizing, and then graphitizing the foam, similar to the process for manufacturing pitch-based carbon fibers. This study was a successful initial attempt to demonstrate that such a foam could be blown into a net-shape as a composite preform or core replacement material.

The original process for fabricating microcellular foams was developed by Colton and Suh [3-4]. They demonstrated microcellular foams could be produced from amorphous polymers by saturating the polymer with a gas and then heating above the glass transition temperature. The sudden release of pressure resulted in a foam of uniformly distributed pores. Dutta, et al. [5] modified this process by saturating with nitrogen under pressure to get better solubility of nitrogen into the pitch. In this process the solubility of nitrogen into the pitch is critical to foam fabrication. Therefore, they first melted the pitch in a steady flow of nitrogen. The pitch was then saturated with nitrogen gas and held for 10 minutes before dropping the pressure and temperature suddenly. The current method of fabricating microcellular foams is a variation of both the Suh and Dutta work. It includes a slight initial pressurization before the pitch is heated. Thus, the pitch is under a slight pressure during heating and is further pressurized at the final temperature.

The preceding work [1] used cylindrical disk foam preforms that were allowed to expand freely. This work was undertaken to determine what processing changes were needed to produce a net shape in the final foams and the effects of those processes on the foam structure.

2. EXPERIMENTAL

2.1 Materials: Foams were processed from AR pitch manufactured by Mitsubishi Gas Chemical Co. The pitch is produced by the catalytic polymerization of naphthalene and supplied in a pellet form. Manufacturing data claims the softening temperature is 239°C and the material is 100 percent anisotropic. The glass transition temperature occurs over a temperature range of 230 to 260°C.

2.2 Processing: The pitch pellets were jet milled into particles of an average size of 1-3 μm. These particles were then pressed into a preform which was placed in a Parr® pressure reactor and heated (under a slight nitrogen pressure) at 3°C/min to the desired final temperature. When the preform reached the final temperature (slightly above its softening point), additional nitrogen was added to obtain the final desired pressure 6.9 MPa (1000 psi), and the entire system was held at these conditions for 15 minutes. Foam results when the pressure is abruptly vented to the atmosphere.

Further processing was required to transform this from a pitch to a carbonaceous foam. After blowing, the foam was quickly removed from the reactor and placed in a 150 to 175°C oven. The foam was then cooled to room temperature where weight and dimensional measurements were taken. The foams were then oxygen stabilized in a forced-air oven until a weight gain of approximately 7 percent had been reached. Samples were then carbonized with a 60°C/hour heating rate to 850°C in a nitrogen environment.

2.3 Net Shape Formation: Three methods were attempted to produce a foam of net shape. The methods are described below and in Table 1. The first method, called "extruded," used a discotic preform that was blown into a foam, and the still hot soft foam was pressed (or extruded) into a mold of rectangular shape. The second method consisted of cutting the

Table 1.   Net-shaped processing: samples and description.

Sample number	Production method	Production description
I	Extruded	Shape formed by pressing mold over still soft foams made with disk preforms
II	Molded	Shape formed by foaming into a mold of the desired shape
III	Free Expansion	Shape cut in preform which is then allowed to expand freely
IV	Free Expansion	Shape cut in preform which is then allowed to expand freely

preform into a rectangular cross section that just fit into a metal mold of the same shape where it was blown into a foam: this method is called "molding." The third, or "free expansion," method started with preforms cut into a rectangular shape; these preforms were allowed to expand freely during the blowing operation.

2.4 Microscopy: The foams were cut into multiple sections to be examined primarily by optical microscopy but also by scanning electron microscopy (SEM). The sections of the samples for optical microscopy were cut according to Figure 1 to allow observation of several different locations and specimen orientations for each of the foamed samples. The samples were vacuum infiltrated with a fluorescently-dyed epoxy potting resin and then polished by ordinary resinographic techniques. The polished plugs were examined with a Nikon Optiphot FXL microscope in both bright-field and fluorescent illumination. Images were obtained using a CCD digital camera. Image-1 image processing software was used to measure the bubble sizes and porosity in the foams. SEMs were obtained digitally using a Leica 360 field emission gun scanning electron microscope operating at 15 kV.

2.5 Density Measurements: Density measurements were made on carbon foam specimens using the Archimedes displacement method with water (ASTM C 373-72 for porous ceramics). Samples were cut from each processing specimen to obtain measurements from top, bottom, inside, and outside. The very outer skin of each sample was removed before measurement.

3. RESULTS AND DISCUSSION

The four samples listed in Table 1 were photographed after foaming and carbonization (see Figure 2). It is immediately obvious that the two freely-expanded samples did not retain their rectangular cross sections but instead expanded into circular cross-section pieces. The foam blocks were classified as "rectangular" for the first two processing methods and "circular" for the freely-expanded processing method. The four samples were examined by optical microscopy as functions of location both from the foam center to outside edge and top to bottom. The viewing orientation was also varied for each of the samples. The orientations of observation were arbitrarily set as "B" to reveal surfaces normal to the foam's long in-plane direction or surfaces normal to the cylinder radial direction; "C" to reveal surfaces normal to the foam's short in-plane direction or surfaces normal to the cylinder hoop direction; and "D" to reveal surfaces normal to the through-the-thickness direction.

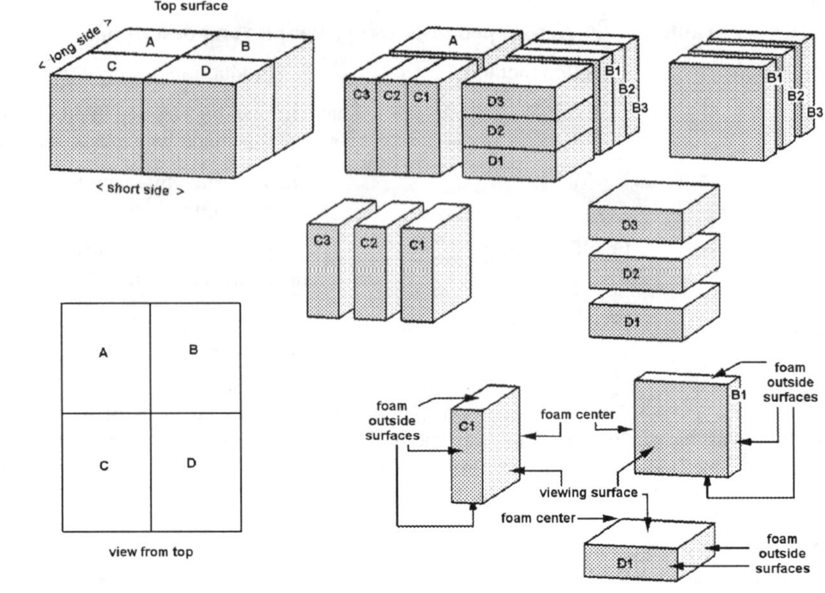

Figure 1. Cutting diagram for the foam samples: a) for rectangular blocks and b) for the cylindrical pieces.

766

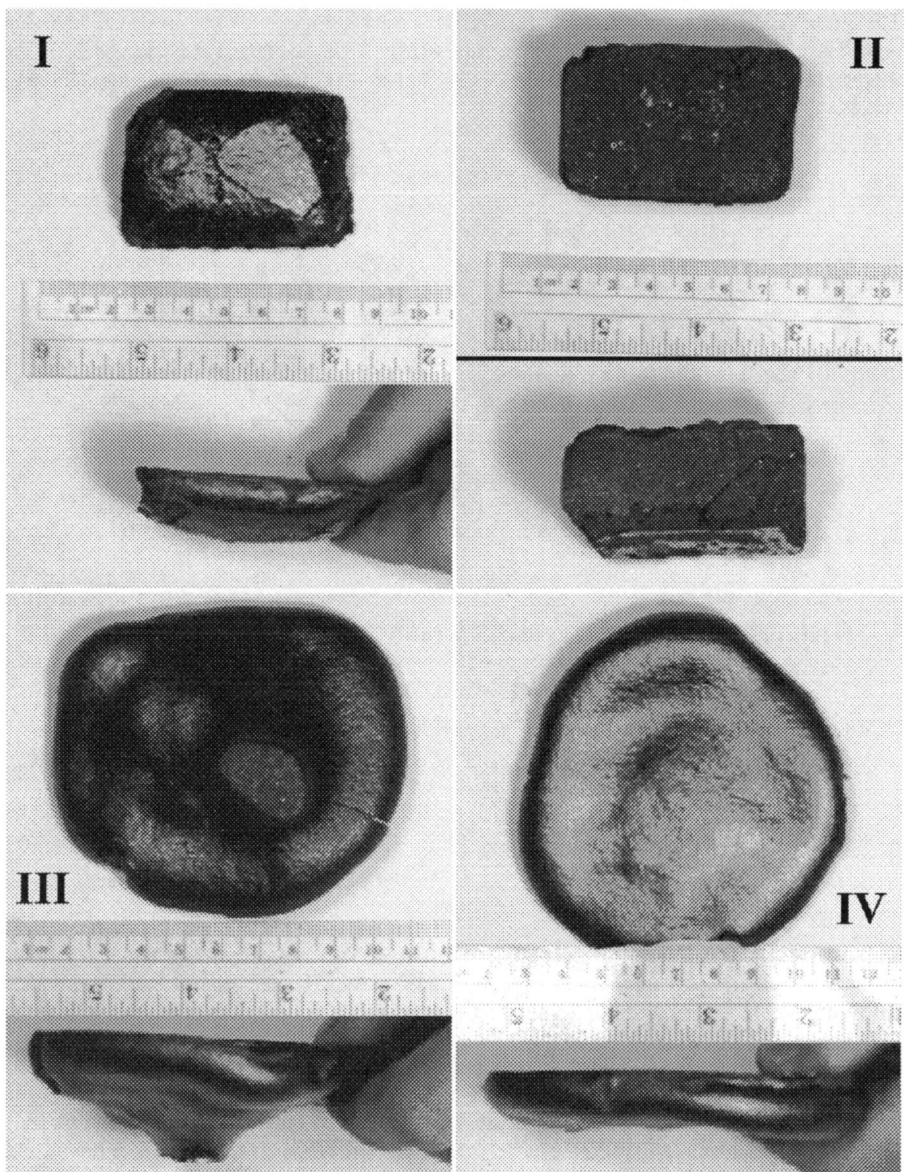

Figure 2.   Photographs of "net-shaped" foams.

Image analysis of the fluorescent images to measure porosity was straightforward; the open-cell bubbles fill with the fluorescent resin and appear yellow against the black carbon struts. There did not appear to be a significant variation in porosity as a function of processing path, position in the foam samples, or orientation of observation. The foams had average porosity values of 81 to 82%. Deviations from the average in any particular view were due more to potting resin infiltration differences than from sample variations. The Archimedes displacement measurements gave porosity values of 87-89%.

The Archimedes "apparent porosity" results are shown in Figure 3. The average "apparent bulk densities" were even less variable with location. The average values for each of the samples are given in Table 2 along with the "apparent porosity" and "apparent density" values.

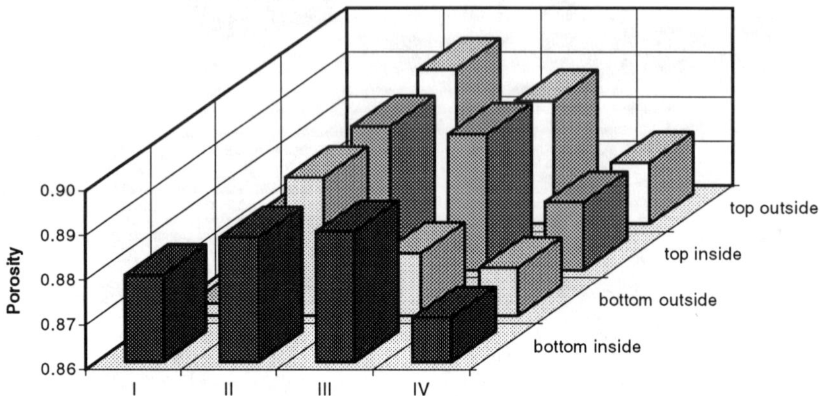

Figure 3.   **Archimedes "apparent porosity" results by sample and location.**

Table 2.   **Archimedes displacement average results.**

Sample number	apparent porosity	apparent bulk density (Mg/m^3)	apparent density (Mg/m^3)
I	0.871	0.241	1.871
II	0.892	0.206	1.911
III	0.885	0.220	1.916
IV	0.872	0.242	1.891

Bubble uniformity can be determined from the bubble size measurements, but can also be evaluated qualitatively from a simple examination of an SEM photo. Figure 4 shows a typical SEM of Foam II. All of the foam samples showed similar uniformity of cell sizes.

Direct measurement of the bubble sizes is not as easy. The usual automated measurements and ASTM D 2856-70 are based on the assumption that each bubble can be separated from its neighbor by a cell wall. Then by measuring the observed bubble diameter or intersection lengths the true average diameter can be calculated. In the open-celled carbon foams, the bubble walls have collapsed (at least partially), making it impossible for the computer to separate cells. The human eye can distinguish the individual bubbles using subtle clues which current PC software is incapable of recognizing.

Measuring the bubble diameters on images is tedious and time consuming, but a good estimate of sizes can be obtained by measuring the number of bubbles in a fixed area. Square images (selected from a larger one from the regions of interest) all the same size were used. The bubbles completely enclosed by the square and those partial bubbles crossing the boundary in only one direction were counted. That is to say one counts the partial bubbles crossing the left edge but not the right edges (and top but not bottom) so that the area covered by the counted bubbles is consistent. By knowing the area of the square, the number

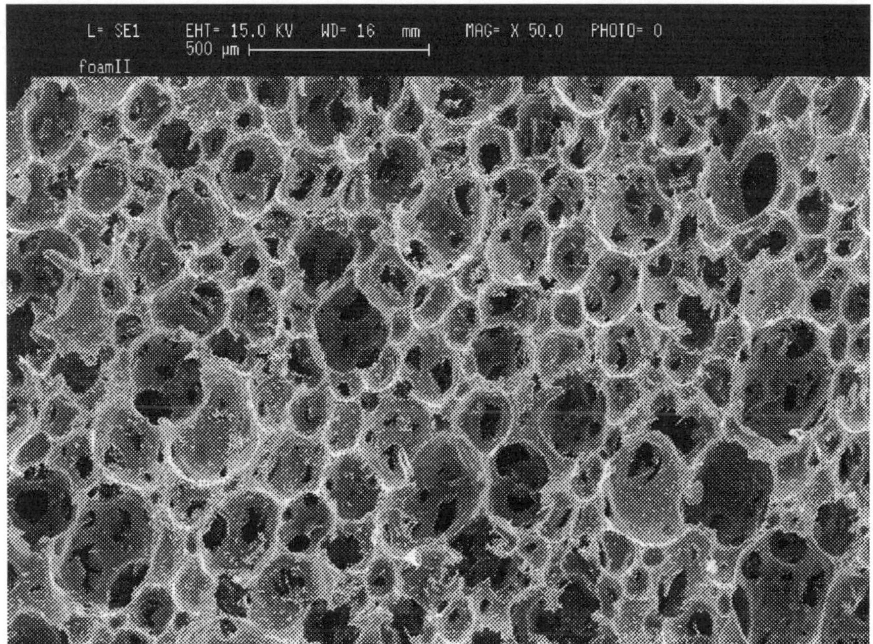

Figure 4.   SEM of Foam II: typical uniformity of all the foam samples.

of bubbles enclosed in that area, and simple geometry, an average diameter can be calculated. This method was used to measure the average cell sizes of the foams.

Figure 5 shows the cell (or bubble) sizes as a function of viewing direction in the center of each specimen. Remember the B samples were cut to view along the long or radial directions

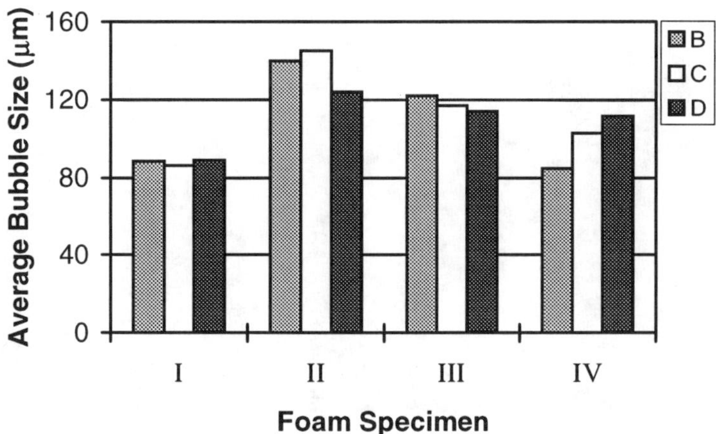

Figure 5.   Average bubble sizes in the center of the foam specimens as a function of viewing direction. B is along the long or radial direction, C is along the short or hoop direction, and D is along the z-axis or through-the-thickness direction.

of the foams, C were cut to view along the short or hoop directions, and D were cut to view along the z-axis or through-the-thickness direction. Basically no major differences were noted within samples which means that the bubbles are isotropic in this part of the foams. The difference in appearance of the largest to smallest bubble sizes is shown in Figure 6.

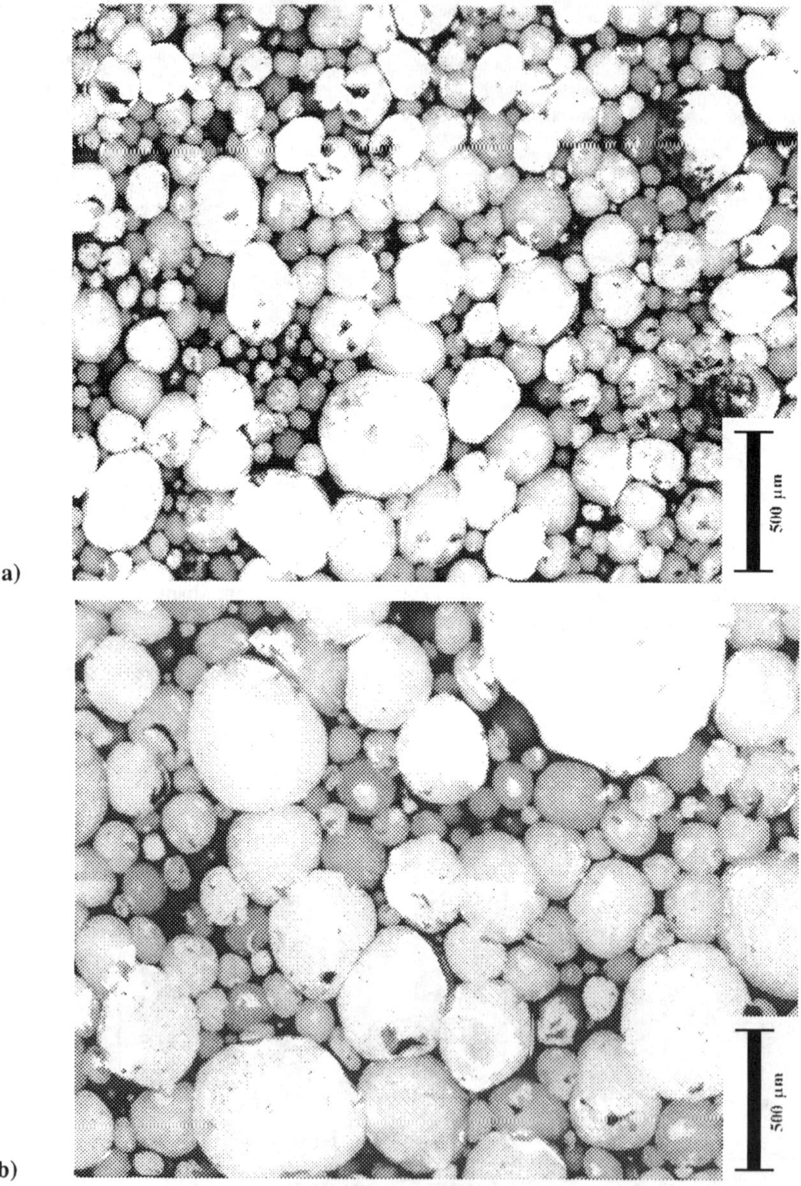

Figure 6.   **Fluorescent micrographs of foams: a) is typical small-celled foam and b) is typical large-celled foam.**

There were some differences among the foam specimens, but this can be seen more easily in Figure 7. This figure shows the average bubble sizes in the central part of the foam

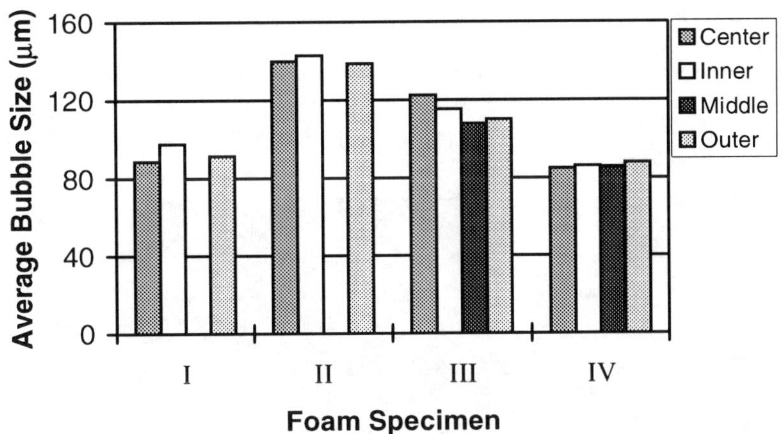

Figure 7. Average bubble sizes in the central part of the foam specimens as a function of LOCATION from the center of the foam to the outer part (but not at the edges).

specimens as a function of location within the foam. The measurements were taken at the center of the foam and then at several other locations towards the outer part, but not at the very edges of the foams. The average bubble size of foams I and IV is slightly more than 80 μm, foam III is 110 to 120 μm, and foam II is about 140 μm.

More interesting are the differences from the center of the foams to the outer edges. Figure 8 shows the size average results. Only foam II appears to show much difference with the edges having a much smaller average size bubble than the center areas. What is not reflected in this

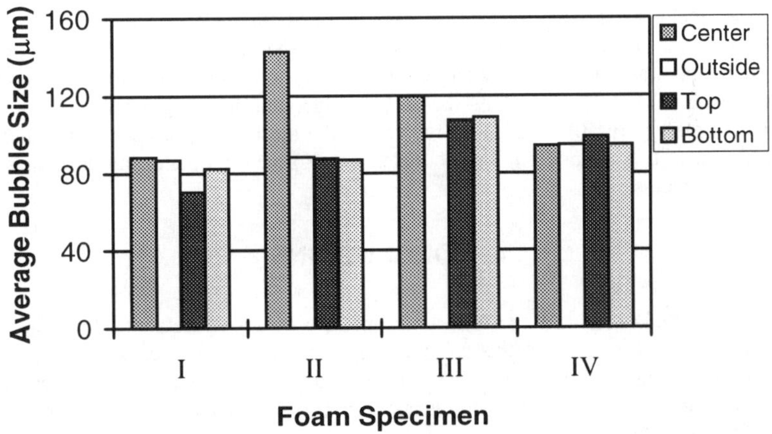

Figure 8. Average bubble sizes in the central part of the foam specimens as a function of POSITION from the center of the foam to the outside edge, top edge and bottom edge.

figure is that the regions picked to count cells deliberately avoided some regions. In foam I there were regions containing collapsed cells, apparently crushed during the extrusion operation. This is shown in Figure 9. Several attempts to produce a foam by extrusion showed that the foam had to be extruded when it was first removed from the blowing chamber and a hot mold used. Since no attempt was made to optimize these conditions, this technique might work without crushing cells if the problems of handling hot pitch and a hot mold together can be solved.

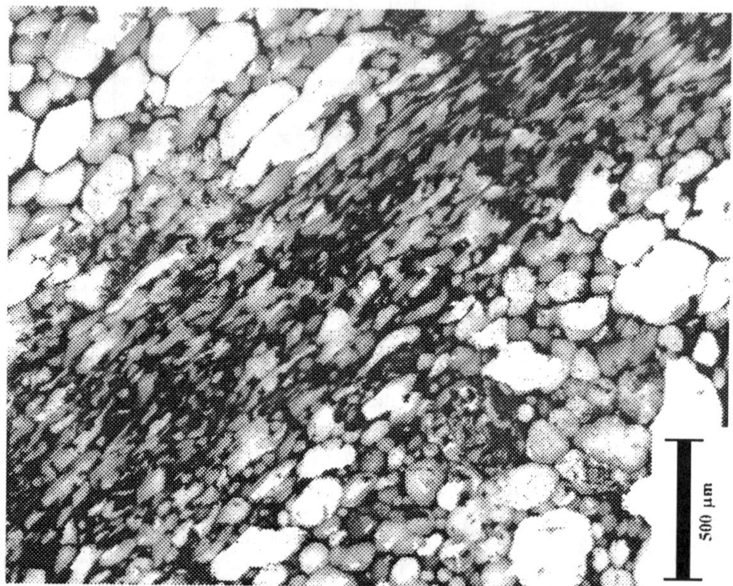

Figure 9.   Example of crushed cells in foam I.

The crushed cells were one of three regions avoided during the cell size measurement operation. A skin, tens of micrometers thick, was on the very outside edge of each foam. The third region was one with very large cells, several millimeters in diameter; these were present in the latter three foams along the bottom edge, and these regions were also avoided (see for example Figure 10). The latter represents the most common form of nonuniformity in these samples. It is unknown if large bubbles were formed in Foam I but collapsed during the extrusion process.

4. CONCLUSIONS

Foaming into a mold produces the best net shape, the final foam retaining the shape of the mold. The extrusion process in this study apparently crushed some cells but this could be dependent on the temperature of the foams and extruding molds which were not examined in detail. Foaming a shaped preform without restrictions results in a circular foam and does not retain the preform shape.

The cell size, porosity and density for all the "net-shape" processing samples were relatively uniform. While the absolute sizes of the cells were somewhat dependent on the shaping method, other studies have shown a much greater dependence of these variations on other processing parameters [1,5,6].

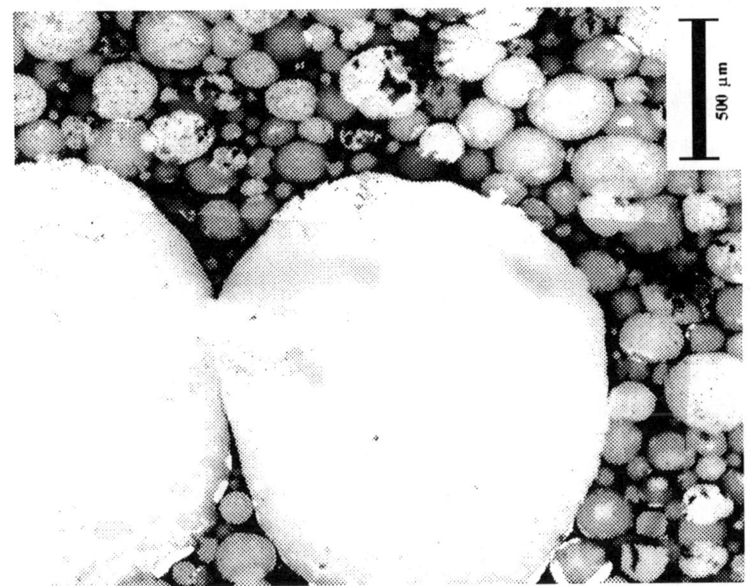

Figure 10.   **Example of occasional very large cells.**

5. ACKNOWLEDGEMENTS

This work was partially supported by Contract number F33615-95-D-5029 and also supported under the U. S. Air Force - Wright Connection summer teacher program.

6. REFERENCES

1.   K. M. Kearns and D. P. Anderson, <u>Proceedings of the ICCM-11</u>, pp. 825 (1997).

2.   R. B. Hall and J. W. Hager, <u>21st Biennial Conference on Carbon Extended Abstracts</u>, pp. 100 (1993).

3.   J. S. Colton and N. P. Suh, <u>Polymer Engineering and Science</u>, <u>27</u> (7), 485 (1987).

4.   J. S. Colton and N. P. Suh, <u>Polymer Engineering and Science</u>, <u>27</u> (7), 500 (1987).

5.   D. Dutta, C. S. Hill, and D. P. Anderson, <u>Novel Forms of Carbon II, MRS Symposium Proceedings</u>, <u>349</u>, 61 (1994).

6.   H. J. Anderson, D. P. Anderson, and K. M. Kearns, <u>23rd Biennial Conference on Carbon Extended Abstracts</u>, <u>II</u>. pp. 4 (1997).

43rd International SAMPE Symposium
May 31-June 4, 1998

EXPERIMENTAL METHODS FOR MEASURING TENSILE AND SHEAR STIFFNESS AND STRENGTH OF GRAPHITIC FOAM

Ajit K. Roy[1], Dave Pullman[2,3], and Kristen M. Kearns[3]

[1]University of Dayton Research Institute
[2]Currently, Hartzell Propeller Inc.
[3]Materials Directorate, Air Force Research Laboratory

ABSTRACT

In an effort to correlate the foam ligament deformation characteristics with the foam bulk properties, it is observed that the compressive test is not a suitable test method due to the premature buckling failure of the foam ligaments. As an alternative, the tensile testing is found to be a desirable test method for this purpose. Experimental techniques for measuring tensile stiffness and strength of open cell foam are described, adapted to a simple specimen configuration. In addition, a fixture to measure the shear stiffness and strength of open cell foam is developed.

KEY WORDS: Graphitic foam, open cell foam, structural foam, porous materials, tensile test method

1. INTRODUCTION

Microcellular graphitic foam is emerging as an ultra lightweight material for many structural applications, such as core materials for a sandwich structure, net shape fabrication of structural components reinforced with fiber preforms, etc. This material macroscopically posses isotropic property which is an added advantage over that of traditional honeycomb materials. The micrograph of a graphitic (open cell) foam microstructure is shown in Figure 1. A representative unit of an open cell foam, based on the minimum surface energy during the foaming process (i.e., the bubble nucleation process), posses a tetrahedral structure of the foam ligaments oriented approximately 109° with each other, see Figure 2. Due to the tetrahedral cell microstructure, the macroscopic properties, such as, foam modulus and strength, are critically influenced by the deformation characteristics of the cell ligaments. Thus, in order to develop a basic understanding of the performance of open cell foam materials, the deformation and failure mechanism of the cell ligaments, thus, critically needs to be studied.

The long term objective of this work is to understand the deformation characteristics of cell ligaments subjected to external loading, and correlate the ligament properties with the processing parameters. An appropriate mechanics model will be used to correlate the ligament properties to the foam bulk (macroscopic) properties to address the long term objective. As an initial step towards the long term objective, the goal for the present study is to develop suitable test methods to measure the bulk (macroscopic) mechanical properties of open cell foam materials. Due to the open porosity of this material, the standard test methods that are used to measure properties of non-porous materials, are not appropriate for measuring the stiffness and strength of foam materials of large porosity (e.g., 80 percent or higher).

2. TEST METHODS

A new generation of graphitic foam (porous) materials is being developed [1-4] with its intended use in structural components as load carrying members. Thus there is a need for developing test methods to characterize the mechanical properties (stiffness and strength) of this material for various loading conditions, such as, compression, tension, and shear. Traditionally, due to ease of loading and its intended use in the past, porous (foam) materials have been generally characterized in compression. As an illustration, the compressive stress-strain behavior of a carbonized phenolic foam with 82 percent porosity is shown in Figure 3. An one-inch cube of the material was subjected to block compression in this case. The initial small slope (up to one percent of the strain) in this stress-stain curve is attributed to excessive deformation of the cell ligaments exposed to the surface. The material appears to carry load up to a strain level of two percent, then beyond that strain level the material fails to carry any further loading, as revealed in the stress-strain curve. The same material, however, in tensile loading appears to carry load much more than the two percent of tensile strain. The apparent failure of the material in compression at about two percent of the strain is attributed to the buckling failure of the cell ligaments. Thus, the cell ligaments are not loaded to its true load carrying capability if the foam is loaded in compression. One of the objectives of this study is to correlate the effects of the processing parameters on the ligament deformation characteristics and its strength. It is known that the processing parameters, especially the graphitization temperature, strongly influence the stiffness and failure strain of the cell ligaments of a graphitic foam. Thus the compressive loading, as discussed above, is not a suitable loading condition to assess the full extent of the ligament deformation mechanism, and the effects of the processing parameters on its failure strain. The tensile loading, in this regard, appears to be a desirable loading condition.

2.1 Tensile Loading Due to high porosity of the material, the tensile coupons prepared from the material could not be directly gripped in the testing frame hydrostatic end grips. Further, in order to apply an uniform displacement on the foam sample subjected to a tensile load, tab materials were bonded to the two ends of the foam sample. Then the tab materials were pin loaded to apply the tensile load to the foam specimen, Figure 4. Specimens of both straight-edge and dogbone configurations were tested to measure the stress-strain behavior. A diamond saw was used to prepare the straight-edge specimens. The dogbone specimens were prepared by using an EDM machining tool. A clip extensometer was initially used to measure strain on the specimen surface. However, due to high porosity of the material, a stable contact of the extensometer clip-points on the specimen surface was not obtained. Thus, the strain applied to the specimen was calculated from the machine head displacement, assuming the amount of strain in the bond between the tab and the foam is negligible compared to that of the material in the gage section. The stress-strain curves of the straight-edge and dogbone specimens are shown in Figures 5 and 6, respectively. The modulus of the material obtained from these two specimen configurations was practically the same. Further, all the specimens tested in these two specimen configurations failed in the gage section. Thus, in view of the simplistic specimen configuration, the straight-edge specimen was considered adequate and suitable for the tensile testing for the graphitic foam. The stress-strain curves of a few graphitic foam specimens, using the straight-edge specimen configuration is shown in Figure 7. The modulus and strength of the material is shown in Table 1.

2.2 Shear Fixture The standard shear test, such as the Iosipescu shear test, is not expected to provide reliable shear properties for high porosity foam. A cylindrical specimen configuration (for example, circular rods) subjected to a axial torsional load is a preferred test for this material to measure its shear properties. A special torsion fixture was designed to measure the shear stiffness and strength of the material, as shown in Figure 8. The twist on the specimen is applied by means of a Worm Gear (60:1 speed reducer) connected to a stepper motor through a 3:1 timing belt. A 100 in-lb torque cell is used to measure the torque applied on the specimen. The twisting angle is calculated from the step angle of the stepper motor. The two ends of the specimen is secured in the fixture by means of set screws.

TABLE 1. Modulus and Strength of the graphitic foam samples tested.

Specimen	Width (in)	Thickness (in)	Gage Length (in)	Modulus Ksi (GPa)	Strength psi (MPa)
1	0.492	0.252	1.5	48.30 (0.333)	236 (1.63)
2	0.492	0.251	1.7	41.72 (0.288)	227 (1.57)
3	0.492	0.251	1.4	44.03 (0.304)	293 (2.02)

3. SUMMARY

The average specific stiffness of the graphitic foam tested above (of density of 0.18 gm/cc) is 4.5 Ksi/(lb/ft3) which is about 40 percent higher than that of a polymeric foam used in many structural applications. Because of its superior specific thermal and electrical conductivity over that of a polymeric foam, it is advantageous to use graphitic foam in applications, such as, thermal planes, heat sink, and electronic packaging. Further, the isotropic property of this material makes it an attractive core materials (over honeycomb core) in sandwich structures subjected to combined loading. In view of its expected use as a structural material, the tensile test method and a fixture for measuring the shear properties are developed. The tensile test method adapted to a simple specimen configuration is discussed.

4. REFERENCES

1. Heather J. Anderson, Kristen M. Kearns, and David P. Anderson, "Microcellular Graphitic Foams", 23rd Biennial Conference on Carbon, July 1997.
2. Kristen M. Kearns, and David P. Anderson, "Microcellular Graphitic Foam Processing", 11th International Conference on Composite Materials, July 1997.
3. David P. Anderson, Kristen M. Kearns, Cindy Tucci, and Gerry Mestemaker, "The Cellular Structure of Net Shaped Pitch-Based Microcellular Carbon-Foams", 43rd International SAMPE Symposium and Exhibition, 1998.

4. Heather J. Anderson, Kristen M. Kearns, and David P. Anderson, "Microcellular Pitch-based Carbon Foam Blown with Helium Gas", _43rd International SAMPE Symposium and Exhibition_, 1998.

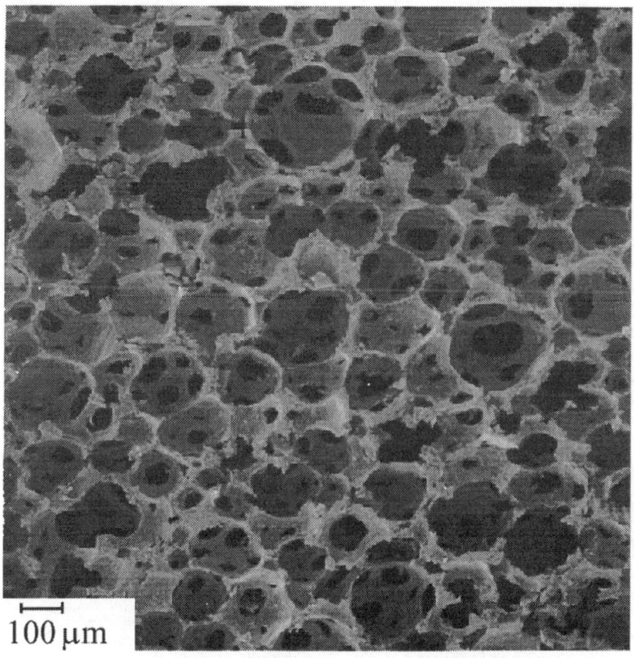

Figure 1. A representative microstructure of a graphitic foam

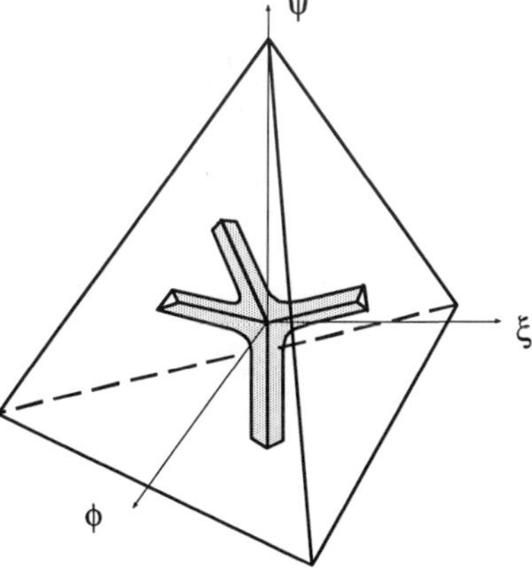

Figure 2. A tetrahedral Representative Volume Element (RVE) of a foam microstructure. The axes of the four ligaments in the RVE are oriented normal to the respective tetrahedral surfaces.

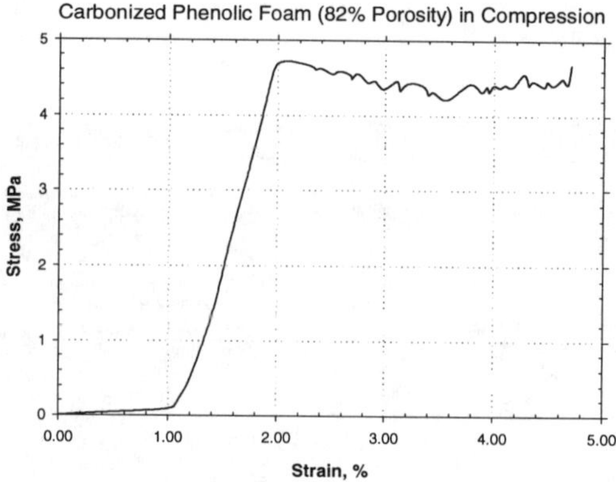

Figure 3. Stress-strain curve of a carbonized phenolic open cell foam subjected to compression.

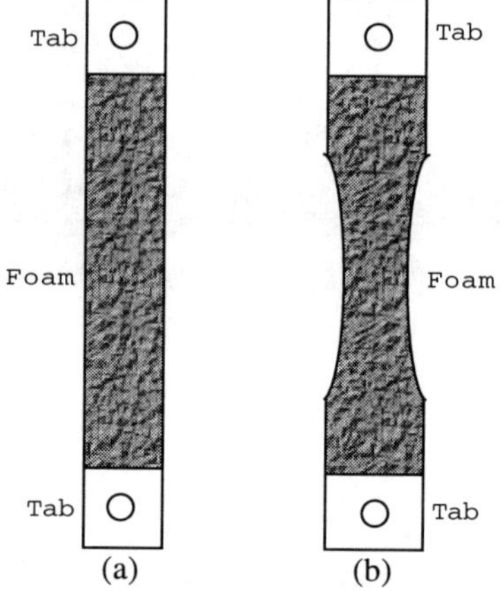

Figure 4. Configuration of the tensile foam specimens. (a) straight-edge specimen, (b) dogbone specimen.

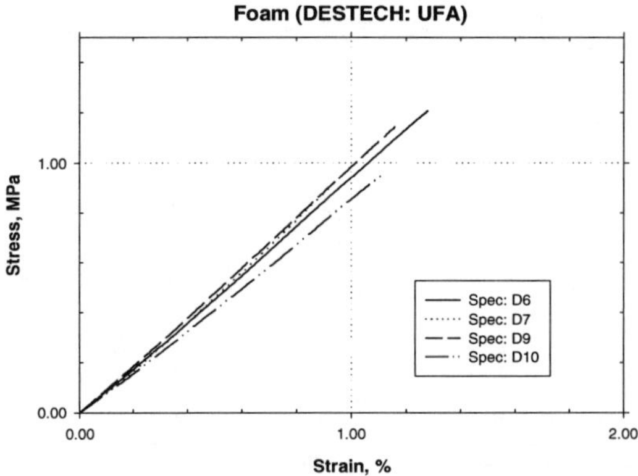

Figure 5. Stress-strain curves of polymeric (Destech: UFA) foam straight-edge specimens subjected to tensile loading.

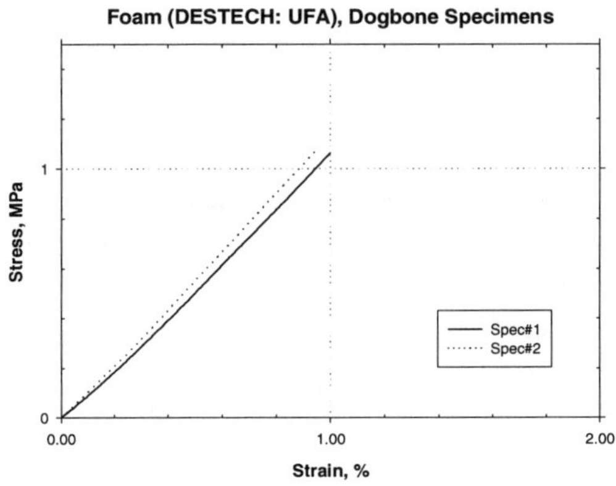

Figure 6. Stress-strain curves of polymeric (Destech: UFA) foam dogbone specimens subjected to tensile loading.

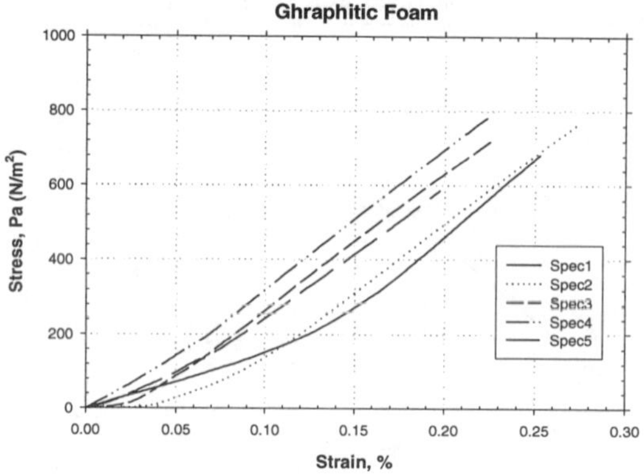

Figure 7. Stress-strain curves of graphitic foam specimens in tensile loading.

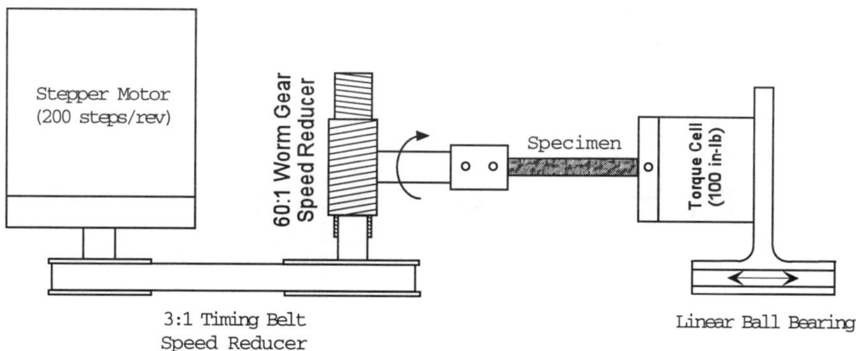

Figure 8. Schematic diagram of the fixture to measure shear properties (stiffness and strength) of foam materials.

PERFORMANCE OF TOUGHENED MODEL EPOXY ADHESIVE SYSTEMS WITH COMPOSITE MATERIALS

Jeremy H. Klug and James C. Seferis
University of Washington/Polymeric Composite Laboratory
Seattle, WA 98195-1750

ABSTRACT

Model epoxy adhesive systems were designed and tested for their bonding ability with carbon fiber composite materials. Silica addition, rubber type, and rubber concentration were varied for the purpose of investigating their influence on intermediate and final properties. The adhesive systems were tested with a carbon fiber prepreg system in both bonded and cocured geometries. Dynamic mechanical analysis and optical microscopy showed that the phase separation of the rubber was significantly influenced by the fumed silica addition. Dispersed phase growth rates were depressed resulting in a lower percentage of toughening domains and in one case the phase separation process was entirely suppressed. Fracture energy, measured by mode I and II critical strain energy release rates, was a strong function of particle size, flexibilized rubber type and content, and residual stress considerations at the interface. There were distinct differences between the bonded and cocured fracture results with the bonded samples generally outperforming the cocured specimens due to an increase in the surface roughness as well as resin flow properties and interfacial stress build-up. Collectively, this work illustrates the importance of phase separation behavior on final part morphology and performance for liquid CTBN toughened epoxy adhesives when used with prepreg material.

KEY WORDS: Epoxy, adhesive, fracture energy

1. INTRODUCTION

Epoxy adhesive systems have been used widely in the aerospace industry because of the wide range of properties and beneficial qualities which they possess. The application of these materials in the aerospace industry includes everything from core splicing adhesive foam to supported film systems used for structural purposes (1). Epoxies are relatively inexpensive when compared to other high temperature adhesives, can be cured with a wide range of materials resulting in considerable shelf and pot life variations, adhere well to most substrates, and provide good solvent resistance (2, 3). They are also inherently less brittle than most other aerospace matrix materials (e.g. phenolics), however, they still often require improvements in their fracture toughness to prevent crack propagation and premature failure during the lifetime.

Liquid reactive carboxyl terminated butadiene acrylonitrile (CTBN) elastomers have been employed in epoxies for several decades as toughening agents (4-11). These materials are frequently added or adducted with the epoxy matrix prior to cure. During the cure cycle, the incompatibility between the rubber and growing epoxy chains forces a phase separation where usually rubber rich domains are created in the continuous epoxy rich domain. However, a small amount of CTBN material usually remains in the continuous phase and flexibilizes the primary phase causing a slight drop in the glass transition temperature (12).

Extensive research has been conducted in the area of CTBN toughening of epoxy matrices including areas such as optimum toughening particle size, toughening mechanisms, optimum adhesive layer thickness, bimodal particle distribution influence, secondary phase formation and growth, and the effect of viscosity on morphological development (4, 6-9, 13-18). Work has also been performed in the area of trying to model the phase separation and growth behavior of the thermosetting systems (19-21).

The analyses of different adhesive characteristics often involve the use of model systems which provide the opportunity of obtaining a fundamental understanding of the influences of chemical composition and physical characteristics. Understanding a material's processing-structure-property interrelationships is a necessary part for material design and manufacturing purposes (22). Model prepreg systems have been previously developed by Seferis and co-workers to investigate commercial prepreg materials (23, 24). These studies have generated valuable information which has allowed for improvements in commercial systems through specialized teams created and composed of material suppliers, customers, manufacturers, and academia (25). Continuing in this manner, model adhesive materials of aerospace quality are currently being produced to accumulate valuable fundamental information.

The purpose of this study is to investigate the bonding characteristics of toughened epoxy systems with composite materials currently used in the aerospace industry. The complex anisotropy and heterogeneity of these materials makes bonding behavior somewhat difficult to predict due to multiple curing reactions, resin flow during cure, interfacial phenomena and interactions, etc. Specifically, this study will investigate the role of flexibilized rubber on the fracture toughness and bonding behavior with both cocured and bonded carbon fiber composite materials. The amount and type of phase separation for CTBN toughened epoxy systems are known to be a very important factors regarding final end-use properties. Collectively, this work will help provide a better understanding of the importance of the rubber phase separation process for liquid CTBN toughened epoxy adhesives when used with composite systems.

2. EXPERIMENTAL

The epoxy materials utilized in the study consisted of a combination of difunctional DGEBA type resins and a tetrafunctional TGMDA resin. Dicyandiamide was used as the curing agent and diuron as a cure accelerator. The tougheners employed included Hycar CTBN 1300x13 and CTBN 1300x8, both liquid modifiers. To increase the viscosity of certain systems and reduce the dispersed phase growth rate, hydrophobic fumed silica (Aerosil R202) was incorporated.

For toughened resins, CTBN material was prereacted with a difunctional epoxy to form an adduct for later incorporation into the base system. The CTBN adduct mixture was combined and mixed at 150°C for one-half hour after which point triphenylphosphine (approximately 0.2 wt%) was added to catalyze the reaction. A high shear mixer was used to blend the system for approximately three minutes and the resin was quenched to room temperature to prevent reaction initiation. Resin films 0.14 mm in thickness were finally prepared.

The different resin formulations were based on manipulation of three variables including silica addition, CTBN toughener type, and toughener level. The resin systems will be referred to by a numbering system and the corresponding differences between them are shown in Table 1.

Resin System	Rubber Type	Rubber Level	Silica Addition
1	x8	5 phr	N
2	x8	5 phr	Y
3	x8	20 phr	N
4	x8	20 phr	Y
5	x13	5 phr	N
6	x13	5 phr	Y
7	x13	20 phr	N
8	x13	20 phr	Y
9	none	n/a	Y
10	none	n/a	N

Table 1. Classification of resin systems

Glass transition information for the model adhesive systems was found using Dynamic Mechanical Analysis (DMA). A TA Instruments 983 DMA module was utilized for the analysis interfaced to a Thermal Analyst 2000 Controller.   These samples were equilibrated at -90°C for five minutes and a 5°C per minute heating ramp was applied until 225°C was reached. Two samples, obtained from bulk unfilmed material, were tested for each adhesive. Sample dimensions were approximately 8 mm by 3.5 mm by 1.5 mm and a frequency of 1 Hz was used with an oscillation amplitude of 0.2 mm.

Fracture specimens were prepared using a carbon fiber 3K-70-PW prepreg material. Both mode I and mode II fracture toughness were tested using the double cantilever (DCB) and end-notch flexure (ENF) methods. Panels were constructed using eighteen plies of prepreg with one layer of adhesive in the midplane and a FEP (fluorinated poply -ethylene-propylene) film was placed in the middle to act as a crack starter. The bonded prepreg panels were cured at 177°C for two hours with a layer of fiberglass peel-ply used to create a fresh surface conducive to bonding. The panels, measuring 33 cm by 8.25 cm by 0.43 cm, were cured at 177°C for two hours using heat-up and cooling rates of 2.8°C per minute and a compaction pressure of 309 kPa.

DCB specimens were tested in a Universal Testing Machine to determine the mode I (tensile) fracture energy with a crosshead speed of 2.54 cm per minute until a final displacement of 6.35 cm was encountered (26, 27) . The energy required to propagate the resulting crack was recorded along with the sample width (approximately 1.27 cm) and crack length increase. Three to four samples were tested for each adhesive to give the reported value.

ENF specimens were tested in the same apparatus to determine the mode II (in-plane shear) fracture energy (26, 27) using a crosshead speed of 0.254 cm per minute. The crack tip was positioned using a 2.5x magnifying apparatus to one inch from both the support and loading pins. The sample was loaded until crack propagation occurred (usually unstable growth). The specimen was then repositioned so that the crack tip was again at the required setting. One sample was tested for each adhesive from which five to six values were obtained.

Optical microscopy was used to investigate the morphology of the cured resins using both the DMA samples (bulk material) and fracture specimens (thin film). Both transmitted light (thin-section microscopy) and reflected light specimens were analyzed. When required, samples were placed in an OsO_4 solution for approximately 18 hours to stain the CTBN regions. An imaging analysis software program was used to calculate representative dispersed phase volume fractions from the photomicrographs by employing a binary imaging contrasting technique and recording characteristic pixel values (28). Scanning electron microscopy (SEM) was also utilized for morphological analysis using an accelerating voltage of 15 kV and a working distance of 20 mm.

3. RESULTS AND DISCUSSION

Dynamic mechanical analysis specimens were tested for the various formulations from neat resin plaques cured at 177°C using a 2.8°C per minute heat-up rate. Most of the toughened epoxy subambient tests showed both low and high temperature transitions, characteristic of two distinct phases present in the materials. The resin systems were characterized for their transition temperature(s) and the results are displayed in Table 2. All T_g values were identified as the peak(s) in the tan δ trace.

RESIN	1	2	3	4	5	6	7	8	9	10
Low Temperature Transition	-56	-44	-44	-38	-17	-60*	-7	-13	-55*	-56*
High Temperature Transition	173	172	172	167	171	168	167	162	176	176

* faint transitions

Table 2. Dynamic mechanical analysis (DMA) results for glass transition temperatures of adhesive formulations

Photomicrographs of the toughened epoxy systems are shown in Figure 1 illustrating the different morphologies observed for the bulk resin systems as influenced by the toughener type, toughener amount, and silica addition. A wide range of microstructures was possible by manipulation of these variables. Resin system 6 did not show any phase separated material under transmitted or reflected visible light. This sample was also examined using SEM and a distinct second phase was not present. The photographic results for sample 8 were manipulated using computer software to more clearly illustrate the phase differentiation. The dispersed phase area or volume percentage was not affected by this procedure.

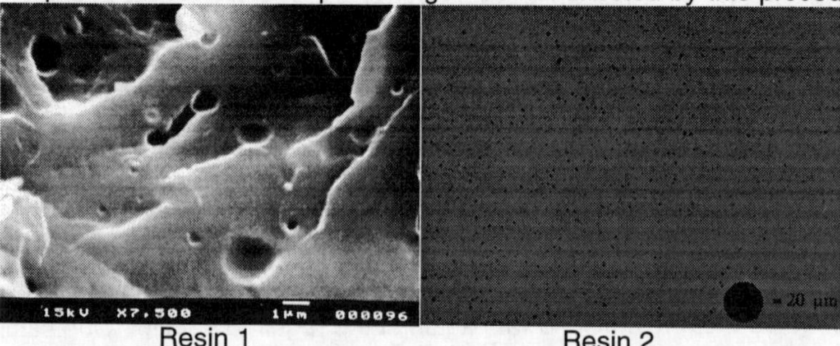

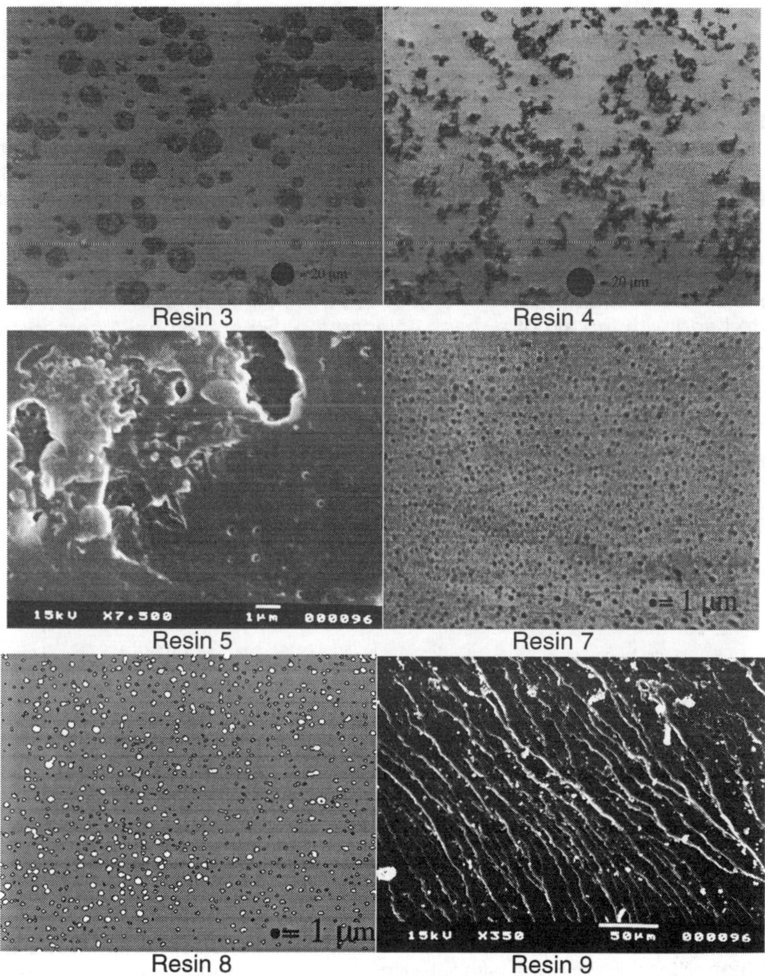

Figure 1. Photomicrographs for model resin systems.

Image analysis was utilized with the photomicrographs of the bulk adhesive systems and the dynamic mechanical analysis results to calculate volume fractions of rubber and epoxy in the different phases. Two of the most common methods used for copolymer compositional analysis are the Fox equation and the Gordon-Taylor equation (29, 30) both of which have been applied for several different modified epoxy combinations with reasonable results (12, 14, 17, 31). The Gordon-Taylor relationship was utilized in this analysis because it has been shown to give slightly better results for flexibilized CTBN blends (versus plasticized blends) (12). This relationship equates the change in the glass transition temperature $\Delta T_{g,E}$ for the epoxy-rich phase with the mass fraction of rubber in this phase, w_r^c, such that

$$\Delta T_{g,E} = \frac{k w_R^c \left(T_{g,E(0)} - T_{g,R(0)} \right)}{1 + w_R^c (k-1)} \tag{1}$$

where $T_{g,E(r)}$ is the observed glass transition temperature for the epoxy-rich phase, $T_{g,E(0)}$ and $T_{g,R(0)}$ are the glass transition temperatures of the neat epoxy and rubber, and k can either be assigned as a fitting parameter or a theoretical value related to the ratio of the thermal expansion jump of the materials at their respective glass transitions. For the present study, the k value was used as a fitting parameter since resin system 6 was completely flexibilized by the CTBN addition (rubber type x13) allowing a direct calculation of the k parameter which was then used for all subsequent systems containing this toughener. The k value for the x8 CTBN systems was scaled from its theoretical value to give a result that was consistent with the ratio of the actual and theoretical values for the x13 CTBN toughener. The volume fraction of rubber in the continuous phase was then calculated and with the aid of a mass balance the dispersed phase composition found. The results of this compositional analysis along with the average dispersed phase particle diameters appear in Table 3.

Resin System	Volume Fraction Rubber in Continuous Phase	Volume Fraction Rubber in Dispersed Phase	Average Spherical Particle Size (μm)
1	0.021	0.81	1.0-4.0
2	0.028	0.66	0.5-1.5
3	0.028	0.82	10-20 and 2-4
4	0.061	0.61	10.0 (non-circular)
5	0.039	0.81	0.5-1.0
6	0.061	0	-
7	0.069	0.97	0.5-1.2
8	0.11	0.87	0.4-1.2
9	0	0	-
10	0	0	-

Table 3. Compositional results for model systems

The behavior of the resin pairs (i.e. with and without silica) was quite consistent for all formulations. The continuous phase (epoxy-rich) transition temperature was slightly higher for the versions without silica except for the first two model systems showing that the dispersed phase nucleation or growth rate during the cure reaction was decreased by the silica addition. This appears reasonable assuming that the mechanism by which the dispersed phase was produced occurred by nucleation and growth rather than spinodal decomposition which has different kinetics. The nucleation and growth mechanism has been supported for similar epoxy-dicyandiamide-CTBN systems and the characteristic bicontinuous phase formation seen for the spinodal decomposition process was not observed during or after cure of resin samples placed on a heated stage viewed under a microscope(19).

Assuming that the nucleation rate was entirely dependent on free energy or interfacial tension between the rubber and epoxy molecules, it would seem more likely that the growth rate of the rubber-rich phase was reduced by the viscosity increase from the silica addition. This resulted in more rubber remaining in the continuous phase after gelation and phase lock-in and a lower T_g which was observed and is also supported by the fact that if the nucleation rate was increased by the silica addition, a reverse in the trend of the epoxy-rich phase T_g values would have been expected, and this was not observed. This growth rate dependence on viscosity has been part of several models used to predict phase separation behavior for toughened epoxy systems and has shown that the diffusional ability of the rubber molecules to join phase separated domains is a direct function of the viscosity (19, 32).

The silica addition also manifested itself in the low temperature transitions. The rubber rich phase was not as pure for the silica systems showing that the increased viscosity (decreased mobility) created a situation where the thermodynamic driving force for phase separation was large enough to possibly overcome the epoxy inclusion in the dispersed phase.

The rubber type dependence on the morphological characteristics was as would be expected. The more polar toughener (x13) was more compatible with the epoxy mixture and phase separation did not occur as early in the curing process as for the x8 (less compatible) toughener. This implies that the nucleation rate was reduced at a given temperature for the more compatible system. Consequently, assuming similar growth rates, the domain sizes should be reduced for the x13 systems and this behavior was indeed observed experimentally.

In terms of compatibility, one would also expect more rubber in the continuous phase of the x13 toughened systems as was seen, provided that phase separation was not complete for both tougheners. The composition of the dispersed phases, however, did not seem to be as strong a function of the toughener type.

The rubber level dependence on the adhesives' morphologies was apparent for both the x8 and x13 systems. An increase in the rubber concentration resulted in a larger volume fraction of dispersed particles, but the concentration of these domains remained virtually constant for each rubber type irrespective of the rubber loading. The exception to this was the low and high rubber levels for the silica resin prepared with x13 CTBN toughener where phase separation wasn't observed for the 5 phr situation.

It should be mentioned before continuing that bulk versus thin film behavior has been shown to be important in regards to certain materials both from a morphological and intrinsic property point of view (33, 34). For this study, bulk materials used for morphological and glass transition information were prepared using thicknesses to closely approximate materials used in the mechanical property test specimens. Microscopy showed that the different resin morphologies were consistent when compared to the thin film composite application materials.

Fracture energy tests, frequently used to characterize toughened resin systems, were performed using the model epoxy resins with carbon fiber 3K-70-PW prepreg. Both mode I and II fracture specimens were prepared and tested. The relationships used for the fracture energy calculations are given by

$$\text{Mode I:} \qquad G_{Ic} = \frac{\Delta E}{\Delta a \cdot w} \qquad (2)$$

$$\text{Mode II:} \qquad G_{IIc} = \frac{9P_c^2 \cdot C \cdot a^2}{2w\left[2l^3 + 3a^3\right]} \qquad (3)$$

where ΔE denotes the energy released in crack growth, a represents the crack length, w the sample width, P_c the maximum load, C the overall material compliance, and l the length to the loading point (26, 27).

Figures 2 and 3 show the fracture results for the specimens cocured with the carbon fiber-epoxy prepreg material. Samples were void free and adhesive thicknesses were on the order of 0.08-0.10 mm. The mode I results for the cocured systems showed somewhat different trends than the mode II results. In mode I, the rubber type played a more significant role in the performance, and the fumed silica addition did not as drastically influence the results. Model systems 1, 3, and 4 were also the only systems to show predominantly cohesive failure during the testing, the other specimens failed at the interface between the resin systems. It appeared that the morphological changes were not as important for resin systems in this testing, an issue that will be discussed in detail later.

The systems toughened with the more compatible CTBN toughener (x13) did not show an increased toughness over the untoughened systems. All x13 systems showed interfacial failure while most of the x8 toughened formulations and to some degree the untoughened resins showed bulk cohesive failure. Adhesive or interfacial failure would cause morphological effects in the bulk resin to be masked since the failure would not occur in the bulk where the toughening particles/phases were contained. Therefore, for systems that failed interfacially these tests might not be a true indication of the value of these toughening domains. It does, however, illustrate the importance of adhesive applications in so far that failure modes can be manipulated with slight changes in chemistry and formulating procedure (35). The particle size also played a role in the mode I results with larger particle systems showed improved results. One possible reason for this is again stress relief during cool down. Larger rubber particles create larger tangential compressive stresses during the cooling period which increases the load required to initiate failure during testing. However, larger particles do not toughen the bulk matrix to the same degree as smaller (0.5-10 μm) particles, provided the failure is cohesive (5, 9, 36).

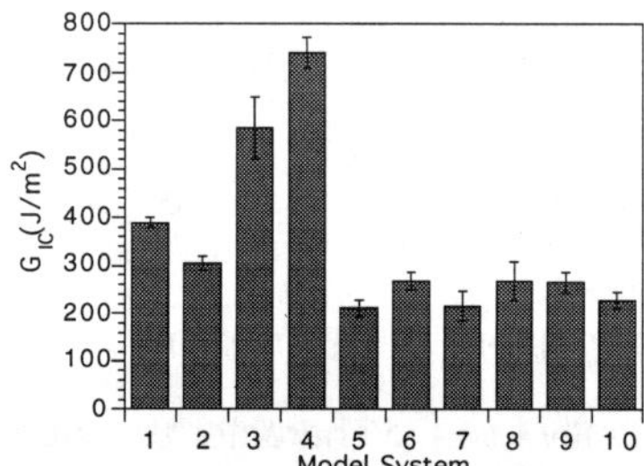

Figure 2. G_{Ic} fracture energy with cocured carbon fiber composite panels

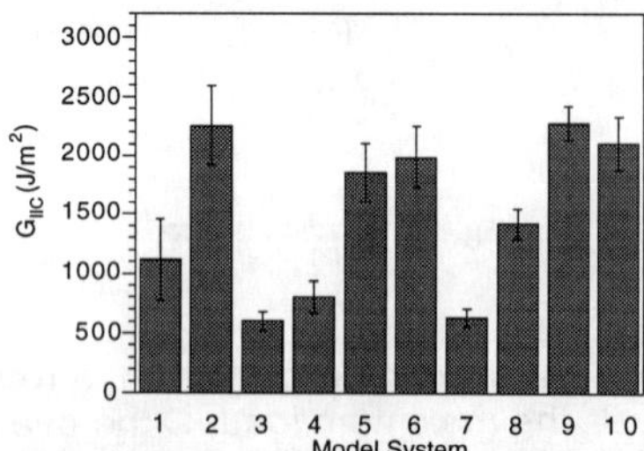

Figure 3. G_{IIc} fracture energy with cocured carbon fiber composite panels

The silica dependence itself did not significantly affect the mode I fracture results for the cocured systems. A very slight increase in the mode I toughness was seen with silica addition for the untoughened resins, possibly a result of a crack pinning mechanism and a slight increase in crack path tortuosity (37). The silica dependence on the fracture toughness by modification of the bulk morphology was evident. The change in the particle size between otherwise identical resin systems was found to slightly influence the mode I fracture toughness. Increased rubber levels were also found to increase the mode I toughness properties as expected provided that the increased rubber content lent itself to phase domain formation and not complete flexibilization.

The mode II fracture cocured samples showed a strong dependence on particle size effects and flexibilized rubber content. The toughest samples in mode II were the untoughened resins followed by those with the smallest dispersed phase inclusions in the continuous phase. The larger the ratio of dispersed phase diameter to adhesive thickness, the larger the stress concentration values for this failure mode and therefore a lower stress/load was required to initiate failure. This idea must also be balanced with the increased continuous phase rubber content which might tend to require less elastic strain energy build-up at the crack tip to initiate crack propagation.

Therefore, the cocured fracture results showed that the manipulation of the nucleation and growth mechanism of dispersed phase toughening by fumed silica addition did influence fracture results. The main mechanism by which this happened was reduced particle size as a consequence of retarded disperse phase growth rates. Mode I results indicated that a switch to interfacial failure by residual stress build-up at the interface greatly inhibited mode I performance for the formulations prepared with the more epoxy compatible toughener. Mode II results showed that retarded growth rates for the dispersed phases were preferred because a smaller phase size lead to increased shear fracture performance.

The bonded fracture results for the model adhesive systems are shown in Figures 4 and 5. The specimens were void free in the bulk and interfacial regions and adhesive thicknesses were on the order of 0.10-0.12 mm.

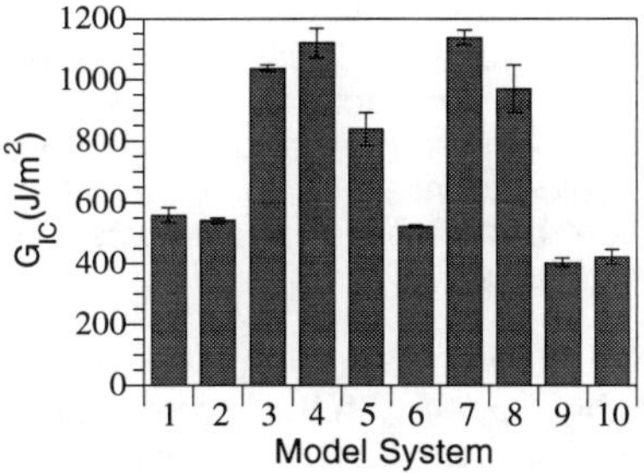

Figure 4. DCB results for bonded composite specimens

The failure modes for the systems also seemed to be a function of these variables, samples 1-4, 6, and 10 showed primarily cohesive failure in mode I while the remainder showed an adhesive/interfacial or a mixed failure mode.

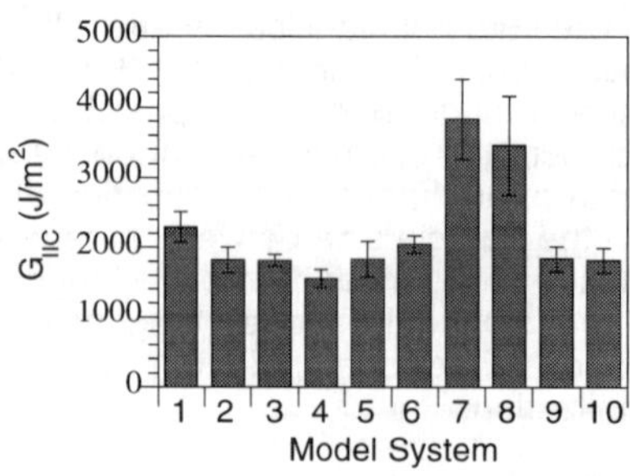

Figure 5. ENF results for bonded composite specimens

The rubber level had a large impact on the fracture energy as might be expected while the rubber type showed to be not as important to the mode I fracture values for bonding situations. However, the rubber type seemed to have a profound effect on the type of failure that was observed. All systems prepared with the more incompatible rubber displayed cohesive failure while a majority of the samples toughened with the more compatible CTBN material showed some adhesive failure. This failure dependence on toughener type was also seen for the cocured mode I samples and suggests that the systems with the larger dispersed phases fail in the bulk of the adhesive (i.e. the crack is deflected from the interface to bulk). This bulk versus interfacial failure for the large particle systems again shows that residual stress relief at the interface can lead to failure mode changes. However, this interfacial failure for the bonded specimens did not result in diminished mode I fracture values as is usually the case. The interfacial failure in mode I testing (specifically for systems 5 through 8) might actually have caused an increase in the perceived fracture energy so that actual mode I values would have been somewhat less. Often interfacial failure between two dissimilar materials is accompanied by mixed mode loading at the crack tip (38). This is a direct consequence of the crack being constrained to a certain path (interface) and often results in artificially high values for double cantilever beam tests.

The modifying effect by the silica on mode I fracture was seen when looking at samples 5 and 6 which just differed in their silica content. The silica version had such diminished nucleation and growth kinetics that a dispersed phase was never formed, instead the CTBN remained flexibilizing the system. Thus, the bulk adhesive possessed a sufficiently low fracture toughness in comparison to a system with lower continuous phase rubber and dispersed rubber particles and the crack propagation never traveled to the interface. The other system, however, had some adhesive failure present but the bulk material was tougher on average and therefore an increased energy was observed.

The ENF results for the bonded composite samples showed that mode II was only affected for the samples highly loaded with the more compatible rubber. These two samples also displayed the most non-linearity in their load-displacement curves prior to catastrophic failure indicating more energy was absorbed by plastic deformation/molecular rearrangement and that the continuous phase rubber content along with the dispersed phase size governed the bonded mode II fracture results. An increased continuous phase rubber content along with smaller dispersed phase particles allowed for the best mode II properties as far as energy absorption before failure was concerned (this combination was only achieved for resin formulations 7 and 8). System 6 had a relatively high continuous phase rubber concentration but did not possess a secondary distinct toughening phase for increased shear fracture performance illustrating the relative importance.

In summary, the bonded fracture results displayed behavior indicating the importance of rubber type, rubber level, and silica content. Mode I toughness showed a strong dependence on rubber level while rubber type and growth retardation (smaller phase size) only manifested themselves in the failure mode and not the absolute fracture values because of possible mixed mode loading. Mode II results for the bonded specimens indicated that a somewhat flexible continuous phase and more importantly a small domain size was desired and consequently the more compatible toughener was favored.

It is apparent after observing Figures 2 through 5 that the variables of silica content, rubber type, and rubber content had a strong influence on the fracture properties of adhesively joined carbon fiber adherends. However, what was not expected was the large discrepancy between cocured and bonded results. The results for the two situations were quite different, with the bonded specimens on average significantly outperforming the cocured samples. This intuitively might not be expected and it should be mentioned that this trend may also be specific to the type of high performance prepreg tested. Regardless of this fact, it brings up several issues which should not be overlooked or discounted when considering bonding and cocuring with adhesive materials.

The bonded versus cocured results indicated that the main mode of adhesion between the two resin systems was not a result of diffusion or molecular entanglement at the interface or interpenetrating networks. These issues have been commonly discussed as the main adhesion and/or toughness contribution mechanisms in polymeric materials, but in this study the diffusion model did not play a significant role (39-42). Mechanical interlocking appeared to have a reasonable influence on the precured panels which were sanded prior to adhesive application by providing on the average increased fracture results. The cocured resin systems on the other hand were smoother on a more microscopic level as a result of hydrostatic like forces developed at the interface between the two liquid-like materials (during cure) meaning that interfacially failed specimens should have slightly higher fracture values because of the larger asperities at the surface providing a greater frictional resistance. An example of this surface roughness for the bonded samples is shown in Figure 6. This also explains for mode I why the bonded interfacially failed specimens showed similar fracture energy values to the cohesively failed bonded samples and why the same behavior wasn't seen for the cocured specimens. Because of the more saw-tooth pattern of the bonded adherends, the shear contribution to the DCB values was significant. However, this reason alone is not enough to fully explain the results since the untoughened bonded resins did not show increased mode II results as they would have if mechanical interlocking was the main factor.

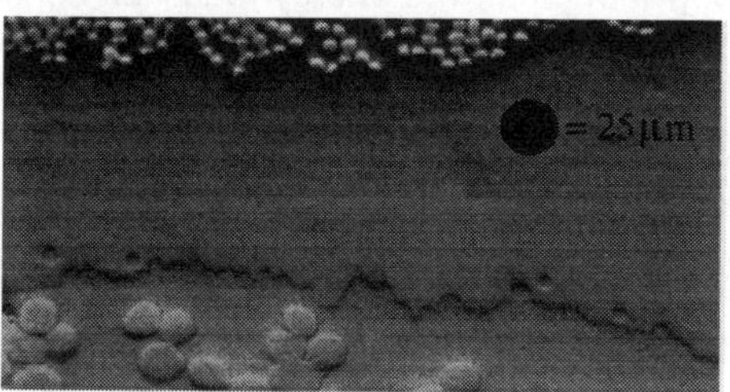

Figure 6. Increased surface roughness characteristic of bonded samples

The G_{IIc} results between the bonded and cocured specimens only showed significant differences for the highly rubberized systems (3,4,7,8) with the smaller particle systems performing best. There was some question concerning the stress relaxation of the various adhesive formulations in mode II testing which might give increased values. Stress relaxation tests were performed on bulk samples for all the adhesive formulations in order to try and simulate the mode II loading rate. The stress decay ratio as a function of initial stress was obtained for the resin systems and the results were a direct function of the rubber remaining in the continuous phase. The spread in the results between the systems with limited flexibilizing rubber and those with more significant levels was only 1.5 to 2.75% (stress percentage reduction over a period of sixty seconds). It seems more that the systems with mixed mode loading performed better by taking advantage of both the adherend roughness and dispersed phase particles.

Primary bonding between resin systems did not appear to provide any adhesive force providing that the resin systems were not poisoned by each other in the cocured case. If primary bonding was to occur, it would have been anticipated in the case of the cocured specimens and an increased interfacial strength realized, however, this was not observed experimentally, in fact inferior numbers were seen in comparison to the bonded structures. The possibility of poisoning was mentioned for the cocuring specimens but this seems remote considering both systems were primarily epoxy (DGEBA) based resins. Different catalysts and accelerators may have been present but should not have detrimentally affected either resin system to the point where it would change failure mechanisms. This was confirmed using differential scanning calorimetry where the different resin systems were put in contact with each other during cure and did not show a decrease in the overall heat of reaction. The value was a weighted average of the individual components as would be expected if there were no interactions.

There are two more arguments that should be made in trying to explain the overall performance increase for the bonded versus cocured specimens. The main ideas have to do with resin flow and preferential migration of the rubberized phase domains to the interface of the dissimilar resin systems and residual stress build-up and dissipation at the interface between the two distinct resin systems.

There were obvious signs of interlaminar/through thickness resin flow in the cocured fracture samples. The flow usually was situated at prepreg-fiber overlap regions (tow interstitial areas) and usually extended from one to three plies in distance from the interface. Despite this resin flow, adhesive layer thicknesses between the cocured and bonded samples were nearly identical, differing on the average by only 5-10% which should not be enough to explain the overall performance differences. Accompanied with the increased resin flow was a higher mobility between and among the resin systems for the cocured samples. This increased mobility along with other free energy reduction considerations is likely what caused some preferential location of the dispersed rubber-rich phases at the interface for the cocured samples. This phenomenon is illustrated in Figure 7 for systems 1 and 3 and might have contributed to the lower average values for cocured samples, especially regarding the mode I results by reducing the interfacial toughness.

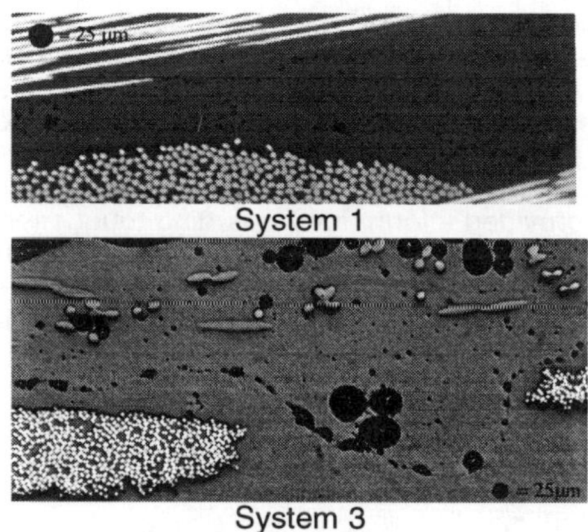

Figure 7. Preferential location of rubber phases at resin interface for cocured specimens

Residual stress relief in polymeric materials is important to the overall optimization of mechanical properties and might have provided some improvement for the bonded samples, however the contribution for the most part would not be sufficient to explain large differences between the bonded and cocured samples. The resin used in the prepreg material was isolated, cured, and stress relaxed at the cure temperature of 177°C for two hours using dynamic mechanical analysis. The results showed a modest decrease of the imposed stress of approximately 20%, however, stresses are again generated in the mechanical samples upon cool down from the "stress-free" cure temperature and the magnitude of the residual stress built from this large ΔT (cure to room temperature) far outweighs any stress relieved in the composite during the bonding operation.

4. CONCLUSIONS

The effect on different properties of CTBN toughener content, type, and fumed silica addition for an epoxy adhesive were studied. Morphological and fracture analyses were conducted for the different model resin systems with wide variations observed.

The resin viscosity increased with silica addition and sufficiently slowed the dispersed phase (rubber-rich) growth rates such that smaller sizes were observed and in one case phase separation was halted. The concentration of rubber in these phases was also decreased with silica addition as determined using dynamic mechanical analysis and microscopy. Fracture samples were prepared with carbon fiber-epoxy prepreg systems under both cocured and bonded situations. Interfacial versus cohesive failure was found to have a strong effect on the fracture energy especially for the bonded specimens. The fracture results showed a definite dependence on the morphological differences that were created with the CTBN type, amount, and silica content. G_{Ic} results favored systems with larger domains which helped to blunt the crack tip during loading while mode II results showed best results for small dispersed phase sizes.

There existed significant differences between the bonded and cocured fracture behavior with the bonded results generally outperforming the cocured. Mechanical interlocking for the bonded samples was determined to have a large positive increase on the fracture samples even for mode I testing provided that interfacial failure was observed. Preferential migration of the dispersed phases to the carbon fiber composite host resin was also observed for the cocured specimens and diminished the overall fracture performance.

Collectively, this work provided information regarding toughened epoxy adhesive systems that helped show the property and morphological dependence on the manipulated flexibilizing and dispersed phase CTBN toughener content as they specifically relate to the joining of carbon-fiber polymer composites.

5. ACKNOWLEDGMENTS

The authors would like to express their thanks to The Boeing Co., SIA Adhesives Inc., and TA Instruments for their support through the Polymeric Composites Laboratory (PCL) consortium.

6. REFERENCES

1.	Boeing, Boeing Material Specification 5-92,
2.	W.G. Potter, Epoxide Resins, Springer-Verlag, New York, 1970 .
3.	C.A. May, Epoxy Resins, Ed. 2, Marcel Dekker, Inc., New York, 1988 .
4.	H. Hsich, Polym. Eng. Sci., 30 (9), 493-510 (1990).
5.	A.J. Kinloch, S.J. Shaw, D.A. Tod and D.L. Hunston, Polymer, 24 , 1341-1353 (1983).
6.	C.B. Bucknall and T. Yoshii, The British Polymer Journal, 10 (3), 53-59 (1978).
7.	W.N. Kim, C.E. Park and C.M. Burns, J. Appl. Polym. Sci., 50 , 1951-1957 (1993).
8.	I.K. Partridge and G.M. Maistros, Plastics, Rubber and Composites Processing and Applications, 23 , 325-330 (1995).
9.	C.K. Riew, E.H. Rowe and A.R. Siebert, R.D. Deanin and A.M. Crugnola ed., Toughness and Brittleness of Plastics, American Chemical Society, Washington D. C., 1976, .
10.	W.A. Romanchick, J.E. Sohn and J.F. Geibel, R.S. Bauer ed., Epoxy Resin Chemistry II, American Chemical Society, Washington D. C., 1983, .
11.	A.R. Siebert, L.L. Tolle and R.S. Drake, Adhesives Age, 7 , (1986).
12.	P. Bussi and H. Ishida, Polymer, 35 (5), 956-966 (1994).
13.	H. Chai, Acta Metallurgica et Materialia, 43 (1), 163-172 (1995).
14.	L.C. Chan, J.K. Gillham, A.J. Kinloch and S.J. Shaw, C.K. Riew and J.K. Gillham ed., Rubber-Modified Thermoset Resins, American Chemical Society, Washington D. C., 1984.
15.	A.J. Kinloch, C.K. Riew ed., Rubber-Toughened Plastics, American Chemical Society, Washington D. C., 1995.
16.	N. Sela, O. Ishai and L. Banks-Solls, Composites, 20 (3), 257-264 (1989).
17.	D. Verchére, H. Sautereau, J.P. Pascault, S.M. Moschair, C.C. Riccardi and R.J.J. Williams, C.K. Riew and A.J. Kinloch ed., Toughened Plastics I: Science and Engineering, American Chemical Society, Washington D. C., 1993.
18.	K. Yamanaka and T. Inoue, J. Mater. Sci., 25 , 241-245 (1990).
19.	J.P. Chen and Y.D. Lee, Polymer, 36 (1), 55-65 (1995).
20.	S.M. Moschiar, C.C. Riccardi, R.J.J. Williams, D. Verchere, H. Sautereau and J.P. Pascault, J. Appl. Polym. Sci., 42 (3), 717-735 (1991).
21.	A. Vazquez, A.J. Rojas, H.E. Adabbo, J. Borrajo and R.J.J. Williams, Polymer, 28 (7), 1156-1164 (1987).

22.   J.C. Seferis, <u>SAMPE J.,</u> <u>24</u> , 6 (1988).
23.   W.J. Lee, J.C. Seferis and D.C. Bonner, <u>SAMPE Qtr.</u>, <u>17</u> , 58 (1986).
24.   M.A. Hoisington and J.C. Seferis, <u>SAMPE Qtr.</u>, <u>24</u> , 10 (1993).
25.   K. Didden, R. Grove, S.-H. Kim, R. McCullough, B. Roundtree, S.-B. Shim, B. Hayes, K.-J. Kim, C. Martin, L. Nguyen, T. Pelton, J. Putnam, D. Renn, C. Watson, Y.T. Shim and J.S. Seferis, <u>Staged Protocol for Global Technology Implementation: Composites with Dual Cure for Manufacturing and Repair</u>, University of Washington Graduate School, 1996.
26.   N.J. Pagano, <u>Interlaminar Response of Composite materials</u>, Ed. Elsevier, New York, 1989 .
27.   J.W. Putnam, <u>Composite Prepreg Process Characteristics</u>, Doctorate Thesis, University of Washington (1996).
28.   W. Rasband, <u>Void analysis was performed on a Model 7100/80 Power MacIntosh using the public domain NIH Image program (written by Wayne Rasband at the U.S. National Institutes of Health and available from the Internet by anonymous ftp from zippy.nimh.nih.gov or on floppy disk from NTIS, 5285 Port Royal Rd., Springfield, VA 22161, part number PB93-504868).</u>,
29.   T.G. Fox, <u>Bulletin of the American Physical Society</u>, <u>Bulletin of the American Physical Society</u>, <u>1</u> (3), 123 (1956).
30.   M. Gordon and J.S. Taylor, <u>Journal of Applied Chemistry</u>, <u>2</u> , 493-500 (1952).
31.   D. Chen, J.P. Pascault, H. Sautereau, R. Ruseckaite and R.J.J. Williams, <u>Polymer International</u>, <u>33</u> , 253-261 (1994).
32.   S.C. Kim, M.B. Ko and W.H. Jo, <u>Polymer</u>, <u>36</u> (11), 2189-2195 (1995).
33.   R.A. Gledhill, A.J. Kinloch, S. Yamani and R.J. Young, <u>Polymer,</u> <u>19</u> (8), 574-582 (1978).
34.   W.D. Bascom, R.L. Cottington, R.L. Jones and P. Peyser, <u>J. Appl. Polym. Sci.,</u> <u>19</u> , 2545-2562 (1975).
35.   J.H. Klug and J.C. Seferis, <u>J. Appl. Polym. Sci.</u>, 1953-1963 , (1997).
36.   F.J. Sultan and F.J. McGarry, <u>Polym. Eng. Sci.</u>, <u>13</u> (1), 29-34 (1973).
37.   Y. Nakamura, M. Yamaguchi, A. Kitayama, M. Okubo and T. Matsumoto, <u>Polymer</u>, <u>32</u> (12), 2221-2229 (1991).
38.   D. Hull and T.W. Clyne, <u>An Introduction to Composite Materials,</u> Cambridge, New York, 1996 .
39.   M.S. High, P.C. Painter and M.M. Coleman, <u>Macromol.</u>, <u>25</u> (2), 797 (1992).
40.   Y.C. Chern, K.H. Hsieh and J.S. Hsu, <u>J. Mater. Sci.</u>, <u>32</u> (13), 3503 (1997).
41.   J.K. Kim and C.D. Han, <u>Polym. Eng. Sci.</u>, <u>31</u> (4), 258-269 (1991).
42.   S.S. Voyutskii and V.L. Vakula, <u>J. Appl. Polym. Sci.</u>, <u>7</u> , 475-491 (1963).

43rd International SAMPE Symposium
May 31-June 4, 1998

EVALUATION OF RETAINED APPARENT SHEAR STRENGTH PROPERTIES IN ADHESIVES BASED ON EPOXY AND CYANATE ESTER CHEMISTRIES

K. Searles [1], G. Tewksbury, B. Tahmasebi, A. Huffstutter
and J. Harvey [2,3‡]

Department of Materials Science and Engineering
Oregon Graduate Institute of Science and Technology
P.O. Box 91000, Portland, OR 97291-1000

ABSTRACT

Tensile lap shear (TLS) testing was employed to evaluate the retention of apparent shear strength properties in two commercially available thermosetting structural film adhesives based on epoxy prepreg and cyanate ester systems. Simple surface preparation techniques were used as opposed to more stringent operations demanded for critical applications. The bonded austenitic stainless steel 316L substrates were subjected to environmental exposures of 24 hour submersion in boiling water and 100 hour thermal aging in air at 121 °C (250 °F) prior to mechanical testing at both room and elevated ((121 °C) [250 °F]) temperatures. Load-displacement response and retention of strength were assessed accordingly, and preliminary results indicate that the epoxy prepreg film adhesive maintains a higher retention of apparent shear strength properties when tested at room temperature, but retention of apparent shear strength in the cyanate ester film adhesive is significantly improved during elevated temperature testing. Finite element analysis verification also indicates that more attention must be given to advanced surface preparation techniques for improved performance of the adhesive systems, especially the cyanate ester formulation.

KEY WORDS: Tensile Lap Shear, Epoxy Prepreg, Cyanate Ester, Film Adhesives.

1. INTRODUCTION

Structural high performance film adhesives such as epoxies or cyanate esters provide a reliable method for joining metallic and non-metallic assemblies used in increasingly hostile environments because of their strength and durability [1]. Notwithstanding, retention of adhesive properties in these environments subsequent to exposure in more demanding damp, elevated temperature conditions is also of significance. As the need for improved resistance to premature failure increases, so does the need for improved surface preparation techniques. In this study, the tensile lap shear performance of two commercially available 121 °C (250 °F) film adhesives is evaluated based on simple surface preparation of 316L stainless steel adherends. Failure behavior and property retention are characterized according to accelerated conditions of 24 hour submersion in boiling water and 100 hour thermal aging at 100% of cure temperature.

---

‡ [1] Intel Corporation, Aloha, OR 97007.

[2] Hewlett Packard Company, Corvallis, OR 97330.

[3] To whom correspondence should be addressed.

2. EXPERIMENTAL PROCEDURE

2.1 Adhesive Selection

The adhesive candidates selected for austenetic 316L rolled stainless steel, metal-to-metal bonding were commerically available in film form and based on either non-woven glass impregnated by bisphenol A/epichlorohydrin epoxy resin chemistry (EX-1548) or non-impregnated, thixotropic and toughened cyanate ester resin chemistry (EX-1516). Both film adhesive types required storage below -12.2 °C (10 °F) free from exposure to contamination, thus avoiding degradation prior to use. The EX-1548 prepreg and EX-1516 systems demanded initial cure temperatures of 113 °C (235 °F) and 121 °C (250 °F), respectively.

2.2 Surface Preparation

A simple surface preparation and conditioning technique was chosen as opposed to pickling (HNO_3/HF/H_2O solution) and etching (HCL/H_3PO_4/HF solution) used for more demanding applications. The 316L stainless steel adherends were surface conditioned with a three-dimensional abrasive and solvent ((CH_3)$_2$CHOH) wiped to remove possible surface film contaminants, followed by a 24 hour bake at 121 °C (250 °F) to remove residual moisture. Subsequent to baking, the adherends were handled with latex gloves to avoid surface oil contamination.

2.3 Adhesive Cure Cycles

Data reported for the EX-1548 and EX-1516 film adhesive TLS strength properties were based on the following cure cycles:

EX-1548 Epoxy Prepreg System:
   a) Ramp oven temperature to 121 °C (250 °F) at a rate of -15.8 °C (3.5 °F).
   b) Hold at 121 °C (250 °F) for 30 minutes minimum.
   c) Cool to below 71 °C (160 °F) at a rate of -13.6 °C (7.5 °F).
   d) Remove and allow additional cooling to room temperature.

EX-1516 Cyanate Ester System:
   a) Ramp oven temperature to 121 °C (250 °F) at a rate of -15.8 °C (3.5 °F).
   b) Hold at 121 °C (250 °F) for 5 hours minimum.
   c) Cool to below 71 °C (160 °F) at a rate of -13.6 °C (7.5 °F).
   d) Remove and allow additional cooling to room temperature.

2.4 Environmental Exposures

The 30 tensile lap shear specimens from each film adhesive system were evaluated for failure strength and property retention according to subjection to the following environmental exposure and testing conditions:

(5) Specimens: controls tested at T_{room}.
(5) Specimens: controls tested at $T_{100\% \, of \, cure}$.
(5) Specimens: subjected to 24 hour submersion in boiling water and tested at T_{room}[§]
(5) Specimens: subjected to 24 hour submersion in boiling water and tested at $T_{100\% \, of \, cure}$.
(5) Specimens: subjected to 100 hour thermal aging at $T_{100\% \, of \, cure}$ and tested at T_{room}.
(5) Specimens: subjected to 100 hour thermal aging at $T_{100\% \, of \, cure}$ and tested at $T_{100\% \, of \, cure}$.

2.5 Mechanical Testing

All room and elevated temperature tensile lap shear tests (Figure 1) were performed in accordance to ASTM D-1002-94 *Standard Test Method for Apparent Shear Strength of Single-Lap-Joint Adhesively Bonded Metal Specimens by Tension Loading* and ASTM D-2295-92 *Standard Test Method for Strength Properties of Adhesives*

[§] Note: $T_{room} = 25.6$ °C (78 °F), $T_{cure} = 121$ °C (250 °F).

in Shear by Tension Loading at Elevated Temperatures. A 90 kN (20 kip) servo-hydraulic Instron testing platform, coupled with an infrared (IR) quartz furnace, was employed to obtain load-displacement failure behavior as a function of temperature and substrate material. This apparatus allowed for either load or stroke control of monotonic and cyclic tensile and compressive loads.

All of the mechanical tests were performed in a manner that permitted the axis of the applied tensile load to coincide with the long axis of the specimen. The wedges in one of the grip assemblies were shimmed by an amount equal to the thickness of the adherend opposite the grip. The specimens were placed in the jaws so that the outer 25.4 mm (1.0 in.) of each end remained in direct contact and load was applied at a constant crosshead displacement rate of 1.3 mm (0.05 in.)/min.

For the elevated temperature testing, a type-K thermocouple was attached to the gage section of the specimen using a capacitive discharge spot welding method. The thermocouple served as the feedback loop for temperature control via an Omega series 9000 CN temperature controller. As the setpoint temperature was approached, the controller would regulate the pulse delay for the IR furnace. Overshoot was nominally less than 2.0% in all cases. Figure 2 depicts the testing and data acquisition system used in determining the failure load-displacement response.

3. RESULTS AND DISCUSSION

The primary objective of this effort was to evaluate the independent performance of two commercially available structural adhesives bonded to stainless steel substrates using tensile lap shear test methods and simple surface preparation techniques as opposed to more difficult methods involving pickling and etching.

3.1 Epoxy Prepreg System

A summary of the tensile lap shear test results and observed failure behavior is presented in Table I. Examination of the control data reveals some scatter, however the strength characteristics are good and compare reasonably well with other available studies using similar adhesive systems. It is surprising that the scatter was higher for the control specimens tested at 121 °C (250 °F), especially since the joint "squeeze out" observed after curing was insignificant. Similarly, the scatter is consistently higher for all of the specimens tested under elevated temperature conditions.

A typically observed superficial cohesive failure (SCOH) is depicted in Figure 3. By using a cold cathode discharge (CCD) camera, appropriate image filters and thresholding, the area fraction for the dominant mode of bond failure was determined. The majority of the epoxy prepreg specimens failed in either a cohesive or superficially cohesive manner. Adhesive failure was only predominantly observed in the 100 hour, thermally aged specimens tested at 100% of curing temperature.

3.2 Cyanate Ester System

The experimental TLS data and observed failure behavior for the cyanate ester system are also summarized in Table I. The scatter and inconsistency are also similar to that observed for the epoxy prepreg system, but what is unanticipated is that the apparent shear strengths at failure are much lower than reported elsewhere [2]. Similar to the CCD imaging methods used for the epoxy prepreg, the cyanate ester (Figure 4) shows significantly less coverage and cohesive failure behavior. In fact, a far larger number of the cyanate ester specimens failed adhesively under all test conditions. This may be explained by the occurrence of "squeeze out" at the adhesive bondline in nearly all of the cyanate ester specimens. Apparently the viscosity of this system becomes much lower during the temperature ramping stage of curing. Clearly, either a fiber supported film and/or an etched surface would most likely result in less "squeeze out", less data scattering and better overall failure characteristics.

3.3 Statistical Evaluation

Figure 5 depicts the mean apparent shear strength and standard deviation for both adhesive systems under control, boiling water submersion and thermal aging conditions. For both adhesive systems, there was a significant reduction in mean strength when exposed to boiling water for a 24 hour duration. The reduction was intensified when these specimens were subjected to elevated temperature testing. Although the majority of the epoxy specimens failed cohesively, the observed adhesive failure in both systems was most likely due to absorption (stiffness changes) and weak interfaces degenerating even more so when submerged.

Both systems responded more favorably to 100 hour thermal aging at 100% of respective cure temperatures, perhaps due to extending curing of the adhesives. The epoxy prepreg adhesive showed an increase of mean TLS strength from 19.6 MPa (2,844.1 psi) at room temperature to 21.6 MPa (3,130.2 psi) at 121 °C (250 °F). The cyanate ester adhesive showed only a marginal decrease from 7.0 MPa (1,005.1 psi) to 6.8 MPa (990.6 psi). However, in both instances the standard deviation increased, indicating more scatter in data.

In reference to Figure 6, the statistical trend was similar with more variation occurring as a function of elevated temperature testing. The only exceptions were the thermally aged cyanate ester specimens tested at room temperature. Clearly, the least amount of confidence can be given to the cyanate ester specimens subjected to boiling water. After observing the joints subsequent to testing, this is not completely expected since several failures were superficially cohesive and not purely adhesive.

3.4 Apparent Shear Property Retention

Shear strength retention in both systems was determined in reference to the control specimens. As shown in Figure 7, the epoxy adhesive retained 74.2% of its shear strength when subjected to boiling water and tested at room temperature. Retention after thermal aging exceeded the mean strength of the room temperature control specimens by 10%, indicating extended curing beyond the initial 30 minute cure time. In both cases involving 24 hour boiling, the adhesive retained only 25-30% of its initial strength.

Similar trends occurred for the cyanate ester thermally aged specimens indicating additional curing. In this case, the retained strength exceeded the elevated temperature control by 24.2%. It is not clear why the trend in retention increased as a function of test temperature after thermal aging.

3.5 FEA Verification

A finite element analysis was performed of single lap joint apparent shear strength in both the epoxy prepreg and cyanate ester systems. The focus was to evaluate the shear strength of these film adhesives assuming perfect surface preparation and hence, bonding conditions. The prescribed displacements for each adhesive model were chosen based on the results obtained for the experimental displacements at failure. As and example, Figure 8 shows typical load-displacement response observed for the epoxy prepreg under all exposures. From the statistical analysis, a mean room temperature, control failure of 19.6 MPa (2,844 psi) corresponded to a mean displacement of 1.4 mm (.055 in.). A prescribed displacement equivalent to the experimental failure displacement led to a maximum numerical apparent shear stress along the cohesive bondline path (Figure 9) of 22.4 MPa (3,245.7 psi). If an assumption is made that the difference between numerical and experimental results indicates efficiency of adhesion, then it is possible that the simple surface preparation technique employed herein was 88% effective for the epoxy system.

The same analogy was also extended to the study of the less than ideal failure strength in the cyanate ester system. A mean room temperature, control failure of 7.0 MPa (1,005 psi) was equivalent to a mean failure displacement of 0.8 mm (.032 in.). This possibly indicated that the same surface preparation technique was only 30% effective, considering that the numerical result was 28.9 MPa (4,191.3 psi). This seems reasonable since experimental results of equal magnitude have been reported previously using advanced techniques such as scale removal via pickling and acid etching. [**]

4. CONCLUSIONS

- The epoxy prepreg adhesive perfomed acceptably when exposed to 24 hour boiling water submersion and 100 hour thermal aging and tested at room temperature. The majority of bondline failures appeared to be cohesive.
- The overall TLS strength of the cyanate ester was quite low compared with published values, however the adhesive property retention at 121 °C (250 °F) after thermal aging was quite favorable.
- Preliminary numerical results and bondline failure examination indicated that simple surface preparation of 316L stainless steel was not applicable to suitable adhesive performance, especially for the cyanate ester formulation.

[**] Material Properties: Cyanate Ester (E_{11}=3.5 GPa, v_{12}=0.30), 316L (E_{11}=200 GPa, v_{12}=0.305)

5. REFERENCES

1. LeGrand, D. S., *Performance Analysis of Tough Film Adhesives*, 26th International SAMPE Technical Conference, 1994: 225.
2. Hamerton, I., Ed.: "Chemistry and Technology of Cyanate Ester Resins", Blackie Academic & Professional, 1994: A.4 340-341.

EX-1548 and EX-1516 are trademarks of Bryte Technologies, Inc.

TLS Strength at Failure (MPa) [psi]							
Adhesive	Adherend	Condition	Temperature (C) [F]	Specimen 1	Specimen 2	Specimen 3	Specimen 4	Specimen 5
Epoxy Prepreg	316L-SS	control	(25.6) [78]	(21.9) [3181.6]	(19.2) [2785.2]	(19.7) [2853.4]	(19.5) [2828]	(17.7) [2572.2]
Epoxy Prepreg	316L-SS	control	(121) [250]	(6.6) [959.1]	(11.5) [1666]	(6.9) [1002]	(7.2) [1049]	(7.3) [1054.8]
Epoxy Prepreg	316L-SS	24 hr boil	(25.6) [78]	(14.1) [2045]	(16.2) [2355.4]	(14.3) [2078.2]	(14.3) [2076.2]	(13.8) [1996.2]
Epoxy Prepreg	316L-SS	24 hr boil	(121) [250]	(1.6) [236.4]	(2.5) [367.2]	(2.2) [318.2]	(2.4) [349.6]	(2.7) [388.6]
Epoxy Prepreg	316L-SS	100 hr exposure	(25.6) [78]	(19.2) [2791]	(22.1) [3195.4]	(19.6) [2840]	(25.3) [3664.2]	(21.8) [3160.2]
Epoxy Prepreg	316L-SS	100 hr exposure	(121) [250]	(1.8) [265.6]	(2.2) [312.6]	(1.8) [265.6]	(0.6) [91.6]	(3.3) [480.6]
Cyanate Ester	316L-SS	control	(25.6) [78]	(7.8) [1134.8]	(5.7) [832]	(7.1) [1025.6]	(5.7) [834]	(8.3) [1199.2]
Cyanate Ester	316L-SS	control	(121) [250]	(4.3) [629]	(2.1) [306.6]	(2.7) [396.4]	(4.4) [632.8]	(4.7) [685.6]
Cyanate Ester	316L-SS	24 hr boil	(25.6) [78]	(1.5) [211]	(1.5 [222.6]	(3.0) [429.6]	(2.7) [398.4]	N/A
Cyanate Ester	316L-SS	24 hr boil	(121) [250]	(0.2) [30]	(1.2) [169.8]	(0.2) [30.5]	(1.8) [261.6]	(1.9) [271.4]
Cyanate Ester	316L-SS	100 hr exposure	(25.6) [78]	(8.6) [1242.2]	(6.9) [1000]	(5.8) [840]	(7.6) [1097.6]	(5.3) [773.4]
Cyanate Ester	316L-SS	100 hr exposure	(121) [250]	(4.9) [716.8]	(5.0) [726.6]	(3.3) [482.4]	(4.3) [621]	(5.1) [744.2]

Observed Failure Behavior							
Adhesive	Adherend	Condition	Temperature (C) [F]	Specimen 1	Specimen 2	Specimen 3	Specimen 4	Specimen 5
Epoxy Prepreg	316L-SS	control	(25.6) [78]	COH	SCOH	SCOH	COH	SCOH
Epoxy Prepreg	316L-SS	control	(121) [250]	SCOH	SCOH	COH	COH	COH
Epoxy Prepreg	316L-SS	24 hr boil	(25.6) [78]	COH	COH	SCOH	COH	COH
Epoxy Prepreg	316L-SS	24 hr boil	(121) [250]	SCOH	SCOH	SCOH	COH	SCOH
Epoxy Prepreg	316L-SS	100 hr exposure	(25.6) [78]	COH	COH	SCOH	COH	SCOH
Epoxy Prepreg	316L-SS	100 hr exposure	(121) [250]	SCOH	ADH	ADH	SCOH	COH
Cyanate Ester	316L-SS	control	(25.6) [78]	COH	ADH	SCOH	SCOH	COH
Cyanate Ester	316L-SS	control	(121) [250]	SCOH	ADH	SCOH	COH	ADH
Cyanate Ester	316L-SS	24 hr boil	(25.6) [78]	ADH	SCOH	SCOH	SCOH	N/A
Cyanate Ester	316L-SS	24 hr boil	(121) [250]	SCOH	SCOH	SCOH	SCOH	SCOH
Cyanate Ester	316L-SS	100 hr exposure	(25.6) [78]	ADH	SCOH	SCOH	ADH	ADH
Cyanate Ester	316L-SS	100 hr exposure	(121) [250]	SCOH	SCOH	ADH	ADH	COH

Notes:
COH denotes strictly cohesive failure.
SCOH denotes superficial cohesive failure.
ADH denotes strictly adhesive failure.

Table I: Experimental results and observed failure behavior in the epoxy prepreg and cyanate ester adhesive systems.

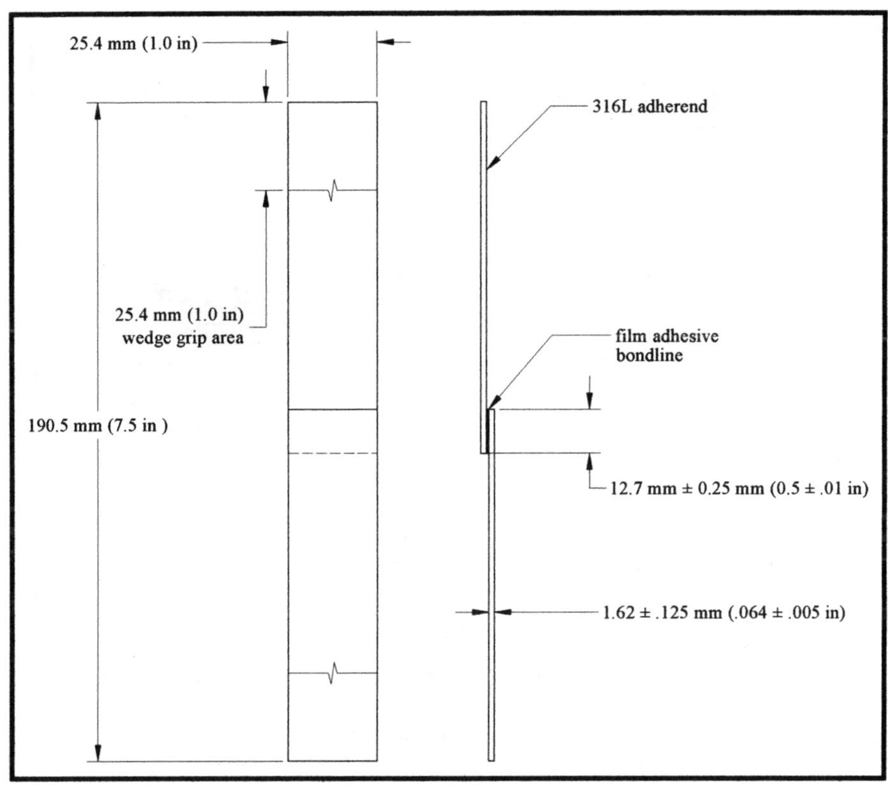

**Figure 1: Depiction and dimensions of tensile lap shear specimens used for
for evaluating the apparent shear performance of the film
adhesives.**

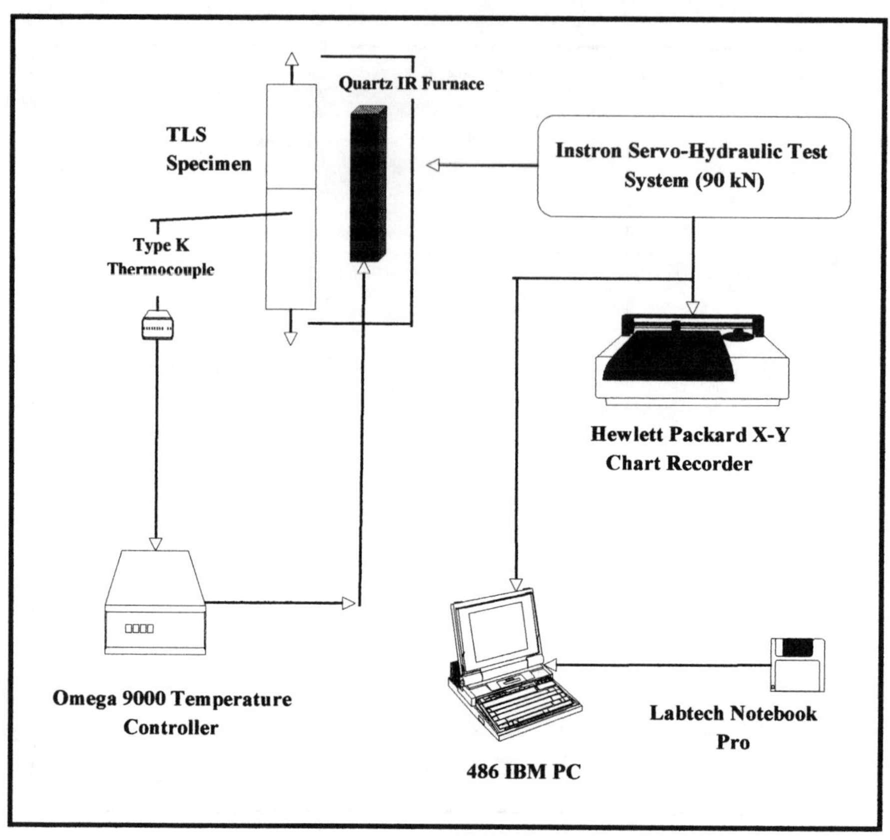

Figure 2: Experimental setup used for room and elevated temperature shear lap joint testing of the film adhesives.

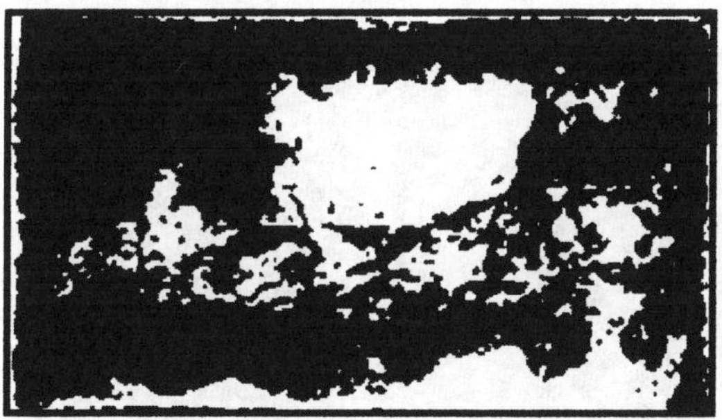

Figure 3: Observed behavior in an epoxy prepreg film adhesive specimen: a) CCD image of SCOH failure, b) calculated failure of 31.4% SCOH failure.

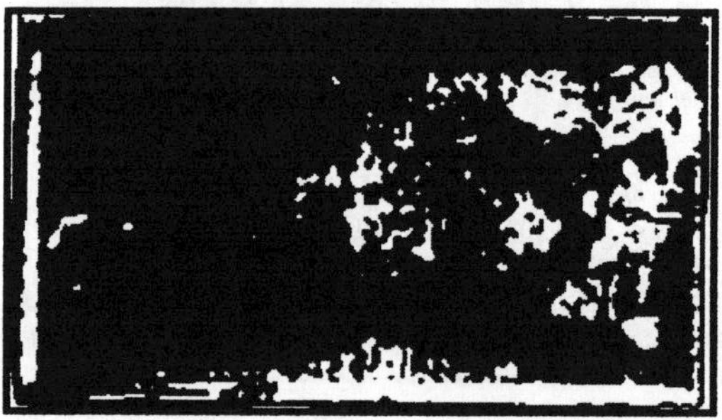

Figure 4: Observed behavior in an cyanate ester film adhesive
specimen: a) CCD image of SCOH failure, b) calculated
failure of 17.2% SCOH failure.

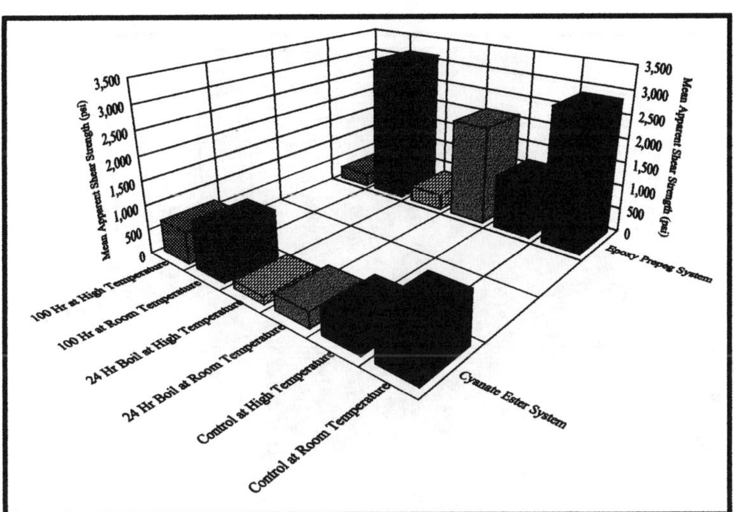

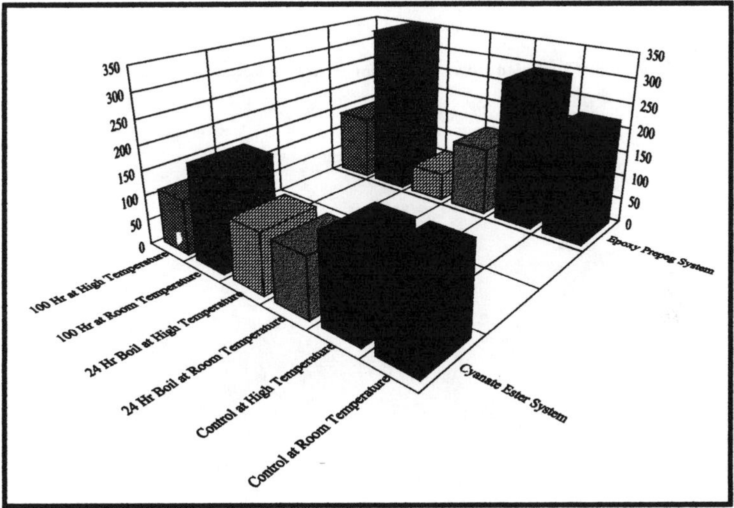

Figure 5: a) Mean strength of both systems and b) standard deviation of strengths with respect to all exposures.

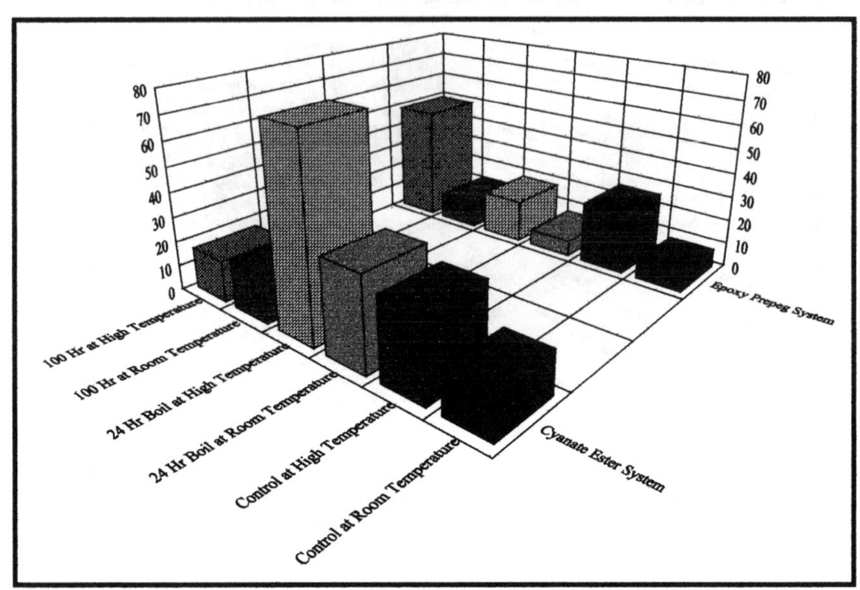

Figure 6: Coefficient of variation for both film adhesive systems.

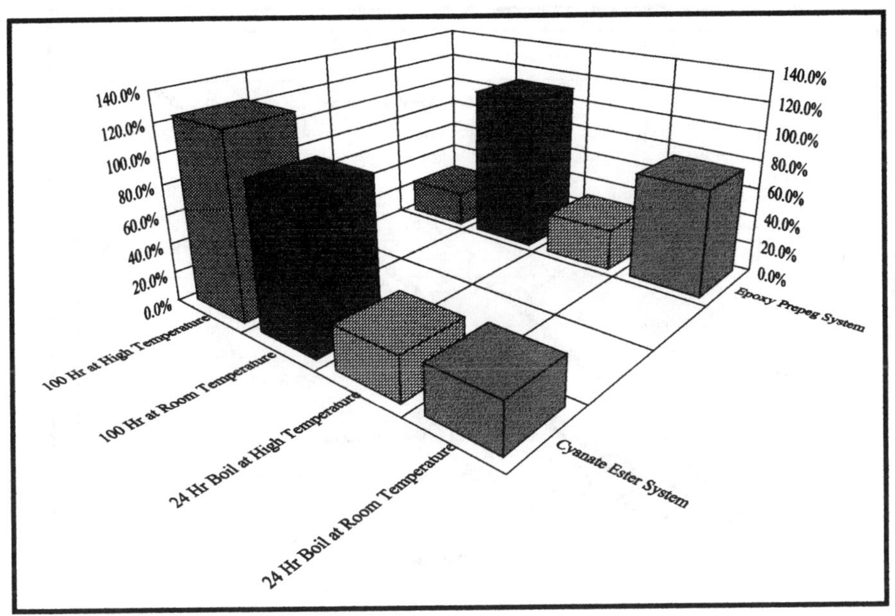

Figure 7: Percent retention of properties for both film adhesive systems.

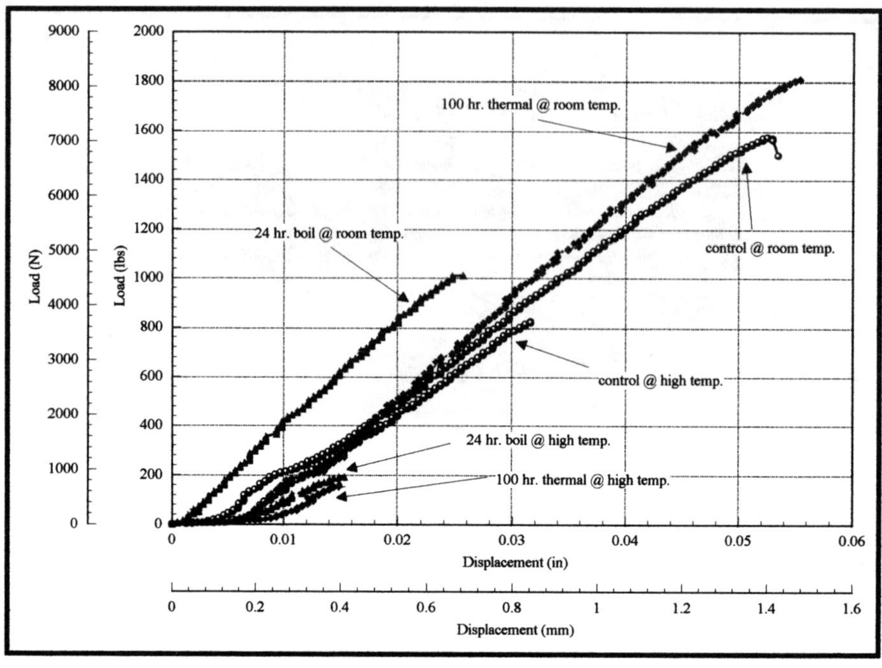

Figure 8: Typical data acquired for the load-displacement response of the epoxy prepreg film adhesive.

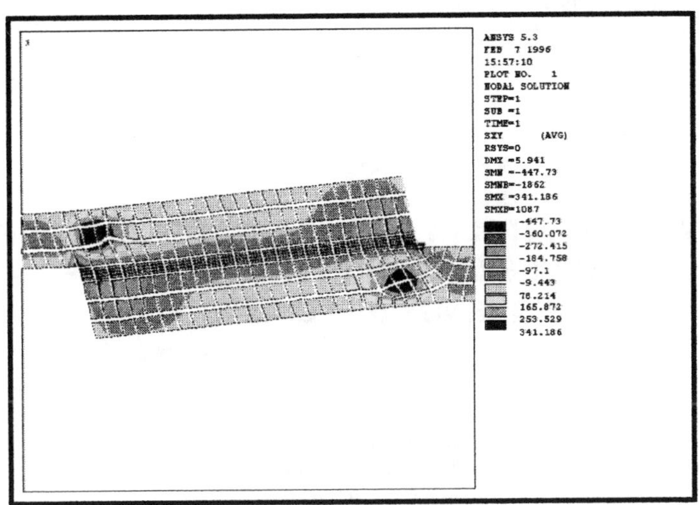

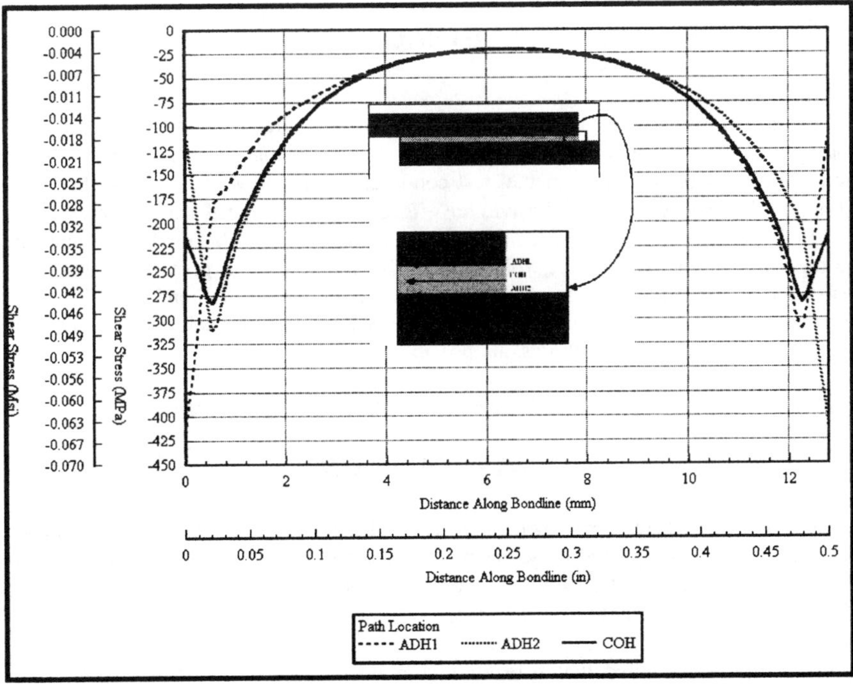

Figure 9: Finite element analysis of epoxy prepreg adhesive bondline to determine apparent strength under ideal adhesion conditions.

43rd International SAMPE Symposium
May 31-June 4, 1998

ENVIRONMENTALLY RESISTANT, CONDUCTIVE ADHESIVE BONDS FOR RADIO FREQUENCY (RF) SHIELDING

John M. Kolyer
Boeing
Guidance, Navigation and Sensors
3370 Miraloma Avenue
Anaheim, CA 92803-3105

ABSTRACT

A conductive bonding system (adhesive, surface finish, and edge sealant) for aluminum-to-aluminum joints in enclosures for Radio Frequency (RF) shielding remains strong and highly conductive under extreme environmental test conditions (salt spray/1,000 hours, then 85°C/1,000 hours). Silver-filled epoxy adhesives are effective, while a nickel-filled epoxy adhesive failed to maintain low electrical resistance. Oxide layers such as chemfilm or chromate etch are too high in electrical resistance, while freshly abraded aluminum gives only temporarily low resistance, presumably because of gradual surface oxidation at the bond line. Therefore, plating is necessary, preferably gold. An epoxy edge sealant excludes salt/moisture, but other materials, e.g., silicones or polysulfides, are possible.

KEY WORDS: Adhesives, Electrically Conductive Materials, Epoxy Resins

1. INTRODUCTION

In a 1993 study (1), electrical resistance increased with thermal aging for a silver-filled epoxy adhesive bonded to nickel-plated or solder-plated surfaces, but resistance with a gold-plated surface remained low even after aging at 85°C/720 hours. The recommendation was to bond silver-filled epoxies only to gold-plated surfaces.

A supplier's article (2) noted that bare aluminum bonded with silver-filled epoxies gave erratic and often high electrical resistance, presumably because of aluminum oxide formation at the interface.

Neither References 1 or 2 included a nickel-filled adhesive or involved salt-spray or high-humidity exposures. One supplier recommended a nickel-filled epoxy adhesive for salt environments without supporting data.

In the absence of detailed available information, the present study was deemed necessary to aid design of RF-protected enclosures and windows. The objective was retention of minimum lap-shear strength of 3.4 MPa (500 psi) and maximum electrical resistance of 16 milliohms/cm^2 bonded area (2.5 milliohms per joint (3) assuming a 1 in.^{2} joint) after salt spray for 1,000 hours followed by 85°C exposure for 1,000 hours.

2. EXPERIMENTAL

2.1 Materials One nickel-filled and five silver-filled epoxy adhesives were tested. Table 1 gives key properties of the three preferred silver-filled adhesives, designated A, B, and C; the others are D and E. A chromate primer (4) was applied before sealing in some cases. Sealants included an anhydride-cured epoxy encapsulant requiring elevated-temperature cure (65°C/24 hours), a thixotropic epoxy adhesive curing at room temperature, a polysulfide (5), and a room-temperature vulcanizing (RTV) silicone (6).

2.2 Lap-Shear Specimens Finger panels of standard dimensions (7), containing four 2.5 x 12.7 cm strips, had an aluminum alloy 2024 core with alloy 1230 (99.3%, minimum, aluminum) surface layers (8).

2.2.1 Surface Treatment Abrasion was by hand-scrubbing with number 320 silicone carbide abrasive paper. Chemfilming (9) was the conductive Class 3. Etching (10) was with sodium dichromate and sulfuric acid. Table 2 gives plating conditions.

2.2.2 Bonding The finger panels were wiped, then sprayed, with acetone and air-dried before adhesive was applied to both surfaces. The overlap was 1.3 cm as prescribed (7), giving a bonded area of 1.3 x 2.5 cm. With two nylon monofilaments (0.23 mm diameter) added as spacers, application of a force of approximately 1300 N resulted in a bond line of approximately 0.15 mm. Adhesives A, B, D, E, and nickel-filled were cured at 71°C/2 hours, adhesive C at 80°C/2 hours.

2.2.3 Sealing The bonded area, including gold plating, was acetone-rinsed and covered with a jacket of sealant approximately 3 mm thick.

2.3 Exposure Conditions Salt spray (salt fog) exposure was at approximately 35°C under standard conditions (11). High-humidity exposure was at 100% relative humidity (RH)/35°C (representing the salt-spray conditions without salt) and at 95% RH/85°C in a sealed vessel over a saturated aqueous solution of potassium sulfate (12). Elevated-temperature exposure was in a circulating-air (forced draft) oven at 85°C ± 2°C.

2.4 Test Methods The same specimen served for periodic electrical testing during exposure and a final strength test. Thus the number of samples needed was reduced, and electrical resistance and mechanical strength could be better correlated.

2.4.1 Lap-Shear Strength Shear in tension was determined with an Instron 1127 testing machine at a strain rate of 1.3 mm/min by the standard method (7).

2.4.2 Electrical Resistance of Bond Clamps were attached to the bonded specimen as shown in Figure 1. The resistance was read, using two leads on a calibrated Hewlett-Packard 4161A or 4262A LCR Meter, between clamps 1 and 2 (reading A), clamps 3 and 4 (reading B), clamps 1 and 4 (reading C), and clamps 2 and 3 (reading D). Since readings C and D include the bond resistance R, algebraic manipulation gives 2R = C+D-A-B, or R = (C+D-A-B)/2. If

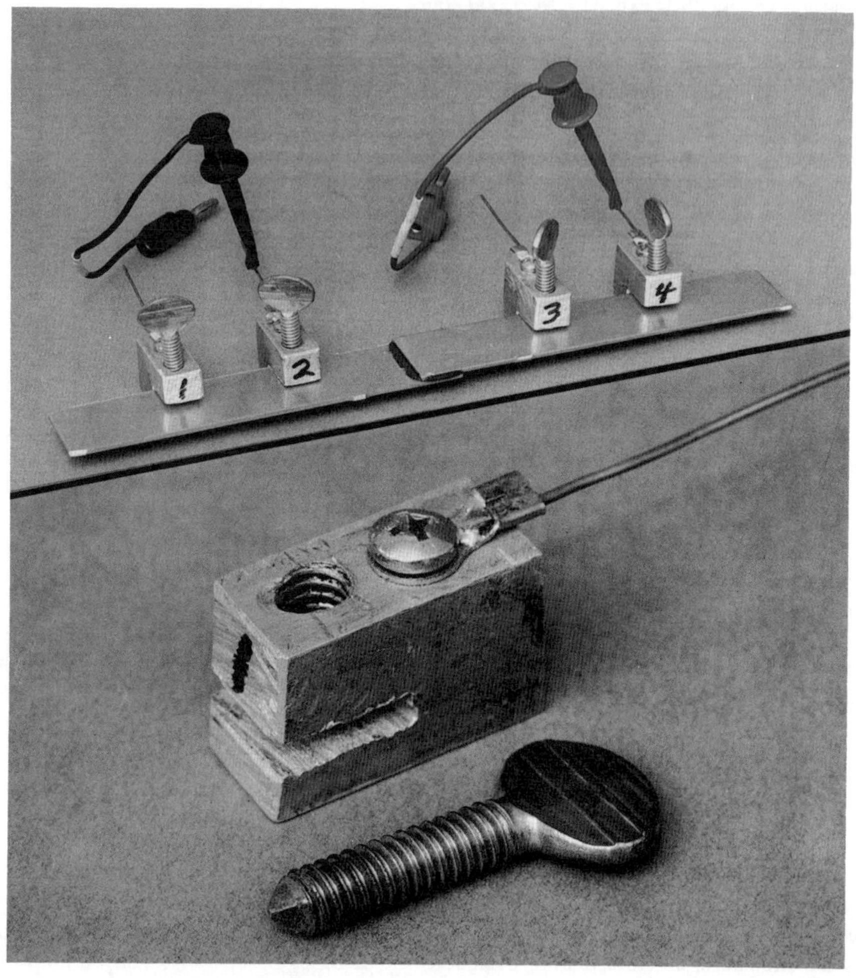

only two leads are used, the resistances of the leads cancel each other. Before attaching the clamps, the points on the 6-32 thumb screws were stroked with No. 320 abrasive paper, and the 18 AWG copper wires (on which clips were fastened to take readings) were polished with No. 600 abrasive paper to remove oxide. Based on observed variability, the calculated estimated error of resistance values at the 95% confidence level was ±2 milliohms for values of 50 milliohms or less and ±5% for higher values. Confirmatory checks were made with known resistance. The target resistance of 16 milliohms/cm.^{2} of bonded area is equivalent to 5 milliohms for the 3.2-cm^2 bonded area of the lap-shear specimens.

3. RESULTS AND DISCUSSION

3.1 Initial Properties Initial properties were determined to rule out bonding systems inadequate even before exposure.

3.1.1 Lap-Shear Strength The goal of 3.4 MPa, minimum, was met with all surface treatments by the nickel-filled adhesive and the preferred epoxy adhesives A, B, and C (Table 1). The other two epoxy adhesives were inadequate on gold plating (2.1 and 2.6 MPa lap-shear versus 3.9-4.9 MPa for the preferred adhesives) and were rejected because gold plating was selected for the preferred bonding system (see below).

Table 1

Preferred Silver-Filled Epoxy Adhesives (Suppliers' Data, Typical Properties)

Designation	Lap-Shear Strength, MPa, alum.-to-alum.	Volume Resistivity, ohm-cm	Advantages
A	5.5	2×10^{-4}	Very high electrical conductivity. Service range to +130°C. Low outgassing.
B	6.9	2×10^{-3}	Room-temperature curing. Service range to +130°C. Easily spread paste.
C	6.9	4×10^{-4}	"Stress-free" for bonding materials with mismatched coefficients of thermal expansion. Flexible at low temperatures (Tg = -20°C, supplier's data). Service range to +130°C. Low outgassing.
NOTE: Tg = glass transition temperature.			

3.1.2 Electrical Resistance All six adhesives tested gave low resistance (0-2 milliohms) against a plated surface or a freshly abraded aluminum surface. The abraded surface showed troughs scooped by the particles on the abrasive paper (Figure 2), and most of the electrical contact may have been with the small area of fresh, bare aluminum exposed at the bottom of these troughs. Consistent with this small area was a limited increase in oxygen/aluminum ratio with time seen by energy dispersive spectroscopy. When the surface was oxidized (as-received, chemfilmed, or etched), only the nickel-filled adhesive gave low resistance, presumably because the hard nickel particles, unlike soft silver particles, penetrated the oxide layer. Thus, this adhesive could be used with chemfilm, which confers a degree of corrosion resistance, under mild conditions but not at elevated temperature as desired (see below).

3.2 Properties after Environmental Exposure Electrical resistance data are presented as graphs (Figures 4-9). For specimens surviving less than 42 days, bond failure occurred within a week after the last data point.

3.2.1 Salt Spray Figure 3 shows typical unsealed, abraded specimens before and after salt-spray exposure. Plated specimens also corroded heavily over the entire surface, starting at

813

Figure 2

Abraded Aluminum Surface, X 1,000

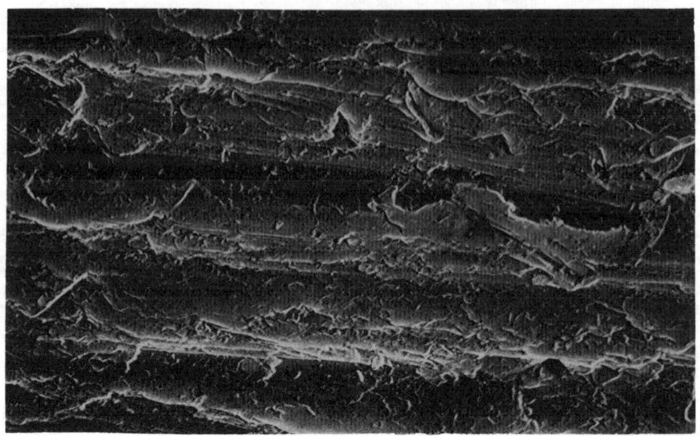

pinholes in the nickel plating. Therefore, samples for edge-sealing were plated only up to 2.5 cm from the end to be bonded (Table 2). As seen in Figure 4, none of the abraded, unsealed specimens survived to be lap-shear tested. The nickel-filled adhesive, recommended by its supplier for salt environments, did retain bond integrity the longest but lost conductivity as rapidly as the silver-filled adhesives. Plating seemed necessary, and in fact with silver or gold plating the resistance stayed low, but all bonds failed in less than 30 days. Next, sealing was tried with bonds with etched aluminum. All four sealants tested were effective, but the room-temperature-curing epoxy adhesive added the most to lap-shear strength. With an abraded surface, this sealant was relatively effective for salt-spray resistance, but subsequent heating caused large rises in resistance (Figure 5), presumably due to thermally accelerated surface oxidation of the aluminum at the bond line. Both plating and sealing were needed, as will be seen.

3.2.2 *High Humidity* As expected, with an abraded surface at 100% RH/35°C (salt spray conditions without salt) visible corrosion was far less and bonds survived longer than in salt spray, but resistance rose rapidly (Figure 6). Adhesives A and B with gold plating maintained low resistance, and the average bond strength after 42 days (1,000 hours) was 1.2 MPa for A and 4.1 MPa for B. At 95% RH/85°C, with gold plating, all bonds failed within seven days for the nickel-filled adhesive and adhesive B, but one bond with adhesive A survived 35 days, and the other three bonds survived 42 days and averaged 1.3 MPa, Again, low resistance was maintained. The conclusion was that edge sealing is needed to exclude moisture even in the absence of salt.

3.2.3 *Elevated Temperature* At 85°C, bonds with an abraded surface and the nickel-filled adhesive or adhesives A or B (D and E were not tested) unacceptably increased in resistance in ambient air (Figure 7), dry air (closed vessel over anhydrous calcium sulfate), or nitrogen (previously purged pressure vessel with dry nitrogen at 0.21 MPa gauge pressure). Surface oxidation under nitrogen might have been caused by oxygen dissolved in the adhesive. Consistent with oxidation causing increased resistance was the retention of low resistance (0-2 milliohms) with a silver- or gold-plated surface for adhesives A and B but not the nickel-filled adhesive (Figure 8). Nickel surfaces oxidize readily, so high resistance may be caused by interfacial resistance between oxide-coated nickel particles and the plating. Again, there

814

Figure 3

Specimens Before and After Salt Spray Exposure

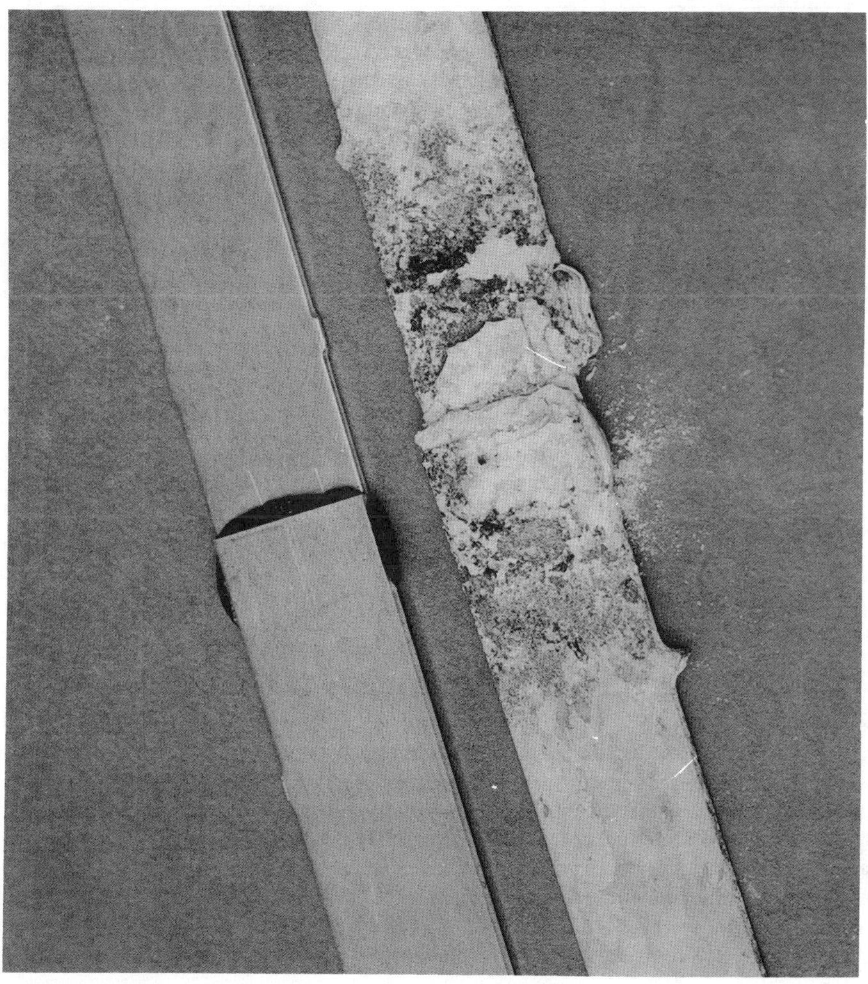

was a rough correlation between increased resistance and decreased lap-shear strength, e.g., 3.9 MPa at 3 milliohms and 1.1 MPa at 8 milliohms for the nickel-filled adhesive on abraded aluminum after 85°C/42 days. All the bonds in Figure 8 for adhesives A and B survived, averaging 5.1 MPa for adhesive A and 3.9 MPa for adhesive B. Adhesive A failed cohesively, as preferred for reliability, whereas adhesive B failed adhesively. One surviving nickel-filled adhesive bond broke on handling, and the other gave 1.7 MPa and failed adhesively.

3.2.4 Salt Spray Followed by Elevated Temperature At this point in the testing, the nickel-filled adhesive was eliminated because it gave increasing resistance at 85°C even on gold plating. For adhesives A and B (and presumably C, which was tested less), plating was

815

Table 2
Plating Procedure

Plating Step	Description	Specification
Degreasing	Cleaning with aqueous alkaline cleaner (Brulin 815GD)	Boeing AA0110-049A
Cleaning	Cleaning with aqueous alkaline cleaner	Boeing AA0109-023AB and AB0210-032A, Type I
Etching, deoxidizing, zincate, copper strike	Preparation of basis material for final plating	Boeing AAA0109-023AB
Nickel plating	Deposition of 13-25 micrometers nickel (electroless) over the entire surface	AMS 2404C (13)
Silver plating	Deposition of 13-25 micrometers silver over the last 5 cm of the end to be bonded	QQ-S-365D (14)
Gold plating	Deposition of 1.3-2.5 micrometers high-purity gold over the last 5 cm of the end to be bonded	MIL-G-45204C (15)

NOTE: Specimens for edge-sealing were nickel-plated and gold-plated only up to 2.5 cm from the end to be bonded, so that the jacket of sealant would cover the whole plated area and salt-induced corrosion could not start at pinholes in the plating.

necessary to survive 85°C with low resistance, and gold was chosen over silver because gold gave longer bond integrity in salt spray and because silver tarnishes. The epoxy adhesive had been selected as the preferred edge sealant, and now it was necessary to demonstrate the effectiveness of the preferred system: adhesive A, B, or C with gold plating and the chosen sealant. Figure 9 shows full success with adhesives A and C, while B gave the required resistance (<5 milliohms in terms of the graph) up to 500 hours (21 days) at 85°C. The final lap-shear strength averaged 10.1 MPa for A (cohesive failure), 4.6 MPa for B (adhesive failure), and 5.2 MPa for C (largely cohesive failure).

3.3 Statistical and Predictive Significance of Results Both electrical resistance and mechanical strength often differed greatly among the four samples of a lap-shear set. Therefore, similar erratic behavior is to be expected for joints in actual hardware, of which the lap-shear bond is a realistic representation. In fact, individually hand-assembled hardware bonds might be even more prone to variation than the lap-shear specimens prepared in sets. Survival of all four specimens of a set under particular environmental conditions is accounted "success" in this work, but of course a fifth specimen might have failed. Obviously, confidence given by passage of only four successive specimens in a context of erratic specimen-to-specimen behavior is limited. Passing a test with four replicates suggests that a high percentage of a thousand replicates would pass, this percentage being the confidence level, but only running more replicates would firm up an estimate of this percentage, e.g., 95%. In conclusion, all that can be said is that passing a severe test with mediocre confidence

816

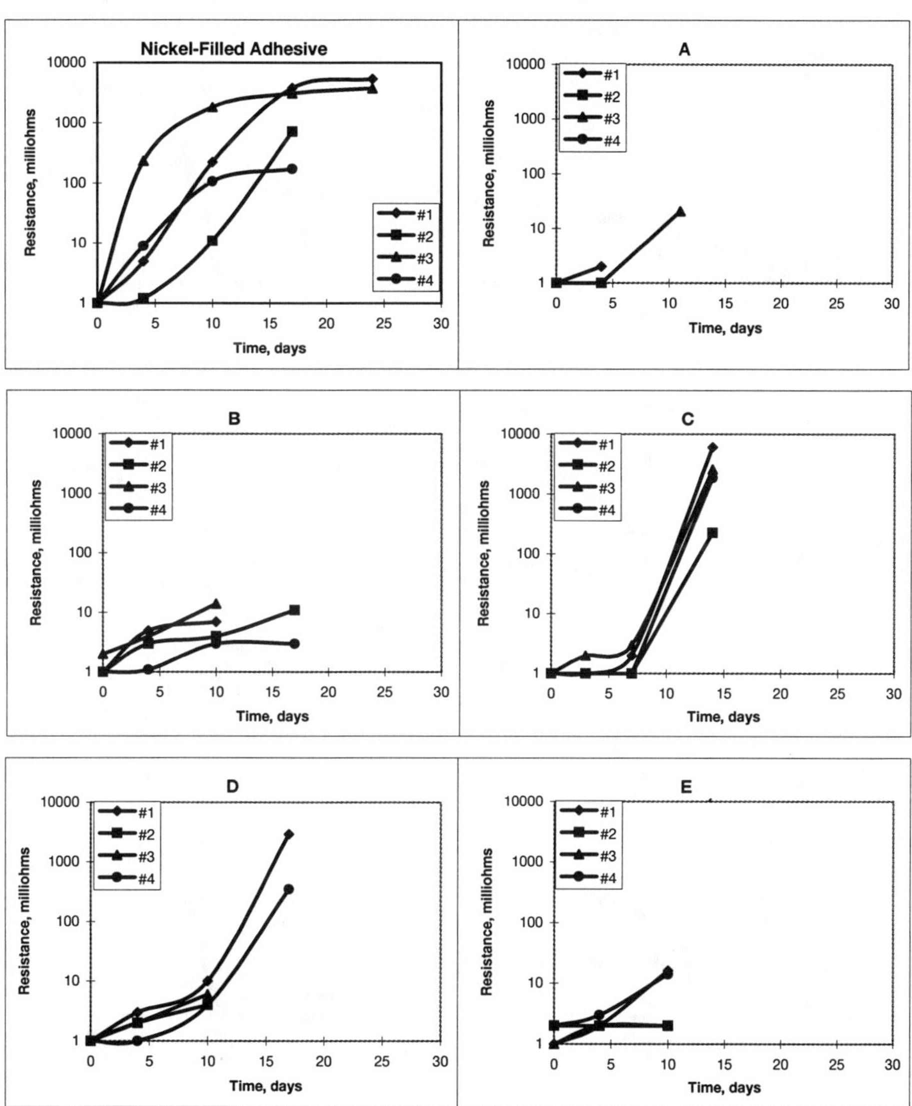

inspires relatively high confidence in performance at milder operating conditions; for example, passing the 85°C/1,000 hour exposure test suggests, but does not guarantee, successful long-term operation of hardware at room temperature.

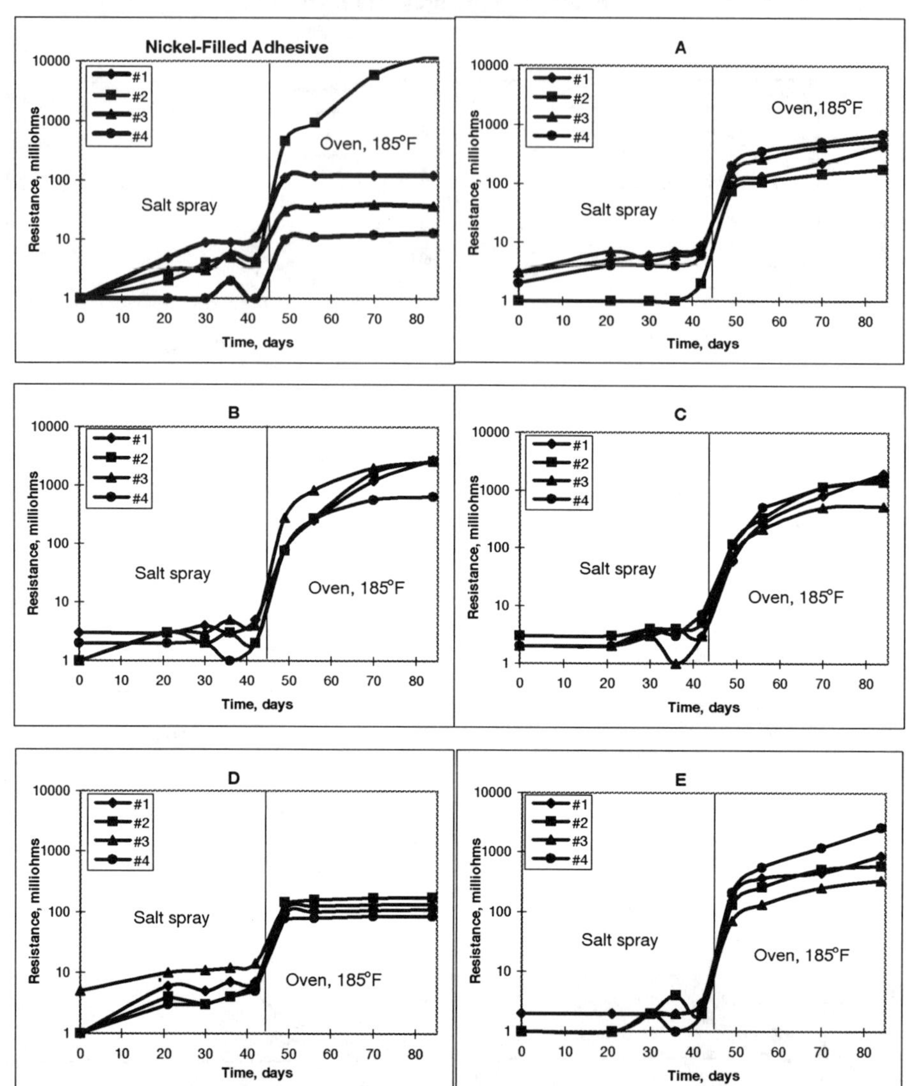

Figure 5. Salt Spray, then 185°F, Abraded, Primed, Sealed with Epoxy

4. CONCLUSION

The objective was an aluminum-to-aluminum conductive bonding system (adhesive, surface finish, and edge sealant) surviving 1,000 hours of salt spray followed by 1,000 hours at 85°C, with retention of mechanical strength and electrical conductivity as stated above. This goal was attained with a silver-filled epoxy, gold plating, and an epoxy adhesive edge sealant.

The two successful adhesives were a high-performance, general purpose product and a low-Tg, "stress free" product suitable for bonding materials with mismatched thermal expansion coefficients.

Figure 6. 100% RH/95°F, Abraded

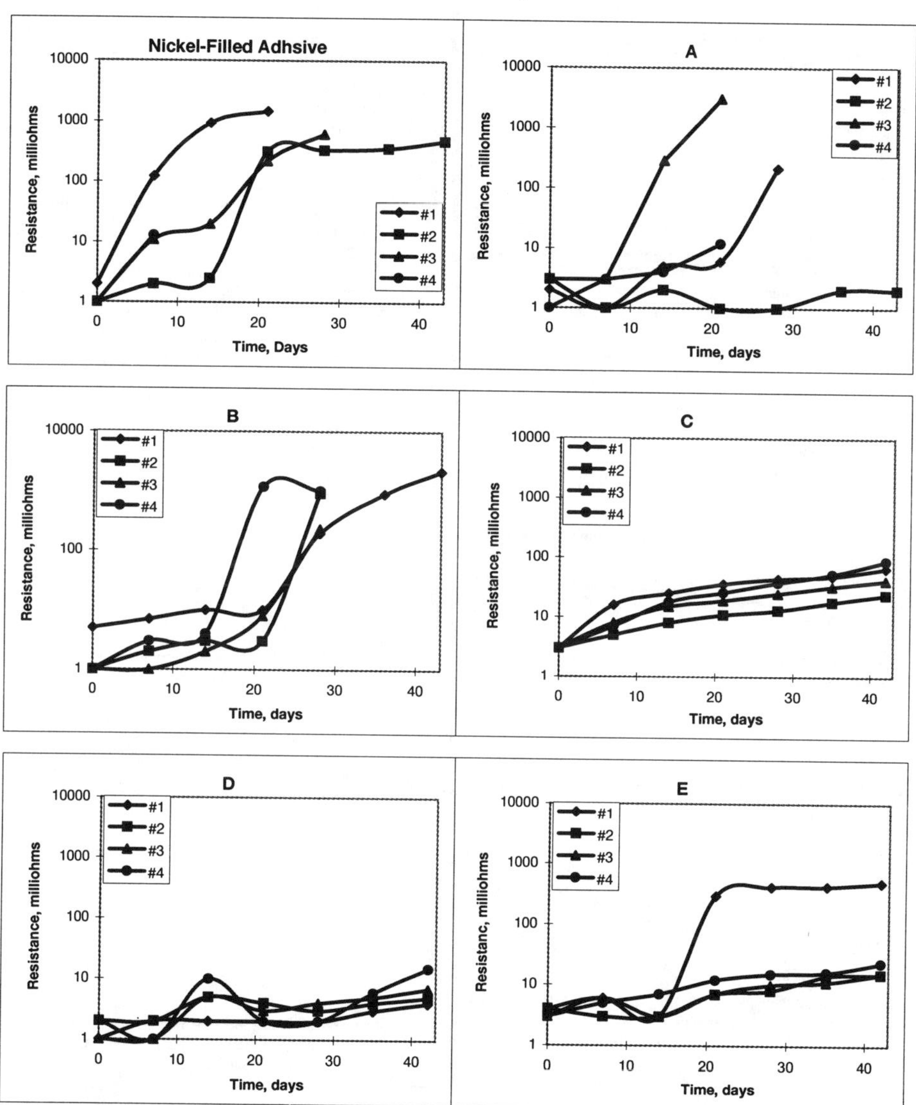

Silver plating is also effective, and many edge sealants, e.g., silicones, polysulfides, polyurethanes, or vapor-deposited coatings, are likely candidates for testing.

Selection of a bonding system will depend on the application and may require real-time testing under anticipated operating conditions for the highest confidence.

5. SUMMARY

The objective was a conductive bonding system (adhesive, surface finish, and edge sealant) for aluminum-to-aluminum joints in enclosures for RF shielding. The bond had to resist salt

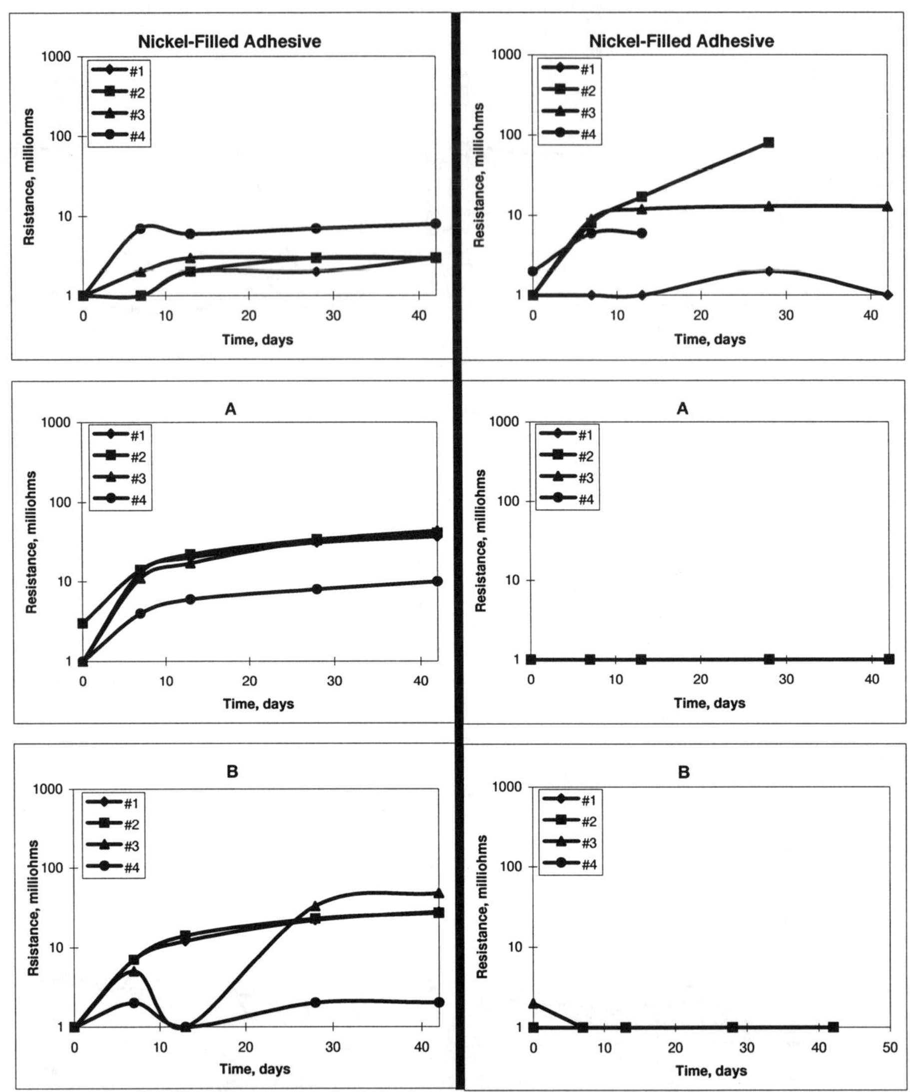

spray for 1,000 hours followed by 85°C exposure for 1,000 hours with retention of adequate mechanical strength (lap-shear >3.4 MPa) and low electrical resistance (<16 milliohms/cm^2 bonded area). The latter goal meets a standard requirement of 2.5 milliohms/joint, with a 1 in.^{2} joint assumed.

A nickel-filled adhesive failed to retain low resistance even against gold plating, presumably because the nickel particles oxidized at the bonding interface. Bare, abraded aluminum with silver-filled epoxies also failed, even when edge-sealed, presumably because of surface oxidation. The successful system comprised silver-filled epoxy, gold plating, and an epoxy edge sealant which excluded salt/moisture and also enhanced mechanical strength.

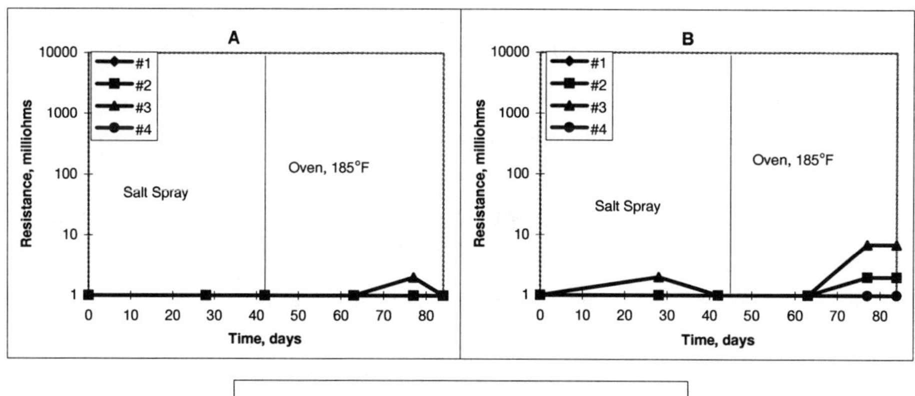

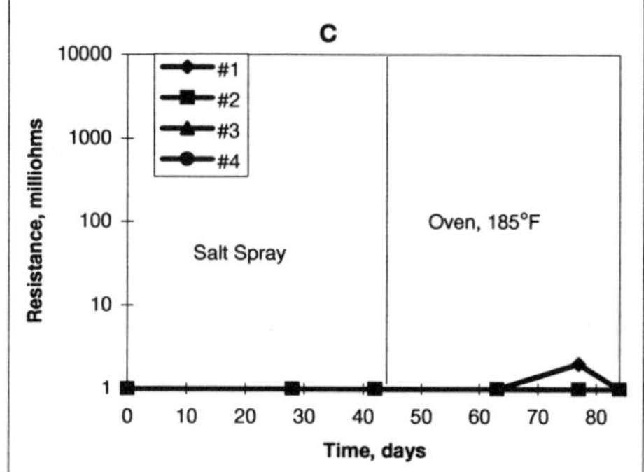

A realistic test method made dual use of lap-shear specimens, representing an actual hardware bond, by determining electrical resistance and finally mechanical strength on the same sample. Four resistance measurements were made through clamps, two on either side of the bond, so that contact resistance canceled out in the calculation. Thus, low-resistance contacts were unnecessary, and corroded/ oxidized samples could be tested quickly with minimum cleanup. Also, dual use of specimens allowed better correlation of strength and resistance data and minimized the number of specimens needed.

6. ACKNOWEDGMENTS

The author thanks Dottie M. Rushton for sample preparation, Dave C. West for plating, George D. Crain for Instron testing, and Arie A. Passchier for computer generation of graphs.

7. REFERENCES

1. Rockwell International study. Proprietary information. Issued 1993.

2. H. S. Kraus, "Peculiar Behavior of Electrically Conductive Adhesives on Aluminum Surfaces," Ablestik Laboratories, Gardena, CA, circa 1980

3. MIL-STD-464, "Electromagnetic Effects Requirements for Systems," paragraph 5.10.3

4. MIL-P-23377G, "Primer Coatings: Epoxy, High-Solids," Type I, Class 1

5. MIL-S-8802F, "Sealing Compound, Temperature-Resistant, Integral Fuel Tanks and Fuel Cell Cavities, High Adhesion," Class B-2

6. MIL-A-46146B, "Adhesives-Sealants, Silicone, RTV, Noncorrosive (for Use with Sensitive Materials and Equipment)"

7. ASTM D 1002-94, "Standard Test Method for Apparent Shear Strength of Single-Lap-Joint Adhesively Bonded Metal Specimens by Tension Loading (Metal-to-Metal)"

8. QQ-A-250/5F, "Aluminum and Aluminum Alloy Plate and Sheet: General Specification for"

9. MIL-C-5541E, "Chemical Conversion Coatings on Aluminum and Aluminum Alloys," Class 3.

10. ASTM D 2651-90, "Standard Guide for Preparation of Metal Surfaces for Adhesive Bonding," procedure for "Sodium Dichromate/Sulfuric Acid Etch" of aluminum

11. ASTM B 117-95, "Standard Procedure for Operating Salt Spray (Fog) Apparatus"

12. ASTM E 104-85, "Standard Practice for Maintaining Constant Relative Humidity by Means of Aqueous Solutions"

13. AMS 2404D, 'Plating, Electroless Nickel"

14. QQ-S-365D, "Silver Plataing, Electrodeposited: General Requirements for"

15. MIL-G-45204C, "Gold Plating, Electrodeposited," Type III

8. BIOGRAPHY

John M. Kolyer is a Senior Engineering Specialist at Boeing in Anaheim, CA. Dr. Kolyer received his Ph.D. in Chemistry from the University of Pennsylvania in 1960 and has been with his present employer for 24 years. His specialties include nonmetallic materials science and electrostatic discharge (ESD) control. He has authored many papers and a book (ESD from A to Z, 2nd Ed., Chapman and Hall, 1996). His web site is http://www.global-eyes.com/jkolyer.

43rd International SAMPE Symposium
May 31-June 4, 1998

CURE PARAMETER EFFECTS ON THE Tg AND CTE OF FLIP CHIP ENCAPSULANTS

Dr. Mark M. Konarski
LOCTITE CORPORATION
ROCKY HILL, CONNECTICUT 06067

ABSTRACT

As the role of direct-chip-attachment increases in the electronics industry, the reliability and performance of COB packaging materials becomes an increasing concern. Although many factors can and do influence component reliability, the biggest determinants of performance are often the glass transition temperature (Tg) and the coefficient of thermal expansion (CTE) of the encapsulant or underfill. The time-temperature relationship during cure of these materials is seen to have a marked effect on both Tg and CTE. This cure cycle effect is compared and discussed for encapsulants with different cure chemistries..

KEY WORDS: Electronics, Encapsulation, Epoxy Resin

1. INTRODUCTION

The glass transition temperature and coefficient of thermal expansion of chip encapsulants have direct effects on semiconductor package performance and reliability.[1] Though they both are usually quoted and accepted as single numerical values, different methods of measurement and sample preparation will provide varying data for the same material. Even within the same sample, the glass transition occurs within a range of temperatures and not as a single point. Factors such as intrachain stiffness, polar forces, and comonomer compatability can affect the size of the glass transition region.[2] Thermal expansion coefficients are affected by polymer chain stiffness, packing, cure shrinkage and filler interactions.[3] Industry drive towards faster line speeds and high production build rates has forced an evolution in flip chip underfill encapsulants. New cure chemistries and development of faster catalysts for the epoxy-anhydride reaction have taken underfill cure times from hours to minutes. Curing at higher temps for less time is another common method employed to decrease the length of cure cycles. Cure cycles have been shown to directly affect thermal properties such as glass transition and the coefficient of thermal expansion. Decisions on material selections for a particular application can often come down to comparisons of these values, and accurate measurement is a key concern.

The purpose of this paper is to provide a summary of the common practices and techniques involved with Tg and CTE.   Methods of measurement are explained and compared. Differences in sample preparation are shown to significantly affect results and suggest concerns about the performance of flip chip packages. A better understanding of the factors affecting these material properties will allow process and design engineers to gain optimum performance from underfill encapsulants and become cognizant of the potential reliability changes that can occur from changes in board-level processing.

2. INSTRUMENTAL MEASUREMENT

The glass transition temperature ("Tg") is the point where a substance changes from a hard-glassy material, to a soft-rubbery one. In monomer or thermoplastic polymers, the transition is from a solid or glass to a flowable liquid.   For crosslinked thermosetting polymers, the transition is to a soft-rubbery composition and tends to occur across a thermal band rather than at a distinct point of temperature. At the glass transition temperature, several easily measurable properties such as volume, dimension, enthalpy, strength and modulus also undergo transitions, and are often used to determine Tg's. Figure 1 schematically portrays the shifts in the glass transition region.

FIGURE 1
PROPERTY CHANGE

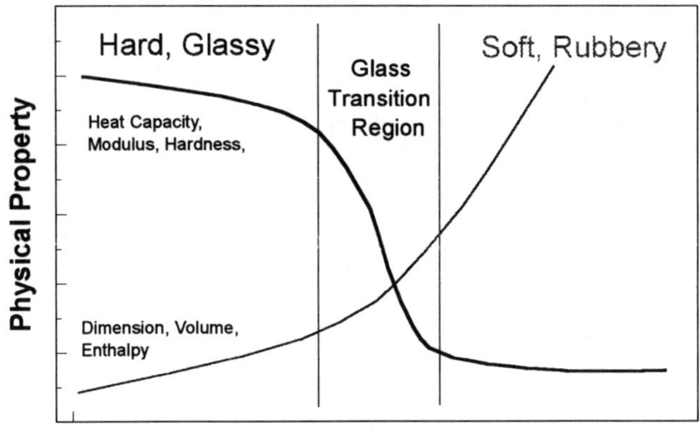

The property shifts at the glass transition temperature of encapsulants can cause severe stress buildup in packaged components leading to premature device failures.[1] Because of this, the Tg of the encapsulant usually places a limit on the use temperature and therefore becomes an important consideration when specifying for a particular application. Chemical composition and extent of cure are the major influenceson Tg, however the method of measurement can vary widely and will affect the data.

The coefficient of thermal expansion is a measure of the fractional change in dimension (usually thickness) per degree rise in temperature. For microelectronics encapsulants, it is often quoted in "ppm/°C" (value x 10^{-6}/°C). Chemical composition, filler loading and cure cycles all affect the value. For typical materials that have non-linear expansion, the specified temperature range will also have an effect on the data, with measurements out closer to the Tg yielding higher values than those quoted across lower temperature as shown in Figure 2.

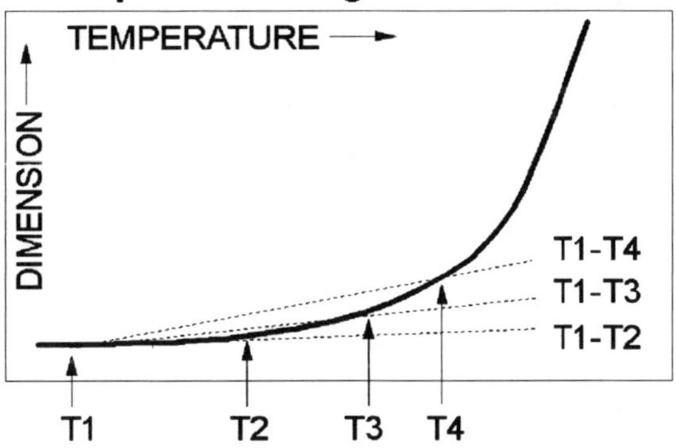

2.1 DSC

Differential scanning calorimetry (DSC) is the quickest and simplest test for Tg. The method requires extremely small samples (typically 5-20 mg) that require no special preparation and material from components on processed boards can be utilized. The method consists of heating the sample in a closely calibrated thermocel where the temperature of the sample is compared to the temperature of a blank reference point within the same cell. Thermodynamic transitions such as melting points and reaction exotherms are easily measured, and the change in heat capacity at the Tg is seen as a shift in the baseline for cured encapsulants as shown in Figure 3.

FIGURE 3
Differential Scanning Calorimetry

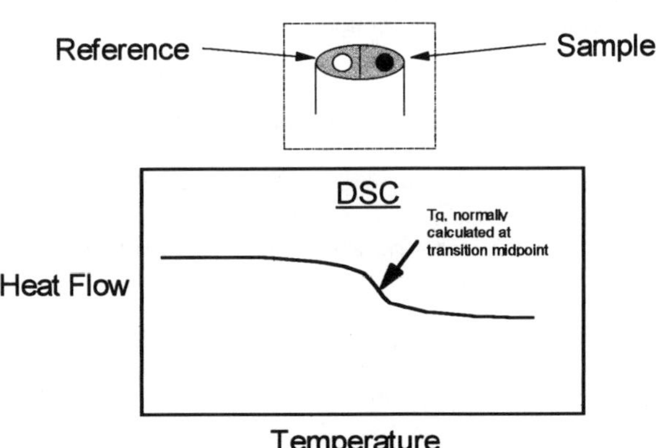

Unfortunately, this fast and convenient method is not universally applicable to all materials. High filler loadings, high crosslink densities, and other thermo-molecular processes can mask the shift due to the Tg and make the transition difficult or impossible to identify.

2.2 TMA

Thermo mechanical analysis (TMA) is the test used to determine thermal expansion coefficients. Since there is a shift to a higher thermal expansion coefficient above the Tg due to changes in molecular free volume[4], the method can also be used to measure the glass transition temperature as shown in Figure 4.

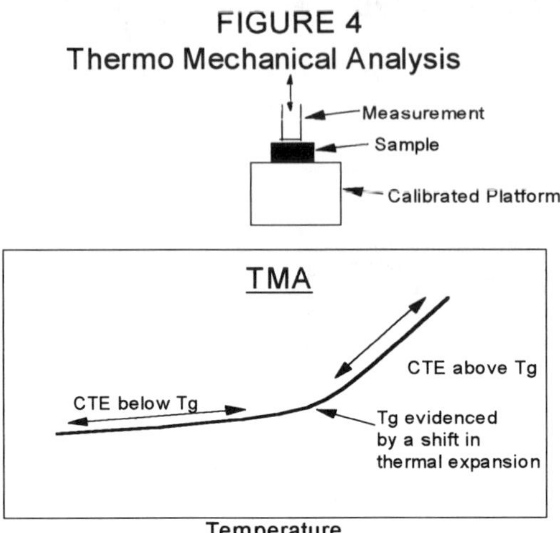

The technique simply consists of heating the sample upon an expansion-calibrated platform, and measuring the dimensional change of the sample with an instrumented probe. The method will also easily follow cure-stress relaxations in and around the glass transition region which sometimes leads to ambiguity in assignment of a specific Tg, and can yield a different value for the same specimen if measured at a different point (for instance the cure stress may be significantly different when measured near the edge of a sample versus its center). The specimens used for CTE/TMA in this paper were all prepared by casting a small disc of net-shape and dimensions. This was done so that any differences are due to the material and its cure, rather than attributed to differences in sample configuration.

2.3 DMA

Dynamic mechanical analysis (DMA) consists of oscillating flexure energy applied to a rectangular bar of the cured encapsulant. The stress that is transfered through the specimen is measured as a function of temperature. Components of material stiffness are separated into a complex modulus and a rubbery modulus. The technique is highly accurate and reproducible although the Tg can be defined in different ways which will have different values as illustrated in Figure 5.

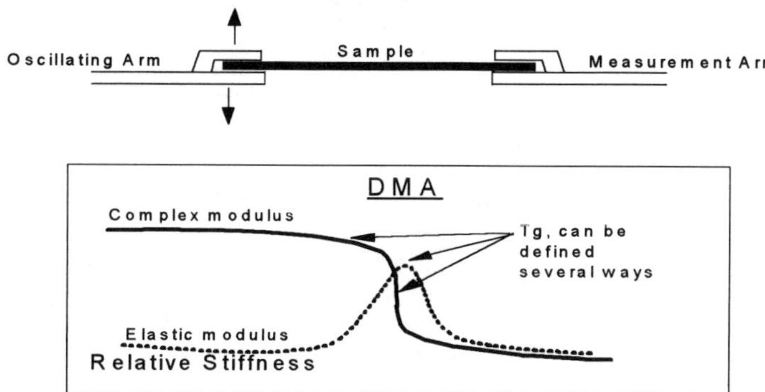

Large samples that must be accurately machined, coupled with longer setups make this test relatively expensive to run.

Each of the methods will produce different data for the same material as shown in the following table.

Instrument Effect on Glass Transition Temperature

INSTRUMENT	GLASS TRANSITION TEMPERATURE
DSC	142°C
TMA	130°C
DMA	137°C (G") 146° (Tan delta)

A single specimen of a developmental epoxy encapsulant was cast and cured for 2 hours at 145°C after gelling at 100°C for one hour. It was then machined into specimens for DSC, TMA and DMA analysis. The results show Tg's ranging from 130°C for the TMA, to 146°C for the DMA measurement.

3. RESULTS AND DISCUSSION

3.1 Chemistry Effects

Chemistry plays an obvious role in determining the glass transition and thermal expansion. Figure 6 depicts a comparison of three underfill chemistries. The samples for this comparison were prepared as unfilled resin castings, so that properties of the different chemistries could be directly compared. The snap-cure underfill has a lower Tg and slightly higher thermal expansion than the more usual epoxy anhydride (unfilled for this comparison). An experimental epoxy-cyanate ester shows the highest Tg and lowest thermal expansion of the three.

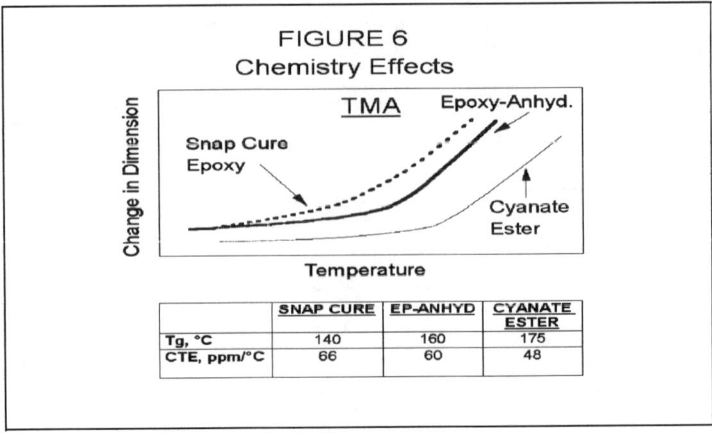

	SNAP CURE	EP-ANHYD	CYANATE ESTER
Tg, °C	140	160	175
CTE, ppm/°C	66	60	48

Figure 7 shows the thermal expansions coefficients for the different materials comprising flip chip packaging at the board level. Thermally induced stress is due to the silicon die being a very low expansion material being mounted to organic composite boards that are an order of magnitude higher. In the unencapsulated package, this stress is focussed at the solder-ball junctions which will fatigue and crack after only a few thermo-cycles if the die is large enough. Underfill encapsulants are used to manage the problem.

FIGURE 7

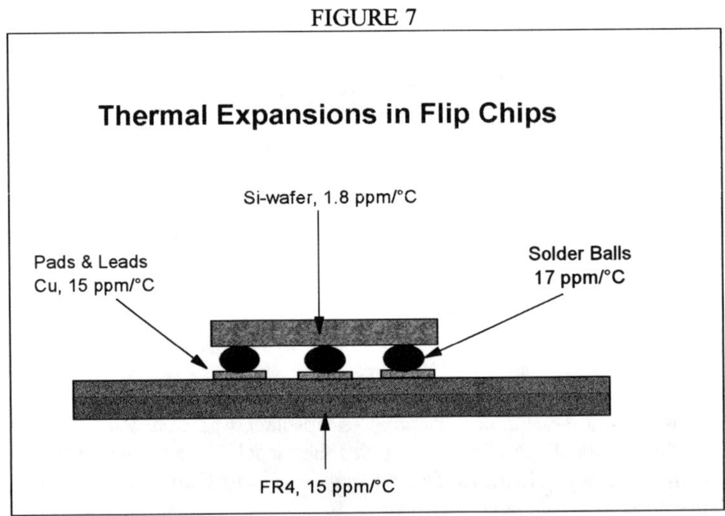

3.2 Filler Loading

Component stress from expansion mismatches between the board and silicon die are dispersed through the underfill encapsulant instead of being concentrated in the solder junctions as in the unencapsulated package. Stress is amplified and compounded when the encapsulant's CTE is high. Lowering the thermal expansion of chip on board encapsulants in order to more closely match expansions of the board, wafer, and solder junctions has been shown to have a direct effect in increasing package reliability. Since all useable encapsulant polymers have expansion coefficients several times larger than the materials to be protected, low expansion fillers are added. The filler loading level has traditionally been as high as possible while retaining the necessary flow and processability. More recently, flip chip underfills have been developed for specific applications where demands for thermal cycling resistance are less severe. In order to achieve extremely rapid flow beneath chips, these underfills are designed with lower filler levels, or no filler at all. The following table shows the effect on CTE.

FILLER LOADING EFFECT ON THERMAL EXPANSION

Underfill	Resin Type	Filler Loading	Cure Time	Tg	CTE
1	A	40%	5 min.	139°C	39 ppm/°C
2	B	40%	20 min.	142°C	37 ppm/°C
3	C	0%	30 min.	160°C	58 ppm/°C
4	C	40%	30 min.	157°C	38 ppm/°C
5	C	60%	30 min.	162°C	27 ppm/°C

Three different epoxy matrices were evaluated. C is a standard epoxy-anhydride, with good Tg and overall properties. It cures in 30 minutes at 150°C. Resin B is also epoxy-anhydride. It has been modified with more rapid cure catalysis and cures in 20 minutes at 150C with a lower Tg. Resin A is a snap curing epoxy formulation with proprietary amine-based cure and a moderate level of filler loading. Concerns about whether a very fast cure chemistry would lead to unusually high cure stress builups appear to be unfounded. At the same filler loadings, resin effects on CTE are somewhat masked, with similar materials all giving similar results.

3.3 Cure Temperature

Cure temperature affects the extent of cure, the rate of gellation and the cure stress which is a function of the gellation temperature. As a result, the cure temperature influences both the coefficient of thermal expansion and the Tg of the encapsulant. As cure temperature increases, crosslink density and degree of cure increase. Since the glass transition is a function of both, the Tg undergoes a corresponding rise until a point where nearly all polymer reaction sites have been used. At that point, the glass transition temperature plateaus with further temperature increase unless thermo-degradation begins. The following table illustrates the point, where increasing the cure temperature from 130°C to 175°C increases the Tg of the encapsulant from 138°C up to 160°C. As the cure temperature is increased beyond 130°C, the CTE also appears to increase indicating some stress buildup. This cure stress is seen to quickly dissipate as the material sees more temperature exposure as evidenced by the lower numbers when the same samples were rerun.

<u>**UNDERFILL CURE TEMPERATURE EFFECT**</u>

Cure Time	Cure Temp	Tg	CTE (ppm/°C)	Rerun CTE (ppm/°C)
30 min.	130°C	138°C	31.4	31.1
30 min.	145°C	150°C	29.0	29.2
30 min.	160°C	162°C	30.1	28.5
30 min.	175°C	160°C	33.2	30.0

3.4 Cure Cycle Time

Older encapsulant technology often recommended intermediate temperature dwells, as a means of managing stress from fairly large cure shrinkages. The gellation of the material at a lower temperature locks in the stress at that point. As the material cooled, the stress would equilabrate at/or near the gellation temperature. The effective temperature delta was therefore decreased as gellation temperatures decreased. Newer underfill encapsulants are engineered for lower cure shrinkage. As a result, there is essentially no measurable difference in the CTE with low temperature cure cycle dwells as shown in the following table.

<u>**STEP-CURE EFFECT ON Tg AND CTE**</u>

Cure Step 1	Final Cure	Tg	CTE (ppm/°C)
none	30′ @ 150°C	160°C	28
60′ @ 100°C	30′ @ 150°C	163°C	27
30′ @ 125°C	30′ @ 150°C	159°C	28

The amount of time at temperature also has an effect. Figure 8 shows the results of different cure times at a cure temperature of 150°C for two different underfill encapsulants.

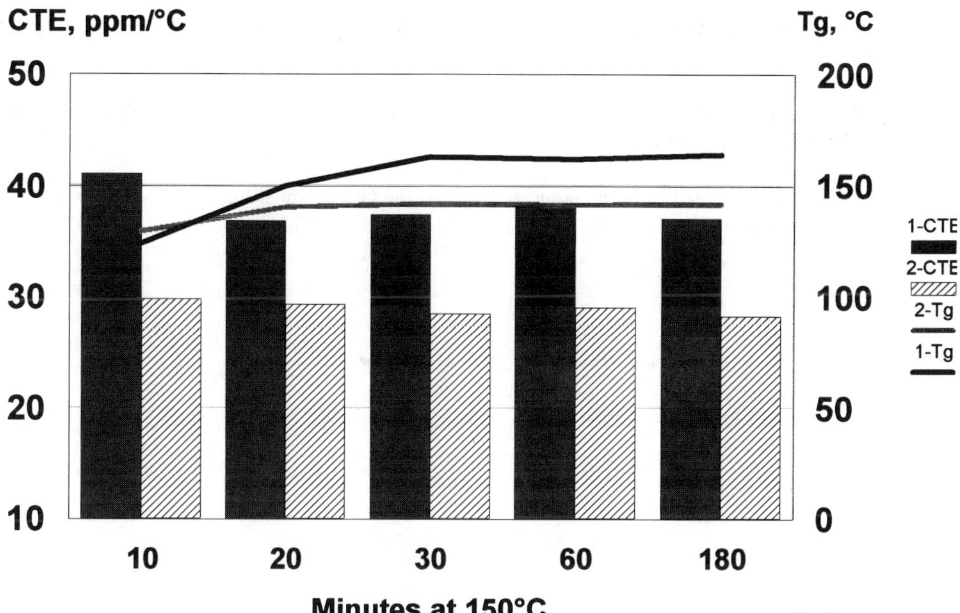

Underfill number "1" is seen to be faster curing than underfill "2", reaching its ultimate Tg in 20 minutes at 150°C versus 30 minutes. CTE for both materials is high when undercured for 10 minutes, dropping to optimum values that then remain relatively constant regardless of further time at temperature.

4. SUMMARY

The glass transition temperature and coefficient of thermal expansion for flip chip underfill encapsulants have been shown to not exist as single numbers, but as ranges of values that can vary depending on measurement techniques and processing methodology. Cure stress is seen to be less of a problem for new underfills with low shrinkage than previous encapsulants. New chemistry yields satisfactory low-stress parts even when cured very rapidly, however designing from data developed under different conditions than those seen at the board level could lead to errors and caution is urged. Cure parameters and measurement techniques can impact the data, and should be noted for any reference to a material's CTE or Tg. Property comparisons should be made on materials that have been processed and tested in similar manners to be meaningful.

5. REFERENCES

[1] Voloshin, A.S. and Tsao, P.H., "Analysis of the Stresses in the Chip's Coating" Proceedings of the 43rd Electronic Component and Technology Conference, (1993).

[2] Roller, Mark B., "The Glass Transition: What's the Point?", J. Coating Technology. (1982), 54(691) pp 33-40.

[3] Fahmy, A.A. and Raga, A.N., "Thermal-Expansion Behavior of Two-Phase Solids", J. Appl. Phys., (1970), 41 pp 5108-5111.

[4] Goldstein, M. "Some thermodynamic Aspects of the Glass Transition: Free Volume, Entropy and Enthalpy Theories", J. Chem. Phys., (1963), 39 pp 3369-3374.

DEVELOPMENT OF CARBON FIBER THERMOPLASTIC COMPOUNDS FOR EMI/RFI APPLICATIONS IN PORTABLE AND WIRELESS ELECTRONICS

Ralph Charbonneau
Product Development Engineer
RTP Company
Winona MN

ABSTRACT

Inherent design freedom and low cost manufacturing have made thermoplastic materials invaluable as housings for portable and wireless electronics. Their use is not problem-free, however, as thermoplastics are transparent to electromagnetic and radio frequency interference (EMI/RFI). Incorporation of metal-coated carbon fibers into thermoplastic materials has been shown to be effective against EMI/RFI. Evaluation of physical and electrical effects of metal-coated carbon fibers in thermoplastics is important in development of EMI/RFI shielding materials. This paper will address both formula and processing in terms of additive type, form and content, and methods of incorporation to achieve an optimized level of protection. An overview of EMI/RFI is given and objectives of EMI testing methods are discussed.

KEY WORDS: EMI/RFI, Electromagnetic Interference, Electromagnetic Shielding, Metal Coated Carbon Fiber.

1. INTRODUCTION

1.1 EMI Concepts   Electromagnetic interference (EMI) is radiation with adverse effects on electrical performance in portable and wireless electronics. While EMI exists across the entire electromagnetic spectrum, from dc electricity at less than 1 Hz to gamma rays above 10^{20} Hz, the great majority of EMI problems are limited to that part of the spectrum between 25 kHz and 10 GHz. This portion is known selectively as the radio

frequency interference (RFI) area and covers radio and audio frequencies. In this paper
the acronym EMI is used to represent both EMI and RFI.

1.2 EMI Problems   Electronic devices are both sources and receptors of EMI. This
fact creates a two-fold problem for manufacturers since both operational integrity of and
emissions from products must be dealt with. In addition, electromagnetic waves have
both a magnetic component and an electric component. The magnetic and electrical wave
components each have a unique EMI effect depending upon frequency and distance from
the radiation source. Resolution of EMI problems necessarily includes consideration of
all these factors.

1.3 EMI Solutions   Securing operational integrity and reducing emissions to meet
applicable standards requires some modification to commonly used thermoplastic
enclosure materials. Thermoplastics are electrical insulators, transparent to electrical
energy, and, therefore, EMI waves pass through them without loss. A range of material
options in two broad categories - surface treatments and composites with conductive
additives - is available for providing EMI shielding to thermoplastic enclosures. This
paper covers composites with conductive additives and, more specifically, composites
with metal-coated carbon fiber additives.

2.   EMI/RFI TESTING/QUALIFICATION

The objective of a test for shielding effectiveness in materials should be determined so
that appropriate value can be obtained. Objectives are usually classified into one of three
categories(1):
- Compliance Testing
- Engineering Evaluation
- Audit/Screening

2.1 Compliance Testing   Compliance testing is a precise and absolute evaluation of a
fully assembled and functioning device. The open field test technique is generally required
for compliance under various agency standards prior to commercialization of most
wireless electronic devices. Free space measurements for both radiated and conducted
emissions are taken in an open field. Such testing requires sophisticated equipment and
must be performed by accredited test labs.

2.2 Engineering Evaluation   The objective of engineering evaluation is to determine
suitability of materials to perform shielding. Less precise than compliance testing,
engineering evaluation still requires a controlled environment for accuracy. Generally, an
open field test site is used for greater accuracy. Equipment is less sophisticated than
required for agency compliance. Common techniques used include dual chamber shielded
box as used in ASTM ES-7-83 (discontinued), transverse electromagnetic cell (TEM-cell),
and coaxial transmission line as in ASTM D 4935-89.

2.3 Audit/Screening   This testing methodology is primarily a screening process, providing a quick look at shielding characteristics of a metal-coated carbon fiber composite. Volume and surface resistivity, microwave reflectance, and attenuation tests on a spectrum analyzer are screening methods that can be used. Results obtained are useful for ranking and screening of candidate shielding composites. Further validation of test results is required through engineering evaluation and compliance levels.

2.3.1 Volume and Surface Resistivity   Volume resistivity is electrical resistance through a unit mass of material and is expressed in units of "ohm-cm." There is a good correlation between volume resistivity and shielding effectiveness of a metal-coated carbon fiber composite. Volume resistivity in the range of 10^{0} to 10^{-3} ohm-cm is generally thought to be acceptable for effective shielding composites (2).

Surface resistivity is electrical resistance over a unit area of material's surface and is expressed in units of "ohms/square." Surface resistivity and shielding effectiveness can be correlated where electrical conductivity is a surface property as in applied conductive films or coatings. In some applications, coating thickness is a variable in shielding effectiveness. However, in the instance of incorporated conductive modifiers, surface resistivity has little value in correlating shielding effectiveness. The value of low surface resistance of plastics in shielding is in providing electrical continuity between components of an assembly.

2.3.2 Microwave Reflectance   A reflectometer measures radio frequency reflectivity of planar material surfaces. The instrument focuses energy at a frequency of 10.525 GHz through a rectangular waveguide aperture onto a target surface. The relative return loss of a sample is compared to a metal plate reference. A determination of the relationship of this reflection coefficient to shielding effectiveness of the material can be made through higher level testing as described under "engineering" or "compliance" levels.

2.3.3 Spectrum Analyzer   A spectrum analyzer with tracking generator and transmitting/receiving probes can be used to screen attenuation values of metal-coated carbon fiber composites. A molded specimen is passed through an electromagnetic field and the reduction in field intensity is measured. Data obtained is attenuation of the incident input signal and not shielding effectiveness of the material. Absolute value of data is useful for ranking and comparison. Determination of the relationship of tested attenuation to shielding effectiveness can be made through higher level testing as described under "engineering" or "compliance" levels.

The key features of testing just described are summarized in Figure 1.

FIGURE 1. EMI Test Methods

Test Method	Field	Specimen	Comments
COMPLIANCE TESTING			
1. Open Field	Far	Device	Used for various agency approvals; sophisticated equipment; high cost; excellent reproducibility.
ENGINEERING EVALUATION			
1. ASTM D4935-89	Far	Doughnut	Modest cost; good reproducibility and correlation; suitable for ranking.
2. TEM Cell	Far	Flat plaque, Enclosure	Modest cost; good reproducibility and correlation; suitable for ranking; tests enclosures.
3. Dual Chamber Box	Near	Flat plaque	Low cost; poor correlation.
AUDIT/SCREENING			
1. Volume Resistivity	N/A	Small part	Low cost; limited correlation.
2. Surface Resistivity	N/A	Small part	Low cost; limited correlation.
3. Microwave Reflectance	Far	Any	Uncertain correlation; in-process tool
4. Spectrum Analyzer	Near	Any	Uncertain correlation; in-process tool

3. SHIELDING MATERIALS

3.1 Shielding Additives   Metal-coated carbon fiber is a composite of metal, typically nickel or copper, and carbon (graphite) fiber. Individual strands of carbon fiber are encapsulated by a thin sheath of metal through electro-plating or vapor deposition processes. The metal coating, typically less than one micron thick, surrounds a carbon fiber that is typically seven microns in diameter. The metal-coated fibers are produced in bundles (tow) of 3,000 to 12,000 filaments. The combination of carbon fiber and metal sheath combines high strength and low weight of carbon fiber with the conductivity and corrosion resistance of metal.

Following application of metal, the coated strands are impregnated with proprietary sizing compositions to enhance compatibility with various thermoplastic polymers. These finished materials are then offered in either continuous or chopped form.

Metal-coated carbon fiber materials have been commercially available since the early '80s and have undergone extensive evaluation as conductive additives in EMI shielding composites.

3.1.1 Sized Nickel-Coated Carbon Fiber   This commercialized product uses nickel as the coating metal, primarily for its corrosion resistance, high conductivity, and relatively low cost. A variety of sizings are available to ease processing into many thermoplastic resins. The product is commonly twenty-five to forty weight percent sizing and sixty to seventy-five weight percent effective nickel-coated carbon fiber. The nickel-to-carbon fiber ratio is variable. For this paper, the ratio was forty-five to fifty weight percent nickel and fifty to fifty-five weight percent carbon fiber. See Figure 2 for details.

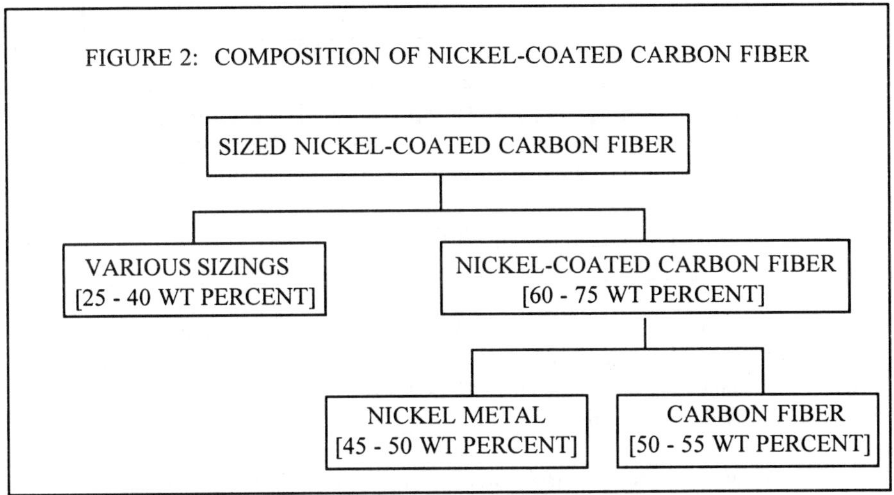

3.1.2 Nickel-Copper-Nickel-Coated Carbon Fiber   This recently commercialized product uses copper, generally in combination with nickel, to provide improved electrical conductivity and greater attenuation of EMI at higher frequencies. The product is supplied with a variety of sizings for many thermoplastic resins. It is twenty-five weight percent sizing and seventy-five weight percent nickel-copper-nickel-coated carbon fiber. The coating on carbon fiber is a tri-layer metal sheath of approximately fifty weight percent metal. The tri-layer is a base of ten weight percent nickel, middle of thirty weight percent copper, and top layer of ten weight percent nickel. See Figure 3 for details.

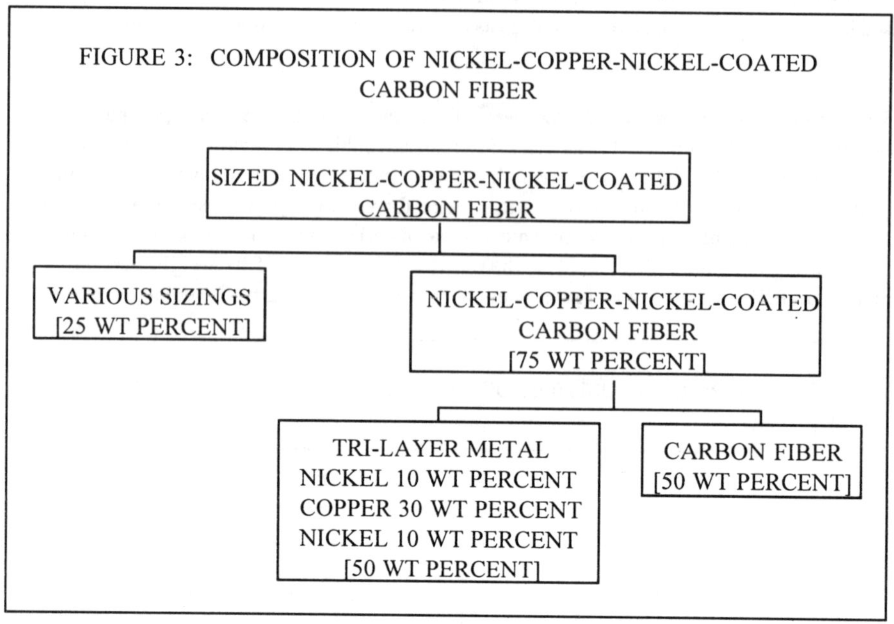

3.2 Shielding Composites   An injection moldable shielding composite can be either of two physical forms - a physical blend of shielding additive and thermoplastic resin or an extruded compound of these components. These are referred to as ***cube blend*** and ***compounded blend***, respectively.

3.2.1 Cube Blend   A cube blend is a molding compound which can be introduced directly to the injection molding machine. It consists of two components - a metal-coated carbon fiber concentrate and a resin matrix consisting of thermoplastic resin and possibly other additives such as flame retardant, wear additive, and pigment. The thermoplastic resin with its additives is extrusion compounded prior to blending with the metal-coated carbon fiber concentrate. Cube blend's advantage is primarily in economics as less metal-coated carbon fiber is needed to achieve a given conductivity and shielding effectiveness. In a cube blend, metal-coated carbon fiber is not subjected to shear of extrusion compounding, resulting in reduced fiber attrition. Additional concentrate can be added during molding if conductivity/shielding falls below specification. Cube blends demand a high level of technical skill for complete dispersion of the metal-coated carbon fibers in molded parts.

3.2.2 Compounded Blend   A compounded blend is a molding compound produced through extrusion compounding of metal-coated carbon fiber, thermoplastic resin, and possibly other additives such as flame retardant, wear additive, and pigment. This compounded blend can also be introduced directly to the injection molding machine. Compounded blend's advantage is in uniformity of molded parts as dispersion of metal-coated carbon fiber is generally complete. Extrusion compounding does cause fiber attrition and loss of conductive/shielding properties, remedied by increased metal-coated carbon fiber content.

4. MATERIAL DATA

Metal-coated carbon fiber was evaluated as an EMI shielding additive in polycarbonate thermoplastic resin. No other modifiers or reinforcements were included in this study. Physical and electrical property comparisons were made between nickel-coated and nickel-copper-nickel-coated carbon fiber composites. Shielding investigations for this paper were performed at the audit/screening level. Effect of shielding additive content was su…d at five, ten, fifteen, and twenty effective weight percent. Both physical forms - cube blend and compounded blend - were included.

Tables 1 through 4 detail physical, electrical, and reflective properties of each material studied.

Table 1. NiCF in Polycarbonate - Cube Blend				
Nickel-Coated Carbon Fiber (wt %)	5	10	15	20
Specific Gravity	1.25	1.28	1.33	1.41
Mold Shrinkage, 3.175 cm thick (%)	0.33	0.22	0.18	0.13
Tensile Strength (MPa)	75	92	105	111
Tensile Modulus (MPa)	4.8	7.0	9.9	10.8
Tensile Elongation (%)	5.0	3.4	2.1	2.0
Flexural Strength (MPa)	127	157	170	176
Flexural Modulus (GPa)	4.5	7.1	9.7	10.2
Izod Impact Strength				
Notched (kJ/m)	0.17	0.13	0.11	0.12
Unnotched (kJ/m)	1.30	0.66	0.56	0.50
Deflection Temperature, 1.81 MPa (deg C)	142	142	145	145
Volume Resistivity (ohm-cm)	3300	5.72	1.06	0.79
Surface Resistivity (ohms/square)	>10E+13	10E+5	10E+5	10E+5
Microwave Reflectivity (%)	85	90	95	97

Table 2. NiCF in Polycarbonate - Compounded Blend				
Nickel-Coated Carbon Fiber (wt %)	5	10	15	20
Specific Gravity	1.24	1.27	1.31	1.40
Mold Shrinkage, 3.175 cm thick (%)	0.31	0.20	0.19	0.14
Tensile Strength (MPa)	76	93	105	110
Tensile Modulus (MPa)	4.7	7.1	9.9	10.7
Tensile Elongation (%)	5.1	3.5	2.2	2.1
Flexural Strength (MPa)	128	159	171	177
Flexural Modulus (GPa)	4.5	7.1	9.6	10.2
Izod Impact Strength				
Notched (kJ/m)	0.17	0.13	0.11	0.09
Unnotched (kJ/m)	1.30	0.50	0.52	0.50
Deflection Temperature, 1.81 MPa (deg C)	142	141	143	145
Volume Resistivity (ohm-cm)	5000	20	3.05	0.95
Surface Resistivity (ohms/square)	>10E+13	10E+8	10E+5	10E+5
Microwave Reflectivity (%)	85	90	95	97

Table 3. NiCuNiCF in Polycarbonate - Cube Blend				
Nickel-Coated Carbon Fiber (wt %)	5	10	15	20
Specific Gravity	1.23	1.25	1.30	1.40
Mold Shrinkage, 3.175 cm thick (%)	0.40	0.22	0.20	0.17
Tensile Strength (MPa)	70	82	98	105
Tensile Modulus (MPa)	4.8	7.0	9.2	10.4
Tensile Elongation (%)	2.0	1.5	1.2	1.1
Flexural Strength (MPa)	100	125	155	160
Flexural Modulus (GPa)	4.0	6.5	8.0	9.5
Izod Impact Strength				
Notched (kJ/m)	0.09	0.10	0.10	0.11
Unnotched (kJ/m)	0.21	0.22	0.25	0.26
Deflection Temperature, 1.81 MPa (deg C)	140	142	145	145
Volume Resistivity (ohm-cm)	1.45	0.16	0.02	0.02
Surface Resistivity (ohms/square)	>10E+13	10E+5	10E+5	10E+5
Microwave Reflectivity (%)	90	95	97	97

Table 4. NiCuNiCF in Polycarbonate - Compounded Blend

Nickel-Coated Carbon Fiber (wt %)	5	10	15	20
Specific Gravity	1.21	1.26	1.31	1.39
Mold Shrinkage, 3.175 cm thick (%)	0.40	0.23	0.19	0.18
Tensile Strength (MPa)	71	83	97	104
Tensile Modulus (MPa)	4.6	7.2	9.4	10.3
Tensile Elongation (%)	2.0	1.5	1.3	1.2
Flexural Strength (MPa)	101	124	144	159
Flexural Modulus (GPa)	3.9	6.4	8.1	9.4
Izod Impact Strength				
Notched (kJ/m)	0.08	0.10	0.10	0.10
Unnotched (kJ/m)	0.21	0.22	0.26	0.25
Deflection Temperature, 1.81 MPa (deg C)	140	142	145	145
Volume Resistivity (ohm-cm)	5.0	2.5	0.50	0.10
Surface Resistivity (ohms/square)	>10E+13	10E+8	10E+5	10E+5
Microwave Reflectivity (%)	87	93	95	97

Figures 4 through 7 complete the test results with data from a spectrum analyzer/tracking generator.

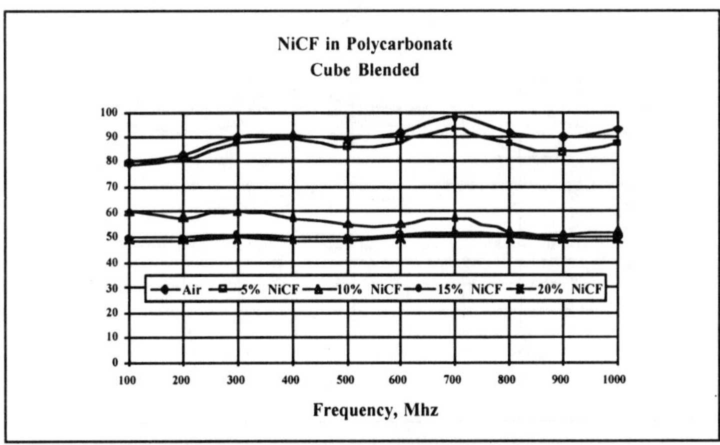

Figure 4. Attenuation of Nickel-Coated
Carbon Fiber in Polycarbonate - Cube Blend

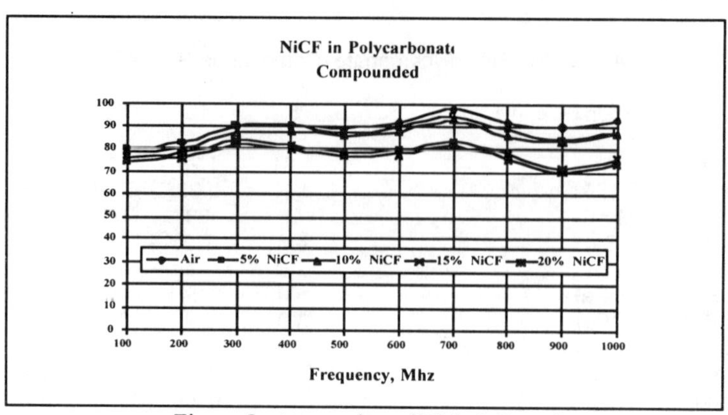

Figure 5. Attenuation of Nickel-Coated
Carbon Fiber in Polycarbonate - Compounded Blend

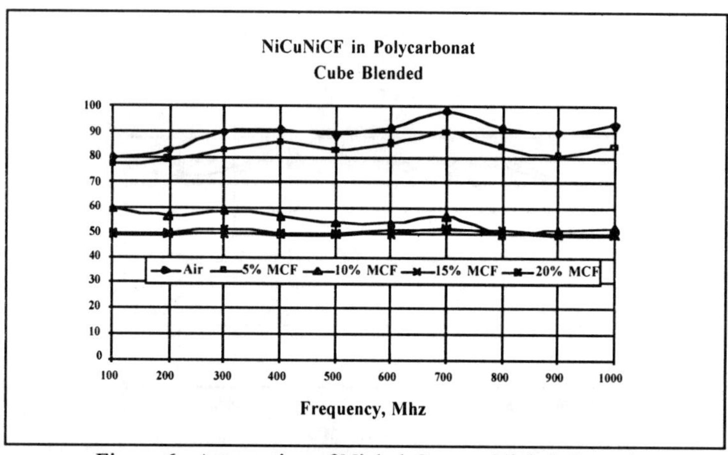

Figure 6. Attenuation of Nickel-Copper-Nickel-Coated
Carbon Fiber in Polycarbonate - Cube Blend

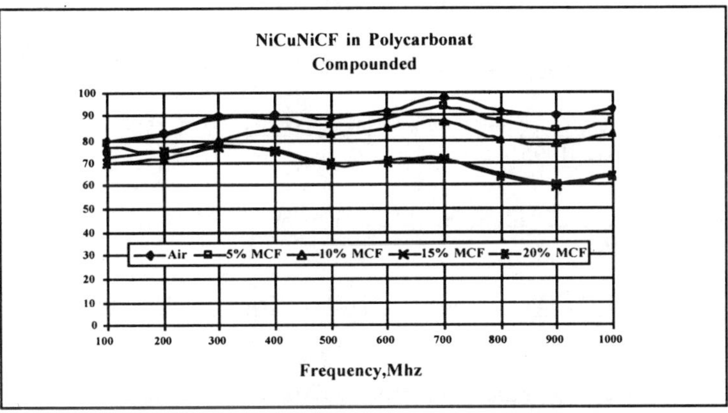

Figure 7. Attenuation of Nickel-Copper-Nickel-Coated
Carbon Fiber in Polycarbonate - Compounded Blend

5. CONCLUSION

Dissemination of data provides insight into the process of product development.
Physical observations are noted including increase in specific gravity, changes in mold
shrink rate, and the expected increases in strength and stiffness of all the products.
Volume resistivity shows improvement in cube blends over compounded blends with
both fiber types and in nickel-copper-nickel tri-layer over nickel alone as coating options.
Minor increase in reflectivity is noted for nickel-copper-nickel blends and insignificant
differences are recorded in reflectivity of cube blends over compounded blends.

Spectrum analyzer scans of noise attenuation show noteworthy differences between
materials. First, attenuation of equivalent concentrations in cube blends is significantly
improved over compounded blends once a threshold concentration of metal-coated carbon
fiber is reached. Second, the attainment of this threshold concentration is at a lower level
in cube blends - between five and ten weight percent in both fiber types - compared to
between ten and fifteen weight percent in compounded blends. Next, the maximization of
attenuation appears to be near fifteen weight percent in cube blends and is not yet reached
at twenty weight percent in compounded blends. Finally, nickel-copper-nickel tri-layer's
stronger attenuation in compounded blends coupled with improved volume resistivity in
both cube and compounded blends suggests some increased value for EMI shielding.

Evaluation of physical and electrical properties of metal-coated carbon fiber in
thermoplastics is important for optimization of formula and processing. An
understanding of test methods and results is crucial to obtain suitable products for
EMI/RFI protection. Metal-coated carbon fiber thermoplastic materials are shown to
provide strong electrical conductivity and EMI attenuation through audit/screening
evaluations. Numerous commercial EMI shielding applications with metal-coated carbon
fiber are cited in the literature (3-5).

6. REFERENCES

1. Parsi, Fariborz, "Electromagnetic Interference and EMI Shielding," August 1997.

2. <u>Electromagnetic Interference Shielding: A Materials Perspective</u>, Tech Trends: Innovation 128, 10-12, 1996.

3. Huang, J.C., <u>EMI Shielding Plastics: A Review</u>, Advances in Polymer Technology, 14(2), 137-150, 1995.

4. Luxon ,Bruce., <u>EMI Shielding Properties of Nickel Coated Graphite Fiber Composites</u>, EMI/RFI Shielding, 1-4, 1984.

5. Markstein, Howard., <u>Proper Shielding Reduces EMI</u>, Electronic Packaging & Protection, 72-78, 1997.

BIBLIOGRAPHY

Ralph Charbonneau, Product Development Engineer and Conductive Materials Process Manager at RTP Company, Winona, MN, has been involved with metal-coated carbon fiber processing since 1993. He also directs RTP Company's R&D efforts at satellite facilities in Dayton, NV, and Ft. Worth, TX.

43rd International SAMPE Symposium
May 31-June 4, 1998

CONDUCTIVE THERMOPLASTICS FOR ESD CONTROL IN PORTABLE AND WIRELESS ELECTRONICS

Larry Rupprecht
RTP Company
Winona, Minnesota 55987

ABSTRACT

The commonly available thermoplastic materials used in portable and wireless electronic devices are susceptible to damage from electrostatic discharge (ESD). Gaining control of static damage follows a logical sequence including identifying source of static and its effects and formulation of static-control conductive thermoplastic materials. This paper explores that sequence through review of static sources and effects, control of static, and introduction of carbon fiber modified products for static-control. Incorporation of carbon fibers into thermoplastic materials can effectively control static while retaining colorability and enhancing physical characteristics in molded components. A selection of static-control conductive thermoplastic products utilizing carbon fiber as an electrically conductive modifier is presented.

KEY WORDS: Carbon Fiber, Conductive Thermoplastic, Electrically Conductive Materials, Electrostatic Discharge, ESD, Thermoplastic Resins/Polymers

1. INTRODUCTION

A multitude of new discoveries in electronics coupled with low-cost manufacturing of both finished goods and their component electronics have led to rapid growth of portable and wireless electronic devices. Low manufacturing costs are derived in part from use of thermoplastic materials. Thermoplastics are generally preferred materials due to their low part-weight, colorability, design freedom, processing ease, and low cost. The commonly available thermoplastic materials are unfortunately susceptible to damaging effects of static electricity, also known as electrostatic discharge (ESD). Uncontrolled static electricity disrupts intended function and renders useless many portable and wireless devices.

2. SOURCE OF STATIC ELECTRICITY

Static electricity is accumulated electrical charge at rest with an overwhelming desire to return to an electrically neutral condition (entropy). Its return to neutral condition can be through three mechanisms - static discharge (rapid), static dissipation (moderate), or static decay (slow).

2.1 Creation of Static Electricity through Triboelectrification   Static electricity is commonly created through contact and separation of electrically dissimilar materials, a process known as triboelectrification. Contact and incidental friction between materials excites each at a molecular level and allows electron transfer from one to the other. Such transfer creates a surplus of electrons on one material and a deficit of electrons on the other. Transferred electrons are isolated when the two materials separate, as when paper is removed from acetate sheet or a shoe sole is pressed and lifted from flooring. Oppositely charged static fields form on the two previously neutral materials.

2.2 Magnitude of Static Charge   The size of a static charge generated through triboelectrification is dependent primarily on four factors:
- material type;
- friction during contact;
- separation speed;
- relative humidity of the near environment.

Each works in concert with the others - if physical and environmental factors are correct, triboelectrification will take place.

2.2.1 Material Type Influence of material type is displayed in a triboelectric series as shown in Figure 1 (1). Materials are positioned in a triboelectric series to show behavior relative to other materials based on propensity for positive or negative triboelectric behavior. Materials widely separated in this series will generate stronger static charges than materials adjacent to each other.

<u>Materials</u>	<u>Polarity</u>
Air, Human Skin, Rabbit Fur	More Positive Charge Tendency
Human Hair, Glass	
Nylon, Wool, Silk	
Aluminum, Paper, Cotton, Steel	
Wood, Amber, Rubber	
Nickel, Copper, Brass	
Silver, Gold, Platinum	
Acetate, Rayon, Polyester, Orlon	
Polyethylene, PVC, Silicon, Teflon	More Negative Charge Tendency

FIGURE 1. TRIBOELECTRIC SERIES

2.2.2 Contact Friction Friction during contact is dependent upon surface texture, contact pressure, and velocity of contact. Smoother surfaces, higher pressures, and faster movements provide more friction and, therefore, greater transfer of electrons.

2.2.3 Separation Speed High separation speed following contact isolates transferred electrons more easily. Slow movement allows opposite charges to neutralize before full separation of the two materials is accomplished.

2.2.4 Relative Humidity Low relative humidity leads to high resistance on material surfaces, hindering charge dissipation or decay until higher voltages are accumulated. High relative humidity provides dissipation or decay at lower voltages, helping to prevent high voltage static accumulation.

3. DAMAGE CAUSED BY STATIC ELECTRICITY

3.1 Particulate Contamination Manufacture of electronic devices and components demands cleanliness not required in many other industries. Particulate material damages mechanical and electrical function at both final product and electronic component levels. At the product level, static-attracted particulate dirt affects temperature, light, and magnetic sensing, prevents mechanical switching, and causes rejection for appearance. At the component level, static-attracted particulates produce electrical shorting within circuitry and destroy silicone chip material through contamination.

3.2 Explosive Discharge A static discharge-generated spark in a flammable or organic dust-filled atmosphere is a likely disaster. Many operations must deal with not only organic dust problems but also with communication and lighting in such dangerous environments. Static-free equipment solutions are mandated.

3.3 Electronic Device Failure Typical damage threshold values for several common devices are shown in Figure 2. These threshold values will decrease as complexity and miniaturization of electronics increases.

Device Type	Typical Threshold Value (volts)
MR Read/Write Component	5 - 100
MOSFET	50 - 200
Transistor	50 - 500
CMOS	100 - 1000

FIGURE 2. TYPICAL DEVICE SENSITIVITY

Figure 3 shows triboelectric charges generated during common activities (1). Serious problems exist when these high triboelectric charges are coupled with low threshold values for electronic devices. From manufacturing and transportation to communication and education, from health care and public safety to just plain living, microelectronics are an integral part of life.

Action	10-20% RH (volts)	50-90% RH (volts)
Walking on synthetic flooring	35,000	1,500
Handling polyethylene bag	20,000	1,200
Rising from synthetic chair	18,000	1,500

FIGURE 3 TYPICAL ELECTROSTATIC CHARGES

4. PREVENTION OF STATIC DAMAGE

Static damage prevention categorized into three disciplines provides direction in developing and implementing effective control plans (2). These three disciplines are -
- Grounding
- Isolation
- Neutralization

4.1 Grounding   Historically, grounding to earth provided a satisfactory level of static protection. Most current microelectronic devices cannot survive a rapid static discharge directly to earth ground. Grounding today is therefore modified to meet needs of sensitive electronics and generally includes slower mechanisms of decay and dissipation. Thus, sensitive devices are sheltered from dangerous voltage spikes through protective materials providing slower charge loss rates. These static dissipative or static decaying materials, in concert with grounded materials, move static charges safely away from electronics to earth ground without voltage spikes (2).

4.2 Isolation   Devices which are themselves insulators cannot be protected by grounding, dissipation, or decay. Protection is provided by isolating them from their environment and allowing only non-charged contact with them. Isolation can be accomplished by eliminating all insulative materials from work areas and by using Faraday cages to provide an electrically neutral environment. Insulative materials can be replaced with dissipative or decay-type materials. Faraday cages can be designed to provide "cradle-to-grave" protection. Static protective work surfaces, ESD garments, static shielding bags and totes, and chip matrix trays are examples of these isolation methods(2)

4.3 Neutralization   Neutralization reduces static charge accumulated on insulative surfaces without actual contact as is needed for grounding. Commercially available ionizers produce a balanced mix of positive and negative ions. When directed across an object to be neutralized, these ions combine with opposite charges and effectively cancel them. Neutralization alone is not a complete preventative measure but it is able to reduce the probability of ESD damage occurring (2).

Gaining knowledge of sources and damage potential of static electricity is the first step in any static protection program. Applying this knowledge to development of static control solutions is the next step. This requires awareness of protective materials and their static control mechanisms. Several conductive thermoplastic compounds are introduced through a discussion of carbon fiber's role in static control and examples of successful static control solutions.

5. CARBON FIBER FOR STATIC CONTROL

5.1 Conductivity of Carbon Fiber   Carbon fiber is inherently more conductive than standard thermoplastic polymers and, while not as conductive as many metals, it is still quite impressive when weight specific resistivity is considered. Table 1 compares mass specific resistivity of carbon fiber and several other materials (3).

TABLE 1. RESISTIVITY		
	Resistivity Micro Ohm-cm	Mass Specific Resistivity* Micro Ohm Gm/Cm^2
H_2O (Deionized)	50,000	50,000
Graphite ($\perp$ to planes)	4,000	8,800
Carbon Fiber ($\cong$ 90% carbonized)	3,000	5,700
Carbon Fiber ($\cong$ 95% carbonized)	1,500	3,800
Mercury (Liquid)	96	1,300
Graphite (Parallel to planes)	40	88
Nickel	6.8	60
Silver	1.6	17
Copper	1.7	15
Aluminum	2.6	7

* Defined as resistance in ohms between ends of a specimen 1 centimeter long and a mass of 1 gram. Units are micro ohm-g/cm^2.
[From W.P. Hettinger, Jr., J.W. Newman, M.D. Kiser, and D.D. Carlos, Ashland Oil Inc., Ashland, KY, "Cost Effective Conductive Carbon Fiber for Composite Applications."]

Earlier discussion of grounding and isolation as preventative measures introduced dissipation and decay of charges (grounding) and Faraday cages (isolation) as means to afford protection. In each of these disciplines, carbon fiber containing thermoplastic compounds are able to provide the desired level of protection by changing the material's resistance. Static control grounding is accomplished by reducing the resistance of surfaces proximate to items facing static damage, allowing static charge to move through or across materials (dissipate) rather than arc through air (discharge). Faraday cages isolate by providing a low resistance (high conductance) enclosure, allowing static charge to move around the in-danger item.

5.2 Threshold Concentration and Percolation Point in a Conductive Compound
Resistance of a carbon fiber modified thermoplastic compound is related to content and
aspect ratio of carbon fiber. B.Webling noted that compound conductivity and modifier
content did not have a linear relationship. Rather, at a characteristic "threshold" modifier
concentration, there was a sudden increase in conductivity with only slight additional
change as concentration increased. Webling described this phenomenon as "percolation."
This percolation point occurs when a modifier's concentration in the compound, inherent
conductivity, and compounded aspect ratio reduce average "inter-particle" separation
distance sufficiently to pass electrons (4).

5.3 Features of Carbon Fiber Modified Thermoplastic Compounds   Carbon fiber
modified thermoplastic static protective materials provide features not found in all other
protective materials. The design freedom of thermoplastics allows physical and electrical
properties to be used as attributes, not features to be overcome in design. Carbon fiber
modified thermoplastic materials eliminate secondary operations usually necessary for
static protection. Carbon fiber's static protective features are permanent and cannot be
removed through abrasion, cleaning and solvents or heat. Carbon fiber modified
thermoplastic materials are also colorable.

5.4 Formulating for Static Control   Generic applications and the carbon fiber filled
products formulated for them are presented next. Carbon fiber as a conductive modifier
and its effect on compound properties in several thermoplastic resins will be discussed.
Conductivity is described by both volume resistivity and surface resistivity. Standard
practice uses surface resistivity of molded articles as the qualifying property in units of
ohms per square.

5.4.1 Integrated Chip (IC) Matrix Tray   Matrix chip trays are used in testing, transport
and storage, and assembly of circuitry for many portable and wireless products. Trays
are often rated by a "bake" temperature for conditioning ICs prior to soldering.
Thermoplastic resins chosen will reflect bake temperature required in addition to part
complexity. Acrylonitrile butadiene styrene (ABS) is used to 105^0C, polycarbonate (PC)
to 125^0C, polysulfone (PSUL) to 150^0C, and polyethersulfone (PES) to 200^0C. In most
instances, milled carbon fiber is compounded at concentrations from fifteen to twenty
percent by weight to achieve surface resistivity of 10^5 to 10^9 ohms/square. The resultant
injection molded tray provides static protection by a combination of grounding and
isolation. During testing and assembly procedures, the IC-filled trays are normally
afforded a contact to ground. During transit and storage, these trays are normally stacked
and enclosed in a full ESD shield. Table 2 provides a property comparison of unmodified
PES resin and carbon fiber filled PES resin.

TABLE 2. IC MATRIX TRAY MATERIALS				
Polyethersulfone (PES)				
			Unmodified	With Milled Carbon Fiber
Specific Gravity	gm/cc		1.37	1.40
Molding Shrinkage	%		0.7	0.7
Izod Impact Strength,				
notched	J/m	(ft.lb/in)	85 (1.6)	54 (1.0)
unnotched	J/m	(ft.lb/in)	640 (12)	534 (10)
Tensile Strength	MPa	(psi)	83 (12000)	103 (15000)
Tensile Elongation	%		50	4
Flexural Strength	MPa	(psi)	127 (18500)	172 (25000)
Flexural Modulus	MPa	(psi)	2550 (370,000)	6200 (900,000)
Deflect Temp, 1.81 MPa	^{0}C	(^{0}F)	205 (400)	210 (410)
Volume Resistivity	Ohm-cm		$>10^{15}$	5000
Surface Resistivity	Ohms/square		$>10^{15}$	10^6
Static Decay Rate	Seconds		Infinity	<2
[Data developed by RTP Company, Winona, MN]				

5.4.2 Conveyor Bearing Bushings Conveyor systems are widely used in manufacturing and assembly plants to transport items through stages of assembly and storage. Thermoplastics are preferred materials in many conveyors for reduced particulation and noise, for flexibility and modularity, and for precision and low cost. Conveyor materials are not always chosen with triboelectric features, humidity, or cleanliness levels of all possible environments considered.

One conveyor manufacturer prevents most static difficulties by building controlled rate grounding into the system. A carbon fiber modified nylon 6-6 (Table 3) material provides static dissipation at a slower rate than the previous all-metal frame. Roller bearings are fitted with injection molded bushings and overmolded sleeves of this material, providing a higher resistance link between moving parts and grounded stationary frame members. Static on conveyed items is then slowly and safely dissipated to ground before accumulating to dangerous levels, thus preventing arcing and high-energy ESD events on the conveyor.

TABLE 3. CONVEYOR BEARING BUSHING				
Polyamide Nylon 6-6				
			Unmodified	With Milled Carbon Fiber
Specific Gravity	gm/cc		1.08	1.14
Molding Shrinkage	%		1.5	0.25
Izod Impact Strength,				
notched	J/m	(ft.lb/in)	800 (15)	130 (2.5)
unnotched	J/m	(ft.lb/in)	1600 (30)	750 (14)
Tensile Strength	MPa	(psi)	50 (7500)	110 (16000)
Tensile Elongation	%		60	4
Flexural Strength	MPa	(psi)	60 (9000)	165 (24000)
Flexural Modulus	MPa	(psi)	1400 (200,000)	5000 (750,000)
Deflect Temp, 1.81 MPa	^{0}C	(^{0}F)	72 (160)	193 (380)
Volume Resistivity	Ohm-cm		$>10^{15}$	1000
Surface Resistivity	Ohms/square		$>10^{15}$	10^5
Static Decay Rate	Seconds		Infinity	<2
[Data developed by RTP Company, Winona, MN]				

5.4.3 Remote Heart Monitor Housing Clinical heart function testing generally stresses patients emotionally, poor performance results, and test validity is questioned. Repeated testing is costly and adds to stress levels. To eliminate emotional responses, patients are fitted with needed sensors which are hard wired to a pocket-sized recording device. Patients remain ambulatory and perform normal activities during a designated time period. Standard heart function information plus time-weighted values are thus obtained without the added emotional stress of testing or clinical environment. Monitor feedback provides greater distinction between serious heart problems and the general stress from clinical environment. Subsequent treatment is more effective.

Monitor housings were molded of precolored ABS thermoplastic, chosen for toughness and resistance to cleaning solutions. Initial trial devices were affected by static charges on patient's body arising from contact with "non-triboelectric compatible" materials through activities such as walking, movement of clothing, and handling paper. Static would cause spikes in data and sometimes destroy circuit elements. Static problem was resolved by compounding carbon fiber into ABS to make the plastic housing function as a Faraday cage and isolate internal circuitry (Table 4). Design freedom of plastics and recycling and color options were retained along with an enhanced resistance to breakage.

TABLE 4. REMOTE HEART MONITOR HOUSING				
Acrylonitrile Butadiene Styrene (ABS)				
			Unmodified	With Milled Carbon Fiber
Specific Gravity	gm/cc		1.04	1.09
Molding Shrinkage	%		0.50	0.20
Izod Impact Strength,				
notched	J/m	(ft.lb/in)	160 (3)	40 (0.8)
unnotched	J/m	(ft.lb/in)	960 (18)	215 (4)
Tensile Strength	MPa	(psi)	40 (6000)	70 (10000)
Tensile Elongation	%		2	1
Flexural Strength	MPa	(psi)	70 (10000)	100 (15000)
Flexural Modulus	MPa	(psi)	2100 (300,000)	5500 (800,000)
Deflect Temp, 1.81 MPa	^{0}C	(^{0}F)	100 (210)	105 (220)
Volume Resistivity	Ohm-cm		$>10^{15}$	1000
Surface Resistivity	Ohms/square		$>10^{15}$	10^4
Static Decay Rate	Seconds		Infinity	<2
[Data developed by RTP Company, Winona, MN]				

6. CONCLUSION

Thermoplastics will remain the preferred materials for portable and wireless electronics. Although naturally insulative, thermoplastic resins can be made static-protective when necessary. Through incorporation of carbon fibers, thermoplastic materials are able to provide static protection through grounding and isolation. The resultant conductive materials are colorable, exhibit enhanced physical properties, and retain desireable design and recycling features of thermoplastics.

7. BIBLIOGRAPHY

Larry Rupprecht, Senior Development Engineer and Conductive Materials Manager at RTP Company, Winona, MN, has been formulating conductive materials for more than twenty years. He has authored numerous articles on static protective and EMI shielding materials.

8. REFERENCES

1.	A. Woltman, ESD - Electrostatic Discharge, "The Charging Process," 1987.

2.	E. Johnson, EOS/ESD Tutorial 1997, pp. S/1/1 - S/1/24.

3.	W.P. Hettinger, Jr., J.W. Newman, M.D. Kiser, and D.D. Carlos, Ashland Oil Inc., Ashland, KY, "Cost Effective Conductive Carbon Fiber for Composite Applications," ACS Meeting, Denver, CO, 1987.

4.	B.Webling, Kunststoffe German Plastics 76, pp. 69-73, (1986).

EMI SHIELDING EFFECTIVENESS OF NICKEL COATED CARBON FIBER AS A LONG FIBER THERMOPLASTIC CONCENTRATE

Malcolm W. K. Rosenow
Business Manager, Fiber Products
Inco Specialty Powder Products

Dr. J. A. E. Bell
Consultant

ABSTRACT

Five different long fiber nickel concentrates containing 60wt% of nickel coated carbon fiber were injection molded into several different thermoplastic polymers. These polymers included, ABS, PA6, PA66, PC, PC/ABS, PEI, and PPS. EMI shielding data is generated for each polymer system, and summarized into one chart independent of a polymer system, with the shielding attenuation limits set within a 95% confidence level. The level of nickel coated carbon fiber was varied from 5, 10, 15, and 20wt% to demonstrate that the maximum shielding attenuation is achieved at 15wt% of nickel coated carbon at 30, 100, 300, and 1,000 MHz. The results show that EMI shielding attenuation ranges from, 80 dB at 30 MHz, to 100 dB at 1,000 MHz., using 15wt% nickel coated carbon fiber. The shielding performance results comparing injection molding the concentrate by dry blending versus compounding the nickel concentrate into a polycarbonate resin, illustrate the need to maintain and preserve the nickel coated carbon fiber aspect ratio.

KEYWORDS: EMI Shielding, Conductive, Nickel Coated Carbon Fibers, Thermoplastic

1. INTRODUCTION

With the introduction of new electronic products, such as laptop computers running at higher frequencies and in smaller packages, the problem of electromagnetic interference (EMI) and thermal management is becoming a challenge to control. The key limitation of plastics in replacing metals is the inherent insulative nature of most plastics. Plastics are therefore vulnerable to electromagnetic interference that limits the ability of the electronic equipment to function satisfactorily in its electromagnetic environment without introducing intolerable disturbances to anything in that environment. The solution is to make the insulative plastic conductive with the introduction of conductive materials, such as, carbon black, stainless steel, carbon fiber and nickel coated carbon fiber.

The above material possibilities define one of the first steps in designing for a conductive injection molded part. Additional factors are taken into consideration for the end-use requirements, such as, physical strength and toughness, flammability rating, anticipated temperature extremes, chemical environment, dimensional tolerances, and color[1].

In addition to the available conductive materials, the performance of the injection molded shielding device will be determined by design. The shape, thickness, relative position of the electronics, type of joints and apertures the injection molded shielding device contains will define its shielding performance[2,3].

If shielding can not be achieved with the design of the electronic board, the solution is to enclose the electromagnetic emissions within a conductive box. Assuming that the emission spectrum of the electronics to be shielded is measured, and subtracting this from the maximum allowable emissions defined by regulatory standards, will yield the frequencies needing attenuation and the degree to which it needs to be attenuated. Once all this is understood, then the selection of the conductive material can take place to make the electronic device including the seams, apertures and injection molded enclosure, electromagnetic compatible (EMC).

Compounded products (short fiber reinforced thermoplastics or SFRTP) using a small fraction of carbon fibers derived from polyacrylonitrile fibers (PAN) will make a polymer partially conductive. Higher fiber percentages, from 30 - 40 weight percent, will provide 40dB of shielding attenuation in the 30 MHz to 1 GHz range.

With the introduction of long fiber reinforced thermoplastic technology (LFRTP), the benefits of long fiber reinforced composites were realized with higher material performance in impact strength and higher flexural strength especially at elevated temperatures over SRFTP. These new thermoplastic composites provide even higher levels of shielding attenuation over the conventional compounded products.

Nickel coated carbon fibers, INCOFIBER™ produced by coating the carbon fibers with nickel by a chemical vapour deposition (CVD) process, combine the advantages of the reinforcing properties with the high conductivity of the nickel coating. Taking this one step further and using a patented long fiber thermoplastic impregnation technology, uniquely different products are available called, INCOSHIELD™ long fiber nickel concentrates. These long fiber nickel concentrates consist of 60% by weight of nickel coated carbon fiber and 40% by weight of a thermoplastic resin. These are subsequently dry blended down with a thermoplastic to a lower fiber level prior to injection molding.

1.1 Technology   The chemical vapour deposition (CVD) process illustrated in Figure 1, coats any unsized carbon fiber. In this process the fiber is pulled through the coater, and nickel is deposited by thermal decomposition of the nickel bearing gas into elemental nickel and carbon monoxide to coat the fibers[4], as shown in Equation 1.

$$Ni(CO)_4 \rightarrow Ni + 4CO \dots\dots\dots (1)$$

This process is not new, having been discovered by Ludwig Mond and Carl Langer in 1889[5], and developed into an industrial process by 1902. This process is the major refining process of

Inco Limited, the world's leading nickel company in North America and Europe. The advanced process technology has been developed with continuous investment in research and development.

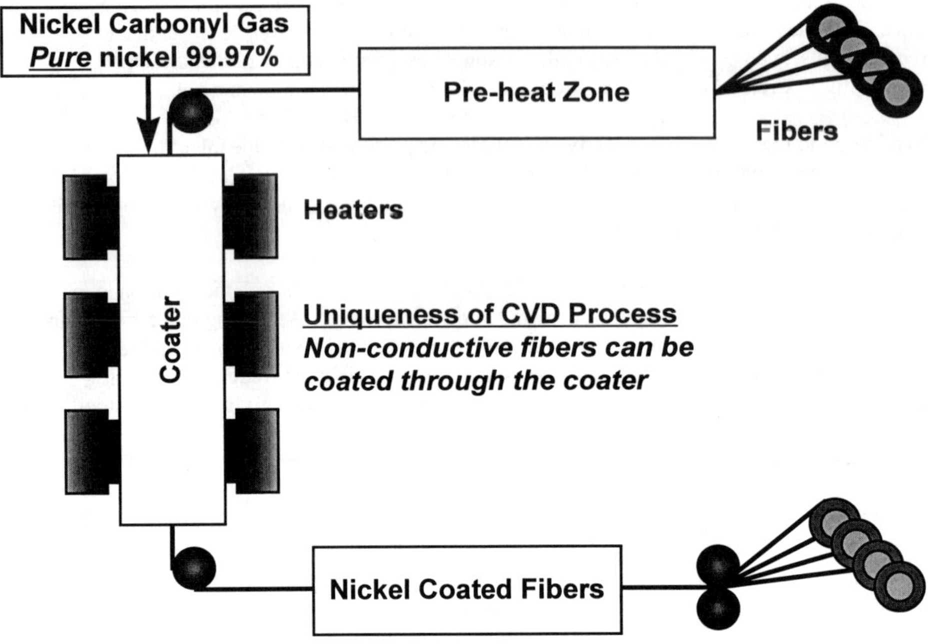

Figure 1 Chemical vapour deposition (CVD) process

The conductivity of each tow is monitored throughout the winding of the finished product, as illustrated in Figure 2. This data is used as a feed back loop to control the level of nickel coating. The fiber can be coated to any desired nickel level with optimized targets of between 42 - 48 wt% for electronic applications[6].

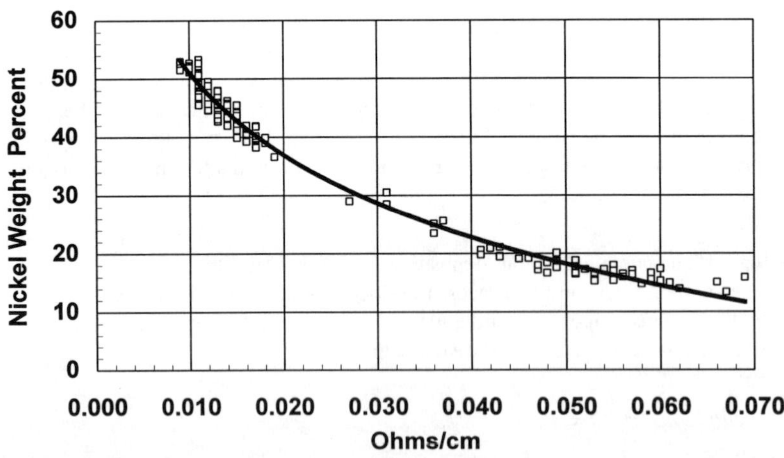

Figure 2 Resistance per unit length for nickel coated carbon fiber as a function of wt% nickel

The gas penetrates the fiber bundle easily to achieve uniform nickel coating throughout the bundle as shown in Figure 3. The uniqueness of the CVD process is that it is not limited to conductive fibers. Non-conductive fibers have been successfully coated[7] and are under evaluation for EMI shielding applications.

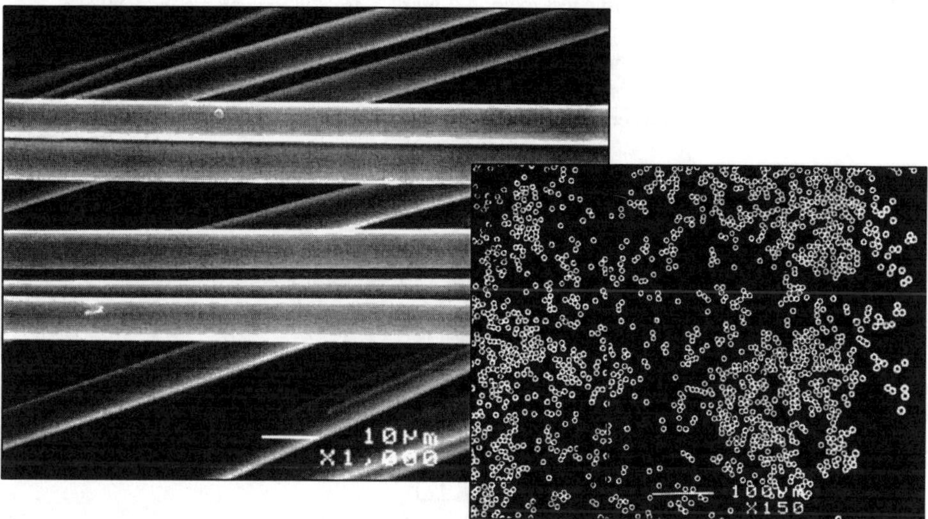

Figure 3 Longitudinal and cross-section on nickel coated carbon fiber

A proprietary thermoplastic process, Figure 4, is then used to fully impregnate the nickel coated fiber bundles with a thermoplastic resin. The material is then cut to a length, generally 6.0 mm, to produce a long fiber nickel concentrate.

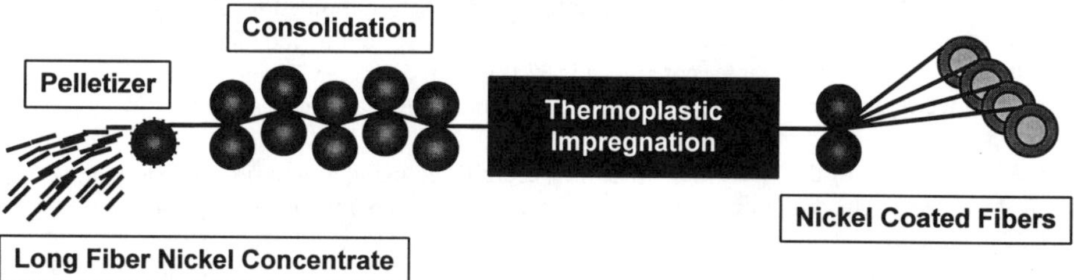

Figure 4 Long fiber thermoplastic process

Various thermoplastic resin systems are available to match the specific applications, unlike the previous technology of electroplating and coating the fibers with a general purpose water based sizing. Commercially available long fiber nickel concentrates are available in the following thermoplastic resins systems; polycarbonate (PC), polyamide 6 (PA6) and polyamide 12 (PA12), polyethylene (PE), polyetherimide (PEI) and polyphenylene sulfide (PPS) and polymethylmethacrylate (PMMA).

The purpose of this paper is to outline the use of this advanced new thermoplastic nickel coated carbon fiber concentrate for EMI shielding in plastics and the properties which can be achieved.

2. Test Method to Measure Shielding Effectiveness

A flanged coaxial transmission line test method, Figure 5, is used to evaluate the EMC (Electromagnetic Compatibility) shielding effectiveness of the injection molded nickel coated carbon fibers in a two dimensional form. This method is based on a variant of the American Society of Testing and Materials, ASTM Method ES7-83, modified by the American National Bureau of Standards, and now encompassed in ASTM D4935-98.

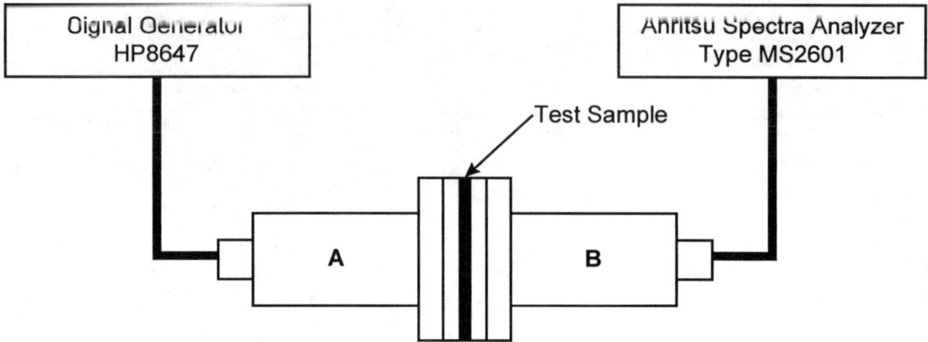

Figure 5 Coaxial transmission line test fixture - ASTM D 4935-89

This test method provides for the measuring of the Electromagnetic Shielding (EM) effectiveness (SE) of a planar material due to a plane-wave, far field wave. The test method is normally valid over a frequency range 30MHz - 1,500MHz. The testing of these samples were over a range of 25MHz - 1,000MHz.

Test plaques, 151 mm x 151 mm x 3.2 mm thick were injection molded for shielding effectiveness testing. Donut specimens are cut with a 133.1 mm outside diameter and an inner annulus of 76.2 mm. The center core piece is 33.0 mm in diameter. In most cases three different specimens of each composition were measured and averaged.

The dynamic range of the test arrangement was established using aluminum sheet in place of the test samples. The dynamic range was in excess of the 110dB to ensure accuracy of the results obtained.

The shielding effectiveness SE(dB) is calculated by the formula given below in Equation 2:

$$SE(dB) = 20logE_1 - 20logE_2 \quad\quad\quad (2)$$

where E_1 is the received signal when the test sample is absent, and E_2 is the received signal when the test sample is present.

Repeatability of the testing was very good, with a maximum variation of $\pm$ 1.0 dB when the test was repeated. Measurement uncertainty due to system errors is in the order of $\pm$ 1.2dB, and errors due to random effects, that is impedance mismatch and measurement drift over a short period, would amount to $\pm$ 1.0 dB. The total measurement error, including operator error, is in the order of $\pm$3.2 dB.

3. Experimental

3.1 Fiber Length Evaluation   The first set of tests evaluated a polycarbonate long fiber nickel concentrated. This was compared to dry blending the material directly in the feed hopper of the injection molding machine with the same material compounded using a twin screw compounder, then injection molded. Both tests made 151 mm x 151 mm x 3.2 mm plaques containing 15 wt% of nickel coated carbon fiber. The shielding effectiveness of the average of 4 samples from each preparation are shown in Figure 6. The twin screw compounded material had a dramatically lower shielding effectiveness of 15 dB compared to the usual 60 dB for the material which was dry blended only.

In order to find out the cause of this dramatic difference, both samples were extracted in dichloromethane on a hot plate with reflux. The length distribution of the fiber is shown for the two samples in Figure 7. Both show break up of the fiber from 6,000 micro meters down to 300 micro meter, however the dry blended material retained longer fibers.

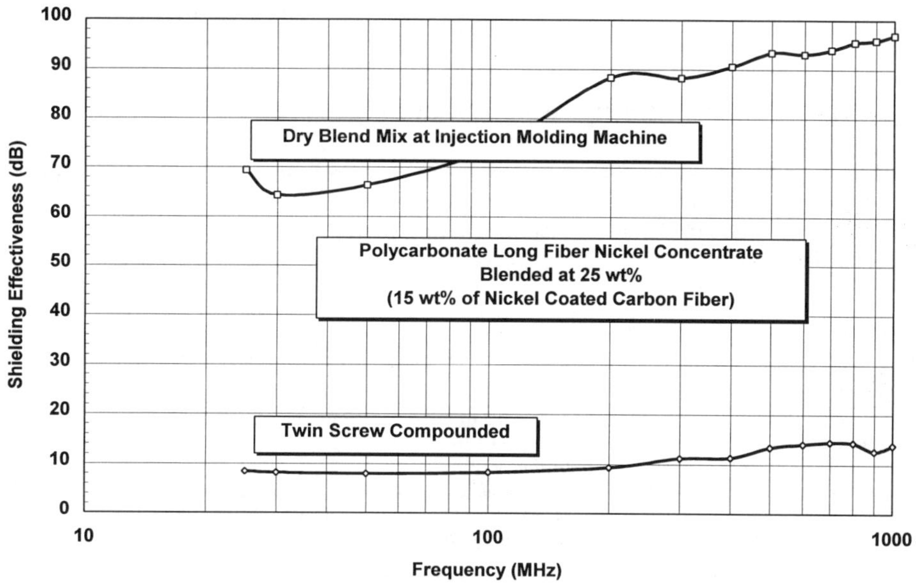

Figure 6 Shielding effectiveness by dry blending vs compounding process

Skillful molding is central to the production of high volume, high quality thermoplastic parts and this is essential when using long fiber reinforced materials[8]. Electrical properties, such as EMI shielding, can be sharply reduced through processes or molding conditions which break or shorten the fiber length.

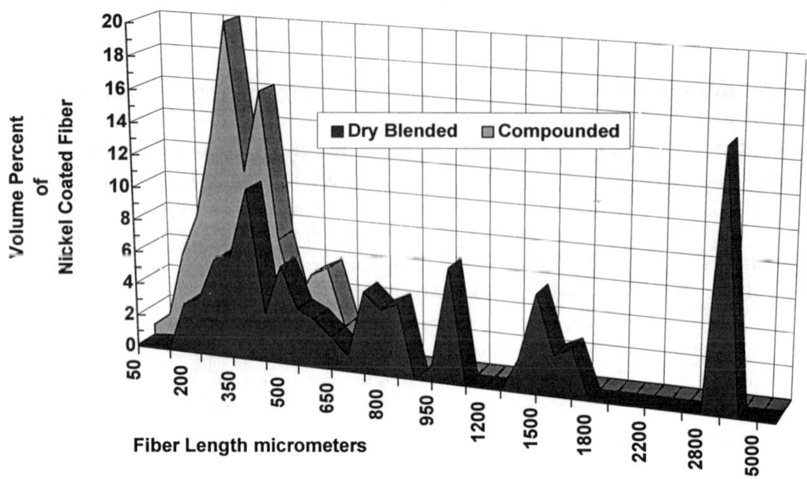

Figure 7 Fiber length distribution of materials dry blended vs compounded

3.1 Shielding Effectiveness   The next series of shielding tests took five different long fiber nickel concentrates containing 60wt% of nickel coated carbon fiber. The materials were injection molded into the following thermoplastic polymers, indicated in Table 1, by dry blending the mixture.

Table 1

Long Fiber Nickel Concentrate		*Blending Polymer*
Polyamide 12 (PA12)	Polyamide 6 (PA6)	Polyamide 6/6 (PA6/6)
Polycarbonate (PC)	Polycarbonate (PC)	
Polyetherimide (PEI)	Polyetherimide (PEI)	
Polymethylmethacrylate (PMMA)	Acrylonitrile-Butadiene-Styrene (ABS)	Polycarbonate/Acrylonitrile-Butadiene-Styrene (PC/ABS)
Polyphenylene sulfide (PPS)	Polyphenylene sulfide (PPS)	

The level of nickel coated carbon fiber was varied from 5, 10, 15, and 20wt% and EMI shielding data was generated for each of the blending polymer systems using two samples per weight percent of nickel coated carbon fiber. A total of seven different blending polymers was used for the five different long fiber nickel concentrates. Data was summarized into one chart independent of a polymer system for each nickel carbon fiber weight percent (see Figures 8, 9, 10 & 11). Confidence level of 95% are illustrated on each chart to indicate the range of EMI shielding data for all the polymers.

Maximum shielding attenuation is achieved at approximately 15wt% of nickel coated carbon at 30, 100, 300, and 1,000 MHz, as shown in Figure 12. EMI shielding attenuation ranges from, 80 dB at 30 MHz, to 100 dB at 1,000 MHz., for 15wt% nickel coated carbon fiber.

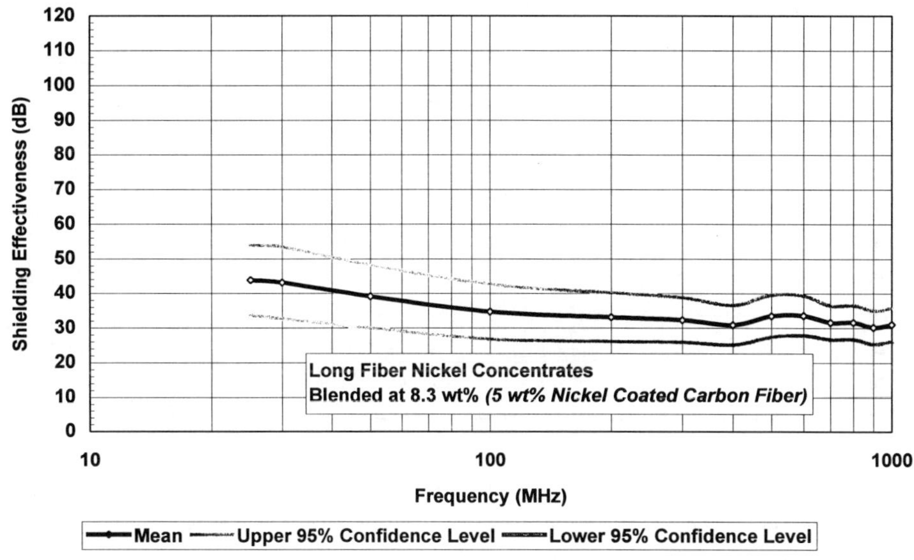

Figure 8 Shielding effectiveness of 5wt% nickel coated carbon fiber independent of the thermoplastic polymer

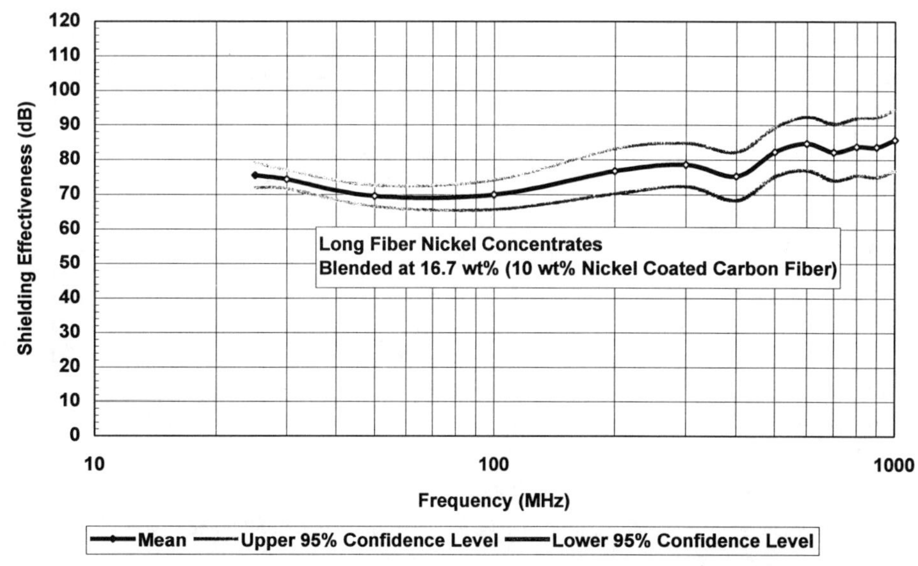

Figure 9 Shielding effectiveness of 10wt% nickel coated carbon fiber independent of the thermoplastic polymer

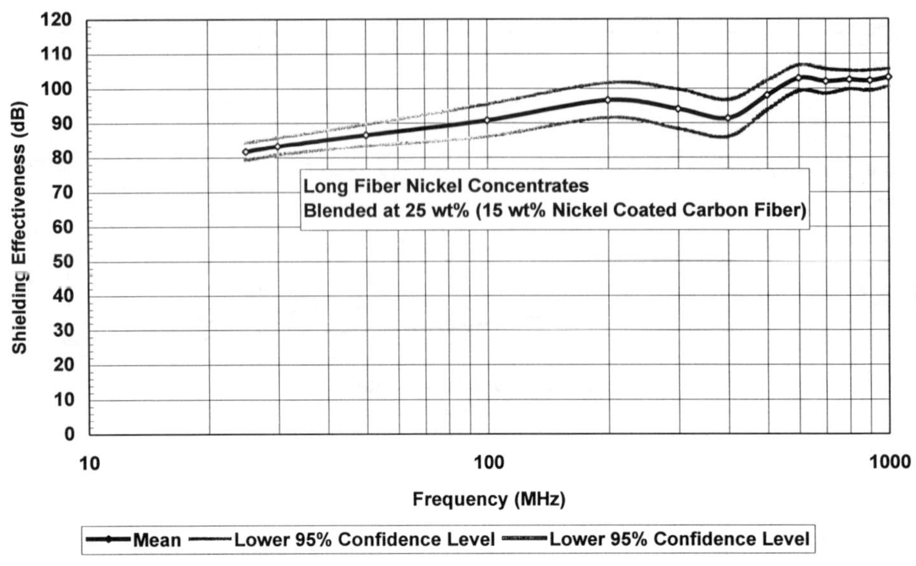

Figure 10 Shielding effectiveness of 15wt% nickel coated carbon fiber independent of the thermoplastic polymer

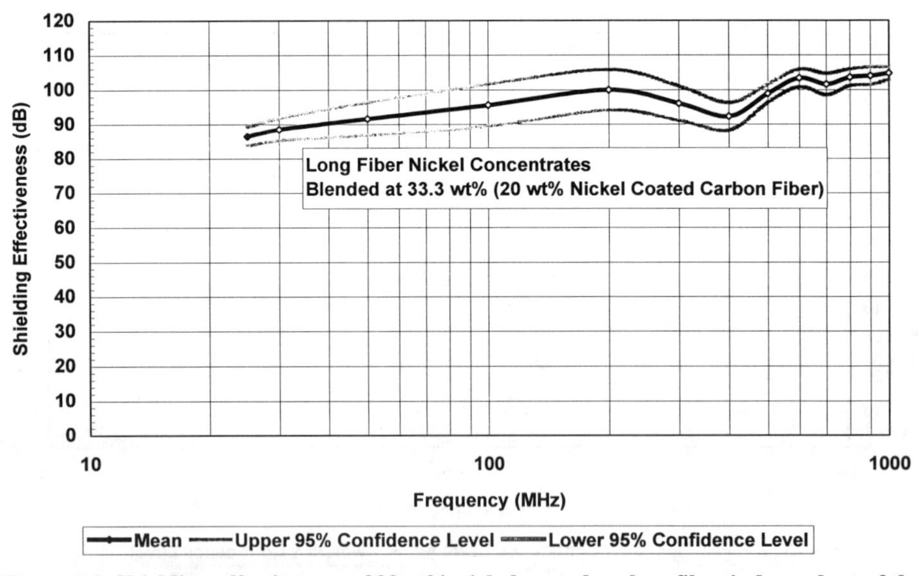

Figure 11 Shielding effectiveness of 20wt% nickel coated carbon fiber independent of the thermoplastic polymer

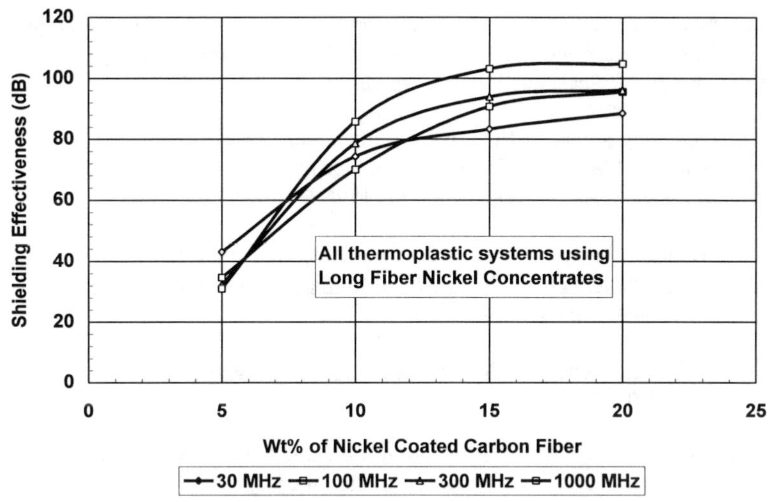

Figure 12 Shielding effectiveness vs nickel coated carbon fiber

4. CONCLUSIONS

A new advanced long fiber nickel concentrate is now available in a variety of thermoplastic resins for EMI shielding applications. Using the same thermoplastic resin as the injection molded part ensures that the physical and electrical properties of the final product are optimized. High levels of shielding effectiveness and physical properties have been obtained.

The most important factor to consider to maximize EMI shielding attenuation or electrical properties is maintaining fiber length. Mechanical and electrical properties can be sharply reduced by conditions which break or shorten the fiber length. Fiber retention can be accomplished through dry blending this new long fiber nickel concentrate, and moreover by using the optimum injection molding parameters

Dry blending at 25 weight percent of this new long fiber nickel concentrate (15 weight percent nickel coated carbon fiber) with a thermoplastic compounded product is recommended for direct use at the injection molding machine to achieve maximum shielding attenuation.

5. REFERENCES

1. Larry Rupprecht and Ray Leaf, "Application-based Selection of Conductive Thermoplastic Materials", <u>ITEM Update 1997</u>, pages 76-84.

2. Chris P. J. H. Borgmans and Steve R. Gerteisen, "Design Considerations for EMI Shielding Using Conductive Plastic Compounds", in the Society of Plastics Engineers <u>Proceedings of the Plastics for Portable Electronics</u>, Las Vegas, NV, January 5th - 6th, 1995.

3. Larry Rupprecht, "Predicting Shielding Effectiveness of Conductive Thermoplastic Materials", in the Society of Plastics Engineers <u>Proceedings of the Plastics for Portable and Wireless Electronics</u>, Phoenix, AZ, November 6th - 8th, 1996.

4. J. A. E. Bell and G. Hansen, "Nickel Coated Fibres for Aerospace Applications", in <u>Proceeding of the 24th International SAMPE</u> Technical Conference, October 20th - 22nd, 1992, Toronto, Ontario, Canada, pages T902-T911.

5. Joseph R. Boldt, Jr., "<u>The Winning of Nickel, Its Geology, Mining, and Extractive Metallurgy</u>", Methuen & Co. Ltd., London.

6. Malcolm W. K. Rosenow and J. A. E. Bell, "Injection Moldable Nickel Coated Carbon Fibre Concentrate for EMI Shielding Applications", in the <u>Proceeding of the SPE 55th Annual Technical Conference, ANTEC'97</u> April 27th - May 2nd, 1997, pages 1492-1498.

7. J. A. E. Bell and G. Hansen, "Properties of Nickel Coated Carbon and Kevlar Fibers Produced by Decomposition of Nickel Carbonyl", in the <u>Proceedings of the 23rd International SAMPE Technical Conference</u>, October 22nd -24th, Kaimesha Lake, New York, pages 1183-1193.

8. Malcolm W. K. Rosenow, "Injection Molding of Long Carbon Fiber Reinforced Thermoplastic Materials", in the <u>Proceeding of the 40th International SAMPE Symposium and Exhibition</u>, May 8th - 11th, 1995, pages 1534-1541

EXCEPTIONAL PROCESSING (EP) TECHNOLOGY PROCESS STUDY FOR INJECTION MOLDED THIN WALL PARTS

Stephen L. Thompson, Kevin R. Quinn
LNP Engineering Plastics Inc.

ABSTRACT

The demand for thin wall applications has driven resin suppliers to formulate materials that will meet the customers' structural requirements, but also be processable at wall thicknesses below 1.5 mm. Low viscosity, fast processing materials are required for such applications. The development of high flow 20% glass filled Polycarbonate resin systems are evaluated to determine if these systems are not only processable, but functional and cost effective. Improvements over typical 20% glass filled Polycarbonate compounds are seen through lower pressures to fill, higher injection speeds, and shorter cycle times, while maintaining part integrity.

KEY WORDS: Thin Wall, Polycarbonate

1. BACKGROUND

Previous designs of computer housings, hand held phones, and other portable communication equipment used thicker walls and fell under the same processing guidelines as typical injection molded parts. It was common for these wall sections to be 4 mm (0.157 in.). Weight reduction was achieved with the use of foaming agents. Currently wall thicknesses in the 1 mm (0.040 in.) ranges are common, and could be considered the norm for thin wall applications. Molders are now looking ahead to 0.80 mm (0.032 in.) and thinner for hand held equipment.

Designing parts with thinner walls means utilizing different processing guidelines. Faster fill times are needed when molding thin wall parts. During injection, regardless of part thickness, a frozen layer or skin is developed along the outside walls, as the material flows into the cavity. The area between these frozen layers is the flow channel. As walls become thinner, the flow channel decreases. This restricted flow means that it is more difficult to fill and pack parts especially when the length to thickness (L:T) ratios are above 100:1. Injecting the material as fast as possible without degradation is critical when processing thin wall parts with a high L:T ratio. Faster injection speeds will help to minimize the skin layer, through shear heating. Injection times below 0.75 sec. are common for wall thicknesses of 1.5 mm (0.060 in.), and 0.40 sec for thicknesses at 1 mm (0.040 in.). To achieve these fast fill times higher injection pressures are needed. In this case study melt pressures up to 240 MPa (35,000 psi) were required to fill the part. High performance injection molding machines are recommended to attain these high speeds and pressures.

As nominal wall thickness decreases, structural integrity becomes a question, hence the use of engineering plastics. The use of fiber reinforcements in a plastic material can substantially increase the strength and stiffness of a part. There are, however, tradeoffs when specifying a new material for an application. When using engineering plastics, especially reinforced compounds, viscosity is a factor that needs to be addressed. Adding reinforcements typically increases the viscosity of a material. Excessive shear rates and degradation is a concern considering the fast injection speeds and higher temperatures needed to push these compounds. Surface appearance at gates and weld lines will be affected by these conditions.

Multiple gates are recommended when the L:T ratio rises above 100:1. Multiple gates will result in weld lines. The weld line integrity of a reinforced material is especially important in thin wall applications. Reinforced plastics with improved flow characteristics are being developed to reduce process temperatures, shear, and surface blemishes. These flow enhanced resins will help minimize the tradeoffs that go along with switching to reinforced materials for thin wall applications.

2. MATERIALS

Four materials, all Polycarbonate (PC) based, were selected for this study. Each material contained 20% short glass fiber (GF) reinforcement. These compounds are differentiated from each other by their base resins and flow enhancement packages. These packages reduce viscosity, increase processability and provide other benefits that will be discussed in this paper. A straight 20% glass filled compound was used as the baseline for the cycle time study. These materials are listed in the order of decreasing viscosity:

Material	**Description**	**Flow Viscosity (300 1/sec)**
Thermocomp® DF-1004	20% GF PC	5214
Thermocomp® DF-1004 EM	20% GF PC	5002
Thermocomp® DF-1004 SM	20% GF PC	3370
Thermocomp® DF-1004 EP	20% GF PC	2319

3. MOLDING

3.1 Mold The tool used in this study was built in conjunction with Moldmasters Limited who was responsible for building the hot runner system for the tool (Figure 1). It contains a 205 mm x 305 mm (8 in. x 12 in.) enclosure (Figure 2) with a nominal wall thickness of 1 mm (0.040 in.). Four gates were used yielding an L:T ratio of 140:1. The four valve gates are 2 mm (0.080 in.) in diameter. In order to hide gate vestiges the gates were placed on the underside of the enclosure. This gate placement required the tool to be fitted with an air operated, A-side ejection system.

One of the goals, when designing the enclosure was to make the part difficult to fill. The L:T ratio is 140:1, and two of the gates are located near holes. Another objective was to make the part design represent real world situations. Multiple valve gates, internal ribs, bosses, gussets, and slotted holes were incorporated into the tool design.

Venting is critical in thin wall applications. The fast injection speeds associated with thin wall molding means that air must escape from the tool quickly. Adequate venting will allow the part to fill out properly and reduce the chances of burning. A 0.038 mm (.0015 in.) perimeter vent is used in the enclosure tool.

3.2 Machine The machine used in this study was a high performance JSW 310 Ton SP unit located at the Triple S research facility. It is equipped to run thin wall tooling applications. The machine has thicker platens to minimize deflection and flashing under high injection pressures. Standard 310 ton machines typically have a shot capacity of 24 oz. (styrene). A 24 oz. barrel would be too large for most thin wall applications, whose shot sizes are minimal. The JSW was equipped with a 13 oz. barrel. A smaller barrel results in shorter residence times, reducing the chance for material degradation. Accumulators provide the extra injection pressure needed for thin wall applications. The JSW injection molding machine is capable of 240 MPa (35,000 psi.) of injection pressure.

3.3 Material Preparation The materials in the study were dried according to the guidelines set in LNP's processing tips for molding fiberglass reinforced thermoplastics bulletin.

4. TRIAL

A cycle time study was set up to determine the processing differences among the four 20% GF PC compounds. Performance was evaluated by finding the optimum cycle for each compound, without sacrificing part quality. Part quality was determined by assessing surface appearance, particularly at the weld lines and gates. Parts were also inspected to assure that there were no cracks on the bosses and/or part corners. When the optimum cycle time for a material was established, fifty samples were molded and the process data was evaluated. From the process data a cycle time cost savings analysis was completed. Mechanical properties of the four compounds were also tested.

To reduce the number of variables associated with the molding process, some of the processing parameters in the cycle time study remained constant. Barrel and manifold temperatures, mold temperature, back pressure, and screw speed were not changed for any of the compounds. See Tables 1 and 4 for the set points of these parameters. The first stage injection pressure was set at 240 MPa (35,000 psi), the machine's maximum melt pressure capability. This would allow the machine to use whatever pressure it needed, within this limit, to fill the part. The hold pressure was set at one half of the first stage injection pressure. Hold pressure tends to have a minimal effect on filling the part since the flow channel is typically frozen after the first stage injection pressure. Setting the hold pressure to one half the first stage injection pressure helped reduce screw bounce-back and did not greatly contribute to packing the part out.

Once a general cycle had been established, the optimum cycle time was achieved through an incremental decrease in cycle time using cooling time, and clamp open-close time. Cooling time was lowered until the parts would fail. Failure occurred in the form of pin penetration, cracking during ejection, or parts not releasing from the tool. The cooling time was then raised until consistent quality parts were being molded.

5. RESULTS

A cycle time study was completed on all four compounds. Figure 3 shows a side by side comparison of the parts. The DF-1004 compound could only be 75% filled, even at a pressure of 240 MPa (35,000 psi). The actual pressure that would be required to fill the part, using this compound, could not be assessed. The DF-1004 compound was evaluated at its maximum fill point, and an optimum cycle was established even though the parts were not completely filled. Table 2 shows the process times for DF-1004. An injection time of 1.8 seconds was required to fill the part. The optimum cycle time for the DF-1004 parts was 20.8 seconds. The DF-1004 EM compound could only be 95% filled under maximum pressure. An optimum cycle time was achieved using the same technique as with the DF-1004 compound. The injection time for DF-1004 EM, as shown in Table 2, was a full second quicker than that of the DF-1004. This resulted in an overall cycle time of 19.8 seconds. Although an optimized cycle had been set-up for these two compounds, it was difficult to truly judge the cycle time capabilities of these materials considering that they did not entirely fill.

The DF-1004 SM and DF-1004 EP materials did fill but processed differently. The DF-1004 SM compound required 213 MPa (30,843 psi.) of melt pressure to fill the part. An injection time of 0.41 seconds was needed to completely fill and pack the enclosure. A cycle time of 17.45 seconds was established and produced consistent parts. A cycle time lower than this resulted in parts sticking to the tool and cracking along the bosses. The surface appearance of the parts looked good. The DF-1004 EP compound was the best performing material in the group. The DF-1004 EP required 194 MPa (28,000 psi) of pressure to fill the enclosure, approximately 10% lower than the DF-1004 SM. The fill time required was 0.36 seconds. The overall cycle time established was 14.5 seconds. The parts still seemed to release from the tool easily at this cycle time. A cooling time lower than this amount resulted in the parts sticking to the tool occasionally. The 14.5 second cycle provided quality parts with no interruption on a consistent basis. A slight surface appearance improvement over the DF-1004 SM was seen with the DF-1004 EP material.

Material properties were tested in accordance with ASTM test methods for each of the Polycarbonate compounds (Table 5). These results show that the switch to flow enhanced reinforced materials does not hinder overall material performance. Special impact testing was conducted on specimens measuring 102 mm x 102 mm (4"x 4"). These samples were taken from the actual enclosure parts and a drop impact test was conducted on all the compounds except the DF-1004. The DF-1004 did not fill far enough to be able to take a test sample. This test applied to the materials in the cycle study at a 1 mm thickness, and show no correlation to the typical ASTM D-3763 drop impact testing. The results show (Table 5) that by switching to the flow enhanced compounds, there is no attributable loss in impact at the 1 mm thickness.

A cost comparison between the DF-1004 EM and DF-1004 EP compounds looked at potential cost savings that could be gained by switching to a flow enhanced reinforced compound. Table 6 shows that the savings in cycle time translates into a 23% cost savings per part.

6. CONCLUSIONS

- Flow enhanced Polycarbonate materials provide:

    - Increased Productivity
    - Faster cycle times
    - Lower injection pressures
    - Lower required melt temperatures
    - Improved mold release

- Part cost savings

- Longer flow lengths (optimum for thin wall applications)

- Lower residual stresses

- Flatter parts

- Improved surface finish

- No loss of impact strength compared to standard Polycarbonate compounds

7. ACKNOWLEDGMENTS

The authors would like to thank Moldmasters Limited for their assistance in the design of the hot runner manifold and construction of the tool. Thanks to Triple S Plastics for the use of their injection molding equipment. Also thanks to Chris Ellis for his work on the mechanical property testing.

8. REFERENCES

1. "Processing Tips For Molding Fiber Glass Reinforced Thermoplastics," LNP Engineering Plastics Inc., Exton, PA, 1992.

TABLE 1

PROCESS TEMPERATURES				
DROP ZONES	**DF-1004**	**DF-1004 EM**	**DF-1004 SM**	**DF-1004 EP**
Zone 1 (°C)	305.0	305.0	305.0	305.0
Zone 2 (°C)	305.0	305.0	305.0	305.0
Zone 3 (°C)	305.0	305.0	305.0	305.0
Zone 4 (°C)	305.0	305.0	305.0	305.0
Zone 5 (°C)	321.0	321.0	321.0	321.0
Zone 6 (°C)	321.0	321.0	321.0	321.0
MELT ZONES	**DF-1004**	**DF-1004 EM**	**DF-1004 SM**	**DF-1004 EP**
Zone 1 (°C)	315.5	315.5	315.5	315.5
Zone 2 (°C)	315.5	315.5	315.5	315.5
Zone 3 (°C)	310.0	310.0	310.0	310.0
Zone 4 (°C)	310.0	310.0	310.0	310.0
Zone 5 (°C)	305.0	305.0	305.0	305.0
MOLD(°C)	**DF-1004**	**DF-1004 EM**	**DF-1004 SM**	**DF-1004 EP**
Core & Cavity	88.0*	88.0	88.0	88.0
Front Plate	60.0	60.0	60.0	60.0

*This was limited by the use of a water heater.

TABLE 2

PROCESS TIMES				
	DF-1004	**DF-1004 EM**	**DF-1004 SM**	**DF-1004 EP**
Inj. Time(sec)	1.80	0.86	0.41	0.36
Hold Time(sec)	0.50	0.50	0.50	0.50
Cycle Time(sec)	20.80	19.80	17.45	14.50

TABLE 3

PROCESS PRESSURES				
	DF-1004	**DF-1004 EM**	**DF-1004 SM**	**DF-1004 EP**
Melt Pressure	236.0 MPa	232.5 MPa	213.0 MPa	194.0 MPa
Hold Pressure	118.0 MPa	116.0 MPa	110.5 MPa	105.5 MPa
Back Pressure	.35 MPa	.35 MPa	.35 MPa	.35 MPa

TABLE 4

OTHER SETTINGS				
	DF-1004	**DF-1004 EM**	**DF-1004 SM**	**DF-1004 EP**
Shot Size	2.50	2.50	2.50	2.50
Decompression	0.1	0.1	0.1	0.1
Screw Recovery	5.0	5.8	6.4	6.0
Screw Speed	50 RPM	50 RPM	50 RPM	50 RPM

TABLE 5

MECHANICAL PROPERTY COMPARISON				
	DF-1004	**DF-1004 EM**	**DF-1004 SM**	**DF-1004 EP**
Tensile Strength, MPa	92.5	99.0	103.0	103.5
Tensile Elongation, %	3.6	3.2	3.0	2.8
Flexural Strength, MPa	160.0	168.0	171.0	170.0
Flexural Modulus, MPa	6027.5	6455.0	6710.0	6752.0
Notched Izod (J/m)	112.0	101.5	85.5	85.5
Unnotched Izod (J/m)	737.0	753.0	721.0	651.5
Falling Dart @ 3.175 mm, J	16.2	11.8	11.1	10.6
Falling Dart @ 1 mm, J	No Data	1.7	1.4	2.0
Spiral Flow @ 0.750 mm	101.5	108.0	127.0	146.0

TABLE 6

COST ANALYSIS		
Materials	*DF-1004 EM*	*DF-1004 EP*
Material Cost ($/lb)	3.89	3.89
Part Weight (lbs)	0.242	0.242
Material Part Cost ($)	0.062	0.062
Machine Cost ($/hr) (310 Ton HP Machine)	50	50
Cycle Time (sec)	19.8	14.5
Production (Parts/hr)	181	248
Machine Cost ($/Part)	0.275	0.200
TOTAL COST ($)	**0.337**	**0.262**

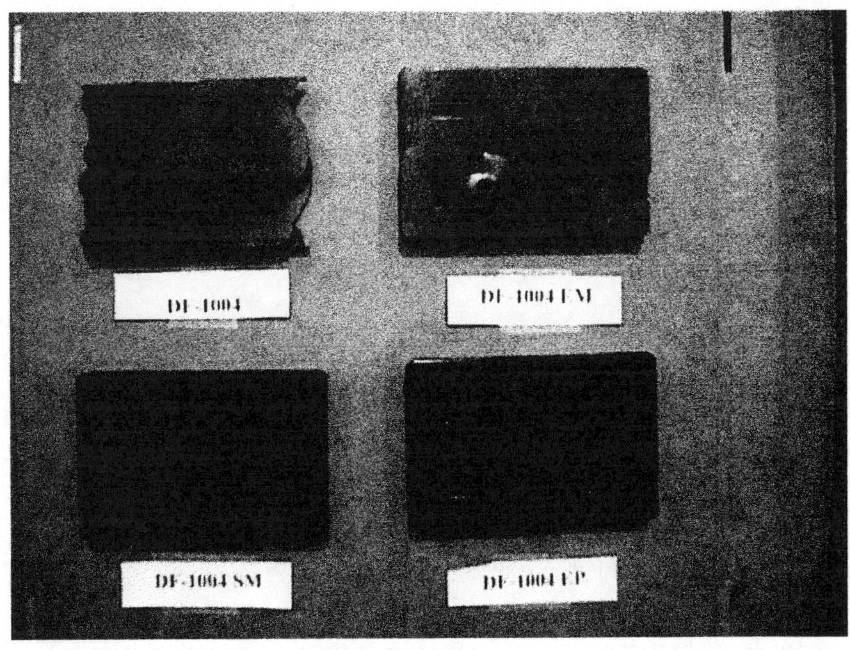

DF-1004
DF-1004 EM
DF-1004 SM
DF-1004 EP

APPLICATIONS FOR COPPER AND NICKEL-COPPER-NICKEL ELECTROPLATED CARBON FIBERS FOR EMI/RFI SHIELDING

Louis G. Morin Jr. and
Robert E. Duvall
Composite Materials, L.L.C.
700 Waverly Ave.
Mamaroneck, New York 10543

ABSTRACT

This presentation identifies and discusses our attempts to remedy the most serious short comings of present day methods of EMI/RFI shielding. The focus is on the use of Copper coated and Nickel-Copper-Nickel coated carbon fiber in high frequency shielding applications. The proposed changes in shielding technology towards on board shielding and back plane shielding of printed circuit assemblies will require a material or combination of materials far more complex and sophisticated than surface shielding methods such as painting, plating or foils. The portable and wireless electronic industries are depending upon faster microprocessor capabilities, which require more sensitive circuitry, which must be protected. The end user, the consumer, is becoming more unforgiving of electronic failure. The next major advances in shielding will require substantial cost savings, design flexibility, increased shielding effectiveness, 100% recyclability, and the least possible exposure to liability.

KEY WORDS: Copper Coated Graphite (CCG), Nickel-Copper-Nickel Coated Graphite (NCNCG), Metal Coated Graphite (MCG)

1. INTRODUCTION

The future needs for composites will be met by the hybridizing of varying fibers and conductive additives in order to be cost effective and meet the specific needs of the OEM's. The old fashioned thought that one fiber, one additive, or one exterior process can achieve this is inherently wrong due to the unique properties of each. The OEM today must be able to request the specific properties he requires for the product and have a conductive additive supplier who can deliver the lowest cost and still be recyclable. Recyclability and exposure to future litigation should be primary concerns in any decision. The purpose here is to discuss the advantages and the deficiencies of processes and products and hence propose practical solutions to the problems.

Since it is obvious that portable and wireless electronics are becoming smaller and lighter, new materials will be required to meet the fast growing needs. Cost is the driving factor, so it is essential that it be thoroughly addressed. It is imperative that the OEM now consider the yield unit cost more than ever before. As the component parts become thinner, the number of parts that can be produced per pound of resin increases dramatically. In costing out a product, it is important that the manufacturer be mindful the economic implications of what is going on. The amount of resin per part is decreasing; the surface area per pound is increasing because walls are becoming thinner; but the per-part costs of any secondary operations that may be required are staying about the same. On a unit cost basis, the production of parts that require secondary operations is becoming more expensive and the production of parts that can be completed in one primary molding operation is becoming less expensive. The economic implications can be dramatic, as Table 1-4 illustrate.

2. CCG and NCNCG FIBERS

The recent invention of a high efficiency copper plating system has pointed the way to the development of two new products. The advancement of copper on carbon fiber marks the breakthrough necessary for electroplated fibers to replace other outdated methods. Copper being half the cost of nickel yet more conductive in both thermal and electrical properties (see Figures 1 & 2) makes it the obvious choice. Carbon fibers electroplated with approximately 30% copper by weight and then flashed with nickel for corrosion resistance and the low contact resistance properties of nickel, now have twice the conductivity of a 60% nickel by weight fiber. This allows lesser amounts of effective fiber per pound to achieve the same shielding effectiveness as other metal coated carbon fibers.

CCG and NCNCG electroplated carbon fibers can achieve another desired benefit for electronic products: enhanced thermal management, which cannot be obtained by other methods. Shielding methods, such as: painting, plating and foils achieve shielding by reflection. Using these methods, the incident EMI signal generated from the circuitry bounces around on the inside continually converting energy to heat. This unintended heat is undesirable and can damage components, leading to intermittent operation or early failure of the device. CCG and NCNCG fibers absorb radiated EMI fields. When these radiated EMI fields strike the fiber reinforced surface, the energy is absorbed and transferred to every fiber. The entire electronic enclosure becomes a thermal conductor drawing the heat away, and protecting the sensitive circuitry. (See Table 5)

The electroplated thickness can be accurately controlled from 2% to 90% by weight depending on the customers' needs. This gives a great amount of design flexibility to the engineer. Custom blends varying fiber lengths and electroplated thickness' can control conductivity and shielding effectiveness by balancing the total number of filaments and their conductivity. For example a 50% nickel coated carbon fiber will produce approximately 520 million ¼" filaments per pound, yet a 30% copper coated carbon fiber will yield approximately 760 million ¼" filaments per pound. Both the aspect ratio of the fiber and the conductivity of the metal used will determine conductivity and shielding effectiveness.

3. HYBRID FIBER CONCENTRATES

The design engineer can now specify the properties required for the product. Conductivity, shielding effectiveness, tensile and modulus can now be created. Through two proprietary systems, pellets are available in a "wire drawn" (similar to long fiber), and in an elongated granule which are compatible with all resins .

3.1 WIRE DRAWN The wire drawn pellet allows for the resin or compounded resin to be extruded with a center "wire" of fibers. This center "wire" can be any combination of CCG, NCNCG, carbon fiber, or fiberglass. This is a pellet now ready for immediate use in injection molding. If so desired other chemical additives are easily added by the compounder to achieve the best possible hybrid.

3.2 ELONGATED GRANULES Produced by this method that can also contain CCG or NCNCG and be made into a hybrid with carbon fiber or fiberglass. The elongated granule operation applies suitable water based binders, compatible with all resins, which allows the compounder and molder the ease of Dry Blend™. Resin pick up varies from 5% to 35% depending on the binder. Uniform dispersion is easily obtained due to the compatible chemistry of the binders.

4. COST of CCG and NCNCG

When calculating the cost of shielding the yield unit cost is still the only way. Permanent molded in reliability as provided with metal coated and unplated carbon fiber is a process requiring no secondary operation. Films or paint to prevent scratching are not necessary. With no secondary operation required, inventory costs are reduced and recyclability is maintained.

5. COST of SILVER/COPPER PAINT AND ELECTROLESS PLATING

There are several factors to consider when calculating the cost of these methods. Silver is a commodity subject to dramatic price fluctuations. Rejects due to scratching, peeling, cracking, flaking and aging of these films are costly. Other added costs include increased inventories, protective layers and gaskets. Painting and plating are unnecessary processes which are subject to continuing and growing environmental regulations and concerns. Costs are great because thin wall parts create a massive amount of surface area per pound. The greatest cost is recycling material the OEM has no immediate use for.

6. LIABILITY ISSUES

More often than ever before, companies are being sued today for improper shielding of their components. General Motors lost a suit and was ordered to pay damages of $1.7 million in September 97, for improper shielding of wiring*. Those who shield by methods which are easily tampered with by consumers put themselves at unnecessary risk. Foils which can be easily removed or paint/plating which are much more easily damaged by the unknowing consumer, will eliminate or impair shielding effectiveness. With CCG or NCNCG you must physically destroy the part to defeat the shielding. Other materials such as fibers that are smaller in length or diameter than 4 microns, some as small as 0.01 microns, fall in the asbestosis range. A major fiberglass manufacturer, an early investor in such products, quickly removed itself from such ownership. It should be difficult to convince your environmental department to approve their use. How will these microscopic fibers be recycled, and if a fire should occur and the filaments go airborne who will be liable?

*Corey Takahashi, <u>The Wall Street Journal</u>, B10, September 8, 1997

7. RECYCLABILITY

Today's world is becoming more focused on the inevitability that companies must learn how to design products that are recyclable. In some European countries for example, landfills are at capacity, and U.S. manufacturers must even recycle the box they send the part over in. If you are not designing parts now that are recyclable, you must prepare to pay dearly for it in the future. Surface finished parts are not recyclable in their existing state and can create severe economic consequences in the future. Carbon fiber or electroplated carbon fiber is 100% recyclable.

8. CONCLUSION

Composites play a very important part in our society. The overall size of the electronic and electrical market in 1996 from data on primary resins sold into it can be conservatively estimated at 536 million pounds in the U.S., 733 million pounds in western Europe and at least 500 million pounds for the rest of the world. This market segment continues to be dominated by conductive coatings with about 1% of it using conductive plastics (plastics using conductive plastics fillers).* Greater market share of conductive fillers is dependent upon innovative new products such as CCG and NCNCG. No other product available today provides decreased costs, recyclablity, molded in reliability and minimizes your liability exposure. The shielding effectiveness is equal to or exceeds other methods and the thermal properties allow for thermal management to prevent excessive internal heating of electronic enclosures. Due to the lower loading required with CCG and NCNCG the greatly reduced unit cost will make this product state of the art in wireless and other shielding applications in the very near future. It is important to remember that molders are not painters or platers. Provide them with a shielding additive that is as easy to add as color, and watch your costs drop, reject rate decrease and your productivity improve.

9. REFERENCES

1. Corey Takahashi, The Wall Street Journal, B10, September 8, 1997
2. Robert D. Leaversuch, Modern Plastics, 51, January 1998

* Robert D. Leaversuch, Modern Plastics, 51, January 1998

Cost of Shielding

How is this calculated?

- **What is the :**

    A) **Weight of the Part**
    B) **Surface Area**
    C) **Cost of Painting / Plating**
    D) **Cost of Foils**
    E) **Cost of Plastic Protectives**
    F) **Cost of Gaskets**
    G) **Cost of Labor**
    H) **Cost of Inventory**
    I) **Cost of Rejects**

Table 2

How can I save $ with COMPMAT?

60 dB Shielding Effectiveness can be achieved with 10% MCG

COMPMAT MCG

EXAMPLE

- 1. MCG at $28.00/LB @ 10%   =   2.80
- 2. Assume a resin cost   =   2.00
- 3. Add line 1 + 2   =   **4.80/Lb**
- 4. Take the weight of your part   =   10 grams
- 5. Determine No. of Parts / Lb   =   45
- 6. Divide the # of Parts by line 3 =   .107

This is your Total Cost per Part   = $ 0.107

This is your Total Cost per Pound   = $ 4.80

Table 3

<u>Painting / Plating</u>

60 dB is achieved but expensive

EXAMPLE

- 1. Assume a resin cost = 2.00/Lb
- 2. Cost of Painting / Plating = .50 ea.
- 3. Take the weight of your part = 10 grams
- 4. Determine No. of Parts / Lb = 45
- 5. Multiply line 2 by line 4 = 22.50
- 6. Add Cost of Gasket, Protective Plastic, Labor, Rejects & Inventory = 1.00
- 7. Add lines 1, 5 & 6 = **25.50**

Table 4

<u>FINISHED COST / LB.</u>

Painting / Plating = $25.50 /lb.
$ 0.56 ea.

COMPMAT MCG = $ 4.80 /lb.
$ 0.107 ea.

COMPARISON OF MEASURED THERMAL AND ELECTRICAL PROPERTIES OF MCG AND AS - 4 FIBERS

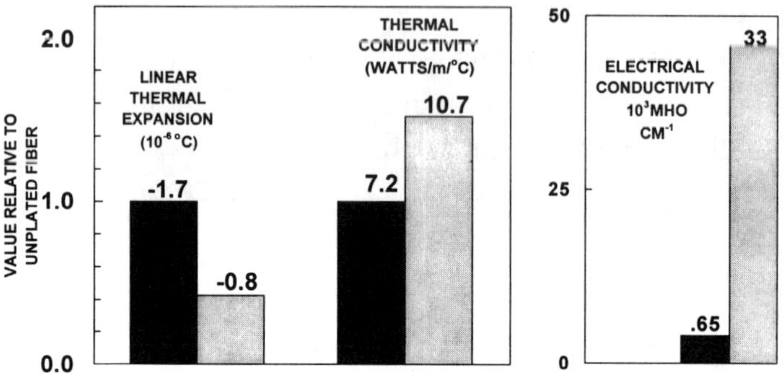

COMPMAT[TM] MCG
Thermal Conductivity

Axial

Carbon Fiber	4.41 Watts/meter/°C
Nickel Coated Graphite	6.40
NCN Coated Graphite	8.28

Transverse

Carbon Fiber	0.95 Watts/meter/°C
Nickel Coated Graphite	0.95
NCN Coated Graphite	1.31

Tested according to ASTM C177.

<u>AS-4 12K</u>

Unplated	34.1 ohms/meter
• Nickel	1.9
• Copper	0.6
• Nickel/Copper/Nickel	0.7

• **Fibers Electroplated to 45% by weight metal**

CARBON-FILLED ENGINEERING THERMOPLASTICS
FOR DEMANDING APPLICATIONS

Jim Fishburn
Wit Bushko
General Electric Corporate Research and Development
Schenectady, NY

Les Goff
General Electric Plastics
Selkirk, NY

Tom Hablitzel
Kevin Andrews
General Electric Plastics
Pittsfield, MA

ABSTRACT

Increasingly stringent material requirements in electronic packaging and business
equipment (internals and externals) applications are pushing the capability of current
engineering thermoplastics. In order to satisfy these fast-changing markets, future
materials must provide advances in design flexibility and conductive/high modulus
performance. These applications also demand levels of material consistency and quality
that set the benchmark for the next generation of engineering thermoplastic materials.

The Custom Engineered Products group of General Electric Plastics specializes in the use
of carbon-based fillers to meet these needs. CEP has utilized carbon-based technologies
(fibers, fibrils, metal-coated fillers) to design a product portfolio focusing on high
modulus, tight tolerance and conductive requirements.

This paper will cover the development of 2 new generations of products containing
carbon-based fillers in GE Plastics resins, in Lexan®, Noryl®, and Ultem®.

INTRODUCTION

GE Plastics has become a leading engineering plastics supplier by using plastics to replace
traditional materials, i.e. wood, metal, and glass. Custom Engineered Products at GE
Plastics has developed materials with unique properties that provide performance in areas
of wear resistance, high modulus, tight tolerance, and static dissipation. Through the use
of GE Plastics resin technology and key additives and filler technologies, CEP is meeting
key customer requirements in these performance areas.

Engineering Thermoplastics (ETP) are incorporated into nearly all of our daily
environment. Computers and information systems are now a routine part of our daily
functions. In the past few years the growth of electronics have generated many
opportunities for the use of unique engineering plastics. One main area of opportunity is

the use of ETP in the management of electrical needs (static dissipation, shielding, etc.) in these growing markets.

Material requirements have become increasingly stringent in the electronics-packaging and business equipment markets. Microprocessor frequency continues to increase and noise from other developing electronics components are driving the need for electrical performance. Increasing heat generation from chips and memory is pushing the need for better materials heat capability as well. Engineering thermoplastics have been playing an ever-increasing role in these applications due to reduced part weight, ability to replace several to many metal parts with one plastic part, and lower over all systems cost. At each turn, thermoplastics and their processing technologies have met critical customer needs. The next opportunity for thermoplastics is in parts requiring electrical properties, ranging from electrostatic dissipation (ESD) to electromagnetic interference/radiofrequency interference (EMI/RFI) shielding.

ESD/EMI requirements are currently being met by the use of either metal components or thermoplastics with metal-based (foils, inserts) components added in secondary assembly steps. Coating technologies have also been employed but are costly and have been suspect to environmental issues. The use of inherently shielding plastics (ISP), the term we will use for conductive plastics (through filler addition), may provide benefits over both technologies. ISP have much greater design flexibility as well as lower weight than metals. Other possible advantages of using ISP may be: the potential for total materials cost savings, cost savings from the elimination of secondary processing steps, and increased resistance to electrical performance decreases through surface defects. As we shall see later, even a seemingly insignificant seam or hole in the shielding material can dominate the shielding characteristics of any part design. With the bulk shielding properties of an ISP, scratches and surface defects are no longer fatal to shielding ability.

Carbon-based filler technologies offer a cost-effective way in which to fulfill electrical needs of ETP for the business equipment and computer markets. The fillers are well designed for such applications providing not only the necessary electrical properties, but also the high modulus needed to maintain rigidity in metal replacing, load bearing parts. GE Plastics materials are currently filling applications that deliver both high modulus and electrical performance. These types of applications work perfectly with carbon-based fillers due to the inherent high modulus of the filler. This is especially important in applications such as paper paths. This same performance, as well as dimensional stability makes carbon-based materials important in electronics packing applications such as microchip transport trays. ISP are also becoming increasingly practical in shielding applications as well. These application could include computer housings, disk drive enclosures, and telecommunications equipment, all rapidly expanding and innovative markets.

GE PLASTICS ROLE

The Custom Engineered Products business in Selkirk, NY has developed a product portfolio designed to meet the demanding performance challenges of tomorrow's applications. General Electric Plastics is a leader in the engineering thermoplastics industry, and the CEP group has at our disposal the entire family of GE engineering thermoplastics resins including Lexan® polycarbonate, Noryl® modified polyphenylene oxide, Ultem® polyetherimide, Cycoloy® engineering blends, etc. The CEP business marries GE Plastic's resin technology with unique filler technologies. As such, the diversified portfolio is categorized into wear-resistance, high modulus, tight tolerance, and ESD/EMI performance. In the development of these products, we have used, among others, carbon-based filler technologies to provide value and performance. Also, the global quality initiatives and processes that GE Plastics employs drives robustness and consistency throughout the product portfolio.

APPLICATION DESIGN

Traditionally, the approach to product development has been strictly: design > electronics > packaging/manufacturing > shielding and secondary operations at the end of the process. At this point, the shielding method and cost of shielding are both dependent upon earlier decisions and can add unnecessary cost. The cost of shielding can be greatly reduced by proactively designing for shielding.

Design for shielding is most effective when taking the shielding requirements into account throughout the product development cycle, including the concept, design, and implementation phases. The concept phase primarily involves understanding design restrictions and limitations, shielding requirements, and shielding method compatibility. Once this information is understood and concepts are generated, we move to the design phase where the concepts are taken to the OEM. Designs can then be modified to suit other needs of the OEM. The designs can also be modified to accommodate various shielding methods. This stage will likely involve multiple iterations with the OEM to ensure shielding and other design concerns are satisfied. The last phase is the implementation of the chosen shielding method. It will also involve other design considerations which can enhance shielding effectiveness.

Once you have a general idea of the "enclosure" or unit, understanding the shape and form factor limitations is key. There may be restrictions due to other standardized components involved or a traditional shape may be required by the OEM. Next, the number and type of seams should be discussed. There are certain openings which will be required (such as the openings for a notebook enclosure). Since efficient aperture design plays such a vital role in improving shielding effectiveness, we should have a good understanding of what type/shape of openings are required, the number of apertures needed, and the location of these openings. If at all possible, the aperture design should be somewhat flexible since this is a key design feature that we can vary in our conceptualizations to enhance shielding.

Next, we should strive to understand the shielding requirements of the enclosure. This includes operating and testing frequencies (many OEM's will test at frequencies higher than their operating frequency), the level of noise generated (if a completely new product, the level of noise can be based on previous/similar designs). Finally, the other customer requirements concerning impact, cosmetics, and other functional needs should be understood as early in the process as possible.

Understanding the shielding requirements of the enclosure is an important step which is often overlooked. The limit on the noise will most likely be dictated by regulations. However, the customer may have even stricter limitations for other functional reasons (i.e., intercomponent interference) so it is important to clarify the driving force behind the customer's limits. Once the limit is known, determining how much noise is generated by the active components is the next step. With both the noise limit and the noise generation, we can now understand the amount of shielding that the enclosure itself must offer (not including the shielding needed for cables and other specific components). It is also important to know whether the unit is tested primarily for ingress (noise coming into the unit) or egress (noise going out of the unit).

These are important questions needed to determine which shielding method will be appropriate for the application.

Once all of the influences on the design are understood, we can conceptualize and take the designs to the OEM to find the concept (or concepts) which best captures their intent. Then, the design can be modified for the possible methods of shielding keeping in mind opportunities for cost out and improved shielding performance.

The trends of the marketplace are playing to the advantages of ISP Materials as the solution of choice. ISP design considerations offers the advantage of

- Elimination of secondary operations (except cosmetic painting)
- High recyclability
- Permits high degree of design flexibility
- Lightweight
- Part Consolidation

The current challenges of ISP materials are the effective filling and orientation of fillers like carbon fiber in corners, curves, and grooves during the injection molding of an enclosure.

The opportunity/advantages of ISP can be utilized with a thorough understanding of a comparison of the methods based on various criteria. Most of these are relatively straightforward. However, there are two areas which often raise some questions:

Recyclability. Paints and Plating must undergo an extensive stripping process to make it suitable for recycling. The run-off (or sludge consisting of copper, nickel, silver) from the stripping process must also be properly disposed of or cleaned since it contains heavy metals that can be toxic. Aluminum (used in vacuum metallization) is not a heavy metal and does not require such an extensive stripping process. Aluminum run-off can be easily disposed of since it does not contain heavy metals. Accessibility to aluminum recycling processes is much better than other metals since aluminum is recycled in many different industries. Also, since vacuum metallization is conducted in a closed process, there are little metal/chemical wastes for disposal.

Low Material Costs. It is very important to remember that many of the cost comparisons found in literature are based on material <u>plus</u> labor cost *estimations*. Since the labor involved will largely be a function of the complexity of the part (masking costs), some of the cost comparisons are not altogether accurate.

MEASUREMENT TECHNIQUES

Accurate evaluation of the shielding capabilities of filled systems is not an obvious task. Widely-used surface resistivity measurements are best suited for highly conductive homogeneous coatings or materials. It is difficult to obtain repeatable results with filled materials because the measurement depends on the contact resistance of the resin-rich surface of samples. Good practice involves use of a four point resistivity measurement technique to reduce the effect of contact resistance and an array of contact points to capture point variation of the surface resistivity. More than 40% variability in bulk resistivity measurements of low-level carbon filled polycarbonate has been observed over the surface of 6x6 inch plaques.

Bulk resistivity measurements are certainly useful in the evaluation of the ESD performance of filled systems, but are fundamentally flawed when it comes to measuring the shielding ability of these composite materials. Resistivity is a good indicator of the ability of the material, even an inhomogeneous one, to dissipate a static charge. Shielding, however, involves the passing of an electromagnetic wave through the bulk of a material.

The fundamental mechanisms of shielding are reflection and absorption from the fillers embedded in the plastic. Each and every fiber in the material participates in this wave scattering process. A direct current-based measurement for material resistivity can only pass through those fibers that make a continuous path between the contact points. Thus a simple relationship used to calculate shielding effectiveness of planar materials based on the resistivity of the sample and its thickness does not hold for filled systems. A modified coaxial transmission line test developed by the National Bureau of Standards can be used to evaluate the shielding effectiveness (SE) of planar materials. Even in this test, in which TEM waves are used, extreme care must be exercised, because the method relies on the capacitative coupling of coaxial flanges. This coupling is difficult to maintain at low frequencies and is affected by even small variations in the thickness distribution of the sample. A lump parameter model of the fixture and the coaxial cables was used to account for the changes in the transmission line characteristics. The method yields good results for the frequency range of 30 to 1500 MHz and sample thicknesses up to about 4 mm.

DESIGN AFFECT ON SHIELDING

Many of the fundamental principles used in designing shielding enclosures with metal or plastics covered with conductive coatings are applicable for the inherently shielding materials. Enclosure features such as venting holes, slots, and grills are subject to the general rule that it is the maximum dimension (L) and not the surface area of the opening that controls the leakage through the opening.

$$SE\ (dB) = 20\ \log \frac{5905.44}{f(MHz)\ L(in)}$$

Increased aperture lengths (L) allow increased wavelengths (lower frequencies) to pass through an enclosure. Apertures and slots should be designed below a critical slot length to reduce EMI emissions.

$$Critical\ Slot\ Length\ (L) = \frac{\lambda}{20}$$

A similar calculation may be made to predict the effect of a series of openings on the shielding effectiveness of an enclosure. In this case the focus should be on the largest opening dimension for the holes in a linear array.

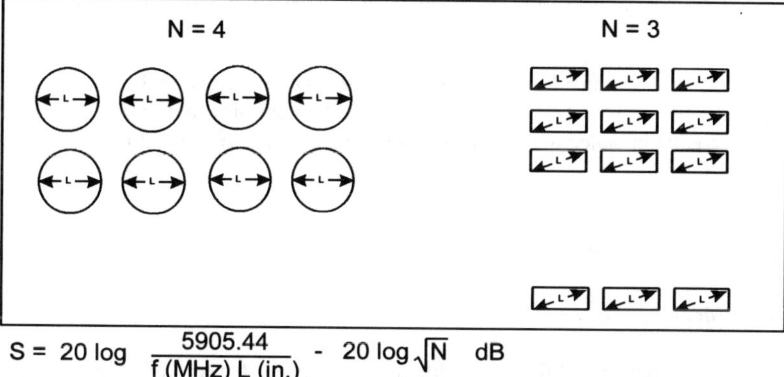

$$S = 20 \log \frac{5905.44}{f\,(MHz)\,L\,(in.)} - 20 \log \sqrt{N} \quad dB$$

In addition to controlling the maximum dimension of an opening, the depth of an opening may be adjusted to effect the shielding performance of the enclosure. This approach, the waveguide technique, is more effective in enclosures made of ISP then of metal sheets or plastics covered with conductive coating.

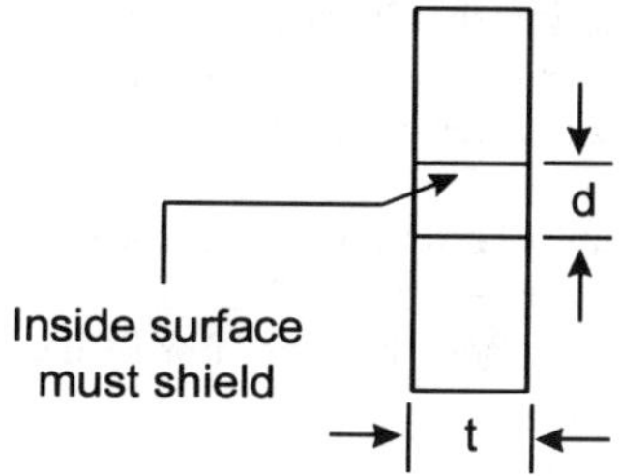

Radiation is attenuated below cutoff frequency (f_c):

f_c (MHz) = 5.9 x 10^9/L (inches)   slot

f_c (MHz) = 6.9 x 10^9/d (inches)   hole

Theoretical attenuation of a wave guide

SE (dB) = 27.2 t/L (slot)   t = depth of opening

         32   t/d (hole)   d = diam. of opening

L = max. linear dimension

Mini-waveguides can be created by adding depth to the openings in the enclosure. The internal walls of such openings are naturally conductive when ISP are used. This is because ISP's conductivity is distributed throughout its volume and not only on the surface of the material.

In contrast to highly conductive metal sheets and coatings, seams made of ISP cannot rely on electrical contact provided by shielding seals. Surface resistance of resin-rich filled materials can be much higher than the bulk conductivity of the material. Low impedance openings can be designed by increasing the capacity of the opening modifications in geometry.

All enhancements should be considered with respect to the overall effect on the complete system. Proper attention to board level design, component positioning and wire management will greatly enhance the opportunity to deliver a cost effect shielded enclosure.

MATERIALS

Carbon-Based Fillers

The various types of carbon-based fillers provide distinct avenues for reaching the stringent requirements of today's applications. However, each type of filler is also unique in the performance that it can impart to a thermoplastic material.

Carbon Black allows one to impart a low level of electrical properties (surface resistivity from 10^{11} to 10^{7} Ohms/sq) without extremely high loadings. This electrical performance is typical of a material designed for ESD applications. The filler will increase the modulus and strength of a resin somewhat. Possible disadvantages with carbon black could be handling at manufacturing, sloughing onto sensitive electrical or magnetic parts, and the inability to produce colors other than shades of black.

Graphite Fibrils can provide the same or increased conductivity (ESD-level) over carbon black, usually at even lower loadings by weight . The fibrils are easily handled and minimally impact the physical performance of the base resin. Again, however fibril-filled materials are also limited in the range of possible colors.

Carbon fibers and metal-coated carbon fibers can provide significantly higher conductivity (lower resistivity) (down to 10^{2} Ohms/sq) and are the most effective carbon-based fillers for use in inherently shielding plastics. They are often used at high loadings to provide the conductivity necessary for shielding. The fibers also provide high modulus to the material. The presence of the anisotropic fillers will modify the processability (stiffness, shrinkage) of the material with respect to an unfilled resin.

Whatever the case, carbon-based fillers provide a rather cost-effective means ESD or EMI performance while each bringing its own blend of physical performance to the material. The strengths of the fillers should be considered when materials are being designed for applications.

ISP

ESD applications have previously been discussed in this paper and include such things as electronics packaging and business equipment components. The main requirement for ESD materials is the ability to dissipate a charge and bleed it to ground safely. This requires a modest level of conductivity in the material. However, in some cases the ability to draw an external charge is undesirable. This could allow the electrocution of the electronics being protected in the first place. Therefore, too much conductivity may also no be desirable.

ESD applications require the ability of the material to dissipate a charge at the part surface, since this is the point of contact with the electronic components or where the triboelectric build-up of energy can occur. Because of this, surface resistivity is often the measurement used to assess ESD performance. Measurement of surface resistivity is quick and provides information about the part surface specifically. A typical ESD material should have a surface resistivity on the order of 10^9 to 10^6 Ohms/sq in order to be of utility. Many carbon-based fillers can provide this level of surface resistivity in an engineering thermoplastic. These include conductive carbon black, graphite fibrils, and carbon fibers or metal-coated carbon fibers.

Figure 1 shows a typical result of surface resistivity measurements from a series of thermoplastic formulations with varying levels of a conductive filler. At low levels of filler loading, the material is insulating. As the filler loading increases, a point is reached where there is a dramatic reduction in the resistivity (increase in conductivity), called the percolation limit. This is the point where conduction begins to take place. The entire drop in resistivity occurs over only a few weight percent of the filler loading. At loadings above the percolation limit, the resistivity is relatively low and fairly constant over a large range of loadings. Depending on the identity of the carbon filler, this percolation threshold will shift either to higher or lower loadings. The relative ranking of the fillers in terms of the lowest to highest loading by weight necessary to reach the percolation threshold goes from graphite fibrils to metal-coated carbon fibers to carbon fibers to carbon black.

Figure 2a shows a comparison of the surface resistivity versus loading (weight %) for carbon fiber and metal-coated carbon fiber in a Noryl® matrix. The measurements were performed with a Pinion meter which has an experimental range of 10^4-$10^{12.5}$ Ohms/sq on injection molded 125mil thick plaques. The fillers were added on compounding in an extruder. The resin is insulative at low levels of each filler. Between 5 and 10% loading by weight we see the characteristic drop in resistivity at the percolation threshold. The two fillers seem to provide rather similar resistivity at all the loadings represented in this Figure. One might expect the more conductive metal-coated carbon fiber to provide lower resistivity at the same loading. This is evident in Figure 2b, which shows the same data represented in loading by volume % rather than weight %. Clearly, the percolation threshold comes at a lower volume % loading of the more conductive metal-coated carbon fiber. At the same weight loading of the two fillers, one would expect less total fibers from a more dense (the metal-coated) material. This may explain the slightly lower Flexural modulus (standard ASTM testing, 250mil thick bars) for metal-coated fibers than carbon fibers at the same weight loading in a Noryl® matrix, Figure 3.

If the bulk resistivity of the samples is measured as opposed to surface resistivity, there is a difference in the performance of the two fillers, Figure 4. In Figure 4a bulk resistivity is shown with respect to weight % of filler while in Figure 4b it is shown with respect to volume % of filler. Measurements were performed on molded Noryl® parts using silver painted electrodes to reduce contact resistance. Resistance values were recorded with a standard multi-meter. The bulk resistivity was calculated from these measurements via:

$$\rho = (R * A) / l$$

where ρ is bulk resistivity, R is the resistance, A is the part cross sectional area, and l is the distance between the electrodes. In Figure 4a we again see that at low loadings the resistivity registers only at the top of the range of the measurement, 10^7 Ohms-cm. The behavior is very similar to that of the surface resistivity until between 10 and 20%, where the resistivity of the carbon fiber material begins to level out. Above this loading, the metal-coated carbon fiber provides lower resistivity than the uncoated fiber. Also, as before, if the data is represented in terms of volume % of filler, the material with the metal-coated carbon fiber reaches the percolation threshold at lower filler loading.

For the two sets of resistivity data presented, the question remains: Which is more representative? As the part weight reduction is one of the distinct advantages of using a thermoplastic over a metal, the weight % representation may be more applicable. If weight reduction were important to the designer or OEM, then one would prefer to get the highest electrical performance from the lowest weight part.

Custom Engineered Products can use all of the above mentioned carbon-based fillers to provide a large range of surface resistivities in engineering thermoplastics depending on filler type and filler loading. With the GE Plastics resin base we can provide the wide range of ESD capability in a large range of heat requirements. Figure 5 shows a plot of use temperature versus surface resistivity. With the employment of the various GE Plastics engineering thermoplastics, we can provide materials that meet any application heat requirements from 100C to 200C. The use of the various carbon-based fillers in different loadings allows us to cover surface resistivities from insulating (10^12 Ohms/sq) to conductive (10^4 Ohms/sq) in all of the resin lines shown. Once the heat requirements and electrical performance have been ascertained, other Critical to Quality (CTQ) variables can be met by applying the correct resin and additive technologies.

EMI shielding applications require the material (or materials package in cases where metal inserts are used) to shield electric or magnetic components from electromagnetic or radio frequency interference. These materials may also serve to protect the public from receiving significant amounts of electromagnetic radiation from increasingly more powerful electronic components. In many cases a conductive material is needed that can absorb and bleed off the incident radiation.

Figure 6 shows a plot of insertion loss (in decibels) with respect to increasing conductive filler loading covering a typical area of performance from the use of carbon-based fillers. The insertion loss value is simply the amount of attenuation that the material provides in the testing geometry:

Insertion Loss(dB) = 20 log (Incident Field Strength /Transmitted Field Strength)

All matrices previously discussed would provide shielding performance in the area designated in this plot. As expected, shielding ability increases as the carbon-based filler loading increases.

Notice that we have a region of shielding performance rather than a line. Differences in the shielding ability of two samples with similar loading of a conductive filler may be due to:

A) The type of carbon-based filler. Each filler has different electrical performance, and would therefore be expected to facilitate shielding differently at each loading.

B) The resin formulation. This could be from the inherent conductivity of the matrix, to other additives that enhance/decrease shielding ability.
C) The processing history of the material. More specifically, the filler attrition and filler dispersion resulting from processing.

The trick, of course, is to design a system with a highly conductive filler that provides- after the rigors of processing- the conductive network that is of utility in the application keeping in mind that fiber attrition and fiber dispersion are important factors in the equation.

Materials that are conductive enough to shield will generally have surface resistivities on the order of 10^2 to 10^5 Ohms/sq. As shown in Figure 5, this is achievable in all of our resins bases. Here again, the choice of the filler comes primarily from the shielding requirements. If two fillers are applicable, then other physical performance requirements can be used to choose the best filler. The matrix resin is usually dictated by performance needs and aesthetic requirements. Whatever the desired combination of these things, the range of carbon-based fillers and resins at the disposal of GE Plastics make it able to supply your resin.

CONCLUSION

In the future, we expect even greater demands to be placed on ISP in ESD/EMI applications. As the market for shielding and electrostatic applications increases, new technologies in ETP will play a major part in meeting the demands. Many of the market requirements will be driven by cost versus performance and design flexibility along with quality and environmental concerns. We also anticipate that regulations and advances in electronics technology will drive the necessity for even higher conductive performance from plastics.

The Custom Engineered Products business, through six sigma quality initiatives, a diverse resin portfolio, and unique resin/filler technologies, is in a position to meet many of these challenges. The GE quality process focuses throughout the ETP chain, starting with the raw materials and finishing with the performance of the finished product. This allows GE Plastics to deliver the maximum performance from our filler technology without additional cost. It also provides a means of designing future products to meet and exceed all customer requirements with an unparalleled quality rigor. Six sigma is also allowing us to reduce our cycle time while increasing customer quality, so that we can provide the best technology at the premium performance/cost balance.

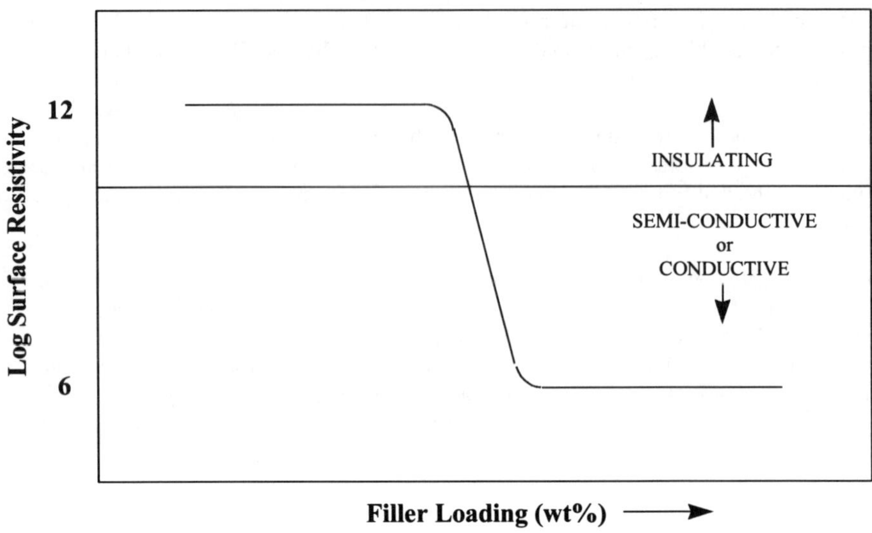

Figure 1: A Typical Surface Resistivity Percolation Curve

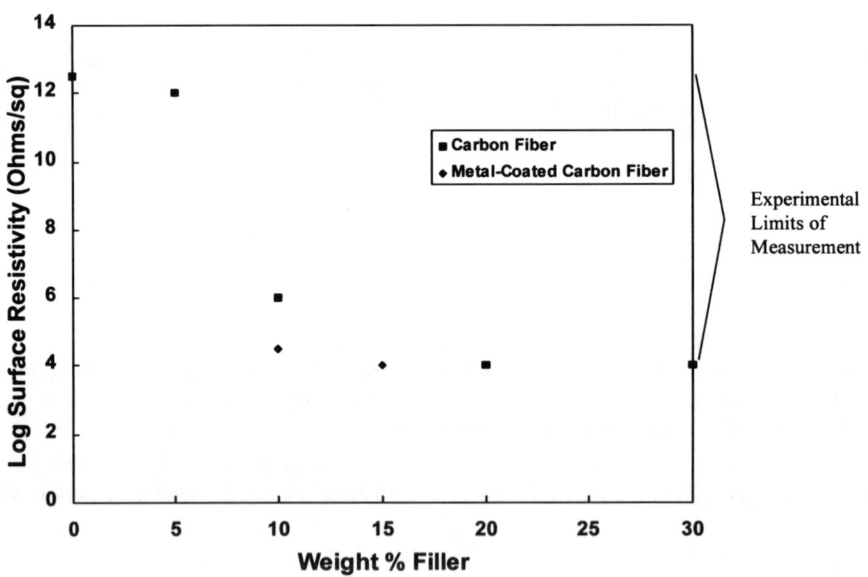

Figure 2a: Surface Resistivity of Carbon Fillers

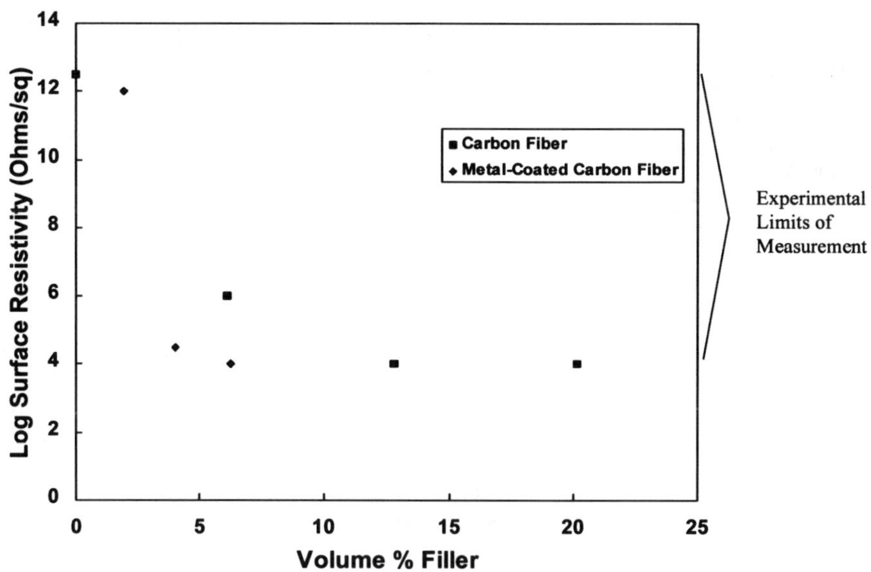

Figure 2b: Surface Resistivity of Carbon Fillers

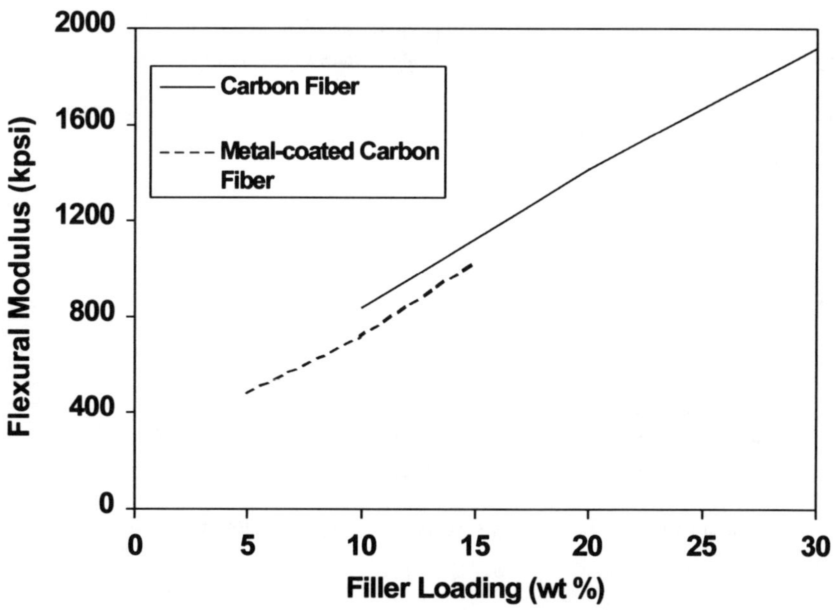

Figure 3: Effect of Filler on Modulus in Noryl

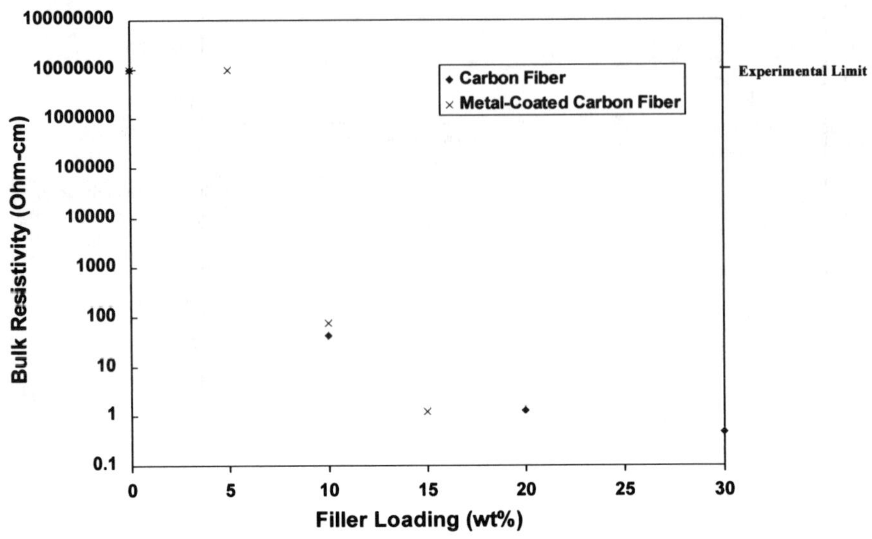

Figure 4a: Bulk Resistivity of Carbon Fillers

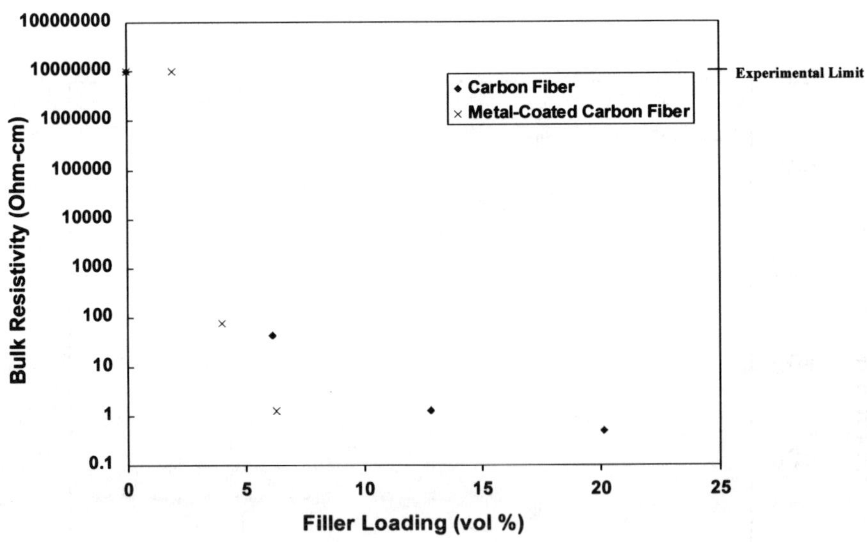

Figure 4b: Bulk Resistivity of Carbon Fillers

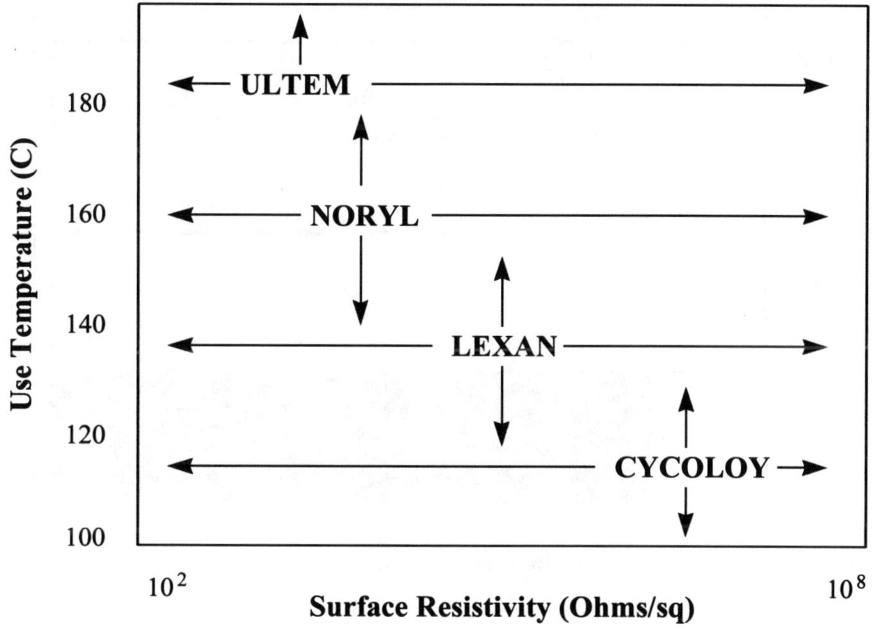

Figure 5: Resin Utility in ESD/EMI

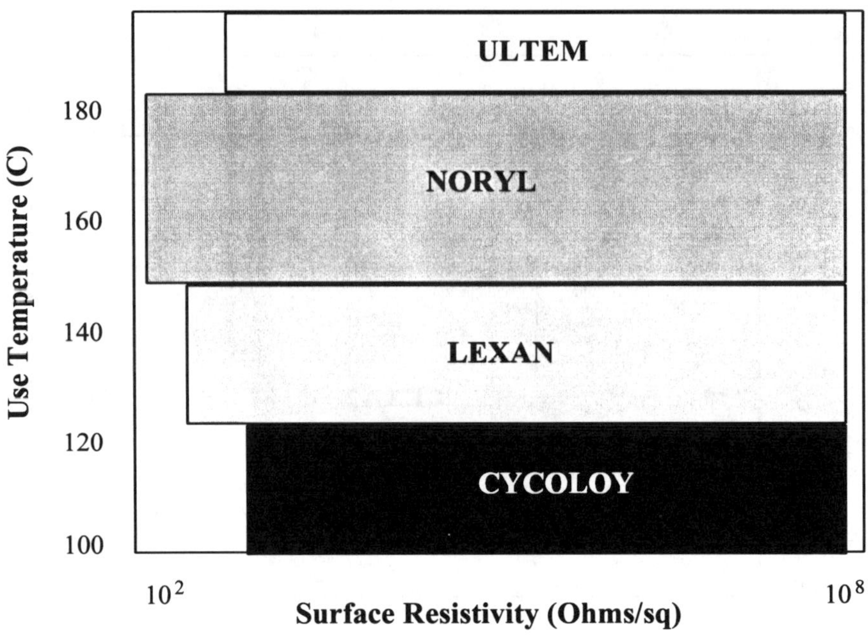

Figure 5: Resin Utility in ESD/EMI

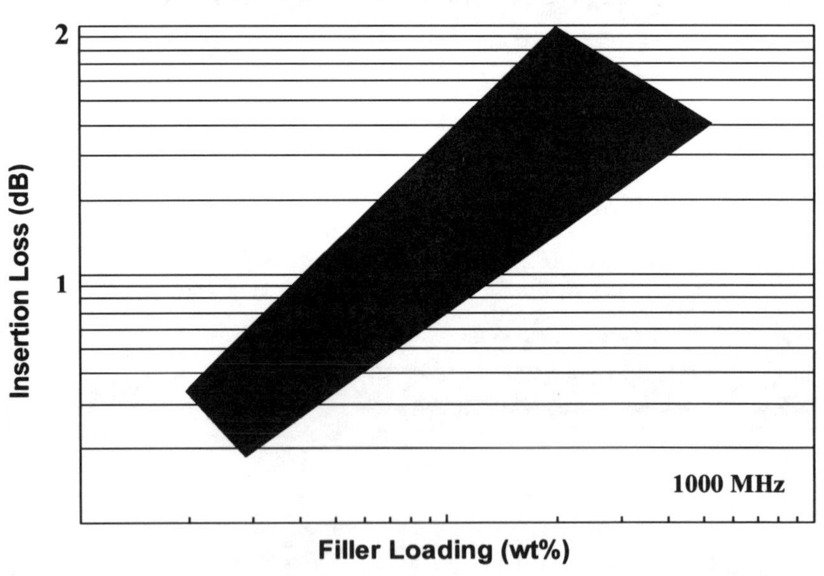

Figure 6: Typical Shielding Range in ISPs

SUCCESSFUL PRODUCTION OF THIN WALL ARTICLES IN LONG-CARBON-FIBRE, FLAME RETARDANT, THERMOPLASTICS

Rik Decoopman
Marketing Manager
VERCO subsidiaries

John Vandaele
Engineering Manager
STRUCTUPLAS N.V.

ABSTRACT

After having supplied the computer industry some years with smc/carbon fibre reinforced compression mouldings, Structuplas, that always has been, first of all, a thermoplastic custom moulder, decided to change the production technique to injection moulding. After three years of hard work, most of the problems are solved. An honest report with recommendations (and thanks) to all parties envolved: raw material suppliers, designers, oem's...

KEY WORDS: Injection Moulding, Applications - Business Equipment, Carbon Fibres.

1. INTRODUCTION

Especially for the computer industry there was and is a high interest in finding companies capable of making thin wall computer[1] housings needing no postoperation for EMI-shielding. For that reason, over the last 5 years, different kinds of technics were developed to achieve this target.

One of those technics, namely the injection moulding of resins, loaded with long carbon fibres[2], will be the subject of this presentation.

_ . _ . _ . _ . _

It is obvious that reducing the wall thickness of a plastic part gives a weight reduction and an enormous cost advantage: both the material cost and the moulding cost are normally reduced.

Unhappily, reducing wall thicknesses, apart from affecting the general properties (mechanical, thermal...) of the part, also goes together with reduced flow paths. Parts can need more gating points, tools can need to be more sturdy and injection machines can need higher clamping forces. For most applications in unfilled, unreinforced plastics, these tool cost increases can be, regarding the reduced part cost, easily accepted.

On the other hand if, as a consequence of the reduction in wall thickness, to upgrade mechanical properties, fillers and/or fibres need to be added, the basic material becomes more expensive and aesthetic problems can arise in gating and weld line areas. Depending on the filler/the fibre, most of these parts need special painting to be aesthetically acceptable.

So for aesthetic reinforced thin wall articles, to be cost effective, due to the needed external finishing, the plastic compound must give a supplementary advantage to the customer...

This can be the case if, as reinforcing help, long-carbon-fibres are used (figure 1/figure 2), which add apart from the needed mechanical properties also inherent shielding properties, normally only realised after expensive, non environmental friendly after treatments as electro less plating or silver plating.

[1]**Thin wall articles**:
        filled/reinforced materials   : wall thickness between 1,0 and 1,5 mm
        unfilled materials   : from 0,8 mm on...

[2]**Long fibre reinforced**: fibres in the pellet min. 5 mm long.
Normally the length of the fibre is equal to the length of the pellet.
Pellets are typically made by pultrusion; their length is around 10 mm.

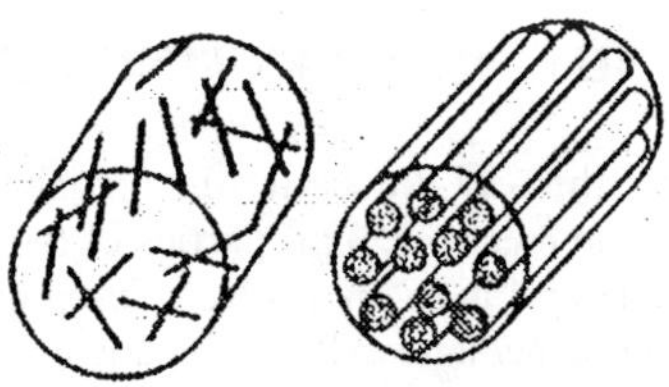

Figure 2 : General comparaison between 30% normal short carbon fibre
and 30% long carbon fibre flame retardant PA.
Although flame retardant: much higher mechanical properties.

899

2. LONG FIBRE COMPOUNDS

EXCELLENT FLOW AND GOOD WETTING OF FIBRES NEEDED

2.1. Basic Resins   The thinner the wall, the better the flowability of the plastic has to be without too much compromising the mechanical properties of the material, also as the flame retardants limit the increase in melt temperature. For filled/reinforced applications, basic resins are as a consequence mostly of crystalline nature: PP / aliphatic as well as partly aromatic PA. When properly processed, the advantages of long fibres to short fibre thermoplastics are essentially: - higher stiffness
- better high temperature performance
- <u>much higher (notched) impact strength, also at low temperatures</u>

2.2. Fibres   Fibres must be chosen in order to give:
- good compatibility with the basic resin.
  Adequate wetting of the fibres is needed: kind of sizing is of major importance.
- good electromagnetic shielding
- normal weight percentages: 30 to 60%

2.3. Flame retardants   Neither red Phosfor, nor PBDFE may be used. No juicing or blooming...

2.4. One pellet materials needed   Some suppliers sell physical mixtures of plastic pellets with carbon fibre bundles.
These kind of materials don't allow:
- vacuum air transport and automatic material line feed. They very easily give segregation and free carbon fibres, which can damage the electric equipment of the injection moulding machines.
- problemless predrying as the fibre bundles can be removed by the drying air and also clug the filters

2.5. Regrind   It can not be denied that, for certain parts, still to be gated by 3 plate runners, regrind can be a problem so:
- re-use only regrind once
- check the impact of the amount of regrind:
  with your raw material supplier: on mechanical and other important properties (flame retardancy...)
  with your customer: on shielding properties.

3. THIN WALL PART DESIGN

GATE MARKS AND WELD LINES IN NON COSMETIC AREAS.
WARPAGE CONTROLLED BY SUITABLE DESIGN.

Be aware that apart from the general design rules:
  Keep it simple
  Wall thickness as constant as possible
  Add stiffness by design through: fitting the internal components
                                      the way of connecting the individual parts
  General draft min. 1°, preferred 1,5°
  Specify general tolerances and only mention in the drawing these dimensions with tolerances,
          which really need to be accurate.
  Notch sensitivity doesn't depend on the material only... but also on design: every change in
          thickness, in stiffness can act as a notch
With this kind of materials sinks are not really the problem, a suitable painting also helps to
mask them.

But special attention has to be given to:

3.1. The gate and weld line marks

3.2. The avoiding of warpage
The number and position of the gates are very important for:
                    - filling
                    - overall aesthetics (weld lines...)
                    - warpage
Whenever possible, the internal tool pressure must be limited to 650 bars, which means that
maximum flow length is, with the actual compounds, only 80-100 mm.
Start the mould fill simulation/mould stress/part stress/part warp simulation to identify problem
areas asap. It is always better to remediate on paper ...

3.3. The avoiding of flash sensible (weak) tool areas   Parting lines must be chosen so that
if flash occurs, its removal doesn't harm the esthetics of the parts...

*And: don't forget that some of the smaller articles used in the assembly, can easily be made
out of the moulding scrap of the big ones...*

4. TOOL DESIGN

TWICE AS STIFF AS USUAL AND EVERYWHERE VENTED

4.1. Steel choice   Most of the plastic materials used are as a consequence of the reinforcement and flame retardancy as well abrasive as corrosive. (Remember corrosion resistance is only obtained with polished surfaces.) For long life times and low maintenance costs: special steels must be used. These steels normally have longer delivery times than standard steels.

4.2. Tool stiffness   Internal pressures of about 650 bars require stiff tools. In 3 plate tools, compromises between tool stiffness and weight of runner have to be made. In case of a too weak tool construction, steel blocks may crack. Calculate all tool components for 1.300 bars internal tool pressure.

4.3. Gating (see also thin wall part design)   Correct gating is of major importance. Standard gating for multiple gates is 3 plate injection. Depending on article design, special hot channels can sometimes be used.

4.4. Venting   Adequate venting is a must: mould deposits (blooming/juicing) must be avoided. As trapped air can give corrosion, a good practise is to vent every feature (boss, rib...). Inserts are always of a big help.

In case of very thin articles it can even be needed to foresee the tool with vacuum suction to remove as much as possible air out of the cavity before injection.

4.5. Tool maintenance   Don't forget to foresee a weekly tool maintenance program.

Comparison of the properties of the main steel groups, showing the most important distinguishing features.

	Wear resistance	Hardness	Toughness	Corrosion resistance	Polish-ability	Texturing properties	Weld-ability	Machin-ability
Case-hardening steels	+++	+++(case)	++ (core)	0	+++	+	0	+++
Prehardened steels	0	0	++	0	++	+++	++	+
Maraging steel	0	++	+++	+	++	++	+++	0
Through-hardenable steels	+++	+++	0	0	+(0)*	++	0	++
Corrosion-resistant steels: Hardened	++	+++	0	++	+++(0)*	+	+(0)*	+

0 to +++ (ascending)* ledeburitic chromium steels

5. MOULDING MACHINE

DON'T DO A SPECIAL JOB ON A STANDARD MACHINE...

5.1. Plasticizing unit   As well as the steel for the tool, the screw and cylinder need to be made out of special steels: resisting the corrosion and the abrasion.
The check valve must be a free flow ring check valve.
In order to prevent too long dwell times: between 60-80% of the injection volume should be used. This requirement can give strange combinations of injection and clamping units as the internal tool pressure is fairly high.

5.2. Fast injection   Due to the thin walls and the fast cooling down of the conductive material, injection speeds are fast and accumulator injection is practically always a must.
Shot weights (120-300 grams) are typical injected under 1,5 seconds.
To guarantee this fast injection, a high injection pressure must be available.

5.3. Clamping unit   With a thickness as thin as 1,2 mm, internal tool pressures are mostly in the region of 65 Mpa what means that for an A4 size article a 400 Tons press is a minimum.
Regarding the fast injection needed, toggle machines are the best choice.

6. MOULDING CONDITIONS

6.1. Predrying   Most materials need predrying: use dry air driers to assure always the material to have the same moisture content what guarantees the same viscosity and same plasticizing stroke = injection volume.

In case scrap can be added, remove all fine "dust" which normally has taken a lot of moisture very quick and also doesn't add anything to the mechanical properties of the part.
Be aware that, if the plastic compound has loose fibres, the filters of your dry air dryers and vacuum hoppers can be very quick choked and drying insufficient, apart from the eventual damage they can do to their electric motors...

6.2. Tool temperatures   Tool temperatures have to be set in a way:
       - internal stresses are low = no warpage after moulding
       - no after crystallisation happens in the ovens while painting
Tool temperatures normally are between 80 and 130 °C depending on the type of plastic and paint.

6.3. Plasticizing for maximum length of fibres   Plasticizing must be optimised to get the fibres at the screw tip still as long as possible. Screw back pressure and speed are dominant factors what can be seen in their impact on the melt temperature.
The real fibre breakage will happen during injection in the runner system and especially in the gate areas.

6.4. Fully automatic cycling is a must   Fully automatic cycling of tools is always needed. In case of 3 plate tools, remove runner as well as part by overhead robot.
Cycle times are short and are fully dependant on needed cooling time to produce a stabilized part that 'll not warp afterwards.

Figure 3:

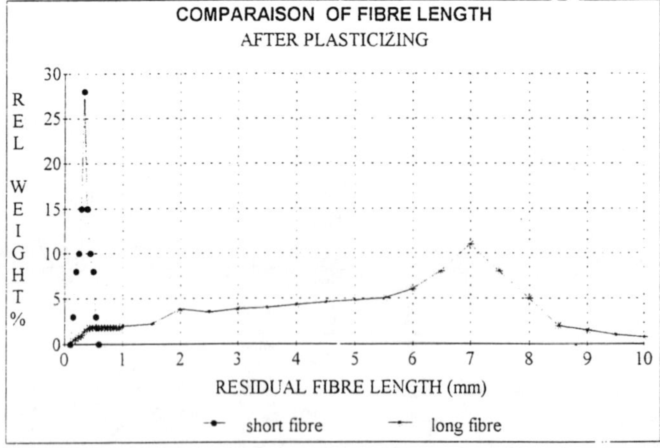

6.5. Operation window

The operation window is quite narrow and defined by

- the shot volume
- the injection time (profile)
- the after pressure time and profile

<table>
<tr>
<td colspan="3">STRUCTUPLAS RIESSELBERG</td>
<td colspan="7" align="center">WORK INSTRUCTION</td>
<td>Doc. Nr.:
WI</td>
<td>766
1.</td>
</tr>
<tr>
<td colspan="3" align="center">CUSTOMER</td>
<td colspan="7" align="center">INJECTION</td>
<td colspan="2">Pag.: 1 / 1</td>
</tr>
<tr>
<td colspan="3" align="center">46 H XXXX</td>
<td colspan="7" align="center">CD BASE XXXXXXXX : WORKING WINDOW</td>
<td colspan="2">M 321 A</td>
</tr>
</table>

FIXED PARAMETERS: DO NOT MODIFY :   7662WVAS.E
MATERIAL SUPPLIER /BATCH : XXXXXXXXXX REF:YYYYYYYYY

REGRIND   : kg   : (AA) +BB

Injection side   :Cilindertemperatures   : °C   : 240   250   250   250   245   40
                  Nozzle temperature   : °C   : 265
                  Screw speed   : tⁱ   : —
                  Screw counter-pressure   : bar   : - —

                  Hydraulic pressure   : bar   : 120 - 140 (result = reading...)
                  Change on after-pressure: mm :   4,5 reference for plastisized volume
                  Cushion   : mm   : 4 mm   (result = reading...)

                                               : mm   : 20   40   70   100   100   100

        Injection profile, mn/sec   : 4,5   : 9   10   25   50   90   130,5

                                               : mm   :   1   10   10   40   40

        After-pressure profile, bar   : 0   :   0   0   0,1   1

                  Cooling time   : sec   : XX
                  After-pressure time step 1   : sec   : Y1
                  Cycle time   : sec   : XX (+/-1)   (result = reading...)
                  Temp strip-off plate   : °C   : 60   BT 1
                  Temp injection half   : °C   : 120   BT 2
Ejection side   :Temp ejection half   : °C   : 120   BT 3

ONLY ADJUST :

        Screw stroke   : mm   : (+/-2)
        Injection time   : sec : (+/-0,1)
        After-pressure, 2nd press/time : bar : (+/-5)   (1-2 sec)

        Stroke   I Injection profile   After-pressure profile
        !xxxxxxxxxxxxxx!   !xxxxxxxxxxxxxxx!   !xxxxxxxxxxxxxxx!
        !xxxxxxxxxxxxxx!   !xxxxxxxxxxxxxxx!   !xxxxxxxxxxxxxxx!
         !   !   !   !   !   !
        AAA   mm   BBB   0,X   Sec   1,Y   CC   bar   DD
        (Burr/sink mark)   (Burr/sink mark/aesthetics)   1   sec   2 (sink mark)
        min   Max   min   Max   min   Max

Goedgekeurd			sampe							
Beoordeeld										
Opgesteld	JVD	JVD	JVD							
Datum	27/2/97	3/3/97	4/2/1998							
Wijz. Nr.	0	1	2	3	4	5	6	7	8	9

7. MOULDED PART QUALITY ASSURANCE

CONSTANT MONITORING IS A MUST

7.1. Material certificates   We all know there can be suddenly a certain shortage of carbon fibres. Please let your supplier check and confirm the availability of the used fibre for the time of the project and ask him to send a certificate about the important characteristics with each batch of material.

7.2. Production checks

<table>
<tr><td colspan="2">STRUCTUPLAS RIESSELBERG</td><td colspan="4" align="center">CONTROL INSTRUCTION</td><td colspan="2">Doc. Nr. :
K V A 7662
28 / 01</td></tr>
<tr><td colspan="2">CUSTOMER</td><td colspan="5" align="center">INJECTION</td><td>Pag. : 1 / 1</td></tr>
<tr><td colspan="2">46HXXXX</td><td colspan="5" align="center">CD BASE XXXXX</td><td>M 321 A</td></tr>
<tr><td colspan="2" align="center">C h e c k   p o i n t s</td><td>Requirement + tolerance</td><td>Number Frequence</td><td>Operator</td><td>Means Reference</td><td colspan="2">Form</td></tr>
<tr><td colspan="2">1. DATUM STAMP</td><td>daily adapted
visual</td><td>first part
type 004
morning</td><td>shift
foreman</td><td>visual</td><td colspan="2">type 004</td></tr>
<tr><td colspan="2">2. FLATNESS MEASUREMENT</td><td>mm MAX under
foot left behind</td><td>1 pce / 2 H</td><td>"</td><td>Flat plate/
gauge
thickness pl</td><td colspan="2">"</td></tr>
<tr><td colspan="2">3. LOOSE FIBRES ?</td><td>NONE</td><td>"</td><td>"</td><td>visual, see
ref. part</td><td colspan="2">"</td></tr>
<tr><td colspan="2">4. ALL GATES OPEN ?</td><td>check runner</td><td>"</td><td>"</td><td>"</td><td colspan="2">"</td></tr>
<tr><td colspan="2">5. CONTOUR LX FILLED ?</td><td>Area 1: filled</td><td>100%</td><td>operator</td><td>"</td><td colspan="2">"</td></tr>
<tr><td colspan="2">6. RIBS LX FILLED ?</td><td>Area 2: filled</td><td>"</td><td>"</td><td>"</td><td colspan="2"></td></tr>
<tr><td colspan="2">7. FLASH (see WVA)</td><td>NONE</td><td>"</td><td>"</td><td>"</td><td colspan="2">"</td></tr>
<tr><td colspan="2">8. DAMAGES</td><td>NONE</td><td>"</td><td>"</td><td>"</td><td colspan="2">"</td></tr>
<tr><td colspan="2">9. OIL - GREASE - DEMOULDING AGENT</td><td>"</td><td>"</td><td>"</td><td>"</td><td colspan="2">"</td></tr>
<tr><td colspan="2">10. DEFORMATION ON BOS ? ?</td><td>NONE</td><td>"</td><td>"</td><td>"</td><td colspan="2">"</td></tr>
<tr><td colspan="2">11. BURNED SPOTS - ROUGH
SURFACE IN VISIBLE AREAS</td><td>NONE</td><td>"</td><td>"</td><td>"</td><td colspan="2">"</td></tr>
<tr><td colspan="2">12. EJECTORS</td><td>NOT VISIBLE ON
VISIBLE SURFACE</td><td>"</td><td>"</td><td>"</td><td colspan="2">"</td></tr>
<tr><td colspan="2">13. SPRUES IN VISIBLE SURFACE</td><td>MAX 0.4
DEEPENED</td><td>"</td><td>"</td><td>"</td><td colspan="2">"</td></tr>
<tr><td colspan="2">14. RESISTANCE MEASUREMENTS
HORIZONTAL - VERTICAL</td><td>MAX 10ohm</td><td>2pcs/new
container</td><td>QC</td><td>ohm meter</td><td colspan="2">see WVA</td></tr>
<tr><td colspan="2">15. FLATNESS REPORT, DETAILED</td><td>3D GEOPACK
7662</td><td>"</td><td>"</td><td>3D machine</td><td colspan="2">"</td></tr>
<tr><td>Approved</td><td></td><td>sampe</td><td></td><td></td><td></td><td></td><td></td></tr>
<tr><td>Reviewed</td><td></td><td></td><td></td><td></td><td></td><td></td><td></td></tr>
<tr><td>Drafted</td><td>ALT</td><td>ALT</td><td>ALT</td><td></td><td></td><td></td><td></td></tr>
<tr><td>Date</td><td>28/02/97</td><td>08/04/97</td><td>10/04/97</td><td></td><td></td><td></td><td></td></tr>
<tr><td>Modif. Nr</td><td>0</td><td>1</td><td>2</td><td>3   4</td><td>5</td><td>6</td><td>7 8 9</td></tr>
</table>

8. FINISHING

NO DISCUSSIONS IF FULL PRODUCTION IS IN ONE HAND

8.1. Degating   Of course, gates are normally placed in areas where no finishing is needed, but this is often, due to the great number of needed gates, normally not the case for all gates.

For the correct removal of these gates, a dedicated milling machine is used to remove the sprue tips.
Remark that already a special gate design is needed to be able to mill the sprue decently. The result is a mark likely to the one of a recessed ejector.

8.2. Deflashing   Choosen tool steel, stiffness, design and accuracy must avoid flash.

8.3. Inserting   Inserting is done on dedicated machines which can insert two parts or at the same time (one type of insert) or two parts after each other with two types of inserts. Machines can be set for ultrasonic or heat inserting.

8.4. Painting   Painting is done in state of the art electrostatic dust poor paint lines. Parts are masked by special masking jigs in non conductive material which doesn't attract the paint. The electrostatic paint system guarantees a good coverage of all areas needed to be painted.

8.5. Tamponprint   Further finishing can be done on several tamponprint machines.

8.6. Assembly   If asked for a lot of small components can be bought and mounted in the basic housings. In this way stock at the OEM can be reduced and unskilled labour can be done at reasonable cost.

9. CONCLUSIONS

The moulding and finishing of thin wall products with long carbon fibre flame retardant compounds is not anymore an adventure and becoming state of the art.
With the experience gained over the last years and with the help of CAE/simulation soft ware, it is possible to estimate the high demands on mould and mouldings. The further enhancement of the EMI shielding properties of the compounds looks to give both, moulding and materials, a bright future.

10. BIOGRAPHICS

Rik Decoopman　：　Director of FARDEC BVBA, Marketing, Sales and Export Consultancy
for the VERCO group of companies

John Vandaele　：　PLASCON N.V., Consultant to the VERCO group of companies.
Experience in thermoforming - extrusion - RIM - injection moulding

11. REFERENCES

1.　Steels for plastic moulding/Edelstahl Witten-Krefeld GmbH Edition: 11/97/c
2.　Faserverkürzung beim Verarbeiten langfasergefüllter Thermoplaste
H J Wolf, Darmstadt　Kunststoffe 83(1993) 1
3.　Thermoplaste mit Rückgrat (Langfaserverstarktes PP und PA)
A Lücke, Frankfurt/Main Kunststoffe 87 (1997) 3
4.　Injection Moldable Nickel Coated Carbon Fibre Concentrates for EMI Shielding
Malcolm W K Rosenow Antec '97/22
5.　LNP Verton Structural compounds
6.　Software can predict fibre orientation
Stephan Kukula　Modern Plastics Int, June 1996/79
7.　Celstran literature
PCI - Winona

RESISTIVITY CONTROL USING WET LAID NONWOVENS

Nigel Walker , Technical Fibre Products Ltd

ABSTRACT

The wet-laid nonwoven process is derived from traditional papermaking. When applied to high performance conductive fibres it produces mats with exceptional planar uniformity, an isotropic fibre arrangement and areal weights as low as $6gm^{-2}$. The use of lightweight nonwoven fabrics allows composite designers to optimise the electrical performance of high cost materials such as carbon, metal or metal coated fibres by placing a precisely tailored amount in the most effective position within a structure.

The nonwoven process is unique in fabric manufacturing techniques because it permits blending of dissimilar fibres in all possible proportions. This property is used to design fabrics with accurately controlled resistivity values. Such fabrics find use in a multitude of applications ranging from static protection through resistive heating to radar signature management .

KEYWORDS: Nonwovens; Electrically conductive materials; Resistive Heating

1 . WET LAID NONWOVENS

Wet laid nonwovens are produced by a modified papermaking process. Papermaking is believed to have been invented in 105 AD in China and is credited to a gentleman named Tsai Lun. Tsai Lun's process involved the mixing of fibres (plant derived cellulose) with water, subsequent draining of the fibre slurry through a mesh, followed by removal of the remaining water by pressing and drying.

Modern papermaking uses essentially the same process but is now a highly sophisticated high volume process capable of producing tens of tonnes per hour.

The term wet laid nonwoven is usually applied when the fibres involved are non cellulose and do not exhibit the strong hydrogen bonds which give conventional papers their high density and strength in the absence of other binding materials.

A common characteristic of all paper-like materials is that they consist of a two-dimensional sheet of short discontinuous fibres, with the fibres randomly arranged within the plane of the sheet.

The wet laid process may be applied to most fibres which have a density more than 1 Mgm^{-3} and a surface which is " wettable" in aqueous media. Advanced fibres which are currently processed via the wet laid route include, carbon, aramid, silicon carbide, metal and metal coated fibres. The fibres used are relatively short, precision chopped to lengths between 3mm and 25mm and may be taken from the heaviest available tows. The fabrics produced are typically lightweight (6 to 50 gm^{-2}) and thin (0.1 to 0.5mm), although the process is also capable of producing heavier and thicker products.

Since the fibres of interest have no inherent bonding mechanism, chemical binders are used to maintain the web structure. It is usual for the binder content to be minimised in order to avoid interfering with the properties of the fibres, typically binder contents will be less than 10% by weight. Binders are usually organic and may be selected from a wide range of thermoplastic and thermosetting polymers including vinyl, epoxy, polyester and phenolic resins ; the choice of binder is determined by compatibility and processing requirements of the matrix resin for the final composite.

Hybrid fabrics may be produced very easily and accurately. Precise quantities of fibre may be mixed in the wet slurry stage and will subsequently be reproduced in the finished fabric. Because there are no geometric constraints any proportion of fibre is possible. The system is not limited to binary mixtures, ternary mixtures are currently established and quaternary systems are certainly possible. The fabrics are true hybrids and may be considered as two or more interpenetrating fibre networks .

Figure 1 : Photomicrograph (*20) of hybrid nonwoven

**Figure 2 : Photomicrograph (*300) of carbon fibre nonwoven
demonstrating binder distribution**

2. RESISTIVITY CONTROL

The electrical resistance of thin planar materials is usually described in terms of surface resistivity, this assumes the thickness to be dimensionless and is a reasonable approximation in the case of materials of less than 0.5 mm thickness.

Surface resistivity measurement is carried out either by measuring the resistance between two electrodes which are separated by a distance equal to their length (1) or by measuring the resistance between to concentric circular electrodes (2) in either case surface resistivity is expressed in ohms per square, $\Omega\square^{-1}$. When measurements of surface resistivity are made on fibrous nonwoven materials it is necessary to describe the pressure applied to the sample when making the measurement.

It has recently been found that for a nickel coated carbon nonwoven, resistance decreases with pressure up to an applied load of approximately 400 kPa, the system remains steady between 400 and 1000 kPa and the resistance begins to increase at higher pressures. This general principle is expected to be true for nonwovens based on other fibre types although the exact pressures at the points of change will vary according to the material.

When a hybrid fabric is produced from a mixture of electrically conductive and non conductive fibres, the surface resistivity of the fabric changes with the fibre proportions in a predictable and controllable manner.

The following table details the results of resistivity measurements for a range of hybrid nonwovens produced from 12mm carbon fibre and 12mm "E" glass fibre, tested with a contact pressure of 200 kPa.

Surface Resistivity $\Omega\square^{-1}$	Carbon content %	Log Resistivity
31500	5	4.5
5200	10	3.72
480	12.5	2.68
263	15	2.42
48.9	17.5	1.69
27	20	1.43
13.2	40	1.12
3.2	100	0.51

Table 1 Surface resistivity change with carbon fibre content

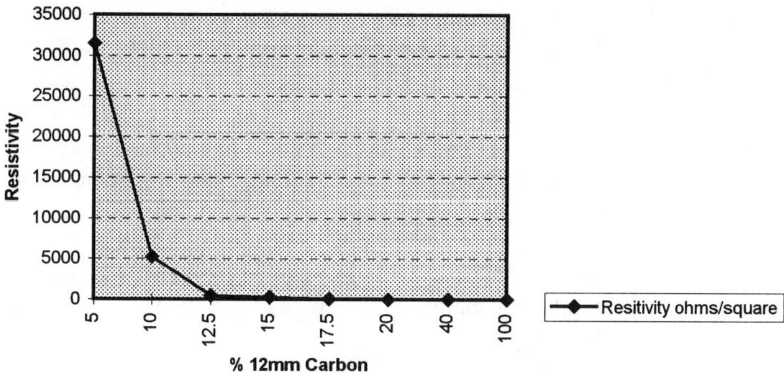

Figure 3 : Resistivity vs carbon content

Whilst the position of the curve changes with the use of alternative fibres, the shape of the curve is characteristic for all nonwovens produced from a mixture of conductive and insulating fibres. In all cases there is a "switching" area, where the network of conductive fibres becomes established, resulting in a very rapid decrease in resistivity, the curve then flattens with subsequent increases in the conductive fibre content producing progressively smaller resistivity changes. The relationship may be better illustrated by plotting the logarithm of the resistivity against the carbon content. (Fig.4).

Extensive studies relating to hybrids of carbon and cellulose (woodpulp) were undertaken by Hoon ,Shelton and Tanner (**3**).These findings should also apply to the materials currently discussed, although the physical differences between carbon and cellulose are likely to be much more significant than those between carbon and glass.

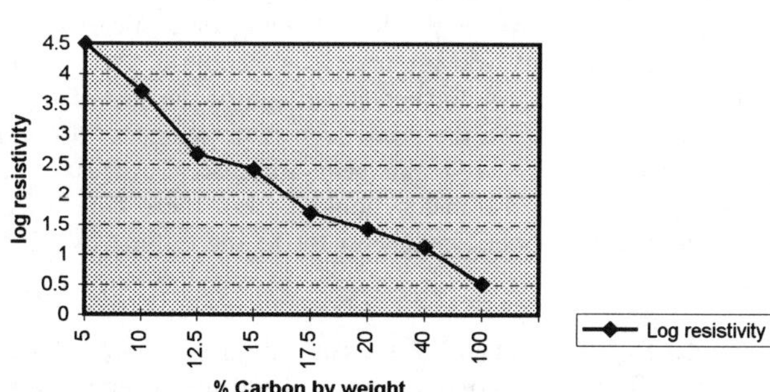

Figure 4: Log Resistivity vs carbon fibre content

The range of fibres which are suitable for nonwoven production include highly conductive materials such as copper or nickel coated carbons and dielectric materials such as quartz and aramids. Given this range of materials it is at least theoretically possible to construct fabrics with resistivity values from 10^{14} $\Omega\square^{-1}$ (100% Quartz) to 0.2 $\Omega\square^{-1}$ (100% Copper coated carbon).

In practice fabrics with surface resistivity values greater than about 10^{9} $\Omega\square^{-1}$ are more usually described in terms of their capacity for absorbing radio frequency energy rather than their surface resistivity.

Most of the applications for fabrics with controlled resistivity described in the remainder of this paper refer to materials which are classified as " conductive" i.e surface resistivity $<10^{5}\Omega\square^{-1}$ (4)(5).

3. STATIC PROTECTION / GROUNDING

In the fields of chemical vessels, pipework and offshore engineering, composite materials are increasingly preferred to metals due to their corrosion resistance, lightweight and ease of fabrication and installation. One of the main restrictions to the increased acceptance of composites is their triboelectric properties, unless treated to produce a conductive surface, most composites have a tendency to collect static charges in use which makes them unsuitable for use in petrochemical or dry powder processing applications.

Conductive nonwoven surfacing veils may be used to overcome this problem very effectively. The use of surfacing veils containing conductive fibres has several advantages over the available alternative methods of surface coating or loading the gel coat with conductive particles

Because the veil forms an integral part of the composite structure there is no risk of loss of conductivity by corrosion , abrasion ,flaking or peeling of the surface layer.
The technique of adding conductive particles to the gel coat also results in a permanently conductive structure however the high loading of particles required to achieve adequate levels of conductivity will often seriously compromise the structural and chemical strength of the gel coat. A further advantage of the use of conductive nonwovens is ease of application , since the nonwoven is presented as a continuous web it is particularly suited to automated and semi-automated processes such as pultrusion ,mandrel wrapping and resin transfer moulding. For these types of process the nonwoven material is usually supplied precision slit to the optimal width for the process

The "ultimate" nonwoven for use in chemical vessels would be one produced from 100% carbon fibre , such veils combine conductivity with excellent chemical resistance and superior mechanical properties. Such fabrics are indeed widely used particularly where hot acid environments are involved or where a high level of conductivity is required.

For larger volume and more cost sensitive applications, a veil produced from carbon combined with either glass or polyester fibre will provide adequate conductivity to ensure safety from static generation or discharge. The surface resistivity of the composite is usually greater than that of the nonwoven, the final measured resistivity is affected by the proximity of the nonwoven to the surface and the pressures used during moulding . Determination of the best fibre blend to achieve a specific resistivity value usually requires a joint prototyping exercise between the nonwoven producer and the composite producer.

In general fabrics for static protective surfaces are selected from the conductive side of the "switching" part of the resistivity / fibre content curve , in practical terms this means that the first choice fabric for static protection is a blend of 20% carbon with either glass or polyester, this fabric combines optimal cost with adequate conductivity.

Static protection by use of conductive nonwovens is by no means confined to thermosetting resin based "composite" systems. Other applications include the combination of nonwovens with vinyl foams to create anti-static bench coverings and the use of conductive nonwovens as a carpet base material. These products are used in the microelectronics industry where static sensitive components require very careful handling.

4. RESISTIVE HEATING

If an electric current is passed through an insulated conductor, energy is lost as heat as a function of resistance. When this principle is used in conjunction with nonwoven fabrics with controlled resistivity it is possible to design a range of lightweight, thin heater elements suitable for a wide variety of applications.

The isotropic nature of the fibres in wet laid nonwovens result in heating elements which rapidly achieve equilibrium with their surroundings and which exhibit a very even heat distribution throughout the plane of the element. This effect is further promoted by the thermal conductivity of the fibres used.

The ability to control the resistivity of the nonwoven allows the " tailoring" of a fabric which matches resistance with available power sources and geometric constraints on the size and shape of the element to achieve the desired heat output.

Elementary physics may be used to predict the power output of these systems:-

Power (Watts) = Voltage^2 / Resistance

So the derived power can be obtained providing the surface resistivity may be expressed as a resistance, this is possible for rectangular elements where the resistance is equal to surface resistivity $\Omega\square^{-1}$ multiplied by the number of squares e.g. for an element 5m * 0.1m with a surface resistivity of $0.5\Omega\square^{-1}$ the effective resistance is 0.5 * 50 = 25 Ω.

4.1 Low power trace heating

Nickel or copper coated carbon fibre nonwovens are the materials of choice for these applications, these fabrics have the lowest available resistivity (in the region 0.2 - 0.5 $\Omega\square^{-1}$) which makes them most suitable for low power, low voltage trace heating systems .

The main practical applications for such heating elements have been in the field of ice protection and portable "comfort" heating. Ice protection systems are required to generate temperatures of less than $50°C$, with rapid heat increase and a very even heat distribution. These systems can easily be incorporated in advanced composite structures without a significant weight penalty and without any redefinition of shape to accommodate the heating element. Such systems are employed to protect aircraft wings and propellers , these are typically powered by a 28V system - so a heater tape as described above (5m * 0.1m) will have a power output of $28^2 / 25 = 31W$, which equates to 62 Wm^{-2}. Smaller elements find use as portable battery powered heaters for body comfort applications , a 6 - 9 V dry cell battery may be used to power small heating elements built into clothing or used as insoles e.g. in Ski boots

4.2 Mid range heater elements

Nonwovens with higher resistivity than pure metal coated fabric are used in applications requiring power densities in the range 300 - 3000 Wm^{-2}. A fabric produced from carbon fibre is highly suited for incorporation in composite structures and , in addition to providing a route for resistive heating may also improve chemical resistance, structural properties or provide EMI shielding.

In order to demonstrate and quantify this application a simple plate heater was constructed by forming a laminate of a 60 gm^{-2} carbon fibre nonwoven, between two outer plies of 900gm^{-2} glass, chopped strand mat, electrical contact was made by two brass electrodes placed in the carbon layer, the whole structure was hand laid with a vinyl ester resin.

The heater panel thus formed was connected to a d.c. power supply with a voltage regulator and tested for surface temperature against varying voltage (fig 5) and surface temperature against time (fig 6).

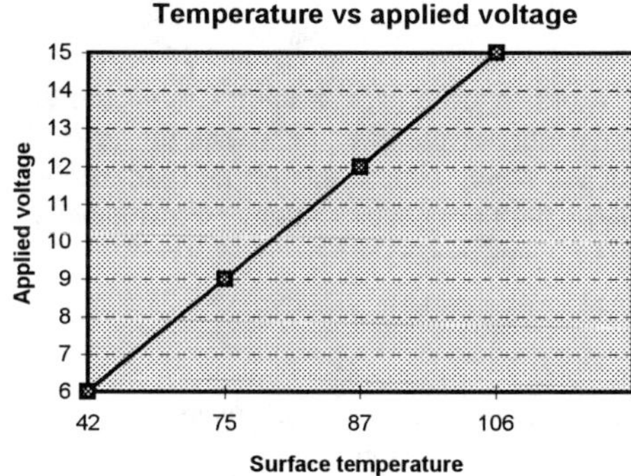

Figure 5: Surface temperature vs applied voltage

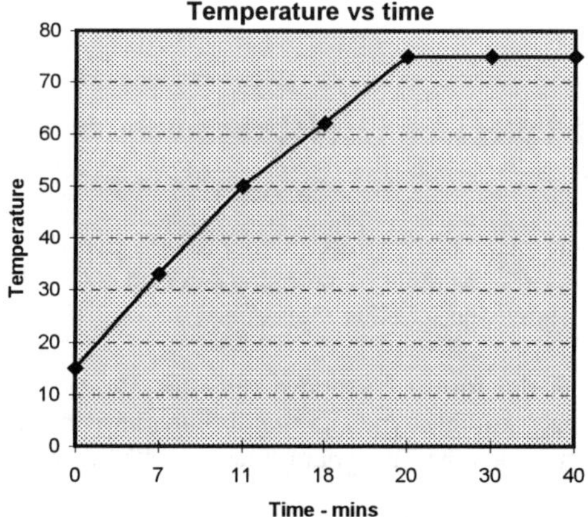

Figure 6: Surface temperature vs time for panel at 9V, 1A.

This type of FRP construction has been used to produce heated moulds, chemical processing vessels and immersion heaters.

A related application which does not involve composites is the production of heater pads in upholstery, particularly in heated car seats.

917

In this application the resilience and durability of the nonwoven as produced becomes a major item of the design. Carbon fibre is too brittle to be used in fabrics subjected to continuous flexing so a metal or metal coated fibre is preferred. This has the further advantage of providing superior conductivity which allows the heater pad to be produced with a large proportion of durable textile fibre whilst still having a resistivity suitable for producing adequate heat from a 12V automotive system.

4.3 Mains powered heaters

Using domestic mains electricity to power heaters made from nonwoven veils with increased resistivity considerably higher power densities may be achieved.

One notable example of this is the design and production of a thin lightweight heater element for use in domestic irons. Since the voltage was fixed at 220V and the size and shape of the element fixed by the design of the iron , a fabric was designed with a resistivity of ~$50\Omega\square^{-1}$ which was then capable of providing a surface temperature of 250 ° C , the power output of this fabric was calculated as being over 10 kWm^{-2}. These levels of temperature and power place particularly high demands on the outer, insulating layers of the heater element. In the domestic iron example both polyimide and poly ether ether ketone (PEEK) were used, the PEEK providing the matrix around the carbon/glass nonwoven and the polyimide providing the outer high temperature insulating layer.

5. EMI / RFI SHIELDING

The use of conductive fibres, carbon or metal coated carbon to provide EMI/RFI shielding is very well documented. (see - for example, papers presented at SAMPE 98, session 3E "Affordable M & P for portable and wireless electronics".)

If the conductive fibres are presented in the form of a nonwoven, this allows the designer of the shielded enclosure to select, the fibre type and the fibre loading and to precisely position the fibre in the composite such that maximum shielding effectiveness is achieved at minimum cost.

Shielding performance against density of carbon fibre (in grams per square metre) is shown in figure 7.

For even higher levels of shielding, or equal shielding with reduced weight penalty, metal coated carbon fibres are used. Nickel is the most widely available metal coating at present but copper and silver are possible where the application demands maximum performance.

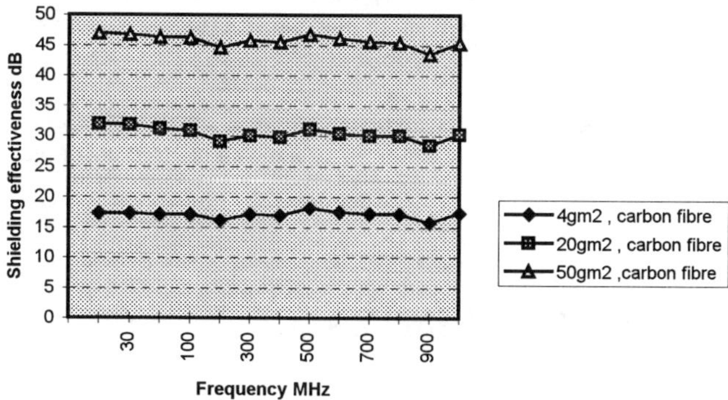

Figure 7 Shielding performance vs carbon loading

Nonwovens are used to best effect in those processes which can make use of their continuous sheet nature. Continuous processes such as SMC moulding and pultrusion can ensure a permanent shielding effect without the need for post production treatment by the incorporation of a nonwoven layer. The ability to cut the material to shape either by hand or by die stamping for volume , makes it suitable for semi continuous processes such as RTM.

Since these nonwovens are paper-like structures, they are also highly suited to wall covering and find applications in architectural shielding, particularly for electronic security and for the creation of electrically "sealed" EMC test chambers.

6. RADAR SIGNATURE MANAGEMENT

At a concentration of less than 5% by weight the conductive fibre network in a nonwoven breaks down and the individual fibres begin to act as dipoles. This property is used as one aspect of defence based radar signature management systems.

The details of this technology is of course restricted, however the property of creating interference with signals at a given wavelength is now finding applications outside the defence area.

Of particular interest is the screening of buildings in radar sensitive areas such as the periphery of airports. Positioning of building without screening is severely restricted because of the risk of "ghost" imaging by reflecting radar from the building surfaces.

Many airports are now insisting that all new buildings are clad with radar neutral materials. The inclusion of a suitable nonwoven in the panel cladding of these buildings offers a simple, low cost solution to this problem.

In contrast to the defence field the frequencies involved are well documented and unlikely to be changed. This means that "narrow band" absorbers can be used ensuring greater simplicity and effectiveness.

7. CONCLUSIONS

The wet laid nonwoven process is unparalleled in its versatility with regard to the materials used in fabric construction.

Hybrids of conductive / insulating fibre allow the development of fabrics with electrical properties " tailored " to specific end uses.

Whilst the properties of these fabrics may be predicted mathematically and from experimental evidence, the complexity of response to processing and the variety of options in fabric design are best addressed by close co-operation between the end -user and fabric producer at the start of the design process

8. REFERENCES

1. BS EN 100 015-1   Basic Specifications: Protection of electrostatic sensitive devices: Part 1 general requirements. BSI Milton Keynes (1991)
2. ANSI/ASTM D 257   Test methods for dc resistance or conductance of insulating materials
3. S.R. Hoon, A. Shelton and B.K. Tanner, J. Materials Science.,20, 3311-3319 (1985).
4. Ministry of Defence (UK) Def Stan 59-98 Handling Procedures for Static Sensitive Devices
5. US Department of Defense - DOD - HDBK-263 Electrostatic Discharge Control Handbook for Protection of Electrical and Electronic Parts Assembly

ON-LINE NDE OF PULTRUDED AIRFRAME STIFFENERS

Roger W. Engelbart and Michael L. VanDernoot
The Boeing Company, St. Louis, MO
Daniel D. Coppens - Anholt Technologies, Inc., Newark, DE
Christian Koppernaes - WeeBee Pultrusions, Beaufort, SC

ABSTRACT

Pultrusion has been demonstrated to be a cost-effective means of fabricating constant cross-section composite structures. However, it has not been widely used in aerospace applications. The goal of this program was to demonstrate equivalent performance of a pultruded, secondarily-bonded stiffener to a hand layup, co-cured stiffener in a secondary structural application. To focus on an actual aircraft application, a hat-stiffened cockpit floor was selected as a basis for design of the pultruded stiffener. As part of the manufacturing effort, on-line nondestructive evaluation was utilized to demonstrate the cost advantage of combining an inspection operation with a continuous fabrication process. The on-line system was based on the Boeing-designed Mobile Automated Scanner (MAUS) III, a portable ultrasonic and eddy current inspection instrument intended for in-service inspection of aircraft. The MAUS was modified to attach to the pultrusion line and provide continuous real-time C-scan data for the fabricated stiffener. The capability of the system to adapt to varying part geometries and provide instantaneous go/no-go indications as well as continuous data was demonstrated.

KEY WORDS: Nondestructive Evaluation (NDE), Pultrusion

1. INTRODUCTION

The conventional hand lay-up, autoclave process of fabricating composite aircraft structures has dominated the industry for many years. Although the process is proven and reliable, it is also known to be very labor intensive and therefore costly. During recent years, considerable effort has been expended investigating alternative processes which have the potential to reduce manufacturing cycle time and expense. Pultrusion offers the capability for continuous fabrication and on-line cure once the set-up has been completed. The process also allows the addition of an in-process inspection operation; quality assurance procedures represent a significant portion of manufacturing costs. On-line inspection not only permits a complete product at the end of the pultrusion line, but also provides the opportunity for real-time feed-back control of process parameters.

2. EXPERIMENTAL

2.1 Component Selection The baseline component on which the pultruded structure was to be modeled had to be based on a typical aircraft application in order to understand the potential of pultrusion to impact actual manufacturing costs. A secondary structure, such as a stiffened door, access panel, or electronics rack, was targeted to reduce the need to meet stringent mechanical properties. The candidate component was to be carbon-reinforced, with a blade or T-stiffener, and a constant cross-section.

Example candidate components were wing and fuselage access panels. Some wing panels are unstiffened, and are glass-reinforced. These could be easily pultruded as ordinary flat panels, but would not be an adequate demonstration of the technology. The fuselage panels were also unstiffened, and were contoured to the outer mold line of the fuselage. While these were of constant cross-section, the panel curvature would have required larger and more expensive tooling for the demonstration.

The selected baseline component was a cockpit floor section, as shown in Figure 2.1-1. The most important design factor for this secondary structure is the force due to cockpit pressurization. Although the stiffeners are of a hat configuration, this geometry proved easier to pultrude than a blade or a T. The floor is constant cross-section and is currently fabricated as a single piece; the stiffeners and the laminate are hand lay-ups, co-cured by autoclave. The floor is later cut into left- and right-hand sections, separated by a splice channel; this is due to the assembly sequences with other cockpit hardware. The stiffener of the left-hand floor was selected as the basis for the pultruded stiffener; this section is small enough to be considered for a future demonstration in pulling a full panel.

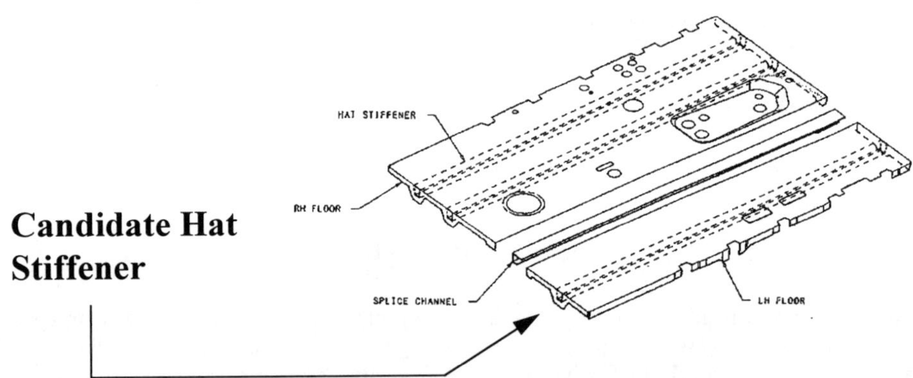

Figure 2.1-1. Baseline Component - Cockpit Floor

2.2 Overview of the Pultrusion Process The pultrusion process, as depicted in Figure 2.2-1, begins with dry reinforcements being continuously pulled from their respective creel systems through alignment guides which form the material into a near net shape of the final product. The preform is then drawn into the die where it is initially compressed in what is termed the fiber compaction region of the die. The compacted preform then enters the injection region of the die, where resin is introduced via mechanical pressure (hydraulic or pneumatic feed) into the profile preform. Injection region temperature and pressure control are critical in order to maintain optimum preform impregnation (i.e., low porosity levels).

The fully impregnated preform is then pulled into the cure region of the die, where the die temperature profile is established to optimize product quality (i.e., minimizing void content, maximizing degree of conversion, maintaining desired surface finish). The profile laminate is then pulled from the die via reciprocating hydraulic clamps. Clamping force and pull speed are process-controlled variables. Finally, the product is cut to length with the use of a fly-away cut-off saw, capable of cutting the product at the instant a signal is transmitted from an on-line limit

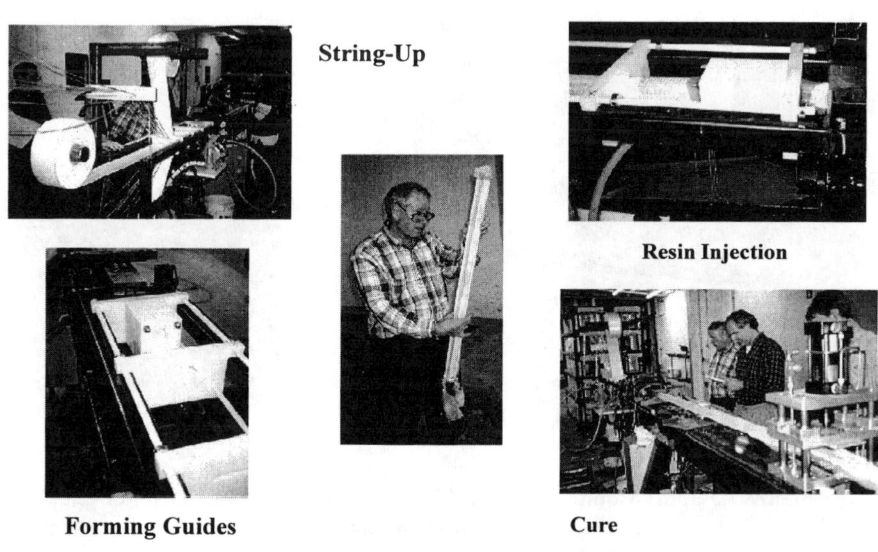

Figure 2.2-1. Pultrusion of Hat Stiffener

2.3 On-Line NDE Post-process nondestructive evaluation is a costly but necessary step in the fabrication of composite aircraft structures. Not only is there set-up and inspection time, but also time to transport the parts to and from a remote inspection facility. The pultrusion process allows the integration of nondestructive evaluation with the fabrication process, effectively reducing inspection time to zero. The on-line NDE task was designed to show

how NDE equipment could be adapted to the processing line to obtain meaningful data during manufacturing. The demonstration was based on the Boeing-designed Mobile Automated Scanner (MAUS) III, a multi-modal instrument which was intended for in-service inspection of aircraft in the field. This device is a portable C-scan inspection system that was developed for rapid inspection of large, complex structures. The system is capable of collecting, imaging, and storing high resolution data in any of the following inspection modes: ultrasonic pulse-echo, ultrasonic resonance, and eddy current. The basic system consists of an electronics case, laptop computer, and scanner, as shown in Figure 2.3-1 a and b. The scanner head was designed to be moved manually across the surface of a structure. At the same time, scanner-mounted transducers experience translational motion with respect to the scan direction. As the scanner moves, strips of inspection data are obtained that range in width from one to eight inches, depending on the number of transducers and the scanner configuration. The laptop computer continuously displays the data in real time and stores it on a removable hard disk. The data can be transferred to optical disk for future evaluation and archiving.

The set-up for the demonstration was performed in the Boeing laboratory using sections of vacuum-assisted resin transfer molded (VARTM) carbon/epoxy stiffeners. A transducer frequency of 2.25 Mhz was selected to allow adequate sound penetration through potentially porous material. Each

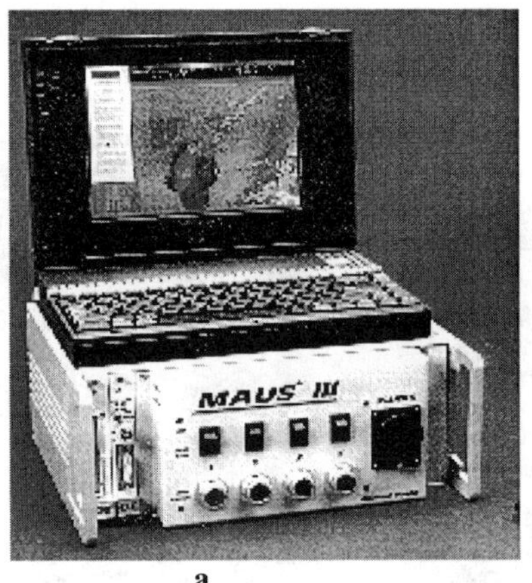
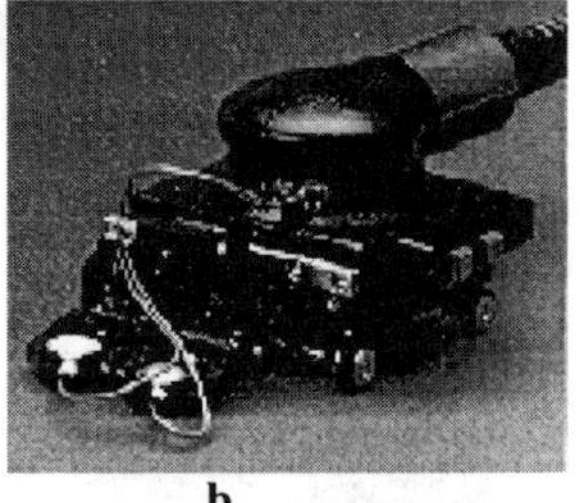

a   b

Figure 2.3-1. (a) Mobile Automated Scanner (MAUS) III Electronics Cage and (b) Scanner

transducer on the scanner normally covers and area two inches wide. The scanner stroke was adjusted to cover the width of the hat, as shown in Figure 2.3-2.

Figure 2.3-2. Laboratory Adjustment of Scanner Stroke

Because the MAUS was designed to move across a stationary surface, it is relatively simple to adapt the system to the pultrusion application in which the scanner is stationary and the surface moves. Additional wheels were mounted on the scanner to accommodate the limited surface of the hat, as shown in Figure 2.3-4. The additional wheels in the rear

Figure 2.3-4. Additional Scanner Wheels for Mounting to Hat Stiffener

were spaced to straddle the stiffener, providing balance to the scanner. The front wheels ride the top of the hat, allowing the system encoders to track the linear distance covered. It was planned to obtain comparative ultrasonic data from the VARTM stiffeners, but the samples proved to be too porous for penetration even at the lower frequency

Figure 2.3-5 illustrates the MAUS set-up on the pultrusion line at WeeBee. The scanner was held in place by a length of aluminum extrusion which in turn was mounted to a video camera tripod. One goal of this demonstration was to show that the MAUS can be

Figure 2.3-5. MAUS On-Line During Pultrusion Run

easily attached to and removed from the line during the process, without disrupting the manufacturing operation. The size, portability, and flexibility of the scanner permit at least temporary attachment to the pultrusion line with materials that are readily available.

During the pultrusion run, the MAUS scanner successfully rode the moving stiffener and oscillated the transducer without and interference from the stiffener surface. Because of difficulties with the process parameters, ultrasonic data was not obtained in-process; four runs were attempted, but none were of sufficient length to permit on-line inspection. The final data was obtained in the laboratory using short sections of the glass stiffeners from the run trials.

Figure 2.3-6 shows a scan segment from one of the stiffener segments. The data is presented in pseudocolor to allow easy visibility of material regions with widely differing ultrasonic attenuation. The stiffener section, although thin, proved to be very

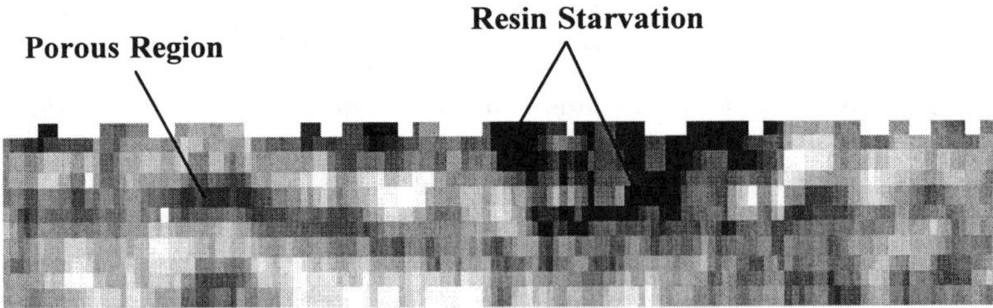

Figure 2.3-6. MAUS C-Scan Data Segment from Glass Stiffener

attenuative, indicating porosity or dryness. This condition was evident from examination of the surface. The scan segment shows that interior of the part may have inadequate resin penetration, as seen by the labeled regions. The light areas represent locations of low attenuation relative to the rest of the part, indicating better resin permeation.

3. CONCLUSIONS

Under this effort, we have identified an aircraft component which is in current service and which could be transitioned from hand lay-up fabrication to continuous fabrication by pultrusion. We have successfully demonstrated the feasibility of incorporating on-line ultrasonic nondestructive evaluation into the process, and we have performed a preliminary cost analysis that the potential of pultrusion to reduce the fabrication cost of the candidate baseline component.

4. ACKNOWLEDGEMENT

This work was performed under a Phas I Small Business Innovation Research (SBIR) program. The program was sponsored by the Naval Air Warfare Center at Patuxent River, MD and managed by the Office of Naval Research.

IN-PROCESS MONITORING OF PRE-STAGED
FIBER PLACEMENT TOWS USING
NUCLEAR MAGNETIC RESONANCE (NMR)

Roger W. Engelbart - The Boeing Company, St. Louis, MO
Dr. Mark Conradi and Dr. R. Dean Stoddard
Washington University, St.Louis, MO

ABSTRACT

Fiber placement usage and applications have increased; however, process control measures have focused mainly on temperature monitoring. Remote infrared measurements are taken at the nip point to try to optimize consolidation and thermocouples are used to control the application of heat to adjust the tack of incoming tows. Although the temperature measurements are accurate, they may not reflect the true condition of the resin with respect to viscosity and cure state. Nuclear magnetic resonance (NMR), which detects the changes in proton spin orientation within a magnetic field, has been shown to be very sensitive to resin viscosity, and therefore, to cure state.

Under the program, IM7/977-3 towpreg material was pre-staged at various degrees of cure in an off-line operation prior to fiber placement. Differential scanning calorimetry (DSC) analyses were performed to verify degree of cure. Towpreg samples in these various cure states were placed in an 8.0 Tesla laboratory magnet and a direct correlation was established between degree of cure and NMR relaxation time. Utilizing a laboratory transport mechanism, lengths of pre-staged towpreg were moved between the poles of the magnet, demonstrating the feasibility of obtaining degree of cure data in-process.

1. INTRODUCTION

Effective monitoring of the degree of cure of pre-staged fiber placement tows requires a technique that responds directly to material conditions, instead of indirect measurement such as temperature or tow speed through the line. Because nuclear magnetic resonance

(NMR) detects the change in alignment of hydrogen nuclei in a magnetic field, the technique is very sensitive to changes in resin due to the curing process.

This paper descibes work that was per formed under the In-Situ Fiber Placement program which was sponsored by the Navy Center of Excellence in Composites Manufacturing Technology. The process monitoring effort was performed in three phases: 1) measurement of the relaxation time T_2 at various temperatures for both 977-3 and 8552 to establish the sensitivity of the parameter to cure; 2) development of a quantitative relationship between degree of cure and T_2; and 3) validation of the capability of NMR to acquire data from moving towpreg in a simulated on-line application.

2. PRINCIPLES OF NMR

The NMR experiments detailed in this report were performed on hydrogen nuclei which are abundant in the epoxy resin. Hydrogen nuclei possess a nuclear spin $I=1/2$ which is oriented randomly before the sample is placed in an external magnetic field. Upon being placed in a magnetic field, the nuclear spins align either parallel or anti-parallel to the external field. Spins which are parallel to the magnetic field have a lower potential energy than those which are anti-parallel to the field; therefore, the preferred alignment is parallel to the external field. The NMR frequency is determined by the strength of the external magnetic field. For the measurements detailed here, the magnetic field strength is 5000 Gauss, yielding an NMR frequency of 21.25 MHz. At this field and at the temperatures used for the NMR measurements (between 80° and 200° C), there is a very small excess fraction (approximately 10^{-5}) of spins in the lower energy state. It is only this small excess in the lower energy state which gives rise to a net magnetization and is the source of NMR signals.

At equilibrium, the magnetization is aligned parallel to the external magnetic field. The NMR experiment is performed by manipulating the magnetization with a radiofrequency pulse, or series of pulses, and observing the magnetization's subsequent evolution. Eventually, the magnetization will return to equilibrium and the whole process can be repeated. The return of the magnetization to equilibrium is characterized by two time constants: T_1, the spin-lattice (or longitudinal) relaxation time, and T_2, the spin-spin (or transverse) relaxation time. T_1 is a measure of how long it takes for the magnetization to return to equilibrium parallel to the external field; T_2 is a measure of how long it takes for the magnetization perpendicular to the magnetic field to decay to zero, its equilibrium value. Both T_1 and T_2 are influenced by motion of the hydrogen atoms. The hydrogen nuclei possess magnetic moments which generate their own small magnetic fields. These fields affect other nearby hydrogen nuclei. The strength of the interaction depends on the relative positions and orientations of the hydrogen nuclei. Motion of the hydrogen nuclei causes the interactions between the hydrogen to fluctuate. It is the fluctuations in the spin-spin interactions which determine both T_1 and T_2. Because the details of the T_1 and T_2 mechanisms differ, the two relaxation rates are sensitive to fluctuations (hydrogen motion) occurring on different time scales. The relaxation time T_2 was chosen as the focus of this study because preliminary work indicated it was more sensitive to hydrogen motions occurring at the temperatures (80° - 200° C) of interest.

3. EXPERIMENTAL

3.1 Characterization of T_2 Sensitivity to Cure   Preliminary studies show that T_2 decreases by up to four orders of magnitude (100 ms - 10 ms) as both the 8552 and 977-3 resins cure. T_2 changes because the rate of hydrogen motion changes as the materials cure. At low degrees of cure there are essentially no cross-links between the short length (low molecular weight) polymer chains; their motion is unhindered and rapid. This rapid hydrogen motion leads to long T_2's (approximately 100 ms at 160° C) for uncuredmaterial. As the materials cure, the chains lengthen and cross-links form between the polymer chains; the severely inhibited motions cause T_2 to be shorter for higher degrees of cure. Due to the large difference in mobility in liquids and solids, the T_2's in the two phases also differ. In liquids, the high mobility leads to a narrow NMR linewidth, or frequency spectrum; this is equivalent to a very long T_2. In solids, the low mobility leads to a broad NMR linewidth and short T_2 .

The range of T_2's for the carbon/epoxy materials demonstrates how much motional rates change as a result of curing. At low degrees of cure the long T_2's show that the epoxy resin is a not-very-viscous liquid. In a liquid the molecules move very rapidly on the timescale of the T_2 measurement. The rapid molecular motion causes the interaction between the hydrogen nuclei to fluctuate rapidly as well. The net effect is motion causes the interaction between the hydrogen nuclei to fluctuate rapidly as well. The net effect is that the interaction between hydrogen nuclei is nearly averaged to zero. Because the NMR linewidth is determined by the distribution of magnetic fields at the hydrogen nuclei, averaging of the interaction between the hydrogen nuclei causes the effective distribution of fields to narrow and the NMR linewidth becomes narrower as well. This effect is known as motional narrowing. Since T_2 is essentially the inverse of the NMR linewidth, a narrower linewidth means a longer T_2.

As curing progresses the growing chain-lengths and denser cross-links hinder the hydrogen motion. The slower the hydrogen motion the less effective the motional narrowing and the broader the NMR linewidth and the smaller is T_2. Eventually, hydrogen motion will become so slow that the atoms are essentially stationary during the T_2 measurement. This is called the rigid-lattice limit. In this limit there is no motional narrowing and the NMR linewidth is broad, meaning T_2 is short. The exact width of the rigid-lattice linewidth is determined by the strength of the interaction between the hydrogen nuclei. The short T_2's seen for degrees of cure as low as 60 or 70 percent show that the epoxy resin is already becoming very rigid on the NMR timescale.
The sample materials used in the preliminary work (monitoring T_2 at constant temperature) were 977-3 and 8552. The samples were prepared by cutting strips of material from the rolls supplied and then folding them several times to form a piece about 6 mm x 6 mm x 1 cm. This piece was then placed into a 8 mm O.D. pyrex tube,

12 cm in length, and the tube was inserted into an NMR probe. The NMR experiments were performed by placing the sample into a heated gas stream. To allow the material to equilibrate at the correct temperature, the NMR experiments were initiated after a 1-2 minute delay. The experiments for this portion of the work were performed in room air; nitrogen purge was not used.

3.2 NMR Response for IM7/977-3 Cure   The NMR T_2 measurements on the 977-3 and 8552 systems demonstrate the sensitivity of T_2 to changes in the rate of hydrogen motion due to curing. Figure 3..2-1 is a plot of T_2^{-1} versus time (note the data is plotted as $1/T_2$, not T_2) for the 977-3 system. The data shown are T_2 measurements performed on six different samples, each allowed to cure at a different temperature. The data show that T_2, or T_2^{-1}, changes by up to four orders of magnitude for curing at the highest temperatures.   The data also show the sensitivity of the kinetics to the curing temperature. Initially, T_2^{-1} is approximately 100 s^{-1} for curing at 82° C and changes only by a factor of 2 after curing for 8 hours. For curing at 201° C the initial relaxation rate is approximately 6 s^{-1} (longer T_2 because the resin is more fluid at the higher temperature) and changes by almost four orders of magnitude in the first 30 minutes of curing. By the time the 977-3 sample has cured for 45 min at 201° C it is essentially rigid on the timescale of the T_2 measurement.

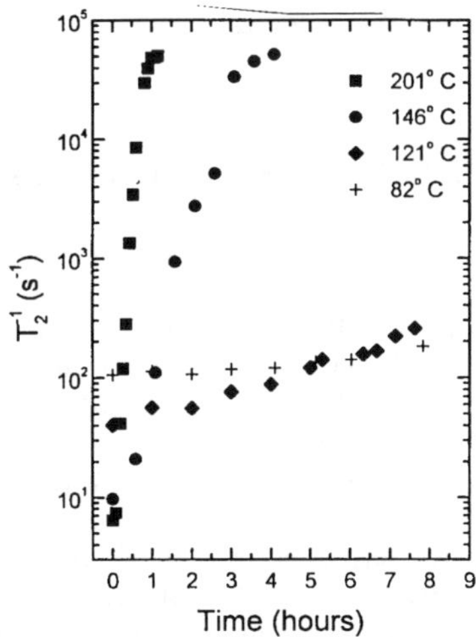

Figure 3.2-1. NMR Relaxation vs. Cure Time, 977-3

3.3 NMR Response for IM7/8552 Cure   Figure 3.3-1 shows similar data for the 8552 system where T_2 is monitored for five samples held at constant temperatures. For the 8552 sample, T_2^{-1} again increases by almost four orders of magnitude during cure at the highest temperatures. The data also show that the onset of the curing reaction for the 8552 system occurs at a lower temperature than for the 977-3 system. By comparing the slopes of T_2^{-1} versus time at approximately 100° C for both samples it can be seen that the slope is steeper for the 8552 sample than for the 977-3, indicating that the reaction is occurring more quickly for the 8552 system.

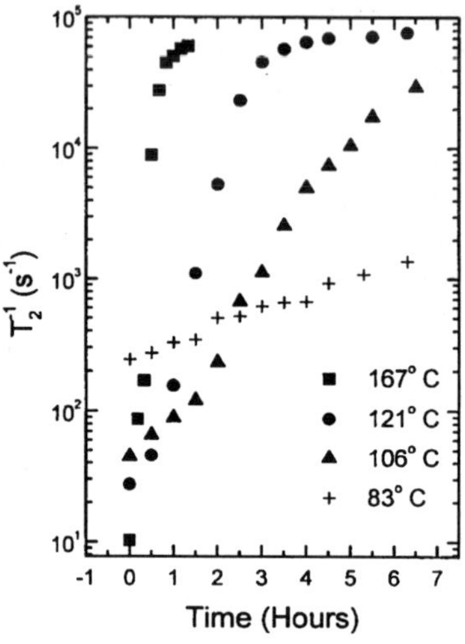

Figure 3.3-1. NMR Relaxation vs Cure Time, 8552

3.4 Static NMR Meaurement-T$_2$ vs Degree of Cure The sample material used in this portion of the work was 977-3. Two separate batches of 977-3 samples were pre-cured by heating in an oven at 200°. Each batch included a complete set of pre-cured samples, as well as one sample which was uncured. The nominal degrees of cure for the different samples were 5, 10, 20, 30, 40, 50, 60 and 70 percent.

The 977-3 material was prepared for curing by cutting strips of the material from the spool. At high degrees of cure the material becomes difficult to cut cleanly. In addition, some epoxy resin flakes off as the material is cut, changing the epoxy/carbon fiber ratio of the sample. To get around this problem, some of the material was pre-cut into squares which fit into the aluminum DSC pans. The rest of the material was folded into NMR samples (see above). The material for both the DSC and NMR measurements were wrapped in aluminum foil for pre-curing together. A separate packet was prepared for each degree of cure. The curing was performed in a Fisher Scientific oven equipped with an access port on top. To prevent oxidation of the material, nitrogen gas was constantly circulated through the oven during curing. Before beginning curing, the oven was pre-heated to 200° C and held there during the course of the curing. Once the oven was at temperature, samples were successively hung in the middle of the oven with a piece of wire for periods of time calculated from a kinetic model of curing. As each sample was removed from the oven it was immediately placed in the cool air above a liquid nitrogen bath to ensure rapid cooling. For storage, the samples were removed from the aluminum foil and placed into glass vials. The samples were then stored in a freezer.

After curing, the degree of cure of each sample was measured by differential scanning calorimetry. The measurements were performed on a Thermal Instruments TA-2970 machine. The samples were prepared for calorimetry by first weighing the aluminum pan to be used. Pieces of 977-3 material were then placed into the pan and it was reweighed to obtain the weight of the sample. The pan was then placed into the apparatus and the measurement begun. The DSC measurement was performed by heating the sample from 50° to 340° C at a rate of 20° C/min. After each measurement the sample pan was discarded and a new one used. The degree of cure for each sample was taken as the average of five individual measurements of degree of cure. To prevent oxidation of the material, the DSC measurements were performed in flowing nitrogen gas.

Once the degree of cure was determined for each sample, T_2 was then measured. The T_2 measurements also included a sample which was left uncured. The NMR measurements were performed on samples which were placed into pyrex tubes, identical to those described above. The NMR probe was pre-heated to the desired temperature and the samples placed into the probe. To allow the NMR samples to equilibrate at the correct temperature, a delay of approximately 1-2 min was used before beginning the T_2 measurement. The measurements at 160° and 140° C were performed on the first batch of sample material. For the 160° C NMR measurements T_2 was taken as the average of T_2 measured on five different samples. For the 140° C measurements T_2 was taken as the average of four measurements for the 20.1% cured sample; otherwise, five measurements were used. The material from the second batch of pre-cured samples were used to make the NMR measurements at 80°, 100° and 120° C. Because of the small amount of additional curing caused by the NMR measurements at these low temperatures, measurements at all three temperatures were performed on the same set of 977-3 samples. The measurements were made in the order 80°, 100°, and then finally, 120° C. For these measurements T_2 was taken as the average of five measurements in all cases.

Figure 3.4-1 is a plot of T_2^{-1} versus degree of cure for the 977-3 system. The T_2 and degree of cure measurements were performed on material which had been heated together to ensure they were identically cured. the data show a correlation between degree of cure as meeasured by differential scanning calorimetry (DSC) and the NMR parameter T_2. this correlation allows degree of cure to be quantitatively determined by a measurement of T_2 at a known temperature. In addition, the data also show that the optimal measuring temperature changes with degree of cure. That is, T_2 measured at $160^{\circ}C$ is most sensitive to degree of cure values from 30 to 70 percent, while T_2 measurements at $80^{\circ}C$ are most sensitive to degree of cure values from 10 to 50 percent.

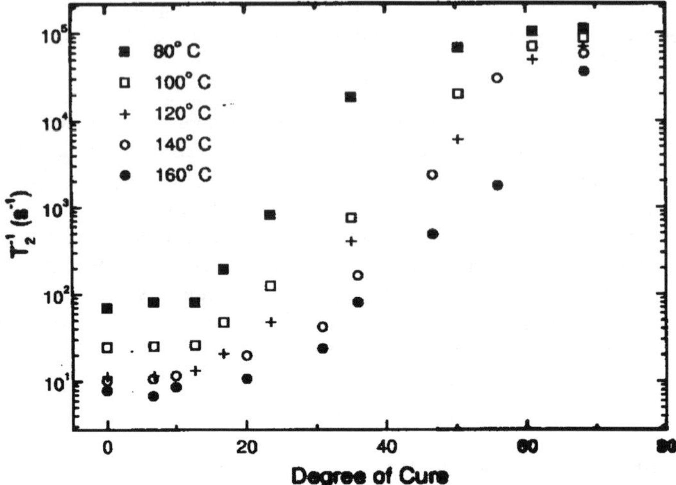

**Figure 3.4-1. Laboratory Correlation Between Degree of Cure
and Spin-Spin Relaxation Time**

The relationship between degree of cure and T_2 was not established for the 8552; at this point in the program it was found that the material supplied had already been exposed to some minimal pre-staging by the vendor. The chance of error was to great to proceed, and the remainder of the effort was focused on the 977-3.

3.5 Dynamic NMR Measurement The third phase of this work was to demonstrate that NMR could be used to determine the DOC of a single tape of moving towpreg. The samples used in this phase of the work were made from the 977-3 material, but were taken from a different roll of material than the samples for the previous work. The two rolls of material came from the same prepreg lot.

The samples were prepared by cutting approximately 10 ft long strips of material from the roll. Five strips of material were cut from the roll. One strip was left uncured while the remaining four pieces of material were nominally cured 10, 20, 30 and 40 %. The samples were cured by sequentially placing them into the pre-heated (200° C) Fisher Scientific oven. For curing, the tapes were looped over a piece of copper wire and suspended inside the oven. To prevent oxidation of the samples nitrogen gas was circulated through the oven during the pre-staging. After pre-staging, the degree of cure

934

of each strip of towpreg was determined by differential scanning calorimetry. The DSC measurements were made on pieces of towpreg which were taken from sections of the tape located approximately 1 foot from both ends of the tape. After removal of the DSC material a strip of tape approximately 8 ft long was left for performing the NMR measurements. The degree of cure of each strip was determined as the average of five separate measurements of degree of cure. The degree of cure of the four pre-staged samples was determined to be 2.1, 21.7, 37.9 and 46.2%.

Following the degree of cure measurements the T_2 of each sample was measured using the NMR probe and a prototype tape transport system , as shown in Figure 5. The set-up is based on pulling the towpreg up through the existing 1.6 inch gap, 5000 Gauss electromagnet and probe. The take-up reel is from an audio tape recorder and is mounted above the magnet. The take-up reel is driven by a 1 rpm motor which translates to a tape transport speed of about 6.3 inches per minute. The supply reel is also from an audio tape recorder and is mounted behind the magnet. Tension in the towpreg is established by a friction clutch on the supply reel consisting of felt pads pressed against the hub of the reel.

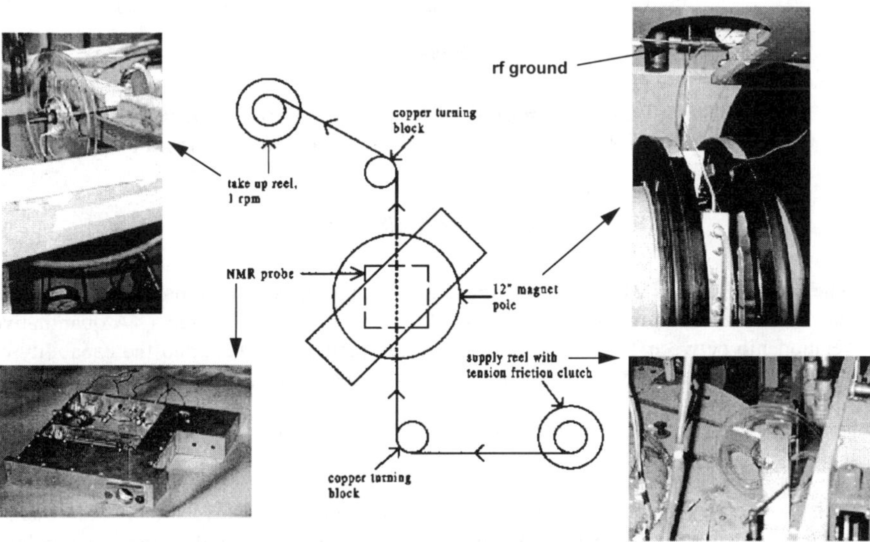

Figure 3.5-1. Laboratory Set-Up for NMR Dynamic Measurements

Figure 3.6-1 is a plot of T_2^{-1} vs. degree of cure for the 977-3 system, with T_2 measured at 80° C. The closed squares represent measurements performed in a standard, static fashion with the carbon/epoxy towpreg contained in a glass sample tube. The open squares represent measurements performed using the modified probe and tape transport system on moving towpreg. The two different measurement techniques are in good agreement, demonstrating the practical applicability of NMR T_2 determinations of degree of cure in the 977-3 carbon/epoxy system.

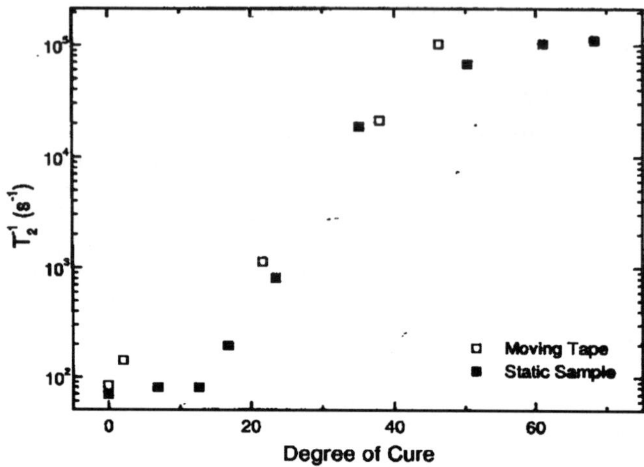

Figure 3.6-1. On-Line Simulation: Correlation Between Degree of Cure and Spin-Spin Relaxation Time for Both Static and Moving Towpreg

4. CONCLUSIONS

The sensitivity of NMR relaxation time T_2 to changes in resin cure has been shown and data for T_2 vs cure time has been collected for two major resin systems. A quantitative relationship between T_2 and degree of cure has been established, and the capability of NMR to acquire degree of cure data in-process has been successfully demonstrated.

5. ACKNOWLEDGEMENT

This work was performed under a program entitled "In-Situ Fiber Placement," sponsored by the Navy Center of Excellence for Composites Manufacturing Technology and monitored by Mr. Bruce Brailsford. The goal of the program was to reduce the autoclave processing necessary for fiber-placed structures by preliminarily advancing the cure state of the tows.

43rd International SAMPE Symposium
May 31-June 4, 1998

MONITORING THE CURE PROCESSES OF POLYMER COMPOSITES AND NEAT RESIN USING AN ULTRASONIC WIRE WAVEGUIDE TECHNOLOGY

Yan Li, Gerald J. Posakony and Suresh M. Menon
XXsys Technologies, Inc.
4619 Viewridge Ave.
San Diego, CA 92123 - 1639

ABSTRACT

In many composite applications, the quality of the manufactured parts or structures is dependent on the cure of the polymer. To assure the quality and the consistency of a composite and to improve the efficiency of the manufacturing processes, it is desirable to have a reliable tool to make in-situ monitoring of the composite while it goes through the cure process. This paper presents an ultrasonic wire waveguide technology that has been developed for the cure monitoring of polymer composites and neat resins. The technology is based on wave propagation in a specially shaped thin wire embedded in a composite. Based on the information extracted from the received ultrasonic waves, the cure of the composite can be monitored nondestructively in real-time. This paper presents experimental results from several resins and composites. It discusses the operation and implementation of this technology as well as applications of this new technology to a variety of composite manufacturing processes.

KEY WORDS: Cure Monitoring, Wire Waveguide, Composite Armor

1. INTRODUCTION

Over the last decade, liquid molding techniques have become more and more popular for manufacturing of composite parts and structures. These techniques are often selected due to the ease of manufacturing automation and the ability to effectively and inexpensively produce large and complex shaped parts and structures. Resin transfer molding (RTM), vacuum assisted RTM (VARTM) and Seemann's composite resin infusion molding process (SCRIMP) are examples of widely applied liquid molding techniques in the composite industry today.

Resin flow and cure play key roles in the liquid molding techniques. Proper resin flow and cure in these processes are crucial to the quality of the composites and manufacturing consistency, particularly when the thickness of the composite is large. As a result, it is highly desirable to have a sensor system that can obtain flow/cure information and use it to achieve intelligent system control during the composite manufacturing process.

There is also a strong need for in-service monitoring and assessment of the integrity, overall health, and damage on the composite structures resulting from normal use, impact, and battles for military applications. In particular, a smart structure which has the capability to detect damage and degradation can be very valuable for ensuring continued integrity of the composite structures.

2. SENSOR DESCRIPTION AND NEAT RESIN EXPERIMENT

An acoustical sensor technology has been developed for the liquid molding composite manufacturing techniques. This technology is based on the wave propagation in a thin metallic wire embedded in the composite part[1-4]. Figure 1 is a schematic of the developed wire waveguide (WWG) sensor. The wire is made of stainless steel and has two

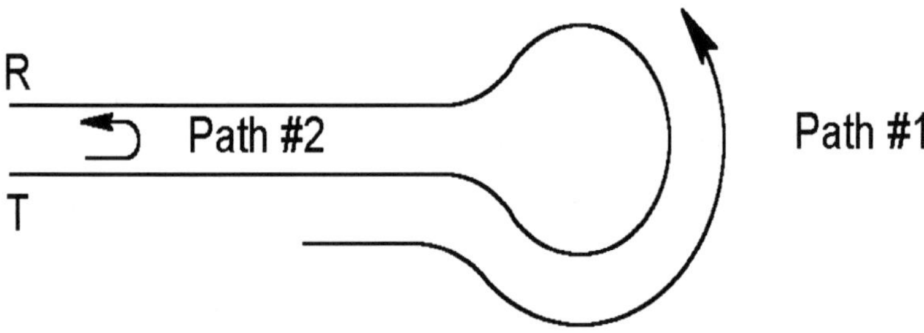

Figure 1. Wire waveguide sensor.

small piezoelectric elements attached to its ends, one is transmitter and the other receiver. The sensor is operated at about 350 kHz and driven with 4 cycles of toneburst.

To demonstrate the sensor operation, an experiment was conducted using a small reusable Teflon tray at the room temperature. The resin used was Shell Epon 828 + Epicure 3234 which was selected for availability and popularity. This experiment is representative to many other neat resin experiments we conducted during the course of the sensor development.

Referring to Figure 1, there are two paths for the ultrasonic wave to propagate from transmitting transducer to the receiving transducer. Path #1 is the route followed by the energy propagated through the wire itself. Path #2 is the route followed by energy

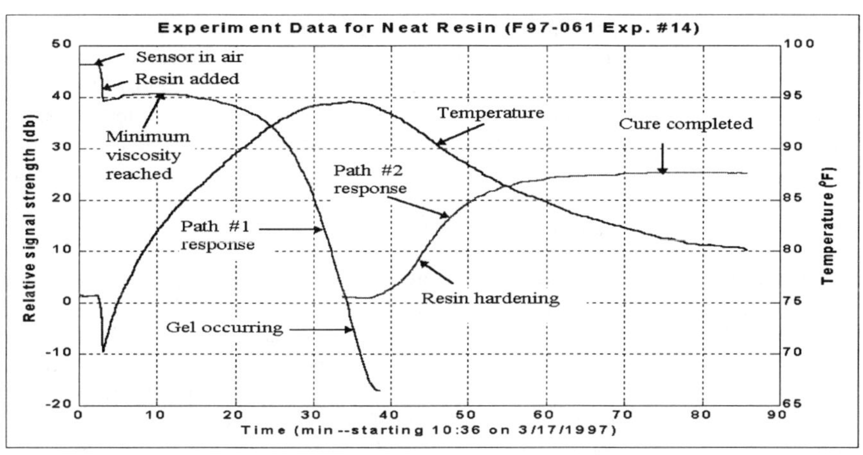

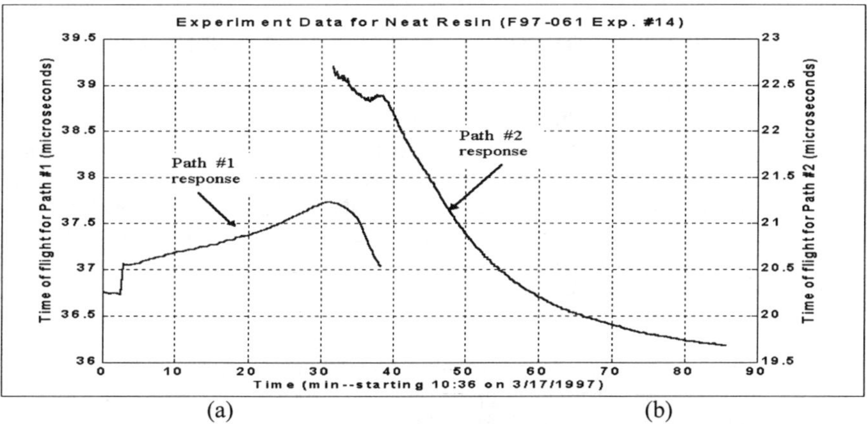

(a)   (b)

Figure 2. Ultrasonic responses of WWG sensor in a neat resin experiment
(a) Amplitude and (b) Time-of-flight.

propagated between the two parallel legs of the wire. These two paths provide the information that describes various stages of a cure process.

The typical ultrasonic amplitude responses from the neat resin experiment are shown in Figure 2a. In addition to the temperature curve obtained from a thermocouple, there are two curves showing the amplitude responses from Paths #1 and #2. Path #1 shows when the resin reached the sensor, the viscosity change and the gelation. Path #2 provides information on the resin hardening through the completion of the cure.

Before the resin was poured into the Teflon tray, ultrasonic energy propagates in the sensor wire. When the resin was poured in, some amount of energy leaked out of the wire,

939

causing a drop in the amplitude. As the curing process progressed, the resin viscosity decreased slightly. This is reflected by the small amount of increase in the transmitted energy in Figure 2a. Before significant gel occurred for the resin, most of the received energy was propagated through Path #1. As the gel progressed, the amount of energy propagated through the wire decreased rapidly. Meanwhile the resin began to harden and provided sufficient support for the ultrasonic energy to propagate via Path #2. Towards the end of cure, the resin reached its viscoelastic asymptote as indicated by the flattening of the amplitude response for Path #2.

Time-of-flight information was also obtained for the experiment and the results are presented in Figure 2b. The response for Path #2 shows an increase in velocity as the resin hardened. Note that the time-of-flight for path #1 reached maximum before the gelation was complete. This seems to be typical in all of the neat resin experiments we conducted in this project. In addition, since Path #2 has a shorter distance, its signal arrives earlier than that of Path #1. In fact, because of our choice of frequency, signals from the two paths can be time-gated to provide separate amplitude responses for the two paths. Typical waveforms for Paths #1 and #2 are shown in Figure 3.

3. WWG IN COMPOSITE MANUFACTURING PROCESSES

3.1 RTM Experiment Several experiments were carried out using the WWG sensor for the RTM process. The RTM mold used in these experiments was a 0.0125m thick 0.2m x 0.2m aluminum mold. After the layers of glass fabric were placed in the mold and sensors embedded in the middle layer, the mold was tighten by eight C clamps. Resin was then injected into the closed mold with thermocouple and sensor leads coming out from the side. The resin was injected from the bottom of the mold with vent ports on four sides as well as at the top. The resin used was Jeffco 1314A and Jeffco 3109B (4:1 ratio) which has a pot life of 15 minutes at the room temperature. Jeffco resin was selected because it was Shell 828 based and readily available.

Typically the injection was done with a cold or lukewarm mold. After the resin was injected, heat was applied to warm up the mold and speed up the cure process. Figure 4 shows typical amplitude and time-of-flight responses from a RTM experiment . The responses are similar to those shown in Figure 2 for the neat resin. The major differences occurred shortly after the resin reached the sensor. There was a quick increase in the amplitude for Path #1. We speculate this is due to the resin lifting the fabrics from the sensor wire. The rise, however, quickly stabilized at a level lower than before the resin reached the sensor. However, we can not explain why the time-of-flight reduced (velocity increased) when resin just reached the sensor at this time. In spite of this, it is clear that the sensor was able to detect when the resin arrived.

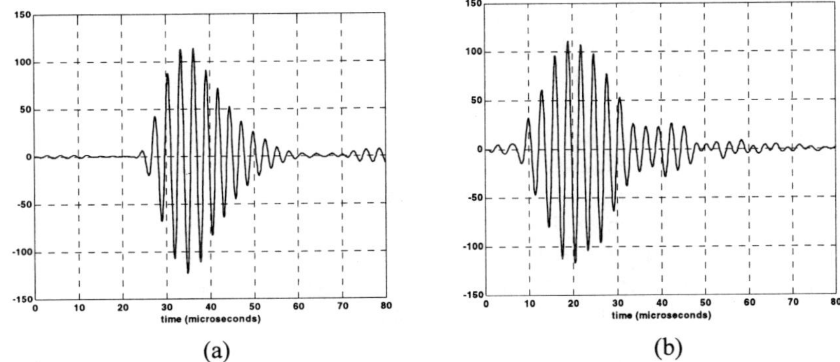

(a)   (b)

Figure 3. WWG waveforms (a) path #1 and (b) path #2.

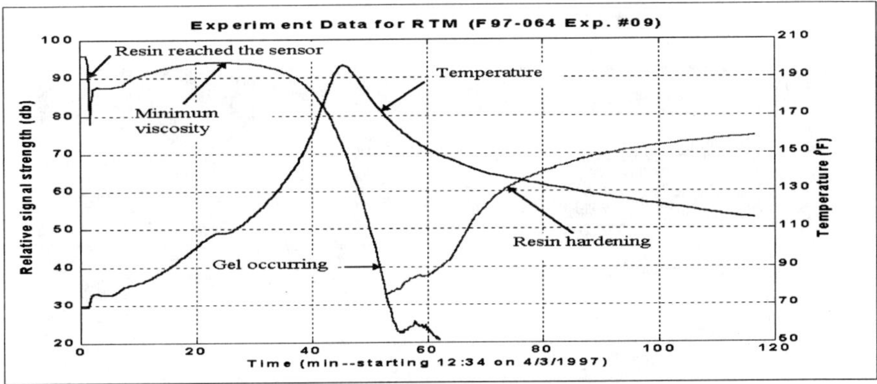

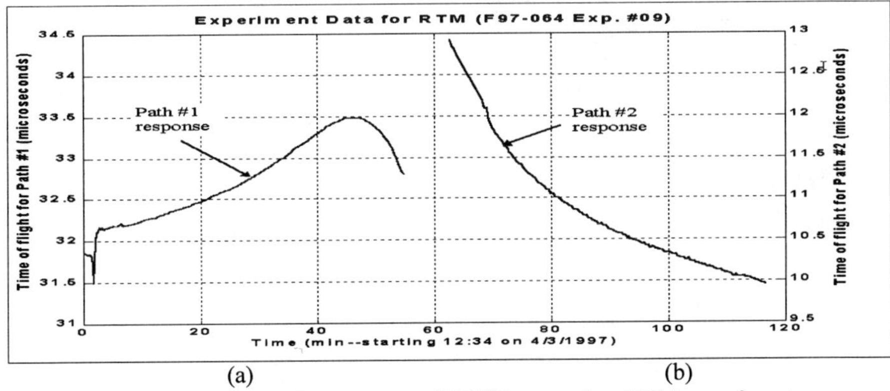

(a)   (b)

Figure 4. Ultrasonic responses of WWG sensor in a RTM experiment
(a) Amplitude and (b) Time-of-flight.

3.2 VARTM/SCRIMP Experiments-Small Mold   To evaluate the effectiveness and applicability of the sensor in the VARTM/ SCRIMP processes, several experiments were carried out using the lower half of the RTM mold. A slight modification was made to the mold such that the entry port for the resin was relocated to the side for convenience purpose. The vacuum was achieved with bagging materials and a vacuum fitting on the surface of the mold. In addition, a controlled heater blanket was placed under the base of the mold. Figure 5 is a photo of the experiment in progress. The resin used was NanYa 128 (Shell 828 based resin) with EAC100NC (Anhydride) plus EAC catalyst (accelerator/ promoter). The mixing ratio was 100:85:3. This resin system has a long pot life (will remain liquid at the room temperature) and low viscosity (~300 cps). This epoxy system was chosen mainly due to the suitability and availability.

Figure 6 shows typical responses obtained from the VATRM/SCRIMP process. Comparing to those in the RTM, the responses were very similar except minor additions of features due to the vacuum process as annotated on the figures.

Figure 5. Photo of a small mold VARTM/SCRIMP experiment.

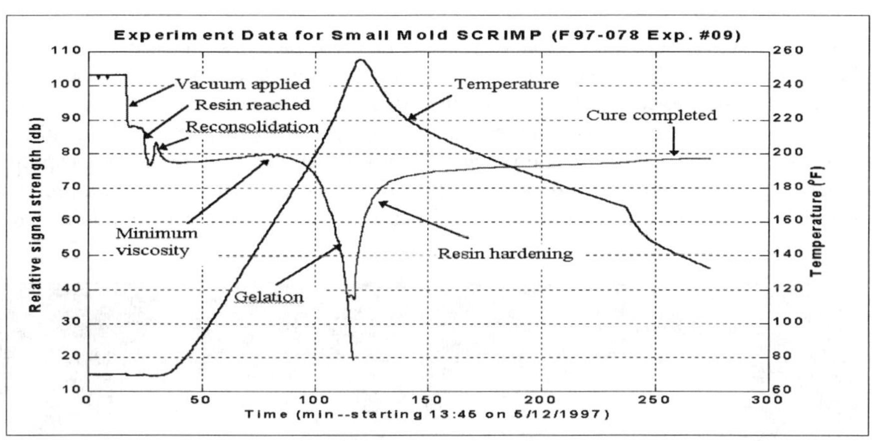

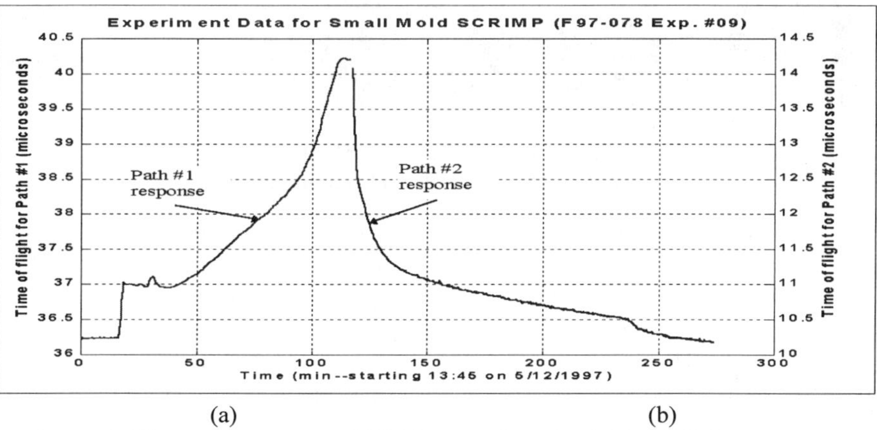

(a)   (b)

Figure 6. Ultrasonic responses of WWG sensor in a VATRM/SCRIMP experiment
(a) Amplitude and (b) Time-of-flight.

3.3 A Large Mold VARTM/SCRIMP Experiment   After the WWG sensor was developed and validated in the small mold SCRIMP, we proceeded to a demonstration in the large mold. The resin used for this was the same as that for the small mold SCRIMP experiment. There were two resin entry ports located on one side of the large mold. A silicone heater blanket was placed under the base of the mold. To speed up the resin infusion process, special resin distribution tubes were used on the edge of the mold. Sensors and thermocouples were installed in the mid-layer of each of the four sections. The mold was not preheated but the heater was turned on about the time the resin was infused. Figure 7 gives two photos showing the sensor placement and resin infusion process during an experiment.

943

Again, the WWG sensors worked very well. Figure 8 includes four sensor responses at four different thickness levels. The responses all have similar characteristics and resemble those in Figure 6 except that it appears the viscosity of the resin did not drop before gel occurs. This may be due to a number of factors, such as the heating of the mold before resin was fully infused or that the vacuum pump we used did not have enough capacity to produce quick second consolidation in the part, etc. In spite of this, we can see clearly when resin reached the sensors, when gel occurred, and when the cure was effectively complete.

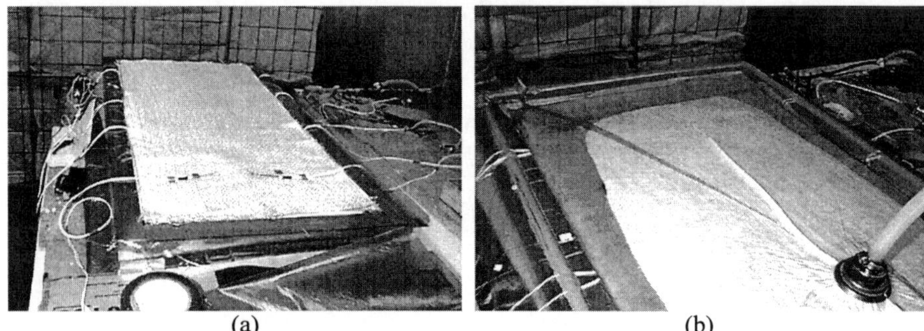

(a)   (b)

Figure 7. Photos of a large mold SCRIMP experiment
(a) Sensor placement and (b) resin infusion.

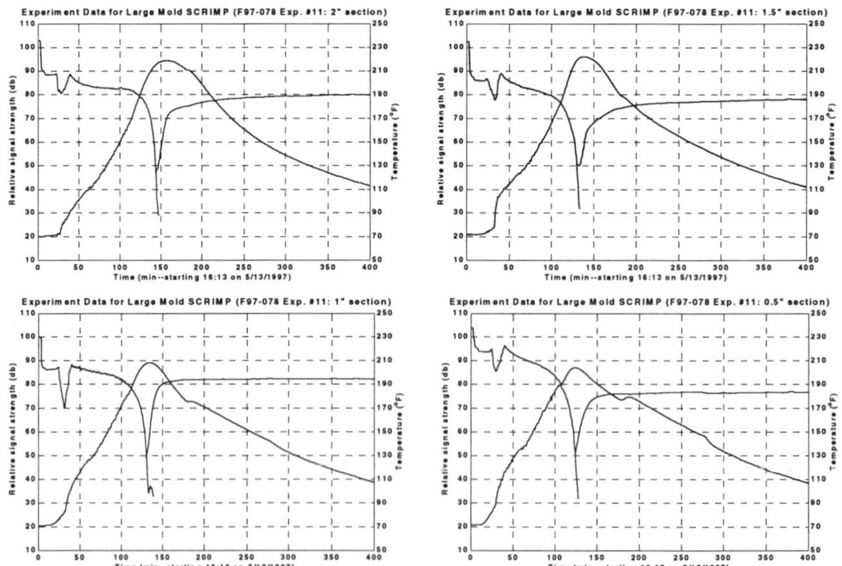

Figure 8. Ultrasonic responses of WWG sensor in a large mold SCRIMP experiment.

4. IMPACT DAMAGE DETECTION EXPERIMENTS

The objective of these experiments is to show that the WWG sensor for the resin flow/cure monitoring can also be used for the measurement of structure vibrational response. Composite parts fabricated from the small mold VARTM/SCRIMP experiments with embedded WWG sensors were used for the study. Two damage detection methods were used in this study. The first method used a film piezo-actuator as a source of vibration. The second method used one embedded WWG sensor as a transmitter and two other embedded WWG sensors as receivers. Both methods demonstrated the damage detection capability.

In both experiments, measurements were made before impact damage was made to the samples. The impact damage was obtained with a 1.8m - 13 kg impact load per SACMA SRM2 (or Boeing BSS7260) standard. The impact head had a diameter of 0.025m, resulting in visible damage. Pictures of impacted parts are shown in Figure 9.

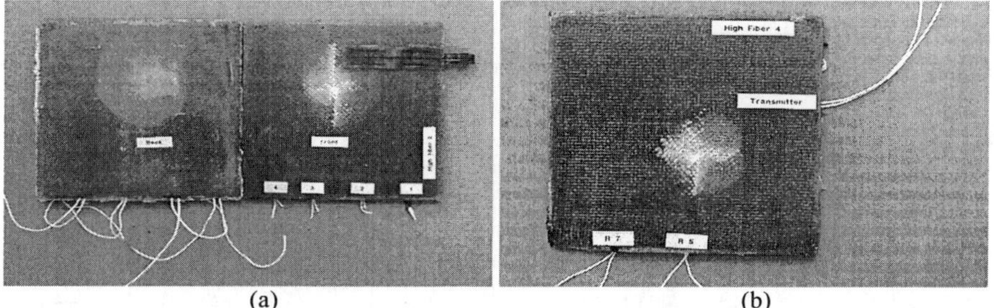

(a)   (b)

Figure 9. Photos of impacted composite samples.

In the vibrational damage detection method, a 0.0125m x 0.05m thin film piezo-actuator was attached to the sample with epoxy. It was driven directly by the HP33120A function generator in continuous wave mode through the frequency range of 0.5 to 200 kHz. Amplitude of resonance signals were measured with two embedded WWG sensors. Table I gives the matrix for the vibration measurements along with the corresponding deviations for a sample. The deviation calculation was made in frequency domain using:

$$deviation = \sqrt{\sum (A - A_0)^2} \bigg/ \sum |A_0| \tag{1}$$

where A and A_0 are the amplitude values for before and after impact. Figure 10 shows the vibrational responses for before and after the impact of the sample. The embedded sensors were able to pick up the differences in the composite resulted from the impact.

In the ultrasonic damage detection experiment, one sensor was used as the transmitter and another two sensors as receivers. We propagated ultrasonic waves in a composite part at 350 kHz. Ultrasonic measurements were made before and after subjecting the composite sample to the same impact level as described before. The impact was purposely done to the perceived wave propagation path. A photo of the part showing relative positions of the

945

sensors and impact location is in Figure 9b. Table II lists the matrix for the ultrasonic measurement and Figure 11 gives the received signals for this experiment. The deviation calculation was made in time domain using Eq. (1). As can be seen in Figure 11, impact introduced changes in the waveform shapes and arrival times of the signals, proving again that the embedded sensors can detect the damage in the composite part.

Table I. Vibration measurement-- % deviation from the first baseline measurement.

WWG sensor	Baseline	Baseline (repeat)	Impacted	Impacted (repeat)
#5	0	0.78	3.23	3.24
#3	0	0.21	3.53	3.55

Table II. Ultrasonic measurement-- % deviation from the first baseline measurement.

WWG sensor	Baseline	Baseline (repeat)	Impacted	Impacted (repeat)
#7	0	0.83	4.24	4.23
#5	0	0.91	6.67	6.30

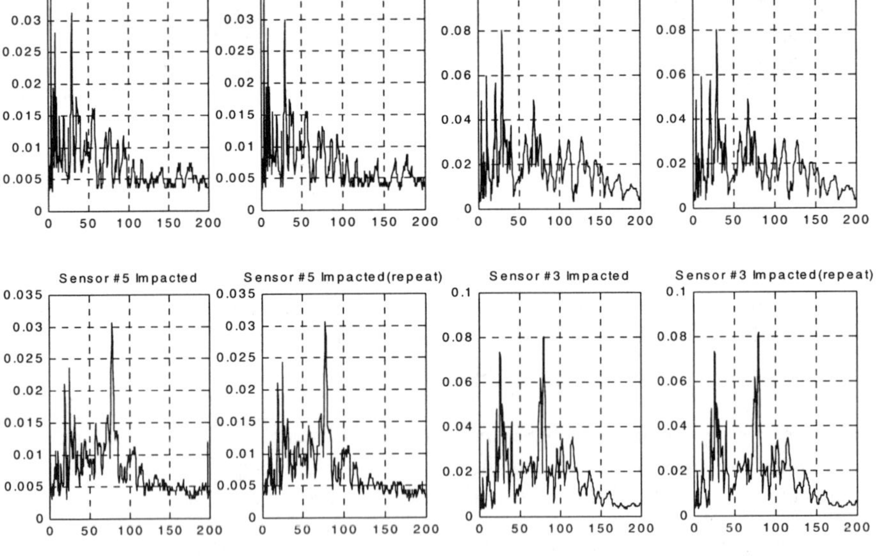

Figure 10. Frequency responses for damage detection – vibration experiment.

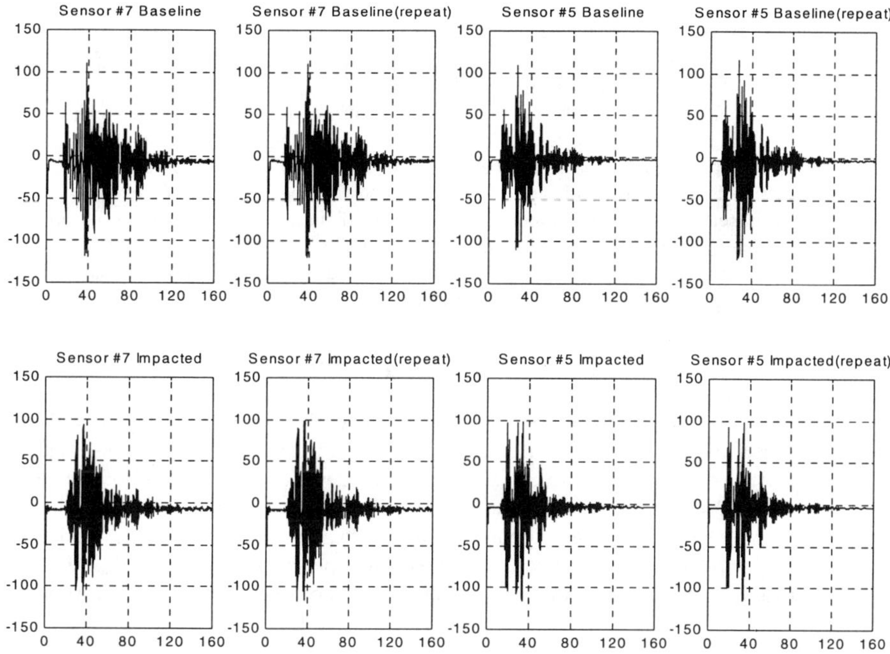

Figure 11. Waveforms for damage detection – ultrasonic experiment.

5. CONCLUSION

We have developed a novel embedded WWG sensor and measurement technology that can be used for flow/cure/damage measurement of composite parts manufactured by liquid molding techniques such as RTM and VARTM/SCRIMP. The measurement can be performed in real time for in-process flow/cure monitoring during fabrication and the same embedded sensor can also be used for damage detection and dynamic vibrational response monitoring after the parts or structures are placed in-service. Experimental results obtained in this project have clearly established the technical feasibility of this new low-cost, rugged, easy to implement ultrasonic WWG sensor technology.

6. ACKNOWLEDGEMENT

This work was supported by an SBIR program funded by the Department of the Army of the United States Department of Defense. Drs. Panagiotis Blanas and Bruce Fink were the program managers. The concept of the WWG sensor was originated from an Advanced Technology Program funded by the National Institute of Standards and Technology of the United States Department of Commerce. Dr. Carol Shutte was the program manager.

7. REFERENCES

1. R. T. Harold and R. Brynsvold, "Strain Measurements inside composite materials using embedded acoustic waveguides", Proceedings for nondestructive evaluation applied to process control of composite fabrication, Oct. 4-5, 1994.
2. Yan Li and Gerald J. Posakony, "Application of ultrasonic wire waveguide technology to cure monitoring of composite jackets for bridge retrofit", Second proceedings for nondestructive evaluation applied to process control of composite fabrication, Oct. 1-2, 1996.
3. Yan Li, "Development and application of an acoustic waveguide technology to in-process cure and in-service dynamic response monitoring of liquid molded composite armor smart structures", Final SBIR phase I report to Department of the Army, July 8, 1997.
4. Yan Li and Gerald J. Posakony, "Measurement of the cure state of a curable materials using an ultrasonic wire waveguide", patent filed, 1997.

8. BIOGRAPHY

Dr. Yan Li -- Principal Engineer. Dr. Yan Li received his Ph.D. degree with research excellence award in ultrasonics from the Department of Engineering Science and Mechanics at Iowa State University. He has been working in the field of nondestructive evaluation (NDE) and sensors since 1987. He has a strong theoretical background in acoustic guided wave propagation as well as practical experience in NDE. He has worked on many projects and published over 30 technical papers in applied NDE. He has made significant scientific contribution to the development of several prototypic instruments.

Dr. Suresh Menon--Senior Research Scientist, XXsys Technologies, Inc. received his Ph.D. in Engineering Science and Mechanics from the Pennsylvania State University. His areas of expertise are theoretical and experimental guided wave (Lamb, Rayleigh, etc.) propagation in different materials especially viscoelastic materials such as various polymers, fiber/polymer composites, metal matrix composites and concrete. His thesis was related to nondestructive evaluation of polymers using ultrasonics, acoustics, optics, pressure and temperature sensors. He has several publications and patents dealing with guided wave propagation in powder injection molded composites, anisotropic polymer composites, and also applications of other sensors for in situ monitoring and control.

Mr. Gerald J. Posakony, P.E., scientific advisor, XXsys Technologies, Inc., has spent the last 35 years in research and development of a first-of-a-kind nondestructive measurement instrumentation. He was manager of the Applied Physics Center of DOE's Pacific Northwest National Laboratory (PNNL) operated by Battelle Memorial Institute. At PNNL he was responsible for the technical and managerial activities of the Automation and Measurement Science, Computer Science and Energy Science Departments. Before going to PNNL, he was Vice President and General Manager of the Research Division of Automation Industries, Inc., in Boulder, CO. His personal research has been in the study of wave propagation and ultrasonic transducers. He is a Fellow in the ASNT, ASTM and AIUM and a member of ASM professional societies. He holds eight patents and is an author of many technical articles in NDE.

43rd International SAMPE Symposium
May 31-June 4, 1998

STRATEGIES FOR SELECTION OF SENSOR SYSTEMS FOR ON-LINE MONITORING AND CONTROL OF THE RESIN TRANSFER MOLDING PROCESS

Karl V. Steiner and Roderic C. Don
University of Delaware, Center for Composite Materials

Suresh G. Advani
University of Delaware, Mechanical Engineering Department

ABSTRACT

Over the last decade, the percentage of composite parts for DoD needs has risen drastically. In the aircraft industry, this figure is likely to increase to levels as high as 65 percent with large sections such as wing spans or fuselages manufactured as single composite components. However, complex manufacturing methods, long cycle times, high material costs and a relatively high number of defects have made this goal elusive. Meanwhile, Resin Transfer Molding (RTM) and the related liquid molding processes continue to be developed as successful technologies for manufacturing complex composite parts with large dimensions. Although process modeling and simulation has advanced rapidly, the manufacturing practice is still dependent on costly trial and error cycles.

A successful integration of on-line sensors is one of the key elements to reducing the cost of these complex structures. An integration of the sensor data with a parallel simulation of the process lays the foundation for interactive process control strategies, which will bridge the gap between manufacturing science and manufacturing practice.

During the RTM process, two steps in process are critical to ensure a defect-free composite part: first, void-free impregnation of the fiber preform in the mold with the injected resin, and second, complete cure of the infiltrated composite structure. Both of these process steps need a sensor system and the science for the prediction of preform permeabilities around edges and corners during impregnation of the resin but kinetic characterization is still in its infancy. The challenge is the implementation of such a system in a cost effective manner.

The paper describes the process-specific selection criteria applied for the investigation of on-line sensing methods for the RTM process. Although the investigation will be focused on RTM as a specific process, the resulting link with a software based sensor for quality analysis will have direct relevance to a variety of other manufacturing processes.

1. INTRODUCTION

The use of advanced materials—in particular, polymeric composites—continues to increase steadily for both military and commercial applications. Among the most promising composites processing techniques for high-quality and low-cost applications are liquid molding processes such as Resin Transfer Molding (RTM) and Vacuum-Assisted Resin Infusion (VARI). The two methods are similar in that they involve placement of a dry fiber preform in a mold, closure and sealing of the mold, injection of a thermosetting polymer resin, and finally curing and removal of the part. In either of these processes, a strong knowledge base regarding the flow of resin through fibrous preforms is required to ensure repeatable, proper filling of the mold and wet-out of the dry fiber preform. In addition, proper curing of the composite part is vital to avoid shrinkage and warpage due to the thermal stresses that may develop as the result of improper heating and cooling.

One of the first considerations during any process development cycle is dedicated to the selection of the most suitable process to fabricate the composite component at hand. RTM has proven to be a versatile and attractive process for high-volume, high-performance, low-cost manufacturing of polymer composites and it is gaining prominence in many areas of the manufacturing industry. In this process, a fibrous preform consisting of reinforcing material is placed in the mold. The mold is then closed, and the inserted preform is compacted depending on clamping pressure and the amount of fibers placed into the mold. In the next stage, a pre-polymer mixed with a catalyst is injected into the mold through gates, impregnating the fiber preform and filling the heated mold. Finally, the resin cures and the composite part can be removed from the mold for final processing steps such as trimming. The typical stages of the RTM process are shown in Figure 1.

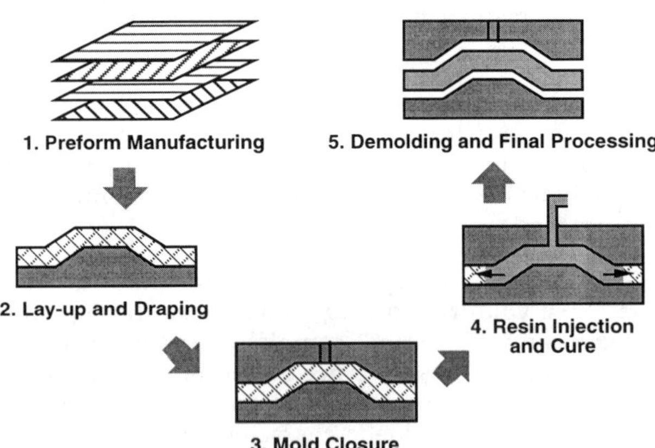

Fig. 1: Five stages of the Resin Transfer Molding process [1]. On-line sensing methods are inserted primarily during the resin injection and cure stage.

2. INTELLIGENT PROCESSING OF RTM

RTM owes its popularity to its near-net-shape manufacturing capabilities and the control the designer has over the orientation of the reinforcing fibers—and thus the ability to tailor the properties of the final part to meet the requirements for the desired application. In order for composite structures to reach their full potential, it is increasingly important to integrate intelligent process control strategies with suitable advanced sensors into an automated, intelligent processing facility to reduce the need for human intervention. To reach this goal, it will be necessary to obtain detailed knowledge about the composite materials properties, an understanding of flow and cure modeling for the RTM process, on-line sensor technologies and their utilization for composites-related applications, integrated process control strategies, and extensive fabrication and production know-how for composite components.

Process models are proving to be important in providing an understanding of the resin impregnation of the fiber preforms and the interrelationships between the fiber preform and resin percolation inside them. This science base forms the foundation for resin and fiber selection, tool design, preform design, and prototype development. However, since the prediction of preform permeabilities and the characterization of cure kinetics are usually between 10 and 15% of the experimentally obtained values and since high-performance parts have practically zero tolerance for defects such as entrapped voids or unsaturated regions, it is essential to monitor and control the manufacturing process on-line in addition to predicting the outcome beforehand [1]. This requires an understanding of not only how the tool design is inherently linked to the manufacturing process but also what critical variables should be monitored and which control strategy should be adopted when the actual process is not performing to expectations.

There are a number of parameters that can vary and influence the RTM process, and some of these critical factors are listed in Table 1.

Table 1: Process Development Factors for the Resin Transfer Molding Process.

RTM Elements	Process Development Factors
Mold	Mold Geometry, Location and Number of Gates and Vents
Preform	Fiber Orientation, Compaction, Permeability, Porosity, Sizing
Resin	Injection (ρ, V, T), Viscosity, Vent Vacuum, Sizing, Temperature, Cure Kinetics, Surface Tension, Conductivity, Heat Capacity
Operation	Injection Pressure, Flow Rate, Clamping Force, Heat Sources
Design Features	Metal Inserts, Foam Core Inserts, Reinforcing Ribs, Stringers
Environment	Resin/Preform Interaction, Resin/Mold Interaction

This allows for several possible combinations of parameters that can be selected and adjusted to simulate the process in a virtual environment. However, each of these simulations is computationally intensive. Therefore, it is not always clear to the user, which approach may lead to the optimal results, since all of these factors contribute to the manufacturing process development cycle for composites. In order to reduce these development costs the following approach has been chosen:

- Process models, which incorporate relevant process physics, must be utilized to optimize the process to meet design and manufacturing constraints.

- Virtual sensors and actuators should be implemented in the process simulations before the actual mold is fabricated and tested. Thus, the number of sensors can be reduced and their locations optimized to enable cost-effective manufacturing.

- A sensor/actuator-based control scheme should be developed to account for the inherent variability of the material and process parameters in the manufacturing process.

- To account for the part-to-part variability that is inevitable on the shop floor, an on-line controller and appropriate sensors should be implemented to enable real-time adjustments to the process during actual part fabrication.

This approach outlines the need for on-line sensors in an RTM environment to enable real-time process control to reduce design costs and improve production reliability and component properties.

3. ON-LINE SENSING ISSUES FOR THE RTM PROCESS

There have been numerous studies conducted and described in the literature that evaluate the use of on-line sensors for process control of composites manufacturing processes [2-9]. In most cases, these studies are focused on a limited set of sensors. In contrast, the goal of this study is to establish guidelines for on-line sensors and to initially consider a broad set of sensors. The selection of the appropriate sensor for any particular composites process is guided by some of the processing considerations described below:

- Is the information gathered by the sensors available in real time in order to be applicable for on-line control purposes? The control-critical time limit includes the times required for acquisition, presentation, and interpretation of the data to enable timely response by the control system.

- Are the sensors reusable if they are placed in the mold or will they have to be replaced for each part since they either remain in the composite part or may get damaged during the production cycle? In many cases, the data acquisition system maybe mounted remotely and low-cost sensors are placed inside or connected to the mold, while the data analysis is conducted externally.

- What is the cost of the sensors–both in initial capital investment and through-out an entire production cycle? For limited runs of specialty components, it may not be feasible to invest in a costly data acquisition system and one might focus on a set of comparatively inexpensive sensors that may have to be replaced with each part produced. However, for higher production runs, the initial investment for a reusable system may negligible when compared with the recurring costs for individual sensors that would have to be replaced with each part.

- Are the sensors suitable for a variety of resins and reinforcing fibers or are there inherent limitations in the selected sensors? In many production scenarios it is appropriate to focus on a specific resin and fiber system; however, the broad applicability of any sensor-based control system may be limited by, for instance, the need for either conductive or non-conductive fibers in the preform.

- Are the sensors adaptable to complex geometries and are they able to take into account the presence of inserted cores or provide information about the fiber orientation and compaction of the preform? Some sensors may not be suitable for complex three-dimensional parts due their physical limitations. Other sensors may not be able to penetrate the composite part in the mold cavity and may be limited to surface measurements only.

- Are the sensors accurate and able to survive harsh environments such as elevated temperatures or pressures outside the originally design envelope? Some sensors are sensitive to elevated temperatures and may be permanently damaged when exposed to excessive heat.

- Is the selected sensor system complex or user-friendly? In some cases, the sensors may be connected easily to the data acquisition system, while other approaches may are time-consuming and may even require a separate calibration procedure for each part in the production cycle.

- And finally, does the sensor provide any multi-functionality besides process monitoring during production? Some embedded sensors have been investigated for their ability to provide additional information for life-cycle monitoring. Such a sensor system would change the cost-benefit analysis for a particular sensor by providing additional and relevant information.

During the RTM process, many of the processing parameters are set, while others can be manipulated to change and optimize the process. Among the set parameters are the designed mold geometry, the type, orientation, and compaction of the fiber preforms, the resulting preform permeability, the resin characteristics, in particular its temperature-dependent viscosity, and the number and location of inlet gates and vents. To control the process, the injection pressure, flow rate, and resin temperature can be varied for each inlet gate. In addition, the temperature of the mold can be varied to change the viscosity of the resin during the filling stage and to selectively accelerate or decelerate the cure cycle.

It is important to select the type of process phenomena that needs to be measured to provide necessary information for process control. The most important parameter during the mold filling stage is the reliable detection of the flow front in the mold cavity–both in plane and through-the-thickness–especially for composite parts with thick sections. Additional sensors may be useful to determine the extent of wet-out and existence of micro voids between the fibers. Pressure and flow-rate sensors may be inserted at each of the inlet gates to monitor the injection of resin, and temperature sensors may be placed in the inlet gates and throughout the mold. Finally, cure sensors enable the in-situ monitoring of the extent of cure throughout the composite structure.

There are several options for the placement of on-line sensors for control during the RTM process, as outlined in Table 2. Sensors could be mounted either in the mold, become part of the composite part, or be used remotely such as in the case of laser-based ultrasound or infrared imaging techniques.

Table 2: Sensor placement options for on-line composites processing.

Sensor Placement	Advantages	Disadvantages
Mounted in Mold	Reusable, non-intrusive in part, production-oriented	may require costly mold modifications
Embedded in Part	provide information about internal state of composite could be used for life-cycle monitoring	expendable, remains in part, cost of sensor is issue sensor may adversely affect part performance
Non-contact	completely non-intrusive, sensing through mold walls	may be limited to controlled environment outside of mold

In addition, sensors can be differentiated based on their sensing mode as follows:

- Distal sensors that provide point measurements at given locations in the mold surface or mold cavity.

- Nodal sensors that provide a distributed sensing option by taking advantage of fibers and sensor wires placed in the mold cavity.

- Averaging sensors that are useful for obtaining averaged bulk properties of selected sections of the composite part.

Distal and nodal sensors have been implemented both in two- and three-dimensional modes, while averaging sensors are limited to one-dimensional representations of two-dimensional measurements.

An evaluation of on-line sensors applicable to the specific needs of the RTM process has led to a selection of the following choices. Table 3 shows a selection of the sensors divided into four categories: flow and wet-out, pressure, temperature, and degree of cure. No assessment of the sensor quality is included at this time since a detailed evaluation will be conducted in the coming months to down-select the most appropriate sensor for each category.

Table 3: Potential sensors to monitor and measure RTM-related effects.

Sensing Method	Flow/Wet-out	Pressure	Temperature	Degree of Cure
Fiber Optics (EFPI Sensor)		X	X	X
Fiber Optics (Bragg Sensor)			X	
Fiber Optics (Evanescent Wave)	X			X
Electromagnetic Conductance and Capacitance	X			X
Ultrasonic Sensors	X			X
Infrared Sensors			X	
Thermocouples, RTD			X	
Pressure Sensitive Distributed Grid	X	X		
Piezo Transducers in Mold Wall or at Inlets	X	X		
Wire Wave Guides	X			

5. SUMMARY

In order to reduce the cost of composites components made with the RTM process, an intelligent processing approach will be required that incorporates process models with on-line sensors and process control. A selection of sensing issues was discussed to aid in the selection of appropriate sensors for on-line process control.

A number of existing sensing methods have been selected that will be further evaluated in laboratory evaluations with several resin and fiber systems. The most promising sensing methods will be down-selected, based primarily on their technical capacity to provide the information required for real-time process control.

Before implementation of the final set of sensors as part of a control system, a cost-benefit analysis will have to be performed that will take into account the different requirements of lab-scale and industrial environments.

6. ACKNOWLEDGEMENTS

The authors gratefully acknowledge the support provided by the Office of Naval Research under Grant #N00014-97-C-0415 for the "Advanced Materials Intelligent Processing Center" at the University of Delaware.

7. REFERENCES

[1] Advani S., M.V. Bruschke and R. Parnas, "Resin Transfer Molding," *Flow and Rheology in Polymeric Composites Manufacturing*, Chapter 12, pp. 465-516, Edited by S. G. Advani, Elsevier Publishers, Amsterdam, 1994.

[2] Hunston, D., et. al, *Assessment of the State-of-the-Art for Process Monitoring Sensors for Polymer Composites*, US Department of Commerce, National Institute of Standards and Technology, NISTIR 4514, 1991.

[3] Steiner, K.V., J.B. Mehl, and M. Lamontia, "An Assessment of Three On-line Void Sensing Methods for In-Situ Manufacturing of Thermoplastic Composites," 29th ISATA International Symposium on Automotive Technology & Automation, Florence, Italy, 1996.

[4] Sincebaugh, P. and K. Meissner, "Intelligent Control of the Resin Transfer Molding Process," NDE Applied to Process Control of Composite Fabrication, NTIAC, St. Louis, MO, 1994.

[5] DuBois, C., S. Malghan, W.O. Ballata, S. Walsh, "Resin Flow Sensing: A Precursor to Successful Composite Control," Second Conference on NDE Applied to Process Control of Composite Fabrication, NTIAC, St. Louis, MO, 1996.

[6] Woerdeman, D.L., and R.S. Parnas, "Cure Monitoring with an Evanescent Wave Fluorescence Sensor," Proceedings of ANTEC '95, Society of Plastics Engineers, Boston, MA, pp. 2805-2811, 1995.

[7] Ganesh, C., et. al, "Predicting Degree-of-Cure of Epoxy Resins With Fiber Optic Sensors and Artificial Neural Networks" 39th International SAMPE Symposium, Anaheim, CA, 1994.

[8] Kranbuehl, D., D. Eichinger, and A. Williamson, "On-line In Situ Control of the Resin Transfer Molding Process," 35th International SAMPE Symposium, Anaheim, CA, 1990.

[9] Fink, B.K., S.M. Walsh, D.C. DeSchepper, and J.W. Gillespie, Jr., "Advances in Resin Transfer Molding Flow Monitoring Using SMARTWeave Sensors," Proceedings of the ASME Materials Division, Vol. 69, No. 2, pp. 999-1016, 1995.

43rd International SAMPE Symposium
May 31-June 4, 1998

FIBER PLACEMENT INSPECTION SYSTEM
AN EXPERIMENTAL APPROACH

W. Augustus Elliott and Reed Hannebaum, PE
Reed Hannebaum Engineering Services
10706 Trenton Ave., St. Louis, MO 63132

ABSTRACT

Detecting and gauging between pairs of edges on robot placed fiber tow gaps where the lack of image contrast makes traditional methods impractical requires a new machine vision approach. A pilot study performed by RHES with Boeing Aircraft Company under the Wright Laboratory program, "Design and Manufacturing of Low Cost Composites - Wing" (DMLCC-W) has addressed illumination and image enhancement, using a CDI light with a cylinder lens image averager. While much work remains in the development of the feature extraction algorithm, the study suggests a direction. The new algorithm will avoid the noise inherent in gradient-based edge detection by employing a two pass system. On the first pass, an imprecise low noise edge detector will find the edge's general neighborhood. Then, a gradient-based scheme will be applied only to the edge neighborhood to find the edge, thus culling the excess noise that could present itself as a false edge.

KEY WORDS: Automation, Machine Vision, Quality Assurance

1. INTRODUCTION

In an airframe component fiber placement application, thorough and constant inspection of many features (i.e. tow gap width) of the product is mandatory to ensure structural stability. However, manual inspection, especially of features like tow gap width, can be time-consuming and costly. If effectively applied, machine vision can drastically improve the inspection costs and manufacturing cycle time for fiber placed aircraft structures.

The design of a robust machine vision solution to this or any other inspection problem involves three (3) major facets:

1. Illumination
2. Image Conditioning
3. Feature Extraction

Each of these must be carefully evaluated to successfully design and implement any vision system. This is especially true with the design of a system for a low contrast inspection such as the gauging of the shadowed gap between two tows of black fiber material. A good illustration of this can be seen in a recent pilot study on the feasibility of automated inspection of the fiber placement process performed by RHES with Boeing Aircraft Company under the Wright Laboratory program, "Design and Manufacturing of Low Cost Composites - Wing" (DMLCC-W).

2. ILLUMINATION

As with any nontrivial vision application, the most important factor in this pilot study was product illumination to best present the feature of interest to the camera. The ideal lighting scheme would obviate the shadows in the gaps, leaving the fiber tows as brightly lit as possible, so as to allow a clear, accurate measurement. Of the standard illumination methods systematically tested on this product, two emerged particularly successful at heightening the contrast.

Low-angle Bright Field Illumination, in which the camera and illuminator are both directed toward a central point at identical angles (see Figure 1), was the first success with creating an acceptable lighting differential or contrast on the black-on-black product. However, the positioning precision and space required to implement this scheme on a robotic fiber placement arm present an array of unnecessary difficulties with mounting, calibration, and the image itself.

Figure 1: Bright Field Illumination

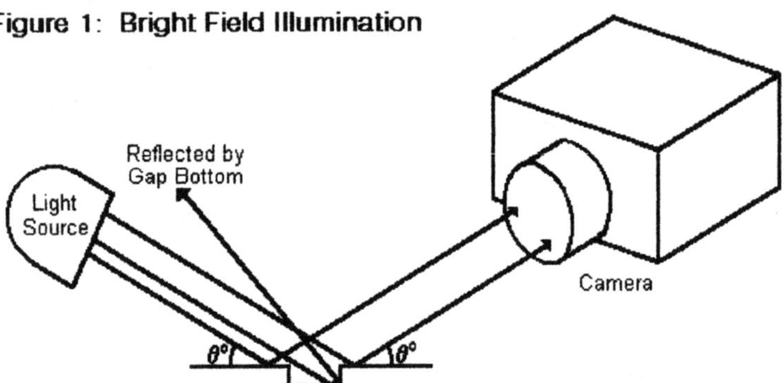

In practice, this method required the camera and light to be mounted along a path of reflection at precise angles with respect to the gap between two fiber tows to produce sufficient contrast in the resulting image. In this tilted position, the camera suffered from an optical distortion condition known as the keystone effect.

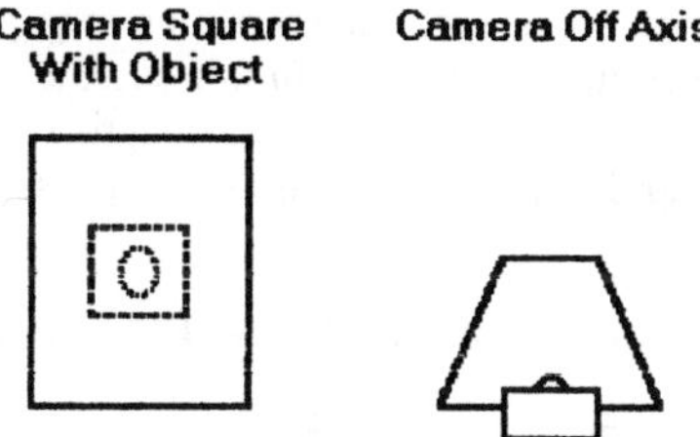

The Keystone Effect is characterized by a foreshortening of the image caused by a linear variation of magnification throughout the image (see Figure 2). An object at one end of the image can appear several times larger or smaller than an identical object at the opposite end. This effect will cause a perfect square to appear trapezoidal or "keystone" shaped. Compensation can be made for this effect, but this type of situation is generally best avoided when possible in gauging applications.

A Continuous Diffuse Illuminator (CDI) was a much better option. The CDI mounts directly to the front of the camera to provide a diffuse omnidirectional source of light (see Figure 3) which highlights the tows, leaving deep, high-contrast shadows in the gaps for measurement. Inherent in this method are the distinct advantages of compactness and lack of distortion, making it better suited to the application than other lighting solutions.

Figure 3: Continuous Diffuse Illuminator

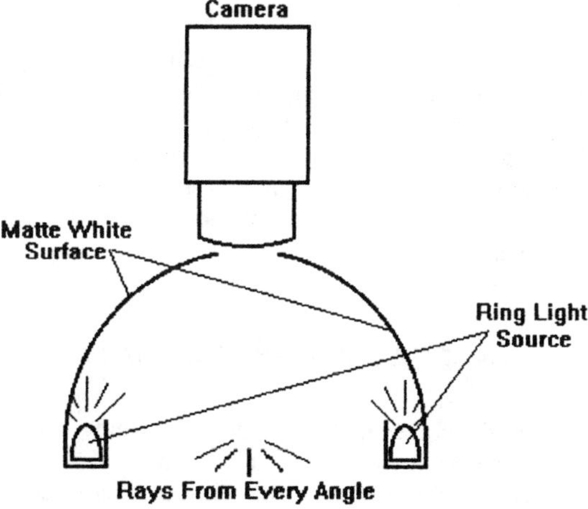

3. IMAGE ENHANCEMENT

Another way to improve conditions for feature extraction is to preprocess or enhance the image to improve the actual feature extraction. In this case, an averaging of the width of the gap in one dimension over its entire on screen length was desirable to lessen the effects of stray fiber strands and lint-like material at the edges of the gaps on the overall gauging process.

Figure 4: Image Averaging in One Dimension

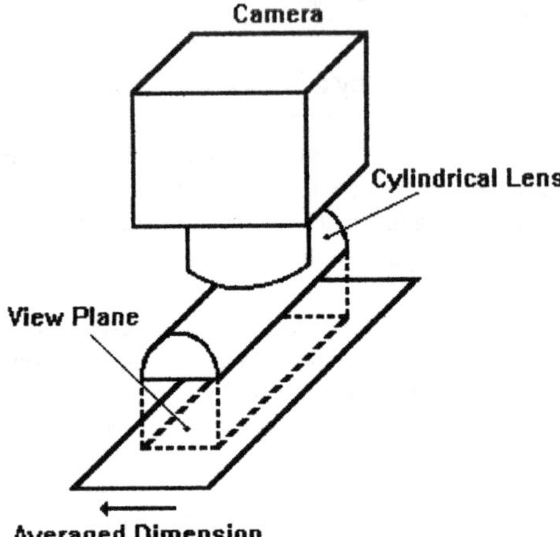

As morphological preprocessing tends to be time consuming on a sequential architecture, an optical implementation of an averaging algorithm proved an effective zero computational overhead alternative. A cylindrical lens (see Figure 4) was inserted into the optical path, blurring the image in the direction parallel to the tows and leaving it unaltered in the dimension of the measurement across the gap. This improved both the consistency and the sharpness of the gap edges sufficiently for the successful application of edge detection and gauging algorithms.

4. FEATURE EXTRACTION

The algorithm used for feature extraction in the pilot study was a one dimensional gradient based edge detection operator (1). It performed a parabolic interpolation to estimate a second derivative of a selected set of colinear pixel values. The advantage of this type of edge detection scheme is the high speed of execution. The disadvantage is the fact that the derivative not only locates the desired turning points of a pixel value curve corresponding to edges, but also exaggerates those sharp and jagged false turning points introduced by various sources of noise in the system.

Commercial vision processors currently utilize two classes of linear edge-type algorithms. One simply counts transitions between "light" and "dark" areas by counting cross points on a threshold line (1,2).   The other calculates the subpixel position of an edge within a transition by taking either a derivative or some other derivative-based operation of the function described by the pixel values along the line segment described by the linear edge detection gauge and finding spikes in the resulting function.   The sharper the spike, the more crisp and defined the edge.

Although the latter is better suited to this application, it is also more susceptible to false edge readings from the striations and random pixel noise.   Pairing and augmenting these two algorithms for the most repeatable measurement will be a fundamental focus of the next phase of this project.

5. FUTURE WORK

Pending funding, a more in-depth study is scheduled for second or third quarter of 1998. With most of the hardware questions now answered, the new focus will be on the development of the new feature extraction algorithm. Using what was learned from the pilot study, an illuminated camera appliance will designed and constructed for the capture of the experimental images.

The question of algorithmic repeatability will then be addressed. To date, stock feature extraction algorithms for caliper-style gauging are not sufficiently dependable with striations present; it is in the design of a new algorithm that the greatest concentration of research effort will be spent.

First, a gradient is computed for the graph of pixel values versus position. This creates a spike at every edge. A tall, sharp spike indicates a high-contrast, easily-discernible edge. Then, a threshold is drawn through the gradient curve, and a distance is measured between the leftmost and rightmost spikes crossing that threshold. If a striation in a tow causes a strong enough spike in the gradient curve to become the leftmost or rightmost crossing spike, the resulting distance measurement is not a valid representation of the gap's width (see Figure 5c).

Figure 5: A Width Gauge
(a) Image Window of Tow Gap

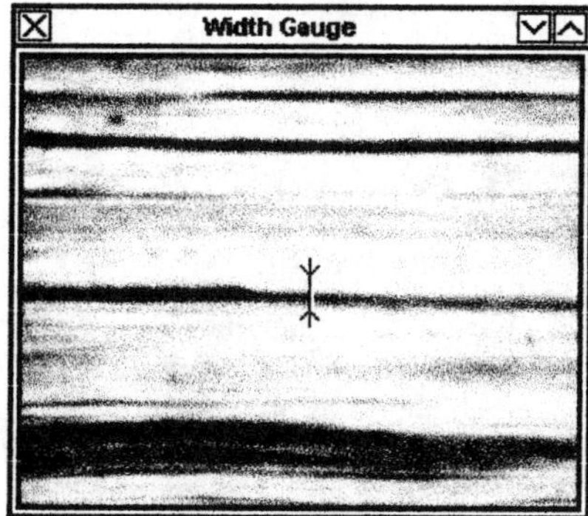

(b) Grayscale Values Along Gauge

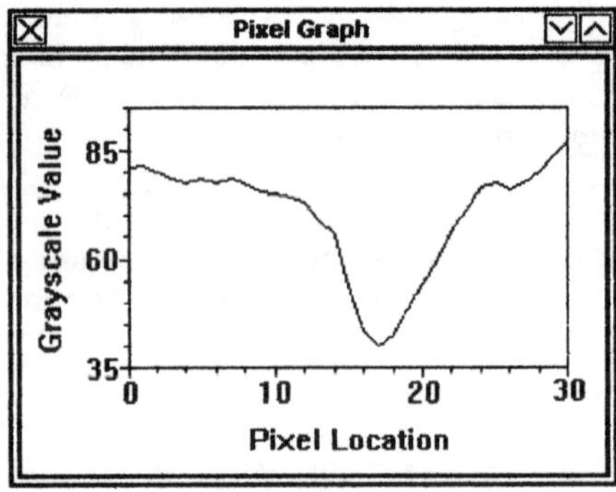

(c) Gradient Strengths Along Gauge

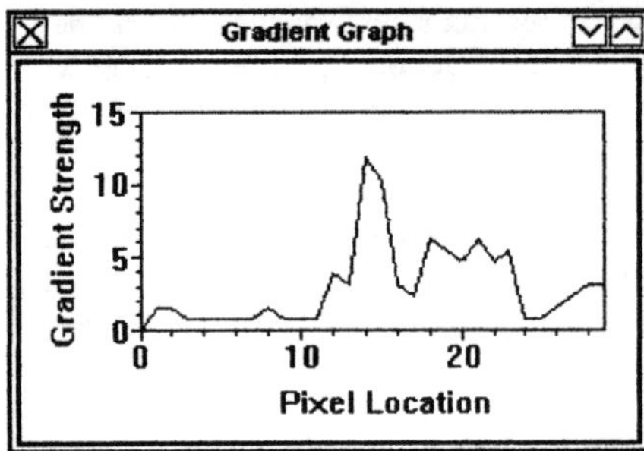

Faced with the limitations of the gradient or derivative based approximations coupled with the extremely high level of noise present in low contrast images, it is clear that a specialized approach is required.

Both gross and acute edge locating algorithms will be utilized in a two pass method. The first pass will employ a non-gradient based, lower accuracy edge detection algorithm (3) to define a narrow neighborhood around the edge. Then, the second pass will utilize a gradient based, high accuracy edge detector only inside the edge neighborhoods to find the true edge locations.

6. CONCLUSIONS

The automated visual inspection of the fiber placement process requires a specialized approach. Lighting, image enhancement, and the feature extraction process are all integral to this approach. It was found that a continuous diffuse illuminator with a cylindrical lens was the best combination of optical hardware to provide a clear image for the feature extraction algorithm. The algorithm will be the focus of the next stage of research, pairing an imprecise low noise edge detection method with a highly precise high noise gradient based algorithm to find the true edges.

7. REFERENCES

1. J. Russ, <u>The Image Processing Handbook</u>, CRC Press, Inc., London,1995, pp. 242-249.

2. E.R. Davies, <u>Machine Vision: Theory Algorithms Practicalities</u>, Academic Press, San Diego, 1997, pp. 103-130.

3. W. Franklin and C. Ray, "Higher isn't Necessarily Better: Visibility Algorithms and Experiments," <u>Advances in {GIS} Research: Sixth International Symposium on Spatial Data Handling</u>, Taylor and Francis Press, Edinburgh, 1994, pp. 751-770.

IN-SITU ULTRASONIC CURE MONITORING SENSORS

B. Boro Djordjevic, B. Milch
Center for Nondestructive Evaluation
Johns Hopkins University
Baltimore, MD 21218

ABSTRACT

Fiber reinforced organic matrix composites applied to large structures require inexpensive cure process control. The focus of this work is the development of simple ultrasonic NDE sensors suitable for manufacturing and in-field use. This sensor is developed around a short wave-guide ultrasonic probe that is embedded in composite for in-situ cure monitoring applications. The sensor responds to changes in resin density and sound velocity during cure and can be quantitatively calibrated for determination of the final cure. The sensor is based on acoustic impedance variance across a material interface and can be utilized over the full cure cycle change of the resin in the composite. Significant advantage of this method is the simplicity of the measurement, low cost of the wave-guide probe and the adaptability of the sensor configuration to various composite-processing environments

KEY WORDS: Composites, Cure, Ultrasonic, Sensors

1. INTRODUCTION

Fabrication of large composite structures with thermosetting polymers requires understanding of the overall cure cycle, including local cure rates and degree of cure completion. There are many analytical tools such as DSC, TMA, or FTIR that characterize cure process (1-2). However, there is no effective manufacturing method to in-situ monitor mechanic changes in the resin during cure process. During the cure, thermoset resin undergoes changes in viscosity, density, and ultrasonic velocity. Cured resin material becomes more rigid and starts to behave as solid with ability to support shear waves with increased stiffness and effective modulus. Ultrasonic waves have been demonstrated as effective way to measure the cure process (3-7). Most of the ultrasonic NDE methods are cumbersome, complex and require advanced signal interpretation. The guided stress wave sensor uses known ultrasonic material coupled into the organic resin and accurately measure signal reflection/transmission due to impedance change in the curing resin.

2. EXPERIMENTAL MEASUREMENTS

Ultrasonic measurements to determine cure process of variety of resin systems were performed using common polymers as wave-guide material. Typical materials are acrylic (PMMA), polystyrene, or polyamide. The selection of wave-guide materials is dependent on expected resin cure behavior, temperature requirements and desired sensitivity to the end cure condition. Figure 1 shows the measurement set-up for ultrasonic cure sensor (UCS) based on short wave-guide probe. Figure 2 illustrates the critical ultrasonic parameters governing signal interactions at probe-resin interface. Although this probe can use non-normal incident angle configurations, the work reported here is based only on simpler, normal incident ultrasonic wave condition. Reflection of the signal at the interface **R** is function of impedance difference across the interfaces. The value of **R** can range from 0 to1. By knowing impedance of the probe one can calculate the impedance of a resin from the level of the signal reflection. The UCS probe can be calibrated by reference to the signal from the known impedance liquid. Thus, the UCS signals can be interpreted quantitatively with the respect to the measured cure condition and the final state of the resin cure.

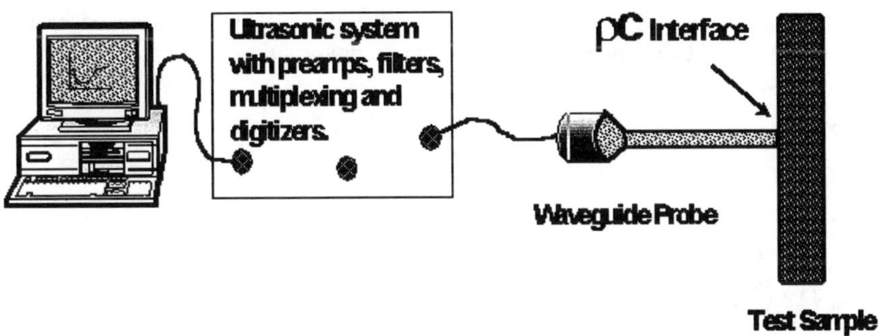

Figure 1. Short wave-guide ultrasonic cure sensor experimental set-up.

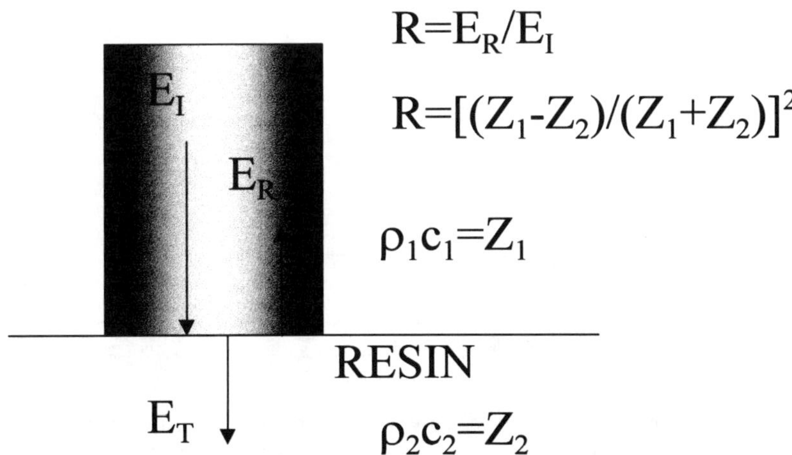

$$R=E_R/E_I$$

$$R=[(Z_1-Z_2)/(Z_1+Z_2)]^2$$

$$\rho_1 c_1 = Z_1$$

$$\rho_2 c_2 = Z_2$$

Figure 2. Summary of the basic principles governing the reflection and transmission of ultrasound used for UCS measurements. **R** reflection coefficient, **Z** is acoustical impedance, ρ is density and **c** is velocity of sound.

Figure 3 shows a typical signal response of the UCS probe during a cure cycle of the resin. The sensor shows all stages of thermoset cure process including gelation, cure, transition stages, and the final cure. The horizontal axis represent time and vertical axis show relative signal from the probe. One should note large ultrasonic signal change and the dynamic signal range plotted on the vertical axis that requires good instrumentation and low amplification noise. Figure 4 is temperature profile of the resin for the same cure test run. The increase in temperature during the exothermic cross-linking process is coincident with response of the UCS probe indicating rapid stiffening of the resin with corresponding increase in acoustical impedance of the material. The UCS probe response is controlled by resin mechanical properties and is related to the change in resin density and elastic constants. Such ultrasonic measurements directly sense mechanical condition of the resin and can predict mechanical behavior of the material. The resin cure process generates specific mechanical changes that are identifiable in the sensor response and can be monitored for process verification and control. The sensor requires one side access and does not depend on distance or time measurements employed by many other ultrasonic cure monitoring methods.

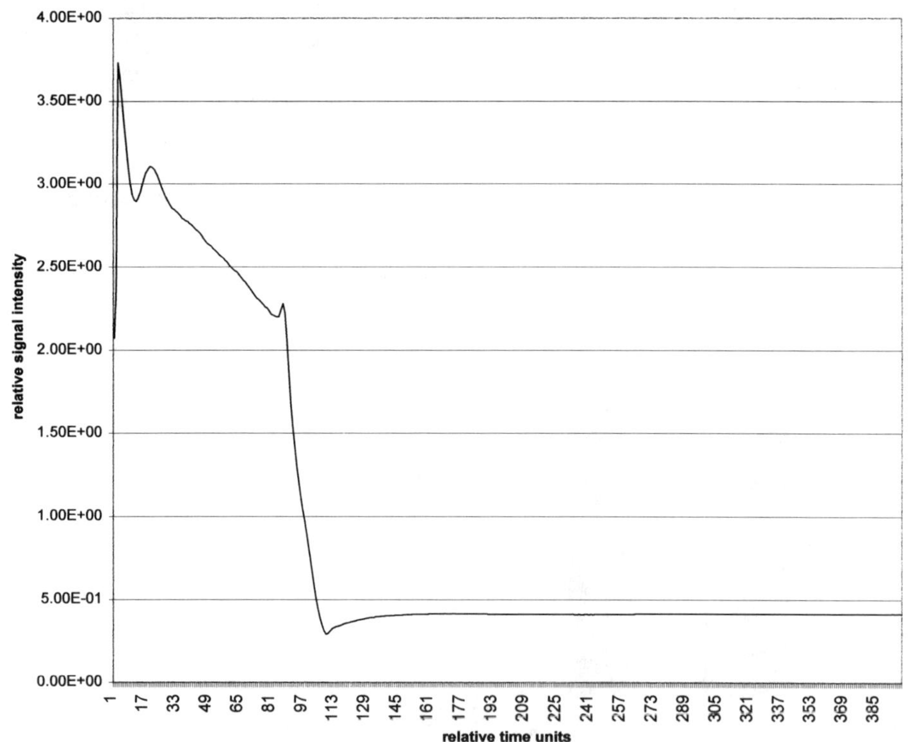

Figure 3. UCS signal response during cure of Polyester resin at room temperature and environment. Note large dynamic range; stable final cure level signal and significant signal changes during different cure stages.

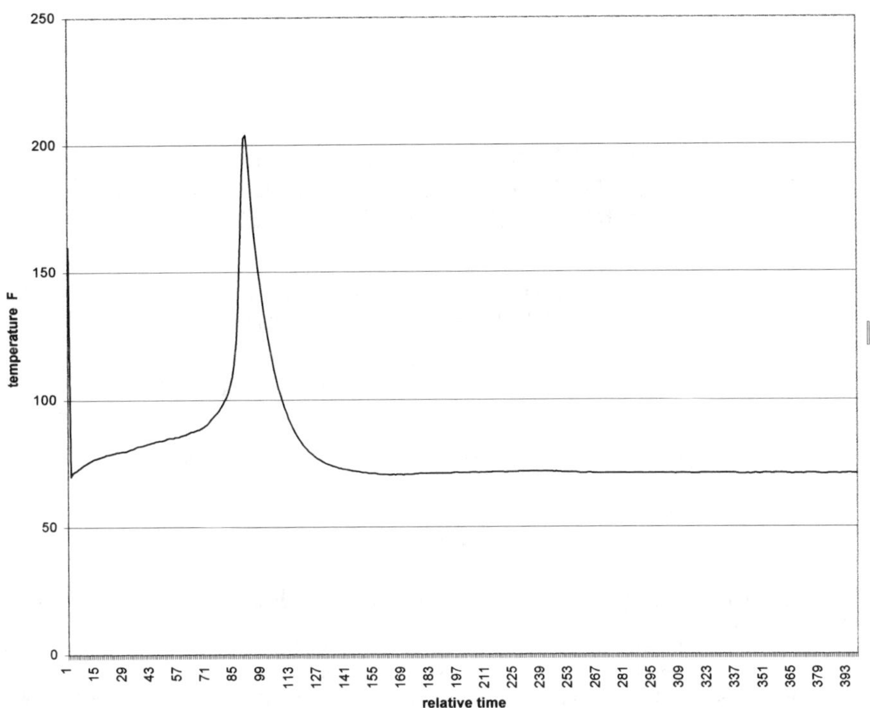

Figure 4. Temperature profile of the cure cycle measured with UCS probe shown in Figure 3.

3. CONCLUSIONS

The UCS probe has been demonstrated to be very sensitive to different stages of a cure cycle and appears to be effective in quantitative determination of the final cure. The benefits of this configuration are based on the simplicity of the probe, low cost of the wave-guide, ability to quantitatively calibrate the signals and ability to configure the probe to variety of resins and processes. The ultrasonic waves are very sensitive to the interface impedance changes and UCS probe is adapted to utilize and maximize this effect.

4. REFERENCES

1. B. J. Hunt, M. I. James, <u>Polymer Characterization</u>, Blackie Academic&Profess. Pub. 1993
2. John Mitchell, <u>Applied Polymer Analysis and Characterization</u>, Vol. I, II, Oxford University Press, 1992
3. R. A. Kline at.al, <u>Rew. Progress in Quantitative Nondestructive Evaluation</u>, Vol 13,Ed. D. O. Thompson, D. E. Chimenti, plenum Press, pp. 2229-2236, 1994
4. G. A. Sofer, E. A. Hauser, J. Polymer Science, <u>18</u>, (6), 611 (1952)
5. E. P. Papadakis, J. Applied Physics, <u>45</u>, (3), 1218 (1974)
6. A. M. Lindrose, Exp. Mechanics, <u>18</u>, (6), 227 (1978)
7. M.J. Ehrlich, et.al., <u>Nondestructive Characterization of Materials VI</u>, Plenum, , pp. 7-12. 1994

EVALUATION AND SELECTION OF REPLACEMENT THERMAL CONTROL MATERIALS FOR THE HUBBLE SPACE TELESCOPE

Jacqueline A. Townsend, Patricia A. Hansen, Mark W. McClendon,
NASA Goddard Space Flight Center, Greenbelt, Maryland, 20771 USA

Joyce A. Dever
NASA Lewis Research Center, Cleveland, Ohio 44135 USA

Jack. J. Triolo
Swales Aerospace, Beltsville, Maryland 20705 USA

ABSTRACT

The mechanical and optical properties of the metallized Teflon® FEP thermal control materials on the Hubble Space Telescope (HST) have degraded over the nearly seven years the telescope has been in orbit. Given the damage to the outer layer of the multi-layer insulation (MLI) that was apparent during the second servicing mission (SM2), the decision was made to replace the outer layer during subsequent servicing missions. A Failure Review Board was established to investigate the damage to the MLI and identify a replacement material. The replacement material had to meet the stringent thermal requirements of the spacecraft and maintain structural integrity for at least ten years.

Ten candidate materials were selected and exposed to ten-year HST-equivalent doses of simulated orbital environments. Samples of the candidates were exposed sequentially to low and high energy electrons and protons, atomic oxygen, x-ray radiation, ultraviolet radiation and thermal cycling. Following the exposures, the mechanical integrity and optical properties of the candidates were investigated using Optical Microscopy, Scanning Electron Microscopy (SEM), and a Laboratory Portable Spectroreflectometer (LPSR). Based on the results of these simulations and analyses, the FRB selected a replacement material and two alternates that showed the highest likelihood of providing the requisite thermal properties and surviving for ten years in orbit.

KEY WORDS: Low Earth Orbit (LEO) Simulated Environments, Teflon® FEP (fluorinated ethylene propylene), Thermal Control Materials

1. INTRODUCTION

The Hubble Space Telescope (HST) was launched into low Earth orbit (LEO) in April 1990 with Multi-Layer Insulation (MLI) blankets on the Light Shield (LS), Forward Shell (FS), and several Equipment Bays (1). The outer layer of these multi-layer blankets was aluminized Teflon® FEP (fluorinated ethylene propylene). Following the First Servicing Mission (SM1) in December 1993, analysis of retrieved MLI blankets revealed that the outer layer was beginning to degrade (3). When astronauts returned for the Second Servicing Mission (SM2) in February 1997, they discovered severe cracking in the outer layer of the MLI blankets on both solar facing and anti-solar facing surfaces of HST. A small specimen of the outer layer was retrieved from the LS region and returned for ground-based analysis (1, 2).

Testing of the MLI specimen that was returned during SM2 revealed that the cracks observed on HST were a form of slow crack growth, which meant that they occurred slowly, under low stress, in the presence of a degrading environmental factor (2, 5). The Teflon® FEP had completely lost plastic deformation capability, indicating that significant chain scission had occurred (2). The material also showed an increased density and crystallinity (2, 9). The solar absorptance of the Teflon® FEP had increased due to bulk changes in the Teflon® FEP and cracking in the vapor deposited aluminum (VDA) backing (2, 7, 11, 13). This damage appeared to be the result of the combination of bulk damage from radiation exposure (electrons and protons) and the nearly 40,000 thermal cycles (-100 to +50 °C) the MLI experienced.

Given the severity of the damage, HST management decided it was likely that repairs to the outer layer would be required during the next servicing mission (SM3) in December, 1999. A Failure Review Board (FRB) was tasked to recommend a replacement material to be deployed during SM3 on the LS that would last through the spacecraft end-of-life (EOL) in 2010. The recommended material was required to maintain structural integrity over the course of ten years and have an EOL solar absorptance over hemispherical emittance ratio (α/ε) of less than 0.28.

In order to find a replacement material, the FRB selected ten promising candidate materials and subjected them to simulations of the HST orbital environment. The effort involved facilities at Boeing Space Systems and three NASA centers: Marshall Space Flight Center (MSFC), Lewis Research Center (LeRC), and Goddard Space Flight Center (GSFC). Following these exposures, the specimens were evaluated at GSFC in terms of crack propagation and morphology and optical properties.

2. MATERIALS

2.1 Candidate Selection Process   To determine which materials should be tested, the FRB assembled a list of seventeen possible replacement materials and established a list of performance criteria. These nine criteria are listed below:

1. Low solar absorptance/thermal emittance ratio $\alpha/\varepsilon \leq 0.28$ at EOL
2. Ability to maintain structural integrity
3. Compatibility with EVA installation
4. Tear resistance
5. Not a source of contamination
6. Commercial availability for SM3 mission
7. Has demonstrated record of long term in-space durability in LEO
8. Suitable to construct a functional outer layer
9. Stowability

As part of the selection process, evaluation criteria with appropriate weighting factors were developed and used in a multiplicative evaluation process to assess each candidate material. The multiplicative evaluation process had been developed external to NASA but used extensively by NASA Lewis Research Center's Electro-Physics Branch for strategic planning and prioritization activities. Board members scored each candidate material according to how well they believed it would meet each of the performance criteria. In doing so, the damage to the current Teflon® FEP material was considered along with the issues specific to each of the various candidate replacement materials. Scores from each board member for each criteria were used in the multiplicative evaluation formula to calculate an overall score for each material. Based on this process, the original list of seventeen possible candidates was pared down to six candidate replacement materials which were then exposed to simulated space environments. Following the simulations, the actual performances of these six materials were evaluated with respect to these same criteria and a final selection was made.

2.2 Candidates   Ten candidate replacement materials were evaluated in simulated low Earth orbit environments. Through this work the numbers below were used to refer to the material.

1. 10 mil Teflon® FEP/VDS/Inconel/non-UV-darkening adhesive/Nomex® scrim
2. 5 mil Teflon® FEP/VDS/Inconel/adhesive/fiberglass scrim/adhesive/2 mil Kapton®
3. 10 mil Teflon® FEP/VDA/non-UV-darkening adhesive/Nomex® scrim
4. 5 mil Teflon® FEP/VDS/non-darkening adhesive/fiberglass scrim/adhesive/2 mil Kapton®
5. 5 mil Teflon® FEP/VDS/Inconel/non-UV-darkening adhesive/Nomex® scrim
6. 5 mil Teflon® FEP /VDA/non-UV-darkening adhesive/Nomex scrim
7. OCLI multi-layer oxide UV blocker/2 mil white Tedlar®
8. 5 mil Teflon® FEP/VDA (the current material)
9. $SiO_2/Al_2O_3/Ag/Al_2O_3$/4 mil stainless steel
0. Proprietary Teflon® FEP/AZ93 White Paint/Kapton®

The first six materials were chosen based on the selection process described in section 2.1. Given the stringent thermal requirements (EOL $\alpha/\varepsilon \leq .28$), the options for candidate replacement materials were limited. Metallized Teflon® FEP met those thermal requirements, and photos taken during SM2 revealed that bonded Teflon® FEP used on HST had maintained its structural integrity. Because of this, several versions of Teflon® FEP/VDA and /VDS (vapor deposited silver) bonded to a scrim were candidates.

Four additional materials were included in the testing for other reasons. The current material (material 8), was included to verify that the test procedure could produce damage similar to that observed in orbit. Material 7 was included because it was used on HST exterior surfaces in other applications, and HST management wanted to anticipate its performance. Materials 9 and 10 were included at the discretion of the FRB Chair. Since the materials chosen through the selection process were so similar, materials 9 and 10 were included so that fundamentally different materials were evaluated in the event that none of the first six was successful.

3. EXPERIMENTAL

Since there was no facility for simultaneous exposure to a LEO-equivalent environment, the specimens were exposed to several environmental factors sequentially. The order of the exposures was designed to cause the maximum damage. Since the cracks in the HST materials were a form of slow crack growth, it was necessary to provide both the environmental factor and low stress in each simulation. In orbit, the stress was most likely associated with the thermal cycling (2). Since the specimens could not be thermal cycled during exposures to other environmental factors, special holders were developed to hold the specimens at constant strain while they were exposed to electrons, protons and AO.

Four sets of the candidates were exposed to electrons and protons at one of two facilities: MSFC or Boeing Space Systems Radiation facility. Following the electron/proton exposure, two sets were exposed to atomic oxygen (AO), and then thermal cycled at GSFC. Two other sets were exposed to x-rays at LeRC and thermal cycling at either LeRC or GSFC. The fluence values for these exposures were based on the estimates of the HST environment found in reference 1 of this volume.

3.1 Specimen Preparation   Samples of the candidates were procured from several different vendors. Specimens with VDA were purchased from Dunmore and were backed with their proprietary, non-UV-darkening, polyester adhesive. Specimens with VDS were purchased from Sheldahl and were backed with their proprietary, non-UV-darkening, polyester adhesive. Material 7 was obtained from GSFC stock, material 8 was supplied by Lockheed Martin Missiles and Space from current stock, and material 9 was manufactured in the Thermal Branch at GSFC. Material 0 was provided by its manufacturer, AZTek. Specimen preparation was done in the Materials Engineering Branch and the Thermal Branch at GSFC.

The full sheet of each candidate was cured according to the manufacturer's specifications. Some were vacuum baked for up to 24 hours; others were received fully cured. Then each sheet was cleansed with an extracted clean-room wipe soaked in analytical-grade isopropyl alcohol. The sheets were then wrapped around a 0.5 cm diameter dowel along two axes to pre-stress the metal backing. Specimens were then cut in five different sizes to accommodate the test fixture at each exposure facility (see Table 1). A microtome blade was used to cut the individual specimens with identical orientation from a single sheet of each candidate material. A new blade was used for each material. Control specimens were cut at the same time and stored in a lab at GSFC. Witness specimens were also cut and traveled with the test specimens to each test site.

TABLE 1: SPECIMEN DIMENSIONS

Specimen Set	Materials	Dimensions (length x width, cm)
M1	Candidates	12.7 x 1.27
M2	Candidates	12.7 x 1.27
M3	5 mil Teflon® FEP/VDA	12.7 x 5.08
B1, B2	Candidates + extras	12.7 x 3.81
B3	5 mil Teflon® FEP/VDA	12.7 x 3.81
G1	Candidates + extras	5.08 x 5.08
L1	Candidates + extras	varied: 15.24 x 12.7 to 25.4 x 20.32

In order to provide a region of stress concentration, each specimen in sets M1, M2, M3, B1, B2 and B3 was sliced through one quarter of its width at a point 5.08 cm from its top. The slices were cut from the inside to the edge of the specimen using a microtome blade. The load used during the radiation exposures was calculated to provide 1000 psi in the net section of the specimen.

Once the specimens were cut they were photographed, and then the test and witness specimens were vacuum baked at 50 °C until the outgassing rate had dropped below the requirements for HST (1.56 x 10^{-9} g/cm^2/hr). Following bakeout, the solar absorptance was measured, and the specimens were hand-carried to the first exposure site.

3.2   Environmental Exposures   Sets of the candidates were exposed to several aspects of the space environment so that the combined effects of the environment could be assessed. In addition to the combined exposures, the effects of thermal cycling and ultraviolet radiation (UV) were evaluated individually. The exposures and set designations are summarized in Table 2. Sample sets were named according to the facility that performed the first exposure. Sets that began with "M" were first exposed at MSFC; "B" sets went to Boeing; "L" sets went to LeRC; and "G" sets remained at GSFC.

TABLE 2: CANDIDATE EXPOSURE SUMMARY

Exposure Set	First Exposure Location	Electron Exposures Duration (years)	Type	Energy (keV)	Proton Energy (keV)	AO (years)	X-ray (years)	Thermal Cycles #	Load	UV (ESH)
M1	MSFC	10	Dose	50 to 500	700	10	-	20,000	taped	-
M2	MSFC	10	Dose	50 to 500	700	-	10	3,200	taped	505
M3	MSFC	6.8	Dose	50 to 500	700	6.8	-	20,000	taped	-
B1	Boeing	10	Fluence	40	40	-	10	1,000	spring	-
B2	Boeing	10	Fluence	40	40	-	-	-	-	-
B3	Boeing	6.8	Fluence	40	40	-	-	-	-	-
L1	LeRC	-	-	-	-	-	-	>1500	mass	-
G1	GSFC	-	-	-	-	-	-	-	-	374
					MSFC		LeRC			GSFC

3.2.1   Combined Environmental Exposures   Six sets of specimens (M1, M2, M3, B1, B2, B3) were exposed sequentially to aspects of the space environment at several different facilities.

3.2.1.1 MSFC Exposure Facilities   Three environmental elements were simulated at MSFC: electrons, protons and atomic oxygen (AO). During each of these exposures, the specimens were mounted to induce the 1000 psi stress described in section 3.1. The procedure and results of this experiment were detailed in reference 12.

971

The electron and proton exposures were completed using their Combined Environmental Effects (CEE) test system. The MSFC staff calculated the dose versus depth profile for HST fluences for each candidate. They then designed a fluence of 50 keV, 220 keV and 500 keV electrons and 700 keV protons which matched that profile as closely as possible. During the CEE exposure, the specimens were under vacuum (5 x 10^{-7} Torr) and were subjected to the electrons of various energies simultaneously and then protons. For all specimens, the exposure times were less than one hour. Specimen sets M1 and M2 were exposed to ten-year HST doses of electrons and protons. Set M3 (the current HST material) was exposed to a SM2-equivalent dose to determine if the observed damage to HST could be duplicated with the test plan (12).

Following electron and proton exposures two sets (M1 and M3) were exposed to atomic oxygen (AO) in their Atomic Oxygen Beam Facility (AOBF). The fluences of the exposures were monitored with control specimens of Kapton® H and pristine Teflon® FEP. The flux was estimated based on measurements of the AO ion current neutralized by the system during a standard run. Set M1 was exposed to a ten-year equivalent HST fluence. Set M3 was exposed to a SM2-equivalent fluence (12).

3.2.1.2 Boeing Exposure Facilities   Three material sets were exposed to electron and proton fluences at Boeing Information, Space and Defense Systems, Radiation Effects Laboratory (14). Rather than matching the dose versus depth profile, the Boeing facility matched the total HST fluence of electrons and protons with 40 keV electrons and 40 keV protons. During the exposure the specimens were under load. Sets B1 and B2 were exposed to ten-year HST fluences. Set B3, the current HST material, was exposed to a SM2-equivalent fluence (14). Following this exposure, no further testing was performed on sets B2 and B3 so that they would be available if any tests or simulations were needed in the future.

3.2.1.3 LeRC Exposure Facilities   Two types of exposures were completed at LeRC: x-ray exposures and thermal cycling. The x-ray exposures were completed in a chamber equipped with an electron gun and an aluminum target. Flux was measured with a photodiode in a standard run, and the flux value was used to calculate the duration of test runs to achieve the desired fluence. Two sets (M2 and B1) were exposed to ten-year HST-equivalent fluences of Al Kα x-rays. Set B1 was then thermal cycled at LeRC.

The LeRC thermal cycling device was comprised of two thermal chambers dwelling at the two temperature limits, -100 and +50 °C. Specimens were held vertically and raised or lowered from one chamber to the other with a mechanical arm. The cycle time, roughly 5 minutes, was driven by the temperature of an exposed thermocouple. The specimens were spring loaded so that they were stressed throughout the cycle (11). Set B1 received 1000 thermal cycles in this chamber.

3.2.1.3 GSFC Exposure Facilities   Rapid thermal cycling and UV exposures were carried out at GSFC. Rapid thermal cycling between -100 °C and +60 °C took place in a modified thermal cycle chamber with a nitrogen purge. Liquid nitrogen vapor and a heat gun were added to the chamber to reduce the period of the cycles to 15 to 20 seconds. Temperatures were monitored with thermocouples taped around the test specimen, and the cycle was driven by a thermocouple affixed with epoxy to a control specimen mounted adjacent to the test specimen (13). Following electron, proton and AO exposures, sets M1 and M3 received 20,000 cycles at GSFC. Set M2 received 3,200 cycles following electron, proton and x-ray exposures.

The GSFC UV exposures were done using a Spectralab X-25 Solar Simulator equipped with a Xenon lamp. The radiation had a minimum wavelength of 180 nm. Following thermal cycling set M2 was exposed to 374 equivalent sun hours (ESH).

3.2.2   Individual Environmental Exposures   In addition to the combined effects, the effects of thermal cycling on larger specimens and UV exposure were evaluated.

3.2.2.1 Large Specimen Thermal Cycling   Since most of the candidate materials were layered with scrim, concerns were raised about the possibility that thermal cycling could result in permanent deformation of the materials due to mismatched coefficients of thermal expansion or creep properties. The specimens used in the combined effects exposures were too small to address these concerns. Considerably larger specimens (set L1) were thermal cycled at LeRC in order to assess the degree of deformation (10).

Specimens were subjected to at least 1500 cycles with a roughly 9 minute period in the thermal chamber described in section 3.2.1.3. Following the cycles the specimens were evaluated for fractional distortion. Fractional distortion was defined as d/h, where h was the height of the suspended specimen and d was the maximum displacement from true vertical (see Figure 1).

FIGURE 1: FRACTIONAL DISTORTION, d/h (10)

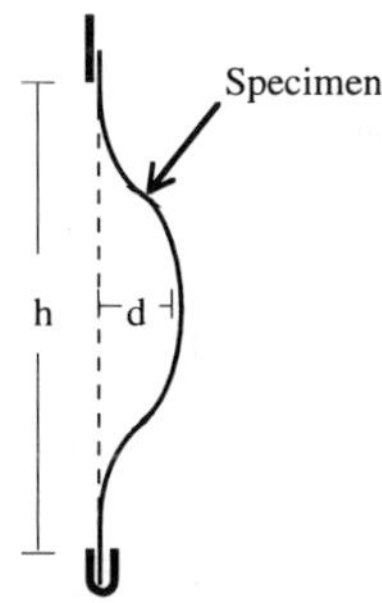

Following thermal cycling, the fractional distortion was calculated for each material. The values are reported in Table 3 (10). The LeRC investigators concluded that shape distortion was a major concern for materials 4 and 7 (10).

TABLE 3: FRACTIONAL DISTORTION (d/h) FROM THERMAL CYCLING (10)

Material #	Initial d/h	Final d/h	Δ	Comments
3	0.020	0.032	0.012	
4	0.008	0.073	0.065	Convex
6	0.020	0.063	0.043	Wavy
7	0.046	0.049	0.003	Concave
8	0.026	0.025	0.001	
9	0.012	0.012	0.000	

3.2.2.2 Ultraviolet Radiation Exposures In order to prove the UV stability of the proprietary adhesives provided by the vendors, the G1 specimens were exposed to ultraviolet radiation and then tested for changes in solar absorptance. The G1 set consisted of two specimens for each candidate material. The specimens were handled vigorously in order to break the metal backing in every specimen so that the UV could reach the adhesive. They were then exposed in the UV chamber (described in section 3.2.1.4) at the beginning of the test plan and remained there as long as possible before the final FRB meeting. Following the meeting, the selected candidate was placed back in the chamber, to determine the maximum value. The pre- and post-exposure solar absorptance values are recorded in Table 4.

4. RESULTS

General observations and absorptance values were recorded before and after each exposure. After the test plan was completed the specimens were sectioned for scanning electron microscope (SEM) analysis of the slice region and the surface.

4.1 General Observations Following the mounting procedure, there appeared to be some evidence of tensile overload at the end of the slice in some specimens. Following electron and proton exposures there were no obvious changes to the specimens, although solar absorptance measurements showed a slight increase (12). Following ten-year AO exposures, the specimens had a matte finish common in AO degradation of materials; this was detected in the solar absorptance measurements (12). No changes were noted after x-ray exposure. Most changes were observed following thermal cycling.

973

TABLE 4: SOLAR ABSORPTANCE PRE- AND POST-UV EXPOSURE

Material	Sample .#	Solar Absorptance After 374 ESH		
		Initial	Post-UV	Δ
10 mil FEP/VDS/Inconel/adhesive/Nomex	1	0.117, 0.094	0.107, 0.094	-0.010, 0
5 mil FEP/VDS/Inconel/adhesive/ fiberglass scrim/adhesive/Kapton	2	0.082, 0.081	0.080, 0.075	-0.002, -0.006
10 mil FEP/VDA/adhesive Nomex	3	0.164, 0.164	0.174, 0.175	0.010, 0.011
5 mil FEP/VDA/adhesive/fiberglass scrim/adhesive/Kapton	4	0.186, 0.187	0.189, 0.194	0.003, 0.007
5 mil FEP/VDS/Inconel/adhesive/Nomex	5	0.095, 0.086	0.083, 0.083	-0.008, -0.003
5 mil FEP/VDA/adhesive/Nomex	6	0.144, 0.137	0.142, 0.143	-0.002, 0.006
5 mil FEP/VDA	8	0.138, 0.135	0.153, 0.143	0.015, 0.008
SiO_2/Al_2O_3/Ag/Al_2O_3/4 mil stainless steel	9	0.074, 0.079	0.079, 0.082	0.005, 0.003
After 1144 ESH				
5 mil FEP/VDA/adhesive/Nomex	6		0.151, 0.155	0.007, 0.180
SiO_2/Al_2O_3/Ag/Al_2O_3/4 mil stainless steel	9		0.094, 0.088	0.020, 0.009

4.1.1   Thermal Cycling   Exposure set B1 (electron, proton, x-ray) was cycled while spring loaded at LeRC, and most of the specimens experienced crack growth. Two specimens (B1.2, B1.4) tore in two along the pre-exposure slice before the 1000 cycles were completed, and specimen B1.8 (the current HST material) had torn most of the way across the width by the end of the cycles. Specimens B1.1 and B1.5 had yellowed regions that seemed to be associated with the adhesive.

Exposure sets M1 and M3 experienced 20,000 thermal cycles at GSFC following electron, proton and AO exposure. The specimens appeared dramatically different following thermal cycling. Before cycling, the surface had a diffuse appearance but still seemed mostly transparant; there was no evidence of yellowing. After cycling the materials were milky and the surfaces were nearly opaque. It is believed that the thermal cycling opened micro-cracks at AO erosion trough sites. This appearance change was detected in the solar absorptance measurements. Some specimens also appeared yellowed at the edges; this seemed to be associated with the adhesive. In addition, most of the specimens exhibited some crack propagation.

Set M2 experienced 3,200 thermal cycles at GSFC following electron, proton and x-ray exposure. Several specimens exhibited crack growth. M2.4 delaminated at the interface between the FEP and the VDA, and the crack propagated most of the width of the specimen.

4.2   Solar Absorptance Measurements   Before and after each exposure and the total exposure, the solar absorptance was measured. In all cases except following x-ray exposure, the measurements were made with a Laboratory Portable Spectroreflectometer (LPSR). Following the x-ray exposure, the absorptance was measured using a UV-Vis-NIR Spectrophotometer (Perkin-Elmer, Lambda-9). The largest changes in solar absorptance occurred during thermal cycling of exposure set M1. Additional increases were noticed following UV exposure. The final solar absorptance values are reported in Table 5. Solar absorptance measurements were not taken prior to thermal cycling for sets M1, M2, M3. The change in solar absorptance recorded in Table 5 was calculated by subtracting the change from all the other exposures from the overall change. Final measurements were not taken for exposure set B1 before the specimens were sectioned, so the values were estimated by summing the change in solar absorptance from each exposure and the initial measurement.

TABLE 5: CHANGE IN SOLAR ABSORPTANCE ($\Delta\alpha$) FOLLOWING
ENVIRONMENTAL EXPOSURES

| Sample | Initial α | $\Delta\alpha$ Following Each Exposure | | | | | Post Test α | Overall $\Delta\alpha$ |
		Charged Particles	Atomic Oxygen	X-ray	Thermal Cycling	Near UV		
M1.1	0.092	-0.001	0.017		0.292*	0.039	0.439	0.347
M1.2	0.076	0	0.022		0.007*	0.068	0.173	0.097
M1.3	0.146	0.007	0.030	Not	0.158*	0.071	0.412	0.266
M1.4	0.167	0	0.040	Exposed	0.119*	-0.006	0.320	0.153
M1.5	0.080	0.004	0.025		0.251*	0.029	0.389	0.309
M1.6	0.138	0	0.033		0.083*	0.050	0.304	0.166
M1.8	0.139	0.007	0.024		0.067*	-0.010	0.227	0.088
M2.1	0.093	0		-0.002	0.115*		0.206	0.113
M2.2	0.079	-0.001		-0.003	0.013*		0.088	0.009
M2.3	0.161	0.004	Not	-0.002	0.049*	Not	0.212	0.051
M2.4	0.174	0	Exposed	-0.003	-0.003*	Exposed	0.168	-0.006
M2.5	0.081	0		-0.003	0.042*		0.093	0.012
M2.6	0.140	0		-0.002	0.011*		0.149	0.009
M2.8	0.133	0.002		-0.002	0.039*		0.172	0.039
M3.1	0.139	0.007	0.040	No Exp.	0.061*	.022	0.269	0.130
B1.1	0.087	0.004		0	0.063		0.150*	
B1.2	0.081	-0.002		-0.001	0.003		0.084*	
B1.3	0.152	0.002		0	-0.005		0.147*	
B1.4	0.178	-0.003	Not	0	0.002	Not	0.180*	
B1.5	0.081	0.002	Exposed	0.003	0.037	Exposed	0.118*	
B1.6	0.135	0.002		0	-0.007		0.128*	
B1.7	0.336	0.004		-	-0.001		0.335*	
B1.8	0.135	0.003		-0.004	0.013		0.185*	
B1.9	0.076	0.001		Not	Not			
B1.0	0.172	0.006		Measured	Measured			

* Values calculated rather than measured (see section 4.2)

4.3 Scanning Electron Microscopy (SEM)   When all of the exposures were completed, the specimens were sectioned, and SEM analysis was performed. The end of the slice region was analyzed to detect propagation and to study the morphology of the crack. Four basic types of fractures were found: tensile overload (TO), slow crack propagation 1 (SC1), slow crack propagation 2 (SC2), and combinations of TO and slow cracking (TO1 or TO2). The features and causes of these fractures are descibed below, and Table 6 summarizes the type of cracking observed by material and exposure set. Crack extent (length) is addressed in Section 4.3.5 and Table 7.

4.3.1 Tensile Overload (TO)   The TO crack surfaces showed plastic deformation with long fibrous tears and thin rippled layers (see Figure 4). This type of failure occurred when the load applied to the specimen, combined with stress concentration at the end of the slice, exceeded the material's ultimate strength while the materials was relatively ductile.

In specimens exposed to AO, the crack surface appeared fibrous. The fibers were oriented in the through-thickness direction and showed little plastic deformation. The AO surface damage appeared to serve as crack-initiating flaws, allowing tensile overload without plastic deformation. The cracks appeared to progress away from the initial notch; the actual crack front tended to move from the AO-exposed surface to the opposite surface.

Evidence of TO was found in specimens B1.1, B1.2, B1.3, B1.4, and B1.7 (see Table 6).

4.3.2 Slow Crack Propagation 1 (SC1)   The SC1 crack surfaces were very flat and perpendicular to the specimen surface. There was very little deformation at the specimen surface. The areas between striations were relatively smooth, and there was no evidence of plastic deformation (see Figure 2). A combination of stress from thermal contraction of the constrained specimen and possible change in material properties during low temperature cycles caused the crack to propagate a short distance. The high temperature portion of the cycle allowed the crack tip to close, therefore the low temperature excursions started with a relatively sharp crack tip. The fracture surface of these cracks most closely resembled those from retrieved HST materials (5).

Evidence of SC1 was found in specimens M1.2, M1.4, M2.3, M2.4, and M2.5 (see Table 6).

4.3.3   Slow Crack Propagation 2 (SC2)   The SC2 crack surfaces were wavy, with some deformation at the specimen surface in the direction perpendicular to the specimen surface. Ductile tearing was observed between and at the crests of wavy striations (see Figure 3). As with the SC1, the crack front progressed during thermal cycles. The tension on the specimen was sufficient to cause some plastic deformation. Because the specimen was always under tension, the crack tip did not close with each high temperature excursion. Therefore, low temperature cycles started with a blunted, stressed crack tip.

Evidence of SC2 was found only found in combination with TO (TO2, see Table 6).

4.3.4   Combination (TO1 or TO2)   The features of the crack were consistent with single tensile overload adjacent to the initial slit and then changed to either SC1 or SC2 described above. This crack occurred when the initial yielding (TO) reduced the stress concentration below that necessary for failure. The crack then progressed as SC1 or SC2 depending upon the conditions (see Figure 4).

Evidence of TO1 or TO2 was found in specimens M1.5, M2.2, M3.1, B1.5, B1.6, and B1.8 (see Table 6).

TABLE 6: CRACK FEATURES BY CANDIDATE AND SET

Material	Sample .#	Crack Type			
		M1 Set	M2 Set	M3 Set	B1 Set
10 mil FEP/VDS/Inconel/adhesive/Nomex	1	None	None		TO
5 mil FEP/VDS/Inconel/adhesive/ fiberglass scrim/adhesive/Kapton	2	SC1	TO2		TO
10 mil FEP/VDA/adhesive Nomex	3	None	SC1		TO
5 mil FEP/VDA/adhesive/fiberglass scrim/adhesive/Kapton	4	SC1	SC1		TO
5 mil FEP/VDS/Inconel/adhesive/Nomex	5	TO1	SC1		TO1
5 mil FEP/VDA/adhesive/Nomex	6	None	None		TO2
OCLI/White Tedlar	7	-	-		TO
5 mil FEP/VDA	8	TO	None	TO1	TO2
SiO2/Al2O3/Ag/Al2O3/Stainless	9	-	-		None
AZ93 White/Kapton	0	-	-		None

4.3.5   Crack Extent   SEM images of the specimens were used to determine how far the cracks propagated from the edge of the pre-exposure slit. Since many of the candidates were layered with different types of scrim, crack length alone was not an effective illustration of the material performance. Table 7 contains descriptive data about the crack propagation in each specimen and each exposure set.

All of the specimens in the B1 exposure set experienced tensile overload. This was most likely caused by the nominal 1000 psi tensile stress during the thermal cycling. Materials in exposure set M2 most frequently experienced slow crack growth similar to the retrieved HST materials.

Specimens with fiberglass scrim (materials 2, 4) experienced the worst cracking. In all test sets these materials showed crack propagation, often accompanied by delamination between the FEP and the VDA or VDS. In the overly-rigorous load conditions of the B1 exposure set, these materials failed completely.

In specimens with Nomex scrim (materials 1, 3, 5, 6) the crack propagation past the pre-exposure slit stopped before the first or second scrim fiber. Often there was evidence of minor delamination between the FEP and the VDA or VDS.

TABLE 7: CRACK FEATURES BY MATERIAL AND EXPOSURE SET

Material Number	Crack Features					
	M1 Set		M2 Set		B1 Set	
	Type	Extent	Type	Extent	Type	Extent
1	None	-	None	-	TO	to next fiber
2	SC1	delam; long	TO2	short, no delam	TO	tore in two
3	None	-	SC1	short of next fiber	TO	to 2nd fiber
4	SC1	long, no delam	SC1	delam, very long	TO	tore in two
5	TO1	delam; to next fiber	SC1	to 2nd fiber	TO1	to next fiber, delam
6	None	-	None	-	TO2	delam, to next fiber
7	-	-	-	-	TO	long micro crack
8	TO	short	None	-	TO2	mixed, 3/4 of width
9	-	-	-	-	None	-
0	-	-	-	-	None	-

5.   SELECTION

The FRB used the candidate selection process described in section 2.1 to rank the candidate materials following the environmental exposures. The materials were evaluated based on their performance with respect to each of the nine factors. Two materials, Proprietary Teflon®/AZ93 White Paint/Kapton® (material 0) and $SiO_2/Al_2O_3/Ag/Al_2O_3$/4 mil stainless steel (material 9) were not considered in this final evaluation. Material 0 was eliminated prior to voting because of problems with particulate contamination and UV darkening. The composite coating on stainless steel (material 9) was not practical for the LS repairs in terms of cost or handling. The remaining candidate replacement materials were ranked as in Table 8.

TABLE 8: FINAL RANKING OF CANDIDATE MATERIALS

Rank	Material Number	Material
1	6	5 mil Teflon® FEP /VDA/adhesive/Nomex scrim
2	3	10 mil Teflon® FEP/VDA/adhesive/Nomex® scrim
3	8	5 mil Teflon® FEP/VDA (the current material)
4	1	10 mil Teflon® FEP/VDS/Inconel/adhesive/Nomex® scrim
5	5	5 mil Teflon® FEP/VDS/Inconel/adhesive/Nomex® scrim
6	2	5 mil Teflon® FEP/VDS/Inconel/adhesive/fiberglass scrim/adhesive/2 mil Kapton®
7	7	OCLI multi-layer oxide UV blocker/2 mil white Tedlar®
8	4	5 mil Teflon® FEP/VDA/adhesive/fiberglass scrim/adhesive/2 mil Kapton®

Material 6, (5 mil Teflon® FEP /VDA/non-UV-darkening adhesive/Nomex scrim), was ranked first and recommended as the replacement material for the new outer layer. However, there was some concern that the absorptance value would increase significantly with more UV exposure. Specimen M2.6 was placed back in the UV chamber at GSFC for additional exposure and absorptance measurements. In the event that the recommended material failed this final test, two alternates that had not been included in this test plan were suggested.

The first alternate was $SiO_2/Al_2O_3/Ag/Al_2O_3$/Kapton®. This material had excellent thermal properties, and the coating proved durable in the electron and proton exposures and the large specimen thermal cycling. The second alternate was unsupported 10 mil Teflon® FEP. This material had the advantage of being commercially available and relatively inexpensive. Also, the 10 mil candidates seemed to perform better in crack resistance than the 5 mil candidates. The project management for HST will make the final selection based on program requirements.

6. CONCLUSION

Based on the nine performance criteria established by the FRB at the beginning of the exposures, material 6 (5 mil Teflon® FEP/VDA/non-UV-darkening adhesive/Nomex scrim) was recommended as the new outer layer for the MLI on the HST LS. The two most important factors in this selection were the optical properties and the structural integrity following the simulated space exposures. Although the absorptance of the selected material did not meet the EOL requirements, it was the best performer among the specimens that maintained structural integrity. Since limited UV exposure was possible during the test plan, UV exposure continues so that the maximum absorptance of this material can be determined before the final decision is made.

Fracture surfaces that resembled those of retrieved HST specimens were observed on several specimens in the M2 exposure set and on two specimens in the M1 exposure set. The highest increase in solar absorptance occured in expsure set M2. Material 1 (10 mil Teflon® FEP/VDS/Inconel/adhesive/Nomex® scrim) had the highest increase in solar absorptance in each exposure set.

7. ACKNOWLEDGMENTS

The staff at the environmental exposure facilities were Dave Edwards (MSFC), Jason Vaughn (MSFC), Dennis Russell (Boeing), Larry Fogdall (Boeing), Kim de Groh (LeRC), Bruce Banks (LeRC) and Wanda Peters (Swales Aerospace). Sample preparation was done by Mary Ayres-Treusdell (GSFC), Tom Zuby (Unisys), and Dave Hughes (Swales Aerospace). Fluence calculations for HST were perfomed by Teri Gregory (Lockheed Martin Technical Operations) and Janet Barth (GSFC).

8. REFERENCES

1) P.A. Hansen, J.A. Townsend, Y. Yoshikawa, J.D. Castro, J.J. Triolo, and W.C. Peters, "Degradation of Hubble Space Telescope Metallized Teflon® FEP Thermal Control Materials", <u>Science of Advanced Materials and Process Engineering Series</u>, <u>43</u>, 000 (1998).

2) J. A. Townsend, P.A. Hansen, J.A. Dever, J.J. Triolo, "Analysis of Retrieved Hubble Space Telescope Thermal Control Materials", <u>Science of Advanced Materials and Process Engineering Series</u>, <u>43</u>, 000 (1998).

3) T. Zuby, K. DeGroh, and D. Smith; "Degradation of FEP Thermal Control Materials Returned from the Hubble Space Telescope," NASA Technical Memorandum 104627, December 1995.

4) A. Milintchouk, M. Van Eesbeek, F. Levadou and T. Harper, "Influence of X-ray Solar Flare Radiation on Degradation of Teflon® in Space" <u>Journal of Spacecraft and Rockets,</u> <u>Vol. 34</u>, No. 4, July-August, 1997.

5) L. Wang, J.A. Townsend, and M. Viens, "Fractography of MLI Teflon® FEP from the HST Second Servicing Mission", <u>Science of Advanced Materials and Process Engineering Series</u>, <u>43</u>, 000 (1998).

6) J.A. Dever, J. A. Townsend, J.R. Gaier and A.I. Jalics, "Synchrotron VUV and Soft X-Ray Radiation Effects on aluminized Teflon® FEP", <u>Science of Advanced Materials and Process Engineering Series</u>, <u>43</u>, 000 (1998).

7) C. He and J.A. Townsend, "Solar Absorptance of the Teflon® FEP Samples Returned from the HST Servicing Missions", <u>Science of Advanced Materials and Process Engineering Series</u>, <u>43</u>, 000 (1998).

8) J.A. Dever, K.K. deGroh, J.A. Townsend, L.L. Wang, "Mechanical Properties Degradation of Teflon® FEP Returned from the Hubble Space Telescope", NASA/TM-1998-206618

9)   J. Dever, "Summary of HST MLI Materials Analysis Results", NASA Internal Memorandum, June 23, 1997.

10)   B. Banks, "Report for 'Potato Chip' Testing of the Candidate HST MLI Replacement Materials", NASA Internal Memorandum, November 21, 1997.

11)   B. Banks, T. Stueber, S. Rutledge, D. Jaworske, W. Peters, "Thermal Cycling-Caused Degradation of Hubble Space Telescope Aluminized FEP Thermal Insulation," AIAA 98-0896, Presented at the 36th Aerospace Sciences Meeting & Exhibit, Reno NV, January 12-15, 1998.

12)   D.L. Edwards, J.A. Vaughn, *Report on Charged Particle and Atomic Oxygen Exposure of Candidate Replacement Materials for the Hubble Space Telescope (HST)*, October 15, 1997.

13)   C. Powers, "Completion of Thermal Cycle Life Tests of HST MLI Samples," NASA Internal Memorandum, January 8, 1998

14)   D. Russell, *Test Report "Radiation Exposure Testing -- HST Thermal Blanket,"* November, 1997

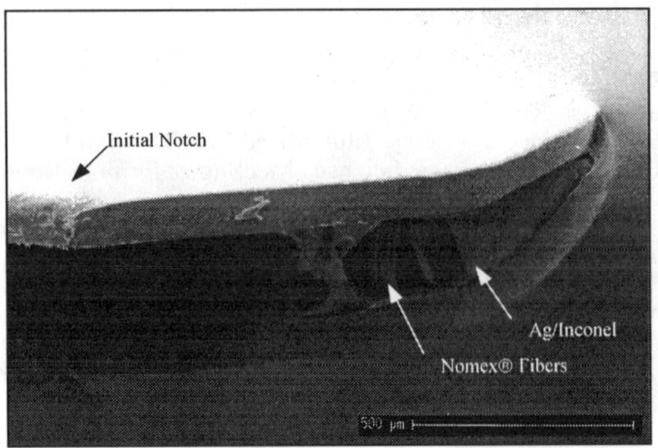

Figure 2a. Specimen M2.5, Entire crack length, including initial razor cut at the far left. Nomex® fibers are revealed behind the crack opening in the center. Silver/Inconel layer is visible at the crack tip to the right.

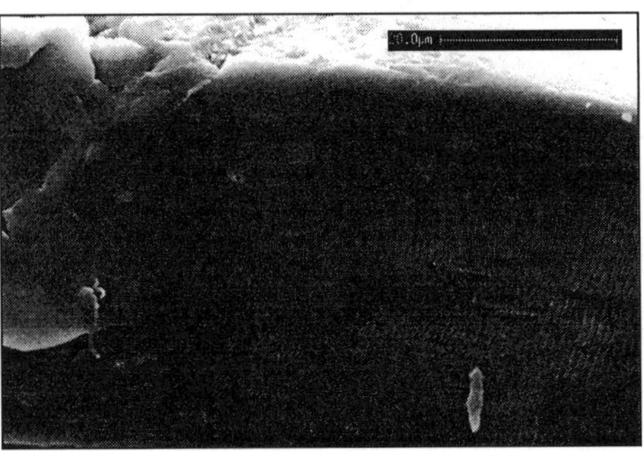

Figure 2b. Specimen M2.5, Initial crack surface, relatively smooth with fine striations.

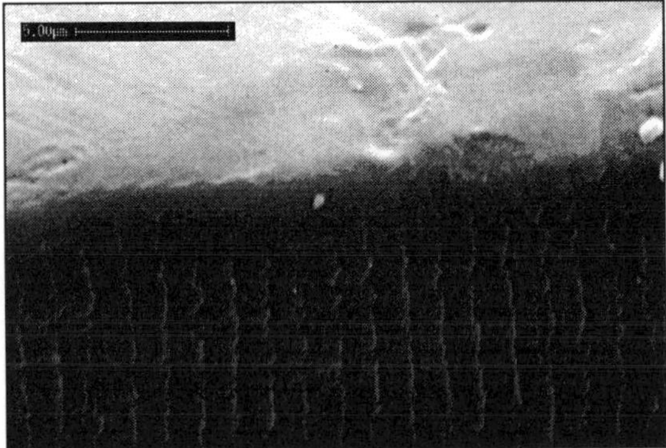

Figure 2c. Specimen 2.5, Detail of striations on the crack surface and the exposed FEP.

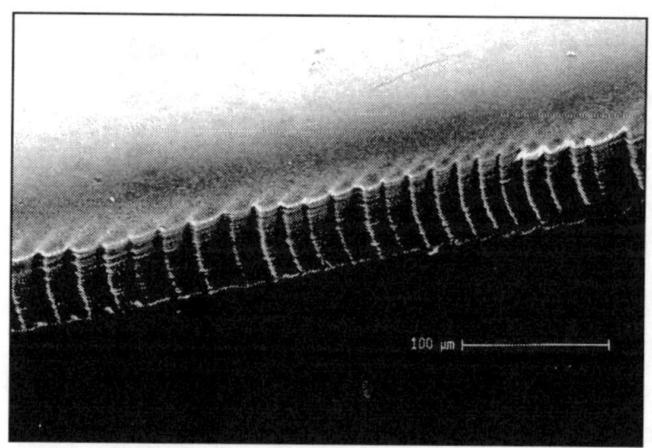

Figure 3a. Specimen B1.8, Five mil FEP with VDA spring loaded during thermal cycling. Shows wave-like striations. Crack propagation was right to left.

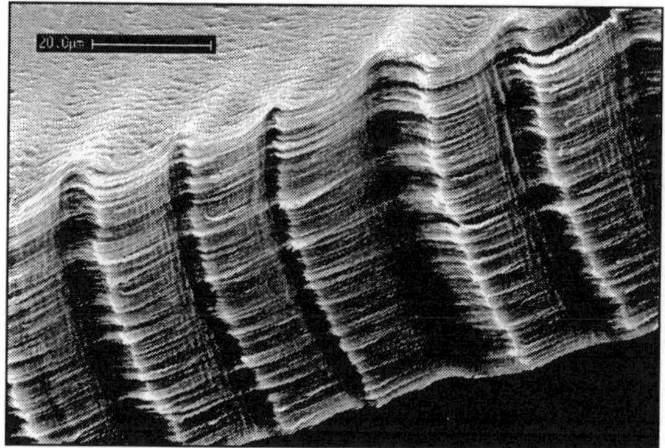

Figure 3b. Detail of the crack surface of specimen B1.8. Crack propagation was right to left.

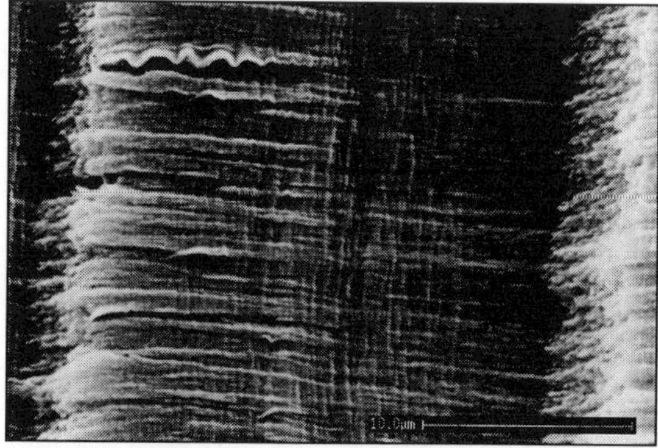

Figure 3c. Detail of wave-like striations showing ductile tearing features, specimen B1.8.

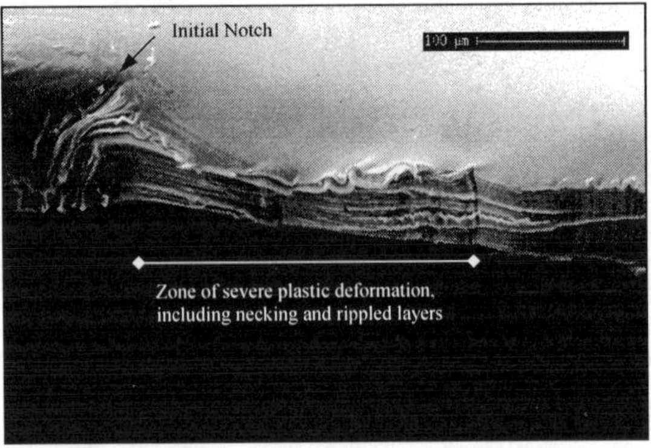

Figure 4a. Specimen B1.6. The initial razor cut is visible at the far left. The crack surface shows significant plastic deformation. Beginning of progressive crack formation is apparent at the far right. Crack propagation was left to right.

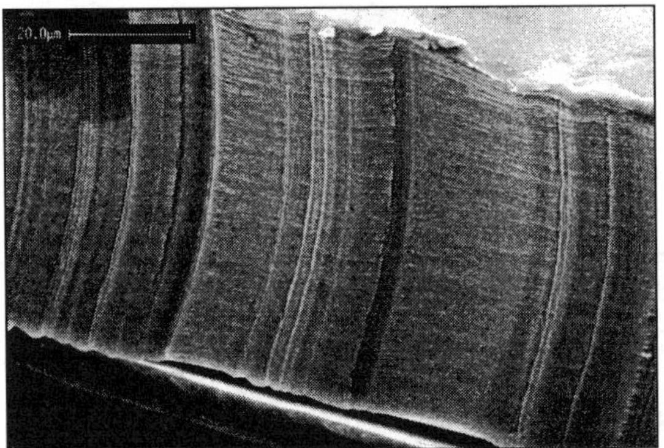

Figure 4b. Specimen B1.6. Midpoint of crack length, crack progression was left to right.

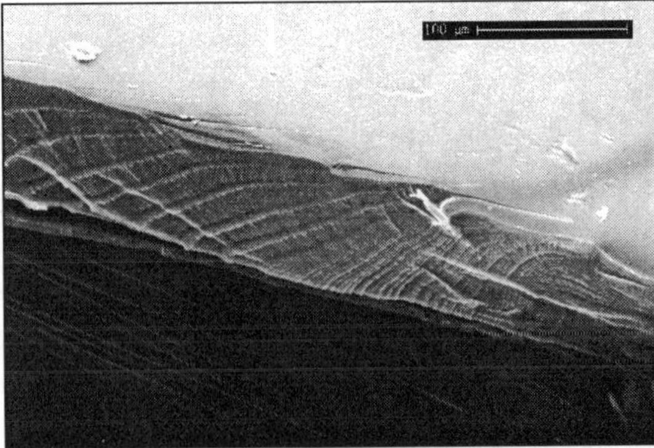

Figure 4c. Specimen B1.6. End of crack. Secondary cracks associated with main crack striations are apparent. Plastic deformation at the extreme right was created during SEM specimen preparation.

43rd International SAMPE Symposium
May 31-June 4, 1998

ATOMIC OXYGEN RESISTANT FIBER FOR BRAIDING

Peter Schuler,Ross Haghighat, Hamid Mojazza
Triton Systems Inc. Chelmsford, MA 01824

ABSTRACT

Organic polymer based threads for applications in Low Earth Orbit (LEO) suffer significant erosion from the combined effects of Atomic Oxygen and Vacuum Ultraviolet radiation which compromise their mechanical and physical properties over time (REF 1, 2, 3). Threads used to stitch Multi Layer Insulation blankets (MLI) have limited lifespans due to these environmental effects and alternative thread materials such as Teflon® coated glass can be extremely difficult to sew. Triton Systems has developed twisted and braided sewing threads made from a unique polymer (TOR™) that are more durable than other competing materials in ground based testing conducted at Marshall Space Flight Center.

This paper discusses the processing of TOR into fibers, the braiding of TOR around an inner Kevlar® core, characterization of the fiber to date, and present and future applications for this product.

KEYWORDS: Atomic Oxygen, Low Earth Orbit, Tether, Braided threads

I. INTRODUCTION

Poly Arylene Ether Benzimidazoles (PAEBI'S) are a new class of polymers invented by Connell, et al [REF 4,5,6,7,8] of NASA and produced under the trade name TOR (**T**riton **O**xygen **R**esistant) by Triton Systems Inc. Chelmsford, MA.   One TOR derivative containing an inorganic oxide link in the polymer backbone, has demonstrated outstanding resistance to erosion by Atomic Oxygen (AO) in ground based simulation tests. Three years of research by Triton has resulted in the development of TOR threads that are far more resistant to Atomic Oxygen erosion in Low Earth Orbit (LEO) than other competing organic polymer materials. The need for longer lasting thread materials is driven by the demand for extended space missions such as the International Space Station (ISSA) and the rising number of satellites deployed in 200 – 500 km orbits where there is a high concentration of Atomic Oxygen.

In 1995, the first of 3 NASA funded programs was started with the goal of developing a space durable thread that would be used to sew Multi Layer Insulation (MIL) blankets together. In the last three years, the polymer synthesis has been scaled up from gram to multipound quantities, fiber spinning has gone from single filament to more than 30 filaments, and pilot scale fiber processing equipment (Figure 1.) has been installed to meet the current demand for thread. In addition, TOR fiber has undergone extensive ground testing as well as qualification and deployment on a long duration space flight.

The evolution of threads from TOR polymer was made in a series of steps, each of which complemented the overall development effort. Our ability to synthesize polymer batches with repeatable results was critical in the early development stages. We identified a spinning process and post processing techniques for this polymer that increased the mechanical properties enough to begin thread manufacturing trials. Work is still underway to optimize and fine tune physical properties of the fiber for specific end uses. Finally, we have identified and are working with end users for these threads products.

Several commercial applications for TOR fiber are being pursued at this time. The primary benefit for these applications is that TOR lasts significantly longer in Low Earth Orbit than other available polymer materials thereby increasing the service life of components made with it.

One of the applications is a high strength sewing thread to stitch insulation blankets used on the International Space Station. This work is being carried out in response to a request from the Johnson Space Flight Center in Houston TX. To meet the high break strength requirements for this thread, TOR fiber is braided around an inner Kevlar core. An alternative application with Marshall Space Flight Center, would use TOR fiber braided around an inner core for a tether deployed in Low Earth Orbit. In this application, the product is not a thread used for stitching, but it has a very similar design. Several different tether designs are being considered for this work, yet all of them face potential failure due to erosion of the polymeric materials without the protection offered by the TOR overbraid.

Figure 1. Pilot Process Fiber Spinning Equipment

2. EXPERIMENTAL

2.1 Chemistry The chemical structure of TOR is shown below in Figure 2. TOR is prepared by the nucleophilic displacement reaction of dihydroxy Benzimidazole monomers with active aromatic dihalides. The TOR polymers are related to polybenzimidazoles that are well known rigid rod polymers used as fibers in textiles, adhesives and foams. The nitrogen atoms they contain allows them to behave like polyimides, such as the well known polymer Kapton[®] (Dupont and Monsanto trademarks), but by combining the Benzimidazole structure with the Arylene ether link and the inorganic oxide link, a polymer was created with higher Tg than Kapton, still retaining the transparency and processability of the polyimides.

2.2 Fiber processing The conversion of TOR polymer into fibers is made using a solution spinning process. Solutions can be prepared with the solvent Dimethyllacetamide (DMAc) in concentrations as high as 35% (solids) but usually in the preferred range of 30% - 32% solids. It is imperative that the solutions be highly filtered and degassed prior to using them as any impurities or gas bubbles in the solution will result in broken and inconsistent filaments. The solution viscosity is in the range of 220,000 – 250,000 cp. Fibers are extruded and pulled through a methanol coagulation bath to remove the DMAc solvent. The diffusion process that occurs in the coagulation bath is highly complex and beyond the scope of this paper, however, it is critical to the production of high quality TOR fibers. The rate at which the diffusion process occurs affects the "as spun" fiber porosity. If the diffusion rate is too rapid, the fiber will be so porous, that they will not "draw" properly in subsequent steps. Conversely, if the diffusion is too slow, the fiber will contain trace amounts of solvent which will "explode" out of the fiber in the post processing steps. The coagulation bath composition, temperature, line speed, and time that the fiber is in the bath are all factors in controlling the diffusion process. Examples of poor quality porous, and high quality dense fiber structures are shown below in Figures 3 and 4.

Experimentation with 3 common spinning methods (shown below in Figure 5.) revealed that the most effective spinning process for making TOR fiber is the dry jet-wet spinning technique. In this process, the individual filaments are extruded into a room temperature air gap before they enter the coagulation bath. As the fiber travels in the air gap, some solvent evaporates, the filament size decreases, and the fiber becomes stronger and more handleable. The height of the air gap must be at least 6 inches to have a significant positive effect on the finished fiber properties. This translates to a 1 – 2 sec time interval in which the fiber is in the air gap.

Post processing of the "as spun" fibers is limited to a heating and stretching of the fiber near the Tg of the polymer, and application of a spin coating to the finished fiber bundle. The post processing converts weak "as spun' fiber to stronger, finer, finished fiber. The fiber tow is drawn across a heated smooth plate between 2 sets of non-heated godet rollers. The draw ratio of these rollers is typically 2X - 3X. In the drawing step, the fiber densifies and undergoes some molecular orientation. An aqueous based finish coating is then applied as the final step in order to keep individual filaments together in the tow.

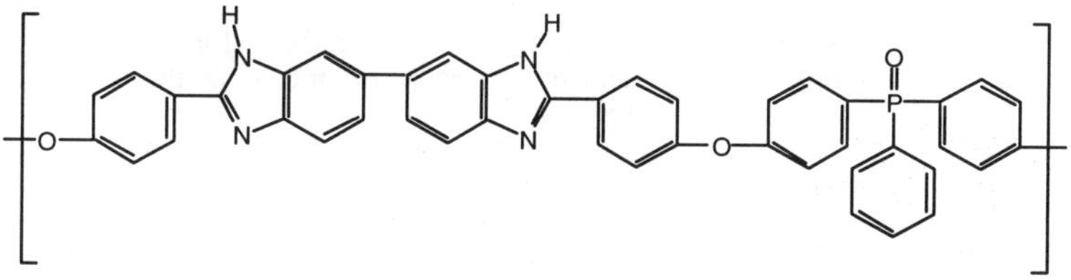

Figure 2. Chemical structure of TOR

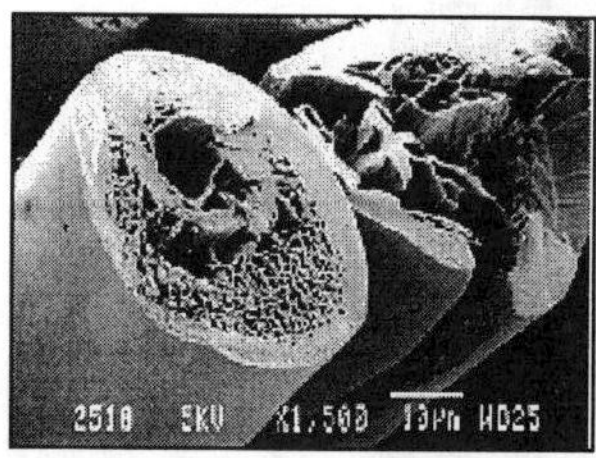

Figure 3 Porous fiber structure

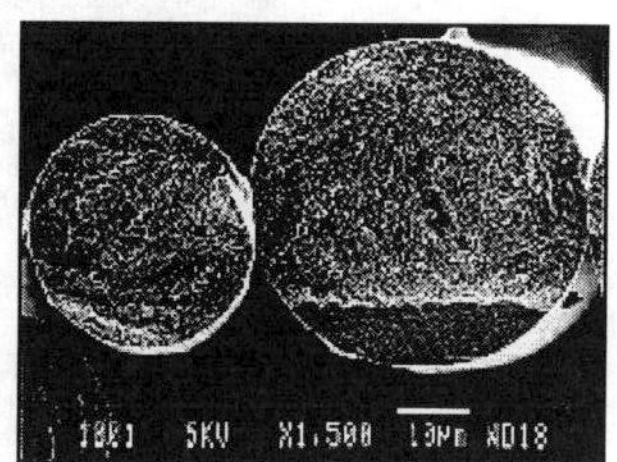

Figure 4 Dense fiber structure

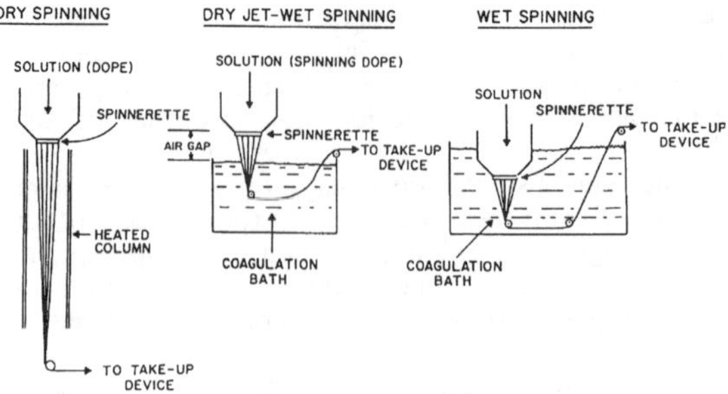

Figure 5. Several solution spinning processes.

2.3 Threads Two types of threads are made from TOR fiber. A twisted thread is available in a range of sizes (D –FF) with low to medium strength properties. A local thread manufacturer has been our partner in doing the actual twisting of the filaments into the finished thread. Generally, these threads are constructed with 3 or 4 filament bundles, depending on the finished size. All of these threads have a final "Z" twist which is common for threads used in a majority of today's commercial sewing machines.

A second type of thread has been developed for applications requiring significantly higher break strengths. This thread is made by braiding TOR fiber over a high strength inner core. In most cases the core is Kevlar, however Spectra or PBO are alternative core materials. This braided thread has been developed to meet 10 – 20 lb break strength needs for a thread with a 0.020" diameter. This would compare in size to a FF sized Nomex thread. We have made braided threads with both 4 and 8 tows wrapped around the inner core. The 8-tow braid provides better protection of the inner core due to the increased density of TOR fiber surrounding it. There is less possibility of the inner core being exposed after the stitching operation and the overall smoothness of the thread is better, which improves the sewability. The elongation properties of high strength materials such as Kevlar and Spectra are generally low and are good candidates for a TOR overbraid. Braiding over a core material with an inherently low elongation is advantageous since it closely matches the elongation properties of TOR. Damage to the overbraid is minimized during the sewing process because both materials yield at the same strain level. If the outer core yields prematurely, substantial damage to the outer braid could occur, compromising the AO protection it provides. We have teamed with Cortland Cable Co. of Cortland NY to do the braiding of these high strength threads.

Figure 6. Standard braiding equipment

3. RESULTS AND DISCUSSIONS

3.1 Environmental Testing Triton Systems has conducted ground based AO testing at the Marshall Space Flight Center environmental effects laboratory under the direction of Mr. Jason Vaughn. Some of the results of those tests are presented below. Figure 7 shows a plot of mass loss due to AO exposure at a fluence level of 1.56×10^{21} atoms/cm^2 for several polymer materials. The results indicate that TOR threads errode significantly less than other candidate organic materials. When atomic oxygen reacts with the TOR polymer surface, a phosphorous rich oxide layer forms on the skin that protects the base polymer from further erosion by atomic oxygen. Figure 8 shows this oxide formation as observed with FTIR and EDS spectra on a sample before and after exposure to atomic oxygen.

Figure 9 shows tensile testing on braided thread before and after exposure to atomic oxygen at the Marshall test facility. After exposure to a fluence level of 2.3×10^{21} atoms/cm^2, there is less than 20% loss of tensile strength. In comparison, other materials in the test (Spectra, Kevlar, Kapton and oriented PTFE) did not have enough integrity to be removed from the test chamber without falling apart. The fact that the braided thread retains a high level of strength after AO exposure makes this material a strong candidate for long duration space flights because it can reduce the number of missions needed to replace insulation materials that have failed at the stitching.

A long exposure flight test of TOR has been completed and the results of that experiment are being generated at the time of this paper being written. In March 1996 samples of TOR fiber were placed on the MIR space station as part of the POSA I flight experiment. These samples passed flammability, toxicity and offgassing qualification tests. The samples were prepared by MSFC and located on tray EOIM-1. The tray was photographed in the deployed position and is shown below in Figure 10. In September 1996, the samples were retrieved and are being evaluated by MSFC. The preliminary indications are that the TOR fiber suffered no visible damage during the mission but additional analysis must be completed. Additional properties of TOR fiber are presented below in Table 1.

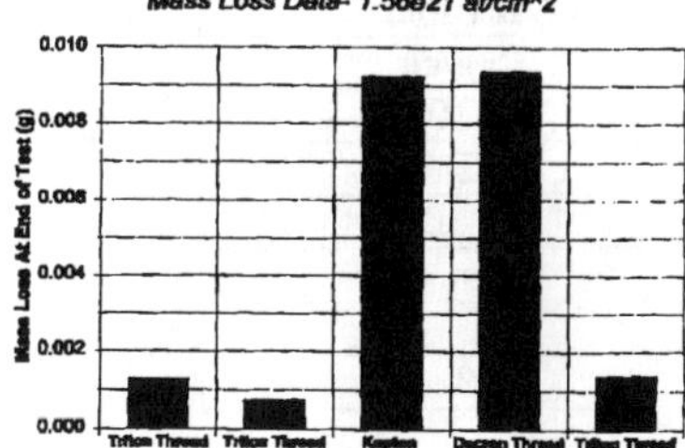

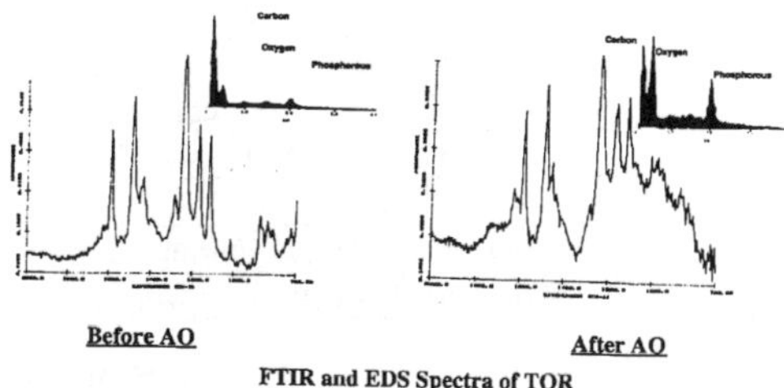

<table>
<tr><td>

Figure 7. AO Test from MSFC

</td><td>

Figure 8. FTIR and EDS spectra of TOR before and after AO exposure

</td></tr>
</table>

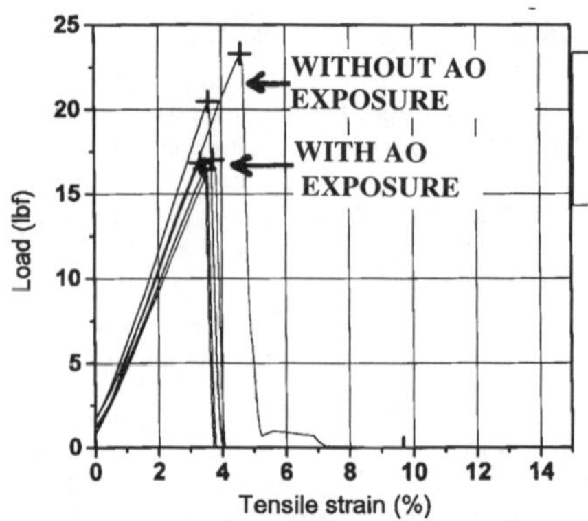

Figure 9. Tensile tests on braided thread before and after AO exposure

Figure 10. TOR fiber samples on POSA I sample tray

Properties	TOR Fiber
Specific gravity	1.35 g/cc
Available denier (per filament)	1 – 10
Tenacity (g/denier) dry	2 – 2.5
Breaking elongation (%)	10 –20
Color	Tan
Volatile condensable material (%)	< 1%
Moisture absorption by weight (%)	5-8

Table 1. TOR fiber Properties

4. SUMMARY AND CONCLUSIONS

Many thread materials considered for applications in Low Earth Orbit suffer from erosion by Atomic Oxygen. This erosion is most damaging at low altitudes (200Km – 400 Km) and affects most organic polymers. As space missions increase in duration, the need to identify, qualify and use more durable materials becomes necessary. Triton Systems has developed TOR polymer for films, resins and fibers that is more resistant to atomic oxygen erosion in ground based simulation testing when compared with other polymeric materials.

Over the past 3 years, the synthesis and processing of TOR has been scaled up to a pilot production level and equipment to manufacture hundreds of pounds of fiber yearly is now operational. Two different thread constructions have been produced from TOR fiber and their mechanical and physical properties have been characterized. The applications for the two threads are distinctly different because of the strength differences. At this time, the braided thread is primarily used in high strength sewing applications (such as MLI blankets) because it will help meet longer lifespan requirements. The lower strength twisted thread is suitable for identification marking.

New applications for the TOR fiber are likely to be braided structures that provide AO protection for internal components. The advancement of tethers for payload transport, power generation and deorbiting use present many requirements for environmental durability, and the TOR AO resistant fiber show promise in their ability to protect the non-metallic components. The construction and future deployment of the International Space Station will also have components that require the use of longer lasting threads. Finally, TOR fiber will be used in the braiding of experimental conduit structures to determine the feasibility of providing environmental protection for existing tool and life support tethers and wiring.

6. REFERENCES

1. "First LDEF Post-Retrieval Symposium Abstracts, June 2 – 8, 1991" NASA CP
2. "LDEF Materials Workshop '91", Nov. 19-22,1991, NASA CP 3162
3. "Third LDEF Post Retrieval Symposium Abstracts", NASA CP 10120 (1993)
4. Proceedings of the American Chemical society, Apr 1994
5. US Patent No. 5,245,044 (Sept. 14, 1993) Connell et al (The United States of America as represented by the Administrator of The National Aeronautics and Space Administration)
6. US Patent No. 5,317,078 (May 31, 1994) Connell et al (The United States of America as represented by the Administrator of The National Aeronautics and Space Administration)
7. US Patent No. 5,412,059 (May 2, 1995) Connell et al (The United States of America as represented by the Administator of The National Aeronautics and Space Administration)
8. US Patent Application No. 08/186,184 (Jan 21, 1994) Connell et al (The United States of America as represented by the Administator of The National Aeronautics and Space Administration)
9. Jason A. Vaughn, "ProSEDs Tether Development, Atomic Oxygen Tests of Non-Conductive Tether Materials" Oct 1997

ACKNOWLEDGMENTS

The authors thank the generous financial support of the National Aeronautics and Space Administration under SBIR contract numbers, NAS8-40146, NAS8-40576

ELECTRICALLY CONDUCTIVE, LOW SOLAR ABSORPTION, ATOMIC OXYGEN RESISTANT FILMS FOR SPACECRAFT THERMAL AND CHARGE CONTROL

John Lennhoff, Marvin Guiles, and George Harris
Triton Systems, Inc. Chelmsford, MA 01824

Jason Vaughn, David Edwards, and James Zwiener
Marshall Space Flight Center, Huntsville, AL 35892

ABSTRACT

There is a significant requirement for an electrically conductive thermal control film for the dissipation of spacecraft charging.   The present technology based upon Indium Tin Oxide (ITO) thin conductive coatings on Teflon or Kapton films suffer from a range of problems, including poor space durability and a tendency to crack. A new blended polymer film developed at Triton Systems combines the properties of an intrinsically electrically conductive polymer with that of an AO resistant polymer. The resulting film, called C-COR for Conductive Colorless Oxygen Resistant, has bulk electrical conductivity adjustable between 10^{-12} and 10^{-8} $(ohm\text{-}cm)^{-1}$, AO resistantance, low absorptance, high emittance, vacuum stability, and good mechanical and UV resistant properties. The C-COR polymer can be used as a coating or as a free standing film. Data will be presented on the details of the space durability, optical, mechanical, and electrical properties of the C-COR film.

KEY WORDS: Electrically Conducting Materials, Polyaryl ether, Space Applications

1. INTRODUCTION

A myriad of radiation impact space vehicles with considerable fluxes and energy in both LEO and GEO. The electric charge build-up resulting from the GEO and polar LEO environments must be dissipated for a variety of reasons, the most significant being the interference to electronic components from capacitive discharge. While Indium tin oxide (ITO) and other surface coatings can provide some surface charge dissipation, the thin surface coatings do not provide bulk charging relief. The ITO layer is readily eroded in some orbital altitudes resulting in an insulating polymer surface being exposed. Fillers, either semi-conducting oxides or metals can produce limited improvements however they significantly alter optical performance of Second Surface Mirror thermal control films. The inorganic or metallic solid fillers also change the physical properties of the polymer films which can result in high spauling, brittle, and high density materials.

The exposure of polymeric materials on spacecraft to the high energy electron environment of space results in the accumulation of secondary electrons and considerable spacecraft charging. When the electrostatic potential from the accumulated electron charge exceeds the dielectric strength of the polymer, a breakdown occurs that can interrupt or damage normal spacecraft function. The charging effect is particularly problematic in high earth orbits, i.e., above 500 km. This charging problem could be eliminated if moderate conductivity could be integrated into the polymer materials. Table 1 lists the values for conductivity and resistance for a range of polymers, metals, and conducting polymers. The regulation of spacecraft charging could be accomplished by materials having resistance of 10^{10} ohm-cm or less (equivalently with conductivity of 10^{-10} (ohm-cm)$^{-1}$ or greater). These materials must also have long-term stability, environmental resistance, and mechanical integrity. Elliptical orbits that involve high GEO, also encounter atomic oxygen (AO) exposure during the LEO portion of orbit. This orbit poses double trouble for spacecraft that will require conductivity and AO resistance.

Table 1. A Listing of the Conductivities and Resistances of some Polymers, Metals, and Conductive Polymers.

Material	Resistance ρ Ohm-cm	Conductivity σ (Ohm-cm)$^{-1}$
FEP-Teflon	10^{18}	10^{-18}
Kapton	10^{17}	10^{-17}
PAE-COR	10^{15}	10^{-15}
Spacecraft Requirement	10^{10}	10^{-10}
C-COR	10^{8}-10^{12}	10^{-12}-10^{-8}
Carbon Powder	1-1000	0.001-1
Conductive Polymers	0.1-1000	10 - 0.001
Polyacetylene	500-1000	0.005 - 0.001
Gold	2.5 x 10^{-10}	4 x 10^{9}
Copper	1.6 x 10^{-10}	6.3 x 10^{9}
Silver	1.5 x 10^{-10}	6.7 x 10^{9}

Triton Systems, Inc. has incorporated a proprietary Conducting Polymer (CP) into a proven space durable and AO resistant polymer called PAE-COR. The marriage of these unique materials, called C-COR, result in a space durable conducting polymer composite thermal blanket. This blanket will provide <u>through thickness</u> conductivity and AO protection. Triton has also incorporated improved VUV resistance for this polymer through inclusion of a fluorescent material into the COR polymer backbone.

Saturated aliphatic organic molecules are electrical insulators, but conjugated organic polymers have some electrical conductivity through the valence bands created by their π electrons. A breakthrough occurred in 1977 when Shirakawa, Louis, MacDiarmid, Chang, and Heeger, and others noted that the conductivity of some conjugated polymers could be dramatically increased by the use of "dopants" or "counter ions" which oxidize or reduce the polymer molecules to provide free conductive paths for electrons. Also, the conductivity of some of these polymers was found to be reversible by reversing the redox reaction - usually by the variation of a potential across the polymer.
Conducting polymers are usually identified by their high aromatic content, such as polyacetylene, polyphenylene, polypyrole, and polyaniline. The polymers typically are doped with a variety of ionic species which assist in the electron transfer and increase conductivity up to a dozen orders of magnitude. These dopants include I_2, Cl_2, AsF_5, BF_4, Na, ClO_4, CF_3SO_3, and more recently organic acids. Many conductive polymers are difficult to process because they do not melt and are not soluble in organic or aqueous solvents. The dopant species can vaporize or oxidize resulting in loss of conductivity or the polymer matrix can decompose, oxidize or be attacked by the dopant during heat processing of conductive polymers. Brittleness is also common among conductive polymers, further inhibiting applications once processed. Organic acids comprise a relatively new type of dopant molecule with demonstrated high thermal stability and induced processing improvements.

This paper summarizes the development and characterization of an electrically conductive, optically transparent, space durable polymer film for spacecraft thermal control. This work was performed under a NASA Marshall Space Flight Center Phase I SBIR Program. The goal of the work was to formulate, fabricate and develop light weight, optically clear, atomic oxygen and VUV stable conductive films for use as SSM's that are static charge dissipaters, called C-COR (for Conductive Oxygen Resistant). Conductivities in the range of 10^{-11} to 10^{-9} S/cm (resistances of 10^{11} to 10^{9} ohm-cm) are desired for this application.

We report here on the synthesis of the required polymers, formulation of the organic acid doped conductive polymer/COR blend, optimization of film processing, and characterization of the resulting films. The polymers synthesized were doped proprietary conductive polymer and poly(arylene ether phenyl phosphine oxide)s, known by Triton as COR (colorless atomic oxygen resistant). In a range of formulations, these polymers were blended in an organic solvent and cast as clear free standing films. These films were characterized for conductivity, optical properties and AO, VUV, and thermal stability, and a database of information regarding conductivity verses formulation was built.

2. EXPERIMENTAL

COR films were blended with the conductive polymer in an organic solvent using proprietary materials and methods. Film casting and curing of C-COR films was performed below the thermal deactivation temperature of the conductive polymer, which is 210 C. C-COR films have been successfully cast and cured, and the process is still being optimized.

The COR polymer was prepared by Triton's modification of the standard Polyarylene ether synthesis method, in a polar aprotic solvent by an aromatic nucleophilic displacement reaction of an activated aryl phenyl phosphine oxide with a bisphenol in the presence of alkali metal carbonates.

Conductivity of the films were measured at Triton Systems using a Keithly electrometer Model 6517A with a Model 8008 resistivity test fixture.

AO exposure of the C-COR films was performed by Jason Vaughn at the NASA MSFC facility. The AO source utilized a fast atom system. Optical properties of the C-COR films were measured for solar absorptance and thermal emittance using an AZ Technology Spectroreflectrometer SPSE and TEMP-2000, respectively.

VUV exposure experiments were performed at the NASA GSFC Thermal Engineering Branch facility by Wanda Peters of Swales Aerospace. An AZ Technology LPSR-200 was used to measure solar absorptance, and a Geir-Dunkle DB-100 was used to measure emittance.

Two C-COR films were tested for vacuum conductivity stability by Jason Vaughn of the Mashall Space Flight Center. The films, with resistances of 2×10^9 and 2×10^{10} ohm-cm were selected for testing. A set of copper rings were vacuum evaporated on the films and leads were soldered to them for measurement of surface conductivity during the test. A sample was placed in the vacuum chamber, connected to the 100 volt power supply, and measurement of the current passed through the film was monitored as the chamber was pumped down to about 10^{-6} Torr. Resistance was calculated from the geometry of the copper rings, the voltage applied and the current measured.

Optical properties of the C-COR films were measured for solar absorptance and thermal emittance using an AZ Technology Spectroreflectrometer SPSE and TEMP-2000, respectively. These measurements were made at NASA/MSFC by Triton staff.

3. RESULTS

Triton Systems carried out a work effort that concentrated on synthesis of the required polymers, formulation of the doped CP/COR blend, optimization of film processing, and characterization of the resulting films. The polymers synthesized were doped Conducting Polymer (CP) and poly(arylene ether phenyl phosphine oxide)s, known by Triton as COR (Colorless atomic Oxygen Resistant). In a range of formulations, these polymers were blended in an acidic organic solvent and cast as clear free standing films. These films were characterized for conductivity and stability, and a database of information regarding conductivity verses formulation was built. Figure 1 shows the relationship between conductive polymer content and film bulk electrical conductivity.

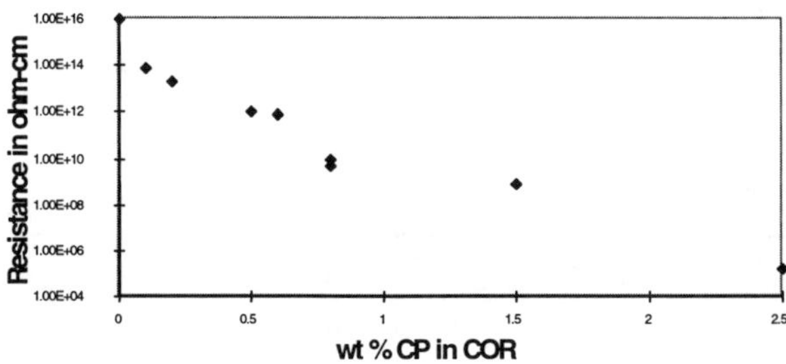

Figure 1 – Conductive Polymer (CP) concentration as a function of Bulk Resistance

These data are presented in Figures 2 and 3 for 1 mil films. A typical space application C-COR film at 10^{10} ohm-cm, 1 mil film would have an α=0.3, ε=0.8 and α/ε= 0.375. Significant reductions in the color of these films are expected as the formulation and methodology matures.

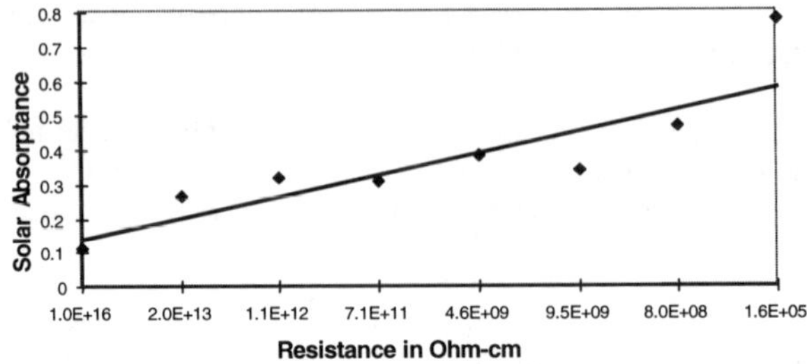

Figure 2 - Solar Absorptance as a function of C-COR Resistance

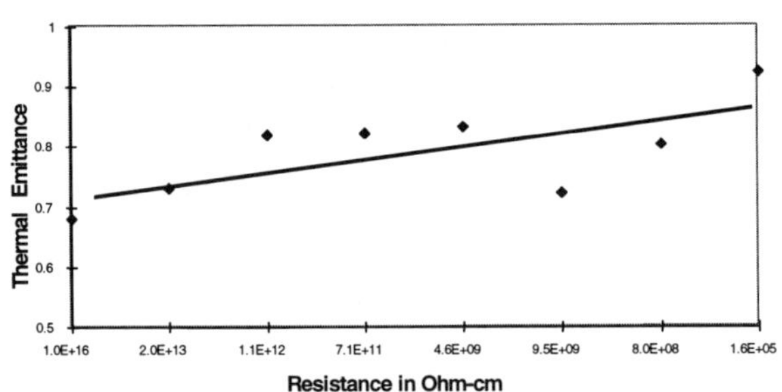

Figure 3 - Thermal Emittance as a function of C-COR Resistance

Figure 2 provides an example of time verses resistance data from one of the films. Many conductive materials depend on hydration, or some other volatile component, to provide conductivity. These types of materials have demonstrated a resistance change of several orders of magnitude from atmospheric to space vacuum. The Triton Systems C-COR materials does not contain any volatile components, and depends upon the polymeric PANi material for its conductivity. The C-COR material demonstrated stable vacuum resistance measurements over the several hour MSFC testing period.

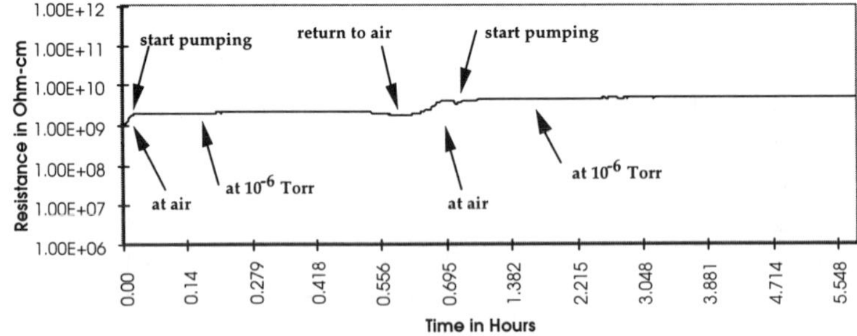

Figure 4 - Vacuum Stability of C-COR Resistance

Figure 5 shows the thermal stability of conductivity, and then the sharp irreversible thermal degradation of the conductivity of a 5.6 % CP blend with COR, as a function of increasing temperature. From Figure 5, CP blend retains its high conductivity of 100 (ohm-cm)$^{-1}$ at temperatures up to T = 200 C. Then at about T = 210 C, the CP blend loses its conductivity, falling to 10^{-8} (ohm-cm)$^{-1}$ at about 210 C, and to 10^{-12} (ohm-cm)$^{-1}$ at higher temperatures.

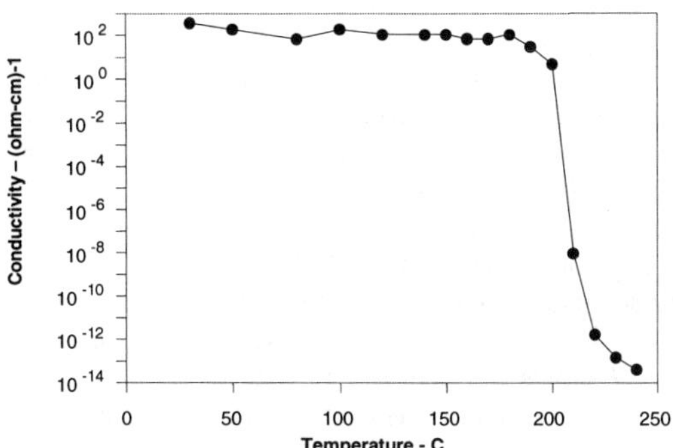

Figure 5 - Conductivity as a function of Temperature for 5.6% CP COR Film

AO resistance was measured at MSFC on a single sample of C-COR film with a conductivity of 10^{-5} S/cm. The erosion data as a function of AO fluence is provided in Figure 6 for the conductive C-COR film, along with COR, FEP Teflon and Kapton insulating polymer films. Table 2 below provides absorptance and emittance data for the beginning and end of the AO exposure test. The EOL AO fluence was 1.8×10^{21} AO/cm^2. The erosion rates for the C-COR are very similar to those for the base COR material. The base COR material has an order of magnitude lower erosion rate compared to FEP Teflon.

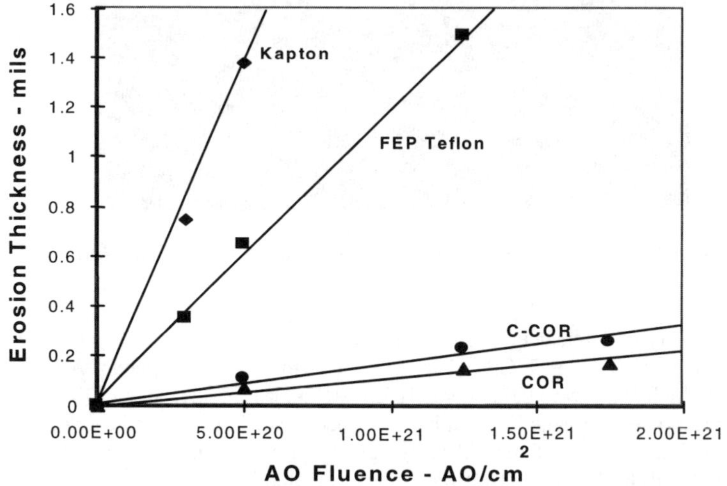

Figure 6 - Erosion Thickness as a function of AO Fluence

Table 2 - Optical Properties of AO Exposed C-COR Film

	α	ε	α/ε
BOL	0.31	0.74	0.42
EOL	0.39	0.75	0.52
Δ	0.08	0.01	0.1

Examination of the C-COR film exposed to AO by scanning electron microscope (SEM) was performed. The SEM uses an electron beam to coerce the emission of secondary electrons from a sample. The secondary electron yield provides a contrast medium for a sample image. This secondary electron yield is coordinated with the position of the scanning electron beam on the sample to provide a 2D image. An insulating sample produces an electron charging phenomenon that defocuses the imaging because of a significantly broadened secondary electron source at the sample surface. Hence, an SEM image can provide details on the conductivity of a sample. An image (2000X) of the AO exposed region in Figure 7 shows a typical AO eroded surface and no evidence of charging or defocusing of the image. The image was acquired at 20 keV and shows no evidence of charging.

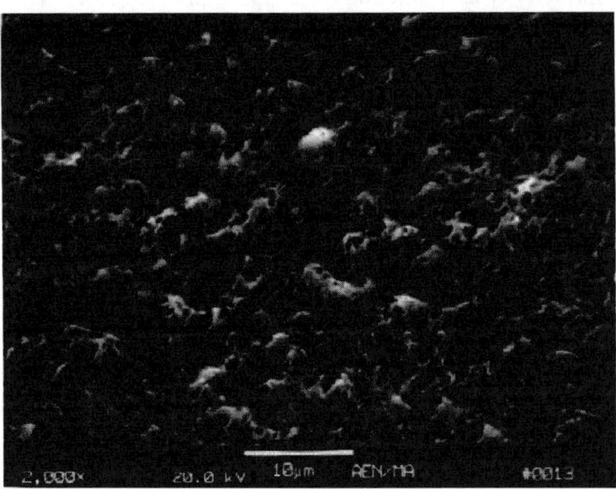

Figure 7- 2000X SEM Image of an AO exposed C-COR Film Sample

The VUV stability of Optical Properties was measured using a C-COR sample with 10^{-5} (ohm-cm)$^{-1}$ conductivity (0.9 % CP). The matrix polymer was a Bisphenol A based COR formulation, which is know to be moderately UV stable. More advanced COR formulation have shown improved UV stability, but have not been tested as the conductive based film. The chemical functionality of the CP is thought to provide some UV stability, as other chemicals with similar functionality are common UV stabilizers. Also the conductivity of the polymer may inhibit certain electron transfer reaction associated with the UV degradation. These concepts are not proven at this time. These UV stability results are provided in the Table 3 below.

Table 3 - Solar Absorptance as a Function of UV Exposure for C-COR

VUV (ESH)	α	$\Delta\alpha$
O	0.282	0
386	0.334	.052
1081	0.351	.061
1500	0.354	.072

Mechanical properties of the base COR polymer were measured and compared to the mechanical properties of the conductive films. A 10^{10} ohm-cm resistance film was measured for mechanical properties and the results are shown in Table 3. Mechanical properties are often a strong function of processing, as has been demonstrated for the base COR material.

Table 3 - Mechanical Properties of COR and 10^{10} ohm-cm C-COR Films

	COR	C-COR (10^{10} ohm-cm)
Tensile Strength (Ksi)	10-12	8-10
Elongation (%)	10-65	4-15
Modulus (Ksi)	350	450

4. CONCLUSIONS

A space durable, optically transparent, electrically conductive polymer film has been developed for use in spacecraft thermal and electric charge management. A range of conductivities have been demonstrated, all of which have excellent AO and vacuum stability, and good UV stability and mechanical properties. The film is an excellent alternative to ITO coated Teflon or Kapton films for charge control in space applications.

5. ACKNOWLEDGEMENTS

This work was performed under NASA Phase I SBIR program Contract # NAS8-97142. The UV exposure and characterization work was performed by Ms. Wanda Peters of Swales Aerospace Corp. working at NASA/GSFC. The authors gratefully acknowledge the efforts of all those who assisted with this program.

CONE CALORIMETER INVESTIGATIONS OF FIRE RESISTANT MATERIALS FOR AIRCRAFT APPLICATIONS

Arthur F. Grand, Ph. D.
Omega Point Laboratories, Inc.
Elmendorf, TX 78112-9784 USA

Edward D. Weil, Ph. D.
Polytechnic University
Brooklyn, NY 11201

ABSTRACT

The "cone calorimeter" (ASTM E1354, ISO 5660) has enormous potential for the characterization of the flammability properties of fire resistant materials intended for aircraft (and other) applications. However, the method is not without its drawbacks, especially for materials of low flammability. Utilization of the results from this test method must be performed with due regard for the limitations of the data for the particular materials of interest. Recent cone calorimeter studies of char-forming thermoplastic materials are discussed. Included in the study were tests on various specimen mounting protocols. Recommendations for the use and interpretation of the results obtainable from the test method are presented.

KEY WORDS: Cone Calorimeter, Heat Release Rate, Flammability

1. INTRODUCTION

1.1 Overall Test Program This study is part of a project by Omega Point Laboratories, Inc., under an SBIR (Small Business Innovation Research) grant from the U. S. Department of Transportation and monitored by the Federal Aviation Administration (FAA).

Polymeric materials are very desirable for aviation use and are in service in many areas, including bulkheads, seats and interior panels. These products have advantages over metals for many applications because of their weight, strength, corrosion resistance, electrical properties, maintenance and ease of manufacture. However, offsetting some of the advantages of polymeric products over metals, one of the most significant disadvantages is flammability.

The fire safety of commercial aircraft has been the subject of periodic activities in the Federal Government. Notable are the Aviation Safety Act of 1988, the U. S. House of Representatives hearings on "Aircraft Cabin Safety and Fire Survivability,"[1] and a U. S. General Accounting Office report[2]. Fires in aircraft pose unique problems for survivability due to the confinement of occupants in a restricted space. Also, the numbers of people involved in any single aircraft fire incident is enough to make the incident newsworthy.

The Federal Aviation Administration (FAA) has been investigating the fire performance of aircraft seats, interiors and baggage compartments (i.e., cargo liners) for many years. The so-called "65/65" rule[3] was developed to address the issue of heat release rate of interiors when subjected to fire exposures comparable to a post-crash fuel fire. An FAA-sponsored conference dealt with current issues and future plans regarding fire resistant aircraft interior materials[4]. Previous SAMPE meetings[5] have addressed fire issues, with participants from the FAA, Navy, Coast Guard, other government agencies, industry, research laboratories and academia.

In response to concerns about flammability issues, the U. S. Navy conducted several programs to characterize the flammability of composite products and to examine "fire barriers" to protect the polymeric substrates[6,7,8,9,10]. In conjunction with the Navy's efforts, the National Institute of Standards and Technology conducted several studies on the flammability of composites[11,12,13,14]. The hazards to people from fire onboard ships and aircraft include the potential for burns, smoke inhalation, and possible loss of life (even if forced to leave the craft). Physical damage to the structure may occur as a result of fire and smoke. The spread of fire must be limited by diligent consideration of the products and materials that potentially could contribute to a fire.

Ignitability, flame spread, smoke release, and other information obtained from fire and flammability tests, are not inherent properties of the products and materials tested; the results may be very dependent on the nature of the test procedure. Specimen size and orientation, external heat flux, type of ignition source, and method of calculation of the data, all influence the results.

1.2 The Cone Calorimeter Test Method The cone calorimeter method (ASTM E1354, ISO 5660) is a state-of-the-art bench scale flammability test procedure with relevance to simulation of larger scale fire scenarios. Its utility for characterizing flammability characteristics of many different products has been demonstrated[15,16] and its accuracy and repeatability are good[17,18]. The apparatus has the potential for being able to provide relevant data for the correlation of laboratory and full-scale test methods[19], and for input into hazard analyses[20]. One of the benefits in using this method for research is the versatility of the apparatus, including various specimen preparation techniques, a range of external heat fluxes, different specimen sizes and orientations, and long exposure times. This permits the researcher considerable freedom in exploring unusual burning characteristics. Typically, a 100 mm x 100 mm specimen, positioned horizontally under a conical heater, is exposed to a radiant heat flux of up to 100 kW/m^2. Obtained from the test run are ignition delay time, peak and average heat release rates, mass loss and mass loss rate, effective heat of combustion and rate of smoke and toxic gas evolution.

1.3 Advantages and Disadvantages of the Cone Test Method for Real Products The cone calorimeter method ("Cone") is not an error-free solution for all types of specimens (nor is any other single fire test protocol). Homogeneous, readily combustible specimens can be evaluated easily and with good repeatability in the Cone. On the other hand, composite materials, materials that burn with extensive char formation, and thermally stable and fire retardant materials all present difficulties in the determination of their flammability properties and in the interpretation of the results[21,22]. Some of the problems typically encountered are described below.

- For difficult to ignite specimens, even high heat flux conditions and the spark ignition source may not be adequate to evaluate the specimens' ignitability.
- Low heat release rates due to thermal stability or char formation may be irregular and less repeatable than for ordinary materials.
- A complication of testing products that do not burn readily is the "edge effect." The standard 100-mm square specimen has a much larger edge-to-surface-area ratio than that of full size panels in actual use. Thus, char forming specimens often evolve flammable gases from underlying material, around the edges of the specimen, even after the surface has developed a substantial char. This edge effect has been discussed previously[23], but has not been totally resolved.
- Thickness of test specimens, and swelling or char formation during the test, can influence the results.
- Testing of small specimens might not reflect some of the effects of full-scale fire testing, such as cracking, char formation and separation of layers.

The cone calorimeter was selected for these studies over the OSU heat release calorimeter (the device required by the FAA for compliance testing of aircraft interiors) because of the ease of setting different heat fluxes (up to 100 kW/m^2), the horizontal specimen orientation and the ready visibility and access to the specimen before and during a test. Also, the Cone permits continuous mass loss measurements and the calculation of an effective heat of combustion. Recently, a study from Omega Point Laboratories compared results from the cone calorimeter and the OSU calorimeter[24].

2. TEST MATERIALS AND PROCEDURES

All of the materials evaluated in this portion of this program were prepared at Polytechnic University. The polymer used was General Electric's aircraft grade ULTEM® 9075, a modified polyetherimide· The gray material, obtained in pellet form, was used as received (although protected from moist air it was not oven-dried as might have been done if mechanical properties had been important for this project). The material is marketed for aircraft use and therefore served as our "baseline" for flammability comparison purposes. It will be abbreviated simply as "PI" in this paper, in order to not imply any direct comparisons of data with any other Ultem products.

The laboratory procedure was as follows: 60 g. of the PI and the desired amount of an additive, if any (the study reported herein is exclusively on results of "blanks") were weighed to the nearest 0.1 g and then processed in a Brabender Plasticorder for 10 minutes at 350° C. The irregular pieces were then molded to a plate of approximately 1.5 mm (1/16") thickness in a Carver press at 103-138 MPa (15,000-20,000 psi) with the platens at 320° C. The molded plate was then cut to a 100 mm x 100 mm square for cone calorimeter testing. The leftover pieces were remolded to a thickness of 3 mm (1/8") and cut into test bars for oxygen index measurements (not included in this paper).

A limited amount of work was also conducted with RADEL R, a polyphenylene sulfone from Amoco, which is currently used in aircraft cabins. This material performed very well in regard to heat release rate, but we found it much more difficult to process in the laboratory equipment available at Polytechnic. It was characterized by limited ignitability at the 50 kW/m^2 heat flux level which was adopted for screening in this project; therefore, it was concluded that it would have been difficult to use this material to screen additives.

Plaques received at Omega Point Laboratories from Polytechnic were conditioned at 23° C and 50 % R.H., in accordance with the cone calorimeter standard test method. Other aspects of this test method were followed, except as noted herein.

Studies were conducted with the specimens prepared in three different specimen configurations, as described below:
1) pan – specimens were wrapped in aluminum foil on all except the exposed face and placed on ceramic fiber insulation ("Kaowool") on the specimen pan, in accordance with the standard ASTM E1354 mounting procedure.
2) pan with wires – specimens were prepared as for the "pan" configuration, and two small stainless steel wires were wrapped around both the specimen and the pan to reduce the effects of specimen warping and char formation. This procedure is used as an alternative to the "wire grid" recommended for intumescent materials in the Cone test standards because the thin wires generally are less intrusive on the formation of char and generally do not adversely affect the behavior of the specimens (the grid can significantly affect fire behavior).
3) edge frame – specimens were wrapped in aluminum foil in a similar fashion to the "pan" configuration, except they were mounted in the edge frame (or "retainer frame"), as described in the ASTM and ISO standards (the ISO standard requires the use of the frame). This configuration protects the edges of the specimen from burning in the early stages of the test run. It is recommended for multi-layer materials, where the cut edge might not reflect the composition of the specimens, and for thicker materials, especially char-forming materials, such as wood.

Experiments were also conducted with a modified test protocol based on that recommended in ISO 5660 and in a paper by Gensous and Grayson[25]. This procedure entails placing the specimen at a distance of 60 mm from the bottom of the cone heater, rather than the usual 25 mm, and is recommended for specimens that intumesce (i.e., develop a large char) under the given heat flux conditions. The additional distance from the cone heater (and from the spark igniter that is attached to the bottom of the cone heater at a distance of approximately 12.5 mm) permits the specimen to intumesce freely without risk of contacting the spark igniter or the heater, or requiring elevation of the cone heater during the test to prevent these occurrences. In these trials, the pan configuration was used; while the referenced study and the ISO test method require the edge frame. Advantages and disadvantages of this procedure will be discussed below.

3. RESULTS AND DISCUSSION

3.1 General Information on Results All of the results presented are based on a 300 s run time, even though these test specimens would have continued to burn (the standard protocol specifies that the run continues until after the specimen has burned out). All of the curves show continued combustion and a non-zero heat release rate at the end of the run. The run times for these specimens were limited to 300 s for the following reasons:

1) Most of the useful information from these materials (e.g., time to ignition and peak heat release rate) is obtained in the early stages of the experiment.
2) Once a char formed, the later stages of burning for these specimens were generally even more irregular than the early stages, due to small flames coming from cracks in the char or around the edges.
3) It is useful to have a fixed run time for comparison of parameters such as total heat release and effective heat of combustion.

The parameters reported from cone calorimeter evaluations include the following:

time to sustained ignition (t_{ig}) – the time until ignition of at least 4 s that does not require the presence of the spark igniter to continue

peak heat release rate (pk. HRR) – the maximum value of HRR during the 300 s; in case of multiple "peaks," the first one is noted

average heat release rate (avg. HRR) – an average of the HRR values over some time period (typically, 60 or 180 s) of time following the time to sustained ignition

total heat release (THR) – the integral of the HRR-time curve for the time of the run

effective heat of combustion (eff. Hc) – the THR divided by the total mass loss

mass loss – loss of mass of the specimen over the time of the run, recorded continuously

avg. specific extinction area (SEA) – a measure of the smoke evolved, normalized to the mass loss of the specimen and computed as an average for the time of the run

smoke release rate (SRR) – another time-dependent measure of smoke evolved, not dependent on mass loss

3.2 Specimen Mounting Alternatives and Interpretation of Results Unfortunately, there is no single specimen mounting configuration (nor any single heat flux) that is suitable for all types of materials and products. Some of the mounting techniques are appropriate for certain types of specimens, such as composites and thick specimens, and can be determined simply by the appearance of the specimen prior to testing. Other specimens may require extensive evaluation of different configurations prior to establishing a protocol. This is impractical in many cases for "standard testing" of a variety of materials. To complicate matters, comparisons of a composite with a homogeneous sample, of an intumescent with a non-intumescent material or of materials with different thicknesses will be difficult, if not impossible. Furthermore, the obvious parameters of time to ignition and peak heat release rate may not necessarily be the best to use for all materials (e.g., these could be more sensitive to specimen mounting configurations). Other choices, such as average HRR or effective heat of combustion, might be preferable. Finally, comparison of smoke values from different materials, especially ones that burn differently due either to their innate characteristics or due to the mounting method, must be considered carefully.

3.3 Test Results – Specimen Mounting The results of evaluating three series of PI "blanks" under the three different specimen preparation configurations (pan only, pan with wires and edge frame) are shown in Table 1 and in Figures 1, 2 and 3. There are clearly some differences in the test results. In addition, these differences are supported by observations. The specimens that were "restrained" either by wires or by the edge frame tended to result in the char breaking more readily than those unrestrained (i.e., in the "pan" configuration). The "break" in the edge frame specimens was a result of the char continuing to grow while the substrate was held down by the edge frame. Splits in the char/substrate created new flaming areas. In the tests with the fine wires, the wires cut into the char causing unburned material to be exposed (this was not readily apparent simply by viewing the burn, but was determined by redoing certain experiments after the heat release curves were obtained). As a result, the restrained specimens either burned more (i.e., higher HRR), as shown in the table and figures, or longer, sometimes with a second peak in the HRR curve (e.g., Fig. 3). While each of these mounting techniques would seem to have some drawbacks for testing specimens; the simplest one, using the pan configuration without wires, appeared to be preferable from a testing point of view. Therefore, subsequent studies were conducted with that configuration.

The repeatability of test results for these materials, in any of the configurations, is not as good as one would like in order to make meaningful comparisons. In particular, it is felt that the differences in times to ignition may be much too dependent on the proximity of the char to the spark igniter and, possibly, on operator techniques. Some of the lack of repeatability (e.g., in HRR values) is a result of the nature of the burning of this type of material. Furthermore, some differences in cone calorimeter results were apparent for different "batches" of test specimens. Therefore, comparisons have been made primarily for series of tests conducted at the same time. These specimens were very thin (1.5 mm), partly to accommodate the need to use as little material as possible during the screening portion of the test program. While similar problems in repeatability occurred with thicker specimen[24], the thinness of these specimens may account for some of the variability in results.

<h3 style="text-align:center">Table 1. Cone Calorimeter Results for PI Under
Three Different Mounting Configurations</h3>

Run Time = 300 s; Heat Flux = 50 kW/m^2

Specimen Config.	Run No.	t_{ig} (s)	Pk HRR (kW/m^2)	THR (kJ)	ML (g)	Eff. Hc (MJ/kg)	60s Avg. HRR (kW/m^2)	180s Avg. HRR (kW/m^2)	SEA (m^2/kg)
No Wires	61	84	73.1	70.9	4.6	15.6	47.4	32.2	297
No Wires	62	90	78.5	52.1	3.4	15.1	50.4	26.4	338
No Wires	63	91	48.5	45.8	2.9	15.9	33.5	22.3	343
	Avg.	88	66.7	56.2	3.6	15.5	43.8	27.0	326
Wires	55	74	139.7	92.4	4.6	20.1	57.0	43.2	317
Wires	56	68	120.4	105.3	5.5	19.3	69.3	48.9	306
Wires	57	94	150.7	95.5	5.0	19.2	79.3	46.9	350
	Avg.	79	136.9	97.7	5.0	19.5	68.5	46.3	324
Edge Frame	58	61	74.0	78.9	4.0	20.0	35.7	39.9	292
Edge Frame	59	66	66.5	75.0	4.2	17.9	26.4	32.3	253
Edge Frame	64	61	83.7	48.7	1.8	26.5	24.1	22.9	322
	Avg.	63	74.7	67.5	3.3	21.5	28.7	31.7	289

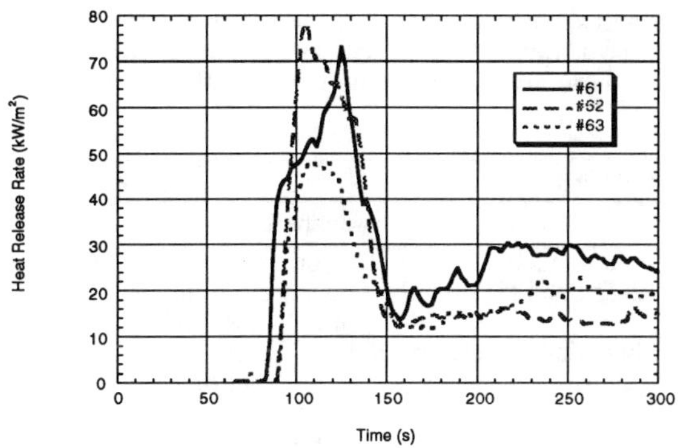

Figure 1. Heat Release Rate for Pan Configuration without Wires

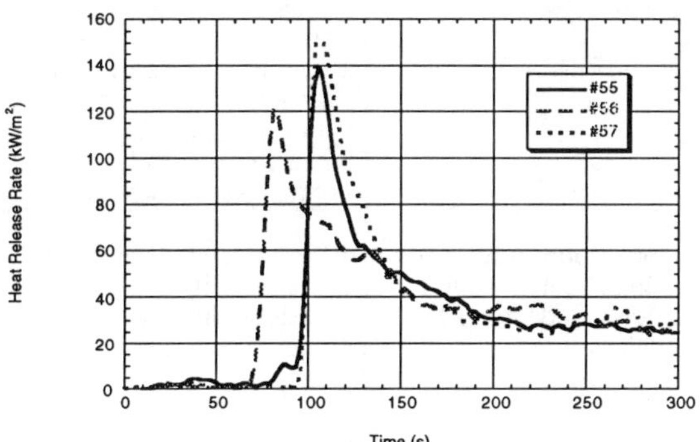

Figure 2. Heat Release Rate for Pan Configuration with Wires

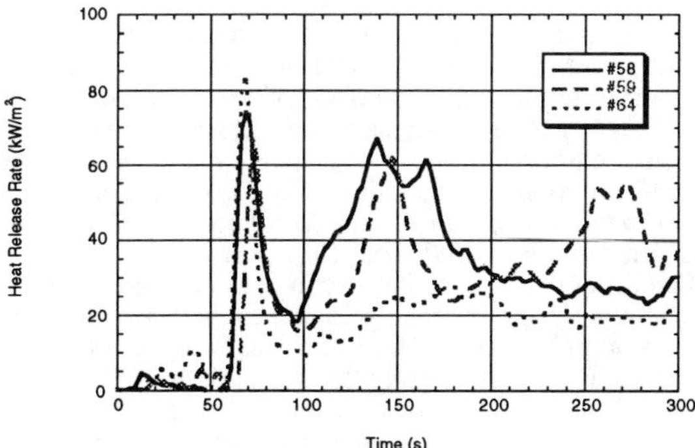

Figure 3. Heat Release Rate for Edge Frame Configuration

3.4 Test Results – Specimen Positioning Additional experiments were tested at two different distances from the cone heater. There are both advantages and disadvantages to testing intumescent materials at the usual 25 mm separation or at the recommended 60 mm separation. At 60 mm, the cone heater was set so that the heat flux at the surface of the specimen was at the desired level. In order to obtain a heat flux of 50 kW/m^2 at 60 mm in our laboratories, the heater was set to an intensity that was approximately 62 kW/m^2 at the usual 25 mm distance. Some of the advantages and disadvantages of the two options are summarized in the chart below (while these are the opinions of the authors, some are the same as those described in reference 25).

Advantages	Disadvantages
25 mm separation	
• standard practice, recommended for most materials in both ASTM and ISO standard test methods • spark igniter at recommended distance from specimen (ca. 12.5 mm)	• intumescent char could touch sparker or cone, resulting in improper ignition and/or mass loss errors • as a result, cone heater assembly must be moved during the run, leading to operator variability • intumescent char is exposed to different heat fluxes as it grows into center of cone heater
60 mm separation	
• no need to move heater during test run • specimen unlikely to touch spark igniter or cone heater • visibility of specimen and char formation	• heat flux must be measured with non-standard setup at the new distance • different heat flux distribution across specimen surface from standard practice • igniter is farther from surface of specimen than usual, possibly leading to non-repeatability or delayed ignition • extra care must be taken against cross-drafts

The results of testing these materials at two different distances from the cone heater are presented in Table 2 and in Figures 4 and 5. The repeatability of the results from this material makes it somewhat difficult to develop firm conclusions; however, the irregularity of results may also be a factor in determining whether or not there is any practical difference between the two specimen exposure distances for this material.

The results at 60 mm separation distance, compared to 25 mm, appear to have somewhat longer times to ignition (but not significantly so) and somewhat higher peak HRR (probably not significant, considering the scatter). The 60 mm tests had higher THR, generally higher mass loss, higher Hc, and higher avg. HRR values (these all seemed to be significantly higher than the results at 25 mm).

These differences in results may be due to differences in heat flux exposure to the surface of the char for the two configurations. Heat flux measurements as a function of distance from the bottom of the cone heater (and across the surface of the specimen) have been reported by Gensous and Grayson[25] and by Paul[26]. These studies illustrate that the heat flux pattern as a function of distance from the bottom of the cone heater is not the same in the 0 - 25 mm range as it is in, say, the 40 - 60 mm range. Due to the unique conical shape of the heater, heat flux exposures of char closer to the heater than 25 mm do not see a much higher heat flux (less than 20 percent increase). On the other hand, heat flux exposures of chars that start at 60 mm separation (with the flux set at that level), actually increase at a greater rate (e.g., the heat flux at 40 mm from the cone heater is approximately 25 percent higher than at 60 mm; at 25 mm, the flux is more than 30 percent higher than at 60 mm). Thus, while the specimens situated 60 mm from the heater have the same nominal starting flux as at 25 mm and have fewer restrictions on char growth, the growing char actually might be exposed to a higher flux than for specimens that start at 25 mm. This observation seems counter-intuitive since the heater appears to be so close to the specimen at 25 mm, and further analysis of these circumstances is recommended.

**Table 2. Cone Calorimeter Results for PI at
Two Different Exposure Distances (25 mm and 60 mm)**

Run Time = 300s; Heat Flux = 50 kW/m^2

Specimen Config.	Run No.	tig (s)	Pk HRR (kW/m^2)	THR (kJ)	ML (g)	Eff. Hc (MJ/kg)	60s Avg. HRR (kW/m^2)	180s Avg. HRR (kW/m^2)
25 mm	230	64	106.7	75.4	4.27	17.7	39.5	36.9
25 mm	231	69	113.2	48.6	2.75	17.7	32.2	19.9
25 mm	232	71	89.3	45.4	2.76	16.5	28.9	20.5
	Avg.	67	110.0	62.0	3.51	17.7	35.9	28.4
60 mm	234	84	122.6	82.4	3.80	21.7	50.4	39.8
60 mm	235	64	141.9	99.3	4.20	23.7	60.2	45.8
60 mm	236	85	101.1	101.9	4.86	21.0	59.1	49.4
	Avg.	74	132.3	90.9	4.00	22.7	55.3	42.8
60 mm + foil	233	80	110.6	78.3	4.25	18.4	65.0	40.4

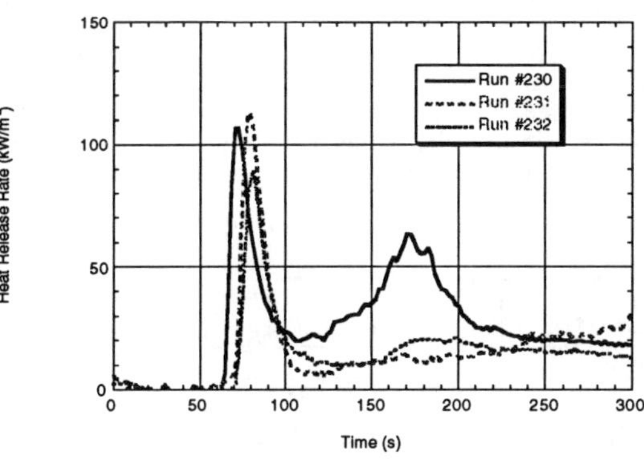

Figure 4. Heat release rate curves at 25 mm separation

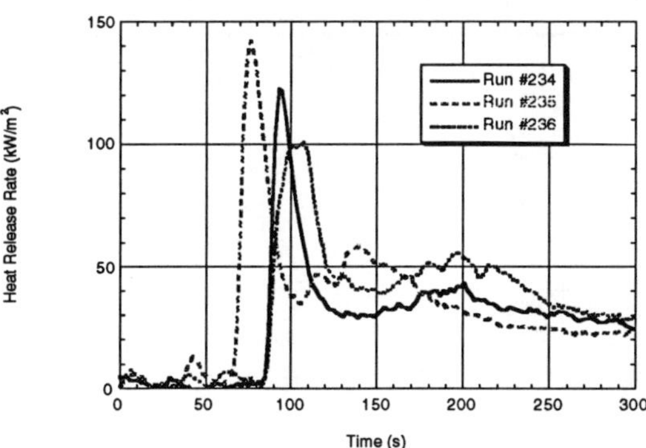

Figure 5. Heat release rate curves at 60 mm separation

Results are also shown in table 2 for a single test conducted at 60 mm distance, but with a tall aluminum foil "boat" that extended up to approximately 25 mm from the cone heater. It has been established (Lukas[27] and at Omega Point Laboratories) that this configuration, with the heat flux set at 25 mm, maintains a constant heat flux all the way to the bottom of the specimen holder (probably due to reflection from the Al foil). Many of the results of this test are intermediate between the other two configurations. The main advantage of this option is that the heat flux can be left as set for the 25 mm distance. The main disadvantage over tests at 60 mm is that the specimen char development is essentially "invisible" due to the tall Al boat.

3.5 Smoke Results Smoke data from the cone calorimeter test method may be expressed either as a time-dependent function or as a number representing total smoke. The former is commonly expressed as "Extinction Coefficient" (analogous to optical density), with units of m^{-1}; while the latter is commonly reported as "Specific Extinction Area" (SEA), a value representing the overall quantity of smoke produced, normalized to specimen mass lost, with units of m^2/kg. An additional calculation, not mentioned in the standard, is "Rate of Smoke Release," with the units of m^2. This may be reported either as a time varying function or as the summation, "Total Smoke Release" (TSR). Table 3 contains SEA and TSR values for the runs conducted at different exposure distances, as an example of smoke results. In this case, the two sets of SEA values are within about 18 percent of one another; whereas the TSR values are different by as much as 30 percent. Figure 6 contains a plot from one run (#231) of RSR and extinction coefficient. Figure 7 contains a plot from the same run of RSR and SEA. It is apparent that the rate of smoke release is similar in pattern to extinction coefficient, and represents the pattern of smoke evolution as a function of time. On the other hand, RSR and SEA produce very different time-dependent plots, the latter being highly dependent on the incremental mass loss obtained during the course of the experiment. Thus, while SEA is the common parameter for reporting smoke values from the cone calorimeter, the RSR and TSR better represent usable values both as time-dependent and as summary functions.

Table 3. Cone Calorimeter Smoke Results

Specimen Config.	Run No.	Avg. SEA (m^2/kg)	TSR (m^2)	CO Yield
25 mm	230	266	1.14	0.17
25 mm	231	289	0.79	0.17
25 mm	232	268	0.74	0.16
	Avg.	278	0.97	0.17
60 mm	234	333	1.27	0.15
60 mm	235	323	1.36	0.17
60 mm	236	298	1.45	0.13
	Avg.	328	1.31	0.16

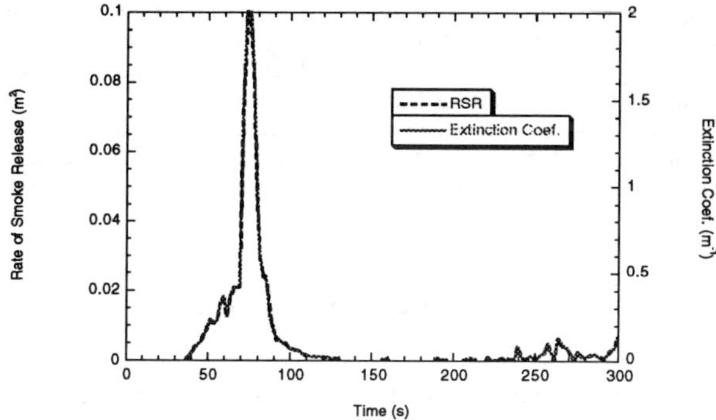

Figure 6. Rate of Smoke Release and Extinction Coefficient

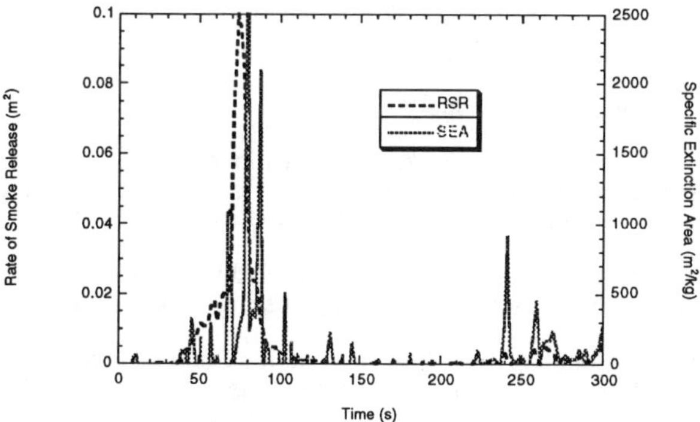

Figure 7. Rate of Smoke Release and Specific Extinction Area

4. CONCLUSIONS AND RECOMMENDATIONS

It has been shown in previous studies that the cone calorimeter test method can be very repeatable for "well-behaved" materials, materials that are homogeneous in their burning behavior and ignite readily under the test conditions. On the other hand, it has been demonstrated in this study that significant problems can arise when dealing with specimens that are not so well-behaved. Intumescent, thermally stable materials, for example, exhibit burning characteristics that may be very subject to influences such as specimen mounting. Furthermore, consideration must be given to positioning intumescent specimens. Such specimens might not exhibit the same degree of repeatability as specimens with a more homogeneous burning characteristics.

Interpretation of cone calorimeter data is an important aspect of utilizing this test method. These data should be analyzed within the context of the test method, and with consideration for the nature of the specimens being evaluated and the reasons for conducting the studies. Smoke data, as well as heat release data, must be carefully evaluated before drawing conclusions.

Additives to aircraft-grade Ultem 9075 as the base resin have been evaluated for comparison to the results shown herein. Tests have been conducted at both 25 mm and at 60 mm separation. The scatter and occasional non-repeatability of the results have made comparisons difficult. A report of these studies is being prepared.

5.0 ACKNOWLEDGMENT

This work was supported by the United States Department of Transportation (DOT) under Contract Number DTRS-57-96-C-00011, to whom grateful acknowledgment is made. The Technical Monitors for this "Phase II" SBIR (Small Business Innovation Research) grant were Dr. Thor Eklund and Mr. Constantine Sarkos, Federal Aviation Administration Technical Center, Atlantic City Airport, NJ. Discussions with Dr. Eklund, Mr. Sarkos, Dr. Richard Lyon and others in the FAA have been very helpful during the conduct of this program. Gratitude is also expressed for the cooperation of the suppliers of the materials used in this part of the program.

6.0 REFERENCES

1. "Aircraft Cabin Safety and Fire Survivability," Hearings before the Government Activities and Transportation Subcommittee of the Committee on Government Operations, United States House of Representatives, U. S. Government Printing Office, Washington, DC 20402, April 11, 1991.

2. "Aviation Safety: Slow Progress in Making Aircraft Cabin Interiors Fireproof," Report to Congressional Requesters, United States General Accounting Office, Gaithersburg, MD 20884, GAO/RCED-93-37, 1993.

3. FAR 25.853 through Ammendment 25-72.

4. R. G. Hill, T. Eklund, and C. P. Sarkos, "International Conference for the Promotion of Advanced Fire Resistant Aircraft Interior Materials," Report No. DOT/FAA/CT-93/3, Federal Aviation Administration, Department of Transportation, February 9-11, 1993.

5. 41st and 42nd International SAMPE Symposiums and Exhibitions, 1996 and 1997, Society for the Advancement of Material and Process Engineering, Vols. 41 and 42.

6. U. Sorathia, T. Dapp and J. Kerr, "Fire Barriers for Composite Structures," DTRC/SME-90-13, David Taylor Research Center, Department of the Navy, February 1990.

7. U. Sorathia, T. Dapp, J. Kerr and J. Wehrle, "Flammability Characteristics of Composites," DTRC/SME-89-90, David Taylor Research Center, Department of the Navy, November 1989.

8. U. Sorathia, "Fire Resistant Barriers for Composites," DTRC/SME-88-28, David Taylor Research Center, Department of the Navy, May 1988.

9. U. Sorathia, "Survey of Resin Matrices for Integrated Deckhouse Technology," DTRC/SME-88-52, David Taylor Research Center, Department of the Navy, August 1988.

10. U. Sorathia, C. M. Rollhauser and W. A. Hughes, "Improved Fire Safety of Composites for Naval Applications," Fire and Materials, Vol. 16, 119-125 (1992).

11. J. E. Brown, J. J. Loftus and R. A. Dipert, "Fire Characteristics of Composite Materials - A Review of the Literature," NBSIR 85-3226, National Bureau of Standards (now National Institute of Standards and Technology), U. S. Department of Commerce, August 1986.

12. J. E. Brown, E. Braun and W. H. Twilley, "Cone Calorimeter Evaluation of the Flammability of Composite Materials," NBSIR 88-3733, National Bureau of Standards (now National Institute of Standards and Technology), U. S. Department of Commerce, March 1988.

13. T. Ohlemiller, and S. Dolan, "Ignition and Lateral Flame Spread Characteristics of Certain Composite Materials," NISTIR 89-4030, National Institute of Standards and Technology (Formerly National Bureau of Standards), U.S. Department of Commerce, January 1989.

14. T. Ohlemiller, "Assessing the Flammability of Composite Materials," NISTIR 89-4032, National Institute of Standards and Technology (Formerly National Bureau of Standards), U.S. Department of Commerce, January 1989.

15. A. F. Grand, "Laboratory Evaluations Using the Cone Calorimeter," Presentation (no written paper) at the Spring Meeting of the Fire Retardant Chemicals Association, San Antonio, March 1994.

16. A. F. Grand, "The Influence of Fire Calorimetry Results on Product Research and Development Activities," Symposium on Fire Calorimetry, 50th Calorimetry Conference, July 23-28, 1995; in DOT/FAA/CT-95/46, M. M. Hirschler and R. E. Lyon, Eds., available from NTIS, Springfield, VA.

17. ASTM Research Report #E05-1008 on "Cone Calorimeter Inter-laboratory Trials," ASTM (January 3, 1990).

18. ISR Research Report from Committee E05 on "Interlaboratory Test Study on ASTM E1354 Standard Test Method for Heat and Visible Smoke Release Rates for Materials and Products Using an Oxygen Consumption Calorimeter," ASTM Institute for Standards Research, 1993.

19. V. Babrauskas, and S. J. Grayson, *Heat Release in Fires*, Elsevier Applied Science, 1992.

20. R. D. Peacock, et al., "An Update Guide for HAZARD I Version 1.2," NISTIR 5410 (May 1994); also "Technical Reference Guide" and "Software User's Guide" for "HAZARD I Fire Hazard Assessment Method" and for "Version 1.1," NIST Handbook 146 (1989 and 1991, respectively)

21. A. F. Grand, "The Use of the Cone Calorimeter to Assess the Effectiveness of Fire Retardant Polymers Under Simulated Real Fire Test Conditions," Interflam '96, Cambridge, England, March 26-28, 1996.

22. A. F. Grand, "Heat Release Calorimetry Evaluations of Fire Retardant Polymer Systems," 42nd International SAMPE Symposium and Exhibition, May 4-8, 1997, Society for the Advancement of Material and Process Engineering, Vol. 42, pp. 1062-1070.

23. T. Kashiwagi, and T. G. Cleary, "Effects of Sample Mounting on Flammability Properties of Intumescent Polymers," Fire Safety J., Vol. 20, , 203–225 (1993).

24. A. F. Grand and J. R. Bundick, "A Comparison of Results between the OSU and Cone Calorimeter Test Methods," Fire and Materials '98, 5th International Conference and Exhibition, February 23-24, 1998, San Antonio.

25. F. Gensous and S. Grayson, "Improved Procedure for Testing Intumescent Materials using the Cone Calorimeter," Interflam '96, Interscience Communications Ltd, England, pp. 977-981.

26. K. T. Paul, "Cone calorimeter: Initial experiences of calibration and use," Fire Safety J., vol. 22, pp. 67-87 (1994).

27. C. Lukas, "Communication: Measurement of heat flux in the cone calorimeter," Fire and Materials, vol. 19, pp. 97-98 (1995).

HEALTH HAZARDS OF CARBON FIBERS FROM BURNING COMPOSITE MATERIALS

Sanjeev Gandhi[†] and Richard E. Lyon
Fire Safety Section AAR-422,
Federal Aviation Administration
Atlantic City Int'l Airport, NJ 08405

[†]Galaxy Scientific Corporation

ABSTRACT

The use of polymer matrix composites continues to increase for primary and secondary structural components in commercial aircraft due to the superior strength to weight ratio of these materials compared to metals. However, there are concerns about the potential health hazards from carbon fibers released from burning composites in aircraft fires. The health risks stem from airborne fibrous dust and micron-size, respirable fibers following aircraft fire or explosion. An extensive review of the literature was conducted to examine the factors contributing to inhalation toxicity of fibrous matter. The toxicological significance of carbon fibers was reviewed in terms of the size and concentration of fibers emitted due to the incineration of composites. The current data derived from animal studies do not reflect high acute toxicity that can be attributed to carbon fibers. Further work is needed to determine if volatile organic vapors have a synergistic effect towards the toxicity of particulates.

KEY WORDS: Carbon Fiber, Polymer Matrix Composites, Combustion Toxicology, Pulmonary Fibrosis, Lung Cancer

1. INTRODUCTION

Polymer structural composites are engineered materials comprised of continuous, high-strength fibers impregnated with a polymer matrix to form a reinforced layer (ply) which is subsequently bonded together with other layers under heat and pressure to form an orthotropic laminate. Continuous carbon fibers are the preferred material for reinforcement in applications where significant reduction in weight is needed along with improved functional performance. Fibers are used alone or in combination with other fibers (hybrids) in the form of continuous fiber fabrics, tapes, and tows or as discontinuous strands. Carbon and graphite fibers are manufactured in a continuous process from materials known as precursor fibers. Polyacrylonitrile (PAN) and pitch are the most commonly used precursors. Typically, carbon fibers have a nominal diameter of

6-8 μm and comprise about 55-60 volume percent of the laminate with the polymer resin being the remainder.

Polymer matrix composites are increasingly replacing metals in commercial aircraft for primary and secondary structural component applications due to their superior specific strength. The usage of lighter-weight structural composites is projected to increase significantly in next-generation commercial aircraft. Figure 1 illustrates the rising trend in the use of polymer composites in commercial airplanes supplied by two leading manufacturers. The percent structural weight is based on the operating empty weight of the aircraft. The Boeing 777 aircraft, with a three-fold increase in the percent weight of structural composites, represents a significant leap from previous generation airplanes. Boeing estimates that the structural weight fraction of polymer composites in their commercial airplanes will increase to about 20 percent over the next 15 years (1).

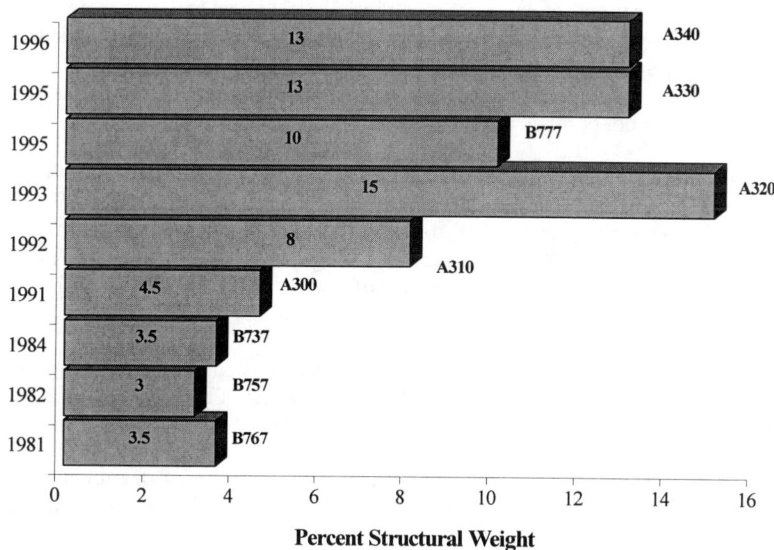

Figure 1. Use of fiber composite in commercial transport aircraft.

However, there are rising concerns about the potential health and safety hazards posed to mishap personnel responding to postcrash aircraft fires due to the carbon fibers released from burning composites. A number of incidents have been reported indicating the irritant and toxic effects of fibrous particulates on firefighting, rescue, and cleanup personnel (2-4). Crewmembers engaged in various activities during and after the aircraft mishap have reported varying degrees of health effects ranging from eye and skin irritation to severe respiratory problems.

Burning of advanced composite materials generates a complex mixture of gases and visible products of incomplete combustion, collectively referred to as smoke. The smoke stream contains a mixture of evolved gases, organic vapors, and solid particles. As the organic resin burns off, the reinforcing fibers are exposed to a turbulent, oxidizing environment that causes them to break up and erode into small microfiber fragments.

Aerosols constitute the visible component of smoke and are comprised of aggregates of solid particles adsorbed with condensed vapors of organic compounds. The toxicity of gases produced during the combustion of polymer materials has been widely studied. Purser has provided a detailed review of the advances over the past three decades in the measurement and health effects of gases (5). In contrast, the toxicity of solid particles including fibers has received little attention. This paper reviews the nature of health hazards posed by carbon fibers released from incineration of composites. In view of the projected increase in the use of composites in commercial aircraft, these hazards are of great concern to firefighters and rescue personnel who work in close proximity to the crash and fire site. Possible health effects are discussed in terms of fiber size and expected inhalation exposure from burning composites.

2. HEALTH HAZARDS FROM FIBROUS MATTER

There are two major routes of exposure involving fibers–dermal and inhalation. Irritation of the skin and eyes resulting from a reaction to sharp, fragmented fibers of a diameter greater than 4-5 μm is a typical response to dermal exposure. The severity of the exposure depends on the fiber size and stiffness. The skin irritation and sensitization effects are however not permanent (6-7).

The inhalation exposure mode, however, poses the greatest potential for human health effects from fibers. Numerous studies have described the causal association between physical dimensions of the fiber and its toxicity. These studies address the health risks due to inhalation of fibrous material via chronic, long-term exposure in the work place environment. Warheit reviewed the toxicology of various mineral and organic fibers including asbestos, rock wool, fiberglass, kevlar, and carbon (7).

The impact on human health due to fiber inhalation is dependent on the fiber deposition and clearance from the pulmonary system. The physical size of the fiber is the most important factor affecting deposition. Any particle is considered a fiber if its aspect ratio, i.e., length to diameter ratio (L/D) is greater than 3 (7). The respirability of a fiber is determined in terms of the aerodynamic diameter D_a, the diameter of an equivalent spherical particle having the same terminal velocity as the fiber. Fibers with diameters smaller than 3 μm and lengths < 80 μm fall in the respirable range and can penetrate deep into the lungs. The bronchial[*] tubes in the lungs have a characteristic branching pattern where each bronchus is divided into two smaller airways. With each division the airways become smaller, resulting in airflow changes along the bronchial path. Thus, fiber deposition along the airways is determined by the fiber diameter and density. Fibers that are not parallel to the flow axis are more likely to be deposited on the airway walls. The larger airway bifurcations are the preferred sites of fiber deposition due to interception. Fibers that are parallel to the airflow can penetrate deeper into the lung's progressively smaller airways. Animal studies with glass fibers have shown that fibers that are small enough to pass through bronchial airways are found deposited in the alveoli tissue (7).

---

[*] Bronchial section consists of six generations of bronchial airways of increasing finer diameter. Respiratory bronchioles are the smallest bronchioles (diameter 0.5 μm) that connect the terminal bronchioles to the alveolar ducts.

Fibers are deposited primarily in the bronchial section. The resulting pathological effects from retained fibers are first observed at this section of the pulmonary system. However, a majority of inhaled fibers is cleared from the respiratory airways over time. There are three modes of fiber clearance, translocation, ingestion due to cellular activity, and dissolution in lung fluids. Translocation is described as movement of the fibers from their initial deposition site to other sites in the lung and into the alveolus. The alveolus is composed of thin walls of epithelial cells that are very sensitive to foreign matter. Also present are pulmonary alveolar macrophages, cells that serve an important function called phagocytosis. A pulmonary self-defense mechanism, macrophages remove bacteria, dead cells, and foreign particles by ingestion. The removal efficiency is limited by the macrophage diameter ($\sim$ 10-15 μm). Fiber and particles with length <10 μm are easily cleared by the macrophages (7-8). Longer fibers ($\sim$20-30 μm), loading of macrophage occurs, slowing the clearance process. The impairment of macrophage clearance functions due to ingestion and buildup is used as an indicator of pathology due to fibers.

The impact of fiber size and duration of residence on the respiratory system has been investigated in numerous studies. Penetration of fibers in the lung causes an inflammatory response called alveolitis. The lung's efforts to repair this damage result in progressive scarring of the lung walls. The interstitial scarring due to deposition of fibrous matter is termed pulmonary fibrosis (8). Fibrosis slows the process of fiber clearance resulting in longer fiber retention. Extended chronic exposure to asbestos is known to result in bronchogenic carcinoma or lung cancer. Epidemiological studies have demonstrated the causal relationship between asbestos exposure over a period of several years and development of lung cancer (8). It is believed that longer fibers with longer retention in the lung are more likely to cause the onset of malignancy. Shorter fibers i.e., < 10 μm are more easily cleared from the lung due to ingestion by the macrophage. The risk of lung cancer increases with the presence of a substantial number of long fibers (>10 μm) with $\approx$ 0.8 μm diameter.

3. TOXICITY OF CARBON FIBERS

The National Aeronautics and Space Administration (NASA) assessed the risk of accidental release and dispersal of carbon fibers from burning composite fires materials following an aircraft crash and explosion (9). A series of small- and large-scale fire tests of composites were conducted to simulate aircraft fires. Fibers were collected using adhesive coated papers and analyzed for fiber count and size distribution by optical and electron microscopy. Microscopic fiber analysis revealed that fiber diameter was significantly reduced to micron size in the fire due to fiber oxidation and fibrillation. Such micron-sized fibers can be inhaled and lead to potential health hazards. The frequency distributions of fiber diameters collected in a burn test compared to the frequency distribution of the original carbon fibers diameter are shown in the Figure 2.

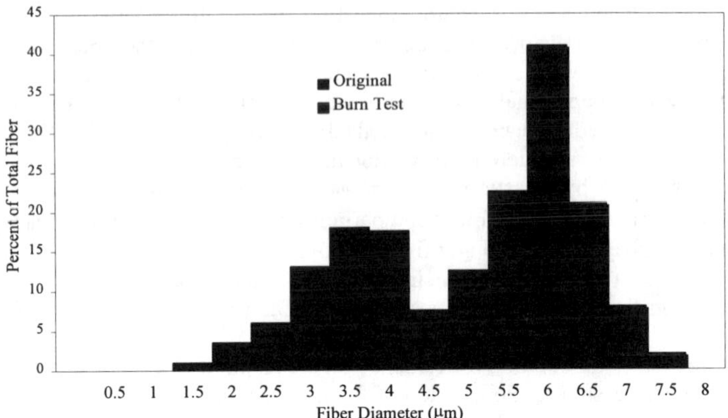

Figure 2. Fiber diameter distribution of original carbon fibers compared with that of fibers released during the burn test.

The collected fibers had a mean diameter of 4 μm versus 6.5 μm for the original fibers. At extreme flame temperatures (> 900 ^{0}C) under oxygen-rich test conditions, large amounts of fibers were completely consumed through oxidation. The overall fiber diameter was reduced drastically in the flame after the matrix was burned away and the fibers were exposed to the flame. Considering the potential health hazards associated inhalation of micron-sized fibers, further efforts were devoted towards physical characterization of these fibers using scanning electron microscopy (SEM).

In a later paper, Seibert presented the results of SEM analysis of respirable fibers (diameter <3 μm, length <80 μm); these fibers constituted less than 24 percent of the total fibers released from burning composites (10). The respirable fibers had an average diameter of 1.5 μm with an average length of 30 μm. Overall, the fiber size spectrum ranged from ≈ 0.5 μm to 5 μm in diameter. Further studies were undertaken to quantify the respirable fiber concentrations and to determine the potential dose from exposure. Fiber concentration was measured by drawing samples from inside the smoke plume with composite material burning inside a 10.7 m diameter pool of jet fuel. All fibers with an L/D ratio greater than 3 were counted and fiber concentrations were determined for a 20-minute burn time. The results indicated an average fiber concentration of 0.14 fibers/cm^3. This is significantly lower than the OSHA mandated criteria for asbestos exposure. Lacking any definitive data on pathological effects, the study concluded that carbon fiber exposure be treated as a nuisance problem similar to fibrous glass or sawdust.

Additional data on the characteristics of the airborne fibers have come from recent aircraft postcrash investigations. Mahar measured the fiber concentration (fibers/cm^3) and aerodynamic diameter of the fibers collected at a military-jet crash site. Mahar reported that there was a significant increase in the particulate levels during cleanup operations with the disturbance of the aircraft wreckage (11). Less than 20 percent of the collected fibers were found to be in respirable range, that is with aerodynamic diameter smaller than 10 μm. The total concentrations of the fibers collected from breathing air zones ranged from 0.02-0.06 fibers/cm^3. The Navy guidelines limit exposure to a time-weighted average of 3.5 fibers/cm^3 of air and a maximum of 10 fibers/cm^3 over a 40-hour workweek.

The U.S. Coast Guard (USCG) recently conducted a series of tests to characterize the graphite fibers emitted from burning graphite/epoxy composites (4). The laboratory-scale tests were conducted at 50 and 75 kW/m^2 using the cone calorimeter equipped with a modified sampling system to maximize the collection of released fibers from the burning composites. The USCG test results are compared to those of NASA and Mahar's crash site investigation in Table 1.

Table 1. Carbon fiber exposure data.

Study	Respirable Fibers (%)	Fiber Size (μm)		Fiber Exposure (f/cm^3)	
		Diameter	Length	Data	NIOSH Standard
NASA (1970s)	<24	1.5	30	0.14	10
Crash Site (1990)	<20	2	7	0.04	10
US Coast Guard (1997)	23	2.5 2.4	50 40	ND	10

To date, no toxicological studies have been conducted to assess the health effects of inhalation of carbon fibers released in fires. No epidemiological or exposure studies could be found on firefighter exposure to fibrous particles. However, few studies have been done on the harmful effects of chronic exposure to airborne carbon fibers and dust in an occupational setting. Martin et al. evaluated the potential health effects of dusts generated during the machining of six carbon fiber-epoxy composites (12). This was the first animal study to focus on the pathological responses caused by mechanically generated aerosol from composites rather than raw carbon fibers. Characterization of the aerosol revealed very few fibers in the generated particles. The fiber diameter ranged between 7-11 μm and there was no evidence of fiber attenuation due to fibrillation. The mean diameter of the respirable fraction of non fibrous particles was 2.7 μm. The particles were introduced via intratracheal injection directly into rat lungs. Invitro studies were also done using rabbit alveolar macrophages. The assessment of pathological damage was based on several end points including cytotoxicity, phagocytosis, and the presence of certain enzymes and proteins in the lavage fluid. The results indicated that two samples from carbon fiber-epoxy composites cured with an aromatic amine agent were significantly more toxic than the controls. The authors concluded that exposure to carbon fiber and composite dust should be treated as more than a nuisance and that further studies with longer-term inhalation exposure were needed to characterize their toxic potential. These findings are consistent with the carcinogenicity of aromatic amines used as curing agents, e.g., 4, 4' methylenedianiline (MDA). OSHA lists MDA as an animal carcinogen and a suspect human carcinogen by any exposure route: ingestion, inhalation, or dermal (4).

Kwan characterized the physical size and aerodynamic properties of particulates generated during laser machining of composites (13). The particle size analysis did not reveal generation of any fiber-like particles (L/D>3). The respirable fraction consisted of spherical particles with D_a <2 μm. Kwan's study did not include animal testing for toxicity. However, chemical extraction of the sampled particulate matter revealed the presence of adsorbed volatile organic compounds with aromatic structure. Most of these chemicals were polycyclic aromatic hydrocarbons (PAH). The toxic effects of these PAH species based on the National Institute of Occupational Safety and Health database ranges from irritation to carcinogenicity. Recent studies at the US Air Force Toxicology Division, support the polycyclic aromatic nature of organic compounds generated during combustion of advanced composite materials (14). Researchers conducted a series of tests using a modified University of Pittsburg apparatus for evaluating combustion products of composites used in military aircraft. A large number (over 90) of chemical species were identified which can be broadly classified as, (i) polycyclic aromatic hydrocarbons, (ii) nitrogen-containing aromatics such as aniline, and (iii) phenol-based organic compounds. Several of these chemicals e.g. aniline, quinoline, and toluidine, are known to induce mutagenic and carcinogenic effects in animals.

Waritz conducted an inhalation study using mechanically generated carbon fiber aerosols derived from polyacrylic (5, 7). Rats were exposed to carbon fiber aerosol for 6 hours/day, 5 days a week, for 16 weeks. The dose was comprised of fibers ~3 μm in diameter, 72 percent of which were 10-60 μm long. This was a single dose study with fiber concentration at 20 mg/m^3 ($\approx$ 40 fibers/cm^3). The rats were sacrificed after 4, 8, 12, and 16 weeks of exposure and after 35 and 80 weeks of recovery. No fibers were found in the lung tissue. Histopathological evaluation did not reveal any fibrosis effects or any changes in pulmonary function response. Warheit conducted a multidose study in which rats were exposed to respirable (1-4 μm diameter), experimental pitch carbon fibers for 1-5 days at concentrations of 50-90 fibers/cm^3 of air (5, 7). The effects in terms of histopathological and pulmonary changes were compared with control rats exposed to experimental PAN-based carbon fibers of averaging 4.4 μm in diameter. The respirable carbon fiber exposure caused only transient, dose-dependent, inflammatory response. Consistent with Waritz, there were no significant pulmonary function changes or evidence of fibrosis in either the pitch- or PAN-based carbon fibers.

The Civil Aviation Authority in the United Kingdom has initiated research to examine the toxicological significance of carbon fibers combined with gaseous products from several structural composites used in a commercial jet aircraft. Results from these studies are not yet available (4).

4. CONCLUSIONS

The structural applications of fiber reinforced polymer composites in commercial transport aircraft structures are increasing steadily and are projected to grow significantly in next-generation aircraft. There are increasing concerns regarding the health hazards posed to firefighters and rescue personnel due to the gaseous, particulate, and fibrous combustion products generated from the incineration of structural aircraft components. During postcrash, spilled fuel fires involving aircraft composites, micron-sized fibers are generated due to oxidation and fibrillation under extreme thermal conditions (temperatures >1000 ^{0}C). These fibers are small enough to be inhaled and deposited in the deep lung region, transporting with them any volatile organic compounds generated in fire. Fiber concentration data from burn tests and crash site investigations suggest that the exposure concentration is much lower than the OSHA mandated limits for asbestos and other silica based fibrous materials. To date, epidemiological studies have not been conducted to assess human exposure to carbon fibers released from burning composites nor have animal studies been published on the role of these fibers in inhalation toxicity. Studies on exposure to respirable sized raw PAN- and pitch-based carbon fibers do not indicate significant adverse health effects such as pulmonary fibrosis. Other studies involving occupational exposure to carbon dust and fibers generated during machining and grinding of composites do not provide adequate evidence of fibrosis or carcinogenic effects.

However, recent studies have identified large number of hazardous chemicals that are associated with particulates emitted during combustion of advanced composites. Several of these chemicals are known to cause mutagenic and carcinogenic effects in animals. Further work is needed to assess the implications of any synergistic or antagonistic interactions between the fibers and volatile organic compounds.

5. REFERENCES

1. U. Sorathia et. al., "A Review of Fire Test Methods and Criteria for Composites," SAMPE J., 33(4), 23, (1997).
2. C. Bickers, "Danger: Toxic Aircraft," Jane's Defense Weekly, 711, (1991).
3. Anon., "Danger: Fibers on Fire," Professional Engineering, 8(11), 10, (1995).
4. S. Gandhi and R. E. Lyon, Health Hazards of Fibers from Burning Composites, FAA Technical Note, in print, 1998.
5. D. A. Purser in P. DiNenno, ed., Toxicity Assessment of Combustion Products, SFPE Handbook of Fire Protection Engineering, 2nd Edition, Society of Fire Protection Engineers, Boston, Massachusetts, 1995, pp. 85-146.
6. V. Vu et. al., "Chronic Inhalation Toxicity and Carcinogenicity Testing of Respirable Fibrous Particles," Regulatory Toxicology and Pharmacology: RTP, 24(3), 202, (1996).
7. D. B. Warheit, Fiber Toxicology, Academy Press, New York, 1993.
8. M. Lippman in D. B. Warheit, ed., Biophysical Factors Affecting Fiber Toxicity, Fiber Toxicology, Academy Press, New York, 1993, pp. 259-303.
9. V. L. Bell, "Potential Release of Fibers from Burning Carbon Composites," NASA N80-29431, Washington D.C., 1980.
10. J. F. Seibert, "Composite Fiber Hazard," Air Force Occupational and Environmental Health Laboratory, AFOEHL Report 90-EI00178MGA, 1990.
11. S. Mahar, "Particulate Exposures from the Investigation and Remediation of a Crash Site of an Aircraft Containing Carbon Composites," American Industrial Hygiene Association J., 51, 459, (1990).
12. T. R. Martin et. al., "Response of the Rat Lung to Respirable Fractions of Composite Fiber-Epoxy Dusts," Environmental Res., 48(1), 57 (1989).
13. J. K. Kwan, "Health Hazard Evaluation of the Post-Curing Phase of Graphite Composite Operations at the Lawrence Livermore National Laboratory," UCRL-LR-104684, Livermore, California, 1990.
14. J. C. Lipscomb et. al., "Combustion Products from Advanced Composite Materials," Drug and Chemical Toxicology, in print, 1998.

LOW-COST FIRE RETARDANT COMPOSITES

W. Kowbel, R. Vaidyanathan, J. R. Patel, and J. C. Withers
MER Corporation, Tucson, AZ 85706

ABSTRACT

Currently used fire-resistant polymer-polymer composites are expensive. In this work, polymeric blends were investigated with the goal of enhanced flame retardation. Subsequently, polymeric fibers were coated with this new polymeric blend, to increase their intrinsic flame retardation. Effect of blending siliconated polymers on the flame-retardation of different resins is reported. Comparison of flame retardation properties of these mixtures was made to standard flame retardation materials.

1. INTRODUCTION

The combustion of polymers takes place in continuous cycles [1]. During burning, the heat generated in the flame is transferred back to the polymer surface which in turn produces volatile polymer fragments or fuel. As a result, more heat is produced, thus perpetuating the cycle. Flame retardancy may be achieved by interrupting this cycle; this could be performed in two ways [1].

- Solid phase inhibition: This involves changes in the polymer substrate, such a polymer cross-linking on the surface forming a carbonaceous char which prevents further burning. Or during burning, the polymer may evolve water, which cools the surface, and increases the amount of energy needed to maintain the flame.
- Vapor phase inhibition: Volatile free-radical inhibitors are released during burning, which inhibit the free-radical reactions taking place in the flame front, and interrupting the flame.

Polymers may be divided into three classes of flame retardant structures, which are: (i) intrinsically flame retardant structures containing either high halogen or aromatic structures as well which have the ability to form char on burning, (ii) less flame retardant materials, which may be made more flame retardant by appropriate chemistry, and (iii) quite flammable polymers, which could be made more flame retardant by the addition of additives [1]. School bus seating materials such as polyurethane fall in the 3rd class of polymer materials, and are therefore, highly flammable.

The choice of a polymer for flame retardancy purposes depends on cost and the importance of flame retardancy in relation to its final use [1]. Intrinsically flame retardant materials are quite expensive. The additive approach of adding organic phosphorus, organic halogen, or organic halogen compounds with antimony oxide could be adopted, with organic phosphorus compounds being the most effective. Improved understanding of the degradation chemistry of these compounds has led to an increased char and alter vapor phase reactions [1]. Yet another approach is to incorporate flame retardant chemical units into the polymer backbone, which would allow the polymer to be more flame retardant as well as be able to maintain the original physical properties of the polymer. Examples include halogen compounds, in the range of 10-25 % by weight of the polymer [1].

Kutty and Nando [2] investigated the effect of adding short Kevlar fibers to polyurethane to form a composite. The flammability of polyurethane polymer was improved by the addition of the Kevlar fibers. It was also observed that the amount of smoke was reduced, while increasing the char yield of the composite as compared to the unfilled polyurethane.

Benrashid and Nelson [3,4] reported that the flammability of polyurethanes was improved by the incorporation of a silicone moiety into the structure of urethane block copolymers. They reported that the block copolymer micro-phase separated, forming a silicone rich surface, which improved the thermal stability, and increased the oxygen index value (which is an indication of the flame retardant nature of the polymer). Additionally, the copolymers showed relatively good resistance to oxygen plasma, and improved flame retardancy compared to non-siliconated, polyether polyurethane block copolymers. The improvement in flame retardancy was related to the increased yield in pyrolytic char and increased resistance of char to oxidation. Benrashid and Nelson [3, 4 and references therein] further reported that siliconated polyurethanes have improved resistance to acids and bases, good mechanical strength, thermal stability, high resistance to moisture, and the potential to be used for coatings. They further concluded that the thermal, surface and flammability characteristics of the polymers are dependent on the ability to develop systems where the silicone content can be differentially varied in the surface and bulk. In particular, systems where the silicone can be maximized at the surface, yet be low in content in the bulk are thought to be good candidate materials for use as coatings, films and insulating plastics. The enhanced fire retardancy of the siliconated surface is thought to be due to the formation of SiO_2 on the surface which insulates and protects the bulk from fire and leads to low smoke formation [3].

This work examines the use of novel polymeric systems for improving the fire retardant properties of existing composite systems.

2. EXPERIMENTAL

2.1 MATERIALS The polymers, resins and their mixtures investigated were: (a) a polysilazane resin, (b) a phenolic resin, (c) a polysiloxane polymer, (d) a poly-dimethyl-siloxane (PDMS), and their mixtures, (e) phenolic resin-PDMS (50/50), (f) phenolic resin-polysiloxane (50/50), (g) polysilazane resin-polysiloxane (25/75). Both Kynol (from American Kynol) and Kevlar (from DuPont) fibers were coated with the above polymeric blends and mixtures.

2.2 MATERIALS CHARACTERIZATION Thermogravimetric analysis (TGA) and differential scanning calorimetry (DSC) analysis were performed using a Perkin-Elmer TG7 system. A 25°C/min heating rate was employed for both the TGA and DSC experiments. Char yield was calculated from the TGA results while T_g was calculated from the DSC data. A V_0

test using a propane gas burner was performed to assess the material fire resistance characteristics.

3. RESULTS

3.1 THERMAL ANALYSIS OF POLYMERS Figures 1-4 show typical TGA plots of different polymeric systems. Figure 1 shows a TGA plot of the phenolic resin. Initial weight loss (wt) below 200°C is attributed to the loss of solvent. A major weight loss after 400°C is attributed to the burn out of the organics. A 60 % wt loss is reached at 650°C. Figure 2 shows a TGA plot of the polysilazane resin. Initial weight loss below 150°C is attributed to loss of the solvent. Only a 15% weight loss is observed at 650°C. In comparison a polysiloxane resin (Figure 3) experienced a 40 % wt loss at 650°C under the same conditions. Thus, the polysilazane type resin offers the best thermal stability. A 25:75 polysilazane/polysiloxane blend, Figure 4, offers improved thermal stability compared to polysiloxane alone (Figure 2). Figures 5 and 6 show the DSC curves of polysilazane and a polysilazane/polysiloxane blend, respectively. It can be seen that the blends of the polysilazene polymer with a polysiloxane polymer greatly enhances the T_g of the blend.

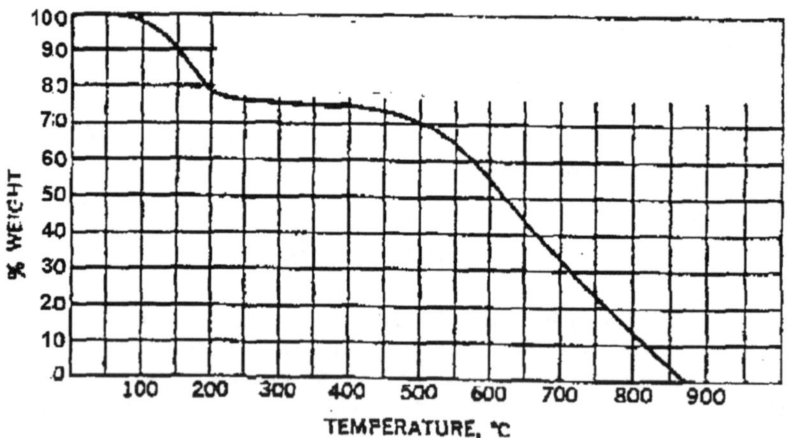

Figure 1.TGA of the phenolic resin.

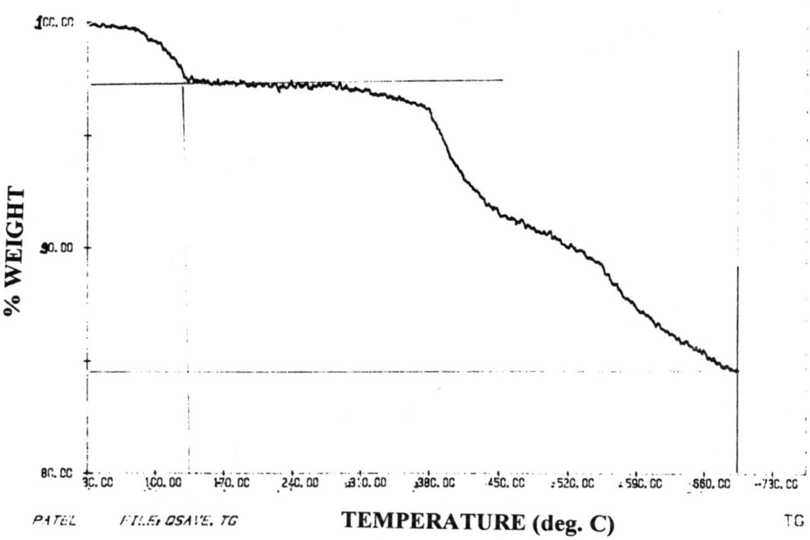

Figure 2. TGA of polysilazane resin

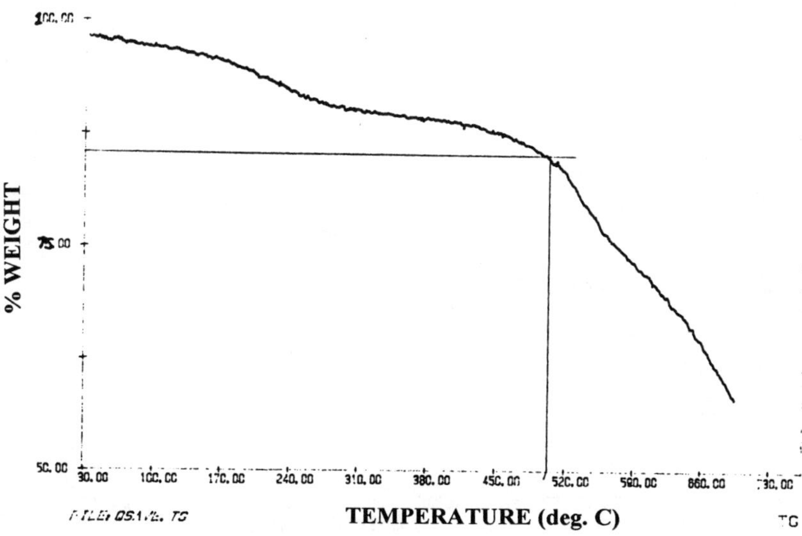

Figure 3. TGA of polysiloxane polymer

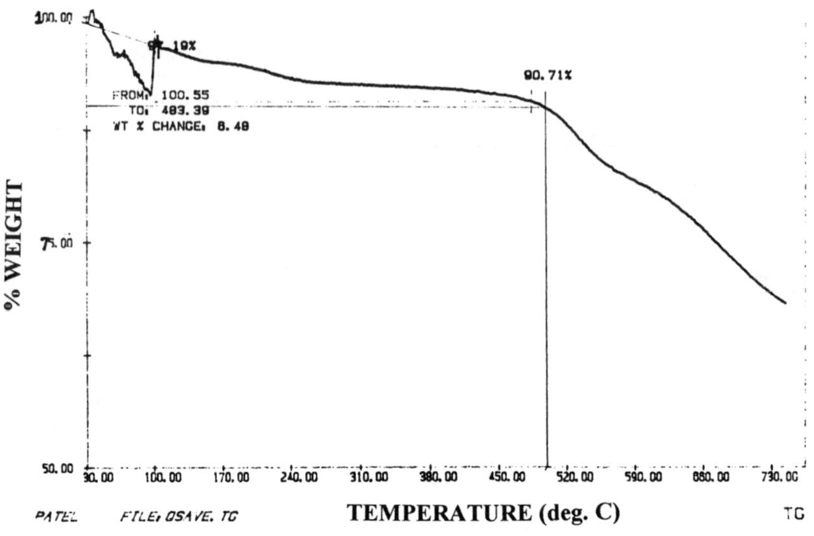

Figure 4. TGA of polysilazane/polysiloxane blend

The TGA and DSC results are summarized in Tables I and II, respectively. The TGA data in Table I lists both the weight loss and char yield at a constant temperature of 500°C. The DSC data is summarized in terms of the glass transition temperature of the resins and its mixtures.

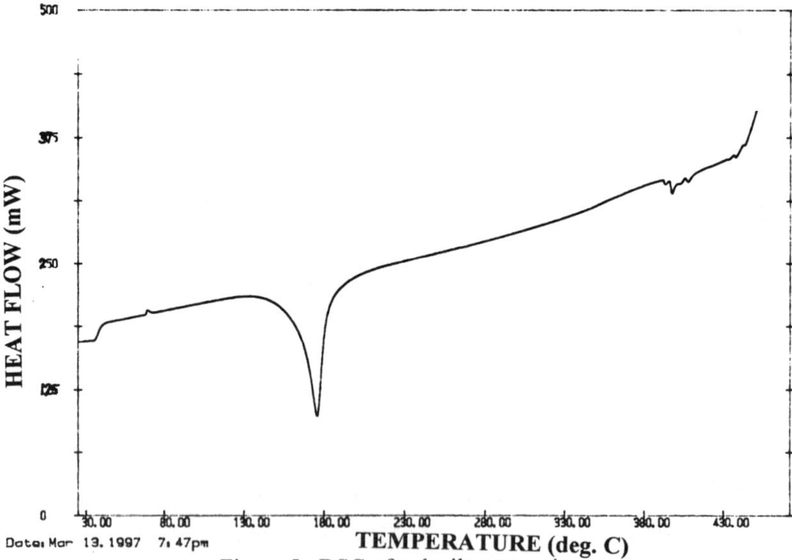

Figure 5. DSC of polysilazane resin

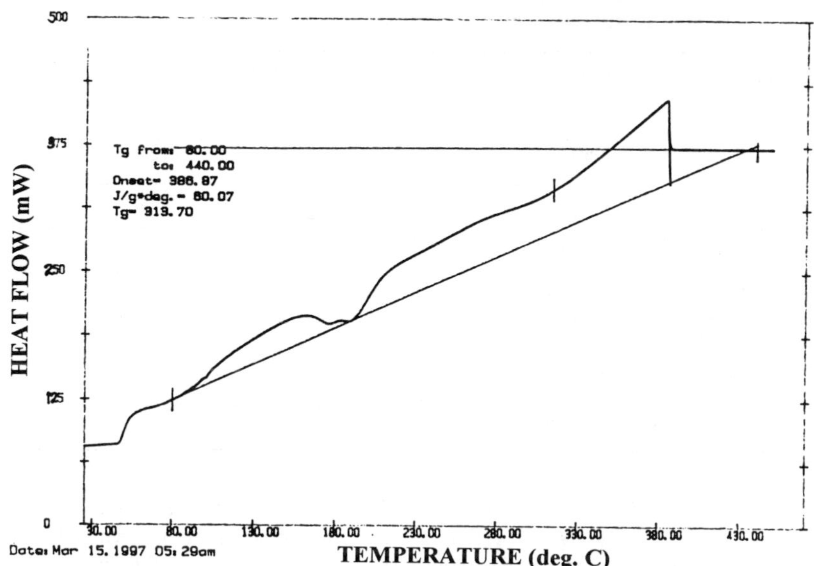

Figure 6. DSC of polysilazane/polysiloxane blend.

Table I. Weight loss and char yield data at 500°C derived from Thermogravimetric analysis (TGA) data for the different resins and mixtures investigated.

#	Resin/mixture	Ratio	Weight loss at 500°C (%)	Char yield at 500°C (%)
1.	Polysilazane resin	100	8.3	91.7
2.	Phenolic resin	100	30.0	70.0
3.	Polysiloxane polymer	100	15.0	85.0
4.	PDMS	100	92.7	7.3
5.	Phenolic resin-PDMS	50:50	62.9	37.1
6.	Phenolic resin-polysiloxane	50:50	29.6	70.4
7.	Polysilazane-polysiloxane	25:75	10.2	89.8

Table II. Glass transition temperature from Differential Scanning Calorimetry (DSC) data for the different resins and mixtures investigated.

#	Resin/mixture	Ratio	Glass transition temperature (°C)
1.	Polysilazane resin	100	159.3
2.	Phenolic resin	100	145.0
3.	Polysiloxane polymer	100	162.7
4.	Phenolic resin-PDMS	50:50	154.0
5.	Phenolic resin-Polysiloxane	50:50	141.8
6.	Polysilazane-polysiloxane	25:75	313.7

It has to be noted that the polysilazane resin evolves ammonia gas when in contact with water or moist surfaces. It was also observed that the amount of ammonia evolved is reduced considerably when the polysiloxane content was increased to 75 % in the mixture. Based on that result, a 25:75 polysilazane-polysiloxane mixture was chosen.

From Table I, it can be seen that the polysilazane resin had a 91.7 % char yield compared to the 70% char yield of the phenolic resin from room temperature to 500°C. Further, the polysiloxane polymer had an 85% char yield up to 500°C. However, PDMS had less than 10 % char yield when heated to 500°C. When PDMS was added to the phenolic resin in equal amounts, the mixture had less than 40 % char yield compared to 70 % for pure phenolic resin alone. Polysiloxane mixtures with either the phenolic resin or the polysilazane resin, showed approximately similar char yields compared to either the polysilazane resin or the phenolic resin alone. This is a significant result, since higher the char yield exhibited by a resin or a mixture, the better its flame retardant properties. Based on that result alone, it appears that polysiloxane mixtures with the phenolic and polysilazane resins would be good candidate polymer mixtures for flame retardation applications. Even though the char yields of the polysiloxane mixtures are slightly lower compared to the pure resins, since polysiloxane is a silicon rich polymer, it could provide better flame retardation in a mixture compared to the pure resins alone.

The results from Table II provide further validation of the results from Table I. It has been reported in the literature that the glass transition temperature or T_g is a good indicator of the flame retardation properties of polymers [5]. A higher T_g corresponds to a better flame retardant polymer. The T_g of the polysilazane, phenolic resins and the polysiloxane polymer are very close to each other. However, when the polysiloxane polymer was added to the polysilazane resin, the T_g value of the mixture was improved by approximately 100 %. In comparison, the addition of polysiloxane did not improve the T_g value of the phenolic resin (see Table II). In fact, a phenolic:PDMS 50:50 mixture showed an improved T_g compared to a phenolic:polysiloxane 50:50 mixture.

3.2 THERMAL ANALYSIS OF POLYMERIC FIBERS Figures 7 and 8 are TGA plots of Kynol and Kevlar fibers, respectively. Kevlar fiber exhibits much better thermal stability, but is much more expensive than the Kynol fiber. Coating of the Kynol fiber with a polysilazane/polysiloxane blend greatly improves its thermal stability, Figure 9. Thus, the use of such polymeric coatings offer a good potential for thermal stability of low-cost polymeric fibers. Thus the polymer fiber/polymer coating composite system offers a good potential for improved fire resistance of composite materials.

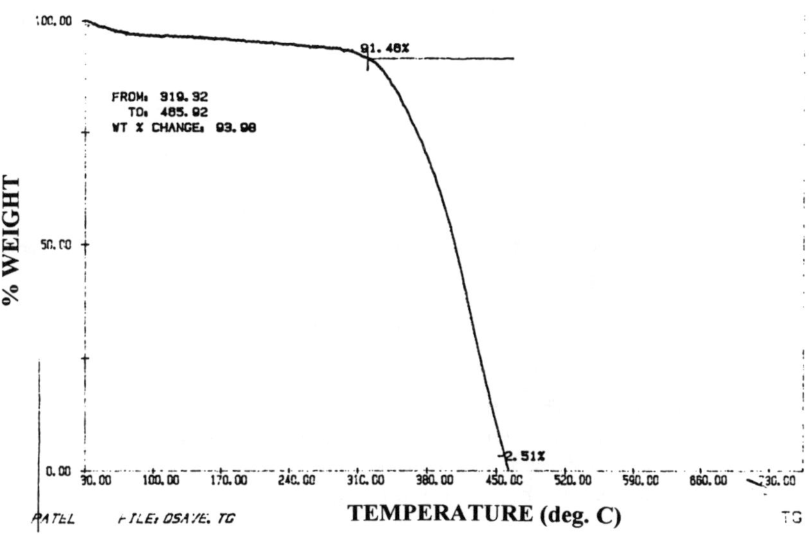

Figure 7. TGA of Kynol fiber.

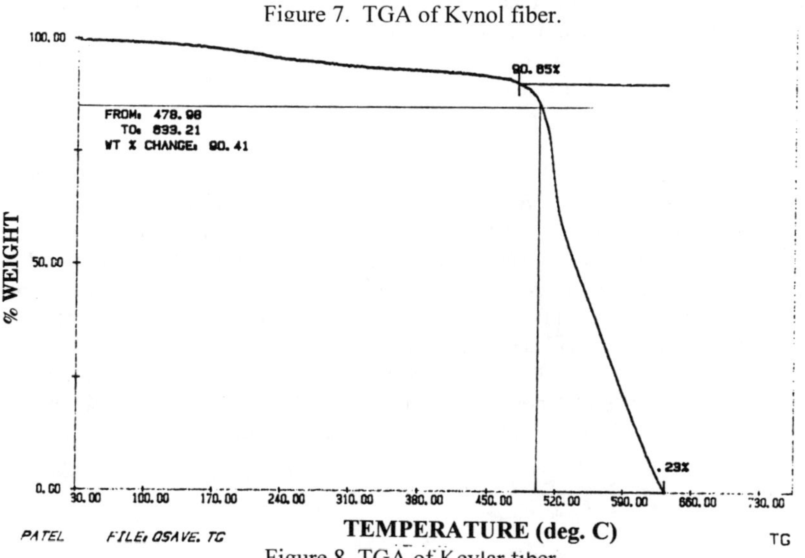

Figure 8. TGA of Kevlar fiber

1025

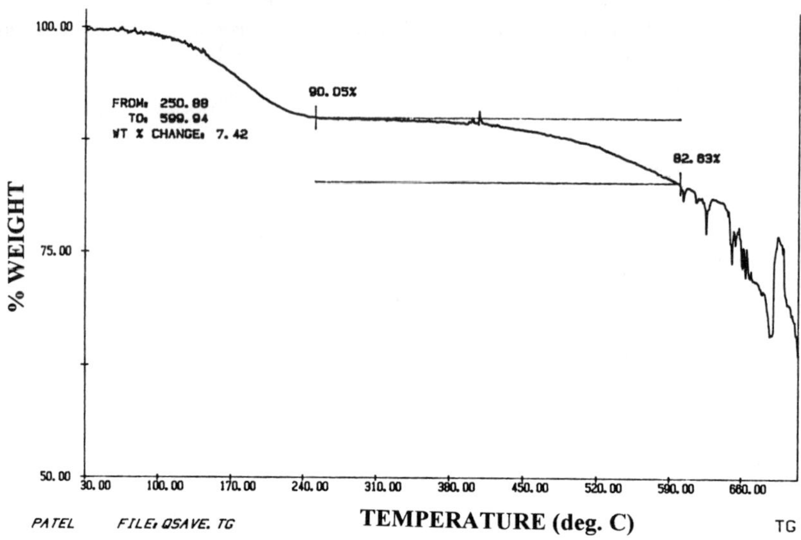

Figure 9. TGA of coated Kynol fiber.

3.3 V-0 TEST RESULTS The V-0 test results are shown in Table III. Recently, it has been suggested that clay addition to Nylon-6 polymer improved the flame retardation considerably [6]. The effect of adding clay particles to the polysilazane and phenolic resins as well as the polysilazane-polysiloxane mixtures was also investigated.

From the results in Table III, it can be seen that the Kevlar fibers are exhibiting better flame resistance compared to Kynol. When only phenolic resin was used as a coating, it did not afford any protection against a sustained flame on both the Kevlar and Kynol fibers. This was also the case when the fibers were coated with a phenolic-PDMS mixture. However, it has to be noted that the Kynol fibers have poor resistance to an open flame compared to a Kynol fabric, which has excellent fire retardant properties.

From Table III, it can be also be seen that polysiloxane and its mixtures with phenolic and polysilazane resins again performed well in resisting a sustained flame. However, the surprising outcome was that a polysilazane-ceramic particle mixture was not at all able to resist a sustained flame, as was suggested by its excellent char yield (90.8%) and high T_g values (319.7°C) from the TGA and DSC results. This indicates that the flame retardant properties are controlled not only by the char yield and T_g values, but by other factors also. At this time, it can only be speculated that these factors may include the silicon content of the polymer/mixture coating, as well as the water of hydration in the additives [7]. For example, recent literature suggests that boric acid may also perform well as an additive for flame retardation [7]. It could be conjectured that since the major difference between the boric acid and the boron nitride is their water of hydration, this could be an ingredient that is needed to give a good flame retardant property as an additive.

Table III. Time to sustain flame for Kevlar and Kynol fibers coated with resin mixtures.

Resin/Mixture	Ratio	Time to sustain flame for Kevlar fibers coated with resin mixtures (sec)	Time to sustain flame for Kynol fibers coated with resin mixtures (sec)
None	NA	10	5
Phenolic	100	10	5
Polysiloxane	100	32	21
Polysilazane	100	36	25
Polysilazane-polysiloxane	25/75	68	56
Polysilazane-ceramic particles	40/60	10	5
Polysilazane-clay particles	40/60	59	53
Phenolic-PDMS	50/50	10	5
Phenolic-polysiloxane	50/50	22	16
Phenolic-clay particles	40/60	38	28
(25% polysilazane-75% polysiloxane)/clay particles	50/50	93	83

The time to sustain a flame in the case of polysilazane and its mixtures with a clay addition was higher than either pure polysilazane resin or its mixtures. As was noted earlier, clay additions have shown much improved flame retardation when added to Nylon-6 polymer [6]. As long as the rheological behavior of polymer mixtures with clay additions can be controlled to be able to spray on to composites or poly urethane foam, this could be an additive worth investigating at a later time. In fact, clay additions to a polysilazane-polysiloxane mixture showed the highest resistance to a sustained flame (see Table III).

4. CONCLUSIONS

The use of a polysiloxane-polysilazane blend as well as the addition of clay particles to a silazane type resin yields good fire resistance. The application of such polymeric based coatings to low-cost Kynol fibers vastly improves their fire resistance.

5. ACKNOWLEDGMENTS

This work was supported by the Department of Transportation, under SBIR contract no. DTRS57-97-C-00080.

6. REFERENCES

1. G. L. Nelson, "Fire and polymers: an overview," in Fire and Polymers II: Materials and Tests for Hazard Prevention, G. L. Nelson, editor, ACS Symposium Series 599, American Chemical Society, Washington, DC, 1995, pp. 1-26.

2. S. K. N. Kutty and G. B. Nando, "Studies on the flammability of short Kevlar® Fiber-Thermoplastic Polyurethane composite," *Journal of Fire Sciences*, 11, pp. 66-79 (1993).

3. R. Benrashid and G. L. Nelson, "Synthesis of new siloxane urethane block copolymers and their properties," *Journal of Polymer Science: Part A: Polymer Chemistry*, 32, pp. 1847-1865 (1994).

4. R. Benrashid and G. L. Nelson, "Flammability improvement of polyurethanes by incorporation of a silicone moiety into the structure of block copolymers," in Fire and Polymers II: Materials and Tests for Hazard Prevention, G. L. Nelson, editor, ACS Symposium Series 599, American Chemical Society, Washington, DC, 1995, pp. 217-235.

5. T. D. Gracik and G. L. Long, "Prediction of thermoplastic flammability by thermogravimetry," *Thermochimica Acta*, **212**, pp. 163-170 (1992).

6. J. W. Gilman, T. Kashiwagi, and J. D. Lichtenhan, "Nanocomposites: a revolutionary new flame retardant approach," *SAMPE Journal*, **33**(4), pp. 40-46 (1997).

7. J. Green, "Mechanisms for flame retardancy and smoke suppression," *Journal of Fire Sciences*, 14, pp. 426-442 (1996).

NON-BURNING SILICONE RESIN COMPOSITE MATERIALS

Timothy C. Chao*, Satyendra K. Sarmah, Ronald P. Boisvert,
Gary T. Burns, Dimitris E. Katsoulis, and William C. Page

Dow Corning Corporation
Midland, Michigan 48686

ABSTRACT

A new experimental silicone resin was prepared and cured. The cured resin was thermally stable and exhibited very low heat realease rates in cone calorimetry experiments. It was fabricated into composite laminates using Nicalon® fabric and glass fabric. The composites were burned in the cone calorimeter under heat flux of 50 kW/m^2. They were exceptionally ignition resistant and had very low peak heat release rates. They generated very little CO and almost no smoke was observed. The composites retained their shape after fire testing and delamination was hardly seen. For mechanical properties, the burned samples retained up to 78% of their initial tensile strengths. In a separate study, solid state ^{29}Si MAS NMR was used to determine cross-linking of hydrosilylation cured silicone resins. The effect of sample size on heat release rate was also investigated. Specimen sizes of 4″x4″x1/4″ and 2″x2″x1/4″ had similar heat release rates. Additionally, TGA/MS was used to study the volatiles generated during pyrolysis of silicone resins. This study further demonstrates that silicone resins are very good materials for fabricating composites with low heat release rate, CO and smoke yields.

KEY WORDS: Silicone Resins, Composite Laminates, Flammability.

1. INTRODUCTION

The first paper of this series [1] examined the flammability of a series of $(PhSiO_{1.5})_x(MeSiO_{1.5})_{0.75-x}(ViMe_2SiO_{0.5})_{0.25}$ resins cured by hydrosilylation, and the effect of fillers on their fire performance and char formation. Resins in this study had peak heat release rates between 100-150 kW/m^2 at an incident heat flux of 50 kW/m^2. Higher phenyl content resins had higher heat release rates, CO and smoke yields. Fillers such as chopped fibers greatly improved the char integrity of the filled resins after fire testing. Even though these results were encouraging, non-burning properties were not observed.

Recently, high temperature organic polymers such as cyanate ester resin [2], phthalonitrile resin [3] and polyphenylene [4] have been reported with low heat release rates. Composites with non-burning properties were made from polymer matrices such as poly(p-phenylenebenzobisoxazole) [5] and Geopolymer, a potassium aluminosilicate [6]. Silicone resins have been known for their low heat release rates [1, 7, 8], however, it appears that no silicone resin has been reported showing non-burning properties.

In an attempt to develop non-burning silicone resin composite materials, a new experimental silicone resin was prepared and fabricated into composite laminates. Both Nicalon® fabric and glass fabric were used to make the laminates by vacuum bagging techniques [9]. The cone calorimetry was used to study their fire performance [10]. The mechanical properties were measured to determine their strength retention after fire testing.

In a seperate study, solid state ^{29}Si MAS NMR was used to study cross-linking of hydrosilylation cured silicone resins. The effect of sample size on the fire performance of silicone composites was also studied. Additionally, TGA/MS data of cured silicone resins was obtained to identify pyrolysis products.

2. EXPERIMENTAL

2.1 Materials. The materials used to synthesize hydrosilylation cured silicone resins were described earlier [1]. For the experimental silicone resin, it was prepared from methyltrichlorosilane (Aldrich). The fabrics used were Nicalon® fabric (Nippon Carbon Company) and glass fabric (Clark-Schwebel).

2.2 Synthesis and Characterization. The synthesis of hydrosilylation cured silicone resin was described earlier [1]. The synthesis of the experimental silicone resin was based on proprietary procedures. Solution ^{29}Si NMR spectra were recorded on a Varian VXR-200 MHz spectrometer operating at 79.46 MHz. A 30 wt % solution in $CDCl_3$ and $Cr(acac)_3$ (0.02 M) was used. Solid state ^{29}Si MAS NMR spectra of cured silicone resins were recorded on the Varian VXR-400 spectrometer. Infrared spectra were recorded on a Perkin Elmer 1600 FTIR spectrometer. Gel permeation chromatography (GPC) data were obtained on a Waters GPC equipped with a 410 differential refractometer detector using THF as an eluent and all values were relative to polystyrene standards. Thermal gravimetric analyses were recorded on a DuPont 592 TGA analyzer. During the TGA analysis a heating rate of 10 °C/min over a temperature range of 25-1000 °C was used. The flow rate was 100 cm^3/min of helium. A Kratos Concept 1H, high performance magnetic sector mass spectrometer was used to collect all real time mass spectral data for gas-phase species evolved from the TGA instrument. Heat release rates were measured on a cone calorimeter (Custom Scientific Instrument) at an incident heat flux of 50 kW/m^2.

2.3 Fabrication of Silicone Resin Composite Laminates. Two composite laminates were fabricated using the experimental silicone resin as a matrix material. One laminate was reinforced with heat-treated 8-harness satin weave, style 7781, E-glass fabric. The other was reinforced with heat treated M-sized ceramic grade 8-harness satin weave Nicalon® fabric. Both fabrics did not contain any sizing. The two prepregs were prepared by dipping the fabrics in 60wt% resin solution in acetone and drying overnight on corrugated aluminum foils in a fume hood at ambient conditions. Care was taken to maintain the integrity of the fabrics during the prepregging process. The resin content in the prepregs were approximately 47.9 wt% (Nicalon®) and 40.5 wt% (glass). The Nicalon® prepreg was cut into 20 pieces (9" x 5") while the glass fabric prepreg was cut into 24 pieces (9" x 4.5"). The two lay-ups, all warps parallel, were placed in the same vacuum bag side by side and cured in an autoclave using a combined debulk and cure cycle. Bleeder cloths were used in the vacuum bag to remove excess resin. The laminates were then postcured at 260°C for 16 hours.

The 20-ply Nicalon® laminate was 0.28" thick and contained approximately 42% fiber by volume. The 24-ply glass laminate was 0.2" thick and contained approximately 56% fiber by volume. Each laminate was then cut into two 4" x 4" pieces. One piece from each laminate was then subjected to fire testing in the cone calorimeter.

2.4 Mechanical Testing. Specimens were cut from both the original and the fire-tested laminates. The specimens were nominally 4" long and 0.5" wide. Tapered end-tabs, 1" long, were bonded to the specimens with epoxy adhesive resulting in 2" gage length. These specimens were subjected to tension tests in the warp direction using an Instron test frame with crosshead speed of 0.1"/min. Mechanical wedge action grips with serrated grip faces were used.

3. RESULTS AND DISCUSSION

3.1 Heat Release Rate. The heat release rate is one of the most important measurements reflecting the flammability of a material [9]. The experimental methyl resin was prepared from methyltrichlorisilane, therefore it was a methyl based silicone resin. The uncured resin had a peak heat release rate of 65 kW/m^2. This decreased to less than 10 kW/m^2 after curing (Figure 1). Both samples were crushed into powder form for fire testing. The resin was fabricated into laminates using Nicalon® fabric and glass fabric. The heat release rates are shown in Figure 2. In both cases, the peak heat release rates were below 10 kW/m^2 at an incident heat flux of 50 kW/m^2. Even with continuous sparking, neither composites ignited during the course of the test (20 minutes). As shown in Figures 3 and 4, the burned samples had the same physical appearance as the original samples except for a very thin silica ash on the top surfaces of the burned samples. In a related study, two slabs (2"x2"x1/4") were prepared using 60 volume % of chopped glass fiber and tested at incident heat flux of 50 kW/m^2 and 75 kW/m^2. Both samples did not ignite and had the peak heat release rates around 15 kW/m^2.

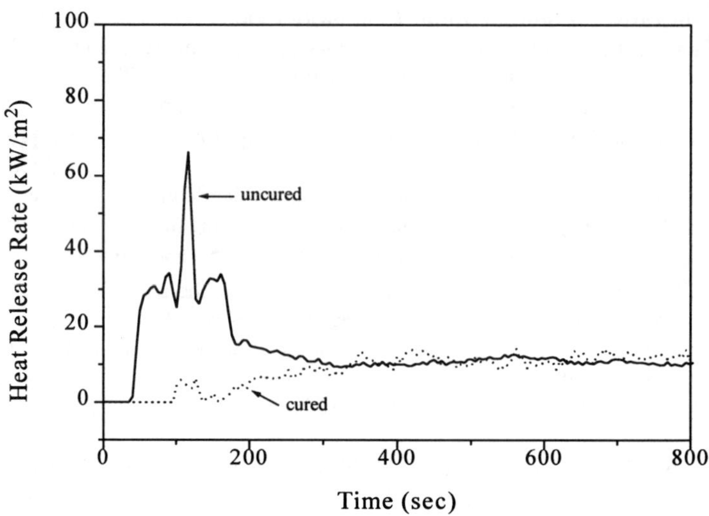

Figure 1. Plot of heat release rates of methyl silicone resin.

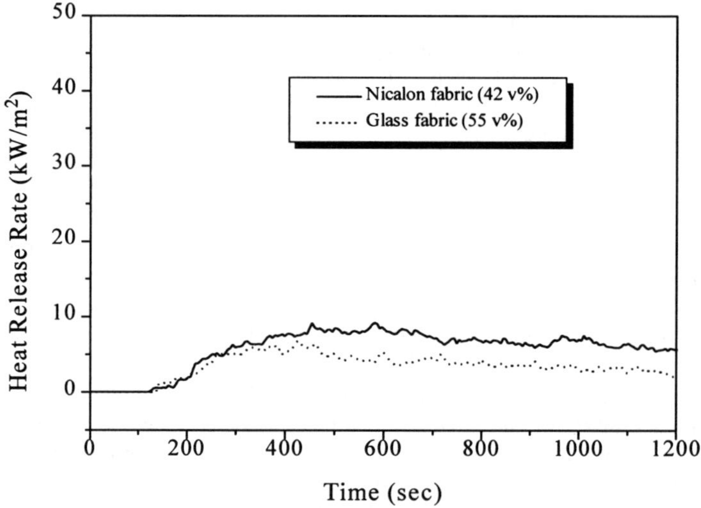

Figure 2. Plot of heat release rates of silicone resin composite laminates.

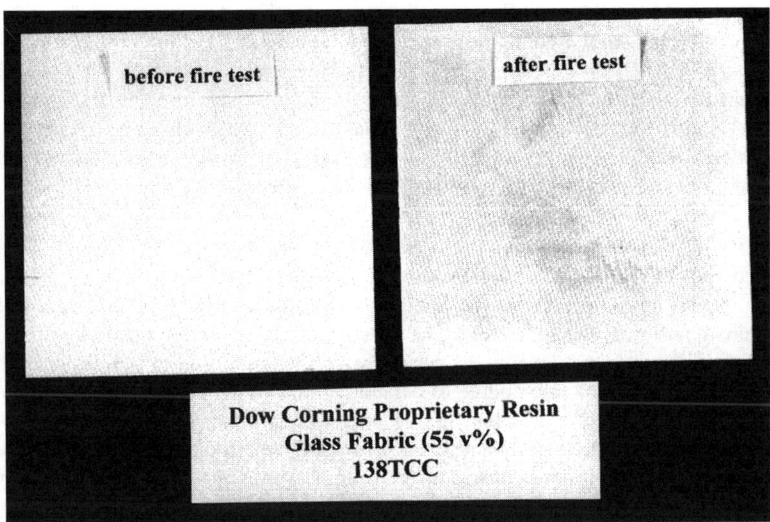

Figure 3. Top view of silicone resin composites before and after fire testing.

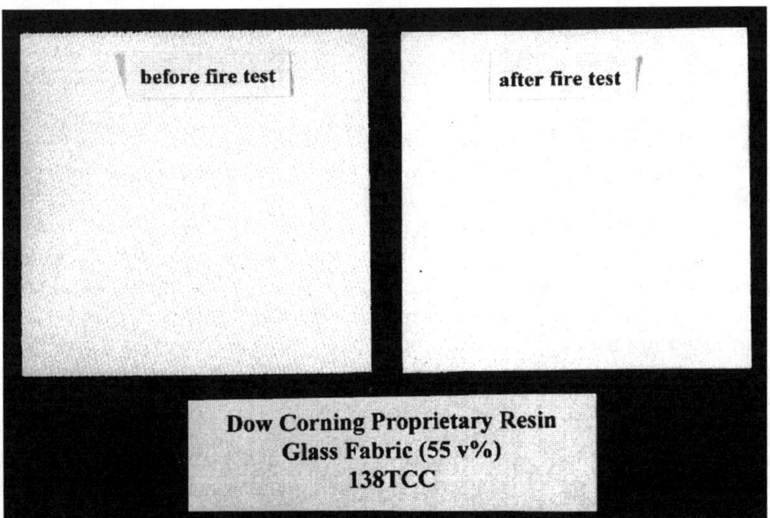

Figure 4. Bottom view of silicone resin composites before and after fire testing.

3.2 Carbon Monoxide and Smoke. Since a methyl based resin was used as the matrix, very low CO and smoke yields were found (Figures 5 and 6). For comparsion, a PMMA standard and a non-brominated epoxy-glass fabric composite (4"x4"x1/8") were burned. Whereas the experimental silicone resin and the PMMA had low CO and smoke yields, the CO yield from the epoxy resin was much higher. In addition, the epoxy's smoke extinction coefficient value reached 13 (not shown). If a brominated epoxy resin was used, the CO and smoke yields would have been higher.

3.3 Mechanical Properties. The mechanical test results are presented in Table 1 for the two composite laminates. Both Nicalon$^{®}$ and glass reinforced composites exhibited similar failure behavior. Both as-made and fire-tested specimens exhibited interlaminar delamination during tensile testing. The matrix material exhibited brittle failure. The top ply of the fire tested specimens, the surface which was exposed to the heater in the cone calorimeter, exhibited some degree of delamination after fire testing. This top ply would separate easily during tensile testing. However, delamination between all plys during tensile testing of both as-made and fire tested specimens indicates poor composite shear properties as well as low toughness of the resin. Figures 7 and 8 show the side view of composites before and after fire testing.

Table 1. Silicone composites tension test results.

Fabric	Tensile Strength (ksi)		Strength Retention (%)
	before fire test	after fire test	
Nicalon$^{®}$	41.3	32.2	78.0
Glass	28.0	13.1	46.8

3.4 Cross-linking Studies of Hydrosilylation Cured Silicone Resins. Earlier solid state ^{29}Si CP/MAS NMR studies [1] showed that ~40% of the original vinyl groups were still present after curing a $(PhSiO_{1.5})_{0.75}(ViMe_2SiO_{0.5})_{0.25}$ resin by hydrosilylation. In that study, tetramethylcyclotetrasiloxane (D_4^H) was used as the crosslinker at a SiH:SiVi molar ratio of 1:1. In order to determine if increasing the SiH content would decrease vinyl content, a higher molar ratio of SiH:SiVi (3.5:1) was used. In addition to the $(PhSiO_{1.5})_{0.75}(ViMe_2SiO_{0.5})_{0.25}$ resin, the methyl analogue $(MeSiO_{1.5})_{0.75}(ViMe_2SiO_{0.5})_{0.25}$ was included in this study. The Pt concentration was 50 ppm and the weight ratio of inhibitor to Pt was 100:1. The samples were cured at 180 °C for 30 minutes followed by a post-cure at 200 °C for 16 hours.

The solid state ^{29}Si MAS NMR spectra are shown in Figure 9, the peak of interest is $ViMe_2SiO_{0.5}$ at -1.5 ppm. The $PhSiO_{1.5}$ and $MeSiO_{1.5}$ peaks are at -79 ppm and -66 ppm, respectively. Two new peaks at 10 ppm and -20 ppm corresponding to the two silicons in the $O_{0.5}SiMe_2CH_2CH_2SiMeO$ crosslink, appear after cure. Residual SiH was not observed.

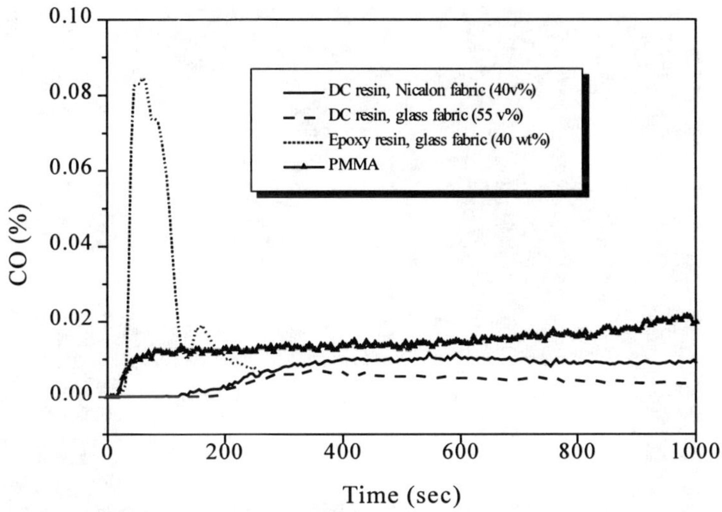

Figure 5. Plot of CO of silicone resin composites, epoxy resin composite and PMMA.

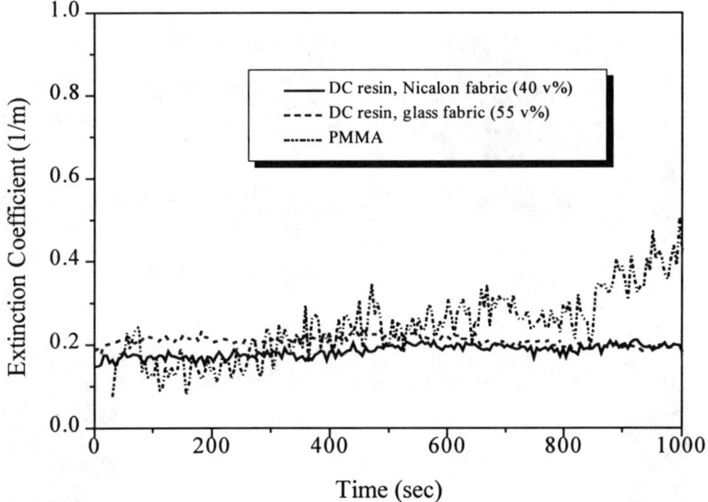

Figure 6. Plot of extinction coefficient of silicone resin composites and PMMA.

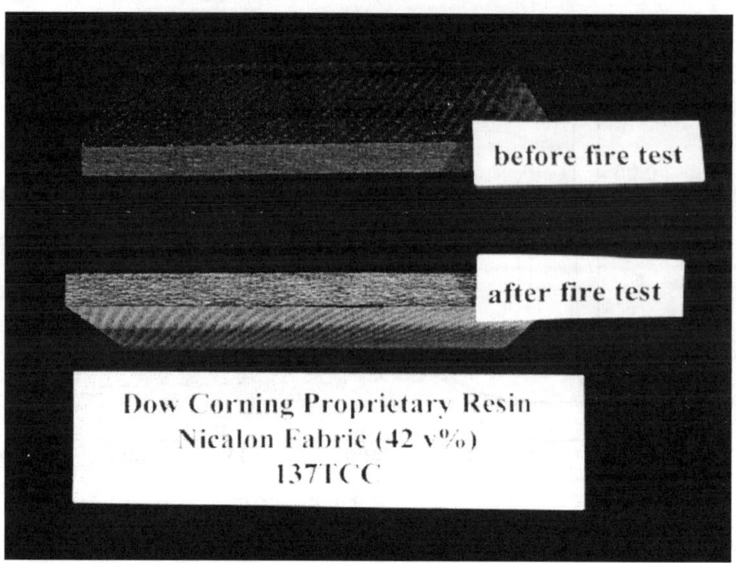

Figure 7. Side view of Nicalon® fabric reinforced silicone resin composite before and after fire testing

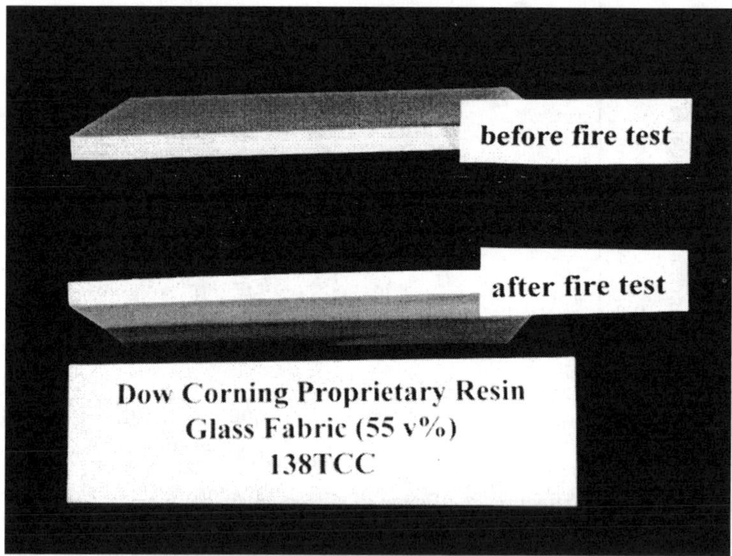

Figure 8. Side view of glass fabric reinforced silicone resin composite before and after fire testing.

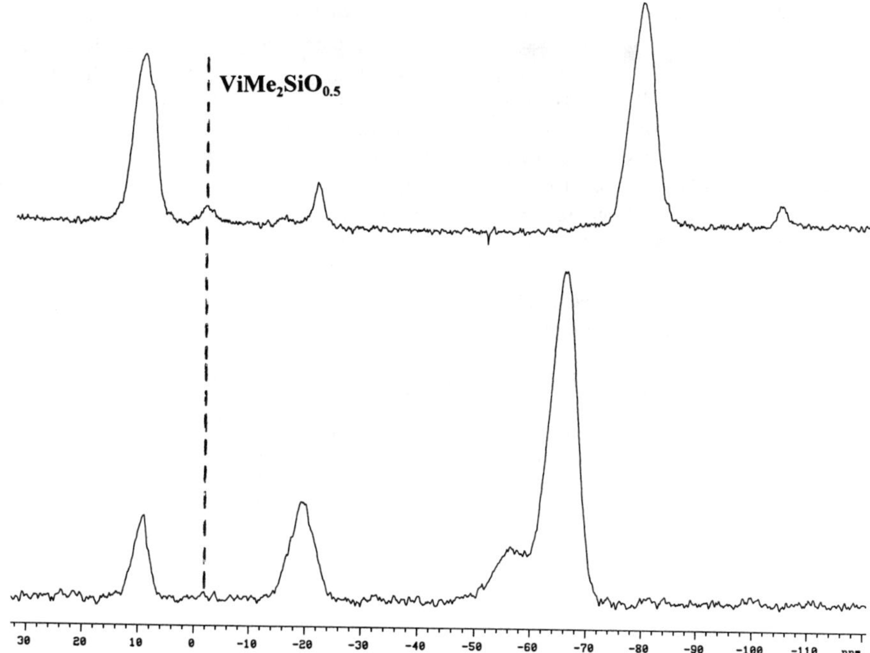

Figure 9. Solid state ^{29}Si MAS NMR of a cured (PhSiO$_{1.5}$)$_{0.75}$(ViMe$_2$SiO$_{0.5}$)$_{0.25}$ resin (top) and (MeSiO$_{1.5}$)$_{0.75}$(ViMe$_2$SiO$_{0.5}$)$_{0.25}$ resin (bottom) using D$_4^H$ as crosslinker (SiH/SiVi=3.5).

For the (PhSiO$_{1.5}$)$_{0.75}$(ViMe$_2$SiO$_{0.5}$)$_{0.25}$ resin, ~10% of the vinyl groups remained after curing while no vinyl groups were observed in the cured (MeSiO$_{1.5}$)$_{0.75}$(ViMe$_2$SiO$_{0.5}$)$_{0.25}$ resin. Clearly, increasing the concentration of the SiH relative to the vinyl concentration in the resin improved the degree of cure for both resins, especially for the methyl resin. The more complete cure for (MeSiO$_{1.5}$)$_{0.75}$(ViMe$_2$SiO$_{0.5}$)$_{0.25}$ resin may be due to a combination of less steric constraints in the methyl resin and the lower viscosity of the resin and crosslinker mixture prior to cure making the SiH groups more accessible to vinyl groups.

Solid state ^{29}Si NMR was recently used to study the hydrosilylation of (MeViSiO)$_4$ and (MeHSiO)$_4$ at two different SiH:SiVi molar ratios [11]. At both ratios, uncross-linked vinyl groups were observed after cure. However, upon heating at 400 °C for 1h under Ar, the vinyl groups were consumed. Even though the system is different, their results suggest that a higher temperature post-cure may result in more complete cure. If true, increasing SiH further may be not necessary to reduce vinyl content.

3.5 The Effect of Sample Size on Heat Release Rate. In order to investigate the relationship between heat release rate and sample size, composites with different sizes were prepared from a $(PhSiO_{1.5})_{0.75}(ViMe_2SiO_{0.5})_{0.25}$ resin filled with different concentrations of milled glass fiber. The results are summarized in Table 2 and Figure 10. A single slab of the 4″x4″x1/4″ sized samples were burned. For the 2″x2″x1/4″ samples, 2-5 slabs were prepared at each fiber loading and the average values are reported. Similar heat release rates were obtained with both the 4″x4″x1/4″ and the 2″x2″x1/4″ slabs at an incident heat flux of 50 kW/m². Similar results were reported from a study using different sized PMMA samples [12].

Table 2. The peak heat release rate data of 4″x4″x1/4″ and 2″x2″x1/4″ samples with different loading of milled glass fiber.

Filler Loading (v%)	Peak Heat Release Rate (kW/m²)	
	4″x4″x1/4″	2″x2″x1/4″
10	202	205
20	190	186
30	145	174
40	135	154
50	115	126

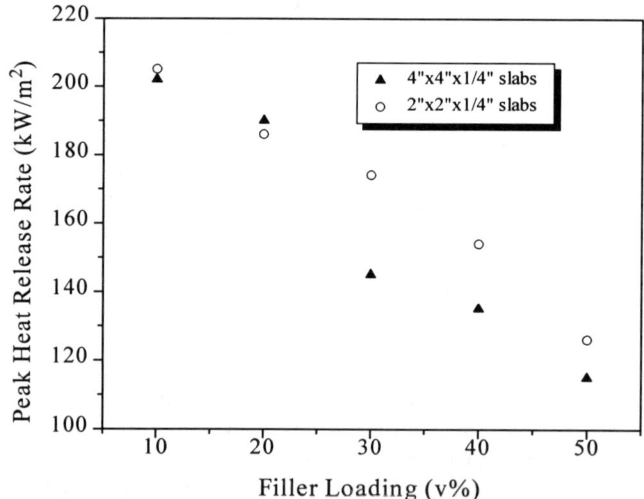

Figure 10. Relationship between the peak heat release rate and the size of samples (4″x4″x1/4″ vs 2″x2″x1/4″) with different loading of milled glass fiber.

3.6 TGA/MS Studies of Silicone Resins. The volatiles generated during pyrolysis of a phenyl resin were analyzed by TGA/MS. Figure 11 shows the TGA curve of a cured $(PhSiO_{1.5})_{0.75}(ViMe_2SiO_{0.5})_{0.25}$ resin heated to 1000 °C in a helium atmosphere at a heating rate of 10 °C/min. The major product was identified as benzene between 480 and 670 °C. In the same temperature range, other materials identified in the effluent stream included ethylene, acetylene and biphenyl. Ethylene and acetylene came from cleavage of vinyl groups, they were not secondary products from the decomposition of benzene. This is verified by checking mass spectrum of benzene, the concentration of decomposition products (*m/e* 27 and 26) in benzene is much lower. Biphenyl was probably the reaction product of aryl fragments of Si-aryl bonds, it has been identified on pyrolysis of poly(methylphenylsilmethylene) recently [13]. The evolution of methane was not observed until 575 °C and it continued until 860 °C.

The TGA of the cured experimental resin clearly shows its higher thermal stability (Figure 12). The minor weight loss around 400 °C was due to the evolution of water. This is probably the product of residual silanol-silanol condensation. The evolution of methane occurred at temperatures around 740 °C.

The results of this study are consistent with earlier TGA/MS studies on silicone resins [14-16]. In general, benzene, toluene, ethylene and acetylene evolve at lower temperatures than methane. Since volatiles are equivalent to fuel in a fire scenario, this study suggests that silicone resins with low phenyl contents should be used as matrices in composites developed for fire applications. This is consistent with the fire testing data collected in this project.

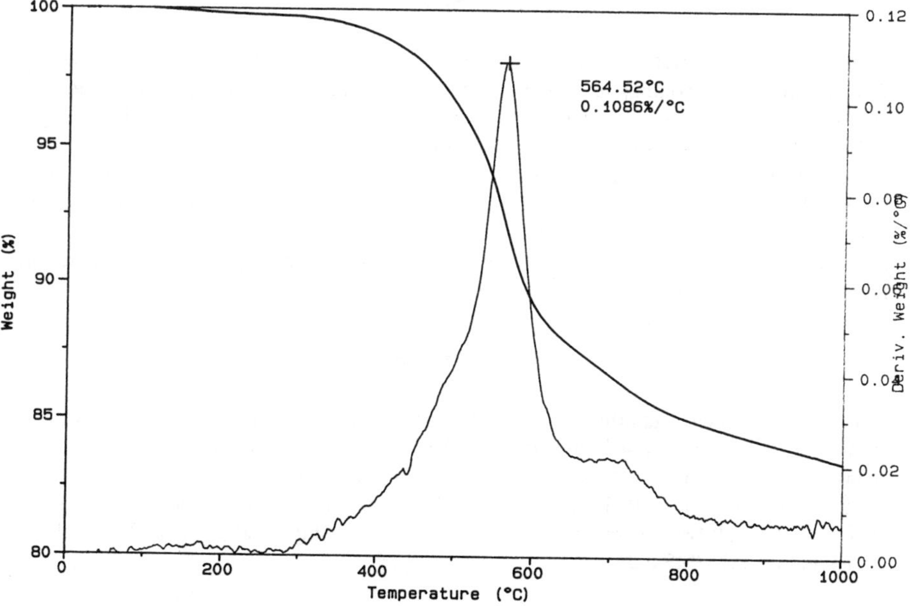

Figure 11. TGA of $(PhSiO_{1.5})_{0.75}(ViMe_2SiO_{0.5})_{0.25}$ resin and $D_4{}^H$ cured at 180 °C for 30 minutes and 200 °C for 16h.

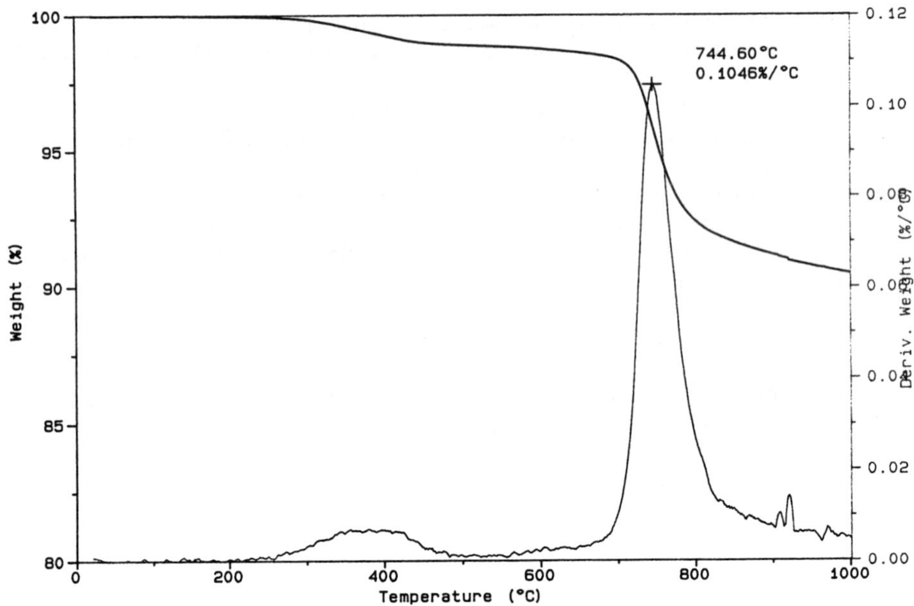

Figure 12. TGA of condensation cured methyl silicone resin.

4. CONCLUSIONS

The new experimental silicone resin used in this research was cross-linked by condensation. The composite laminates had non-burning properties and generated low carbon monoxide and smoke. The laminates retained up to 78% of their tensile strength after fire testing. Increasing the SiH:SiVi ratio in hydrosilylation cured resins improved the extent of cure was shown by solid state ^{29}Si MAS NMR. Peak heat release rates were independent of sample size in a series of silicone resins studied by cone calorimetry. TGA/MS identified pyrolysis products and supported cone calorimetry results. This study demonstrates that silicone resins are good candidates for fabricating next generation non-burning composite materials.

5. ACKNOWLEDGMENTS

This work has been supported by Federal Aviation Administration under Contract No. 95-G-05. We gratefully acknowledge Dr. Richard E. Lyon for his support. We are thankful to Prof. James E. McGrath of Virginia Tech and Dr. Forrest O. Stark of Dow Corning Corporation for their suggestions. We thank Dr. Sanlin Hu for acquiring solid state ^{29}Si MAS NMR spectra and Tom Sanders, Dan Filsinger and Dr. Ron Tecklenburg for collecting TGA/MS data.

6. REFERENCES

1. Chao, T. C., Burns, G. T., Katsoulis, D. E., Page, W. C., <u>Proc. 42nd Int. SAMPE</u>, <u>42</u>, 1355 (1997).
2. Rodriquez-Arnold, J., Arnold, F. E., Lyon, R. E. <u>Recent Adv. Flame Retard. Polym. Mater.</u>, <u>6</u>, 80 (1995).
3. Sastri, S. B., Armistead, J. P., Keller, T. M., Sorathia, U., <u>Proc. 42nd Int. SAMPE</u>, <u>42</u>, 1032 (1997).
4. Trimmer, M. S., Isomaki, M. R., Ulman, M., <u>Proc. 42nd Int. SAMPE</u>, <u>42</u>, 1315 (1997).
5. Kim, P. K., Pierini, P., Wessling, R., <u>J. Fire Sci.</u>, <u>11</u>, 296 (1993).
6. Forden, A. J., Balaguru, P., Lyon, R. E., <u>Proc. 41st Int. SAMPE</u>, <u>41</u>, 24 (1996).
7. Buch, R. R., <u>Fire Safety J.</u>, <u>17</u>, 1 (1991).
8. Rand, P. B., <u>Proc. SPIE-Int. Soc. Opt. Eng.</u>, <u>2934</u>, 104 (1997).
9. Mallick, P. K., <u>Fiber-Reinforced Composites</u>, 2nd ed. Marcel Dekker, Inc., New York, (1993).
10. Babrauskas, V., Grayson, S. eds, <u>Heat Release in Fires</u>, Elsevier Applied Science, (1992).
11. Michalczyk, M. J., Farneth, W. E., Vega, A. J., <u>Chem. Mater.</u>, <u>5</u>, 1687 (1993).
12. Tu, K.-M., <u>Proc. 23rd Int. Conf. Fire Safety</u>, <u>23</u>, 38 (1997).
13. Ogawa, T., Murakami, M., <u>Chem. Mater.</u>, <u>8</u>, 1260 (1996).
14. Burns, G. T., Taylor, R. B., Xu Y., Zangvil, A., Zank, G. A., <u>Chem. Mater.</u>, <u>4</u>, 1313 (1992).
15. Wilson, A. M., Zank, G., Eguchi, K., Xing, W., Yates, B., Dahn, J. R., <u>Chem. Mater.</u>, <u>9</u>, 1601 (1997).
16. Wilson, A. M., Zank, G., Eguchi, K., Xing, W., Yates, B., Dahn, J. R., <u>Chem. Mater.</u>, <u>9</u>, 2139 (1997).

EVALUATION OF THERMAL DEGRADATION
OF THE
RESIN-FIBER INTERFACE
IN
GRAPHITE FIBER REINFORCED LAMINATES
USING
ULTRASONIC SPECTROSCOPY

Alan J. Lesser, Gregory T. Schueneman, Terry R. Hobbs,
Polymer Science & Engineering Department
University of Massachusetts
Amherst, MA 01003

ABSTRACT

In certain fire situations, a structural or load-bearing component made from a PMC may be exposed to excessive thermal loads that thermally degrade the matrix below the surface. In this paper, we report the results of a study to assess the utility of ultrasonic spectroscopy as a means of assessing the residual characteristics of PMCs exposed to excessive thermal loads. Herein, we show that the attenuation measured in the power spectra of ultrasonic energy correlates with degradation in composite performance on graphite fiber epoxy matrix composites exposed to thermal degradation. Attenuation in the power spectra occurs primarily by eroding the highest frequencies first and progressing to the lower frequencies as the degradation progresses. Hence, if this method is to be considered practical as a non-destructive method for evaluation of thermal degradation. Mode I fracture toughness tests also correlate with thermal exposure showing lower values nearer the heat source. However, fractographic investigations of the surfaces using scanning electron microscopy (SEM) showed no evidence of interfacial failure between the fiber and resin for this prepreg. This suggests that the degradation is not necessarily at the interface and occurs in the more resin rich regions as well.

KEYWORDS: Thermal Degradation, Ultrasonic Spectroscopy, Delamination Resistance

1. INTRODUCTION

Polymer matrix composites (PMCs) are commonplace materials in many industries including commercial aircraft, marine, offshore oil platforms, and automotive applications. Over recent years, a safety concern has been raised regarding the use of these materials in situations where potential fire could erupt due to their general susceptibility degrade, decompose, or even combust in certain conditions.

In certain fire situations, a structural or load-bearing component made from a PMC may be exposed to excessive thermal loads that may only char the surface of the material and thermally degrade the matrix below the surface. This thermal degradation may, in turn, dramatically affect the mechanical integrity long before combustion of the matrix occurs. Also, in controlled fire situations, structural or load-bearing components may be exposed to excessive temperatures for short periods of time and show little or no apparent signs of degradation. In these situations, questions arise with regard to what the residual strength is and what non-destructive methods are available today to assess its residual strength. In this paper, we report the results of a study to assess the utility of ultrasonic spectroscopy as a means of assessing the residual characteristics of PMCs exposed to excessive thermal loads.

The use of ultrasound in non-destructive testing has become commonplace in many industries. Currently, ultrasonic B-scans and C-scans are routinely used to image defects in metal, ceramic, and polymeric materials. These methods are useful for mapping larger defects and cracks but are limited in the minimum size which can be detected.

However technological advancements over the recent years in ultrasonic, data acquisition, and processing equipment have enabled the implementation of a number refined spectroscopic methods to investigate the microstructural characteristics of materials. Recently, several studies have reported results from heterogeneity induced ultrasonic attenuation (1-5). These methods have been successfully applied to porous metals (2), to polymers and liquid crystalline systems (3), to porous epoxy (4), and to polymer latex emulsions (5) and colloid suspensions (6).

In 1978 O'Donnell et. al. (1) explored the existence of general relationships between ultrasonic attenuation and phase velocity (i.e., dispersion). In their study, they presented local forms of the general Kramer's-Kronig relationships under the assumptions that the are sufficiently small and do not change rapidly over the frequency range of interest. The local forms relate the attenuation to phase velocity as described over a definite (as opposed to indefinite range) which is of great practical use for application with ultrasonic transducers with a finite range of frequencies. They then illustrated the suitability of these relationships on hemoglobin solutions.

By the mid to late 80's, characteristic power spectra for a number of semicrystalline and amorphous polymers had been measured (3) including PE, PTFE, PMMA, and PC. Comparative studies have also been done to assess changes in the power spectra resulting from controlled heterogeneities including 0.3 -0.5 mm bubbles in PC, and changes resulting from nematic to isotropic phase transitions in liquid crystalline systems (3). These studies were conducted in the 5 MHz range.

In 1986, (2) average pore size down to 100 μm was detected porous aluminum alloy casting. These studies were conducted over frequencies ranging between 1 and 20 MHz. Similar studies were conducted on porous epoxy (4) in using a similar range of frequencies reported in reference 2. However, due to the slower wave speed in the epoxy media, attenuation due to scattering could be detected from pores of average size less than 60 μm. Most recently, these methods have been applied to characterize the particle size in latex emulsions (5) and colloid suspensions (7).

In this paper we investigate the potential of this method to detect mechanical degradation of PMCs exposed to excessive thermal loads. Specifically, we are interested in mechanical degradation that occurs at or near the fiber-resin interface in graphite reinforced epoxy based thermoset composite laminates. Herein, we report results from an investigation whereby unidirectional laminates were exposed to excessive thermal loads predominately along the fiber with the sole purpose to preferentially thermally degrade the polymer near at or near this interface. Ultrasonic excess attenuation from these laminates are then compared to other mechanical and analytical tests to characterize the residual strength of the composites after exposure to the excessive thermal loads.

2. EXPERIMENTAL

2.1   Laminate Fabrication   The laminates for this study were fabricated from a commercially available prepreg, R922 provided to us by Hexcell Corporation. The prepreg is made with a G30-500 graphite fiber and an amine cured epoxy-based matrix. Unidirectional panels of 30 plies total thickness were fabricated and cured in accordance with the manufacturer's recommendations. The panels were fabricated in a hand lay-up fashion with a 0.7 mm Teflon spacer inserted midway through the lay-up sequence along one edge of the panel. The panels were then compression molded with a final cure stage of 2.5 hours @ 176 C and 840 KPa pressure. Double cantilever beam (DCB) tests specimens were then machined from each panel to a total length of 20 cm and a width of 25 cm. The notch from produced from the Teflon spacer measured approximately 12 mm.

2.2   Ultrasonic Spectroscopy   Selected DCB specimens were submerged in water and scanned in a backscatter (pulse/echo) mode with a Panametrics Inc. 20 MHz focused transducer. The transducer used has a 3 mm diameter element and a focal point 19 mm from its face. A Panametrics Model 5601A/TT pulser/receiver was used to drive the transducer. The signal from the pulser/receiver was digitized at a rate of 1GHz with a Sonix STR81G PC based digitizer. A schematic of the acoustic equipment is shown in Figure 1.

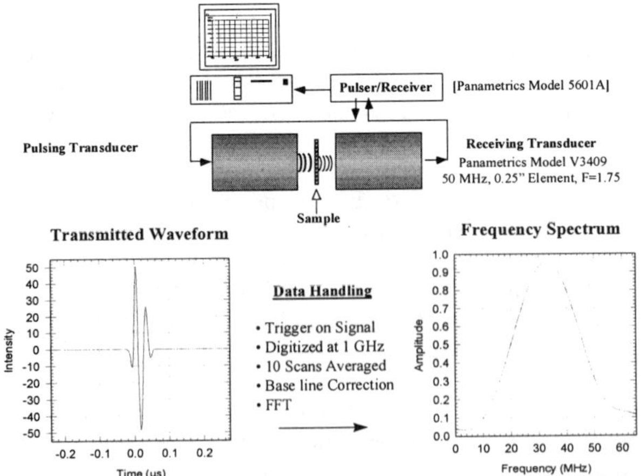

Figure 1: Schematic of the Acoustic Spectroscopy Setup

Panametrics Inc. Multiscan software was used to set the trigger threshold and gate the echoes returning from the sample. The software controls two stepper motors that move the transducer in the X and Y directions via Panametrics P1399 position match generator and MEI motion controller. Before scanning, the specimens were brushed while submerged to remove surface bubbles. Three A-scans were collected and averaged every 0.25 mm along the length of the sample and digitally stored as a B-scan. B-scans were collected every 2.5 mm along the width of the sample. The B-scans were collected to form a three-dimensional data array for each DCB specimen. The 3D array can be separated into A-scans at selected points in the specimen, B-scans along the specimen length or width of the specimen, and C-scans to map specific depths across the entire sample.

The standard technique of ultrasonic phase spectroscopy (see Figure 1) utilizes two types of time domain traces for measurement of the attenuation coefficient and dispersion. The first is a reference trace typically acquired from a water-only-path. Subsequently, the trace obtained from the specimen. In our case we utilize the use the specimen traces from samples before exposure as the reference traces and traces from the heat exposed samples to calculate the attenuation.

The average power spectra for each specimen were obtained by first averaging the A-scan traces along the centerline length of the DCB specimen. Next the baseline was calculated and subtracted from the entire data set. The data were then Fourier transformed and presented as the power spectra for each specimen.

After the specimens were exposed to an excessive heat source, individual A-scans at specified locations away from the heat source were processed individually to obtain the power spectra for that specific location. Next, the average power spectra for that sample was then used to calculate the attenuation caused by the exposure to the heat source.

2.3 Heat Exposure   Thermogravimetric analyses were conducted on samples of the laminates to determine an approximate decomposition temperature. A 15 mg sample was heated at a rate of 10 C /min. This test indicated the onset of degradation for this material to be in the range of 360 – 375 C.

After the DCB specimens were ultrasonically scanned to obtain the average power spectra, an edge containing fiber ends (either the notch side or the opposing side) was polished. Thermocouples were then placed at the polished end, the opposing end and midway along the length of each specimen. The specimens were then wrapped with glass wool to act as a thermal insulation and the polished edge was exposed to a heat source for a pre-specified length of time.

Two different heat exposures were conducted for this study. The first series involved submerging the polished tip of the specimens into a 12 mm deep silicone oil bath. The oil bath was maintained at a temperature of 205 C and the specimens were exposed for a period of 28 days. The steady-state temperature measured at the mid-point averaged 130 C and the temperature at the opposing end averaged approximately 90 C. For these experiments, the notched ends were submerged into the oil bath.

The second type of heat exposure is similar the first case except that sand replaced the silicone oil and was preheated to 450 C. The edges of the DCB specimens were submerged into the heated sand for a period of 90 minutes. During this time the temperature at the midpoint of the specimens reached 314 C and at the far end reached. The ends opposite the notched ends were polished and directly exposed to the heat source in this set of experiments.

This was necessary since severe decomposition of the composite prohibited further preparation of the DCB specimens (i.e., application of the hinges) if the notched ends were exposed.

2.4 Mode I Energy Release Rate Measurement (G_{1c})   The DCB test specimens were prepared for testing by first adhering hinges to the opposing faces near the end of the notch tip for gripping, and coating the sides of the specimens with a white brittle paint. The test specimens were carefully loaded to pre-crack them before the actual testing started. During this initial loading, any debris bridging the notch surfaces was cut away with a razor blade. The Mode I tests were conducted at a crosshead speed of 2 mm/min. Crack lengths were measured optically during the testing and verified/corrected after inspection of the fracture surfaces after the testing was complete.

The energy release rates were determined using the compliance method (7) using measured values of critical load for that crack length together with the specimen compliance at the specified crack lengths. In this approach, the measured compliance is used to calculate the energy release rates using Equation 1:

$$G_{1C} = \frac{P^2}{2b}\frac{\partial C}{\partial a}$$

1

where G_{1C} is the energy release rate (fracture energy), P is the applied load, C is the specimen compliance, a is the crack length, and b is the specimen width. Figure 2 is a plot of the measured calibration versus crack length curve for the DCB test specimens. The curve fit is a third order polynomial with a Pearson's correlation coefficient of 0.990.

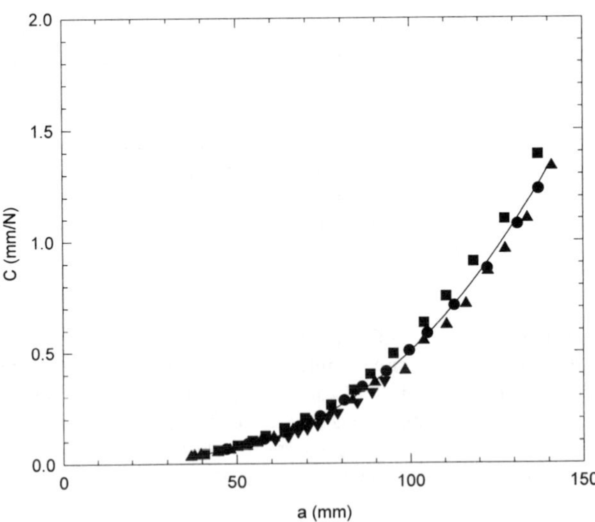

Figure 2: Plot of measured compliance curve for Mode I energy release rate experiments.

3. DISCUSSION OF RESULTS

Each DCB specimen was ultrasonically scanned as detailed in the experimental in Section 2. Note that although a 20 MHz transducer was used the peak frequency from the reflected energy occurred at approximately 7 MHz. This is shown typically for a specimen in Figure 3. Note that the power spectra was windowed from the back reflection off the back wall of the specimen to maximize response from the attenuation from the sample. If the sample were removed and the time trace was analyzed without the DCB specimen, the center of energy would be closer to the 20 MHz transducer rating. This implies that the unidirectional panel is adsorbing a significant amount of the higher frequency sound.

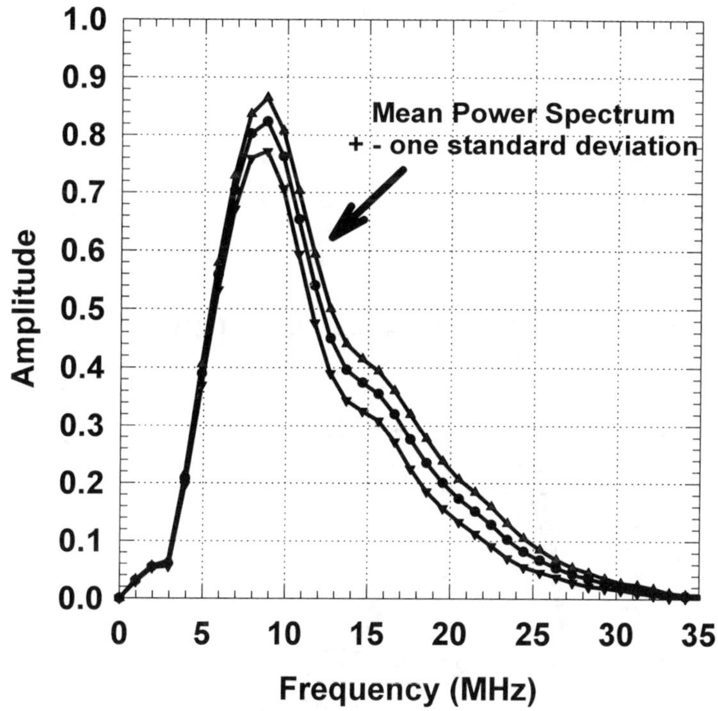

Figure 3: Average power spectra for un-exposed DCB test specimen.

A series of DCB test specimens were then exposed to a high temperature heat source. The heat source was applied purposely applied to an edge with exposed fiber to promote maximum conductivity along the length of the specimen and to accentuate fiber-resin interfacial degradation. Only one end of the specimen was exposed to the heat source to promote a steady-state gradient of temperature along the length of the specimen. Consequently, it was anticipated that a single specimen exposed in this fashion would contain a similar gradient of resin and fiber-resin interfacial degradation that could be correlated with ultrasonic and fracture toughness measurements.

DCB specimens were exposed to two different conditions (see Section 2). One series of specimens were exposed to at one end to a temperature of 205 C for a period of 28 days and a second set was exposed to a temperature of 450 C for a period of 90 minutes.

After exposure the DCB specimens were again ultrasonically scanned as described in Section 2. However, in these scans the power spectra were not averaged along the length of the specimen, as was the case for the sample before exposure. Instead individual power spectra were recorded and plotted at selected positions along the length of the exposed specimens. These spectra are shown typically for the test specimens exposed to 205 C for 28 days in Figure 4. The plots in Figure 4 are labeled with numbers that denote the measurement distance away from the heat source. Figure 4 shows that there is a clear reduction in the power spectra from the back wall that becomes more pronounced as the measurements are taken nearer the heat source. This, of course, indicates that additional attenuation occurs in the composite as a consequence of the heat exposure.

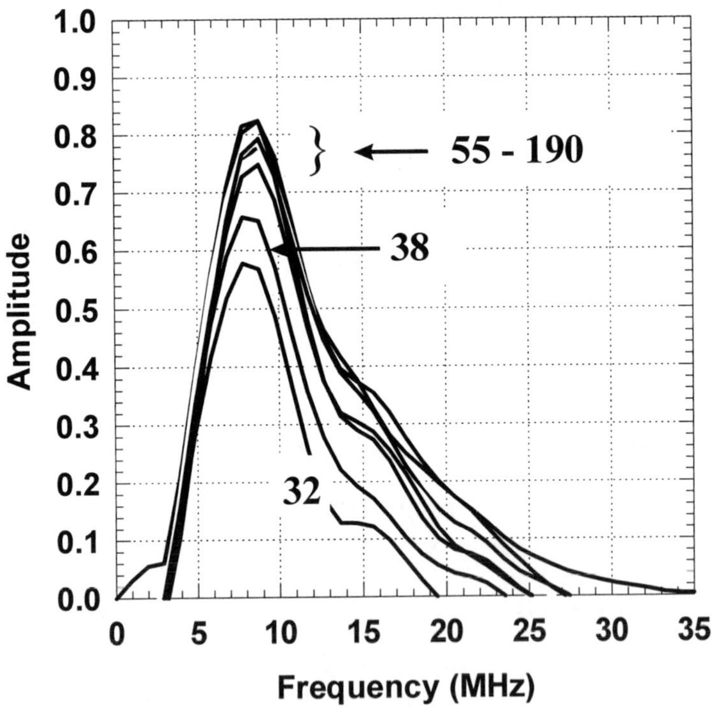

Figure 4: Power spectra for the DCB specimen exposed to 205 C heat source.

The attenuation (absorption), A, of ultrasonic energy due to the heat treatment on the DCB specimen can be calculated by:

$$A(f) = A_w - \ln\left(\frac{P}{P_o}\right)$$

□ 2

where A_w is the attenuation coefficient of water, P_o is the power spectrum intensity before exposure (Figure 3) and P is the intensity after exposure (Figure 4). The absorption spectra for the sample heat aged at 205 C are presented in Figure 5.

It is interesting to note from Figure 5 that total attenuation occurs over the entire length of the sample at frequencies greater than approximately 20 MHz. However, as measurements are taken closer to the heat source, significantly more acoustic energy is attenuated at lower frequencies. The amount of attenuation continues to increase to the point where, at 32 mm away from the heat source, nearly 40% of the energy is attenuated at 5 MHz, and all acoustic energy is attenuated above 13 MHz.

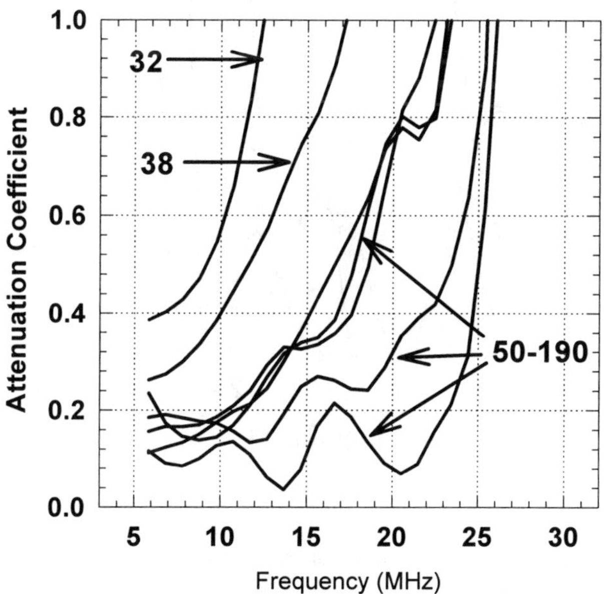

Figure 5 : Ultrasonic Attenuation for sample heat aged at 205 C for 28 days

Figure 6 presents the attenuation spectra for the samples exposed to 450 C at 90 minutes. In this specimen complete attenuation was measured on the sample at all distances less than 45 mm from the heat source indicating severe degradation. At 50 mm, it can be seen that the degradation was still significant enough that over the frequency range of the transducer, a minimum of 80% attenuation occurred. At 55 mm, the attenuation was still significant but at 60 to 150 mm the attenuation was almost the same as the unexposed sample. When comparing this result to the sample exposed at 205 C (see Fig. 5), one can see that Figure 6 suggest that there is a higher gradient of degradation in the sample exposed to the higher heat source, which is in line with expectations.

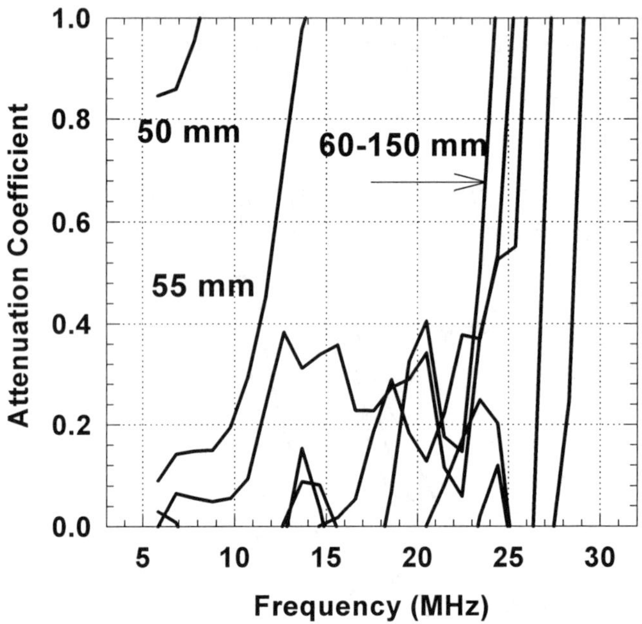

Figure 6: Ultrasonic Attenuation for sample heat aged at 450 C for 90 minutes.

Mode I, critical energy release tests (G_{1c}) were conducted on the DCB test specimens to evaluate if any degradation in the resin or resin/fiber interface occurred as a consequence of the heat loading. Under normal conditions, the energy critical energy release rate becomes independent of crack length and is ascribed the value of G_{1c}. As a result, a typical test on DCB specimens usually reaches a steady state condition corresponding to the G_{1c} value. However, if the fracture energy changes along the specimen's length, as we might expect from our heat treated samples, then a steady state value will not be reached.

Figure 7 shows the G_{1c} results for a control (unexposed) and one exposed to the heat treatment. Note first that control specimen produced a G_{1c} value of 130 J/mm2. This is 18% higher than that reported by the manufacturer (8). Nonetheless, this value is in the range of that reported by the manufacturer and differences might be attributed to differences in fiber volume. More interestingly, Figure 7 shows that the fracture energy measured on the heat-treated samples were significantly reduced to levels ranging between 60 - 80 J/mm2. Also, the fracture energies on the heat-treated sample are at their lowest value nearest the heat source and increase gradually as measurements are made further away from the heat source.

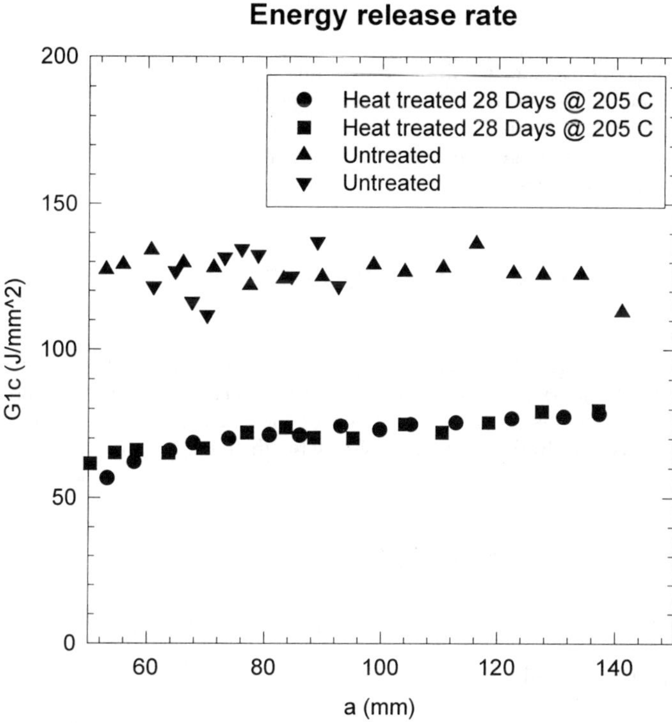

Figure 7: G_{1c} fracture energy measurements on DCB control and heat-treated (205 C for 28 days).

Scanning electron micrographs were also taken of the surface to identify if any evidence of interfacial failure was responsible for the decrease in the measured fracture energy and increase in acoustic attenuation of the test specimens. Unfortunately, no direct evidence was obvious from this investigation so the micrographs are not included in this publication.

4. CONCLUSIONS

This study has shown that the attenuation measured in the power spectra of ultrasonic energy correlates with degradation in composite performance on graphite fiber epoxy matrix composites exposed to thermal degradation. Attenuation in the power spectra occurs primarily by eroding the highest frequencies first and progressing to the lower frequencies as the degradation progresses. The range of frequencies we observed most sensitive are between 5-25 MHz, but we believe this to be closely related to the frequency response of the transducer. However, at the lower frequencies (i.e., < 5 MHz) visual degradation is also evident. Hence, if this method is to be considered practical as a non-destructive method for evaluation thermal degradation, frequencies greater that 5 are suggested.

Mode I fracture toughness tests also correlated with thermal exposure showing lower values nearer the heat source. However, fractographic investigations of the surfaces using scanning electron microscopy (SEM) showed no evidence of interfacial failure between the fiber and resin for this prepreg. This suggests that the degradation is not necessarily at the interface and occurs in the more resin rich regions as well.

5. ACKNOWLEDGEMENTS

The authors would like to gratefully acknowledge the Federal Aviation Administration for their financial support and Hexcel Corporation for the materials used in this study. The authors would like to also thank Richard E. Lyon of the FAA, Shaw M. Lee of Hexcell Corporation, and Steven R. Berbube of Panametrics for their many thoughtful discussions during this work.

5. REFERENCES

1.	M. O'Donnell, E. T. Jaynes, and J. G. Miller, J. Acoust. Soc. Am. 63 (6), 1935 (1978)

2.	L. Alder, J. H. Rose, and C. Mobley, J.Appl. Phys., 2 (15), 336 (1986).

3.	K. Matsushige, H. Okabe, S. Shichijyo, and T. Takemura, Japanese J. Appl. Phys., 24 34, (1985).

4.	M. S. Hughes, S. M. Handley, and J. G. Miller, IEEE Journal, Ultrasonics Symposium, 9, 1041 (1987).

5.	C. Verdier, and M. Piau, J. Acoust. Soc. Am. 101 (4), 1868 (1997).

6.	T. Matsuoka, et. al., Japanese J. Appl. Phys. 36, 2972 (1997)

7.	J. G. Williams, A. J. Kinloch, J. Mater. Sci. Letters, 8, 125 (1989)

8.	Shaw Ming Lee, Hexcel Corp. private communication

FLAMMABILITY STUDIES OF POLYMER LAYERED SILICATE NANOCOMPOSITES

Jeffrey W. Gilman*, Takashi Kashiwagi, James E. T. Brown, Sergei Lomakin[†]
National Institute of Standards and Technology[ʃ] , Gaithersburg, MD

Emmanuel P. Giannelis, Evangelos Manias
Cornell University
Ithaca, NY

ABSTRACT

Polymer layered silicate (PLS) nanocomposites are materials with unique properties when compared to conventional filled polymers (1). For example, the mechanical properties of a nylon-6 layered-silicate nanocomposite, with a silicate mass fraction of only 5 %, show excellent improvement over those for the pure nylon-6. The nanocomposite exhibits a 40 % higher tensile strength, 68 % greater tensile modulus, 60 % higher flexural strength, and a 126 % increased flexural modulus. The heat distortion temperature (HDT) is increased from 65° C to 152° C (2). Previously, we reported on the flammability properties of *delaminated* nylon-6 layered silicate nanocomposites and *intercalated* polymer layered-silicate nanocomposites prepared from polystyrene, PS, and polypropylene-graft-maleic anhydride, PP-g-MA (3,4). Here, we will briefly review these results and report on our initial studies of the flammability of <u>thermoset PLS nanocomposites:</u> *intercalated* vinyl ester silicate and *intercalated* epoxy silicate nanocomposites.

1. INTRODUCTION

In the pursuit of improved approaches to fire retarding polymers a wide variety of concerns must be addressed, in addition to the flammability issues. For commodity polymers the low cost of these materials requires that the fire retardant (FR) approach also be of low cost. This limits solutions to the problem primarily to additive type approaches. These additives must be low cost and easily processed with the polymer. In addition, any additive must not excessively degrade the other performance properties of the polymer, and it must not create environmental problems in terms of recycling or disposal of the final product.

Polymer layered silicate (PLS) nanocomposites are hybrid organic polymer - inorganic materials with unique properties when compared to conventional filled polymers (1). In some cases increased thermal stability, an important property for improving flammability performance, is also accompanied by a doubling of tensile modulus and strength, as well as decreased gas permeability, increased solvent resistance, and higher heat distortion temperature. Methods to prepare PLS nanocomposites have been developed by several groups over the last decade (2, 5, 6, 7, 8, 9, 10).

KEY WORDS: Nanocomposite, Flammability, Silicate.

[†] NIST Guest Researcher from the Russian Academy of Sciences.
[ʃ] This work was carried out by the National institute of Standards and Technology (NIST), an agency of the U. S. government, and by statute is not subject to copyright in the United States.

In general these methods achieve molecular level incorporation of layered silicate (e.g., monmorillonite) into the polymer by addition of a modified silicate; during the polymerization (*in situ*), or to a solvent swollen polymer, or to the polymer melt (1). Two terms (*intercalated* and *delaminated*) are used to describe the two general classes of nano-morphology that can be prepared. The *intercalated* structure, where the extended polymer chains are inserted into the gallery space between the individual silicate layers (see Figure 1). These are well ordered multi-layered structures. The *delaminated* (or *exfoliated*) structures result when the individual silicate layers are well dispersed in the organic polymer. The interlayer spacing (20 nm -200 nm) is on the order of the radius of gyration of the polymer. The silicate layers in a *delaminated* structure are not as well ordered as in a *intercalated* structure.

Nylon-6 and polycaprolactone silicate-nanocomposites have been prepared through *in situ* ring opening polymerization of monomers. This process requires some re-development of the polymer manufacturing process, and although nylon-6 silicate nanocomposite is commercially available, this approach may not be as attractive for other polymers. Melt intercalation of the polymer directly into the layered silicate was recently developed. In this process the appropriately modified layered silicate (cation exchanged montmorillonite) and the polymer are combined in the melt to form the nanocomposite (1). This process is of course most advantageous for thermoplastic polymers. For thermosetting resins the *in situ* polymerization method is still suitable, and the appearance of several publications (5, 10, 11) and patents (12, 13, 14) demonstrates that new thermoset silicate-nanocomposites with unique properties can be prepared. In the case of an unsaturated polyester thermoset resin, the *intercalated* nanocomposite was shown to have a higher Young's modulus than both the pure polymer and the *delaminated* nanocomposite (10).

2. BACKGROUND

Thermal analysis (TGA) of several different PLS nanocomposite resin systems has revealed the intriguing result that *intercalated* nanocomposites are more thermally stable than the *delaminated* nanocomposites (4, 9, 10). Molecular dynamics simulations of the thermal degradation of a series of polypropylene (PP)/graphite nanocomposites recently found that the most pronounced stabilization of the polymer occured where the graphite layers were separated by 3 nm; that is, had an *intercalated* structure (15). Previously, we reported on the flammability properties of several thermoplastic polymer nanocomposites; *delaminated* nylon-6 layered silicate nanocomposites and *intercalated* polymer layered-silicate nanocomposites prepared from polystyrene, PS, and polypropylene-graft-maleic anhydride, PP-g-MA (3, 4, 9). The flammability data for nylon-6, PS and PP-g-MA, and new data on nylon-12 is shown in Table 1. The data shows that both the peak and average heat release rates (HRR) were reduced significantly for *intercalated* and *delaminated* PLS nanocomposites with low mass fraction (2 % to 5 %) of silicate. The HRR plots for nylon-6 and nylon-6 silicate-nanocomposite (mass fraction 5 %) at 35 kW/m^2 heat flux are shown in Figure 2, and are typical of those found for all the PLS nanocomposites in Table 1. The nylon-6 nanocomposite has a 63 % lower HRR than the pure nylon-6. Furthermore, for the PS silicate nanocomposite the magnitude of improvement in flammability performance is comparable to that found for PS flame retarded using a total mass fraction of 30 % of decabromodiphenyl oxide, DBDPO, and antimony trioxide, Sb_2O_3, (see Table 1). This is accomplished without as much of an increase in the soot (SEA) or CO yields. The data also indicates that the rate of mass loss during combustion of the PLS nanocomposite is significantly reduced from the values observed for the pure polymers (see Figure 3). Since the heat of combustion, SEA and carbon monoxide yields are unchanged this suggests that the source of the improved flammability properties of these materials is due to differences in condensed phase decomposition processes and not to a gas phase effect.

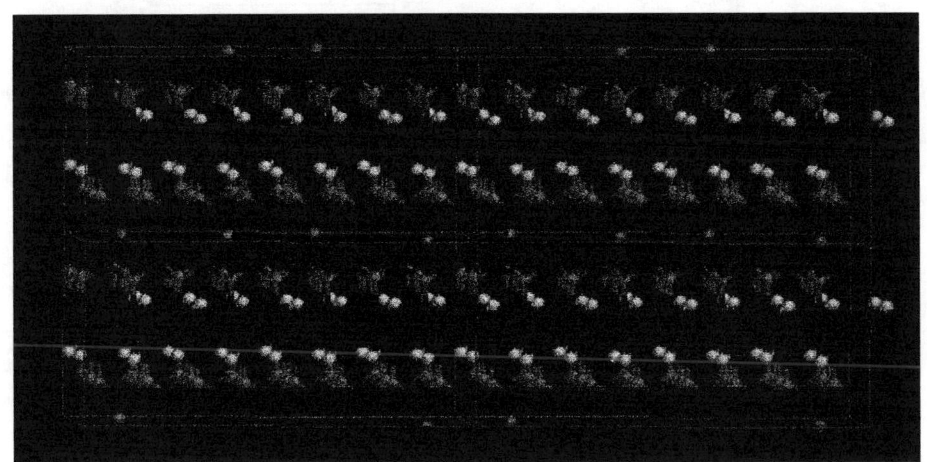

Figure 1. Molecular representation of sodium montmorillonite, showing two aluminosilicate layers with the Na$^+$ cations in the interlayer gap or gallery (1.14 nm layer-to-layer spacing).

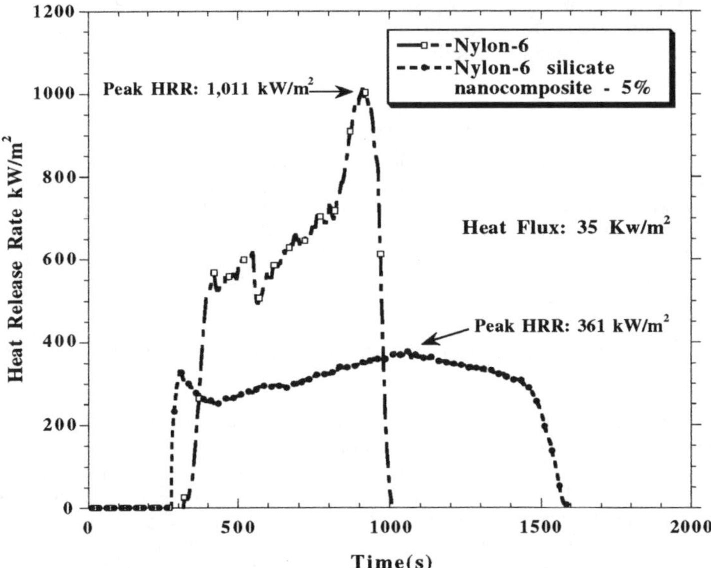

Figure 2. Comparison of the Heat Release Rate (HRR) plot for nylon-6, nylon-6 silicate-nanocomposite (mass fraction 5 %) at 35 kW/m^2 heat flux, showing a 63 % reduction in HRR for the nanocomposite.

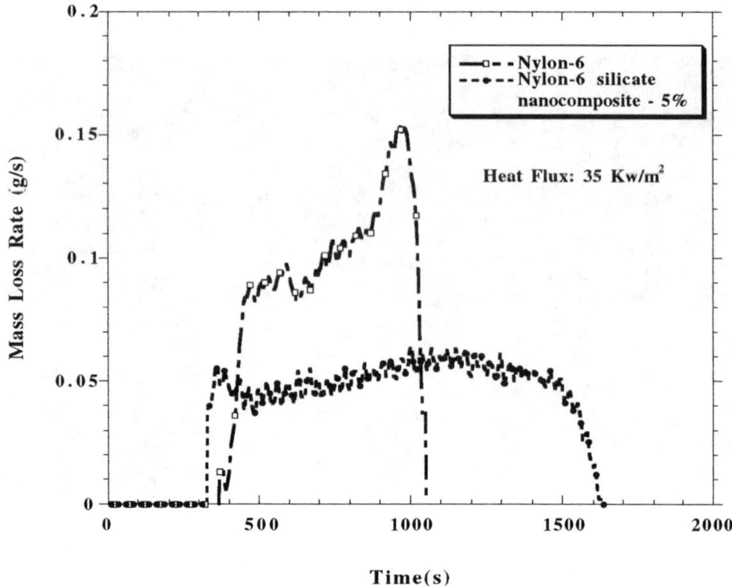

Figure 3. The mass loss rate data for nylon-6, nylon-6 silicate-nanocomposites (5 %). The curves closely resemble the HRR curves (Figure 2), indicating that the reduction in HRR for the nanocomposites is primarily due to the reduced mass loss rate and the resulting lower fuel feed rate to the gas phase.

3. EXPERIMENTAL

Evaluations of flammability were done using the Cone Calorimeter (16). The tests were done at an incident heat flux of 35 kW/m² using the cone heater. A heat flux of 35 kW/m² represents a typical small-fire scenario (17). Peak heat release rate, mass loss rate and specific extinction area (SEA) data, measured at 35 kW/m², are reproducible to within ± 10 %. The carbon monoxide and heat of combustion data are reproducible to within ± 15 %. The uncertainties for the Cone calorimeter are based on the uncertainties observed while evaluating the thousands of samples combusted to date. The cone data reported here is the average of two or three replicated experiments. Cone samples for thermoplastic polymers were prepared by compression molding the samples (25g - 55 g) into 75 mm diameter disks, 10 mm-15 mm thick, using a press with a heated mold.

The vinyl ester (18) samples were prepared at room temperature (25 °C) by mixing the resin (modified bisphenol-A epoxy based vinyl ester, a combination of a nitrile rubber and bisphenol-A epoxy based vinyl ester, mass fraction of 58 % in styrene - Derakane 8084, and bis-A/novolac epoxy based vinyl ester, a combination of bisphenol-A epoxy based vinyl ester and novolac epoxy based vinyl ester, mass fraction of 67 % in styrene - Derakane 441, both from Dow Chemical Co., see Figure 6) with the initiator (2-butanone peroxide, mass fraction of 1.25 %) the cobalt catalyst (mass fraction of 0.3 %, Cobalt napthenate, mass fraction of 6 % in mineral spirits, OMG Americas Inc.) and an organically modified silicate (OMS), dimethyl ditallow ammonium montmorillonite, Closite-15A, Southern Clay Products, Inc.) using an overhead mechanical stirrer for 5 minutes. The mixtures were poured into aluminum sample dishes (75 mm diameter x 15 mm depth) and cured at room temperature for 12 h and then post-cured at 70° C for 8 h.

The epoxy samples of diglycidyl ether bisphenol-A based epoxy (DGEBA, DER 331, Dow Chemical Co.) were prepared using a previously published procedure (11). The mixtures were poured into aluminum sample dishes (75 mm diameter x 15 mm depth) and cured at room temperature for 18 h and then post-cured at 100° C for 3 h. Methylenedianiline (MDA) and benzyldimethylamine (BDMA) were used as curing agents. The OMS used was dimethyl ditallow ammonium montmorillonite.

Nylon-6 and nylon-6 silicate nanocomposites (silicate mass fraction of 2 % and 5 %) nylon-12 and nylon-12 silicate nanocomposites (silicate mass fraction of 2 %) were obtained from UBE industries and were used as received.

Preparation of PS- silicate -nanocomposite (silicate mass fraction of 3 %) was accomplished by melt blending PS (Styron 6127, Dow Chemical Co.) with dimethyl dioctadecyl ammonium-exchanged montmorillonite (1). This yields a nanocomposite with the *intercalated* structure. The intergallery spacing, by X-ray diffraction, XRD, is 3.1 nm (2θ = 2.6°). The *immiscible* PS-silicate mix, where the silicate is only mixed in at the primary-particle size scale (~5 µm), is prepared under the same melt blending conditions except the alkyl ammonium used to compatibilize the montmorillonite has only one octadecyl R group instead of two. This renders the ion exchanged montmorillonite slightly less organophilic and intercalation does not occur.

Preparation of PP-g-MA- silicate -nanocomposite (silicate mass fraction of 5 %) by melt blending was accomplished by pressing the PP-g-MA (mixed with the dimethyl ditallow ammonium-exchanged montmorillonite, Closite 15A, at 160 °C for 30 minutes using a Carver press, followed by heating in a vacuum oven for several hours at 160 °C. This yields a nanocomposite with the *intercalated* structure. The intergallery spacing, by XRD analysis, is 3.6 nm. PP-g-MA (m.p. 152° C) was purchased from Aldrich and contains a mass fraction of 0.6 % maleic anhydride. It has a melt index of 115 g / 600 s and a Mw of ~ 10K, Mn ~ 5K.

X-ray diffraction (XRD) spectra were collected on a Phillips diffractometer using Cu Kα radiation, (λ = 0.1505945 nm). Powder samples were ground to a particle size of less than 40 µm. Solid polymer-silicate monoliths were typically 14 mm by 14 mm with a 2 mm thickness.

For the transmission electron microscopy (TEM), the char was broken into small pieces, embedded in an epoxy resin (Epofix), and cured overnight at room temperature. Ultra-thin sections were prepared with a 45° diamond knife at room temperature using a DuPont-Sorvall 6000 ultramicrotome. Thin sections (nominally 50 nm-70 nm) were floated onto water and mounted on 200-mesh carbon-coated copper grids. Bright-field TEM images were obtained with a Philips 400T microscope operating at 120 kV, utilizing low-dose techniques.

4. RESULTS AND DISCUSSION

Epoxy and vinyl ester resin systems represent a large fraction of the commercial thermoset polymer, coating, and composite markets (19, 20). A reduction in the flammability of these inherently highly flammable materials should increase their use. For the purpose of determining the flammability properties of <u>thermoset</u> PLS nanocomposites, especially the *intercalated* type, we have prepared and studied two *intercalated* vinyl ester nanocomposites and two *intercalated* epoxy nanocomposites.

4.1. Characterization of Nanocomposites by XRD. The microstructure of the epoxy and vinyl ester nanocomposites was characterized using XRD. The XRD patterns which reveal the *intercalated* structure of the modified-bisphenol-A (MBA) vinyl ester silicate nanocomposite and the bis-A/novolac (BAN) vinyl ester silicate nanocomposite are shown in Figure 4. The interlayer spacing for the MBA nanocomposite is 4.8 nm which represents a 1.5 nm expansion of the interlayer spacing of the silicate layers present in the original organically modified montmorillonite. The interlayer spacing for the BAN nanocomposite is 6.2 nm which represents a 2.9 nm expansion of the interlayer spacing of the silicate layers. The different degree of *intercalation* for these two vinyl ester systems is most likely due to the different structures of the components of the formulations, shown in Figure 6, and the resulting polarity and conformational mobility differences.

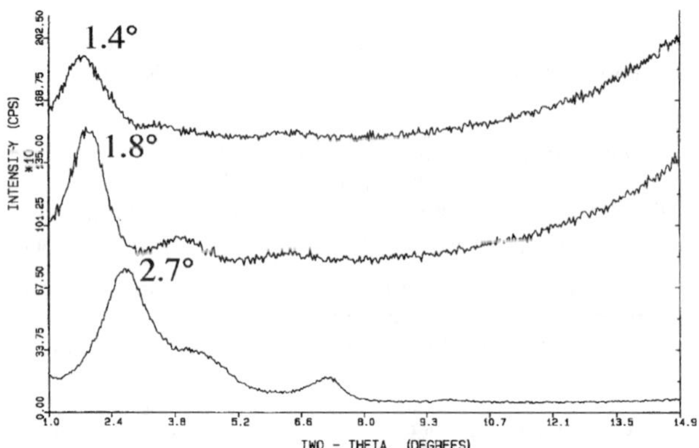

Figure 4. XRD patterns of: pure OMS, dimethyl ditallow ammonium- montmorillonite (bottom, $2\theta = 2.7°$, layer spacing 3.3 nm); cured modified-bisphenol-A epoxy based vinyl ester silicate nanocomposite (middle, 6 % silicate, $2\theta = 1.8°$, layer spacing 4.8 nm) and cured bis-A/novolac epoxy based vinyl ester silicate nanocomposite (top, 6 % silicate, $2\theta = 1.4°$, layer spacing 6.2 nm).

The XRD characterization of the DGEBA epoxy silicate nanocomposites, cured with either MDA or BDMA, shown in Figure 5, confirms *intercalated* structures, with interlayer spacings of 3.5 nm and 4.3 nm, respectively. These results parallel previously published work where MDA/DGEBA epoxy silicate nanocomposite exhibited less expansion of the silicate spacing than the BDMA/DGEBA epoxy silicate nanocomposite (11). It was proposed that the MDA, a diamine, could bridge the silicate layers and thereby prevent complete delamination. The BDMA, a monoamine, would not have this limitation. In the previous system however, the BDMA/DGEBA nanocomposite was fully *delaminated* whereas it is only *intercalated* here. The reason for this difference lies in the different alkyl ammonium montmorillonites used. In the previous study where the BDMA/DGEBA nanocomposite *delaminated,* the OMS used was bis(hydroxyethyl) methyl tallow montmorillonite. The hydroxyethyl groups in this OMS react with the oxirane rings of the DGEBA aiding in the dispersion of the silicate in the resin. The resulting epoxy network contains chemical bonds to the alkyl ammonium cation and therefore an ionic tether to the silicate (11). The strength of this ionic tether between the polymer and the silicate was found to have a direct relationship to the tensile strength of the nylon-6 nanocomposites (21). We used an OMS without reactive functionality (dimethyl ditallow ammonium montmorillonite). This results in a cured DGEBA nanocomposites without an ionic tether to the silicate for both the MDA and BDMA cured epoxies. It should be noted that since the MDA contains four active N-H hydrogens it can react directly as a crosslinking agent with DGEBA. The BDMA contains no active N-H, and it acts only as a catalyst for the homopolymerization of DGEBA. This results in formation of a purely polyether network for the BDMA/DGEBA epoxies (11, 19).

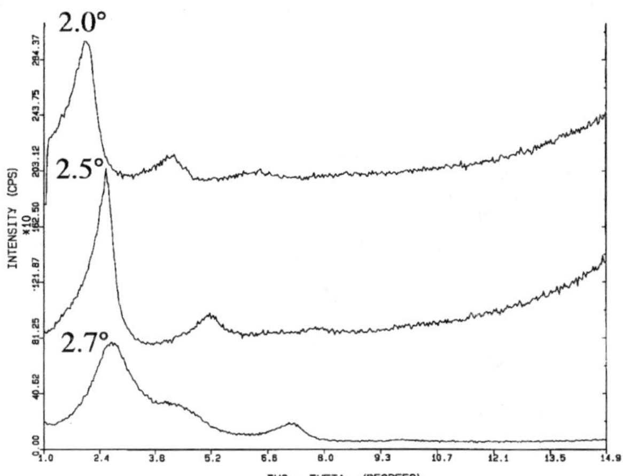

Figure 5. XRD patterns of: pure organic modified silicate, dimethyl ditallow ammonium-montmorillonite (bottom, 2q = 2.7°, layer spacing 3.3 nm); MDA cured epoxy silicate nanocomposite (middle, 6 % silicate, 2q = 2.5°, layer spacing 3.5 nm) and BDMA cured epoxy silicate nanocomposite (top, 6 % silicate, 2q = 2.0°, layer spacing 4.3 nm).

4.2. Epoxy and Vinyl Ester Flammability Properties. The results of combustion of the vinyl esters and their respective nanocomposite versions in the Cone calorimeter are shown in Table 2. Just as with the thermoplastic PLS nanocomposites both the peak and average heat release rates (HRR) were significantly improved for these *intercalated* vinyl ester nanocomposites with low mass fraction (6 %) of silicate. Furthermore, the primary difference (aside from HRR) between the flammability properties of the pure vinyl esters and the nanocomposites is the mass loss rates (MLR). The heat of combustion (H_c) soot (SEA) and carbon monoxide yields are unchanged.

The results of combustion of the MDA/DGEBA and BDMA/DGEBA epoxies and their respective nanocomposite versions in the Cone calorimeter are shown in Table 3. The data shows essentially the same behavior as for the vinyl ester nanocomposites, about a 40 % reduction in peak HRR, average HRR, and average MLR. The heat of combustion (H_c), soot (SEA) and carbon monoxide yields are unchanged. The HRR plots for DGEBA/MDA and the DGEBA/MDA silicate nanocomposite (6%) are shown in Figure 7. This plot is representative of the HRR behavior for both the epoxies and vinyl esters. Although ignition times in the cone calorimeter are accompanied by somewhat large uncertainties (± 25%) shorter ignition times are generally observed for the PLS nanocomposites. This may be caused by the low thermal stability of the quaternary alkyl ammonium cation contained in the organic modified montmorillonite used to prepare the PLS nanocomposites.

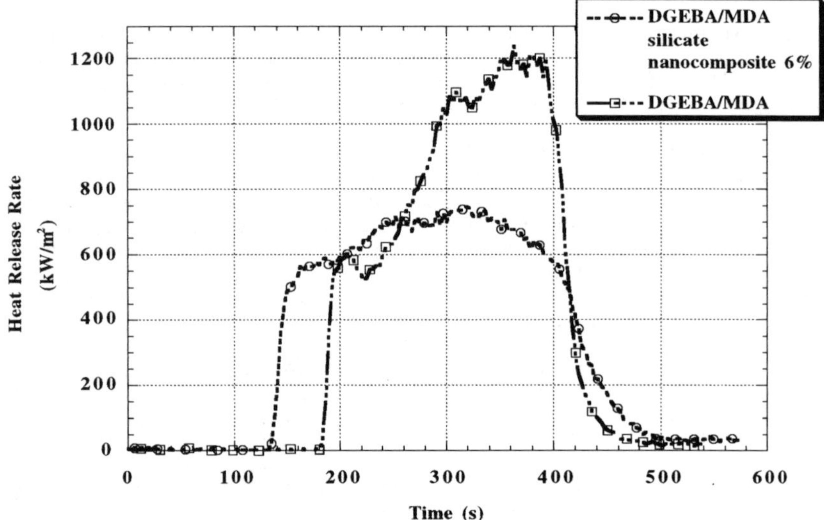

bisphenol-A epoxy vinyl ester resin

novolac epoxy vinyl ester resin

Figure 6. Vinyl ester resin chemical structures.

Figure 7. Comparison of the Heat Release Rate (HRR) plots for DGEBA/MDA and DGEBA/MDA silicate nanocomposite (mass fraction 6 %) at 35 kW/m² heat flux, showing a 40 % reduction in the peak HRR for the nanocomposite.

Table 1. Cone Calorimeter Data

Sample (*structure*)	Residue Yield (%) ± 0.5	Peak HRR (Δ%) (kW/m²)	Mean HRR (Δ %) (kW/m²)	Mean H_c (MJ/kg)	Mean SEA (m²/kg)	Mean CO yield (kg/kg)
Nylon-6	1	1,010	603	27	197	0.01
Nylon-6 silicate-nanocomposite 2% *delaminated*	3	686 (32%)	390 (35%)	27	271	0.01
Nylon-6 silicate-nanocomposite 5% *delaminated*	6	378 (63%)	304 (50%)	27	296	0.02
Nylon-12	0	1710	846	40	387	0.02
Nylon-12 silicate-nanocomposite 2% *delaminated*	2	1060 (38%)	719 (15%)	40	435	0.02
P S	0	1,120	703	29	1,460	0.09
PS silicate- mix 3% *immiscible*	3	1,080	715	29	1,840	0.09
PS silicate-nanocomposite 3% *intercalated*	4	567 (48%)	444 (38%)	27	1,730	0.08
P S w/ DBDPO/Sb_2O_3 30%	3	491 (56%)	318 (54%)	11	2,580	0.14
PP-g-MA	0	2,030	861	38	756	0.04
PP-g-MA silicate nanocomposite 5% *intercalated*	8	922 (54%)	651 (24%)	37	994	0.05

Heat flux : 35 kW/m², H_c : Heat of combustion, SEA : Specific Extinction Area

Table 2. Cone calorimeter data for MBA and BAN vinyl esters

Sample	Residue Yield (%) ± 0.5	Peak HRR (Δ%) (kW/m²)	Mean HRR (Δ %) (kW/m²)	Mean MLR (g/s m²)	Mean H_c (MJ/kg)	Mean SEA (m²/kg)	Mean CO yield (kg/kg)
Mod-Bis-A Vinyl Ester	0	879	598	26	23	1360	0.06
Mod-Bis-A Vinyl Ester**	8	656 (25%)	365 (39%)	18 (30%)	20	1300	0.06
Bis-A /Novolac Vinyl Ester	2	977	628	29	21	1380	0.06
Bis-A /Novolac Vinyl Ester **	9	596 (39%)	352 (44%)	18 (39%)	20	1400	0.06

Heat flux : 35 kW/m², H_c : Heat of combustion, SEA : Specific Extinction Area, MLR : Mass Loss Rate

** : 6 % silicate *intercalated* nanocomposite

Table 3. Cone calorimeter data for DGEBA/MDA and DGEBA/BDMA epoxies.

Sample	Residue Yield (%) ± 0.5	Peak HRR (Δ%) (kW/m²)	Mean HRR (Δ %) (kW/m²)	Mean MLR (g/s m²)	Mean H_c (MJ/kg)	Mean SEA (m²/kg)	Mean CO yield (kg/kg)
Epoxy resin DGEBA/MDA	11	1296	767	36	26	1340	0.07
Epoxy resin DGEBA/MDA**	19	773 (40%)	540 (29%)	24 (33%)	26	1480	0.06
Epoxy resin DGEBA/BDMA	3	1336	775	34	28	1260	0.06
Epoxy resin DGEBA/BDMA**	10	769 (42%)	509 (35%)	21 (38%)	30	1330	0.06

Heat flux : 35 kW/m², Hc : Heat of combustion, SEA : Specific Extinction Area, MLR : Mass Loss Rate

** : 6 % silicate *intercalated* nanocomposite

The data suggests that the source of the improved flammability properties of these materials is due to differences in condensed phase decomposition processes and not to a gas phase effect. As stated above these flammability properties are characteristic of both the thermoplastic and thermoset PLS nanocomposites.

4.2. Char Formation and Characterization. Comparison of the residue yields for thermoplastics and thermosets reveals little improvement in the carbonaceous char yields, once the presence of the silicate in the residue is accounted for. This was somewhat surprising since other studies of the thermal reactions in layered organic-silicate intercalates, at 400 °C, reported formation of carbonaceous-silicate residues and other condensation and crosslinking type reaction products (22). These data indicate that, although the mechanism of flame retardancy may be very similar for each of the systems studied, it is not via retention of a large fraction of carbonaceous char in the condensed phase. A recent study in our laboratory of the nylon-6 silicate nanocomposite pyrolysis process, in a nitrogen atmosphere, using a radiative gasification apparatus (shown in Figure 8) revealed that the reduction in MLR does not occur until the sample surface is partially cover with char (4). The MLR data for these N_2 gasification experiments are shown in Figure 9 . Visual observation of the pyrolysis shows that at 180 s when the MLR for the nylon-6 silicate nanocomposite slows, compared to the pure nylon-6, the surface of the nanocomposite is over 50 % covered by char.

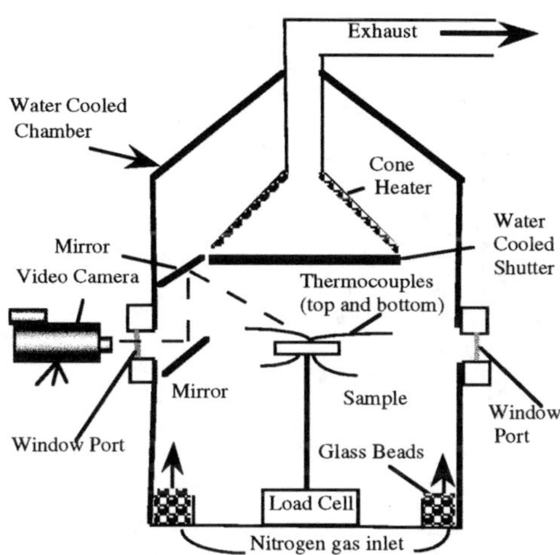

Figure 8. A schematic of the radiative gasification apparatus (1 m diameter, 2 m height). The gasification apparatus allows pyrolysis, in a nitrogen atmosphere, of samples identical to those used in the Cone calorimeter.

TEM of a section of the combustion char from the nylon-6 silicate-nanocomposite (5 %) is shown in Figure 10. A multilayered silicate structure is seen after combustion, with the darker, 1 nm thick, silicate sheets forming a large array of fairly even layers. This was the primary morphology seen in the TEM of the char, however, some voids were also present. The delaminated hybrid structure, appears to collapses during combustion. The nanocomposite structure present in the resulting char appears to enhance the performance of the char through reinforcement of the char layer. This multilayered silicate structure may act as an excellent insulator and mass transport barrier, slowing the escape of the volatile products generated as the nylon-6 decomposes (4). XRD analysis of chars from combustion of nylon-6, and the two DGEBA nanocomposites (Figure 11) shows that the interlayer spacing of the chars is the same, 1.3 nm, independent of the chemical structure (thermoplastic polyamide, thermoset aromatic diamine cured epoxy or thermoset homopolymerized epoxy) or nano-structure (*delaminated* or *intercalate*) of the nanocomposite.

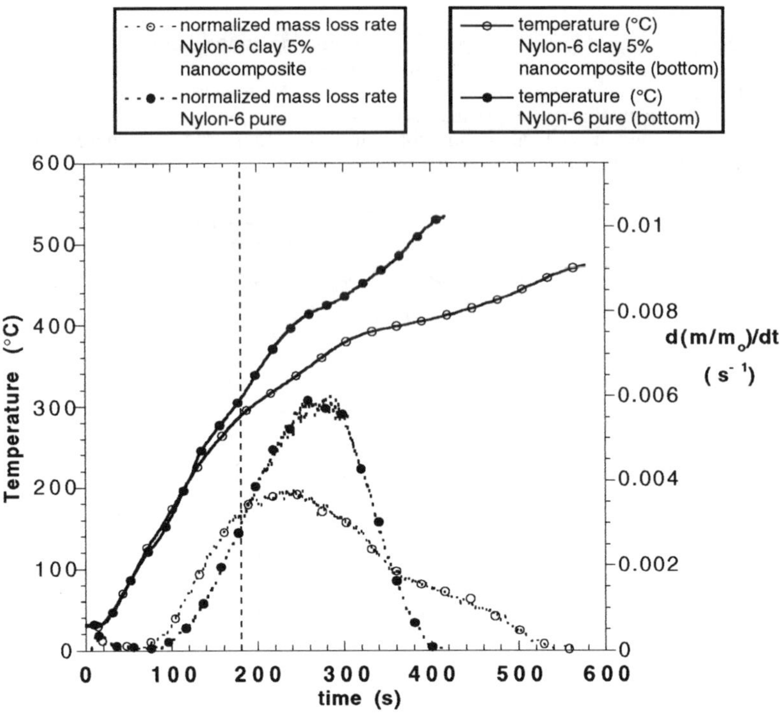

Figure 9. Normalized Mass loss rate and temperature versus time plots for the gasification experiments for nylon-6 and nylon-6 silicate (5%) nanocomposite in a N_2 atmosphere. All samples were exposed to a flux of 40 kW/m^2 in a N_2 atmosphere. The mass loss rate curves begin to differ at 180 seconds when the surface of the nanocomposite sample is partially covered by char. The insulating effect of the char can be seen in the bottom-surface thermocouple data for the nanocomposite.

Figure 10. TEM of a section of the combustion char from the nylon-6 silicate-nanocomposite (5 %) showing the carbon-silicate (1 nm thick, dark bands) multilayered structure. This layer may act as an insulator and a mass transport barrier.

5. FUTURE WORK

The future focus of this project will be on the continued development of a fundamental understanding of the fire retardant mechanism of polymer layered silicate nanocomposites. The effect of nano-structure on the flammability properties of a given nanocomposite system will be investigated. This work will be done within the Flammability of Nanocomposites Consortium.

6. CONCLUSIONS

The flammability properties of <u>thermoplastic</u> and <u>thermoset</u> polymer layered silicate nanocomposites are reported. The peak and average heat release rate (HRR) are reduced by 40 % to 60% in *delaminated* and *intercalated* nanocomposites containing a silicate mass fraction of only 2 % to 6 %. Not only is this a very promising new method for flame retarding polymers, but it does not have the usual drawbacks associated with other additives. That is, the physical properties are not degraded by the additive (silicate); instead they are greatly improved. Furthermore, this system does not increase the carbon monoxide or soot produced during the combustion. The nanocomposite structure of the char appears to enhance the performance of the char layer. This layer may act as an insulator and a mass transport barrier slowing the escape of the volatile products generated as the polymer decomposes.

7. ACKNOWLEDGMENTS

The authors would like to thank Dr. R. Lyon and the Federal Aviation Administration for partial funding of this work, through Interagency Agreement DTFA0003-92-Z-0018. We would also like to thank Mr. Michael Smith for Cone Calorimeter analysis, Ms. Lori Brassel for preparation of vinyl ester and epoxy samples, Dr. Catheryn Jackson and Dr. Henri Chanzy for TEM analysis of the char samples, Dr. M. Nyden for the montmorillonite structure, and Dr. James Cline for use of XRD facilities.

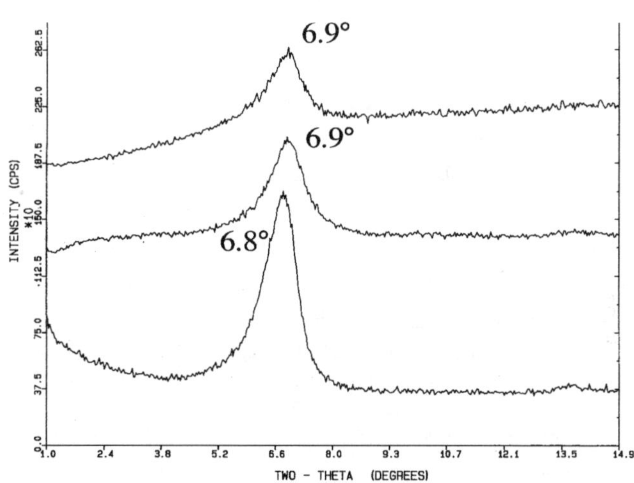

Figure 11. Char XRD patterns of silicate nanocomposites: nylon-6 (<u>bottom</u>), DGEBA/BDMA (<u>middle</u>), and DGEBA/MDA (<u>top</u>). All chars have a interlayer spacing of ~1.3 nm, which is confirmed by TEM.

8. REFERENCES

1. E. P. Giannelis, *Advanced Materials*, <u>8</u>, (1) 29 (1996).

2. Y. Kojima, A. Usuki, M. Kawasumi, A. Okada, Y. Fukushima, T. Kurauchi, and O. Kamigaito, J. Mater. Res. <u>8</u>, 1185 (1993).

3. J. W. Gilman, T. Kashiwagi, J. D. Lichtenhan, SAMPE Journal, , 33, no. 4, 40 (1997).

4 . J. W. Gilman, T. Kashiwagi, S. Lomakin, E. P. Giannelis, E. Manias, J. D. Lichtenhan, P. Jones in Fire Retardancy of Polymers : the Use of Intumescence, G. Camino, M. Le Bras, S. Bourbigot & R. DeLobel eds., The Royal Society of Chemistry, Cambridge (1998) in press.

5. M.S. Wang, T. J. Pinnavaia, Chem. Mater. 6, 468 (1994).

6. A. Usuki, A. Okada, T. Kurauchi, J. App. Poly. Sci. 63, 137 (1997).

7. A. Usuki, Y. Kojima, M. Kawasumi, A. Okada, Y. Fukushima, T. Kurauchi, and O. Kamigaito, *J. Mater. Res.* 8, 1179 (1993).

8. Y. Kojima, A. Usuki, M. Kawasumi, A. Okada, Y. Fukushima, T. Kurauchi, and O. Kamigaito, *J. Mater. Res.* 8, 1185 (1993).

9. J. Lee, T. Takekoshi and E. Gannelis, Mat. Res. Soc. Symp. Proc. vol 457, p. 513, (1997).

10. J. Lee and E. Giannelis, Polymer Preprints, vol. 38, p. 688, (1997).

11. P.B. Messersmith, E. P. Gianellis, Chem. Mater. 6, 1719 (1994).

12. S. Miyanaga, Y. Tsunoda, Japanese Patent JP 09,255,747, 1996.

13. A. Usuki, et.al., US Patent 4,889,885, 1989.

14. E. N. Kresge, D. J. Lohse, US patent 5,665,183, 1997.

15. M. Nyden, J. W. Gilman, Comp. and Theor. Polym. Sci. in press (1998).

16. V. Babrauskas, R. Peacock, *Fire Safety Journal*, 18, 255 (1992).

17. V. Babrauskas, *Fire and Materials*, 19, 243 (1995).

18. Certain commercial equipment, instruments, materials, services or companies are identified in this paper in order to specify adequately the experimental procedure. This in no way implies endorsement or recommendation by NIST.

19. R. S. Bauer, ed. Epoxy Resin Chemistry II, ACS Symposium Series 221, American Chemical Society, Washington, D.C. 1983.

20. C. D. Dudgeon in Composites: Engineered Materials Handbook, vol 1, sec. 2, 90, ASM International, , Metals Park, OH, 1987.

21. A. Usuki et. al., J. Appl. Poly. Sci., 55, 119 (1995).

22. J. M. Thomas, "Intercalation Chemistry", Chapter 3, p. 55, Academic Press, Inc., London, 1982.

FIRE SAFETY OF COMPOSITES IN THE U.S. NAVY

Usman Sorathia, John Ness, Michael Blum
Carderock Division, Naval Surface Warfare Center
West Bethesda, MD 20817-5700

ABSTRACT

Composite materials for marine applications offer the benefit of weight savings, corrosion resistance, and reduced life cycle costs. The composite structures used in marine applications tend to be large, complex, and thick. As such, the use of low temperature non-autoclave cure resins is desirable. The U.S. Navy is presently using fire retarded (brominated) vinyl ester resin for some topside composite structures. These composites are produced by vacuum assisted resin transfer molding. Brominated vinyl ester resin generates dense heavy smoke with high yields of carbon monoxide. Acid gases such as hydrogen bromide are also produced. Several alternative resins with and without non halogenated fire retardants have been recently evaluated by NSWCCD in small scale fire test methods. These included cone calorimeter testing at three different fluxes of 25, 50, and 75 kW/m^2. The summary of results, including smoke production rate and carbon monoxide yield, is presented for different vinyl ester resins with and without additives. Brominated epoxy vinyl ester marked #1168 exhibited lower heat release rates, but significantly higher smoke generation and CO yield when compared with non brominated vinyl esters. Of the additives studied, aluminum trihydrate (15 phr, #1196) shows 20 and 25% decrease in peak heat release rates, 24 and 13% decrease in average heat release rates, and 27 and 24% decrease in average mass loss rates at radiant heat fluxes of 50 and 75 kW/m^2 respectively.

KEY WORDS: Low Smoke Composites, Vinyl Esters, Fire Retardants.

INTRODUCTION

During the past ten years, there has been a resurgence of interest in the development and application of composites for both primary and secondary load-bearing structures of Navy ships such as lightweight foundations, deckhouses, masts; for machinery components such as composite piping, valves, centrifugal pumps and heat exchangers; and for auxiliary or support items such as gratings, stanchions, vent screens, ventilation ducts, etc. A recent notable large composite application is Advanced Enclosed Mast/Sensor (AEM/S) System which has been installed on USS RADFORD. This renewed interest in composite materials is driven by fleet needs to reduce maintenance, save weight, increase covertness and provide affordable alternatives to metallic components with lower life cycle costs [1]. Technical issues which have limited composite use on board Navy warships include total acquisition costs (materials, fabrication and installation), an adequate performance database and knowledge of design allowables, and the combustible nature (and, hence, the fire, smoke and toxicity threat) of polymer matrix systems.

The use of composites inside Naval submarines is now covered by MIL-STD-2031 (SH), Fire and Toxicity Test Methods and Qualification Procedure for Composite Material Systems Used in Hull, Machinery, and Structural Applications Inside Naval Submarines [2]. Two guiding criteria were established for the use of composite systems aboard Navy vessels [3]. The first is that the composite system will not be the fire source, i.e., it will be sufficiently fire resistant not to be a source of spontaneous combustion. The second is that ignition of the composite system will be delayed until the crew can respond to the primary fire source, i.e., the composite system will not result in rapid spreading of the fire. MIL-STD-2031 contains test methods and requirements for flammability characteristics such as Oxygen Index (ASTM D2863 modified), Smoke Obscuration (ASTM E662), Flame Spread (ASTM E162), Cone Calorimeter heat release rates (ASTM E1354), Burn Through Test (resistance to flame penetration), Navy Quarter scale test (flashover potential), N-gas smoke toxicity (presence of super toxic materials), large-scale open environment fire test (burning and extinguishing characteristics in normal atmospheres), large-scale pressurizable fire test (burning and extinguishing characteristics in pressurized atmospheres), etc. This selection of small and intermediate scale tests was considered to be the most indicative for the conditions to which the materials would be exposed aboard submarines.

The Navy currently has no specific fire related standard for composite structures in surface ships. Recently, NAVSEA 03G has provided guidance and requirements for use of composite materials in selected applications such as vent screens, catwalk grating, composite louvers, ventilation ducting, etc. The flammability requirements for surface ships are different than the submarines. In addition to material flammability concerns such as smoke generation, flame spread and flashover, the critical issue in surface ship fires is the residual strength of composite structures at elevated temperatures for a period of 30-60 minutes [4].

LOW SMOKE COMPOSITE MATERIALS

The U.S. Navy is currently using brominated vinyl ester matrix resin with glass reinforcement for composite applications in topside surface ship structures. This matrix resin was selected due to its good corrosion resistance and toughness. Bromine is an effective flame retardant, especially when combined with antimony oxide. Bromination of vinyl ester resin imparts fire retardancy as manifested by flame spread and lower heat release rates[5]. However, this fire retardant system functions primarily in the gas phase causing incomplete combustion. As such, brominated resins produce dense smoke, increase in the yield of carbon monoxide, and hydrogen bromide. Recent interest in the use of organic matrix composite materials in U.S. Navy submarines and ships has generated the requirement for significant improvement in the flammability performance of these materials including reduction in the amount of smoke, carbon monoxide, and corrosive combustion products. New fire retardant approaches for organic matrix composite materials are needed to address the smoke issues and to further reduce the flammability of these composites.

NSWCCD has undertaken a Research & Development effort which will result in the identification, optimization, and subsequent scale up of low smoke producing matrix resins with non halogenated fire retardants which can be processed by non auto-clave VARTM process. Use of this non-halogenated fire retardant system will produce composite materials with improved flammability properties and comparable or better mechanical and environmental properties than brominated resins.

STUDY OF VINYL ESTER RESINS

In the first stage, several glass reinforced vinyl ester composites were fabricated using vacuum assisted resin transfer molding process. These included brominated bisphenol A epoxy vinyl ester resin (1168), non brominated epoxy vinyl ester resin (1167), epoxy novolac vinyl ester resin (1169), and bisphenol A epoxy vinyl ester resin (1170). The composites were made to a thickness of 0.25 inch with a fiber content of about 70% by weight using 24 oz/yd^2 glass woven roving. The resins were cured at ambient temperatures by the use of cobalt naphthenate, dimethyl aniline, and methyl ethyl ketone peroxide as promoters and catalyst. The composites were further post cured at 160°F for 6 hrs. The composite materials were evaluated for tensile (ASTM D 638), compression (ASTM D 695), and Iosipescu shear (ASTM D 5379M), smoke generation (ASTM E 662), and heat release rates in cone calorimeter (ASTM E1354) respectively. The data for mechanical characterization is shown in Table 1. The data from smoke chamber and cone calorimeter are shown in Tables 2 and 3 respectively. Figures 1, 2 and 3 show smoke production rates, carbon monoxide yields, and the heat release rates for these glass reinforced composites with different vinyl ester resins.

Table 2 includes smoke data from cone calorimeter such as specific extinction area and smoke production rate. All testing was performed in the horizontal orientation. For each panel at any given flux, three coupons were tested and the results presented are the averages of three coupons. Specific extinction area (SEA, m^2/kg) is defined as the smoke which is produced per unit mass of material being volatilized, i.e. k V_s / $\dot{m}$, where k is the extinction coefficient (1/m), V_s is the standard volume flow rate of air (m^3/s), $\dot{m}$ is the mass loss rate of sample, (kg/s). S.E.A. is usually adopted to express the extent of the contribution to the smoke generation. Smoke production rate is given as the product of specific extinction area and the mass loss rate.

Table 1: Mechanical characterization of different glass/vinyl ester (VARTM) composites.

Test	Tensile (ASTM D 638)		Compression (ASTM D 695)			Iosipescu Shear (ASTM D 5379M)	
	Modulus Gpa	Strength Mpa	Modulus Gpa	Strength MPa	Poisson Ratio	Modulus Gpa	Strength MPa
1167	25.9	441	30.7	368	0.159	4.1	95
1168	27.5	432	27.4	324	0.157	3.7	89
1169	21.7	330	29.0	309	0.171	4.4	101
1170	20.6	385	27.4	218	0.173	4.6	82

Smoke Production Rate

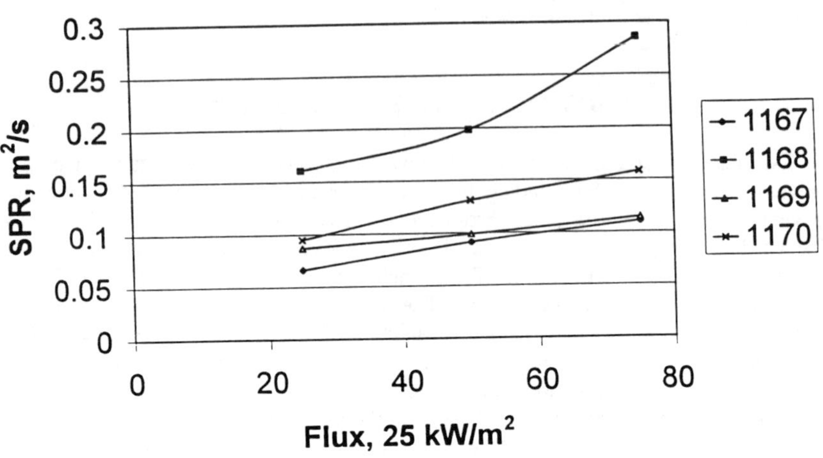

Figure 1: Smoke Production Rate Versus Heat Flux.

CO Yield

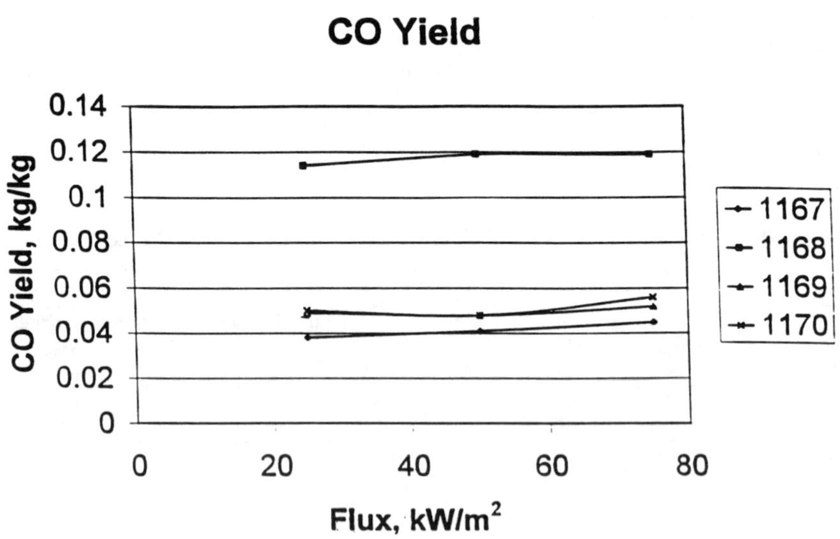

Figure 2: Carbon Monoxide Yield Versus Heat Flux.

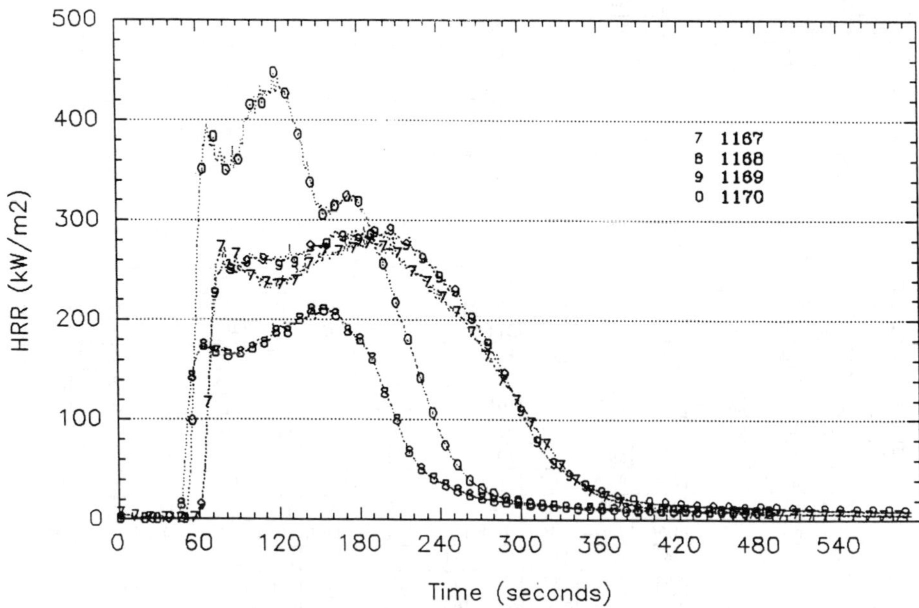

Figure 3: Heat Release Rate versus Time (@75 kW/m²)

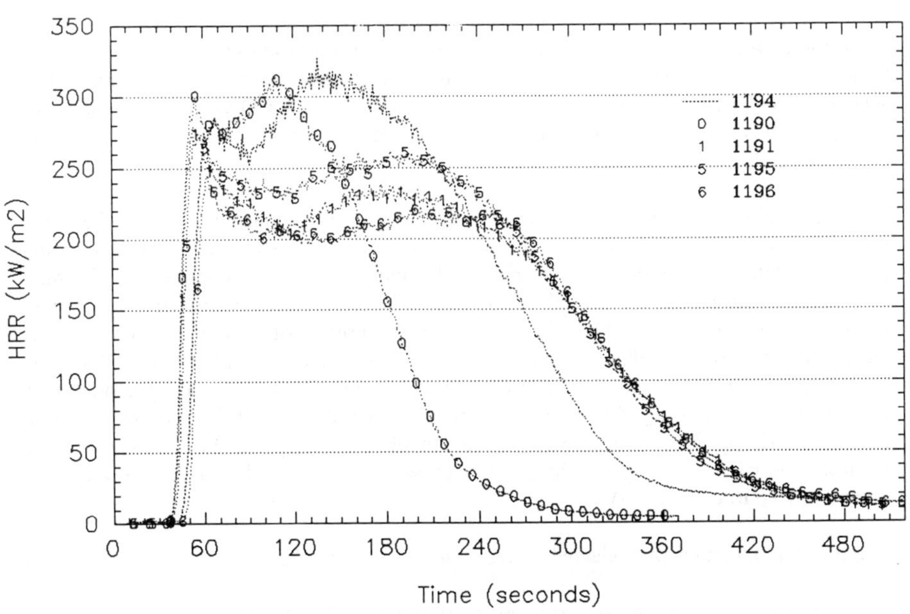

Figure 4: Heat Release Rate versus Time (@75 kW/m²)

Table 2: Smoke data (ASTM E662) for different glass/vinyl ester composites.

Test (Flaming)	1167	1168	1169	1170
Dmax, Flaming	173	593	217	197
Ds(300s), Flaming	103	503	154	185
CO, ppm	300	800	200	300
CO_2, %v	2	0.5	2	2
HCL, ppm	ND	Tr	ND	ND
HCN, ppm	2	2	Tr	Tr

Table 3: Heat release and smoke data (ASTM E1354) for different glass/vinyl ester composites.

Test	1167			1168			1169			1170		
Flux, kW/m^2	25	50	75	25	50	75	25	50	75	25	50	75
Time to Ignition, sec	320	85	42	214	52	29	302	85	42	259	75	36
Peak HRR, kW/m^2	308	276	281	147	152	217	342	302	303	356	348	432
Avg. HRR, kW/m^2	106	116	128	70	60	89	136	129	138	132	128	168
Avg. HRR (180s), kW/m^2	187	203	215	105	112	158	223	226	229	236	248	300
Avg. HRR (300s), kW/m^2	180	184	190	92	86	108	211	198	203	190	179	202
Total HRR, MJ/m^2	64	59	59	29	28	33	68	62	63	58	55	61
Avg. MLR, $g/s.m^2$	7.94	9.15	11.2	11.99	13.02	18.2	10.9	12.15	13.1	10.4	12.76	15
Avg. SEA, m^2 / kg	836	999	986	1341	1524	1569	796	815	872	914	1027	1050
Avg. SPR, m^2/s	.0663	.0914	.1104	.1608	.1984	.2856	.0868	.0990	.1142	.0950	.1310	.1575
Avg. Eff. HOC, MJ/kg	24	22	21	10	11	11	22	21	21	24	23	24
Peak CO, ppm	293	250	327	777	753	1030	410	330	407	452	350	495
Avg. CO yield, kg/kg	0.038	0.041	0.045	0.114	0.119	0.119	0.049	0.048	0.052	0.050	0.048	0.056

STUDY OF VINYL ESTER RESIN WITH DIFFERENT ADDITIVES

From the data presented in the previous section, non brominated epoxy vinyl ester resin based composite marked #1167 was selected for further study with additives. The selection was based on the basis of low smoke generation, low CO yield, and comparable or superior mechanical properties. However, low smoke properties for this resin are also accompanied by higher heat release rates.

As mentioned previously, composite structures used in Navy applications tend to be large, complex, and thick. Navy prefers the use of vacuum assisted resin transfer molding process to fabricate composite structures as this method produces composites with high fiber content, up to 70% by weight. However, all the additives evaluated in this study could not be processed by VARTM due to powder additives not flowing through the glass fabric with the resin and filtering out. The composite panels with additives produced by VARTM contained patches of dry and resin starved areas. Hence, the additive study was conducted by processing the composites first by hand lay up followed by vacuum bagging. This produced composites with about 60% by weight of fiber content. As a control, the #1167 was also fabricated without any additive by similar process for comparative purposes (1194).

The fire retardant additives selected in this study are described below:

1. **Enhanced Char Forming Proprietary Additive (10 phr, 1191):** Char formation (carbons stay in the condensed phase) is recognized as an excellent method of reducing flammability. Char formation reduces the amount of volatile polymer pyrolysis fragments, or fuel, available for burning in the gas phase; this in turn reduces

the amount of heat released and fed back to the polymer surface. The char also insulates the underlying polymer and acts as a mass transport barrier delaying the volatilization of fuel. A distinct advantage of purely char enhancing flame retardant approach is that the generation rates of smoke and corrosive combustion products are significantly reduced compared to the pure polymers and halogenated flame retardants. The char enhancing additive was mixed with epoxy vinyl ester resin prior to making a composite with glass fiber by hand lay up and followed by vacuum bagging.

2. **Polymer Layered Silicate Nano-composites (10 phr, 1195):** Polymer nanocomposites represent an alternative to conventionally (macroscopically) filled polymers [6]. Recent studies of Nylon 6 with silicate dispersed or layered in the polymer matrix at the nano scale (5%) show increase in the heat distortion temperature by up to 100°C, and reduction in the peak heat release rates by up to 63% at heat fluxes of 50 kW/m^2 [7]. Also, unlike other flame retardant approaches for polymers which tend to reduce mechanical properties, the silicate-nanocomposites have shown some improvement in mechanical properties. Silicate material was mixed with modified vinyl ester resin at 85°C for four hours with reflux condenser to prevent volatilization of styrene. Upon cooling the mix, the modified vinyl ester-silicate layered nanocomposite with glass woven roving was fabricated by hand lay up followed by vacuum bagging.

3. **Aluminum trihydrate (15 phr, 1196):** Aluminum tri hydrate is widely used as flame retardant in unsaturated thermosetting resins. ATH works by endothermically releasing water at the fire exposure temperatures. Usually, ATH is used at high loading levels. However, due to the limitations of VARTM process preferred for large scale fabrication, only 15 phr aluminum trihydrate was used in this study. The composite with glass fiber was fabricated by hand lay up followed by vacuum bagging.

4. **Siloxane Powder Additive (10 phr, 1190):** The siloxane powdered additives are a combination of polydimethyl siloxane and fumed silica. Recent studies [8] have found that siloxane powdered additive 4-7081, when used at a loading level of 5 wt.% with polycarbonate and ethylene vinyl acetate copolymer, reduced their peak values of HRR by 60% and 51% respectively. The additive is available in different sizings. In this study, siloxane powdered additive 4-7081 was used at a loading level of 10 phr. The additive was mixed with the liquid vinyl ester at room temperature prior to hand lay up followed by vacuum bagging.

The composite panels with different additives were evaluated by cone calorimeter at fluxes of 25, 50, and 75 kW/m^2. All testing was performed in the horizontal orientation. For each panel at any given flux, three coupons were tested and results presented are the averages of three coupons. Table 4 presents the results on heat release rates and smoke generation. Figure 4 shows the heat release rates for these composites.

RESULTS AND DISCUSSION

Cone calorimeter evaluation of the fire retardant study has provided data with respect to peak and average heat release rates, smoke production rates, and CO yields for vinyl ester resins with and without additives. Evaluation was conducted at three different fluxes to simulate small and large fires.

Table 4: Heat release and smoke data (ASTM E1354) for glass/vinyl ester composites with additives.

Test	1194 (Control)			1190			1191			1195			1196		
Flux, kW/m^2	25	50	75	25	50	75	25	50	75	25	50	75	25	50	75
Time to Ignition, sec	323	94	43	240	74	37	346	89	38	334	85	37	386	99	45
Peak HRR, kW/m^2	337	313	325	309	307	311	317	279	293	332	283	283	303	249	245
Avg. HRR, kW/m^2	139	190	171	133	161	169	141	151	170	139	162	164	121	144	149
AHRR (180s), kW/m^2	213	245	236	195	215	215	194	182	199	201	198	192	172	167	177
AHRR (300s), kW/m^2	190	202	200	146			191	188	199	198	191	199	165	167	180
Total HRR, MJ/m^2	62	62	61	44	43	42	71	69	66	79	64	69	59	56	60
Avg. Mass LR, g/s.m^2	9.08	10.89	11.98	9.52	10.99	12.36	8.03	8.81	10.40	7.84	8.96	10.29	6.84	7.99	9.06
Avg. SEA, m^2 / kg	822	1086	1111	902	1185	1231	785	902	1061	802	998	1120	726	852	988
Avg. SPR, m^2/s,	.0746	.1183	.1331	.0859	.1302	.1522	.0630	.0795	.1103	.0629	.0894	.1152	.0497	.0681	.0895
Avg. Eff. HOC, MJ/kg	24	23	22	23	23	22	24	22	21	24	24	23	23	23	23
Peak CO, ppm	273	300	388	300	303	370	272	285	303	283	280	308	280	238	240
Avg. CO yield, kg/kg	.046	0.047	.051	.049	0.048	0.049	0.047	0.047	0.043	.0409	0.044	0.048	.036	.037	.037

In the first step, brominated epoxy vinyl ester marked #1168 exhibited lower heat release rates, but significantly higher smoke generation and CO yield when compared with non brominated vinyl esters. It also resulted in lower time to ignition. The effect of bromination is thus evident which causes incomplete combustion and is also reflected in lower effective heat of combustion. However, at the high flux of 75 kW/m^2, results indicate significant increase in mass loss rate which may be partially explained by the increase in peak heat release rate and total heat release, but may be accompanied by increased production of hydrogen bromide. Smoke chamber data also indicated high smoke and carbon monoxide generation for brominated vinyl ester. Non brominated epoxy vinyl ester marked # 1167 was selected for further study due to lower smoke generation and CO yields. This was also accompanied by higher peak and average heat release rates. This vinyl ester selection was also based on the evaluation of tensile, compressive, and shear properties evaluation conducted concurrently with small scale fire testing.

In the second step, epoxy vinyl ester marked #1167 was studied in conjunction with selected fire retardant additives. The selection of additives was made on the basis of their potential to lower heat release rates. The incorporation of additives could not be accomplished using preferred vacuum assisted resin transfer molding due to the particle sizes of the additives selected and their inability to flow through the glass woven roving to fully wet the reinforcement. In some cases, the additive appeared to filter out and resulted in non homogeneous composite panels. No attempt was made in this study to optimize the VARTM process, or optimize the particle size of the additives. It was assumed that comparative screening of the additives could be accomplished by hand lay up followed by consolidation of the composite panels with vacuum bagging.

Results are shown in Table 4 and also in Figure 4. All additives were effective in reducing peak heat release rates to some extent. Of the additives studied, aluminum trihydrate (#1196) shows 20 and 25% decrease in peak heat release rates at radiant heat fluxes of 50 and 75 kW/m^2 respectively; and 24 and 13% decrease in average heat release rates at 50 and 75 kW/m^2 respectively. Also, average mass loss rates were decreased by 27 and 24% at 50 and 75 kW/m^2 respectively.

Future work will include the compatibility of different sizings of siloxane powder additive, optimization of VARTM to process glass reinforced composites with additives, and the effect of additive and corresponding loading levels on mechanical properties of glass reinforced composite materials.

SUMMARY

Composite materials for marine applications offer the benefit of weight savings, corrosion resistance, and reduced life cycle costs. The composite structures used in marine applications tend to be large, complex, and thick. As such, the use of low temperature non-autoclave cure resins is desirable. The U.S. Navy is presently using fire retarded (brominated) vinyl ester resin for some topside composite structures. These composites are produced by vacuum assisted resin transfer molding. Brominated vinyl ester resin generates dense heavy smoke with carbon monoxide and hydrogen bromide. Several alternative resins with and without non halogenated fire retardants are being evaluated by NSWCCD for potential use in shipboard applications.

Results indicate that non brominated epoxy vinyl ester (#1167) possesses superior balance of mechanical properties and low smoke generation with lower yields of carbon monoxide. However, this improvement in smoke and CO generation is accompanied by higher peak heat release rates.

Several fire retardant additives were evaluated in conjunction with non brominated epoxy vinyl ester (#1167). These included enhanced char forming additive, siloxane powder additive, silicate nanocomposite, and aluminum tri hydrate. Of the additives studied, aluminum trihydrate (15 phr, #1196) shows 20 and 25% decrease in peak heat release rates, 24 and 13% decrease in average heat release rates, and 27 and 24% decrease in average mass loss rates at radiant heat fluxes of 50 and 75 kW/m^2 respectively.

ACKNOWLEDGMENTS

This work was sponsored by the Office of Naval research (ONR 332) as part of the Seaborne Structural Materials Project for Composites, 034S60, PE 602234N, Work Unit 1-6430-705. The ONR sponsor is Mr. James Kelly.

REFERENCES

1. Caplan, I., "Marine Composites - Learning Lessons Along The Way," First International Workshop on Composite Materials for Offshore Operations, University of Houston, Texas, 26-28 Oct. 1993.

2. MIL-STD-2031 (SH), "Fire and Toxicity Test Methods and Qualification Procedure for Composite Material Systems Used in Hull, Machinery, and Structural Applications Inside Naval Submarines" (Feb 1991).

3. DeMarco, R.A., "Composite Applications at Sea: Fire Related Issues," 36th International SAMPE Symposium, Vol. 36 (Apr 1991).

4. U. Sorathia and Timothy Dapp, "Structural Performance of Glass/vinyl Ester Composites at Elevated Temperatures", SAMPE Journal, Vol. 33, No. 4, July/August 1997.

5. Sorathia, U. and C.P. Beck, "Fire-Screening Results of Polymers and Composites," Proceedings of Improved Fire and Smoke Resistant Materials for Commercial Aircraft Interiors, National Research Council, Publication NMAB-477-2, National Academy Press, Washington, DC (1995).

6. E. P. Giannelis, Advanced Materials, $\underline{8}$, 29, (1996).

7. Jeffrey W. Gilman, Takashi Kashiwagi, Joseph D. Lichtenhan, "Nanocomposites: A Revolutionary New Flame Retardant Approach", SAMPE Journal, Vol. 33, No. 4, July/August 1997.

8. Pape, P.G. and Romenesko, D.J, " The Seventh Annual BCC Conference on Flame Retardancy: Recent Advances in Flame Retardancy of Polymeric Materials", Stamford, Connecticut, May 1996.

Subject Index

Page

Subject Index

Page

Ageing

Analysis/Testing

Subject Index

Page

Subject Index

Page

Subject Index

Page

Subject Index

Page

Subject Index

                                                    Page

Equipment and Machinery

Fabrics

Failure Analysis

Fasteners

Subject Index

Page

Subject Index

Page

Subject Index

Page

Subject Index

Subject Index

Page

Page

Repair

Resin Transfer Molding

Retrofit/Rehabilitation

Subject Index

Page

Subject Index

Subject Index

Page

Subject Index

Page

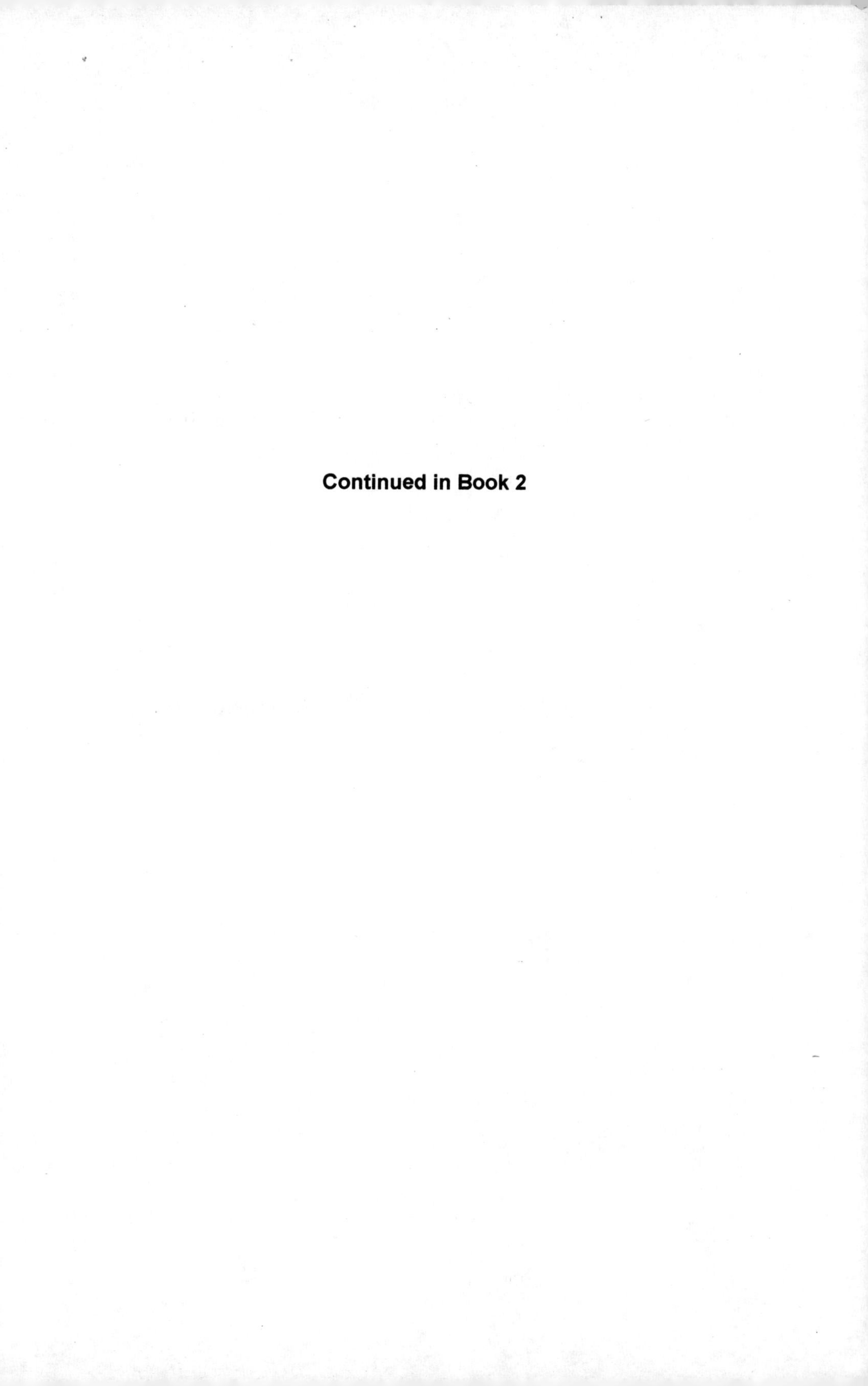

Continued in Book 2